Geophysical Monograph Series

Geophysical Monograph Series

169 **Magnetospheric ULF Waves: Synthesis and New Directions** *Kazue Takahashi, Peter J. Chi, Richard E. Denton, and Robert L. Lysak (Eds.)*
170 **Earthquakes: Radiated Energy and the Physics of Faulting** *Rachel Abercrombie, Art McGarr, Hiroo Kanamori, and Giulio Di Toro (Eds.)*
171 **Subsurface Hydrology: Data Integration for Properties and Processes** *David W. Hyndman, Frederick D. Day-Lewis, and Kamini Singha (Eds.)*
172 **Volcanism and Subduction: The Kamchatka Region** *John Eichelberger, Evgenii Gordeev, Minoru Kasahara, Pavel Izbekov, and Johnathan Lees (Eds.)*
173 **Ocean Circulation: Mechanisms and Impacts—Past and Future Changes of Meridional Overturning** *Andreas Schmittner, John C. H. Chiang, and Sidney R. Hemming (Eds.)*
174 **Post-Perovskite: The Last Mantle Phase Transition** *Kei Hirose, John Brodholt, Thorne Lay, and David Yuen (Eds.)*
175 **A Continental Plate Boundary: Tectonics at South Island, New Zealand** *David Okaya, Tim Stem, and Fred Davey (Eds.)*
176 **Exploring Venus as a Terrestrial Planet** *Larry W. Esposito, Ellen R. Stofan, and Thomas E. Cravens (Eds.)*
177 **Ocean Modeling in an Eddying Regime** *Matthew Hecht and Hiroyasu Hasumi (Eds.)*
178 **Magma to Microbe: Modeling Hydrothermal Processes at Oceanic Spreading Centers** *Robert P. Lowell, Jeffrey S. Seewald, Anna Metaxas, and Michael R. Perfit (Eds.)*
179 **Active Tectonics and Seismic Potential of Alaska** *Jeffrey T. Freymueller, Peter J. Haeussler, Robert L. Wesson, and Göran Ekström (Eds.)*
180 **Arctic Sea Ice Decline: Observations, Projections, Mechanisms, and Implications** *Eric T. DeWeaver, Cecilia M. Bitz, and L.-Bruno Tremblay (Eds.)*
181 **Midlatitude Ionospheric Dynamics and Disturbances** *Paul M. Kintner, Jr., Anthea J. Coster, Tim Fuller-Rowell, Anthony J. Mannucci, Michael Mendillo, and Roderick Heelis (Eds.)*
182 **The Stromboli Volcano: An Integrated Study of the 2002–2003 Eruption** *Sonia Calvari, Salvatore Inguaggiato, Giuseppe Puglisi, Maurizio Ripepe, and Mauro Rosi (Eds.)*
183 **Carbon Sequestration and Its Role in the Global Carbon Cycle** *Brian J. McPherson and Eric T. Sundquist (Eds.)*
184 **Carbon Cycling in Northern Peatlands** *Andrew J. Baird, Lisa R. Belyea, Xavier Comas, A. S. Reeve, and Lee D. Slater (Eds.)*
185 **Indian Ocean Biogeochemical Processes and Ecological Variability** *Jerry D. Wiggert, Raleigh R. Hood, S. Wajih A. Naqvi, Kenneth H. Brink, and Sharon L. Smith (Eds.)*
186 **Amazonia and Global Change** *Michael Keller, Mercedes Bustamante, John Gash, and Pedro Silva Dias(Eds.)*
187 **Surface Ocean–Lower Atmosphere Processes** *Corinne Le Quèrè and Eric S. Saltzman (Eds.)*
188 **Diversity of Hydrothermal Systems on Slow Spreading Ocean Ridges** *Peter A. Rona, Colin W. Devey, Jérôme Dyment, and Bramley J. Murton (Eds.)*
189 **Climate Dynamics: Why Does Climate Vary?** *De-Zheng Sun and Frank Bryan (Eds.)*
190 **The Stratosphere: Dynamics, Transport, and Chemistry** *L. M. Polvani, A. H. Sobel, and D. W. Waugh (Eds.)*
191 **Rainfall: State of the Science** *Firat Y. Testik and Mekonnen Gebremichael (Eds.)*
192 **Antarctic Subglacial Aquatic Environments** *Martin J. Siegert, Mahlon C. Kennicut II, and Robert A. Bindschadler*
193 **Abrupt Climate Change: Mechanisms, Patterns, and Impacts** *Harunur Rashid, Leonid Polyak, and Ellen Mosley-Thompson (Eds.)*
194 **Stream Restoration in Dynamic Fluvial Systems: Scientific Approaches, Analyses, and Tools** *Andrew Simon, Sean J. Bennett, and Janine M. Castro (Eds.)*
195 **Monitoring and Modeling the Deepwater Horizon Oil Spill: A Record-Breaking Enterprise** *Yonggang Liu, Amy MacFadyen, Zhen-Gang Ji, and Robert H. Weisberg (Eds.)*
196 **Extreme Events and Natural Hazards: The Complexity Perspective** *A. Surjalal Sharma, Armin Bunde, Vijay P. Dimri, and Daniel N. Baker (Eds.)*
197 **Auroral Phenomenology and Magnetospheric Processes: Earth and Other Planets** *Andreas Keiling, Eric Donovan, Fran Bagenal, and Tomas Karlsson (Eds.)*
198 **Climates, Landscapes, and Civilizations** *Liviu Giosan, Dorian Q. Fuller, Kathleen Nicoll, Rowan K. Flad, and Peter D. Clift (Eds.)*
199 **Dynamics of the Earth's Radiation Belts and Inner Magnetosphere** *Danny Summers, Ian R. Mann, Daniel N. Baker, Michael Schulz (Eds.)*
200 **Lagrangian Modeling of the Atmosphere** *John Lin (Ed.)*
201 **Modeling the Ionosphere-Thermosphere** *Jospeph D. Huba, Robert W. Schunk, and George V. Khazanov (Eds.)*
202 **The Mediterranean Sea: Temporal Variability and Spatial Patterns** *Gian Luca Eusebi Borzelli, Miroslav Gačić, Piero Lionello, Paola Malanotte-Rizzoli (Eds.)*
203 **Future Earth - Advancing Civic Understanding of the Anthropocene** *Diana Dalbotten, Gillian Roehrig, Patrick Hamilton (Eds.)*
204 **The Galápagos: A Natural Laboratory for the Earth Sciences** *Karen S. Harpp, Eric Mittelstaedt, Noémi d'Ozouville, David W. Graham (Eds.)*
205 **Modeling Atmospheric and Oceanic Flows: Insights from Laboratory Experiments and Numerical Simulations** *Thomas von Larcher, Paul D. Williams (Eds.)*
206 **Remote Sensing of the Terrestrial Water Cycle** *Venkat Lakshmi (Ed.)*
207 **Magnetotails in the Solar System** *Andreas Keiling, Caitríona Jackman, Peter Delamere (Eds.)*

Geophysical Monograph 208

Hawaiian Volcanoes

From Source to Surface

Rebecca Carey
Valérie Cayol
Michael Poland
Dominique Weis
Editors

This Work is a co-publication between the American Geophysical Union and John Wiley and Sons, Inc.

AGU
American Geophysical Union™

WILEY

This Work is a co-publication between the American Geophysical Union and John Wiley & Sons, Inc.

Published by John Wiley & Sons, Inc., Hoboken, New Jersey
Published simultaneously in Canada

Library of Congress Cataloging-in-Publication Data

ISBN: 978-1-118-87204-8

Cover image: Halema'uma'u Lava Lake, Kīlauea Volcano, Hawai'i Volcanoes National Park, Hawai'i, USA
(Samuel Holt, University of Tasmania, Australia, December 3, 2013).

Back cover logo: AGU Chapman Conference, 2012, "Hawaiian Volcanoes, From Source to Surface," designed by Ken Hon, University of Hawai'i, Hilo

Printed in the United States of America

10 9 8 7 6 5 4 3 2 1

CONTENTS

CONTRIBUTORS

Richard M. Allen
Department of Earth and Planetary Science, University of California, Berkeley, California, USA

Kyle R. Anderson
Hawaiian Volcano Observatory, U.S. Geological Survey, Hawai'i National Park, Hawaii, USA

Maxim D. Ballmer
Department of Geology and Geophysics, University of Hawai'i at Mānoa, Honolulu, Hawaii, USA
and
Now at Earth-Life Science Institute, Tokyo Institute of Technology, Meguro, Tokyo, Japan

Elspeth Barnes
Pacific Center for Isotopic and Geochemical Research, Department of Earth, Ocean, and Atmospheric Sciences, University of British Columbia, Vancouver, British Columbia, Canada

Pascale Besson
Institut de Physique du Globe de Paris, UMR CNRS 7154, Sorbonne Paris Cité, Université Paris Diderot, Paris, France

Rebecca Carey
School of Physical Sciences and Centre of Excellence in Ore Deposits, University of Tasmania, Hobart, Tasmania, Australia

Katharine V. Cashman
School of Earth Sciences, University of Bristol, Bristol, United Kingdom

Valérie Cayol
Laboratoire Magmas et Volcans, Université de Lyon, CNRS-UJM-UBP-IRD, Saint Etienne, France

Cheng Cheng
Department of Earth and Planetary Science, University of California, Berkeley, California, USA

David A. Clague
Monterey Bay Aquarium Research Institute, Moss Landing, California, USA

Harmony V. Colella
Department of Earth Sciences, University of California at Riverside, Riverside, California, USA
and
Department of Geology, Miami University at Ohio Oxford, Ohio, USA
and
Now at School of Earth and Space Exploration, Arizona State University, Tempe, Arizona, USA

Brian L. Cousens
Department of Earth Sciences, Carleton University, Ottawa, Ontario, Canada

Hannah R. Dietterich
Department of Geological Sciences, University of Oregon, Eugene, Oregon, USA
and
Now at Volcano Science Center, U.S. Geological Survey, Menlo Park, California, USA

James H. Dieterich
Department of Earth Sciences, University of California at Riverside, Riverside, California, USA

Andrea Di Muro
Observatoire Volcanologique du Piton de la Fournaise, Institut de Physique du Globe de Paris, La Plaine des Cafres, La Réunion, France

Marie Edmonds
Department of Earth Sciences, University of Cambridge, Cambridge, United Kingdom

Tamar Elias
Hawaiian Volcano Observatory, U.S. Geological Survey, Hawai'i National Park, Hawaii, USA

Todd Ericksen
Hawai'i Institute of Geophysics and Planetology, University of Hawai'i at Mānoa, Honolulu, Hawaii, USA

Sarah Fagents
Hawai'i Institute of Geophysics and Planetology, University of Hawai'i at Mānoa, Honolulu, Hawaii, USA

Valérie Ferrazzini
Observatoire Volcanologique du Piton de la Fournaise, Institut de Physique du Globe de Paris, La Plaine des Cafres, La Réunion, France

Michael O. Garcia
Department of Geology and Geophysics, University of Hawai'i at Mānoa, Honolulu, Hawaii, USA

Christine Garofalo
Institut de Physique du Globe de Paris, UMR CNRS 7154, Sorbonne Paris Cité, Université Paris Diderot, Paris, France

Yves Gaudemer
Institut de Physique du Globe de Paris, UMR CNRS 7154, Sorbonne Paris Cité, Université Paris Diderot, Paris, France

Diane Hanano
Pacific Center for Isotopic and Geochemical Research, Department of Earth, Ocean, and Atmospheric Sciences, University of British Columbia, Vancouver, British Columbia, Canada

Andrew J. L. Harris
Laboratoire Magmas et Volcans, Université Blaise Pascal, Clermont Ferrand, France

Lauren Harrison
Pacific Center for Isotopic and Geochemical Research, Department of Earth, Ocean, and Atmospheric Sciences, University of British Columbia, Vancouver, British Columbia, Canada

Christina Heliker
Hawaiian Volcano Observatory, U.S. Geological Survey, Hawai'i National Park, Hawaii, USA

Rosalind T. Helz
U.S. Geological Survey, Reston, Virginia, USA

Richard P. Hoblitt
Cascades Volcano Observatory, U.S. Geological Survey, Vancouver, Washington, USA

Bruce F. Houghton
Department of Geology and Geophysics, University of Hawai'i at Mānoa, Honolulu, Hawaii, USA

Garrett Ito
Department of Geology and Geophysics, University of Hawai'i at Mānoa, Honolulu, Hawaii, USA

Scot K. Izuka
Pacific Islands Water Science Center, U.S. Geological Survey, Honolulu, Hawaii, USA

Jessica H. Johnson
Hawaiian Volcano Observatory, U.S. Geological Survey, Hawai'i National Park, Hawaii, USA
and
Center for the Study of Active Volcanoes, University of Hawai'i, Hilo, Hawaii, USA
and
School of Earth Sciences, University of Bristol, Bristol, United Kingdom
and
Now at School of Environmental Sciences, University of East Anglia, Norwich, United Kingdom

Benjamin H. Mackey
Department of Geological Sciences, University of Canterbury, Christchurch, New Zealand

John Maclennan
Department of Earth Sciences, University of Cambridge, Cambridge, United Kingdom

Larry G. Mastin
Cascades Volcano Observatory, U.S. Geological Survey, Vancouver, Washington, USA

Michael May
Department of Geology, University of Otago, Dunedin, New Zealand

Nicole Métrich
Institut de Physique du Globe de Paris, UMR CNRS 7154, Sorbonne Paris Cité, Université Paris Diderot, Paris, France

Asta Miklius
Hawaiian Volcano Observatory, U.S. Geological Survey, Hawai'i National Park, Hawaii, USA

Emily K. Montgomery-Brown
California Volcano Observatory, U.S. Geological Survey, Menlo Park, California, USA

Tim R. Orr
Hawaiian Volcano Observatory, U.S. Geological Survey, Hawai'i National Park, Hawaii, USA
and
Department of Geology and Geophysics, University of Hawai'i at Mānoa, Honolulu, Hawaii, USA

Carolyn Parcheta
Department of Geology and Geophysics, University of Hawai'i at Mānoa, Honolulu, Hawaii, USA
and
Now at Jet Propulsion Laboratory, Pasadena, California, USA

Matthew R. Patrick
Hawaiian Volcano Observatory, U.S. Geological Survey, Hawai'i National Park, Hawaii, USA

Aline Peltier
Observatoire Volcanologique du Piton de la Fournaise, Institut de Physique du Globe de Paris, La Plaine des Cafres, La Réunion, France

Michael Poland
Hawaiian Volcano Observatory, U.S. Geological Survey, Hawai'i National Park, Hawaii, USA

Rob W. Porritt
Department of Earth and Planetary Science, University of California, Berkeley, California, USA
and
Now at Department of Earth Science, University of Southern California, Los Angeles, California, USA

J. Michael Rhodes
Department of Geosciences, University of Massachusetts, Amherst, Massachusetts, USA

Diana C. Roman
Department of Terrestrial Magnetism, Carnegie Institution for Science, Washington, D.C., USA

Timothy R. Rose
National Museum of Natural History, Smithsonian Institution, Washington, D.C., USA

Michael C. Rowe
School of Environment, University of Auckland, Auckland, New Zealand

Scott K. Rowland
Department of Geology and Geophysics, University of Hawai'i at Mānoa, Honolulu, Hawaii, USA

David R. Sherrod
Cascades Volcano Observatory, U.S. Geological Survey, Vancouver, Washington, USA

Isobel Sides
Department of Earth Sciences, University of Cambridge, Cambridge, United Kingdom

S. Adam Soule
Geology and Geophysics Department, Woods Hole Oceanographic Institution, Woods Hole, Massachusetts, USA

Thomas Staudacher
Observatoire Volcanologique du Piton de la Fournaise, Institut de Physique du Globe de Paris, La Plaine des Cafres, La Réunion, France

A. Jeff Sutton
Hawaiian Volcano Observatory, U.S. Geological Survey, Hawai'i National Park, Hawaii, USA

Donald A. Swanson
Hawaiian Volcano Observatory, U.S. Geological Survey, Hawai'i National Park, Hawaii, USA

Lauren Swavely
Department of Geology and Geophysics, University of Hawai'i at Mānoa, Honolulu, Hawaii, USA

Weston A. Thelen
Hawaiian Volcano Observatory, U.S. Geological Survey, Hawai'i National Park, Hawaii, USA

Carl R. Thornber
Cascades Volcano Observatory, U.S. Geological Survey, Vancouver, Washington, USA

Sylvie Vergniolle
Institut de Physique du Globe de Paris, UMR CNRS 7154, Sorbonne Paris Cité, Université Paris Diderot, Paris, France

Benoit Villemant
Institut de Sciences de la Terre de Paris, Université Pierre et Marie Curie, Paris, France

Dominique Weis
Pacific Center for Isotopic and Geochemical Research, Department of Earth, Ocean, and Atmospheric Sciences, University of British Columbia, Vancouver, British Columbia, Canada

Kelly M. Wooten
Hawaiian Volcano Observatory, U.S. Geological Survey, Hawai'i National Park, Hawaii, USA

and

Department of Geological and Mining Engineering and Sciences, Michigan Technological University, Houghton, Michigan, USA

PREFACE

In addition to being a popular tourist destination and a source of inspiration for artists, writers, and explorers, Hawaiian volcanoes provide essential data for the foundation of many tenets of Earth science. During one of the first scientific expeditions to Hawai'i, in 1840, American geologist James Dwight Dana recognized that volcanoes like Kīlauea and Mauna Loa should "share equally ... the attention of investigators" with the much-better-known (at that time) Italian volcanoes of Vesuvius and Etna. Since Dana's first visit to the islands, Hawai'i has enjoyed a standing of one of the best places in the world to study volcanoes. Indeed, research conducted in the archipelago has had a profound impact on the development of not just volcanology but all of Earth science.

Frequent eruptive activity and ease of access have established Hawai'i as an excellent natural laboratory, which is why Dr. Thomas Jaggar chose to build a facility for continuous observation of earthquakes and volcanoes, today's Hawaiian Volcano Observatory, at the summit of Kīlauea Volcano in 1912. From that time forward, Hawaiian volcanoes have served as testing grounds for models of mantle plumes, as sites for investigations of magma ascent, storage, transport, and eruption, and as analogs for other volcanoes on Earth and across the solar system. Hawai'i is the type location of hotspot volcanism and is a key element in models of global tectonics and mantle structure. In addition, activity of Hawaiian volcanoes has spurred technological developments that have led to outstanding advances in geophysical methods.

Since the start of the new millennium, the volcanoes of Hawai'i have been the targets of numerous major multidisciplinary studies, encompassing disciplines from chemistry to physics, depths from the core-mantle boundary to the atmosphere, and environments from submarine to subaerial. Drilling studies have recovered several kilometers of core representing hundreds of thousands of years of previously hidden volcanic history, which sheds light on models of magma supply and hotspot dynamics; submarine mapping and sampling has provided constraints on the age and physiochemical evolution of individual volcanoes; large-scale submarine/subaerial seismic deployments have imaged the deep roots of Hawai'i; and unprecedented eruptive activity coupled with modern geophysical/geological/geochemical instrumentation and techniques has elucidated the characteristics of shallow magma storage and transport and the mechanisms of eruption.

These and other recent scientific developments motivated the American Geophysical Union Chapman Conference "Hawaiian Volcanoes, From Source to Surface," held in Waikoloa Beach on the Island of Hawai'i during 20–24 August 2012, the centennial year of the Hawaiian Volcano Observatory. The goal of the conference was to highlight the current understanding of how Hawaiian volcanoes work and to stimulate multidisciplinary discussions about future directions for studies of basaltic volcanoes in Hawai'i and elsewhere. About 180 scientists from 12 countries were in attendance, including approximately 40 students who represented the next generation of Hawai'i volcano researchers.

This monograph, which represents a sampling of the themes presented at the Chapman Conference, describes the current knowledge of Hawaiian volcanism, from its origin within Earth to its physical and chemical manifestations at the surface. While most of the research articles contained in the pages that follow are specific to Hawai'i, the lessons learned are applicable to a range of volcano types and locations and should therefore be of broad interest to the geoscience community.

An important outcome of the Chapman Conference was the recognition that better communication and coordination are needed among researchers who focus on different disciplines and aspects of how Hawaiian volcanoes work. Our intent in organizing this volume was therefore to parallel the structure of the conference, that is, to trace a packet of magma from its point of origin within the mantle, through ascent into subvolcanic storage reservoirs and plumbing systems, transport toward the surface, and finally to eruption. We eschew disciplinary bounds in favor of this more integrated approach, which recognizes, for example, that the gas plume emanating from Kīlauea's summit is ultimately controlled by the chemistry of the mantle and that most of the processes occurring at depth can only be inferred from measurements made at the surface.

The volume opens with a reflection on how studies of Hawaiian volcanoes and their deep roots have helped shape our understanding of Earth processes, from the deep mantle to the atmosphere. The 22 chapters that follow begin within the mantle and gradually ascend toward the surface, providing insights into the chemical and physical manifestations of Hawaiian volcanism along the way. The monograph closes with a discussion of the major questions that continue to surround Hawaiian volcanism, based not only on the preceding chapters but also on breakout discussions from the Chapman Conference. As such, that chapter represents something of a "white paper" that details critical research questions that are currently preventing improved understanding of how Hawaiian volcanoes work and that should be targets for future study.

We would like to extend our appreciation to the many people and organizations that made the Chapman Conference a memorable and successful event, first and foremost to the attendees that participated in discussions and shared the results of their research. We are grateful for the time and energy that the other meeting organizers dedicated to the project: Paul Okubo (U.S. Geological Survey), Ken Hon (University of Hawai'i at Hilo), Simon Carn (Michigan Technological University), Helge Gonnermann (Rice University), and Scott Rowland (University of Hawai'i at Mānoa). Funding was generously made available by the National Science Foundation, U.S. Geological Survey, and International Association for Volcanology and Chemistry of the Earth's Interior, and the American Geophysical Union meetings staff provided exceptional service in support of conference operations. Finally, we thank the authors and reviewers of the chapters that follow for their enthusiasm and patience as this monograph was assembled. We are confident that it meets the goal of providing a benchmark for understanding Hawai'i's volcanoes and that it will serve as an important resource and guide for future studies in years to come.

Rebecca Carey
University of Tasmania

Valérie Cayol
Centre National de Recherche Scientifique

Michael Poland
U.S. Geological Survey

Dominique Weis
University of British Columbia

ABOUT THE COMPANION WEBSITE

This book is accompanied by a companion website:

www.wiley.com/go/Carey/Hawaiian_Volcanoes

The website includes figures, tables and appendices from book chapters 4, 5, 7, 8, 12, 18, 20 and videos from chapter 18.

1

How and Why Hawaiian Volcanism Has Become Pivotal to Our Understanding of Volcanoes from Their Source to the Surface

Michael O. Garcia

ABSTRACT

Hawai'i is a superb venue for volcanological research because of the high frequency and easy access to eruptions and its unique geological setting in the central Pacific far from plate boundaries or continents. Studies of Hawaiian volcanoes and their deep roots have helped shape our understanding of Earth processes from the deep mantle to the atmosphere. Since the creation of the Hawaiian Volcano Observatory in 1912, research on Hawaiian volcanoes has led to many fundamental discoveries. A few of these discoveries spanning the range of topics discussed in this volume are reviewed here: the role of Hawaiian geology in understanding mantle-derived magmas; the importance of helium isotopes in determining the nature and structure of the mantle; the physical controls on lava flow type, emplacement, and how far they will travel; and Hawaiian submarine geology, including the newest volcano (Lō'ihi) and the giant Nu'uanu landslide. A critical knowledge level has been achieved about Hawaiian volcanoes allowing scientists to test hypotheses that address essential issues about the way Earth works. Working on Hawaiian volcanoes is a humbling experience not only due to their grandeur but also because with new knowledge we realize how much is still not well understood.

1.1. INTRODUCTION

This volume, *Hawaiian Volcanoes: From Source to Surface,* is the third American Geophysical Union monograph devoted to Hawaiian volcanism. The other two are: monograph 92, *Mauna Loa Revealed: Structure, Composition, History and Hazards* [*Rhodes and Lockwood*, 1995], and monograph 128, *Hawaiian Volcanoes: Deep Water Perspectives* [*Takahashi et al.*, 2002]. This is a remarkable testament to the importance and interest of Hawaiian volcanism to the geosciences community. Why has research on Hawaiian volcanism warranted this attention?

In this chapter, a personal perspective is given on the influences that may have led to the focus of many early 20th century studies on Hawaiian volcanoes. A review is presented of four broad areas of discovery that have been made about fundamental magmatic and other geological processes based on studies of Hawaiian volcanoes. These discoveries have led to Hawai'i becoming the "type example" for many magmatic and volcanic phenomena. Studies of Hawaiian volcanoes and their deep roots have and are continuing to shape our understanding of Earth processes from the deep mantle to the atmosphere (source to surface).

Three questions will be addressed below: (1) What attracted early 20th century geoscientists to Hawai'i and made it a preferred site for studying volcanism? (2) Why has research on Hawaiian volcanism been so important to our overall understanding of magma generation, evolution, and eruption? (3) What are some of the noteworthy but perhaps not-so-obvious discoveries that have come from studying Hawaiian volcanoes since the founding of the Hawaiian Volcano Observatory (HVO) in 1912? These discoveries include the origin of magma series and the melting history of Hawaiian volcanoes, primitive noble gases (helium), the undersea volcano Lō'ihi, the giant submarine Nu'uanu landslide, and lava flow dynamics and emplacement. Many other exciting topics related

Deptartment of Geology and Geophysics, University of Hawai'i at Mānoa, Honolulu, Hawaii, USA

Hawaiian Volcanoes: From Source to Surface, Geophysical Monograph 208. First Edition.
Edited by Rebecca Carey, Valérie Cayol, Michael Poland, and Dominique Weis.

Companion Website: www.wiley.com/go/Carey/Hawaiian_Volcanoes

to research on Hawaiian volcanoes are not included in this review, although some are discussed in this volume.

1.2. WHAT HAS ATTRACTED VOLCANOLOGISTS TO HAWAI'I?

The allure of active volcanism has long beckoned scientists to the Hawaiian Islands. What they found has brought them back and, in some cases (G. A. Macdonald), led them to stay. Many factors have contributed to Hawai'i becoming a premier venue for studying volcanic processes: (1) Eruptions are frequent, mostly quiescent and easily accessible [e.g., *Dana*, 1890; *Macdonald et al.*, 1983]. (2) The tropical setting allows year-round observations. (3) Excellent publicity about Hawaiian volcanoes by newspaper and book writers (e.g., Mark Twain and Isabella Bird), artists who painted majestic scenes of the volcanoes (Figure 1.1), and national lobbying efforts by citizens of Hawai'i who were interested in volcanology. (4) Hawai'i is in the United States (since its annexation in 1898 [*Daws*, 1968]) and has for several decades had frequent, direct air service from many major Pacific and North American cities. (5) Hawaiians and other local residents have great *aloha* for visitors.

The Hawaiian Islands have been called "the loveliest fleet of islands that lies anchored in any ocean" [Mark Twain, in *Frear*, 1947]. Thus, despite the considerable distance of the islands from any continent (e.g., ~3800 km from California), Hawai'i has drawn volcanologists and geologists for more than 170 years. The first geologist to visit Hawai'i was James D. Dana, who became one of the giants of American geology and its first volcanologist [*Appleman*, 1987]. He arrived with the U.S. Exploring Expedition in 1840 after a long sea voyage from Fiji and just following a spectacular eruption on Kīlauea's East Rift Zone [*Appleman*, 1987]. The chief naturalist on this expedition was the noted American entomologist and artist Titian Peale, who produced many sketches and beautiful paintings of Kīlauea Volcano. His work is part of what has become known as the "Volcano School," a group of artists who painted dramatic nocturnal scenes of Kīlauea and Mauna Loa volcanoes. The French painter Jules Tavernier (1844–1889)] was arguably the most important member of the Volcano School. His painting of a fiery night at Kīlauea in the 1880s stirs the souls of geologists and other volcanophiles (Figure 1.1). The school included an international array of artists: Ernst Christmas (Australia, 1863–1918), Constance Cumming (Scotland, 1837–1924), Charles Furneaux (America, 1835–1913), D. Howard Hitchcock (America, 1861–1943), and Ogura Itoh (Japan, 1870–1940). Their paintings brought the beauty and excitement of active Hawaiian volcanoes to the attention of many across the United States and around the world.

Before his first visit to Hawai'i, Dana visited other volcanic areas, including the frequently active Italian volcanoes (Stromboli and Vesuvius), which were both erupting during his visit [*Dana*, 1835]. He also explored

Figure 1.1 Painting of Kīlauea Volcano entitled "Volcano at Night," on canvas by Jules Tavernier, circa 1885–1889. Used by permission from the Honolulu Museum of Art.

other oceanic Islands, including the Cape Verdes, Tahiti, Fiji and, Samoa, prior to arriving in Hawai'i. Dana made a second visit to Hawai'i and Italy in preparation for his insightful and well-illustrated book *Characteristics of Volcanoes with Contributions of Facts and Principles from the Hawaiian Islands* [*Dana*, 1890]. In this book, he compared and contrasted the features and activity of Hawaiian volcanoes with those of Vesuvius and Etna. Although the Italian volcanoes were better known in the 19th century, Dana felt that Italy and Hawai'i should share equally in the attention of scientific investigators. *Dana* [1890] promoted visiting the Hawaiian Islands by noting that only a two-week voyage was required from New York (three weeks from Europe) to visit the "great, open, free-working craters of Hawai'i." He also noted the usually quiet way Hawaiian volcanoes send forth lava streams 30–50 km long, creating a "peculiarly instructive field for the student of volcanic science, as well as attractive to the lover of the marvelous."

The early missionaries also played an important role in making the presence of Hawai'i's active volcanoes known and exciting to the western world. Reverends A. Bishop, C. S. Stewart, W. Ellis, J. Goodrich, and T. Coan published books and/or articles on their observations of eruptions of Kīlauea and Mauna Loa volcanoes. Most notable was Reverend Coan, who was an avid eruption watcher and good friend of James Dana. The two corresponded regularly. Dana, who was an editor for the *American Journal of Science*, published "with comments" several of Coan's reports on eruptions of Kīlauea and Mauna Loa in the journal [e.g., *Dana*, 1850]. Coan made many fundamental observations about volcanic processes, including the importance of lava tubes (which he called "pyroducts") for transporting lava ~20 km from its vent during Mauna Loa's 1843 eruption. He observed this eruption after hiking for four days through the jungle. Dana disagreed with this report and published his opinion in the same article, arguing the eruption must have been fed by a long fissure [*Dana*, 1850]. The eyewitness (Coan) was correct about the importance of lava tubes in transporting fluid lava 10 km or more from its primary vent. Several other reports by *Coan* [1856, 1869, 1880] were also published in the *American Journal of Science* (without comment) after notable eruptions of Kīlauea or Mauna Loa.

The books and articles by Dana, missionaries and others, and the paintings of the Volcano School artists undoubtedly inspired many other pioneering volcanologists and geologists to visit and report on Hawai'i's active volcanoes and their lavas, including *Daly* [1911], *Hitchcock* [1911], *Jaggar* [1912], *Perret* [1913], *Cross* [1915], and *Powers* [1915]. These early studies laid the foundation and were a motivation for subsequent work, including the diverse studies that are presented in this monograph. One insightful comment in the preface to *Dana*'s [1890] book is still true today: "But much remains to be learned from the further study of the Hawaiian volcanoes." The chapters in this monograph do much to increase our knowledge of how volcanoes work, although they also raise many questions for future research.

The establishment of the HVO in 1912 was a critical milestone in the eventual emergence of Hawaiian volcanism as a classic example for many volcanic processes. After the catastrophic eruptions of Mt. Pelée (Lesser Antilles) in 1902 and Krakatau (Indonesia) in 1883, where tens of thousands of people were killed, there was a heightened awareness that a volcano observatory was needed to monitor and learn more about volcanoes. In 1909, Thomas Jaggar and Reginald Daly, both professors at the Massachusetts Institute of Technology, visited Hawai'i and developed the idea of establishing a permanent site for a volcano observatory on Kīlauea [*Wright et al.*, 1992]. In 1911, Jaggar recruited Frank Perret, an entrepreneur and inventor (he worked for Edison as a teenager and received a patent for his electric fan) and a self-taught volcanologist, to begin the work of establishing an observatory. The new observatory was permanently established during the following year and involved the help of many other scientists, including Fusakichi Omori, the Japanese pioneer in seismology whom Jaggar visited after his 1909 trip to Kīlauea [*Apple*, 1987], and gas sampling and analysis experts E. Shepard and A. Day, from the Carnegie Institution [e.g., *Shepard*, 1925]. For more information about HVO, see the new compilation by *Tilling et al.* [2014] as well as the pictorial history of Hawaiian volcanism by *Wright et al.* [1992].

The creation of Hawai'i National Park in 1916 (the name was changed to Hawai'i Volcanoes National Park in 1961) was another notable event that brought attention to the active volcanoes in Hawai'i. Jaggar was a vocal proponent in the creation of the new park. He gave guided tours of Kīlauea to visiting U.S. Congress members, spoke to many groups about the need for the park, and even went to Washington, D.C., to testify. In his testimony, *Jaggar* [1916] emphasized that "There is no place on the globe more favorable for the systematic study of volcanology and the relations of local earthquakes to volcanoes as in Hawai'i, and for this reason alone, if for no other, it would be appropriate to set aside a national park in this wonderland of volcanic activity, where the earth's primitive processes are at work, making new land and adding new gases to the atmosphere." Within eight months of his testimony, President Wilson signed the bill creating Hawai'i National Park, the 12th national park in the United States [*Wright et al.*, 1992].

Several key additional factors were critical for the emergence of Hawai'i as a premier venue for studying magmatic processes. Hawai'i's distant location from any continent or plate boundary has led to its selection as a site for testing many hypotheses, including the anatomy of hotspot volcanoes via deep scientific drilling to 3.5 km [*Garcia et al.*, 2007; *Stolper et al.*, 2009] and whether mantle plumes extend into the deep mantle [*Wolfe et al.*, 2009; *Weis et al.*, 2011]. In comparison to other mantle plumes, the Hawaiian mantle plume is considered the hottest [*Sleep*, 1990]. The high temperature of the plume is

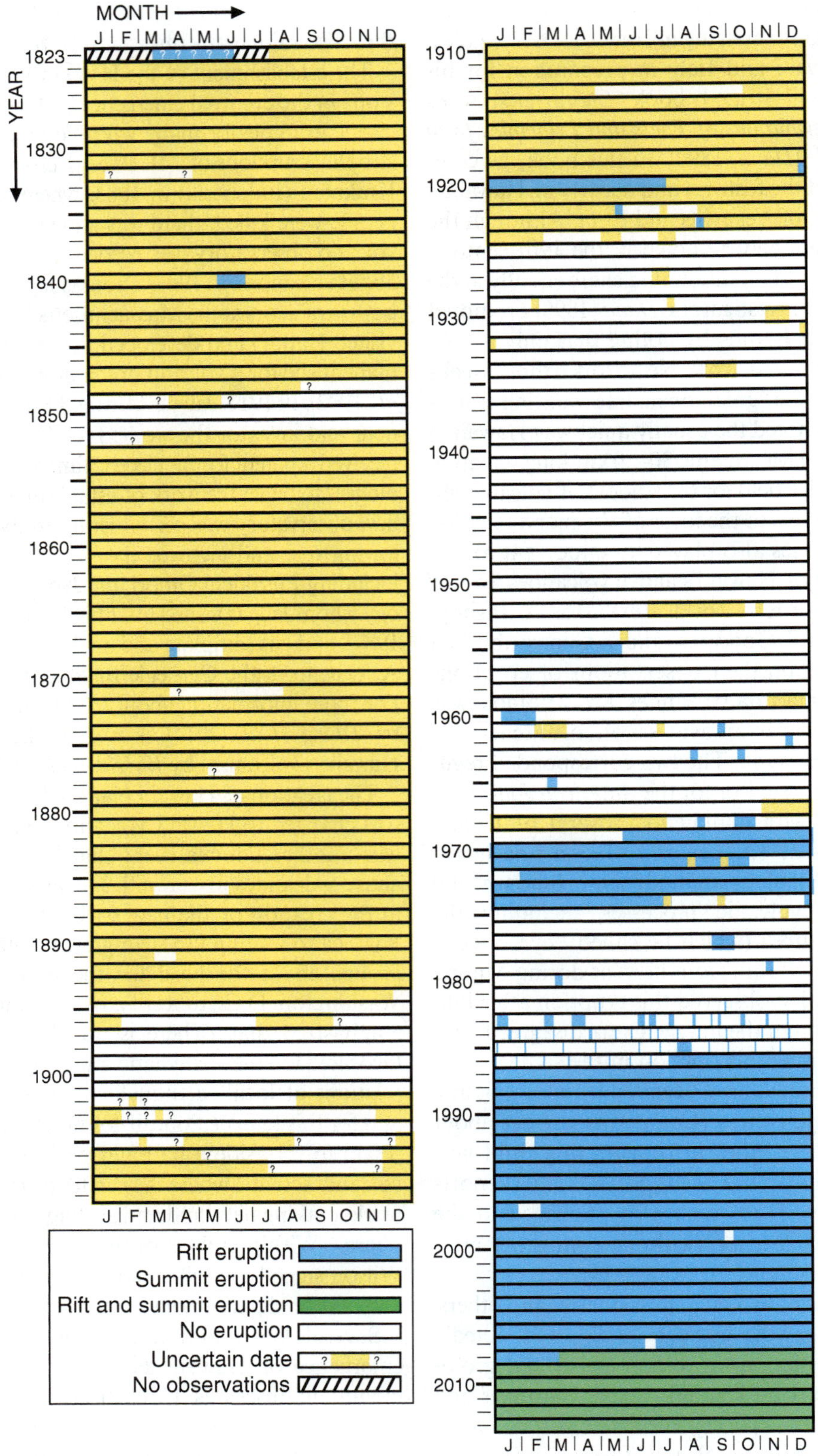

Figure 1.2 Graphic summary of the historical eruptive activity at Kīlauea Volcano from the first written observations in August 1823 to December 2013. Activity is subdivided by location (summit shown in gold and rift zones in blue). Eruptive activity since March 2008 (shown in green) has been nearly continuous both at the summit in Halema'uma'u Crater and in and around the Pu'u 'Ō'ō vent along the East Rift Zone. Modified from *Macdonald and Eaton* (1964) and *Garcia et al.* [2003], based on descriptions by *Dana* [1890], *Brigham* [1909], *Fiske et al.* [1987], *Bevens et al.* [1988], *Macdonald et al.* [1983], *Heliker and Mattox* [2003], and T. Orr (2013, personal communication).

undoubtedly responsible for Hawai'i's frequent eruptions. Since westerners arrived in Hawai'i and began recording eruptive activities ~190 years ago, Kīlauea has been prolific (Figure 1.2), which has promoted volcanological research on and tourism to the Hawaiian Islands.

1.3. HOW HAVE STUDIES OF HAWAIIAN VOLCANOES INFLUENCED OUR KNOWLEDGE OF MAGMATIC AND VOLCANIC PROCESSES?

Many discoveries have been made about fundamental Earth processes as a result of research on Hawaiian volcanoes. Rather than provide a list of these discoveries, four examples are discussed below to give a flavor of their diversity and impact on our understanding of geological phenomena. Obvious examples that are well documented elsewhere are not included (e.g., mantle plumes and volcano monitoring; see Chapter 24). For some of these examples given below, I have participated in the research (melting history and marine geology), whereas others I have observed with interest (noble gases and lava flows).

1.3.1. Origin of Magma Series and Melting History of Hawaiian Volcanoes

One fundamental geological controversy in which studies of Hawaiian lavas played an early and pivotal role had to do with the origin of magma series. This story is superbly told by *McBirney* [1993] and is summarized here. Field work on the Eocene Scottish volcanic centers in the 19th century led to the realization that alkalic basaltic rocks are the dominant rock type, whereas tholeiitic rocks are minor in abundance and mostly differentiated [*Bailey et al.*, 1924]. This observation led to the hypothesis that alkaline magmas were primary and mantle-derived, and the later tholeiitic lavas were formed as a result of contamination of alkaline magmas by continental crust (siliceous metamorphic rocks and granites [*McBirney*, 1993]). Debate on this issue raged until *Lacroix* [1928] reported that the huge shield volcanoes in Hawai'i are made of tholeiitic lavas with only minor alkalic lavas, and no continental crust is available for contamination to form tholeiitic lavas. This story was refined by *Macdonald* [1963] based on extensive field studies and collaboration with T. Katsura for geochemical work [e.g., *Macdonald and Katsura*, 1964]. They showed that on the subaerial Hawaiian shield volcanoes, alkalic lavas were found only as late-stage caps, typically comprising ~1 vol% of the volcano, with the bulk of the volcano being tholeiitic [*Macdonald and Abbott*, 1977]. Laboratory work that used observations from Hawai'i showed that both types of magma can be primary and that depth and extent of melting are the determining variables in controlling whether alkalic (deep and low degree of melting) or tholeiitic magma (shallower and higher degree of melting) is produced [e.g., *Green and Ringwood*, 1967].

The Hawaiian magma evolution story changed dramatically with the discovery of an active volcano south of the Island of Hawai'i (Lō'ihi; Figure 1.3), which is told in more detail below (Section 1.3.3.1). Lō'ihi was found to have both tholeiitic and alkalic lavas [*Moore et al.*, 1982]. Detailed sampling via submersible led to the recognition that the tholeiites form only a thin veneer overlying a base of weakly to strongly alkalic lavas (see summary by *Garcia et al.* [2006a] and references cited therein). This critical observation led to a petrogenetic melting model for Hawaiian volcanoes [*Frey et al.*, 1990]. This model, revised with new ages and volumes (Figure 1.4), is consistent with numerical modeling studies of Hawaiian volcanoes [e.g., *Ribe and Christensen*, 1999]. It provides a context for explaining the observed variations in rock type during the growth of Hawaiian volcanoes. The model is simplistic and does not account for large, short-term variations in magma supply rate (e.g., an order of magnitude for historical Kīlauea [*Pietruszka and Garcia*, 1999]). Low magma supply and infrequent eruption at the initial and end stages of growth of Hawaiian volcanoes are correlated with low degrees of melting and eruption of alkaline lavas. Higher magma supply and more frequent eruptions occur when the volcano is centered over the hotspot and tholeiitic lavas are erupted, producing >90% of the volcano (Figure 1.4). This model was refined for the postshield stage using a combination of field work, theoretical petrology, and geochemistry by Fred Frey (with colleagues and students). They developed a conceptual model that related the reduced magma supply during the waning stages of volcanism to the eruption of nearly aphyric, differentiated lavas (hawaiite-trachyte). They hypothesized that the conduit supplying magma to the shallow reservoir system of the volcano solidified, leading to ponding of alkalic basaltic magma at or near the MOHO where it underwent crystal fractionation to produce lower density (buoyant) magma [*Frey et al.*, 1991]. The Hawaiian evolutionary growth model (Figure 1.4) has been used to explain magma-type variations on other oceanic islands, including the Canaries [*Hoernle and Schmincke*, 1993], Easter Island [*Haase et al.*, 1997], and Réunion [*Albarede et al.*, 1997], as well as continental basalt sequences around the world [e.g., *Xing et al.*, 2011; *Lease et al.*, 2008].

1.3.2. Noble Gases–Helium Isotopes

The study of noble gases has greatly enhanced our understanding of the chemical heterogeneity and structure of Earth's mantle and the origin of Earth's atmosphere [e.g., *Graham*, 2002; *Moreira*, 2013]. Noble gases have been inert to biological and chemical reactions during Earth's history, making them superb tracers of mantle processes [e.g., *Moreira and Kurz*, 2013]. Ocean island basalts have played a key role in noble gas research because their lavas avoid the contaminating effects of continental crust, allowing them to serve as excellent windows into the mantle. The preferred sites for trapping

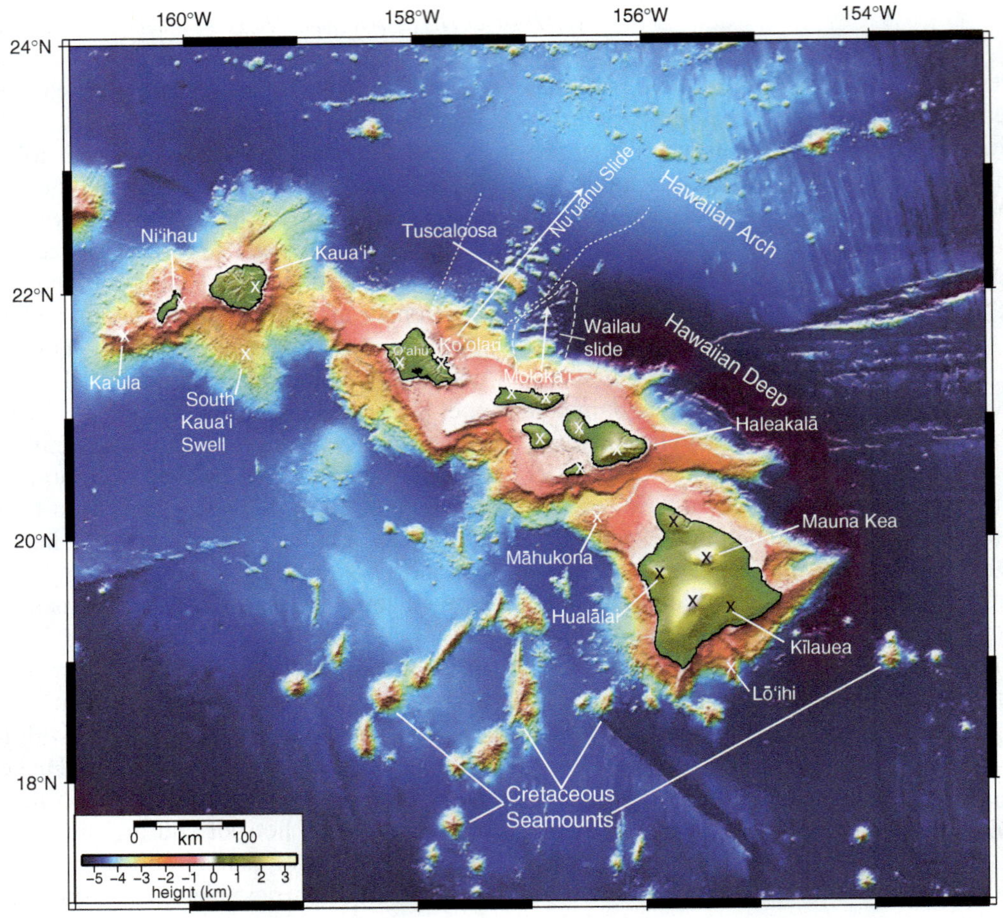

Figure 1.3 Bathymetric map of the Hawaiian Islands region with features discussed in the text labeled. Summits or centers of volcanoes are labeled with an "x." The Hawaiian Deep is the dark blue area surrounding the islands. The Hawaiian Swell is the light blue area to the north of the Deep. The base map was created by Garrett Ito and used by permission.

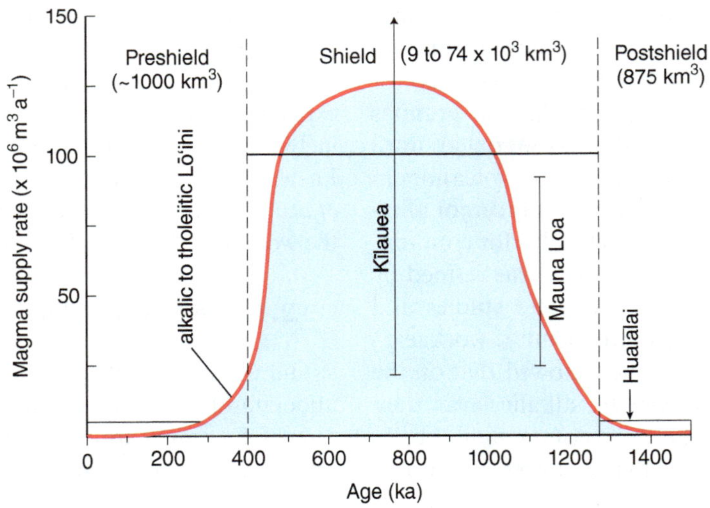

Figure 1.4 Growth history model for a Hawaiian shield volcano. This composite model is based on volume and age estimates for the preshield stage [*Garcia et al.*, 1995; *Guillou et al.*, 1997], shield [*Robinson and Eakins*, 2006; *Frey et al.*, 1990], and postshield [*Frey et al.*, 1990]. Magma supply rate estimates (vertical bars) are from *Pietruszka and Garcia* [1999] and *Poland et al.* [2012] for Kīlauea, *Wanless et al.* [2006] for Mauna Loa, and *Moore et al.*, [1987] for Hualalai.

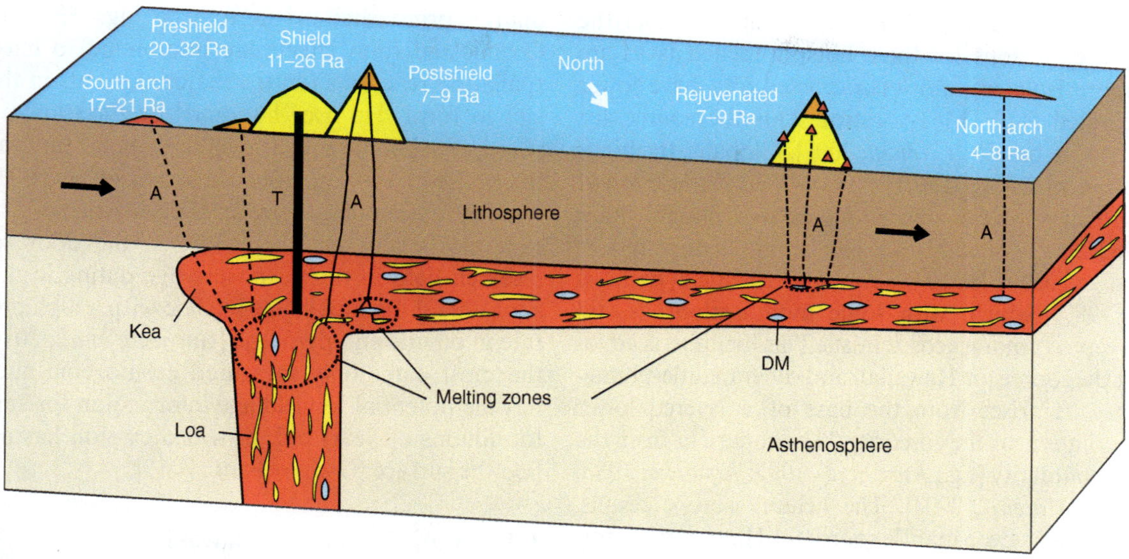

Figure 1.5 Schematic cross section of the Hawaiian mantle plume drawn through the Kea side of the Hawaiian Islands. The plume contains at least three distinct components: the plume matrix is considered the Kea component. The other two dominant components are the Loa (yellow blobs) and depleted mantle (DM, blue blobs). Overlying the schematic are He isotope values ($^3He/^4He$) given in Ra (atmospheric ratio). Arrows show direction of movement of the lithosphere. The South Arch alkaline lavas have relatively high $^3He/^4He$ values. Lō'ihi lavas have the highest $^3He/^4He$ values. Downstream from Lō'ihi, $^3He/^4He$ values decrease, becoming identical to those of mid-ocean ridge basalt (MORB) for the postshield and rejuvenated stages and the North Arch (A, alkalic, T, tholeiite). Schematic after *Garcia et al.* [2010].

noble gases within basalts are the glassy rinds of submarine lavas or melt inclusions in olivine and pyroxene phenocrysts [e.g., *Lupton*, 1983]. Among the noble gases, helium isotopes have received the most attention in the study of mantle heterogeneity because excess ^3He is an indicator of primitive mantle reservoirs and is ubiquitously present in oceanic lavas [*Graham*, 2002]. The other isotope of helium, ^4He, is mostly radiogenic in origin, produced primarily from the decay of U and Th. Helium has a short atmospheric residence time (~1 Myr) and is thought to be lost from the slab prior to subduction [*Lupton*, 1983]. Thus, its concentration is lowered but the $^3He/^4He$ ratio of the mantle is not substantially changed by subduction [*Gonnermann and Mukhopadhyay*, 2009].

A wealth of He isotope data has been collected over the last 35 years from ocean island basalts since the discovery of elevated $^3He/^4He$ ratios in midocean ridge and back-arc basin basalts (9–12 times the atmospheric ratio, Ra [*Krylov et al.*, 1974; *Lupton and Craig*, 1975]). The first indication that Hawaiian magmas have higher $^3He/^4He$ ratios than midocean ridge basalt came from analysis of fumarole gas from Kīlauea, which yielded a $^3He/^4He$ ratio of 15 Ra [*Craig and Lupton*, 1976]. The early noble gas studies on Hawaiian rocks examined phenocrysts in lavas from Kīlauea and Haleakalā volcanoes and mantle xenoliths from Salt Lake Crater on O'ahu and Hualālai volcano [e.g., *Kaneoka and Takaoka*, 1978; 1980; *Kyser and Rison*, 1982]. They found elevated $^3He/^4He$ ratios in the lavas, which implies a primitive source for Hawaiian magmas, and that the xenoliths were not genetically related to Hawaiian tholeiites.

The discovery and sampling of Lō'ihi Seamount led to a frenzy of noble gas work on Hawaiian rocks after the first reported helium measurements revealed very high $^3He/^4He$ ratios (up to 32 Ra [*Kurz et al.*, 1982]). Subsequent studies of Lō'ihi samples showed a wide range of $^3He/^4He$ ratios (20–32 Ra [*Kurz et al.*, 1983; *Kaneoka*, 1983; *Rison and Craig*, 1983]) with no apparent correlation with age (e.g., lavas from the 1996 Lō'ihi eruption yielded $^3He/^4He$ ratios of 26 Ra, midway in the range [*Garcia et al.*, 1998]). The wide range in $^3He/^4He$ ratios showed that the Lō'ihi's source is heterogeneous, which is consistent with results for other radiogenic isotopes [*Garcia et al.*, 2006a]. Studies of older Hawaiian volcanoes generally yielded lower ratios (mostly <26 Ra) and an overall trend of decreasing $^3He/^4He$ ratios with decreasing age during shield and continuing into postshield volcanism [e.g., *Kurz et al.*, 1987, 1995, 2004] (Figure 1.5). The postshield transitional lavas from Māhukona are an exception to this temporal trend, with their relatively high $^3He/^4He$ ratios (16–21 Ra) and Lō'ihi-like Pb isotope ratios. These $^3He/^4He$ high ratios were interpreted to indicate a Lō'ihi source component for some Māhukona rocks [*Garcia et al.*, 2012].

Basalts from many other oceanic islands also show elevated $^3He/^4He$ ratios (>15 Ra; Samoa, Galápagos, Juan Fernandez, Pitcairn, Society Islands, Azores [see references in *Hofmann et al.*, 2011]). These results indicate that the mantle source regions for these hotspot-related lavas are deeper, more isolated, and less degassed and have lower time-integrated (U + Th)/^{3}He ratios than the source for midocean ridge basalts [*Graham*, 2002]. Globally, the highest and most primitive ^{3}He/^{4}He ratios (>30 Ra) are found only in Lō'ihi and Icelandic plume-derived basalts [*Kurz et al.*, 1982; *Stuart et al.*, 2003]. The helium isotope results for Hawaiian basalts led to the two-layer mantle hypothesis [*Allègre*, 1987], which, with some modifications, is still widely accepted among geoscientists. This hypothesis advocates that the source for Hawaiian and other mantle-plume-related basalts arises from the base of a layered lower mantle, perhaps near the core/mantle boundary or from the 670 km discontinuity [e.g., *Kurz et al.*, 1982; *Kaneoka*, 1983; *Allegre and Moreira*, 2004]. The helium isotope results, which advocated a deep mantle source for Hawaiian basalts, spurred seismologists to look for and find seismic evidence of a deep plume under Hawai'i [e.g., *Wolfe et al.*, 2009].

Another surprising discovery was the presence of relatively high ^{3}He/^{4}He ratios (17 and 21 Ra) in alkali lavas from the South Arch volcanic field upstream from the Hawaiian plume [*Hanyu et al.*, 2005]. In contrast, alkalic lavas from the postshield and rejuvenation stages of Hawaiian volcanism have values typical of MORB (7–9 Ra; Figure 1.5). These results and the previous work on shield tholeiitic lavas led to the realization that there is a strong asymmetry in the distribution of ^{3}He/^{4}He ratios in Hawaiian basalts, with lavas erupted ahead of the main shield phases of volcanism having consistently higher values (Figure 1.5). The Lō'ihi and South Arch samples also have high volatile contents (especially CO_2), which prompted the hypothesis that metasomatic fluids play an important role in carrying high ^{3}He/^{4}He fluids to the upstream side of the Hawaiian plume [*Dixon and Clague*, 2001; *Hanyu et al.*, 2005; *Hofmann et al.*, 2011]. These fluids are thought to form during incipient melting in the plume.

Science sometimes leaps forward from serendipitous events. The study of helium isotopes in Haleakalā lava and minerals is a fascinating example that ties in with Hawaiian mythology. Haleakalā is the Hawaiian "house of the sun." Early Hawaiians applied the name to the summit area of this volcano, where the demigod Maui snared the sun and forced it to slow its journey across the sky [*Westervelt*, 1910; *Pukui et al.*, 1974]. It is in this area where basalts yielded the highest ^{3}He/^{4}He values ever reported in Hawai'i (34–37 Ra). These values were assumed to be a primordial signature [*Kaneoka and Takaoka*, 1978]. Subsequent work on a suite of lavas drilled into outcrops at the summit of the volcano showed that the ^{3}He/^{4}He decreased with depth in the cores and that the heated olivine had higher ratios (especially at lower temperatures) than olivine crushed in vacuo (>60 vs. 8 Ra [*Kurz*, 1986; *Kurz et al.*, 1987]). These results were interpreted to indicate a cosmogenic rather than a primordial origin for the high ^{3}He/^{4}He ratios. Cosmic radiation effects are interpreted to have caused spallation reactions (on the major elements in the minerals) to form ^{3}He over hundreds of thousands of years [*Kurz*, 1986] in the cold arid climate (polar tundra zone) at the house of the sun, ~3050 m above sea level. Haleakalā was the first reported terrestrial occurrence of in situ cosmogenic helium and helped open the door to a new research field in rock exposure age dating [e.g., *Granger et al.*, 2013]. Being a stable nuclide with a high production rate in olivine and pyroxene [*Goehring et al.*, 2010], ^{3}He is the most commonly measured cosmogenic nuclide and has the potential to yield age information for surfaces up to millions of years old, provided erosion has not modified the surface [*Granger et al.*, 2013].

1.3.3. Marine Studies of Hawai'i

One of Earth's last frontiers is beneath its oceans, as witnessed by the discoveries that are continuing to be made during expeditions to the flanks and seafloor around the Hawaiian Islands (Figure 1.3). Thomas Jaggar was a visionary in advocating for marine research. Part of his rationale for construction of a new observatory in Hawai'i was its unique position in the central Pacific, making it favorable for the study of the deep-sea floor [*Jaggar*, 1913]. Jaggar was greatly influenced by the work of *Dana* [1890], which included one of the first bathymetric maps of the Pacific. Although based on very limited soundings by the U.S. Hydrographic Office, *Dana* [1890] showed the Hawaiian chain extended to Kure Island at ~29°N and that a depression surrounds and is within 65 km of the Hawaiian Islands (highlighted by the 3000 fathom contour on his map). Unlike other depressions in the North Pacific off Japan and the Aleutian Islands, the Hawaiian depression was interpreted by *Dana* [1890] as a possible consequence of gravitational pressure related to the nearby volcanoes with the amount of subsidence related to volcano size. Given the limited bathymetric data and geological knowledge at the time, it is amazing that Dana was able to define this feature, albeit crudely, and to offer a reasonable explanation for its origin. This depression is now called the Hawaiian Deep and it is paired with an arch. Both features are thought to be related to lithospheric flexure caused by the rapid and voluminous loading of the Pacific Plate by Hawaiian volcanism [e.g., *Jackson and Wright*, 1970; *Bianco et al.*, 2005]. The Hawaiian Arch is superimposed on the broader uplift (swell) related to the Hawaiian plume (Figure 1.3). The Hawaiian Islands are one of foremost locations to study the structure and dynamics of mantle plumes using features related to the swell. For example, ~200 km downstream from the vertical axis of the Hawaiian plume, swell topography was used to estimate the plume's excess temperature (400 K), radius

(50–70 km), and upper mantle viscosity (10^{20} and 3×10^{20} Pa s [*Zhong and Watts*, 2002]). The Hawaiian Swell has also been interpreted to be a result of crustal underplating by intrusion of Hawaiian magma, suggesting that the swell is partially supported by shallow chemical buoyancy [*Leahy et al.*, 2010].

Following the development of the echo sounding method in the 1920s and its use by the U.S. Navy to survey the oceans, a bathymetric map of the North Pacific basin was published illustrating the basic features of the Hawaiian chain [*U.S. Navy*, 1939]. A more detailed version of this map that focused on the flanks of the Hawaiian Islands was presented by *Stearns* [1946]. It included many interesting features that were not discussed, including the dramatic rift zones of Haleakalā, Kīlauea, and West Moloka'i volcanoes and the submarine volcanoes Lō'ihi and Māhukona on the south and northwest flanks of the Island of Hawai'i. These features are evident on a current bathymetric map of the Hawaiian Islands (Figure 1.3). Lō'ihi and Māhukona were "rediscovered" in the 1980s. For more on the history of marine expeditions in Hawaiian waters prior to 2000 (including the Gloria surveys that provided acoustic backscatter maps of the U.S. Exclusive Economic Zone 370 km offshore from Hawai'i), see *Moore and Clague* [2002].

Many discoveries have been made on the flanks and seafloor surrounding the Hawaiian Islands. Two of these discoveries warrant special attention and are highlighted here: Lō'ihi Seamount and the Nu'uanu landslide (Figure 1.3). Lō'ihi Seamount is thought to represent the youngest volcano in the Hawaiian chain (*Moore et al.*, 1982), whereas the Nu'uanu landslide was a catastrophic debris avalanche that originated on the submarine flanks of Ko'olau Volcano [*Moore*, 1964]. J. G. Moore was the lead author on both studies. He is a pioneer and the foremost expert on the marine geology of Hawaiian waters. Although his work started with and continued mapping granites in the Sierra Nevada Mountains of California [*Moore*, 2000], he has done more during the last 50 years to illuminate the marine geology of Hawai'i than anyone else. His tenure at HVO as scientist-in-charge in 1962–1964 may have served as a springboard for his later marine studies.

1.3.3.1. Lō'ihi Seamount

Located ~35 km south of the Island of Hawai'i (Figure 1.3), Lō'ihi first became a known bathymetric feature in 1940 on a U.S. Coast and Geodetic chart (no. 4115). No attention was drawn to the seamount until a large earthquake swarm occurred south of the Island of Hawai'i in 1952. The initial thought was the swarm might be an active submarine volcano [*Macdonald*, 1952]. No tremor was recorded and the epicenter locations were aligned in an east-west band ~10 km north of Lō'ihi [*Klein*, 1982]. These results led *Macdonald* [1952] to conclude that the swarm was related to faulting and not an active volcano. A bathymetric survey of the seafloor southeast of the Island of Hawai'i in 1954 revealed more details about Lō'ihi and four other seamounts, which were all given Hawaiian names [*Emery*, 1955]. Because of its elongate shape, the name Lō'ihi (meaning "long" [*Puki and Elbert*, 1984]) was recommended by some Hawaiian scholars. Lō'ihi is one of many seamounts surrounding the Hawaiian Islands (Figure 1.3). Rocks from some of these other seamounts were dated by K-Ar methods yielding Cretaceous ages [*Dymond and Windom*, 1968], which led to Lō'ihi being considered as an "older volcanic feature" [*Moore and Fiske*, 1969].

Renewed interest in Lō'ihi followed two earthquake swarms south of the Island of Hawai'i in 1971–1972 and 1975. The swarms were interpreted to be related to volcanic activity, and the improved HVO seismic network placed the epicenters beneath Lō'ihi Seamount [*Klein*, 1982]. This prompted a marine expedition to the seamount in 1978 to determine if it was an active volcano. Photographs of its summit area showed young-looking lavas. Subsequent dredging of Lō'ihi obtained fresh, glassy lavas [*Moore et al.*, 1982]. These and other rock samples were extensively analyzed geochemically [see *Garcia et al.*, 2006a, for a summary]. The source for its lavas is now considered one of the end members for the geochemical variation in Hawaiian shield volcano lava for major elements and Pb isotopes [e.g., *Weis et al.*, 2011; *Jackson et al.*, 2012].

Many bathymetric surveys were made of Lō'ihi to better understand its formation and to make comparisons after seismic and volcanic events. The height of Lō'ihi is estimated to be ~3.5 km above its basement [*Garcia et al.*, 1995], making it comparable in height to Etna Volcano. Thus, Lō'ihi is a respectable-sized volcano (by continental standards), although it is dwarfed by its larger neighbor Mauna Loa (~13 km tall). Two ~300 m deep pit craters were discovered at the summit of Lō'ihi during early swath mapping surveys [*Malahoff et al.*, 1982; *Fornari et al.*, 1988]. The Alvin and Pisces V submersibles were used to collect samples in stratigraphic order from these pit craters and from the dissected east flank of Lō'ihi, revealing that the volcano is dominantly alkalic with a thin veneer of tholeiitic lavas [*Garcia et al.*, 1995]. The alkalic lavas yielded unspiked K-Ar ages up to 102 ± 13 ka [*Guillou et al.*, 1997]. Reconstruction of Lō'ihi's history using these ages indicates it is considerably older than expected, possibly 400 ka (Figure 1.4). Another exciting discovery was the presence of excess ^{3}He, methane, and microorganisms in the water column above Lō'ihi, suggesting that the volcano has hydrothermal activity [*Malahoff et al.*, 1982]. This was confirmed during Alvin dives in 1987, which found warm springs (15–30°C [*Karl et al.*, 1988]).

The largest swarm of earthquakes ever observed at a Hawaiian volcano occurred at Lō'ihi Seamount in July

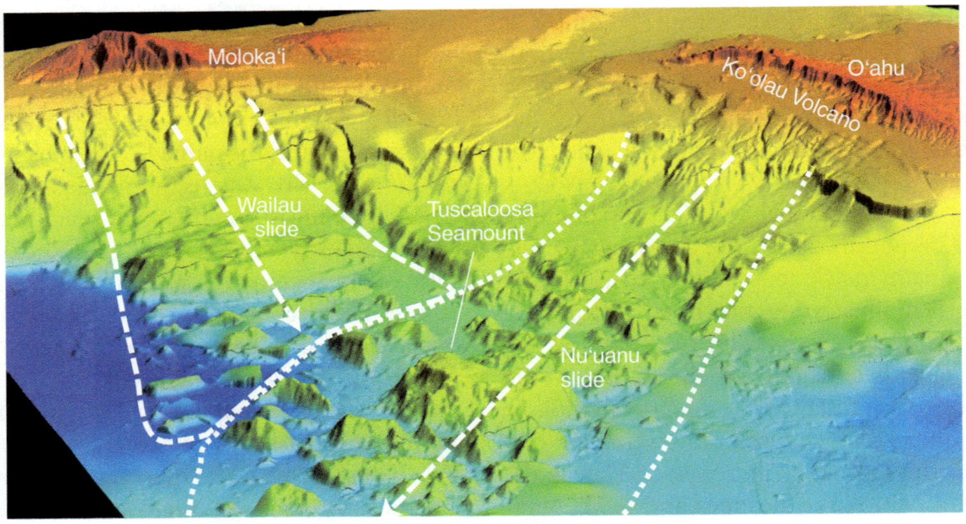

Figure 1.6 Shaded relief map showing a three-dimensional perspective view of the Nu'uanu and Wailau landslide debris. White lines are the boundaries of the debris fields with arrows showing the direction of motion away from the flanks of shield volcanoes on O'ahu and Moloka'i. Map was constructed using the bathymetric data of *Smith et al.* [2002]. Figure created by Greg Moore and used by permission.

and early August 1996 [*Caplan-Auerbach and Duennebier*, 2001]. During or following the earthquake swarm, a new large summit pit crater similar in size to the two other Lō'ihi craters formed [*Lō'ihi Science Team*, 1997]. Intense hydrothermal plume activity was found in the water above the new crater [*Lō'ihi Science Team*, 1997]. Hydrothermal fluids from the vents had temperatures up to 200°C and contained polysulfide minerals (wurtzite, pyrrhotite, and chalcopyrite) signifying even higher fluid temperatures (>250°C [*Davis and Clague*, 1998]). Research is continuing on Lō'ihi to document the hydrothermal system and the associated biological communities.

Samples of shiny basalt collected after the earthquake swarm were dated using the short-lived isotope ^{210}Po (138 day half-life). Their ages indicate eruption during the first half of 1996, prior to the earthquake swarm [*Garcia et al.*, 1998]. Petrological studies of the new lavas suggest that they crystallized at moderate depths (8–9 km), which is about 1 km below the hypocenters for earthquakes from the 1996 swarm [*Garcia et al.*, 1998; *Caplan-Auerbach and Duennebier*, 2001]. Taken together, the petrological and seismic evidence imply that Lō'ihi's current magma chamber is considerably deeper (≥4 km) than the shallow magma reservoirs (2–4 km) associated with the nearby active shield volcanoes Kīlauea and Mauna Loa.

1.3.3.2. Nu'uanu Landslide

High sea cliffs are common around the Hawaiian Islands (e.g., north coasts of Kaua'i and East Moloka'i). *Dana* [1890] proposed that these cliffs were the products of faulting. Others suggested that many of the cliffs were formed by marine erosion [*Hitchcock*, 1900; *Wentworth*, 1927] or the headwalls of giant landslides [*Stearns and Macdonald*, 1946]. The debate about these cliffs and the possibility they are related to landslides continued for decades. The Geological LOng-Range Incined Asdic (GLORIA) surveys (1986–89) documented the ubiquitous distribution of landslides surrounding the Hawaiian Islands [*Normark et al.*, 1993]. At least 68 major landslides (more than 20 km long) were recognized between Midway and the Island of Hawai'i [*Moore et al.*, 1994]). The largest is the Nu'uanu slide off the northeast flank of O'ahu [*Normark et al.*, 1993]. The full extent of the slide (Figure 1.6) was not appreciated until the detailed swath mapping by Japan Marine Science and Technology Center (JAMSTEC) during the cooperative Japan-U.S. program in 1998–1999. This survey covered an area of ~50,000 km^2 off the north flanks of the islands of O'ahu and Moloka'i [*Smith et al.*, 2002]. Both islands are sources of a major landslide (Figure 1.6). The JAMSTEC Hawai'i program included submersible diving to observe and collect samples from the blocks associated with the slides [*Yokose*, 2002]. The geochemistry of most samples collected from the Nu'uanu slide debris is identical to the distinctive composition of lavas from Ko'olau volcano on the east side of O'ahu (e.g., high SiO_2, low CaO and FeO), supporting the argument that Ko'olau was the main source of the blocks northeast of O'ahu [*Shinozaki et al.*, 2002]. The lavas from blocks off the north coast of Moloka'i (which are aligned parallel to the island) have compositions unlike that of Ko'olau and are thought to be part of the Wailau landslide from East Moloka'i [*Shinozaki et al.*, 2002].

Four piston cores (each ~7 m long with one or more volcanic sand or silt layers) were taken during the JAMSTEC program with the goal of documenting the timing of the slides using paleomagnetics and paleontology. The volcanic sand layers are thought to be related to major landslides. One core was taken from the flat summit of Tuscaloosa Seamount, the largest block in the Nu'uanu slide (30 km long, 17 km wide, and >2 km thick). The core has a >26 cm thick, black sand layer with glass compositions identical to those of East Moloka'i lavas and unlike those of Ko'olau Volcano [*Sherman et al.*, 2002]. The sediments just above the sand yielded a fossil age of 1.77–2.0 Ma, which places a minimum age for both the older Nu'uanu and younger Wailau slides [*Sherman et al.*, 2002]. Unfortunately, the JAMSTEC piston cores were unable to delineate the age of the Nu'uanu slide.

Serendipity occurred during the planning for International Ocean Drilling Program (IODP) Leg 200, which was designed to develop an ocean floor observatory along a deep-sea cable between California and Hawai'i. Time (42 h) was allocated to drill up to 100 m into the sediments on the crest of the Hawaiian Arch (260 km from O'ahu) to determine the depositional history, age, thickness of, and hazards associated with the Nu'uanu slide. Unexpectedly, indurated sediments were encountered at ~13 m in the hole, slowing the drilling and resulting in only a 41 m deep hole. To our astonishment, eight distinct volcanic sand layers (the lower seven of which have glass with Ko'olau compositions) were recovered with mixed or Eocene radiolaria [*Garcia et al.*, 2006b]. Both features are common in deep-sea turbidite layers [*Garcia and Hull*, 1994]. The upper seven sand layers were probably deposited between 1.0 and 2.15 Ma based on paleomagnetics [*Garcia et al.*, 2006b]. The age of the lower sand, which is >5 m thick, is unknown. Four of the sand layers are >1.5 m thick. It is unclear which (if any) were related to the Nu'uanu slide, given that many deeper turbidite layers were detected in seismic profiles across the arch [*Garcia et al.*, 2006b]. The identification of numerous turbidite layers in both the seismic profiles and piston cores indicates that debris from many landslides were able to travel >260 km from O'ahu and climb up the 500 m high Hawaiian Arch to be deposited at this site. Thus, landslides from Hawaiian volcanoes are common and pose a more serious threat than previously assumed. How big is that threat?

The detailed JAMSTEC bathymetry map was used to estimate the Nu'uanu landslide volume, to reconstruct Ko'olau volcano to its prelandslide size, and to evaluate the hazards associated with the landslide. The volume of the submarine portion of the slide is estimated at $2–3 \times 10^3$ km^3, making it one of the largest slides on Earth [*Satake et al.*, 2002]. The debris avalanche field is ~50 km wide and extends for ~150 km from Ko'olau (Figure 1.6). If the debris field was caused by a single landslide, modeling shows that the tsunami height it generated would have been >100 m along the north side of O'ahu and Moloka'i, up to 70 m along the southern California coast, where it was directed, and 10–40 m along the coasts of Oregon and Washington [*Satake et al.*, 2002]. Thus, Hawaiian landslides have the potential to be one of the largest natural disasters on Earth.

The documentation of giant landslides around the Hawaiian Islands resulted in the search for similar features around other oceanic islands. It was no surprise that large landslides were discovered around many other oceanic islands (e.g., Canary Islands, Cape Verde Islands, Réunion, Tristan da Cunha [*Holcomb and Searle*, 1991; *Masson et al.*, 2002; *Ancochea et al.*, 2010]). Large landslides around oceanic islands had been suspected for many decades where gaps were identified in the original shape of the islands [e.g., *Daly*, 1924; *Fairbridge*, 1950]. The availablilty of swath mapping to survey up to 10 km wide sections of the deep ocean floor made it possible to confirm the presence of the submarine landslides in various climates and tectonic settings [*Holcomb and Searle*, 1991]. These results indicate that many oceanic volcanoes have been substantially reduced in size by landslides and that significant risks are posed by these volcanoes. However, the frequency of such collapses on Hawaiian volcanoes seems to be low (1 per 20 ka [*Garcia et al.*, 2006b]) compared to the risks posed by other natural hazards (earthquakes, fires, and floods).

1.3.4. Dynamics and Emplacement of Lava Flows

Studies of Hawaiian volcanoes have been fundamental to our evolving understanding of the dynamics and emplacement of lavas flows. For example, the classic terms that are now universally used for describing two principal flow types are of Hawaiian origin. *Dutton* [1884] was the first to introduce to the geological community the terms "pāhoehoe" and "'a'ā," which were used by Hawaiians to describe these two distinctive flow types. Although these terms were considered "barbarous" by some geologists at the time, they were quickly adopted by most geologists [*Macdonald*, 1972]. Pāhoehoe is characterized by a smooth, billowy, rolling, or ropey surface, whereas 'a'ā has a rough, jagged, spinose, and clinkery surface. These two common flow types were studied extensively by *Macdonald* [1953, 1967], primarily using Hawaiian examples. He noted that both viscosity and the amount of internal disturbance owing to flowage were critical in determining flow type.

Macdonald's work was extended by others using Hawaiian examples to better understand the internal mechanics of lava flows. *Peterson and Tilling* [1980] evaluated the causes for the transition of basaltic lava from pāhoehoe to ʻaʻā (e.g., volumetric flow rate, flow dimensions, slope, and momentum) as well as the internal mechanics within lava flows and how they create the observed flow features. *Peterson and Tilling* [1980] assumed that nearly all Hawaiian basaltic lava erupts as pāhoehoe. *Rowland and Walker* [1990] showed, using examples from recent Mauna Loa and Kīlauea eruptions, that the volumetric discharge rate plays a key control in determining flow type, with ʻaʻā forming when flow rates are higher than >5–10 m^3/s, whereas at lower rates pāhoehoe flows are erupted. The lavas from the ongoing eruption of Kīlauea demonstrate this effect beautifully. ʻAʻā was produced mostly during the high-fountaining, high-effusion-rate episodes at the Puʻu ʻŌʻō vent from 1983 to 1986, whereas pāhoehoe flows were prevalent from 1986 to 1992 during the low-effusion and low- to no-fountaining eruption at the Kupaianaha satellite vent [*Rowland and Walker*, 1990]. In most cases, discharge rate is primarily controlled by processes within the volcano (e.g., relaxation of a stressed, shallow magma chamber and degassing [*Rowland and Walker*, 1990]). Through extensive field work, George Walker and his students documented the varieties of pāhoehoe flow types [pipe-vesicle bearing (P type), spongy (S type) and toothpaste], the distribution of crystals within flows, and their internal and external flow structures [e.g., *Walker*, 1987, 1989; *Rowland and Walker*, 1987, 1988; *Wilmouth and Walker*, 1993]. These studies led to better understanding of the controls of viscosity on lava flow dynamics and processes of degassing.

Studies of Kīlauea lavas were pivotal in the recognition of the importance of inflation to the emplacement of pāhoehoe sheet flows. Sheet flows, which are a major crust-forming component in flood basalts, continental rifts, and along fast-spreading midocean ridges [e.g., *Ballard et al.*, 1979; *Macdonald et al.*, 1989; *Tolan et al.*, 1989], are emplaced without channels. *Macdonald* [1953] was probably the first to apply the term "inflation" to pāhoehoe flow growth for features he observed in Hawaiian lavas. The concept was more fully documented by *Walker* [1991] in describing historical lavas within Kīlauea Caldera. *Hon et al.* [1994] quantified the effects of inflation while monitoring Kīlauea's active flow field, showing that a 20 cm thick flow lobe was able to expand into a many meter thick sheet over a period of weeks. This concept of slow but relentless growth was applied to continental flood basalt provinces to radically change ideas for the method and timing of flood basalt emplacement. *Self et al.* [1996, 1998], *Thordarson and Self* [1998], and *Vye-Brown et al.* [2013] proposed that inflation of these voluminous flows [~1000 km^3] occurred over periods of decades, instead of days to weeks, as previously suggested, which reduces the potential environmental impact of flood basalt eruptions [*Self et al.*, 1998]. The inflation model has extraterrestrial applications in helping to explain emplacement of voluminous flows on Mars and Io [*Keszthelyi et al.*, 2000, 2006].

Lava flows present substantial hazards to communities around the world. Civil defense managers need to know how far a channelized lava flow can travel before it solidifies. Knowing the answer to this question is imperative for assessing the hazard during an eruption. The FLOWGO model was developed to answer this question. FLOWGO is a kinematic, self-adaptive, numerical model to describe the down-flow thermal and rheological evolution of channel-contained lava [*Harris and Rowland*, 2001]. It offers a means of analyzing lava flow thermo-rheological relationships; identifying important factors that determine how far a channel-fed flow can extend; assessing lava flow hazards; and reconstructing flow regimes for prehistoric, unobserved, or remote flows. FLOWGO considers flow velocities, widths, surface crust parameters, heat budget, cooling rate, absolute temperature, crystallinity, viscosity, and yield strength in assessing the distance a channelized lava flow will travel. The results of the model were field tested using Mauna Loa, Kīlauea, and Etna eruptions, where vent locations, eruption durations, effusion rates, and channel dimensions were known, achieving good results [*Harris and Rowland*, 2001]. The model was then applied to assessing the risk from future eruptions of Mauna Loa, a volcano that sent a flow into the city of Hilo (Figure 1.3) in 1881 and another that reached within 6 km of the city in 1984. The modeling results indicated that topography alone cannot be used to derive potential inundation distances from channel-fed flows [*Rowland et al.*, 2005]. For example, Hilo, which sits in a topographic low, is relatively safe from all but the highest effective-effusion rate channel-fed flows, especially if northeast rift eruptions continue to be produced from vents at elevations >2500 m. However, if a tube-fed eruption occurs, such as the one in 1881, the city will be threatened again. The FLOWGO model has been exported to other volcanic regions, including the Galápagos and Mt. Cameroon [*Rowland et al.*, 2003; *Wantim et al.*, 2013].

Lava from the 1984 eruption of Mauna Loa has been instrumental in understanding the effects of microphenocrysts on flow dynamics. During this eruption, the microphenocryst content of lava near the vent increased from 0.5 to 30 vol% without any concurrent change in either bulk magma composition or eruption temperature (1140 ± 3°C [*Lipman et al.*, 1985]). This change was attributed to undercooling of the magma by 20–30°C from degassing of volatiles [*Lipman et al.*, 1985]. The undercooling and crystallization increased the effective viscosity of the magma, which probably caused volumetric flow rates to decrease and led to stagnation of the flow [*Lipman et al.*, 1985]. The increase in crystallinity helped save the city of Hilo from being inundated by lava. These effects are broadly similar to those inferred to result from degassing during eruption and downslope flow of lava on Etna volcano [*Sparks and Pinkerton*, 1978]. The importance of microphenocrysts in controlling lava effective viscosity has been applied to understanding magma emplacement for a wide range of rock types in diverse settings, including Mount St. Helens dacite [*Cashman*, 1992], Inyo dome rhyolite [*Swanson et al.*, 1989], Mt. Etna basalt [*Armienti et al.*, 1994] and Mt. Fuji basalt [*Ishibashi*, 2009].

1.4. SUMMARY

Studies of Hawaiian volcanoes from their source to surface have been pivotal in our quest to better understand many Earth processes. Volcanologists from all over the world have been and are continuing to be attracted to Hawai'i for research. The reasons for this attraction include the high frequency of eruptions, and year-round and easy access as well as Hawai'i's relatively uncomplicated geological setting in the center of the Pacific Plate. People are familiar with Hawai'i because it is a favorite tourist destination. Thus, research done on Hawai'i is of broad appeal. More importantly, a critical level of knowledge has been achieved about Hawaiian volcanoes since the founding of the HVO in 1912. Therefore, scientists are drawn to Hawai'i to test hypotheses and to address fundamental issues about the way Earth works.

As a result of research on Hawai'i over the last 100 years, tremendous progress has been made in our understanding of how magma is generated, what happens to it during ascent, and how it is transported on the surface as lava flows. The recognition that Hawai'i's giant shield volcanoes consist almost entirely (90%–95%) of tholeiitic lava reversed early notions that tholeiitic magma was formed by contamination of alkaline magma by continental crust. This recognition coupled with work on midocean ridge basalts led to a resurgence in experimental petrology that showed that the degree and depth of melting are critical for determining the composition of parental magmas (tholeiitic vs. alkaline). These results were combined with field and geochemical studies to place the evolution of Hawaiian volcanoes into a platetectonic framework. The discovery of excess ^{3}He in Hawaiian basalts unleashed an explosion of work on noble gases in oceanic basalts. Studies of noble gases in Hawaiian lavas led to the recognition of a distinct, relatively primitive, and undegassed reservoir deep within the mantle. These results prompted a revolution in models for the structure and heterogeneity of Earth's mantle.

The underwater geology around the Hawaiian Islands has revealed many new insights into geological processes, as predicted by *Jaggar* [1913]. One example is Lō'ihi Volcano, which was thought to be a Cretaceous seamount until two earthquake swarms in the 1970s led to marine expeditions that found fresh, glassy lavas and active hydrothermal springs. Lō'ihi is now considered the youngest member of the Hawaiian-Emperor chain. In 1996, this infant volcano was the site of the largest earthquake swarm ever recorded on a Hawaiian volcano following a small eruption.

The enormous size, relatively steep submarine slopes, and rapid growth of Hawaiian volcanoes cause them to become gravitationally unstable and collapse. Dozens of giant landslides, some with debris extending more than 200 km from their source, have been recognized along the Hawaiian ridge and around other oceanic volcanoes. These landslides are thought to have produced colossal tsunamis. Thus, landslides from oceanic volcanoes pose a major risk to populations bordering Earth's oceans. However, the frequency of these landslides appears to be low (at least in Hawai'i) with a possible reoccurrence interval of every 20,000 yr.

Hawai'i is the classic site for two characteristic lava types with Hawaiian names: 'a'ā and pāhoehoe. The frequent and changing styles of eruption of Mauna Loa and Kīlauea volcanoes have provided superb natural laboratories for evaluating the dynamics of lava flows and how they are emplaced. Extrusion rate is a dominant factor in determining whether 'a'ā or pāhoehoe flows are formed during a Hawaiian eruption. When pāhoehoe lava is erupted, the transition to 'a'ā is governed by volumetic flow rate, flow dimensions, slope, crystallinity, and momentum as well as the internal mechanics.

As shown by the excellent chapters on diverse topics in this monograph, much has been learned about the processes related to Hawaiian volcanism since the work of *Dana* [1890]. Nevertheless, after working on Hawaiian volcanism for the last 38 years, it has become apparent to me how little is really known about volcanoes. I have more questions now than when I started my research. Nevertheless, the next decade of study on Hawaiian volcanoes will be exciting as more interdisciplinary work is done and new techniques are developed to probe Earth's interior leading to more discoveries. Regardless of any discoveries, we will continue to be in awe of the power and beauty of the fireworks from the Hawaiian volcano goddess Pele.

ACKNOWLEDGMENTS

Mahalo nui loa to Michael Poland and the other editors of this monograph (Rebecca Carey, Dominique Weis, and Valérie Cayol) for inviting me to write this review chapter. The chapter reflects my nearly four decades of experience working on Hawaiian volcanoes with many colleagues (e.g., J. Michael Rhodes, Dominique

Weis, Fred Frey, Marc Norman, Garrett Ito, Ed Stolper, and Dave Muenow) and students (e.g., Aaron Pietruszka, R. Michael Easton, Dorsey Wanless, Jared Marske, Steve Spengler, and Howard West). Conversations with and comments by Michael Poland, Scott Rowland, Mark Kurz, and Fred Frey improved the manuscript for this chapter. I thank Nancy Hulbirt, Garrett Ito, and Greg Moore for providing Figures 1.2, 1.3, and 1.6, respectively, and Pauline Sugino of the Honolulu Museum of Art for permission to use the image of the Jules Tavernier painting, which is a gift of Mrs. E. Faxon Bishop, 1959 (2562.1). Constructive reviews by Don Swanson, Karen Harpp, and Helge Gonnerman improved the manuscript markedly and are greatly appreciated. This work was supported by a grant from the National Science Foundation (EAR-1118741) and it is SOEST Contrib. no. 9224.

REFERENCES

Albarede, F., B. Luais, G. Fitton, M. Semet, E. Kaminski, B. J. Upton, P. Bachelery, and J.-L. Cheminee (1997), The geochemical regimes of Piton de la Fournaise volcano [Reunion] during the last 530,000 years, *J. Petrol.*, *39*, 171–201.

Allègre, C. J. (1987), Isotope geodynamics, *Earth Planet. Sci. Lett.*, *86*, 175–203.

Allègre, C. J., and M. Moreira (2004), Rare gas systematics and the origin of oceanic islands: The key role of entrainment at the 670 km boundary layer, *Earth Planet. Sci. Lett.*, *228*, 85–92.

Ancochea, E., M. J. Huertas, F. Hernán, and J. L. Brändle (2010), Volcanic evolution of São Vicente, Cape Verde Islands: The Praia Grande landslide, *J. Volc. Geotherm. Res.*, *198*, 143–157.

Apple, R. A. (1987) Thomas A. Jaggar, Jr. and the Hawaiian Volcano Observatory, in *Volcanism in Hawaii*, Prof. Pap. 1350, edited by R. W. Decker et al., pp. 1619–1644, U.S. Geol. Surv., Reston, Va.

Appleman, D. E. (1987), James D. Dana and the origins of Hawaiian volcanology: The U.S. Exploring Expedition in Hawaii, 1840–41, in *Volcanism in Hawaii*, Prof. Pap. 1350, edited by R. W. Decker et al., pp. 1607–1618, U.S. Geol. Surv., Reston, Va.

Armienti, P., M. T. Pareschi, F. Innocenti, and M. Pompilio (1994), Effects of magma storage and ascent on the kinetics of crystal growth. The case history of the 1991–93 Mt. Etna eruption, *Contrib. Mineral. Petrol.*, *115*, 402–414.

Bailey, E. B., T. C. Clough, W. B. Wright, J. E. Richey, and G. V. Wilson (1924), Tertiary and post-Tertiary geology of Mull, Loch Aline and Oban, *Mem. Geol. Surv. Scotland*, p. 445.

Ballard, R. D., T. R. Holcomb, and T. H. van Andel (1979), The Galapagos Rift at 86°W: 3. Sheet flows, collapse pits, and lava lakes of the rift valley, *J. Geophys. Res.*, *84*, 5407–5422.

Bevens, D., T. J. Takahashi, and T. L. Wright (Eds.) (1988), *The Early Serial Publications of the Hawaiian Volcano Observatory*, Hawaii Natural History Assoc., vols. 1–3, Hawaii National Park, Hawaii.

Bianco, T., G. Ito, J. Becker, and M. O. Garcia (2005), Secondary Hawaiian volcanism formed by flexural arch decompression, *Geochem. Geophys. Geosyst.*, *6*, Q08009, doi:10.1029/2005GC000945.

Brigham, W. T. (1909), *The Volcanoes of Kilauea and Mauna Loa on the Island of Hawaii, Their Variously Recorded History to the Present Time. Memoirs of the Bernice Pauahi Bishop Museum 2*, Bishop Museum Press, Honolulu, Hawaii. (Reprinted 1974, Kraus Reprint Company, Millwood, N.Y.)

Caplan-Auerbach, J., and F. K. Duennebier (2001), Seismicity and velocity structure of Loihi Seamount from the 1996 earthquake swarm, *Bull. Seismol. Soc. Am.*, *91*, 178–190.

Cashman, K. V. (1992), Groundmass crystallization of Mount St. Helens dacite, 1980–1986: A tool for interpreting shallow magmatic processes, *Contrib. Mineral. Petrol.*, *109*, 431–449.

Coan, T. (1856), On the eruption at Hawaii, *Am. J. Sci.*, 2d ser., *22*, 240–243.

Coan, T. (1869), Notes on the recent volcanic disturbances of Hawaii, *Am. J. Sci.*, 2d ser., *47*, 89–98.

Coan, T. (1880), Recent activity of Mauna Loa and Kilauea, *Am. J. Sci.*, 3d ser., *20*, 71–72.

Craig, H., and J. E. Lupton (1976), Primordial neon, helium, and hydrogren in oceanic basalts, *Earth. Planet. Sci. Lett.*, *31*, 369–385.

Cross, W. (1915), *Lavas of Hawaii and Their relations*, Prof. Pap., 88, U.S. Geol. Surv., Reston, Va.

Daly, R. A. (1911), The nature of volcanic action, *Proc. Am. Acad. Arts Sci.*, *47*, 47–122.

Daly, R. A. (1924), *The Geology of American Samoa*, Publ. 340, pp. 95–143, Carnegie Inst. Wash., Washington, D. C.

Dana, J. D. (1835), On the condition of Vesuvius in July, 1834, *Am. J. Sci.*, *27*, 281–288.

Dana, J. D. (1850), On the volcanic activity of Hawaii, *Am. J. Sci.*, ser. 2, *29*, 235–244.

Dana, J. D. (1890), *Characteristics of Volcanoes with Contributions of Facts and Principles from the Hawaiian Islands*, Dodd, Mead and Co., New York.

Davis, A. S., and D. A. Clague (1998), Changes in the hydrothermal system at Loihi Seamount after the formation of Pele's pit in 1996, *Geology*, *26*, 399–402.

Daws, G. (1968), *Shoal of Time: A History of the Hawaiian Islands*, Macmillan, New York.

Dixon, J. E., and D. A. Clague (2001), Volatiles in basaltic glasses from Loihi Seamount, Hawaii: Evidence for a relatively dry plume component, *J. Petrol.*, *42*, 627–654.

Dutton, C. E. (1884), *Hawaiian Volcanoes*, 4th Annual Rept., pp. 75–219, U.S. Geol. Surv, Reston, Va.

Dymond, J., and H. L. Windom (1968), Cretaceous K–Ar ages from Pacific Ocean seamounts, *Earth Planet. Sci. Lett.*, *4*, 47–52.

Emery, K. O. (1955), Submarine topography south of Hawaii, *Pac. Sci.*, *9*, 286–291.

Fairbridge, R. W. (1950), Landslide patterns on oceanic volcanoes and atolls, *Geograph. J.*, *115*, 84–119.

Fiske, R. S., T. Simkin, and E. A. Nielsen, E. A. (Eds.) (1987), *The Volcano Letter*, Smithsonian Inst. Press, Washington, D. C.

Fornari, D. J., M. O. Garcia, R. C. Tyce, and D. G. Gallo (1988), Morphology and structure of Loihi seamount based on seabeam sonar mapping, *J. Geophys. Res.*, *93*, 15227–15238.

Frear, W. F. (1947), *Mark Twain and Hawaii*, Lakeside Press, Chicago, Ill.

Frey, F., W. Wise, M. O. Garcia, H. West, and S. T. Kwon (1990), Evolution of Mauna Kea volcano, Hawaii: Petrologic and geochemical constraints on postshield volcanism, *J. Geophys. Res.*, *95*, 1271–1300.

Frey, F. A., M. O. Garcia, W. S. Wise, A. Kennedy, P. Gurriet, and F. Albarede (1991), The evolution of Mauna Kea volcano, Hawaii: Petrogenesis of tholeiitic and alkalic basalts, *J. Geophys. Res.*, *96*, 14,347–14,375.

Garcia, M. O., and D. Hull (1994), Turbidites from giant Hawaiian landslides: Results from Ocean Drilling Program Site 842, *Geology*, *22*, 159–162.

Garcia, M. O., D. J. P. Foss, W. B. West, and J. J. Mahoney (1995), Geochemical and isotopic evolution of Loihi Volcano, Hawaii, *J. Petrol.*, *26*, 1647–1674.

Garcia, M. O., K. H. Rubin, M. D. Norman, J. M. Rhodes, D. W. Graham, D. Muenow, and K. Spencer (1998), Petrology and geochronology of basalt breccia from the 1996 earthquake swarm of Loihi Seamount, Hawaii: Magmatic history of its 1996 eruption, *Bull. Volcanol.*, *59*, 577–592.

Garcia, M. O., J. Caplan-Auerbach, E. H. De Carlo, M. D. Kurz, and N. Becker (2006a), Geology, geochemistry and earthquake history of Loihi seamount, Hawaii's youngest volcano, *Chemie der Erde*, *66*, 81–108.

Garcia, M. O., S. B. Sherman, G. F. Moore, G. Acton, R. Goll., I. Popova-Goll, and J. Natland (2006b), Frequent landslides from Koʻolau Volcano: Results from ODP site 1223, *J. Volcanol. Geotherm. Res.*, *151*, 251–268.

Garcia, M. O., E. H. Haskins, E. Stolper, and M. Baker (2007), Stratigraphy of the Hawaiian Scientific Drilling Project: Anatomy of a Hawaiian volcano, *Geochem, Geophys, Geosyst.*, *8*, Q02G20, doi:10.1029/2006GC001379.

Garcia, M. O., L. Swinnard, D. Weis, A. R. Greene, T. Tagami, H. Sano, and C. E. Gandy (2010), Petrology, geochemistry and geochronology of Kauaʻi lavas over 4.5 Ma: Implications for the origin of rejuvenated volcanism and the evolution of the Hawaiian plume, *J. Petrol.*, *51*, 1507–1540.

Garcia, M. O., D. Weis, D. Hanano, G. Apuzen-Ito, A. Flinders, and M. Kurz (2012), Age, geology, geophysics and geochemistry of Mahukona volcano, Hawaii, *Bull. Volcanal.*, *74*, 1445–1463.

Goehring, B. M., M. D. Kurz, G. Balco, J. M. Schaefer, J. Licciardi, and N. Lifton (2010), A reevaluation of in situ cosmogenic ^{3}He production rates, *Quat. Geochron.*, *5*, 410–418.

Gonnermann, H. M., and S. Mukhopadhyay (2009), Preserving noble gases in a convecting mantle, *Nature*, *459*, 560–564.

Graham, D. W. (2002), Noble gas isotope geochemistry of mid-ocean ridge and ocean island basalts: Characterization of mantle source reservoirs, *Rev. Mineral. Geochem.*, *47*, 247–318, Mineral. Soc. Am., Washington, D.C.

Granger, D. E., N. A. Lifton, and J. K. Willenbring (2013), A cosmic trip: 25 years of cosmogenic nuclides in geology, *Geol. Soc. Am. Bull.*, *125*, 1379–1402.

Green, D. H., and A. E. Ringwood (1967), The genesis of basaltic magmas, *Contrib. Mineral. Petrol.*, *15*, 103–190.

Guillou, H., M. O. Garcia, and L. Turpin, L. (1997), Unspiked K-Ar dating of young volcanic rocks from the Loihi and Pitcairn seamounts, *J. Volcanol. Geotherm. Res.*, *78*, 239–250.

Haase, K. M., P. Stoffers, and C. D. Garbe-Schonberg (1997), The petrogenetic evolution of lavas from Easter Island and neighbouring seamounts, near-ridge hotspot volcanoes in the SE Pacific, *J. Petrol.*, *38*, 785–813.

Hanyu, T., D. A. Clague, I. Kaneoka, T. J. Dunai, and G. R. Davies (2005), Noble gas systematics of submarine alkalic lavas near the Hawaiian hotspot, *Chem. Geol.*, *214*, 135–155.

Harris, A. J. L., and S. K. Rowland (2001), FLOWGO: A kinematic thermo-rheological model for lava flowing in a channel, *Bull. Volcanol.*, *63*, 20–44.

Heliker, C., and T. N. Mattox (2003), The first two decades of the Puu Oo-Kupaianaha eruption; chronology and selected bibliography.in *The Puu Oo-Kupaianaha Eruption of Kilauea Volcano, Hawaii: The First 20 Years*, Prof. Pap. 1676, edited by C. Heliker et al., pp. 121–136, U.S. Geol. Surv., Reston, Va.

Hitchcock, C. H. (1900), The geology of Oahu, *Geol. Soc. Am. Bull.*, *11*, 23–35.

Hitchcock, C. H. (1911), *Hawaii and Its Volcanoes*, 2nd ed. with supplement, Hawaii Gazette, Honolulu.

Hoernle, K., and H. Schmincke (1993), The role of partial melting in the 15-Ma geochemical evolution of Gran Canaria: A blob model for the Canary hotspot, *J. Petrol.*, *34*, 599–626.

Hofmann, A. W., C. G. Farnetani, M. Spiegelman, and C. Class (2011), Displaced helium and carbon in the Hawaiian plume, *Earth Planet. Sci. Lett.*, *312*, 226–236.

Holcomb, R. T., and R. C. Searle (1991), Large landslides from oceanic volcanoes, *Marine Geotech.*, *10*, 19–32.

Hon, K., J. Kauahikaua, R. Denlinger, and K. Mackay (1994), Emplacement and inflation of pahoehoe sheet flows: Observations and measurements of active lava flows on Kilauea Volcano, Hawaii, *Geol. Soc. Am. Bull.*, *106*, 351–370.

Ishibashi, H. (2009), Non-Newtonian behavior of plagioclase-bearing basaltic magma: Subliquidus viscosity measurement of the 1707 basalt of Fuji volcano, Japan, *J. Volcanol. Geotherm. Res.*, *181*, 78–88.

Jackson, E. D., and T. L. Wright (1970), Xenoliths in the Honolulu Volcanic Series, *J. Petrol.*, *11*, 405–430.

Jackson, M. G., D. Weis, and S. Huang (2012), Major element variations in Hawaiian shield lavas: Source features and perspectives from global ocean island basalt (OIB) ststematic, *Geochem. Geophys. Geosyst.*, *13*, Q09009, doi:10.1029/2012GC004268.

Jaggar, T. A. (1912), *Report of the Hawaiian Volcano Observatory*, Soc. of Arts of Mass. Inst. of Technol., Boston.

Jaggar, T. A. (1913), Scientific work on Hawaiian volcanoes, Hawaiian Gazette, 1913, reprinted in Bevens, D., T. J. Takahashi, and T. L. Wright (1988), *The Early Serial Publications of the Hawaiian Volcano Observatory*, *1*, 478–491, Hawaii Natural History Assoc., Hawaii.

Jaggar, T. A. (1916), A bill [H.R. 9525] to establish a national park in the Territory of Hawaii: Hearing before the committee on public lands, 3, Feb. 1916, Gov. Print. Ofc.,Washington, D.C.,

Kaneoka, I. (1983), Noble gas constraints on the layered structure of the mantle, *Nature, 302*, 698–700.

Kaneoka, I., and N. Takaoka (1978), Excess 129Xe and high 3He/4He ratios in olivine phenocrysts of Kapuho lava, and xenolithic dunites from Hawaii, *Earth Planet. Sci. Lett.*, *39*, 382–386.

Kaneoka, I., and N. Takaoka (1980), Rare gas isotopes in Hawaiian ultramafic nodules and volcanic rocks: Constraints on genetic relationships, *Science*, *208*,1366–1368.

Karl, D. M., G. M. McMurtry, A. Malahoff, and M. O. Garcia (1988), Loihi Seamount, Hawaii: A mid-plate volcano with a distinctive hydrothermal system, *Nature*, *335*, 532–535.

Keszthelyi, L., A. S. McEwen, and T. Thordarson (2000), Terrestrial analogs and thermal models for Martian flood lavas, *J. Geophys. Res.*, *105*[E6], 15,027–15,049.

Keszthelyi, L. P., S. Self, and T. Thordarson (2006), Flood lavas on Earth, Io, and Mars, *J. Geol. Soc. Lond.*, *163*, 253–264.

Klein, F. W. (1982), Earthquakes at Loihi submarine volcano and the Hawaiian hot spot, *J. Geophys. Res.*, *87*, 7719–7726.

Krylov, A., B. A. Mamyrin, L. A. Khabarin, T. I. Mazina, and Y. I. Silin (1974), Helium isotopes in ocean floor bedrock, *Geochem. Intern.*, *11*, 839–844.

Kurz, M. D. (1986), In situ production of terrestrial cosmogenic helium and some applications to geochronology, *Geochim. Cosmochim. Acta*, *50*, 2855–2862.

Kurz, M. D., W. J. Jenkins, and S. R. Hart (1982), Helium isotopic systematics of oceanic islands and mantle heterogeneity, *Nature*, *297*, 43–47.

Kurz, M. D., W. J. Jenkins, S. Hart, and D. Clague (1983), Helium isotopic variations in Loihi Seamount and the Island of Hawaii. *Earth Planet. Sci. Lett.*, *66*, 388–406.

Kurz, M. D., M. O. Garcia, F. A. Frey, and P. A. O'Brien (1987), Temporal helium isotopic variations within Hawaiian volcanoes: Basalts from Mauna Loa and Haleakala, *Geochim. Cosmochim. Acta*, *51*, 2905–2914.

Kurz, M. D., T. C. Kenna, D. Kammer, J. M. Rhodes, and M. O. Garcia (1995), Isotopic evolution of Mauna Loa volcano: A view from the submarine southwest rift, in *Mauna Loa Volcano Revealed*, Geophys. Monogr. Ser. 92, edited by J. M. Rhodes and J. P. Lockwood, pp. 289–306, AGU, Washington, D. C.

Kurz, M. D., J. Curtice, D. E. Lott III, and A. Solow (2004), Rapid helium isotopic variability in Mauna Kea shield lavas from the Hawaiian scientific drilling project, *Geochem. Geophys. Geosyst.*,*5*(4),QC4G14,doi:10.1029/2002GC000439.

Kyser, T. K., and W. Rison (1982), Systematics of rare gas isotopes in basic lavas and ultramafic xenoliths, *J. Geophys. Res.*, *87*, 5611–5630.

Lacroix, A. (1928), La composition minéralogique et chimique des roches éruptives et particulièrement des laves mésozoiques et plus récentes de la Chine orientale, *Bull. Geol. Soc. China*, *7*, 13–59.

Leahy, G. M., J. A. Collins, C. J. Wolfe, G. Laske, and S. C. Solomon (2010), Underplating of the Hawaiian Swell: Evidence from teleseismic receiver functions, *Geophys. J. Int.*,*183*, 313–329

Lease, N. A., A. Abdel Rahman, and M. Fattah (2008), The Euphrates volcanic field, northeastern Syria: petrogenesis of Cenozoic basanites and alkali basalts, *Geol. Mag*, *145*, 685–701.

Lipman, P. W., N. G. Banks, and J. M. Rhodes (1985), Degassing-induced crystallization of basaltic magma and effects on lava rheology, *Nature*, *317*, 604–607.

Lō'ihi Science Team (1997), Researchers rapidly respond to submarine activity at Loihi volcano, Hawaii, *EOS Trans. AGU*, *78*, 229–233.

Lupton, J. E. (1983), Terrestrial inert gases: Isotope tracer studies and clues to primordial components in the mantle, *Ann. Rev. Earth. Planet. Sci.*, *11*, 371–414.

Lupton, J. E., and H. Craig (1975), Excess ^{3}He in oceanic basalts: Evidence for terrestrial primordial helium, *Earth Planet. Sci. Lett.*, *26*, 133–139.

Macdonald, G. A. (1952), The South Hawaii earthquakes of March and April, 1952, The Volcano Letter 515, reprinted in *The Volcano Letter*, edited by R. S. Fiske et al., 1987, Natl. Mus. Nat. Hist., Smithsonian Inst. Press, Washington, D. C.

Macdonald, G. A. (1953), Pahoehoe, aa and block lava, *Am. J. Sci.*, *251*, 169–191.

Macdonald, G. A. (1963), Relative abundance of intermediate members of the oceanic basalts: A discussion, *J. Geophys. Res.*, *68*, 5100–5102.

Macdonald, G. A. (1967), Forms and structures of extrusive basaltic rocks, in *The Poldervaart Treatise on Rocks of Basaltic Composition*, edited by H. H. Hess and A. Poldervaart, pp. 1–61, Interscience Publishers, New York.

Macdonald, G. A. (1972), *Volcanoes*, Prentice Hall, Englewood Cliffs, N. J.

Macdonald, G. A., and A. T. Abbott (1977), *Volcanoes in the Sea: The Geology of Hawaii*, Univ. of Hawaii Press, Honolulu.

Macdonald, G. A., and J. P. Eaton (1964), Hawaiian volcanoes during 1955, *U. S. Geol. Surv. Bull.*, *1171*, p. 170.

Macdonald, G. A., and T. Katsura (1964), Chemical composition of the Hawaiian lavas, *Geol. Soc. Am. Mem.*, *116*, 477–522.

Macdonald, G. A., A. T. Abbott, and F. L. Peterson (1983), *Volcanoes in the Sea: The Geology of Hawaii*, Univ. of Hawaii Press, Honolulu.

Macdonald, K. C., R. Haymon, and A. Shor (1989), A 220 km^2 recently erupted lava field on the East Pacific Rise near lat 8°S, *Geology*, *17*, 212–216.

Malahoff, A., G. M. McMurtry, J. C. Wiltshire, and Y. Hsueh-Wen (1982), Geology and chemistry of hydrothermal deposits from active submarine volcano Loihi, Hawaii, *Nature*, *298*, 234–239.

Masson, D. G., A. B. Watts, M. J. R. Gee, R. Urgeles, N. C. Mitchell, T. P. Le Bas, and M. Canals (2002), Slope failures on the flanks of the western Canary Islands, *Earth Sci. Rev.*, *57*, 1–35.

McBirney, A. R. (1993), *Igneous Petrology*, Jones and Bartlett, Boston.

Moore, J. G. (1964), *Giant Submarine Landslides on the Hawaiian Ridge*, Prof. Pap. 501-D, pp. D95–98, U.S. Geol. Surv., Reston, Va.

Moore, J. G. (2000), *Exploring the Highest Sierra*, Stanford University Press, Stanfrod, Calif.

Moore, J. G., and D. A. Clague (2002), Mapping the Nuuanu and Wailau landslides in Hawaii, in *Hawaiian Volcanoes: Deep Underwater Perspectives*, edited by E. Takahashi et al., Geophys. Monogr. 128, pp. 223–244, AGU, Washington, D. C.

Moore, J. G., and R. S. Fiske (1969), Volcanic substructure inferred from the dredge samples and ocean-bottom photographs, Hawaii, *Geol. Soc. Am. Bull.*, *80*, 1191–1202.

Moore, J. G., D. A. Clague, and W. R. Normark (1982), Diverse basalt types from Loihi seamount, Hawaii, *Geology*, *10*, 88–92.

Moore, J. G., W. R. Normark, and R. T. Holcomb (1994), Giant Hawaiian landslides, *Annu. Rev. Earth Planet. Sci.*, *22*, 119–144.

Moore, R. B., D. A. Clague, M. Rubin, and W. Bohrson (1987), Hualalai volcano: A preliminary summary of geologic, petrologic, and geophysical data., In: *Volcanism in Hawaii*, Decker, R. W., edited by R. W. Decker, T. L. Wright, and P. H. Stauffer, Prof. Pap. 1350, pp. 571–585, U.S. Geol. Surv., Reston, Va.

Moreira, M. A. (2013), Noble gas constraints on the origin and evolution of Earth's volatiles, *Geochem. Perspect.*, *2*, 299–403.

Moreira, M. A., and M. D. Kurz (2013), Noble gases as tracers of mantle processes and magmatic degassing, in *The Noble Gases as Geochemical Tracers, Advances in Isotope Geochemistry*, edited by P. Burnard, Springer-Verlag, Berlin, Heidelberg.

Normark, W. R., J. G. Moore, and M. E. Torresan (1993), Giant volcano-related landslides and the development of the Hawaiian Islands, *U.S. Geol. Surv. Bull.*, *2002*, 184–196.

Perret, F. A. (1913), Volcanic research at Kilauea in the summer of 1911, *Am. J. Sci.*, 4th ser., *36*, 475–483.

Peterson, D. W., and R. I. Tilling (1980), Transition of basaltic lava from pahoehoe to aa, Kilauea Volcano, Hawaii: Field observations and key factors, *J. Volcanol. Geotherm. Res.*, *7*, 271–293.

Pietruszka, A. P., and M. O. Garcia (1999), A rapid fluctuation in the mantle source and melting history of Kilauea Volcano inferred from the geochemistry of its historical summit lavas (1790–1982), *J. Petrol.*, *40*, 1321–1342.

Poland, M. P., A. Miklius, A. J. Sutton, and C. R. Thornber (2012), A mantle-driven surge in magma supply to Kīlauea Volcano during 2003–2007, *Nature Geosci.*, *5*(4), 295–300, doi:10.1038/ngeo1426.

Powers, S. (1915), Hawaii's great volcanoes and the study of them, *Bull. Am. Geograph. Soc.*, *47*, 577–583.

Pukui, M. K., and S. H. Elbert, (1984), *Hawaiian Dictionary*, Univ. of Hawaii Press, Honolulu.

Pukui, M. K., S. H. Elbert, and E. T. Mookini (1974), *Place Names of Hawaii*, Univ. of Hawaii Press, Honolulu.

Rhodes, J. M., and J. P. Lockwood, (Eds.) (1995), *Mauna Loa Revealed: Structure, Composition, History and Hazards*, Geophys. Monogr. 92, AGU, Washington, D. C.

Ribe, N. M., and U. R. Christensen (1999), The dynamical origin of Hawaiian volcanism, *Earth Planet. Sci. Lett.*, *171*, 517–531.

Rison, W., and H. Craig (1983), Helium isotopes and mantle volatiles in Loihi Seamount and Hawaiian Island basalts and xenoliths, *Earth Planet. Sci. Lett.*, *66*, 407-426.

Robinson, J. E., and B. W. Eakins (2006), Calculated volumes of individual shield volcanoes at the young end of the Hawaiian ridge, *J. Volcanol. Geotherm. Res.*, *151*, 309–317.

Rowland, S. K., and G. P. L. Walker (1987), Toothpaste lava: Characteristics and origin of a lava structural type transitional between pahoehoe and a'a, *Bull. Volcanol.*, *49*, 631–641.

Rowland, S. K., and G. P. L. Walker (1988), Mafic-crystal distributions, viscosities, and lava structures of some Hawaiian lava flows, *J. Volcanol. Geotherm. Res.*, *35*, 55–66.

Rowland, S. K., and G. P. L. Walker (1990), Pahoehoe and a'a in Hawaii: Volumetric flow rate controls the lava structure, *Bull. Volcanol.*, *52*, 615–628.

Rowland, S. K., A. J. L. Harris, M. J. Wooster, H. Garbeil, P. J. Mouginis-Mark, F. Amelung, and L. Wilson (2003), Volumetric characteristics of lava flows from interferometric radar and multispectral satellite data: The 1995 Fernandina and 1998 Cerro Azul eruptions in the western Galápagos, *Bull. Volcanol.*, *65*, 311–330.

Rowland, S. K., H. Garbeil, and A. J. L. Harris (2005), Lava channel lengths and hazards on Mauna Loa determined from thermal and downslope modeling with FLOWGO, *Bull. Volcanol.*, *67*, 634–647.

Satake, K., J. R. Smith, and K. Shinozaki (2002), Volume estimate and tsunami modeling for the Nu'uanu and Wailau landslides, in *Hawaiian Volcanoes: Deep Underwater Perspectives*, edited by E. Takahashi et al., Geophys. Monogr. 128, pp. 333–348, AGU, Washington, D. C.

Self, S., T. Thordarson, L. Keszthelyi, G. P. L. Walker, K. Hon, M. T. Murphy, P. Long, and S. Finnemore (1996), A new model for the emplacement of Columbia River basalts as large, inflated pahoehoe lava flow fields, *Geophys. Res. Lett.*, *23*, 2689–2692.

Self, S., L. Keszthelyi, and T. Thordarson (1998), The importance of pahoehoe, *Ann. Rev. Earth Planet. Sci.*, *26*, 81–110.

Shepard, E. S. (1925), The analysis of gases obtained from volcanoes and from rocks, *J. Geol.*, *33*, 289–370.

Sherman, S. B., M. O. Garcia, and E. Takahashi (2002), Geochemistry of volcanic glasses from piston cores taken north of O'ahu and Moloka'i islands, Hawaii, in *Hawaiian Volcanoes: Deep Underwater Perspectives*, edited by E. Takahashi et al., Geophys. Monogr. 128, pp. 263–277, AGU, Washington, D. C.

Shinozaki, K., Z. Y. Ren, and E. Takahashi (2002), Geochemical and petrological characteristics of Nuuanu and Wailau landslide blocks, in *Hawaiian Volcanoes: Deep Underwater Perspectives*, edited by E. Takahashi et al., Geophys. Monogr. *128*, pp. 297–310, AGU, Washington, D. C.

Sleep, N. H. (1990), Hotspots and mantle plumes: Some phenomenology, *J. Geophys. Res.*, *95*, 6715–6736.

Smith, J. R., K. Satake, and K. Suyehiro (2002), Deepwater multibeam sonar surveys along the southeastern Hawaiian ridge: Guide to the CD-ROM, in *Hawaiian Volcanoes: Deep Underwater Perspectives*, edited by E. Takahashi et al., Geophys. Monogr. *128*, pp. 333–348, AGU, Washington, D. C.

Sparks, R. S. J., and H. Pinkerton, (1978), The effect of degassing on the rheology of basaltic lava, *Nature*, *276*, 385–386.

Stearns, H. T. (1946), Geology of the Hawaiian Islands, *Hawaii Div. Hydrogr. Bull.*, *8*, 112.

Stearns, H., and G. A. Macdonald (1946), Geology and groundwater resources of Hawaii, *Hawaii Div. Hydrogr. Bull. 9*, 363.

Stolper, E., D. J. DePaolo, and D. M. Thomas (2009), Deep drilling into a mantle plume volcano: The Hawaii Scientific Drilling Project, *Sci. Drilling*, *7*, 1–14.

Stuart, F. M., S. Lass-Evans, J. G. Fitton, and R. M. Ellam (2003), High $^{3}He/^{4}He$ ratios in picritic basalts from Baffin Island and the role of a mixed reservoir in mantle plumes, *Nature*, *424*, 57–59.

Swanson, S., M. T. Nancy, H. R. Westrich, and J. C. Eichelberger (1989), Crystallization history of Obsidian Dome, Inyo Domes, California, *Bull. Volcanol.*, *51*, 161–176.

Takahashi, E., P. W. Lipman, M. O. Garcia, J. Naka, J., and S. Aramaki (Eds.) (2002), *Hawaiian Volcanoes: Deep Water Perspectives*, Geophys. Monogr. *128*, AGU, Washington, D. C.

Thordarson, T., and S. Self, (1998), The Roza flow of the Columbia River Basalt Group: A gigantic pahoehoe flow field, *J. Geophys. Res.*, *103*(B11), 27,411–27,445.

Tilling, R. I., J. P. Kauahikaua, S. R. Brantley, and C. A. Neal (2014), The Hawaiian Volcano Observatory—a natural laboratory for studying basaltic volcanism, Chapter 1 of Poland, M. P., T. J. Takahashi, and C. M. Landowski, (eds.), Characteristics of Hawaiian Volcanoes, *U.S. Geol. Surv. Prof. Pap. 1801*, pp. 1–64, doi:10.3133/pp18011.

Tolan, T. L., S. P. Reidel, M. H. Beeson, J. L. Anderson, K. R. Fecht, and D. A. Swanson (1989), Revisions to the extent and volume of the Columbia River Basalt Group, *Geol. Soc. Am. Spec. Pap*, *239*, 1–20.

U.S. Navy (1939), Bathymetric Chart of the North Pacific Ocean, Hydrographic Office, no. 5486, U.S. Navy, Washington, D. C.

Vye-Brown, C., S. Self, and T. L. Barry (2013), Architecture and emplacement of flood basalt flow fields: Case studies from the Columbia River Basalt Group, NW USA, *Bull. Volcanol.*, *75*, 697–718, doi:10.1007/s00445-013-0697-2.

Walker, G. P. L. (1987), Pipe vesicles in Hawaiian basaltic lavas: Their origin and potential as paleoslope indicators, *Geology*, *15*, 84–87.

Walker, G. P. L. (1989), Spongy pahoehoe in Hawaii: A study of vesicle-distribution patterns in basalt and their significance, *Bull. Volcanol.*, *51*, 199–209.

Walker, G. P. L. (1991), Structure, and origin by injection under surface crust, of tumuli, "lava rises," "lava-rise pits," and "lava-inflation clefts" in Hawaii, *Bull. Volcanol.*, *53*, 546–558.

Wanless, V. D., M. O. Garcia, J. M. Rhodes, D. Weis, M. Norman, D. J. Fornari, M. Kurz, and H. Guillou (2006), Submarine radial vents on Mauna Loa Volcano, Hawaii, *Geochem. Geophys. Geosyst. (*G3*)*, 7, Q05001, doi:10.1029/2005GC001086.

Wantim, M. N., M. Kervyn, G. G. J. Ernst, M. A. del Marmol, C. E. Suh, and P. Jacobs (2013), Numerical experiments on the dynamics of channelised lava flows at Mount Cameroon volcano with the FLOWGO thermo-rheological model, *J. Volcanol. Geotherm. Res.*, *253*, 35–53.

Weis, D., M. O. Garcia, J. M. Rhodes, M. Jellinek, and J. S. Scoates (2011), Role of the deep mantle in generating the compositional asymmetry of the Hawaiian mantle plume, *Nature Geosci.*, *4*, 831–838.

Wentworth, C. K. (1927), Estimates of marine and fluvial erosion in Hawaii, *J. Geol.*, *1*, 117–133.

Westervelt, W. D. (1910), *Legends of Maui: A demi god of Polynesia and his mother Hina*, Hawaiian Gazette, Co., Honolulu.

Wilmouth, R. A., and G. P. L. Walker (1993), P-type and S-type pahoehoe: A study of vesicle distribution patterns in Hawaiian lava flows, *J. Volcanol. Geotherm. Res.*, *55*, 129–142.

Wolfe, C. J., S. C. Solomon, G. Laske, J. A. Collins, R. S. Detrick, J. A. Orcutt, D. Bercovici, and E. H. Hauri (2009), Mantle shear-wave velocity structure beneath the Hawaiian hot spot, *Science*, *326*, 1388–1390.

Wright, T. L., T. J. Takahashi, and J. D. Griggs (1992), *Hawai'i Volcano Watch; A Pictorial History, 1779–1991*, University of Hawai'i Press and Hawai'i Natural History Association, Honolulu.

Xing, Y., S.- L. Yang, H.-L. Chen, Z.- Q. Chen, Z.-L. Li, G. E. Batt, and Y.-Q. Li (2011), Permian flood basalts from the Tarim Basin, Northwest China: SHRIMP zircon U–Pb dating and geochemical characteristics, *Gondwana Res.*, *20*, 485–497.

Yokose, H. (2002), Landslides on the windward flanks of Oahu and Molokai, Hawaii, in *Hawaiian Volcanoes: Deep Underwater Perspectives*, edited by E. Takahashi et al., Geophys. Monogr. 128, pp. 245–261, AGU, Washington, D. C.

Zhong, S., and A. B. Watts (2002), Constraints on the dynamics of mantle plumes from uplift of the Hawaiian Islands, *Earth Planet. Sci. Lett.*, *203*, 105–116.

2

Seismic Constraints on a Double-Layered Asymmetric Whole-Mantle Plume Beneath Hawai'i

Cheng Cheng[1], Richard M. Allen[1], Rob W. Porritt[1,2], and Maxim D. Ballmer[3,4]

ABSTRACT

It is generally accepted that mantle plumes are responsible for hotspot chains and as such provide insight to mantle convection processes. Among all the hotspots, the Hawaiian chain is a characteristic example that has been extensively explored. However, many questions remain. If a plume does exist beneath the Hawaiian chain, what is the shape, size, and orientation of the plume conduit? To what extent can the seismic structure of the plume be mapped? Can we see a continuous plume conduit extending from the lower to the upper mantle? At what depth do melting processes occur? Here, we combine constraints from three data sets (body waves, ballistic surface waves, and ambient noise) to create 3D images of the velocity structure beneath the Hawaiian Islands from a depth of ~800 km to the surface. We use data from the Hawaiian Plume Lithosphere Undersea Melt Experiment (PLUME), which was a network of four-component broadband ocean bottom seismometers that had a network aperture of ~1000 km. Our multiphase 3D model results indicate there is a large deep-rooted low-velocity anomaly rising from the lower mantle. At transition zone depths the conduit is located to the southeast of Hawai'i. A 2% S-wave anomaly is observed in the core of the plume conduit around 700 km depth, which, once corrected for damping effects, suggests a 200–250°C temperature anomaly assuming a thermal plume. In the upper mantle, there is a horizontal plume "pancake" at shallow depths beneath the oceanic lithosphere, and there is also a second horizontal low-velocity layer in the 250 to 410 km depth range beneath the island chain. This second layer is only revealed after surface wave phase velocity data are incorporated into the inversion scheme to improve the constraints on the structure in the upper ~200 km. We suggest this feature is a deep eclogite pool (DEP), an interpretation consistent with geodynamic modeling [*Ballmer et al.*, 2013]. The model also shows reduced lithospheric velocities compared to the typical ~100 Myr old lithosphere, implying lithospheric rejuvenation by the plume. In addition, a shallow (~20 km) low-velocity anomaly is observed southeast of the Island of Hawai'i. This suggests a newly modified lithosphere, as might be expected in the location of an emerging new island in the Hawaiian chain.

2.1. INTRODUCTION AND MOTIVATION

The Hawaiian Islands are an ideal place to study intraplate hotspots. Many researchers consider it to be a case example of a deep-rooted whole-mantle plume [*Morgan*, 1971]. While a plume origin is broadly accepted, there is an ongoing debate about the morphology of the plume system, including the depth of origin, and the direction from which the plume originates if it is not vertical. The structure of the plume in the upper mantle and how it interacts with the overriding lithosphere of the

[1]*Department of Earth and Planetary Science, University of California, Berkeley, California, USA*

[2]*Department of Earth Science, University of Southern California, Los Angeles, California, USA*

[3]*Department of Geology and Geophysics, University of Hawai'i at Mānoa, Honolulu, Hawaii, USA*

[4]*Earth-Life Science Institute, Tokyo Institute of Technology, Meguro, Tokyo, Japan*

Hawaiian Volcanoes: From Source to Surface, Geophysical Monograph 208. First Edition.
Edited by Rebecca Carey, Valérie Cayol, Michael Poland, and Dominique Weis.

Companion Website: www.wiley.com/go/Carey/Hawaiian_Volcanoes

Pacific Plate are also open questions with a variety of geochemical interpretations [e.g., *Lassiter et al.*, 1996; *Abouchami*, 2005; *Huang et al.*, 2011; *Weis et al.*, 2011] and geodynamical models attempting to predict the possible interactions [e.g., *Detrick and Crough*, 1978; *Monnereau et al.*, 1993; *Farnetani and Hofmann*, 2009, 2010; *Farnetani et al.*, 2012; *Rychert et al.*, 2013].

Seismic imaging techniques provide a powerful mechanism to constrain the 3D structure and origin of the island chain. There are several regional seismic studies of Hawai'i that are based on onshore station data [*Woods and Okal*, 1996; *Priestley and Tilmann*, 1999; *Tilmann et al.*, 2001] or offshore station data [*Wolfe et al.*, 2009, 2011; *Laske et al.*, 2007, 2011]. Studies that rely exclusively on on shore recorded data have a limited aperture (width) of the seismic array and the poor ray path coverage makes it impossible to fully assess the deeper mantle structure. More recent regional studies instead make use of the offshore deployment of seismometers during the Plume and Lithosphere Undersea Melt Experiment (PLUME), which increases the aperture and thereby constrains structure over a wider area at shallow depth and also deeper into the lower mantle.

For example, *Wolfe et al.* [2009, 2011] use P- and S-wave arrivals from teleseismic earthquakes to image mantle structure to great depths and conclude that the plume stem extends into the lower mantle, with an origin southeast of Hawai'i. SS precursor observations are consistent with this result [*Schmerr and Garnero*, 2006; *Schmerr et al.*, 2010]. However, imaging with inverse scattering of SS waves has been interpreted to suggest the presence of an 800 to 2000 kilometer wide thermal anomaly in and immediately below the transition zone 1000 km west of Hawai'i [*Cao et al.*, 2011]. The conclusion drawn is that hot material does not rise from the lower mantle through a narrow vertical plume but instead accumulates near the base of the transition zone before being entrained into flow toward Hawai'i. *Cao et al.* [2011] also find a thinned transition zone southeast of Hawai'i but disregard it as the estimated excess temperature is too low and is lower than the 300–400°C estimated excess temperature for the anomaly to the west. *Laske et al.* [2011] analyze Rayleigh waves recorded across the PLUME network at frequencies between 10 and 50 mHz, thereby constraining structure in the upper 100–200 km. Their study reveals lithospheric rejuvenation within an area likely confined to within 150 km of the island chain.

In an effort to better constrain the 3D structure of the upper mantle beneath Hawai'i, we combine body and surface wave observations using a joint inversion scheme [*Obrebski et al.*, 2011] and a finite frequency kernel approach. Our approach uses teleseismic body wave travel time measurements and surface wave phase velocity information from ballistic surface waves and ambient noise cross-correlation measurements. All constraints are jointly inverted to obtain a multiphase tomographic shear wave velocity model. The resulting model constrains structure from the surface down to ~800 km depth. It is simultaneously consistent with all the seismic observations, meaning that it takes advantage of the surface wave constraints to resolve shallow (<200 km) structure, while being consistent with teleseismic traveltimes that are able to constrain deeper structure.

2.2. DATA AND METHOD

The PLUME experiment included a large network of four-component broadband ocean bottom seismometers (OBSs) occupying more than 70 sites and having an overall aperture of more than 1000 km [*Laske et al.*, 2009]. PLUME was designed as a tool to determine the deep-mantle seismic velocity structure beneath the Hawaiian hotspot island chain. PLUME was a two-year deployment that ran from January 2005 through June 2007. The first stage of this deployment was a 500 km wide OBS network with interstation spacing of ~80 km, where data were recorded continuously from January 2005 through January 2006. In the second stage of the deployment, the OBSs occupied a 1000 km wide region with station spacing of about 220 km from April 2006 through June 2007. In this two-phase deployment, data from the broadband sensors had undetermined orientations.

The first step in our processing is orienting the PLUME OBS horizontal components using teleseismic P-wave particle motions. Generally, our method produces stable and reliable orientations (average standard deviation is about 6°) over a wide range of earthquake back azimuths. To maximize the signal-to-noise ratio (SNR), we began by applying a 0.04–0.1 Hz bandpass filter and measured ~1100 P-wave relative arrival times on vertical component data. These P-wave measurements were used to determine orientations. We compared our estimated orientations with those determined using surface waves and found them to be consistent. Once these data were rotated to radial and transverse orientations, ~750 S-wave relative arrival times (including direct S and SKS phases) were determined from the SV component via multi channel cross correlation. Of these, we selected 75 events distributed in as wide a range of back azimuth directions as possible (Figure 2.1), restricting the data to events with epicentral distances greater than 30° and M_w >5.5. As part of the waveform-by-waveform quality control, arrivals were picked manually using the Antelope dbpick software. This software has an interface for viewing waveform data and the ability to pick arrival times and provides markers that are then used as a starting point for the cross-correlation step. We use a multi channel least squares cross-correlation approach [*VanDecar and Crosson*, 1990] that results in a relative travel time delay data set. We select only the highest quality data based on

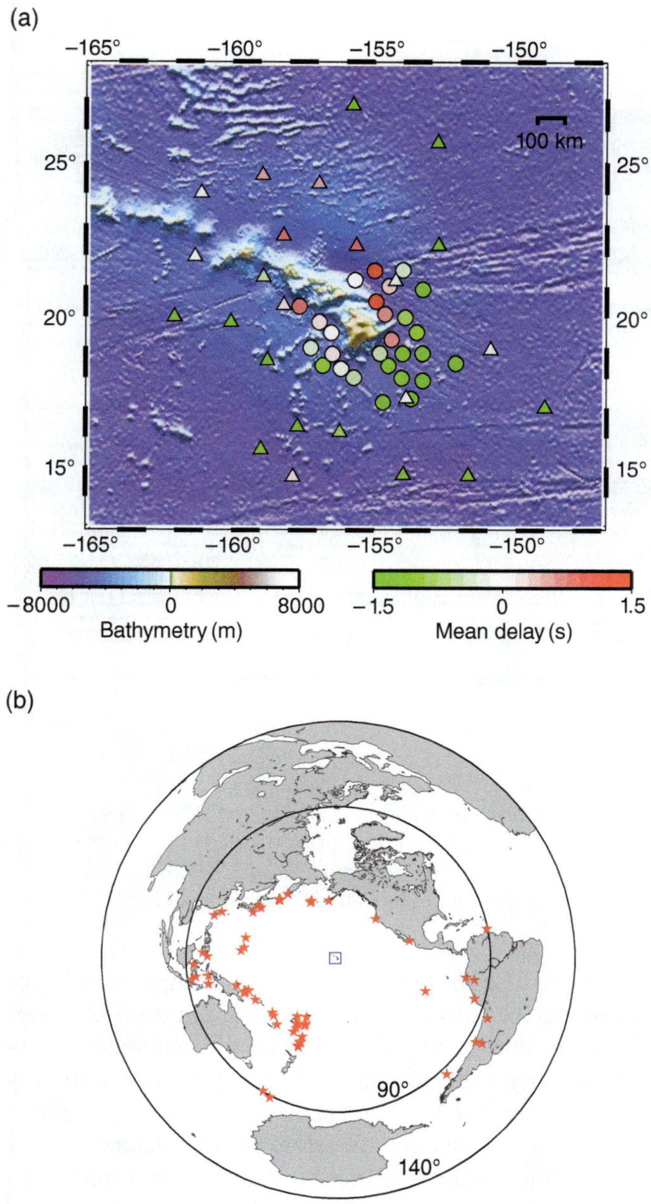

Figure 2.1 Map of the study area. (a) Ocean bathymetry showing seismometer locations. Stations deployed in the first year are indicated by circles and those deployed in the second year are marked by triangles. The station colors indicate the mean body wave delays (measured at 0.04–0.1 Hz). Only stations that successfully recorded data are shown. (b) Map of earthquakes (red stars) used in this study and our study location (blue box). Black circles are 90° and 140° from the study region.

the standard deviation of the cross-correlation-derived delay times to make sure that our body wave data set contains reliable shear arrivals [*Obrebski et al.*, 2011].

The surface wave phase data we use here come from two different sources. The first is ambient noise cross-correlation measurements in the period band of 10–25. Due to the relatively high noise environment of OBS data, a linear stack of ~1 year of ambient noise empirical Green's function results is still rather noisy, which makes accurate measurements of surface wave phase velocity difficult. To reduce this problem, we apply a time-frequency domain phase weighted stacking (tf-PWS) method [*Schimmel and Gallart*, 2007; *Schimmel et al.*, 2011], which efficiently increases the SNR. The tf-PWS is an extension of the phase-weighted stack method that is a non linear stack where each sample of a linear stack is weighted by an amplitude-unbiased coherence measure. The idea here is leveraging the time-frequency phase stack, which is based on the time-frequency decomposition of each trace obtained through the S transform. The results before and after applying the tf-PWS differ significantly (Figure 2.2) and the number of visible dispersion curves within each period band after implementing the tf-PWS is greatly increased. From our time-frequency analysis we observe two wave trends with different travel times (Figure 2.2). The T1 phase is the surface wave energy and the T2 phase is the acoustic wave propagating in the water.

The second source of phase velocity measurements comes from ballistic surface waves. These are the direct surface wave energy as opposed to scattered energy, including ambient noise. We use a two-plane wave tomography method [*Forsyth and Li,* 2005; *Yang and Forsyth,* 2006a, 2006b] in the period band of 25–100. Different from the traditional two-station one-plane-wave method, our method uses the amplitude and phase information simultaneously and the interference of two plane waves to model each incoming teleseismic wavefield. This approach can account for the scattering and multipathing caused by lateral heterogeneities and was developed to image regional-scale structures with network apertures typically up to 1000 km, for example, PLUME. This method has been applied successfully in various regions with a similar network configuration as ours [e.g., *Yang and Forsyth,* 2006b; *Yang et al.*, 2008]. Using the same methodology, we derive phase velocity at a variety of period bands from 25 to 100 s.

To simultaneously invert the phase velocity constraints with the body wave relative travel time constraints, we must determine phase velocity anomaly constraints. This is achieved by subtracting the phase velocities calculated for a background model from the absolute phase velocities. We explored the use of several background models and compared the resulting velocity structure at different crustal depths. We found only slight differences in the models indicating that the choice of background model used does not significantly alter our results. Given this, we used the global average ocean Preliminary Reference Earth Model (PREM) model as the background model.

Following *Obrebski et al.* [2011] we create a joint matrix of body wave relative travel time anomalies and the surface wave phase velocity anomalies to use in a joint inversion. The model space extends from 165°W to 145°W and 12°N to 28°N and to a depth of 1000 km. The model grid

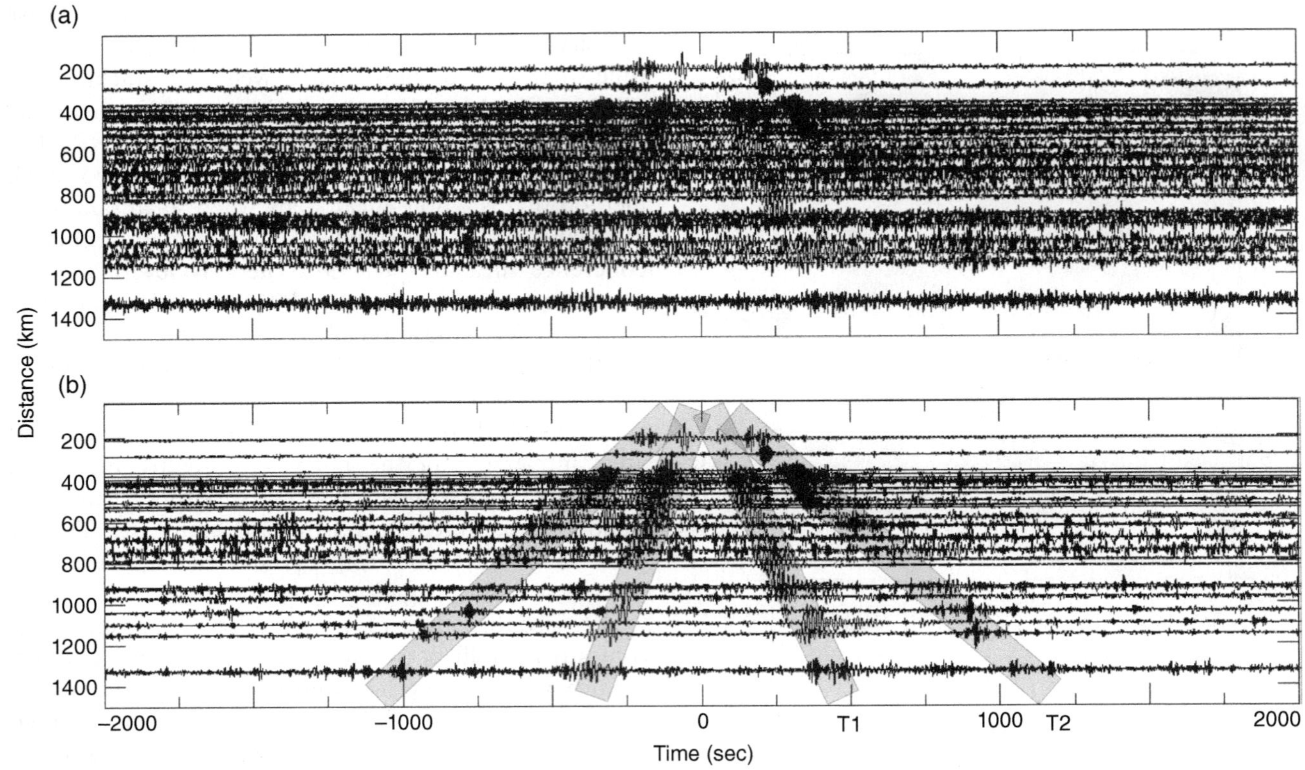

Figure 2.2 Seismic record sections. (a) Record section of the ambient noise cross-correlation between station PL41 and other stations derived using a traditional linear stacking method. (b) The record section shown in (a) except derived using a phase-weighted stacking method [*Schimmel and Gallart*, 2007; *Schimmel* et al., 2011]. Boxed phases, labeled T1 and T2, are surface waves and acoustic waves propagating in the water column, respectively.

includes 33 nodes in both the horizontal and vertical directions, yielding grid spacings of ~30 and ~55 km in the vertical and horizontal directions, respectively. The relative body wave delays are inverted using finite-frequency sensitivity kernels that account for the frequency-dependent width of the region to which body waves are sensitive and also accounts for wavefront healing effects. Our tomographic method uses paraxial kernel theory to calculate the Born approximation forward scattering sensitivity kernels for teleseismic arrival times [*Dahlen et al.*, 2000; *Hung et al.*, 2000, 2004]. The surface wave matrix is made of relative phase velocities estimated for 15 frequencies (10, 12, 15, 18, 20, 22, and 24 s, contributed from ambient noise; 25, 29, 33, 40, 50, 66, 83, and 100 s, contributed from ballistic surface waves) at each node, which constrain the velocity structure from 0 to 300 km depth. To weight the body wave and surface wave constraints in the joint matrix, we use the same weighting scheme as *Julia et al.* [2000] and *Obrebski et al.* [2011]. They define the parameter p in their weighting formula, which allows for a manual tuning of the relative contribution of each data set. After experimentation with various values of p, we settled on using a value of 0.7 as optimal. The sensitivity of very shallow (<60 km) velocity structure to the body waves is also reduced through a ramp parameter. This is set to zero at the surface and increases linearly to 1 at 60 km depth. This is a multiplicative factor applied to the body wave kernels to prevent the body waves from introducing very short wavelength (one station spacing) velocity anomalies at shallow depths where the ray paths are vertical and there is no resolution. The shallowest part of our model is therefore entirely determined by surface wave data constraints. Station terms and event corrections are also included in the inversion. Our inversion requires damping and uses least-squares QR (LSQR) [*Paige and Saunders*, 1982] to iterate to a final model. We also apply a smoothing factor to the model space. To choose the inversion damping parameter, we examine the residual misfit curves as a function of the model norm and find 0.2 is the best damping factor for our inversion scheme.

2.3. IMAGING RESULTS

As a first reference model we produce a smoothed 3D model derived using inversion of only the SV body wave travel times, which we refer to as HW13-SV (Figures 2.3, 2.4b, and 2.4e). The structure is very similar to the teleseismic body wave result of *Wolfe et al.* [2009], as we would expect. We observe a fairly continuous low-velocity feature extending from the surface down through the transition zone and into the deep mantle (Figures 2.3 and 2.4b). We

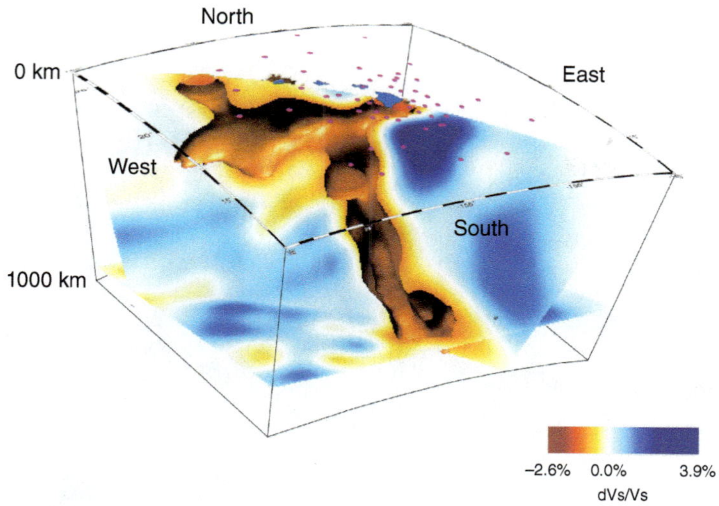

Figure 2.3 A 3D view of our body-wave-only S-wave velocity anomaly model (HW13-SV) for the mantle beneath Hawai'i. Warm colors indicate low-velocity anomaly and cool colors indicate high-velocity anomaly. The value of the isosurface is –1.5%. The model reveals a low-velocity region elongated in the sub vertical direction from the base of the model extending up into the uppermost mantle. This feature is several-hundred kilometers wide and dips to the southeast. We interpret it as the plume conduit. In the upper mantle the low-velocity anomalies are predominantly horizontal and oriented parallel to the island chain. We interpret this as the plume "head," i.e., material that is being dragged along with and beneath the Pacific lithosphere.

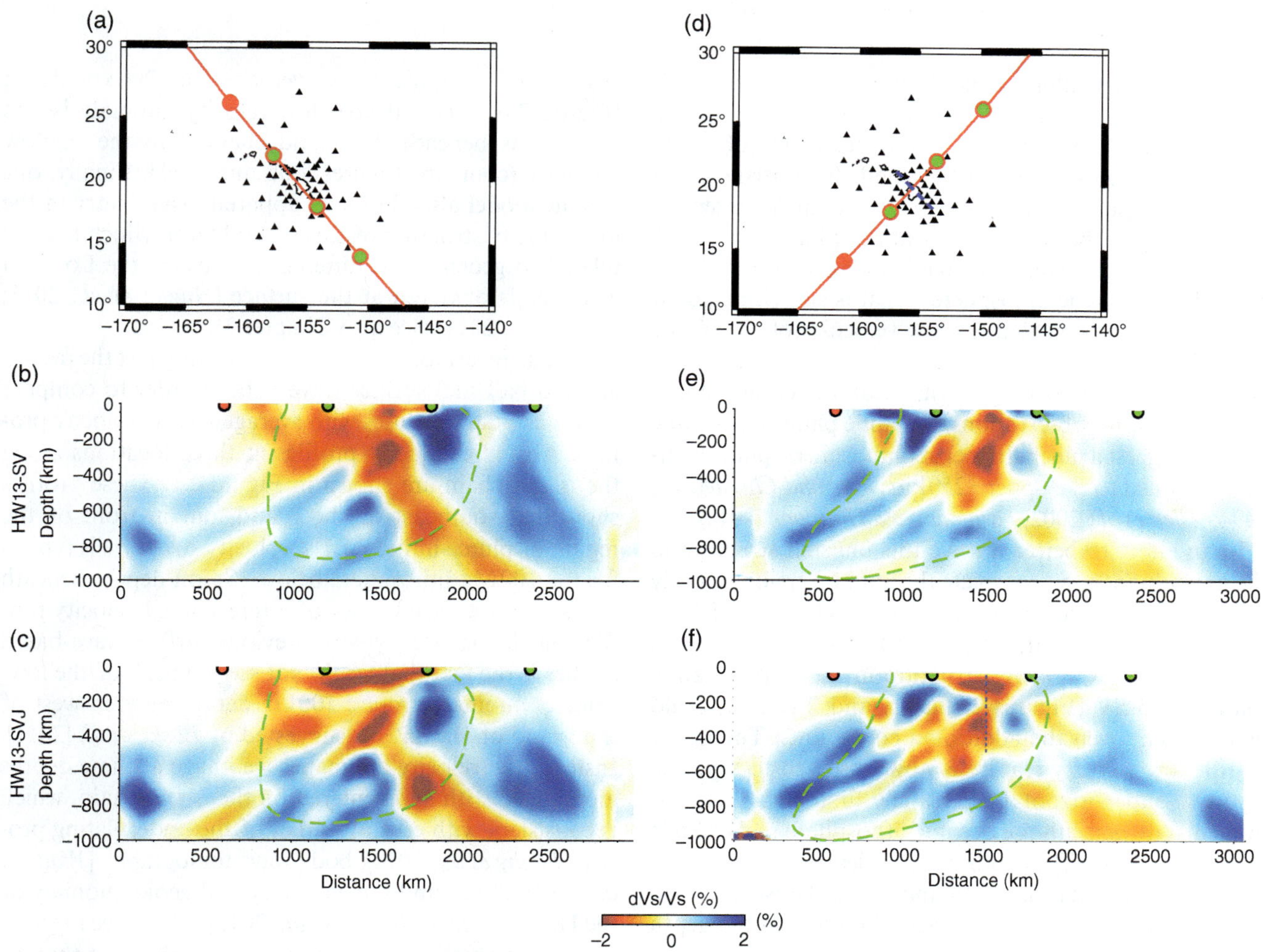

Figure 2.4 Vertical cross sections through the HW13 models: (b, c) cross sections parallel to the Pacific Plate motion; (e, f) cross sections perpendicular to the plate motion. The locations and orientation of the cross sections along with the distribution of stations are shown in (a,d). Results (b) and (e) are from the body-wave-only inversion (HW13-SV); (c) and (f) are from the joint ambient noise, surface wave and body wave inversion (HW13-SVJ). The dashed green contours encompass areas with highest teleseismic body wave ray coverage. The vertical dashed blue line in (d) is beneath the center of the island chain.

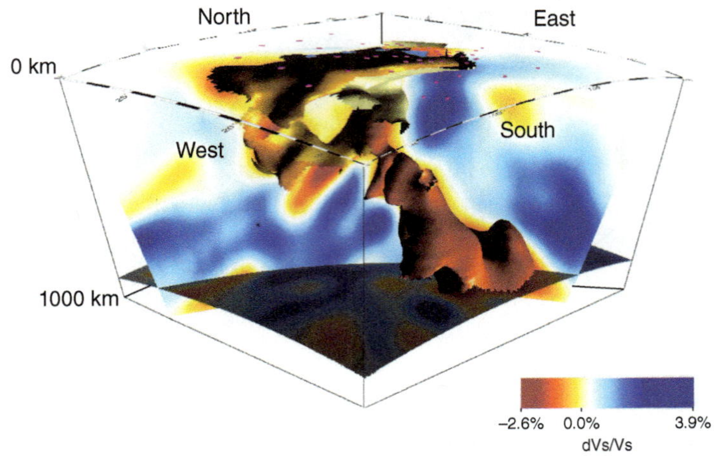

Figure 2.5 A 3D view of our body wave and surface wave joint S-wave velocity anomaly model (HW13-SVJ). This is shown with a similar perspective to the 3D image of HW13-SV in Figure 2.3. The model shows what is interpreted as the plume conduit extending from the lower mantle southeast of Hawai'i up into the upper mantle beneath Hawai'i. In the upper mantle two subhorizontal plume pancakes are observed. The shallow pancake is clearest and extends beneath the lithosphere to a depth of 150 km. The second pancake is less continuous but forms a subhorizontal layer in the 250–400 km depth range. We interpret this as the DEP. This feature is revealed in the joint inversion due to the improved resolution in the upper mantle.

interpret this as a plume conduit, the main stem of which comes from the mantle to the southeast of the surface islands. In the upper mantle, low velocities are observed over a wider region extending the length of the island chain and from the surface down to ~400 km depth. We interpret this low-velocity body as the present-day plume "pancake," i.e., the low-velocity material that is spreading horizontally beneath the oceanic lithosphere. What is surprising about this image is the fact that the low velocities of the plume pancake extend so deep, an observation also noted by *Wolfe et al.* [2009]. Both geodynamical models and previous tomographic observations of mantle plumes in oceanic settings suggest a much thinner sublithospheric plume pancake confined to the upper ~250 km [*Ribe and Christensen*, 1994; *Allen et al.*, 2002; *Farnetani and Hofmann*, 2010].

In an effort to better resolve this unusual structure in the upper mantle, we complete the joint inversion of body and surface wave constraints, which we refer to as HW13-SVJ (Figures 2.4c, 2.4f, 2.5, and 2.6). This model retrieves a more complex velocity structure in the upper mantle than HW13-SV (Figures 2.3, 2.4b, and 2.4c), as would be expected given the additional constraints. The upper mantle low-velocity feature is now observed to separate into two sub horizontal layers. The first is shallow and immediately beneath the oceanic lithosphere and extends to ~150 km depth. The depth extent is clearest in Figure 2.4c and its lateral extent beneath and along the island chain is shown in Figure 2.6b. The second layer is in the depth range of ~250–400 km and is somewhat less continuous in the horizontal direction than the shallow layer (Figures 2.4c, 2.5, and 2.6d), but is distinct from and not continuous with the shallow layer, as illustrated by the absence of significant low velocities at 200 km depth (Figure 2.6c). This deeper low-velocity anomaly is less continuous beneath the island chain than the shallow anomaly (compare Figures 2.6b and 2.6d). Finally, our seismic model also shows an apparent asymmetry in the low-velocity structure of this second layer, which may be related to geochemical differences between the Loa and Kea trends observed at the surface [*Huang et al.*, 2011; *Weis et al.*, 2011] (see also Chapter 3).

We also invert for absolute velocity using just the (ambient and ballistic) surface wave data in order to compare these absolute velocities with average Pacific velocity profiles. Absolute velocity profiles at three locations along the island chain are shown in Figure 2.7. These profiles show that southeast of the island, out in front of the region of plume influence, the velocity structure is typical for the 100 Ma lithosphere below ~75 km depth. Beneath the islands we see evidence of a rejuvenated velocity profile. One inconsistency with previous surface-wave-based studies is the fact that we do not see the peak of the low-velocity anomaly around 100 km depth centered west of the Island of Hawai'i, as suggested by *Laske et al.* [2011]. Instead, the velocity anomalies we image at these depths are aligned with and centered on the island chain, which is consistent with crustal studies of the underplating process [*Leahy et al.*, 2010], body wave tomography [*Wolfe et al.*, 2009, 2011], and the geometry and geoid anomaly of the Hawaiian Swell [*Cadio et al.*, 2012]. We do see the low-velocity anomalies extending further to the west than to the east, which is consistent with a recent receiver function study suggesting that the melt path to the Island of Hawai'i originates to the west [*Rychert et al.*, 2013].

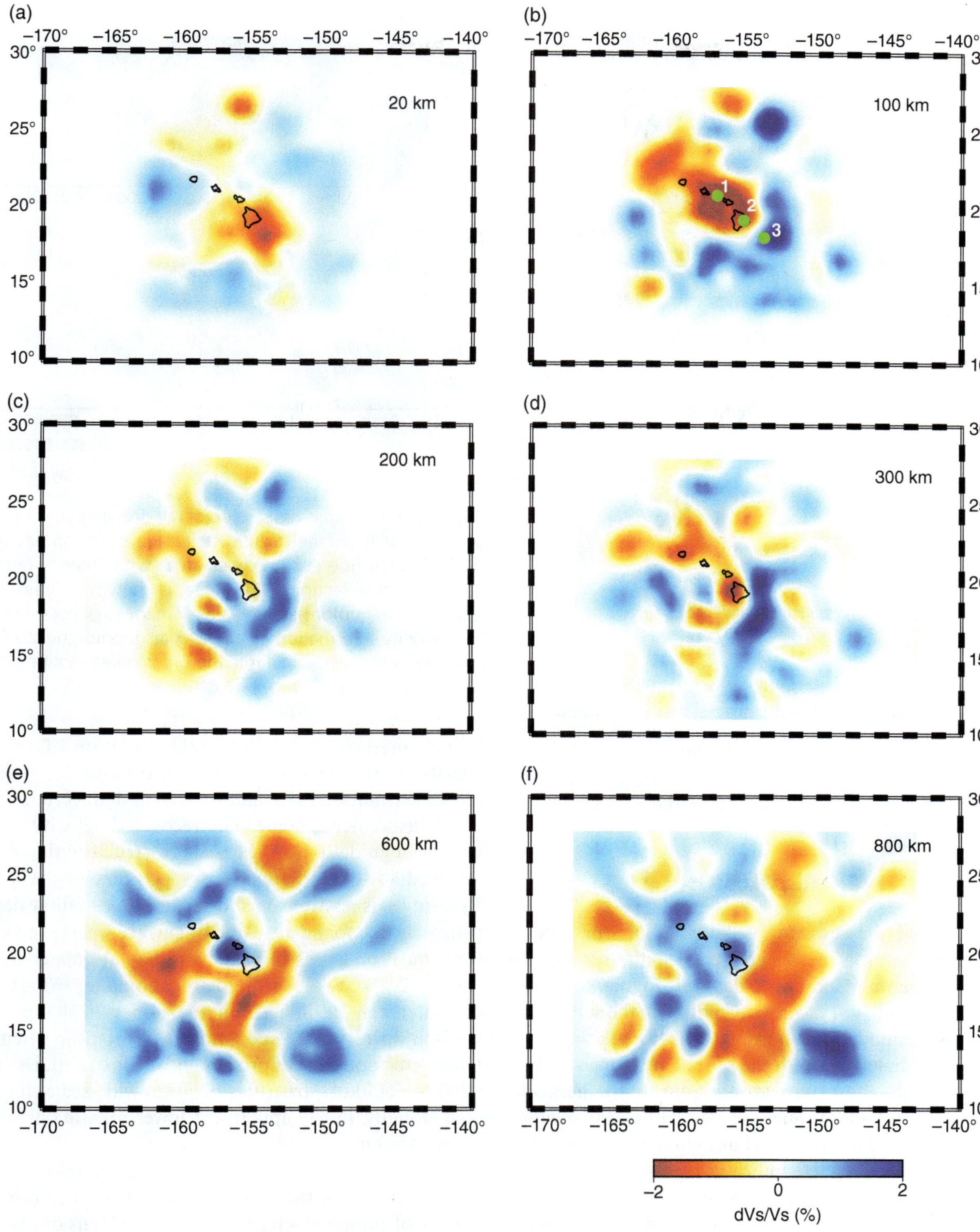

Figure 2.6 Maps of shear wave velocity perturbations in HW13-SVJ at (a) 20, (b) 100, (c) 200, (d) 300, (e) 600 and (f) 800 km depth. Numbered green dots in subpanel (b) are the locations of low-velocity zone 1, low-velocity zone 2, and high-velocity zone 3 in Figure 2.7.

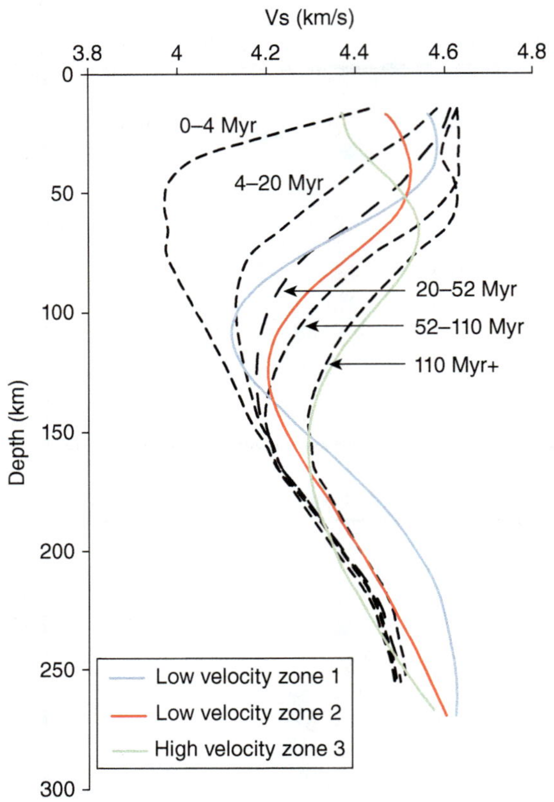

Figure 2.7 Upper mantle absolute velocity profile derived from ambient noise and surface wave inversions. The dashed lines are average mantle velocity profiles for the Pacific lithosphere with various ages [*Nishimura and Forsyth*, 1989]. The colored lines show the velocity profiles at various locations along the length of the island chain and are indicated in Figure 2.6b.

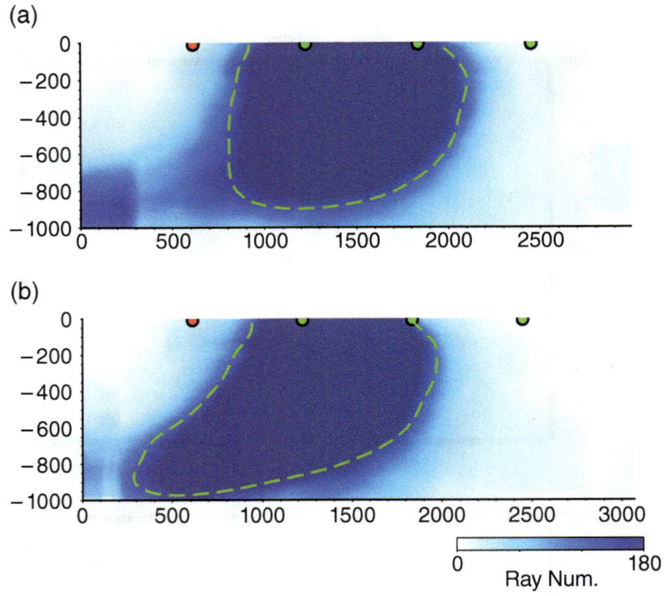

Figure 2.8 Vertical cross section of ray density maps. (a) Cross section that is parallel to the Pacific Plate motion where the dashed green line encompasses areas with "high" ray coverage. This cross section is the same as shown in Figures 2.5b and 2.5c. The color bar shows how many rays contributed to the velocity information at a particular model grid point. (b) Cross section perpendicular to the Pacific Plate motion and the same as shown in Figures 2.5e and 2.5f.

2.4. RESOLUTION

We explored the resolving capabilities of our models using several approaches. First, we show the ray density from the teleseismic body wave data set. Figure 2.8 shows the region with highest ray density plotted in the same cross sections as the model shown in Figure 2.4. The coverage is densest immediately beneath the seismic network and to a depth similar to the maximum aperture of the network. The green dashed line outlines the region of dense coverage and is also plotted on the model in Figure 2.4. Ray density is by no means a complete indicator of resolution (see the following additional tests) and crossing rays are needed to resolve structure. However, it is an indicator of where there are many ray paths constraining the velocity structure.

We next conduct checkerboard resolution tests using synthetic velocity models consisting of alternating high- and low-velocity boxes. These synthetic velocity models are used to generate synthetic data, that are then inverted to assess the ability of the data set to recover the synthetic models. Again, these types of tests are not perfect and do make the assumption that the finite frequency sensitivity kernels used are a "perfect" representation of the true sensitivity of the constraints. We do add 5% noise to the constraints to simulate errors and uncertainties in the measurements made. The recovered velocity models provide a guide to how the scale of structure that can be resolved varies as a function of position in the model. Resolution is possible on shorter scales at shallow depths, primarily due to the inclusion of surface waves, and structures must be larger to be constrained at greater depths. Figure 2.9 shows that velocity anomalies ~200 km in diameter can be imaged in the upper ~250 km where surface waves provide constraints. At depths where the teleseismic body waves provide most constraint, i.e., from ~300 to ~600 km, structures ~350 km wide are well recovered. Finally, in the uppermost lower mantle only structures ~450 km wide are recovered.

Finally, we construct a suite of synthetic velocity models to further test the ability of our data set to resolve a variety of input velocity structures. We focus on the ability of the data set to constrain vertical plumelike anomalies at a variety of depths. Figure 2.10 shows the input and the recovered velocity structure for tests with a vertically oriented low-velocity anomaly 200 km wide and with a peak velocity anomaly of −4%. In the first test (Figures 2.10a, and 2.10e) the anomaly extends from 0 to 300 km. The recovered velocity structure successfully captures the input structure in that the strongest low-velocity

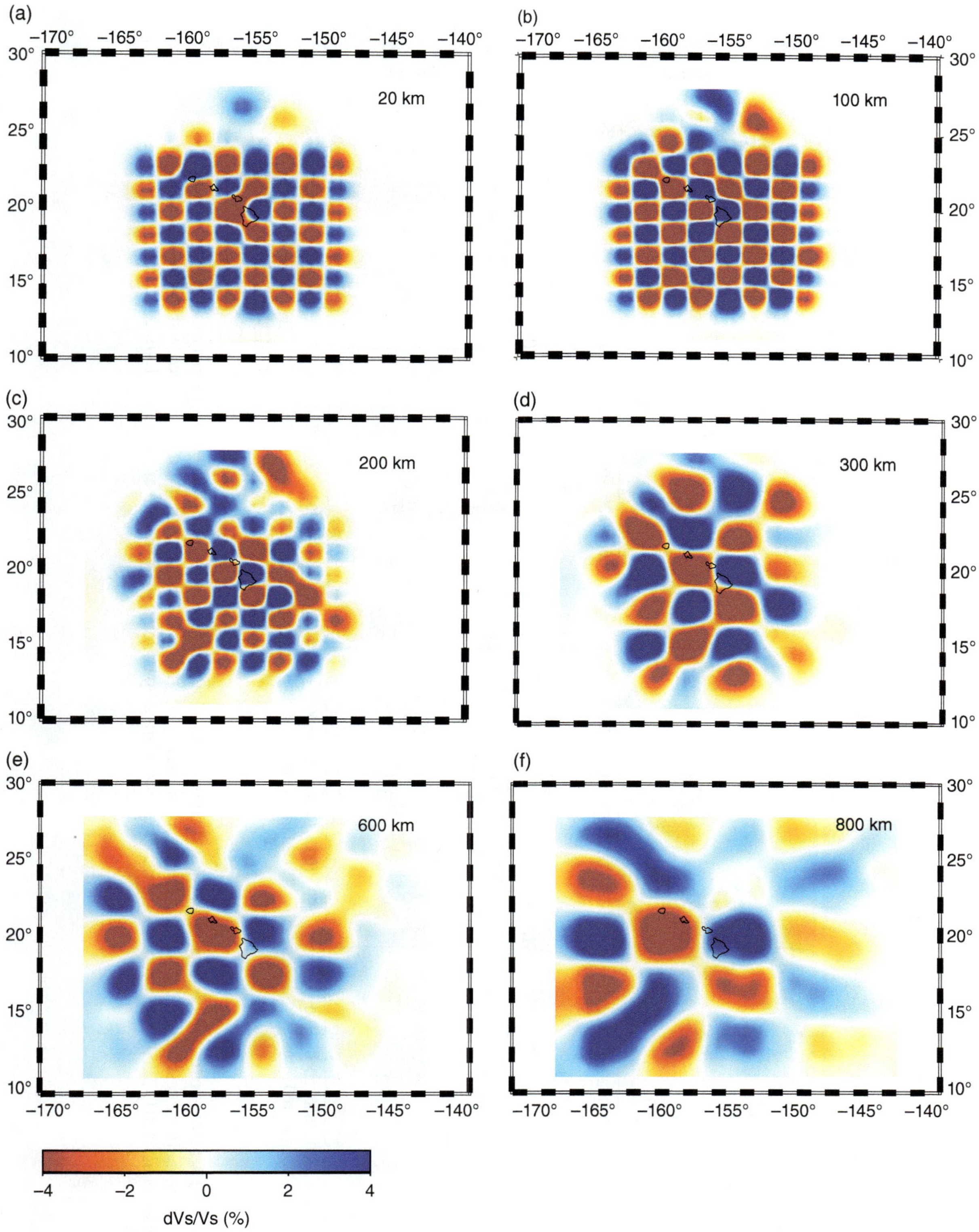

Figure 2.9 Checkerboard resolution tests for input (synthetic) velocities anomalies of ±4%. Horizontal slices through the recovered velocity anomalies are shown at depths of 20, 100, 200, 300, 600, and 800 km.

anomaly also extends from 0 to 300 km depth and has approximately the same width. The amplitude of the anomaly also approaches the input −4% at most depths. Ray path smearing effects are also seen with high- and low-velocity anomalies radiating down and outward from the input velocity structure. However, these smearing effects are all low amplitude with the largest streaking anomalies only being ~0.5%. Based on this test, we

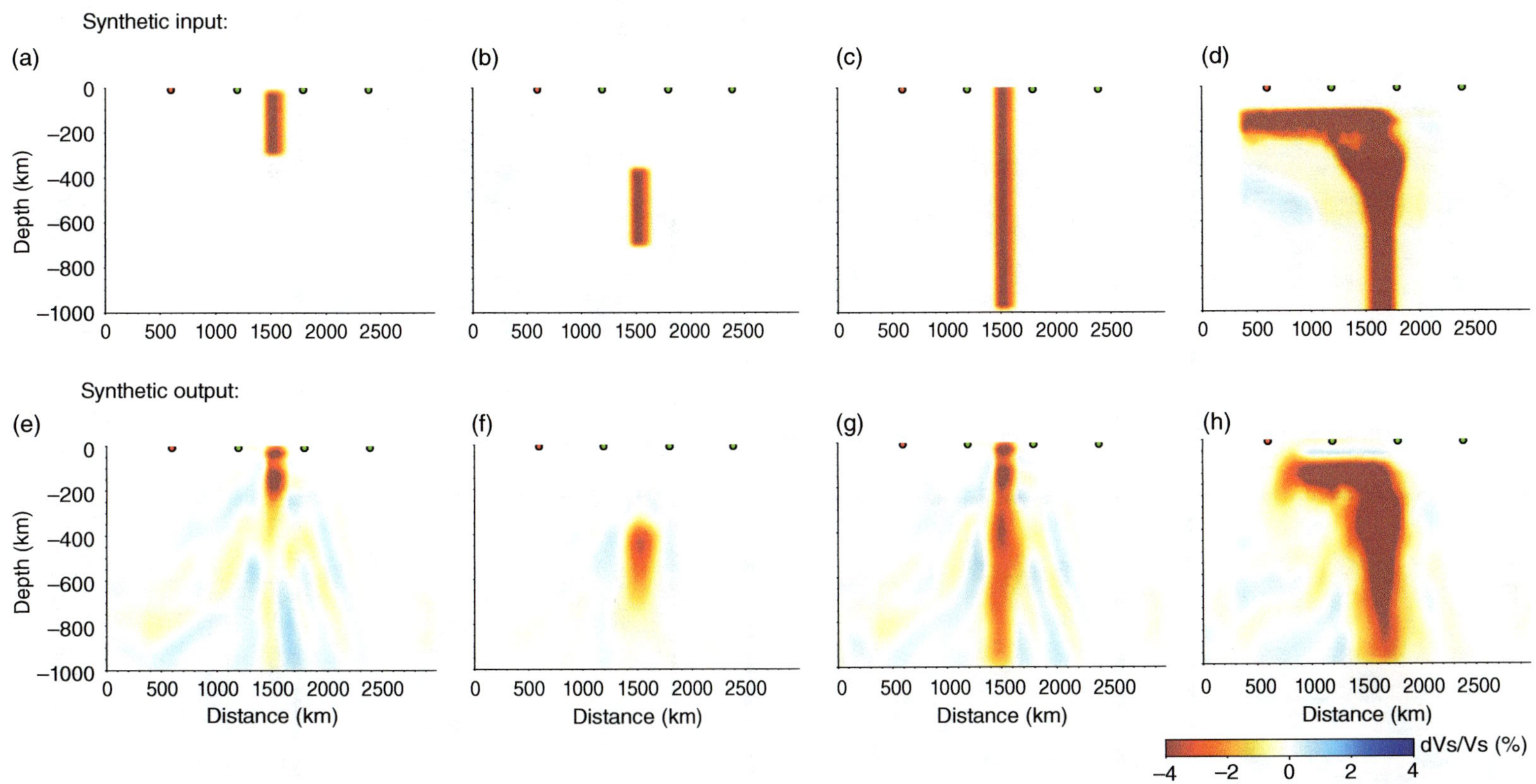

Figure 2.10 Synthetic resolution tests: (a–d) vertical cross sections through four synthetic input velocity models; (e–h) recovered velocity structure. These tests include vertical columns in various depth ranges (a–c; e–g) and a test of the model structure expected for *Ballmer et al.'s* [2013] geodynamic model (d, h). The sections are all parallel to plate motion as in Figures 2.4b, 2.4c, and 2.8a.

conclude that our model would show a strong velocity anomaly constrained to the upper 300 km if the true anomaly structure was only in the upper 300 km. Next we explore the possibility that the low-velocity anomaly is only 350–650 km deep (Figures 2.10b and 2.10f). Again the recovered velocity structure is largely constrained to the same depth range with less smearing than the first test. The amplitude recovery is reduced to −2% to −3%, meaning that we only recover 50%–75% of the velocity anomaly. Next we use a low-velocity anomaly that extends throughout the full depth range of the model (Figures 2.10c and 2.10 g) and find that this is also well recovered with only low-amplitude anomalies resulting from smearing of velocities along downward going ray paths similar to what was observed in the first test. The amplitude of the velocity anomalies in the input plume stem also decreases with depth in a similar fashion to the first two tests.

In the final test we use a synthetic velocity structure based on the "two-layer" geodynamic model for mantle flow beneath Hawai'i developed by *Ballmer et al.* [2013]. We estimated the absolute velocities from the predicted temperature, density, and lithostatic pressure at each point of the quasi-steady-state model using the relationships of *Faul and Jackson* [2005] and *Jackson and Faul* [2010] with specifications reported in *Ballmer et al.* [2013]. We then subtract the 1D reference velocity model to determine velocity anomalies for input into our synthetic test. The recovered structure is good beneath the seismic array, including the variable size of the low-velocity anomaly in the upper mantle associated with Ballmer's two-layer structure. These tests leave us confident that the upper mantle structure is well resolved in our seismic inversion.

2.5. DISCUSSION

2.5.1. Structure and Origin of Plume Conduit

Our imaging results suggest that a subvertically oriented low-velocity anomaly exists beneath Hawai'i that extends through the upper mantle and down into the uppermost lower mantle. We interpret this feature as being indicative of a whole-mantle-plume source for the Hawaiian island chain. This structure and interpretation is similar to the conclusion of *Wolfe et al.* [2009]. Both results reveal low velocities within the mantle transition zone and in the topmost lower mantle, suggesting there is a deep source region for the Hawaiian plume. A thermal boundary layer near the base of the transition zone has been suggested as a candidate source region for some plumes [e.g., *Cserepes and Yuen*, 2000]. However, our image does not support this hypothesis as we see no evidence of a broadening of the low-velocity anomaly beneath the 660 km discontinuity, as would be expected in that case.

We interpret our model to indicate that the plume conduit has an origin southeast of the Island of Hawai'i. As shown by Figures 2.4 and 2.6, there are several locations in the uppermost lower mantle that have low velocities in our model. However, we interpret the anomaly to the southeast to be the plume conduit for three reasons. (1) The low velocities in the southeast quadrant are larger in size (lateral extent) and amplitude (see Figure 2.6f) than any other quadrant, reaching 2% when others are closer to 1%. (2) The smaller low-velocity anomalies seen in other quadrants are outside the region of dense ray coverage, i.e. outside the green dashed lines in Figure 2.4. (3) The anomalies extending southeast are the most continuous (Figures 2.4 and 2.5). It is also worth noting that the teleseismic ray coverage is not dominated by rays coming from the southeast (Figure 2.1b) so the plume conduit that we image/interpret is not due to a particularly dense ray bundle coming from that direction. Our observation that the plume origin is to the southeast is also consistent with geodynamic arguments that mantle convection in general, and the fast-moving Pacific Plate in particular, shear and tilt the plume conduit in the upper and lower mantle [*Richards and Griffiths*, 1988; *Steinberger and O'Connell*, 1998; *Steinberger et al.*, 2004; *Farnetani and Hofmann*, 2010].

Several previous studies have explored topography of the transition zone discontinuities in the region beneath Hawai'i in an effort to identify anomalies that may be related to the source of the islands. *Shen et al.* [2003] observe transition zone thinning across a broad zone beneath the island chain, which they interpreted as reflecting higher-than-normal temperatures beneath the region due to the passage of plume material through the transition zone. In contrast, *Li et al.* [2000] and *Wölbern et al.* [2006] suggest that the transition zone is thinned ~200 km west of Hawai'i. In a study of SS precursors, *Cao et al.* [2011] argue that ponded plume material sits beneath the transition zone several degrees west of Hawai'i. Our imaging results do not support these observations, instead suggesting that the plume conduit is intersecting the transition zone ~200 km southeast of Hawai'i.

A number of factors can cause velocity heterogeneity in the mantle, including variations in temperature (with contributions from both anharmonic and anelastic components [e.g., *Karato*, 1993]), water content [*Karato*, 2003], melt [e.g., *Hammond and Humphreys*, 2000], grain size [*Faul and Jackson*, 2005; *Jackson and Faul*, 2010], and bulk composition. The last factor includes depletion by melt extraction [*Jordan*, 1979], although recent studies suggest that the effects of depletion on P- and S-wave velocities and V_p/V_s ratios may be minor [*Schutt and Lesher*, 2006; *Afonso et al.*, 2010]. In the uppermost lower mantle (~700 km depth) the S wave velocity anomalies are around 2% in our model. As smoothing and damping are part of the inversion scheme, the velocity anomalies we obtain are minimum estimates [*Allen et al.*, 2002]. From our synthetic tests (Figure 2.10) we find that we recover about 50%–75% of the anomaly at this depth, which implies the real velocity anomalies in the plume conduit likely exceed ~3% [*Allen and Tromp*, 2005]. These velocity anomalies imply an excess temperature of 200–250°C [*Schilling*, 1991; *Allen et al.*, 2002], consistent with petrological estimates [*Herzberg et al.*, 2007].

2.5.2. Upper Mantle Structure and Double-Layered Plume

The low-velocity structure that we image in the upper mantle is in disagreement with the classic version of plume theory. The classic plume model [e.g., *Morgan*, 1972; *Ribe and Christensen*, 1994; *Farnetani and Hofmann*, 2010] involves a near-vertical plume conduit rising through the entire mantle and spreading at the base of the lithosphere to form a pancake of hot material. The main hotspot forms where the conduit feeds the pancake, which is dragged away by plate motion, to the northwest in the case of Hawai'i. However, instead of a single shallow plume pancake, we image two distinct horizontal low-velocity layers in the upper mantle. One layer is at <150 km depth, and the second is in the ~250–400 km depth range. Whereas this study is the first to image two separate layers, it is consistent with a previous study from *Wolfe et al.* [2009], who have recovered one broad low-velocity body that extends from the base of the lithosphere to ~400 km depth.

Our imaged model structure is consistent with a recent geodynamic model [*Ballmer et al.*, 2013] developed in an effort to explain the thick low-velocity body imaged by *Wolfe et al.* [2009]. While the model was developed simply to explain the existence of a plume pancake to ~400 km depth beneath the island chain, it also suggests a double layer to the plume pancake, as we observe. In Ballmer et al.'s model the plume is composed of 85% peridotite and 15% chemically dense eclogite. As the plume material rises through the upper mantle, it generates a deep eclogitic pool (DEP) at 300–410 km depth, from which a shallow upwelling rises further to feed the pancake [*Ballmer et al.*, 2013]. The compositions used in the geodynamic model are based on geochemical constraints for Hawaiian lava composition, which is thought to contain mafic materials such as eclogites, in addition to peridotite [*Hauri et al.*, 1996; *Sobolev et al.*, 2005; *Herzberg*, 2011; *Pietruszka et al.*, 2013]. These mafic materials likely originate from subducted oceanic crust that is entrained by the plume and are denser than peridotite throughout the upper mantle. The ascent of a plume rich in eclogite is thus controlled by a competition between negative compositional and positive thermal buoyancy. In addition, it is affected by phase transitions in the upper mantle that modulate the densities of eclogite and peridotite to cause a maximum of the excess density of eclogite at 300–410 km depth [*Aoki and Takahashi*, 2004]. It is this excess density maximum that causes the plume to stall and to form a DEP [*Ballmer et al.*, 2013]. When the material rising out of the DEP crosses the coesite-stishovite phase transition at 300 km depth [*Aoki and Takahashi*, 2004], the density of eclogite sharply decreases, thereby inducing rapid upwelling in a narrow plume conduit. The predicted contrast between a wide DEP and a narrow shallow plume [*Ballmer et al.*, 2013] compares well with our observations (Figures 2.5 and 2.6).

2.5.3. Plume Interaction with Lithosphere

The age of the Pacific Plate around the Hawaiian Islands is about 100 Ma [*Müller et al.*, 2008]. Figure 2.7 shows the typical velocity profiles for oceanic plates of different ages derived from surface wave measurements [*Nishimura and Forsyth*, 1989]. We determine the absolute velocity based on our surface wave constraints and compare them to the typical profiles at three different locations that span regions in both the low- and high-velocity portions of the lithosphere (green dots in Figure 2.6b). Location 1 is halfway along the island chain, which is in the middle of the strong low-velocity anomaly at ~100 km depth. Location 2 is at the southwest coast of the Island of Hawai'i and on the edge of the strong low-velocity anomaly. Location 3 is further southwest and "upstream" of the plume, outside the low-velocity region of the plume.

Compared to the typical 100 Ma lithosphere velocity profile that would be expected beneath Hawai'i, the velocity at location 1 is most reduced. The velocity reduction extends from the surface to ~140 km, producing a velocity profile that is more similar to the typical profile for the 20–52 Ma lithosphere. The velocity at location 2 is also reduced to ~140 km depth, but to a lesser extent. These observations are consistent with the concept of lithospheric rejuvenation [e.g., *Detrick and Crough* 1978; *Von Herzen et al.*, 1989; *Monnereau et al.*, 1993; *Li et al.*, 2004]. Location 1 has been rejuvenated to a greater extent than location 2, as would be expected given that location 1 has been under the influence of the lithosphere-modifying plume for a longer time than location 2. Our observations are consistent with those of *Laske et al.* [2011]. We suggest rejuvenation of the lithosphere should be understood primarily as a velocity reduction in the lithosphere and not necessarily a mechanical thinning of the lithosphere.

Location 3 is 150 km southeast of the Island of Hawai'i and is upstream of the plume. We therefore expect little or no rejuvenation effect. Below 75 km depth the velocity profile at location 3 is very similar to the typical profile for the 100 Ma lithosphere (Figure 2.7). However, the velocity is

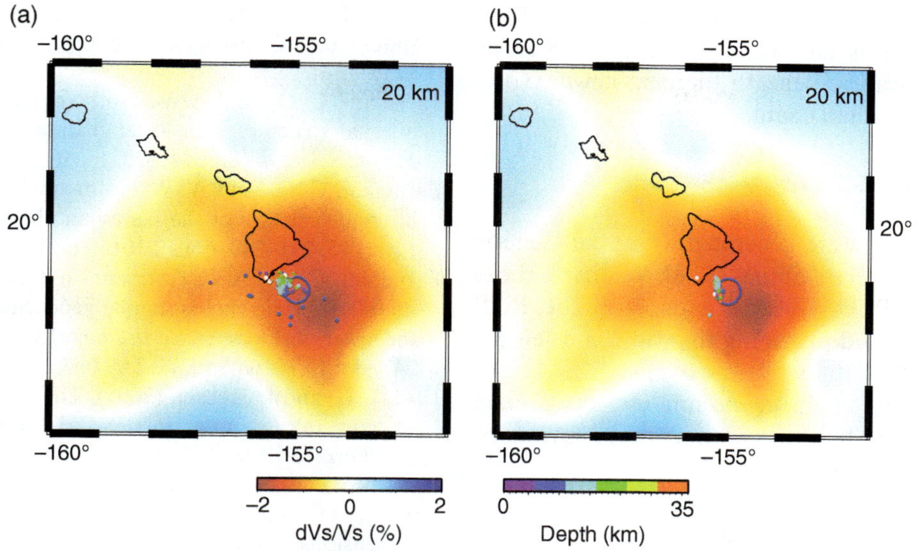

Figure 2.11 Map view of HW13-SVJ at 20 km depth with the location of earthquake swarms from (a) 1971 and (b) 1996 and the location of the Lō'ihi Seamount (blue circle). The lower right color scale indicates earthquake depth and the lower left color scale indicates velocity perturbation in percent.

significantly reduced at shallower depths. The shallow low-velocity zone is also imaged in HW13-SVJ (Figures 2.6a and 2.11). We are able to image this low-velocity anomaly at 20 km depth due to the inclusion of the higher-frequency constraints from ambient seismic noise (down to 0.1 Hz), which have wavelengths of ~45 km. These wavelengths are still too long to image magma chambers, so these shallow low velocities instead indicate a broad region of the lithosphere that must be modified to cause the reduction in velocity. The location of the shallow low-velocity anomaly is centered between earthquake swarms in the upper 30 km and the location of the Lō'ihi Seamount (Figure 2.11). We therefore propose that we are imaging evidence of shallow lithospheric modification in response to the arrival of the next volcano in the island chain.

2.6. SUMMARY

We derive a 3D model (HW13-SVJ) of the Hawaiian plume extending from the lower mantle to the surface using a multiconstraint seismic tomography method that uses the PLUME OBS data. Checkerboard resolution tests and the synthetic tests suggest that the primary features in our model are robust to 800 km depth. Our results show that under the Hawaiian hotspot a continuous low-velocity anomaly extends from the roots of the volcanoes downward into the lower mantle. The plume origin is southeast of the island chain, which is consistent with what mantle flow models predict. The geometry of this low-velocity body is consistent with a whole-mantle plume feeding the hotspot in this region. The 2% low-velocity anomaly in the lower mantle implies a 200–250°C temperature anomaly once corrected for damping and assuming a purely thermal plume. These modeled deep features are similar to prior body wave tomography results [*Wolfe et al.*, 2009, 2011].

The addition of surface wave phase velocity constraints from earthquakes and ambient seismic noise provides additional constraints in the upper mantle, including the lithosphere and crust, which are jointly inverted with the body wave constraints. In the upper mantle, the structure of the low-velocity plume deviates substantially from the classic plume pancake model. Our model suggests the presence of two sub horizontal low-velocity layers in the upper mantle. The first is immediately below the oceanic lithosphere, as expected. The second anomaly is directly above the 410 km boundary and extends up to ~250 km depth. This secondary tomographic feature is consistent with geodynamic models of plumes with high eclogite content [*Ballmer et al.*, 2013]. They predict a low-velocity layer immediately above the 410 discontinuity produced by the accumulation of eclogite. The model also suggests a lateral heterogeneity inside this low-velocity layer that may be related to the observed geochemistry asymmetry at the surface (Chapter 3).

At more shallow, lithospheric depths, our model shows lower velocities than expected for the ~100 Ma oceanic lithosphere. Instead the velocity profiles are more similar to the Pacific lithosphere with a mantle age of 20–50 Ma. This is consistent with previous observations and interpretations of a rejuvenated lithosphere beneath the Island of Hawai'i [*Laske et al.*, 2011]. Finally, at the shallowest depth

(~20 km), we see a low-velocity anomaly southeast of the Island of Hawai'i that is likely indicative of the process of lithospheric modification beneath the next newly evolving island in the Hawaiian chain.

ACKNOWLEDGMENTS

Most of the figures were generated using the Generic Mapping Tool [*Wessel and Smith*, 1998]. We thank those involved with the PLUME experiment and the IRIS DMC for collecting and providing the wideband seismic data we used in this study. We also acknowledge the careful reviews and helpful comments of Cinzia Farnetani, Bernhard Steinberger, anonymous reviewers, and Dominique Weis as editor.

REFERENCES

Abouchami, W. (2005), Lead isotopes reveal bilateral asymmetry and vertical continuity in the Hawaiian mantle plume, *Nature, 434*, 851–856.

Afonso, J. C., J. Ranalli, M. Fernàndez, W. Griffin, S. Y. O'Reilly, and U. Faul, (2010), On the Vp/Vs–Mg# correlation in mantle peridotites: Implications for the identification of thermal and compositional anomalies in the upper mantle, *Earth Planet. Sci. Lett. 289*, 606–618.

Allen, R. M., et al. (2002), Imaging the mantle beneath Iceland using integrated seismological techniques, *J. Geophys. Res.*, *107*, doi:10.1029/2001JB000595.

Allen, R. M., and J. Tromp (2005), Resolution of regional seismic models: Squeezing the Iceland anomaly, *Geophys. J. Int.*, *161*, 373–386 doi:10.1111/j.1365-246X.2005 .02600.x.

Aoki, I., and E. Takahashi, (2004). Density of MORB eclogite in the upper mantle, *Phys. Earth Planet. Int.*, *143–144*, 129–143.

Ballmer, M., G. Ito, C. Wolfe, and S. Solomon, (2013). Double-layering of a thermochemical Hawaiian plume in the upper mantle, *Earth Planet. Sci. Lett.*, *376*, 155–164, doi:10.1016/j.epsl.2013.06.022.

Cadio, C., M. D. Ballmer, I. Panet, M. Diament, and N. Ribe (2012), New constraints on the origin of the Hawaiian swell from wavelet analysis of the geoid to topography ratio, *Earth Planet. Science Lett.*, *359*, 40–54.

Cao, Q., R.D. Van der Hilst, M.V. De Hoop, and S.-H. Shim (2011), Seismic imaging of transition zone discontinuities suggests hot mantle west of Hawaii, *Science*, *332*, 1068, doi:10.1126/science.1202731.

Cheng, C., Allen, R. M., Porritt, R. W., Ballmer, M. D., 0000 Seismic constraints on a double-layered asymmetric whole-mantle plume beneath Hawaii, in: R. Carey, D. Weis, M. Poland, (Eds), *Hawaiian Volcanism: From Source to Surface*, AGU Monograph, this volume.

Cserepes, L. and D. A. Yuen (2000). On the possibility of a second kind of mantle plume, *Earth Planet. Sci. Lett. 183*, 61–71.

Dahlen, F. A., S. H. Hung, and G. Nolet (2000). Frechet kernels for finite-frequency traveltimes—I. Theory, *Geophys. J. Int.*, *141*, 157–174.

Detrick, R. S. and S. T. Crough (1978). Island subsidence, hot spots, and lithospheric thinning, *J. Geophys. Res.*, *83*, 1236–1244.

Farnetani, C. G., and A. W. Hofmann (2009), Dynamics and internal structure of a lower mantle plume conduit, *Earth Planet. Sci. Lett. 282*, 314–322, doi:10.1016/j.epsl.2009.03.035.

Farnetani, C. G., and A. W. Hofmann (2010), Dynamics and internal structure of the Hawaiian plume, *Earth Planet. Sci. Lett. 295*, 231–240, doi:10.1016/j.epsl.2010.04.005.

Farnetani, C. G., A.W. Hofmann and C. Class (2012), How double volcanic chains sample geochemical anomalies from the lowermost mantle, *Earth Planet. Sci. Lett.*, *359–360*, 240–247, doi:10.1016/j.epsl.2012.09.057.

Faul, U.H. and I. Jackson (2005), The seismological signature of temperature and grain size variations in the upper mantle, *Earth Planet. Sci. Lett.*, *234*, 119–134, doi:10.1016/j.epsl.2005.02.008.

Forsyth, D. W., and A. Li (2005), Array-analysis of two-dimensional variations in surface wave velocity and azimuthal anisotropy in the presence of multipathing interference, in Seismic Earth: Array Analysis of Broadband Seismograms, Geophys. Monogr. 157, edited by A. Levander and G. Nolet, pp. 81–97, AGU, Washington, D. C.

Hammond, W. C., and E. D. Humphreys (2000), Upper mantle seismic wave velocity: Effects of realistic partial melt geometries, *J. Geophys. Res. 105*, 10,975–10,986.

Hauri, E. H., J. C. Lassiter, and D. J. DePaolo (1996), Osmium isotope systematics of drilled lavas from Mauna Loa, Hawaii, *J. Geophys. Res.*, *101*(B5), 11,793–11,806, doi:10.1029/95JB03346.

Herzberg, C., P. D. Asimow, N. Arndt, Y. L. Niu, C. M. Lesher, J. G. Fitton, M. J. Cheadle, and A. D. Saunders (2007), Temperatures in ambient mantle and plumes: constraints from basalts, picrites, and komatiites, *Geochem. Geophys. Geosyst.*, *8*, 34.

Herzberg, C., (2011), Identification of source lithology in the Hawaiian and Canary islands: Implications for origins, *J. Petrol.*, *52*, 113–146.

Huang, S., P. S. Hall, and M. G. Jackson (2011), Geochemical zoning of volcanic chains associated with Pacific hotspots, *Nature Geosci.*, *4*, 874–878.

Hung, S. H., F. A. Dahlen and G. Nolet (2000), Fréchet kernels for finite-frequency traveltimes—II. Examples, *Geophys. J. Int.*, *141*, 175–203.

Hung, S. H., Y. Shen and L.Y. Chiao (2004), Imaging seismic velocity structure beneath the Iceland hot spot: A finite frequency approach, *J. Geophys. Res.*, *109*, B08305, doi:10.1029/2003JB002889.

Jackson, I., and U. H. Faul (2010) Grainsize–sensitive viscoelastic relaxation in olivine: Towards a robust laboratory–based model for seismological application, *Phys. Earth Planet. Int.*, *183*, 151–163, doi:10.1016/j.pepi.2010.09.005.

Jordan, T. H., (1979), Mineralogies, densities, and seismic velocities of garnet lherzolites and their geophysical significance, in: The Mantle Sample: Inclusions in Kimberlites and Other Volcanic, edited by F. R. Boyd and H. O. A. Meyer, pp. 1–14, AGU, Washington, D.C.

Julia, J., C. J. Ammon, R. B. Herrmann, and A. M. Correig (2000), Joint in-version of receiver function and surface wave dispersion observations, *Geophys. J. Int.*, *143*, 99–112.

Karato, S. (1993), Importance of anelasticity in the interpretation of seismic tomography, *Geophys. Res. Lett. 20*, 1623–1626.

Karato, S. (2003), Mapping water content in the upper mantle in Inside the Subduction Factory. Geophys. Monogr.*138*, edited by J. M. Eiler, pp. 135–152, AGU, Washington, D. C.

Laske, G., J. Phipps Morgan, and J. A. Orcutt (2007), The Hawaiian SWELL pilot experiment: evidence for lithosphere rejuvenation from ocean bottom surface wave data, in Plates, Plumes and Planetary Processes, Spec. Pap. 430, edited by G. R. Foulger and D. M. Jurdy, pp. 209–233, Geol. Soci. of Am., Boulder, Colo.

Laske, G., J. A. Collins, C. J. Wolfe, S. C. Solomon, R.S. Detrick, J. A. Orcutt, D. Bercovici, and E.H. Hauri (2009), Probing the Hawaiian hot spot with new broadband ocean bottom instruments, *EOS Trans. Am. Geophys. Union.*, *90*, 362–363.

Laske, G., A. Markee, J. A. Orcutt, C. J. Wolfe, J. A. Collins, S. C. Solomon, R.S. Detrick, D. Bercovici, and E. H. Hauri (2011), Asymmetric shallow mantle structure beneath the Hawaiian Swell—Evidence from Rayleigh waves recorded by the PLUME network, *Geophys. J. Int.*, *187*, 1725–1742.

Lassiter, J. C., D. J. DePaolo and M. Tatsumoto (1996), Isotopic evolution of Mauna Kea volcano: Results from the initial phase of the Hawaiian Scientific Drilling Project, *J. Geophys. Res, 101*, 11,769–11,780.

Leahy, G. M., J. A. Collins, C. J. Wolfe, G. Laske, and S. C. Solomon (2010), Underplating of the Hawaiian Swell: Evidence from teleseismic receiver functions, *Geophys. J. Int. 183*, 313–329, doi:10.1111/j.1365-246X.2010.04720.x

Li, X., R. Kind, K. Priestley, S. Sobolev, and F. Tilmann (2000). Mapping the Hawaiian plume conduit with converted seismic waves, *Nature, 405* (2000), 938–941.

Li, X., R. Kind, X. Yuan, I. Wölbern, and W. Hanka (2004), Rejuvenation of the lithosphere by the Hawaiian plume, *Nature*, *427* (2004), 827–829.

Monnereau, M., M. Rabinowicz, and E. Arquis (1993), Mechanical erosion and reheating of the lithosphere: A numerical model for hotspot swells, *J. Geophys. Res.*, *98*, 809–823.

Morgan, W. J. (1971), Convective plumes in the lower mantle, *Nature*, v. 230, pp. 42–43.

Morgan, W. J. (1972), Deep mantle convection plumes and plate motions, Bulletin of the American Association Petroleum Geology, v. 56, pp. 203–213.

Müller, R. D., M. Sdrolias, G. Gaina, and W. R. Roest (2008), Age, spreading rates, and spreading asymmetry of the world's ocean crust. *Geochem. Geophys. Geosyst.*, *9*, Q04006, doi:10.1029/2007GC001743.

Nishimura, C.E. and D.W. Forsyth (1989). The anisotropic structure of the upper mantle in the Pacific, *Geophys. J. Int.*, *96*, 203–229.

Obrebski, M., R. M. Allen, F. Pollitz, and S.-H. Hung (2011), Lithosphere-asthenosphere interaction beneath the western United States from the joint inversion of body-wave traveltimes and surface-wave phase velocities, *Geophys. J. Int.*, *185*, 1003–1021, doi:10.1111/j.1365-246X.2011.04990.x.

Paige, C. C. and M. A. Saunders (1982), LSQR: An algorithm for sparse linear equations and sparse least squares, *TOMS*, *8*, 43–71.

Pietruszka, A. J.M.D. Norman, M. O. Garcia, J. P. Marske, and D. E. Burns (2013), Chemical heterogeneity in the Hawaiian mantle plume from the alteration and dehydration of recycled oceanic crust, *Earth Planet. Sci. Lett.*, *361*, 298–309.

Priestley, K., and F. Tilmann (1999), Shear-wave structure of the lithosphere above the Hawaiian hot spot from two-station Rayleigh wave phase velocity measurements, *Geophys. Res. Lett.*, *26*, 1493–1496.

Ribe, N. M. and U. R. Christensen (1994), Three-dimensional modeling of plume-lithosphere interaction, *J. Geophys. Res.*, *99*, 669–682.

Richards, M. A. and R. W. Griffiths (1988), Deflection of plumes by mantle shear flow: experimental results and a simple theory. *Geophys. J. Int. 94*, 367–376.

Rychert, A. C., G. Laske, N. Harmon and P. Shearer (2013). Seismic imaging of melt in a displaced Hawaiian plume, *Nature Geosci.*, *6*, 657–660, doi:10.1038/ngeo1878.

Schilling, J. G. (1991), Fluxes and excess temperatures of mantle plumes inferred from their interaction with migrating midocean ridges, *Nature*, *352*, 397–403.

Schimmel, M., and J. Gallart (2007), Frequency-dependent phase coherence for noise suppression in seismic array data, *J. Geophys. Res.*, *112*, B04303, doi:10.1029/2006JB004680.

Schimmel, M., E. Stutzmann, and J. Gallart (2011), Using instantaneous phase coherence for signal extraction from ambient noise data at a local to a global scale, *Geophys. J. Int.*, *184*, 494–506, doi:10.1111/j.1365-246X.2010.04861.x.

Schmerr, N., and E. Garnero (2006), Investigation of upper mantle discontinuity structure beneath the central Pacific using SS precursors, *J. Geophys. Res, 111*, doi:10.1029/2005JB004197.

Schmerr, N., E. Garnero, and A. McNamara (2010), Deep mantle plumes and convective upwelling beneath the Pacific Ocean, *Earth Planet. Sci. Lett.*, *294*,143–151.

Schutt, D. L. and C. E. Lesher (2006), Effects of melt depletion on the density and seismic velocity of garnet and spinel lherzolite, *J. Geophys. Res.*, *111*, B05401, doi:10.1029/2003JB002950.

Shen, Y., C. J. Wolfe and S. C. Solomon (2003), Seismic evidence for a mid-mantle discontinuity beneath Hawaii and Iceland, *Earth Planet. Sci. Lett.*, *214*, 143–151.

Sobolev, A. V., A. W. Hofmann, S. V. Sobolev, and I. K. Nikogosian (2005), An olivine-free mantle source for Hawaiian shield basalts, *Nature*, *434*, 590–597.

Steinberger, B., and R. J. O'Connell (1998), Advection of plumes in mantle flow: implications on hotspot motion, mantle viscosity and plume distribution, *Geophys. J. Int.*, *132*, 412–434.

Steinberger, B., R. Sutherland and R. J. O'Connell (2004), Prediction of Emperor-Hawaii seamount locations from a revised model of plate motion and mantle flow, *Nature, 430*, 167–173.

Tilmann, F. J., H. M. Benz, K. F. Priestley and P.G. Okubo (2001). P-wave velocity structure of the uppermost mantle beneath Hawaii from traveltime tomography, *Geophys. J. Int.*, *146*, 594–606.

VanDecar, J. C. and R. S. Crosson (1990), Determination of teleseismic relative phase arrival times using multi-channel cross-correlation and least squares, *Bull. seism. Soc. Am.*, *80*, 150–169.

Von Herzen, R. P., M. J. Cordery, R. S. Detrick, and C. Fang (1989), Heat flow and the thermal origin of hot spot swells: The Hawaiian Swell revisited, *J. Geophys. Res.*, *94*, 13,783–13,799.

Weis, D., M. O. Garcia, J. M. Rhodes, M. Jellinek, J.S. Scoates (2011), Role of the deep mantle in generating the compositional asymmetry of the Hawaiian mantle plume, *Nature Geosci. 4*, 831–838.

Wessel, P. and W. H. F. Smith (1998), New, improved version of the Generic Mapping Tools released, *EOS Trans. AGU, 79*, 579.

Wölbern, I., A. Jacob, T. Blake, R. Kind, X. Li (2006), Deep origin of the Hawaiian tilted plume conduit derived from receiver functions, *Geophys. J. Int., 166*, 767–781.

Wolfe, C. J., S. C. Solomon, G. Laske, J. A. Collins, R. S. Detrick, J. A. Orcutt, D. Bercovici, and E. H. Hauri (2009), Mantle shear-wave velocity structure beneath the Hawaiian hot spot, *Science*, *326*, 1388–1390, doi:10.1126/science.1180165.

Wolfe, C. J., S. C. Solomon, G. Laske, J. A. Collins, R. S. Detrick, J. A. Orcutt, D. Bercovici, and E. H. Hauri (2011), Mantle P-wave velocity structure beneath the Hawaiian hot spot, *Earth planet. Sci. Lett.*, *303*, 267–280.

Woods, M. T. and E.A. Okal (1996), Rayleigh-wave dispersion along the Hawaiian Swell: A test of lithospheric thinning by thermal rejuvenation at a hotspot, *Geophys. J. Int.*, *125*, 325–339.

Yang, Y., and D. W. Forsyth (2006a), Regional tomographic inversion of amplitude and phase of Rayleigh waves with 2-D sensitivity kernels, *Geophys. J. Int.*, *166*, 1148–1160.

Yang, Y., and D. W. Forsyth (2006b), Rayleigh wave phase velocities, small-scale convection and azimuthal anisotropy beneath southern California, *J. Geophys. Res.*, *111*, B07306, doi:10.1029/2005JB004180.

Yang, Y., A. Li, and M. H. Ritzwoller (2008), Crustal and uppermost mantle structure in southern Africa revealed from ambient noise and teleseismic tomography, *Geophys. J. Int.*, *174*(1), 235–248, doi:10.1111/j.1365-246X.2008.03779.x.

3

Asymmetric Dynamical Behavior of Thermochemical Plumes and Implications for Hawaiian Lava Composition

Maxim D. Ballmer[1,2], Garrett Ito[1], and Cheng Cheng[3]

ABSTRACT

The Hawaiian Kea and Loa volcano trends have commonly been interpreted as directly reflecting a compositional zonation within the Hawaiian plume stem, inherited from the lowermost mantle. As this zonation is often associated with variations in mafic material and as such materials, especially intrinsically dense eclogites, impact mantle flow, this study aims to characterize the ascent and melting of bilaterally zoned thermochemical plumes. Our geodynamic models predict that plumes bearing $\gtrsim$12% eclogite tend to stagnate as a deep eclogitic pool (DEP) in the mid-upper mantle where phase changes lead to a maximum in eclogite excess density. This behavior can explain recent seismic-tomography results and predicts thermal asymmetry of material rising out of the DEP to feed the hotspot. Thermal asymmetry is caused by the effects of ambient mantle flow or plume stem zonation on DEP dynamics and ultimately boosts peridotite melting on the melting zone's hotter side. This hotter side is hence less dominated by melting of mafic materials, despite being fed by equally or *more* such materials than the cooler side. These results suggest that the Kea side of the Hawaiian plume is equally or more eclogitic than the Loa side, opposite to previous interpretations. Care should thus be taken in mapping geographical variations in lava chemistry into the deep mantle.

3.1. INTRODUCTION

Better constraints on the make-up of the deep mantle are needed to advance our understanding of the differentiation and secular evolution of our planet [cf. *Boyet and Carlson*, 2005; *Labrosse et al.*, 2007]. In the framework of mantle plume theory, hotspot lavas provide one of the most important means of probing the composition of the deep mantle. For example, the global-scale DUPAL anomaly among hotspot lavas [*Dupre and Allegre*, 1983] has been associated with the seismically imaged large low shear wave velocity provinces (LLSVPs) [*Castillo et al.*, 1998; *Wen et al.*, 2001; *Wen*, 2006]. These LLSVPs are compositionally dense structures in the deep mantle [*Ishii and Tromp*, 1999; *Masters et al.*, 2000; *Mosca et al.*, 2012], from the margins of which most mantle plumes are inferred to rise [*Burke and Torsvik*, 2004; *Torsvik et al.*, 2006; 2010]. Regional geographical variations in ocean-island basalt geochemistry have also been interpreted as reflecting lower mantle compositional heterogeneity in the Pacific [e.g., *Konter et al.*, 2008].

Similarly, local-scale geographical variations of lava geochemistry within single hotspot chains, such as within the Marquesas, Society, and Samoa chains, are interpreted as reflecting heterogeneity in the deep mantle [*Workman et al.*, 2004; *Huang et al.*, 2011; *Chauvel et al.*, 2012; *Payne et al.*, 2013]. Another example that has recently attracted a great deal of interest is the bilateral geochemical asymmetry of Hawaiian volcanism (Figure 3.1). The Hawaiian Islands have long been interpreted to be a double chain of volcanoes [*Jackson et al.*, 1972], of which

[1]*Department of Geology and Geophysics, University of Hawai'i at Mānoa, Honolulu, Hawaii, USA*
[2]*Earth-Life Science Institute, Tokyo Institute of Technology, Meguro, Tokyo, Japan*
[3]*Department of Earth and Planetary Sciences, University of California, Berkeley, California, USA*

Hawaiian Volcanoes: From Source to Surface, Geophysical Monograph 208. First Edition.
Edited by Rebecca Carey, Valérie Cayol, Michael Poland, and Dominique Weis.

Companion Website: www.wiley.com/go/Carey/Hawaiian_Volcanoes

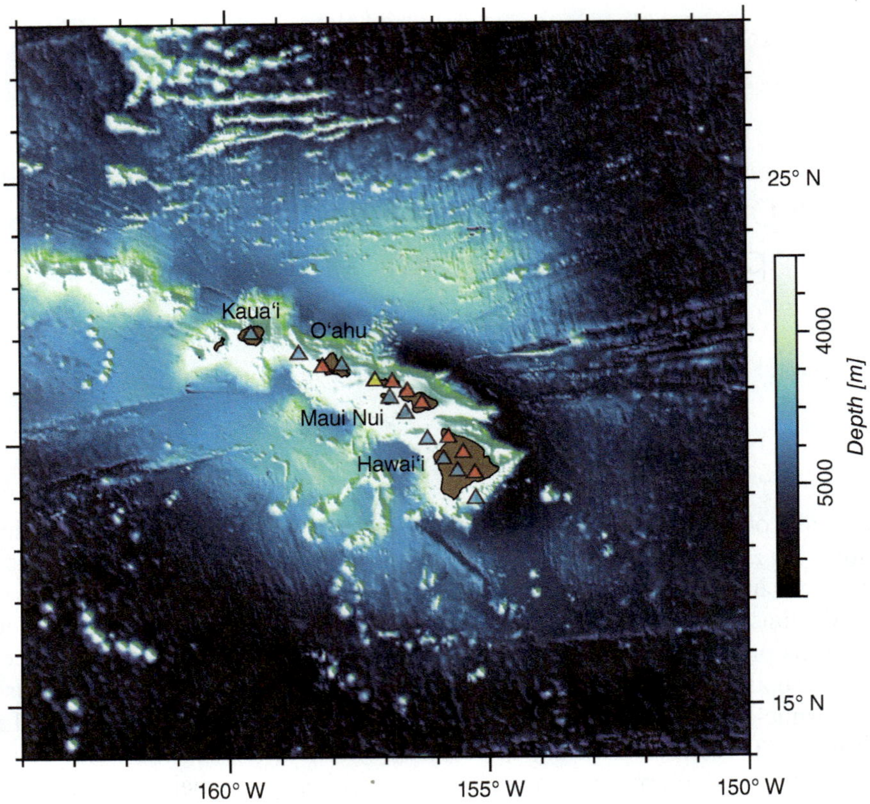

Figure 3.1 Map view of Hawaiian volcanism and bathymetry. Red and blue triangles denote sites of Kea-type and Loa-type shield-stage lavas (distinguished by their $^{208}Pb^{*}/^{206}Pb^{*}$ compositions [*Weis et al.*, 2011]), respectively. Classifications of Wai'anae and Ka'ena ridge volcanoes are based on *Sinton et al.* [2014]. Yellow triangles mark volcanoes that straddle the dividing line in $^{208}Pb^{*}/^{206}Pb^{*}$ as defined by *Abouchami et al.* [2005].

the prevalent morphological trends, the southwestern "Loa" and the northeastern "Kea" trend, display distinct geochemical signatures [*Frey and Rhodes*, 1993; *Kurz et al.*, 1996; *Abouchami et al.*, 2005; *Weis et al.*, 2011]. This distinction has been interpreted as revealing a bilateral compositional zonation of the deep plume conduit [e.g., *Abouchami et al.*, 2005]. *Huang et al.* [2011] and *Weis et al.* [2011] argue that the Hawaiian plume, as apparently rising from the margin of a LLSVP, entrains compositionally distinct materials into either sides of the plume stem.

The possibility that entrainment of distinct materials may be expressed in hotspot-lava composition requires that plumes rise with little lateral material exchange or rotation across their conduits (about a vertical axis). In classical theory, plumes are described as upwellings that rise near vertically through the entire mantle to support magmatism, purely driven by thermal buoyancy [*Morgan*, 1972; *Griffiths and Campbell*, 1990; *Sleep*, 1990; *Ribe and Christensen*, 1994, 1999]. Based on such purely thermal plume models (i.e., with density being independent of composition), *Farnetani and Hofmann* [2009, 2010] demonstrate that patterns of deep-mantle composition can indeed be directly reflected by geographic variations in hotspot-lava geochemistry. However, the Hawaiian plume is thought to carry compositionally dense, mafic materials (e.g., entrained from the LLSVP or a slab graveyard), thereby potentially behaving more complexly than a classical plume [e.g., *Farnetani and Samuel*, 2005], particularly if these materials are initially distributed nonuniformly, or even bilaterally asymmetrically, within the deep part of the plume stem.

Evidence for the presence of mafic materials (e.g., pyroxenite or eclogite) in the source of Hawaiian lavas comes from major-element, trace element, as well as isotopic data [*Hauri*, 1996; *Huang and Frey*, 2005; *Sobolev et al.*, 2005, 2007; *Huang et al.*, 2007; *Herzberg*, 2011; *Jackson et al.*, 2012; *Pietruszka et al.*, 2013]. These materials are inferred to be more abundant in the source of Loa than of Kea volcanoes and are likely to contribute to the geochemical distinction between the two trends [*Sobolev et al.*, 2005; *Greene et al.*, 2010; *Jackson et al.*, 2012].

Eclogites are significantly denser than peridotites and thus strongly affect plume ascent. Analogue and numerical geodynamic models show that such eclogitic or "thermochemical" plumes display asymmetric and time-dependent behavior [*Davaille*, 1999; *Farnetani and Samuel*, 2005; *Lin and van Keken*, 2005; *Kumagai et al.*,

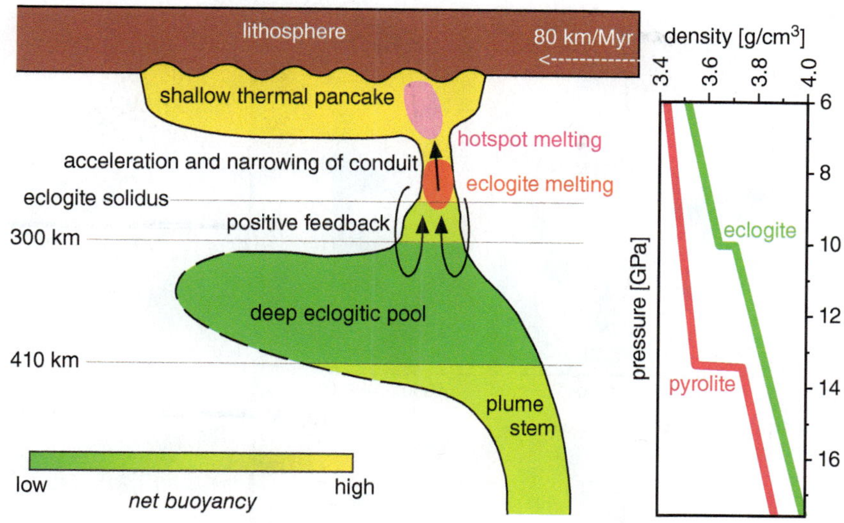

Figure 3.2 Conceptual figure (not to scale) of a double-layered thermochemical plume. The peak in excess density of eclogite relative to pyrolite (and also relative to peridotite; see left panel as reproduced from *Aoki and Takahashi* [2004]) can induce stagnation of a thermochemical plume to a deep eclogitic pool (DEP) in the depth range of ~300 to ~410 km. Such a DEP, in combination with the shallow thermal pancake ponding beneath the lithosphere, places a large volume of hot and seismically slow material in the upper mantle, which is required by seismic tomography. The reduction of negative compositional density due to the coesite-stishovite phase transition at ~300 km depth as well as due to eclogite melting starting at ~250 km depth pulls eclogitic material out of the DEP in a positive-feedback mechanism.

2008], much in contrast to classical plume behavior, which so far has inspired the interpretations of geochemical asymmetry of Hawaiian lavas. This complex behavior is caused by the competition between diffusive, positive thermal buoyancy and nondiffusive, negative compositional buoyancy in thermochemical plumes. Phase transitions can account for additional complications [e.g., *Farnetani and Samuel*, 2005]. For example, *Ballmer et al.* [2013] show that a peak in the excess density of eclogite relative to peridotite due to phase changes in the olivine and quartz systems [*Aoki and Takahashi*, 2004] can cause thermochemical (i.e., eclogitic) plumes to pool in a broad layer at depths of 300–410 km depth (deep eclogitic pool, DEP) before rising to feed the hotspot (Figure 3.2).

Evidence for such complex dynamical behavior beneath the Hawaiian hotspot is provided by recent seismic tomography. Body wave tomography, based on data collected during the regional Plume and Lithosphere Undersea Melt Experiment (PLUME), images a thick (~400 km) asymmetric body of slow seismic velocities beneath the hotspot [*Wolfe et al.*, 2009, 2011]. These seismic models disagree with classical plume theory, which predicts a much thinner (<100 km) and symmetric layer ("pancake") to pond beneath the lithosphere. The presence of a large volume of low-velocity materials is further supported by a recent joint inversion of body waves, surface waves, and ambient noise recorded by PLUME (see Chapter 2). In addition to body wave tomography alone, these joint inversions resolve two layers of seismically slow material in the upper mantle: one beneath the lithosphere (at ~100 km depth) and another one contained in the depth range of ~250 to ~450 km (Figure 3.3; see Chapter 2). Seismic resolution tests indicate that these seismic constraints are consistent with the presence of a double-layered (DEP and pancake) thermochemical plume beneath the Hawaiian hotspot [*Ballmer et al.*, 2013] (see Chapter 2).

In this study, we explore the dynamics of bilaterally asymmetric thermochemical plumes in which one side of the plume carries a higher fraction of compositionally dense materials (such as eclogite) than the other side. By predicting geographical patterns of volcanism due to the melting of mafic and ultramafic source lithologies, we investigate the link between the deep compositional zonation and geographical variations in lava composition. Our results show that thermochemical plume dynamics in the upper mantle can create geographical patterns of lava composition independent of a zonation in the deep-plume conduit. They can veil, or even apparently reverse, any deep bilateral zonation.

3.2. METHODS AND MODEL DESCRIPTION

We use a Cartesian version of the finite element code Citcom [*Moresi et al.*, 1996] to model thermochemical plume dynamics. The model domain of length 5280 km (x direction), width 3300 km (y direction), and height 660 km

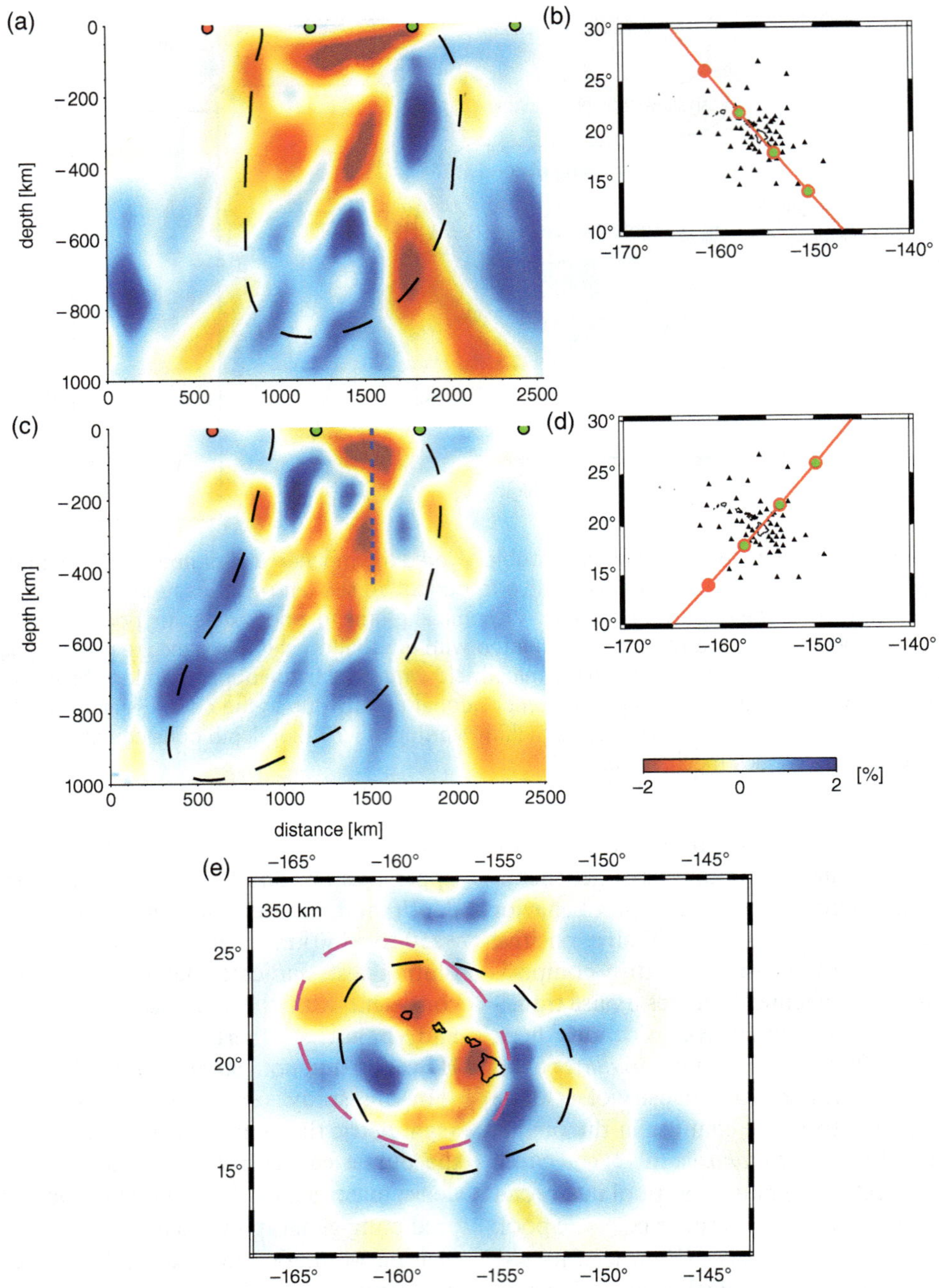

Figure 3.3 Cross sections through the seismic model HW13-SVJ of Cheng et al. (see Chapter 2) produced from a joint inversion of shear wave and surface wave (teleseismic as well as ambient noise) travel times. The location of the vertical cross sections in panels (a) and (c) are denoted in the insets in panels (b) and (d) as red lines, respectively (with red and green dots as geographical markers). Black triangles in (b) and (d) show the seismic network used for the inversion. The blue dashed line in (c) approximately separates the Kea and Loa volcano subchains. (e) The horizontal cross section is located at the same depth (350 km) as the cross sections through the synthetics shown in Figure 3.9. Ray coverage is highest within the thin black dashed lines (cf. Figure 2.6 in Chapter 2). A wide body of slow shear wave velocities at typical DEP depths is segmented by fast structure toward the west and is surrounded by the pink dashed ellipse. The seismic images show complex structure, crudely consistent with (a) a double-layered plume in the upper mantle and a slanted stem rooted in the lower mantle, (c) a shallow plume that is asymmetric about any axis parallel to plate motion, and (e) a DEP with relatively complex structure. Panels (a) through (d) are taken from Figure 2.4 with minor modifications.

Table 3.1 Key Plume Parameters and Average Densities of the Plume Core for All Cases.

					Average Plume-Core Net Buoyancy	
	r_P	r_E	$\Phi_{ECL,NE}$	$\Phi_{ECL,SW}$	At 300–410 km Depth	Elsewhere
Case S1	75 km	90 km	15%	15%	−9.7 kg/m³	6.8 kg/m³
Case S2	75 km	100 km	15%	15%	−10.7 kg/m³	5.8 kg/m³
Case Z1	65 km	78 km	8%	10%	3.5 kg/m³	13.4 kg/m³
Case Z2	70 km	84 km	10%	12%	−0.9 kg/m³	11.2 kg/m³
Case Z3	75 km	90 km	12%	15%	−6.4 kg/m³	8.4 kg/m³
Case Z4	70 km	84 km	11%	15%	−5.3 kg/m³	9.0 kg/m³
Case Z5	75 km	90 km	10%	16%	−5.3 kg/m³	9.0 kg/m³

Note: Radii of the plume's thermal and compositional anomalies (r_P and r_E) as well as initial eclogite contents on either side of the plume ($\Phi_{ECL,NE}$ and $\Phi_{ECL,SW}$) are given. Positive and negative net buoyancies refer to plume-core material being less and more dense than the ambient nonplume mantle, respectively. Cases S1 and S2 are analogous to cases A and B by *Ballmer et al.* [2013], respectively.

(z direction) is discretized in 768 × 512 × 96 finite elements with the smallest elements (i.e., 4.5 × 4.5 × 4.5 km) located in the asthenosphere near the hotspot. This fine resolution is needed to accurately model melting processes and predict magma compositions. A horizontal velocity in the x direction of $v_{plate} = 80$ km/Myr is imposed at the top boundary to simulate Pacific Plate motion. All boundaries, except for the front ($x = 0$) and back ($x = 5280$ km) sides are closed to in- and outflow. At a distance of 3135 km from the front side (i.e., inflow boundary), 2145 km from the back side (i.e., outflow boundary), and 1650 km from both lateral side boundaries, a Gaussian thermal anomaly of half-width r_P (see Table 3.1) and amplitude of +300 K is imposed to supply a plume at the base of the model. Upward flow of plume material into the model domain (i.e., from the lower into the upper mantle) is only allowed within a radius of $4.8r_P$ from the plume center, where the bottom boundary is locally open to vertical flow.

In the core of the plume, a cylinder of radius r_E is taken to contain eclogite (Figure 3.4). The radius of this cylinder, r_E, is $1.2r_P$ in most models, except for case S2, in which $r_E = 100\text{ km} \approx 1.33r_P$ (cf. Table 3.1). The imposed initial volume percent of eclogite within the cylinder is $\Phi_{ECL,SW}$ on the southwestern (SW) side of the plume and $\Phi_{ECL,NE}$ on the northeastern (NE) side, both varied for different models between 8% and 16% (cf. Figure 3.4). With plate motion parallel to the x direction, the y direction is NE to SW. In symmetric cases S1 and S2, the imposed distribution of eclogite is uniform ($\Phi_{ECL,SW} = \Phi_{ECL,NE}$, Figure 3.4). For the other models (cases Z1–Z5), a compositional bilateral zonation in the deep plume stem is simulated by imposing $\Phi_{ECL,SW} > \Phi_{ECL,NE}$. This aspect of our model setup is novel compared to any thermochemical plume study to date, including *Ballmer et al.* [2013]. We note that the end-member situations of purely axisymmetric (S1 and S2) versus bilaterally asymmetric (Z1–Z5) compositional structure are idealizations of what is probably more complex in reality.

In contrast to the eclogitic core ($r \le r_E$) of the plume, the relatively cool plume outskirts ($r > r_E$) with excess temperatures <~110 K (case S2: <~87 K) as well as the cooler ambient mantle contain no eclogite. Both the plume and the ambient mantle are dominated by peridotitic lithologies: They initially contain $\Phi_{HP} = 20\%$ hydrous peridotite in addition to the percentage $\Phi_{DP} = 100\% - \Phi_{HP} - \Phi_{ECL,SW}$ of dry peridotite. The remainder of the mantle consists of a refractory lithology that does not melt beneath the hotspot (cf. *Stracke et al.*, 2011). Accordingly, the initial volume fraction of the refractory lithology is $\Phi_{RL} = \Phi_{ECL,SW}$ in the ambient mantle, $\Phi_{RL} = 0\%$ on the SW side of the deep plume stem, and $\Phi_{RL} = \Phi_{ECL,SW} - \Phi_{ECL,NE}$ on the NE side. Composition is advected with the flow, evolves due to melting and hybridization (see next paragraph), and is steadily replenished (and set to initial conditions) where material enters the model domain at the front side of the model box and the base of the plume.

Including eclogite in the models impacts the results through both the associated variations in the mantle density due to solid-phase changes and melting (Figure 3.2) and the expression of mafic material in the magmas. To account for the effects of phases changes in the quartz and olivine systems, the excess density of eclogite relative to peridotite is defined to be 220 kg/m³ in the depth range 300–410 km and 110 kg/m³ elsewhere, based on the laboratory experimental results of *Aoki and Takahashi* [2004]. For eclogite melting, we use *Yasuda et al.*'s [1994] empirically derived parameterization. As such, the models predict eclogite to begin melting first at a depth of ~260 km and continue to melt to a maximum allowable extent of 60% at a depth of ~150 km within the hottest core of the plume [*Sobolev et al.*, 2005, 2011]. We assume that once any eclogitic melt is formed, it instantaneously reacts with the ambient peridotite in a one-to-one fashion to

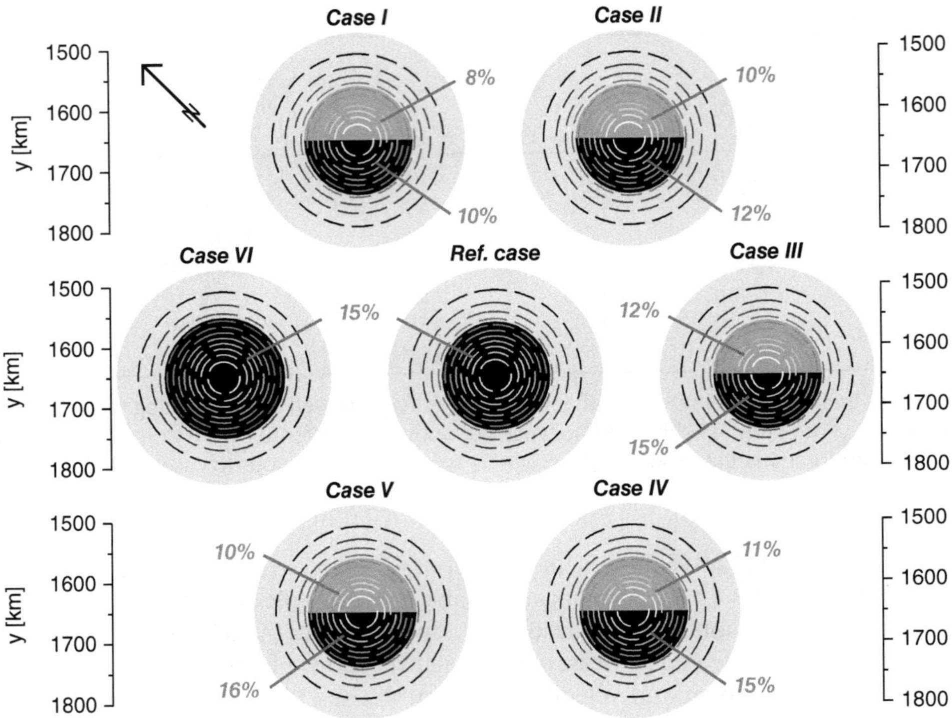

Figure 3.4 Imposed eclogite content at the base of the deep plume stem for all cases. The eclogite-bearing core of the plume is colored black to dark grey, and the surrounding noneclogitic, but still warm outskirt of the plume stem is light grey. Eclogite contents on the two sides of the plume core are labeled. Dashed lines are contours of imposed potential temperature from 1375°C (outside, black) to 1625°C (inside, white) in 25°C increments. In each model, the center of the plume at the base of the box is at $x = 3135$ km and $y = 1650$ km. The black arrow in the top-left corner points northward.

form bimineralic, olivine-free pyroxenite [*Yaxley and Green*, 1998; *Sobolev et al.*, 2005, 2007; *Herzberg*, 2011]. In the hotspot melting zone at about 110–200 km depth, both peridotite and pyroxenite melt, and do so according to the parameterizations of *Hirschmann* [2000] and *Pertermann and Hirschmann* [2003], respectively. Further details of the melting parameterization, as well as of the mantle density and rheology treatments are reported in the Appendix. Model parameters are listed in Table 3.2.

Following *Ballmer et al.* [2010, 2011, 2013], the predicted volumetric fraction X_{PX} of pyroxenite-derived to the total (i.e., pyroxenite + peridotite derived) volume of volcanism is the quantity used to characterize lava composition. This quantity is computed by assuming all magmas rise vertically and mix perfectly before reaching the surface. To predict the average composition of shield-stage volcanism on the SW and NE sides of the hotspot (i.e., $X_{PX,SW}$ and $X_{PX,NE}$, respectively), we integrate over two adjacent, rectangular strips on the surface. Each strip is 80 km long in the x direction (i.e., parallel to plate motion) and 63.4 km wide in the y direction; thus the total width (in y) of the rectangular zone, from which extracted magmas are focused to form the double volcano chain, is 126.8 km. To capture main shield-stage volcanism (of duration $\tau_{shield} \approx 1$ Myr) along the two (i.e., Kea and Loa) volcano trends the two strips (of length 80 km $= \tau_{shield}/v_{plate}$) are positioned to maximize the total volcanic flux between them with similar fluxes from each strip. Accordingly, the boundary between the two strips nearly bisects the absolute peak in vertical magma flux from below and defines the central axis of the model hotspot chain. It is this axis that is used to define bilateral symmetry or asymmetry of plume parameters and volcanism.

3.3. RESULTS

The models are relevant to the Hawaiian hotspot in that their predictions are consistent with key observations. In all our models (cf. Table 3.1), a hot mantle plume rises through the upper mantle to support localized hotspot (i.e., shield-stage) volcanism, focused over an area ~100 km long and ~75 km wide, consistent with geological constraints. The plume thermal anomaly imposed at the base of the model of +300 K agrees well with chemical geothermometry [*Herzberg et al.*, 2007]. From case to case, the predicted average volcanic flux varies between

Table 3.2 Notation (see also section A.2).

Parameter	Symbol	Value
Reference ambient mantle temperature	T_m	1350°C
Plume peak thermal anomaly	ΔT_{plume}	300 K
Adiabatic gradient	γ	0.3 K/km
Heat capacity	c_P	$1250\,J \cdot kg^{-1} \cdot K^{-1}$
Thermal diffusivity	κ	$10^{-6}\,m^2/s$
Thermal expansivity	α	$3 \times 10^{-5}\,K^{-1}$
Latent heat of melt	L	560 kJ/kg
Critical porosity in peridotite	φ_C	0.40%
Critical porosity in pyroxenite	$\varphi_{C,PYX}$	10%
Residual porosity in peridotite	φ_R	0.40%
Residual porosity in pyroxenite	$\varphi_{R,PYX}$	5%
Water partitioning coefficient		0.01
Bulk water content in ambient mantle	c_0	310.64 ppm
Water content below which peridotite behaves like dry peridotite[a]	c_{dry}	6 ppm
Effective mantle viscosity	η_{eff}	$1.6 \times 10^{19}\,Pa \cdot s$
Activation energy[b,c]	E^*	300 kJ/mol
Activation volume	V^*	$5 \times 10^{-6}\,m^3/mol$
Dehydration stiffening coefficient[a]	ξ	100
Melt lubrication exponent[d]	ζ	−40
Mantle density	ρ_0	$3300\,kg/m^3$
Excess density of eclogite	$\Delta\rho_{ECL}$	$110–220\,kg/m^3$§
Density change with depletion in peridotite	$\Delta\rho_F$	$-165\,kg/m^3$
Magma density	ρ_M	$2800\,kg/m^3$
Depth of box	D	660 km
Top velocity boundary condition	v_{plate}	80 km/Myr
Gravity acceleration	G	$9.8\,m/s^2$
Ideal gas constant	R	$8.314\,Jmol^{-1}\,K^{-1}$

[a]*Hirth and Kohlstedt* [1996, 2003].
[b]*Karato and Wu* [1993].
[c]*Hirth* [2002].
[d]*Kohlstedt and Zimmerman* [1996].

~130,000 and ~190,000 km^3/Myr (or ~4.1 and ~6.0 m^3/s; see Table 3.3), a range that is within that of published measurements [*van Ark and Lin*, 2004; *Vidal and Bonneville*, 2004; *Robinson and Eakins*, 2006]. The modeled plumes also dynamically push up the plate to support hotspot swells of heights $870\,m \leq h_{swell} \leq 960\,m$ and widths $1130\,km \leq w_{swell} \leq 1340\,km$, similar to observations [*Wessel*, 1993; *Crosby and McKenzie*, 2009] (also see arch-shaped region of shallow seafloor around Hawai'i in Figure 3.1). As the amount of mafic materials carried by the plume varies from case to case, we modulate the plume radius to maintain good agreement with these key hotspot characteristics (Table 3.1).

One result common to all of the models is that mafic lithologies heavily contribute to hotspot volcanism. Any eclogite carried by the plume, of which the initial contents in our models vary between 8% and 16% from case to case (and from side to side of the plume), melts to a maximum degree of 60% in the depth range of about 150–260 km to react with ambient peridotite (in a 1:1 fashion) and to form pyroxenite [*Yaxley and Green*, 1998; *Sobolev et al.*, 2005, 2007]. The resulting moderate fractions of pyroxenite (i.e., 9.6%–19.2% of the mantle) dominate magmatism, contributing $X_{PX} > 50\%$ ($66\% < X_{PX} < 83\%$) to the model hotspot lavas, even though the source consists mainly of peridotite (with volume fractions of 74.4%–87.2%). This dominance is caused by the much higher isobaric melt productivities in pyroxenite than in peridotite and is reinforced by consumption of latent heat by the first at the expense of the latter [*Phipps Morgan*, 2001; *Katz and Rudge*, 2011]. While such a dominance of melting mafic materials at the Hawaiian hotspot is indeed supported by many recent studies [*Hauri*, 1996; *Huang and Frey*, 2005; *Sobolev et al.*, 2005, 2007; *Huang et al.*, 2007; *Herzberg*, 2011; *Jackson et al.*, 2012; *Pietruszka et al.*, 2013], other studies favor a dominance of peridotitic melting [cf. *Putirka et al.*, 2011]. We also note that our predictions in terms of X_{PX} are upper bounds, as incomplete wall-rock

Table 3.3 Key Model Predictions for All Cases.

	h_{swell}	w_{swell}	Volcanic Flux[a]	$X_{PX,SW}$	$X_{PX,NE}$	$X_{PX,SW} - X_{PX,NE}$
Case S1	920 m	1200 km	5.4 m³/s	83.56%	82.71%	0.85%
Case S2[b]	732 m	1100 km	3.4 m³/s	87.77%	83.55%	4.22%
	(65 m)	(31 km)	(0.6 m³/s)	1.93%	2.34%	1.55%
Case Z1	870 m	1130 km	4.1 m³/s	70.49%	65.25%	5.24%
Case Z2	940 m	1200 km	6.0 m³/s	76.22%	73.54%	–2.68%
Case Z3	940 m	1290 km	6.5 m³/s	81.32%	82.52%	–1.20%
Case Z4	890 m	1190 km	5.5 m³/s	66.08%	66.10%	–0.02%
Case Z5	960 m	1340 km	5.6 m³/s	70.09%	79.04%	–8.95%

Note: Swell height, h_{swell}; swell width, w_{swell}; contribution of mafic lavas in volcanism on the southwestern and northeastern sides of the hotspot, $X_{PX,SW}$ and $X_{PX,NE}$.
[a]Volcanic fluxes reported are average total volcanic fluxes including secondary volcanism occurring well away from the hotspot.
[b]In case S2, the reported key model predictions strongly vary over model time; the reported values are averages with standard deviations reported in parentheses.

reactions in the deep eclogite melting zone can produce mafic lithologies with lower productivities than simulated in this study (e.g., websterites instead of pyroxenites [*Mallik and Dasgupta*, 2012]).

3.3.1. Double Layering of Plume Material in Upper Mantle

The presence of dense eclogite leads to thermochemical plumes that are wider and rise more sluggishly compared to noneclogitic (i.e., thermal or "classical") plumes with similar thermal buoyancy fluxes. Moreover, the doubling of the excess density of eclogite in the depth range of 300–410 km due to the effects of phase transitions in the quartz and olivine systems [*Aoki and Takahashi*, 2004] can profoundly modify plume morphologies and upwelling dynamics (cf. *Ballmer et al.*, 2013). If enough eclogite is present in the deep plume conduit (see below), the rise of the plume partially stalls in this depth range to form a layer of hot material hundreds of kilometers wide and ~150 km thick (Figure 3.5), which we refer to as the deep eclogitic pool.

For example, in cases S1 and S2, the eclogitic plume core (originally 15% eclogite) has a net (i.e., thermal plus compositional) buoyancy of −3.3 kg/m³ to about −23 kg/m³ within the critical depth range of 410–300 km, imaged as the blue region in Figures 3.5b,e (buoyancy is defined as the density deficit relative to the ambient mantle at the reference temperature). Accordingly, this material tends to sink back down into the transition zone. However, it is supported from below by the deep plume conduit (red colors at the bottom of Figures 3.5b,e), which in the transition zone is almost entirely positively buoyant (i.e., by up to ~+13.2 kg/m³) and is sustained on its sides by the noneclogitic plume outskirts (with a buoyancy of up to ~+10 kg/m³; red sheath around the blue DEP in Figures 3.5b,c,e,f). Therefore, the eclogitic core spreads laterally (but not indefinitely) to form a wide DEP (Figures 3.5a–f). As material continues to feed the DEP from below, the DEP inflates to cross the 300 km deep coesite-stishovite phase transition, where the material instantaneously regains a positive net buoyancy (Figures 3.5c,f). The resulting increase in upwelling rate above this phase transition sets up a local minimum in dynamic pressure that guides the (not necessarily vertical) rise of material through the DEP. This local minimum is enhanced by the consumption of eclogite due to melting, occurring in the depth range of about 150–260 km. This process strongly enhances the net buoyancy. Consequently, the shallow plume conduit is narrower and rises more rapidly (by rates of up to ~1 m/yr) than the deep plume stem and ponds beneath the lithosphere as a thin (<100 km) pancake that supports the hotspot swell.

Similarly, compositionally zoned cases Z3–Z5 with average eclogite contents >12% display a strong double layering of plume material with a wide DEP (Figures 3.5 m–u). In these cases, material from the more eclogite rich SW sides of the plumes primarily inflates the DEP as shown in Figures 3.6d–f by contours of initial eclogite content (i.e., eclogite content with any modifications due to melting removed). The materials from the eclogite-poor NE sides of the plumes instead rise relatively rapidly through and contribute relatively little volume to the DEP as they largely remain positively buoyant (by up to +3.3 kg/m³ (case Z3) and +7.7 kg/m³ (case Z5)).

In contrast, in compositionally zoned cases Z1 and Z2 with average eclogite contents <12%, most of the plume-core material remains positively or near neutrally buoyant even in the depth range of 300–410 km. The plumes therefore rise without stalling significantly to form a DEP (Figures 3.5 g–l). Just a portion of the cooler outer plume core in case Z2, particularly on the SW side, becomes

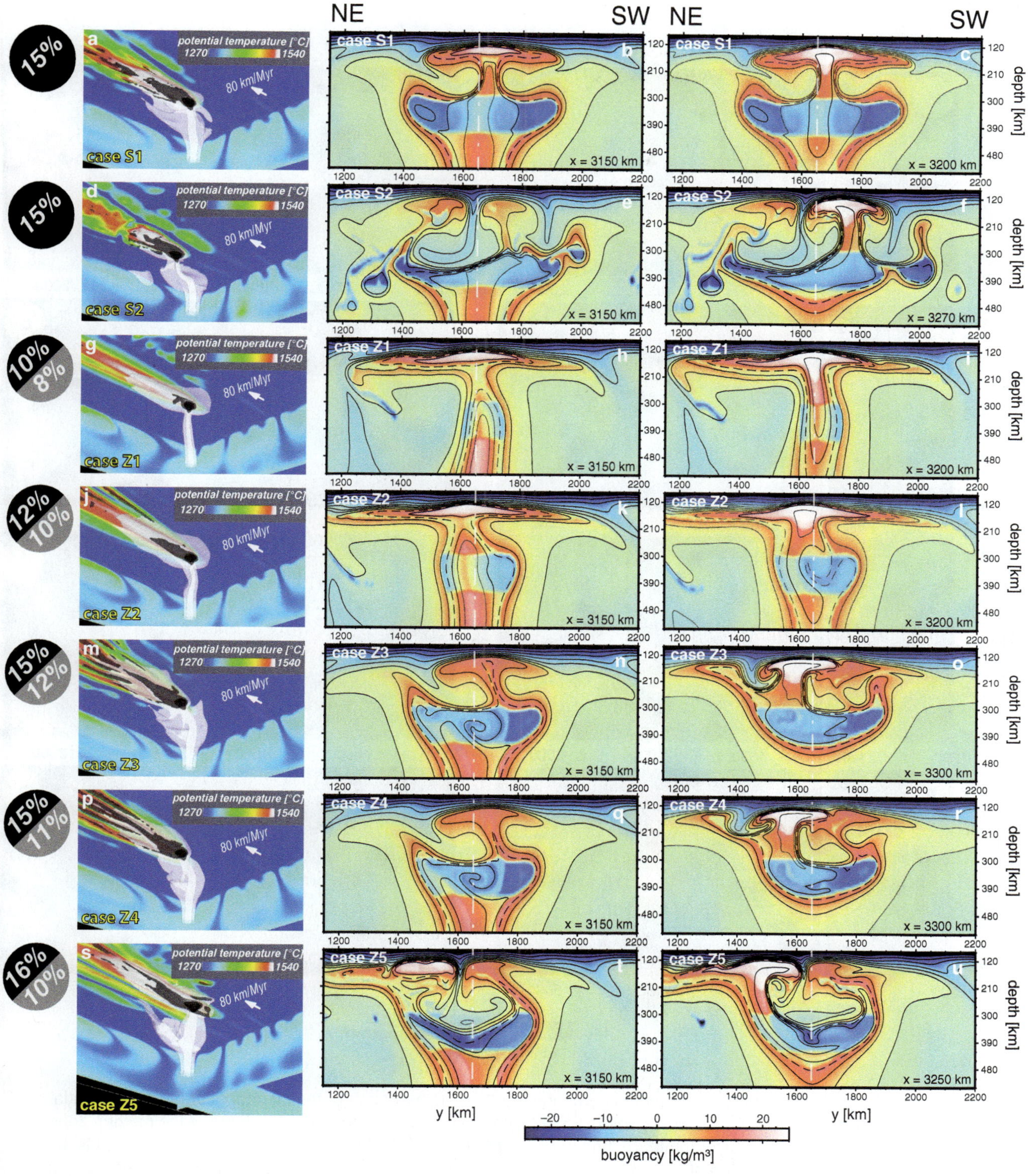

Figure 3.5 Temperature, melting, and net buoyancy of all cases (rows as annotated). Left column shows perspective view from below of potential temperature (colors) with the 1550 and 1620 °C isosurfaces shown in translucent white and solid white, respectively. Grey and black isosurfaces denote sites of minor and major magma generation, respectively. Middle and right columns show cross sections of net buoyancy (colored) perpendicular to plate motion. Net buoyancy is the sum of thermal and compositional contributions, expressed as density deficit relative to the ambient mantle at the reference potential temperature. Dashed-dotted lines mark the symmetry axis of the model's initial and boundary conditions. Middle column (h–n) shows cross sections through the hottest part of the deep plume conduit and DEP. Right column shows cross sections, slightly further downstream, through the hottest part of the shallow plume conduit (i.e., close to the peak hotspot melt production). Contours denote potential temperature in 50 °C increments with the 1500 °C isotherm dashed. Disk insets schematically show the initial distribution of eclogite, similar to Figure 3.4.

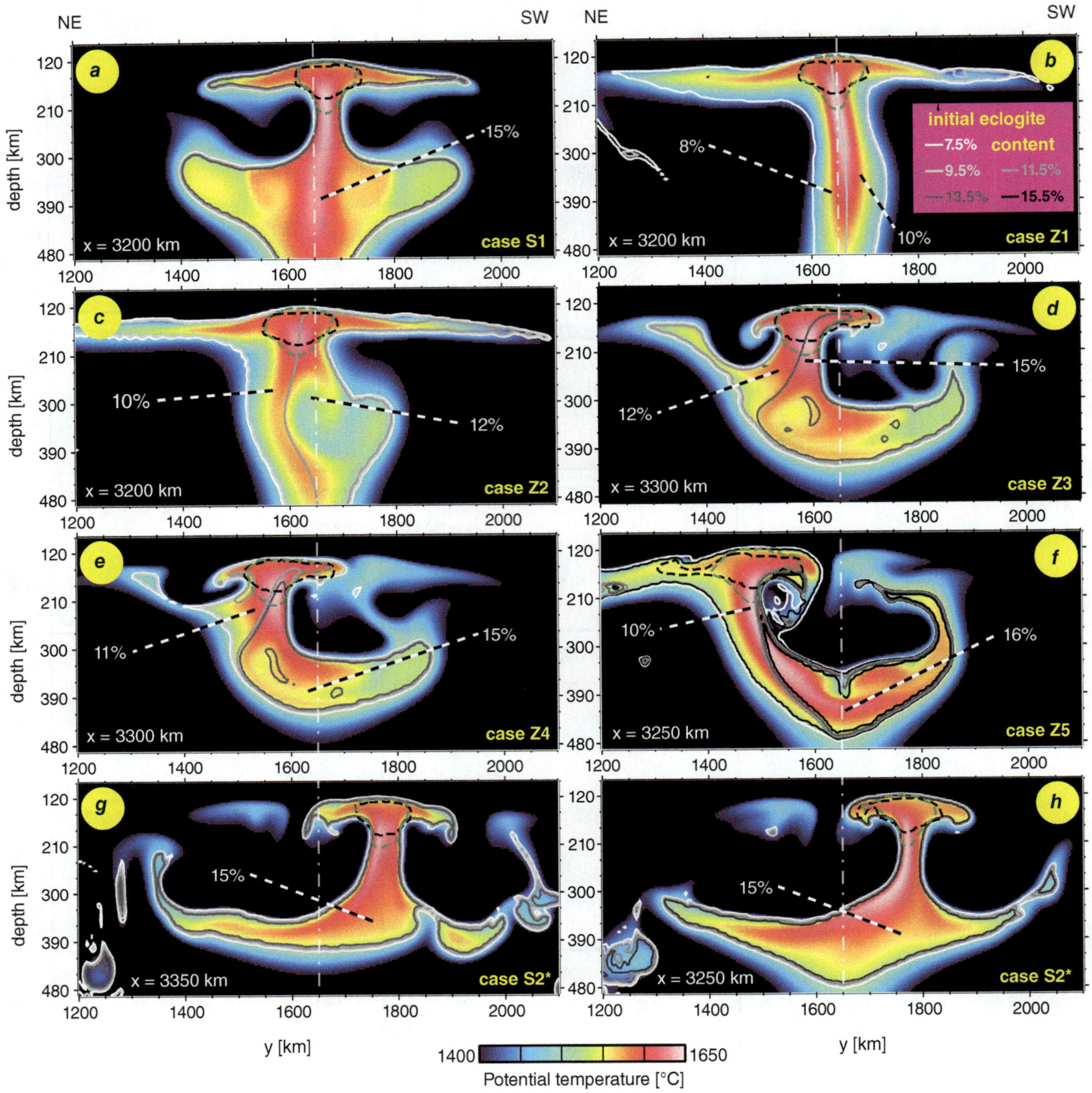

Figure 3.6 Vertical cross sections oriented perpendicular to plate motion of potential temperature (colors), initial eclogite content (white-to-black solid contours, see legend in upper right corner), and shape of the melting zone (dashed green contours for pyroxenite, dashed black for peridotite) for all cases. The initial eclogite content is the original content at the position of the material at the base of the model box, from where it has been advected (i.e., with the effects of eclogite melting removed). As shown by the solid contours running more or less vertically along the plume axis in the compositionally zoned cases Z1–Z5, the two sides of the plume do not physically mix with each other. In cases S2 and Z3–Z5, temperatures across the melting zones are asymmetric in this cross section, and the degrees of peridotite melting are higher (not shown) and the depth of the solid (of both lithologies but of peridotitic lithologies in particular) are deeper on the hotter side of the hotspot melting zone. Pyroxenite melts to maximum degrees on both sides. Dashed-dotted lines mark the symmetry axis of the model's initial and boundary conditions. (*) For case S2, two different snapshots are shown, one of which (g) corresponds to the snapshot shown in Figure 3.8i, the other one of which (h) corresponds to that in Figure 3.8j.

marginally negatively buoyant, leading to only a slight (~50%) widening of the plume conduit in the mid-upper mantle (Figure 3.5k). The plume dynamics in cases Z1 and Z2 are indeed much like that of a classical thermal plume (Figures 3.5g–l).

In summary, average initial plume-core eclogite contents >12% appear to be required to form a DEP in plumes with parameters that are realistic for the Hawaiian plume (see above). As the dynamical behavior is governed by the competition of thermal and compositional buoyancy forces, smaller such contents would be required for other plumes with smaller excess temperatures [*Herzberg et al.*, 2007]. For example, a content of >6% is likely to be sufficient to induce formation of a DEP for a perhaps more common plume with a peak excess temperature of ~150 K. The estimated critical eclogite contents would have to be slightly corrected upward (by ~3%) if the ambient mantle itself contained eclogite (~3%), as is indicated by studies of midocean ridge basalts that suggest a content of ~5% pyroxenite in the upper mantle [*Hirschmann and Stolper*, 1996]. Considering this possibility, the eclogite contents modeled here can be regarded as excess contents relative to the ambient mantle.

3.3.2. Asymmetry in Plume Behavior and Melting

Our models display complex and asymmetric structures not only at the depths of the DEP (i.e., 410–300 km) but also above this depth range (Figure 3.6). These asymmetries are seen in the overall morphology of the plume as a whole as well as in the temperatures and eclogite contents across the shallow plume conduit and plume pancake (Section 3.3.2.1). The structure of temperature and initial eclogite content (i.e., eclogite content imposed in the deep plume stem) within the melting zone are then found to produce bilateral asymmetries in the mafic contribution to volcanism (Section 3.3.2.2). Despite these asymmetries, the current models generally reproduce the main results of purely thermal plume models [*Farnetani and Hofmann*, 2009, 2010; *Farnetani et al.*, 2012] that materials originating from distinct sides of the deep plume stem mostly remain on their sides all the way to the surface (evident in contours of initial eclogite content, Figure 3.6). However, the added effects of compositional buoyancy as well as melting behavior of peridotitic vs. mafic material in our models lead to different conclusions about the relationship between magma and source compositions.

3.3.2.1. Thermochemical Plume Dynamics, Temperature, and Composition

In case Z3, strong asymmetry in temperature rises out of a compositionally zoned deep plume stem and DEP (Figure 3.6). The higher imposed eclogite content on the SW versus the NE side of the deep plume core ($\Phi_{ECL,SW} = 15\%$, $\Phi_{ECL,NE} = 12\%$) leads to a higher average excess density on the SW side, a density contrast that doubles in the depth range of 300–410 km where the DEP forms. This situation has three important consequences: First, the NE side of the DEP (with a near-neutral average buoyancy of about $-0.9\,kg/m^3$) is preferentially entrained into the shallow plume, whereas the SW side of the DEP (with a strongly negative average buoyancy of about $-9.7\,kg/m^3$) displays more extensive pooling and lateral spreading (see contours in Figure 3.6d). Second, the more eclogitic material from the SW side of the deep plume stem fills over half of the DEP (imaged as broad orange-to-yellow area in Figure 3.7m), pushing the less eclogitic material as well as the whole shallow plume conduit to the NE (dashed-dotted line in Figure 3.6d is symmetry axis of the model setup for reference). Third, the less hot, outer plume-core material from the SW side of the deep plume stem fails to rise through the depth range of 300–410 km (green colors on the right-hand side of Figure 3.6d; dark blue colors Figures 3.5n,o). Therefore, the materials that enter the shallow plume from the (initially more eclogitic) SW side are, on average, hotter than those on the (initially less eclogitic) NE side. The variations in initial eclogite content in the DEP thus create a "thermal filter" for the temperatures entering the shallow plume. The resulting thermal asymmetry in the shallow plume conduit is visualized in Figures 3.7m–o as an offset between the peak in temperature (bull's eye of dark grey contours; black cross) and the focus in mantle upwelling (bull's eye of red contours; red cross). As the focus of upwelling lies very close to the peak hotspot volcanic flux, which defines the hotspot location, case Z3 predicts bilateral asymmetry with higher temperatures on the SW side of the hotspot melting zone (green and black dashed lines encircle the pyroxenite + peridotite melting zone in Figure 3.6d with colors highlighting the thermal asymmetry).

Increasing the contrast in initial eclogite contents of the deep plume stem (from case Z3 to cases Z4 and Z5) boosts the asymmetry in the make-up and temperature distribution of the shallow plume conduit. The modeled plumes in cases Z4 and Z5 carry average eclogite contents (Table 3.1) that are similar to those of case Z3 and therefore generally behave similarly to that in case Z3. However, they display greater cross-chain variations in initial eclogite contents (Table 3.1). As the thermal filter effect is greater, the proportion of the less eclogitic (NE) relative to the more eclogitic (SW) material within as well as the thermal asymmetry across the shallow plume conduit increases from cases Z3 to Z4 and Z4 to Z5 (preponderance of green and blue colors increases from Figure 3.7o to Figure 3.7r to Figure 3.7u). The asymmetry of the overall morphology of the plume also increases with

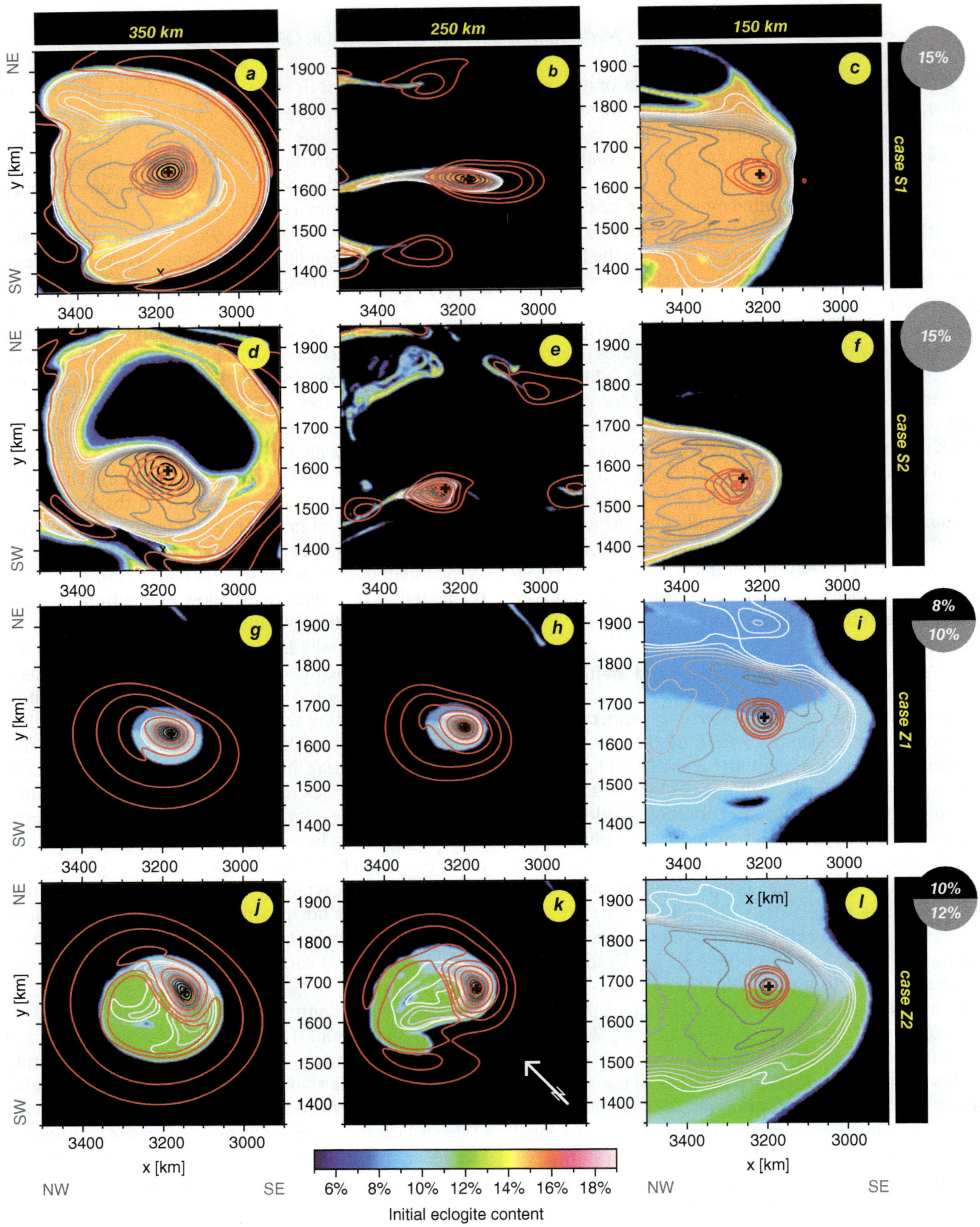

Figure 3.7 Plume upwelling rates, temperatures, and initial eclogite contents (colors with black denoting 0%) for all cases (rows as annotated) in horizontal cross sections at 150, 250, and 350 km depth (columns as annotated). Black-to-white contours denote mantle potential temperature in 10 °C increments between 1520 °C (white) and 1640 °C (black). Red contours denote plume upwelling rates on a log-scale (spaced by factor $\log_{10}(0.2)$). Disk insets (top) schematically show the initial distribution of eclogite, similar to Figure 3.4. The black and red crosses (as well as the bull's eyes of the grey-to-black and of the red contours) mark the peak in plume temperature and the focus of most vigorous upwelling, respectively. Whereas for cases S1 and Z1 and Z2, the centers of these bull's eyes fall on top of each other, they are displaced from each other for cases S2 and Z3 and Z5. Such a displacement identifies an asymmetry in temperature structure across the ascending plume. For reference, the deep plume stem at the bottom of the box (i.e., 660 km depth) is centered at $x = 3135$ km and $y = 1650$ km; plate motion is from right to left by rates of 80 km/Myr.

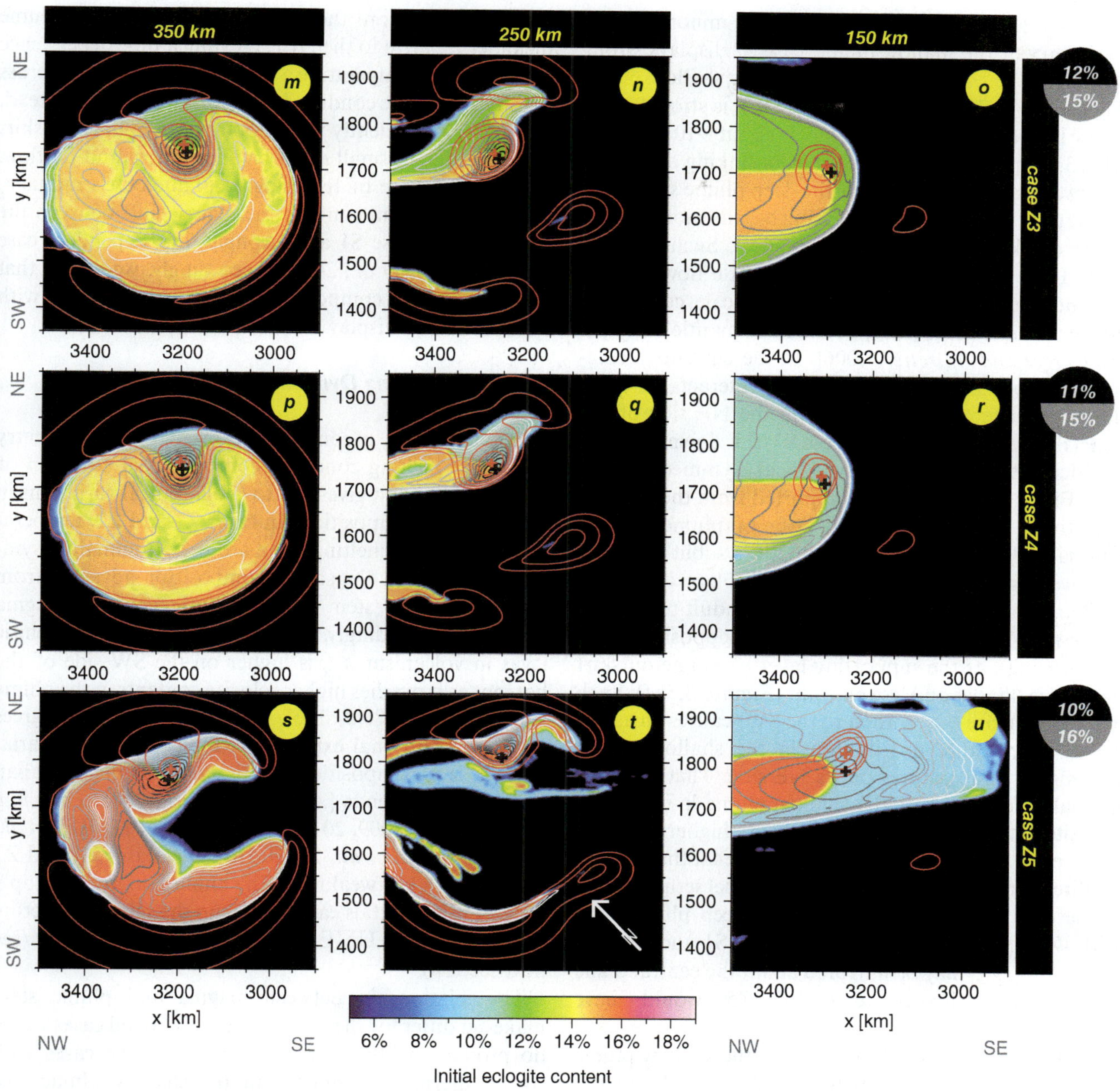

Figure 3.7 Cont'd

greater initial eclogite contrast (Figures 3.5m–u), as the less eclogitic, NE material, which predominantly feeds the shallow plume, is pushed increasingly to the margins of the DEP (Figure 3.7). Accordingly, the offset between the peak temperature and the peak upwelling rate increases from cases Z3 (~15 km) to Z5 (~40 km).

In contrast to cases Z3–Z5, cases Z1 and Z2 do not display significant thermal asymmetry of the shallow plume (Figure 3.7) despite a deep-rooted compositional asymmetry that differs only slightly from that in case Z3 (In cases Z1, Z2, and Z3, $\Phi_{ECL,SW}$ equals $1.25\Phi_{ECL,NE}$, $1.2\Phi_{ECL,NE}$, and $1.25\Phi_{ECL,NE}$, respectively; cf. Figure 3.4). The main cause for the lack of the asymmetry is the lack of a large DEP in cases Z1 and Z2. Accordingly, plume material rises rapidly through the depth range of 300–410 km, the thermal filter effect (as discussed above for cases Z3–Z5) is negligible, and any thermal asymmetry of the shallow plume conduit relative to the peak upwelling rate (and hence hotspot center) is minimal (Figure 3.7).

For a nonzoned plume, any thermal asymmetry across the shallow plume conduit would have to be explained by entirely different mechanisms than those described above, as the make-up of the deep plume stem in these cases is axisymmetric (with 15% eclogite throughout the plume core). While case S1 displays only very minor thermal asymmetry of the shallow plume, case S2 displays strong asymmetry, across both the DEP and the shallow plume (Figures 3.5–3.7). In both cases, the DEP is strongly negatively buoyant (average buoyancy is about $-10\,kg/m^3$; cf. Table 3.1) and is again prevented from sinking down into the lower mantle by the rise of the deep plume stem from below (and is contained laterally by the noneclogitic plume outskirts rising around its sides). Such a quasi-stable DEP is sensitive to ambient mantle flow such as that induced by sublithospheric small-scale convection [*Ballmer et al.*, 2007, 2011] or ambient mantle wind [cf. *Steinberger and Antretter*, 2006]. In the current calculations, a sublithospheric downwelling interacts with the plume pancake and leans down on the NE side of the DEP (Figures 3.5 and 3.6). In case S1, this downwelling is insufficient to induce strong dynamical asymmetry within the DEP, as the DEP is well supported from the sides by a relatively thick and buoyant layer of plume-outskirt material (Figures 3.5b,c). In case S2, the sublithospheric downwelling is instead sufficient to displace the whole DEP and to cause the shallow plume conduit to rise off-center (i.e., from the DEP on the side opposite to the downwelling), as the supporting layer of plume-outskirt material is thinner and less buoyant (Figures 3.5e,f). Such an off-center rise (i.e., shifted to the SW) causes an asymmetric temperature distribution in the shallow plume with higher temperatures on its NE side. That said, the thermal gradient across the shallow plume in case S2 is opposite to that in cases Z3–Z5 (i.e., with higher temperatures on the SW side). We note, however, that the SW and NE directions are conventions of our model setup defined by the distribution of eclogite in the deep plume stem (with $\Phi_{ECL,SW} \geq \Phi_{ECL,NE}$) and that models S1 and S2 with $\Phi_{ECL,SW} = \Phi_{ECL,NE}$ may be mirrored about the central plane at $y = 1650$ (mirrored cases Z1–Z5 would have $\Phi_{ECL,SW} < \Phi_{ECL,NE}$).

The temperature distribution across the shallow plume in case S2 is not only asymmetric but also time dependent (Figures 3.6g,h). Such time dependence is related to episodic extraction of hot material out of the DEP. Any extraction tends to increase the net density of the DEP. In case S2, this increase is sufficient to cause the DEP to spread laterally and sink slightly, a process that chokes the flux of material through the DEP's roof. Once outflow is smaller than inflow from below, the DEP reinflates until it is thick enough for material to rise above the coesite-stishovite phase transition at 300 km depth to generate another shallow-plume pulse (also see Movie S3 by *Ballmer et al.*, [2013]). The process begins again, and resulting shallow-plume pulsations provide a dynamical explanation for the variations in total volcanic flux documented along the Hawaiian-Emperor chain [*van Ark and Lin*, 2004; *Vidal and Bonneville*, 2004; *Ballmer et al.*, 2013]). In addition, the location of the shallow plume meanders relative to the DEP, creating a time dependence of the average temperature and thermal asymmetry across the shallow plume conduit. In cases S1 and Z1–Z5 instead, the DEP is sufficiently well supported by plume-outskirt material (due to small r_E) to balance inflow and outflow. Thus, an increase of the eclogitic plume core's radius r_E from 90 to 100 km controls the regime shift between the pseudosteady case S1 and strongly time dependent case S2 [cf. *Ballmer et al.*, 2013]. Accordingly, we expect that double-layered, compositionally zoned plumes with $r_E \approx 100\,km$ also display shallow-plume pulsations.

3.3.2.2. Melting Dynamics and Its Expression in Volcanism

In our models, a combination of thermal asymmetry across the melting zone and compositional zonation of the plume conduit gives rise to geographical variations in hotspot-lava composition. In cases Z1 and Z2, temperatures across the melting zone are only minimally asymmetric; therefore, the source composition advected from the deep plume stem predominantly controls magma make-up. Accordingly, the contribution of pyroxenitic lavas in volcanism X_{PX} is higher on the SW side of the hotspot as it overlies higher eclogite contents in the plume stem (Figures 3.8c,d, Table 3.3). This prediction agrees with the traditional explanation for geographical variations in lava composition, particularly for the Hawaiian Kea and Loa trends [*Abouchami et al.*, 2005; *Farnetani and Hofmann*, 2009, 2010; *Huang et al.*, 2011; *Weis et al.*, 2011; *Farnetani et al.*, 2012]. However, cases Z1 and Z2 suffer from a key weakness: The plumes do not develop a broad DEP and thus cannot explain the seismic structure imaged by the PLUME regional seismic tomography (see next section).

The relationship between magma and plume stem make-up differ for the compositionally zoned cases which do produce a DEP (Z3–Z5). These are the cases with strong thermal asymmetry in the shallow plume (see Section 3.3.2.1), and results show that peridotite melting starts deeper (Figures 3.6d–h) and reaches higher degrees on the hotter side of the shallow plume conduit than on its cooler side. The effect of thermal asymmetry on pyroxenite melting is much smaller, as pyroxenite reaches maximum degrees of melting on both sides. These melting dynamics systematically reduce X_{PX} on the hotter (and more eclogitic) SW side of the hotspot (Figures 3.8e–g) and accordingly trade off with the effects of source composition in terms of X_{PX}.

In case Z4, for example, the shallow thermal asymmetry consequently leads to a more-or-less symmetric distribution of X_{PX} across the main melting zone, thus completely obscuring the bilateral asymmetry in the composition of the plume conduit (Figure 3.8f, Table 3.3).

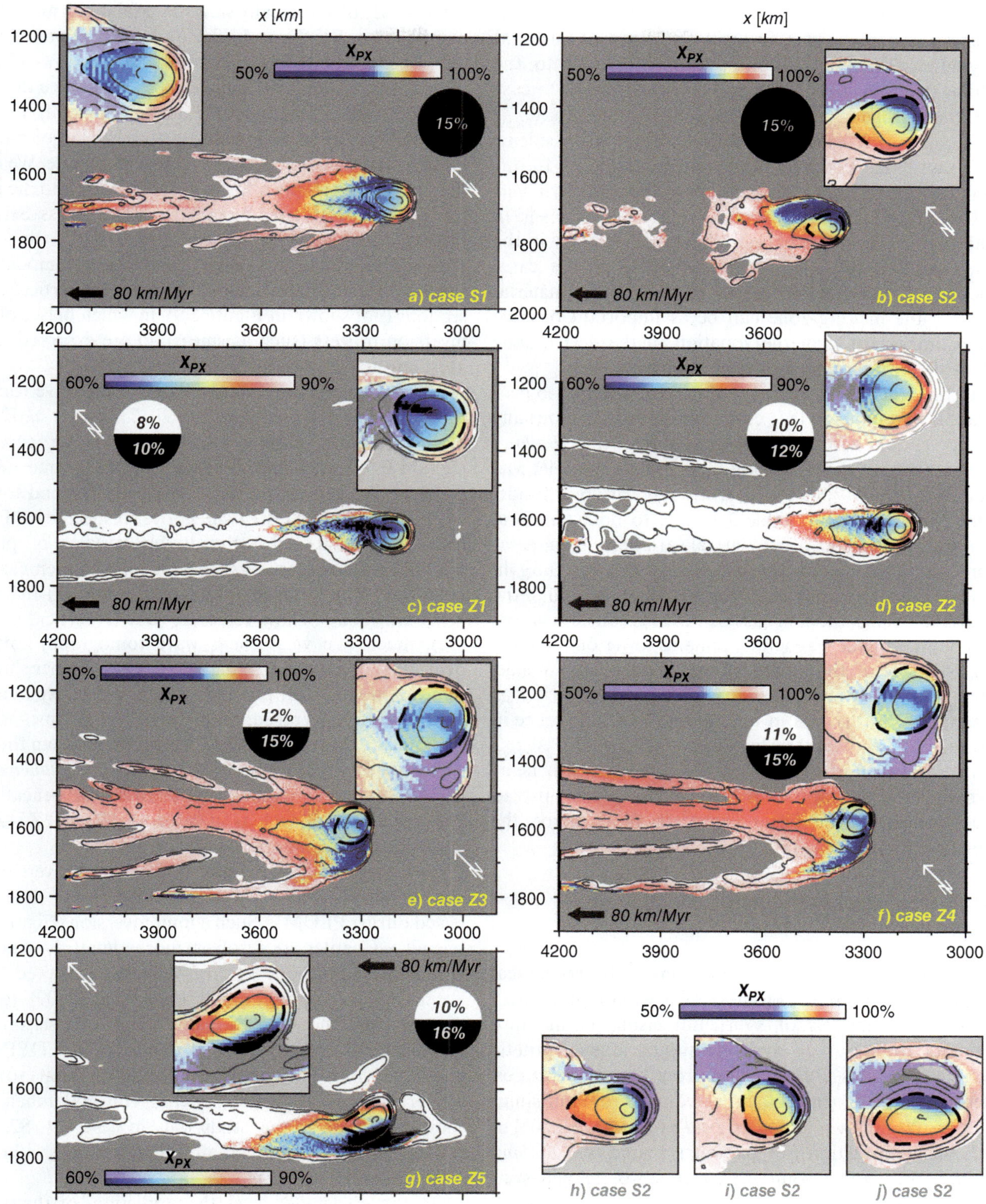

Figure 3.8 Geographical distribution of volcanism for all cases. Dark-grey contours show the flux of volcanism with solid contours at 0.1, 1, and 10 $km^3 km^{-2} Myr^{-1}$ from outside to inside. Colors display X_{PX} (note the difference in scales for panels (c,d) and (g) compared to the other panels). Insets zoom into the hotspot region (nonshaded and inside the black dashed line). The edge lengths of the insets is 200 km, corresponding to a zoom factor of 2:1. White arrows mark the model's northern direction. For time-dependent case S2, only one snapshot is fully shown (b), but zooms into the hotspot region of three additional snapshots are also given (h–j). The snapshot shown in (i) corresponds to that in Figure 3.6 g; the snapshot shown in (j) corresponds to that in Figure 3.6 h. Averages of X_{PX} for each side of the hotspot are reported in Table 3.3.

In cases Z3 and Z5, the effect of thermal asymmetry on melting dynamics even reverses the sense of bilateral asymmetry in magma composition compared to the plume's compositional zonation (Figures 3.8e,g). In case Z3, the average X_{PX} is greater on the NE side with $X_{PX,SW} - X_{PX,NW} = -1.2\%$ despite the fact that it is the SW side of the plume that contains more eclogite initially ($\Phi_{ECL,SW} - \Phi_{ECL,NE} = +3\%$). Case Z5 predicts an even greater reversal in asymmetry at the surface ($X_{PX,SW} - X_{PX,NE} \approx -9\%$), although it has the strongest bilateral asymmetry in the deep conduit ($\Phi_{ECL,SW} - \Phi_{ECL,NE} = +6\%$) of the cases modeled. Thus, the thermal asymmetry of the material feeding the melting zone can be as important or even more important than the zonation in the deep plume stem in controlling the geographic patterns of X_{PX}.

In fact, thermal asymmetry alone can give rise to geographic variations in lava composition even without any compositional zoning in the deep plume stem [*Ballmer et al.*, 2013]. In case S2, both $X_{PX,SW}$ (~83.4% to ~91.5%) and $X_{PX,NE}$ (~79.0% to ~90.8%) strongly vary over model time (Figures 3.8h–j; Table 3.3), due to both temporal variations in the flux of and lateral variations in temperature within the shallow plume conduit (Figures 3.6g–h; also see last paragraph of Section 3.3.2.1). Although the initial eclogite content in case S2 remains constant across the plume core, X_{PX} generally remains lower on the (hotter) NE side of the hotspot than the (cooler) SW side, except for a very few snapshots with near-symmetric X_{PX}, one of which is shown in Figure 3.8h. The difference in average X_{PX} between the two sides (i.e., $X_{PX,SW} - X_{PX,NE}$) ranges between ~0.6% and ~6.8%. Case S1 with near-symmetric temperatures in the shallow plume instead does not display significant asymmetry in X_{PX} across the hotspot (Figure 3.8a, Table 3.3).

3.3.3. Implications of Seismic Constraints on Plume Dynamics

To compare our model predictions with geophysical constraints from seismic tomography, we compute three-dimensional models of synthetic seismic shear wave velocity from our geodynamic simulations. Synthetic seismic velocities are calculated from the temperatures, densities, and eclogite contents predicted from our simulations. We use the method of *Faul and Jackson* [2005] to do this computation (for parameters used, see *Ballmer et al.* [2013]), and account for the effects of eclogite content on seismic velocities according to *Xu et al.* [2008]. For comparison, we also predict a synthetic seismic velocity model for a purely thermal plume (i.e., based on the reference model of *Ballmer et al.* [2011]).

Figure 3.9 shows horizontal cross sections through the synthetic velocity models at 350 km depth and elucidates that the models fall into two categories: first, models with a wide body of seismically slow material in the mid-upper mantle (cases S1, S2, and Z3–Z5) due to the presence of a DEP; second, models without such a wide low-velocity body (cases Z1 and Z2), similar to the reference thermal (i.e., classical) plume model (Figure 3.9a). Only plumes from the first category are able to stabilize large volumes of seismically slow material in the upper mantle. We note that Figure 3.9 shows a forward model of synthetic seismic velocities unfiltered by the inversion process. Seismic filtering (i.e., by a resolution test) would instead be required to directly compare these velocity models to seismic tomography images, particularly for vertical cross sections (not shown in Figure 3.9), in which biases of the inversion process (such as smearing) are expected to be most strongly expressed.

Using the inversion matrix of the shear wave tomography model of *Wolfe et al.* [2009], *Ballmer et al.* [2013] performed such seismic resolution tests for case S1 (Figure 3.9b) and the reference thermal plume model (Figure 3.9a), representative cases for the first and second (with and without DEP) categories, respectively. These tests demonstrated that the double layering of plume material as in our case S1 (i.e., analogous to their case A) is expected to be imaged as one thick (~400 km) and wide (~500 km) body of low velocities (due to vertical smearing), much like what shear wave tomography imaged beneath Hawai'i [*Wolfe et al.*, 2009]. They further showed that our case S1 can provide a much better explanation for the station-averaged, shear wave arrival times measured during the seismic PLUME experiment than the reference thermal plume model (originally by *Ballmer et al.* [2011]). Accordingly, manifestation of a DEP beneath the Hawaiian hotspot such as in cases S1, S2, and Z3–Z5 is demanded by the seismic constraints.

This interpretation is supported by joint inversion of surface wave, shear wave and ambient noise data collected during PLUME. Such a joint inversion (Figure 3.3; see also Chapter 2) resolves more detailed structure within the large body of low velocities retrieved from shear-wave tomography alone [*Wolfe et al.*, 2009]. It suggests the presence of two distinct layers of seismically slow material at the predicted depths of the DEP and plume pancake. This geophysical evidence points toward a thermochemical plume beneath Hawai'i, much like those portrayed in this study (i.e., in cases S1, S2, and Z3–Z5), and away from a more classical plume without a DEP (cases Z1 and Z2).

On a more detailed level, the predictions of the subset of our models with a DEP also appear to be in good agreement with the specific structures resolved by joint tomography. In terms of its lateral extent, the predicted body of seismically slow material (i.e., DEP) in a subset of our models (especially, S1, S2, and Z3–Z5) is consistent with observations (dashed ellipses in Figures 3.9

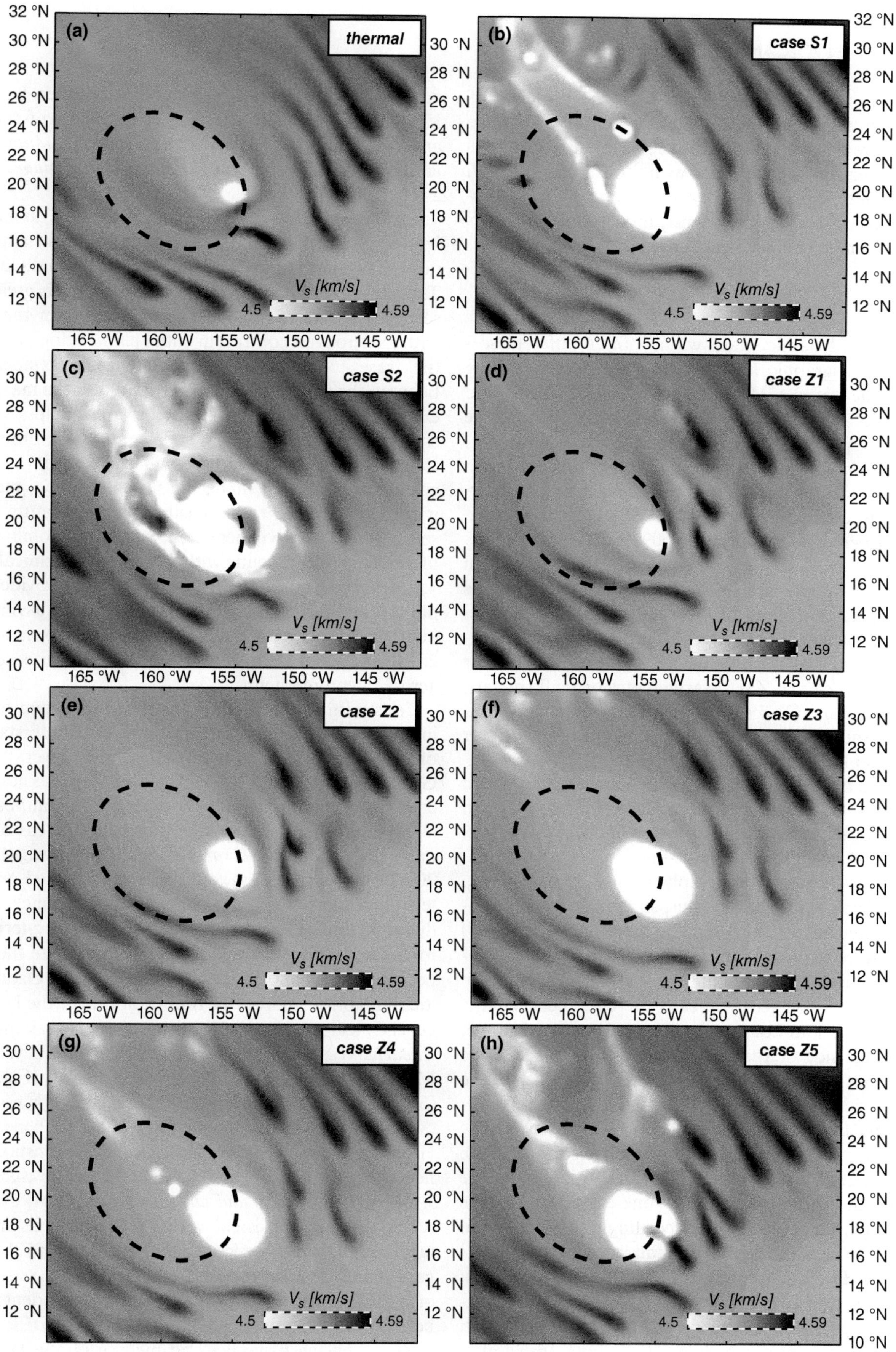

Figure 3.9 Synthetic seismic shear wave velocities at 350 km depth (shades of grey). Shown are horizontal cross sections through three-dimensional synthetic models computed from the predicted temperatures, densities, and eclogite contents of (a) a thermal plume model [*Ballmer et al.*, 2011], (b,c) cases S1 and S2, and (d–h) cases Z1–Z5. In interpolating the three-dimensional model to the cross section shown, synthetic seismic velocities are averaged over a depth interval of 50 km. The dashed ellipse is analogous to that shown in Figure 3.3e and thus denotes the approximate size and location of the body of low seismic wave speeds imaged by joint seismic tomography (see Chapter 2).

and 3.3e), albeit displaced eastward. The presence of short-wavelength, seismically fast structures within the DEP, as predicted by cases S2 and Z5 (Figures 3.9c,h), is also generally consistent with observations (cf. Figure 3.3e). These first-order agreements between predictions of simplified geodynamic models and geophysical constraints are indeed encouraging.

3.4. DISCUSSION

In our geodynamic study, we find three different scenarios for geographic variations in the materials melting beneath the hotspot: (1) Cases without a DEP (Z1 and Z2) predict a direct geometrical relationship between the geographic pattern of mafic content in volcanism (X_{PX}) and the zonation of the deep plume conduit, similar to the traditional explanation for the geochemical differences between the Loa and Kea trends [*Abouchami et al.*, 2005; *Farnetani and Hofmann*, 2009, 2010; *Huang et al.*, 2011; *Weis et al.*, 2011; *Farnetani et al.*, 2012]). (2) In contrast, compositionally zoned cases with a DEP (cases Z3–Z5) predict a nonintuitive relationship, as thermal asymmetry across the shallow plume conduit boosts peridotite melting on the more eclogitic side to veil (Z4) or even reverse (Z3 and Z5) the asymmetry of mafic materials in volcanism. (3) Finally, thermal asymmetry is sufficient to generate a geographic asymmetry in X_{PX} even for nonzoned plumes (case S2).

Of these three scenarios, only scenarios 2 and 3, which include a DEP, are applicable to Hawai'i. Only they can account for the stabilization of large volumes of hot, seismically slow material in the mid-upper mantle, as is indicated by the PLUME tomography results [cf. *Wolfe et al.*, 2009; *Ballmer et al.*, 2013]. Thus, shallow thermal asymmetry rising out of the DEP (as in scenarios 2 and 3) appears to be as or more important than a deep compositional zonation of the plume (as only in scenario 2) in giving rise to the geochemical difference between the Hawaiian Kea and Loa trends. In both scenarios 2 and 3, the low-X_{PX} side of the melting zone is predicted to be hotter (thus peridotite melts to higher extents) than the high-X_{PX} side. As Kea-type lavas are thought to be influenced less by mafic melting than Loa-type lavas (i.e., $X_{PX,SW} > X_{PX,NE}$ [*Sobolev et al.*, 2005; *Greene et al.*, 2010; *Herzberg*, 2011; *Jackson et al.*, 2012]), we infer that the Kea side of the mantle melting zone is hotter than the Loa side [cf. *Xu et al.*, 2014]. Accordingly, the distinction between the Kea and Loa trends can arise out of either a nonzoned Hawaiian plume (scenario 3) [cf. *Ballmer et al.*, 2011, 2013] or a zoned plume with greater eclogite contents on the low-X_{PX} Kea side of the deep plume stem (scenario 2), not on the high-X_{PX} Loa side as previously inferred [e.g., *Sobolev et al.*, 2005]. In either case, the model results in combination with the seismic evidence argue against a direct geographic link between Loa and Kea lava geochemistry and the composition of the SW and NE sides of the deep plume stem [cf. *Abouchami et al.*, 2005; *Farnetani and Hofmann*, 2010; *Huang et al.*, 2011; *Weis et al.*, 2011].

Future studies should be designed to distinguish between scenarios 2 and 3 and hence to test whether the deep Hawaiian plume is indeed compositionally zoned. Thermal asymmetry across the melting zone is greater in scenario 2 than in scenario 3 and greatest in the cases with the strongest contrast in initial plume eclogite content (i.e., case Z5 in this study). Differences in temperature between the two sides control variations in the maximum extent of peridotite melting across the hotspot, which should be testable using major elements and various trace elements. The challenge here will be to distinguish the contributions from mafic versus peridotitic melting, which are expected to be sensitive to temperature themselves. Such a test will hence likely require a combination of geodynamic modeling with targeted geochemical analysis. Geophysical evidence may be able to provide complementary constraints on the amplitude of asymmetry in temperature across the melting zone [cf. *Rychert et al.*, 2013] and thus ultimately on that in eclogite content across the plume stem.

Constraining the make-up of the plume stem feeding into the DEP will serve to improve our understanding of the behavior of the plume in the lower mantle. Geodynamic models have predicted complex thermochemical plume dynamics for a range of compositions involving plume pulsations on timescales of tens of Myr or longer [e.g., *Lin and van Keken*, 2005; *Kumagai et al.*, 2008]. Such pulsations would be superimposed on midterm pulsations rising out of the DEP itself, such as predicted by our study (e.g., case S2). These long-term and midterm pulsations ultimately act to modulate the compositions and temperatures across the melting zone and are thus able to address the time evolution of the Loa/Kea geochemical asymmetry, which may or may not [cf. *Abouchami et al.*, 2005] persist up the chain beyond the Moloka'i Fracture Zone or even switch polarity (Figure 3.1 [*Sinton et al.*, 2014]). We also note that there is no direct evidence for or against the existence of a DEP beneath the Hawaiian hotspot tens of Myr ago. While crustal thickness variations along the Hawaiian-Emperor chain [*van Ark and Lin*, 2004; *Vidal and Bonneville*, 2004] on timescales that are consistent with DEP pulsations may be taken as indirect evidence [cf. *Ballmer et al.*, 2013], future geodynamic studies should assess the long-term stability of a DEP, especially with strongly time-dependent plume behavior in the lower mantle feeding into the DEP.

Resolving the dynamics and melting of the Hawaiian plume in space and time is indeed needed to improve our understanding of the mantle as a whole. For example, it may provide insight into the compositions of the materials making up the deep lower mantle as well as the dynamical processes by which the Hawaiian plume may or may not entrain LLSVP material [cf. *Steinberger and Torsvik*, 2012]. Studying other plumes in addition to the Hawaiian plume may provide constraints on the heat and material fluxes across the mantle as well as the possible origins of the LLSVPs. However, our results suggest that high-resolution seismic images of these other plumes would be required to assess the origin of any local-scale geographic variations in lava geochemistry across the related hotspots.

3.5. CONCLUSIONS

In this study, we examine the upper mantle dynamics of mantle flow and melting of thermochemical plumes with a bilateral zonation in the content of mafic materials rising out of the lower mantle. The predictions of our geodynamic simulations can be grouped within three different scenarios:

1. Thermochemical plumes with a bilateral asymmetry in eclogite content, and average such contents of <12%, behave similar to classical thermal plumes with little mixing between the two sides of the plume and a direct relationship between source and lava composition, consistent with previous findings [*Farnetani and Hofmann*, 2009, 2010]. However, these plumes cannot account for the stabilization of large volumes of hot material in the mid-upper mantle, as is evident for the Hawaiian plume by seismic tomography.

2. In contrast, compositionally zoned plumes with average eclogite contents of >12% intermittently stagnate in the depth range of about 300–410 km to form a DEP, and hence can reconcile seismic constraints. This pooling is due to a relative density maximum of eclogite at these depths, which in turn only allows the hottest material to enter the shallow plume on the more eclogitic side of the plume. The resulting thermal asymmetry across the melting zone gives rise to more vigorous peridotite melting on the initially more eclogitic side, therefore reducing the expression of mafic materials in volcanism X_{PX} to the point that it is similar to or even lower than on the other (i.e., initially less eclogitic) side.

3. Moreover, compositionally nonzoned plumes that rise asymmetrically out of a DEP (e.g., due to ambient mantle flow) can also sustain a thermal gradient across the melting zone and thus geographic variations in X_{PX} across the hotspot.

These model predictions (i.e., scenarios 2 and 3) have important implications for Hawaiian plume dynamics and the composition of the deep mantle. Asymmetry in X_{PX} across the hotspot, as evident by the geochemical distinction between the Kea and Loa trends [e.g., *Sobolev et al.*, 2005], is an explicit indication not necessarily of a (bilateral) compositional zonation of the deep plume stem, but rather of a thermal gradient across the hotspot melting zone with higher temperatures on the Kea side. If a zonation in the content of mafic materials indeed exists, our results suggest that it is the NE Kea side of the deep plume stem that contains more eclogite than the SW Loa side, contrary to previous interpretations. Our findings put into question previous efforts of mapping the deep mantle from geographical variations in hotspot lava composition.

APPENDIX: MELTING, DENSITY, AND RHEOLOGY PARAMETERIZATIONS

3A.1. Melting Parameterization

The solidi versus depth functions and the rate change in melt fraction per increment of pressure release of eclogite, pyroxenite, and peridotite are parameterized using polynomial regressions to experimental data [*Yasuda et al.*, 1994; *Hirschmann*, 2000; *Pertermann and Hirschmann*, 2003]. We assume the rate of heat consumption by melting is proportional to the latent heat $L = 560$ kJ/kg and melting rate. However, the processes of eclogite melting and subsequent hybridization of the ambient peridotite to form pyroxenite together are assumed to neither consume nor release latent heat. As soon as the volume fraction of melt in a given lithology exceeds a critical porosity (i.e., $\phi_C = 0.4\%$ in peridotite and $\phi_{C,PYX} = 10\%$ in pyroxenite), magmas are instantaneously extracted and taken to feed volcanism [cf. *McKenzie*, 1985; *Schmeling*, 2006]; thereafter, the solid maintains residual porosities of $\phi_R = 0.4\%$ and $\phi_{R,PYX} = 5\%$. This treatment is equivalent to assuming that the timescale of melt extraction is much smaller than that of mantle flow [*Kelemen et al.*, 1997]. The porosities applied are bracketed by estimates [*Yaxley and Green*, 1998; *Faul*, 2001; *Stracke et al.*, 2006].

Compared to that of dry peridotite, the melting temperature of hydrous peridotite is reduced with increasing water content in the liquid, c_L (i.e., by $43c_L$ K/wt% [*Katz et al.*, 2003]). The liquid water concentration decreases during melting as an inverse function of depletion (i.e., extent of previous or ongoing melting) of hydrous peridotite, F_{HP}:

$$c_L = \frac{c}{D_{H_2O}} = \frac{c_O}{F_{HP} + D_{H_2O}\left(1 - F_{HP}\right)} \quad (3.1)$$

for $F_{HP} < \phi_C$ and

$$c_L = \frac{c}{D_{H_2O}} = c_O \frac{\left(1 - \frac{F_{HP} - \varphi_C}{1 - \varphi_C}\right)^{\frac{1}{\varphi_C\left(1 - D_{H_2O}\right) + D_{H_2O}} - 1}}{\varphi_C + D_{H_2O}\left(1 - \varphi_C\right)} \quad (3.2)$$

for $F_{HP} \geq \phi_C$ [*Zou*, 1998] with c the water content in the solid and $D_{H_2O} = 0.01$ the partitioning coefficient of water.

3A.2. Density Parameterization

Mantle density varies as a function of temperature, composition, and melt content:

$$\rho = \rho_0 - \alpha(T - T_0) + (\rho_M - \rho_0)\varphi + \Delta\rho_F(F_{DP}\Phi_{DP} + F_{HP}\Phi_{ECL}) + \Delta\rho_{ECL}\Phi_{ECL} \quad (3.3)$$

$$\text{with} \quad \varphi = \varphi_{PYX}\Phi_{PYX} + \varphi_{DP}\Phi_{DP} + \varphi_{HP}\Phi_{HP} \quad (3.4)$$

and subscripts PYX, ECL, HP, and DP referring to the components pyroxenite, eclogite, hydrous peridotite, and dry peridotite, respectively; Φ_{XY} is the volume fraction of a component; ϕ_{XY} the is porosity in a component; F_{XY} is the depletion of a component; T is temperature; T_0 is reference temperature; ρ is density; ρ_0 is reference density; α is thermal expansivity; ϕ is bulk porosity; ρ_M is density of melt; $\Delta\rho_F$ is density change with depletion; and $\Delta\rho_{ECL}$ is excess density of eclogite. The densities of both dry and hydrated peridotite decrease as depletion increases during progressive melting, since heavy minerals and oxides are preferentially consumed during melting [*Schutt and Lesher*, 2006]. In contrast, we assume the densities of eclogite and pyroxenite to be independent of depletion (due to their relatively small volume fraction and uniform mineralogy, respectively). The $\Delta\rho_{ECL}$ was fixed at 220 kg/m³ in the depth range 300–410 km and was 110 kg/m³ elsewhere [*Aoki and Takahashi*, 2004]. We also ignore the effect of the refractory lithology on bulk density (i.e., by setting $\Delta\rho_{RL} \equiv 0$), assuming a negligible intrinsic negative density anomaly of this minor component. As the volume fraction of the refractory component $\Phi_{RL} = \Phi_{ECL,SW} - \Phi_{ECL}$ (except for where eclogite melting has taken place) with $\Phi_{ECL,SW}$ the initial volume fraction of eclogite on the SW side of the plume stem (i.e., a constant for each calculation), $\Delta\rho_{RL}$ also largely trades off with $\Delta\rho_{ECL}$.

3A.3. Rheology Parameterization

We apply a Newtonian rheology to model convection in the mantle. We neglect the stress dependence of mantle viscosity η and focus on the effects of composition, temperature, and melt. While stress-dependent dislocation creep is relevant for plume-lithosphere interaction in the asthenosphere [*Asaadi et al.*, 2011], stress-independent diffusion creep is thought to be the dominant mechanism of deformation in the lower upper mantle and at the depths of the DEP [*Karato and Wu*, 1993]. We chose an activation energy $E^* = 300$ kJ/mol, which is appropriate for modeling diffusion creep [*Karato and Wu*, 1993; *Hirth*, 2002]. Finally, we account for the effects of dehydration in olivine [*Hirth and Kohlstedt*, 1996] and melt retention on rheology [*Kohlstedt and Zimmerman*, 1996]. Whereas melt retention lubricates the mantle, dehydration stiffening during progressive melting is thought to be dominant [*Karato*, 1986]. We parameterize mantle rheology such that hydrous peridotite stiffens by a factor of $\xi = 100$ as it dehydrates from $c = 100$ ppm to $c = c_{dry}$ [*Hirth and Kohlstedt*, 1996]. Mantle rheology is instead taken to be insensitive to the abundance of olivine-free (eclogite, pyroxenite) lithologies, since deformation of a fine-scale (i.e., compared to the length scale of convection) assemblage is predominantly compensated by the weakest lithology (i.e., peridotite) as long as it forms the matrix. Substituting equation (1) or (2) into equation (5) thus leaves with a rheology dependent on the depletion in hydrous peridotite F_{HP} and melt content φ, in addition to the dependence on temperature T and depth z:

$$\eta = \eta_m \exp\left(\frac{E^* + \rho_m g z V^*}{RT} - \frac{E^*}{RT_m}\right) \frac{\left[\left(c - c_{dry}\right)\frac{\xi}{\xi - 1}\right]^{\Phi_{HP}}}{\exp(\zeta\varphi)} \quad (3.5)$$

with η_m the reference mantle viscosity and all other parameters reported in Table 3.2. The effective mantle viscosity $\eta_{eff} = 1.6 \times 10^{19}$ Pa · s defines the minimum viscosity of the asthenosphere in a mantle column well away from the hotspot, typically reached at ~160 km depth.

ACKNOWLEDGMENTS

M.D.B. and G.I. were supported by NSF grant EAR-1141938. Calculations were performed on NSF's TeraGrid (XSEDE) using allocation TG-EAR120012 as well as at the Hawai'i Open Supercomputing Center. We thank associate editor Mike Poland and two anonymous reviewers for comments that helped to improve the manuscript.

REFERENCES

Abouchami, W., A. W. Hofmann, S. J. G. Galer, F. A. Frey, J. Eisele, and M. Feigenson (2005), Lead isotopes reveal bilateral asymmetry and vertical continuity in the Hawaiian mantle plume, *Nature*, *434*, 851–856.

Aoki, I., and E. Takahashi, (2004), Density of MORB eclogite in the upper mantle, *Phys. Earth Planet. Int.*, *143*, 129–143.

Asaadi, N., N. M. Ribe, and F. Sobouti (2011). Inferring nonlinear mantle rheology from the shape of the Hawaiian swell, *Nature*, *473*, 501–504.

Ballmer, M. D., J. van Hunen, G. Ito, P. J. Tackley, and T. A. Bianco (2007), Non-hotspot volcano chains originating from small-scale sublithospheric convection, *Geophys. Res. Lett. 34*, doi:10.1029/2007GL031636

Ballmer, M. D., G. Ito, J. van Hunen, and P. J. Tackley (2010), Small-scale sublithospheric convection reconciles geochemistry and geochronology of "Superplume" volcanism in the western and south Pacific, *Earth Planet. Sci. Lett.*, *290*, 224–232.

Ballmer, M. D., G. Ito, J. van Hunen, and P. J. Tackley (2011), Spatial and temporal variability in Hawaiian hotspot volcanism induced by small-scale convection, *Nature Geosci.*, *4*, 457–460.

Ballmer, M. D., G. Ito, C. J. Wolfe, and S. C. Solomon (2013), Double layering of a thermochemichal plume in the upper mantle beneath Hawaii, *Earth Planet. Sci. Lett.*, *376*, 155–164.

Boyet, M., and R. W. Carlson (2005), Nd-142 evidence for early (>4.53 Ga) global differentiation of the silicate Earth, *Science*, *309*, 576–581.

Burke, K., and T. H. Torsvik (2004) Derivation of Large Igneous Provinces of the past 200 million years from long-term heterogeneities in the deep mantle, *Earth Planet. Sci. Lett.*, *227*, 531–538.

Castillo, P. R., J. H. Natland, Y. L. Niu, and P. F. Lonsdale (1998), Sr, Nd and Pb isotopic variation along the Pacific-Antarctic risecrest, 53–57 degrees S: Implications for the composition and dynamics of the South Pacific upper mantle, *Earth Planet. Sci. Lett.*, *154*, 109–125.

Chauvel, C., R. C. Maury, S. Blais, E. Lewin, H. Guillou, G. Guille, P. Rossi, and M.-A. Gutscher (2012), The size of plume heterogeneities constrained by Marquesas isotopic stripes, *Geochem. Geophys. Geosyst.*, *13*, Q07005.

Crosby, A. G., and D. McKenzie (2009), An analysis of young ocean depth, gravity and global residual topography, *Geophys. J. Int.*, *178*, 1198–1219.

Davaille, A. (1999), Simultaneous generation of hotspots and superswells by convection in a heterogeneous planetary mantle, *Nature*, *402*, 756–760.

Dupre, B., and C. J. Allegre (1983), Pb-Sr variation in Indian Ocean basalts and mixing phenomena, *Nature*, *303*, 142–146.

Farnetani, C. G., and A. W. Hofmann (2009), Dynamics and internal structure of a lower mantle plume conduit, *Earth Planet. Sci. Lett.*, *282*, 314–322.

Farnetani, C. G., and A. W. Hofmann (2010), Dynamics and internal structure of the Hawaiian plume, *Earth Planet. Sci. Lett.*, *295*, 231–240.

Farnetani, C. G., and H. Samuel (2005), Beyond the thermal plume paradigm, *Geophys. Res. Lett.*, *32*, L07311, doi:10.1029/2005GL022360.

Farnetani, C. G., A. W. Hofmann, and C. Class (2012) How double volcanic chains sample geochemical anomalies from the lowermost mantle, *Earth Planet. Sci. Lett.*, *359*, 240–247.

Faul, U. (2001), Melt retention and segregation beneath mid-ocean ridges, *Nature, 410*, 920–923.

Faul, U. H., and I. Jackson (2005), The seismological signature of temperature and grain size variations in the upper mantle, *Earth Planet. Sci. Lett.*, *234*, 119–134.

Frey, F. A., and J. M. Rhodes (1993), Intershield geochemical differences among Hawaiian volcanoes: Implications for source compositions, melting process and magma ascent paths, *Phil. Trans. R. Soc. Lond. A*, *342*, 121–136.

Greene, A. R., M. O. Garcia, D. Weis, G. Ito, M. Kuga, J. Robinson, and S. Yamasaki (2010), Low-productivity Hawaiian volcanism between Kaua'i and O'ahu, *Geochem. Geophys. Geosyst.*, *11*, Q0AC08.

Griffiths, R. W., and I. H. Campbell (1990) Stirring and structure In mantle starting plumes, *Earth Planet. Sci. Lett.*, *99*, 66–78.

Hauri, E. H. (1996), Major element variability in the Hawaiian mantle plume, *Nature, 382*, 415–419.

Herzberg, C. (2011), Identification of source lithology in the Hawaiian and Canary Islands: Implications for origins, *J. Petrol.*, *52*, 113–146.

Herzberg, C., P. D. Asimow, N. Arndt, Y. L. Niu, C. M. Lesher, J. G. Fitton, M. J. Cheadle, and A. D. Saunders (2007), Temperatures in ambient mantle and plumes: Constraints from basalts, picrites, and komatiites, *Geochem. Geophys. Geosyst. 8*, 34.

Hirschmann, M. M. (2000), Mantle solidus: Experimental constraints and the effects of peridotite composition, *Geochem. Geophys. Geosyst.*, *1*, doi: 10.1029/2000gc000070.

Hirschmann, M. M., and E. M. Stolper (1996), A possible role for garnet pyroxenite in the origin of the "garnet signature" in MORB, *Contrib. Mineral. Petrol.*, *124*, 185–208.

Hirth, G. (2002), Laboratory constraints on the rheology of the upper mantle, *Plastic Deformation of Minerals and Rocks*, *51*, 97–120.

Hirth, G., and D. L. Kohlstedt (1996), Water in the oceanic upper mantle: Implications for rheology, melt extraction and the evolution of the lithosphere, *Earth Planet. Sci. Lett.*, *144*, 93–108.

Hirth, G., and D. L. Kohlstedt (2003), Rheology of the upper mantle and mantle wedge: A view from the experimentalists, in *Inside the subduction factory*, edited by J. Eiler, pp. 83–105, AGU, Washington, D. C.

Huang, S., M. Humayun, and F. A. Frey (2007), Iron/manganese ratio and manganese content in shield lavas from Ko'olau Volcano, Hawai'i, *Geochim. Cosmochim. Acta*, *71*, 4557–4569.

Huang, S., P.S. Hall, and M. G. Jackson (2011), Geochemical zoning of volcanic chains associated with Pacific hotspots, *Nature Geosci.*, *4*, 874–878.

Huang, S. C., and F. A. Frey (2005), Recycled oceanic crust in the Hawaiian Plume: Evidence from temporal geochemical variations within the Koolau Shield, *Contrib. Mineral. Petrol.*, *149*, 556–575.

Ishii, M., and J. Tromp (1999), Normal-mode and free-air gravity constraints on lateral variations in velocity and density of Earth's mantle, *Science, 285*, 1231–1235.

Jackson, E. D., E. A. Silver, and G. B. Dalrymple (1972), Hawaiian-Emperor chain and its relation to Cenozoic circumpacific tectonics, *Geol. Soc. Am. Bull.*, *83*, 601–618.

Jackson, M. G., D. Weis, and S. Huang (2012), Major element variations in Hawaiian shield lavas: Source features and perspectives from global ocean island basalt (OIB) systematics, *Geochem. Geophys. Geosyst.*, (*13*), Q09009, doi: 10.1029/2012GC004268.

Karato, S. (1986), Does partial melting reduce the creep strength of the upper mantle, *Nature*, *319*, 309–310.

Karato, S., and P. Wu (1993), Rheology of the upper mantle: A synthesis, *Science*, *260*, 771–778.

Katz, R. F., and J. F. Rudge (2011), The energetics of melting fertile heterogeneities within the depleted mantle, *Geochem. Geophys. Geosyst.*, *12*, doi: 10.1029/2011GC003834

Katz, R. F., M. Spiegelman, and C. H. Langmuir (2003), A new parameterization of hydrous mantle melting, *Geochem. Geophys. Geosyst.*, *4*, 1073, doi: 10.1029/2002GC000433.

Kelemen, P. B., G. Hirth, N. Shimizu, M. Spiegelman, and H. J. B. Dick (1997), A review of melt migration processes in the adiabatically upwelling mantle beneath oceanic spreading ridges, *Phil. Trans. R. Soc. Lond. Ser. Mathematicalphys. and Eng. Sci.*, *355*, 283–318.

Kohlstedt, D. L., and M. E. Zimmerman (1996), Rheology of partially molten mantle rocks, *Ann. Rev. Earth Planet. Sci.*, *24*, 41–62.

Konter, J. G., B. B. Hanan, J. Blichert-Toft, A. A. P. Koppers, T. Plank, and H. Staudigel (2008), One hundred million years of mantle geochemical history suggest the retiring of mantle plumes is premature, *Earth Planet. Sci. Lett.*, *275*, 285–295.

Kumagai, I., A. Davaille, K. Kurita, and E. Stutzmann (2008), Mantle plumes: Thin, fat, successful, or failing? Constraints to explain hot spot volcanism through time and space, *Geophys. Res. Lett.*, *35*, L16301.

Kurz, M. D., T. C. Kenna, J. C. Lassiter, and D. J. DePaolo (1996), Helium isotopic evolution of Mauna Kea volcano: First results from the 1-km drill core, *J. Geophys. Res. Solid Earth.*, *101*, 11781–11791.

Labrosse, S., J. W. Hernlund, and N. Coltice (2007), A crystallizing dense magma ocean at the base of the Earth's mantle, *Nature*, *450*, 866–869.

Lin, S. C., and P. E. van Keken (2005), Multiple volcanic episodes of flood basalts caused by thermochemical mantle plumes, *Nature*, *436*, 250–252.

Mallik, A., and R. Dasgupta (2012), Reaction between MORB-eclogite derived melts and fertile peridotite and generation of ocean island basalts, *Earth Planet. Sci. Lett.*, *329*, 97–108.

Masters, G., G. Laske, H. Bolton, and A. Dziewonski (2000), The relative behavior of shear velocity, bulk sound speed, and compressional velocity in the mantle: Implications for chemical and thermal structure, in *Geophysical Monograph on Mineral Physics and Seismic Tomography from the Atomic to the Global Scale*, edited by S. Karato et al., pp. 63–87, AGU, Washington, D. C.

McKenzie, D. (1985) The extraction of magma from the crust and mantle, *Earth Planet. Sci. Lett.*, *74*, 81–91.

Moresi, L., S. Zhong, and M. Gurnis (1996), The accuracy of finite element solutions of Stokes' flow with strongly varying viscosity, *Phys. Earth Planet. Int.*, *97*, 83–94.

Morgan, W. J. (1972), Plate motions and deep mantle convection, *Geol. Soc. Am. Memoir*, *132*, 7–22.

Mosca, I., L. Cobden, A. Deuss, J. Ritsema, and J. Trampert (2012), Seismic and mineralogical structures of the lower mantle from probabilistic tomography, *J. Geophys. Res*, *117*, B06304.

Payne, J. A., M. G. Jackson, and P. S. Hall (2013), Parallel volcano trends and geochemical asymmetry of the Society Islands hotspot track, *Geology*, *41*, 19–22.

Pertermann, M., and M. M. Hirschmann (2003), Partial melting experiments on a MORB-like pyroxenite between 2 and 3 GPa: Constraints on the presence of pyroxenite in basalt source regions from solidus location and melting rate, *J. Geophys. Res.*, *108*, 2125.

Phipps Morgan, J. (2001) Thermodynamics of pressure release melting of a veined plum pudding mantle, *Geochem. Geophys. Geosyst.*, *2*, 2000GC000049.

Pietruszka, A. J., M. D. Norman, M. O. Garcia, J. P. Marske, and D. H. Burns (2013), Chemical heterogeneity in the Hawaiian mantle plume from the alteration and dehydration of recycled oceanic crust, *Earth Planet. Sci. Lett.*, *361*, 298–309.

Putirka, K., F. J. Ryerson, M. Perfit, and W. I. Ridley (2011), Mineralogy and composition of the oceanic mantle, *J. Petrol.*, *52*, 279–313.

Ribe, N. M., and U. R. Christensen (1994), Three-dimensional modeling of plume-lithosphere interaction, *J. Geophys. Res.*, *99*, 669–682.

Ribe, N. M. and U. R. Christensen (1999), The dynamical origin of Hawaiian volcanism, *Earth Planet. Sci. Lett.*, *171*, 517–531.

Robinson, J. E., and B. W. Eakins (2006), Calculated volumes of individual shield volcanoes at the young end of the Hawaiian ridge, *J. Volcanol. Geotherm. Res.*, *151*, 309–317.

Rychert, C. A., G. Laske, N. Harmon, and P. M. Shearer (2013), Seismic imaging of melt in a displaced Hawaiian plume, *Nature Geosci.*, *6*, 657–660.

Schmeling, H. (2006), A model of episodic melt extraction for plumes, *J. Geophys. Res.*, *111*, B03202, doi:03210.01029/0200 4JB003423.

Schutt, D. L., and C. E. Lesher (2006), Effects of melt depletion on the density and seismic velocity of garnet and spinel lherzolite, *J. Geophys. Res.*, *111*, B05401.

Sinton, J., D. E. Eason, M. Tardona, D. G. Pyle, I. van der Zander, H. Guillou, A. Flinders, D. A. Clague, and J. J. Mahoney (2014), Ka'ena Volcano: A precursor volcano of the island of O'ahu, Hawai'i, *Geological Society of America Bulletin*.

Sleep, N. H. (1990), Hotspots and mantle plumes: Some phenomenology, *J. Geophys. Res.*, *95*, 6715–6736.

Sobolev, A. V., A. W. Hofmann, S. V. Sobolev, and I. K. Nikogosian (2005), An olivine-free mantle source of Hawaiian shield basalts, *Nature*, *434*, 590–597.

Sobolev, A. V., et al. (2007), The amount of recycled crust in sources of mantle-derived melts, *Science*, *316*, 412–417.

Sobolev, S. V., A.V. Sobolev, D. V. Kuzmin, N. A. Krivolutskaya, A. G. Petrunin, N. T. Arndt, V. A. Radko, and Y. R. Vasiliev (2011) Linking mantle plumes, large igneous provinces and environmental catastrophes, *Nature*, *477*, 312–316.

Steinberger, B., and M. Antretter (2006), Conduit diameter and buoyant rising speed of mantle plumes: Implications for the motion of hot spots and shape of plume conduits, *Geochem. Geophys. Geosyst.*, *7*, Q11018.

Steinberger, B., and T. H. Torsvik (2012), A geodynamic model of plumes from the margins of Large Low Shear Velocity Provinces, *Geochem. Geophys. Geosyst.*, *13*, Q01w09.

Stracke, A., B. Bourdon, and D. McKenzie (2006), Melt extraction in the Earth's mantle: Constraints from U-Th-Pa-Ra studies in oceanic basalts, *Earth Planet. Sci. Lett.*, *244*, 97–112.

Stracke, A., J. E. Snow, E. Hellebrand, A. von der Handt, B. Bourdon, K. Birbaum, and D. Günther (2011), Abyssal peridotite Hf isotopes identify extreme mantle depletion, *Earth Planet. Sci. Lett.*, *308*, 359–368.

Torsvik, T. H., M. A. Smethurst, K. Burke, and B. Steinberger (2006), Large igneous provinces generated from the margins of the large low-velocity provinces in the deep mantle, *Geophys. J. Int.*, *167*, 1447–1460.

Torsvik, T. H., K. Burke, B. Steinberger, S.J. Webb, and L. D. Ashwal (2010), Diamonds sampled by plumes from the core-mantle boundary, *Nature*, *466*, 352–355.

van Ark, E., and J. Lin (2004), Time variation in igneous volume flux of the Hawaii-Emperor hot spot seamount chain, *J. Geophys. Res.*, *109*, B11401

Vidal, V., and A. Bonneville (2004), Variations of the Hawaiian hot spot activity revealed by variations in the magma production rate, *J. Geophys. Res.*, *109*, B03104.

Weis, D., M. O. Garcia, J. M. Rhodes, M. Jellinek, and J. S. Scoates (2011), Role of the deep mantle in generating the compositional asymmetry of the Hawaiian mantle plume, *Nature Geosci.*, *4*, 831–838.

Wen, L. (2006), A compositional anomaly at the Earth's core-mantle boundary as an anchor to the relatively slowly moving surface hotspots and as source to the DUPAL anomaly, *Earth Planet. Sci. Lett.*, *246*, 138–148.

Wen, L. X., P. Silver, D. James, and R. Kuehnel (2001), Seismic evidence for a thermo-chemical boundary at the base of the Earth's mantle, *Earth Planet. Sci. Lett.*, *189*, 141–153.

Wessel, P. (1993), Observational constraints on models of the Hawaiian hot-spot swell, *J. Geophys. Res.*, *98*, 16,095–16,104.

Wolfe, C. J., S. C. Solomon, G. Laske, J. A. Collins, R. S. Detrick, J. A. Orcutt, D. Bercovici, E. and H. Hauri (2009), Mantle shear-wave velocity structure beneath the Hawaiian hot spot, *Science, 326*, 1388–1390.

Wolfe, C. J., S.C. Solomon, G. Laske, J. A. Collins, R. S. Detrick, J. A. Orcutt, D. Bercovici, and E. H. Hauri (2011), Mantle P-wave velocity structure beneath the Hawaiian hotspot, *Earth Planet Sci. Lett.*, doi:10.1016/j.epsl.2011.1001.1004.

Workman, R. K., S. R. Hart, M. Jackson, M. Regelous, K. A. Farley, J. Blusztajn, M. Kurz, and H. Staudigel (2004), Recycled metasomatized lithosphere as the origin of the enriched mantle II (EM2) end-member: Evidence from the Samoan volcanic chain, *Geochem. Geophys. Geosyst.*, *5*, doi: 10.1029/2003GC000623.

Xu, G., S. Huang, F. A. Frey, J. Blichert-Toft, W. Abouchami, D. A. Clague, B. Cousens, J. G. Moore, and M. H. Beeson (2014), The distribution of geochemical heterogeneities in the source of Hawaiian Shield lavas as revealed by a transect across the strike of the Loa and Kea spatial trends: East Molokai to West Molokai to Penguin Bank, *Geochem. Cosmochem. Acta*, *132*, 214–237.

Xu, W., C. Lithgow-Bertelloni, L. Stixrude, and J. Ritsema, (2008), The effect of bulk composition and temperature on mantle seismic structure, *Earth Planet. Sci. Lett.*, *275*, 70–79.

Yasuda, A., T. Fujii, and K. Kurita (1994), Melting phase-relations of an anhydrous midocean ridge basalt from 3 to 20 GPa: Implications for the behavior of subducted oceanic-crust in the mantle, *J. Geophys. Res.*, *99*, 9401–9414.

Yaxley, G. M. and D. H. Green (1998), Reactions between eclogite and peridotite: Mantle refertilisation by subduction of oceanic crust, *Schweizerische Mineralogische und Petrographische Mitteilungen*, *78*, 243–255.

Zou, H. B. (1998) Trace element fractionation during modal and nonmodal dynamic melting and open-system melting: A mathematical treatment, *Geochem. Cosmochim. Acta*, *62*, 1937–1945.

4

Major-Element and Isotopic Variations in Mauna Loa Magmas over 600 ka: Implications for Magma Generation and Source Lithology as Mauna Loa Transits the Hawaiian Plume

J. Michael Rhodes

ABSTRACT

This chapter covers about 600 ka of Mauna Loa's magmatic history, contrasting the lack of major-element variability with significant isotopic diversity. There has been little change in the major-element parental magma compositions over this time period. In contrast, there is considerable variation in isotopic ratios, implying varying contributions of Loa and Lō'ihi plume components over time. There is, however, no clear age-related trend. Recent studies have questioned the ubiquity of peridotite as the source of Hawaiian shield-building tholeiites. They argue, instead, for the presence of recycled crustal components in the discrete form of eclogite/pyroxenite. The lack of a clear correlation between isotopes and major elements, especially SiO_2, suggests that there is no need for this eclogite/pyroxenite component in the source. A peridotite fertilized by melts from recycled crustal components is sufficient. The uniform but high SiO_2 content of Mauna Loa magmas coupled with varying isotopic compositions and a garnet trace element signature is consistent with the formation of these magmas through the reaction of a low-SiO_2 parental magma (3–4 GPa) with buoyant, depleted, residual harzburgite at shallow levels in the plume.

4.1. INTRODUCTION

Subaerial samples from Mauna Loa volcano span about 120 ka of its magmatic history and include historical lavas (1843–1984 [*Rhodes*, 1989; *Rhodes and Hart*, 1995]), prehistoric ^{14}C-dated lavas (0.2–36 ka [*Lockwood*, 1995; *Trusdell and Lockwood*, 2014a,b,c]), and samples from the Hawai'i Scientific Drilling Project (0.3–120 ka [*Rhodes*, 1996; *Rhodes and Vollinger*, 2004; *Sharp et al.*, 1996; *Sharp and Renne*, 2005]. Recently, *Jicha et al.* [2012] documented $^{40}Ar/^{39}Ar$ ages of lavas (59 to around 600–700 ka) from Mauna Loa's submarine southwest rift zone and from the 1600 m high Kahuku landslide scarp that dissects the submarine Southwest Rift Zone [*Garcia et al.*, 1995]. In short, we now have around 600–700 ka of magmatic history for Mauna Loa volcano: a time span exceeding that obtained for the magmatic record of Mauna Kea volcano by the Hawai'i Scientific Drilling Project [*Sharp and Renne*, 2005; *Garcia et al.*, 2007; *Stolper et al.*, 1996, 2009].

Given a 9–10 cm/yr northwesterly movement of the Pacific Plate, during this 600–700 ka magmatic record, Mauna Loa will have traversed about 60 km across the Hawaiian plume and, according to various plume models, should have sampled magmas generated from different parts of the plume. In this chapter I explore some of the major-element and isotopic compositional changes in Mauna Loa lavas over this 600–700 ka time span and examine what these changes (or lack of them) can tell us about melting and magma generation within the plume, the nature of the source lithology (eclogite/pyroxenite vs. peridotite), and the heterogeneous nature of the plume. Of particular importance is the observation that the major-element composition of inferred parental magmas changes very little over this period in the volcano's

Department of Geosciences, University of Massachusetts, Amherst, Massachusetts, USA

Hawaiian Volcanoes: From Source to Surface, Geophysical Monograph 208. First Edition.
Edited by Rebecca Carey, Valérie Cayol, Michael Poland, and Dominique Weis.

Companion Website: www.wiley.com/go/Carey/Hawaiian_Volcanoes

magmatic history, in contrast with the isotopic variation, which is significant, particularly in older submarine lavas.

4.2. COMPOSITIONAL VARIATION IN MAUNA LOA LAVAS

4.2.1. Sampling and Ages

Historical Eruptions (N = 165) There are 33 recorded eruptions of Mauna Loa between 1843 and 1984 [*Macdonald and Abbott*, 1970; *Lockwood and Lipman*, 1987; *Rhodes and Hart*, 1995]. All of these, with the exception of two radial vent eruptions on the northwest (NW) flank in 1859 and 1877, have occurred in the summit caldera or along the Northeast Rift Zone or Southwest Rift Zone. Only one of these, 1984, was sampled systematically during an active eruption [*Rhodes*, 1989].

^{14}C-Dated Prehistoric Eruptions (N = 205) Most Mauna Loa surface flows are younger than 36 ka. Ages for these lavas have been obtained by ^{14}C dating of charcoal excavated from beneath the flows [*Lockwood and Lipman*, 1987; *Lockwood*, 1995; *Trusdell and Lockwood*, 2014a,b,c]. Again, most eruptions originated from the summit caldera and from the two rift zones. In addition, there have been intermittent periods of high magma supply when the summit caldera has overflowed, producing flows down both the NW and southeast (SE) flanks of the volcano [*Lockwood*, 1995]. There were also occasional radial vent eruptions on the NW flank during this same period [*Trusdell and Lockwood*, 2014a,b,c]. Locations of these samples, together with samples of the historical eruptions, are shown by *Trusdell and Lockwood* [2014a,b,c].

Hawai'i Scientific Drilling Lavas (N = 47) The Hawai'i Scientific Drilling Project (HSDP) drilled two holes on the outskirts of Hilo, penetrating and sampling about 245 m of sub aerial Mauna Loa lavas before entering Mauna Kea lavas [*Stolper et al.*, 2009]. They are inferred to be long-traveled flows from the Northeast Rift Zone. In places the flows are interlayered with sediments and coral indicating that lava accumulation kept pace with subsidence (~2.3 mm/yr [*Moore et al.*, 1990]) of Mauna Loa. The ages of these HSDP samples range from 0.3 to 120 ka and are interpolated from several age measurements in the core of the HSDP pilot hole [*Beeson, et al.*, 1996; *Sharp et al.*, 1996; *Sharp and Renne*, 2005]. The relationship between depth (m) and age (ka) can be approximated by

$$\text{Age} \sim D \times 1.065 - D^2 \times 0.00128 + D^3 \times (8.66 \times 10^{-6}),$$

where D is depth (m).

Submarine Lavas (N = 425) Submarine pillow lavas have been sampled by dredging and submersible dives from submarine slopes on the western flank of the volcano; from deposits of the South Kona submarine landslide and radial vents on the western flank of the volcano; from the submarine Southwest Rift Zone and from the submarine portion of the Kahuku landslide scarp that dissects the Southwest Rift Zone [*Garcia et al.*, 1989, 1995; *Davis et al.*, 2003; *Wanless et al.*, 2006a,b]. *Jicha et al.* [2012] provide $^{40}Ar/^{39}Ar$ ages of lavas sampled from the Southwest Rift Zone and from the submarine Kahuku landslide scarp that range in age from 59 to around 700. At around 300–400 ka the lava accumulation rate declines dramatically from around 18–20 mm/yr to about 1–2 mm/yr. Consequently, an inferred age for undated lavas on the southwest rift zone can be obtained from the age-vs.-water-depth (mbsl) relationship in Table 1 of *Jicha et al.* [2012] as follows

$\text{Age} \sim D \times 1.065 - 465.8$ for depths shallower than 860 mbsl.

$\text{Age} \sim D \times 0.0526 + 401.9$ for depths greater than 860 mbsl.

The ages of the other submarine lavas are poorly constrained. Samples of the submarine slopes on the western flank of the volcano are probably very young, possibly contemporary with subaerial lavas (<36 ka?). The submarine radial vents are also considered to be young (<47 ka) based on flow morphology and the thickness of Mn coating [*Wanless et al.*, 2006a]. The South Kona landslide is thought to have occurred around 200 ka [*Lipman et al.*, 1988; *Moore et al.*, 1989]. An $^{40}Ar/^{39}Ar$ age of 247 ± 50 ka for a sample of the landslide debris [*Jicha et al.*, 2012] provides an indication of the age of this material and a maximum age for this event. Some samples of the landslide deposit have anomalous isotopic ratios [*Weis et al.*, 2011] that are only found below a sample dated at 460 ± 36 ka in the Kahuku landslide scarp [*Jicha et al.*, 2012], indicating that some of the landslide debris may be over 400 ka. Locations of the submarine lavas are shown by *Garcia et al.* [1995], *Davis et al.* [2003], *Wanless et al.* [2006a,b], *Weis et al.* [2011], and *Jicha et al.* [2012].

The major-element abundances in these lavas were measured by X-ray fluorescence analysis following procedures documented by *Rhodes* [1996] and *Rhodes and Vollinger* [2004]. The data are provided in Supplementary Appendix A. The isotopic data are from *Kurz and Kammer* [1991], *Kurz et al.* [1995], *Rhodes and Hart* [1995], *Wanless et al.* [2006a,b], *Marske et al.* [2007], and *Weis et al.* [2011].

4.2.2. Major Elements

It has long been recognized that the major-element composition of Mauna Loa lavas is distinct from that of the neighboring volcanoes of Kīlauea, Mauna Kea, and

Lō'ihi [e.g., *Wright*, 1971; *Garcia et al.*, 1989; *Rhodes et al.*, 1989; *Frey and Rhodes*, 1993; *Hauri*, 1996]. There is a tendency for SiO_2 to be higher and TiO_2 and CaO to be lower at a given MgO content in Mauna Loa lavas relative to the other volcanoes. Total alkali content ($Na_2O + K_2O$) is also slightly lower in MgO-normalized Mauna Loa lavas, as are incompatible trace element abundances and ratios (e.g., Nd, La, Ce, Ba, Zr).

4.2.2.1. Olivine Control Trends

Mauna Loa lavas plot on extremely well-defined compositional trends extending from as low as 5.7% to as much as 33.2% MgO (Figure 4.1). These linear trends converge toward olivine compositions, indicative of the strong influence of olivine accumulation and fractionation on these so-called olivine control trends [e.g., *Powers*, 1955; *Murata and Richter*, 1966; *Wright*, 1971; *Wilkinson and Hensel*, 1988; *Clague et al.*, 1995; *Rhodes*, 1995]. The exceptions to these trends include lavas that are multiply saturated with clinopyroxene and plagioclase and have evolved beyond olivine control (MgO < 6.8%), some rarer lavas that have fractionated clinopyroxene or plagioclase, and the enigmatic alkalic submarine radial vent lavas [*Wanless et al.*, 2006b]. According to *Wanless et al.* [2006b], the low-SiO_2 and high incompatible element content of these alkalic lavas are a consequence of extensive clinopyroxene fractionation at moderate depth, not from low degrees of melting.

The prevailing interpretation of olivine control trends is that they are a consequence of combined olivine fractionation and accumulation. The strong linear nature of the trends suggests, however, that they result largely from mixing and less from fractionation. For example, *Clague et al.* [1995] attribute the linear trends in the whole-rock data from Kīlauea's submarine Puna Ridge to entrainment

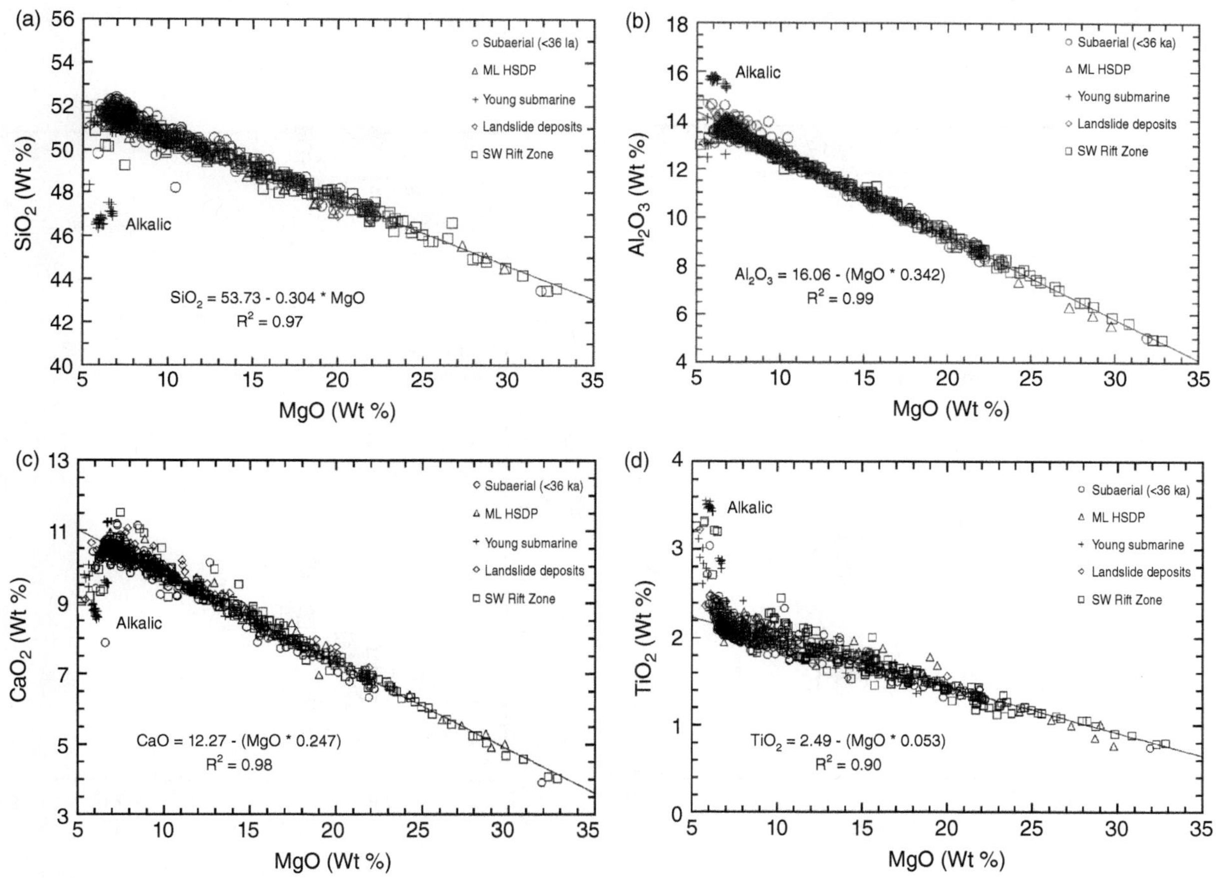

Figure 4.1 Variation with MgO of (a) SiO_2; (b) Al_2O_3; (c) CaO; and (d) TiO_2 for Mauna Loa lavas. For simplicity the lavas have been subdivided into subaerial lavas (<36 ka); subaerial lavas from the Hawai'i Scientific Drilling Project (1.5–120 ka); young submarine lavas, including both radial vent and slope lavas from the submarine western flank (probably <50 ka); submarine landslide deposits (>200 ka); and lavas from the submarine Southwest Rift Zone and the submarine section of the Kahuku landslide scarp (59 to >600 ka (but mostly >400 ka)). The data used to construct the olivine control regression lines are screened for samples that are altered ($K_2O/P_2O_5 < 1.0$) and have undergone fractionation beyond olivine control (MgO < 6.8%).

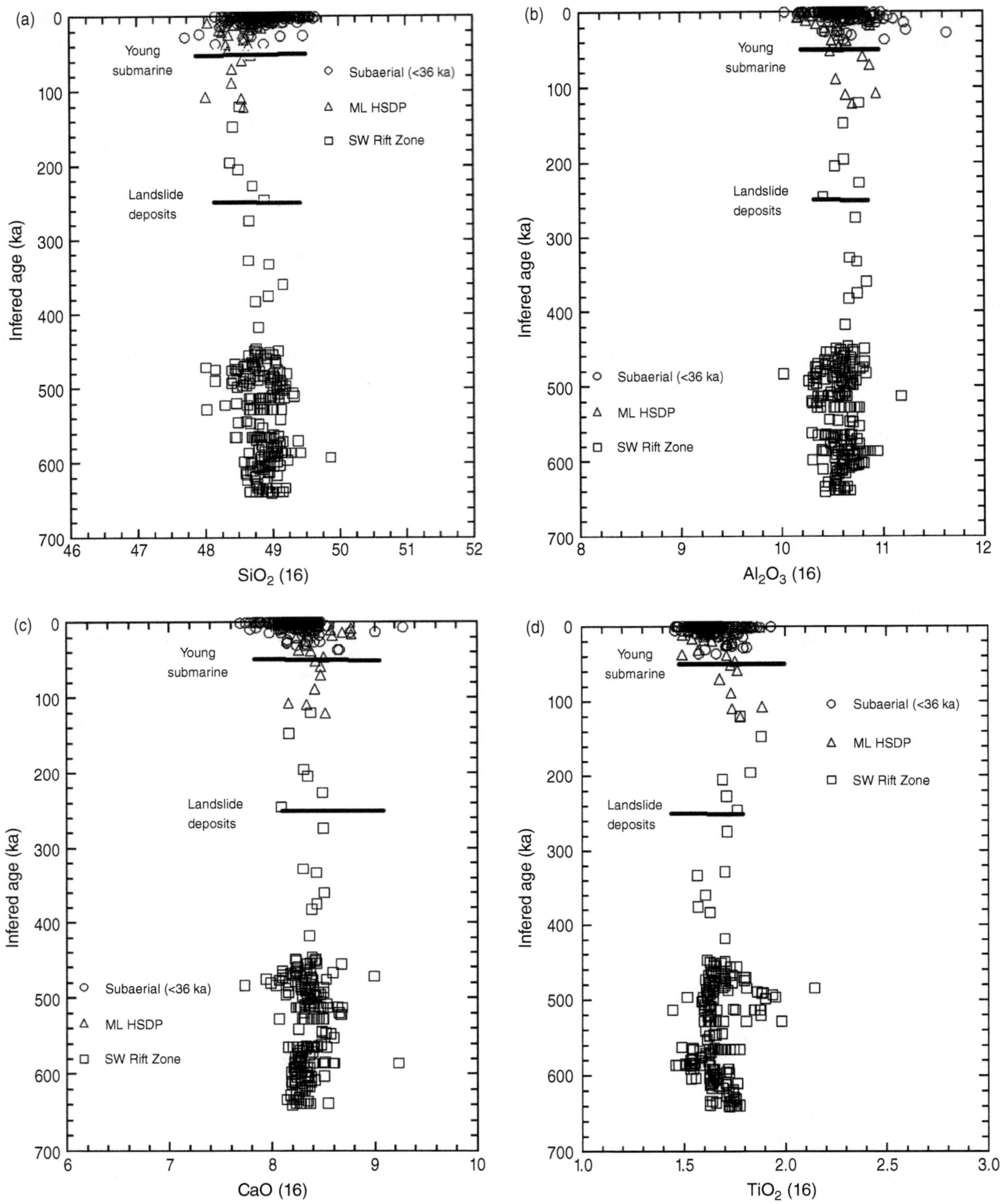

Figure 4.2 Plots of normalized (a) SiO_2; (b) Al_2O_3; (c) CaO; and (d) TiO_2 for Mauna Loa lavas with inferred age (see text for details). The data are normalized to 16% MgO using the olivine control regression equations from Figure 4.1. The data are screened to exclude altered samples or those that have fractionated beyond olivine control. Symbols are as in Figure 4.1. Because of a lack of age information, data for young submarine lavas (slope and radial vents) and landslide deposits are shown as thick horizontal lines at 50 and 250 ka, respectively.

of an olivine cumulate mush (average composition $Fo_{87.9}$) into a later, unrelated, differentiated, low-MgO (6.5–7.0%) magma. *Rhodes and Vollinger* [2004] argue that similar olivine control trends in Mauna Kea lavas result from three-component mixing between an olivine cumulate mush, an evolved multiply saturated magma (MgO ~7%), and parental picritic magmas (13%–18% MgO). This is the preferred interpretation for these Mauna Loa olivine control trends, where it can be shown that the picritic parental magmas plot on or very close to the olivine-evolved magma whole-rock mixing trend [*Rhodes and Garcia*, 2008; *Rhodes*, in preparation].

An important aspect of these olivine control trends is that the oxides are remarkably uniform at a given MgO content, irrespective of the age or the location of the sample (Figure 4.1). This also applies to those oxides that characteristically vary from volcano to volcano (e.g., SiO_2, CaO, TiO_2), especially volcanoes on either of the "Loa" or "Kea" trends [e.g., *Frey and Rhodes*, 1993; *Hauri*, 1996; *Jackson et al.*, 2012]. For example, if altered lavas (loss on ignition > 1%; $K_2O/P_2O_5 < 1.0\%$) and those that have fractionated beyond olivine control (MgO < 6.8%) are excluded, then

$$SiO_2 = 53.73 - 0.304 \times MgO\ (R^2 = 0.97)$$
$$TiO_2 = 2.49 - 0.053 \times MgO\ (R^2 = 0.90)$$
$$CaO = 12.27 - 0.247 \times MgO\ (R^2 = 0.98)$$
$$Al_2O_3 = 16.06 - 0.342 \times MgO\ (R^2 = 0.99)$$

The implications are startling. At a given MgO content, there has been little or no change in the major-element compositions of Mauna Loa's lavas over about 600 ka of magmatic history. This is illustrated in Figure 4.2, where the MgO-normalized content of these four oxides are plotted against their inferred age. Samples that have fractionated beyond olivine control or are inferred to have fractionated clinopyroxene are excluded from this figure. The data are normalized to 16% MgO using the regression slopes given above. This approach is preferred to the more common method of adding or subtracting small increments of equilibrium olivine to melt compositions in order to mimic olivine fractionation. This is because mixing between an olivine mush and low-MgO magmas (~7%) has overprinted the olivine fractionation that produced the low-MgO magmas and is the dominant process creating these trends. Only samples with inferred ages are plotted in this figure. Because of the uncertainty in the ages of the submarine slope and radial vent lavas and the submarine landslide deposits, they are shown schematically as horizontal lines at 50 and 250 ka, respectively. Note that they fall within the range for young subaerial and old submarine lavas. There is an obvious dearth of data between 120 and 400 ka because, as noted earlier, the lava accumulation rate on the submarine southwest rift zone declined dramatically from around 18–20 mm/yr to around 1–2 mm/yr [*Jicha et al.*, 2012] at around 300–400 ka.

This uniformity in major elements does not extend to MgO itself. There is a clear distinction between younger subaerial lavas and older submarine lavas (Figure 4.3). The younger subaerial lavas (<120 ka) are dominated by low-MgO lavas (7–8% MgO) at the end of the olivine control trend. These are the so-called reservoir lavas of *Rhodes* [1989] that are inferred to result from steady state mixing and replenishment of a magma body perched at the pigeonite-melt reaction point at the end of the olivine fractionation trend [*Rhodes*, 1989; *Rhodes and Hart*, 1995]. Picritic lavas (MgO > 12%), although present, are relatively rare. By contrast, the older submarine lavas (>120 ka) contain a much greater proportion of picritic lavas. Low-MgO lavas (7–8% MgO), although present, are less abundant (Figure 4.3). This could imply a dramatic shift in parental magmas from picritic to basaltic around 120 ka (see below) or, more plausibly, reflect the accumulation of olivine deeper in the magmatic plumbing system [*Garcia et al.*, 1995; *Rhodes*, 1995].

4.2.2.2. MgO-FeO Relationships

The variation of FeO with MgO, together with information on olivine compositions, provides insights into olivine accumulation and parental magma compositions (Figure 4.4). There is a tight cluster of lavas at the low MgO end of the FeO-vs.-MgO olivine control trend (MgO ~7%; FeO ~9.8%). Lavas with significantly higher FeO values but similar MgO contents are evolved and multiply saturated and have fractionated clinopyroxene and plagioclase beyond olivine control. These evolved lavas are also higher in TiO_2 and K_2O and lower in CaO (Figure 4.1) as a consequence of multiphase fractionation. The overall MgO-FeO trend is one of broadly subhorizontal to slightly increasing FeO content with increase in MgO. This variation in FeO with MgO is consistent with the crystallization and accumulation of olivine (Fo_{88-91}) into the low-MgO magma (Figure 4.4). Clearly, olivine (Fo_{88-91}) has not crystallized from a low-MgO magma with around 7% MgO and 9.8% FeO. A picritic parental melt is required. Assuming a $KD_{Fe\text{-}Mg}$ of 0.32 [*Putirka*, 2005] for partitioning between olivine and Mauna Loa magmas, lavas crystallizing Fo_{88-91} olivine will have FeO/MgO around 0.76–0.55 and contain between 13% and 19% MgO (Figure 4.4). Using a higher $KD_{Fe\text{-}Mg}$ of 0.34, as recently proposed by *Matzen et al.* [2011], would result in even higher MgO contents.

Although picritic glasses have not been identified on Mauna Loa [*Davis et al.*, 2003], they are present on Kīlauea's submarine East Rift Zone [*Clague et al.*, 1991, 1995] and in some of Kīlauea's prehistoric explosive eruptions (see Chapter 7). It could be argued that because

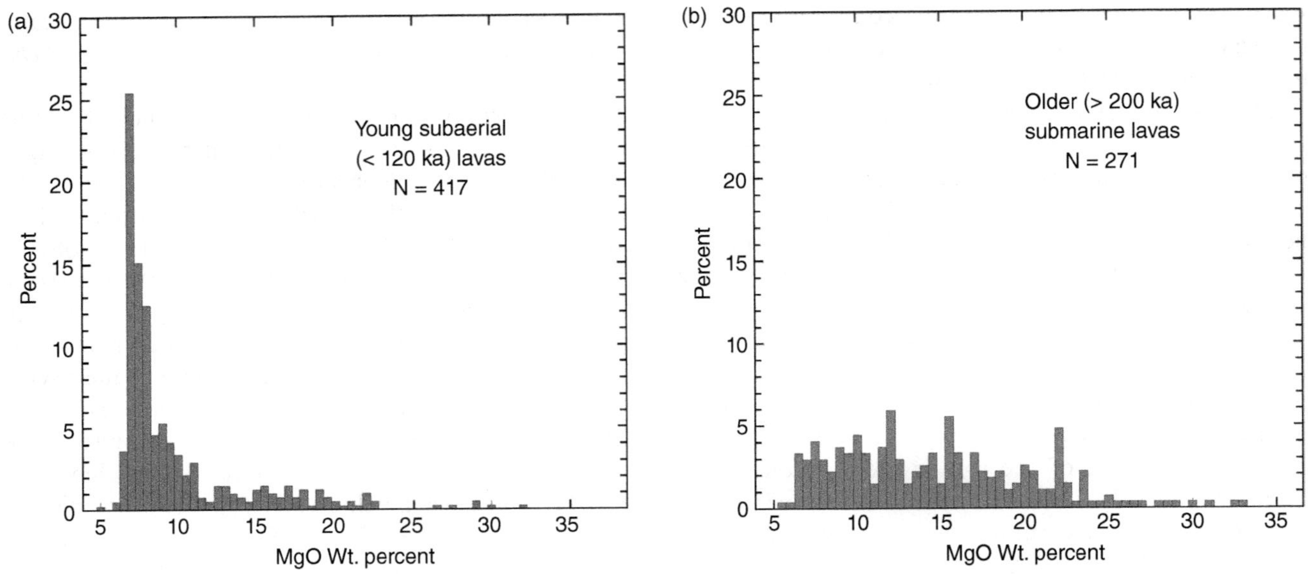

Figure 4.3 Frequency histograms for MgO in (a) subaerial lavas (<120 ka) and (b) older submarine lavas (>200 ka).

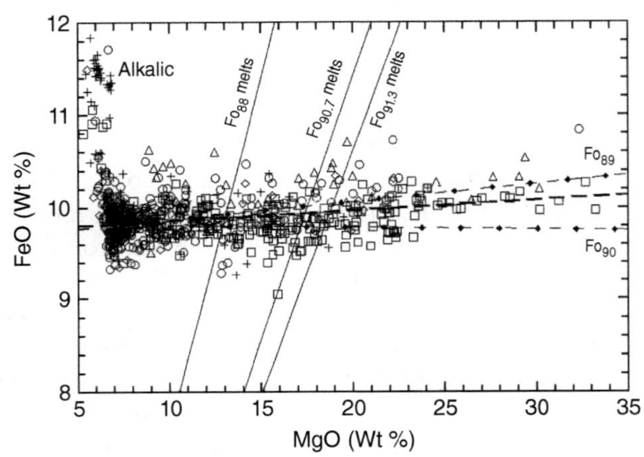

Figure 4.4 FeO vs. MgO for Mauna Loa lavas. Symbols are as in Figure 4.1. FeO is calculated from total iron as Fe_2O_3 on the assumption that the oxygen fugacity of the lavas was close to magnetite-wüstite (MW) [*Rhodes and Vollinger*, 2005]. The olivine control regression line (thick dashed line) is compared with olivine addition lines constructed by adding Fo_{89} and Fo_{90} olivine to a melt with 7.0% MgO and 9.8% FeO. The three oblique lines represent the FeO/MgO of melts in equilibrium with Fo_{88}, $Fo_{90.7}$, and $Fo_{91.3}$, assuming a $KD_{Fe\text{-}Mg}$ of 0.32 [*Putirka*, 2005] The intersection of these lines with the olivine control regression trend provide an estimate of the MgO and FeO content of parental magmas.

these olivine control trends are dominated by olivine accumulation into a low-MgO magma, there is no a priori reason why the parental magmas should fall on these whole-rock trends. *Rhodes and Garcia* [2008], however, show that by incremental addition of equilibrium olivine to submarine Mauna Loa glasses that have not undergone clinopyroxene or plagioclase crystallization, melts in equilibrium with Fo_{88-91} olivine do indeed fall on the whole-rock olivine control trend, close to the predicted compositions. Therefore, using the intersection of the FeO/MgO ratio of melts in equilibrium with critical olivine compositions with the whole-rock trend is a valid alternative to adding small increments of equilibrium olivine to evolved melt compositions in order to estimate parental magma compositions.

The most common forsterite content of olivine cores in relatively young (<120 ka) subaerial Mauna Loa lavas varies from Fo_{87-91}, with a modal value of Fo_{88} and a maximum value of $Fo_{90.7}$ [*Wilkinson and Hensel*, 1988; *Rhodes*, 1995; *Baker et al.*, 1996; *Garcia*, 1996]. In contrast, olivine cores in older submarine lavas (~400 ka) tend to be more forsteritic with most common values ranging from Fo_{88-91}, a modal value of Fo_{90} and a maximum value of $Fo_{91.3}$ [*Garcia et al.*, 1995; *Garcia*, 1996]. If these differences are significant, it might imply that older picritic Mauna Loa magmas with around 19% MgO were slightly hotter and were produced by slightly more melting than younger magmas with around 17% MgO (Figure 4.4). Taking the maximum forsterite content of $Fo_{91.3}$ and using a modification of the classic olivine-melt geothermometer [*Roeder and Emslie*, 1970; *Putirka*, 2005] and a $KD_{Fe/Mg}$ of 0.32, the estimated MgO content of the most primitive submarine Mauna Loa magma is 18.8% MgO (Figure 4.4). For the younger magmas in equilibrium with $Fo_{90.7}$, the estimated MgO content is 17.4% MgO. The inferred major-element compositions of parental magmas erupted prior to ~200 ka are compared with those erupted after ~200 ka in Table 4.1. The implications are that over 600 ka

Table 4.1 Estimated compositions of Mauna Loa parental magmas.

	Young Lavas (<200 ka)	Old Lavas (>200 ka)
SiO_2	48.44	48.01
TiO_2	1.58	1.50
Al_2O_3	10.10	9.62
Fe_2O_3*	12.27	12.06
MnO	0.17	0.17
MgO	17.40	18.80
CaO	7.96	7.61
Na_2O	1.66	1.58
K_2O	0.27	0.25
P_2O_5	0.17	0.16
Total	100.01	99.76
Mg value	0.75	0.77

Note: Fe_2O_3* is total Fe expressed as Fe_2O_3. The Mg value is Mg/(Mg + Fe) after adjusting Fe^{3+}/total Fe to 0.1.

the MgO content of the parental magma, and therefore its temperature, may have declined slightly. Alternatively, it may be that the magma supply and recharge was higher in the past, precluding extensive fractionation in shallow magma reservoirs.

4.2.3. Isotopic Ratios of Sr, Nd, and Pb

Early studies of the isotopic ratios of Sr, Nd, and Pb in young (<36 ka) subaerial Mauna Loa lavas showed that they were relatively uniform with typical values around 0.70388 ± 0.00012, 0.51287 ± 0.00006, and 18.12 ± 0.07, respectively [*O'Nions et al.*, 1977; *Tatsumoto*, 1978; *Kurz and Kammer*, 1991; *Kurz et al.*, 1995; *Rhodes and Hart*, 1995]. These values plot in an intermediate position on the well-defined "Hawaiian array" [e.g., *Staudigal et al.*, 1984; *West et al.*, 1987; *Weis et al.*, 2011] extending from Ko'olau lavas at one extreme and Kīlauea and Lō'ihi lavas at the other. Despite this relative uniformity, *Rhodes and Hart* [1995] identified systematic temporal variation in historic lavas that mimicked variations in incompatible element ratios that were related to eruption rates.

More recent studies of older lavas from the Mauna Loa section of HSDP and samples recovered from Mauna Loa's submarine Southwest Rift Zone reveal greater isotopic diversity [e.g., *Kurz et al.*, 1995; *DePaolo et al.*, 2001; *Blichert-Toft et al.*, 2003; *Abouchami et al.*, 2005]. The HSDP data (1.5–120 ka) extend the range in Sr, Pb, and Nd isotopes to both slightly higher and lower values. Deeper samples, inferred to be older than 40 ka, all have $^{3}He/^{4}He$ greater than 12, in contrast with younger samples, which typically range between 8 and 10 [*DePaolo et al.*, 2001]. All submarine lavas have higher $^{3}He/^{4}He$ ranging from 13 to 20 [*Kurz et al.*, 1995] and include samples with lower $^{87}Sr/^{86}Sr$ and higher $^{206}Pb/^{204}Pb$ and $^{143}Nd/^{144}Nd$ that tend toward values more typical of Kīlauea and Lō'ihi volcanoes. These same samples also have low Zr/Nd (<12) more typical of Kīlauea and Lō'ihi than Mauna Loa. *Kurz et al.* [1995] and *DePaolo et al.* [2001] suggest that these apparent trends in isotopic ratios with relative age reflect progressive sampling by Mauna Loa as it traverses a radially zoned plume.

The recent review by *Weis et al.* [2011] on the isotopic characteristics of the Hawaiian mantle plume contains a great deal (120 samples) of new, high-precision, isotopic data for Mauna Loa. These include historical and ^{14}C-dated subaerial lavas together with submarine samples obtained by dredging from the Southwest Rift Zone, landslide deposits, western slope lavas, and radial vents [*Garcia et al.*, 1989; *Davis et al.*, 2003; *Wanless et al.*, 2006a,b] and samples recovered from the submarine extension of the Kahuku landslide scarp by the submersibles Pisces V and Jason II [*Garcia et al.*, 1995; *Weis et al.*, 2011]. A brief description of the location of these samples is given in the supplementary information of *Weis et al.* [2011]. These data, in conjunction with the $^{40}Ar/^{39}Ar$ dating of the submarine lavas [*Jicha et al.*, 2012], allow us to place the isotopic evolution of Mauna Loa magmas over the past 600 ka in a much clearer perspective.

Compared with the young subaerial lavas (<36 ka), there is greater isotopic diversity in the older submarine lavas (59 to ~600 ka [*Jicha et al.*, 2012]). The ratio $^{87}Sr/^{86}Sr$ ranges from 0.7035 to 0.7040, $^{206}Pb/^{204}Pb$ from 18.03 to 18.39, and $^{143}Nd/^{144}Nd$ from 0.51283 to 0.51297 (Figure 4.5). Most submarine lavas, however, have Sr, Nd, and Pb isotopic ratios that are indistinguishable from much younger subaerial lavas (Figure 4.5). Therefore, in contrast with prior claims [e.g., *Kurz et al.*, 1995; *DePaolo et al.*, 2001], there is no systematic change in these isotopic ratios with age (Figure 4.6). There are old (~400 ka) lavas that are isotopically indistinguishable from young subaerial lavas. Most importantly, as shown clearly by the Pb isotopic data (Figure 4.5c), almost all of the Mauna Loa data plot on the Loa side of the Kea-Loa boundary defined by *Abouchami et al.* [2005]. Only three young (~1 ka) subaerial lavas, with $^{206}Pb^{*}/^{204}Pb^{*} < 0.95$, plot on the Kea side of this boundary [*Marske et al.*, 2007]. The overall trend in Pb isotopic ratios is from the main cluster of young (<36 ka) subaerial lavas towards Lō'ihi lava compositions. Several samples from the South Kona landslide deposit overlap with Lō'ihi compositions (Figure 4.5). One of these samples has been dated at 247 ± 50 ka [*Jicha et al.*, 2012]. Several samples with unusually high $^{208}Pb/^{204}Pb$ relative to $^{206}Pb/^{204}Pb$ and therefore relatively high $^{206}Pb^{*}/^{204}Pb^{*}$ [*Weis et al.*, 2011] are present in the Kahuku submarine landslide scarp and in dredge samples from

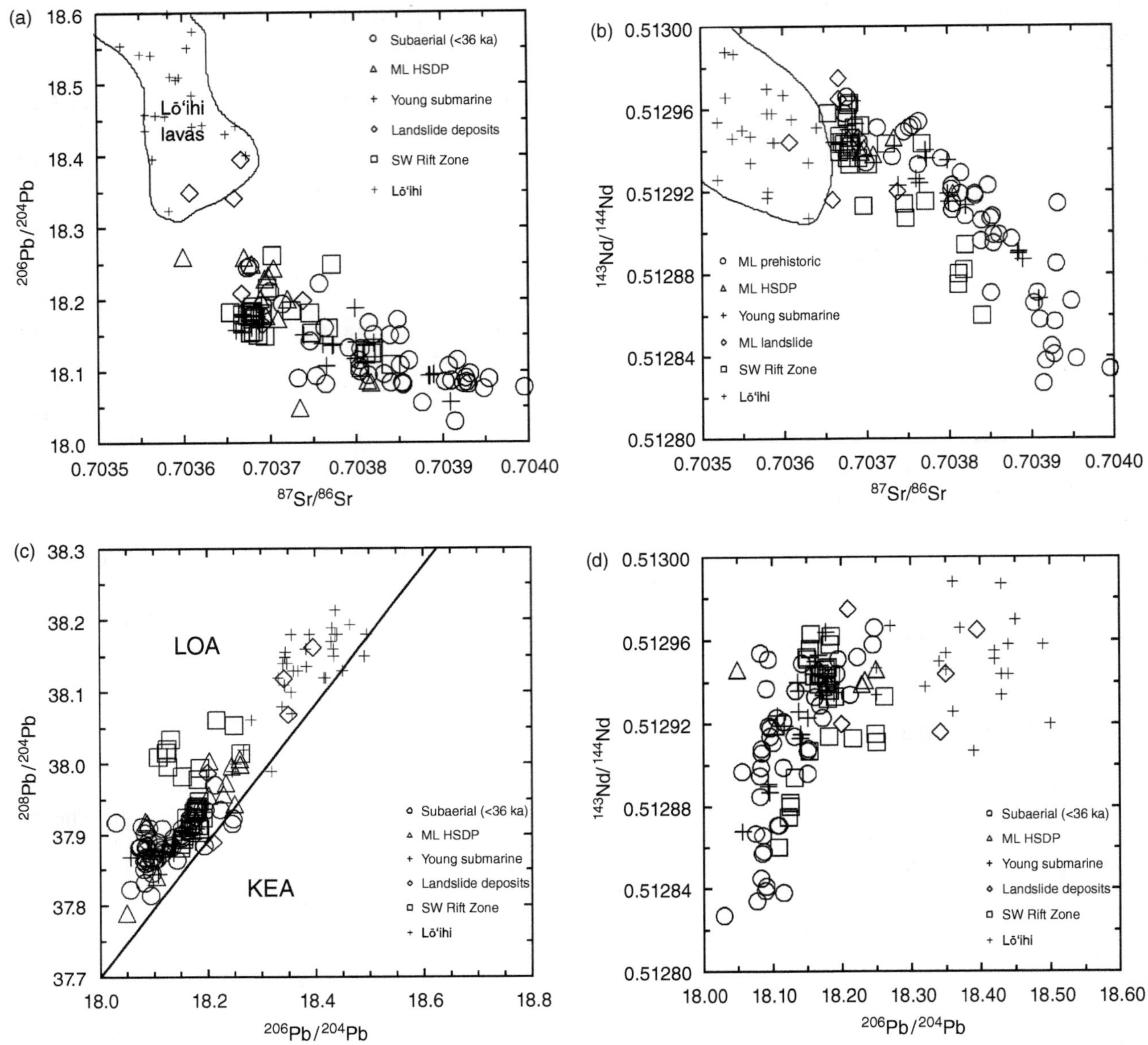

Figure 4.5 Plots of radiogenic isotope data for Mauna Loa lavas. The symbols are as in Figure 4.1. The data are from *Kurz and Kammer,* 1991; *Kurz et al.,* 1995; *Rhodes and Hart,* 1995; *Blichert-Toft et al.,* 2003; *Abouchami et al.,* 2005; *Wanless et al.,* 2006a,b; *Marske et al.,* 2007; and *Weis et al.,* 2011. Data for Lō'ihi shield-building tholeiites are shown for comparison (data from the Georoc database compilation of *Jackson et al.,* 2012). The solid black line separating Loa trend lavas from Kea trend lavas in (c) is from *Abouchami et al.* [2005].

the submarine Southwest Rift Zone (Figure 4.5). All of these samples are below a sample in the scarp dated at 461 ± 36 ka [*Jicha et al.*, 2012], indicating that these lavas with elevated $^{208}Pb/^{204}Pb$ relative to $^{206}Pb/^{204}Pb$ were erupted prior to ~460 ka. Importantly, the overall isotopic variations in Mauna Loa lavas correspond to about 43% of the entire shield-stage Hawaiian array for $^{87}Sr/^{86}Sr$ and 36% of the array for both $^{206}Pb/^{204}Pb$ and $^{143}Nd/^{144}Nd$.

4.3. DISCUSSION

4.3.1. Covariation of Major Elements and Isotopes

The isotopic variations in Hawaiian shield-building lavas is typically described in terms of three end-member plume components: a Kea component, best exemplified by Mauna Kea lavas; a Loa component, illustrated by lavas from the Makapu'u stage of Ko'olau volcano

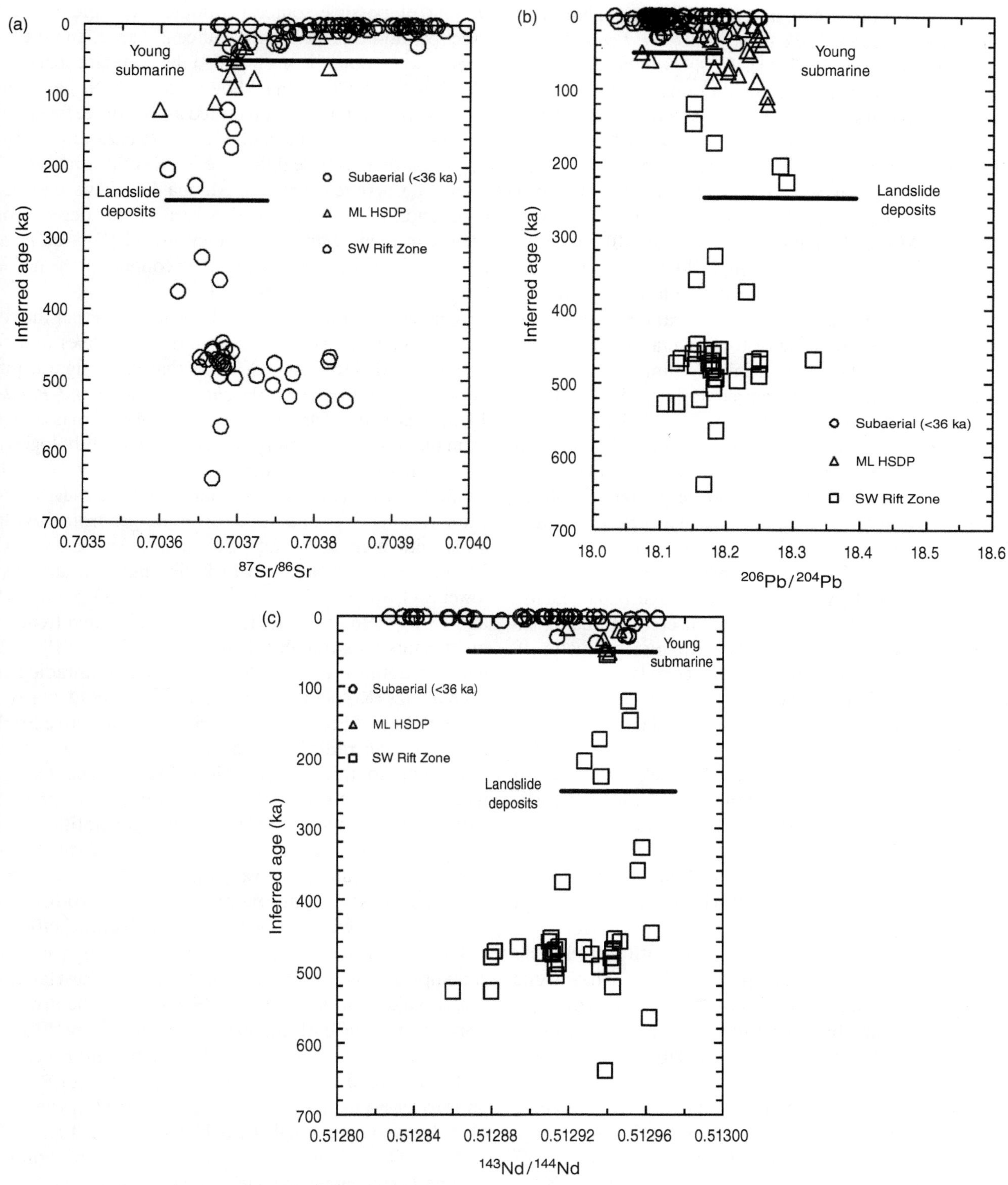

Figure 4.6 Plots of radiogenic isotope data for Mauna Loa lavas with inferred age (see text for details). The isotope data are from the same sources listed in Figure 4.5. As in Figure 4.2, data for young submarine lavas (slope and radial vents) and landslide deposits are shown as thick horizontal lines at 50 and 250 ka, respectively.

[*Haskin and Garcia*, 2004]; and a Lōʻihi component [e.g., *Staudigal et al.*, 1984; *Stille et al.*, 1986; *West et al.*, 1987; *Hauri*, 1996; *Weis et al.*, 2011]. As shown above, Mauna Loa lavas contain varying contributions of the Loa and Lōʻihi components, with the Lōʻihi component tending to be more prevalent in some older lavas (>240 ka) and the Loa component more common in younger lavas (<36 ka). The Kea source component that characterizes Kīlauea and Mauna Kea magmas appears to be of only minor influence in Mauna Loa magmas. *Weis et al.* [2011] identify an additional component with high $^{208}Pb/^{204}Pb$ relative to $^{206}Pb/^{204}Pb$ in old (>460 ka) Mauna Loa lavas that has not been identified at other Hawaiian volcanoes. A growing consensus suggests that the isotopic signature of the Loa component reflects substantial amounts of recycled oceanic crust plus minor sediments [e.g., *Hauri*, 1996; *Blichert-Toft et al.*, 1999; *Huang and Frey*, 2005; *Fekiacova et al.*, 2007; *Jackson et al.*, 2012]. What is of critical importance is whether or not this recycled crustal component is intrinsic to a source peridotite perhaps fertilized by crustal component melts or whether a petrologically distinct, crustal-derived lithology such as eclogite or pyroxenite is involved in the melting process. *Hauri* [1996], for example, was the first to suggest that dacitic magmas, produced from melting discrete domains of eclogite within the plume, mix with picritic melts of peridotite to produce the broadly correlated range in SiO_2 and isotopic characteristics of Hawaiian magmas. Others [e.g., *Takahashi and Nakajima*, 2002; *Huang and Frey*, 2005; *Fekiacova et al.*, 2007; *Jackson et al.*, 2012] have proposed variations on this model, especially to account for the high SiO_2 content and characteristic isotopic signature of Makapuʻu-stage Koʻolau lavas [*Haskin and Garcia*, 2004]. *Sobolev et al.* [2005, 2007] have introduced a more complex model in an attempt to explain the high Ni content of Hawaiian olivines and their host magmas. In this model silicic melts from discrete eclogite/pyroxenite domains react with surrounding peridotite to produce secondary pyroxenite. They argue that subsequent melting of differing proportions of these secondary pyroxenite and peridotite source materials is responsible for differences in the compositions of Hawaiian magmas and for their correlated isotopic and major-element compositional characteristics. *Herzberg* [2006, 2011] goes even further arguing that most Hawaiian magmas, with the exception of the enigmatic high-CaO Mauna Kea lavas [*Stolper et al.*, 2004; *Rhodes et al.*, 2012] and possibly some Lōʻihi tholeiites, are derived solely through melting of pyroxenite. Others [e.g., *Putirka et al.*, 2011; *Rhodes et al.*, 2012] question the role of Ni abundances in olivines and lavas to distinguish between a pyroxenite or peridotite source for Hawaiian magmas. They argue that, given the uncertainties in the partitioning of Ni between olivine and melt, and the large range in the Ni content of possible source peridotites, it is premature to reject peridotite as the mantle source of Hawaiian magmas. This second point is exemplified by a recent paper by *Herzberg et al.* [2013] that shows that high-Ni peridotites (2360 ppm vs. 1960 pm) are required as the source for certain plume-related magmas. Additionally, *Rhodes et al.* [2012] show that there is little difference in Ni concentrations at a given MgO content between Mauna Loa lavas and both high- and low-SiO_2 Mauna Kea lavas. These observations are inconsistent with the Sobolev model [*Sobolev et al.*, 2005, 2007], which predicts higher Ni contents in the higher-SiO_2 lavas relative to lower-SiO_2 lavas.

What is paramount in all of these models is that the correlated trends between Sr, Nd, and Pb isotopes and the major elements (SiO_2, CaO, TiO_2) that differ significantly between volcanoes [e.g., *Wright*, 1971; *Frey and Rhodes*, 1993; *Hauri*, 1996] are a consequence of magmas derived from melting varying proportions of discrete lithologies of eclogite/pyroxenite and peridotite. *Jackson et al.* [2012] review this correlation in Hawaiian shield-stage lavas. They show that the Hawaiian data array ranges from Koʻolau lavas with high SiO_2 and $^{87}Sr/^{86}Sr$ and low CaO, TiO_2, $^{206}Pb/^{204}Pb$, and $^{143}Nd/^{144}Nd$ to Lōʻihi and Kea lavas with lower SiO_2 and $^{87}Sr/^{86}Sr$ and higher CaO, TiO_2, $^{206}Pb/^{204}Pb$, and $^{143}Nd/^{144}Nd$. They attribute these correlated trends to melting mixtures of high-SiO_2 eclogite (with low $^{206}Pb/^{204}Pb$) with peridotite as a source for high-SiO_2 and unradiogenic Pb magmas (Koʻolau) and a mixture of low-SiO_2 pyroxenite (with high $^{206}Pb/^{204}Pb$) and peridotite as a source for the low-SiO_2 high radiogenic Pb magmas (Lōʻihi and Kea). According to this interpretation, Mauna Loa magmas within this range should, therefore, contain contributions from both eclogite/pyroxenite and peridotite source components. Importantly, on the other hand, melting of peridotite, fertilized by varying amounts of melts of recycled crustal components, will not produce the correlation between major-element and isotopic ratios.

The isotopic variation of Mauna Loa lavas is significant, encompassing about 58% of the entire array for Hawaiian shield-building lavas for $^{208}Pb/^{204}Pb$, 43% of the array for $^{87}Sr/^{86}Sr$, and 36% of the array for both $^{206}Pb/^{204}Pb$ and $^{143}Nd/^{144}Nd$ [*Weis et al.*, 2011]. Although there is no clear systematic trend with age, there is a tendency for younger lavas to have a higher Loa component contribution and for older lavas to have a higher Lōʻihi component (Figures 4.5 and 4.6). One might infer from this isotopic variation that Mauna Loa magmas include high but varying proportions of melt derived from an eclogite/pyroxenite source. If so, we should expect to find comparable variation in major-element abundances, especially for SiO_2, CaO, and TiO_2, and significant covariation between these oxides and isotopic ratios [e.g., *Hauri*, 1996; *Jackson et al.*, 2012]. In fact, the data are more complex. As was shown earlier, there is remarkable overlap in these oxides in olivine control trends throughout the entire 600 ka of Mauna Loa's recorded magmatic history (Figure 4.1). This point is further emphasized by the fact that when these oxides are normalized to a constant MgO content (16%), they are very uniform and there are no obvious differences throughout the 600 ka magmatic record (Figure 4.2). An MgO content of 16% has no petrogenetic connotation in this chapter, but is adopted for comparison with the analysis of *Jackson et al.* [2012], who use a parental magma with 16% MgO, in equilibrium with Fo_{90}, to evaluate the correlation of major elements and isotopic ratios in Hawaiian shield lavas. Even though MgO-normalized CaO shows little variation with inferred age (Figure 4.2), it tends to mimic the overall trends for $^{87}Sr/^{86}Sr$ and $^{206}Pb/^{204}Pb$ (Figures 4.7a and 4.7b) in Hawaiian shield lavas [*Jackson et al.*, 2012]. There is no such correlation, however, with $^{208}Pb^{*}/^{206}Pb^{*}$, in contrast with the overall negative correlation observed for Hawaiian shield lavas (Figure 4.7c). There is no distinct correlation of

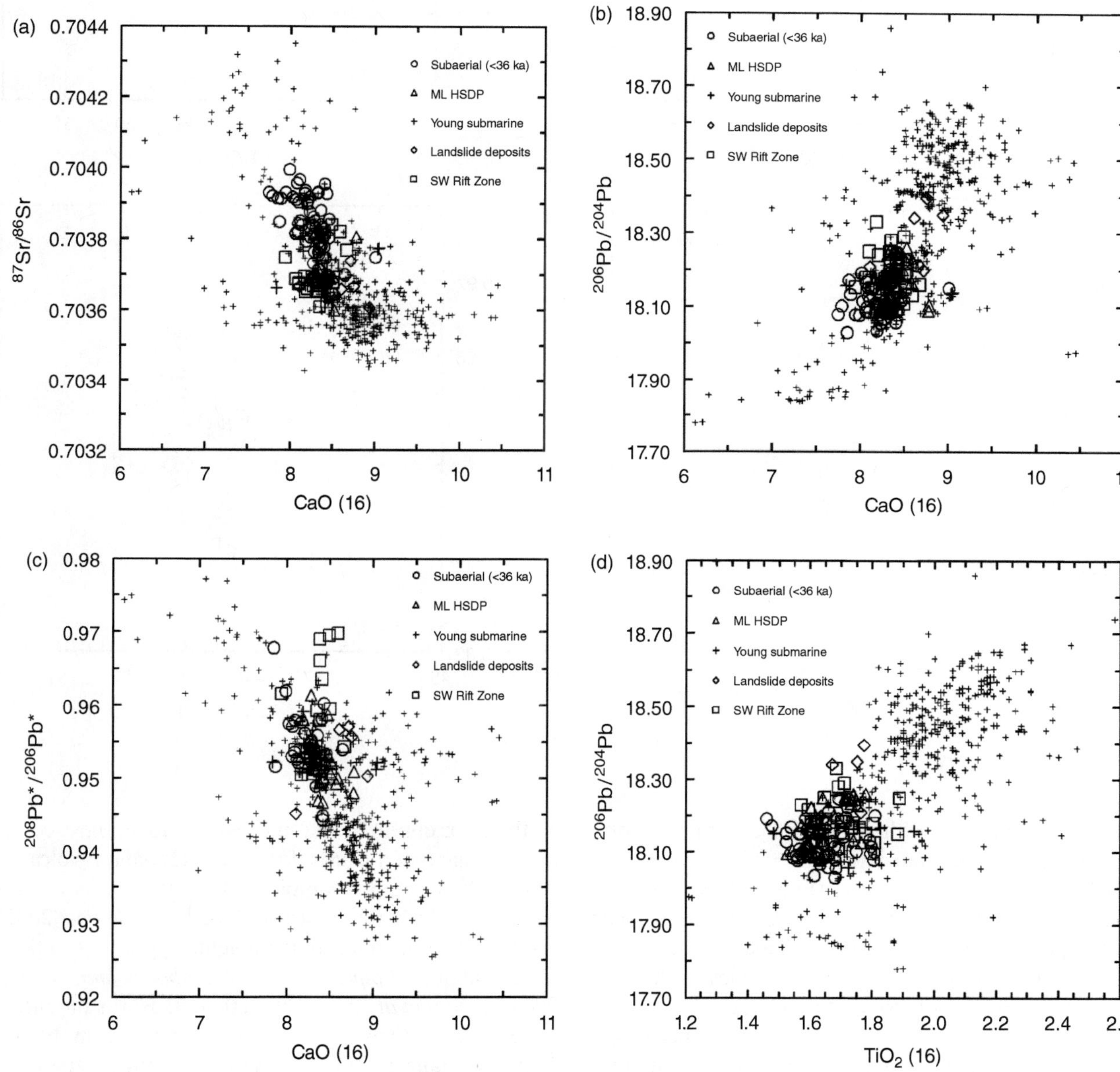

Figure 4.7 Plots of MgO-normalized CaO, TiO_2 and SiO_2 versus $^{87}Sr/^{86}Sr$, $^{206}Pb/^{204}Pb$, $^{206}Pb^{*}/^{204}Pb^{*}$, and $^{143}Nd/^{144}Nd$ for Mauna Loa lavas. The isotopic data are from the same sources listed in Figure 4.5. Following *Jackson et al.* [2012], the data have been screened for samples that are altered ($K_2O/P_2O_5 < 1.0$) or have undergone fractionation beyond olivine control (MgO < 6.8%). Shown for comparison is the data compilation of *Jackson et al.* [2012] for all Hawaiian shield-building tholeiites (small black crosses).

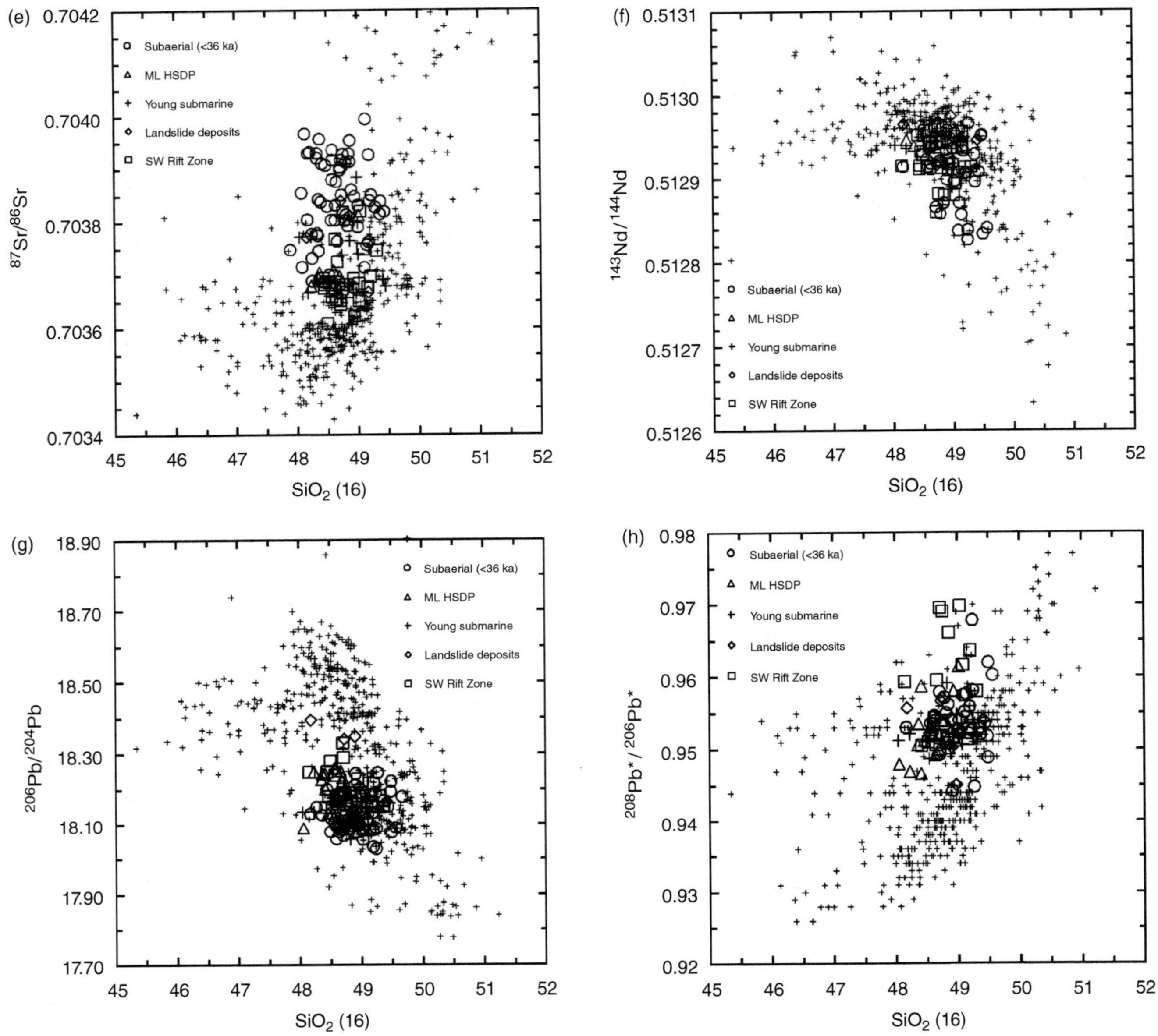

Figure 4.7 Cont'd

MgO-normalized TiO_2 with $^{206}Pb/^{204}Pb$ or other isotopic ratios, but the data do fall in a cluster along the overall trends for Hawaiian shield lavas (Figure 4.7d). The real problem is with SiO_2. Normalized SiO_2 is invariant, with the consequence that there is no correlation of MgO-normalized SiO_2 with the diagnostic isotopes $^{206}Pb/^{204}Pb$, $^{208}Pb^{*}/^{206}Pb^{*}$, $^{87}Sr/^{86}Sr$, or $^{143}Nd/^{144}Nd$ (Figures 4.7e–h). In short, the test fails and there is no compelling need for a lithologically discrete eclogite/pyroxenite component in the source of Mauna Loa magmas.

Given the overall correlation between major elements and isotopic ratios claimed for Hawaiian shield-building tholeiites [e.g., *Hauri*, 1996; *Jackson et al.*, 2012], one might ask whether there is something unique about Mauna Loa. With the exception of Mauna Kea, we do not have comparable long-term records for other Hawaiian volcanoes. Mauna Kea is more complicated because in its recorded ~320 ka shield-building history [*Sharp and Renne*, 2005] it erupted three distinct major magma types with different SiO_2 contents [*Stolper et al.*, 2004; *Rhodes and Vollinger*, 2004; *Rhodes et al.*, 2012]. Nonetheless, as was the case for Mauna Loa, there is no obvious correlation between MgO-normalized SiO_2 and isotopic ratios in any of the three magma types (Figure 4.8). This is particularly critical for the high-SiO_2 lavas. They are the most abundant and were erupted throughout the entire shield-building stage and encompass the entire range of $^{206}Pb/^{204}Pb$, $^{87}Sr/^{86}Sr$, and $^{143}Nd/^{144}Nd$ in Mauna Kea shield-stage lavas

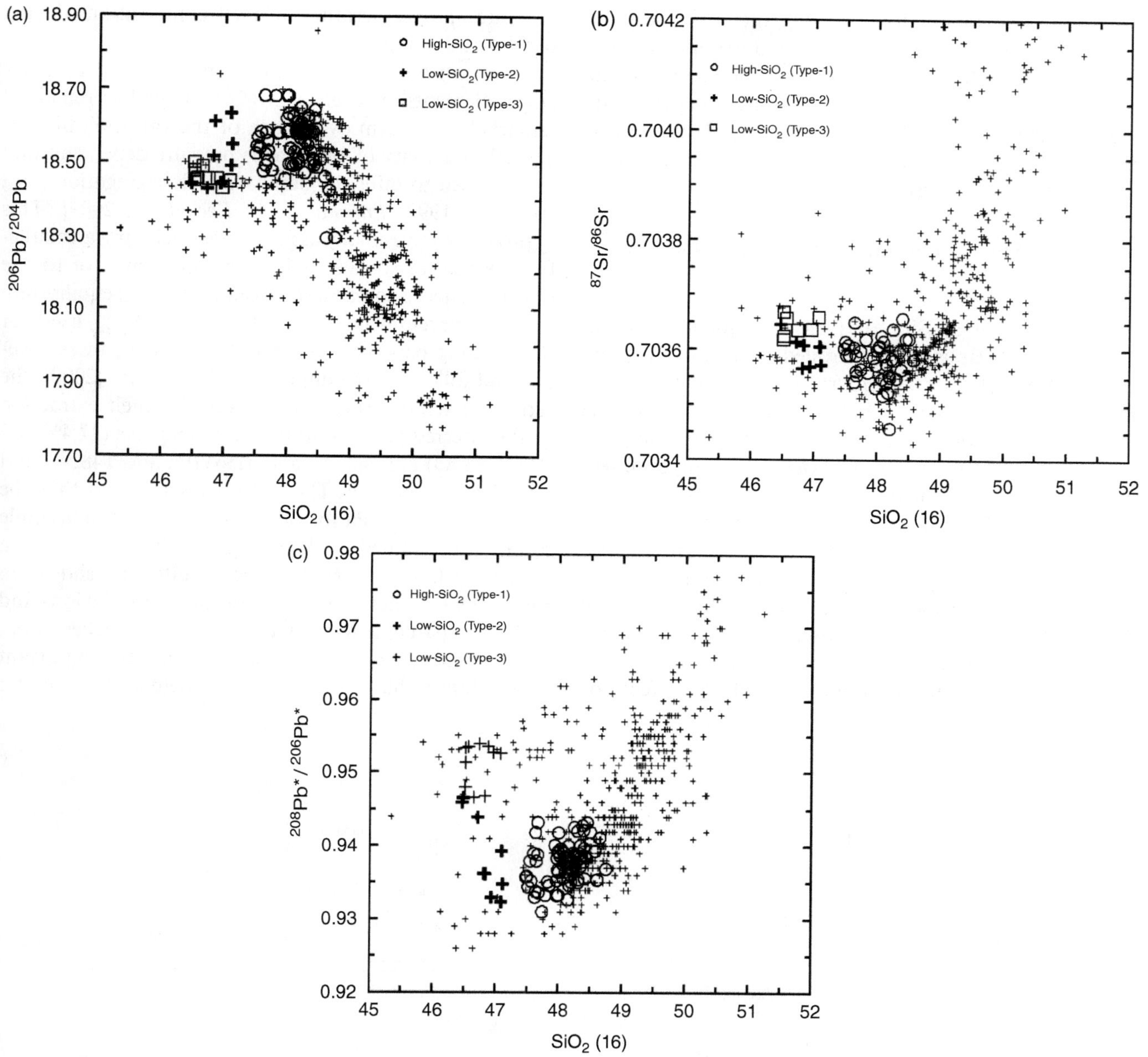

Figure 4.8 Plots of MgO-normalized SiO_2 versus $^{206}Pb/^{204}Pb$, $^{87}Sr/^{86}Sr$, and $^{206}Pb^*/^{204}Pb^*$ for Mauna Kea shield-building tholeiites. The Mauna Kea tholeiites are divided into three magma types according to the classification schemes of *Rhodes and Vollinger* [2004] and *Stolper et al.* [2004]. Major-element data are from *Rhodes and Vollinger* [2004] and *Rhodes et al.* [2012]. Isotopic data are from *Eisele et al.*, 2003; *Blichert-Toft et al.*, 2003; *Abouchami et al.*, 2005; *Bryce and DePaolo*, 2005; and *Blichert-Toft and Albarède*, 2009. The data have been screened for samples that are altered ($K_2O/P_2O_5 < 1.0$) or have undergone fractionation beyond olivine control (MgO < 6.8%). Shown for comparison is the data compilation of *Jackson et al.* [2012] for all Hawaiian shield-building tholeiites (small black crosses).

populating both the Lo-8 and Mid-8 Pb isotopic trends [*Eisele et al.*, 2003; *Abouchami et al.*, 2005; *Blichert-Toft and Albarède*, 2009; *Rhodes et al.*, 2012; *Nobre Silva et al.*, 2013]. Why are CaO and TiO_2 possibly consistent with the presence of a discrete eclogite/pyroxenite component in the source of Mauna Loa (and Mauna Kea) magmas whereas SiO_2 is not? Experimental melting of an inferred Mauna Loa parental magma with about 18% MgO [*Maaloe*, 2004] shows that the parental melt was in equilibrium with a harzburgite residue, not lherzolite. Therefore, melting was extensive and both CaO and TiO_2 would have partitioned almost completely into the melt. This tells us nothing of the source lithology, and therefore the weak correlation between CaO and TiO_2 and isotopic ratios is consistent with both melting of an eclogite/pyroxenite source and a peridotite fertilized by eclogite/pyroxenite melts. This is not the case for SiO_2, which is dependent on the pressure and temperature of melt segregation in addition to the lithology of the source.

4.3.2. Why Such Uniformity in SiO_2 in Mauna Loa Magmas?

If the content of SiO_2 and other major elements in Mauna Loa lavas is not controlled by melting varying contributions of an eclogite/pyroxenite component in the plume peridotite source, what is the explanation for the remarkable uniformity of MgO-normalized compositions of these lavas for over 600 ka? It has long been recognized that experimental melting of peridotite can produce wide variations in major-element melt compositions, from alkalic basalts to picrites, and that the composition is controlled by both the extent of melting and the pressure (depth) of melt segregation [e.g., *Kushiro*, 1996; *Walter*, 1998; *Wagner and Grove*, 1998; *Longhi*, 2002]. *Stolper et al.* [2004, Figure 13] and *Putirka et al.* [2011, Figure 6] provide useful summaries of the effects of pressure and temperature on melt compositions during peridotite melting.

As shown above, the SiO_2 content, and that of other major-elements, at a given MgO content has remained remarkably uniform over much of the volcano's history. The silica activity of magmas is pressure dependent and can be used to infer the depth of melt segregation [e.g., *Albarède*, 1992; *Putirka*, 2008; *Lee et al.*, 2009]. This approach involves many assumptions, including a volatile-free peridotite source. It is extremely sensitive to the olivine forsterite composition assumed to be in equilibrium with the parental magma and to the $KD_{Fe/Mg}$ used for partitioning between olivine and melt. With these caveats, and following the approach of *Lee et al.* [2009], the estimated temperature and pressure for melt extraction of the inferred parental Mauna Loa magmas (17.4% and 18.8% MgO) are 1441°C and 1.56 GPa and 1482°C and 1.91 GPa, respectively. These estimates are lower than the 1550°C and 2.05GPa for olivine-orthopyroxene multiple saturation obtained by *Maaloe* [2004] for a Mauna Loa picrite with 18.2% MgO. The results are shown in Figure 4.9 together with similar estimates for the low- and high-SiO_2 parental Mauna Kea magmas [*Stolper et al.*, 2004; *Rhodes and Vollinger*, 2004]. Again, the important conclusion is that the melting parameters (*T*, *P*, *F*) for

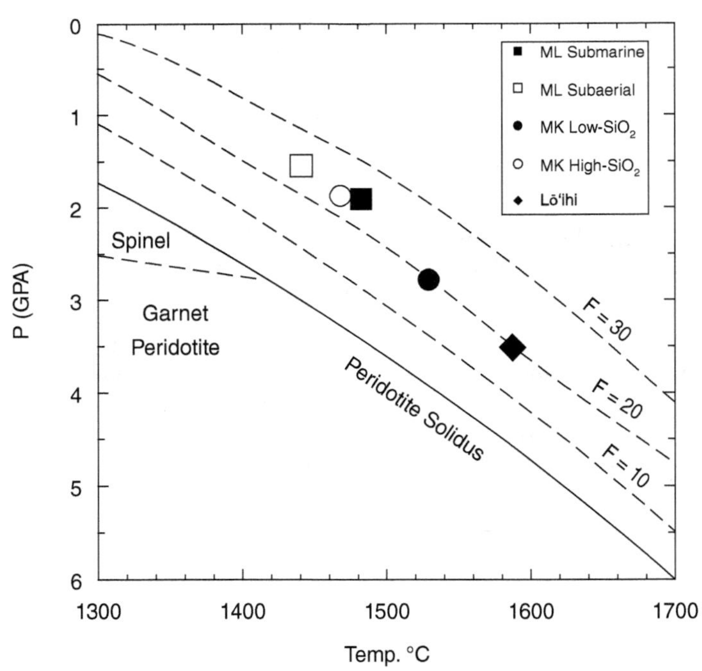

Figure 4.9 Temperature and pressure calculated for several inferred Mauna Loa, Mauna Kea, and Lō'ihi parental magmas following the approach of *Lee et al.* [2009]. In these calculations $KD_{Fe\text{-}Mg}$ is 0.32 and the oxygen fugacity is magnetite-wüstite (MW). The Mauna Loa subaerial and submarine parental magmas are assumed to be in equilibrium with $Fo_{90.7}$ and $Fo_{91.3}$, respectively; the Mauna Kea parental magmas with $Fo_{90.7}$, and the Lō'ihi magma with, $Fo_{91.3}$. The peridotite solidus and melt fraction isopleths are from *Lee et al.* [2009].

Mauna Loa magmas have not changed significantly over 600 ka. How can this be?

The first problem is that these inferences do not fit well within the concept of a thermally zoned plume [e.g., *Watson and Mckenzie*, 1991; *Ribe and Christensen*, 1999; *Farnetani and Hofmann*, 2010]. Assuming a northwestward Pacific Plate motion of 9–10 cm/yr, Mauna Loa should have traversed about 60 km of the Hawaiian plume over 600 ka. According to current plume models, this should have resulted in varying amounts of melt generation at different depths within the plume, producing different magma compositions. Some of these differences will be averaged out because of pooling from a broad (~25 km? [*DePaolo and Stolper*, 1996]) magma capture zone, but nonetheless some variation should be evident. Similarly, one would expect that aggregate melts produced by polybaric melting in an ascending plume should exhibit some differences over 600 ka. The almost constant major-element compositions inferred for parental Mauna Loa magmas are inconsistent with these interpretations. Could this imply that there is no significant decreasing radial thermal gradient from the plume center but, instead, the core of the plume is thermally uniform, perhaps with a sharp thermal gradient toward the margin?

Second, the trace element data for Mauna Loa lavas [e.g., *Budahn and Schmitt*, 1985; *Garcia et al.*, 1995; *Rhodes and Hart*, 1995; *Norman and Garcia*, 1999] require equilibrium with residual garnet to explain the rare earth element patterns and the almost constant abundances of the heavy rare earths [*Hofmann et al.*, 1984]. The estimated pressures and depths of the inferred Mauna Loa parental magmas (MgO ~17.4%–18.8%) are too shallow to be in equilibrium with garnet (>2.3 GPa [*Lee et al.*, 2009]) near the peridotite solidus (Figure 4.9). This is a pervasive problem in Hawaiian petrology [e.g., *Eggins*, 1992; *Wagner and Grove*, 1998; *Takahashi and Nakajima*, 2002, *Stolper et al.*, 2004]. Experimental melting of basalts and peridotites [see syntheses by *Stolper et al.*, 2004; *Putirka et al.*, 2011] show that high-SiO_2 parental magmas, with high enough MgO to crystallize Fo_{88-91}, result from melt segregation at around 1.5–2 GPa and are in equilibrium with harzburgite. That is much too shallow (50–65 km) to be in equilibrium with garnet lherzolite (>80 km). By contrast, there is no such problem for low-SiO_2 parental magmas. They result from melt segregation at around 3–4 GPa, corresponding to mantle depths of 95–125 km, and are in equilibrium with garnet lherzolite.

Third, the inferred segregation depths are well within the lithosphere. There are compelling isotopic arguments, especially Os isotopes, that shield-building Hawaiian tholeiites were not produced by melting the Cretaceous lithosphere overlying the Hawaiian plume [e.g., *Hauri et al.*, 1996; *Lassiter and Hauri*, 1996; *Blichert-Toft et al.*, 1999; *Abouchami et al.*, 2000; *Fekiacova et al.*, 2007]. There is debate concerning the thickness of the lithosphere above the Hawaiian plume and its relationship to the Hawaiian Swell. One model attributes the buoyancy of the swell to the erosion and thinning of the lithosphere by the hot Hawaiian plume [e.g., *Detrick and Crough*, 1978; *Liu and Chase*, 1989; *Moore and Schubert*, 1997]. The other attributes it to accumulation of buoyant melt-depleted plume material beneath a normal-thickness lithosphere [e.g., *Sleep*, 1990; *Phipps Morgan et al.*, 1995]. Recent seismological estimates of the thickness of the lithosphere beneath Hawai'i range from around 88 to 110 km [*Priestly and Tilmann*, 1999; *Li et al.*, 2004]. There are three possibilities, therefore: The depth estimates for melt segregation are unrealistic; the lithosphere has been thinned and the seismic resolution is insufficient to distinguish between lithosphere- and buoyant-depleted plume harzburgite beneath the lithosphere; or the plume-derived melt and associated residual harzburgite intrude deep into the lithosphere. [*Phipps Morgan et al.*, 1995; *Wagner and Grove*, 1998].

4.3.3. Ascending Peridotite Melt-Harzburgite Interaction

Experimental melting of peridotites and basalts indicate that high-SiO_2 basaltic Hawaiian tholeiitic magmas are produced at low pressures (<2.2 GPa), corresponding to depths of less than 71 km within the lithosphere, and that they are in equilibrium with a harzburgite residue. They are not in equilibrium with garnet lherzolite at any pressure. Conversely, trace element data for these magmas require deeper melt segregation from a garnet lherzolite source. In order to explain this conundrum, a number of authors have appealed to melt interaction between ascending melt and overlying depleted harzburgite [e.g., *Eggins*, 1992; *Wagner and Grove*, 1998; *Stolper et al.*, 2004]. The basic concept is that low-SiO_2 melts, produced by melting garnet lherzolite at depths of 90–130 km within the plume, interact, on ascent, with overlying depleted residual harzburgite (the residue of prior melt extraction within the plume). The reaction involves the dissolution of orthopyroxene from the harzburgite, compensated for by the crystallization of olivine from the melt. *Wagner and Grove* [1997] applied this model to derive a primitive Kīlauea tholeiite magma [*Clague et al.*, 1995] from an aggregate of low-SiO_2 experimental melts. Similarly, *Stolper et al.* [2004] showed that the major-element compositions of the ubiquitous high-SiO_2 Mauna Kea tholeiites could be produced from a low-SiO_2 Mauna Kea tholeiite by dissolution of orthopyroxene plus clinopyroxene and crystallization of olivine. The distinct isotopic differences between these two magma types preclude this model, but it is plausible that a low-SiO_2 tholeiite with comparable isotopic characteristics (the type 2 lavas of *Rhodes and Vollinger*, 2004) was the parent of the high-SiO_2 Mauna Kea tholeiites. As noted by *Stolper et al.* [2004], if melt-harzburgite interaction is a prevailing process in the Hawaiian plume, then low-SiO_2 magmas will be the exception.

The uniform high-SiO_2 content of Mauna Loa magmas over ~600 ka, and the uniformity of most other major-elements, can most readily be explained by the interaction of a low-SiO_2 melt and plume-derived residual harzburgite. There are no low-SiO_2 tholeiitic lavas within the entire ~600 ka Mauna Loa record, so it is difficult to infer a primitive magma composition. Perhaps it is similar to low-SiO_2 Lō'ihi tholeiites, since Lō'ihi is on the Loa trend and is close to the leading edge of the plume and therefore its magmas have probably had less chance to interact with a lherzolite residue. Additional support is provided by the existence of old (>200 ka), high-SiO_2 content Mauna Loa lavas with characteristic Lō'ihi isotopic ratios (Figure 4.5). A critical requirement of this model is that magmas should retain the isotopic characteristics, and trace element garnet signature of the low-SiO_2 parental magma. *Wagner and Grove* [1998] showed that the reaction of a low-SiO_2 melt with harzburgite to produce a Kīlauea parental magma had little influence on the garnet diagnostic La/Yb. The only effect was to slightly decrease the abundance of both elements. Using the reactants from Table 4 of *Wagner and Grove* [1998]) and typical concentrations of Sr, Nd, and Pb in primitive Mauna Loa lavas [*Norman and Garcia*, 1999], AFC calculations [*DePaolo*, 1981] show only slight change in $^{87}Sr/^{86}Sr$, $^{143}Nd/^{144}Nd$, and $^{206}Pb/^{204}Pb$ irrespective of the ratios assumed for the reactant harzburgite.

Herzberg [2006, 2011] has proposed that only the enigmatic high-CaO Mauna Kea lavas [*Stolper et al.*, 2004; *Rhodes et al.*, 2012] and possibly primitive Lō'ihi tholeiites are peridotite melts. He infers that all other Hawaiian tholeiites are melts of pyroxenite. The thrust of the argument, in addition to high Ni in olivines and melts (see earlier) is that, with the exception of these high-CaO lavas and some Lō'ihi lavas, all other inferred parental Hawaiian tholeiites are lower in CaO than the experimental peridotite melts of *Walter* [1998]. Additionally, peridotite melts should plot on the olivine side of the thermal divide between garnet and orthopyroxene in the system olivine-quartz-CATS (Ca Tschermak's pyroxene) at 3–4 Gpa. All high-SiO_2 primitive and parental melts (including Kīlauea and Mauna Kea as well as Mauna Loa) plot on the silica side of this thermal divide at these pressures and are therefore, according to Herzberg, melts of recycled oceanic crust in the form of eclogite/pyroxenite. Reaction of a peridotite-derived Lō'ihi-like primitive melt with a harzburgite residue as it ascends to shallower levels in the plume provides a solution to this dilemma. Co-crystallization of olivine with resorption of orthopyroxene from the harzburgite will increase the SiO_2 content of the melt as its composition moves away from olivine across the CATS-orthopyroxene join. Additionally, this process will decrease the CaO and TiO_2 content of the melt, both of which tend to be lower in Loa trend lavas than in those of the Kea trend. In short, if the melt-harzburgite reaction is a viable process, there is no need for discrete pyroxenite/eclogite lithologies in the Hawaiian plume source.

Given estimates of 10%–30% melt extraction to account for the composition of Hawaiian shield-building tholeiites, it follows that most of the ascending plume is residual harzburgite with perhaps minor lherzolite. The most likely location for extensive interaction of low-SiO_2 primitive melts with this harzburgite residue is at the top of the plume [*Eggins*, 1992; *Wagner and Grove*, 1998] as the buoyant residual harzburgite is dragged horizontally by the overlying lithosphere [*Rhodes and Hart*, 1995; *Ribe and Christensen*, 1999]. Providing orthopyroxene is not exhausted by the process, the SiO_2 content of the magma will be fixed by its saturation with olivine and orthopyroxene [*Stolper et al.*, 2004], resulting in an ideal situation for producing an almost invariant SiO_2 magma content over a long period of time. If initial melting at depth to produce the low-SiO_2 primitive magma involves varying proportions of the Loa and Lō'ihi source components in fertilized peridotite, we will end up with lavas with varying isotopic ratios but almost uniform SiO_2, as is the case for Mauna Loa. A similar scenario probably applies to the high-SiO_2 lavas of Mauna Kea [*Stolper et al.*, 2004; *Rhodes and Vollinger*, 2004]. These lavas occur throughout the entire shield-building stage but belong to both the Lo-8 and Mid-8 Pb isotopic trends with widely differing $^{206}Pb/^{204}Pb$ [e.g., *Blichert-Toft and Albarède*, 2009; *Rhodes et al.*, 2012; *Nobre Silver et al.*, 2013]. The associated but less common low-SiO_2 lavas of Mauna Kea [*Stolper et al.*, 2004; *Rhodes and Vollinger*, 2004] may have avoided reaction with the harzburgite by flow through olivine-lined channels formed by prior melt-harzburgite interaction [e.g., *Kelemen et al.*, 1995; *Stolper et al.*, 2004].

4.4. CONCLUSIONS

1. Recent $^{40}Ar/^{39}Ar$ dating of Mauna Loa's submarine lavas has extended our knowledge of the eruptive history of the volcano from the 0–120 ka subaerial record to over 600 ka. Although less continuous, this exceeds the record availabe for Mauna Kea volcano from the Hawai'i Scientific Drilling Project. During this 600 ka period, lavas that have not fractionated beyond olivine control follow exceedingly well-defined olivine control trends. This indicates that the major-element composition of Mauna Loa parental magmas has remained remarkably constant over 600 ka. There is a suggestion, however, from olivine phenocryst compositions that parental,

young (<120 ka) subaerial magmas may have been slightly lower in MgO (17.4%) than older (>200 ka) submarine lavas (18.8% MgO) and therefore, perhaps, a little cooler.

2. In contrast with the major-element data, the isotopic data for Pb, Sr, and Nd are highly variable, encompassing around 58% of the entire Hawaiian array for $^{208}Pb/^{204}Pb$, 43% of the array for $^{87}Sr/^{86}Sr$, and 36% of the array for both $^{206}Pb/^{204}Pb$ and $^{143}Nd/^{144}Nd$. There is a tendency for subaerial lavas to have higher $^{87}Sr/^{86}Sr$ and lower $^{206}Pb/^{204}Pb$ and $^{143}Nd/^{144}Nd$ than older submarine lavas. It must be stressed, however, that the isotopic range of older submarine lavas are considerable. Some have isotopic ratios comparable to historical lavas; others have isotopic ratios that overlap with those of Lō'ihi. There is therefore no clear trend with age. Despite the isotopic diversity, indicative of varying proportions of Loa and Lō'ihi plume components, the major-elements have typical Mauna Loa compositions.

3. During its recorded 600 ka eruptive history Mauna Loa should have traversed about 60 km of the Hawaiian plume. Current plume models and petrological melting experiments would lead one to expect a variety of major-element magma compositions reflecting a radial decrease in temperature and corresponding differences in the depth of melt segregation. Similarly, if the source components of Hawaiian magmas are a witches' brew of peridotite and recycled crustal eclogite/peridotite as recent models propose, then again we should expect a diversity of magma compositions, reflecting varying proportions of these components. This is not the case; therefore, other explanations should be sought.

4. There is an overall consensus that the broad (but not always entirely convincing) correlation between the isotopic ratios of Sr, Pb, and Nd with major-element oxides such as SiO_2, CaO, and TiO_2 in Hawaiian shield-building magmas is a consequence of melting recycled crustal components in the Hawaiian plume. What is not agreed upon is whether or not this recycled crustal component is present as discrete lithological domains of eclogite/pyroxenite or whether it is present in the form of peridotite "fertilized" by melts of these crustal components. Opinion encompasses the entire spectrum from "no peridotite" to "only peridotite." The uniformity in the major-element composition of Mauna Loa magmas over 600 ka, coupled with a lack of correlation between MgO-normalized SiO_2 and isotopic ratios, suggests that the source of these magmas is peridotite, not varying proportions of peridotite and eclogite/pyroxenite.

5. The paradox of the uniformity in major-element compositions, coupled with considerable isotopic diversity, over Mauna Loa's recorded history can be explained by reaction of a low-SiO_2 parental magma with a buoyant, depleted harzburgite residue from prior melting within the plume, as the magma ascends through the upper regions of the plume. This accounts for the high, and constant, SiO_2 content of the magmas, the retention of the trace element garnet signature of the parental magma, and its variable isotopic characteristics reflecting the contribution of varying source components. There is no need for a discrete eclogite/pyroxenite source component. Ironically, however, even if it existed, its presence would be overprinted by the melt-harzburgite reaction! A major problem with this model, which also applies to all models attempting to explain high-SiO_2 Hawaiian tholeiites, is that the inferred depth of melt segregation is too shallow and within the lithosphere. Either the inferred depths are in error, the lithosphere over the Hawaiian plume is eroded and thinner than seismological models suggest, or plume-derived melts and associated residual harzburgite intrude deep into the lithosphere.

SUPPLEMENTARY DIGITAL DATA

The following data items are available online at www.wiley.com/go/Carey/Hawaiian_Volcanoes

Appendix 4A. Major element analyses of Mauna Loa lavas.

ACKNOWLEDGMENTS

This work was funded by the National Science Foundation. Special thanks to Peter Lipman, Jack Lockwood, and Frank Trusdell for encouragement, guidance, and friendship during my long fascination with Mauna Loa. Also, to Mike Garcia for initiating the submarine studies of Mauna Loa. Pete Dawson kept the instruments running and Mike Vollinger did the XRF analyses. Thanks also to Matt Jackson and Claude Herzberg for perceptive and thorough reviews and to my wife Claudia for careful editing. They are not to blame for the final product!

REFERENCES

Abouchami, W., S. J .G. Galer, and A. W., Hofmann (2000), High precision lead isotope systematics of lavas from the Hawaiian Scientific Drilling Project, *Chem. Geol., 169*, 187–209.

Abouchami, W., A. W. Hofmann, S. J. G. Galer, F. A. Frey, J. Eisele and M. Feigenson (2005), Pb isotopes reveal bilateral asymmetry in the Hawaiian plume, *Nature, 434*, 851–856.

Albarède, F. (1992), How deep do common basaltic magmas form and differentiate? *J. Geophys. Res., 97*, 10,997–11,009.

Baker, M. B., S. Alves, and E. M. Stolper (1996), Petrology and petrography of the Hawaii Scientific Drilling Project Lavas: Inferences from olivine phenocryst abundance and compositions, *J. Geophs. Res., 101*, 11,715–11,727.

Beeson, M. H., D. A. Clague and J. P. Lockwood (1996), Origin and depositional environment of clastic deposits in the Hilo drill hole, Hawaii, *Geophys. Res., 101*, 11,617–11,629.

Blichert-Toft, J., and F. Albarède (2009), Mixing of isotopic heterogeneities in the Mauna Kea plume Conduit, *Earth Planet. Sci. Lett., 282*, 190–200.

Blichert-Toft, J., F. A. Frey and F. Albarède (1999), Hf isotope evidence for pelagic sediments in the source of Hawaiian basalts, *Science, 285*, 879–882.

Blichert-Toft, J., D. Weis, C. Maerschalk and F. Albarède (2003), Hawaiian hotspot dynamics as inferred from the Hf and Pb isotope evolution of Mauna Kea volcano, *Geochem. Geophys. Geosyst., 4,* doi:10.1029/2002GC000340.

Bryce, J., and D. DePaolo (2005), Geochemical structure of the Hawaiian plume: Sr, Nd and Os isotopes in the 2.8 km HSDP-2 section of Mauna Kea volcano, *Geochem. Geophys. Geosyst., 6,* doi:10.1029/2004GC000809.

Budahn, J. R., and R. A. Schmitt (1985), Petrogenetic modeling of Hawaiian tholeiitic basalts: A geochemical approach, *Geochim. Cosmochim. Acta, 49*, 67–87.

Clague, D. A., W. S. Weber and J. E. Dixon (1991), Picritic glasses from Hawaii, *Nature, 353*, 553–556.

Clague, D. A., J. G. Moore, J. E. Dixon and W. B. Friesen (1995), Petrology of submarine lavas from Kilauea's Puna Ridge, Hawaii, *J. Petrol., 36*, 299–349.

Davis, M. G., M. O. Garcia and P. Wallace (2003), Volatiles in glasses from Mauna Loa volcano, Hawaii: Implications for magma degassing and contamination, and growth of Hawaiian volcanoes, *Contrib. Mineral. Petrol., 144*, 570–591.

DePaolo, D. J. (1981) , Trace element and isotopic effects of combined wallrock assimilation and fractional crystallization, *Earth. Planet Sci Lett., 53*, 189–202.

DePaolo, D. J.,and E. M. Stolper (1996), Models of Hawaiian volcano growth and plume structure: Implications of results from the Hawaii Scientific Drilling Project, *J. Geophys Res., 101*, 11,643–11,654.

DePaolo, D. J., J. G. Bryce, A. Dodson, D. A. Shuster, and B. M. Kennedy (2001), Isotopic evolution of Mauna Loa and the chemical structure of the Hawaiian plume, *Geochem. Geophys. Geosyst., 2,* 1044, doi:10.1029/2000GC000139.

Detrick, R. S., and S. T. Crough (1978), Island subsidence, hotspots, and lithospheric thinning, *J. Geophys. Res., 83*, 1236–1244.

Eggins, S. M. (1992), Petrogenesis of Hawaiian tholeiites. 2. Aspects of dynamic melt segregation, *Contrib. Mineral. Petrol., 110*, 387–397.

Eisele, J., W. Abouchami, S. J. G. Galer and A. W. Hofmann (2003), The 320 ky Pb isotope evolution of the Mauna Kea lavas recorded in the HSDP-2 drill core, *Geochem. Geophys. Geosyst., 4,* doi:10.1029/2002GC000339.

Farnetani, C. G., and A. W. Hofmann (2010), Dynamics and internal structure of the Hawaiian plume, *Earth Planet Sci. Lett., 296*, 231–240.

Fekiacova, Z., W. Abouchami, S. J. G. Galer, M. O. Garcia and A. W. Hofmann (2007), Origin and temporal evolution of Koolau Volcano, Hawaii: Inferences from isotopic data on the Koolau Scientific Drilling Project (KSDP), the Honolulu Volcanics and ODP Site 843, *Earth Planet. Sci. Lett., 261*, 65–83.

Frey, F. A., and J. M. Rhodes (1993), Inter-shield geochemical differences among Hawaiian volcanoes: Implications for source compositions, melting processes and magma ascent paths, *Philos. Trans. R. Soc. Lond. Part A, 342,* 121–136.

Garcia, M. O. (1996), Petrography and olivine and glass chemistry of lavas from the Hawaii Scientific Drilling Project, *J. Geophys. Res., 101*, 11,701–11,713.

Garcia, M. O., K. Muenow, J. Aggrey and J. O'Neil (1989), Major element, volatile, and stable isotope geochemistry of Hawaiian submarine tholeiitic glasses, *J. Geophys. Res., 94*, 10,526–10,538.

Garcia, M. O., T. Hulsebosch, and J. M. Rhodes (1995), Olivine-rich submarine basalts from the Southwest Rift Zone of Mauna Loa volcano: Implications for magmatic processes and geochemical evolution, in *Mauna Loa Revealed: Structure, Composition, History, and Hazards,* Geophys. Monogr. *92*, edited by J. M. Rhodes and J. P. Lockwood, pp. 219–239, AGU, Washington, D. C.

Garcia, M. O., E. H. Haskins, E. M. Stolper and M. Baker (2007), Stratigraphy of the Hawaii Scientific Drilling Project core (HSDP2): Anatomy of a Hawaiian shield volcano, *Geochem. Geophys. Geosyst., 8,* doi:10.1029/2006GC001379.

Haskin, E. H., and M. O. Garcia (2004), Scientific drilling reveals geochemical heterogeneity within the Koolau shield, Hawaii, *Contrib. Mineral. Petrol., 147*, 162–188.

Hauri, E. H. (1996), Major element variability in the Hawaiian mantle plume, *Nature, 283,* 415–419.

Hauri, E. H., J. C. Lassiter and D. J. DePaolo (1996), Osmium isotope systematics of drilled lavas from Mauna Loa, Hawaii, *J. Geophys. Res., 101,* 11,793–11,806.

Herzberg, C. (2006), Petrology and thermal structure of the Hawaiian plume from Mauna Kea volcano, *Nature, 444*, 605–609.

Herzberg, C. (2011), Identification of source lithology in the Hawaaian and Canary Islands: Implications for origins, *J. Petrol., 52*, 113–146.

Herzberg, C., P. D. Asimov, D. A. Ionov, C. Vidito, M. G. Jackson and D. Geist (2013), Nickel and helium evidence for melt above the core-mantle boundary, *Nature, 493*, 393–397.

Hofmann, A. W., M. D. Feigenson and I. Raczek (1984), Case studies on the origin of basalts III. Petrogenesis of the Mauna Ulu eruption, Kilauea, 1969–1971, *Contrib. Mineral. Petrol., 88*, 24–35.

Huang, S., and F. A. Frey (2005), Recycled oceanic crust in the Hawaiian Plume: Evidence from temporal geochemical variations within the Koolau shield, *Contrib. Mineral. Petrol., 149*, 556–575, doi:10.1007/s00410-005-0664-9.

Jackson, M. G., D. Weis and S. Huang (2012), Major element variation in Hawaiian shield lavas: Source features and perspectives from global ocean island (OIB), *Geochem. Geophys. Geosyst., 13*, doi:10.1029/2012GC004268.

Jicha, B. R., J. M. Rhodes, B. S. Singer and M. O. Garcia (2012), $^{40}Ar/^{39}Ar$ geochronology of submarine Mauna Loa volcano, Hawaii,*J. Geophys. Res., 117*,B09204,doi:10,1029/2012JB009373.

Kelemen, P. B., N. Shimizu and V. J. M. Salters (1995), Extraction of mid-ocean ridge basalt from the upwelling mantle by focused flow of melt in dunite channels, *Nature, 375*, 747–753.

Kurz, M. D., and D. P. Kammer (1991), Isotopic evolution of Mauna Loa volcano, *Earth. Planet. Sci. Lett., 103*, 257–269.

Kurz, M. D., T. C. Kenna, D. P. Kammer, J. M. Rhodes and M. O. Garcia (1995), Isotopic evolution of Mauna Loa volcano; a view from the submarine southwest rift zone, in

Mauna Loa Revealed: Structure, Composition, History and Hazards, Geophys. Monogr. 92, edited by J. M. Rhodes and J. P. Lockwood, AGU, Washington, D. C.

Kushiro, I. (1996), Partial melting of a fertile mantle peridotite at high pressures: An experimental study using diamond aggregates, in *Reading the Isotopic Code,* Geophys. Monogr. 95, edited by A. Basu and S. Hart, AGU, Washington, D. C.

Lassiter, J. C., and E. H. Hauri (1996), Osmium isotope variations in Hawaiian lavas: Evidence for recycled oceanic lithosphere in the Hawaiian plume, *Earth Planet. Sci. Lett., 164,* 483–496.

Lee, C.-T., P. Luffi, T. Plank, H. Dalton and W. P. Leeman (2009), Constraints on the depth and temperatures of basaltic magma generation on Earth and other terrestrial planets using a new barometer for mafic magmas, *Earth Planet. Sci. Lett., 279,* 20–33.

Li, X., R. Kind, X. Yuan, I. Wölbern and I. Hanka (2004), Rejuvenation of the lithosphere by the Hawaiian plume, *Nature, 427,* 827–829.

Lipman, P. W., W. P. Normark, J. G. Moore, J. B. Wilson and C. E. Gutmacher (1988), The giant Alika debris slide, Mauna Loa, Hawaii, *J. Geophys. Res. 93,* 4279–4299.

Liu, G. M., and C. G. Chase (1989) , Evolution of midplate hotspot swells: Numerical solutions, *J. Geophys. Res., 94,* 5571–5584.

Lockwood, J. P. (1995), Mauna Loa eruptive history: The preliminary radiocarbon record, in *Mauna Loa Revealed: Structure, Composition, History and Hazards,* Geophys. Monogr. 92, edited by J. M. Rhodes and J. P. Lockwood, pp. 45–80, AGU, Washington, D. C.

Lockwood, J. P., and P. W. Lipman (1987), Holocene eruptive history of Mauna Loa volcano, *U. S. Geol. Surv. Prof. Pap., 1350,* 509–535.

Longhi, J. (2002), Some phase equilibrium systematics of lherzolite melting, *Geochem. Geophys. Geosyst., 1020,* doi: 10,1029/2001GC000204.

Maaloe, S. (2004), The PT-phase relations of an MgO-rich Hawaiian tholeiite: The composition of primary Hawaiian tholeiites, *Contrib. Mineral. Petrol., 148,* 236–246.

Macdonald, G. A. and A. T. Abbott (1970), *Volcanoes in the Sea,* University of Hawaii Press, Honolulu, p.441.

Marske, J. P., A. J. Pietruszka, D. Weis, M. O. Garcia and J. M. Rhodes (2007), Rapid passage of a small-scale mantle heterogeneity through the melting regions of Kilauea and Mauna Loa volcanoes, *Earth Planet. Sci. Lett., 259,* 34–50.

Matzen, A. K., M. B. Baker, J. R. Beckett and E. M. Stolper (2011), Fe-Mg partitioning between olivine and high-magnesian melts and the nature of Hawaiian parental liquids, *J. Petrol. 52,* 1243–1263.

Moore, J. G., D. A. Clague, R. T. Holcomb, P. W. Lipman, W. R. Normark and M. E. Torresan (1989), Prodigious submarine landslides on the Hawaiian Ridge, *J. Geophys. Res., 94,* 17,465–17,484.

Moore, J. G., W. R. Normark and B. J. Szabo (1990), Reef growth and volcanism on the submarine southwest rift zone of Mauna Loa, Hawaii, *Bull. Volcanol., 52,* 375–380.

Moore, W., and G. Schubert (1997), Lithospheric thinning and chemical buoyancy beneath the Hawaiian Swell, *Geophys. Res. Lett., 24,* 1287–1290.

Murata, K. J., and D. H. Richter (1966), Chemistry of the lavas of the 1959–60 eruption of Kilauea volcano, Hawaii, Prof. Pap. 357-A, U.S. Geol. Surv., Reston, VA.

Nobre Silva, I. G., D. Weis and J. S. Scoates (2013), Isotopic systematics of the early Mauna Kea shield phase and insight into the deep mantle beneath the Pacific Ocean, *Geophys. Geochem. Geosyst., 14,* 659–676.

Norman, M. D., and M. O. Garcia (1999), Primitive magmas and source characteristics of the Hawaiian plume: Petrology and geochemistry of shield picrites, *Earth Planet. Sci. Lett., 168,* 27–44.

O'Nions, R. K., P. J. Hamilton and N. M. Evensen (1977), Variations in 143Nd/144Nd and $^{87}Sr/^{86}Sr$ ratios in oceanic basalts, *Earth Planet. Sci. Lett., 34,* 13–22.

Phipps Morgan, J., W. J. Morgan and E. Price (1995), Hotspot melting generates both hotspot volcanism and hotspot swell? *J. Geophys. Res., 100,* 8046–8062.

Powers, H. A. (1955), Composition and origin of basaltic magmas on the Hawaiian islands, *Geochem. Cosmochem. Acta, 7,* 77–107.

Priestly, K., and F. Tilmann (1999), Shear-wave structure of the lithosphere above the Hawaiian hot spot from two-station Rayleigh wave phase velocity measurements, *Geophys. Res. Lett., 26,* 1493–1496.

Putirka, K. D. (2005), Mantle potential temperatures at Hawaii, Iceland, and the mid-ocean ridge system, as inferred from olivine phenocrysts: Evidence for thermally driven mantle plumes, *Geochem. Geophys. Geosyst., 6,* Q05L08, doi:10.1029/2005GC000915434.

Putirka, K. D. (2008), Thermometers and barometers for volcanic systems, in *Minerals, Inclusions and Volcanic Processes, Reviews in Mineralogy & Geochemistry, vol. 69,* edited by K. D. Putirka and F. J. Tepley III, *Min. Soc. Amer.,* Chantilly, VA.

Putirka, K., F. J. Ryerson, M. Perfitt and W. I. Ridley (2011), Mineralogy and composition of the oceanic mantle, *J. Petrol., 52,* 279–313.

Rhodes, J. M. (1989), Geochemistry of the 1984 Mauna Loa eruption: Implications for magma storage and supply, *J. Geophys. Res., 93,* 4453–4466.

Rhodes, J. M. (1995), The 1852 and 1868 Mauna Loa picrite eruptions: Clues to parental magma compositions and the magmatic plumbing system, in *Mauna Loa Revealed: Structure, Composition, History and Hazards,* Geophys. Monogr. *92,* edited by J. M. Rhodes and J. P. Lockwood, pp. 241–262, AGU, Washington, D. C.

Rhodes, J. M. (1996), Geochemical stratigraphy of lava flows sampled by the Hawaii Scientific Drilling Project, *Geophys. Res., 101,* 11,729–11,746.

Rhodes, J. M., and M. O. Garcia (2008), Olivine-control trends for Hawaiian magmas and their consequences for magma temperatures and the plume hypothesis (poster), IAVCEII Meet., Reyjavik, Iceland.

Rhodes, J. M., and S. R. Hart (1995), Episodic trace element and isotopic variations in historical Mauna Loa lavas: Implications for magma and plume dynamics, in *Mauna Loa Revealed: Structure, Composition, History and Hazards,* Geophys. Monogr. 92, edited by J. M. Rhodes and J. P. Lockwood, pp. 263–288, AGU, Washington, D. C.

Rhodes, J. M., and M. J. Vollinger (2004), Composition of basaltic lavas sampled by phase-2 of the Hawaii Scientific Drilling Project: Geochemical stratigraphy and magma types, *Geochem. Geophys. Geosyst., 5*, Q03G13, doi:10.1029/2002GC000434.

Rhodes, J. M., and M. J. Vollinger (2005), Ferric/ferrous ratios in 1984 Mauna Loa lavas: A contribution to understanding the oxidation state of Hawaiian magmas, *Contrib. Mineral. Petrol., 149*, 666–674.

Rhodes, J. M., K. P. Wenz, C. A. Neal, J. W. Sparks and J. P. Lockwood (1989), Geochemical evidence for invasion of Kilauea's plumbing system by Mauna Loa magmas, *Nature, 337*, 257–260.

Rhodes, J. M., S. Huang, F. A. Frey, M. Pringle and G. Xu (2012), Compositional diversity of Mauna Kea shield lavas recovered by the Hawaii Scientific Drilling Project: Inferences on source lithology, magma supply, and the role of multiple volcanoes, *Geochem. Geophys. Geosyst., 13*, Q03014, doi:10.1029/2011GC003812.

Ribe, N. M., and U. R. Christenson (1999), The dynamic origin of Hawaiian volcanism, *Earth Planet. Sci. Lett., 171*, 517–531.

Roeder, P. L., and R. F. Emslie (1970), Olivine-liquid equilibrium, *Contrib. Mineral. Petrol., 29*, 275–289.

Sharp, W. D., and P. R. Renne (2005), The $^{40}Ar/^{39}Ar$ dating of core recovered by the Hawaii Scientific Drilling Project (phase 2), Hilo, Hawaii, *Geochem. Geophys. Geosyst., 6*, Q04G17, doi:10.1029/2004GC000846.

Sharp, W. D., B. D. Turrin, P. R. Renne and M. A. Lanphere (1996), The $^{40}Ar/^{39}Ar$ and K/Ar dating of lavas from the Hilo 1-km core hole, Hawaii Scientific Drilling Project, *J. Geophys. Res., 101*, 11,607–11,616.

Sleep, N. H. (1990), Hotspots and mantle plumes: Some phenomenology, *J. Geophys. Res., 95*, 6715–6736.

Sobolev, A. V., A. W. Hofmann, S. V. Sobolev and I. K. Nikogosian (2005), An olivine free mantle source of Hawaiian shield basalts, *Nature, 434*, 590–597.

Sobolev, A. V., et al. (2007), The amount of recycled crust in source of mantle-derived melts, *Nature, 316*, 412–417.

Staudigal, H., A. Zindler, S. R. Hart, T. Leslie, C.-Y. Chen and D. Clague (1984), The isotopic systematics of a juvenile intraplate volcano: Pb, Nd, and Sr isotope ratios from Loihi seamount, Hawaii, *Earth Planet Sci. Lett., 69*, 13–29.

Stille, P., D. M. Unruh and M. Tatsumoto (1986), Pb, Sr, Nd and Hf isotopic constraints on the origin of Hawaiian basalts and evidence for a unique mantle source, *Geochim. Cosmochim. Acta, 50*, 2303–2319.

Stolper, E. M., D. J. DePaolo, and D. M. Thomas (1996), Introduction to special session: Hawaii Scientific Drilling Project, *J. Geophys. Res., 101*, 11593–11598.

Stolper, E., S. Sherman, M. O. Garcia, M. Baker and C. Seaman (2004), Glass in the submarine section of the HSDP2 drill core, Hilo, Hawaii, *Geochem. Geophys. Geosyst.,* 5, Q07G15, doi:10.1029/2003GC000553.

Stolper, E. M., D. J. DePaolo and D. M. Thomas (2009) , Deep drilling into a mantle plume volcano: The Hawaii Scientific Drilling Project, *Sci. Drill., 7*, 4–14.

Takahasi, E., and K. Nakajima (2002), Melting processes in the Hawaiian plume: An experimental study, in *Hawaiian Volcanoes: Deep Underwater perspectives,* Geophys. Monogr. 128, edited by E. Takahashi et al., pp. 403–418, AGU, Washington, D. C.

Tatsumoto, M. (1978), Isotopic composition of lead in oceanic basalt and its implication to mantle evolution, *Earth Planet. Sci. Lett., 38,* 13–29.

Trusdell, F. A., and J. P. Lockwood (2014a) Geologic map of the northeast flank of Mauna Loa volcano, Island of Hawaii, Hawaii, U.S. Geol. Surv. SIM 2932-A, scale 1:50,000, in press.

Trusdell, F. A., and J. P. Lockwood (2014b), Geologic map of the central-southeast flank of Mauna Loa volcano, Island of Hawai'i, Hawaii, U.S. Geol. Surv. SIM 2932-B, scale 1:50,000, in press.

Trusdell, F. A., and J. P. Lockwood (2014c), Geologic map of southern Mauna Loa volcano, Island of Hawai'i, Hawaii, U.S. Geol. Surv. SIM 2932-C, scale 1:50,000, in press.

Wagner, T. P., and T. L. Grove (1998), Melt/harzburgite reaction in the petrogenesis of tholeiitic magma from Kilauea Volcano, Hawaii, *Contrib. Mineral. Petrol., 131*, 1–12.

Walter, M. J. (1998), Melting of garnet peridotite and the origin of komatiite and depleted lithosphere, *J. Petrol., 39*, 29–60.

Wanless, V. D., M. O. Garcia, F. A. Trusdell, J. M. Rhodes, M. D. Norman, D. Weis, D. J. Fornari, M. D. Kurz and H. Guillou (2006a), Submarine radial vents on Mauna Loa volcano, Hawaii, *Geochem. Geophys. Geosyst., 5*, Q05001, doi:10.1029/2005GC001086.

Wanless, V. D., M. O. Garcia, J. M. Rhodes, D. Weis and Norman M. D. (2006b), Shield stage alkali volcanism on Mauna Loa volcano, Hawaii, *J. Volcanol. Geotherm. Res., 151*, 141–155.

Watson, S., and D. McKenzie (1991), Melt generation by plumes: A study of Hawaiian volcanism, *J. Petrol., 32*, 501–537.

Weis, D., M. O. Garcia, J. M. Rhodes, M., Jellinek and J. S. Scoates (2011), Role of the deep mantle in generating the compositional asymmetry of the Hawaiian mantle plume, *Nature Geosci., 4*, 831–838.

West, H. B., D. C. Gerlach, W. P. Leeman and M. O. Garcia (1987), Isotopic constraints on the on the origin of Hawaiian lavas from the Maui volcanic complex, *Hawaii, Nature, 330*, 216–220.

Wilkinson, J. F. G., and H. D. Hensel (1988), The petrology of some picrites from Mauna Loa and Kilaues, Hawaii, *Contrib. Mineral. Petrol., 98*, 326–345.

Wright, T. L. (1971), Chemistry of Kilauea and Mauna Loa lava in space and time, *Prof. Pap. 735*, U.S. Geol. Surv., Reston, Va.

5

Lithium Isotopic Signature of Hawaiian Basalts

Lauren Harrison, Dominique Weis, Diane Hanano, and Elspeth Barnes

ABSTRACT

Recycling of oceanic crust and sediment is a common mechanism to account for the presence of chemical heterogeneities observed in oceanic island basalts (OIBs). Because of the sizeable fractionation of lithium isotopes in low-temperature environments, lithium serves as a tracer for recycled material in OIB sources. In this study, we analyzed 88 samples of Hawaiian basalt from all volcanic stages and 10 samples of altered oceanic crust from Ocean Drilling Program (ODP) Site 843 for lithium isotopes. The measured range of lithium isotopes is δ^7Li = 0.8‰–5.7‰. Correlations of lithium isotopes with radiogenic isotopes indicate lithium isotopes may be used to trace mantle sources in Hawaiian lavas. Loa trend shield volcanoes appear to show lower δ^7Li, differentiating between the Loa and Kea geochemical trends. Similarly, postshield lavas have lower δ^7Li than shield lavas. In Hawaiian basalts, lithium isotopes help distinguish between Loa source components: Ko'olau Makapu'u shield stage lavas may have between 1% and 5% of a carbonate input and Hualālai postshield and shield lavas may reflect incorporation of subduction eroded lower continental crust. Comparison of this data set with worldwide OIB published lithium isotopic data indicates that the lithium isotopic system behaves systematically on a mantlewide scale.

5.1. INTRODUCTION

Hawaiian volcanism delineates into two distinctive geographical volcanic trends, the Kea and the Loa trends, which are also different in their geochemical and isotopic characteristics [*Jackson et al.*, 1972; *Tatsumoto*, 1978; *Abouchami et al.*, 2005; *Huang et al.*, 2011b; *Weis et al.*, 2011]. Traditionally, the geochemical differences between Loa and Kea trend shield lavas have been attributed to mixing of different components in the source of the Hawaiian mantle plume, i.e., an enriched end-member (Ko'olau) characterized by the presence of recycled oceanic crust ± sediment, a more depleted end member (Kea) with high ε_{Nd}, ε_{Hf}, and $^{206}Pb/^{204}Pb$, and a more undegassed component defined by high $^3He/^4He$ (Lō'ihi) [*Kurz et al.*, 1983; *Stille et al.*, 1986; *Frey et al.*, 1994; *Hauri*, 1996; *Tanaka et al.*, 2008; *Jackson et al.*, 2012]. Furthermore, some postshield and rejuvenated samples require a fourth and more isotopically depleted source to account for observed isotopic variation [*Huang et al.*, 2005; *Xu et al.*, 2005, 2007; *Fekiacova et al.*, 2007; *Hanano et al.*, 2010]. Although many studies have focused on resolving, characterizing, and determining the proportions these different sources contribute to Hawaiian volcanism during the stages of a volcano's life, many questions still remain that up to now radiogenic isotopic systems such as Pb, Hf, Nd, and Sr have not fully answered. In this chapter, we use the lithium stable isotopic system to obtain additional insights on the source characteristics of Hawaiian lavas.

Geochemically, lithium is a moderately incompatible element and is therefore concentrated in the continental crust. The two isotopes of lithium, ^{6}Li and ^{7}Li, are fractionated in low-temperature aqueous environments

Pacific Center for Isotopic and Geochemical Research, Department of Earth, Ocean, and Atmospheric Sciences, University of British Columbia, Vancouver, British Columbia, Canada

Hawaiian Volcanoes: From Source to Surface, Geophysical Monograph 208. First Edition.
Edited by Rebecca Carey, Valérie Cayol, Michael Poland, and Dominique Weis.

Companion Website: www.wiley.com/go/Carey/Hawaiian_Volcanoes

because ^{7}Li preferentially partitions into the fluid phase. Consequently different Earth reservoirs have unique signatures in lithium isotopes [e.g., *Tang et al.*, 2010, and references therein]. The flux of lithium between these reservoirs is largely controlled by interactions of solid geological materials with hydrous fluids and the temperatures of those fluids; i.e., at metamorphic temperature lithium is preferentially extracted from mineral phases, while at hydrothermal temperature lithium is added through incorporation in secondary minerals [*Chan et al.*, 1992; *Huh et al.*, 2004; *Gao et al.*, 2012]. Thus lithium primarily fluxes into oceanic crust from seawater as it is hydrothermally altered, and this signature is in turn radically altered by prograde metamorphism and dewatering during subduction [*Zack et al.*, 2003; *Elliott et al.*, 2004; *Tomascak*, 2004; *Tang et al.*, 2010]. Through this complex cycling, lithium isotopes have the potential to be a unique tracer for the presence of altered oceanic crust and sediments in the sources of oceanic island basalts such as Hawai'i [*Zack et al.*, 2003; *Elliott et al.*, 2004, 2006; *Kobayashi et al.*, 2004; *Ryan and Kyle*, 2004; *Tomascak*, 2004; *Nishio et al.*, 2005; *Chan et al.*, 2009; *Vlastélic et al.*, 2009; *Tang et al.*, 2010; *Krienitz et al.*, 2012].

Recently, however, the application of lithium isotopes as tracers of recycling processes has been questioned for several reasons. First, the discovery of lithium diffusion at small scale (mineral-matrix) that results in lithium isotopic heterogeneities has led to concerns that the original source lithium isotopic signature may be obliterated during transport and cooling of melts [*Richter et al.*, 2003; *Seitz et al.*, 2004; *Lundstrom et al.*, 2005; *Jeffcoate et al.*, 2007; *Parkinson et al.*, 2007]. Diffusion also presents the possibility of homogenization of lithium isotopes during residence time in the mantle, further complicating the use of lithium as a source tracer [*Halama et al.*, 2008]. Second, a lack of predictable behavior of lithium isotopes, including the absence of a significantly heavy lithium isotopic signature in arc lavas [*Chan et al.*, 1999, 2001, 2002b], a lack of correlation of lithium isotopes with other isotopic systems [*Chan and Frey*, 2003], or the lack of significant variation in lithium isotopic signature from mid-ocean ridge basalt (MORB) mantle in other mantle-derived melts [*Chan and Frey*, 2003; *Tomascak et al.*, 2008], has called into question the robustness and systematics of the processes that control the lithium isotopic system. Evaluating the practicality of lithium isotopes as a mantle tracer in well-characterized Hawaiian lavas is a major aim of this study.

This study uses lithium isotopes to (1) corroborate the usefulness of this isotopic system in oceanic island basalt studies [*Chan et al.*, 2009; *Vlastélic et al.*, 2009; *Krienitz et al.*, 2012]; (2) further characterize Hawaiian plume source characteristics for preshield, shield, postshield, and rejuvenated lavas; and (3) assess the prevalence and impact of recycled materials in creating observed mantle heterogeneities in Hawai'i and especially in the difference between shield-stage Loa and Kea trend lavas. After controlling for different degrees of source melting, fractionation of lithium isotopes during crystal fractionation, and perturbation of lithium in samples by alteration, we have found relative differences in lithium isotopes between Kea and Loa trend volcanoes and between the shield and postshield volcanic stages. We also show that lithium isotopes in Hawaiian lavas correlate with Pb, Hf, Nd, and Sr radiogenic isotopic systems, indicating that the lithium isotopic signature of erupted lavas can be used as a tracer of the mantle source in Hawaiian lavas. An analysis of lithium isotopes and various potential geochemical indices of sediment recycling (La/Nb, Sr/Nb, $^{87}Sr/^{86}Sr$, and $^{143}Nd/^{144}Nd$) suggest the presence of a heterogeneous sedimentary contribution in the composition of Loa trend volcanoes. Trace element and isotopic data indicate that this sedimentary source may be ancient carbonate for the Makapu'u section of Ko'olau volcano and lower continental crust for Hualālai [*Willbold and Stracke*, 2006; *Huang et al.*, 2009, 2011a]. Finally, a comparison of this Hawaiian data with published lithium isotopic data for other oceanic island basalts (OIBs) shows systematic behavior of lithium in the framework of traditionally defined mantle end members [high-μ (HIMU), enriched mantle one (EM-I), enriched mantle two (EM-II), and PREvalent MAntle (PREMA); *White*, 1985; *Zindler and Hart*, 1986; *Hofmann*, 2003; *White*, 2010; *Krienitz et al.*, 2012] where Hawai'i has traditionally been classified as an EM-I-type oceanic island.

5.2. SAMPLES AND METHODS

The aim of this study is a wide-scale investigation of the lithium isotopic signatures present in Hawai'i to both characterize Hawai'i in the context of other oceanic islands and to test for the presence of recycled material in the source of Loa trend volcanoes. With this aim in mind, samples spanning the greatest compositional, geographical, and temporal range as well as different volcanic stages were analyzed to constrain the range of lithium isotopic signatures in Hawaiian lavas (Figure 5.1).

Lithium isotopic ratios were measured on 88 samples of Hawaiian lavas, including preshield, shield, postshield, and rejuvenated volcanic stages and a temporal range from less than a century old to ~4.3 Ma [Kaua'i; *Garcia et al.*, 2010]. Published descriptions and localities of many of these samples are provided in the references listed in Table 5.1. In addition, 10 samples of ~110 Ma Pacific altered upper oceanic crust from ODP Site 843 (~320 km west of the Island of Hawai'i; see Figure 5.1 [*King et al.*, 1993; *Fekiacova et al.*, 2007]) were analyzed for lithium isotopes. Also included in the figures and discussion is the Mauna Loa, Mauna Kea, and Ko'olau lithium isotopic data of *Chan and Frey* [2003]. For this chapter, the term

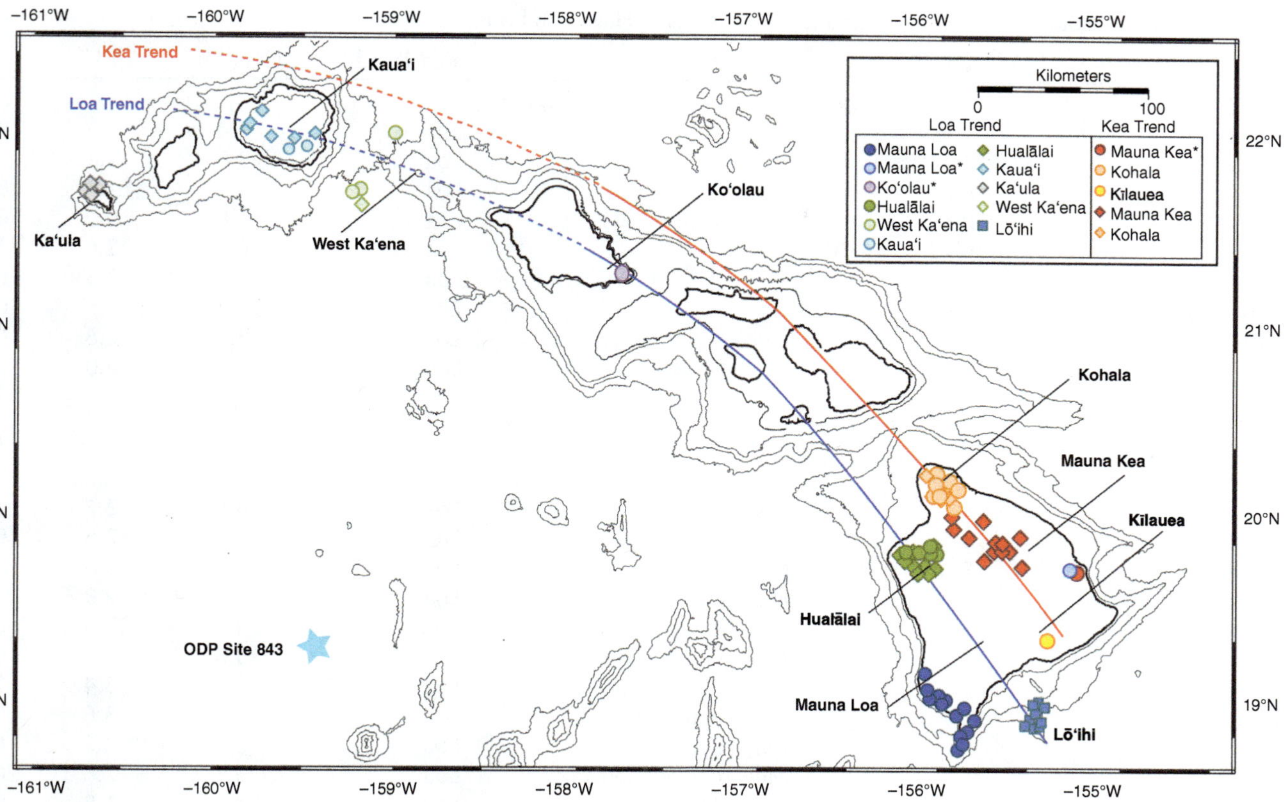

Figure 5.1 Map of Hawai'i and sample locations. The basemap for this figure is the Main Hawaiian Islands Chart 750-001 Version 17 (released January 2011) downloaded from the Hawai'i Mapping Research Group of the School of Ocean and Earth Science and Technology at the University of Hawai'i at Mānoa's website (Main Hawaiian Islands Multibeam Bathymetry Synthesis: www.soest.hawaii.edu). Approximate locations of samples are indicated by symbols that are also used in the figures that follow [*Feigenson et al.*, 1983; *Clague*, 1987; *Garcia et al.*, 1989, 1993, 1995, 1998, 2010; *Chan and Frey*, 2003; *Greene et al.*, 2010; *Hanano et al.*, 2010; *Weis et al.*, 2011, in prep. *Jicha et al.*, 2012; J. M. Rhodes, unpublished (Hualālai postshield samples)]. Approximate locations of the Loa and Kea geochemical trends also shown as a blue and red line, respectively. Symbols for this and all following figures are divided into circles for shield lavas, squares for preshield lavas, and diamonds for postshield and rejuvenated lavas. Data noted in legend with an asterisk are from *Chan and Frey* [2003].

Ko'olau refers to the Makapu'u section of Ko'olau volcano. This data set compliments existing Pb-Hf-Nd-Sr isotopic, major and trace element data from several sources (see Table 5A.1 for a full compilation). In the cases where high-precision trace element data did not exist (Hualālai, Kohala, and Mauna Loa shield samples; Lō'ihi preshield samples; and Ka'ula and Kaua'i shield, postshield, and rejuvenated samples), a full characterization was carried out at the Pacific Center for Isotopic and Geochemical Research (PCIGR) at the University of British Columbia.

Samples were crushed and powdered following several different techniques (please see references in Table 5A.1 for a full description). Whole-rock powders were analyzed for lithium isotopic compositions and concentrations at the PCIGR. For lithium isotope analysis, whole-rock powders were leached in 0.05 N HCl for one hour at room temperature, a procedure that has been shown to have no or limited effect on primary magmatic lithium isotopic ratios but that is necessary to remove possible contamination from a heavy, posteruption lithium isotopic signature contained in secondary clays [*Huh et al.*, 2004; *Vlastélic et al.*, 2009]. Our results show that this heavy alteration signature may be effectively removed by leaching (Figure 5.2). However, for 9 out of 10 altered oceanic crust samples from ODP Site 843 leaching did not change the lithium isotopic signature beyond analytical error of unleached samples. Furthermore, leaching did not change the lithium isotopic signature of samples outside of two standard deviations (2SD) of average East Pacific Rise MORB (Figure 5.2).

After leaching an aggressive two-step combination HF and $HClO_4$ digestion was carried out following method C of *Yokoyama et al.* [1999] to minimize the presence of insoluble fluorides. The two ion exchange column procedure of *Jeffcoate et al.* [2004] was used to separate and purify

Table 5.1 Lithium isotopic compositions and concentrations of Hawaiian lavas.

Sample	Volcanic Stage	Volcanic Series	Loa or Kea Trend	Li (ppm)	δ^7Li	2σ
Lō'ihi						
ATV-1-4	PreS		Loa	4.4	5.5	0.4
186-3[b]	PreS		Loa	3.0	4.2	0.0
1801-9[b]	PreS		Loa	4.3	1.7	0.3
1804-17[b]	PreS		Loa	5.8	0.9	0.4
P287-1[b]	PreS		Loa	6.1	2.6	1.1
158-10[b]	PreS		Loa	5.3	3.5	0.2
158-1dup[b]	PreS		Loa	5.8	2.6	0.7
P286-14[b]	PreS		Loa	4.1	2.8	0.5
1804-21[b]	PreS		Loa	4.4	2.0	0.8
Kīlauea						
KIL-93	S		Kea	4.4	3.7	1.3
Mauna Loa						
ML-63[d]	S	Historic	Loa	4.8	2.7	0.3
ML-190[d]	S	Historic	Loa	5.3	2.6	0.3
ML-95[d]	S	Historic	Loa	4.7	3.0	0.3
ML-41[d]	S	Historic	Loa	4.5	2.9	0.3
ML-91[d]	S	Historic	Loa	4.2	4.0	0.4
J2-019-08[d]	S	Mile High Section	Loa	3.8	3.1	0.1
J2-019-08dup[d]	S	Mile High Section	Loa		3.4	0.5
J2-019-19[d]	S	Mile High Section	Loa	4.7	3.6	0.2
J2-019-01[d]	S	Mile High Section	Loa	5.4	4.5	0.7
J2-020-14[d]	S	Mile High Section	Loa	5.1	2.6	0.7
184-7[d]	S	Pisces Dive	Loa	4.4	1.8	0.3
L91-55[d]	S	Prehistoric	Loa	4.7	2.8	0.7
L00-503[d]	S	Prehistoric	Loa	4.0	3.7	0.5
L00-003[d]	S	Prehistoric	Loa	4.6	2.8	0.6
L00-506[d]	S	Prehistoric	Loa	4.2	3.5	0.9
L87-235[d]	S	Prehistoric	Loa	4.5	3.0	0.3
L00-014[d]	S	Prehistoric	Loa	4.4	3.3	0.4
Mauna Kea						
02AMK-1[a]	PS	Laupāhoehoe	Kea	10.0	5.2	0.2
02AMK-2[a]	PS	Hāmākua	Kea		3.2	0.3
02AMK-3[a]	PS	Hāmākua	Kea	4.0	3.6	0.3
02AMK-4[a]	PS	Laupāhoehoe	Kea	10.0	5.0	0.3
02AMK-5[a]	PS	Hāmākua	Kea	4.0	4.1	0.6
02AMK-6[a]	PS	Laupāhoehoe	Kea	10.9	4.4	0.2
02AMK-7[a]	PS	Hāmākua	Kea	7.4	3.5	1.0
02AMK-8[a]	PS	Laupāhoehoe	Kea	8.1	4.7	0.8
02AMK-10[a]	PS	Laupāhoehoe	Kea	8.3	2.7	0.5
02AMK-11[a]	PS	Laupāhoehoe	Kea	10.5	3.7	0.6
02AMK-12[a]	PS	Hāmākua	Kea	10.5	3.4	0.9
02AMK-13[a]	PS	Hāmākua	Kea	3.6	3.0	0.4
Hualālai						
02AHU-1[a]	PS	Hualālai	Loa	6.3	2.0	0.4
02AHU-2[a]	PS	Hualālai	Loa	4.8	2.5	0.3
02AHU-3[a]	PS	Hualālai	Loa	9.0	2.9	0.9
02AHU-4[a]	PS	Hualālai	Loa	7.4	2.6	0.1
02AHU-4dup[a]	PS	Hualālai	Loa		3.0	0.4
02AHU-5[a]	PS	Hualālai	Loa	7.3	2.2	0.4
02AHU-6[a]	PS	Hualālai	Loa	8.7	1.3	0.2
02AHU-7[a]	PS	Hualālai	Loa	10.6	3.1	0.1
02AHU-8[a]	PS	Hualālai	Loa	32.9	1.9	0.8
02AHU-9[a]	PS	Hualālai	Loa	5.6	0.8	0.3

Table 5.1 Cont'd

Sample	Volcanic Stage	Volcanic Series	Loa or Kea Trend	Li (ppm)	δ^7Li	2σ
02AHU-10[a]	PS	Hualālai	Loa	5.6	1.9	1.0
02AHU-11[a]	PS	Hualālai	Loa	6.0	3.5	0.5
02AHU-13[a]	PS	Hualālai	Loa	6.5	1.5	0.3
KK-9-2[i]	S		Loa	4.7	3.4	0.8
KK-9-2 dup[i]	S		Loa		3.9	0.4
KK-11-2[i]	S		Loa	5.9	3.1	0.5
KK-11-5[i]	S		Loa	3.2	3.3	0.2
KK-13-1[i]	S		Loa	3.5	3.5	1.0
KK-14-5[i]	S		Loa	3.9	2.4	0.6
KK-78-12-11-1[i]	S		Loa	3.9	3.7	0.3
HU-01[j]	PS	Hualālai	Loa	9.0	2.7	0.9
HU-02[j]	PS	Hualālai	Loa	7.7	3.0	0.1
Kohala						
02AKA-2[a]	PS	Hāwī	Kea	8.9	2.6	0.6
02AKA-1[a]	PS	Hāwī	Kea	9.1	3.0	0.2
02AKA-3[a]	PS	Hāwī	Kea	9.1	3.5	0.9
02AKA-4[a]	PS	Hāwī	Kea	14.0	4.0	0.4
KH-14[c]	PS	Hāwī	Kea	5.3	2.1	0.5
KH-11[c]	PS	Hāwī	Kea	6.5	3.0	0.3
KH-11dup[c]	PS	Hāwī	Kea		3.2	0.7
02AKA-5[a]	LS	Polulū	Kea	8.8	1.8	0.2
02AKA-6[a]	LS	Polulū	Kea	6.1	1.9	0.1
02AKA-7[a]	LS	Polulū	Kea	7.2	2.0	0.3
MG-2B[c]	LS	Polulū	Kea	7.2	3.1	0.3
KO3-6[k]	LS-PS		Kea	5.5	3.3	0.4
MG-5A[c]	LS	Polulū	Kea	5.1	4.4	0.8
MG-4A[c]	LS	Polulū	Kea	4.1	3.8	1.0
KO1-27[k]	LS		Kea	4.9	3.1	0.6
Ko'olau						
USGS standard	S		Loa	3.8	4.3	0.9
West Ka'ena						
J2-296-01[e]	PS-Rej	Jason2 Dive 296	Loa	6.1	4.3	0.3
J2-305-14[e]	LS	Jason2 Dive 305	Loa	8.2	4.4	0.6
J2-305-14dup[e]	LS	Jason2 Dive 305	Loa		5.0	0.7
J2-306-15[e]	LS	Jason2 Dive 306	Loa	4.6	5.2	1.1
J2-306-27[e]	LS	Jason2 Dive 306	Loa	6.3	3.2	0.3
Kaua'i						
KV03-5[f]	PS	Waimea Canyon	Loa	5.1	4.2	0.3
KV03-5dup[f]	PS	Waimea Canyon	Loa		3.6	0.2
KV05-10[f]	PS	Waimea Canyon	Loa	4.4	3.3	0.1
KV03-15[f]	Rej	Kōloa Volcanics	Loa	6.8	4.1	0.1
KV03-17[f]	Rej	Kōloa Volcanics	Loa	6.6	3.3	0.3
GA-650[f]	PS	Waimea Canyon	Loa	5.1	2.5	0.0
GA-565[f]	PS	Waimea Canyon	Loa	4.1	2.2	0.5
KV04-19[f]	S-LS	Waimea Canyon	Loa	5.3	3.1	0.3
KV04-21[f]	S-LS	Waimea Canyon	Loa	6.1	2.5	1.2
Ka'ula						
J2-300-10[g]	Rej	Jason2 Dive 300		7.2	4.2	0.4
J2-300-20[g]	Rej	Jason2 Dive 300		4.1	3.1	0.4
J2-300-30[g]	Rej	Jason2 Dive 300		4.4	4.1	0.7
J2-300-41[g]	Rej	Jason2 Dive 300		3.8	3.3	0.5
J2-300-41dup[g]	Rej	Jason2 Dive 300			3.7	1.0

(Continued)

Table 5.1 Cont'd

Sample	Volcanic Stage	Volcanic Series	Loa or Kea Trend	Li (ppm)	δ^7Li	2σ
ODP Site 843: Altered Oceanic Crust						
001R 01W 28-30[h]	AOC	Leg 136 0843B-			2.1	0.6
004R 02W 29-31[h]	AOC	Leg 136 0843B-			3.5	0.4
001R 01W 37-39[h]	AOC	Leg 136 0843B-			3.3	1.3
001R 01W 56-58[h]	AOC	Leg 136 0843B-			1.7	0.9
004R 01W 83-85[h]	AOC	Leg 136 0843B-			4.0	1.0
001R 06W 84-86[h]	AOC	Leg 136 0843B-			3.3	0.2
001R 03W 94-96[h]	AOC	Leg 136 0843B-			1.6	0.6
002R 01W 105-106[h]	AOC	Leg 136 0843B-			3.1	0.7
001R 02W 122-124[h]	AOC	Leg 136 0843B-			3.8	0.5
001R 02W 122-124dup[h]	AOC	Leg 136 0843B-			4.3	0.8
003R 02W 8-12[h]	AOC	Leg 136 0843A-			3.5	1.3

Note: PreS = Preshield; S = Shield; LS = Late Shield; PS = Postshield; Rej = Rejuvenated; AOC = Altered Oceanic Crust.
Supporting Data for the above samples (major and trace elements; radiogenic isotopes) comes from:
[a] *Hanano et al.*, 2010.
[b] D. Weis, unpub; *Harrison et al.*, in prep.; *Garcia et al.*, 1993, 1995, 1998
[c] D. Weis, unpub; *Harrison et al.*, in prep.; *Feigenson et al.*, 1983
[d] *Weis et al.*, 2011.
[e] *Greene et al.*, 2010.
[f] *Garcia et al.*, 2010.
[g] *Weis et al.*, in prep.
[h] *King et al.*, 1993; *Fekiacova et al.*, 2007.
[i] *D. Weis*, unpub.; *Harrison et al., in prep.; Clague*, 1987.
[j] *D. Weis*, unpub.; *Harrison et al., in prep.; J. M. Rhodes*, unpub.
[k] *D. Weis*, unpub.; *Harrison et al., in prep.; Garcia et al.*, 1989.

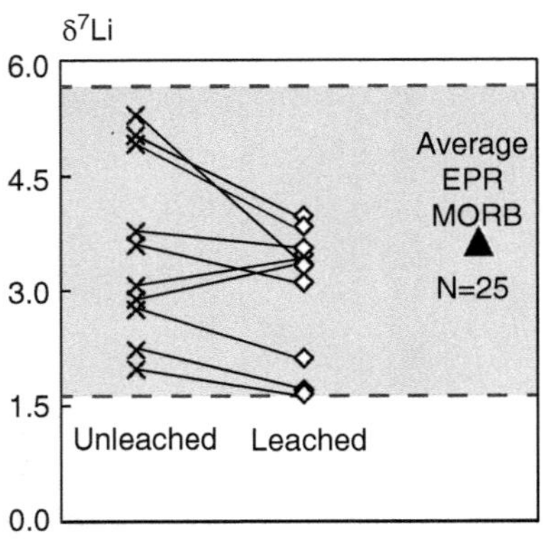

Figure 5.2 ODP Site 843 leaching results. Average East Pacific Rise MORB is calculated from *Elliott et al.* [2006] and *Tomascak et al.* [2008]. The grey field bounded by dashed lines indicates 2SD of average EPR MORB. Leached and unleached sample pairs are linked. The lithium isotopic signature of 8 out of 10 samples lightens with gentle leaching.

lithium, with greater than 98.5% recovery. Total procedural blanks contained 0–0.2 ng lithium, negligible in comparison to the amount of lithium present in the samples (on average ~45 ng). Lithium isotopes were analyzed on a muticollector inductively coupled plasma mass spectrometer (MC-ICP-MS; Nu 021 Nu Instruments Ltd.) using the conventional blank-standard-sample-standard-blank bracketing technique and employing the National Institute of Standard and Technology (NIST) lithium carbonate standard L-SVEC. Each sample was analyzed a minimum of three times to statistically assess precision. During a typical three day analytical session, the average ^{7}Li/^{6}Li value for the L-SVEC standard was 13.95±0.06 (2 standard deviation, *N*=242). Lithium isotopic ratios are reported here as per mil variation relative to L-SVEC based on the equation below [*Coplen*, 1995; *Carignan et al.*, 2004]:

$$\delta^7\text{Li} = \left[\left(\frac{\text{Li}^7/\text{Li}^6{}_{\text{sample}}}{\text{Li}^7/\text{Li}^6{}_{\text{L-SVEC}}}\right) - 1\right] \times 1000$$

Repeated analyses of U.S. Geological Survey (USGS) reference material BHVO-2, a Hawaiian basalt, resulted in an average δ^7Li=3.2 ± 2.5‰ (2SD; *N*=55), slightly lower than the published range (4.0‰–5.5‰; see compilation in Table 5A.2 and *Brant et al.* [2012]). This may be a result of our removal through leaching of the heavy lithium isotopic signature imparted to BHVO-2 by deposition of marine aerosols, a procedure only one other study carried out [*Vlastélic et al.*, 2009]. Analysis of unleached BHVO-2 resulted in δ^7Li=4.3 ± 0.4‰ (2SD, *N*=6), confirming this hypothesis. In-house reference materials Kil-93 (Hawaiian basalt from the 1993 Kīlauea eruption, chilled) and the synthetic Li carbonate Puratronic® (Alfa Aesar 99.999% Li_2CO_3, Lot #23267) were reproducible over the three year course of this study, with values of 3.3 ± 1.7‰ (2SD, *N*=77) and 80.9 ± 1.5‰ (2SD, *N*=232), respectively. Reproducibility on individual days of analysis was much better, i.e., 3.5 ± 0.6‰ (2SD, *N*=4) for Kil-93 and 80.8 ± 0.5‰ (2SD, *N*=10) for Puratronic®. All samples analyzed have 2SD equal to or less than 1.3 per mil.

Trace element concentrations of samples without existing data were measured at the PCIGR using a Thermo Finnigan Element2 high-resolution ICP-MS and following the methods reported by *Pretorius et al.* [2006] and *Carpentier et al.* [2013]. Powdered samples were digested in a mixture of 10:1 HNO_3-HF with no prior leaching. After digestion, samples were fluxed in HCl to eliminate insoluble fluorides, diluted 5000 times, and analyzed on the Element2 using a 1 ppb In spike as an internal standard and the USGS basaltic standard BCR-2 for external calibration. BHVO-2 was analyzed with all samples, and resulting values were within 2SD of published values for all elements [*Weis et al.*, 2006; *Chauvel et al.*, 2011; *Carpentier et al.*, 2013, and references therein]. Full procedural duplicates and replicate measurements showed excellent agreement, and most elements were analyzed with one relative standard deviation (RSD) of less than 5%.

Finally, to place these results in the context of the worldwide geochemical signature of the oceanic mantle, we have compiled existing lithium isotopic data for other oceanic island basalts along with supporting major, trace element, and Pb-Nd-Sr data for these samples (Table 5A.3). All literature radiogenic isotope data are normalized to the same standard values to ensure comparability [*Weis et al.*, 2011]. Nd isotopes are normalized to a $^{143}Nd/^{144}Nd$ La Jolla value of 0.511858; Sr isotopes to a $^{87}Sr/^{86}Sr$ SRM 987 value of 0.710248; Pb isotopes to a SRM 981 values of 16.9405 for $^{206}Pb/^{204}Pb$, 15.4963 for $^{207}Pb/^{204}Pb$, and 36.7219 for $^{208}Pb/^{204}Pb$; and Hf to a $^{176}Hf/^{177}Hf$ JMC 475 value of 0.282160 [*Weis et al.*, 2006, 2007].

5.3. RESULTS

Lithium isotopic compositions and concentrations are reported in Table 5.1. A compilation of supporting published data for these samples (major and trace element data; Pb, Nd, Hf, and Sr isotopic ratios) is provided in Table 5A.1 in the Appendix.

Shield samples for all volcanoes in this study are tholeiitic basalts. The postshield and rejuvenated samples are mostly alkalic basalts and range in composition from foidite to trachyte. All of the preshield samples from Lōʻihi are tholeiitic basalt. Lithium concentrations of all samples in this data set range from 3 to 14 ppm and increase with decreasing weight percent MgO, reflecting the moderately incompatible behavior of lithium. The ratio of lithium to elements of similar compatibilities (for example, Y, Dy, and Yb) does not show any systematic variation versus an incompatible element (Nb). This suggests the partition coefficient of lithium does not change significantly between the different samples of Hawaiian lavas analyzed in this study [*Chan and Frey*, 2003].

5.3.1. Alteration Control

It is important to have stringent controls on the degree of alteration of analyzed samples because of the propensity of lithium to fractionate in low-temperature aqueous environments [*Chan et al.*, 1992; *Huh et al.*, 2004; *Marschall et al.*, 2007]. In addition, interaction of basaltic rocks with seawater ($\delta^7Li \approx 31$‰ [*Jeffcoate et al.*, 2004, and references therein]) will enrich the rock in lithium and increase the lithium isotopic signature to heavier values [*Chan et al.*, 1992, 2002a]. The alteration index of *Wright* [1971], shown in Figure 5.3a, was used to eliminate samples that are significantly altered [*Frey et al.*, 1990]. Most samples plot within the range of magmatic K/Rb (380–630 [*Chan and Frey*, 2003]) and have K_2O/P_2O_5 greater than unity. Variations outside of this window may still be included if the cause for the variation is from differentiation and not alteration. On a plot of Rb versus K_2O (Figure 5.3b), most samples form coherent arrays. Because the extremely high K/Rb values of Kauaʻi postshield sample GA-650 are coupled with a normal lithium isotopic signature and a tight linear correlation on Figure 5.3b, interaction with a lithium-poor fluid, such as fresh groundwater, is likely and probably did not impact the lithium isotopic signature [*Chan and Frey*, 2003]. Samples with abnormal lithium isotopic signatures and K/Rb>750 were removed from all following plots and discussion, along with Hawaiʻi Scientific Drilling Project (HSDP) samples with $K_2O/P_2O_5<0.9$ [*Chan and Frey*, 2003].

5.3.2. Effect of Crystal Fractionation

Major processes with the potential to disturb mantle source lithium isotopic signatures are solid state diffusion of lithium between mantle minerals, diffusion of lithium between minerals and melt, and isotopic fractionation during crystal fractionation [*Chan et al.*, 2009]. Understanding how these processes impact measured lithium isotopic values is key to the application of lithium isotopes as a mantle tracer. Analysis of Kīlauea Iki whole rocks found that lithium isotopes are not fractionated by crystallization of olivine, plagioclase, clinopyroxene, and Fe-Ti minerals [*Tomascak et al.*, 1999]. Subsequent studies have, however, found a significant isotopic disequilibrium between olivine and clinopyroxene phenocrysts and the groundmass of basaltic rocks [*Seitz et al.*, 2004; *Beck et al.*, 2006; *Magna et al.*, 2006; *Jeffcoate et al.*, 2007; *Parkinson et al.*, 2007]. ^6Li diffuses ~3.35% faster than ^7Li into clinopyroxene, with the result that clinopyroxene usually records the lowest

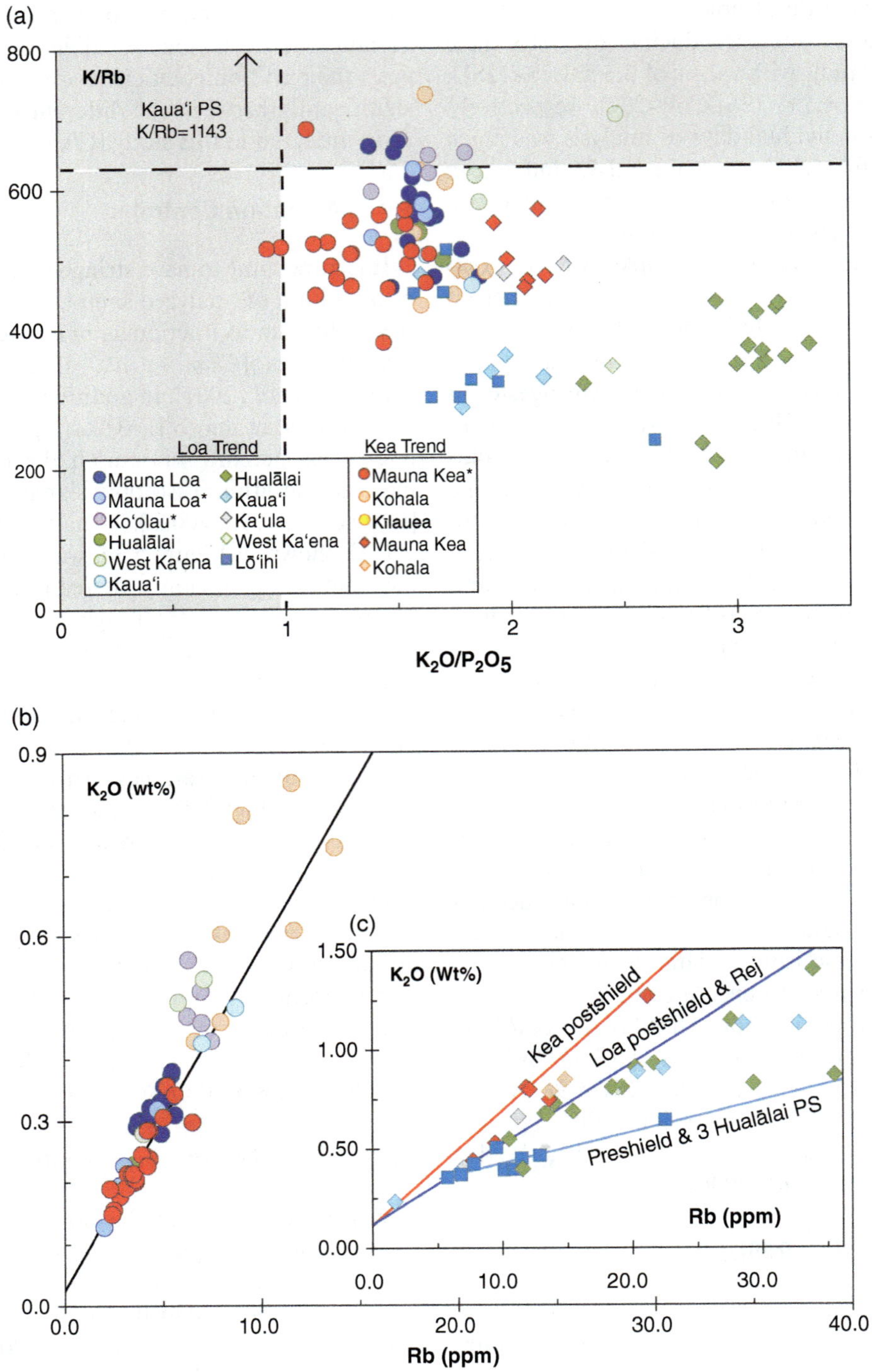

Figure 5.3 Alteration indices of Hawaiian basalts. (a) K/Rb versus K_2O/P_2O_5. Samples with anomalous δ^7Li signatures and K/Rb>750 and $K_2O/P_2O_5<0.9$ (not shown) were removed from figures and discussion. (b) Weight percent K_2O versus ppm Rb for shield lavas. Coherent trends in these diagrams indicate that variation is due to melting or crystallization processes rather than alteration, which would not have a linear signature in this plot. All shield samples fall along one trend (black line). (c) Weight percent K_2O versus ppm Rb for preshield, postshield, and rejuvenated lavas. Here the data define three separate trends: red line for Kea trend postshield lavas; dark blue line for Loa trend postshield lavas and rejuvenated lavas; light blue line for preshield Lō'ihi lavas and three Hualalai postshield lavas. Data noted in legend with an asterisk are from *Chan and Frey* [2003]. Symbols are as in Figure 5.1.

δ^7Li signatures in a rock sample [*Richter et al.*, 2003; *Coogan et al.*, 2005]. In terms of whole-rock δ^7Li measurements, isotopic disequilibrium presents a measurable problem only if clinopyroxene is fractionated *Chan et al.*, 2009; *Vlastélic et al.*, 2009].

In our sample set, none of the shield samples show correlations of lithium isotopes with MgO, indicating that crystal fractionation at low pressure does not fractionate lithium isotopes for these samples. On the other hand, some of the postshield lavas from Kohala, Mauna Kea, and Hualālai show evidence of clinopyroxene fractionation, as exhibited by the decrease of Sc below 5 wt% MgO for these samples (Figure 5.4a) [*Wolfe et al.*, 1997; *Hanano et al.*, 2010]. Clinopyroxene fractionation measurably impacts lithium isotopic signatures as seen in Ba/Y and Nb/Y versus δ^7Li plots (Figures 5.4b and 5.4c). Customarily these trace element ratios are used as a proxy for the degree of partial melting. High Ba/Y or Nb/Y indicate high degrees of partial melting as a result of decreases in residual mantle clinopyroxene abundances [*Huang et al.*, 2011a]. However, the positive slope in Figures 5.4b and 5.4c suggests that residual clinopyroxene (with preferentially lower δ^7Li compositions than the rock matrix and other phenocrysts [*Jeffcoate et al.*, 2007; *Parkinson et al.*, 2007]) left in the source mantle by undergoing less partial melting would decrease the lithium isotopic signature of the melt rather than increase it as expected. The explanation for this unexpectedly heavy lithium isotopic signature is due not to source residual clinopyroxene but instead to clinopyroxene fractionation after separation of melts from the mantle source and before eruption. As Sc is compatible in clinopyroxene but not olivine, it can be used as a proxy for samples that have fractionated out clinopyroxene (see Figure 5.4a). When samples with Sc concentrations below 10 ppm are removed from this data set, the correlation between Ba/Y, Nb/Y, CaO, and MgO versus δ^7Li for postshield samples disappears; therefore, the low-Sc samples will not be considered in the discussion on source processes below. Finally, a correlation of Kaua'i postshield lavas δ^7Li with Nb/Y, Ba/Y, and MgO is a well-documented feature of these rocks that reflects a heterogeneous mantle source rather than perturbation of lithium isotopes by nonsource processes [*Garcia et al.*, 2010]. Coincidentally, the relationship between clinopyroxene fractionation and lithium isotopic signature suggests that isotopic disequilibrium between phenocrysts and groundmass in basaltic rocks is a result of

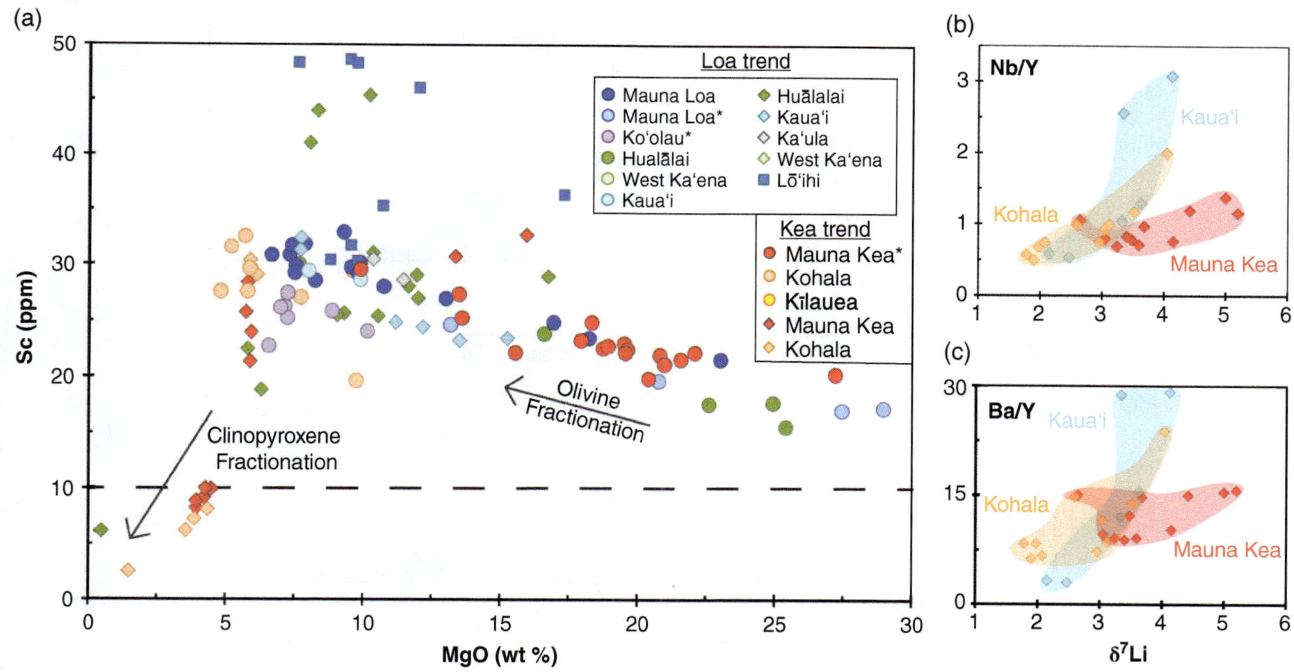

Figure 5.4 (a) Plot of Sc (ppm) versus weight percent MgO. The inverse correlation for the majority of data on this plot arises from the concentration of incompatible Sc during olivine fractionation, and the inflection point, which occurs in samples below 5 wt% MgO, reveals clinopyroxene fractionation [*Vlastélic et al.*, 2009]. Samples below 10 ppm Sc were removed from the discussion to control for the effects of clinopyroxene fractionation (as seen in Nb/Y and Ba/Y versus δ^7Li in (b) and (c)) on lithium isotopic signature. (b) Plot of Nb/Y versus δ^7Li and (c) plot of Ba/Y versus δ^7Li for postshield lavas from Kaua'i, Kohala, and Mauna Kea. Ba/Y and Nb/Y usually increase with decreasing amounts of partial melt [*Huang et al.*, 2011a]. Here the positive correlation results from clinopyroxene fractionation affecting the lithium isotopic signatures (see discussion in text). When samples with less than 10 ppm Sc are removed, the positive correlation in these plots disappears. Data noted in legend with an asterisk are from *Chan and Frey* [2003]. Symbols are as in Figure 5.1.

equilibrium or kinetic fractionation during magmatic differentiation rather than solid state diffusion of lithium.

5.3.3. Lithium Isotopic Signature

5.3.3.1. Shield Lavas

Shield stage lavas range from δ^7Li values of 1.8‰ to 5.7‰. The Loa and Kea geochemical trends are characterized by systematic differences in lithium isotopes, with Kea trend shield lavas appreciably heavier (25 of 31 samples with δ^7Li above 3.5‰) than Loa trend shield volcanics (only 19 of 36 above 3.5‰) (Figure 5.5). When known outlier volcano Ko'olau (see Ko'olau section in discussion below) is removed from consideration, the difference between Loa and Kea shields becomes more apparent (i.e., only 13 of 31 Loa trend shield samples above 3.5‰; Figure 5.5). This result is contrary to the findings of *Chan and Frey* [2003], who did not see a significant difference between Loa and Kea trend volcanoes when studying Mauna Loa, Mauna Kea, and Ko'olau shield lavas (N = 54). This discrepancy may arise from the fact that this study expanded to more volcanoes and focuses on volcanoes shown to have wide-ranging compositions [*Weis et al.*, 2011]. Loa trend volcanoes are also more heterogeneous in lithium isotopes than Kea trend volcanoes (Figure 5.5). Mauna Kea shield samples [from *Chan and Frey*, 2003] are separated into low- and

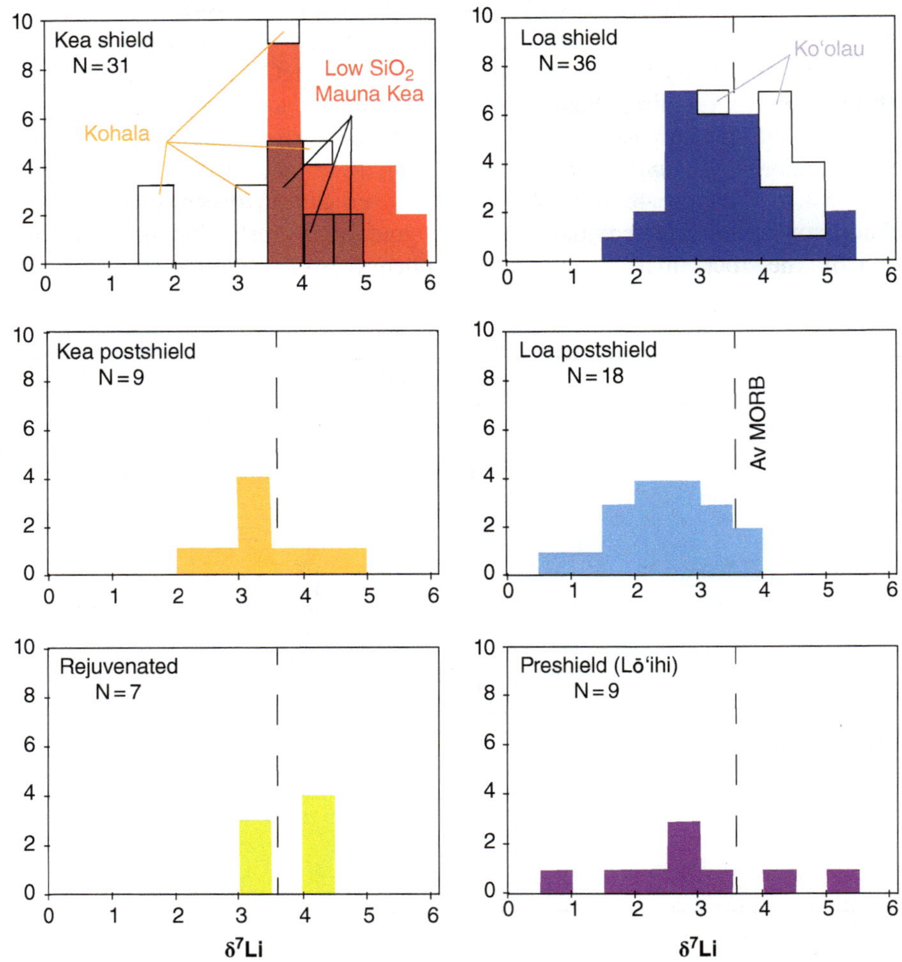

Figure 5.5 Histograms of lithium isotopic data of this study combined with Mauna Kea and the Makapu'u section of Ko'olau volcano data from *Chan and Frey* [2003]. The y axis on each plot indicates the number of samples within 0.5‰ intervals of δ^7Li variation. Kea trend shield lavas are displaced to higher average δ^7Li than Loa trend shield lavas. Loa trend shield lavas have more heterogeneous lithium isotopic signatures than Kea trend shield lavas. For both Kea and Loa trend volcanics, postshield lavas are lower in average δ^7Li than shield lavas. The dashed line in each histogram indicates the average $\delta^7\text{Li} \approx 3.7 \pm 1$‰ (1SD) of MORB [calculated from *Moriguti and Nakamura*, 1998; *Elliott et al.*, 2006; *Tomascak et al.*, 2008; *Nishio et al.*, 2007; and *Chan et al.*, 2009]. Lavas from the Makapu'u section of Ko'olau volcano are distinguished from the rest of Loa trend lavas because the heavier lithium isotopic signature of this volcano is unique from the rest of the Loa trend volcanoes (see discussion). Also note that Kohala volcano, not Mauna Kea, exhibits the lowest lithium isotopic signature of Kea trend shield lavas. Low SiO_2 Mauna Kea lavas are distinguished in dark red and are based on the classification of *Rhodes and Vollinger* [2004].

high-SiO_2 groups based on the characterization of *Rhodes and Vollinger* [2004]. Generally, higher-SiO_2 Mauna Kea shield samples also exhibit heavier lithium isotopic signatures (Figure 5.5) [*Chan and Frey*, 2003].

5.3.3.2. Postshield Lavas

Postshield samples have systematically lighter lithium isotopic signatures than shield lavas (average of all postshields: δ^7Li=2.8‰, *N*=27; average all shields: δ^7Li=3.8‰, *N*=67) (Figure 5.5). On the Island of Hawai'i, the transition from shield to postshield on Mauna Kea and Hualālai is accompanied by a lightening in lithium isotopic signature (Figures 5.7c and 5.7d). This agrees with the Pb-Pb linear trends observed by *Hanano et al.* [2010] for the shield to postshield transitions of these volcanoes with Mauna Kea and Hualālai becoming less radiogenic with time. Finally, it is important to note that postshield samples from Hualālai extend the Hawaiian lithium isotopic signature to the lightest measured whole-rock values for all oceanic islands analyzed to date and appear to be sampling an end member in lithium isotopic space (Figures 5.7b, 5.8b, and 5.9b). Hualālai postshield volcanics range from δ^7Li=0.8‰ to δ^7Li=3.5‰ (*N*=14) with almost three quarters of samples below δ^7Li=2.6‰. Hualālai postshield lavas also have among the lowest ^{206}Pb/^{204}Pb observed in Hawai'i [*Hanano et al.*, 2010; *Weis et al.*, in prep.]. It is important to note that the source of this light lithium isotopic signature is not due to clinopyroxene fractionation, as samples exhibiting signs of this process have not been included in the data set (Figure 5.4).

5.3.3.3. Rejuvenated and Preshield Lavas

Of all the volcanic stages, rejuvenated and preshield samples represent the most and least homogeneous stages, respectively, in lithium isotopic signatures. Rejuvenated samples have an average δ^7Li=3.78±0.99 (2SD, *N*=7), the most homogeneous volcanic stage sampled in this study. It is also important to note that the remarkable similarity of rejuvenated samples from Ka'ula, Kaua'i, and West Ka'ena in lithium isotopes is similar to the homogeneity of Hawaiian rejuvenated samples in general as observed in trace elements and radiogenic isotopes [*Yang et al.*, 2003; *Frey et al.*, 2005; *Fekiacova et al.*, 2007; *Garcia et al.*, 2010; *Weis et al.*, in prep.]. Preshield samples from Lō'ihi are in the range δ^7Li=0.9‰–5.5‰ (*N*=9), by far the largest variation for a volcano observed in the 88 samples analyzed.

5.3.4. Lithium Isotopes of Hawai'i Versus Global Oceanic Island Basalts

Globally, oceanic island basalts range from δ^7Li=2.5‰ in the Foundation Seamount Chain to δ^7Li=8.0‰ on Rurutu Volcano in the Cook-Austral Islands [*Vlastélic et al.*, 2009; *Krienitz et al.*, 2012]. Hawaiian data (this study), particularly from Hualālai and Lō'ihi volcanoes, extend the global range of values down to δ^7Li=0.8‰. On the Li/Y plot (Figure 5.6), it is interesting to note that most OIBs plot above the average MORB value of 0.20 ± 0.04 [*Sun and McDonough*, 1989; *Bouman et al.*, 2004; *Krienitz et al.*, 2012; *Gale et al.*, 2013] and some below. Hualālai postshield volcanics are characterized by high Li/Y, much

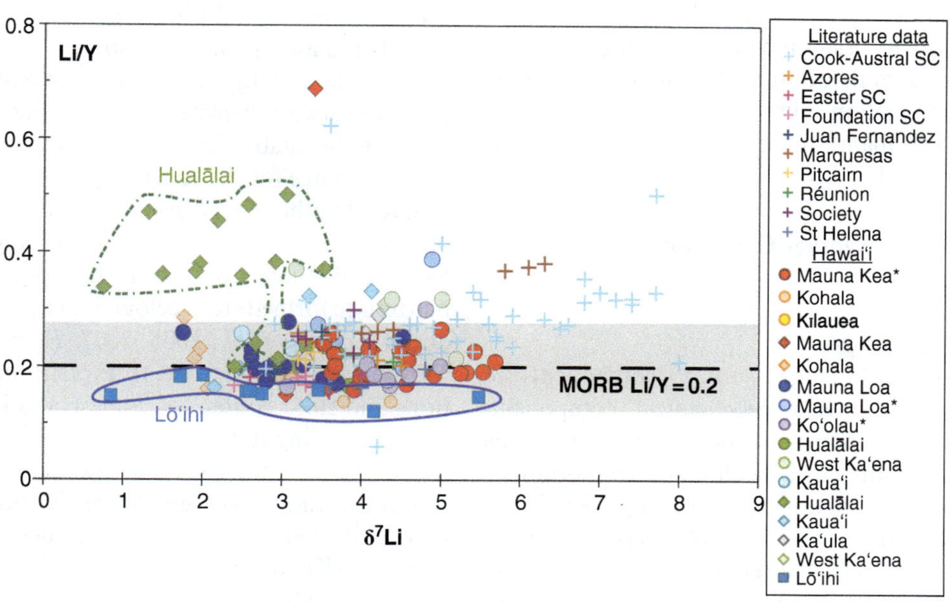

Figure 5.6 Li/Y versus δ^7Li for compiled oceanic island basalts. The heavy dark line denotes the MORB Li/Y value of 0.2 ± 0.04, which is calculated from N-MORB values of *Sun and McDonough* [1989] and *Gale et al.* [2013; *Bouman et al.*, 2004; *Krienitz et al.*, 2012]. The grey field indicates 2SD of MORB Li/Y. SC is shorthand for seamount chain. Data sources in Table 5A.3. Symbols are as in Figure 5.1.

higher than the 0.2 of MORB and systematically higher than any other oceanic island basalt. Lōʻihi samples are among the lowest Li/Y values while HIMU oceanic islands have consistently higher values.

5.4. DISCUSSION

5.4.1. Lithium Isotopic Signature in Hawaiian Lavas: A Source Signature?

The variation observed in lithium isotopes in this study is small and generally cluster around average MORB ($\delta^7Li \approx 3.7‰$ [*Moriguti and Nakamura*, 1998; *Elliott et al.*, 2006; *Tomascak et al.*, 2008; *Nishio et al.*, 2007; *Chan et al.*, 2009]). However, lithium isotopic signatures correlate systematically with other trace elements and radiogenic isotopes, implying that lithium isotopes do behave systematically enough to be used as a geochemical tracer.

Hawaiian volcanoes that are geochemically distinct in Pb-Hf-Nd-Sr isotopes are also distinct in lithium isotopes (Figures 5.7, 5.8, and 5.9; Hf isotopes not shown). Kea trend volcanoes are systematically higher in δ^7Li, $^{206}Pb/^{204}Pb$, $^{207}Pb/^{204}Pb$, $^{208}Pb/^{204}Pb$, and $^{143}Nd/^{144}Nd$ and lower in $^{87}Sr/^{86}Sr$ than Loa trend volcanoes. Because this trend holds for both shield and postshield lavas, it is more likely that variations in lithium isotopes arise from mantle heterogeneities rather than from nonlinear diffusion, melting, or crystallization processes. Although Koʻolau is distinctly a Hawaiian end-member in radiogenic isotopic systems (Pb, Hf, Nd, Sr), in lithium isotopes it seems that Hualālai and Mauna Kea form the extremes of an isotopic mixing array that can be resolved in lithium isotopes as well as Pb, Hf, Nd, and Sr isotopes. In fact, the covariation between lithium and radiogenic isotopes suggests that lithium can preserve old, low-temperature fractionation signatures throughout long residence times in the mantle (up to two billion years according to calculations by *Vlastélic et al.* [2009]).

5.4.2. Hawaiian End-Member Components in Lithium Isotopes

5.4.2.1. The Kea Component

The depleted shield-stage source component in Hawaiian lavas is the Kea geochemical component. Although the Kea component is most dominant in Kea trend volcanoes, it contributes to all Hawaiian volcanism and has been shown to be a long-lived component in the Hawaiian mantle plume (>80 Myr [*Blichert-Toft et al.*, 2003; *Regelous et al.*, 2003; *Frey et al.*, 2005; *Portnyagin et al.*, 2012; *Tanaka et al.*, 2008; *Jackson et al.*, 2012]). Although there is still debate on the lithology of the source of Hawaiian shield volcanoes, we shall first consider a strictly peridotitic source for Kea trend volcanoes. This peridotitic source can originate in the upper mantle as either entrained asthenosphere or entrained lithosphere or from the deep mantle as either plume matrix or recycled oceanic lithosphere [*Chen and Frey*, 1985; *Eiler et al.*, 1996; *Hauri*, 1996; *Lassiter et al.*, 1996; *Lassiter and Hauri*, 1998; *Weis et al.*, 2011; *Jackson et al.*, 2012; *Nobre-Silva et al.*, 2013]. The lithium budget of peridotite is controlled by olivine, a mineral that is remarkably homogeneous with respect to the lithium isotopic signature [*Seitz et al.*, 2004; *Magna et al.*, 2006; *Jeffcoate et al.*, 2007]. The range of observed lithium isotopic signature of olivine in fertile peridotitic mantle xenoliths with minor to no metasomatism or melt extracted is $\delta^7Li = 3.0‰–5.2‰$ with an average bulk composition of $\delta^7Li \approx 3.5‰$ [*Jeffcoate et al.*, 2007]. This composition is both near average MORB and a common composition for Kea trend samples (Figure 5.5). Furthermore, the range observed by *Tomascak et al.* [2008], for global MORB extends to $\delta^7Li = 5.6‰$ very near the heaviest Kea trend volcanic lithium isotopic signature of $\delta^7Li = 5.7‰$ (Mauna Kea [*Chan and Frey*, 2003]). Thus, although the average of all Kea trend volcanoes ($\delta^7Li \approx 4.3‰$) is heavier than average MORB ($\delta^7Li \approx 3.7‰$), a depleted peridotite similar to MORB may explain all observed Kea trend variation, as can fertile, previously unmelted peridotite. There is the possibility that the slightly heavier average of Kea trend volcanism relative to average MORB is due to incorporation of very small amounts of a HIMU-type mantle, which is characterized by heavier lithium isotopic signatures in the range of $\delta^7Li = 2.8‰–8.0‰$ [*Chan et al.*, 2009; *Vlastélic et al.*, 2009; *Krienitz et al.*, 2012]. However, because the Kea component lithium isotopic signature is so near average MORB and the composition of prevalent mantle (PREMA; see discussion below and Figure 5.11), it is more likely that it records the average lithium isotopic signature of peridotitic mantle. Finally, because lithium isotopes cannot distinguish between depleted, recycled, and fertile peridotite, they are unsuitable for distinguishing between a deep or shallow source for the Kea component in Hawaiʻi.

Several authors have put forth a different source model of Hawaiian shield volcanoes, including those from the Kea trend, that is a mixture of both peridotite and a pyroxenite formed from eclogitic melts reacting with peridotite [*Sobolev et al.*, 2005, 2007; *Herzberg*, 2006, 2011]. This model was proposed to explain high Ni concentrations and silica content of some Hawaiian parental magmas. According to the model, Hawaiian lavas with higher weight percent SiO_2 such as Koʻolau and Mauna Loa reflect the addition of greater proportions of this hybrid pyroxenite source than volcanoes such as Lōʻihi, Mauna Kea, and Kīlauea [*Sobolev et al.*, 2005]. As discussed in the sections above, peridotite and pyroxenite theoretically should produce different lithium isotopic signatures in the resulting melt due to lower δ^7Li in clinopyroxene [*Richter et al.*, 2003; *Jeffcoate et al.*, 2007; *Parkinson et al.*, 2007]. Thus lithium isotopes may provide some insight in this

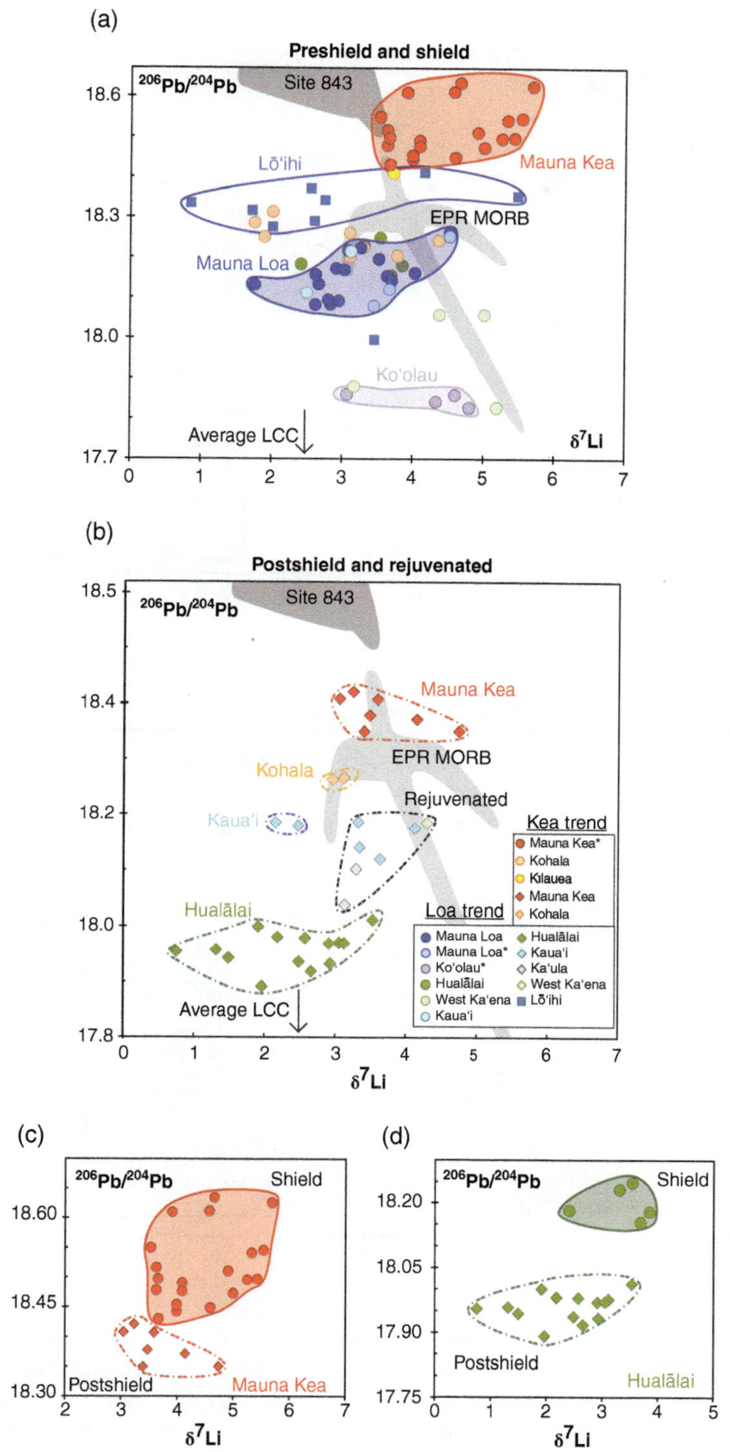

Figure 5.7 Comparison of the Pb and Li isotopic characteristics for preshield, shield, postshield, and rejuvenated volcanism on Hawai'i. (a) Shield and preshield lavas. Aside from Ko'olau and West Ka'ena (see discussion in text), Hawaiian shield lavas exhibit a positive correlation. (b) Postshield and rejuvenated lavas. Hualālai postshield samples extend to the lowest δ^7Li values on Hawai'i. For plots (a) and (b) the field for EPR MORB is compiled data from *Elliott et al.* [2006] and *Tomascak et al.* [2008]. ODP Site 843 basalts are noted by the dark grey field. Plots (c) and (d) show the transition from shield to postshield in $^{206}Pb/^{204}Pb$ versus δ^7Li for Mauna Kea and Hualālai. Note different vertical and horizontal scales for (c) and (d). The lithium isotopic composition lightens with the transition from shield to postshield volcanic stage at these individual volcanoes. Data noted in legend with an asterisk are from *Chan and Frey* [2003]. The composition of lower continental crust is indicated by the arrow and is from *Teng et al.* [2008] and *Rudnick and Goldstein* [1990]. Symbols are as in Figure 5.1.

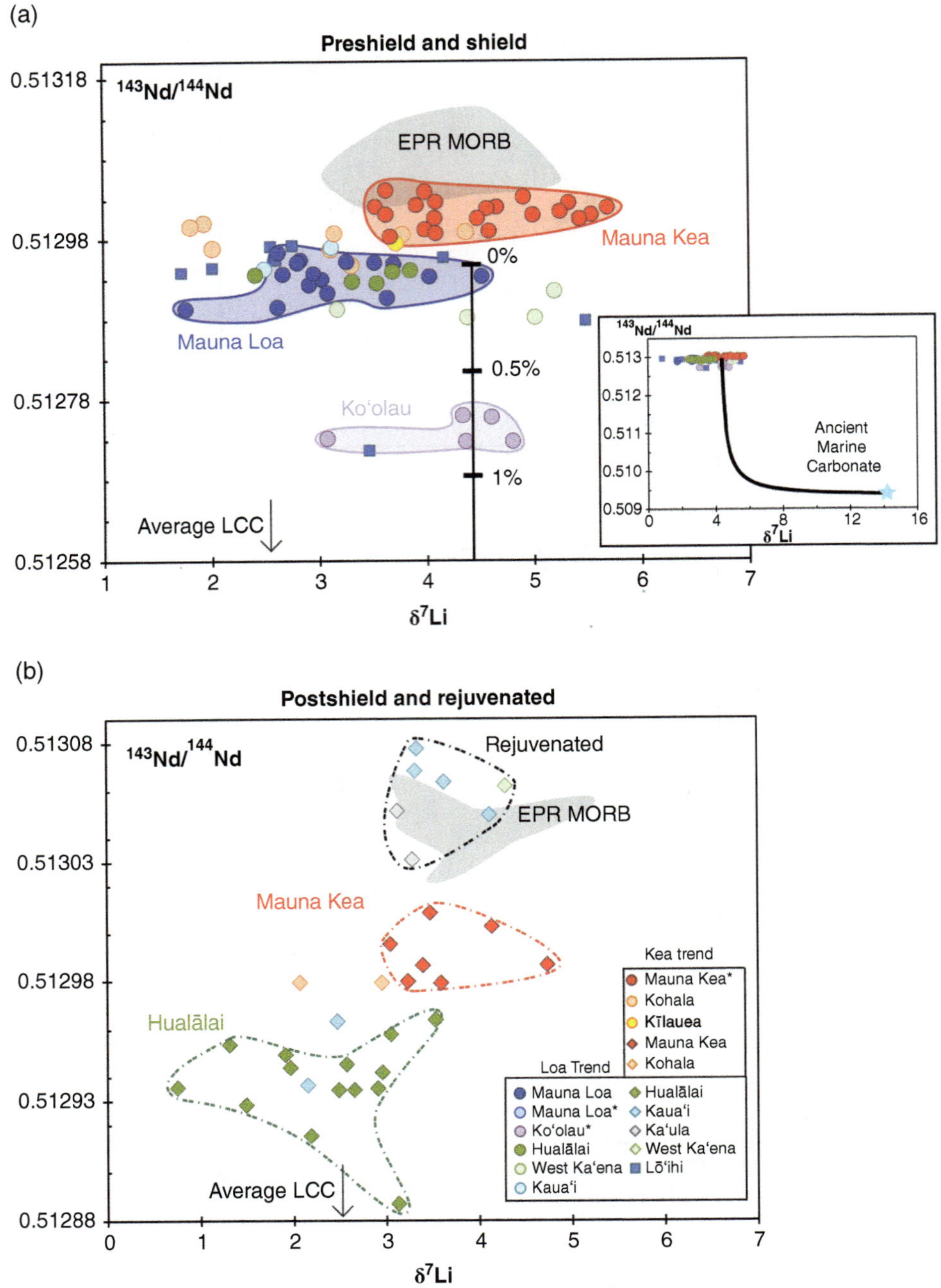

Figure 5.8 $^{143}Nd/^{144}Nd$ versus δ^7Li for (a) shield and preshield volcanics and (b) postshield and rejuvenated volcanics. Note different vertical and horizontal scales for (a) and (b). The line in (a) illustrates the mixing of an ancient marine carbonate with a plume component [*Huang et al.*, 2011a] (see Table 5.2 for model parameters). Tick marks are at intervals of 0.5%. The maximum amount of ancient marine carbonate in Ko'olau lavas based on Nd isotopes is 1%. Inset to (a) shows a larger-scaled view of the plotted data and the full mixing line to an ancient marine carbonate end member (see text for discussion). The field for EPR MORB is compiled data from *Elliott et al.* [2006] and *Tomascak et al.* [2008]. Data noted in legend with an asterisk are from *Chan and Frey* [2003]. The composition of lower continental crust is indicated by the arrow and is from *Teng et al.* [2008] and *Stosch and Lugmair* [1984]. Symbols are as in Figure 5.1.

debate. First, consider Mauna Kea, proposed to include ~45% of this hybrid pyroxenitic source [*Sobolev et al.*, 2005]. If an increase in the amount of SiO_2 in Mauna Kea lavas is due to a greater addition of a pyroxenitic melt in these lavas, they should also exhibit lower lithium isotopic signature. However low-SiO_2 Mauna Kea lavas have lower lithium isotopic signature relative to the higher-SiO_2 lavas (Figure 5.5). The lithium isotopic signature between Mauna Loa and Ko'olau also conflicts with expected results. According to *Sobolev et al.* [2005] there is

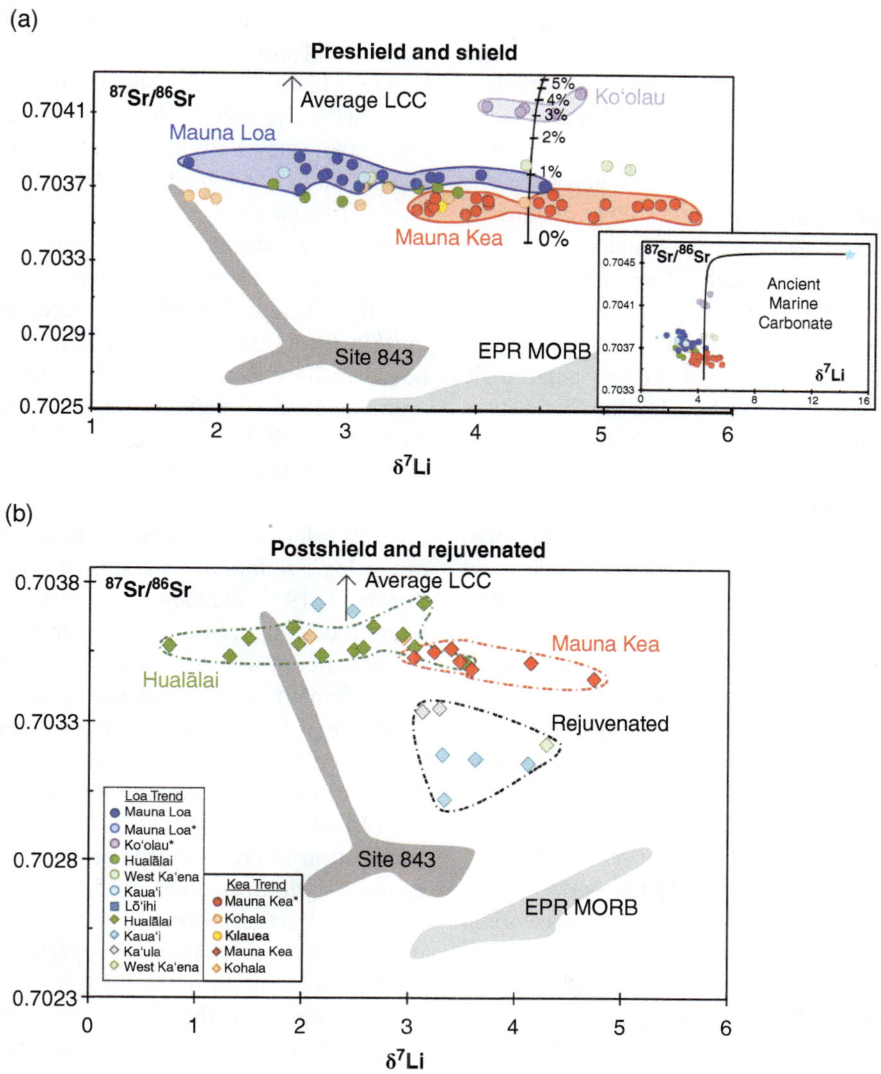

Figure 5.9 $^{87}Sr/^{86}Sr$ versus δ^7Li for (a) shield and preshield volcanics and (b) postshield and rejuvenated volcanics. Note different vertical and horizontal scales for (a) and (b). The line in (a) illustrates the mixing of an ancient marine carbonate with a plume component [*Huang et al.*, 2011a] (see Table 5.2 for model parameters). Tick marks are at intervals of 1%. The maximum amount of ancient marine carbonate in Ko'olau lavas based on Sr isotopes is 5%. Inset to (a) shows a larger-scaled view of the plotted data and the full mixing line to an ancient marine carbonate end member (see text for discussion). For plots (a) and (b) the field for EPR MORB is compiled data from *Elliott et al.* [2006] and *Tomascak et al.* [2008] and does not overlap with Hawaiian data. ODP Site 843 basalts are noted by the dark grey field. Data noted in legend with an asterisk are from *Chan and Frey* [2003]. The composition of lower continental crust is indicated by the arrow and is from *Teng et al.* [2008] and *Stosch and Lugmair* [1984]. Symbols are as in Figure 5.1.

approximately 60% hybrid pyroxenite in Mauna Loa and 80% in Ko'olau, supported by the higher weight percent SiO_2 in Ko'olau lavas. A higher percentage of pyroxenite in the source of a volcano should produce lighter lithium isotopic signatures, a relationship not observed in this study as Ko'olau exhibits the heaviest lithium isotopic signature of any Loa trend volcano (Figure 5.5). Thus lithium isotopes do not support the addition of a hybrid pyroxenite in the source of Hawaiian shield stage magmas.

5.4.2.2. Lō'ihi: Heterogeneity in Lithium Isotopes

The substantial heterogeneity observed in δ^7Li for Lō'ihi is perhaps not extraordinary when considering the overall anomalous and heterogeneous character of this volcano [*Moore et al.*, 1982; *Garcia et al.*, 1993, 1995, 1998, 2006]. Lō'ihi is a very young submarine volcano (5–102 ka [*Guillou et al.*, 1997]) and is the only Hawaiian volcano that has been extensively sampled during the preshield stage and whose chemistry may be controlled by the first melting of plume heterogeneities with different solidi [*Huang et al.*,

2005]. The initial stages of melting of a new, previously unmelted portion of the mantle source along with the lower magma production of the preshield stage (characterized by mostly alkalic lavas of a lower percentage melt) presents the opportunity to resolve smaller scale mantle heterogeneities. In the shield stage, such heterogeneities may be fully exhausted from the mantle source or may be overwhelmed by the greater percentage melts and larger magma chambers of the shield stage [*Garcia et al.*, 2010, and references therein]. This observation implies that the lithium isotopic heterogeneity observed on Lōʻihi volcano provides a unique look at the source of the very small scale heterogeneities that are not sampled during the shield stage (e.g., smaller scale than the heterogeneity observed in shield stage Mauna Loa and Kīlauea by *Marske et al.* [2007]). Conversely, the heterogeneity of Lōʻihi lithium isotopic signatures may be an artifact of its dynamic hydrothermal system and frequent summit collapses that have been shown to contaminate some Lōʻihi samples with a seawater-derived component [*Kent et al.*, 1999; *Garcia et al.*, 2006]. Incorporation of varying amounts of seawater-altered rocks into Lōʻihi's summit magma chamber would create substantial lithium isotopic heterogeneities, although this mechanism would result in heavier lithium isotopic signatures and cannot account for the extremely light lithium isotopic signatures found in some Lōʻihi samples (Figure 5.5).

5.4.2.3. Hualālai: Light Lithium End Member

Hualālai volcano is identified here and in previous work as a perplexing outlier among Loa trend volcanoes, with the lowest documented $^{206}Pb/^{204}Pb$ at a given $^{208}Pb/^{204}Pb$ for postshield lavas [*Hanano et al.*, 2010]. Hualālai also exhibits the highest Li/Y ratio of almost any oceanic island basalt (Figure 5.6). Li and Y have very similar partition coefficients in basaltic systems, so variations in these elements relative to one another are unlikely to be related to partial melting or fractional crystallization [*Ryan and Langmuir*, 1987]. Because Y is essentially a fluid-immobile element, variations in Li/Y are thus usually attributed to the geochemical behavior of lithium during subduction and, more specifically, to the differences in lithium geochemical behavior based on both types of material being subducted and depth level in the subducting package. Lithium is a fluid-mobile element that enters fluids as subducted materials are dewatered, leaving the slab both depleted in lithium and with a lighter lithium isotopic signature by up to 3‰ [*Bouman et al.*, 2004; *Marschall et al.*, 2007; *Tang et al.*, 2010]. If this process is incomplete or heterogeneous, there is the potential to introduce a seawater-lithium enriched reservoir into the deep mantle through recycling oceanic crust [*Chan et al.*, 2009; *Vlastélic et al.*, 2009; *Krienitz et al.*, 2012]. Hualālai postshield samples may sample such a source. Conversely, the low Li/Y source exhibited in lavas from Lōʻihi relative to MORB may arise from a subducted source in which much of the lithium was removed during subduction.

Hualālai postshield lavas also exhibit the most EM-I-like signature of any volcano or volcanic stage on Hawaiʻi (see Figure 5.11 below). This signature includes low $^{206}Pb/^{204}Pb$ and $^{87}Sr/^{86}Sr$ as well as high abundances of Li, Ba, and Rb and depletions of Pb relative to Ce and Nd [*Lustrino and Dallai*, 2003; *Nishio et al.*, 2004; *Willbold and Stracke*, 2006; *Hanano et al.*, 2010]. It has also been shown that ultramafic xenoliths with EM-I-type geochemical characteristics from southwestern Japan and eastern Russia also have light lithium isotopic signatures, down to $\delta^7Li \approx -17‰$ [*Nishio et al.*, 2004] (it should be noted that this value may be lower than the actual EM-I mantle source due to measurements solely on clinopyroxene in this study). Although *Nishio et al.* [2004] attributed the light lithium isotopic signature of EM-I mantle to a mixture of the upper, most highly altered layer of oceanic crust with pelagic sediments, the low values of La/Nb of Hualālai postshield lavas rule out pelagic sediment as a source of these lavas [*Blichert-Toft et al.*, 1999; *Hanano et al.*, 2010]. Instead, a mixture of altered oceanic crust and either subduction eroded or delaminated lower continental crust may account for the trace element, radiogenic isotopes, and lithium isotopic signature of Hualālai postshield volcanics (see Figures 5.7, 5.8, and 5.9) [*Willbold and Stracke*, 2006]. Lower continental crust has an average lithium isotopic composition of $\delta^7Li \approx 2.5‰$ [*Teng et al.*, 2008] that when mixed with subduction-altered oceanic crust may be light enough to explain signatures down to $\delta^7Li = 0.8‰$, as observed for Hualālai. Furthermore, because mantle lithosphere, which is difficult to separate from overlying lower continental crust in delamination scenarios, carries a distinct osmium isotopic signature that is not observed in oceanic island basalts, it is more likely that the lower continental crust in the source of Hualālai was subduction eroded rather than delaminated [*White*, 2010]. Finally, although the lithium isotopic signature of the lower continental crust ($\delta^7Li \approx 2.5‰$) [*Teng et al.*, 2008] and pelagic sediment ($\delta^7Li \approx 2.25‰$) [*Bouman et al.*, 2004] are too similar to distinguish at the current level of analytical precision, we propose that radiogenic isotopic and trace element data imply that the source of heterogeneity in Hualālai postshield samples is more likely the presence of a small amount of recycled lower continental crust rather than pelagic sediments.

5.4.2.4. Lithium Isotopes of Makapuʻu Section of Koʻolau Volcano: A Sedimentary Component

Previous authors have attributed a light lithium isotopic signature in mantle melts to melting of recycled oceanic crust and/or sediments [*Brooker et al.*, 2004; *Tomascak*, 2004; *Elliott et al.*, 2004, 2006; *Kobayashi et al.*, 2004; *Tang et al.*, 2010; *Krienitz et al.*, 2012]. A recycled sedimentary source has been proposed for some Hawaiian lavas based on evidence from hafnium, osmium, lead, calcium, and oxygen isotopes as well as trace and major elements [*Frey et al.*, 1994; *Roden et al.*, 1994; *Eiler et al.*, 1996; *Lassiter and Hauri*, 1998; *Blichert-Toft et al.*, 1999; *Jackson et al.*, 1999, 2012; *Huang and Frey*, 2005; *Huang et al.*, 2009, 2011b]. Recycled altered oceanic crust is also cited as a source of mantle heterogeneity for Hawai'i, oceanic island basalts, and MORB [*Zindler and Hart*, 1986; *Rehkämper, and Hofmann*, 1997; *Hofmann*, 2003; *Salters et al.*, 2006; *Jackson et al.*, 2012; *Gale et al.*, 2013]. The average lithium isotopic signatures for shield lavas from Loa trend volcanoes correlate with geochemical indices of sediment input, i.e., La/Nb and Sr/Nb (Figure 5.10) [*Huang and Frey*, 2005]. The volcanoes at either end of this mixing array are Ko'olau and Lō'ihi with Hualālai and Mauna Loa as intermediates (a trend also observed for these volcanoes in Pb-Hf-Nd-Sr isotope systematics [*Frey et al.*, 1994; *Tanaka et al.*, 2008]). Kea trend volcanoes plot off of these trends, suggesting Kea trend volcanoes do not have a recycled sedimentary input

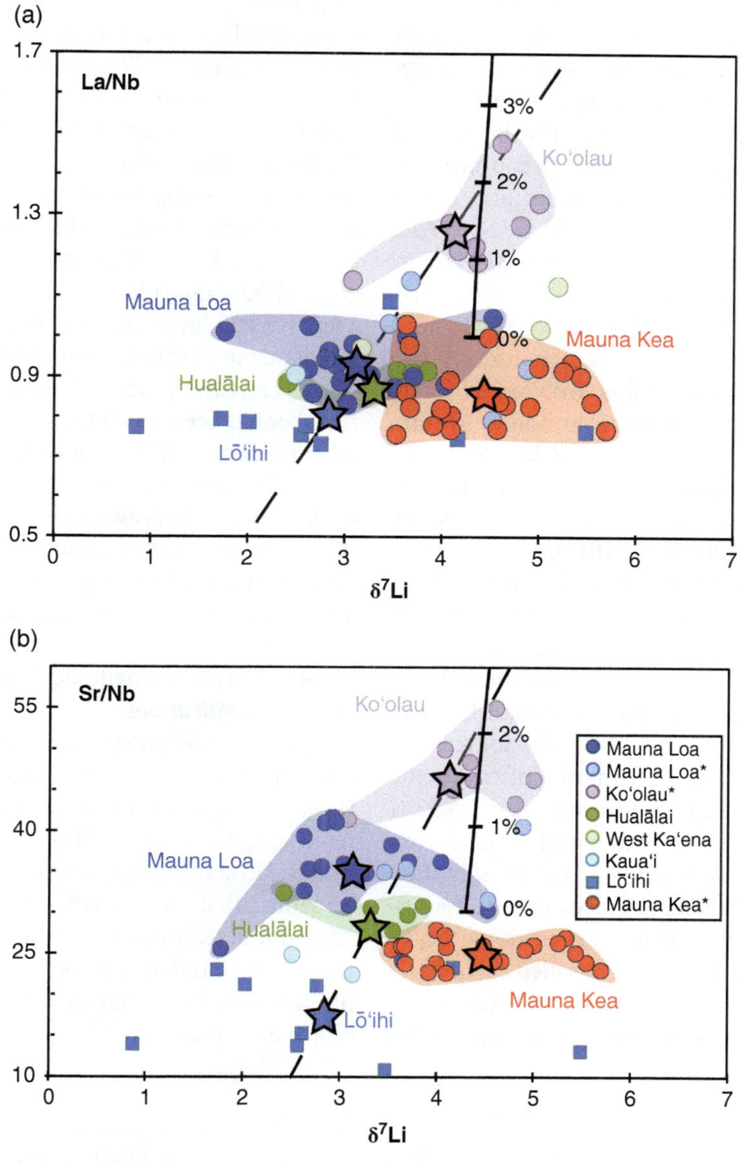

Figure 5.10 Plots of (a) La/Nb and (b) Sr/Nb versus δ^7Li showing shield data from Mauna Kea, Hualālai, Mauna Loa, Kaua'i, Lō'ihi, Ko'olau, and West Ka'ena. Averages of each volcano are marked with a black-outlined star of the same color. The dashed lines show linear regressions of Loa trend volcano averages. There are correlations on each of these plots between average Lō'ihi at low La/Nb and Sr/Nb and Ko'olau at high La/Nb and Sr/Nb. Averages for Mauna Loa and Hualālai plot between Ko'olau and Lō'ihi. Mauna Kea plots entirely off of the trend and is thus not affected by this mixing relationship. The solid black line in each plot illustrates the mixing of an ancient marine carbonate with a plume component [*Huang et al.*, 2011a] (see Table 5.2 for model parameters). Tick marks are at intervals of 1%. The maximum amount of marine carbonate in the Makapu'u section of Ko'olau based on La/Nb and Sr/Nb is 3%–4%. Data noted in legend with an asterisk are from *Chan and Frey* [2003]. Symbols are as in Figure 5.1.

[*Jackson et al.*, 2012] and instead samples the average deep Pacific mantle [*Nobre-Silva et al.*, 2013].

Based on Figure 5.10, the lithium isotopic signature of Loa trend shield lavas plots on a mixing array with Lō'ihi and Hualālai comprising the lightest lithium isotopic signature and Ko'olau the heaviest. We believe this reflects a continuum that reflects the heterogeneity of the Loa trend source between the recycling of a carbonate sedimentary input and a lower crustal input. In several papers, *Huang and Frey* [2005] and *Huang et al.* [2009, 2011a] have argued for the presence of ancient recycled carbonate in the source of Ko'olau Volcano based on major and trace element abundances as well as radiogenic and calcium isotopic signatures of these volcanoes. Lithium isotopes seem to agree with this hypothesis for several reasons. The lavas from the Makapu'u section of Ko'olau Volcano, an end-member volcano in many trace element, radiogenic and stable isotopic plots [e.g. *Lassiter and Hauri*, 1998; *Tanaka et al.*, 2002, 2008, *Fekiacova et al.*, 2007; *Huang et al.*, 2011a; *Jackson et al.*, 2012; *Teng et al.*, 2013], plot consistently off-trend in Sr, Nd, and Pb isotopic plots versus lithium because of their heavier lithium isotopic signature relative to other Loa trend volcanoes. Thus Ko'olau is likely sampling an end member that is not common in Mauna Loa, Hualālai, and Kaua'i (Figures 5.7, 5.8, and 5.9), also supported by the much steeper slope of Ko'olau relative to every other Hawaiian volcano in Pb isotopic space [*Abouchami et al.*, 2005; *Weis et al.*, 2011]. Furthermore, pelagic sediments have extremely high lithium abundances (up to 57.3 ppm) and lithium isotopic signatures between $\delta^7Li = 1.7‰$ and $\delta^7Li = 4.8‰$ with an average signature of 2.25‰ [*Bouman et al.*, 2004]. Thus the relatively heavy lithium isotopic signature of the Makapu'u section of Ko'olau volcano precludes pelagic sediment as a major source.

In contrast, the lithium isotopic signature of carbonate sediments varies between $\delta^7Li = 6.4‰$ and $\delta^7Li = 23.33‰$ and is controlled by both equilibrium with the fluids they are in contact with (usually a heavy seawater signature ≈32‰) and sedimentary diagenetic processes (e.g., recrystallization of silica and carbonate in the presence of heavy seawater) [*Chan et al.*, 2006]. Ancient recycled carbonates have higher $^{87}Sr/^{86}Sr$ [*Huang and Frey*, 2005; *Ray et al.*, 2002], and marine carbonates in general have high Sr/Nb and low Th/La [*Plank and Langmuir*, 1998], characteristics also exhibited by Ko'olau lavas. The low Lu/Hf values of Ko'olau relative to other Loa trend volcanoes has been used to argue against a modern carbonate source, where the lower concentrations of Hf relative to Lu drive the trace element ratios to slightly higher values. However, ancient carbonates may have had a lower Lu/Hf than modern carbonates. In pre-Phanerozoic time, a high-rare earth element (REE) scavenging phosphate component was likely absent from subducting carbonate-rich sediments due to the lack of abundant fish and shell debris in oceanic sediments before the Cambrian Explosion [*Plank and Langmuir*, 1998; *Willbold and Stracke*, 2006]. Contrary to *Huang and Frey* [2005], we propose a largely phosphate-poor carbonate component in the source of Makapu'u section Ko'olau lavas. Based on a simple two-component mixing model [*Langmuir et al.*, 1978; *Huang et al.*, 2011a] (see model input parameters in Table 5.2), we calculate roughly 1%–5% carbonate in the Makapu'u shield stage of Ko'olau volcano. It is interesting to note that some West Ka'ena lavas are sampling a similar end member as Ko'olau, but in smaller proportions (Figures 5.7a, 5.8a, and 5.9a). Thus the proposed ancient recycled carbonate end member has been present in the Loa trend Hawaiian mantle from at least West Ka'ena to Ko'olau, roughly 1–3 million years [*Doell and Dalrymple*, 1973; *Greene et al.*, 2010; *Huang et al.*, 2011a]. The presence of recycled material in the source of Loa trend volcanoes and the absence of this material in Kea trend volcanoes may be the main reasons behind Loa and Kea trend differences in lithium isotopes.

5.4.2.5. Lithium Isotopic Heterogeneity in Loa Trend Volcanoes

Loa trend volcanoes are considerably more heterogeneous in lithium isotopes than Kea trend volcanoes (Figure 5.5), as also observed in radiogenic isotopic systems [e.g., *Weis et al.*, 2011]. We have also proposed above that the Loa trend volcanic source contains recycled components that are carbonate-rich sediments in the Makapu'u section of Ko'olau volcano and lower continental crust and altered oceanic crust rich (with a lighter lithium isotopic signature) in Hualālai. The greater heterogeneity observed in Loa trend volcanoes and especially

Table 5.2 Geochemical model parameters for Figures 5.8, 5.9, and 5.10.

	δ^7Li	Li (ppm)	$^{143}Nd/^{144}Nd$	Nd (ppm)	$^{87}Sr/^{86}Sr$	Sr (ppm)	La (ppm)	Nb (ppm)	Th (ppm)
Hawaiian plume	4.4[a]	1.6[c]	0.512946[a]	1.354[c]	0.7034[e]	18.21[e]	0.687[c]	0.713[c]	0.085[c]
Ancient carbonate	14.5[b]	0.8[d]	0.50938[d]	10.13[d]	0.7046[e]	800[e]	14.22[d]	0.79[d]	0.29[d]

[a]Average of all Mauna Loa and Mauna Kea shield stage analyses of samples in this study.
[b]Heavy end-member carbonate of *Chan et al.* [2006], chosen because subducted carbonates have likely undergone diagenetic recrystallization, a process found to result in heavier lithium isotopic signatures.
[c]Primitive mantle values from *Sun and McDonough* [1989].
[d]Guatemala carbonate (Deep Sea Drilling Program (DSDP) Site 495) of *Plank and Langmuir* [1998].
[e]*Huang et al.* [2011a].

Lōʻihi is therefore likely due to the fact that the lithium isotopic signature of subducting materials is incredibly heterogeneous. For example, the range of lithium isotopic signatures in altered oceanic crust is large (δ^7Li = −2.29‰–13.81‰) [*Gao et al.*, 2012] and highly dependent on both the water rock ratio (which changes with depth in the crust) and the highly channelized and flow-dependent nature of hydrothermal systems. In addition, there is a large range in the possible lithium isotopic signature of subducting sediment that may have a large impact on recycled lithium signatures. The four main types of sediment (carbonate-rich, pelagic clay and ooze-rich, terrigeneous, and volcaniclastic) all have very different lithium isotopic signatures (δ^7Li values of 13.0‰–23.3‰, 1.3‰–14.5‰, −1.7‰–2.5‰, and ≈ 6‰, respectively) [*Bouman et al.*, 2004]. The type of sediment recycled, different dehydration behaviors and processes, and the spatial locations and mixing of each type of oceanic crust and/or sediment in the source could impact the sampled heterogeneity of oceanic island basalts. Finally, the lower mantle near the core-mantle boundary, described by some as a "slab graveyard," may be sampling many different types of materials subducted over presumably millions of years, reflecting the long-term evolution of elemental and isotopic signatures of continental crust, seawater, and biogenic activity [*Plank and Langmuir*, 1998; *Vervoort and Blichert-Toft*, 1999; *Ray et al.*, 2002; *Valley et al.*, 2005]. The lithium isotopic signature of such a mélange of material is surely not uniform [*Stracke et al.*, 2003], and the heterogeneity observed in the Loa trend volcanoes in Hawaiʻi possibly reflects this.

5.4.3. Why Are Postshield Lavas Lighter in Lithium Isotopic Signature Than Shield Lavas?

The transition from shield to postshield stage in Hawaiian volcanoes is accompanied by changes in lava alkalinity, trace elements, and isotopic ratios [*Feigenson et al.*, 1983; *Clague and Dalrymple*, 1987; *Frey et al.*, 1990; *Lassiter et al.*, 1996; *Xu et al.*, 2005, 2007; *Hanano et al.*, 2010], a relationship that seems to hold true in lithium isotopes. Because the trace element abundances of shield and postshield lavas typically fall along common fractionation lines [see *Garcia et al.*, 2010, Figure 4, and *Weis et al.*, in prep.], the changes in lava geochemistry most likely stem from changes in degree of partial melt, magma supply, and depth of melt segregation rather than a wholesale change in source as is required for rejuvenated lavas [*Frey et al.*, 1990; *Garcia et al.*, 2010; *Hanano et al.*, 2010; *Weis et al.*, in prep.]. *Jeffcoate et al.* [2007] found that the lithium isotopic signatures of clinopyroxene in fertile peridotite xenoliths were heavier than clinopyroxene in more depleted xenoliths. In addition, ^{7}Li will preferentially partition into the mobile fluid phase in any system [*Chan et al.*, 1992; *Tomascak*, 2004, and references therein; *Chan et al.*, 2006]. Thus any previous depletion of the postshield source rock by shield-stage melting should systematically lower the lithium isotopic signature of postshield magmas. This effect is quite small, however, as no fractionation is observed between olivine and melt and the fractionation factor between clinopyroxene and olivine is on the order of 0.9985 [*Tomascak et al.*, 1999; *Chan and Frey*, 2003; *Jeffcoate et al.*, 2007]. Consequently, variable source depletion cannot fully explain the observed reduction in lithium isotopic signatures of postshield lavas. However, the lightening of lithium isotopic signal could result from the lower degrees of partial melting that characterize postshield volcanism relative to shield volcanism. Pyroxenite has a lower solidus than peridotite and will thus be the first mantle lithology to melt [*Hirschmann and Stolper*, 1996; *Pertermann and Hirschmann*, 2003; *Lambart et al.*, 2012]. If any pyroxenite was not fully melted during the shield stage, or exist as vertical "stringers" of heterogeneity in the Hawaiian mantle plume that are consistently migrating into the plume melting zone [*Blichert-Toft et al.*, 2003; *Marske et al.*, 2007; *Hanano et al.*, 2010], then the smaller degrees of melting associated with postshield volcanism would result in a higher proportion of the clinopyroxene-rich source in the melt. Higher modal melting of clinopyroxene, a mineral with a preferentially light lithium isotopic signal, would result in a lighter lithium isotopic signature in the resulting erupted whole rock [*Seitz et al.*, 2004; *Magna et al.*, 2006; *Jeffcoate et al.*, 2007]. This effect may be magnified by the higher concentrations of lithium in mantle pyroxenite relative to fertile peridotite [*Seitz and Woodland*, 2000]. The ≈32% higher lithium concentrations of postshield relative to shield lavas support the possibility of a greater role for pyroxenite source in postshield lavas. The signature of the transition from shield to postshield in lithium isotopes remains ambiguous, but preferential partial melting of clinopyroxene accompanied by smaller melt fractions would lighten the lithium isotopic signatures of postshield lavas.

5.4.4. Is MORB-Related Lithosphere or Asthenosphere Assimilated Into the Hawaiian Plume?

The proposed participation of oceanic lithosphere and crust in Hawaiian volcanism is a topic of continued discussion [*Chen and Frey*, 1985; *Gaffney et al.*, 2004; *Frey et al.*, 2005; *Fekiacova et al.*, 2007; *Garcia et al.*, 2010; *Hanano et al.*, 2010]. In Nd and Pb isotopes, most rejuvenated lavas overlap with East Pacific Rise lavas (Figures 5.8b, and 5.9b), although binary mixing trends among Hawaiian lavas do not intersect MORB fields [*Abouchami et al.*, 2000; *Fekiacova et al.*, 2007; *Hanano et al.*, 2010]. There are several lines of evidence indicating that the oceanic crust and lithosphere are not entrained and melted to form Hawaiian basalts. First, rejuvenated lavas do not overlap with MORB in Sr isotopes (Figure 5.9a). Second, ODP Site 843 lavas, which are more representative of low-temperature altered

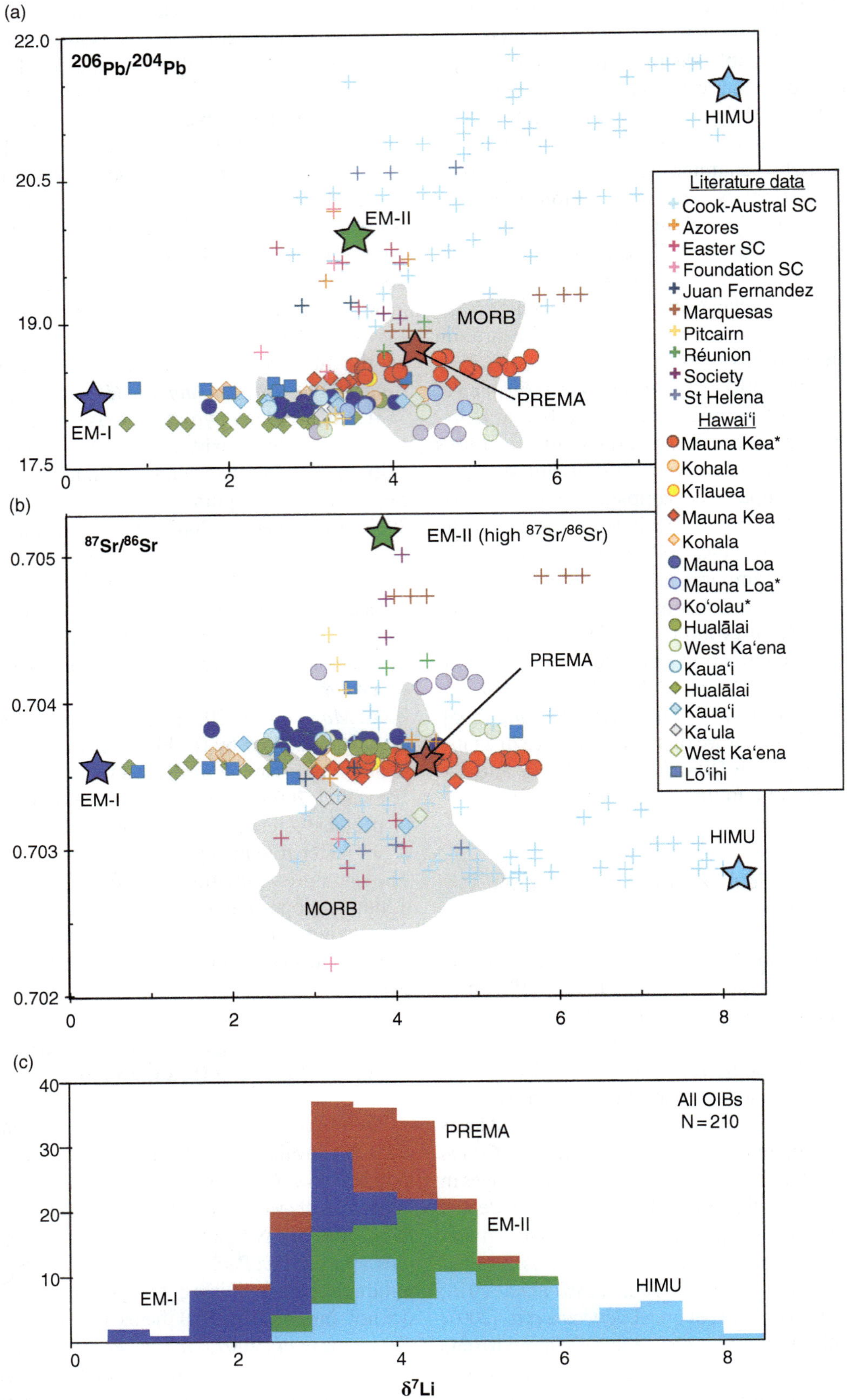

Figure 5.11 Cont'd

oceanic crust that may be entrained and melted in Hawai'i than fresh MORB, do not overlap with the Hawaiian shield or postshield data in Pb, Sr, or Li isotopes (Figures 5.7 and 5.9), and do not follow the same trend in Nd isotopes (Figure 5.8). Although a detailed analysis of the sources of rejuvenated volcanism is beyond the scope of this chapter, lithium isotopes indicate there is not a significant proportion of "young" oceanic crust and/or lithosphere in the source material of any stage of Hawaiian volcanism.

5.4.5. Global Lithium Systematics

In the lead-lithium isotopic space (Figure 5.11a), there is a general positive trend for all oceanic island basalts that have been analyzed for lithium isotopes. The continuity of this trend among data from various sources and for oceanic islands from all oceans (and versus strontium isotopes in Figure 5.11b) indicates mantlewide coherent behavior of lithium isotopes. Although not as neatly constrained as with radiogenic isotopes, data frequency is now sufficient to identify and classify traditional mantle reservoirs in lithium isotopic space. The mantle end members EM-I, EM-II, HIMU, and PREMA [*Zindler and Hart*, 1986; *Hofmann*, 2003] show defined fields (Figures 5.11a, 5.11b, and 5.11c). PREMA is the hub of variations with δ^7Li ≈ 4.5‰, and this composition also corresponds to the Kea end member (see *Nobre-Silva et al.* [2013] for a discussion of a Kea-like prevalent mantle). HIMU is characterized by a lithium isotopic signature δ^7Li in the range of ≈ 2.5‰–8.5‰, EM-II of ≈ 2.5‰–6.0‰, and EM-I of ≈0.5‰–4.5‰ (Figure 5.11c). *Krienitz et al.* [2012] also found that large-scale OIB mantle reservoirs (HIMU, EMI, EMII [*Zindler and Hart*, 1986]) are characterized by unique lithium isotopic signatures. Our lithium isotopic signatures of mantle end members are within the range described by Krienitz et al. with the exception that Hualālai extends the EM-I mantle source down to δ^7Li ≈ 0.5‰ and the HIMU mantle source shows a greater range due to the inclusion of lithium isotopic data from the Cook-Austral seamount chain [*Chan et al.*, 2009; *Vlastélic et al.*, 2009].

5.5. CONCLUSION

The application of lithium isotopes to Hawai'i provides an additional perspective into the source components of the different stages of Hawaiian volcanism that is not offered by radiogenic isotopes. Our study of Hawaiian lavas using this isotopic system has led to the following interpretations:

1. Correlations of lithium isotopes with Pb, Hf, Nd, and Sr isotopes and trace elements indicate that lithium can be used as a source tracer if the data are filtered to remove the effects of fractional crystallization of clinopyroxene and any postmagmatic alteration.
2. Lithium isotopic signatures of 88 Hawaiian basalts extend the range observed to date in Hawai'i in this isotopic system, varying from δ^7Li = 0.8‰–5.7‰. This is also the first study that has identified a difference between the Kea and Loa geochemical trends in lithium isotopes, reflecting the presence of different source components for each of these trends.
3. Postshield lavas are generally lighter than shield stage lavas in lithium isotopes, a result of smaller degrees of partial melting of a clinopyroxene-bearing source, which has a lighter lithium isotopic signature. This may suggest that pyroxenite heterogeneities in the Hawaiian mantle plume exist as vertical stringers rather than discrete blobs.
4. The lithium isotopic signature near PREMA of Kea trend lavas suggests that the source of this component is similar to normal peridotitic mantle.
5. The enriched end member in radiogenic isotopic systems and trace elements is Makapu'u Ko'olau, while the light end member in lithium isotopic space is Hualālai. The unique heavy lithium isotopic signature of Ko'olau may be explained by incorporation of 1%–5% ancient carbonate in the source.
6. The cause of the heterogeneity observed in lithium isotopes for Loa trend volcanoes is not directly resolvable due to the extreme variation and overlap in the lithium isotopic signature of different possible types of recycled material (i.e., marine sediments, oceanic crust, lower continental crust) as well as uncertainties in how homogeneous subduction dewatering processes and mantle mixing of subducted

Figure 5.11 Cont'd. A compilation of oceanic island basalts with lithium isotopic data versus (a) $^{206}Pb/^{204}Pb$ and (b) $^{87}Sr/^{86}Sr$ and a histogram of all OIB lithium isotopic data (c). The MORB field is compiled data from *Moriguti and Nakamura* [1998], *Elliott et al.* [2006], *Tomascak et al.* [2008] *Nishio et al.* [2007], and *Chan et al.* [2009]. All Pb isotopes have been normalized to a SRM 981 $^{206}Pb/^{204}Pb$ value of 16.9405 and all Sr isotopes have been normalized to a SRM 987 $^{87}Sr/^{86}Sr$ value of 0.710248 [*Weis et al.*, 2011]. Symbols for Hawaiian data from this study and *Chan and Frey* [2003] are the same as in previous diagrams. SC is shorthand for seamount chain. Stars represent extreme compositions of the mantle end members rather than the entire range. The OIB histogram (c) includes all compiled literature data (Table 5A.3) and Hawaiian data from this study. The *y* axis on this plot indicates the number of samples within 0.5‰ intervals of δ^7Li variation. In general, HIMU is characterized by lithium isotopic signature in the range of δ^7Li ≈ 2.5‰–8.5‰, EM-II of δ^7Li ≈ 2.5‰–6.0‰, and EM-I of δ^7Li ≈ 0.5‰–4.5‰. The classification of compiled samples into groups shown in this diagram (HIMU, EM-I, EM-II, PREMA) is a general sorting only as many of the compiled samples are classified as a mix of two end members (see Table 5A.3).

material are for lithium isotopes. Thus we simply put forth that the light lithium isotopic signature observed in Hualālai postshield samples may arise from the presence of a mix of altered oceanic crust and subduction eroded lower continental crust, but a more detailed major and trace element and isotopic analysis would be necessary to fully support this.

SUPPLEMENTARY DIGITAL DATA

The following data items are available online at www.wiley.com/go/Carey/Hawaiian_Volcanoes

Appendix 5.1. Major and trace element and isotope ratios from this study and Chan and Frey, 2003.

Appendix 5.2. Literature compilation of BHVO-2 lithium isotopic signature.

Appendix 5.3. Compilation of oceanic island basalt geochemical data.

ACKNOWLEDGMENTS

Our thanks go to Jane Barling, Kathy Gordon, and Vivian Lai for assistance analyzing lithium isotopes and trace element compositions. We are also grateful to M. Garcia, M. Rhodes, and M. Feigenson for some of the samples used in this study and to the Integrated Ocean Drilling Program for providing ODP Site 843 samples. Numerous discussions with Emily Mullen contributed greatly to the manuscript, as did constructive reviews from Shichun Huang and an anonymous reviewer. This work was supported through D. Weis' NSERC Discovery Grant, a University of British Columbia Four Year Fellowship to L. Harrison, and a Postdoctoral Fellowship at the University of British Columbia to E. Barnes.

REFERENCES

Abouchami, W., S. J. G. Galer, and A. W. Hofmann (2000), High precision lead isotope systematics of lavas from the Hawaiian Scientific Drilling Project, *Chem. Geol.*, *169*, 187–209, doi:10.1016/S0009.2541(00).00328-4.

Abouchami, W., A. W. Hofmann, S. J. G. Galer, F. A. Frey, J. Eisele, and M. Feigenson (2005), Lead isotopes reveal bilateral asymmetry and vertical continuity in the Hawaiian mantle plume, *Nature*, *434*(7035), 851–856, doi:10.1038/nature03402.

Beck, P., M. Chaussidon, J. A. Barrat, P. Gillet, and M. Bohn (2006), Diffusion induced Li isotopic fractionation during the cooling of magmatic rocks: The case of pyroxene phenocrysts from nakhlite meteorites, *Geochim. Cosmochim. Acta*, *70*, 4813–4825, doi:10.1016/j.gca.2006.07.025.

Blichert-Toft, J., F. A. Frey, and F. Albarède (1999), Hf isotope evidence for pelagic sediments in the source of Hawaiian basalts, *Science*, *285*, 879–882, doi:10.1126/science.285.5429.879.

Blichert-Toft, J., D. Weis, C. Maerschalk, A. Agranier, and F. Albarède (2003), Hawaiian hot spot dynamics as inferred from the Hf and Pb isotope evolution of Mauna Kea volcano, *Geochem. Geophys. Geosys.*, *4*(2), 8704, doi:10.1029/2002GC000340.

Bouman, C., T. Elliott, and P. Z. Vroon (2004), Lithium inputs to subduction zones, *Chem. Geol.*, *212*, 59–79, doi:10.1016/j.chemgeo.2004.08.004.

Brant, C., L. A. Coogan, K. M. Gillis, W. E. Seyfried, N. J. Pester, and J. Spence (2012), Lithium and Li-isotopes in young altered upper oceanic crust from the East Pacific Rise, *Geochim. Cosmochim. Acta*, *96*, 272–293, doi:10.1016/j.gca.2012.08.025.

Brooker, R. A., R. H. James, and J. D. Blundy (2004), Trace elements and Li isotope systematics in Zabargad peridotites: Evidence of ancient subduction processes in the Red Sea mantle, *Chem. Geol.*, *212*, 179–204, doi:10.1016/j.chemgeo.2004.08.007.

Carignan, J., D. Cardinal, A. Eisenhauer, A. Galy, M. Rehkämper, F. Wombacher, and N. Vigier (2004), A reflection on Mg, Cd, Ca, Li, and Si isotopic measurements and related reference materials, *Geostand. Geoanal. Res.*, *28*(1), 139–148, doi:10.1111/j.1751-908X.2004.tb01050.x.

Carpentier, M., D. Weis, and C. Chauvel (2013), Large U loss during weathering of upper continental crust: The sedimentary record, *Chem. Geol.*, *340*, 91–104, doi:10.1016/j.chemgeo.2012.12.016.

Chan, L. H., and F. A. Frey, (2003), Lithium isotope geochemistry of the Hawaiian plume: Results from the Hawaii Scientific Drilling Project and Koolau Volcano, *Geochem. Geophys. Geosys.*, *4*(3), 8707, doi:10.1029/2002GC000365.

Chan, L. H., J. M. Edmond, G. Thompson, and K. Gillis (1992), Lithium isotopic composition of submarine basalts: Implications for the lithium cycle in the oceans, *Earth Planet. Sci. Lett.*, *108*, 151–160, doi:10.1016/0012-821X(92)90067-6.

Chan, L. H., W. P. Leeman, and C.F. You (1999), Lithium isotopic composition of Central American volcanic arc lavas: Implications for modification of subarc mantle by slab-derived fluids, *Chem. Geol.*, *160*, 255–280, doi:10.1016/S0009-2541(99)00101-1.

Chan, L. H., W. P. Leeman, and S. Tonarini (2001), Lithium isotope compositions of South Sandwich Arc and Southwest Washington Cascades: A comparative study of arc processes, *EOS Trans*, AGU, 82, F1293 (Abstract).

Chan, L. H., J. C. Alt, and D. A. H. Teagle (2002a), Lithium and lithium isotope profiles through the upper oceanic crust: A study of seawater-basalt exchange at ODP sites 504B and 896A, *Earth Planet. Sci. Lett.*, *201*, 187–201, doi:10.1016/S0012-821X(02)00707-0.

Chan, L. H., W. P. Leeman, and C. F. You (2002b), Lithium isotopic composition of Central American volcanic arc lavas: Implications for modification of subarc mantle by slab-derived fluids: Correction, *Chem. Geol.*, *182*(2–4), 293–300, doi:10.1016/S0009-2541(01)00298-4.

Chan, L. H., W. P. Leeman, and T. Plank (2006), Lithium isotopic composition of marine sediments, *Geochem. Geophys. Geosys.*, *7*(6), Q06005, doi:10.1029/2005GC001202.

Chan, L. H., J. C. Lassiter, E. H. Hauri, S. R. Hart, and J. Blusztajn (2009), Lithium isotope systematics of lavas from the Cook-Austral Islands: Constraints on the origin of HIMU mantle, *Earth Planet. Sci. Lett.*, *277*, 433–442, doi:10.1016/j.epsl.2008.11.009.

Chauvel, C., S. Bureau, and C. Poggi (2011), Comprehensive chemical and isotopic analyses of basalt and sediment reference materials, *J. Geostand. Geoanal. Res.*, *35*(1), 125–143, doi:10.1111/j.1751-908X.2010.00086.x.

Chen, C.-Y., and F. A. Frey (1985), Trace element and isotopic geochemistry of lavas from Haleakala Volcano, East Maui, Hawaii: Implications for the origin of Hawaiian basalts, *J. Geophys. Res.*, *90*(B10), 8743–8768, doi:10.1029/JB090iB10p08743.

Clague, D. A. (1987), Hawaiian xenolith populations, magma supply rates, and development of magma chambers, *Bull. Volcanol.*, *49*, 577–587, doi:10.1007/BF01079963.

Clague, D. A., and G. B. Dalrymple (1987), The Hawaiian-Emperor volcanic chain, Part 1: Geologic evolution, *in* Decker, R. W., Wright, T. L. and P. H. Stauffer, eds., *Volcanism in Hawaii, Prof. Pap. 1350, U.S. Geol. Surv.*, *1*, 5–54, Washington DC.

Coogan, L. A., S. A. Kasemann, and S. Chakraborty (2005), Rates of hydrothermal cooling of new oceanic upper crust derived from lithium-geospeedometry, *Earth Planet. Sci. Lett.*, *240*(2), 415–424, doi:10.1016/j.epsl.2005.09.020.

Coplen, T. B. (1995), Atomic weights of the elements, *Pure Appl. Chem.*, *68*, 2339–2359.

Doell, R. R., and G. B. Dalrymple (1973), Potassium-argon ages and paleomagnetism of the Waianae and Koolau volcanic series, Oahu, Hawaii, *Geol. Soc. Am. Bull.*, *8*, 1217–1242, doi:10.1130/0016-7606(1973)84<1217:PAAPOT>2.0.CO;2.

Eiler, J. M., K. A. Farley, J. W. Valley, A. W. Hofmann, and E. M. Stolper (1996), Oxygen isotope constraints on the sources of Hawaiian volcanism, *Earth Planet. Sci. Lett.*, *144*, 453–468, doi:10.1016/S0012-821X(96)00170-7.

Elliott, T., A. Jeffcoate, and C. Bouman (2004), The terrestrial Li isotope cycle: Light-weight constraints on mantle convection, *Earth Planet. Sci. Lett.*, *220*(3–4), 231–245, doi:10.1016/S0012-821X(04)00096-2.

Elliott, T., A. Thomas, A. Jeffcoate, and Y. L. Niu (2006), Lithium isotope evidence for subduction-enriched mantle in the source of mid-ocean-ridge-basalts, *Nature*, *443*, 565–568, doi:10.1038/nature05144.

Feigenson, M. D., A. W. Hofmann, and F. J. Spera (1983), Case studies on the origin of basalt: II The transition from tholeiitic to alkalic volcanism on Kohala volcano, Hawaii, *Contrib. Mineral. Petrol.*, *84*, 390–405.

Fekiacova, Z., W. Abouchami, S. J. G. Galer, M. O. Garcia, and A.W. Hofmann (2007), Origin and temporal evolution of Ko'olau Volcano, Hawai'i: Inferences from isotope data on the Ko'olau scientific drilling project (KSDP), the Honolulu volcanics and ODP Site 843, *Earth Planet. Sci. Lett.*, *261*, 65–83, doi:10.1016/j.epsl.2007.06.005.

Frey, F. A., W. S. Wise, M. O. Garcia, H. West, S.-T. Kwon, and A. Kennedy (1990), Evolution of Mauna Kea volcano, Hawaii: Petrologic and geochemical constraints on post-shield volcanism, *J. Geophys. Res.*, *95*(B2), 1271–1300, doi:10.1029/JB095iB02p01271.

Frey, F. A., M. O. Garcia, and M. F. Roden (1994), Geochemical characteristics of Koolau Volcano: Implications of intershield geochemical differences among Hawaiian volcanoes, *Geochim. Cosmochim. Acta*, *58*(5), 1441–1462, doi:10.1016/0016-7037(94)90548-7.

Frey, F. A., S. Huang, J. Blichert-Toft, M. Regelous, and M. Boyet (2005), Origin of depleted components in basalt related to the Hawaiian hot spot: Evidence from isotopic and incompatible element ratios, *Geochem. Geophys. Geosys.*, *6*(1), Q02L07, doi:10.1029/2004GC00757.

Gaffney, A., B. K. Nelson, and J. Blichert-Toft (2004), Geochemical constraints on the role of oceanic lithosphere in intra-volcano heterogeneity at West Maui, Hawaii, *J. Petrol.*, *45*(8), 1663–1687, doi:10.1093/petrology/egh029.

Gale, A., C. A. Dalton, C. H. Langmuir, Y. Sum, and J.-G. Schilling (2013), The mean composition of ocean ridge basalts, *Geochem. Geophys. Geosys.*, *14*, 489–518, doi:10.1029/2012GC004334.

Gao, Y., et al. (2012), Downhole variation of lithium and oxygen isotopic compositions of oceanic crust at East Pacific Rise, ODP site 1256, *Geochem. Geophys. Geosys.*, *13*(10), Q10001, doi:10.1029/2012GC004207.

Garcia, M. O., D. W. Muenow, and K. E. Aggrey (1989), Major element, volatile, and stable isotope geochemistry of Hawaiian submarine tholeiitic glasses, *J. Geophys. Res.*, *94*(B8), 10525–10538, doi: 10.1029/JB094iB08p10525.

Garcia, M. O., B. A. Jorgenson, J. J. Mahoney, E. Ito, and A. J. Irving (1993), An evaluation of temporal geochemical evolution of Loihi Summit Lavas: Results from Alvin Submersible Dives, *J. Geophys. Res.*, *98*(B1), 537–550, doi:10.1029/92JB01707.

Garcia, M. O., D. J. P. Foss, H. B. West, and J. J. Mahoney (1995), Geochemical and isotopic evolution of Loihi Volcano, Hawaii, *J. Petrol.*, *36*(6), 1647–1671.

Garcia, M. O., K. H. Rubin, M. D. Norman, J. M. Rhodes, D. W. Graham, D. W. Muenow, and K. Spencer (1998), Petrology and geochronology of basalt breccia from the 1996 earthquake swarm of Loihi seamount, Hawaii: Magmatic history of its 1996 eruption, *Bull. Volcanol.*, *59*, 577–592, doi:10.1007/s004450050211.

Garcia, M., J. Caplan-Auerbach, E. H. De Carlo, M. D. Kurz, and N. Becker (2006), Geology, geochemistry and earthquake history of Lo'ihi Seamount, Hawai'i's youngest volcano, *Chem. Erde*, *66*, 81–108.

Garcia, M. O., L. Swinnard, D. Weis, A. R. Greene, T. Tagami, H. Sano, and C.E. Gandy (2010), Petrology, geochemistry, and geochronology of Kaua'i lavas over 4.5 Myr: Implications for the origin of rejuvenated volcanism and the evolution of the Hawaiian Plume, *J. Petrol.*, *51(7)*, 1507–1540, doi:10.1093/petrology/egq027.

Greene, A., M. O. Garcia, D. Weis, G. Ito, M. Kuga, J. Robinson, and S. Yamasaki (2010), Low-productivity Hawaiian volcanism between Kaua'i and O'ahu, *Geochem. Geophys. Geosys.*, *11*, Q0AC08, doi:10.1029/2010GC003233.

Guillou, H., M. O. Garcia, and L. Turpin (1997), Unspiked K-Ar dating of young volcanic rocks from Loihi and Pitcairn hot spot seamounts, *J. Volcanol. Geotherm. Res.*, *78*, 239–249, doi:10.1016/S0377-0273(97)00012-7.

Halama, R., W. F. McDonough, R. L. Rudnik, and K. Bell (2008), Tracking the lithium isotopic evolution of the mantle using carbonatites, *Earth Planet. Sci. Lett.*, *265*, 726–742, doi:10.1016/j.epsl.2007.11.007.

Hanano, D., D. Weis, J. S. Scoates, S. Aciego, and D. J. Depaolo (2010), Horizontal and vertical zoning of heterogeneities in the Hawaiian mantle plume from the geochemistry of consecutive postshield volcano pairs: Kohala-Mahukona and Mauna Kea–Hualalai, *Geochem. Geophys. Geosys.*, *11*(1), Q01004, doi:10.1029/2009GC002782.

Harrison, L., D. Weis, and M. O. Garcia (in prep.), What's different about Hualalai, Kohala and Loihi, *Geological Society of America Abstracts with Programs*, *46*(6), 298.

Hauri, E. H. (1996), Major-element variability in the Hawaiian mantle plume, *Nature*, *382*, 415–419, doi:10.1038/382415a0.

Herzberg, C. (2006), Petrology and thermal structure of the Hawaiian plume from Mauna Kea volcano, *Nature*, *444*, 605–609, doi:10.1038/nature05254.

Herzberg, C. (2011), Identification of source lithology in the Hawaiian and Canary Islands: Implications for origins, *J. Petrol.*, *52*(1), 113–146, doi:10.1093/petrology/egq075.

Hirschmann, M. M., and E. M. Stolper (1996), A possible role for garnet pyroxenite in the origin of the "garnet signature" in MORB, *Contrib. Mineral. Petrol.*, *124*, 185–208, doi:10.1007/s004100050184.

Hofmann, A. W. (2003), Sampling mantle heterogeneity through oceanic basalts: Isotopes and trace elements, *in* R. W. Carlson, ed., The Mantle and the Core, *in* H. D. Holland and K. K. Turekian, eds.,Treatise of Geochemistry Vol 2, 61–102, Elsevier-Pergamon, Oxford, doi:10.1016/B0-08-043751-6/02123-X.

Huang, S., and F. A. Frey (2005), Recycled oceanic crust in the Hawaiian plume: Evidence from temporal geochemical variations within Koolau Shield, *Contrib. Mineral. Petrol.*, *149*, 556–575, doi:10.1007/s00410-005-0664-9.

Huang, S., M. Regelous, T. Thordarson, and F. A. Frey (2005), Petrogenesis of lavas from Detroit Seamount: Geochemical differences between Emperor Chain and Hawaiian volcanoes, *Geochem. Geophys. Geosyst.*, *6*(1), Q01L06, doi:10.1029/2004GC000756.

Huang, S., W. Abouchami, J. Blichert-Toft, D. A. Clague, B. L. Cousens, F. A. Frey, and M. Humayun (2009), Ancient carbonate sedimentary signature in the Hawaiian plume: Evidence from Mahukona volcano, Hawaii, *Geochem. Geophys. Geosyst.*, *10*(8), Q08002, doi:10.1029/2009GC002418.

Huang, S., J. Farkaš, and S. B. Jacobsen (2011a), Stable calcium isotopic compositions of Hawaiian shield lavas: Evidence for recycling of ancient marine carbonates into the mantle, *Geochim. Cosmochim. Acta*, *75*, 4987–4997, doi:10.1016/j.gca.2011.06.010.

Huang, S., P. S. Hall, and M. G. Jackson (2011b), Geochemical zoning of volcanic chains associated with Pacific hotspots, *Nature Geosci.*, *4*(12), 874–878, doi:10.1038/ngeo1263.

Huh, Y., L. H. Chan, and O. A. Chadwick (2004), Behavior of lithium and its isotopes during weathering of Hawaiian basalt, *Geochem. Geophys. Geosyst.*, *5*(9), Q09002, doi:10.1029/2004GC00729.

Jackson, E. D., E. L. I. A. Silver, and G. B. Dalrymple (1972), Hawaiian-Emperor chain and its relation to Cenozoic circum-pacific tectonics, *Bull. Geol. Soc. Am.*, *83*(3), 601, doi:10.1130/0016-7606(1972)83[601:HCAIRT]2.0.CO;2.

Jackson, M. C., F. A. Frey, M. O. Garcia, and R. A. Wilmoth (1999), Geology and geochemistry of basaltic lava flows and dikes from the Trans-Koolau tunnel, Oahu, Hawaii, *Bull. Volcanol.*, *60*, 381–401, doi:10.1007/s004450050239.

Jackson, M. G., D. Weis, and S. Huang (2012), Major element variations in Hawaiian shield lavas: Source features and perspectives from global ocean island basalt (OIB) systematics, *Geochem. Geophys. Geosyst.*, *13*(9), Q09009, doi:10.1029/2012GC004268.

Jeffcoate, A. B., T. E. Elliott, A. Thomas, and C. Bouman (2004), Precise/small sample size determinations of lithium isotopic compositions of geological reference materials and modern seawater by MC-ICP-MS, *Geostand. Geoanal. Res.*, *28*(1), 161–172, doi:10.1111/j.1751-908X.2004.tb01053.x.

Jeffcoate, A. B., T. Elliott, S. A. Kasemann, D. Ionov, K. Cooper, and R. Brooker (2007), Li isotope fractionation in peridotites and mafic melts, *Geochim. Cosmochim. Acta*, *71*(1), 202–218, doi:10.1016/j.gca.2006.06.1611.

Jicha, B. R., J. M. Rhodes, B. S. Singer, and M.O. Garcia (2012), $^{40}Ar/^{39}Ar$ geochronology of submarine Mauna Loa volcano, Hawaii, *J. Geophys. Res.*, *117*, doi:1 0.1029/2012JB009373.

Kent, A. R., D. A. Clague, M. Honda, E. M. Stolper, I. D. Hutcheon, and M. D. Norman (1999), Widespread assimilation of a seawater-derived component at Loihi Seamount, Hawaii, *Geochim. Cosmochim. Acta*, *63*(18), 2749–2761, doi:10.1016/S0016-7037(99)00215-X.

King, A. J., D. G. Waggoner, and M.O. Garcia (1993), Geochemistry and petrology of basalts from Leg 136, Central Pacific Ocean, *Proc. Ocean Drilling Prog. Sci. Results*, *136*, 107–118, doi:10.2973/odp.proc.sr.136.211.1993.

Kobayashi, K., R. Tanaka, T. Moriguti, K. Shimizu, and E. Nakamura (2004), Lithium, boron, and lead isotope systematics of glass inclusions in olivines from Hawaiian lavas: Evidence for recycled components in the Hawaiian plume, *Chem. Geol.*, *212*, 143–161, doi:10.1016/j.chemgeo.2004.08.050.

Krienitz, M.-S., C.-D. Garbe-Schönberg, R. L. Romer, A. Meixner, K. M. Haase, and N. A. Stroncik (2012), Lithium isotope variations in ocean island basalts—Implications for the development of mantle heterogeneity, *J. Petrol.*, *53*(11), 2333–2347, doi:10.1093/petrology/egs052.

Kurz, M. D., W. J. Jenkins, S. R. Hart, and D. Clague (1983), Helium isotopic variations in volcanic rocks from Loihi Seamount and the Island of Hawaii, *Earth Planet. Sci. Lett.*, *66*, 388–406, doi:10.1016/0012-821X(83)90154-1.

Lambart, S., S. Laporte, A. Provost, and P. Schiano (2012), Fate of pyroxenite-derived melts in the peridotitic mantle: Thermodynamic and experimental constraints, *J. Petrol.*, *53*(3), 451–476, doi:10.1093/petrology/egr068.

Langmuir, C. H., R. D. Vocke, G. N. Hanson, and S. R. Hart (1978), A general mixing equation with applications to Icelandic basalts, *Earth Planet. Sci. Lett.*, *37*, 380–392, doi:10.1016/0012-821X(78)90053-5.

Lassiter, J. C., and E. H. Hauri (1998), Osmium-isotope variations in Hawaiian lavas: Evidence for recycled oceanic lithosphere in the Hawaiian plume, *Earth Planet. Sci. Lett.*, *164*, 483–496, doi:10.1016/s0012-821x(98)00240-4.

Lassiter, J. C., D. J. DePaolo, and M. Tatsumoto (1996), Isotopic evolution of Mauna Kea volcano: Results from the initial phase of the Hawaii Scientific Drilling Project, *J. Geophys. Res.*, *101*, 11,769–11,780.

Lundstrom, C. C., M. Chaussidon, A. T. Hsui, P. Kelemen, and M. Zimmerman (2005), Observations of Li isotopic variations in the Trinity Ophiolite: Evidence for isotopic fractionation by diffusion during mantle melting, *Geochim. Cosmochim. Acta*, *69*(3), 753–751, doi:10.1016/j.gca.2004.08.004.

Lustrino, M., and L. Dallai (2003), On the origin of EM-I end-member, *Neues Jahrbuch für Mineralogie*, *179*, 85–100, doi:10.1127/0077-7757/2003/0179-0085.

Magna, T., U. Wiechert, and A. N. Halliday (2006), New constraints on the lithium isotope compositions of the Moon and terrestrial planets, *Earth Planet. Sci. Lett.*, *243*, 336–353, doi:10.1016/j.epsl.2006.01.005.

Marschall, H. R., P. A. E. Pogge von Strandmann, H. M. Seitz, T. Elliott, and Y. Niu (2007), The lithium isotopic composition of orogenic eclogites and deep subducted slabs,

Earth Planet. Sci. Lett., *262*(3–4), 563-580, doi:10.1016/j.epsl.2007.08.005.

Marske, J. P., A. J. Pietruszka, D. Weis, M. O. Garica, and J. M. Rhodes (2007), Rapid passage of a small-scale mantle heterogeneity through the melting regions of Kilauea and Mauna Loa Volcanoes, *Earth Planet. Sci. Lett.*, *259*, 34–50, doi:10.1016/j.epsl.2007.04.026.

Moore, J. G., D. A. Clague, and W. R. Normark (1982), Diverse basalt types from Loihi seamount, Hawaii, *Geology*, *10*, 88–92, doi:10.1130/0091-7613(1982)10<88:DBTFLS>2.0CO;2.

Moriguti, T., and E. Nakamura (1998), Across-arc variation of Li isotopes in lavas and implications for crust/mantle recycling at subduction zones, *Earth Planet. Sci. Lett.*, *163*, 167–174, doi:10.1016/S0012-821X(98)00184-8.

Nishio, Y., S. Nakai, J. Yamamoto, H. Sumino, T. Matsumoto, V. S. Prikhod'ko, and S. Arai (2004), Lithium isotopic systematics of the mantle-derived ultramafic xenoliths: Implications for EM1 origin, *Earth Planet. Sci. Lett.*, *217*, 245–261, doi:10.1016/S0012-821X(03)00606-X.

Nishio, Y., S. Nakai, T. Kogiso, and H. G. Barsczus (2005), Lithium, strontium, and neodymium isotopic compositions of oceanic island basalts in the Polynesian region: Constraints on a Polynesian HIMU origin, *Geochem. J.*, *39*, 91–103.

Nishio, Y., S. Nakai, T. Ishii, and Y. Sano (2007), Isotope systematics of Li, Sr, Nd, and volatiles in Indian Ocean MORBs of the Rodrigues Triple Junction: Constraints on the origin of the DUPAL anomaly, *Geochim. Cosmochim. Acta*, *71*, 745–759, doi:10.1016/j.gca.2006.10.004.

Nobre-Silva, I., D. Weis, and J. S. Scoates (2013), Isotopic systematics of the early Mauna Kea shield phase and insight into the deep mantle beneath the Pacific Ocean, *Geochem. Geophys. Geosyst.*, *14*, 659–676, doi:10.1002/ggge.20047.

Parkinson, I. J., S. J. Hammond, R. H. James, and N. W. Rogers (2007), High-temperature lithium isotope fractionation: Insights from lithium isotope diffusion in magmatic systems, *Earth Planet. Sci. Lett.*, *257*(3–4), 609–621, doi:10.1016/j.epsl.2007.03.023.

Pertermann, M., and M. M. Hirschmann (2003), Anhydrous partial melting experiments on MORB-like eclogite: Phase relations, phase compositions and mineral-melt partitioning of major elements at 2–3 GPa, *J. Petrol.*, *44*(12), 2173–2201, doi:10.1093/petrology/egg074.

Plank, T., and C. H. Langmuir (1998), The chemical composition of subducting sediment and its consequences for the crust and mantle, *Chem. Geol.*, *145*, 325–394, doi:10.1016/S0009-2541(97)00150-2.

Portnyagin, M., D. Savelyev, K. Hoernle, F. Hauff, and D. Garbe-Schönberg (2012), Mid-Cretaceous Hawaiian tholeiites preserved in Kamchatka, *Geology*, *36*(11), 903–906, doi:10.1130/G25171A.1.

Pretorius, W., D. Weis, G. Williams, D. Hanano, B. Kieffer, and J. S. Scoates (2006), Complete trace elemental characterization of granitoid (USGS G-2, GSP-2) reference materials by high resolution inductively coupled plasma-mass spectrometry, *Geostand. Geoanal. Res.*, *30*(1), 39–54, doi:10.1111/j.1751-908X.2006.tb00910.x.

Ray, J. S., M. W. Martin, J. Veizer, and S. A. Bowring (2002), U-Pb zircon dating and Sr isotope systematics of the Vindhyan Supergroup, India, *Geology*, *30*(2), 131–134, doi:10.1130/0091-7613(2002)030<0131:UPZDAS>2.0.CO;2.

Regelous, M., A. W. Hofmann, W. Abouchami, and S. J. G. Galer (2003), Geochemistry of lavas from the Emperor Seamounts, and the geochemical evolution of Hawaiian magmatism from 85 to 42 Ma, *J. Petrol.*, *44*(1), 113–140, doi:10.1093/petrology/44.1.113.

Rehkämper, M. and A. W. Hofmann (1997), Recycled ocean crust and sediment in Indian Ocean MORB, *Earth Planet. Sci. Lett.*, *147*, 93−106, doi:10.1016/S0012-821X(97)00009-5.

Rhodes, J. M., and M. J. Vollinger (2004), Composition of basaltic lavas sampled by phase-2 of the Hawaii Scientific Drilling Project: Geochemical stratigraphy and magma types, *Geochem. Geophys. Geosys.*, *5*(3), doi:10.1029/2002GC000434.

Richter, F. M., A. M. Davis, D. J. DePaolo, and E. B. Watson (2003), Isotope fractionation by chemical diffusion between molten basalt and rhyolite, *Geochim. Cosmochim. Acta*, *67*(20), 3905–3923, doi:10.1016/S0016-7037(03)00174-1.

Roden, M. F., T. Trull, S. R. Hart, and F. A. Frey (1994), New He, Nd, Pb and Sr isotopic constraints on the constitution of the Hawaiian plume: Results from Koolau Volcano, Oahu, Hawaii, USA, *Geochim. Cosmochim. Acta*, *58*(5), 1431–1440, doi:10.1016/0016-7037(94)90547-9.

Rudnick, R. L., and S. L. Goldstein (1990), The Pb isotopic composition of lower crustal xenoliths and the evolution of lower crustal Pb, *Earth Planet. Sci. Lett.*, *98*, 192–207, doi:10.1016/0012-821x(90)90059-7.

Ryan, J. G., and P. R. Kyle (2004), Lithium abundance and lithium isotope variations in mantle sources: Insights from intraplate volcanic rocks from Ross Island and Marie Byrd Land (Antarctica) and other oceanic islands, *Chem. Geol.*, *212*(1–2), 125–142, doi:10.1016/j.chemgeo.2004.08.006.

Ryan, J. G., and C. H. Langmuir (1987), The systematics of lithium abundances in young volcanic rocks, *Geochim. Cosmochim. Acta*, *51*, 1727–1741, doi:10.1016/0016-7037(87)90351-6.

Salters, V. J. M., J. Blichert-Toft, Z. Fekiacova, A. Sachi-Kocher, and M. Bizimis (2006), Isotope and trace element evidence for depleted lithosphere in the source of enriched Ko'olau basalts, *Contrib. Mineral. Petrol.*, *151*, 297–312, doi:10.1007/s00410-005-0059-y.

Seitz, H.-M., and A. B. Woodland (2000), The distribution of lithium in peridotitic and pyroxenitic mantle lithologies—an indicator of magmatic and metasomatic processes, *Chem. Geol*, *166*, 47–64, doi:10.1016/S0009-2541(99)00184-9.

Seitz, H.-M., G. P. Brey, Y. Lahaye, S. Durali, and S. Weyer (2004), Lithium isotopic signatures of peridotite xenoliths and isotopic fractionation at high temperature between olivine and pyroxenes, *Chem. Geol*, *212*, 163–177, doi:10.1016/j.chemgeo.2004.08.009.

Sobolev, A. V., A. W. Hofmann, S. V. Sobolev, and I. K. Nikogosian (2005), An olivine-free mantle source of Hawaiian shield basalts, *Nature*, *434*, 590–597, doi:10.1038/nature03411.

Sobolev, A.V., et al. (2007), The amount of recycled crust in sources of mantle-derived melts, *Science*, *316*, 412–417, doi:10.1126/science.1138113.

Stille, P., D. M. Unruh, and M. Tatsumoto (1986), Pb, Sr, Nd, and Hf isotopic constraints on the origin of Hawaiian basalts and evidence for a unique mantle source, *Geochim. Cosmochim. Acta*, *50*, 2303–2319, doi:10.1016/0016-7037(86)90084-0.

Stosch, H.-G., and G. W. Lugmair (1984), Evolution of the lower continental crust: Granulite facies xenoliths from the Eifel, *West Germany, Nature*, *311*, 368–370, doi:10.1038/311368a0.

Stracke, A., M. Bizimis, and V. J. M. Salters (2003), Recycling oceanic crust: Quantitative constraints, *Geochem. Geophys. Geosys.*, *4*(3), 8003, doi:10.1029/2001GC000223.

Sun, S.-s. and W. F. McDonough (1989), Chemical and isotopic systematics of oceanic basalts: implications for mantle composition and processes, *Geol. Soc. London Sp. Pub.*, *42*, doi:10.1144/GSL.SP.1989.042.01.19.

Tanaka, R., E. Nakamura, and E. Takahashi (2002), Geochemical evolution of Koolau Volcano, Hawaii, in Geophys. Monogr. 128, pp. 311–332, doi:10.1029/GM128p0311.

Tanaka, R., A. Makishima, and E. Nakamura (2008), Hawaiian double volcanic chain triggered by and episodic involvement of recycled material: Constraints from temporal Sr-Nd-Hf-Pb isotopic trend of the Loa-type volcanoes, *Earth Planet. Sci. Lett.*, *265*, 450–465, doi:10.1016/j.epsl.2007.10.035.

Tang, Y.-J., H.-F. Zhang, and J.-F. Ying (2010), A brief review of isotopically light Li—A feature of the enriched mantle?, *Geol. Rev.*, *52*(9), 964–976, doi:10.1080/00206810903211385.

Tatsumoto, M. (1978), Isotopic composition of lead in oceanic basalt and its implication to mantle evolution, *Earth Planet. Sci. Lett.*, *38*, 63–87, doi:10.1016/0012-821X(78)90126-7.

Teng, F., N. Dauphas, S. Huang, and B. Marty (2013), Iron isotopic systematics of oceanic baslats, *Geochim. Cosmochim. Acta*, *107*, 12–26, doi:10.1016/j.gca.2012.12.027.

Teng, F.-Z., R. L. Rudnick, W. F. McDonough, S. Gao, P. B. Tomascak, and Y. Liu (2008), Lithium isotopic composition and concentration of the deep continental crust, *Chem. Geol.*, *255*, 47–59, doi:10.1016/j.chemgeo.2008.06.009.

Tomascak, P. B. (2004), Developments in the understanding and application of lithium isotopes in the earth and planetary sciences, *Rev. Mineral. Geochem.*, *55*(1), 153, doi:10.2138/gsrmg.55.1.153.

Tomascak, P. B., F. Tera, R. T. Helz, and R. J. Walker (1999), The absence of lithium isotope fractionation during basalt differentiation: New measurements by multicollector sector ICP-MS, *Geochim. Cosmochim. Acta*, *63*(6), 907–910, doi:10.1016/S0016-7037(98)00318-4.

Tomascak, P. B., C. H. Langmuir, P. J. Roux, and S. B. Shirey (2008), Lithium isotopes in global mid-ocean ridge basalts, *Geochim. Cosmochim. Acta*, *72*(6), 1626–1637, doi:10.1016/j.gca.2007.12.021.

Valley, J. W., et al. (2005), 4.4 Billion years of crustal maturation: Oxygen isotope ratios of magmatic zircon, *Contrib. Mineral. Petrol.*, *150*, 561–580, doi:10.1007/s00410-005-0025-8.

Vervoort, J. D., and J. Blichert-Toft (1999), Evolution of the depleted mantle: Hf isotope evidence from juvenile rocks through time, *Geochim. Cosmochim. Acta*, *63*, 533–556, doi:10.1016/S0016-7037(98)00274-9.

Vlastélic, I., K. Koga, C. Chauvel, G. Jaques, and P. Télouk (2009), Survival of lithium isotopic heterogeneities in the mantle supported by HIMU-lavas from Rurutu Island, Austral Chain, *Earth Planet. Sci. Lett.*, *286*(3–4), 456–466, doi:10.1016/j.epsl.2009.07.013.

Weis, D., et al. (2006), High-precision isotopic characterization of USGS reference materials by TIMS and MC-ICP-MS, *Geochem. Geophys. Geosyst.*, *7*(8), Q08006, doi:10.1029/2006GC001283.

Weis, D., B. Kieffer, D. Hanano, I. Nobre Silva, J. Barling, W. Pretorius, C. Maerschalk, and N. Mattielli (2007), Hf isotope compositions of U.S. Geological Survey reference materials, *Geochem. Geophys. Geosyst.*, *8*(6), Q06006, doi:10.1029/2006GC001283.

Weis, D., M. O. Garcia, J. M. Rhodes, M. Jellinek, and J. S. Scoates (2011), Role of the deep mantle in generating the compositional asymmetry of the Hawaiian mantle plume, *Nature Geosci.*, *4*(12), 831–838, doi:10.1038/ngeo1328.

Weis et al., (in prep.), What do post-shield and rejuvenated lavas tell us about the Hawaiian mantle Plume? Ka'ula Island, in preparation.

White, W. M. (1985), Sources of oceanic basalts: Radiogenic isotopic evidence, *Geology*, *13*, 115–118, doi:10.1130/0091-7613(1985)13<115:SOOBRI>2.0.CO;2.

White, W. M. (2010), Oceanic island basalts and mantle plumes: The geochemical perspective, *Annu. Rev. Earth Planet. Sci.*, *38*, 133–160, doi:10.1146/annurev-earth-040809-152450.

Willbold, M., and A. Stracke (2006), Trace element composition of mantle end-members: Implications for recycling of oceanic and upper and lower continental crust, *Geochem. Geophys. Geosys.*, *7*(4), Q04004, doi:10.1029/2005GC001005.

Wolfe, E. W., W. S. Wise, and G. B. Dalrymple (1997), The geology and petrology of Mauna Kea volcano, Hawaii—A study of post-shield volcanism, Prof. Pap. 1557, U.S. Geol. Surv., Menlo Park, CA.

Wright, T. L. (1971), Chemistry of Kilauea and Mauna Loa lava in space and time, Prof. Pap. 735, U.S. Geol. Surv., Washington, D.C.

Xu, G. F., F. A. Frey, D. A. Clague, D. Weis, and M. H. Beeson (2005), East Molokai and other Kea-trend volcanoes: Magmatic processes and sources as they migrate away from the Hawaiian hot spot, *Geochem. Geophys. Geosyst.*, *6*, Q05008, doi:10.1029/2004GC000830.

Xu, G. F., F. A. Frey, D. A. Clague, W. Abouchami, J. Blichert-Toft, B. Cousens, and M. Weisler (2007), Geochemical characteristics of West Molokai shield- and post-shield- stage lavas: Constraints on Hawaiian plume models, *Geochem. Geophys. Geosyst.*, *8*, Q08G21, doi: 10.1029/2006GC001554.

Yang, H. J., F. A. Frey, and D. A. Clague (2003), Constraints on the source components of lavas forming the Hawaiian North Arch and Honolulu Volcanics, *J. Petrol.*, *44*, 603–627, doi:10.1093/petrology/44.4.603.

Yokoyama, T., A. Makishima, and E. Nakamura (1999), Evaluation of the coprecipitation of incompatible trace elements with fluoride during silicate rock dissolution by acid digestion, *Chem. Geol.*, *157*(3–4), 175–187, doi:10.1016/S0009-2541(98)00206-X.

Zack, T., P. B. Tomascak, R. L. Rudnick, C. Dalpe, and W. F. McDonough (2003), Extremely light Li in orogenic eclogites: The role of isotope fractionation during dehydration in subducted oceanic crust, *Earth Planet. Sci. Lett.*, *208*(3–4), 279–290, doi:10.1016/S0012-821X(03)00035-9.

Zindler, A., and S. Hart (1986), Chemical geodynamics, *Annu. Rev. Earth Planet. Sci.*, *14*, 493–571, doi:10.1146/annurev.ea.14.050186.002425.

6

Onset of Rejuvenated-Stage Volcanism and the Formation of Līhu'e Basin: Kaua'i Events That Occurred 3–4 Million Years Ago

David R. Sherrod[1], Scot K. Izuka[2], and Brian L. Cousens[3]

ABSTRACT

For ocean-island volcanoes, an understanding of rejuvenated-stage volcanism requires knowing the age of onset, duration of the volcanic episode, and length of quiescence that preceded the rejuvenated pulse. On the Island of Kaua'i, cuttings from two lava flows intercepted in the Hanamā'ulu well (Līhu'e basin) and assigned to the Kōloa Volcanics on basis of major-element, trace-element, and isotope geochemistry have yielded $^{40}Ar/^{39}Ar$ ages of 3.11 and 3.22 Ma. A deeper lava, with age 3.42 Ma, is tentatively assigned to the Kōloa. These ages partly fill a 1 Myr gap in the previously published ages for the Kōloa Volcanics and suggest that rejuvenated-stage volcanism on Kaua'i has persisted for 3.2 Myr or longer. The quiescent period may have been as short as 0.2–0.3 Myr, possibly briefer.

The eastern third of Kaua'i subsided as much as 1.3 km along an island-long north-south fault between 4.7 and 3.4 Myr ago. The fault marks a stunted episode of extension, inasmuch as displacement diminishes rapidly from the inland area north and south toward the coast and the fault lacks expression on the island's submarine flanks. The Līhu'e basin formed sometime thereafter: The new ages are from lava flows that bracket shallow marine sediment, so the basin had subsided below sea level before 3 Ma. Though poorly constrained, rates of subsidence are in the range of those from Lāna'i, which today sits at a distance from the hotspot that is similar to the paleogeographic position of Kaua'i 3 Ma.

6.1. INTRODUCTION

Samples from drill holes in the Hawaiian Islands provide rare opportunities to assess lava accumulation rates, downhole compositional changes, and the ages of key horizons. In basins, drill holes may sample a more complete depositional record than is preserved in surface exposures. Such is the case for Kaua'i's Līhu'e basin, which is unique in the islands as an inland basin that experienced one or more marine incursions interrupted by lava flows. Subsequent infilling of the Līhu'e basin by late-stage lava flows has resulted in one of the thickest accumulations of rejuvenated-stage volcanic rocks on the Island of Kaua'i [*Gandy*, 2006].

The rejuvenated stage of volcanism is particularly well represented on Kaua'i (Figure 6.1). The Kōloa Volcanics on Kaua'i is the most extensive rejuvenated-stage formation in Hawai'i, covering more than a third of the island's surface and having an estimated volume of 58 km^3 [*Garcia et al.*, 2010]. The ability to fully assess the age range of the Kōloa Volcanics has been hampered, however, in part

[1] *Cascades Volcano Observatory, U.S. Geological Survey, Vancouver, Washington, USA*

[2] *Pacific Islands Water Science Center, U.S. Geological Survey, Honolulu, Hawaii, USA*

[3] *Department of Earth Sciences, Carleton University, Ottawa, Ontario, Canada*

Hawaiian Volcanoes: From Source to Surface, Geophysical Monograph 208. First Edition.
Edited by Rebecca Carey, Valérie Cayol, Michael Poland, and Dominique Weis.

Companion Website: www.wiley.com/go/Carey/Hawaiian_Volcanoes

because the exposed, oldest part of the Kōloa is fragmentary in its preservation and also because the boundary between lowermost Kōloa and underlying postshield alkalic lava is ill defined in the absence of intracanyon relations, substantial weathering contrasts, or angular discordance where the Kōloa Volcanics mantles postshield alkalic lava.

Here we report three new radiometric ages and whole-rock geochemical and isotopic analyses interpreted to be from the Kōloa Volcanics in the deep subsurface of the Līhu'e basin (Figure 6.1). Samples for analysis came from cuttings recovered during the drilling of the Hanamā'ulu water-monitoring well, which is sited where the Kōloa Volcanics extends to a depth of more than 220 m below sea level (305 m downhole depth) [*Reiners et al.*, 1999]. The samples came from near the bottom of the section of Kōloa Volcanics penetrated by the well. The lava flows are interlayered with marine sediment that was deposited in shallow water but now lies as deep as 185 m below sea level [*Izuka and Resig*, 2008]. The lava flows thus provide

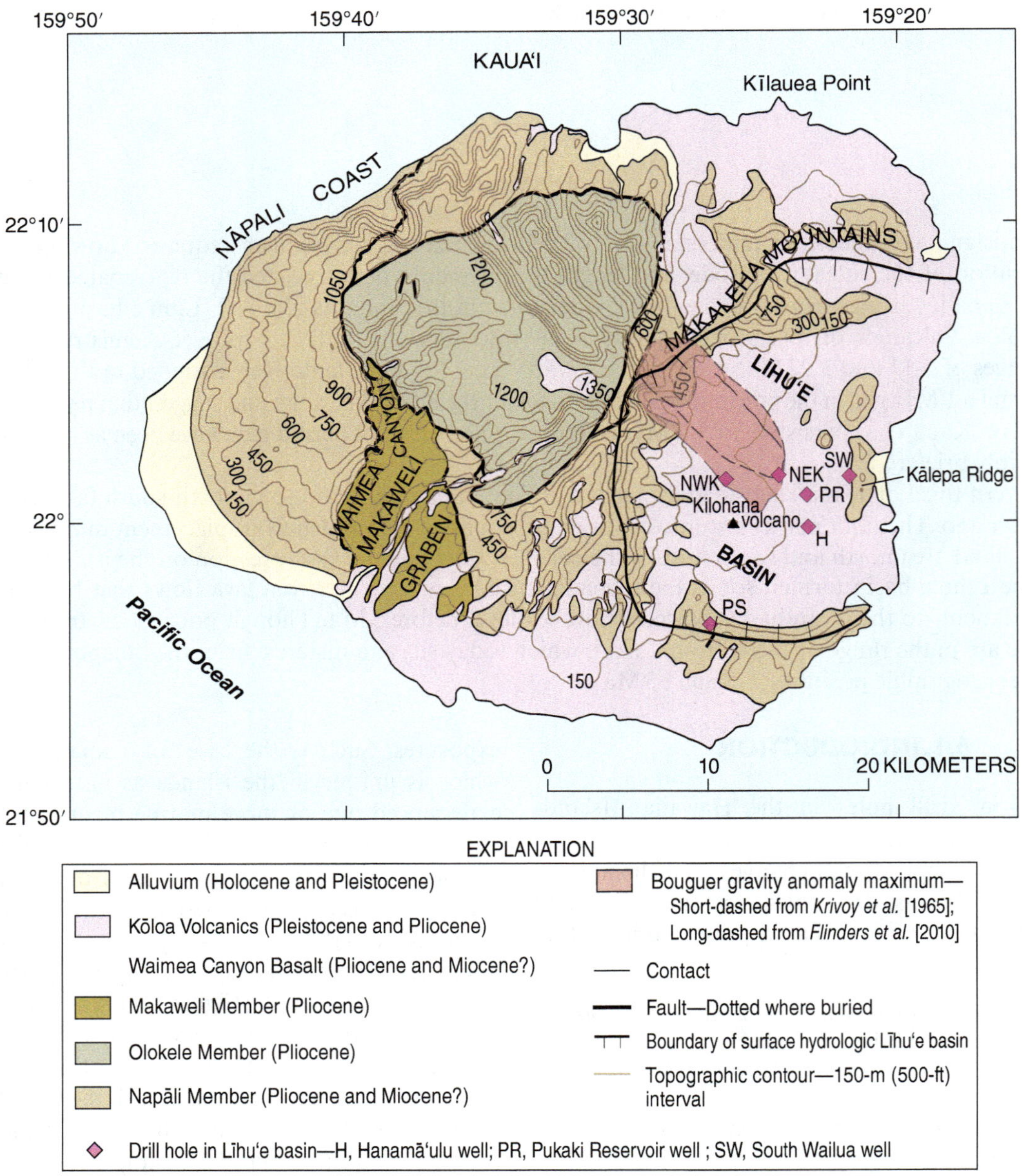

Figure 6.1 Simplified geologic map showing the Līhu'e basin and distribution of Kōloa Volcanics, Kaua'i. Generalized from *Macdonald et al.* [1960]. Shown for reference is area bounding the maximum values for complete Bouguer gravity anomaly, corresponding to 340 mGal contour of *Krivoy et al.* [1965] or 360 mGal contour of *Flinders et al.* [2010]. Hachured line is Līhu'e hydrologic basin of some studies [e.g., *Gingerich and Izuka*, 1997b], not to be confused with the Līhu'e geomorphic and structural basin described herein.

bracketing ages for the formation of ancient shorelines in the Līhu'e basin, constraints on the age of basin formation, and information useful for estimating accumulation rates for the Kōloa Volcanics, subsidence rates for Kaua'i, and tectonics of the Hawaiian Island chain.

6.2. SETTING

The following geologic synopsis is drawn chiefly from *Macdonald et al.* [1960], *Langenheim and Clague* [1987], and *Sherrod et al.* [2007]. Kaua'i was built by midoceanic hotspot shield and postshield volcanism between about 5 and 4 Ma. Stratigraphically, the Waimea Canyon Basalt (Figure 6.2) includes tholeiite lava flows of the shield stage that form the bulk of Kaua'i plus sparse deposits of hawaiite and mugearite in late-shield or postshield time (Figure 6.1). Faulting and erosion of the Kaua'i shield created large depressions that were partly filled with lava flows from late-shield, postshield, and rejuvenated-stage volcanism. The rejuvenated-stage Kōloa Volcanics, chiefly lava flows, overlies the Waimea Canyon Basalt unconformably to disconformably; the Kōloa consists primarily of undersaturated alkalic mafic volcanic rocks and lesser interbedded sedimentary rocks. Rejuvenated-stage volcanism lingered into the Pleistocene. Indeed, given (1) such a protracted history, (2) 30 Kōloa ages in the past 2.5 Ma [*Garcia et al.*, 2010], (3) no gap between ages greater than about 300,000 years, and (4) youngest lava flows only 0.15–0.22 Ma in age [*Garcia et al.*, 2010], future Kōloa eruptions remain a possibility.

6.2.1. Līhu'e Basin

The Līhu'e basin is a semicircular depression on eastern Kaua'i (Figure. 6.1). It is encircled on the west, north, and south by mountains and ridges formed by the shield-stage lava flows of the Waimea Canyon Basalt. The rejuvenated-stage Kōloa Volcanics fills much of the basin. The Waimea Canyon Basalt emerges as isolated small hills and ridges surrounded by Kōloa Volcanics within the basin. The Kōloa Volcanics also covers extensive areas to the north and south of the Līhu'e basin (Figure 6.1).

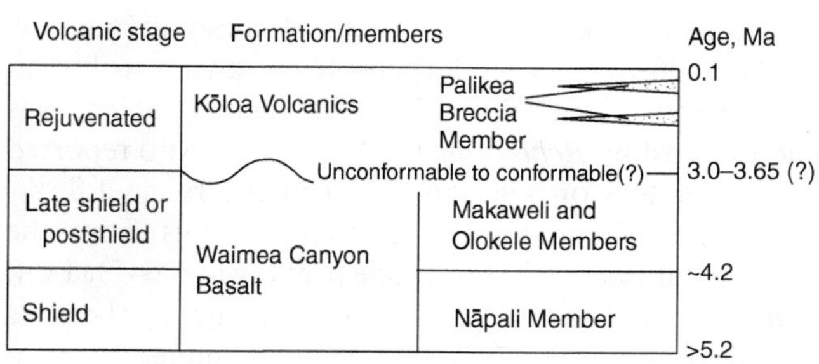

Figure 6.2 Stratigraphic column showing Kaua'i volcanic formations and their approximate age range. Synthesized from stratigraphic interpretation of *Macdonald et al.* [1960], with names revised in accordance with *Langenheim and Clague* [1987].

The formation of the basin has been an enigma, attributed to processes as disparate as stream erosion [*Stearns*, 1967, 1985], caldera collapse [*Macdonald et al.*, 1960; *Holcomb et al.*, 1997], and landslide [*Reiners et al.*, 1999; *Sherrod et al.*, 2007]. Exploratory drilling indicates that in some parts of the basin the contact between the underlying shield volcano and the overlying basin-filling sediment and rejuvenated-stage volcanic rocks is 200–300 m below present sea level [*Gingerich and Izuka*, 1997a, b; *Izuka and Gingerich*, 1997a, b, c, d; *Reiners et al.*, 1999].

6.2.2. Hanamā'ulu Well

The Hanamā'ulu well is one of six deep exploratory wells drilled by the U.S. Geological Survey in 1995–1997 as part of a water resources assessment project in cooperation with the County of Kaua'i Department of Water. The well, located at latitude 21.99447°N, longitude 159.38544°W, was drilled to a depth of 305 m from an altitude of 83 m above sea level [*Izuka and Gingerich*, 1997a] (Figure 6.3). During drilling, cuttings were collected across 1.5 m intervals. Macroscopic, petrographic, geochemical, and paleontological analysis of the cuttings showed that the Hanamā'ulu well penetrated 220 m of Kōloa Volcanics, thin to thick residual soil and alluvial deposits, and two intervening layers (65 m total) of shallow-marine sediment [*Izuka and Gingerich*, 1997a; *Reiners et al.*, 1999; *Izuka and Resig*, 2008] (Figure 6.3). Neither pillow lava nor hyaloclastite is reported in the drill log. The section of Kōloa Volcanics penetrated by the well includes the Palikea Breccia Member, which comprises debris flows shed from shield-stage lavas in the highlands to the west [*Reiners et al.*, 1999]. The two marine layers are also present in the Pukaki Reservoir well, 2 km north of the Hanamā'ulu well (Figure 6.1) [*Izuka and Resig*, 2008]. A generalized stratigraphic section of the Līhu'e basin, relying on the five deep monitoring wells, shows thickening and thinning of the Kōloa Volcanics across the basin, with the thickest sequence in the Hanamā'ulu well [*Reiners et al.*, 1999, their Figure 5; see also *Gandy*, 2006, his cross section, Figure 14A].

6.2.3. Methods

Samples for this study were selected from three depths to bracket the two marine sedimentary layers in the Hanamā'ulu well (Figure 6.3). Initial selection of sample bags targeted those with cuttings of mostly uniform lava

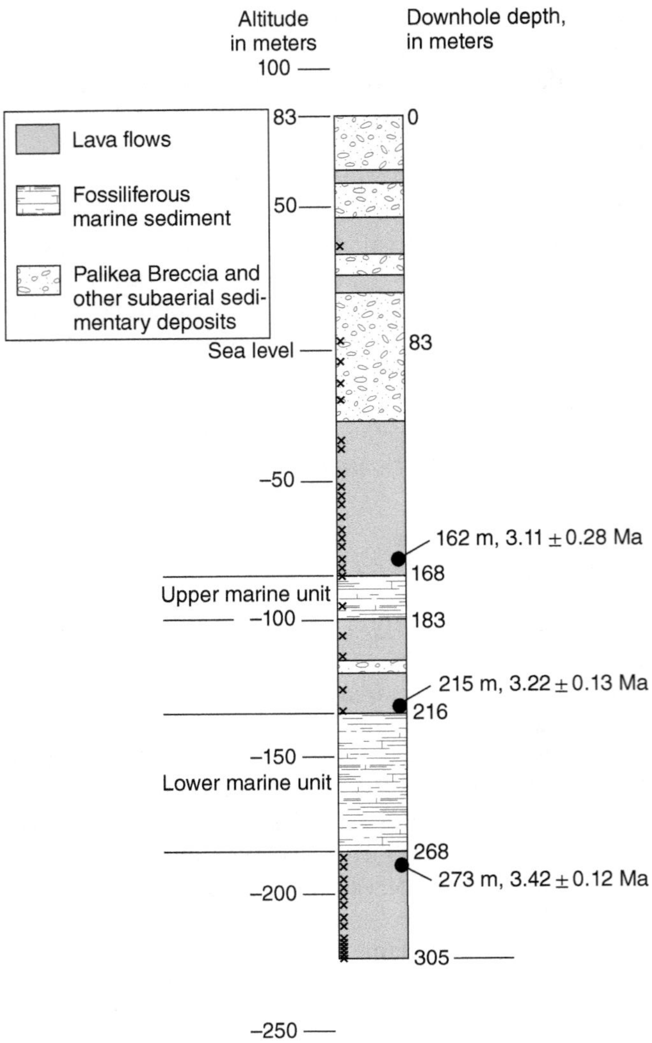

Figure 6.3 Schematic section through Hanamā'ulu well [from *Izuka and Resig*, 2008]. Thick wedge of Palikea Breccia Member of the Waimea Canyon Basalt at altitude 19 to −28 m is based on interpretation by *Reiners et al.* [1999]. New ages from drill cuttings are from samples (filled circles) above, between, and below the two fossiliferous marine sedimentary units. Sample depth for previously published chemical analyses shown by X's [*Reiners and Nelson*, 1998].

chips to avoid sedimentary interbeds. The cuttings were washed and sieved into size-sorted fractions, then hand picked under a binocular microscope to achieve further sample homogeneity. The representative lithology for each depth interval corresponds to that of the most common chips in the large-diameter fraction (8–16 mm); petrographic descriptions appear in an appendix. Hand picking proceeded through successively finer fractions until enough mass (150–200 g) was obtained for $^{40}Ar/^{39}Ar$ age determinations. These sample weights required processing material as fine as 1–2 mm, which formed 20%–35% of each sample submitted for dating. Sample numbers indicate the depth interval, in feet, as is customary for U.S. drilling operations.

Samples were processed by John Huard in the noble gas mass spectrometry laboratory in the College of Earth, Ocean, and Atmospheric Sciences at Oregon State University (http://geochronology.coas.oregonstate.edu/). After crushing, sieving, washing, and drying, mineral separates were prepared by passing each sample through a Frantz magnetic separator. The effect in basaltic lava is to create a feldspar-enriched sample, with much of the iron-bearing material (olivine, clinopyroxene, glass, and oxide minerals) pulled aside during the magnetic separation. Though commonly referred to as groundmass separates, samples so processed include phenocrysts and microphenocrysts that cannot be resolved from finer-grained minerals after grinding. The resulting material, chiefly plagioclase, underwent mild acid cleaning (HNO_3, HF), ultrasonic wash, drying, and a final hand picking under a binocular microscope. After irradiation, samples were heated incrementally and the extracted gas analyzed using a Mass Analyser Products model 215–50 rare gas mass spectrometer. The laboratory dating standard is FCT-3 (biotite), which is assumed to have the same age as FCTs sanidine from the same rock, namely 28.03 ± 0.18 Ma [*Renne et al.*, 1994], cited here with 2σ error.

For whole-rock geochemistry, a few large chips (4–16 mm) amassing about 5 g total for each sample were submitted to the Washington State University Geoanalytical Laboratories (http://www.sees.wsu.edu/Geolab/equipment/xrf.html). Major- and trace-element analyses (Table 6.1) were determined by X-ray fluorescence spectroscopy using an automated Thermo-ARL spectrometer. Analytical precision for major elements is better than 0.5% (http://www.sees.wsu.edu/Geolab/note/xrfprecision.html). For the trace elements, precision is better than 10% and commonly in the range 3–5% except for those elements in concentrations less than 10 ppm (Pb, Th, U), in which case the precision may be as poor as 50% of reported value.

The totals for the major-element analyses are low (91.76%–96.76%; Table 6.1), probably owing to hydration in the subsurface setting. The same problem was encountered by *Reiners and Nelson* [1998], who reported values for loss on ignition (LOI) in the range 1.85%–9.34% for 38 analyses of similar cuttings from the Hanamā'ulu well, with the mode peaking at 6%–7% LOI. Hydration and ensuing alteration affect mostly the glass and ferromagnesian minerals in our samples, components that are largely removed by magnetic separation during preparation of the feldspar concentrates used for dating.

A subset of the whole-rock powders was selected for Pb, Sr, and Nd isotopic analysis (Table 6.2) utilizing

Table 6.1 Whole-rock major and trace-element chemical analyses from dated samples of Kōloa Volcanics in Hanamā'ulu well, Kaua'i.

Sample no.	HTZ-530/535	HTZ-705/710	HTZ-895/900
Formation	Kōloa Volcanics	Kōloa Volcanics	Kōloa Volcanics
Age (Ma)	3.11 ±0.28	3.22 ±0.13	3.42 ±0.12
Rock type	Picrobasalt	Basanite	Thol. basalt
Major-Element Analyses (wt%)			
SiO_2	40.51	41.30	41.99
TiO_2	2.585	2.461	1.642
Al_2O_3	9.72	10.46	9.94
FeO*	12.76	12.80	12.31
MnO	0.193	0.198	0.169
CaO	11.90	11.42	10.12
MgO	14.92	14.12	14.10
K_2O	0.53	0.78	0.38
Na_2O	1.98	2.65	0.93
P_2O_5	0.549	0.573	0.180
Total	95.64	96.76	91.76
Trace-Element Analyses (ppm)			
Ni	413	347	386
Cr	581	404	998
Sc	24	23	33
V	302	292	283
Ba	549	562	193
Rb	27	20	8
Sr	775	733	227
Zr	177	144	81
Y	22	23	15
Nb	48.3	42.8	16.3
Ga	17	18	14
Cu	82	78	66
Zn	115	123	97
Pb	2	2	0
La	34	35	11
Ce	74	74	24
Th	5	6	3
Nd	40	35	14
U	2	1	0

Note: Analyses by X-ray fluorescence at Geoanalytical Labs, Washington State University, Pullman, WA. Sample numbers indicate downhole depth, in feet. Well location 159.38544°W, 21.99447°N, WGS84.

Table 6.2 Pb, Sr, and Nd isotope data for Hanamā'ulu drillhole cuttings, Līhu'e basin, Kaua'i.[a]

Sample	$^{206}Pb/^{204}Pb$	±	$^{207}Pb/^{204}Pb$	±	$^{208}Pb/^{204}Pb$	±	$^{87}Sr/^{86}Sr$	±	$^{143}Nd/^{144}Nd$	±
HTZ-530	18.205	0.002	15.437	0.002	37.922	0.005	0.703319	0.000008	0.513005	0.000009
HTZ-705	—[b]		—[b]		—[b]		0.703089	0.000005	0.513067	0.000007
HTZ-895	18.114	0.005	15.446	0.005	37.763	0.012	0.703145	0.000010	0.513078	0.000008

[a] Samples analyzed at Carleton University, Ontario, Canada, by thermal ionization mass spectrometry using a Triton system. Column chemistry methods, spectrometer fractionation coefficients, and analytical measurements of standards are described in text. Error is 2σ for all isotope analyses.

[b] No run for HTZ-705 Pb isotopic ratios.

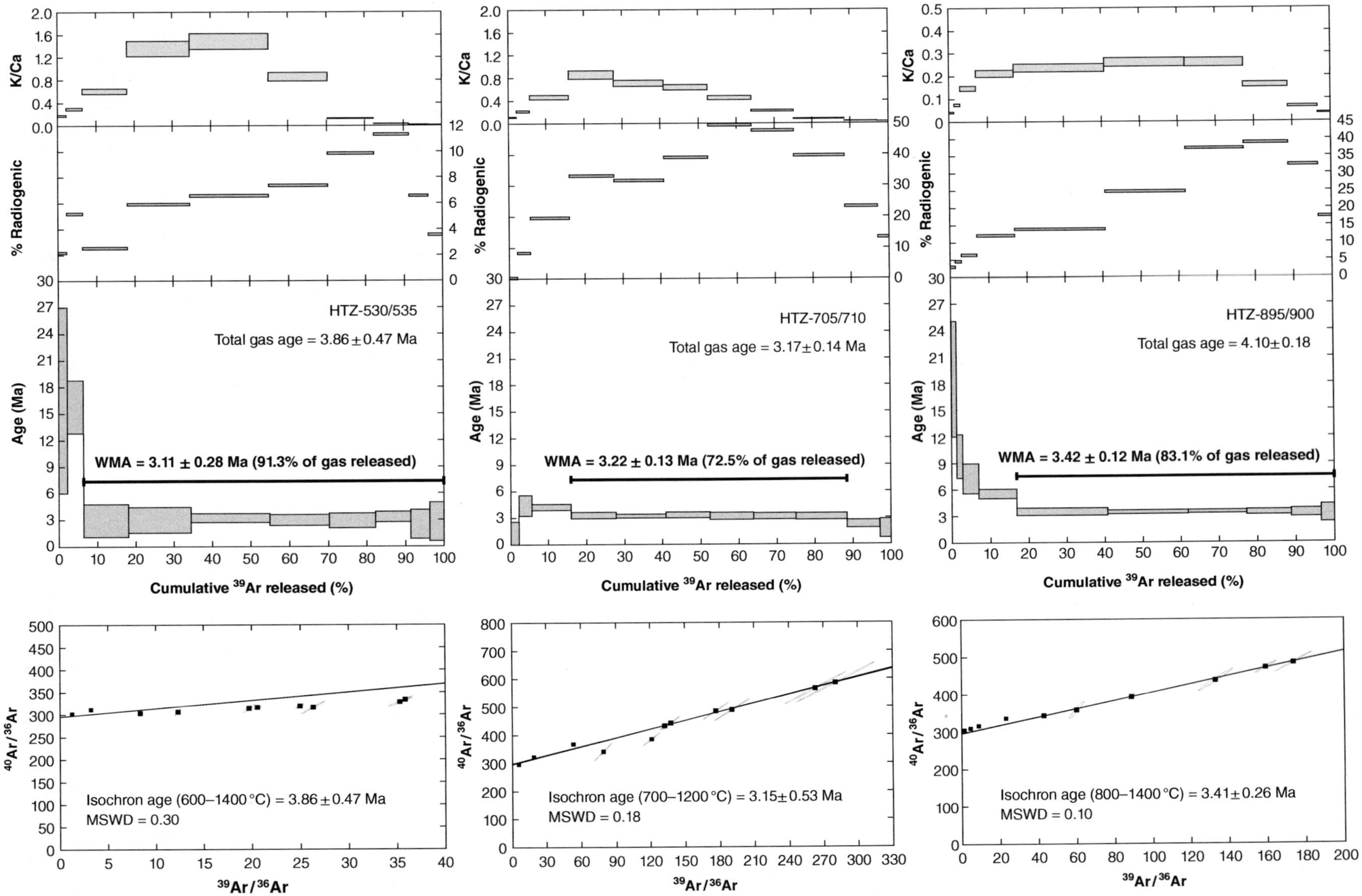

Figure 6.4 Argon plateau and isochron diagrams for dated samples from the Hanamā'ulu well, Kaua'i (WMA, weighted mean age; MSWD, mean square of weighted deviates).

a ThermoFisher Triton T1 thermal ionization mass spectrometer at Carleton University. Lead is separated in Bio-Rad 10 mL polyethylene columns and Dowex AG1-8X anion resin, using 1 N HBR to elute other elements and 6 N HCl to elute Pb. All Pb mass spectrometer runs are corrected for fractionation using NBS981. The average ratios measured for NBS981 are $^{206}Pb/^{204}Pb = 16.890 \pm 0.009$, $^{207}Pb/^{204}Pb = 15.426 \pm 0.009$, and $^{208}Pb/^{204}Pb = 36.494 \pm 0.031$, based on 30 runs between September 2008 and July 2013. The fractionation correction is +0.13%/amu [on the basis of values by *Todt et al.*, 1996]. Strontium is separated using a 14 mL Bio-Rad borosilicate glass chromatography column containing 3.0 mL of Dowex AG50-X8 cation resin. Sr is eluted using 2.5 N HCl, and the rare earth elements (REEs) are then eluted using 6 N HCl. The Sr isotope ratios are normalized to $^{86}Sr/^{88}Sr = 0.11940$. Two Sr standards are run at Carleton University, NIST SRM987 ($^{87}Sr/^{86}Sr = 0.710239 \pm 14$, N = 40, September 2008 to July 2013) and the Eimer and Amend (E&A) $SrCO_3$ ($^{87}Sr/^{86}Sr = 0.708012 \pm 15$, N = 18, September 2008 to July 2013). The REEs are dissolved in 0.26 N HCl and loaded into an Eichrom chromatographic column containing a 2 cm high bed of Teflon powder coated with HDEHP (di(2-ethylhexyl) orthophosphoric acid [*Richard et al.*, 1976]). Neodymium is eluted using 0.26 N HCl. The Nd isotope ratios are normalized to $^{146}Nd/^{144}Nd = 0.72190$. Thirty runs of an internal Nd metal standard yield $^{143}Nd/^{144}Nd = 0.511823 \pm 12$ (N = 40, September 2008 to July 2013), equivalent to a value for the La Jolla standard of 0.511852. All quoted uncertainties are 2-sigma (2σ) standard deviations of the mean.

6.3. AGES

All three samples produced satisfactory age data (Figure 6.4 and Table 6.3, plateau ages, 2σ error for all ages cited herein). The ages are in correct stratigraphic stacking order, which furthers our confidence in them. From oldest (deepest) to youngest (shallow) they are 3.42 ± 0.12 Ma at 273 m depth below ground surface (HTZ-895/900), 3.22 ± 0.13 Ma at 215 m depth (HTZ-705/710), and 3.11 ± 0.28 Ma at 162 m depth (HTZ-530/535). The ages overlap at 2σ uncertainty, so statistically they provide only some broad age range for the entire span of strata encompassed by the samples. Regardless, the intervening gravel and marine beds likely indicate the passage of thousands or hundreds of thousands of years during the deposition of 111 m of lava flows and sedimentary deposits, consistent with the spread of ages.

Plateau ages are favored in this report because, of the several types of ages derived from $^{40}Ar/^{39}Ar$ systematics, the data meet the commonly accepted criteria for a meaningful incremental-heating age [*McDougall and Harrison*, 1999, p. 32], namely (1) a well-defined plateau (horizontal age spectrum with no significant slope) comprising more than 50% of ^{39}Ar released, (2) a well-defined isochron for plateau gas fractions, (3) concordant plateau and isochron ages, and (4) an $^{40}Ar/^{36}Ar$ isochron intercept not significantly different from 295.5. For the Hanamā'ulu well samples, each plateau age calculation comprises at least 70% of ^{39}Ar released.

6.4. GEOCHEMISTRY AS GUIDE TO STRATIGRAPHIC ASSIGNMENT

But are they all Kōloa Volcanics ages? As originally defined, the rejuvenated stage ensues after a lengthy quiescence in volcanic activity at Hawaiian volcanoes [e.g., *Stearns and Macdonald*, 1942; *Macdonald et al.*, 1960]. Subsequent dating, however, has shown the quiescence ranges from 2.0 to as little as 0.6 Myr [*Garcia et al.*, 2010; *Clague and Sherrod*, 2014]. Because outcrop-scale stratigraphic relations cannot be observed in cuttings, the stratigraphic assignment of the rocks analyzed for ages in our study must rely on geochemistry.

On some Hawaiian volcanoes, geochemical characteristics clearly distinguish rejuvenated-stage rocks from those of the shield and postshield stages: rejuvenated-stage rocks commonly are more alkalic than those of the

Table 6.3 Radiometric age data by $^{40}Ar/^{39}Ar$ analysis for Hanamā'ulu well samples.

	Plateau Age			Total Gas	Isochron Age			
Sample	%^{39}Ar (steps)	Age (Ma)	Error (2σ)	Age (Ma)	Age (Ma)	Error (2σ)	MSWD	$^{40}Ar/^{36}Ar_i$
HTZ-530/535	73 (8 of 10)	**3.11**	**0.28**	3.86 ± 0.94	3.21	1.58	0.30	294.7 ± 5.6
HTZ-705/710	72 (6 of 11)	**3.22**	**0.13**	3.17 ± 0.28	3.15	1.06	0.18	299.0 ± 27.6
HTZ-895/900	83 (6 of 10)	**3.42**	**0.12**	4.10 ± 0.36	3.41	0.52	0.10	296.1 ± 7.9

Note: All samples are groundmass separates, mostly feldspar. Preferred age is plateau age, shown in bold. Well location 159.38544°W, 21.99447°N, WGS84. Samples are cuttings; sample number indicates downhole depth interval sampled, in feet.

preceding stages. The distinction is less clear on Kauaʻi, however. Although an alkali-silica plot of published geochemical data for the rejuvenated-stage Kōloa Volcanics describes a compositional field that is for the most part more alkalic than the shield-stage Nāpali Member of the Waimea Canyon Basalt, the fields overlap slightly (Figure 6.5). The geochemical data of our two uppermost dated samples plot within the rejuvenated-stage field defined by other published analyses from the Kōloa Volcanics. Lava flows above the lower marine layer are almost entirely picrobasalt [terminology of *Le Bas et al.*, 1986], as was recognized previously by *Reiners and Nelson* [1998] on the basis of mineral modes in thin section. The assignment of these two samples to the Kōloa is thus supported by the geochemical data and consistent with previous geochemical and petrographic studies.

The chemical analysis of our lowest dated sample lies in the overlap area and thus could be interpreted as either Kōloa Volcanics or Waimea Canyon Basalt. *Reiners and Nelson* [1998] and *Reiners et al.* [1999] concluded, on the basis of geochemistry and petrology, that rocks at the bottom of the Hanamāʻulu well belong to the Kōloa Volcanics, but the distinction is not compelling on the alkali-silica plot.

Trace-element variation diagrams better distinguish the rejuvenated-stage volcanic rocks from underlying strata, but they too suggest a diffuse boundary between rejuvenated and late-shield or postshield alkalic lava flows (for example, log(Zr/Y) vs. log(Nb/Y), Figure 6.6 [data from *Clague and Dalrymple*, 1988; *Maaløe et al.*, 1992; *Mukhopadhyay et al.*, 2003; *Garcia et al.*, 2010]). Late-shield or postshield analyses (Makaweli Member of Waimea Canyon Basalt) plot within and between the fields defined by published analyses for rejuvenated- and shield-stage lava flows on element ratio diagrams. The elements Zr, Y, and Nb are among the group thought to be relatively immobile during alteration and low-grade metamorphism. Other trace-element variation diagrams yield similar results: Alkalic

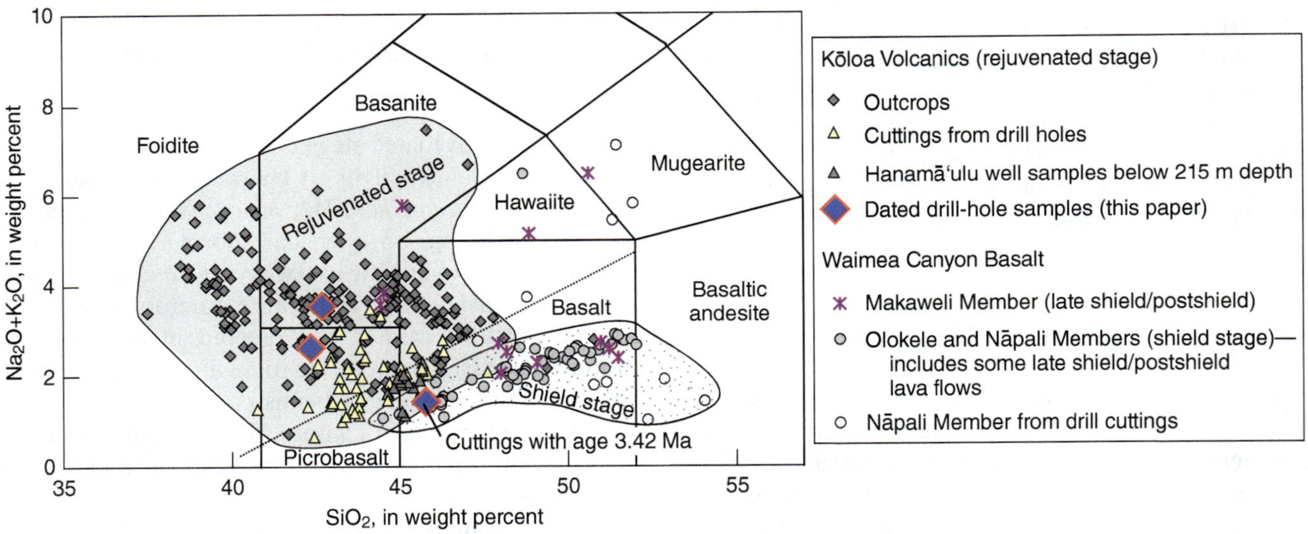

Figure 6.5 Alkali-silica (Na_2O+K_2O versus SiO_2) diagram for all Kauaʻi rock analyses. Rock classification grid fields from *Le Bas et al.* [1986]; shown dashed is boundary separating tholeiitic and alkalic basalt [*Macdonald and Katsura*, 1964]. New analyses from Table 6.1; all analyses are normalized to 100%, volatile free. Wide range of composition for Makaweli Member (of Waimea Canyon Basalt) corresponds to geochemical variation that characterizes late-shield to postshield transition. Drill cuttings from pre-Kōloa strata are part of Waimea Canyon Basalt but were not assigned to specific members within it [*Reiners and Nelson*, 1998]; these also show a wide range of composition that likely results from random sampling through the late-shield/postshield sequence of lava flows. Kauaʻi chemical data referenced in order of number of samples: *Garcia et al.* [2010], 93 analyses; *Clague and Dalrymple* [1988], 55 analyses; *Reiners et al.* [1999], 54 analyses; *Reiners and Nelson* [1998], 46 analyses; *Mukhopadhyay et al.* [2003], 35 analyses; *Palmiter* [1975], 29 analyses; *Maaløe et al.* [1992], 18 analyses; *Macdonald et al.* [1960], 14 analyses; *Garcia* [1993], 14 analyses; *Macdonald and Katsura* [1964], 12 analyses; *Macdonald* [1968], 11 analyses; *Feigenson* [1984], 11 analyses; *Cross* [1915], 3 analyses; *Kay and Gast* [1973], 3 analyses; this chapter (Table 6.1, 3 analyses); and *Washington and Keyes* [1926], 1 analysis.

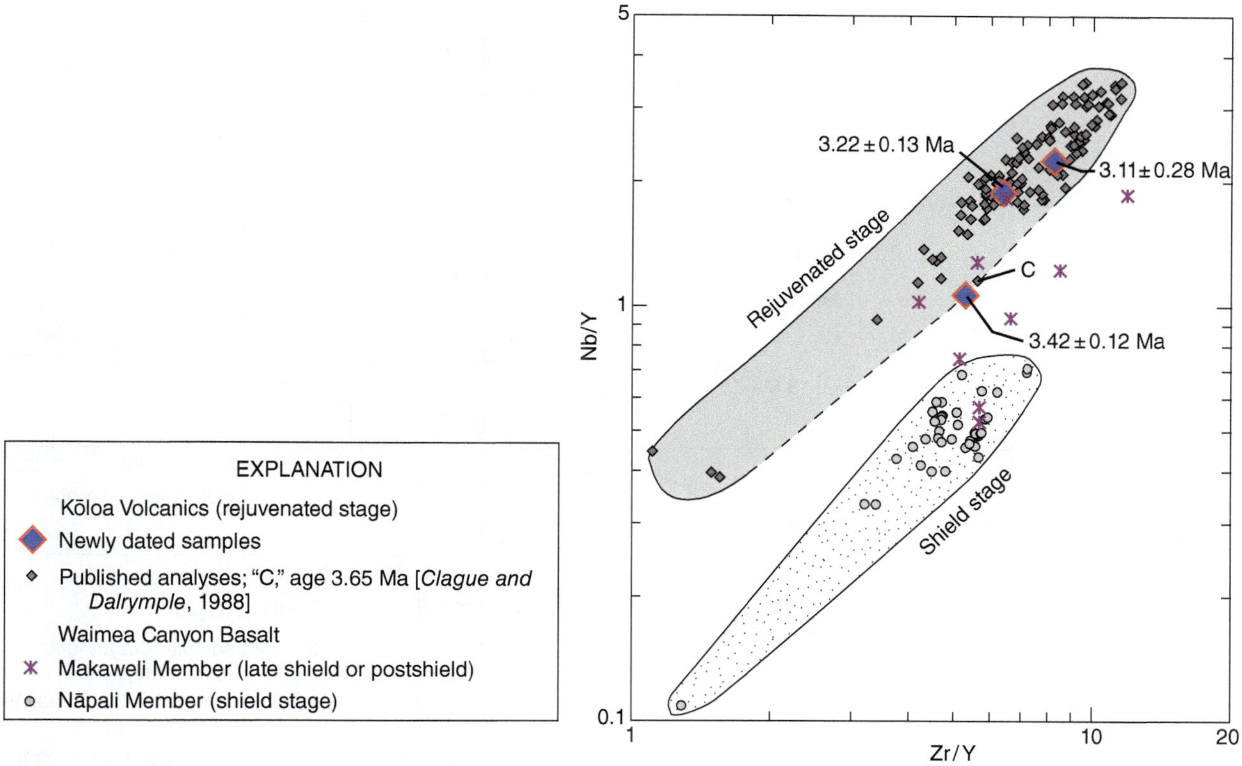

Figure 6.6 Nb/Y and Zr/Y variations for Kaua'i lava flows. Data from *Clague and Dalrymple* [1988], *Maaløe et al.* [1992], *Garcia et al.* [2010], and this work. Ages from Table 6.3.

rejuvenated-stage lava and tholeiitic shield-stage lava can be resolved graphically, but the late-shield or postshield lava flows, which range from tholeiitic to alkalic [*Garcia et al.*, 2010], lie between and transgress the various boundaries. Our younger two samples plot squarely in the rejuvenated-stage field, whereas the older sample (HTZ-895) falls near the boundary and might be interpreted as postshield alkalic lava of the Waimea Canyon Basalt.

Isotopic ratios for Sr and Nd were obtained to further characterize the newly dated samples (Figure 6.7). For Kaua'i in general, isotopic ratios decrease for $^{87}Sr/^{86}Sr$ and increase for $^{143}Nd/^{144}Nd$ upsection from shield-stage tholeiite to more alkalic postshield and rejuvenated-stage lava flows, as was shown by *Clague and Dalrymple* [1988] and *Garcia et al.* [2010]. But those ratios fail to distinguish the alkalic postshield lava from rejuvenated-stage lava (Figure 6.7a). The new data are similar to the findings of *Reiners and Nelson* [1998] from the Hanamā'ulu well (Figure 6.7b); for example, the deepest samples have some of the lowest $^{87}Sr/^{86}Sr$ values among Kaua'i lava flows, a feature that favors their interpretation as rejuvenated-stage lava.

6.5. AGE RANGE OF KŌLOA VOLCANICS

Previously published ages from the Kōloa Volcanics are, with one exception, younger than about 2.6 Ma (Figure 6.8) [*McDougall*, 1964, 1979; *Clague and Dalrymple*, 1988; *Hearty et al.*, 2005; *Garcia et al.*, 2010]. The exception is an age of 3.65 ± 0.06 Ma obtained by *Clague and Dalrymple* [1988] from a basanitic cobble in conglomerate beneath a lava flow in west-central Kaua'i. They assigned the clast to the Kōloa Volcanics on the basis of its isotope, major-, and trace-element geochemistry, but a 1 Myr gap separates the 3.65 ± 0.06 Ma age from the next youngest Kōloa age (2.59 ± 0.12 Ma). This gap persisted through more extensive dating of surface outcrops of Kōloa Volcanics, which led *Garcia et al.* [2010] to conclude that the 3.65 ± 0.06 Ma age is not from the Kōloa Volcanics. The inception of rejuvenated-stage volcanism has thus been a matter of debate, with onset either about 3.65 Ma [*Clague and Dalrymple*, 1988] or 2.6 Ma [*Garcia et al.*, 2010]. The controversy has implications for the timing of major geologic events in the history of Kaua'i and the Hawaiian Archipelago, the duration of rejuvenated-stage volcanism, and theories on the genesis of rejuvenated-stage magmatism on Kaua'i.

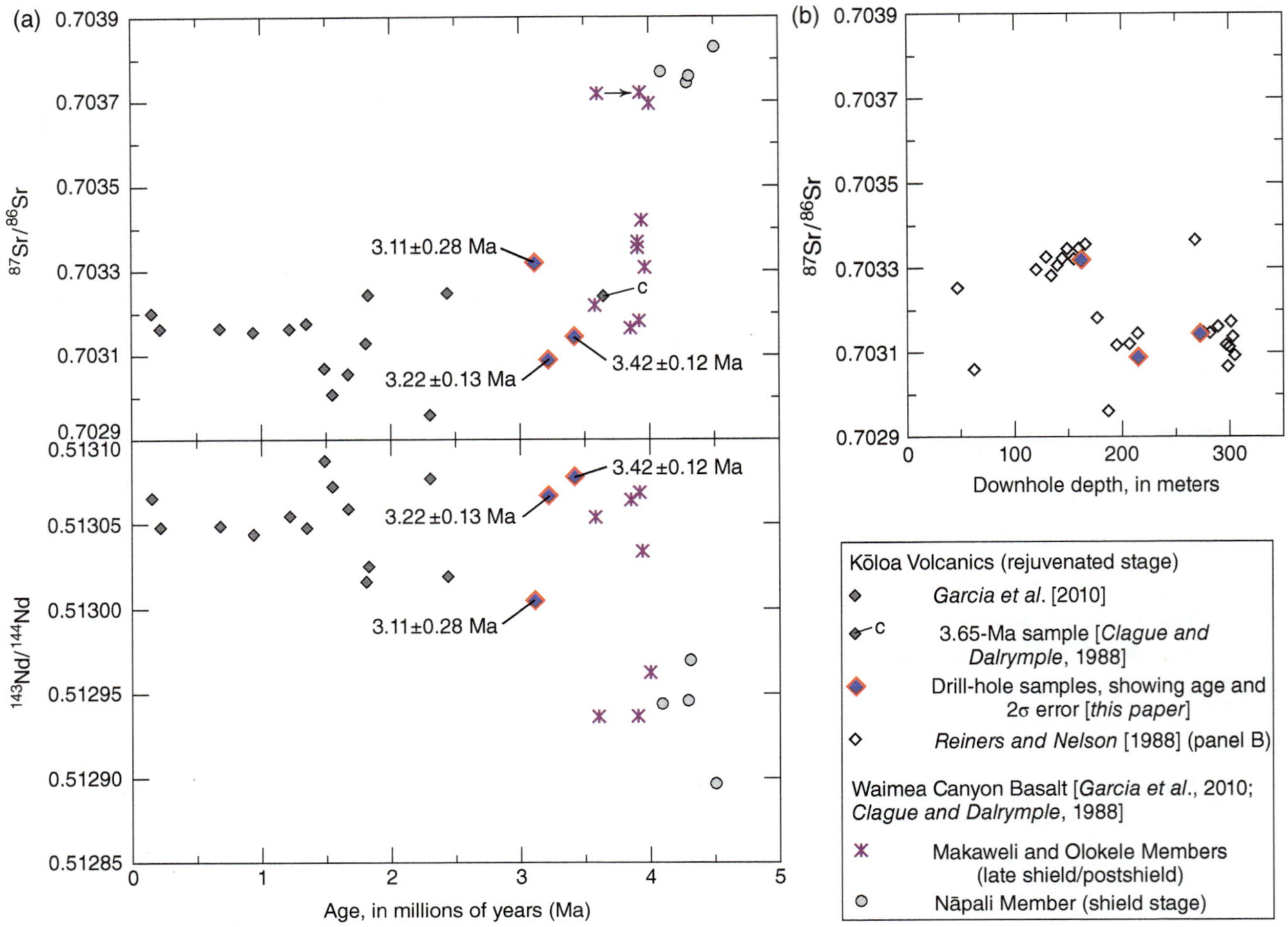

Figure 6.7 (a) Strontium and neodymium isotope ratios for dated lava flows on Kaua'i. Modified from *Garcia et al.* [2010] by addition of new analyses from cuttings in Hanamā'ulu well (Table 6.2). Ages from Table 6.3. Shield-stage lava and tholeiitic part of postshield-stage lava form fields distinct from the more alkalic rocks of postshield and rejuvenated stages, which are similar in their isotopic ratios. (b) Strontium isotope ratios from Hanamā'ulu well by downhole depth. Our new isotopic data match the findings of *Reiners and Nelson* [1998], with the deepest samples having some of the lowest $^{87}Sr/^{86}Sr$ values among Kaua'i lava flows.

The ages from this study partly fill the 1 Myr gap in the previously published ages for the Kōloa Volcanics (Figure 6.8). The chemistry and ages of at least the upper two Hanamā'ulu well samples indicate that eruption of the Kōloa Volcanics began at least 3 Myr ago. These findings leave open the possibility that the previously determined age of 3.65 ± 0.06 Ma [*Clague and Dalrymple*, 1988] was properly assigned to the Kōloa Volcanics and that rejuvenated-stage volcanism on Kaua'i persisted for about 3.5 Myr. In comparison, the Honolulu Volcanics (rejuvenated-stage deposits on O'ahu) erupted over a period of only about 0.7–0.8 Myr [*Ozawa et al.*, 2005].

Neither the initiation (timing) of Kōloa Volcanics eruptions nor the gap between postshield- and rejuvenated-stage volcanism is revealed by the new ages reported here. But with the onset pushed back to at least 3 Ma, probably to 3.4 Ma, and possibly even earlier, the interval between postshield- and rejuvenated-stage volcanism may be as little as 0.2–0.3 Myr, briefest in the island chain, if borne out. The distinction has importance when evaluating the models that explain rejuvenated volcanism [e.g., *Garcia et al.*, 2010, and references therein].

The ages from our study indicate that the gap in previously published ages may result simply from the difficulty in finding oldest candidates for dating and not because of a fundamental pause in rejuvenated-stage volcanic activity. The new ages in this chapter come from samples in only the lower half of the well, so it is likely that more of the gap could be filled by dating samples from shallower stratigraphic levels. The oldest

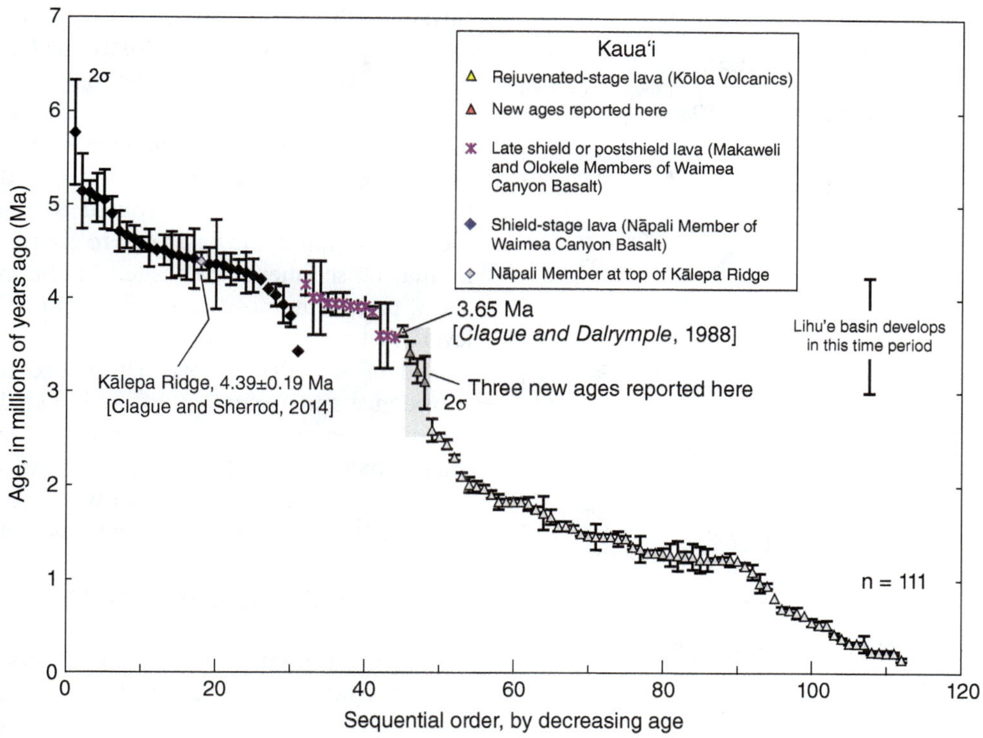

Figure 6.8 All Kaua'i radiometric ages grouped stratigraphically and then ordered sequentially. Data from *Evernden et al.* [1964], *McDougall* [1964, 1979], *Clague and Dalrymple* [1988], *Hearty et al.* [2005], *Garcia et al.* [2010], *Clague and Sherrod* [2014], and our new data. Labeled is the previously published 3.65 Ma age, which is controversial in its stratigraphic assignment, as discussed in text. Data published before 1977 relied on older decay constants and are recalculated here with new decay constants [*Steiger and Jäger*, 1977] for consistency. The data of *Evernden et al.* [1964] are insufficient to allow recalculation, so the method of *Dalrymple* [1979] was used, which leads to ages about 2% older than originally reported.

dated sample from the well, 274 m depth, is underlain by another 25–30 m of lava flows. If all of the Hanamā'ulu well is in Kōloa Volcanics, as suggested by *Reiners and Nelson* [1998] or *Reiners et al.* [1999], then Kōloa ages older than those reported here might be expected.

6.6. RATES OF KŌLOA STRATIGRAPHIC ACCUMULATION, TIMING OF LĪHU'E BASIN FORMATION, AND RATES OF ISLAND SUBSIDENCE

Independent of stratigraphic arguments, the new ages from this study give insight about the formation of the Līhu'e basin. The site of the Hanamā'ulu well must have been near sea level to allow deposition of the lower shallow marine layer (Figure 6.3); therefore, the Līhu'e basin must have formed before deposition of the lower marine unit. The two dated lava flows above the lower marine layer provide the minimum age of the basin. Their weighted mean age is 3.20 ± 0.12 Ma, so the basin formed before 3 Ma.

The age depth data can be converted to rates of stratigraphic accumulation during the time about 3 Ma. The data form a linear trend that indicates stratigraphic accumulation (lava flows and sediment) at a rate 0.3–0.4 mm per year (Figure 6.9). The analytical uncertainty, however, allows a wide range, from as little as 0.1 to 10 mm per year. The faster rate assumes that the 111 m of strata, including those of two marine incursions, might have been deposited in as little as 10,000 years, but it seems an unrealistically rapid accumulation rate because the presence of reefs in the marine deposits and marine sediment more than 30 m thick would require a longer period of sediment accumulation. Also, the lower marine unit shows a facies transition from reef to estuary, which indicates shoreline progradation by sediment deposition, likely over a lengthy period [*Izuka and Resig*, 2008].

Following the initial subsidence that formed the Līhu'e basin, the site continued to subside, as indicated by the deposition of the shallow marine layers and intervening lava flows. We refer to this phase as "island subsidence," although sea-level changes may have contributed to the

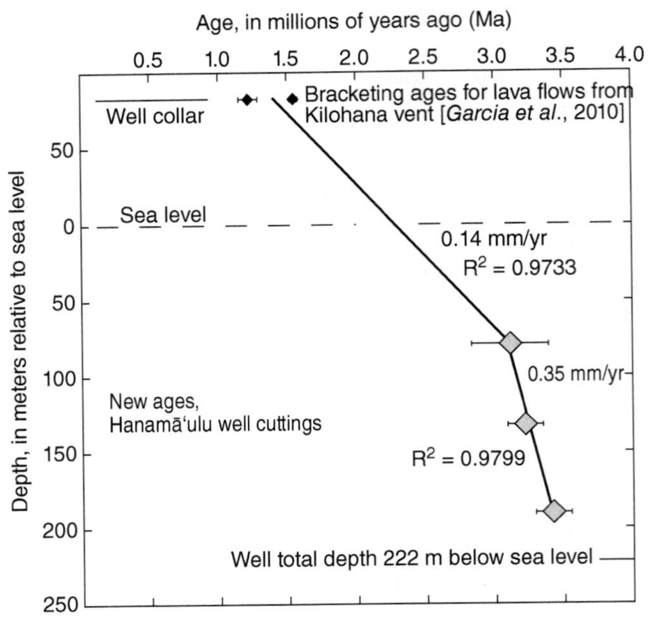

Figure 6.9 Age-depth plot showing stratigraphic accumulation rates. Ages from Table 6.3.

apparent subsidence. The simplest interpretation for a stratigraphic sequence that includes two marine layers and an intervening 30 m sequence of lava flows is that the accumulation rate roughly matched the island subsidence rate.

For the Hawaiian Archipelago, the connection between island building and subsidence depends on distance from hotspot and the depression of oceanic lithosphere by the load of the nearby large young volcanoes. Subsidence can be as great as 2–3 mm per year at the shoreline of a young large island [*Moore*, 1987; *Ludwig et al.*, 1991]. A substantially lower rate would have characterized Kaua'i at 3 Ma, when its geographic setting relative to the Hawaiian hotspot would have been comparable to the position of Lāna'i today (Figure 6.10); that is, about 250 km from the hotspot and about 200 km from the point of greatest subsidence. For Lāna'i, the rate of subsidence has been low during the past 30,000 years, in the range +0.1 and –0.4 mm per year on the basis of drowned reef evidence [*Webster et al.*, 2006]. These rates are comparable to the island subsidence rates for Kaua'i 3 Myr ago, computed from the age depth data in this study.

A lower averaged stratigraphic accumulation rate characterizes the time after 3 Ma. To establish the youngest time point, we take ages from surface lava flows in the area of the Hanamā'ulu well. These, from Kilohana, a small rejuvenated-stage shield volcano, are between 1.55 ± 0.03 and 1.22 ± 0.07 Ma [*Garcia et al.*, 2010, p. 1515]. The sample altitudes are between 50 and 130 m, so we use the 83 m altitude at the Hanamā'ulu well collar (Figure 6.9). The age depth curve connecting the Kilohana surface with the shallowest of the samples in the well indicate stratigraphic accumulation at a time-averaged rate of 0.14 mm per year (Figure 6.9).

The accumulation rate during the 1.5 Myr period following 3 Ma probably has more variation (higher and lower rates) than the long-term averaged rate. Nor can the rate be definitively attributed to island subsidence rates, even though the basin had to continue subsiding for its marine sediments, deposited in shallow water about 3 Ma, to become positioned 86–185 m below present sea level. The time between 3 and 1.5 Ma spans a complex period of Kaua'i's history. The island underwent both subsidence and flexural uplift as it edged outward from the Hawaiian moat, arriving finally (1.5 Ma) in a hotspot-relative paleogeographic position similar to that of O'ahu today (Figure 6.10). The island was also subject to eustatic sea-level fluctuations during this period.

6.7. HOW DID THE LĪHU'E BASIN FORM?

Postulated origins for the Līhu'e basin include stream erosion, caldera collapse, and landslide. The stream erosion origin [*Stearns*, 1967] was offered before the thickness and composition of the basin fill was known. It was refuted on geomorphic grounds by G. A. Macdonald [*Stearns*, 1967, p. 108], who emphasized that the general trend of the Kālepa-Nonou ridge is nearly across the supposed course of the ancestral stream valleys, or roughly at right angles to the trend of ridges that would have formed by the erosive action of consequent streams. Erosion may have modified the basin extensively, but it was not the primary cause of the basin's formation [*Macdonald et al.*, 1960].

The caldera hypothesis presumes magma withdrawal. It is based partly on the roughly circular aspect of the basin. As explained by *Macdonald et al.* [1960, p. 96], both the Līhu'e depression and another graben, the Makaweli, were "related to a general decrease of pressure in the underlying magma reservoir as the end of the [Kaua'i shield] volcanic cycle drew near." Calderas are common features of Hawaiian volcanoes in the preshield and shield stage, but none are known to have developed during postshield or rejuvenated-stage volcanism such as characterized by Kaua'i in the time 3–3.5 Ma. (A caldera origin cannot be wholly dismissed; see Section 6.7.3.)

We propose instead a structural origin unrelated to reduction in magmatic pressure for the formation of the basin. The basic data set to interpret is a structure contour map drawn on the top of the Waimea Canyon Basalt (Figure 6.11a). In the Līhu'e basin the contoured surface corresponds roughly to the contact between the Waimea Canyon Basalt and Kōloa Volcanics, a depiction that relies heavily on the drill-hole data from the late 1990s. The contoured surface within the basin is roughly what

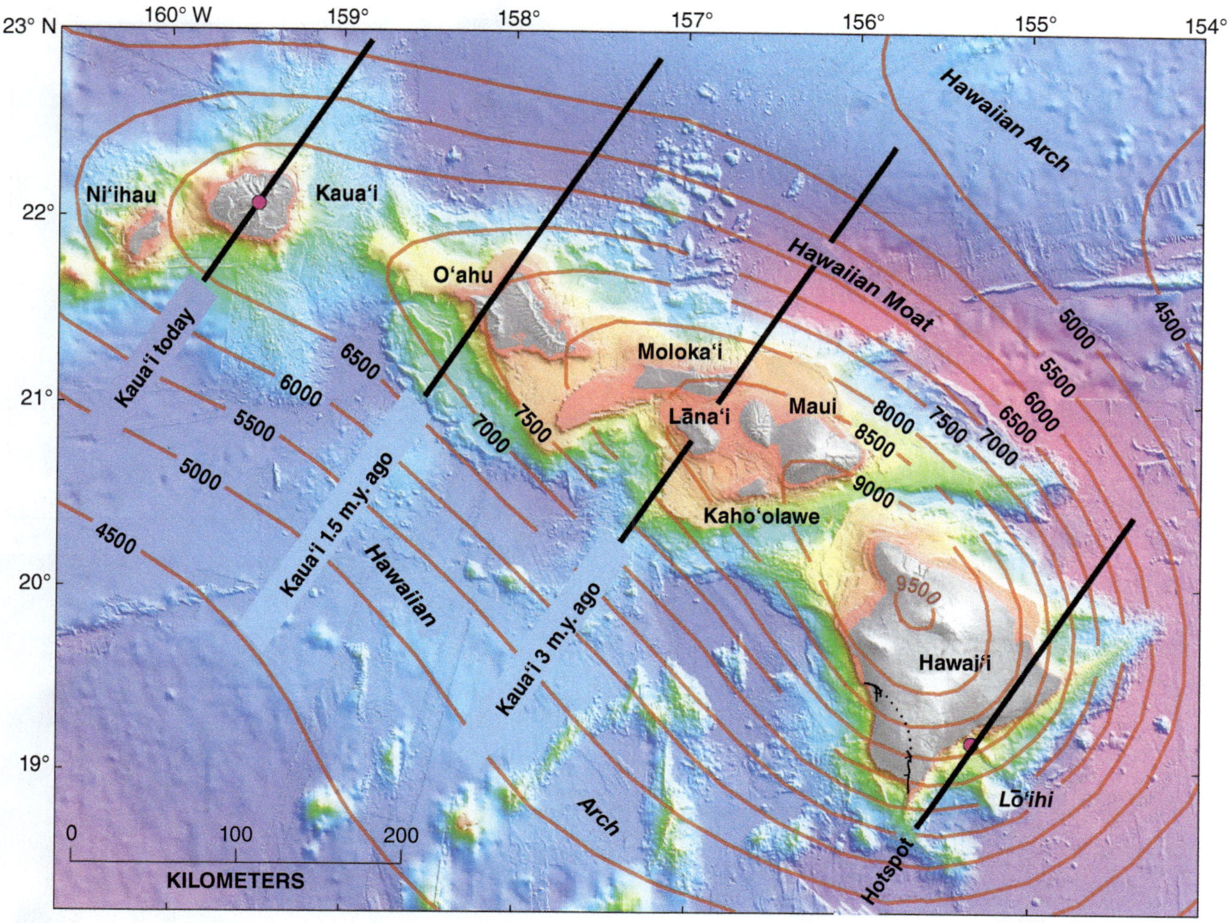

Figure 6.10 Major Hawaiian Islands with black chron lines showing the distance of Kaua'i from the Hawaiian hotspot at 3, 1.5, and 0 Ma, assuming lithospheric translation of 100 km/Myr [*Simkin et al.*, 2006]. At 3.0 Ma, soon after the onset of Kōloa volcanism, Kaua'i was at a position similar to that of Lāna'i today with respect to subsidence rate resulting from some large-island load closer to the hotspot. At 1.5 Ma, Kaua'i was at a position similar to O'ahu today. By comparison, flexural uplift likely affected Kaua'i's subsidence history between 3 and 1.5 Ma. Red lines are contours showing depth in meters below sea level to top of oceanic crust [*Watts and ten Brink*, 1989], which is depressed because of the load of young volcanoes and high heat flow in area of Hawaiian hotspot. Base from *Eakins et al.* [2003].

would have been exposed between about 3 and 3.5 Ma, to judge from the new ages reported here. Our map for the Līhu'e basin is consistent with that drawn by *Gandy* [2006] for calculating Kōloa Volcanics thickness.

More difficult is interpreting the ~3 Ma configuration across the mountainous terrain of Kaua'i, where the Kōloa Volcanics unit is scattered or absent, where the base of the Kōloa can be as young as a few hundred thousand years, and where erosion of the past 3 Myr has greatly modified the top of the Waimea Canyon Basalt compared to buried surfaces. Broad topographic surfaces in central and western Kaua'i are likely the vestige of the top of the Kaua'i shield. In the northern and eastern parts of the island where erosion is more severe, topographic contours were first traced, and then the structure contours were drawn to connect points of equal altitude along ridgelines so that contour lines enclose the greatest breadth of area.

The structure contours in the center of the island are shown abutting a northward striking fault, discussed more fully in the next section. Presumably strata of the Waimea Canyon Basalt are truncated at this structure. If not a fault, then a northward trending monocline is required to bridge the high (western) and low (eastern) surface altitudes at the top of the Waimea Canyon Basalt.

6.7.1. West Side of Līhu'e Basin: An Island-Long Escarpment

A prominent feature of the structure contour map is its similarity to topographic contrasts across the island. All of east Kaua'i, including the Līhu'e basin,

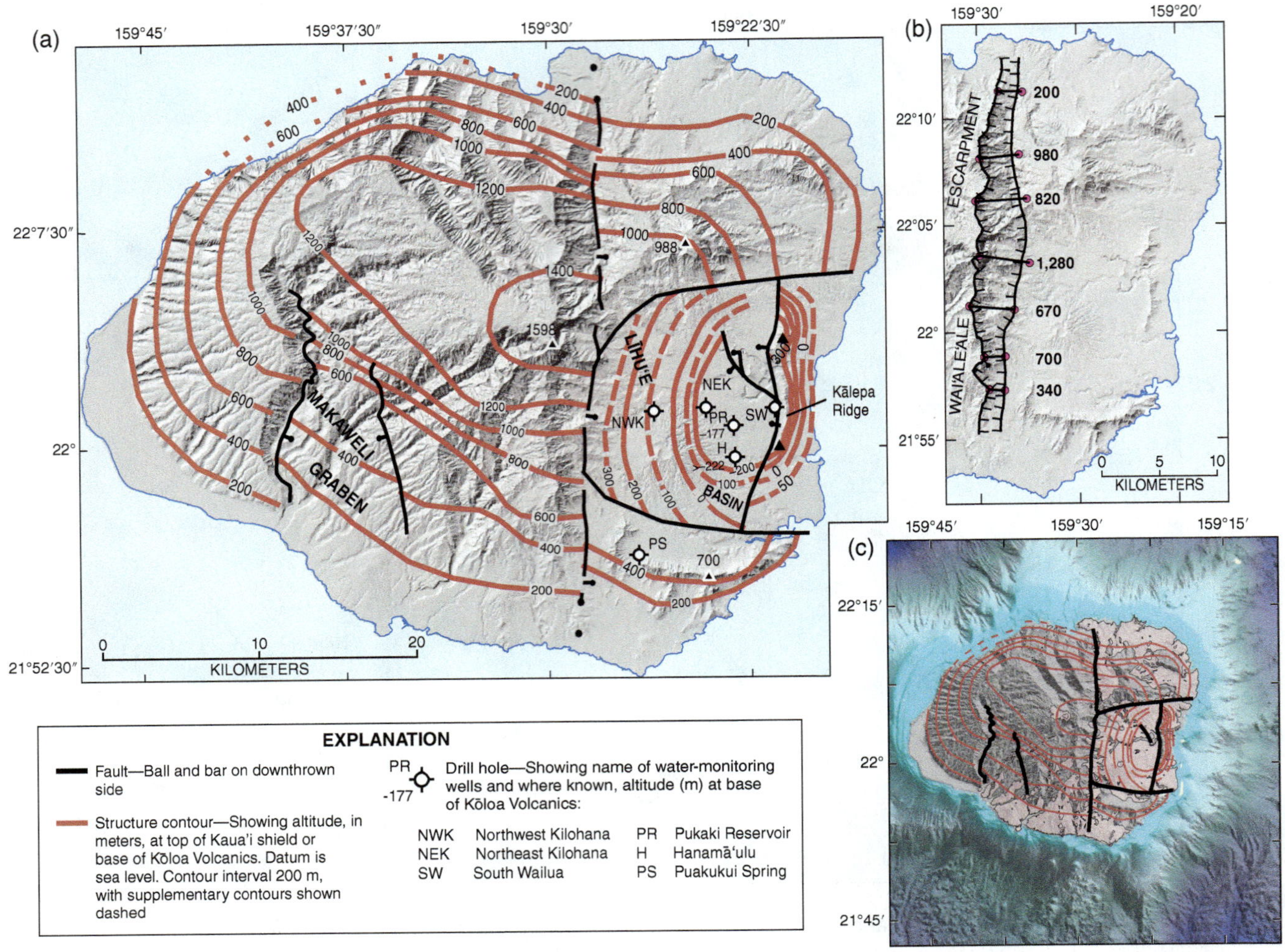

Figure 6.11 Major structural features on Island of Kaua'i. (a) Structure contour diagram. Contours depict altitude, in meters relative to present sea level, at base of Kōloa Volcanics or top of Waimea Canyon Basalt. Two faults on west side of island bound the Makaweli graben [*Macdonald et al.*, 1960]. New interpretations here are major north-south fault and structures in Līhu'e basin. Hillshade relief map from State of Hawaii Office of Planning [http://planning.hawaii.gov/gis/download-gis-data/]. (b) Depiction of altitude contrast across steep zone that separates topographically higher western Kaua'i from lower eastern Kaua'i. Bold line on left is traverse that follows the north-south crestline of the island (Wai'ale'ale escarpment); line on right approximates the break in slope at the foot of the escarpment. Numbers indicate topographic difference, in meters, between high and low paired points along the two traverses. Between them lies a pali (cliff or steep area), probably an ancient fault escarpment. (c) Bathymetric shaded-relief map showing submarine flanks of Kaua'i. Offset on proposed north-south fault diminishes from island center toward shoreline, where it loses expression completely. Similarly, Makaweli graben-bounding faults lack expression on Kaua'i's submarine flanks. Bathymetric data are multibeam synthesis from University of Hawai'i's School of Ocean and Earth Science and Technology (http://www.soest.hawaii.edu/HMRG/multibeam/index.php).

lies substantially lower topographically than the higher western part of the island. The boundary zone can be mapped by plotting altitude contrasts between points along the edge of the highland and those at the foot of the slope that divides western and eastern parts (Figure 6.11b). The difference is as great as 1280 m near the center of the island, diminishing north and south to about 200 m before losing expression at the coastal plain. The zone of altitude contrast is roughly linear, even though deeply fretted by erosion.

On the other Hawaiian Islands, similar topographic contrasts have formed chiefly as headwalls to major landslides. Examples include the fault-bounded Hilina Pali on the south flank of Hawai'i; the Kahuku-Kealakekua fault system on the southwest slope of Mauna Loa; and the seacliffs that bound the north side of East Moloka'i.

A notable exception is Nu'uanu Pali on O'ahu, the foot of which has a stratigraphically deep (600 m) exposure of the Kailua caldera and dike complex of the Ko'olau volcano; thus, Nu'uanu Pali is thought to be entirely erosional [*Stearns and Vaksvik*, 1935; *Macdonald et al.*, 1983].

For Kaua'i, it is difficult to explain such an areally extensive topographic contrast without invoking a structural origin. The evidence for stratigraphic offset is imperfect, however. No mapped stratigraphic unit is sufficiently extensive to measure displacement of shield-stage strata of the Waimea Canyon Basalt across the island-long escarpment. Radiometric ages from the Waimea Canyon Basalt, as a surrogate for stratigraphic position, are sparse. No rocks have been dated along the crest of the topographic escarpment. In eastern Kaua'i, an age of 4.39 ± 0.19 Ma, from the summit of Kālepa Ridge [*Clague and Sherrod*, 2014], is toward the younger reach of shield-stage ages (Figure 6.5). This age correlation is the best available line of evidence that lava flows at the top of the Waimea Canyon Basalt in the lower, eastern part of the island are probably at about the same stratigraphic position as those near the top of the higher, western part.

6.7.2. Origin of Escarpment

We propose that the 30 km long, eastward-facing steep slope (pali) on Kaua'i originated as a headwall escarpment for a major gravity-driven or mobile-flank fault that ceased motion before the east side of the island dropped completely below sea level. By that interpretation, the east-central part of the island has sagged nearly 1300 m relative to the western half. The Līhu'e basin's west side is coincident with the headwall escarpment, but the basin itself lies wholly within the area of the large downdropped eastern block. Previous geologic literature largely ignores the origin of this major escarpment. It may have been hinted at by *Macdonald et al.* [1960, p. 24], who reconciled anomalous dips of lava flows on Kālepa Ridge (generally 10°–12° east shown on a map by *Macdonald et al.* [1960]) by flattening of the dip to the west, owing to a "sagging of that portion of the shield, as a precursor to the greater collapse that appears to have contributed to the formation of the Lihue basin."

The fault lacks expression on the north and south submarine flanks of Kaua'i (Figure 6.11c), which is in agreement with its diminished offset north and south from the central area (Figure 6.11a). Kōloa Volcanics lava flows have coated the coastal plain and submarine slopes in the past 2–3 Myr [*Clague and Dalrymple*, 1988; *Garcia et al.*, 2010], further masking evidence for the fault. Similarly, the Makaweli graben, in the western part of the island, loses expression before reaching the coastal plain (Figure 6.11a [*Macdonald et al.*, 1960]). Conceivably the Makaweli graben formed by the same extensional process that allowed the eastern part of the island to subside.

Structural geometry of mobile-flank faults, in cross section, may be likened to the clay-cake experiments pioneered by Hans Cloos in the 1920s [cited and illustrated by *Mandl*, 1988]. As an island's flank extends, grabens may develop owing to opposed dips of two fault sets: main and antithetic. These are depicted by way of cross sections for Kaua'i and the clay-cake model (Figure 6.12). In terms of island geography, the main fault dips toward the unbuttressed seaward slope, whereas the antithetic fault is oppositely inclined, dipping back toward the island's center. These faults cut down deeper than the floor of the graben, which forms a disconformable depositional surface. Kālepa Ridge is the eastside footwall to the graben. The west flank of Kālepa Ridge must be steep, because the South Wailua well, drilled close to it (Figure 6.1), penetrated the Waimea Canyon Basalt only near the bottom of the well [*Izuka and Gingerich*, 1997c; *Reiners et al.*, 1999]. Our cross section is similar to that of *Reiners et al.* [1999, their Figure 5].

The steep normal faults may root into a décollement near or above the ocean floor-island interface or they terminate in a diffuse zone of compensation within the island (Figure 6.12a). The island's eastern submarine flank shows no bathymetric evidence for deep structural disturbance, such as an eastward directed thrust (Figure 6.11c); but if the décollement was at the ocean-floor interface, its east-flank expression might easily be obscured by sedimentation in the time since 3 Ma, given the small total strain across the island-long fault or Līhu'e basin (<5% across breadth of section in Figure 6.12b). A similar explanation applies to a diffuse zone of compensation.

By this interpretation, formation of the Līhu'e basin was synchronous with or followed soon after displacement along the island-long mobile-flank fault. During that episode, lava flows at Kālepa Ridge and stratigraphically similar strata in the graben block were downdropped from lava flows of similar age at the top of the flank-fault footwall, a stratigraphic hypothesis that can be tested by additional dating along the Wai'ale'ale escarpment. This line of reasoning suggests that the Līhu'e basin formed sometime between 4.39 ± 0.19 Ma (age from top of Kālepa Ridge) and 3 Ma (age of lava flows atop marine sediment in the basin) (Table 6.4).

Two examples elsewhere along the island chain may provide useful analogues for fault systems that cut through the summit regions of Hawaiian shield volcanoes. On Lāna'i, a 16 km long, northwest trending system of normal faults transects the center of the island [*Stearns*, 1940], creating a topographic escarpment with 600 m of relief. The faults lose expression before reaching the shoreline. This fault system has about one-half the length and one-half the

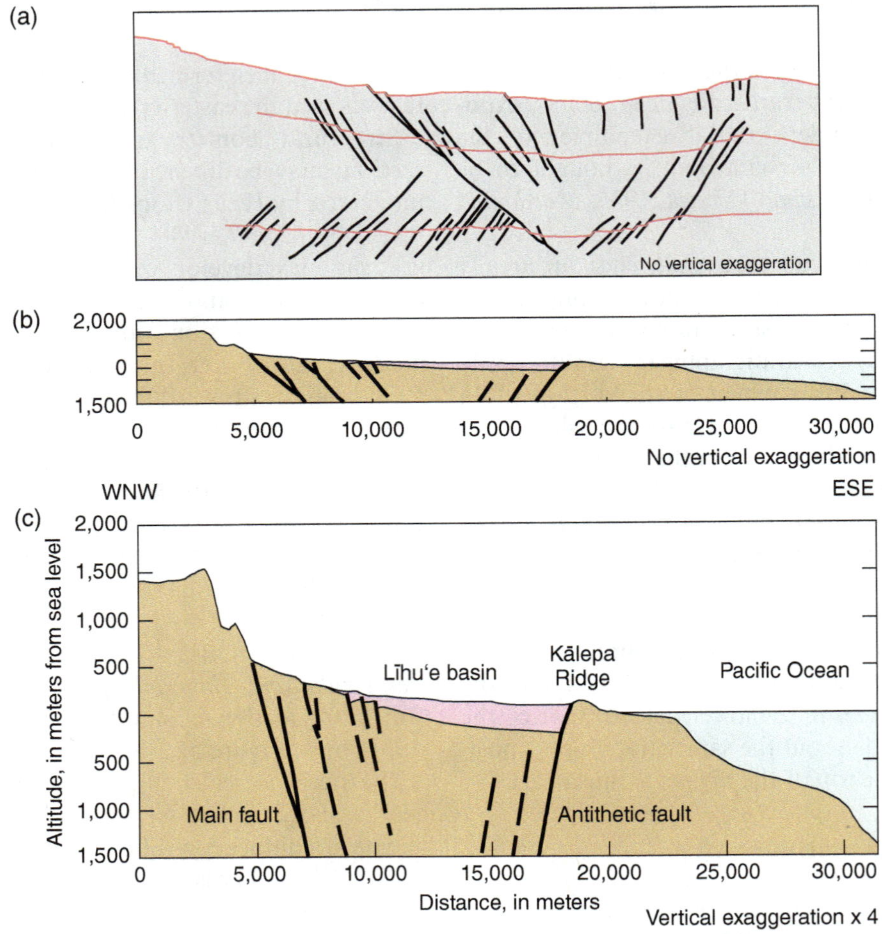

Figure 6.12 Cross section of eastern Kaua'i showing depth of Līhu'e basin. (a) Cross-sectional tracing of deformation and structure resulting from clay cake extended by stretching of rubber base [reproduced from *Mandl*, 1988]. (b) Cross section from highland of western Kaua'i to eastern coast; no vertical exaggeration. Fault pattern from panel (a). (c), Same as panel (b) with vertical exaggeration.

Table 6.4 Geologic history of eastern Kaua'i.

Period (million years ago)	Events	Island Subsidence Rate
From 3.11 to present	**At Hanamā'ulu well site, lava flows of Kōloa Volcanics bury upper marine unit; Palikea Breccia Member of Kōloa Volcanics deposited; continued eruption of Kōloa Volcanics and formation of Kilohana volcano. Time-averaged stratigraphic accumulation rate 0.14 mm/y**	**At least 95 m over 3.11 Myr, but includes periods of subsidence, uplift, and global sea-level fluctuations**
From 3.22 to 3.11	**Deposition of upper marine unit at Hanamā'ulu well site**	**0.3–0.4 mm/y**[a]
3.22	**Lava flows from Kōloa Volcanics bury lower marine unit at Hanamā'ulu well site**	**0.3–0.4 mm/y**[a]
	Major stream drainage in Līhu'e basin pushed to north [*Izuka and Resig*, 2008]	
From 3.42 to 3.22	**Deposition of lower marine unit at Hanamā'ulu well site**	**0.3–0.4 mm/y**[a]
From 4.4 to 3 (?)	**Faulting along island-long escarpment (Figure 6.11); formation of Līhu'e basin by landsliding or faulting**	**0.3–0.4 mm/y**[a]
3.65 [*Clague and Dalrymple*, 1988]	Earliest eruption of Kōloa Volcanics [*Clague and Dalrymple*, 1988]	**0.3–0.4 mm/y**[a]
From 5.5 to 3.5 [*Sherrod et al.*, 2007]	Formation of Kaua'i shield, Olokele depression (caldera?), Makaweli graben; deposition of Nāpali, Makaweli, and Olokele (and Haupu?) Members of the Waimea Canyon Basalt [*Macdonald et al.*, 1960; *Sherrod et al.*, 2007]	2–3 mm/y, by analogy to Island of Hawai'i today [*Moore*, 1987; *Ludwig et al.*, 1991]

Note: Boldface indicates values and events interpreted on the basis of data in this study; other data and interpretations from cited sources.

[a] Similar rates are described for Lāna'i today [*Webster et al.*, 2006].

escarpment height of the proposed Kaua'i fault. The Lāna'i structural feature was interpreted by *Moore et al.* [1989] as the head of the Clark debris avalanche, that is, the upper limit of the region of mass failure.

On Kohala (Island of Hawai'i), a northwest trending system of normal faults 13 km long has created a summit graben along the top of the volcano [*Wolfe and Morris*, 1996]. The fault zone is upslope of the offshore Pololū debris avalanche, and the summit structural depression was interpreted as a pull-apart graben developed at the head of the landslide [*Moore et al.*, 1989].

6.7.3. Circular Aspect of Līhu'e Basin

The Līhu'e topographic basin presents a semicircular shape on shaded relief maps. As described by *Macdonald et al.* [1960, p. 94], "The circular shape of the basin immediately suggests its origin by caldera collapse."

That the Līhu'e basin may once have been near the summit region of the Kaua'i volcano has some credence in islandwide gravity data [*Flinders et al.*, 2010], which have maximum values in the northern part of the basin (Figure 6.1). Gravity maxima on Hawaiian shield volcanoes commonly correspond to crystallized summit magma chambers. The isogal lines for gravity maxima extend northwest beyond the basin, crossing the island-long escarpment [*Krivoy et al.*, 1965; *Flinders et al.*, 2010]. But to ascribe Līhu'e basin origin to caldera formation ignores the oblique trend of the gravity maximum relative to the basin outline. Also, the Līhu'e basin would be large for a hotspot island caldera, 15 km across. A comparison, by diameter, of Hawaiian Island and other Pacific island calderas suggests that 5 km is a more representative breadth according to *Flinders et al.* [2010], who concluded that the summit of the Kaua'i volcano has likely been stripped away by extensive mass wasting or erosion.

If the Līhu'e structural feature originated as Kaua'i's shield-stage caldera, it must have remained intact even as the eastern third of the island subsided along the island-long north-south fault. Indeed, the fault would have edged the west side of the caldera. A caldera origin removes the need for the tear faults shown in Figure 6.11, which are required to form a box-shaped graben. Instead, existing ring faults would have undergone renewed displacement as the (now-postshield) basin floor subsided below sea level. Steep to vertical ring faults are thought to characterize calderas, so no additional extensional strain would have been required as the Līhu'e basin became a marine sediment-holding basin. Thereafter, the basin's depositional history would be the same by caldera or noncaldera interpretation.

The question remains: Do curvilinear structural features necessarily rule out the graben subsidence model suggested here? An answer comes from younger examples on the Island of Hawai'i. There, curvilinear fault traces characterize some of the lesser headwalls associated with major flank-failure faults such as the Kahuku-Kealakekua fault system on the west side of Mauna Loa or the Hilina Pali (fault) on south flank of Kīlauea [*Wolfe and Morris*, 1996; *Trusdell et al.*, 2006]. Small faults that diverge from the Kealakekua fault delineate a semicircular feature similar in scale (mapwise) to the breadth of the Līhu'e basin.

6.8. SUMMARY

Cuttings from three lava flows in the lower part of a drill hole in the Līhu'e basin yield ages of 3.42, 3.22, and 3.11 Ma, in correct stratigraphic order. Geochemically the two younger lava flows are part of the Kōloa Volcanics, and the oldest lava may well be Kōloa; thus the ages fill a gap in the dating archive and record a history that reaches back nearly to a previously reported Kōloa age, 3.65 Ma [*Clague and Dalrymple*, 1988].

The geologic history of eastern Kaua'i includes movement along a heretofore undescribed island-long northward striking fault that has lowered the eastern third of Kaua'i by hundreds of meters sometime between 4.4 and 3 Ma. Shallow-water marine sedimentary rocks are interbedded within the 3.42–3.11 Ma sequence of lava flows described in this study, so the Līhu'e basin formed by 3 Ma or earlier.

Stratigraphic accumulation rates in the Līhu'e basin, from age-depth relations, were about 0.3–0.4 mm/yr from 3.42 to 3.11 Ma. (Rates could be as low as 0.1 mm/yr to as much as 10 mm/yr when the 2σ error of the ages is considered.) These rates probably correspond to island subsidence rates because two shallow-water marine layers are interbedded in that part of the volcanic sequence. The rates are similar to those of Lāna'i today, a volcano that sits at a distance from the Hawaiian hotspot similar to that occupied by Kaua'i at 3 Ma in its history.

ACKNOWLEDGMENTS

Drill cuttings used in this study were obtained from wells drilled by the U.S. Geological Survey in cooperation with the County of Kaua'i Department of Water. Discussions about Hawaiian Island structure with R. Denlinger helped to clarify our own thoughts. The manuscript was reviewed by M. A. Coombs, M. O. Garcia, S. Yamasaki, and two anonymous reviewers.

APPENDIX: SAMPLE DESCRIPTIONS

HTZ-530/535 Slightly porphyritic picrobasalt, fine grained with sparse (<1%) olivine phenocrysts to 1 mm. Sugary equigranular groundmass (plagioclase + pyroxene).

HTZ-705/710 Slightly porphyritic basanite, olivine to 1 mm as phenocryst in sugary fine-grained groundmass. Pyroxene phenocrysts to 1 mm, <<1%.

HTZ-895/900 Clinopyroxene-bearing picritic basalt, with olivine phenocrysts ≤1 mm, 10–15% and cpx ~1 mm, 5%.

REFERENCES

Clague, D. A., and G. B. Dalrymple (1988), Age and petrology of alkalic postshield and rejuvenated-stage lava from Kauai, Hawaii, *Contrib. Mineral. Petrol.*, *99*, 202–218.

Clague, D. A., and D. R. Sherrod (2014), Growth and degradation of Hawaiian volcanoes, Chapter 3 of Poland, M. P., T. J. Takahashi, and C. M. Landowski, (eds.), *Characteristics of Hawaiian Volcanoes*, U.S. Geol. Surv. Prof. Pap. 1801, pp. 97–146, doi:10.3133/pp18013.

Cross, W. (1915), *Lavas of Hawaii and Their Relations*, Prof. Pap. 88, U.S. Geol. Surv., 97 pp.

Dalrymple, G. B. (1979), Critical tables for conversion of K-Ar ages from old to new constants, *Geology*, *7*(11), 558–560.

Eakins, B. W., J. E. Robinson, T. Kanamatsu, J. Naka, J. R. Smith, E. Takahashi, and D. A. Clague (2003), Hawaii's volcanoes revealed, U.S. Geol. Surv. Geol. Invest. map I–2809, scale about 1:850,000, available at http://geopubs.wr.usgs.gov/i-map/i2809.

Evernden, J. F., D. E. Savage, G. H. Curtis, and G. T. James (1964), Potassium-argon dates and the Cenozoic mammalian chronology of North America, *Am. J. Sci.*, *262*, 145–198.

Feigenson, M.D. (1984), Geochemistry of Kauai volcanics and a mixing model for the origin of Hawaiian alkali basalts, *Contrib. Mineral. Petrol.*, *87*(2), 109–119.

Flinders, A. F., G. Ito, and M. O. Garcia (2010), Gravity anomalies of the Northern Hawaiian Islands; implications on the shield evolutions of Kauai and Niihau, *J. Geophys. Res.*, *115*, B08412, 15 pp., doi:10.1029/2009JB006877.

Gandy, C. E. (2006), Volume and petrologic characteristics of the Koloa Volcanics, Kaua'i, M.S. thesis, Dept. of Geol. and Geophys., Univ. of Hawai'i at Mānoa, Honolulu, O'ahu, Hawai'i, 107 pp.

Garcia, M. O. (1993), Pliocene-Pleistocene volcanic sands from Site 842: Products of giant landslides, in *Proc. Ocean Drill. Prog., Sci. Res.*, *136*, edited by R. H. Wilkens, J. Firth, and J. Bender, pp. 53–63, College Station, Texas.

Garcia, M. O., L. Swinnard, D. Weis, A. R. Greene, T. Tagami, H. Sano, and C. E. Gandy (2010), Petrology, geochemistry and geochronology of Kaua'i lavas over 4.5 Myr: Implications for the origin of rejuvenated volcanism and the evolution of the Hawaiian plume, *J. Petrol.*, *51*(7), 1507–1540, doi:10.1093/petrology/egq027.

Gingerich, S.B., and S.K. Izuka (1997a), Construction, geologic log, and aquifer tests of the Puakukui Springs Monitor Well (State Well 2-5626-01), Lihue, Kauai, Hawaii, Open-file Rep. 97–38, U.S. Geol. Surv., 18 pp, Reston, Virginia.

Gingerich, S. B., and S. K. Izuka (1997b), Construction, geologic log, and aquifer tests of the Northwest Kilohana monitor well (State Well 2–0126–01), Lihue, Kauai, Hawaii, Open-file Rep. 97–40, U.S. Geol. Surv., 20 pp, Reston, Virginia.

Hearty, P. J., D. B. Karner, P. R. Renne, S. L. Olson, and S. Fletcher (2005), $^{40}Ar/^{39}Ar$ age of a young rejuvenation basalt flow: Implications for the duration of volcanism and the timing of carbonate platform development during the Quaternary on Kaua'i, Hawaiian Islands, *New Zealand J. Geol. Geophys.*, *48*, 199–211.

Holcomb, R. T., P. W. Reiners, B. K. Nelson, and N.-L. Sawyer (1997), Evidence for two shield volcanoes exposed on the island of Kauai, Hawaii, *Geology*, *25*(9), 811–814, doi:10.1130/0091-7613(1997)025<0811:EFTSVE>2.3.CO;2.

Izuka, S. K., and S. B. Gingerich (1997a), Construction, geologic log, and aquifer tests of the Hanamaulu monitor well (State Well 2–5923–08), Lihue, Kauai, Hawaii, Open-file Rep. 97–36, U.S. Geol. Surv., 23 pp, Reston, Virginia.

Izuka, S. K., and S. B. Gingerich (1997b), Construction, geologic log, and aquifer tests of the Northeast Kilohana monitor well (State Well 2–0124–01), Lihue, Kauai, Hawaii, Open-file Rep. 97–37, U.S. Geol. Surv., 21 pp, Reston, Virginia.

Izuka, S. K., and S. B. Gingerich (1997c), Construction and geologic log of the South Wailua monitor well (State Well 2–0121–01), Lihue, Kauai, Hawaii, Open-file Rep. 97–38, U.S. Geol. Surv., 14 pp, Reston, Virginia.

Izuka, S. K., and S. B. Gingerich (1997d), Construction, geologic log, and aquifer tests of the Pukaki Reservoir monitor well (State Well 2–0023–01), Lihue, Kauai, Hawaii, Open-file Rep. 97–41, U.S. Geol. Surv., 22 pp, Reston, Virginia.

Izuka, S. K., and J. M. Resig (2008), Evidence of late Pliocene-early Pleistocene marine environments in the deep subsurface of the Lihue basin, Kauai, Hawaii, *Palaios*, *23*, 442–451, doi:10.2110/palo.2007.p07-042r.

Kay, R. W., and P. W. Gast (1973), The rare earth content and origin of alkali-rich basalts, *J. Geol.*, *81*(6), 653–682.

Krivoy, H. L., M. Baker, Jr., and E. E. Moe (1965), A reconnaissance gravity survey of the Island of Kauai, Hawaii, *Pacific Sci.*, *19*(3), 354–358.

Le Bas, M. J., R. W. Le Maitre, A. Streckeisen, and B. A. Zanettin (1986), Chemical classification of volcanic rocks based on the total alkali-silica diagram, *J. Petrol.*, *27*(3), 745–750.

Langenheim, V. A. M., and D. A. Clague (1987), The Hawaiian-Emperor volcanic chain; part II. Stratigraphic framework of volcanic rocks of the Hawaiian Islands, in *Volcanism in Hawaii*, Prof. Pap. 1350, vol. 1 edited by R. W. Decker et al., pp. 55–84, U.S. Geol. Surv., Reston, Virginia.

Ludwig, K. R., B. J. Szabo, J. G. Moore, and K. R. Simmons (1991), Crustal subsidence rates off Hawaii determined from $^{234}U/^{238}U$ ages of drowned coral reefs, *Geology*, *19*(2), 171–174, doi:10.1130/0091-7613(1991)019<0171:CSROHD>2.3.CO;2.

Maaløe, S., D. James, P. Smedley, S. Petersen, and L.B. Garmann (1992), The Koloa volcanic suite of Kauai, Hawaii, *J. Petrol.*, *33*(4), 761–784.

Macdonald, G. A. (1968), Composition and origin of Hawaiian lavas, *Geol. Soc. Am. Mem.*, *116*, 477–522.

Macdonald, G. A., and T. Katsura (1964), Chemical composition of Hawaiian lavas, *J. Petrol.*, *5*(1), 82–133.

Macdonald, G. A., D. A. Davis, and D. C. Cox (1960), Geology and ground-water resources of the Island of Kauai, Hawaii, *Hawaii Div. Hydrogr. Bull., 13*, 212 pp.

Macdonald, G. A., A. T. Abbott, and F. L. Peterson (1983), *Volcanoes in the Sea,* 2nd ed., Univ. of Hawaii Press, Honolulu.

Mandl, G. (1988), *Mechanics of Tectonic Faulting: Models and Basic Concepts*, Elsevier, Amsterdam.

McDougall, I. (1964), Potassium-argon ages from lavas of the Hawaiian Islands, *Geol. Soc. Am. Bull., 75*(2), 107–127.

McDougall, I. (1979), Age of shield-building volcanism of Kauai and linear migration of volcanism in the Hawaiian island chain, *Earth Planet. Sci. Lett., 46*, 31–42.

McDougall, I., and T. M. Harrison (1999), *Geochronology and Thermochronology by the $^{40}Ar/^{39}Ar$ Method,* 2nd ed., Oxford Univ. Press, Oxford.

Moore, J. G. (1987), Subsidence of the Hawaiian ridge, in *Volcanism in Hawaii*, Prof. Pap. 1350, vol. 1, edited by R.W. Decker et al., pp. 85–100, U.S. Geol. Surv., Reston, Virginia.

Moore, J. G., D. A. Clague, R. T. Holcomb, P. W. Lipman, W. R. Normark, and M. E. Torresan (1989), Prodigious submarine landslides on the Hawaiian ridge, *J. Geophys. Res., 94*(B12), 17,465–17,484.

Mukhopadhyay, S., J. C. Lassiter, K. A. Farley, and S. W. Bogue (2003), Geochemistry of Kauai shield-stage lavas: Implications for the chemical evolution of the Hawaiian plume, *Geochem., Geophys., Geosyst., 4*(1), doi:10.1029/2002GC000342.

Ozawa, A., T. Tagami, and M. O. Garcia (2005), Unspiked K-Ar dating of the Honolulu rejuvenated and Koolau shield volcanism on Oahu, Hawaii, *Earth Planet. Sci. Lett., 232*, 1–11.

Palmiter, D. B. (1975), Geology of the Koloa Volcanic Series of the south coast of Kauai, Hawaii, M.S. thesis, Dept. of Geol. and Geophys., Univ. of Hawai'i at Mānoa, Honolulu, O'ahu, Hawai'i, 87 pp.

Reiners, P. W., and B. K. Nelson (1998), Temporal-compositional-isotopic trends in rejuvenated-stage magmas of Kauai, Hawaii, and implications for mantle melting processes, *Geochim. Cosmochim. Acta, 62*(13), 2347–2368.

Reiners, P. W., B. K. Nelson, and S. K. Izuka (1999), Structural and petrologic evolution of the Lihue basin and eastern Kauai, Hawaii, *Geol. Soc. Am. Bull., 111*(5), 674–685.

Renne, P. R., A. L. Deino, R. C. Walter, B. D. Turrin, C. C. Swisher, T. A. Becker, G. H. Curtis, W. D. Sharp, and A.-R. Jaouni (1994), Intercalibration of astronomical and radioisotopic time, *Geology, 22*(9):783–786.

Richard, P., N. Shimizu, and C. J. Allègre (1976), $^{143}Nd/^{146}Nd$, a natural tracer: An application to oceanic basalts, *Earth Planet. Sci. Lett., 31*(2), 269–278, doi:10.1016/0012-821X(76)90219-3.

Sherrod, D. R., J. M. Sinton, S. E. Watkins, and K. M. Brunt (2007), Geologic map of the State of Hawaii, Open-file Rep. 2007–1089, U.S. Geol. Surv., 83 pp, Reston, Virginia.

Simkin, T., R. I. Tilling, P. R. Vogt, S. H. Kirby, P. Kimberly, and D. B. Stewart (2006), This dynamic planet: World map of volcanoes, earthquakes, impact craters, and plate tectonics, 3rd ed., Geologic Inv. Map I-2800, scale 1:30,000,000, U.S. Geol. Surv., Reston, Virginia.

Stearns, H. T. (1940), Geology and ground-water resources of the islands of Lanai and Kahoolawe, Hawaii, *Haw. Div. Hydrogr. Bull., 6*, 1–60.

Stearns, H. T. (1967), Geology of the Hawaiian Islands, 2nd printing, with seven-page supplement by G. A. Macdonald, *Haw. Div. Hydrogr. Bull., 8*, 112 pp.

Stearns, H. T. (1985), *Geology of the State of Hawaii,* 2nd ed., Pacific Books, Palo Alto, Calif.

Stearns, H. T., and G. A. Macdonald (1942), Geology and ground-water resources of the island of Maui, *Haw. Div. Hydrogr. Bull., 7*, 344 pp.

Stearns, H. T., and K. N. Vaksvik (1935), Geology and ground-water resources of the island of Oahu, Hawaii, *Haw. Div. Hydrogr. Bull., 1*, 479 pp.

Steiger, R. H., and E. Jäger (1977), Convention on the use of decay constants in geo- and cosmochronology, *Earth Planet. Sci. Lett., 36*, 359–362.

Todt, W., R. A. Cliff, A. Hanser, and A. W. Hofmann (1996), Evaluation of a ^{202}Pb–^{205}Pb double spike for high-precision lead isotope analysis, in *Earth Processes: Reading the Isotopic Code,* Geophys. Monogr., *95*, edited by A. Basu and S. Hart, pp. 429–437, AGU, Washington, D.C., doi:10.1029/GM095p0429.

Trusdell, F. A., E. W. Wolfe, and J. Morris (2006), Digital database of the geologic map of the Island of Hawai'i, U.S. Geol. Surv. Data Series 144, version 1.0, available at http://pubs.usgs.gov/ds/2005/144/, accessed March 2013.

Washington, H. S., and M. G. Keyes (1926), Petrology of the Hawaiian Islands: V, the Leeward Islands, *Am. J. Sci., 12*, 336–352.

Watts, A. B., and U. S. ten Brink (1989), Crustal structure, flexure, and subsidence of the Hawaiian Islands, *J. Geophys. Res., 94*, 10,473–10,500.

Webster, J. M., et al. (2006), Drowned coralline algal dominated deposits off Lanai, Hawaii; carbonate accretion and vertical tectonics over the last 30 ka, *Marine Geol., 225*(1–4), 223–246, doi:10.1016/j.margeo.2005.08.002.

Wolfe, E. W., and J. Morris (1996), Geologic map of the Island of Hawaii, Misc. Invest. Map I–2524-A, scale 1:100,000, U.S. Geol. Surv., Reston, Virginia.

7

Evidence for Large Compositional Ranges in Coeval Melts Erupted from Kīlauea's Summit Reservoir

Rosalind T. Helz[1], David A. Clague[2], Larry G. Mastin[3], and Timothy R. Rose[4]

ABSTRACT

Petrologic observations on Kīlauea's lavas include abundant microprobe analyses of glasses, which show the range of melts available in Kīlauea's summit reservoir over time. During the past two centuries, compositions of melts erupted within the caldera have been limited to MgO = 6.3–7.5 wt%. Extracaldera lavas of the 1959, 1971, and 1974 eruptions contain melts with up to 10.2, 8.9, and 9.2 wt% MgO, respectively, and the 1924 tephra contains juvenile Pele's tears with up to 9.1 wt% MgO. Melt compositions from explosive deposits at Kīlauea, including the Keanakāko'i (A.D. 1500–1800), Kulanaokuaiki (A.D. 400–1000), and Pāhala (10–25 ka) tephra units, show large ranges of MgO contents. The range of melt MgO is 6.5–11.0 wt% for the Keanakāko'i; the Kulanaokuaiki extends to 12.5% MgO and the Pāhala Ash includes rare shards with 13–14.5% MgO. The frequency distributions for MgO in the Keanakāko'i and Kulanaokuaiki glasses are bimodal, suggesting preferential magma storage at two different depths. Kīlauea's summit reservoir contains melts ranging from 6.5 to at least 11.0 wt% MgO, and such melts were available for sampling near instantaneously and repeatedly over centuries. More magnesian melts are inferred to have risen directly from greater depth.

7.1. INTRODUCTION

Kīlauea Volcano is one of the most active and best studied volcanoes in the world. Early observers of the lavas produced during summit activity noted that the compositional range was quite limited and olivine phenocryst content usually low [e.g., *Powers*, 1955], though occasionally more olivine rich flows were produced. Before the electron microprobe came into common use, glass compositions were rarely obtained; consequently even olivine-rich lavas such as those of the 1959 eruption [*Murata and Richter*, 1966] were assumed to be mixtures of relatively differentiated melt (MgO ~ 7.0 wt%) with varying amounts of olivine [*Wright*, 1973]. The persistently olivine poor, low-bulk MgO character of the lavas has also led to models showing Kīlauea's summit reservoir as relatively small, containing a single batch of uniform magma at any given time [*Pietruska and Garcia*, 1999].

Subsequent work on scoria from the 1959 eruption showed that glass composition was quite variable (ranging from 6.4 to 10.2 wt% MgO) and that MgO content tended to increase as bulk MgO (olivine phenocryst) content increased [*Helz*, 1987; see also appendix to *Helz*, 2009]. Recently obtained glass analyses presented by *Helz et al.* [2014b] show that the 1959 eruption was not unique in this respect. Three additional eruptions at Kīlauea in the twentieth century have produced coeval melts with MgO ranging from 6.5 to 9.0 wt% MgO, including the 1971 and 1974 effusive eruptions (locations

[1] *U.S. Geological Survey, Reston, Virginia, USA*
[2] *Monterey Bay Aquarium Research Institute, Moss Landing, California, USA*
[3] *Cascades Volcano Observatory, U.S. Geological Survey, Vancouver, Washington, USA*
[4] *National Museum of Natural History, Smithsonian Institution, Washington, D.C., USA*

Hawaiian Volcanoes: From Source to Surface, Geophysical Monograph 208. First Edition.
Edited by Rebecca Carey, Valérie Cayol, Michael Poland, and Dominique Weis.

Companion Website: www.wiley.com/go/Carey/Hawaiian_Volcanoes

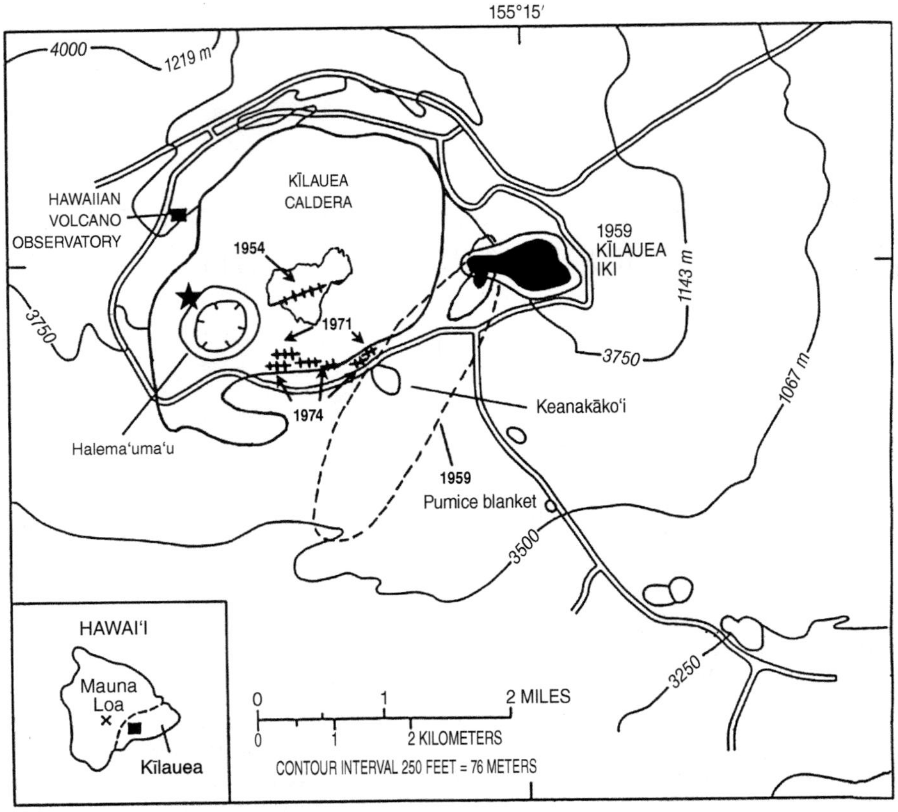

Figure 7.1 Summit area of Kīlauea Volcano showing the locations of the twentieth-century eruptions and fissures discussed in this chapter. The underlying map and information on the 1954 and 1959 eruptions are based on *Richter et al.* [1970, Figure 3]. The locations of the 14 August 1971 and the 19–20 July 1974 fissure systems, described by *Duffield et al.* [1982] and *Lockwood et al.* [1999], respectively, are from *Neal and Lockwood* [2003]. The star marks the location where the 1924 tephra samples were collected.

shown in Figure 7.1), plus the 1924 explosive eruption of Halema'uma'u. Data on the composition of glasses in Kīlauea tephras, presented below, show still wider ranges in MgO content, commonly within individual thin layers of tephra. These results offer some insight into the thermal structure of the present reservoir and into previous states of the system.

7.2. CONTEXT AND DEFINITIONS

At the outset, it is important to distinguish Kīlauea's thermal reservoir (as defined by the dramatic graphics of *Ryan* [1988]) from the smaller magma bodies that are contained within this larger volume of hot, aseismic material. The thermal reservoir is large and (so far as is known) continuous. The magma bodies within it must be smaller and are assumed to be segmented, based on previous work [e.g., *Fiske and Kinoshita*, 1969; *Ryan et al.*, 1983], as discussed by *Helz et al.* [2014b].

The relationship between the thermal reservoir and the melt bodies within it constitute a feedback loop: The magma bodies provide the heat that keeps the thermal reservoir hot enough to be aseismic. However, the position of a given melt body within the reservoir and its size determine how hot it can remain in storage. By looking at variations in the MgO content of well-quenched summit glasses, this chapter documents that a considerable range of coeval melt compositions may occur within the thermal reservoir. The discussion first will focus on the twentieth-century eruptions and then will consider briefly the MgO variation in several older tephra units for which glass analyses are available [*Helz et al.*, 2014a, 2014b].

We focus on MgO variation for two reasons: (1) It is the oxide that varies the most and hence offers the best look at variations in melt composition within Kīlauea's summit reservoir and (2) the MgO content of glasses in equilibrium with olivine is a proxy for temperature [*Helz and Thornber*, 1987]. Therefore differences in MgO content offer insight into the relative temperature and hence the relative residence time and/or physical position of the various coeval melts within the thermal reservoir.

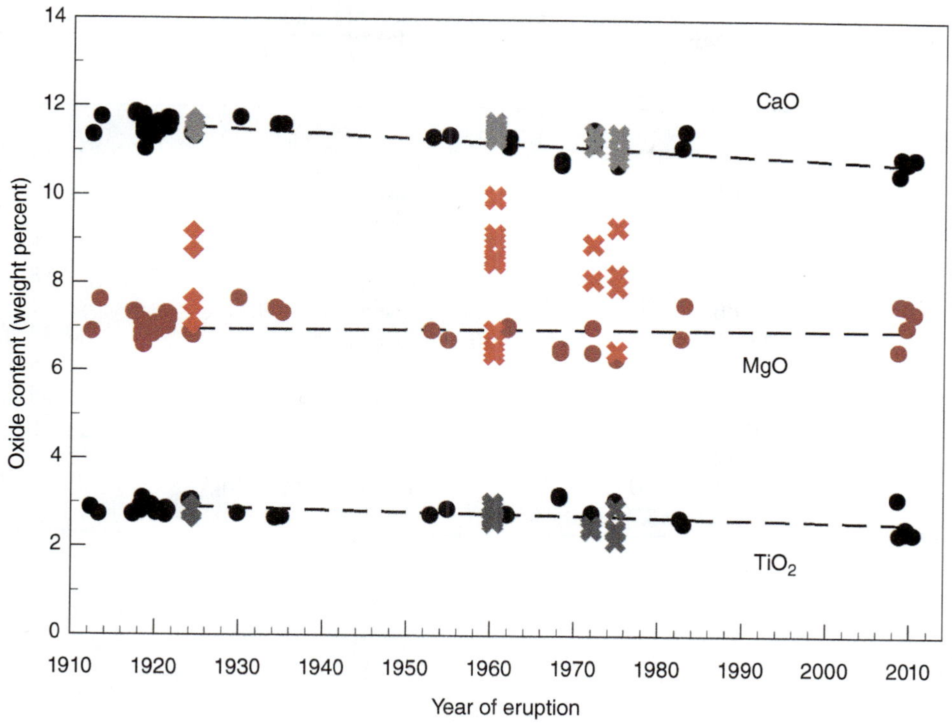

Figure 7.2 Composition of glasses erupted at or near the summit of Kīlauea from 1912 to 2010. Intracaldera glasses from effusive eruptions are shown as circles; extracaldera glasses from 1959, 1971, and 1974 are shown as crosses; and Pele's tears from the 1924 eruption are shown as diamonds. Data for most 1952–1982 eruptions are from *Helz et al.* [1995] and *Wright and Helz* [1996]; data for 1959 are from *Helz* [2009]; data for 1912–1924 and 1971–1974 are by *Helz et al.* [2014b] and Table 7.1 of this chapter; data for 2008–2010 are from C. R. Thornber (unpublished). The dashed lines show least-squares fits to the long-term trends in CaO, MgO, and TiO_2 in the effusive glasses from 1924 to 2010.

7.3. GLASS COMPOSITIONS AT KĪLAUEA ERUPTED DURING THE TWENTIETH CENTURY

Figure 7.2 shows the range of compositions of glasses in scoria and spatter erupted at Kīlauea's summit from 1912 to 2010. By restricting samples investigated to fragmental ejecta, we largely avoid the effects of post eruptive crystallization and minimize those of near-surface cooling. This is important, as earlier results [Helz et al., 1995], notably on the 1984 Mauna Loa eruption, show clearly that MgO in glasses from flow samples run lower than those in spatter.

Oxides shown here include CaO, MgO, and TiO_2; more complete oxide data are shown in Figure 3 of *Helz et al.* [2014b]. It is clear from the figure that most oxides have a very limited range of compositions, with MgO having the largest range of these three oxides. Two sets of glassy samples have MgO contents extending above the general trend. First are glasses from the 1959, 1971, and 1974 eruptions, which surfaced along or just outside the boundary of the Kīlauea Caldera; the locations of the relevant 1971 and 1974 fissures are shown in Figure 7.1. The other high-MgO samples are among a collection of fresh, unbroken, glassy Pele's tears from two samples of the juvenile component of the 1924 tephra (star in Figure 7.1; see analyses in Table 7.1).

The relative frequency of occurrence of different MgO contents in these summit glasses is shown in two histograms in Figure 7.3. This separation into two histograms is necessary because of inconsistent levels of sampling of the twentieth-century Kīlauea summit eruptions: Only the 1959, 1971, and 1974 eruptions, the two very brief 1982 eruptions, and the ongoing summit activity have been extensively sampled, making it difficult to combine all data in one histogram. The top histogram (Figure 7.3a), though not an ideal data base, does have a strong peak at 6.5–7.0 wt% MgO, accurately reflecting the preponderance of low-MgO melts in summit eruptions. The 1959 distribution (Figure 7.3b), with a conspicuous peak at 8–9 wt% MgO, emphasizes the unique character of this eruption in the twentieth–century period.

In evaluating the data in Figure 7.3a, one critical observation is that the full range of melts observed for the 1971 and 1974 fissures was erupted almost simultaneously, as the fissures were active for only a few hours [*Duffield et al.*, 1982; *Lockwood et al.*, 1999]. On 14 August 1971, the line of vents shown in Figure 7.1 erupted melts with

Table 7.1 Glass compositions of intact Pele's tears from 1924 eruption of Kīlauea Volcano.

Sample no.	F04-16-7	F04-16-7	F04-16-7
No. of points	28	2	3
Glass compositions			
SiO_2	49.61	50.92	49.33
TiO_2	3.01	2.57	2.71
Al_2O_3	13.59	13.47	12.77
FeO	11.24	10.68	11.18
MnO	0.17	0.16	0.18
MgO	6.86	7.49	9.13
CaO	11.47	11.31	11.32
Na_2O	2.30	2.14	2.05
K_2O	0.58	0.48	0.54
P_2O_5	0.29	0.26	0.26
Sum	99.12	99.48	99.47
Comments on glass	Average of 10 tears with MgO = 6.57%–7.05%	Tear with higher MgO	Tear with highest MgO

Analyst: T. R. Rose.

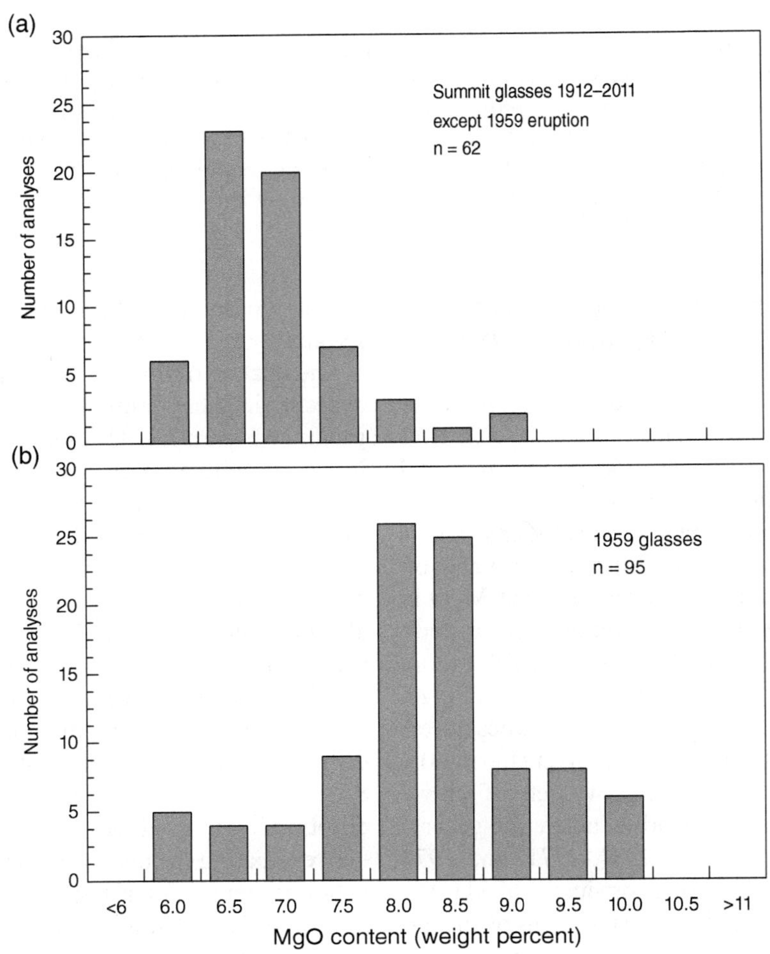

Figure 7.3 Frequency distribution of MgO contents in glasses erupted at Kīlauea's summit between 1912 and 2010. Data in (a) as in Figure 7.2; data in (b) are from *Helz* [2009] and R. T. Helz (unpublished).

6.4%–7.0% MgO on the caldera floor and 8.1%–8.9% MgO on the caldera rim, immediately outside the bounding faults, for a temperature contrast of over 40°C [*Helz and Thornber*, 1987]. On 19–20 July 1974 a parallel line of vents (Figure 7.1) erupted lavas with 6.28% MgO on the caldera floor and up to 8.18%–9.24% MgO next to and just beyond the caldera boundary, for a temperature contrast of 60°C. The easternmost of the 1974 fissures shown was the source of the unusually fluid lava that achieved hyperelevation flow, as described by *Heslop et al.* [1989]. It is possible that this behavior was a consequence of its relatively high melt temperature, which would result in lower viscosity for the lava.

The fresh, unbroken glassy Pele's tears from the 1924 eruption were all found together in two small samples and, given the nature of the eruption [*Jaggar and Finch*, 1924], would also have been erupted within hours or a few days of each other. Their glass MgO contents range from 6.5% to 9.13% (Table 7.1), again showing clearly that glasses with this range of MgO contents were available for near-simultaneous eruption from Kīlauea's summit thermal reservoir.

The 1959 eruption is unique among Kīlauea's historic eruptions. Its MgO peak falls at 8%–9% MgO, equal to the small collection of higher-MgO glasses from the 1971, 1974, and 1924 eruptions. However, the distribution extends to melts with MgO > 9.0%. It has long been known that the 1959 eruption was mixed magma [*Wright*, 1973]. The melts with MgO > 9.0% are rich in the 1959E component, identified by *Helz* [1987] as the juvenile component of the eruption. That is, this component is the magma tracked from depth to the surface by *Eaton and Murata* [1960], the arrival of which preceded the 1959 eruption. It was intruded into the erupting body three times (in phases 1, 4–5, and 8; see discussions by *Helz* [1987]; see also *Eaton et al.* [1987]).

However, Figure 7.3b shows that this was not the dominant thermal component of the eruption. Rather, the most frequent MgO content (8.0%–9.0%) of the eruption is that seen in the late phase 1 and phase 2 scoriae, which consist almost entirely of the stored (1959W) component [*Wright*, 1973]. The full range of glass MgO contents in most 1959W-rich samples varies from 6.35–6.99 wt% (in Iki-1 and Iki-9) to 8.5–8.6 wt% (in Iki-44 and Iki-7; see analyses by *Helz* [1987]). The 1959W-rich samples were erupted over two weeks during phases 1, 2, and 3 of the eruption (14-29 November 1959). Thus, unlike the 1971, 1974, and 1924 material, they were not erupted within hours of each other: Extensive real-time observation of the eruption [*Richter et al.*, 1970] and sample collection [*Murata and Richter*, 1966] show that only one melt composition was being erupted at any given time. However, the entire range of melt MgO content for both magmatic components was erupted over the first week of the eruption.

7.4. DISCUSSION OF RESULTS FOR TWENTIETH-CENTURY ERUPTIONS

The clearest instances of simultaneous eruption of melts of varying MgO content (and hence temperature) are those from the 1971 and 1974 fissures, described above. The 1971 summit eruption began shortly after the first period of activity at Mauna Ulu shutdown [*Duffield et al.*, 1982] and the lavas observed at this and several other fissures have compositions that can be expressed in terms of magma components defined for the Mauna Ulu lavas [*Wright et al.*, 1975]. However, the 14 August 1971 fissure system, which crosses the caldera boundary near Keanakāko'i Crater (Figure 7.1), is the only one with higher MgO contents, in both glass and whole-rock compositions. Similarly, the 1974 summit eruption followed the end of Mauna Ulu activity in 1974; again, the lavas from the fissure discussed here can be expressed in terms of Mauna Ulu components in the latest of those lavas [*Wright and Tilling*, 1980]. As in 1971, only this one fissure (among the many that erupted in the summit area in 1974; see especially *Neal and Lockwood* [2003], for all of the vents of the 1971 and 1974 summit eruption) produced lavas with higher bulk and glass MgO contents. Not only did the various eruptive vents on the caldera floor and in Halema'uma'u produce only low-MgO lavas, but there is no jump in MgO content where the fissures cross the caldera boundary toward Kīlauea's Southwest Rift Zone.

The composition of olivine phenocrysts and microphenocrysts shows a pattern similar to glass MgO content: Those in lavas erupted from Halema'uma'u or on the caldera floor are relatively Fe rich with compositions mostly in the range $Fo_{82-83.5}$ [*Baker*, 1987; *Rowe et al.*, 2009; *Thornber et al.*, 2010] (R. T. Helz, unpublished data), while the samples with higher glass and bulk MgO content contain some olivine with higher forsterite, much of it in the range Fo_{86-87}, typical of other olivine-rich Kīlauea lavas [*Wright*, 1971; *Helz*, 1987; *Helz et al.*, 2014b]. The most magnesian olivine core observed ($Fo_{88.5}$) is a sulfide-bearing olivine, which resembles the class 5 olivines in the 1959 scoria and in Kīlauea Iki lava lake [*Helz*, 1987].

The glassy Pele's tears from the samples of 1924 tephra are found within two small hand specimens, so they must have been erupted within hours or days of each other. The example of the 1959 eruption, from a separate line of activity north and west of the 1971 and 1974 fissure systems, is more complex. This zone extends from Halema'uma'u along the 1954 eruptive fissure (Figure 7.1) to the inferred source of the Ailā'au lavas, a topographic high near the east end of the Kīlauea Iki pit crater [*Clague et al.*, 1999; *Holcomb*, 1987]. Historic activity along this line includes the eruptions of 1832, 1868, 1954, 1959, and April 1982 [see also discussion by *Poland et al.*, 2014].

The lowest-MgO representatives of the 1959W magmatic component are identical to the lavas and glasses of the 1954 summit eruption (Figure 7.1), so it can be inferred that this component had been stored somewhere in the summit reservoir for at least 5 years, prior to being erupted in 1959. However, the dominant volume of the 1959W magma was considerably hotter, with glass MgO = 8%–9%, not the 6.5%–7.0% seen in the 1954 lavas and the earliest 1959 lavas. As discussed by *Helz et al.* [2014b], the eruption of 1959W melts with two distinct ranges of MgO suggests that there were two different bodies of this magma type in storage, prior to the 1959 eruption.

Estimates of depth for subsurface phenomena are notoriously imprecise, and this is true for magma bodies as well. However, most observers would expect the common, relatively low MgO summit lavas to have emerged from shallow storage beneath the caldera floor. Indeed, many samples included in Figures 7.2 and 7.3 were collected from active lava lakes at Kīlauea's summit. These include the samples from 1912–1932 collected by Jaggar and those from the 1967–1968 eruption [*Kinoshita et al.*, 1969]. In the latter case, the ascending magma was tracked from a reservoir estimated to lie at a depth of 2–3 km to the surface [*Fiske and Kinoshita*, 1969] over a period of 22 months. Similarly, the spatter erupted in the ongoing eruption in Halema'uma'u is believed to derive from a sill lying at a depth of roughly 1 km [*Johnson et al.*, 2010; see also *Poland et al.*, 2014].

The rare, more magnesian melts, which have appeared at three different locations, most likely come from somewhat greater depths. For the 1971 and 1974 eruptions, the onset of summit activity was attributed to a breakdown of the magmatic route out to Mauna Ulu, with the backed-up magma being erupted at the summit instead [*Duffield et al.*, 1982; *Lockwood et al.*, 1999]. The high-MgO melts and lavas in these two events were erupted approximately above the point where magma leaves the summit reservoir to enter Kīlauea's East Rift Zone. It seems reasonable to infer that they rose from where the East Rift Zone branches off from the summit reservoir, a depth estimated to be 2–4 km [*Ryan*, 1988; *Poland et al.*, 2014].

The more magnesian part of the stored component of the 1959 eruption could have risen from similar depth below the surface but lying along the alternative "Kīlauea Iki" trend of *Poland et al.* [2014]. The most magnesian melts in the 1924 tephra, however, presumably came from a more centrally located body. Regardless of the exact depth of these three magma bodies, it is reasonably clear that (1) they all had similar MgO contents and (2) they came from different locations, at least in plan view. They provide evidence for the repeated (or continuous) presence of melts with up to 9% MgO over a broad area of the Kīlauea's thermal reservoir throughout much of the twentieth century.

7.5. COMPOSITION OF GLASSES IN OLDER KĪLAUEA TEPHRA

In recent years it has been recognized that Kīlauea has a considerable history of explosive activity, and several tephra units have been mapped, described, and dated. These include (1) the Keanakāko'i Tephra of *Swanson et al.* [2012], erupted between A.D. 1500 and 1800, after formation of the Kīlauea Caldera; (2) the Kulanaokuaiki Tephra of *Fiske et al.* [2009], estimated to have been erupted between A.D. 400 and 1000; and (3) the Pāhala Ash of *Easton* [1987], estimated to be 10–25 ka old. Glass data from these tephra units were reviewed briefly by *Helz et al.* [2014b] and are discussed in more detail here.

The tephra units are complex mixtures of crystal fragments and lithic material in addition to the glasses that are assumed to represent quenched magmatic liquids. Furthermore, the glasses may be extensively altered and there is no way to reconstruct the crystallinity of the magma. Therefore, whole-rock chemistry, which has been the backbone of petrologic evaluation of lava compositions at Kīlauea [*Helz et al.*, 2014b], is of limited value in studying the tephra. A better approach to studying their chemistry is to obtain microbeam analyses of fresh glasses and crystals within the tephra. The glass compositions discussed here are from fresh glassy shards and Pele's tears within the various tephra units. Some of the lithic fragments include interstitial glass, but such glasses are not included in the figures presented below, as their association with the eruption is presumed to be accidental.

In the Keanakāko'i and Kulanaokuaiki units discussed here, each sample has come from a "layer" discernible in the field. The nature and origin of these layers, their thickness, and the time needed to produce them are (with some important exceptions) not specified in the references above. The Pāhala section studied by *Clague et al.* [1995a] showed little layering.

7.5.1. Glasses from Keanakāko'i Tephra

The Keanakāko'i Tephra, which forms much of the upper surface around Kīlauea's caldera, formed during a series of explosive eruptions, of magmatic, phreatomagmatic, and phreatic character [*McPhie et al.*, 1990], erupted over three centuries [*Swanson et al.*, 2012]. The Keanakāko'i deposits are of both fall and pyroclastic density current origin. The lower part of the Keanakāko'i is dominantly vitric and the upper part dominantly lithic. The dominantly vitric section is as thick as 8 m and the dominantly lithic section is 4–5 m.

The Keanakāko'i samples discussed here were first described by *Mastin et al.* [2004]; they are from the dominantly vitric part of the section. They do not constitute a complete section of the Keanakāko'i as the upper scoria layers [*Swanson et al.*, 2012] and the uppermost "golden pumice" of *Sharp et al.* [1987] are not included. The samples

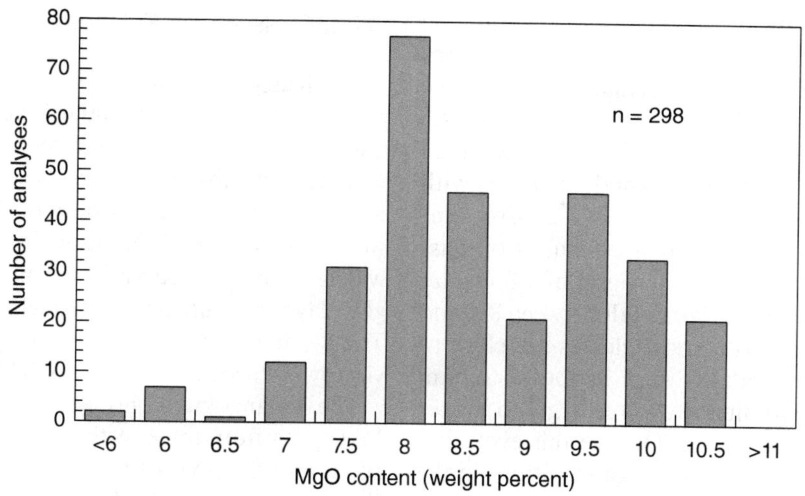

Figure 7.4 Frequency distribution of the MgO content in glasses from the Keanakāko'i samples of *Mastin et al.* [2004]; data are in the Appendix, also in *Helz et al.* [2014a].

record most of the interval during which juvenile component tephra predominated (ca. A.D. 1500–1700; *Swanson et al.* [2012]), though less completely than other sections. There is at least one time break, probably lasting no more than several decades, within this interval [*Swanson et al.*, 2012].

The frequency of occurrence of MgO contents for the Sand Wash and southwest fissure sections of *Mastin et al.* [2004] is shown in Figure 7.4; the results of *Garcia et al.* [2011] on a thicker section of the Keanakāko'i Tephra are similar and extend the chemical characterization of the Keanakāko'i.

The frequency distribution in Figure 7.4 is conspicuously bimodal, with peaks at 8% and 10% MgO; a single glass shard with MgO = 10.98% is the most magnesian found. Figure 7.5 shows the same data plotted by their vertical position, based on Table 2 of *Mastin et al.* [2004]. The most striking feature of these data is that, of the 28 individual layers shown, two-thirds contain glasses with a range of MgO content exceeding 2% by weight, corresponding to a temperature range of 40°C, using the glass geothermometer of *Helz and Thornber* [1987]; the absolute temperatures are not cited because the water content of the tephra may be appreciably higher than that in the calibration experiments. This range is equivalent to the range of MgO observed in most twentieth-century eruptions (Figure 7.3a) and is about half of the 75°C range of quenching temperatures observed in the 1959 eruption (Figure7.3b). Analyses from two adjacent layers (411 and 409 from unit IIA1), with a narrow and wide range in MgO content, respectively, are shown in the Appendix table 7A.1.

Given the fragmental nature of the deposits, the extensive ranges in MgO content within individual samples might reflect secondary reworking. The distribution of analyses in Figure 7.5, however, makes this unlikely for two reasons. First, there are nine layers having narrow

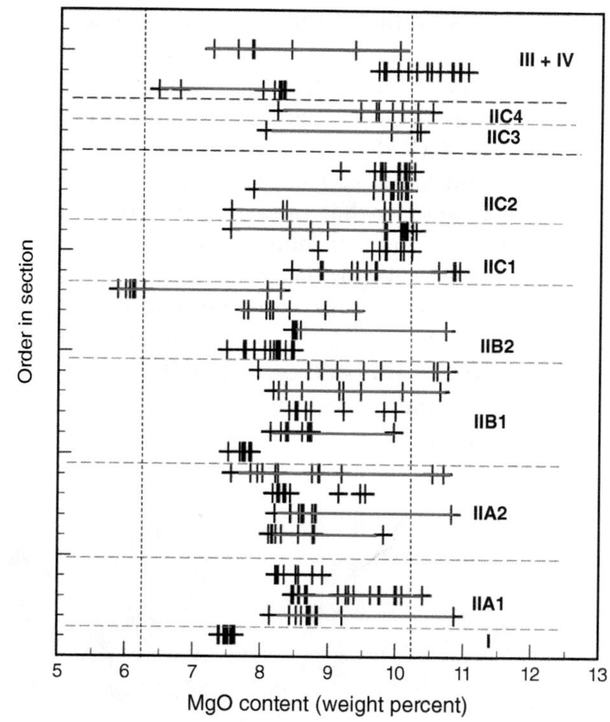

Figure 7.5 Stratigraphic distribution of MgO in samples from the Keanakāko'i Tephra, plotted at arbitrary vertical intervals. This composite section includes all data from samples studied by *Mastin et al.* [2004] using the stratigraphic sequence and identifications given in Table 1 of that paper. The tephra sections are about 4 m thick. Vertical lines show the range of MgO contents seen in the 1959 scoria (about 75°C using the glass geothermometer of *Helz and Thornber* [1987]) for comparison. The red horizontal lines indicate 19 layers where the range of MgO in the layer exceeds 2 wt%. There are eight layers with tighter MgO ranges that occur throughout the two sections (excluding unit I, which is the basal reticulate from a different section of the deposit).

ranges in MgO, seven of them intercalated within the stratigraphically continuous sections. Their positions within the stratigraphic units are irregular, and they have both low and high glass MgO contents. It seems unlikely that they alone could have been unaffected by extensive reworking of the rest of the layers. Second, the layers with wide-ranging MgO contents exhibit fairly conspicuous stratigraphic control on where the densest cluster of glass data falls: In units IIA and IIB (stratigraphy of *Decker and Christiansen* [1984]), most glasses fall between 8% and 9% MgO, whereas in unit IIC, most glasses are close to 10% MgO. Thus the bimodality seen in the histogram reflects a shift from predominantly lower MgO to higher MgO glasses in the younger units. Despite this evolution through time, the total range of glass compositions in the Keanakākoʻi is almost constant over most of the section.

There is one conspicuous exception: Figures 7.4 and 7.5 include glasses from a differentiated layer that lie outside the observed usual range. Photomicrographs of scoria from this differentiated layer are shown in Figure 7.6, where it can be seen that the highly differentiated glasses (most with MgO < 6.5%) are not small pools of interstitial glass in a crystalline matrix but make up most of the scoria. This layer also contains a subordinate component of scoria with MgO > 8.0%, an example of which is also shown. The contrast in vesicularity plus the gap in glass compositions suggests that this layer includes material from two different magma bodies. *Mastin et al.* [2004] tentatively correlated this layer as scoria layer 6 of *McPhie et al.* [1990], but it is now thought to be a separate layer (D. A. Swanson, personal communication). It falls near the boundary between units IIB and IIC, but it is not known whether it marks a significant hiatus at that stratigraphic position.

7.5.2. Glasses from Kulanaokuaiki Tephra

The Kulanaokuaiki Tephra of *Fiske et al.* [2009] erupted between A.D. 400 and 1000. Like the Keanakākoʻi, this unit includes deposits from a series of explosive eruptions, with considerable time gaps between events. Unlike the Keanakākoʻi, surface exposures of this unit are limited: The work of *Fiske et al.* [2009] depended on correlations between widely separated sections based on the position of a unique glass layer with uncommonly high TiO_2 and K_2O contents. The Kulanaokuaiki Tephra is the upper subunit of the previously undivided Uēkahuna Ash [*Dzurisin et al.*, 1995].

The Kulanaokuaiki histogram in Figure 7.7 shows glass MgO data from three sections: Tree Molds, the Uēkahuna Bluff, and the South Flank [*Fiske et al.*, 2009]. As for the Keanakākoʻi data, the histogram for the Kulanaokuaiki glasses is irregular, with a well-defined peak at 7% MgO and a broad high at 8.5%–9.5% MgO. However, the Keanakākoʻi has few glasses with MgO < 7.0%, while the Kulanaokuaiki 2 and 3 units contain abundant differentiated glasses. Another difference is that the Keanakākoʻi contains no glasses with MgO > 11%, while MgO in the Kulanaokuaiki glasses extends to 12%–12.5% MgO [*Rose et al.*, 2011].

Figures 7.8a and 7.8b show the vertical distribution of MgO contents for two of the three sections included in Figure 7.7. The Tree Molds section (Figure 7.8a) is the most complete section of the Kulanaokuaiki Tephra known and is 65 cm thick. The analyzed part of the Uēkahuna Bluff composite section is somewhat thicker (~2 m), with the tephra below the Kulanaokuaiki 2 marker unit not yet analyzed. As for the Keanakākoʻi, the Appendix includes two individual layers (F02-26-10a and F02-26-9a), both from the Uēkahuna Bluff section, with narrow- and wide-ranging MgO contents, respectively.

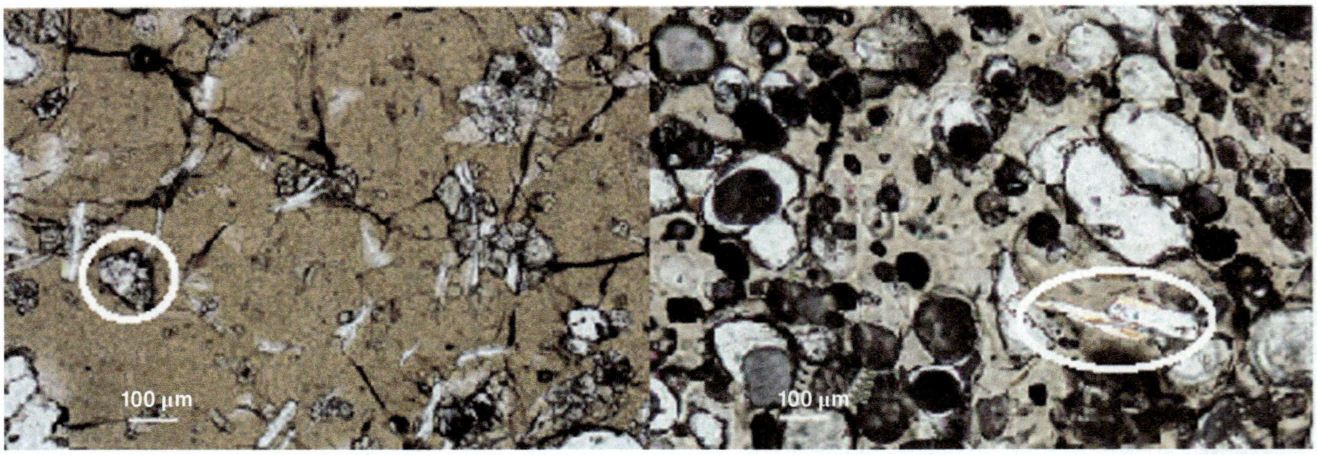

Figure 7.6 Photomicrographs of glassy scoria from sample 514 of *Mastin et al.* [2004]. The sample on the left is representative of the scoria with MgO < 6.5% and contains scattered clusters of augite (greenish) + plagioclase (white), plus olivine(circled), in a matrix of brown glass. This same layer also contains scoria with glass MgO > 8.0%, with rare olivine crystals (circled) in highly vesicular pale brown glass, as shown on the right.

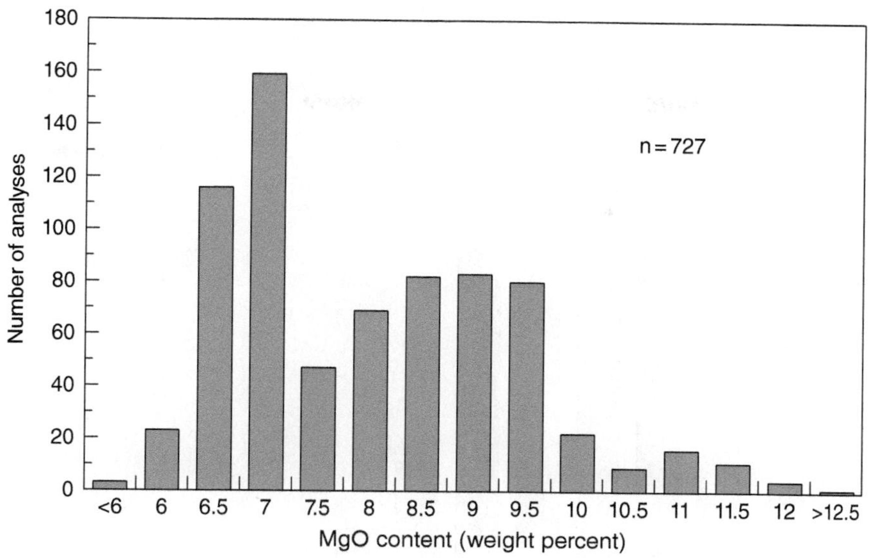

Figure 7.7 Frequency distribution of the MgO content in glasses for the Tree Molds, Uēkahuna Bluff, and South Flank sections through the Kulanaokuaiki Tephra of *Fiske et al.* [2009]. Data are in the Appendix, also in *Helz et al.* [2014a].

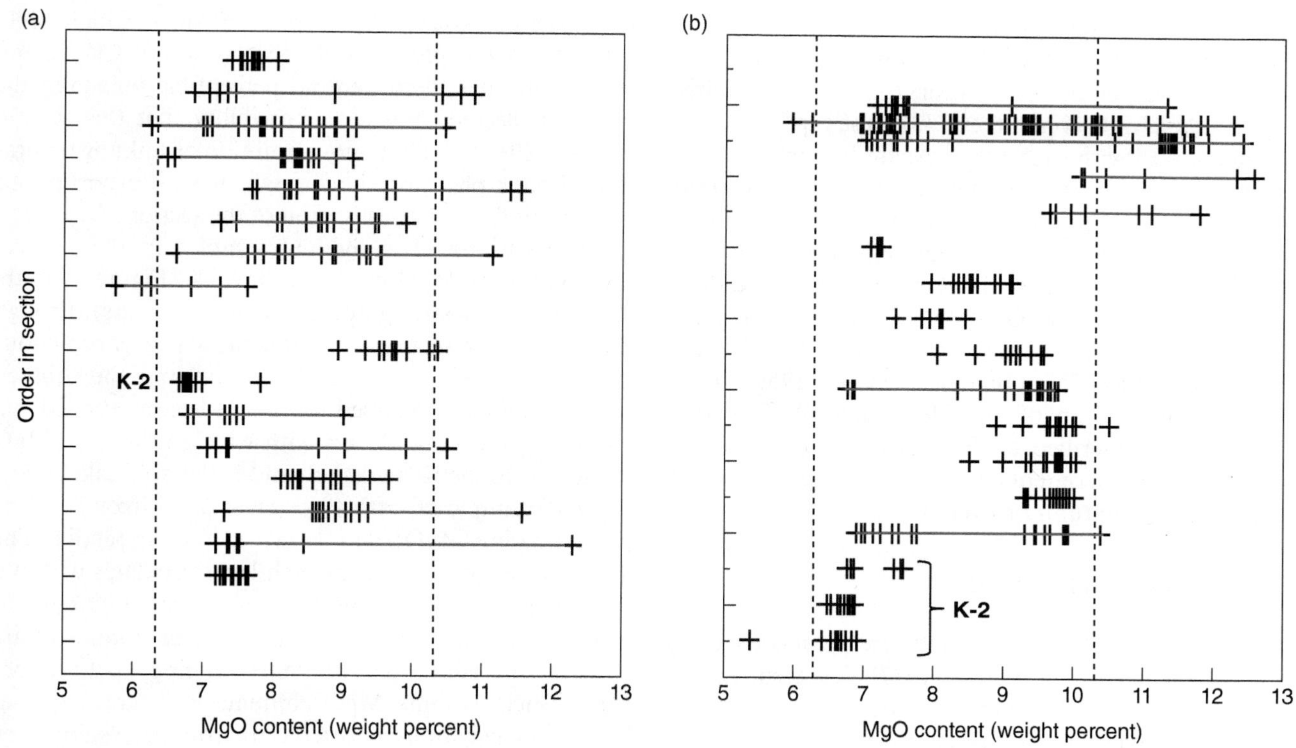

Figure 7.8 MgO variation in two sections of the Kulanaokuaiki Tephra of *Fiske et al.* [2009], which was erupted over the period A.D. 400–1000. The Tree Molds section (a) is 65 cm thick; the analyzed part of the Uēkahuna Bluff section (b), which is a composite of two adjacent sections in the Uēkahuna Bluff, is 2 m thick. The high-Ti,K Kulanaokuaiki 2 marker horizon is labeled K-2 in both sections. Layers with MgO ranges > 2.0% are marked in red; there are layers with narrower ranges in MgO content, distributed throughout both sections. Some layers have MgO ranges that exceed the entire range seen in the 1959 eruption, which is shown by the vertical dashed lines.

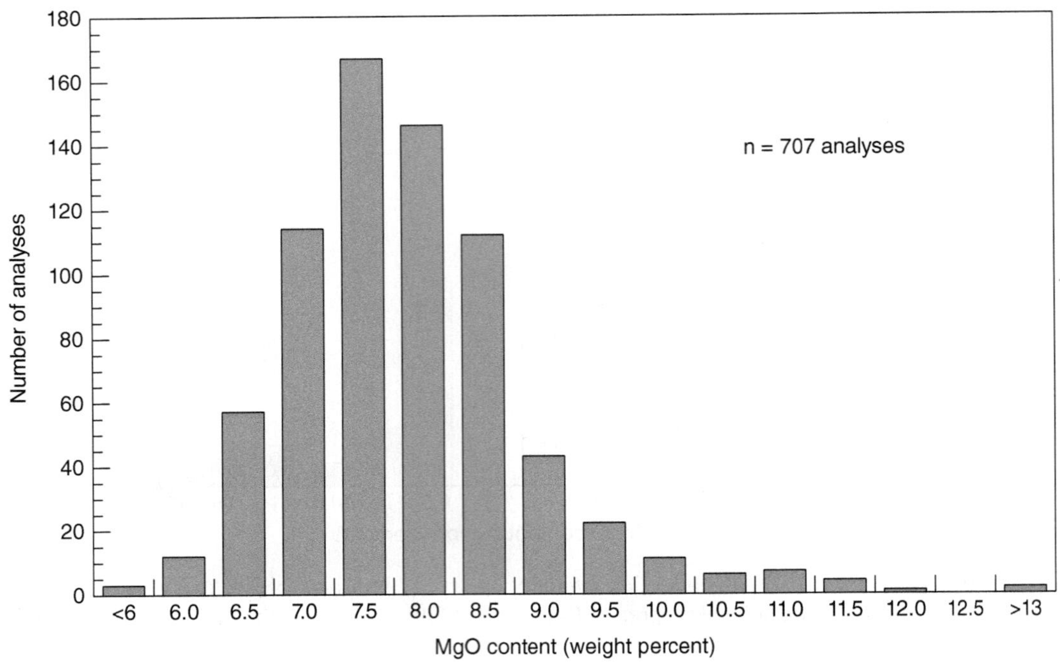

Figure 7.9 Frequency distribution of MgO content in glasses from the Pāhala Ash of *Easton* [1987]. This section of the tephra is found at the top of Hilina Pali [*Clague et al.*, 1995a]. Data are in the Appendix, also in *Helz et al.* [2014a].

In comparing these sections with the Keanakāko'i, one sees again that (1) the glass MgO contents range widely in many layers, (2) there are also layers that exhibit a limited range of glass MgO content, and (3) these types of layers are intercalated with each other. Again it seems unlikely that the ranges can be attributed to secondary reworking but appear to be a primary feature of the layers, though with the caveat that the time represented by individual layers is unknown. Because the Kulanaokuaiki includes a wider range of glass MgO contents, the ranges in some layers not only exceed those in the Keanakāko'i but are broader than the 75°C range seen in the 1959 scoria. One layer at the top of the Uēkahuna Bluff section (Figure 7.8b), described by *Rose et al.* [2011], contains glasses with MgO contents corresponding to a range of quenching temperatures of over 100°C.

7.5.3. Glasses from Pāhala Ash

The oldest tephra unit considered here is a single section from the Pāhala Ash of *Easton* [1987]. The term Pāhala Ash has been applied to many ash deposits on Kīlauea and Mauna Loa. The data discussed here are from one particular section at the top of Hilina Pali, sampled by and reported on briefly by *Clague et al.* [1995a], with an age of 10–25 ka; the relationship of this section to other parts of the Pāhala Ash is unknown. As for the previous two tephra units, the MgO data are presented first as a histogram (Figure 7.9) and then by height through the section collected (Figure 7.10). Unlike the previous two examples, the distribution of MgO contents in the Pāhala is almost Gaussian, with a long tail on the high-MgO side. This shape, plus the field observation of Clague and co-workers that there were no perceptible breaks in the section collected, raises the possibility that this section was deposited in a continuous eruption of unknown duration, though proof of either single or multiple eruptions must await further investigation of the section.

The locations of individual samples in the vertical distribution of MgO contents shown in Figure 7.10 are plotted at their stratigraphic height (in meters) through this 26 m thick section. The thickness is a conspicuous difference between the Pāhala section and the thinner, younger tephra discussed above. However, the Pāhala contains intercalated layers with wide and narrow MgO ranges, as do the Keanakāko'i and Kulanaokuaiki deposits; one example of each type (layer 53 (narrow) and 54 (wide-ranging MgO)) are shown in the Appendix. The main cluster of analyses in each layer oscillates within a fairly narrow range, while the maximum MgO content increases upward. The uppermost layer contains rare glass shards with MgO contents ranging up to 14.5% MgO. Such extreme MgO contents in Kīlauea glasses have previously been reported only from deep submarine glasses [*Clague et al.*, 1991, 1995b].

7.6. DISCUSSION OF RESULTS FOR TEPHRA GLASS COMPOSITIONS

The tephra data show that Kīlauea has erupted melts of varying MgO content, even within the time encompassed by individual thin layers (hours to weeks, based on

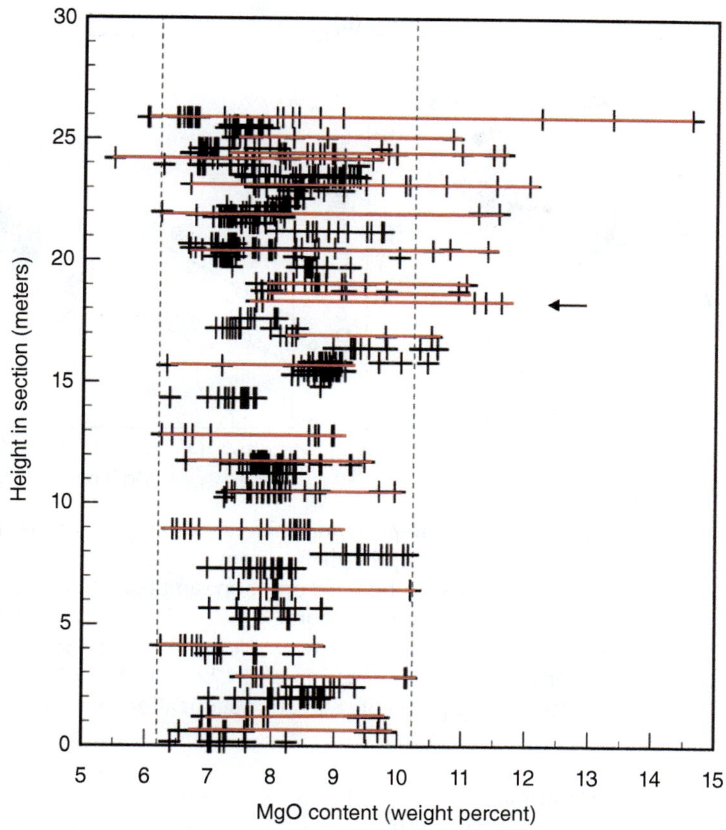

Figure 7.10 Stratigraphic distribution of MgO content of glasses in the Hilina Pali section of the Pāhala Ash. The section is 26 m thick and contains no discernable hiatus. Layers highlighted in red have MgO ranges > 2.0%, as in Figures 7.5 and 7.8. There are at least 21 such layers in the Pāhala and many more that have ranges close to that. The arrow marks the position of layer 37, at a height of 18.26 m. Vertical dashed lines as in Figures 7.5 and 7.8.

observations of historic eruptions), occasionally exceeding the entire range of MgO contents observed in glasses erupted in the twentieth century. Two important questions to be considered are: How are these highly variable glasses produced and to what extent could they have existed in Kīlauea's thermal reservoir, which lies between 2 and 6 km below the surface [*Ryan*, 1988]? As is the case for bulk lava compositions at Kīlauea [*Wright*, 1971], the tephra melts can be readily divided into "differentiated" compositions, that is, melts that were crystallizing the multiphase assemblage olivine + augite ± plagioclase prior to eruption, and "primitive" compositions, that is, melts that were crystallizing olivine (±chromite) only prior to eruption.

7.6.1. Differentiated Melts (MgO < 7.5%) in Tephra Deposits

The relationship between glass CaO and MgO is shown in Figure 7.11 for all tephra glasses discussed above to show the effects of crystallization of olivine, augite, and plagioclase on glass compositions. In Kīlauea lavas, augite begins to crystallize just below ~7.5% MgO, while plagioclase first crystallizes at ~7.0% MgO at low pressure and water contents [e.g., *Helz and Thornber,* 1987]. The differentiated Keanakāko'i and Kulanaokuaiki glasses (Figure. 7.11; see also Table 7.2, colums 11, 12, 15, and 16) show an abrupt downward trend in CaO at about 7.0% MgO, as would be expected if significant amounts of plagioclase were crystallizing. This is consistent with the petrography of the differentiated glass in Keanakāko'i sample 514, which contains abundant laths of plagioclase, usually in clusters with augite (Figure 7.6a).

What is the significance of these layers that contain differentiated glass almost exclusively (see Figures 7.5 and 7.8)? First, the fact that plagioclase has begun to crystallize in these melts at the same MgO content as it does in near-surface conditions means that either these melts were stored at low pressures in Kīlauea's summit reservoir prior to eruption or they were unusually dry, as the presence of H_2O drastically suppresses the incoming of plagioclase relative to the mafic phases. Their appearance as distinct layers in these two explosive units is curious, therefore, as neither possibility is consistent with the melts themselves being the source of the explosions. In particular, it seems unlikely that these dry, differentiated melts were the main carrier magma for the extremely explosive Kulanaokuaiki 3 eruption described by *Fiske et al.* [2011].

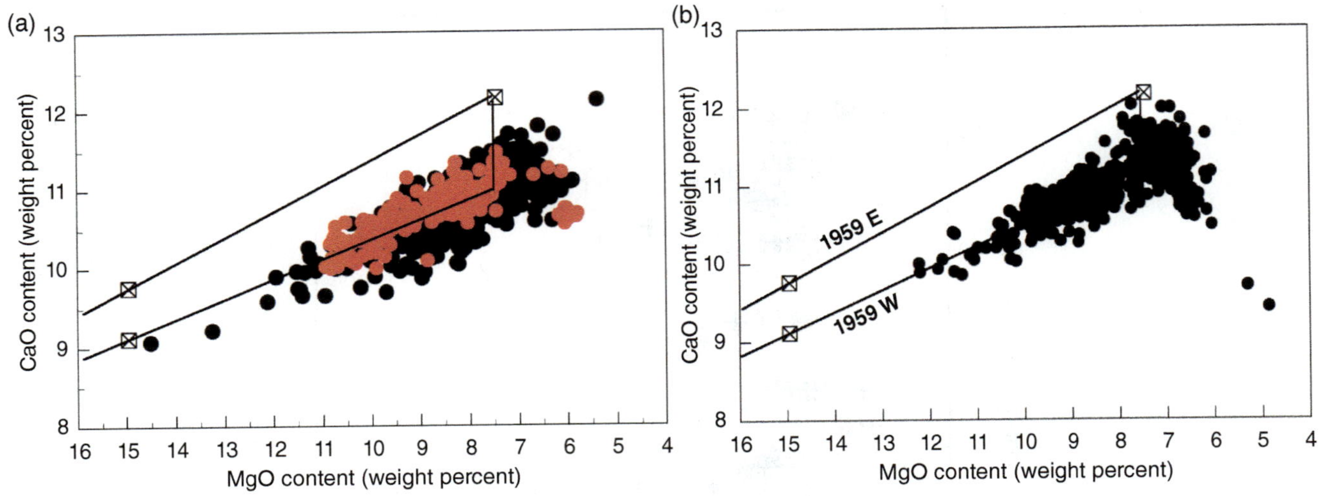

Figure 7.11 CaO vs. MgO in glasses from the three tephra units discussed in this chapter; (a) Includes the Pāhala glasses (black) and Keanakāko'i glasses (red); (b) glasses from the Kulanaokuaiki Tephra. The outlined box shows the range of whole-rock compositions for the 1959 eruption samples [*Murata and Richter*, 1966] and early drill core from Kīlauea Iki lava lake [*Richter and Moore*, 1966], with the juvenile (1959E) component being higher in CaO and the stored (1959W) component lower [*Helz*, 1987]. The sides of the field show the orientation of olivine control lines for Fo_{86-87}. The termination of the box at MgO = 7.5 wt% marks the end of olivine-only variation. Glasses with lower MgO contents reflect crystallization of olivine + cpx ± plagioclase.

Second, these melts (see Figure 7.6a) appear to have been extracted from within a semicoherent three-phase crystal mush, as was suggested for the Kulanaokuaiki 3 melts by *Rose et al.* [2000]. The MgO content of glasses in these differentiated tephra layers is similar to that of the melts that migrated spontaneously from the lower mush zone in Kīlauea Iki lava lake, in response to unstable density gradients in the melt column [*Helz et al.*, 1989]. Such self-segregation, whether before or during eruption, would help explain why the differentiated tephra layers often have a limited range of melt MgO content and also relatively low crystal contents.

The character and occurrence of rare differentiated melts in the Pāhala samples differ from the younger materials discussed above. First, the Pāhala low-MgO glass shards are not segregated into layers dominated by such compositions (Figure 7.10). Second, there is no roll-over in CaO at ~7.0% MgO: On the contrary, the data array simply stops, with the exception of one Pāhala glass (59-2) shard that plots at 5.5% MgO and 12.1% CaO (Figure 7.11a). Lastly, the differentiated Pāhala glasses contain only a little plagioclase, which looks like it precipitated as a quench phase after fragmentation (Figures 7.12c,d) or none at all (Figure 7.13e). Taken together, these observations suggest that the uppermost part of the Pāhala source was stored deeper and/or had a higher H_2O content than the source bodies for the younger differentiated melts. This is consistent with the high sulfur contents of many Pāhala glasses (Table 7.2), and with the apparent size of the eruption, as 26 m of ash implies that its parent bodies were much larger than the sources of the relatively small individual eruptions that make up the Keanakāko'i and Kulanaokuaiki tephras.

7.6.2. Primitive Melts (MgO > 7.5% MgO) in Tephra Deposits

The vast bulk of the glasses found in the tephra units are "primitive" melts [as defined by *Wright*, 1971], crystallizing only olivine ± chromite. As can be seen in Figure 7.11, the compositional arrays extend roughly parallel to the olivine control lines that bound the whole-rock composition for the 1959 eruption. Their slopes should tend to steepen at MgO contents below 9.0% MgO, as the olivine crystallizing from the melt becomes more iron rich than Fo_{86-87}; this is particularly evident in the array of Kulanaokuaiki glasses. Questions to be considered here are (1) how was this vast range of glass MgO contents produced and (2) what part of this range can be considered to have been stored in Kīlauea's thermal reservoir, at 2–6 km, prior to eruption?

Possible answers to the first question include (1) the range of melts results in part by rapid crystallization of olivine during transport and quenching of the glasses, (2) some of the range of melts was produced by last-minute mixing of disparate magmas, shortly before fragmentation and eruption, and (3) the range of glasses represents a range of preexisting melts sampled in a few hours, days, or weeks during explosive eruptions.

Table 7.2 Glass compositions, some with compositions of coexisting olivine, from selected tephra samples.

Tephra unit	Pāhala 1	Pāhala 2	Pāhala 3	Pāhala 4	Pāhala 5	Pāhala 6	Pāhala 7	Kulanaokuaiki 8	Kulanaokuaiki 9	Kulanaokuaki 10
Shard identification no.	66-1	66-7	66-10	66-10	37-4	66-5	65-6	F9-7-3-35	S9-18-2-21	F02-26-10a (79–97)
No. of points	7	7	1	2	4	3	4	3	12	19
Photomicrograph	13a		13b	13b	13c		13e		13f	
SiO_2	48.92	49.32	49.33	50.33	50.27	50.37	50.74	49.77	49.37	49.92
TiO_2	1.92	2.12	2.17	2.58	2.21	2.52	2.63	2.15	2.07	2.47
Al_2O_3	11.17	11.52	11.56	13.08	12.79	13.04	13.96	11.87	11.71	12.83
Cr_2O_3	0.17	0.14	0.15	0.04	0.08	0.07	0.03	—	—	—
FeO	11.41	11.25	11.27	10.88	10.89	11.20	11.20	11.36	11.74	11.07
MnO	0.17	0.18	0.16	0.18	0.17	0.16	0.18	0.18	0.18	0.18
MgO	14.56	13.30	12.78	9.08	9.45	9.25	6.27	12.54	12.49	9.72
CaO	9.01	9.37	9.58	10.23	10.50	10.27	11.17	9.72	9.52	10.69
Na_2O	1.77	1.64	1.88	2.12	2.06	2.04	1.98	1.98	1.94	2.12
K_2O	0.33	0.32	0.34	0.43	0.39	0.44	0.46	0.38	0.37	0.48
P_2O_5	0.18	0.19	0.22	0.26	0.21	0.26	0.27	0.22	0.21	0.30
Sum	99.61	99.35	99.44	99.21	99.02	99.62	98.89	100.17	99.60	99.78
Sulfur (ppm) in glass	1082	1112	1100	1220	297	1268	1220	—	144	—
Olivine (Fo in mol percent)			90.0 (core) 89.5 (rim)	ditto	90.2 (core) 86.1 (rim)	84.7 (core) 88.1 (rim)	81.5		88	
Comments on glass	Highest MgO in unit	Next highest MgO	Highest MgO in 66-10	Next to olivine	Next to olivine			Highest MgO in unit	Next highest MgO	Mid range MgO

Table 7.2 Cont'd

Tephra unit	Kulanaokuaiki	Kulanaokuaiki	Keanakākoʻi	Keanakākoʻi	Keanakākoʻi	Keanakākoʻi
	11	12	13	14	15	16
Shard identification no.	F02-26-7 (31–40)	F02-26-7 (41–49)	602-3	411 (1–10)	514-9	514-10
No. of points	10	9	1	15	5	4
Photomicrograph					6a	6b
Glass compositions						
SiO_2	49.77	51.85	50.26	51.36	51.07	50.34
TiO_2	3.18	2.76	2.21	2.36	2.95	2.35
Al_2O_3	14.10	14.11	12.41	13.51	13.38	13.53
Cr_2O_3	—	—	—	—	0.03	0.04
FeO	10.96	10.95	11.80	11.39	11.98	11.20
MnO	0.17	0.21	0.15	0.23	0.20	0.17
MgO	6.67	6.66	10.98	8.45	6.31	8.20
CaO	11.47	10.83	10.01	10.83	10.62	10.60
Na_2O	2.41	1.84	2.03	1.83	2.33	2.05
K_2O	0.77	0.48	0.33	0.43	0.49	0.45
P_2O_5	0.43	0.27	0.18	0.16	0.25	0.24
Sum	99.93	99.96	100.36	100.55	99.61	99.17
Sulfur (ppm) in glass	—	—	140	361	220	334
Olivine (Fo in mol percent)					77.3	84.8
Comments on glass	High Ti,K glass in K-2 layer	Low-Ti,K glass in same K-2 layer	Highest MgO in unit	Midrange MgO	Low-MgO glass	Higher-MgO glass

Note: Dashes indicate that no determination was made.

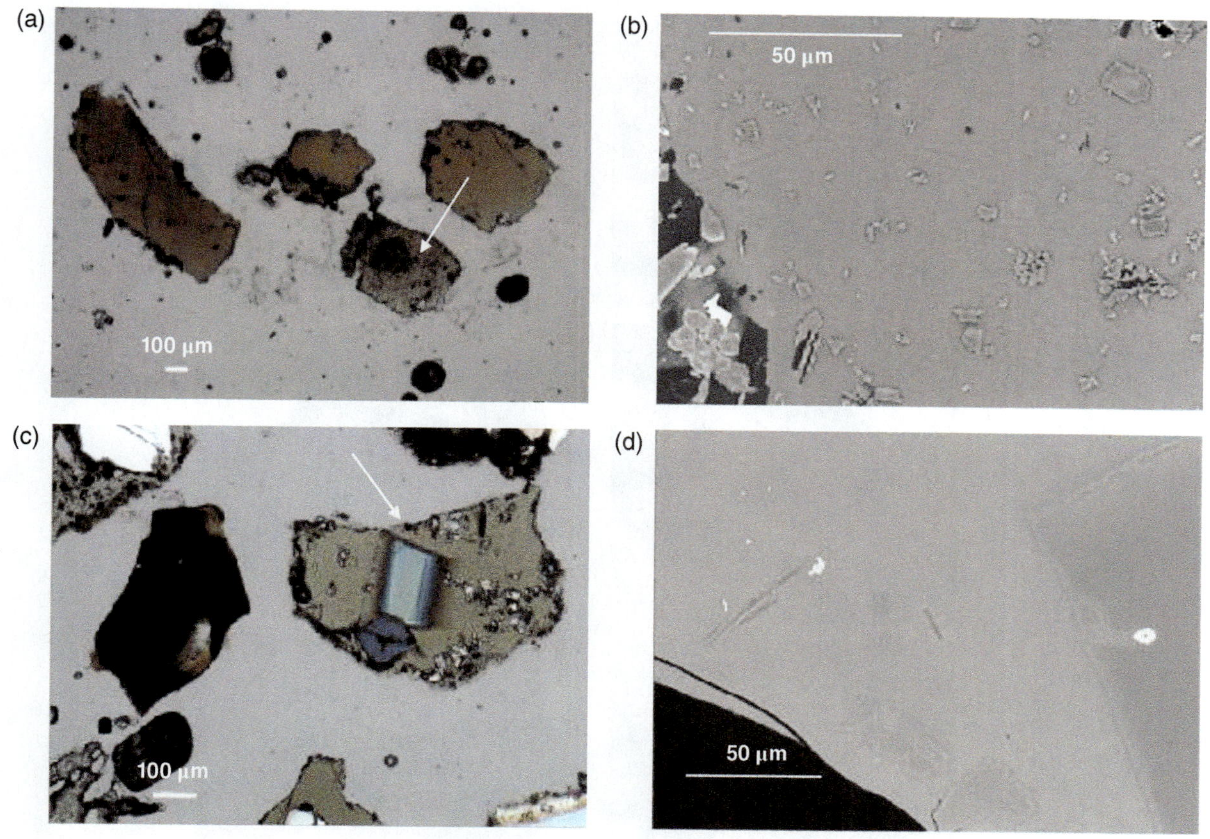

Figure 7.12 Photomicrographs (nicols partly crossed) and scanning electron microscope (SEM) images of shards with highly differentiated glasses from the upper part of the Pāhala Ash: (a) four glass shards in sample 65 (height 25.81 m). (c) two contrasting glass shards in sample 59 (height 24.1 m). The dark brown glass in the shard to the left in (c) has MgO > 9.0%, over 3% higher than the differentiated glass to the right. The arrows in (a) and (c) point to areas in the shards that are shown at higher magnification in the backscattered electron (BSE) images (b, d). In each case, quench-textured crystals are present, but plagioclase (dark gray/black crystals) is scarce in the glass. Shard 65-14 (a, b) includes a microgabbroic clot containing cpx + plag + ilmenite, seen at the lower left in (b).

7.6.3. Behavior of High-MgO Glasses During Quenching

The upper end of the range of concentration of MgO in the tephra glasses increases with increasing age: The maximum MgO seen in the Keanakāko'i is ~11.1%, in the Kulanaokuaiki is 12.5%, and in the Pāhala reaches 14.5% (see selected analyses in Table 7.2). Given these MgO levels, it is reasonable to consider whether some of the range of glass compositions might be produced by imperfect quenching of such highly magnesian melts.

The effects of the growth of olivine crystals during the quench can be seen most clearly in the relatively coarse shards present in the Pāhala Ash. The sample mounts consist mostly of submillimeter shards (Figures 7.12 and 7.13); while some are glass only and others are broken fragments of olivine, there are many that contain both olivine and glass. Table 7.2 presents data for coexisting glass and olivine compositions from several fragments in the Pāhala, one unit in the Kulanaokuaiki [*Rose et al.*, 2011], and sample 514 from the Keanakākoi Tephra.

A reconnaissance study of multiphase (glass + olivine ± chromite) shards found in the uppermost parts of the Pāhala shows that olivine ranges in composition from Fo_{90+} to $Fo_{81.5}$ (see Table 7.2, also photomicrographs in Figures 7.13b, c, and e), varying in parallel with melt MgO contents. The covariance of olivine and melt compositions demonstrates that the extreme melt ranges were not produced by quenching problems but mostly reflect preeruptive melt and olivine compositions. Glass and olivine data from the uppermost layer in the Kulanaokuaiki and sample 514 from the Keanakākoi (Table 7.2) fall in this compositional continuum as well, suggesting that they also record preeruptive components.

The effects of crystal growth on the quench can be evaluated from selected shards. The first, 66-10 (Figure 7.13b), contains three euhedral olivines ($Fo_{90.0-89.6}$) and strongly zoned glass (MgO = 12.78%–9.08%), with MgO decreasing near the olivine crystals. Shard 37-4, shown in Figure 7.13c, is much lower in the Pāhala section (Figure 7.10) but has similar olivine and glass rim compositions (Table 7.2); this suggests that highly magnesian glasses may occur at that

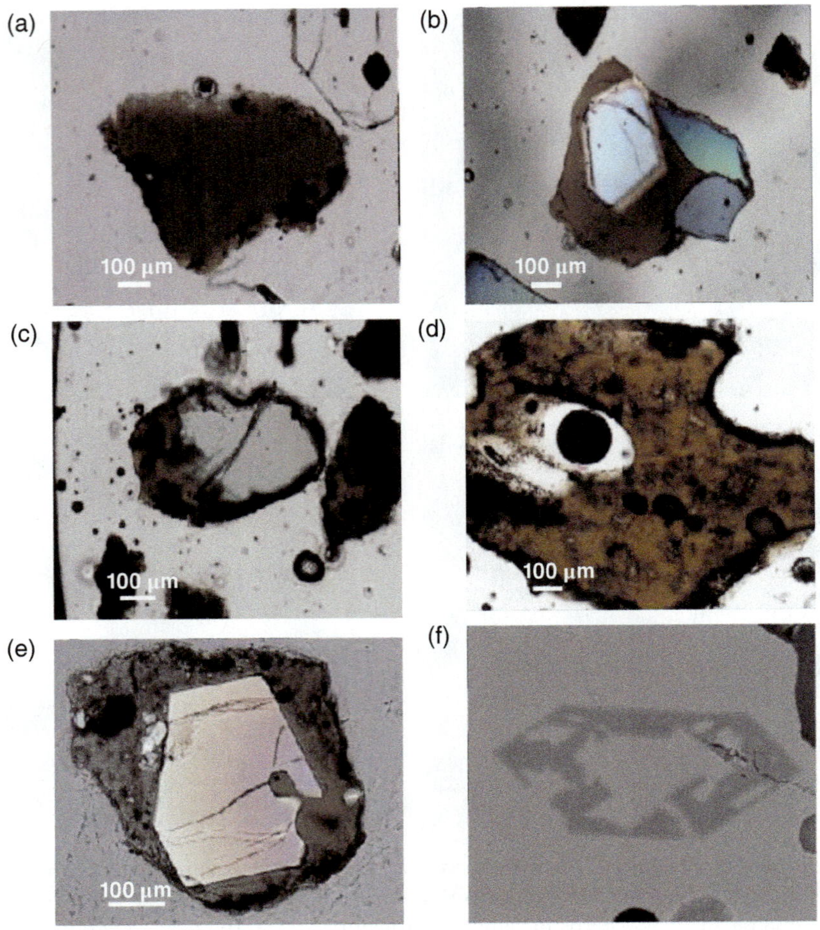

Figure 7.13 Photomicrographs of glass shards ± olivine in Pāhala samples 66 (height = 25.84 m), 65 (height 25.81 m), and 37 (height 18.26 m), (a) shard 66-1 is clear dark glass with MgO = 14.6% by weight; (b) shard 66-10 is dark glass (MgO variable) containing three Fo_{90} olivines; (c) shard 37-4 contains Fo_{90} olivine with a rim of dark glass; (d) shard 37-1,2 shows pervasive tiny quench olivine in the glass; (e) shard 65-6 contains $Fo_{81.5}$ and differentiated glass, with small cpx. Panel (f) is an SEM image of a 100 μm long olivine (S9-18-2-29) in the uppermost Kulanaokuaiki, showing rapid growth habit. Compositions are given in Table 7.2.

level (18.28 m) as well, though none is present in the sample mount. However, these quenching effects are very local, extending over distances of <100 μm, so most glass shards in the Pāhala or elsewhere should not be significantly affected by this style of quench overgrowth of olivine.

A second style of quench phase precipitation can be seen in Figure 7.13d, in which tiny olivine ± augite crystals nucleate throughout the body of the melt. Growth of such crystallites will lower the MgO content; however, the resulting glass is uniform in composition, presumably because the crystallites are dispersed throughout the body of the melt. The quench crystallites seen in Figure 7.12 similarly have only minor influence on glass composition. Hence neither style of growth of crystals during the quench would be expected to affect the bulk of melts erupted, especially that in the Pāhala section discussed here. However, the highly magnesian melts tend to be preserved in the uppermost units, perhaps reflecting the fact that they are more vulnerable to complete alteration and/or devitrification than most glasses. Their survival also suggests that these layers were cooled to ambient temperatures prior to deposition.

7.6.4. Evidence for Magma Mixing

Multiphase shards from the uppermost Pāhala samples contain evidence for late-stage mixing of thermally contrasting magmas. One example (shard 66-5, see data in Table 7.2) contains olivine ($Fo_{84.5}$) with a thin more forsteritic rim ($Fo_{88.1}$) in a glass with an average MgO content of 9.25%; this rim would have been produced when a relatively differentiated part of the Pāhala magma body was invaded by much hotter or more oxidizing melt. Other strongly zoned olivines (both normal and reverse zoned) have been observed in the Pāhala samples, though few have been analyzed. More extensive study of compound olivine + glass shards, though beyond the scope of this chapter, would shed light on how extensive such late-stage, preeruptive magma mixing was in the Pāhala magma body. However, though mixing may increase the number of shards with glass of intermediate composition, it cannot explain the composition of glasses in shards with unzoned olivines. Furthermore, the end-member glasses (and olivines) must in any case exist both before mixing and before eruption. Thus the wide-ranging MgO contents seen in so many tephra layers were most probably produced by near-instantaneous sampling of a wide range of preexisting melts found within Kīlauea's plumbing.

7.6.5. Compositional Limits on Melts in Kīlauea's Summit Reservoir

One question that remains is: Can the entire range of melts found in the tephra units have been stored in Kīlauea's summit reservoir? One curious feature of the variation of MgO with stratigraphic position in the Keanakāko'i (Figure 7.5) is that so many layers stop at 10.5%–11.0% MgO. Similarly, glass compositions in the lower two-thirds of the Pāhala section stop at 10.5%–11.0% MgO. Does this pattern imply that there may be a maximum MgO content for melts that have been in storage in the Kīlauea's thermal summit reservoir for any appreciable time?

It is widely known that the average olivine composition in olivine-rich lavas at Kīlauea has been close to Fo_{86-87} for many centuries, as reviewed by *Helz et al.* [2014b]. The slopes of the glass arrays in Figure 7.11 confirm that these tephra crystallized olivine in this range. The study of *Roeder and Emslie* [1970] on the exchange of Mg and Fe between olivine and melt shows that Kīlauea melts with ~10%–11% MgO will (at typical FeO and Fe_2O_3 contents) be in equilibrium with these commonest of olivine compositions. But where do these melts reside?

Kīlauea lavas that are rich in olivine carry it not in single crystals but in loose aggregates, as described by *Helz* [1987] and particularly *Schwindinger and Anderson* [1989]. This latter study found that virtually all olivine in the 1959 eruption scoria occurred in clusters, typically with four to six grains per cluster. These are loose aggregates with appreciable void space (distinct from the tight dunitic clots of mildly deformed olivines also found in the 1959 lavas [*Helz*, 1987]), and are found in most olivine-rich Kīlauea lavas. A possible explanation for the widespread occurrence of layers with minor amounts of melt with MgO = 10.5%–11.0% in the Keanakāko'i and Pāhala deposits (Figures 7.5 and 7.10) is that the eruptions have entrained loosely aggregated olivine-rich material from near the base of the summit reservoir and have (inevitably) sampled this interstitial liquid in the process.

Further exploration of this concept is beyond the scope of this chapter. However, one important implication of this model is that melts with MgO > 11.0%, found locally in the Kulanaokuaiki and Pāhala deposits, were probably not stored in Kīlauea's summit reservoir; rather they came directly from greater depths, spending little or no time at the storage levels characteristic of the current summit reservoir.

7.7. TEMPORAL AND SPATIAL CONSTRAINTS ON MELT AND MINERAL COMPOSITIONS IN THE SUMMIT RESERVOIR

The complex variations in MgO content of glasses reported above offer some insight into the behavior of Kīlauea's magmatic plumbing, in particular the relationship between melt temperature and residence time and/or depth of storage. Two important points are (1) the system is very active, so that consistency of melt and mineral compositions over extended periods implies near steady state behavior, and (2) storage times for individual melts or magmas appear to be short in comparison.

Observations of twentieth-century eruptions offer the most direct indication on time constraints on magma storage and differentiation. Distinct magma batches in Kīlauea's summit reservoir have succeeded each other about every 2–7 years during the twentieth century [*Helz et al.*, 2014b]. One of these, the 1967–1968 magma batch, was tracked by *Fiske and Kinoshita* [1969] from a depth of ~3 km to the surface over a period of 22 months. The magma arrived having cooled to glass MgO contents of 6.4%, at bulk MgO contents of 7.4–7.6 wt% MgO. This example and the many other intracaldera lavas with similar melt and bulk compositions show that even large batches of magma that reach shallow storage at normal ascent rates differentiate on the way.

The relationship between the magma of the 1954 summit eruption and the stored component of the 1959 eruption is also relevant. The first part of this distinctive magma batch appeared at the surface in 1954, with bulk MgO content of 7.2–7.3 wt% and glass MgO contents of 6.7–7.0 wt%. The second part, erupted in 1959 just outside the caldera, was dominated by glasses with 8–9 wt% MgO. The higher MgO content observed in melts from the later eruption requires either deeper storage of the magma body or later intrusion time or both. However, as ascent to the uppermost part of the thermal reservoir at normal rates generally leads to differentiation to low-MgO contents, then storage at greater depth seems indicated.

The persistence of Fo_{86-87} as the average olivine phenocryst composition in Kīlauea lavas (and its inferred dominance in the tephras, as indicated in Figures 7.11a,b) also imposes some constraints on the system. Evaluation of olivine crystallization and reequilibration in the most olivine-rich parts of Kīlauea Iki lava lake shows that olivine composition shifted from the original Fo_{86-87} to ~Fo_{80} in less than two decades during closed-system cooling and crystallization [*Helz et al.*, 2014b]. The centuries-long persistence of Fo_{86-87} as the average olivine composition in Kīlauea magmas thus implies steady resupply of melt to the site of this accumulated olivine, hypothesized to be near the base of the thermal reservoir.

Observations on the older tephras are compatible with the time scales observed for twentieth-century magmas. For example, the available data on coexisting melt and olivine compositions show that their parental bodies tracked equilibrium fairly closely over a wide range of melt compositions. This suggests that their ascent rates and/or residence times (prior to the onset of eruption) were not critically different from those of twentieth-century magma batches. Similarly, the time scale over which melt MgO in the Keanakāko'i Tephra reaches but does not exceed MgO = 10.5%–11.0% is two centuries [*Swanson et al.*, 2012]. This is the same order of magnitude as the period over which Kīlauea lavas contain Fo_{86-87} as the average olivine phenocryst composition, supporting the idea that these two observations may be related parts of a single process. To summarize: Steady emplacement of magma, rising slowly enough to maintain olivine-melt equilibrium, appears to be the rule within Kīlauea's summit reservoir prior to eruption. This implies that the cause for the wide ranges of MgO contents observed in tephra glasses is to be found in the processes of the explosive eruptions themselves.

7.8. SUMMARY

This chapter has focused on documenting the variation of MgO in fresh glasses erupted from Kīlauea's summit during the twentieth century and over previous centuries to millenia as well. The purpose of the overview is to evaluate how much variation there has been in melts stored in Kīlauea's plumbing at any given time and over time. The results are summarized in Table 7.3, which shows where the four examples reviewed here fall relative to each other in glass MgO content.

In interpreting the patterns in Table 7.3, we assume that ascent rates for individual bodies of magma within the thermal reservoir are broadly similar prior to the onset of eruption, as was discussed above. Within a given column in Table 7.3, therefore, depth of storage should increase downward: That is, because more magnesian melts are hotter, they should have come from greater depths than more differentiated but coeval melts. Thus temperature and pressure tend to increase going down each column. However, because of possible differences in water content among these units and documentable differences in sulfur

Table 7.3 Cross-walk summary of variations in MgO content in Kīlauea glasses for twentieth-century eruptions, Keanakāko'i and Kulanaokuaiki Tephras, and Pāhala Ash.

MgO Level in Glass	Twentieth Century	Keanakāko'i	Kulanaokuaiki	Pāhala
6.0–6.5%	Minor intracaldera effusive, earliest 1959	Rare	Rare	Uncommon
6.5–7.0%	**Peak intracaldera effusive,** minor 1959	Rare	K-2, 3 only	Minor
7.0–7.5%	**Peak intracaldera effusive,** minor 1959	Rare	**K-2, 3 plus peak K-1,4,5**	**Peak**
7.5–8.0%	Minor	**Peak IIA, IIB**	Shallow saddle	**Peak**
8.0–8.5%	**Peak 1959,** present 1924, present 1971, 1974 (extracaldera)	**Peak IIA, IIB**	**Peak K-1,4,5**	**Peak**
8.5–9.0%	**Peak 1959,** present 1924 tephra 1971 (extracaldera)	**Peak IIA, IIB**	**Peak K-1,4,5**	**Peak**
9.0–9.5%	Present 1959	Saddle	**Peak K-1,4,5**	Minor
9.5–10.0%	Present 1959	**Peak IIC1-3**	**Peak K-1,4,5**	Minor
10.0–10.5%	Present 1959	**Peak IIC1-3**	Some K-1,4,5	Persistent trace
10.5–11.0%	Inferred 1959	Less common but persistent	Minor K-1,4,5	Persistent trace
Melts (Glasses) from Below Summit Reservoir				
11–12%	Absent	Absent?	Minor	Trace
Above 12%	Absent	Absent	Trace, up to 12.5% MgO	Trace, up to 14.5% MgO

The bold face indicates the locations of peaks in the MgO histograms (Figures 7.3, 7.4, 7.7, and 7.9) and emphasizes when two peaks exist instead of one.

content (see Table 7.2), neither melt temperature nor depth of storage may be the same as one moves from one column to the next.

Nevertheless there are points of resemblance among several of these examples. The most differentiated parts of the Kulanaokuaiki deposits, for example, are very similar to twentieth-century glasses in the same range. Both frequency distributions have a dominant peak at 6.5%–7.5% MgO, with glasses crystallizing olivine + cpx + plagioclase, and a lesser cluster of olivine-only glasses at higher (though not identical) MgO contents.

By contrast, with the exception of sample 514, the Keanakākoʻi data are shifted to distinctly higher glass MgO contents: This alone suggests that little of the Keanakākoʻi was erupted from shallow storage, in agreement with previous work [*Mastin et al.*, 2004; *Swanson et al.*, 2012]. Also, the Keanakākoʻi glasses are more strongly bimodal than the Kulanaokuaiki glasses and those from the present (twentieth-century) system. The tendency toward bimodal frequency distribution in three of four instances raises the possibility that there are usually two preferred levels of magma storage within the summit reservoir, though their absolute depth may vary from one period to the next.

The Pāhala Ash is distinct from the other three examples, not only in its age and size, but in having a single-mode Gaussian distribution of glass MgO content. It is conceivable that this 26 m thick section was produced during one long but essentially continuous eruption. If so, the preexisting collection of magma bodies was more voluminous than typical for Kīlauea, and petrologic evidence, including observed sulfur contents, suggests that even the most differentiated parts were stored deeper than similarly low-MgO melts in the present thermal reservoir system. Given the presence of rare near-primary melts with ~14.5% MgO, it is tempting to speculate that the Pāhala eruption was triggered in part by destabilization of the preexisting magmatic system by intrusion of this very hot liquid.

7.9. CONCLUSIONS

We have looked at variations in MgO content in glasses erupted from Kīlauea's summit. Specific observations and conclusions are:

1. Four twentieth-century eruptions include coevally stored melts with MgO ranging from 7% to 9% MgO, a wider range than traditionally recognized as being present in the summit reservoir. During each eruption, these melts were erupted within hours or days of each other. The time covered (1924–1974) suggests that this range of melts is commonly present in the existing reservoir.

2. Individual layers in the Keanakākoʻi and much of the Pāhala section reported on here have minor amounts of glass with MgO contents of up to 10.5%–11.0%. Such melts would be in equilibrium with Fo_{86-87}, the dominant olivine phenocryst composition in subaerial Kīlauea lavas of all ages. Minor sampling of such melts, hypothesized to occur as interstitial liquid in olivine cumulates at the base of the summit reservoir, is consistent with their widespread occurrence in the Keanakākoʻi and Pāhala deposits. It also suggests an approximate upper MgO content for melt stored in Kīlauea's thermal reservoir for a period of time.

3. Older explosive eruptions contain glasses with higher MgO content (up to 12.5% in the Kulanaokuaiki Tephra and 14.5% in the Pāhala). Such melts are inferred to have not been stored for significant time in the summit reservoir but to have come from deeper in the system. They will have encountered stored magma as they moved upward and may have contributed to triggering explosive eruptions.

4. Thin individual tephra layers with glasses having MgO ranges greater than 2% are common in all tephra sections analyzed. This feature appears to be primary and has implications for the eruption process, including range of depth of excavation, depth of fragmentation, and extent of magma mixing and quenching prior to deposition.

5. In some cases, the distribution of MgO content is bimodal, suggesting that there have been two different levels of preferred magma storage at certain times.

Examination of the variation of MgO content in fresh glasses provides a simple but effective approach to evaluating the range of melts present in Kīlauea's thermal reservoir and for comparing complex series of explosive eruptions with each other. Because melt MgO varies linearly with temperature, this approach will allow evaluation of the thermal structure of Kīlauea's plumbing at the time of the activity. A further profitable line of inquiry will be to look at the composition of olivines coexisting with melt in these tephra in order to constrain ascent rates and storage times of individual magma bodies and to elucidate the extent of magma mixing prior to fragmentation and eruption. The data that form the basis of this contribution should serve as a basis for further work on issues relating to both magma storage and explosive eruptions at Kīlauea.

SUPPLEMENTARY DIGITAL DATA

The following data items are available online at www.wiley.com/go/Carey/Hawaiian_Volcanoes

Table 7A.1. Glass compositions from the Keanakākoʻi, Kulanaokuaiki and Pāhala tephra units.

ACKNOWLEDGMENTS

The Pāhala glasses were analyzed by Alice Davis and those from the Keanakākoʻi Tephra by Mel Beeson, who are thanked for their contributions. The Pāhala Ash section was sampled by DAC with assistance from Roger Denlinger and LGM and several volunteers at HVO during the 1990s. We also thank reviewers Margaret Mangan, Don Swanson, Michael Rowe, Dominique Weis, and an anonymous reviewer for helpful comments.

REFERENCES

Baker, N. A. (1987), A petrologic study of the 1982 summit lavas of Kīlauea Volcano, Hawaii: Honolulu, Senior Honors thesis, Univ. of Hawai'i at Mānoa, 61 p.

Clague, D. A., W. S. Weber, and J. E. Dixon (1991), Picritic glasses from Hawaii, *Nature*, 353(6344), 553–556, doi:10.1038/353553a0.

Clague, D. A., M. H. Beeson, R. P. Denlinger, and L.G. Mastin (1995a), Ancient ash deposits and calderas at Kīlauea Volcano, *Eos Trans. AGU, 76*(46, suppl.), p. F666, Abstract no. V22C-03.

Clague, D. A., J. G. Moore, J. E. Dixon, and W. G. Friesen (1995b), Petrology of submarine lavas from Kīlauea's Puna Ridge, Hawaii, *J. Petrol.*, *36*(2), 299–349, doi:10.1093/petrology/36.2.299.

Clague, D. A., J. T. Hagstrum, D. E. Champion, and M. H. Beeson (1999), Kīlauea summit overflows; their ages and distribution in the Puna District, Hawai'i, *Bull. Volcanol.*, *6*(6), 363–381, doi:10.1007/s004450050279.

Decker, R. W., and R. L. Christiansen (1984), Explosive eruptions at Kīlauea Volcano Hawaii, in Explosive Volcanism: Inception, Evolution, and Hazards, edited by National Research Council et al., pp. 122–132, National Acad. Press, Washington, D. C.

Duffield, W. A., R. L. Christiansen, R. Y. Koyanagi, and D. W. Peterson (1982), Storage, migration, and eruption of magma at Kīlauea Volcano Hawaii, 1971–1972, *J. Volcanol. Geotherm. Res.*, *13*(3–4), 273–307, doi:10.1016/0377-0273(82)90054-3.

Dzurisin, D., J. P. Lockwood, T. J. Casadevall, and M. Rubin (1995), The Uwēkahuna Ash Member of the Puna Basalt: Product of violent phreatomagmatic eruptions at Kīlauea Volcano, Hawaii, between 2800 and 2100 ^{14}C years ago, *J. Volcanol. Geotherm. Res.*, *66*, 163–184.

Easton, R. M. (1987), Stratigraphy of Kīlauea Volcano, in *Volcanism in Hawaii*, edited by R. W. Decker et al., vol. 1, pp. 243–260, U.S. Geol. Surv. Prof. Pap. 1350.

Eaton, J. P., and K. J. Murata (1960), How volcanoes grow, *Science*, *132*(3432), 925–938, doi:10.1126/science.132.3432.925.

Eaton, J. P., D. H. Richter, and H. L. Krivoy, (1987), Cycling of magma between the summit reservoir and Kīlauea Iki lava lake during the 1959 eruption of Kīlauea Volcano, in *Volcanism in Hawaii*, edited by R. W. Decker et al., vol. *2*, pp. 1307–1335, U.S. Geol. Surv. Prof. Pap. 1350.

Fiske, R. S., and W. T. Kinoshita (1969), Inflation of Kīlauea Volcano prior to its 1967–1968 eruption, *Science*, *165*(3891) 341–349, doi:10.1126/science.165.3891.341.

Fiske, R. S., T. R. Rose, D. A. Swanson, D. E. Champion, and J.P. McGeehin (2009), Kulanaokuaiki Tephra (ca. A.D. 400–1000); newly recognized evidence for highly explosive eruptions at Kīlauea Volcano, Hawai'I, *Geol. Soc. Am. Bull.*, *121*(5–6), 712–728, doi:10.1130/B26327.1.

Fiske, R. S., T. R. Rose, and D. A. Swanson (2011), Diatreme-like eruption at Kīlauea: The Kulanaokuaiki-3 tephra (~900 CE), *Eos Trans. AGU*, 2011 Fall Meet., Abstract V33E-06

Garcia, M. O., A. E. Mucek, and D. A. Swanson (2011), Geochemistry of glass and olivine from Keanakāko'i tephra at Kīlauea Volcano, Hawai`I, *Eos Trans. AGU,* 2011 Fall Meet., Abstract V41A-2480.

Helz, R.T. (1987), Diverse olivine types in lavas of the 1959 eruption of Kīlauea Volcano, and their bearing on eruption dynamics, in *Volcanism in Hawaii*, edited by R. W. Decker et al., vol. *1*, pp. 691–722, U.S. Geol. Surv. Prof. Pap. 1350.

Helz, R.T. (2009), Processes active in mafic magma chambers; the example of Kīlauea Iki lava lake, Hawaii, *Lithos*, *111*(1–2), 37–46, with electronic supplement, doi:10.1016/j.lithos.2008.11.007.

Helz, R. T., and C. R. Thornber (1987), Geothermometry of Kīlauea Iki lava lake, Hawaii, *Bull. Volcanol.*, *49*(5), 651–668, doi:10.1007/BF01080357.

Helz, R. T., H. Kirschenbaum, and J. W. Marinenko (1989), Diapiric transfer of melt in Kīlauea Iki lava lake, Hawaii: A quick, efficient process of igneous differentiation, *Geol. Soc. Am. Bull.*, *101*(4), 578–594, doi:10.1130/0016-7606(1989)101<0578:DTOMIK>2.3.CO;2.

Helz, R. T., N. G. Banks, C. Heliker, C. A. Neal, and E. W. Wolfe (1995), Comparative geothermometry of recent Hawaiian eruptions, *J. Geophys. Res.*, *100*, 17,637–17,657, doi:10.1029/95JB01309.

Helz, R. T., D. A. Clague, L. G. Mastin, and T. R. Rose (2014a), Electron microprobe analyses of glasses from Kīlauea tephra units, U.S. Geol. Surv. Open-File Rep. 2014-1090, 24p. plus 2 appendices in separate files, dx.doi.org/10.3133/ofr20141090.

Helz, R. T., D. A. Clague, T. W. Sisson, and C. R. Thornber, (2014b), Petrologic insights into basaltic volcanism at historically active Hawaiian volcanoes, Chapter 6 of Poland, M. P., T. J. Takahashi, and C. M. Landowski, (eds.), *Characteristics of Hawaiian Volcanoes*, U.S. Geol. Surv. Prof. Pap. 1801, pp. 237–292, doi:10.3133/pp18016.

Heslop, S. E., L. Wilson, H. Pinkerton, and J. W. Head III (1989), Dynamics of a confined lava flow on Kīlauea volcano, Hawaii, *Bull. Volcanol.*, *51*, 415–432.

Holcomb, R.T. (1987), Eruptive history and long-term behavior of Kīlauea Volcano, in *Volcanism in Hawaii*, edited by R. W. Decker et al., vol. *1*, pp. 261–350, U.S. Geol. Surv. Prof. Pap. 1350.

Jaggar, T. A., Jr., and R. H. Finch (1924), The explosive eruption of Kīlauea in Hawaii, 1924, *Am. J. Sci.* Ser. 5, *8*(47), *353*–374, doi:10.2475/ajs.s5-8.47.353.

Johnson, D. J., A. A. Eggers, M. Bagnardi, M. Battaglia, M. P. Poland, and A. Miklius (2010), Shallow magma accumulation at Kīlauea Volcano, Hawai'i, revealed by microgravity surveys, *Geology*, *38*(12), 1139–1142; doi:10.1130/G31323.1.

Kinoshita, W. T., R. Y. Koyanagi, T. L. Wright, and R. S. Fiske (1969), Kīlauea Volcano; the 1967–68 summit eruption, *Science*, *166*(3904), 459–468, doi:10.1126/science.166.3904.459.

Lockwood, J. P., R. I. Tilling, R. T. Holcomb, F. Klein, A. T. Okamura, and D. W. Peterson (1999), Magma migration and resupply during the 1974 summit eruptions of Kīlauea Volcano, Hawai'i, U.S. Geol. Surv. Prof. Pap. 1613, pp. 1–37.

Mastin, L. G., R. L. Christiansen, C. Thornber, J. Lowenstern, and M. Beeson (2004), What makes hydromagmatic eruptions violent? Some insights from the Keanakāko'i Ash, Kīlauea Volcano Hawai'I, *J. Volcanol. Geotherm. Res.*, *137*(1–3), 115–131, doi:10.1016/j.jvolgeores.2004.05.015.

McPhie, J., G. P. L. Walker, and R. L. Christiansen (1990), Phreatomagmatic and phreatic fall and surge deposits from explosions at Kīlauea Volcano, Hawaii, 1790 A.D.; Keanakāko'i Ash Member, *Bull. Volcanol.*, *52*(5), *334*–354, doi:10.1007/BF00302047.

Murata, K. J., and D. H. Richter (1966), The 1959–60 eruption of Kīlauea Volcano, Hawaii; chemistry of the lavas, U.S. Geol. Surv. Prof. Pap. 537-A, pp. A1–A26.

Neal, C. A. and Lockwood, J. P., 2003, Geologic Map of the Summit Region of Kīlauea Volcano, Hawaii, U.S. Geol. Surv. Geol. Inv. Series I-2759.

Pietruszka, A. J., and M. O. Garcia (1999), The size and shape of Kīlauea Volcano's summit magma storage reservoir; a geochemical probe: *Earth Planet. Sci. Lett.*, *167* (3–4), 311–320, doi:10.1016/S0012-821X(99)00036-9.

Poland, M. P., A. Miklius, and E. K. Montgomery-Brown (2014), Magma supply, storage, and transport at shield-stage Hawaiian volcanoes, Chapter 5 of Poland, M. P., T. J. Takahashi, and C. M. Landowski, (eds.), *Characteristics of Hawaiian Volcanoes*, U.S. Geol. Surv. Prof. Pap. 1801, pp. 179–234, doi:10.3133/pp18015.

Powers, H. A. (1955), Composition and origin of basaltic magma of the Hawaiian Islands: *Geochim. Cosmochim. Acta*, *7*(1–2) 77–107, doi:10.1016/0016-7037(55)90047-8.

Richter, D. H., and J. G. Moore (1966), The 1959–60 eruption of Kīlauea Volcano, Hawaii; petrology of the Kīlauea Iki lava lake, U.S. Geol. Surv. Prof. Pap. 537-B, pp. B1–B26.

Richter, D. H., J. P. Eaton, J. Murata, W. U. Ault, and H. L. Krivoy (1970), The 1959–60 eruption of Kīlauea Volcano, Hawaii; chronological narrative, U.S. Geol. Surv. Prof. Pap. 537-E, pp. E1–E73.

Roeder, P. L., and R. F. Emslie (1970), Olivine-liquid equilibrium, *Contrib. Mineral. Petrol.*, *29*(4), 275–289, doi:10.1007/BF00371276.

Rose, T. R., R. S. Fiske, and D. A. Swanson (2000), The Kulanaokuaiki tephra: The glassy components of unit 3, *Eos Trans. AGU*, *81*, Fall Meet. Suppl., Abstract V11B-14.

Rose, T. R., R. S. Fiske, and D. A. Swanson (2011), High-MgO vitric ash in Upper Kulanaokuaiki Tephra, Kīlauea Volcano, Hawai'i; a preliminary description, Abstract V41A-2481 presented at 2011 Fall Meeting, AGU, San Francisco, California, 5–9 Dec.

Rowe, M. C., C. R. Thornber, and T. R. Orr (2009), Kīlauea 2008: Primitive components and degassing recorded by olivine-hosted melt inclusions and matrix glasses, *Goldschmidt Conf. Abstr. Geochim. Cosmochim. Acta*, *73*(13, Suppl. 1), A1126.

Ryan, M. P. (1988), The mechanics and three-dimensional internal structure of active magmatic systems; Kīlauea Volcano, Hawaii, *J. Geophys. Res.*, *93*(B5), 4213–4248, doi:10.1029/JB093iB05p04213.

Ryan, M. P., J. Y. K. Blevins, A. T. Okamura, and R. Y. Koyanagi (1983), Magma reservoir subsidence mechanics; theoretical summary and application to Kīlauea Volcano, Hawaii, *J. Geophys. Res.*, *88*(B5), 4147–4181, doi:10.1029/JB088iB05p04147.

Schwindinger, K. R., and A. T. Anderson, Jr. (1989), Synneusis of Kīlauea Iki olivines, *Contrib. Mineral. Petrol.*, *103*(2), 187–198, doi:10.1007/BF00378504.

Sharp, R. P., D. Dzurisin, and M. C. Malin (1987), An early 19th century reticulite pumice from Kīlauea Volcano, in Volcanism in Hawaii, edited by R. W. Decker et al., vol. 1, pp. 395–404, U.S. Geol. Surv. Prof. Pap.1350.

Swanson, D. A., T. R. Rose, R. S. Fiske, and J. P. McGeehin (2012), Keanakāko'i Tephra produced by 300 years of explosive eruptions following collapse of Kīlauea's caldera in about 1500 CE, J. Volcanol. Geotherm. Res., *215–216*, 8–25, doi.org/10.1016/j.jvolgeores.2011.11.009.

Thornber, C.R. M. C. Rowe, D. T. Adams, J. C. Rea, and T. R. Orr (2010), Application of microbeam techniques to identifying and assessing comagmatic mixing between summit and rift eruptions at Kīlauea Volcano, *Eos Trans. AGU*, 2011 Fall Meet., Abstract V43F-1.

Wright, T. L. (1971), Chemistry of Kīlauea and Mauna Loa lava in space and time, U.S. Geol. Surv. Prof. Pap. 735, pp. 1–40.

Wright, T. L. (1973), Magma mixing as illustrated by the 1959 eruption, Kīlauea Volcano, Hawaii, *Geol. Soc. Am. Bull.*, *84*(3), 849–858, doi:10.1130/0016-7606(1973).

Wright, T. L. and R. T. Helz (1996), Differentiation and magma mixing on Kīlauea's East Rift Zone: A further look at the eruptions of 1955 and 1960. Part II. The 1960 lavas, *Bull. Volcanol.*, *57*, 602–630.

Wright, T. L., and R. I. Tilling (1980), Chemical variations in Kīlauea eruptions 1971–1974, *Am. J. Sci.*, *280-A* (The Jackson Volume), Pt. 1, 777–793.

Wright, T. L., D. A. Swanson, and W. A. Duffield (1975), Chemical compositions of Kīlauea east-rift lava, 1968–1971, *J. Petrol.*, *16*(1), 110–133, doi:10.1093/petrology/16.1.110.

8

Petrologic Testament to Changes in Shallow Magma Storage and Transport During 30+ Years of Recharge and Eruption at Kīlauea Volcano, Hawai'i

Carl R. Thornber[1], Tim R. Orr[2], Christina Heliker[2], and Richard P. Hoblitt[1]

ABSTRACT

Petrologic monitoring of Kīlauea Volcano from January 1983 to October 2013 has yielded an extensive record of glass, phenocryst, melt inclusion, and bulk-lava chemistry from well-quenched lava. When correlated with 30+ years of geophysical and geologic monitoring, petrologic details testify to physical maturation of summit-to-rift magma plumbing associated with sporadic intrusion and prolonged magmatic overpressurization. Changes through time in bulk-lava major- and trace-element compositions, along with glass thermometry, record shifts in the dynamic balance of fractionation, mixing, and assimilation processes inherent to magma storage and transport during near-continuous recharge and eruption. Phenocryst composition, morphology, and texture, along with the sulfur content of melt inclusions, constrain coupled changes in eruption behavior and geochemistry to processes occurring in the shallow magmatic system.

For the first 17 years of eruption, magma was steadily tapped from a summit reservoir at 1–4 km depth and circulating between 1180 and 1200°C. Furthermore, magma cooled another 30°C while flowing through the 18 km long rift conduit, before erupting olivine-spinel-phyric lava at temperatures of 1150–1170°C in a pattern linked with edifice deformation, vent formation, eruptive vigor, and presumably the flux of magma into and out of the summit reservoir. During 2000–2001, a fundamental change in steady state eruption petrology to that of relatively low-temperature, low-MgO, olivine(-spinel)-clinopyroxene-plagioclase-phyric lava points to a physical transformation of the shallow volcano plumbing uprift of the vent. Preeruptive comagmatic mixing between hotter and cooler magma is documented by resorption, overgrowth, and compositional zonation in a mixed population of phenocrysts grown at higher and lower temperatures. Large variations of sulfur (50 to >1000 ppm) in melt inclusions within individual phenocrysts and among phenocrysts in most samples provide an unequivocal glimpse of rapid crystal growth amid sulfur degassing at <30 MPa in a turbulent preeruptive environment. We speculate that, during the last decade, one or more shallow open-system reservoirs developed along the conduit between the summit and Pu'u 'Ō'ō and now serve to buffer the magmatic throughput associated with ongoing recharge and eruption.

Lava with identical trace-element signatures erupted simultaneously at the summit and at Pu'u 'Ō'ō from 2008 to 2013 confirms magmatic continuity between the vents. Complementary changes in compositions of matrix glasses, phenocrysts, and melt inclusions of summit tephra are mirrored by similar changes in contemporaneous rift lava at eruption temperatures 20–35°C lower than those at the summit. Petrologic parameters measured at opposite ends of the shallow magmatic plumbing system are both correlated with summit deformation, demonstrating that effects of summit magma chamber pressurization are translated throughout interconnected magma pathways in the shallow edifice.

[1] *Cascades Volcano Observatory, U.S. Geological Survey, Vancouver, Washington, USA*
[2] *Hawaiian Volcano Observatory, U.S. Geological Survey, Hawai'i National Park, Hawaii, USA*

Hawaiian Volcanoes: From Source to Surface, Geophysical Monograph 208. First Edition.
Edited by Rebecca Carey, Valérie Cayol, Michael Poland, and Dominique Weis.

Companion Website: www.wiley.com/go/Carey/Hawaiian_Volcanoes

8.1. INTRODUCTION

The most prolific era of shield building at Kīlauea Volcano in over half a millenniium was heralded by small, hours-long eruptions at its summit in April and September 1982. In January 1983, a soon-to-be robust linkage was established between magma fed and stored beneath the summit since 1982 and lava subsequently erupted on the East Rift Zone (ERZ; Figure 8.1). By 2013, the ensuing ERZ eruption had added approximately 4.0 km^3 of olivine tholeiite basalt to Kīlauea's south flank (see Chapter 18, this volume). This estimated dense-rock-equivalent (DRE) volume almost certainly represents only a portion of the magma transferred into the shallow Kīlauea edifice since the eruption began. It is unclear how the magmatic plumbing beneath the summit and along the ERZ has evolved while accommodating more than three decades of magmatically driven edifice deformation. Over 30+ years of eruption, and regardless of either "active" or "passive" controls [e.g., *Lundgren et al.*, 2013], magmatic intrusions along the upper and middle ERZ are likely to have significantly increased both the complexity and capacity of interconnected magma pathways and storage chambers along the rift. The evolution of the shallow magmatic plumbing of Kīlauea Volcano during this eruptive era is revealed by changes in the petrology and geochemistry of erupted lava. In this chapter, we assess these changes through analysis of the petrologic data compiled on freshly quenched lava and tephra samples collected throughout the 1983–2013 ERZ eruption as well as those from the 2008–2013 summit eruption. Our primary objective in routine eruption sampling and analysis is to contribute a petrologic perspective toward a better understanding of one of the most thoroughly monitored and researched volcanoes in the world. In this effort, we merely touch upon the interpretive value of a unique and very large treasure trove of petrologic data on Kīlauea bulk lava and its constituent phenocrysts and glasses. This wealth of information is ripe with potential for further modeling of dynamic geochemical changes associated with Kīlauea magmatism and constitutes a substantial and perhaps archetypical database for natural low-pressure phase relations of olivine-tholeiitic basalts in general.

Heliker and Mattox [2003] and *Orr et al.* (Chapter 18) give chronologic narratives of Kīlauea's ERZ eruption from 1983 to 2013, including the episode nomenclature used herein. *Thornber* [2003] gives an interpretive summary of the petrology and geochemistry of samples collected from ERZ vents through September 2001. During the 1983–2001 interval, 2.3 km^3 of new lava was erupted at an average annual rate of 0.12 km^3 (DRE). In the decade that followed, Kīlauea's ERZ eruption was bolstered by a significant surge of magma input, as evidenced by summit inflation and increased gas emissions from 2003 to 2007 [*Poland et al.*, 2008, 2012]. The resulting magmatic overpressurization of the shallow summit reservoir system preceded and probably led to the onset of continuous summit eruption in March 2008 [*Wilson et al.*, 2008; *Wooten et al.*, 2009; *Poland et al.*, 2009, 2012]. Between 2001 and 2013, another 1.7 km^3 of lava vented from within or near Puʻu ʻŌʻō, disrupted only briefly by two conduit-breaching eruptions. The first was in June 2007 at Kanenuiohamo, and the other was the Kamoamoa fissure eruption in March 2011 (Figure 8.1). Despite significant disparity in their petrologic traits, as reported herein, both of these small-volume eruptions are linked to magmatic overpressurization of the shallow edifice.

In addition to data for 1983–2001 ERZ samples interpreted previously by *Thornber* [2001, 2003] and *Thornber et al.* [2003b], we use data on samples collected and analyzed since 2001, including ERZ and summit eruption products, to develop an interpretive overview of 1983–2013 eruption petrology. These additional 12 years of petrologic data, in conjunction with new and improved geophysical and observational data, have enhanced our perception of the dynamic igneous environment sustained within the Kīlauea edifice during prolonged magma recharge and eruption. As established for the first two decades of eruption [*Thornber*, 2003], both subtle and sudden changes in the temperature, chemistry, and mineralogy of ERZ lava have continued to correspond to fluxing of magma into Kīlauea and inherent deformation of the summit and ERZ. During 2008–2013, the petrology of simultaneously erupted summit and ERZ lava give a clearer picture of how and where magma is recharged, stored, and transported within Kīlauea's edifice before erupting and how the shallow eruptive plumbing system has transformed in response to repeated intrusions and magmatic overpressurization.

8.2. KĪLAUEA ERUPTION SAMPLES AND PETROLOGIC DATA SOURCES

Thornber et al. [2003c] present major-element geochemical data for 1002 fresh lava samples erupted from ERZ vents during January 1983 to September 2001, including 795 samples with bulk-lava major-element analysis by wavelength dispersive X-ray fluorescence (WDXRF) and 652 averaged matrix glass compositions (~6520 analyses) compiled, in part, from earlier reports cited therein. Bulk lava and glass trace elements along with Nd, Sr, and Pb isotopes are published separately for a representative subsuite of the 1983–2001 samples [*Thornber et al.*, 2003a]. Also incorporated in this long-term assessment of Kīlauea rift-lava petrology are compositional data for phenocrysts and melt inclusions in pre-2001 samples that were previously published by *Thornber* [2001] and *Thornber et al.* [2003b].

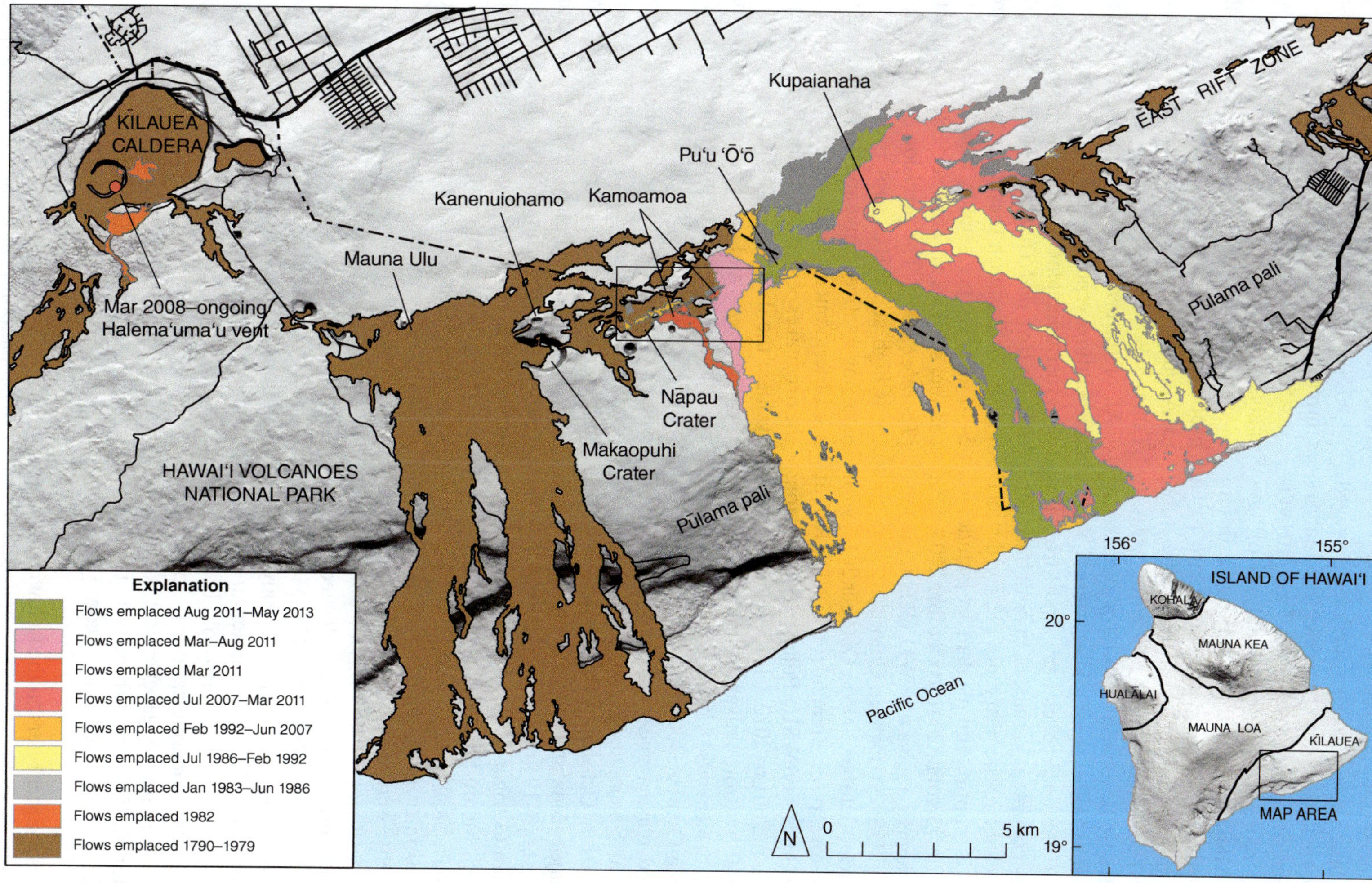

Figure 8.1 Map of summit and upper-middle ERZ of Kīlauea Volcano. Recent and historic lava flows at the summit, along the upper Southwest Rift Zone, and along the upper to middle ERZ are distinguished in the map key.

ERZ bulk geochemistry and microbeam data for 769 lava samples are presented as averages for specific intervals of eruption between October 2001 and November 2013 in Supplementary Tables 8.1s and 8.2s. Supplementary Data Table 8.1s also includes averaged 2001 to 2013 bulk-rock major and trace element analyses of USGS Geochemical Standards BHVO-1 and BHVO-2, in addition to representative bulk analyses of lava throughout the March 6–9, 2011 Kamoamoa fissure eruption. The post-2001 data reviewed here are for a range of sample morphologies (i.e., tears, spatter, tube, and surface flows) collected as near to venting areas as possible. Such variations, related to those particulars of surface emplacement, affect sporadic scattering in the temporal trends of sample geochemistry. Apart from this scatter, temporally consistent variations in both bulk-lava and glass major-element chemistry are clearly distinct. The post-2001 ERZ eruption sample data set includes:

- 309 samples from October 2001 to October 2006, with bulk-lava major-element analysis by WDXRF at the U.S. Geological Survey (USGS), Denver, Colorado;
- 261 samples from October 2006 to October 2013, with bulk-lava major and trace elements by WDXRF at Washington State University (WSU), Pullman, Washington;
- 101 samples from 2001 to 2012, with 36-element long-count instrumental neutron activation analysis (INAA) (including rare earth elements), USGS, Denver, Colorado;
- 8 samples from October 2006 to October 2013, with Induction coupled plasma mass spectrometer (ICPMS) analyses of rare earth elements at WSU;
- 453 samples from October 2001 to October 2013, with averaged major-element electron microprobe analyses of matrix glasses (5–10 analyses per sample); 84 samples collected from October 2001 to October 2004 were analyzed at USGS, Menlo Park, California, and the rest were done at USGS, Denver, Colorado.

More than 7500 microprobe analyses of phenocrysts and melt inclusions, were mostly completed at USGS, Denver, Colorado, except for 250 olivine and 228 melt inclusion analyses on 51 samples collected between January 1999 and October 2004 at USGS, Menlo Park, and 101 phenocryst and 105 inclusion analyses on 10 samples taken over the 2000–2001 interval that were performed at the University of Bristol, U.K. (Fiona Couperthwaite, written communication, October 2009). Averaged and representative geochemical and microbeam analyses for ERZ samples from 2001 to October 2013 are available in Supplementary Data Tables 8.1s, 8.2 s, and 8.3 s.

In addition, this chapter refers to the following data on 165 Kīlauea summit tephra samples, collected from March 2008 to August 2013:

- 54 bulk tephra samples analyzed using WDXRF for major and trace elements at WSU;
- 20 bulk tephra samples analyzed for rare earth elements using ICPMS, 14 at WSU and 6 at USGS, Denver;
- 6 bulk tephra samples analyzed for 36 elements by INAA at USGS, Denver;
- >1500 microprobe analyses of matrix glass in 200 samples;
- >500 microprobe analyses of phenocrysts and ~100 analyses of melt inclusions.

Averaged and representative geochemical and microprobe analyses for summit tephra samples from 2008 to August 2013 are available in Supplementary Data Tables 8.4s, 8.5s, and 8.6s.

All of the eruption temperatures and inferred liquidus temperatures reported here are determined using the MgO glass thermometer calibrated using Kīlauea starting materials by *Helz and Thornber* [1987] (T°C = 20.1 wt% MgO + 1014°C). Eruption thermometry values are averages of 5–10 analyses per sample, and distal lava tube or near-channel samples are vent corrected for 0.9°C/km cooling [*Thornber*, 2001]. The precision of MgO-glass thermometry is high at ±0.1°C, so it provides a precise means of assessing relative temperature changes, but its accuracy as a geothermometer is no better than the ±10°C stated for the original experimental calibration. Calculations used to evaluate the distribution of ferrous iron and magnesium between olivine and surrounding glass are made assuming an equilibrium distribution coefficient (K_D(Fe-Mg)$^{olv/liq}$ = (XFeO/XMgO)^{olv} (XMgO/XFeO)^{liq}) of 0.30 (±0.03, where X = mole percent), established for Kīlauea ERZ basalt by *Roeder and Emslie* [1970]. This value is preferred here over more recent determinations [e.g., *Marzten et al.*, 2011, and others summarized therein] because it was precisely determined using Kīlauea ERZ 1960 starting material in low-pressure, oxygen fugacity (fO_2)-controlled experiments at temperatures of 1150–1250°C that are directly applicable to current eruption products. Based upon repeated measurements in 1998 ERZ lava samples [*Thornber*, 2001; *Roeder et al.*, 2003], a Fe^{3+}/Fe^{2+} ratio in Kīlauea glasses is assumed such that FeO^{liq} /FeO$_t^{liq}$ = 0.9. For distal tube samples, a corrected olivine-glass distribution coefficient (K^*_D(Fe-Mg)$^{olv/liq}$) uses olivine core compositions and averaged matrix glass compositions that are distance corrected for cooling. This amounts to a K_D correction of 0.002 per km of tube distance.

8.3. AN OVERVIEW OF 1983–2013 KĪLAUEA ERZ LAVA PETROLOGY

8.3.1. MgO Systematics of Prolonged ERZ Eruption

The temporal variation of weight percent magnesia (MgO) in matrix glasses and bulk lava tracks 30+ years of fluctuations in the eruption temperature and the "primitive character" of magma erupted from Kīlauea's

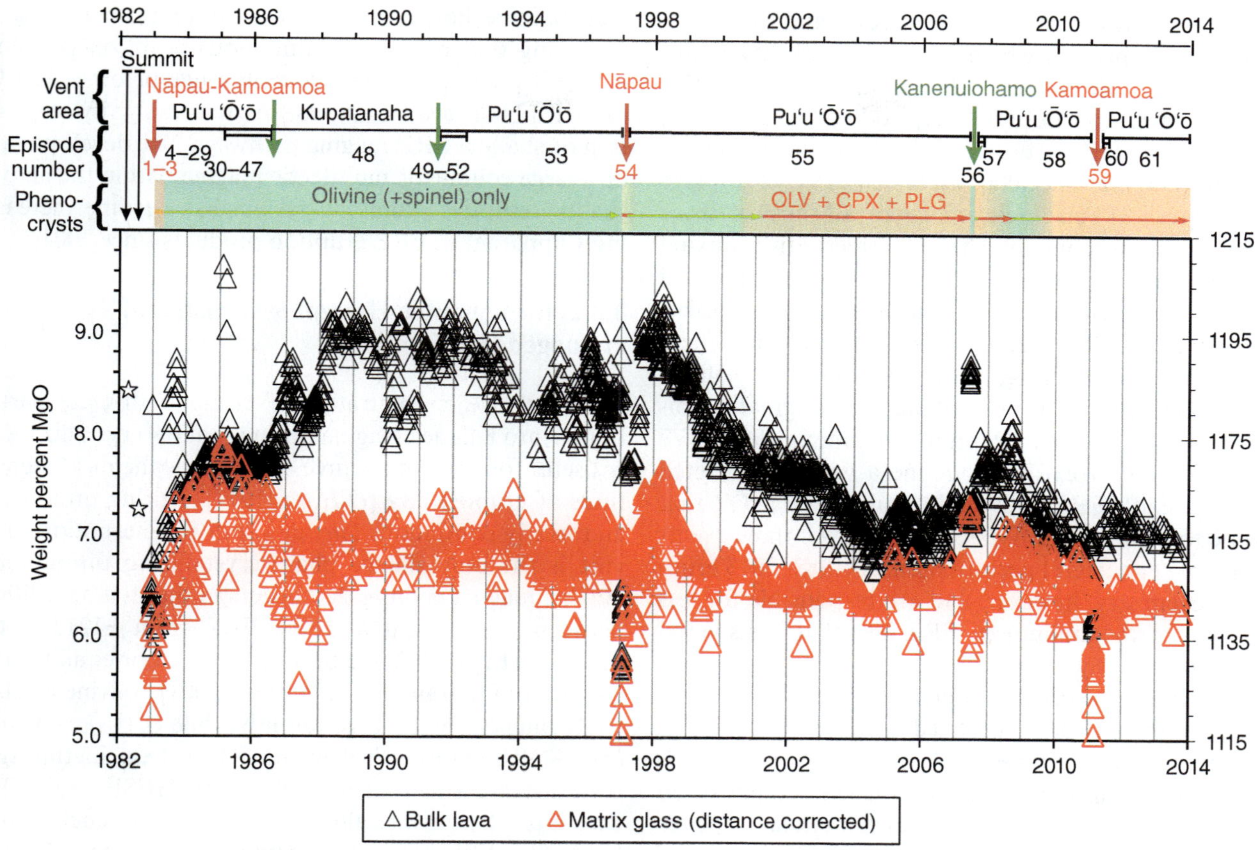

Figure 8.2 Temporal variation of MgO weight percent and MgO thermometry of 1983–2013 ERZ lava. Data are shown for bulk-lava analyses (black triangles) and average matrix glasses (red triangles). Glass temperatures are vent corrected by adding 0.9°C or 0.05 wt% MgO glass per kilometer of tube distance [*Thornber*, 2001]. The timeline at the top is annotated with vent locations (shown in Figure 8.1), episode numbers [*Heliker and Mattox*, 2003] (see also Chapter 18), and phenocryst mineralogy: olivine (OLV), clinopyroxene (CPX), and plagioclase (PLG). Downward pointing arrows indicate changes in vent location. Red arrows signify brief eruptions of low-temperature mixed magmas during episodes 1–3, 54, and 59. The 1982 summit bulk lava compositions are shown for reference (open stars) [*Garcia et al.*, 2003].

ERZ (Figure 8.2). The MgO systematics thus provide a meaningful context for reviewing short- and long-term changes in ERZ lava that are linked, along with edifice deformation, vent formation, and eruptive vigor, to the flux of magma into and out of the shallow magma plumbing system.

Beginning in mid-1983, bulk-lava MgO and corrected glass temperatures started rising and falling in tandem as lava vented continuously at or near Pu'u 'Ō'ō. The coupled changes persisted through shifts in the eruptive vent to Kupaianaha and then back again to Pu'u 'Ō'ō. As indicated at the top of Figure 8.2, lava that flowed steadily from primary vents during the first 18.5 years of eruption contained 3–4 vol% of olivine phenocrysts with trace amounts of chromian spinel. Synchronous changes of MgO in bulk lava and matrix glasses during this prolonged "olivine-only" eruptive interval reveal the repetition of a limited range of olivine-saturated magmatic conditions that, as constrained in time and space by geodetic measurements, were efficiently transmitted from the summit region to the ERZ vent [*Thornber*, 2003]. The prevalence of near-equilibrium olivine crystallization from melts equivalent to bulk-lava compositions and without significant loss or accumulation of olivine during transit along the length of the ERZ conduit is demonstrated by olivine-liquid relations in well-quenched near-vent samples from 1994 through 1998 (episodes 53 and 55) [*Thornber*, 2001]. That study suggests most crystals formed within a two-day period in magma tapped from a summit reservoir circulating between 1180 and 1200°C, cooled 30°C while flowing through the ERZ conduit, and erupted at temperatures of 1150–1170°C.

Distinctly low MgO values are recorded for three brief eruptive intervals, the first being the initial fissure eruptions of January 1983 (episodes 1–3) that extended 7.5 km downrift from the west side of Nāpau Crater, past the eventual Pu'u 'Ō'ō vent. Both the second and third low-MgO eruptive episodes, which occurred on 29–30 January 1997, in Nāpau Crater (episode 54) and 5 March 2011, between Nāpau and Pu'u 'Ō'ō (episode 59; Figure 8.1), were conduit-breaching eruptions uprift of the main Pu'u 'Ō'ō vent [*Owen et al.*, 2000; *Lundgren et al.*, 2013] (see also Chapter 18) Petrologic studies of all three of these eruptive periods confirm shallow preeruptive mixing between hotter, presumably summit-derived magma and cooler magma isolated for decades in shallow dikes [*Garcia et al.*, 1989, 1992; *Thornber et al.*, 2003b]. These three suddenly occurring, short-lived and low-volume episodes indicate the survival of shallow magmatic remnants from the nearby eruptions spanning 1955–1977 that, by their very existence, define a weakened stretch along the middle ERZ between Nāpau and Pu'u 'Ō'ō that is prone to rupture and fissure eruption. All three events caused only ephemeral perturbations in 1983–2013 eruption geochemistry, however, and are responsible for less than 0.1% of the amount of ERZ lava that has spewed nearly continuously since 1983.

Aside from these brief disruptive events, MgO in ERZ lava issued throughout prolonged eruptive intervals has cycled up and down between 6.8 and 10 wt% (Figure 8.2) over periods of months to years. This bulk composition range encompasses that of historical summit lava [*Garcia et al.*, 2003] and defines an overall temperature span of 60°C [*Helz and Thornber*, 1987] for olivine-saturated magma conditions persistent in the shallow edifice during prolonged recharge and eruption.

Thornber [2003] interpreted the cyclic variations of bulk lava MgO between observed limits as the geochemical expression of shifts in geostatic balance between two end-member magma components within a magmatically pressurized shallow plumbing system. The upper MgO limit of 10 wt% represents the shallow recharge component at ~1215°C, after it sheds high-Fo olivine that crystallizes from primitive magma deeper within the edifice [*Clague and Denlinger*, 1994]. The other end member, at 6.8–7.0 wt% MgO and ~1155°C, is the most common among historically erupted Kīlauea magma compositions. This is a low-pressure multitectic composition for olivine tholeiite magma at which clinopyroxene and plagioclase crystallize together with olivine (±spinel) [*Thompson and Tilley*, 1969; *Helz and Thornber*, 1987]. As pointed out by other studies of olivine-basalt systematics [*O'Hara*, 1977; *O'Hara and Matthews*, 1981; *Defant and Nielson*, 1990; *Rhodes and Hart*, 1995], the "buffering" effects of open-system replenishment of a persistent magma reservoir explain the lack of differentiation below the multiphase cotectic. In open-system reservoirs, this cooler phenocryst-bearing magma persists in a five-phase invariant condition at which the formation and dissolution of phenocrysts are sustained by near-continuous magmatic recharge. We can also infer from corollary reasoning that prolonged continuous eruption will not persist if preeruptive magma cools much below ~1150°C (6.8 wt% bulk-lava MgO), as this would signify stagnation of shallow ERZ magma pathways. The development of a large volume of multitectic magma within the active shallow eruptive plumbing system has become increasingly apparent in ERZ eruption products since 2001.

8.3.2. Incompatible-Element Systematics of Prolonged ERZ Eruption

The relative concentration of elements that are partitioned into Kīlauea magma during shallow crystallization is useful for tracking source-distinct geochemical signatures of erupted lava [*O'Hara*, 1977]. During prolonged and nearly continuous (*steady state*) ERZ eruption, olivine is the only silicate phase crystallizing throughout 90% of magma cooling that occurs in the shallow edifice between the summit and the ERZ from ~1215°C to ~1155°C (Figure 8.2, see Section 8.3.1.). For equilibrium growth during transport [*Thornber*, 2001] , olivine is relatively impervious to large-ion lithophile (LILs) elements (e.g., K, Rb,) rare earth elements (REEs) (e.g., La through Yb), and high-field-strength elements (HFSEs) (e.g., Ti, Zr, Nb, Y). Mineral/melt distribution coefficients ($D_i^{min/melt} = X_i^{min}/X_i^{melt}$) determined for these olivine-incompatible elements (i) in Kīlauea ERZ tholeiite have $D_i^{olv/melt}$ values less than 0.01 [*Norman et al.*, 2005]. Additionally, Sc, Ca, and Al are olivine incompatible with $D_i^{olv/melt}$ values of <0.1, <0.04, and <0.002, respectively.

Throughout steady state ERZ eruptive intervals (episodes 48, 53, 55, 58, and 61; Figure 8.2), bulk-lava concentrations of olivine-incompatible elements vary inversely with MgO. As observed for bulk-lava CaO variation between 10 and 7 wt% MgO (Figure 8.3), incompatible-element concentrations of steady state ERZ lava define a liquid line of descent (LLD) affected by varying amounts of olivine fractionation within the shallow edifice. Furthermore, this olivine-saturated LLD shifted progressively toward lower concentrations of olivine-incompatible elements from 1985 to 1995 (steady state episodes 48, 53, and 55) [*Thornber*, 2003], where it has remained throughout subsequent steady state ERZ eruption (episodes 58 and 61, as of 2013) (Figure 8.3).

Depletion of incompatible elements in 1985–1995 ERZ magma prior to shallow olivine crystallization is also evident from a steady decline of their olivine-normalized concentrations [*Thornber*, 2003] and by their relative concentrations, portrayed as ratios, such as CaO/Al2O3, La/Yb, and Zr/Y, which are unaffected by olivine fractionation (Figures 8.4a, b). This trend extrapolates back in time to relatively enriched 1982 summit lava compositions and is broadly consistent

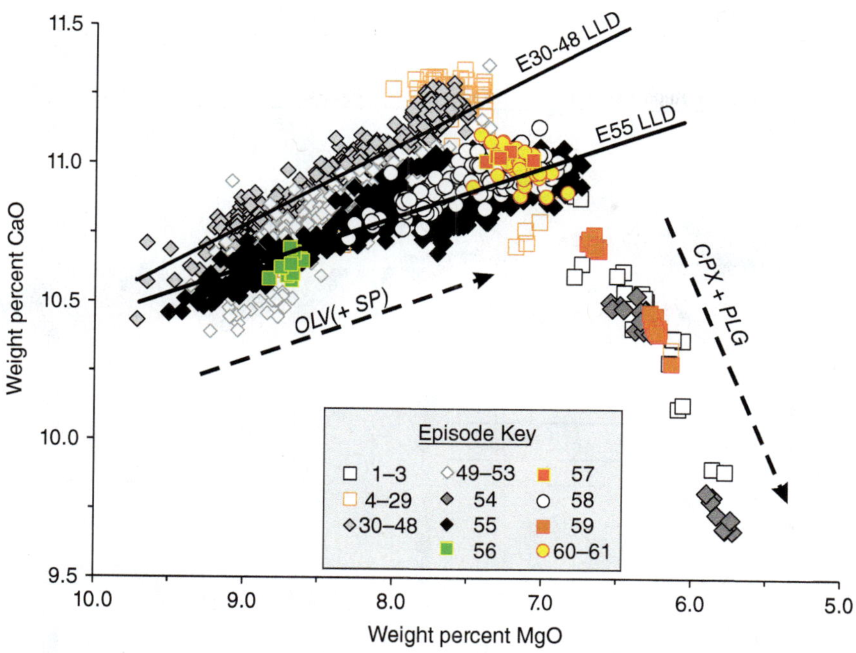

Figure 8.3 CaO versus MgO in 1983–2013 ERZ lava. Symbols shown in legend are for sequential eruptive episodes as indicated in Figure 8.2. The arrows point to relative changes in magma composition resulting from low-pressure fractional crystallization of (1) olivine and spinel (OLV+SP) and (2) clinopyroxene and plagioclase (CPX+PLG). Between 10 and 7 wt% MgO, incompatible-element variation is consistent with olivine-only fractionation. Linear-fit regression of near-continuous 1985–1995 olivine-phyric eruption products (episodes 30–48) define an olivine-controlled liquid line of descent (E30-48 LLD) that is distinctly richer in Ca than the trend for 1998–2001 (episode 55) olivine-only lava (E55 LLD). Aside from the 2011 Kamoamoa eruption (episode 59), all post-2001 multiphyric lower-MgO ERZ lava through 2013 (episodes 57, 58, 60, and 61), along with MgO-rich lava erupted at Kanenuiohamo in 2007 (episode 56) and olivine-phyric Puʻu ʻŌʻō lava from 2008–2009 (episode 58), lie along the same incompatible-element "depleted" E55LLD. The cusp at 6.8–7 wt% MgO is the five-phase multitectic point for shallow Kīlauea magma. ERZ lava with <6.8% MgO (episodes 1–3, 54, and 59) is depleted in moderately incompatible elements (e.g., Ca and Sc), while highly incompatible elements (e.g., LIL, REE, and HFSE) are variably enriched relative to olivine-controlled compositions [compare Figure 2 in *Thornber,* 2003).

with relative variations of Pb isotopes reported by *Garcia et al.* [1996, 2000, 2003], *Pietruszka and Garcia* [1999a], *Thornber et al.* [2003a], and *Greene et al.* [2014]. Previous and ongoing studies of geochemical trends in ERZ eruption products by *Hoffman et al.* [1984], *Garcia et al.* [1992, 1996, and 2000], *Pietruszka et al.* [2006], *Marske et al.* [2008], and *Greene et al.* [2014] attribute decadal trends, as well as shorter-term variations, directly to rapidly changing melting components or processes in the mantle that are then preserved during transport through the upper mantle, through the lithosphere, and into and out of a magma-laden volcanic edifice. In contrast, *Thornber* [2003] explains this decadal change in the parental signature of ERZ lava in terms of progressive assimilation and flushing of older magma residing within the shallow summit during prolonged magmatic recharge and eruption. In that paper, an inherent linkage between summit deflation and the long-term decline in La/Yb in ERZ lava (Figure 8.4c) is demonstrated by close agreement between geodetic modeling of summit reservoir shrinkage and a dynamic mixing model for ERZ La/Yb changes since the 1982 summit eruption. The model assumes that the "end-product" composition of 2001 ERZ lava matches that of the uniformly low La/Yb magma introduced to the shallow summit reservoir at 0.1 km^3 per year since 1982. Both geophysical and geochemical models yielded summit reservoir volume reduction of 2% per year from 1982 to 2001. This result implies that a diminishing proportion of older, incompatible-element enriched and isotopically distinct magma resident beneath the summit was progressively flushed out through the ERZ by a post-1982 influx of new magma derived from a uniform mantle source. The hypothesis that incompatible-element depleted magma completely occupied pathways throughout the actively fed shallow magmatic plumbing system by 2001 [*Thornber*, 2003] has now been substantiated by the geochemical parity between ERZ and summit 2008–2013 eruption products (discussed in Section 8.7 and Chapter 20).

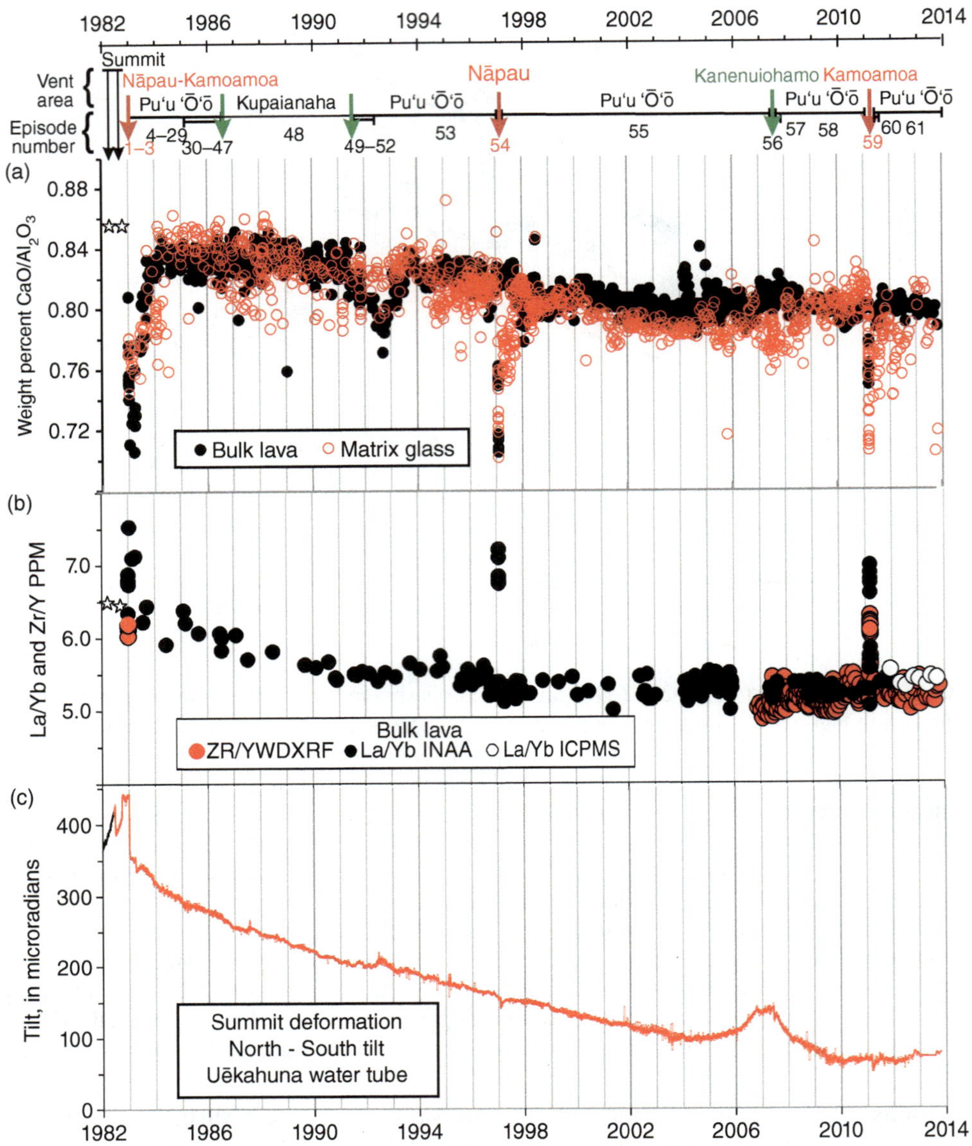

Figure 8.4 Temporal variation of bulk-lava incompatible-element ratios and summit tilt during 1982–2013. (a) CaO/Al_2O_3 weight percent in bulk lava and matrix glasses. (b) La/Yb and Zr/Y in bulk-lava samples. ICPMS values for La/Yb were down corrected by 3% based on INAA values for duplicate unknowns and replicate analyses of BHVO-1,2. Stars denote uncorrected data for April and September 1982 summit lava, from *Garcia et al.* [2003] (major elements, in part (a) and *Pietruzska and Garcia* [1999a] (trace elements). (c) Uēkahuna north-south water-tube tilt data shows long-term Kīlauea summit deflation (negative change) for this interval and inflation (positive change) associated with a 2003–2007 surge in magma supply [*Poland et al.*, 2012].

8.4. CHRONOLOGIC EVOLUTION OF POST-2001 ERZ VOLCANIC AND MAGMATIC CONDITIONS

8.4.1. Late Episode 55 (2001–2007): Unexpected Changes During a Magma Supply Surge

In March 1998, after 14 months of increasingly heightened eruptive activity that followed recovery of Pu'u 'Ō'ō vents after the Nāpau Crater fissure eruption in 1997 (episode 54, Figure 8.1) [*Thornber et al.*, 2003b; *Owen et al.*, 2000], ERZ lava achieved its most primitive state since high-fountaining episodes 30–31 in 1985, erupting near-equilibrium olivine-phyric lava with 9.5 wt% bulk MgO at 1168°C (Figure 8.2). This hallmark was followed by a seven-year decline in bulk MgO and eruption temperature to the lowest sustainable eruption levels of 6.8 wt% and 1140°C in November 2004. On 12 September 1999, amidst this systemic decline in MgO, an intrusion into the upper ERZ, west of Mauna Ulu (Figure 8.1) displaced magma from both the summit reservoir and the

Pu'u 'Ō'ō eruption conduit [*Cervelli et al.*, 2002]. About a year later, between September 2000 and April 2001, a fundamental and long-term shift in the petrologic character of rift lava became apparent. During that interval, steady state Pu'u 'Ō'ō vent products gradually transformed from sparsely olivine-spinel phyric to distinctly multiphyric, with phenocrysts and glomerocrysts of olivine (±spinel) + clinopyroxene ± plagioclase (the chemistry and petrology of phenocrysts in ERZ lava is presented and further discussed in relation to variations of host chemistry in Section 8.5.).

The onset of persistently low MgO, multiphyric lava effusion at Pu'u 'Ō'ō coincides with a summit dilation from April 2001 to June 2002 that was attributed to magmatic "backup" of the ERZ [*Cervelli and Miklius*, 2003; *Miklius*, 2005]. From 2001 through 2003, after the phenocryst mineralogy changed from olivine only to multiphyric, eruption temperatures stagnated within a few degrees of 1145°C, then declined and bottomed out in November 2004 at 1140°C and 6.8 wt% bulk MgO. This seemingly stalled preeruptive condition, at the lowest sustainable eruption temperature, persisted despite the well-documented 2003–2007 surge of new magma into the shallow edifice and also despite a corresponding increase in lava effusion rates during this interval [*Poland et al.*, 2012]. Other than a brief pulse of hotter lava (1153°C) in early 2005, bulk MgO contents remained consistently low, averaging 7.14 wt% ± 0.16 (1σ) in 2005 and 7.10 wt% ± 0.16 (1σ) throughout 2006. Vent tephra and upper elevation lava flows were issued with 3–5 vol% percent of a multiphyric phenocryst assemblage at consistently low eruption temperatures of ~1145°C ±1°C (1σ) throughout the 2003–2007 surge and until the June 2007 eruption at Kanenuiohamo (episode 56). This persistent low-temperature eruptive condition at Pu'u 'Ō'ō is interpreted by *Poland et al.* [2012] to have resulted from magmatic pressurization of the edifice during a 2003–2007 surge in magma supply that forcibly flushed cooler magma from marginal zones or pockets in the plexus of open-system magmatic pathways along the ERZ conduit.

8.4.2. Episode 56 (17–19 June 2007): A Glimpse of Summit-Derived Magma Uprift of Pu'u 'Ō'ō

Four years of surge-related edifice inflation culminated on 17–19 June 2007 with a forceful magma intrusion into the shallow uppermost ERZ between Pauahi Crater and Makaopuhi Crater that caused another collapse of the Pu'u 'Ō'ō Crater floor (as also occurred in January 1997 and September 1999) followed by a small eruption on the east flank of the prehistoric Kanenuiohamo cone (Figure 8.1) [*Poland et al.*, 2008; *Montgomery-Brown et al.*, 2010]. This brief event (episode 56) occurred early on 19 June, discharging ~1500 m^3 of the hottest olivine-only rift lava erupted since 1998 (Figure 8.2). Bulk MgO was 8.6–8.8 wt%, and air-quenched spatter preserved glass temperatures as high as 1162°C. An incompatible-element signature that is indistinguishable from episode 55 magma (Figures 8.3 and 8.4) confirms a comagmatic relation with previous and subsequent ERZ eruption products. Erupted at 15°C hotter than the late-episode 55 multiphyric magmas, the episode 56 lava is hypothesized to be the summit-derived component in post-2001 Pu'u 'Ō'ō multiphyric lava by *Thornber et al.* [2007]. This concept is borne out by similar petrologic traits of tephra from the 2008–2013 Kīlauea summit eruption reported herein and by Rowe et al. in Chapter 20.

8.4.3. Episode 58 (2007–2011): Latent Surge Translation Along the ERZ

Following episode 56, lava reappeared in Pu'u 'Ō'ō in early July 2007 (episode 57) with the same low-MgO (~7.1%) multiphyric character as late-episode 55 lava. In late July, fissures extended 1.5 km downrift from Pu'u 'Ō'ō (episode 58) and, after extrusion localized from the easternmost vent, eruption temperature steadily climbed for a year, while bulk compositions migrated back up the same olivine-controlled episode 55 LLD (Figures 8.2 and 8.3). In August 2008, five months after the onset of Kīlauea's summit eruption in March 2008 (discussed in Section 8.7.), bulk MgO and temperature of ERZ lava peaked at 8.2 wt% bulk MgO and 1157°C. At about that time, eruption petrology retransitioned to olivine only for another 10½ months, while wt% MgO steadily declined (Figure 8.2). The August 2008 timing of the MgO maxima, approximately one year after the upper ERZ intrusion and eruption at Kanenuiohamo, is suggestive of the timing for translation of surge-related magma through the overpressurized magma pathways and pockets in the shallow rift zone to Pu'u 'Ō'ō.

By June 2009, bulk MgO and eruption temperature of rift lava had again fallen below 7.3 wt% and 1150°C, and clinopyroxene and plagioclase phenocrysts reappeared along with olivine. The ERZ lava effusion rate waned throughout 2009–2010 as the MgO downtrend gradually bottomed out near 7.0 wt% and 1140°C. From late 2010 until the Kamoamoa fissure eruption (episode 59) in March 2011, there was a slight ERZ MgO and temperature increase. *Lundgren et al.* [2013] and *Orr et al.* (Chapter 18) attribute the episode 59 conduit breech to magmatic overpressure of the plumbing system. This condition probably resulted from a back-up in the rift plumbing system due to conduit constriction during the preceding low-temperature, low-output interval from the main eruptive vent just east of Pu'u 'Ō'ō.

8.4.4. Episode 59 (5–9 March 2011): Another Ephemeral Perturbation of ERZ Lava Output

The 5–9 March 2011 Kamoamoa fissure eruption (episode 59), uprift of Puʻu ʻŌʻō, (Figure 8.1), put another glitch in the 30-year time line of eruption MgO systematics (Figure 8.2). About $2.7 \times 10^6 m^3$ of lava issued from two clusters of closely spaced eruptive fissures. The fissure vents extended 2.5 km from the east rim of Nāpau Crater with a 0.36 km gap between simultaneously active east and west fissure groups (Figure 8.1) (Chapter 18). Episode 59 eruptive fissures were overlapping or proximally en echelon with fissures that opened in 1983 (episode 1) and 1997 (episode 54) (Figure 8.5). Each of these three sets of fissure eruptions produced a similar low range of 6.1–6.7 wt% bulk MgO lava at similarly low eruption temperatures of 1130–1145°C (Figures 8.2 and 8.3), and all are geochemically distinguished from steady state ERZ lava by a range of relatively enriched incompatible element signatures (e.g., La/Yb; Figure 8.4b). In each case, geodetic modeling suggests magma was displaced from beneath Makaopuhi Crater, diking upward toward the same shallow zone near Nāpau Crater [*Okamura et al.*, 1988; *Owen et al.*, 2000; *Lundgren et al.*, 2013].

Episodes 54 and 59 were both accompanied by magma withdrawal from beneath Puʻu ʻŌʻō and the summit, as magma from both ends of the shallow plumbing system was suddenly drawn toward the zone of rift rupture. It is apparent from geophysical similarities of both eruptions that comparable geochemical traits of episode 59 are the likely result of preeruptive magma hybridization, as demonstrated for episode 54 by *Thornber et al.* [2003b]. In that study, details of phenocryst-liquid relations, along with bulk geochemistry models for combined fractionation and mixing, indicate shallow preeruptive mixing between "open-system," near-cotectic magma and discrete, fractionated 1963 and 1968 magma pods, spatially correlated with nearby 1963 and 1968 eruptive fissures. This conclusion is reinforced by U-series radionuclide disequilibria studies on episode 54 lava [*Girard et al.*, 2012], which indicate a "decades-old" component in episode 54 lava as well as in episodes 1–3 lava.

Recurrent fissure eruptions extending across the region from Nāpau Crater to the present Puʻu ʻŌʻō vent throughout the 1960s attest to prolific shallow dike intrusions in this zone (Figure 8.5). Considering the eruptive history of this segment of the rift, albeit without such elaborate substantiation as for episode 54, multigenerational magma mixing in 2011 Kamoamoa lava is clearly apparent from bulk-lava incompatible-element ratios (e.g., La/Yb, Zr/Y, Nb/Y; Figures 8.4 and 8.6) that are relatively unaffected by low-temperature pyroxene and plagioclase fractionation of Kīlauea ERZ magma. All episode 59 bulk-lava compositions fall on a mixing

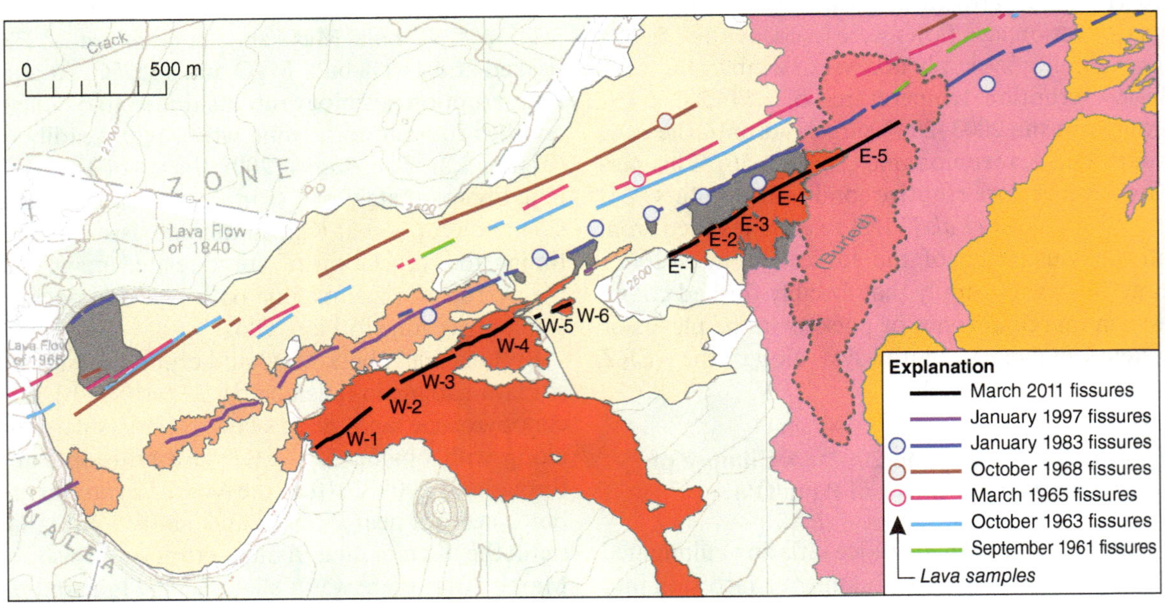

Figure 8.5 Detailed map of Nāpau Crater–Kamoamoa area. The map (area outlined in Figure 8.1) highlights fissure vents and lava flows from recent eruptions. The 5–9 March 2011 Kamoamoa eruption (episode 59) is emphasized with red flow areas and black lines for west and east fissure vent sequences, designated as W-1-6 and E-1-5, respectively, by *Orr et al.* (Chapter 18, this volume). Other eruptions shown here include January 1997 (episode 54, purple lines and dark tan areas); January 1983 (episode 1, blue lines and dark gray areas); 1961, 1963, 1965, and 1968 (light tan areas and green light blue, pink, and brown lines, respectively); 1992–2011 Puʻu ʻŌʻō lava flows (orange area). Locations for 1983, 1965, and 1968 samples used in this study are shown with circles.

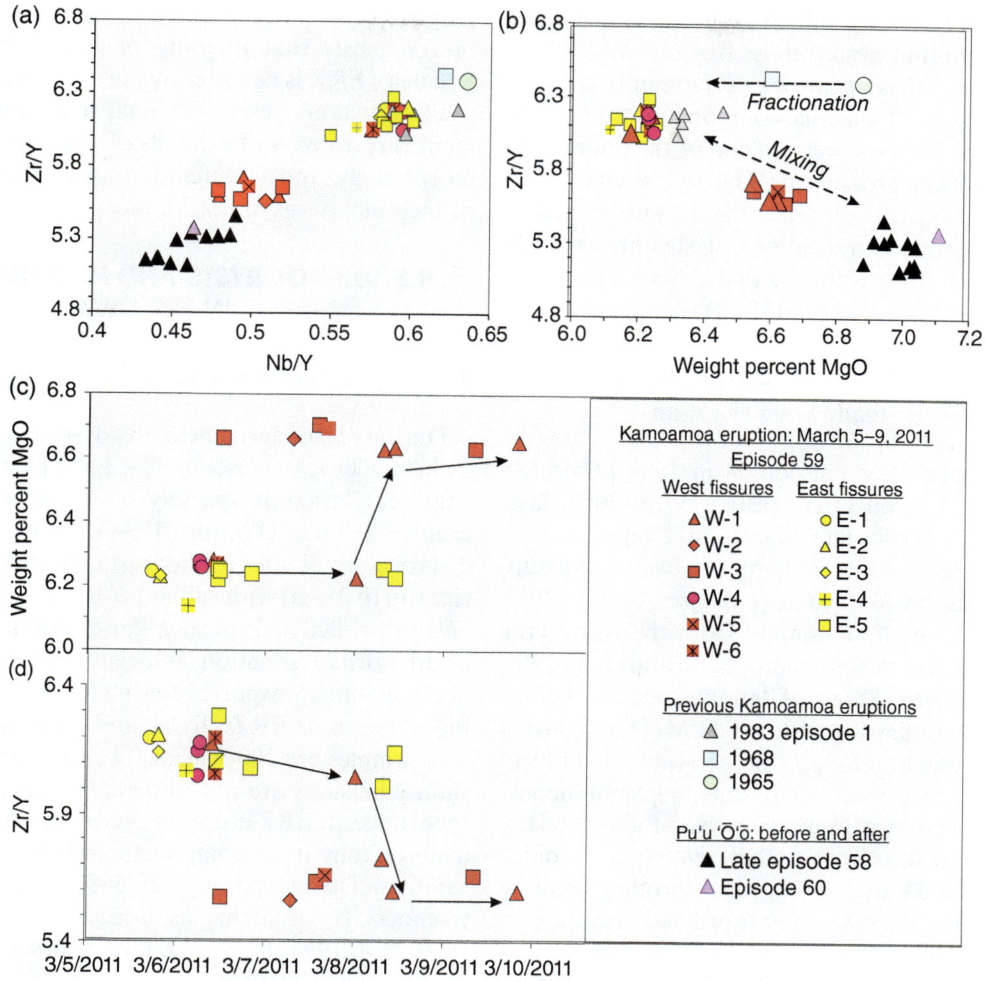

Figure 8.6 Bulk-lava MgO and incompatible-element relations of 5–9 March 2011 Kamoamoa fissure eruption (episode 59): (a) Zr/Y versus Nb/Y (b) Zr/Y versus MgO, showing relative paths of fractionation and mixing (c) MgO versus time; and (d) Zr/Y versus time, with vectors indicating temporal progression of the westernmost and longest-lived vent (W-1) that issued the most flow volume. Symbols in the legend specify each of the spatially consecutive west and east fissure segments shown in Figure 8.5; W-1–6 (red symbols) and E-1–5 (yellow symbols). Data for Pu'u 'Ō'ō lava erupted from October 2010 to February 2011 (episode 58, black filled triangles) and again in June 2011 (episode 60, purple filled triangle) are presented for comparison, along with chemistry of 1965, 1968 and 1983 Kamoamoa lava. Episode 59 data shown here are for the sequence of samples representing almost every activity interval for each fissure segment, as specified by *Orr et al.* (Chapter 8, Figure 18.6) and in supplementary data (Table 8.1a).

line between recent and subsequent Pu'u 'Ō'ō lava and the fractionated equivalents of newly analyzed samples from nearby, en echelon fissure vents active in 1965, 1968, and 1983 (episode 1) (Figures 8.5 and 8.6a,b). Circumstantial evidence of physical proximity and geochemical signature strongly suggests that episode 1 products were themselves mixed with fractionated 1960s magma, and it stands to reason that fractionated hybrid leftovers from 1983 dike magma were a component in 2011 mixed magmas. It is possible that third-generation (or higher) rehybridized magma, mixed and refractionated repeatedly, has been left again in episode 59 feeder dikes beneath Kamoamoa.

Compositional variations in space and time during the four-days of the Kamoamoa fissure eruption reveal a progression of magma mixing and displacement of 1960s dike residua with new magma diverted from Pu'u 'Ō'ō through an uprift breach in the ERZ conduit (Figure 8.6). During initial March 5 pulses from the E-1 to E-3 fissures (Figure 8.5), the predominant lava source is fractionated 1960s dike residua. On March 6, the W-3 vent issued a brief pulse of a hotter hybrid mixture that preceded ejection of cooler evolved lava from E-3, E-4, E-5, W-4, W-5, and W-6. Subsequently, March 7–8 eruptions from the easternmost fissures (E-4, E5) yielded only high-Zr/Y, low-MgO lava, while higher MgO and lower Zr/Y characterized March 7–9 products from the longer-lived and more voluminous westernmost fissures (W-1, W-2, W-3). Collectively, these spatial and temporal correlations indicate stagnant magma components were more efficiently flushed from the western end of the fissure system as the eruption progressed. The interpretation of flushing of the stagnant component from west to east is consistent with geodetic indications of steadily expanding surface dike volume from the first day to the end of the eruption [*Lundgren et al.*, 2013].

8.4.5. Episodes 60 and 61 (2011–2013): Resumption of Persistently Buffered Steady State Eruption

Just as happened after episode 54 in 1997 [*Thornber et al.*, 2003b] and again after episode 56 in 2007, lava returned to Puʻu ʻŌʻō in late March 2011 (episode 60), when the previously active magma pathway repressurized following its uprift breach [*Lundgren et al.*, 2013] (Chapter 18). The earliest samples from the rising lava lake in Puʻu ʻŌʻō's crater were multiphyric and chemically indistinguishable from those of late episode 58 (before the 5 March 2011 collapse), with a bulk MgO of 7.1 wt% and a similar low MgO-glass temperature of 1140°C (Figure 8.2). The return of steady state, depleted incompatible-element signature in newly arrived Puʻu ʻŌʻō lava (Figures 8.4b and 8.6) implies that remnants of older magma encountered during passage through shallow Kamoamoa feeder dikes were *not* translated into the primary conduit feeding steady state ERZ eruption.

The Puʻu ʻŌʻō lava lake steadily rose to the point of overflow in August 2011, when lava broke through the lower western flank of the cone and the crater collapsed again. As the crater subsequently refilled, there was a steady decline in the already low eruption temperatures, and bulk lava MgO varied erratically between <6.8 and 7.5 wt% (Figure 8.2). The crater refilled to overflowing again by late September 2011,when a persistent vent was established through a fissure on the upper east flank of Puʻu ʻŌʻō (episode 61) (Chapter 18). The erratic wt% bulk MgO variations that characterized both episode 60 and crater-filling stages of early episode 61 are attributed to abnormal phenocryst content (4–8 vol%) probably associated with recycling at the vent.

From September 2011 through October 2013, steady effusion from the Puʻu ʻŌʻō vent yielded lava with an average bulk MgO of 7.14 wt% (±0.10 (1σ) for 31 samples) at a consistently low temperature of 1143°C (±2°C (1σ) for 39 samples) (Figure 8.2). Vent tephra and upper elevation lava flows remained consistently multiphyric with 3–5 vol% phenocrysts. Episode 61 incompatible trace-element signatures have retained a consistent "depleted" signature (e.g., Zr/Y = 5.23 ± 0.10 (1σ) for 43 samples) (Figure 8.4b). Continuance of a steady state eruptive condition that hovers near multitectic composition suggests that ongoing magma throughput along Kīlauea's ERZ is persistently buffered by an open-system shallow magma reservoir. Such a preeruptive environment is revealed in the details of phenocryst mineralogy, morphology, and texture, along with the sulfur content of melt inclusions in ERZ lava.

8.5. PHENOCRYSTS AND MELT INCLUSIONS IN ERZ LAVA

8.5.1. Petrographic Characteristics

During prolonged near-steady-state eruption from 1986 to 2000 (i.e., episodes 48–53 and part of episode 55), near-vent lava was sparsely olivine-spinel-phyric. Lava erupted at Puʻu ʻŌʻō from 1994 through 1998 contained 2–4 vol% of olivine phenocrysts, usually 0.5–0.7 mm in size (up to 2 mm) with subhedral to euhedral morphology [*Thornber*, 2001]. Typically, these were unzoned or with slight normal zonation associated with growth during cooling with transport. Melt inclusions are a very common feature of ERZ olivine and most in well-quenched lava samples are 100% glass. The CO_2- and H_2O-bearing fluid inclusions are not observed in phenocrysts of any type in recent ERZ eruption products. Chromian spinel is a ubiquitous trace component (~0.05 vol%) of rift lava samples. The morphology of 5–50 μm size spinel microphenocrysts, occurring as inclusions in olivine or free floating in 1994–1998 ERZ lava, is described in detail by *Roeder et al.* [2003].

Between September 2000 and April 2001, as bulk MgO dropped below 7.5 wt% and steady state eruption temperatures to ~1145°C, olivine-spinel-phyric eruption petrology gradually changed to olivine (+spinel)-clinopyroxene-plagioclase phyric. By mid-2001, lava flowing from the Puʻu ʻŌʻō vent typically contained 3–5 vol% of this multiphyric assemblage.

With the exceptions of MgO-rich, olivine-only lava briefly erupted at Kanenuiohamo in June 2007 (Figures 8.7a,b) and again at Puʻu ʻŌʻō from July 2008 to June 2009 (Figures 8.7c,d), all post-2001 samples of ERZ vent spatter, tube flows, and surface flows originating from vents proximal to and within Puʻu ʻŌʻō's crater have been multiphyric. Petrography of the March 2011 Kamoamoa eruption (episode 59) remains to be reported in detail, but the range of phenocryst associations of the mixed multiphyric assemblage in that eruption is not dissimilar to that of Puʻu ʻŌʻō lava both before and after the event (late episode 58 and episode 60). All samples examined, including those with the most geochemically enriched episode 59 magma mixtures, lack the distinctive large xenocrystic remnants found in 1997 Nāpau Crater lava (episode 54) [*Thornber et al.*, 2003b].

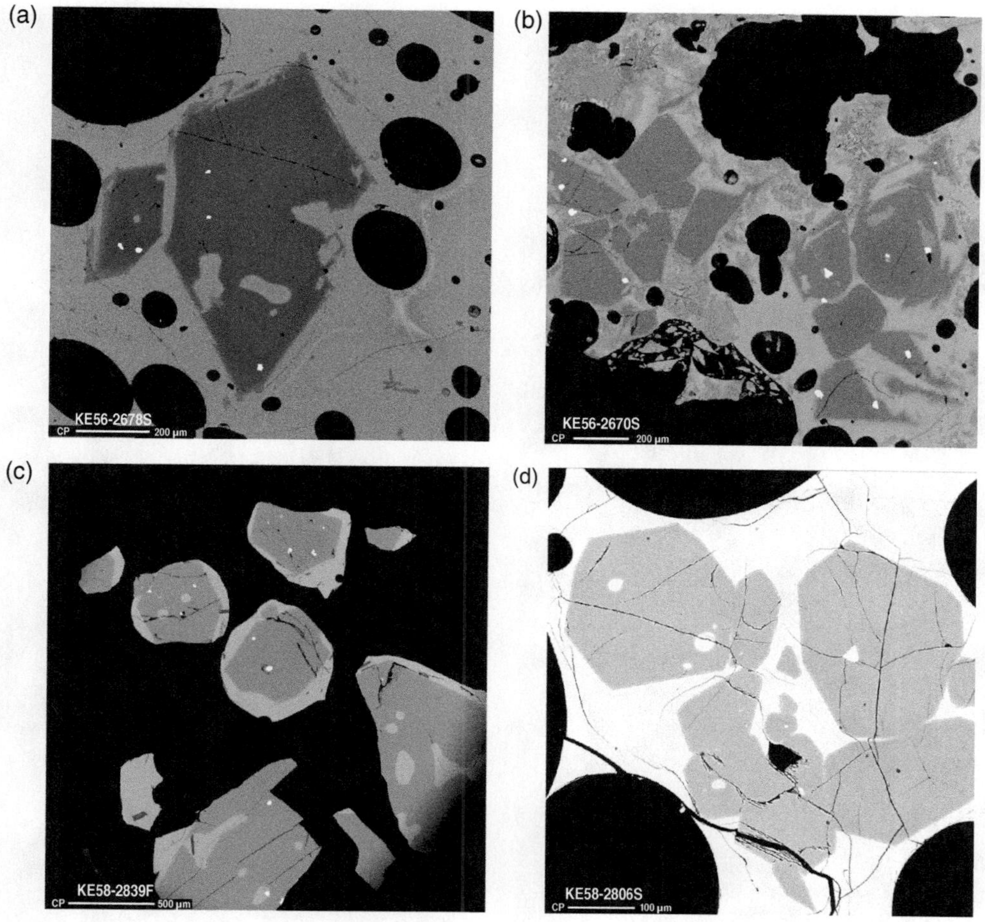

Figure 8.7 Backscattered electron images of olivine-spinel-phyric ("olivine-only") ERZ lava erupted in June 2007 and 2008–2009 (Figure 8.2). Lighter and darker shades of gray are an inversed function of averaged atomic number and contrast levels vary between images. The mostly uniform light gray in the matrix is basalt glass and darker gray olivine contains melt inclusions comprised of light gray glass. The bright white specks are microphenocryts of chromian spinel. Black background areas of mounting epoxy outline vesicles and lava fragments. (a) KE56-2678S, 19 June 2007 (200 μm bar scale); Kanenuiohamo fissure-vent spatter. $Fo_{83.4}$ olivine phenocrysts with skeletal overgrowth of $Fo_{81.0}$, matching calculated equilibrium olivine compositions for bulk lava and host glass of $Fo_{83.6}$ and $Fo_{81.0}$, respectively. MgO-glass temperature is 1161°C and melt inclusion sulfur concentration ranges from 889 to 949 ppm. (b) KE56-2670S, 19 June 2007 (200 μm bar scale); Kanenuiohamo fissure-vent spatter with euhedral and skeletal $Fo_{83.4}$-$Fo_{83.1}$ olivine phenocrysts. The average olivine phenocryst composition in this sample is $Fo_{83.5}$; the same as the equilibrium value calculated for the bulk-lava composition (8.7 wt% MgO). (c) KE58-2839F, 7 May 2009 (500 μm bar scale); Pu'u 'Ō'ō lava flow from a tube breakout at the upper end of Royal Gardens subdivision, 5.5 km east of the vent. Among olivine phenocrysts hand-picked for this grain mount, most are unzoned with near $Fo_{80.0}$ compositions, but one (top center) has core-to-rim zoning of $Fo_{81.5-79.6}$. Equilibrium olivine is calculated as $Fo_{81.1}$ for bulk lava and $Fo_{80.0}$ for the matrix glass with distance corrected MgO temperature of 1148°C. A range of sulfur concentration is measured in melt inclusions: 871 ppm in $Fo_{81.5}$ (top center); 780, 422 ppm in $Fo_{80.2}$ (upper left); and 247, 229, and 156 in $Fo_{80.8-80.1}$ (bottom left). (d) E58-2806S, 13 July 2008 (100 μm bar scale); Pu'u 'Ō'ō spatter from the east spillway vent. A cluster of subhedral olivine phenocrysts in glass with an MgO temperature of 1152°C. $Fo_{81.4}$ olivine core compositions compare to $Fo_{82.0}$ in equilibrium with bulk lava and $Fo_{81.0}$ for host glass. Sulfur concentrations in the two largest melt inclusions of the upper-left olivine crystal are 1005 (left side) and 865 ppm (right side).

Olivine phenocrysts in steady-state multiphyric lava typically range from 0.3 to 0.7 mm in size and either occur as free-floating crystals or are intergrown with augite and plagioclase in open-textured, 0.5–5 mm glomerophyric clusters (microgabbro), (Figures 8.8a, c–f, j –l). Olivine crystals, while typically without obvious compositional zonation, occasionally exhibit normal zoning of the outermost 20–10 μm (Figure 8.8c) and,

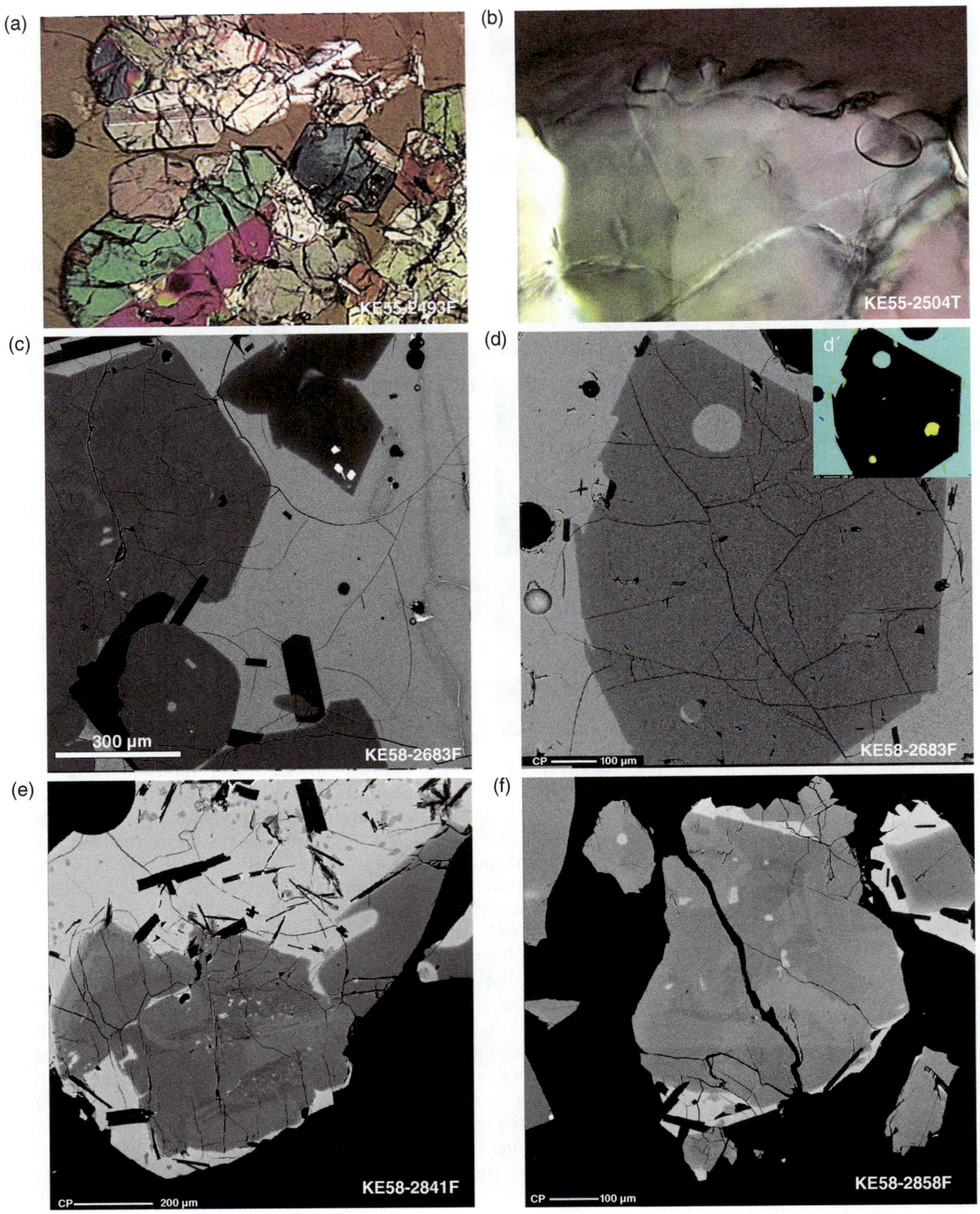

Figure 8.8 Images of multiphyric post-2000 ERZ lava samples. In backscattered electron imagery, olivine is usually a darker gray than clinopyroxene. Darkest gray (or black) plagioclase occurs as blocky crystals, acicular laths, and euhedral, rounded, or irregular inclusions in olvine and augite. (a) KE55-2493S, 3 July 2005 (field of view is 2 mm); spatter from a vent on the south flank of Pu'u 'Ō'ō. Transmitted polarized-light image showing a typical, loosely aggregated, microgabbroic cluster of augite, plagioclase, and olivine phenocrysts in translucent brown glass. (b) KE55-2504T, 19 August 2005 (field of view is 250 µm); tephra from a vent on the east side in Pu'u 'Ō'ō Crater. Transmitted light image looking down the edge of a resorbed augite phenocryst in contact with host glass.

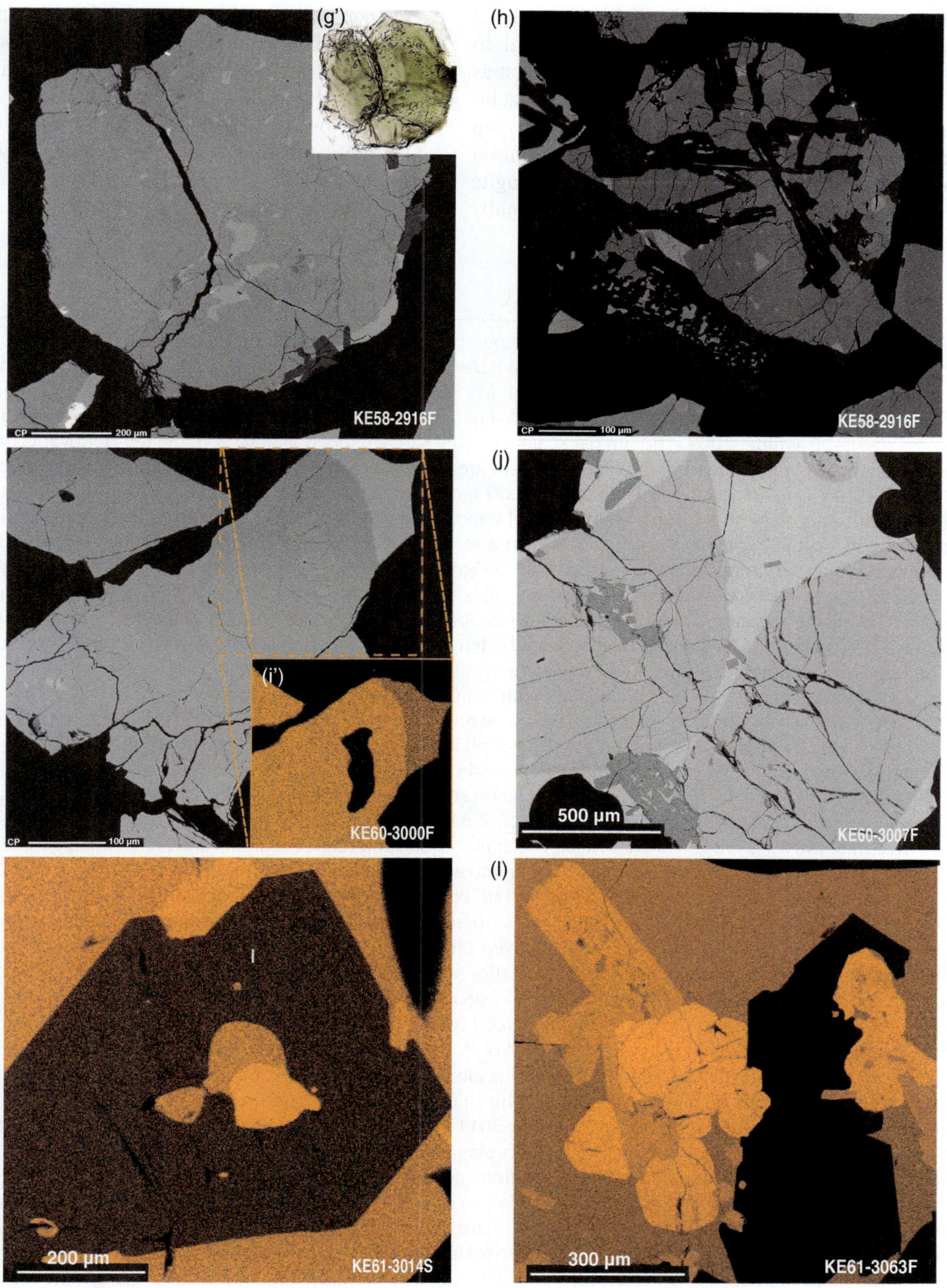

Figure 8.8 (Cont'd) Cuspate or hummocky edge textures are commonly displayed by augite phenocrysts that are dissolving in magma before eruption; such textures are observed in mineral dissolution experiments by *Thornber and Huebner* [1985]. (c) KE58-2683F, 28 August 2007 (300 μm bar scale); lava pond overflow sample, 785 m from Pu'u 'Ō'ō west-flank vent with a mixture of high- and low-temperature phenocrysts. An olivine cluster (top center) includes spinel (white) and normally zoned Mg-rich olivine with core-rim compositions of $Fo_{83.9-79.9}$ (darker to lighter gray). A large subhedral to rounded augite (left side) ($Mg\#_{83.7-84.7}$; $Wo_{36.4-39.5}$; 2.1–2.5 wt% Al_2O_3) is intergrown with $An_{71.8-72.0}$ plagioclase and a relatively Fe–rich olivine of $Fo_{79.5-79.9}$ composition (bottom left center).

more rarely, reversed zoning. Within and between ERZ samples, olivine morphology ranges from euhedral to hopperlike to subhedral, rounded, and embayed. Just as before 2001, many olivine crystals contain melt inclusions large enough for routine microbeam analysis. In multiphyric lava, olivine phenocrysts frequently contain inclusions of well-rounded plagioclase and augite (Figures 8.8c, j, k), and olivine remnants are occasionally overgrown by augite (Figure 8.8i) or plagioclase. Chromian spinel is more conspicuously associated with free-floating olivine (Figure 8.8c) and less apparent in microgabbro glomerocrysts or in olivine that contains plagioclase or augite (Figures 8.8d–f, j –l).

Distinctly emerald-green augite phenocrysts, up to 3mm in size but commonly 0.5–1mm (Figure 8.8g), began to appear intermittently with olivine in 2000 and

Figure 8.8 (Cont'd) Melt inclusions in lower-temperature olivine have sulfur concentrations of 308 and 192 ppm. (d) KE58-2683F, same sample as (c) (100 μm bar scale). Euhedral $Fo_{79.5}$ olivine includes resorbed remnants of augite crystals ($Mg\#_{82.0}$, $Wo_{40.6}$, 2.9 wt% Al_2O_3). (d′) Same image area as (d) in which augite and glass inclusions are more clearly distinguished in an X-ray map highlighting calcium concentration in orange. (e) KE58-2841F, 4 June, 2009 (100 μm bar scale); a frothy shelly pāhoehoe flow from the top of Royal Gardens subdivision, 5.5 km southeast of the Pu'u 'Ō'ō vent area (200 μm bar scale). An already low eruption temperature of ~1145°C (extrapolated to back to the vent) facilitates substantial development of posteruptive augite (Mg# 79.0) and plagioclase (An 82–79) microphenocrysts and phenocryst rim growth. The rounded olivine phenocrysts on the right ($Fo_{80.7}$) crystallized from a melt at 1155°C (7.0 wt% MgO). The large clinopyroxene phenocryst has a compositionally mottled core ($Mg\#_{83.2-81.3}$, $Wo_{32.6-40.8}$ 1.8–3.8 wt% Al_2O_3) with abundant irregular shaped melt inclusions, reflecting a complex growth history at more than 15°C cooler than adjacent higher Fo olivine remnants. (f) KE58-2858F, 16 October 2009 (100 μm bar scale); pāhoehoe breakout at the base of Royal Gardens. The augite fragment on the left ($Mg\#_{82.8-83.9}$) has a glass inclusion with 913 ppm sulfur. The large euhedral augite in the center has a compositionally mottled core ($Mg\#_{83-82}$, $Wo_{38.2-40.9}$, 2.0–2.8 wt% Al_2O_3) and irregular shaped melt inclusions with sulfur concentrations ranging from 490 to 1000 ppm. The small euhedral olivine on the right has a $Fo_{80.5}$ core and $Fo_{78.5}$ outer rim. (g) KE58-2916F, 10 November 2010 (300 μm bar scale); pāhoehoe flow on the coastal plain. A single augite phenocryst with mottled zoning and abundant irregular shaped melt inclusions is emerald green in transmitted light (as shown in (g′). Augite is intergrown with plagioclase ($An_{71.7}$, top right). Darker shaded augite ($Mg\#_{83.5}$, $Wo_{40.4}$, 3.0 wt% Al_2O3) is distinct from zones of lighter shade augite ($Mg\#_{84.5}$, $Wo_{38.0}$, 2.6 wt% Al_2O_3). Small dark and irregular shaped inclusions are pigeonite ($Mg\#_{81.5}$, $Wo_{4.0}$, 1.0 wt% Al_2O_3). Melt inclusion sulfur concentrations are 440 ppm (largest central inclusion), 446 ppm (top center); 589 ppm (upper left) and 344 (lower left). (h) KE58-2916F, 10 November 2010 (100 μm bar scale); same sample as (g). Ophitic to subophitic intergrowth of plagioclase ($An_{67.8}$, bottom left) and augite ($Mg\#_{81.4-82.9}$, $Wo_{38.4-39.6}$, 3.1–2.1 wt% Al_2O_3). Abundant melt inclusions in both phases have sulfur concentrations of 275 –390 ppm. Subhedral and embayed olivine phenocrysts picked from this sample (not shown) have $Fo_{80.6}$ and $Fo_{79.8}$ core compositions, the latter with melt inclusions containing 307–508 ppm sulfur. Calculated olivine for the bulk lava with 7.0 wt% MgO is $Fo_{80.4}$, and the matrix glass has an MgO temperature of 1148°C with equilibrium olivine of $Fo_{79.8}$. (i) KE60-3000F, 29 June 2011 (100 μm bar scale); pond overflow in southeast Pu'u 'Ō'ō crater. Augite ($Mg\#_{83.25}$, $Wo_{38.1}$, 2.1 wt% Al_2O_3) overgrown on the remnants of olivine ($Fo_{79.4}$, upper right). The irregular lathe of olivine is clearly seen as the dark area in the X-ray map (i′) showing calcium concentrations highlighted in orange. The sulfur content of augite melt-inclusions (lower left) ranges from 319 to 458 ppm. (j) KE60-3007F; 11 August 2011 (500 μm bar scale); overflow from Pu'u 'Ō'ō west flank vent. Olivine ($Fo_{79.8}$) phenocrysts intergrown with plagioclase and including plagioclase along with melt containing 362 ppm sulfur. Plagioclase ($An_{71.7}$, bottom center) has inclusions with 245, 319, 448 ppm sulfur. Equilibrium olivine is calculated as $Fo_{81.3}$ for bulk-lava and $Fo_{79.3}$ for matrix glass. Other olivine phenocrysts in this sample with similar composition ($Fo_{78.8-79.5}$) have melt inclusions with 455, 676, and 834 ppm sulfur. (k) KE61-3014F, 12 September 2011 (100 μm bar scale); vent spatter on the northeast rim of Pu'u 'Ō'ō cone. X-ray map of Ca showing a small euhedral olvine ($Fo_{79.4}$, black) intergrown with plagioclase ($An_{73.9}$, intermediate orange) that has overgrown the resorbed remnants of an augite phenocryst ($Mg\#_{81.5}$, $Wo_{38.9}$, 3.0 wt% Al_2O_3, bright orange). Sulfur in each of two melt inclusions is 284 ppm (right side) and 229 ppm (left side, with augite) This sample has a low matrix glass temperature of 1142°C with calculated $Fo_{78.9}$ at equilibrium. (l) KE61-3063F, 11 June 2013 (400 μm bar scale); helicopter-dip sample of Pu'u 'Ō'ō lava lake in notheast side of the crater. X-ray map of Ca showing a microgabbroic phenocryst cluster comprising olivine $Fo_{79.8-79.3}$ (black), augite (bright orange), and plagioclase (intermediate orange); melt inclusions in augite (upper right) contain 352 and 241 ppm S; plagioclase in olivine is $An_{78.7}$ and the larger $An_{75.6}$ phenocryst contains abundant inclusions with sulfur ranging from 532 to 1003 ppm.

then appeared regularly with plagioclase and olivine by spring 2001. Initially, pyroxenes were nearly all well-rounded crystals with embayed or finely scalloped edges (Figure 8.8b) indicative of crystal dissolution shortly before eruption [*Thornber and Huebner*, 1985]. After prolonged eruption at lower temperatures, plagioclase became more prevalent and the range of morphology among coexisting augite phenocrysts extended to both embayed and rounded to crisp, euhedral crystals, both as free-floating crystals and in microgabbro clusters (Figures 8.8a–l). Many Pu‘u ‘Ō‘ō pyroxenes are riddled with both melt and mineral inclusions and also display smooth "hour-glass" zonation as well as much more chaotic zonation patterns (Figures 8.8c, e, f, g, i). Such features are consistent with rapid pyroxene growth in a dynamic crystallization environment.

Plagioclase is present as acicular to blocky phenocrysts with occasional rounded or irregular faces, ranging in size from microlitic (<100 μm) to ~1 mm and occurring as free-floating crystals interspersed within or adjacent to variably consolidated microgabbro clusters (Figures 8.8a, h, l). Smaller plagioclase crystals are commonly included by clinopyroxene and olivine (Figures 8.8j, i). Plagioclase phenocrysts commonly bear melt inclusions and, like augite, some exhibit mottled or complex compositional zonation associated with abundant melt inclusions.

For the duration of episode 61 until October 2013, the petrography of Pu‘u ‘Ō‘ō multiphyric lava continued to portray a dynamic preeruptive environment where growth, dissolution, and regrowth of phenocrysts served to buffer low-pressure multitectic conditions prior to eruption.

8.5.2. Olivine

Electron microprobe analyses of olivine phenocrysts in ERZ lava have been conducted regularly from 1994 to 2014 (Figure 8.9). Other than the occasional appearance of relatively Mg rich olivine phenocrysts that are also cognate to ERZ lava [*Thornber*, 2001], the forsterite content of olivine (Fo = 100 (XMgO/(XMgO/XFeO)), where X is the molar proportion) consistently ranges between the upper and lower limits regulated by equilibrium crystallization over the temperature range defined by bulk-lava and host-glass compositions (Figure 8.9a). The inherent mimicking of typical olivine compositions and those of bulk lava and host glasses throughout the ERZ eruption is strikingly apparent when comparing the running averages for measured Fo contents of all olivine cores and Fo values calculated to be in equilibrium with bulk and vent-corrected glass compositions (Figure 8.9b). Among 357 analyzed olivine phenocrysts in 328 near-vent samples collected from 1994 to 1998 (late episode 53 to early episode 55) [*Thornber*, 2001], 88% are nearly in equilibrium with matrix glasses and range in composition from $Fo_{81.5}$ to $Fo_{80.5}$ (Figure 8.10). From 1998 to 2001 (throughout much of episode 55), averaged olivine phenocryst compositions become less forsteritic with time, displaying an overall decline from Fo_{83} to $Fo_{79.5}$ that tracks midway between averages of values predicted from host bulk-lava and glass compositions. This persistent relation points to efficient ERZ delivery of magma tapped from beneath the summit at a range of temperatures (defined by bulk-lava MgO) and containing mostly equilibrium olivine phenocrysts grown during cooling en route to Pu‘u ‘Ō‘ō [*Thornber*, 2001]. A correlation between prolonged summit deflation (Figure 8.9c) and a steady decline in olivine-controlled MgO systematics of ERZ lava (Figure 8.9b) suggests that ERZ magma was being tapped from a progressively cooler summit reservoir of diminishing volume. However, a running average of olivine Fo variation in ERZ lava does not portray changes in an evolving mix of higher Fo and lower Fo phenocrysts that were grown in host magmas at different temperatures and then mixed together before eruption.

Commensurate with the onset of persistent clinopyroxene and plagioclase phenocrysts in steady state ERZ lava, the declining track of averaged olivine Fo content stalls and flexes upward from $Fo_{80.5}$ to $Fo_{81.5}$ between 2001 and 2003 before falling to $Fo_{79.5}$ in 2004 (Figure 8.9b). The same persistent range of averaged olivine core compositions ($Fo_{81.5\text{–}80.5}$) are nearly in equilibrium with matrix glasses in 1994–1997 olivine-only ERZ lava. However, in 2001–2003 multiphyric lava, bulk MgO is much lower (Figure 8.2), and averaged olivine core compositions are distinctly more forsteritic than olivine that is in equilibrium with host glass (i.e., $K^{*}_{D}(\text{Fe-Mg})^{\text{olv/host}} << 0.3$) (Figure 8.9b). This difference is due to comagmatic mixing. Beginning in 2001, a new $Fo_{79\text{–}80.5}$ population, crystallizing with clinopyroxene and plagioclase in magma buffered at ~1150 –1145°C, is mixed with $Fo_{83\text{–}80.5}$ derived from 1180–1150°C magma (MgO thermometry values of olivine-equilibrated Kīlauea melts). The mixed-magma effects on the olivine phenocryst population are evident from histograms showing populations for ranges of olivine core compositions in steady state ERZ lava before and after 2001 (Figure 8.10). After bottoming out at $Fo_{79.5}$ in 2004, the average Fo content of olivine in multiphyric lava climbed to Fo_{82} in 2005, coincident with a brief pulse of hotter lava (1153°C) in early 2005. This average number reflects the mixed population of near-equilibrium olivine ($Fo_{80.5\text{–}81.5}$) with a significant number of Mg-rich olivine phenocrysts (Fo_{84}-FO_{82}) during this interval (Figures 8.9a and 8.10).

Relatively hot and primitive Kanenuiohamo lava (19 June 2007; episode 56) yielded $Fo_{85.3}$ olivine phenocrysts (the most forsteritic ERZ olivine since 1998) along with near-equilibrium crystals ranging from Fo_{84} to Fo_{83} (Figures 8.7a, b, 8.9a, and 8.10). Episode 56 phenocrysts

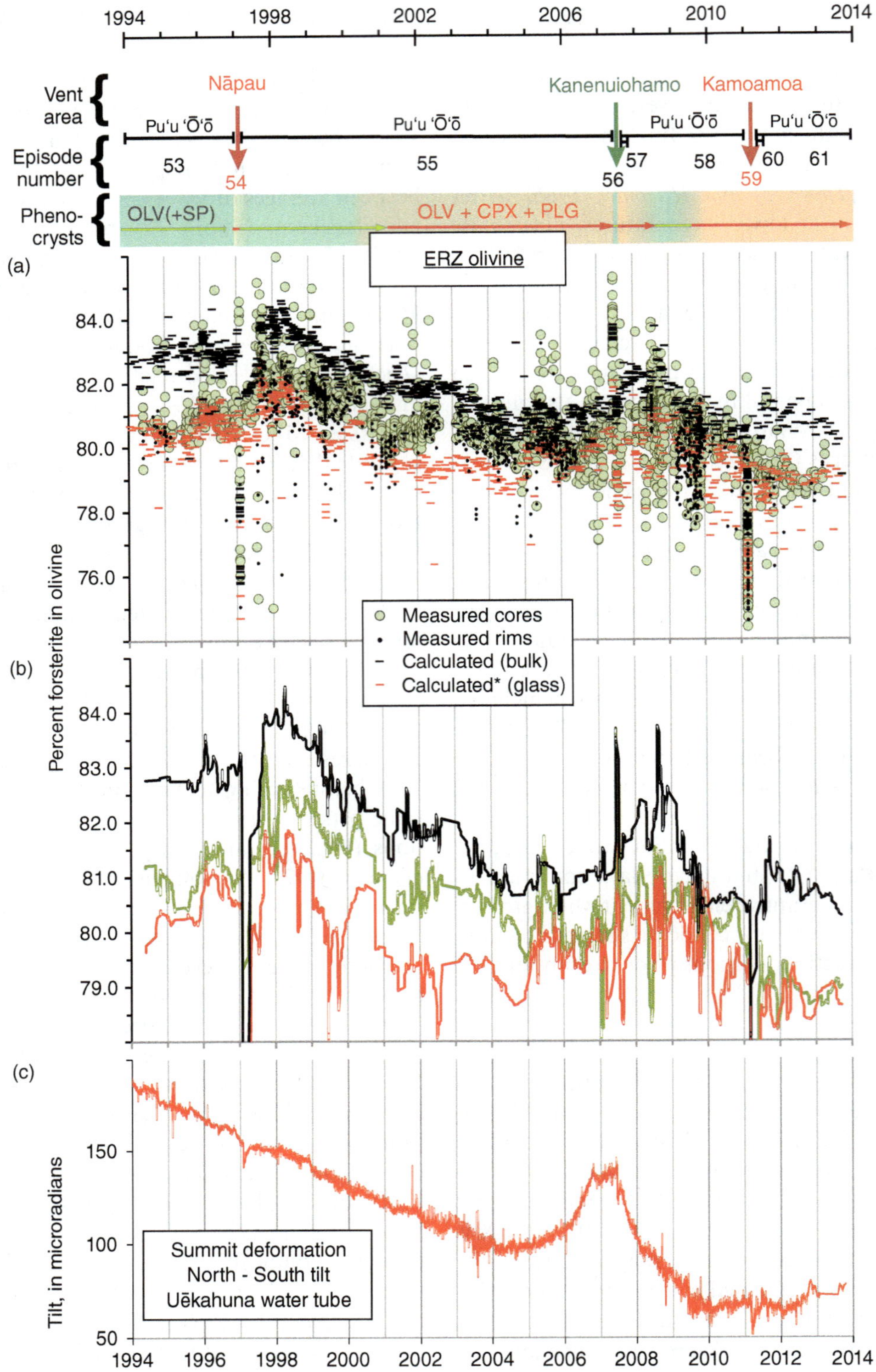

Figure 8.9 Temporal variation of measured and calculated forsterite content of olivine in 1994–October 2013 ERZ lava. The timeline at the top is annotated with vent locations (shown in Figure 8.1), episode numbers, and phenocryst mineralogy: olivine (OLV), spinel (SP), clinopyroxene (CPX), and plagioclase (PLG). (a) Measured phenocryst cores (green-filled circles) and rims (small black-filled circles) are compared with equilibrium values calculated from bulk-lava compositions (black dashes) (K_D (Fe-Mg)$^{olv/liq}$ = 0.30) and from matrix glasses (red dashes), corrected for cooling during transport from the vent (K^*_D = 0.30 + 0.002/km) (as explained in Section 8.2.0). A large population of olivine microphenocrysts is included in the episode 59 (March 2011) data shown here. (b) Running average of every 10 consecutive olivine core analyses (green line) is compared with the running averages of values calculated to be in equilibrium with bulk lava (black line) and with matrix glass compositions (red line). (c) Uēkahuna north-south water tube tilt data show the long-term Kīlauea summit deflation (negative change) from 1994 to 2003–2004 and inflation (positive change) associated with a 2003–2007 surge in magma supply [*Poland et al.*, 2012].

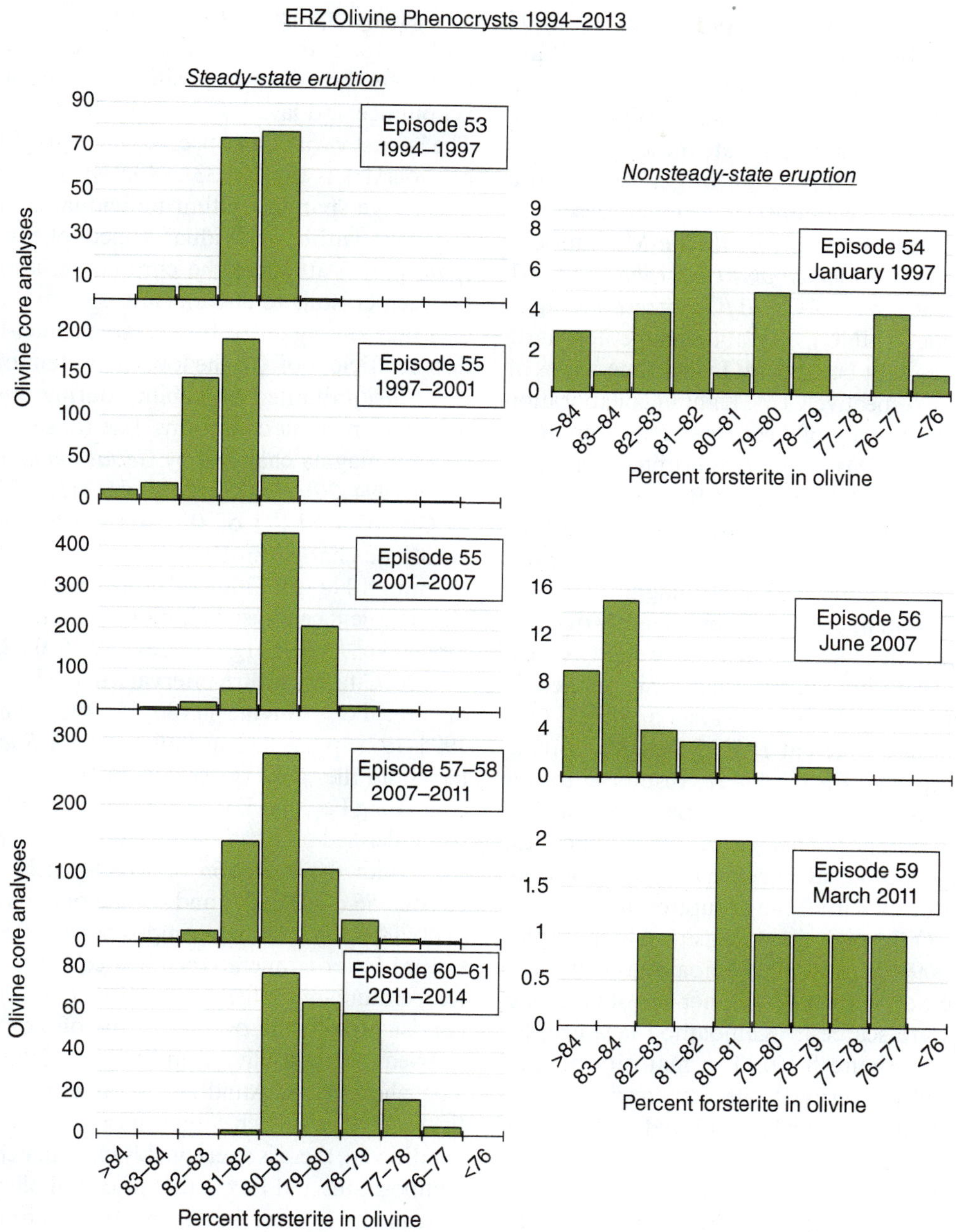

Figure 8.10 Histograms showing the number of analyses within ranges of olivine core compositions between $Fo_{84–76}$ analyzed for sequential intervals of ERZ eruption from 1994 to October 2013. (Left) Olivine in lava issued during prolonged near-continuous effusion (*steady state eruption*) from primary vents near Pu'u 'Ō'ō (late episode 53, episodes 55, 57, 58, 60, and 61). After the onset of multiphyric lava conditions in 2001 (during episode 55), the higher-Fo olivine ($Fo_{>81–84}$) gradually diminishes or disappears, while the relative abundance of lower-Fo olivine ($Fo_{<81–76}$) increases. (Right) Olivine in lava issued from intermittent short-lived fissure vents (*non–steady state eruption*) uprift of Pu'u 'Ō'ō (episodes 54, 56, and 59). Data for each event are presented in sequence with the progression of steady state intervals on the left. The histogram for episode 59 (right bottom) shows compositions of only seven olivine phenocrysts (>100 µm in size) that were observed amid many zoned $Fo_{77–76}$ olivine microphenocrysts (<100 µm) analyzed in samples from this event (Figure 8.9a).

provide an in situ glimpse of the Mg-richest summit-derived component of the comagmatic mixture erupted as Pu'u 'Ō'ō lava before and after the Kanenuiohamo event.

In accord with other petrologic characteristics of lava erupted at Pu'u 'Ō'ō during July 2007 (episode 57 and early episode 58), relatively low Fo olivine, nearly in equilibrium with host glasses, prevailed over a wide mixture of phenocryst core compositions. Higher-Fo olivine phenocrysts are rare and exhibit distinct zonation. One such $Fo_{83.9}$ olivine, plucked from the vigorous outflow of fissure vents east of Pu'u 'Ō'ō on 28 August 2007 (KE58-2683 F), exhibits gradational zonation over 45 μm to an outermost rim composition of $Fo_{79.9}$, which is in equilibrium with the host melt composition (Figure 8.8c). This information is sufficient to estimate the length of time that the crystal of high-temperature origin resided in a lower-temperature melt using diffusion models described by *Martin et al.* [2008] and applying the Fe-Mg inter diffusional coefficients of *Dohmen and Chakraborty* [2007]. Calculations done at $T = 1160°C$ and fO_2 corresponding to nickel/nickel oxide (NNO), fayalite/magnetite-quartz (FMQ), and one log unit below FMQ yield time-scales of 43, 69 and 90 days, respectively (M. Kahl in collaboration with D. Morgan, University of Leeds, personal communication, October 2013). This olivine was erupted exactly 69 days after the 17 June 2007 upper ERZ intrusion and Kanenuiohamo eruption of Fo_{84}-bearing lava, suggesting that this phenocryst could be a remnant of rift-reservoir injection at the height of the 2003–2007 magma surge.

The average Fo content of olivine rose steadily over the next year, mirroring the steady increase in weight percent bulk MgO and eruption temperature. This peaked in August 2008, with $Fo_{83.5}$–Fo_{82} phenocrysts in olivine-(spinel-) phyric lava that erupted at 1157°C (Figures 8.7c, d) perhaps indicating the ERZ's latent response to the 2003–2007 magma surge defined, in part, by summit inflation (Figure 8.9c) [*Poland et al.*, 2012]. For the remainder of episode 58, average olivine compositions lowered synchronously with falling eruption temperature and bulk-lava MgO through the retransition to multiphyric lava in June 2009 and until the Kamoamoa eruption in March 2011 (see Section 8.4.4). Higher-Fo olivine phenocrysts (Fo_{82-81}) are scarce in Kamoamoa lava erupted during episode 59 (Figure 8.10). The majority of analyzed episode 59 olivine crystals were microphenocrysts (<100 μm), and many of those displayed complex or reversed zoning characteristics. Persistently lower Fo olivine (Fo_{80-78}) in multiphyric Pu'u 'Ō'ō lava returned in episode 60 and has persisted throughout episode 61. By October 2013, sustained temperatures of average olivine crystallization (≤1140°C) are close to matching the equilibration temperatures of host glasses (Figure 8.9b), and there is a corresponding shift toward prevalence of euhedral morphology of lower-Fo olivine (Figures 8.8j, k, l). In addition, high-forsterite olivine crystals have been scarce in episode 61 (Figures 8.9a and 8.10), indicating diminishing influence of a more primitive component to comagmatic mixing prior to eruption.

8.5.3. Cr-Spinel

The chemistry of chromian-spinel in 1994–2013 ERZ lava provides additional perspective on long-term changes in eruption petrology. *Roeder et al.* [2003] present over 300 analyses of spinel in 13 representative 1994–1998 ERZ vent samples and found them to define a narrow range within the broader compositional array of 1959 Kīlauea Iki pumice and lava lake samples by *Scowen et al.* [1991]. The largest variability observed in 1994–2013 ERZ microphenocrysts is that of 33.5–49.9 wt% Cr_2O_3. Not only is that range spanned within individual samples, it is also observed within individual zoned phenocrysts. *Roeder et al.* [2003] attribute the chrome variations within and between crystals to circumstances of primary crystallization that are regulated by the large spinel-liquid distribution coefficient of Cr, the low Cr content of the melt, and its diffusion-limited availability during crystal growth in turbulent magma conditions, like those proposed for the summit magma chamber by *Gerlach et al.* [2002].

For 1994–2013 ERZ lava, spinel compositional variations in $Cr^{3+}/(Cr^{3+}+Al^{3+})$ of 0.73–0.61 are complimented by antithetic changes in $Fe^{3+}/(Fe^{3+}+Cr^{3+}+Al^{3+})$ of 0.10–0.22 (Figures 8.11a–d). This oxidation-sensitive parameter regulates trivalent cation substitution over a range of melt-equilibrated $Fe^{2+}/(Fe^{2+}+Mg^{2+})$ values of 0.46–0.58, corresponding to a 30°C temperature interval affected by crystallization of Fo_{83}–$Fo_{80.5}$ olivine (1180–1150°C). The bulk of the 1994–1998 spinel data, including data of *Roeder et al.* [2003], lie within the range delineated by "equipotential" lines for coexisiting Fo_{83} and Fo_{81} olivine at 1180 and 1160°C, respectively, defined by *Roeder et al.* [2003] (Figure 8.11a). In contrast, with the exception of MgO-rich lava issued during episode 56 (June 2007) and olivine-only lava in 2008–2009, post-2001 ERZ spinel compositions reflect a clear shift toward lower-temperature spinel equilibration with lower-temperature multiphyric melts (Figure 8.11b). This trend is consistent with the overall decline of Fo content of both near-equilibrium olivine and of occasional higher-Fo olivine phenocrysts. Amid an overall scattering toward relatively Cr poor, Fe rich compositions (related to local growth conditions), there is a reasonably close match between spinel compositions and Fo content of coeval olivine phenocrysts. A reasonably close match between running averages of $(Cr^{3+}/(Cr^{3+}+Al^{3+}))/(Fe^{2+}/(Fe^{2+}+Mg^{2+}))$ in spinel with running averages of olivine phenocryst core compositions (Figure 8.12a) emphasizes the systematic dependency of spinel compositions on olivine-controlled glass and bulk-lava MgO systematics (Figure 8.9b).

8.5.4. Augite

The pyroxene compositional data presented for ERZ eruption samples are culled from a considerably larger data set on the basis of stringent stoichiometric criteria such that analyses presented are limited to those with cation totals within 0.5% of 4.00 (i.e., ±0.02) in an augite unit formula $(Ca,R^{2+})(R^{2+},R^{3+},Ti^{4+})(Si,Al)_2O_6)$ [*Cameron and Papike*, 1980]. This refined criteria help validate

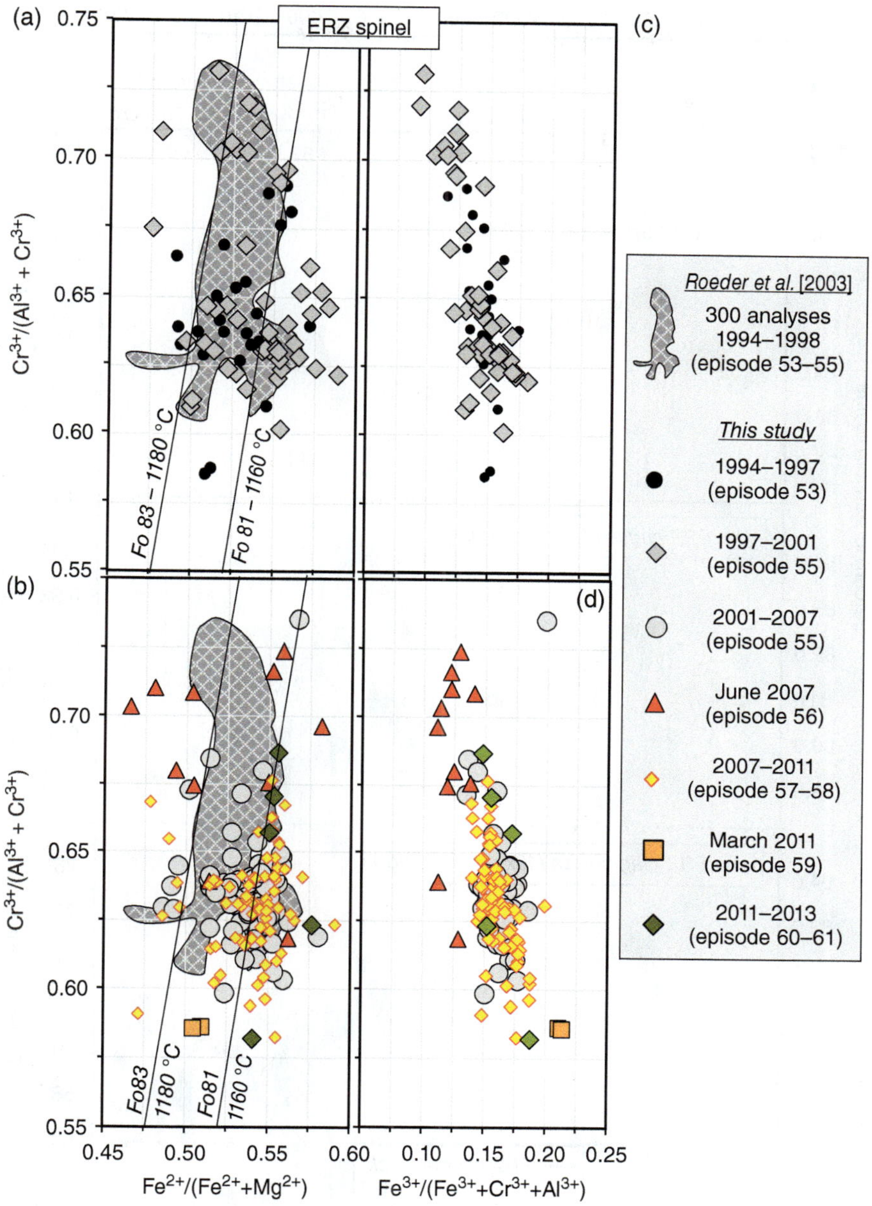

Figure 8.11 Spinel compositions in samples of ERZ lava from 1994 to October 2013. Cations are calculated on the basis of four oxygens, with Fe^{3+} and Fe^{2+} assigned according to stoichiometry and charge balance. Symbols shown in legend are for sequential eruptive episodes as indicated in Figure 8.9. (a, b) $Cr^{3+}/(Cr^{3+}+Al^{3+})$ versus $Fe^{2+}/(Fe^{2+}+Mg^{2+})$ for 1994–2001 and 2001–2013, respectively. Stippled area represents the field of ~300 spinel compositions for 1994–1998 ERZ lava reported by *Roeder et al.* [2003]. The diagonal lines are "equipotential" lines from *Roeder et al.* [2003] delineating spinel compositions in equilibrium with Fo_{83} and Fo_{81} olivine at 1180 and 1160°C. (c, d) $Cr^{3+}/(Cr^{3+}+Al^{3+})$ versus $Fe^{3+}/Fe^{3+}+Cr^{3+}+Al^{3+})$ for 1994–2001 and 2001–2013, respectively.

pyroxene analyses with relatively high aluminum and titanium concentrations and also those of metastable subcalcic compositions observed not only in the microphenocryst (<100 μm) population and as phenocryst rim overgrowths but also in some phenocryst interiors.

The compositional array of clinopyroxene phenocryst (>100 μm) core compositions in 2000–October 2013 Puʻu ʻŌʻō lava (from episodes 55, 57, 58, 60, and 61) is established by 1007 carefully scrutinized analyses in 224 samples. In accordance with accepted pyroxene nomenclature using normalized Ca^{2+}-Mg^{2+}-Fe_t^{2+}componentry (percent wollastonite-enstatite-ferrosilite, assuming all iron is ferrous; *Ross and Huebner* [1979]), the observed range of phenocryst core compositions of Wo_{40-35}, En_{57-45}, and Fs_{22-16} are classified as "augite" (Figure 8.13a). While the compositional range is slightly skewed toward a subcalcic

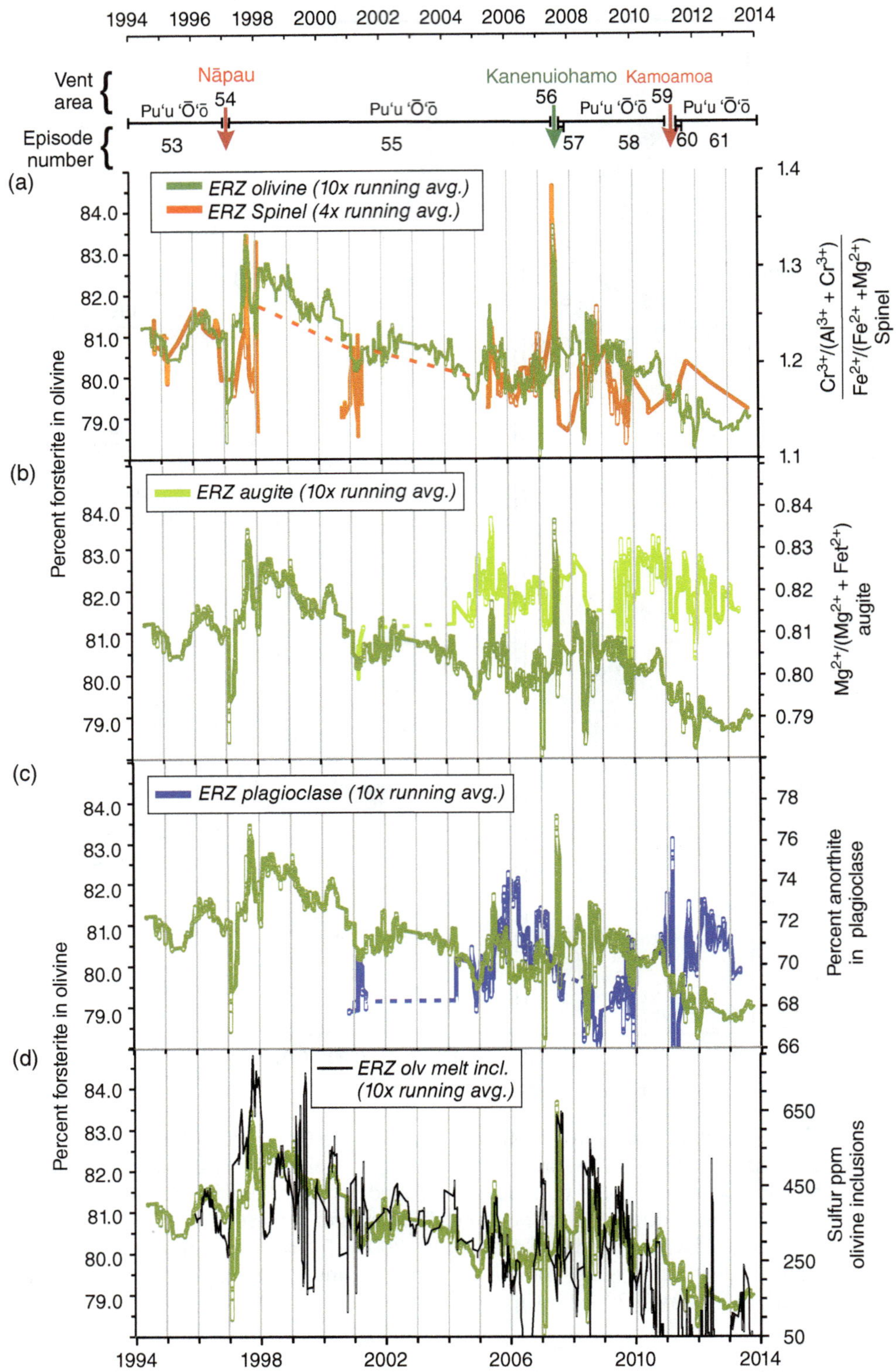

Figure 8.12 Composition comparisons for olivine, spinel, augite, and plagioclase in ERZ lava from 1994 to October 2013. The running average Fo content of every 10 consecutive olivine core analyses (dark green line, left axis) from Figure 8.9b is compared with: (a) spinel, as the running average of $(Cr^{3+}/(Cr^{3+}+Al^{3+}))/(Fe^{2+}/(Fe^{2+}+Mg^{2+}))$ for every four consecutive analyses (orange line, right axis); (b) augite, as the running average of $(Mg^{2+}/(Fe_t^{2+}+Mg^{2+})$ for every 10 core analyses (light green line, right axis); and (c) plagioclase, as the running average An content for every 10 analyses (core and rims) (blue line, right axis). (d) Sulfur in olivine melt inclusions, as the running average of 10 analyses in parts per million. Dashed lines indicate intervals with no data.

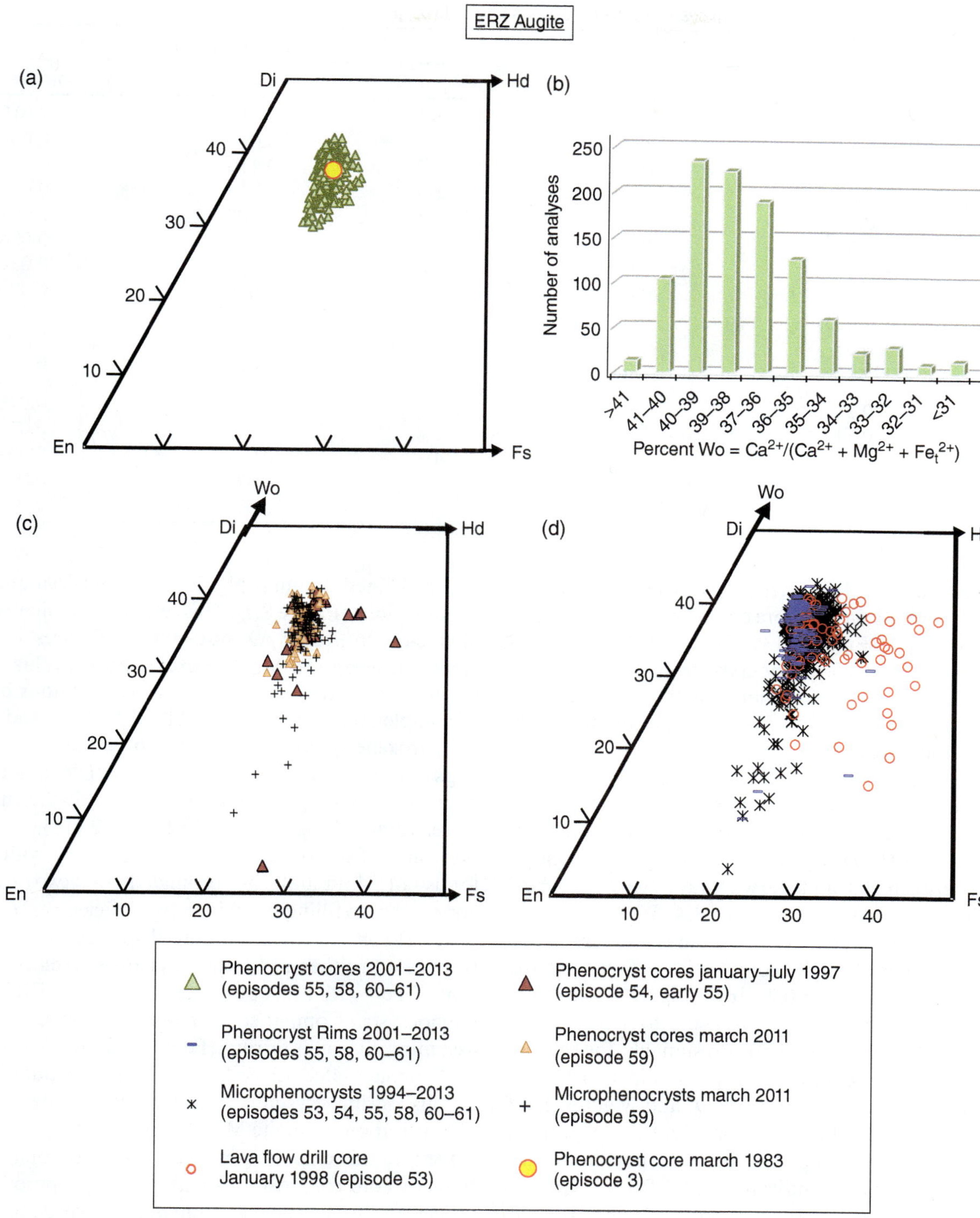

Figure 8.13 Pyroxenes in 2000–October 2013 Kīlauea ERZ lava. (a) Augite pheoncryst cores (green triangles) in steady state multiphyric lava plotted in the left half of the enstatite (En)–diopside (Di)–hedenbergite (Hd)–ferrosilite (Fs) pyroxene quadrilateral. (b) Histogram showing the number of analyses within wollastonite (Wo) ranges for augite core compositions. (c) Compositions of pyroxene phenocrysts and microphenocrysts in hybrid lava from 1997 Nāpau (episode 54) and 2011 Kamoamoa (episode 59) eruptions. (d) Compositions of posteruptive pyroxene microphenocrysts and phenocrysts in episodes 53–58 coastal lava flows (asterisks) and a 1998 active lava flow drill core [*Thornber et al.*, 2004] (open red circles).

Table 8.1 Average of 1,001 augite analyses from 2000–2013 Kīlauea ERZ lava.

	Weight Percent Oxides			Cations Based on 6 Oxygens	
	Average	Standard Deviation		Avarage	Standard Deviation
SiO_2	52.2	0.97	Si^{4+}	1.918	0.019
TiO_2	0.64	0.20	Ti^{4+}	0.018	0.005
Al_2O_3	2.5	0.51	Al_t^{3+}	0.110	0.022
FeO_t	7.0	0.57	Fe_t	0.214	0.018
MnO	0.2	0.03	Mn^{2+}	0.005	0.001
MgO	17.8	0.65	Mg^{2+}	0.977	0.034
CaO	18.5	0.91	Ca^{2+}	0.730	0.036
Na_2O	0.23	0.04	Na^+	0.016	0.003
K_2O	0.00	0.01	K^+	0.000	0.000
Cr_2O_3	0.68	0.18	Cr^{3+}	0.020	0.005
Total	99.76	1.06		4.007	0.011
Wo	0.38	0.02	$Al_{(IV)}$	0.082	0.019
En	0.51	0.01	$Al_{(VI)}$	0.029	0.013
Fs	0.11	0.01	Fe^{3+}	0.022	0.034
		$Mg^{2+}/(Fe_t+Mg^{2+})$		0.82	0.01
		$Mg^{2+}/(Fe^{2+}+Mg^{2+})$		0.83	0.02

population, most ERZ phenocryst core analyses are close to the overall average core composition of $Wo_{38(\pm1\sigma=2)}En_{51(\pm1\sigma=1)}Fs_{11(\pm1\sigma=1)}$ (Figure 8.13b, Table 8.1). The compositional range of pyroxene phenocrysts from episode 54 and 59 mixed-magma eruptions is similar to the range observed for post-2000 Pu'u 'Ō'ō lava, with the exception of rare Fe-rich orthopyroxene found in an episode 54 sample (Figure 8.13c).

ERZ augite phenocrysts incorporate a variable range of basaltic components other than Wo-En-Fs, referred to as "OTHERS" by *Huebner* [1980]. Most have Al_2O_3 concentrations between 2.0 and 3.0 wt% (average of 2.5 ± 0.5 (1σ)). However, a colossal range of 1.4–5.2 wt% Al_2O_3 is observed in the entire phenocryst-core array (Figure 8.14a). The TiO_2 content of augite cores varies between 0.2 and 1.4 wt%, averaging 0.6 (±0.2 (1σ)) (Figure 8.14b), and Cr_2O_3 concentration ranges from <0.1 to 1.2 wt% with an average of 0.7 (±0.20 (1σ)). Presumably for reasons described above for chromium dispersion in spinel, there is no clear correlation between Cr_2O_3 and MgO in ERZ augite phenocrysts. In fact, consistently sympathetic covariations are not apparent between any individual components among significant OTHERS in Kīlauea augite phenocrysts. There is a broad inverse relation between the collective abundance of Al^{3+}, Ti^{4+}, Fe_t^{2+}, and Na^+ versus that of Mg^{2+} + Cr^{3+} (Figure 8.14c), which serves as a fractionation index related to augite paragenesis [*Thornber*, 1994].

ERZ augite phenocrysts spanning the entire array of observed compositions exhibit both resorbed and euhedral morphologies (described above and in Figure 8.8). This observation predicates simultaneous growth and dissolution within an overall MgO temperature range that is defined by multi phyric ERZ bulk-lava and glass compositions (Figure 8.2). This relation is similar to ERZ olivine-controlled MgO systematics (Figures 8.9a, b). However, using $(Fe\text{-}Mg)^{cpx/liq}$ exchange to evaluate equilibrium augite thermometry is, at best, tenuous because of complexities introduced by Al^{3+}, Ti^{4+}, Cr^{3+}, and Fe^{3+} in the pyroxene lattice, not the least of which is Fe^{3+}. The Mg# ($=Mg^{2+}/(Mg^{2+}+Fe_t^{2+})$) of Pu'u 'Ō'ō augite phenocryst cores ranges from 75 to 85, with the majority closer to an average value of 82.1 (±1.2 (1σ)). The proportion of Fe^{3+} to Fe_t^{2+} for augite analyses (with cation totals of $4.0\pm.02$), as estimated by two different approaches to filling stoichiometric deficiencies [*Lindsley*, 1983; *Droop*, 1987], is broadly dispersed over a range from zero to either 40% or 60%. There is no clear correlation of Fe^{3+} with other augite components in ERZ augite phenocrysts. Comparison of augite *Mg#*'s calculated assuming all iron is ferrous (Fe_t^{2+}) versus those using true Fe^{2+}, compensated for Fe^{3+}, infers variable oxidation conditions during crystallization that contribute to a full 10% variation of augite Mg# (Figure 8.14d).

Despite this oxidation-induced Mg# variability, variations of core compositions with time are approximately bracketed by the calculated range of "near-equilibrium" augite Mg# derived from host glasses and bulk melts with assumed $FeO/FeO_t=0.9$ and using a distribution coefficient of K_D $(Fe\text{-}Mg)^{cpx/liq}=((FeO_t/MgO)^{cpx}$ $(MgO/FeO)^{liq}=0.26$ (Figure 8.15a). This value, here defined by Kīlauea augite-host compositions, is well within the range of 0.28 (±0.08 (1σ)) established by *Putirka* [2008] using a combined average of 1245 published experimental observations. A closely matched pattern of augite Mg# running averages with running averages of olivine Fo content

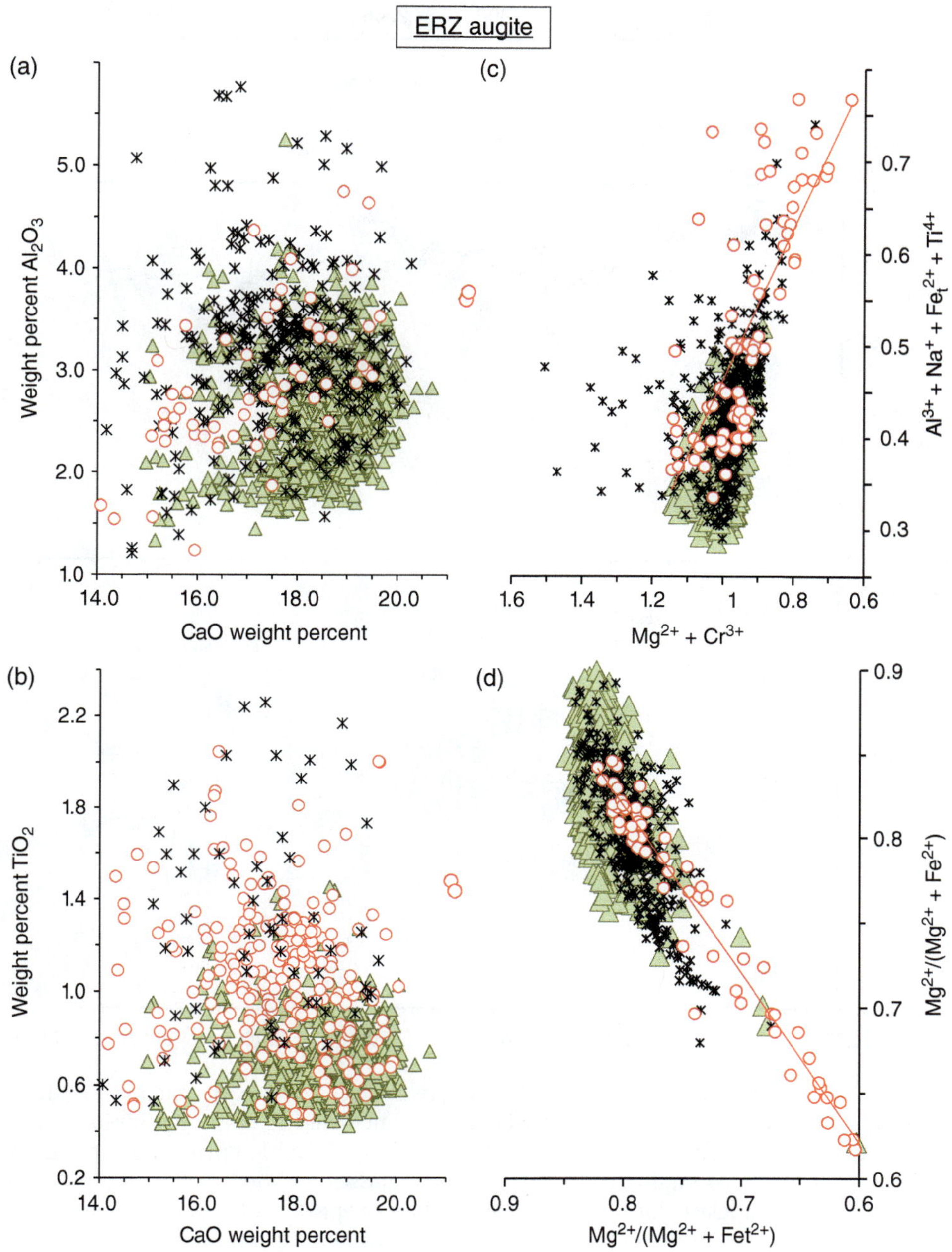

Figure 8.14 Compostional variation of significant minor components in 2000–October 2013 ERZ augite phenocryst cores (green triangles) is compared with posteruptive clinopyroxene compositions reported for olivine-spinel phyric, episode 53–58 coastal lava flows (asterisks), including a 1998 (episode 55) lava flow drill core [*Thornber et al.*, 2004] (open red circles). (a) Weight percent Al_2O_3 variation with CaO; (b) weight percent TiO_2 variation with CaO; (c) $Al^{3+} + Ti^{4+} + Fe_t^{2+} + Na^+$ (cations per 6 oxygens) variation with that of $Mg^{2+} + Cr^{3+}$ (*inverted x axis*); and (d) $Mg^{2+}/(Mg^{2+} + Fe^{2+})$ (cations per six oxygens) without an Fe^{3+} component is compared with $Mg\# = Mg^{2+}/(Mg^{2+} + Fe_t^{2+})$, assuming all iron is ferrous (*inverted x axis*). The strong linear correlation of drill-core pyroxenes (red line) suggests a consistent Fe^{3+}/Fe^{2+} ratio is maintained in the lava flow during posteruptive cooling and fractionation.

(Figure 8.12b) further emphasizes the systematic dependency of augite Mg# on olivine-controlled glass and bulk-lava MgO systematics (Figure 8.9b).

The bulk-magma MgO overprint on average Pu‘u ‘Ō‘ō augite phenocryst compositions points to the broader systematic control of prevalent preeruptive magma conditions. However, such simplification of pyroxene-liquid systematics disguises a record of kinetic disarray apparent from broad compositional diversity of augite phenocrysts within and between lava samples. In fact, much

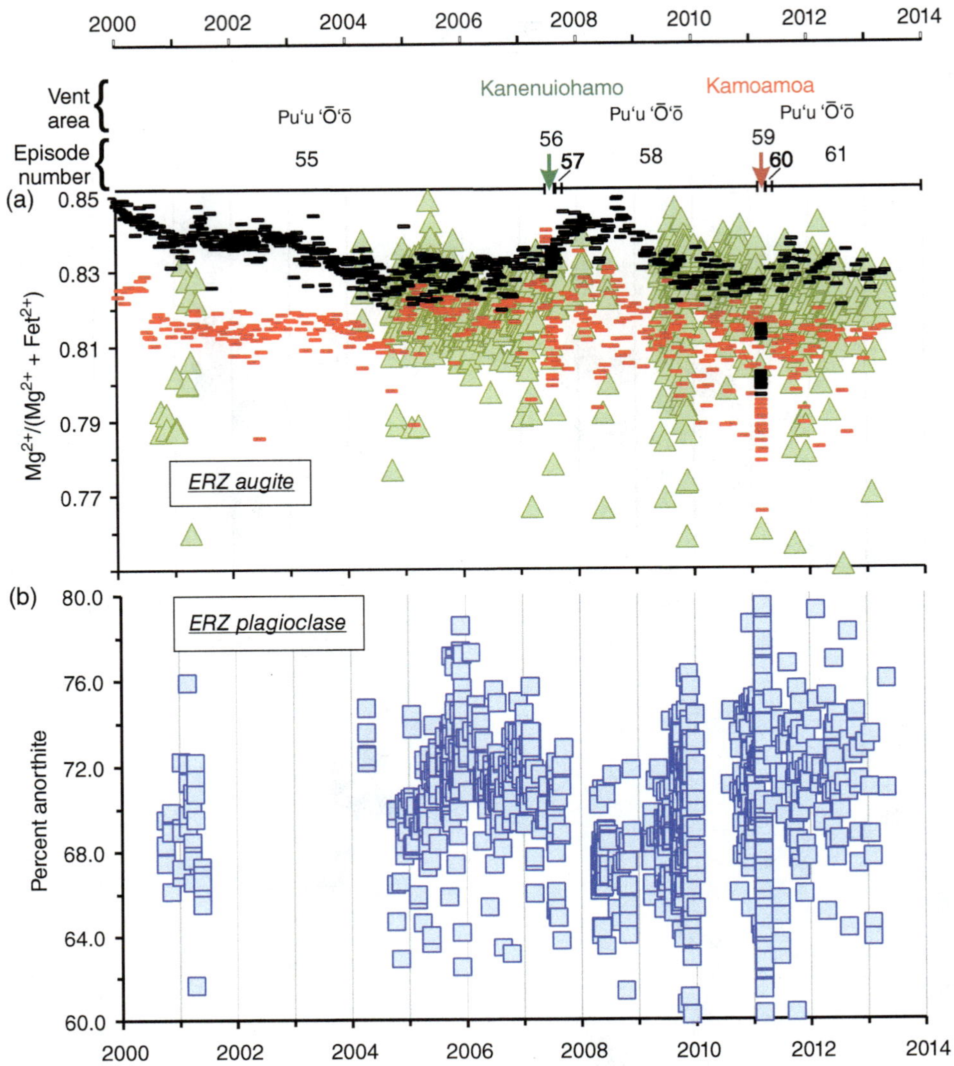

Figure 8.15 Temporal variation during 2000–2013 of augite and plagioclase crystals in ERZ lava. (a) Augite phenocryst cores (green triangles) are compared with values calculated to be in equilibrium with glass (red dashes) and bulk-lava compositions (black dashes). Mg# = $Mg^{2+}/(Mg^{2+} + Fe_t^{2+})$ in pyroxene is calculated without consideration of Fe^{3+}. Calculated equilibrium values for host glasses and bulk liquids are made assuming a $K_D(\text{Fe-Mg})^{cpx/liq} = 0.26$. (b) Plagioclase cores and rims (blue squares); episode 59 data include microphenocrysts. There is a data gap from 2001 to 2004 for pyroxene and plagioclase in ERZ lava.

of the entire range of core compositions is repeatedly observed within single phenocrysts of augite exhibiting complex zoning, abundant melt inclusions, and occasional pigeonite microphenocryst inclusions (e.g., Figures 8.8e, f, g). As pointed out by *Huebner* [1980], anomalously high concentrations of aluminum and titanium in augite can be attributed to growth during rapid cooling [e.g., *Huebner and Turnock*, 1980; *Grove and Bense*, 1977]. Evidence that some Pu'u 'Ō'ō phenocryst growth rates occasionally outpace the crystal/liquid mobility of calcium, aluminum, and titanium is provided by comparing the compositions of quickly grown, posteruptive augite in recent ERZ lava (Figures 8.13d and 8.14). Clinopyroxene microphenocrysts (<100 μm) grown during surface transport of distal or slowly quenched flows or in slowly quenched near-vent samples, as well as pyroxene in drill core from a naturally cooled active lava flow when near-vent samples were olivine-spinel phyric [*Thornber et al.*, 2004], all accommodate a much higher range of Al_2O_3 and TiO_2 than ERZ augite phenocrysts, and many have metastable (nonequilibrated) subcalcic compositions, similar to pigeonite inclusions in phenocrysts. The fact that posteruptive pyroxene compositions completely overlap with those of magmatic augite phenocrysts is, in itself, a testament to shallow preeruptive conditions that are conducive to rapid growth during undercooling.

This comparison of pre- and posteruptive augite in recent ERZ lava helps to demonstrate that augite phenocrysts form at low pressures corresponding to shallow depths within the eruption plumbing system. This conclusion is consistent with landmark studies of the mineralogy of Kīlauea olivine tholeiite crystallization at the Makaopuhi and Kīlauea Iki lava lakes [*Evans and Moore*, 1968; *Helz*, 1980] and low-pressure phase relations from 1 atm laboratory experiments [*Thompson and Tilley*, 1969; *Helz and Thornber*, 1987].

In stark contrast to our conclusion of a low-pressure genesis of Kīlauea ERZ augite phenocytsts, *Putirka* [1997, 2008] interprets the Puʻu ʻŌʻō augite-liquid equilibrium to have been established prior to transport from mantle depths. The Puʻu ʻŌʻō augite analysis of an episode 3 sample (KE3-88) featured by *Putirka* [2008] to exemplify equilibration with bulk melt at 5.1 kbar is (within error) the same as a well-constrained average of more recent Puʻu ʻŌʻō augite phenocryst compositions (Figure 8.13a, Table 8.1). Based upon what we know of the shallow plumbing origin of the ERZ multitectic phenocryst assemblage, combined with persistent evidence for kinetic disequilibrium associated with rapid growth of ERZ augite phenocrysts, it appears that deep magma transport theories should not be justified solely on the basis of geobarometry using typical augite phenocrysts in olivine tholeiite lava associated with shield-building eruptions. In particular, petrogenetic interpretations specifically invoking deep-sourced pyroxene phenocrysts in recent Kīlauea ERZ eruptions are, at best, circumspect (e.g., *Putirka* [1997, 2008] for 1983 Puʻu ʻŌʻō lava and *Vinet and Higgins* [2010] for 1969–1974 Mauna Ulu lava).

8.5.5. Plagioclase

An_{60} to An_{80} is the compositional range established for plagioclase in 2000–October 2013 ERZ lava from a series of ~1200 analyses, where An is the percentage of anorthite (An, $CaAl_2Si_2O_8$) in the feldspar ternary system with albite (Ab, $NaAlSi_3O_8$) and orthoclase (Or, $KAlSi_3O_8$). The unsorted array of ERZ plagioclase data presented here, while mostly comprised of phenocryst core compositions interspersed with rim analyses, also include plagioclase overgrown by augite or olivine and ~250 microphenocryst compositions with a broad compositional range from episode 59 (Figure 8.15b). The average ERZ plagioclase composition is $An_{70.2\ (\pm 1\sigma\ 3.4)}$ $Ab_{29.3(\pm 1\sigma\ 3.3)}$ $Or_{0.5\ (\pm 1\sigma\ 0.2)}$. Plagioclase in ERZ lava is restricted to the highest-temperature portion within a much broader array of feldspar compositions defined by Kīlauea low-pressure crystallization in the Makaopuhi lava lake by *Evans and Moore* [1968] (Figure 8.16).

ERZ plagioclase phenocrysts commonly occur without compositional zonation from core to rim. However, both normal-zoned (with higher An cores and lower An rims) and reverse-zoned plagioclase phenocrysts are not uncommon. Furthermore, irregular and complexly mottled zonation is a noteworthy feature among plagioclase phenocrysts with abundant melt inclusions. As with chemically mottled and inclusion-riddled augites, these plagioclase grains are a likely testament of kinetic disequilibria inherent to a dynamic crystallization environment. These relations justify further scrutiny of plagioclase data to assess subtleties of multi phyric compositional relations. There is no clear correlation between running averages of Fo content in ERZ olivine and An percent measured for unsorted plagioclase data in 2000–October 2013 lava (Figure 8.12c).

8.5.6. Melt Inclusions in ERZ Lava

Melt inclusions in ERZ olivine phenocrysts have been routinely analyzed by microprobe since 1994 [*Thornber*, 2001; *Thornber et al.*, 2003b]. Melt inclusions in ERZ augite and plagioclase phenocrysts, along with glasses trapped in open-textured microgabbroic clusters, have been routinely analyzed since 2005. In all, more than 1550 melt inclusion analyses are compiled herein for 1994–October 2013 ERZ lava (Figure 8.17). In addition, Chapter 20, by *Rowe et al.*, presents a detailed study of major and trace-element signatures of olivine melt inclusions from Kīlauea lava samples erupted during 2008. That separate study provides confirmation of the inherently cognate olivine-liquid relations of steady state ERZ lava inferred by *Thornber* [2001]. In the following discussion, we focus on sulfur, the most variable melt inclusion component within and between all types of Puʻu ʻŌʻō phenocrysts. The track of magmatic sulfur variation preserved in ERZ melt inclusions provides a temporal log of preeruptive sulfur dioxide (SO_2) degassing associated with changes in the dynamics of shallow magma transport and storage in the edifice.

As observed in the paired olivine inclusion data for 2008 olivine-spinel-phyric Puʻu ʻŌʻō samples (Chapter 20), higher magmatic sulfur concentrations correlate with more forsteritic hosts that crystallized at higher temperatures. A more poignant case for a strong correlation between magmatic sulfur and olivine-controlled MgO systematics of 1994–2013 ERZ lava (Figure 8.9b) is demonstrated by comparing running averages of olivine Fo content with those of sulfur in olivine melt inclusions (Figure 8.12d).

An increased frequency of melt inclusions with higher sulfur concentrations of 500–1000 ppm is associated with a vigorous, high-temperature eruption interval in 1998 and again during the height of surge-influenced Puʻu ʻŌʻō compositions in 2008 (Figures 8.9c, 8.12d, and 8.17a). Intermittent appearances of Mg-rich olivine phenocrysts

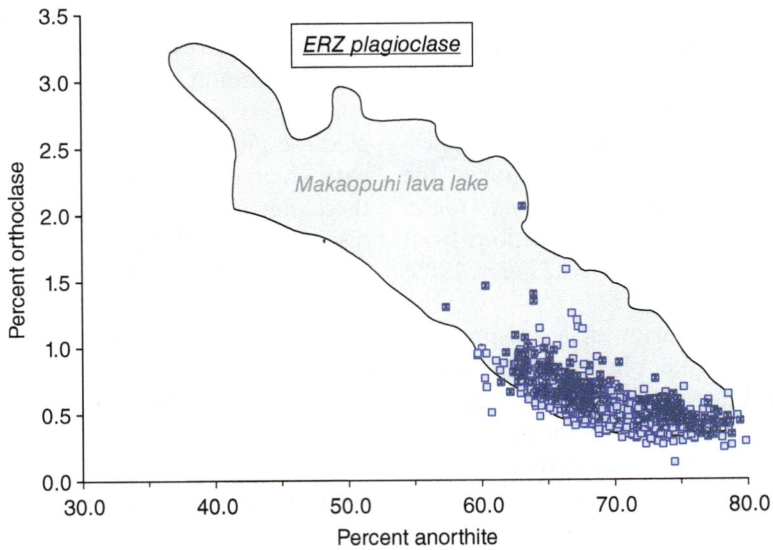

Figure 8.16 Range of anorthite (An) and orthoclase (Or) components of plagioclase in 2000–October 2013 ERZ lava compared to that of Makaopuhi lava lake samples reported by *Evans and Moore* [1968]. Phenocryst core and rim analyses in episodes 58, 60, and 61 Pu'u 'Ō'ō lava (blue squares) are plotted with microphenocrysts and phenocrysts in lava erupted in March 2011 at Kamoamoa during episode 59 (distinguished by asterisks).

($>Fo_{82}$) with inclusions containing >500 ppm S are persistent in low-MgO, post-2000 multiphyric Pu'u 'Ō'ō lava (episodes 55, 58, 60, and 61), but the typical lower-temperature olivine phenocrysts ($<Fo_{80.5}$) have melt inclusions with 250–450 ppm S. This intermediate range of magmatic sulfur is also most commonly trapped within and between coprecipitated phenocrysts of augite and plagioclase in post-2000 ERZ lava (Figures 8.17a, b). Despite the persistent appearance of higher-Fo olivine phenocrysts grown in hotter, S-rich magma, most ERZ lava phenocrysts grow amid storage and transport conditions that are conducive to persistent partial SO_2 degassing. The consistently broad dispersion of high to low sulfur (~1000 to <50 ppm) trapped by coexisting olivine ± clinopyroxene ± plagioclase phenocrysts in nearly every carefully examined multiphyric ERZ lava sample attests to crystallization amid SO_2 degassing in a dynamic shallow regime. Of particular significance in this regard is a similarly high range of magmatic sulfur that is frequently observed within single phenocrysts of inclusion-riddled and chaotically zoned augite and plagioclase (Figures 8.8e–l).

This evidence for heterogeneity of magmatic sulfur in the preeruptive environment documents the evolution of SO_2 as well as a redox reaction within host melts that is significant enough to affect Fe^{3+}/Fe^{2+} and, consequently, Mg/Fe^{2+} ratios of phenocrysts and melts. As elaborated in the comment exchange between *Gerlach* [2004] and *Roeder et al.* [2004], crystal-glass thermometers and oxygen barometers that utilize carefully determined ferric/ferrous ratios on 1994–1998 Pu'u 'Ō'ō lava [*Poustovetov and Roeder*, 2001] consistently indicate that erupted lavas are reduced (~1 log unit fO_2) relative to well-established magmatic fO_2 values from Kīlauea gas fumaroles [*Gerlach*, 1993], a conclusion that is substantiated by *Burgisser and Scalliet* [2007]. The range from 1000–< 50 to ~1000 ppm (~0.1 wt%) of magmatic sulfur (~0.1 wt%) commonly trapped within samples, and even within individual phenocrysts, accounts for production of 0.5 wt% Fe_2O_3, changing the Fe^{3+}/Fe^{2+} from 0.162 to 0.109 and thus affecting a 1% increase in forsterite values, assuming $K_D(\text{Fe-Mg})^{olv/liq}$ of *Roeder and Emslie* [1970]. Also, wide ranges of estimated Fe^{3+} among Pu'u 'Ō'ō augite phenocrysts (Table 8.1) is inherently coupled to large variations of magmatic sulfur and provides convincing testament to the prevalence of redox disequilibria in a dynamic and perhaps turbulent preeruptive environment.

8.6. SUMMARY OF EVIDENCE FOR SHALLOW OPEN-SYSTEM-RESERVOIR MIXING ALONG THE ERZ

The variations of 1994–2013 ERZ phenocryst compositions and melt inclusion sulfur portrayed here, together with salient petrographic characteristics of ERZ lava described above, provide a framework for interpreting volcanic processes responsible for coupled changes in eruptive behavior and lava chemistry. A complete lack of CO_2- and H_2O-bearing fluid inclusions among the entire ERZ phenocryst suite reported on here is a testament to crystallization within the shallow edifice. This "evidence of absence" indicates that ERZ phenocrysts formed in partly degassed magma, expunged of carbon dioxide

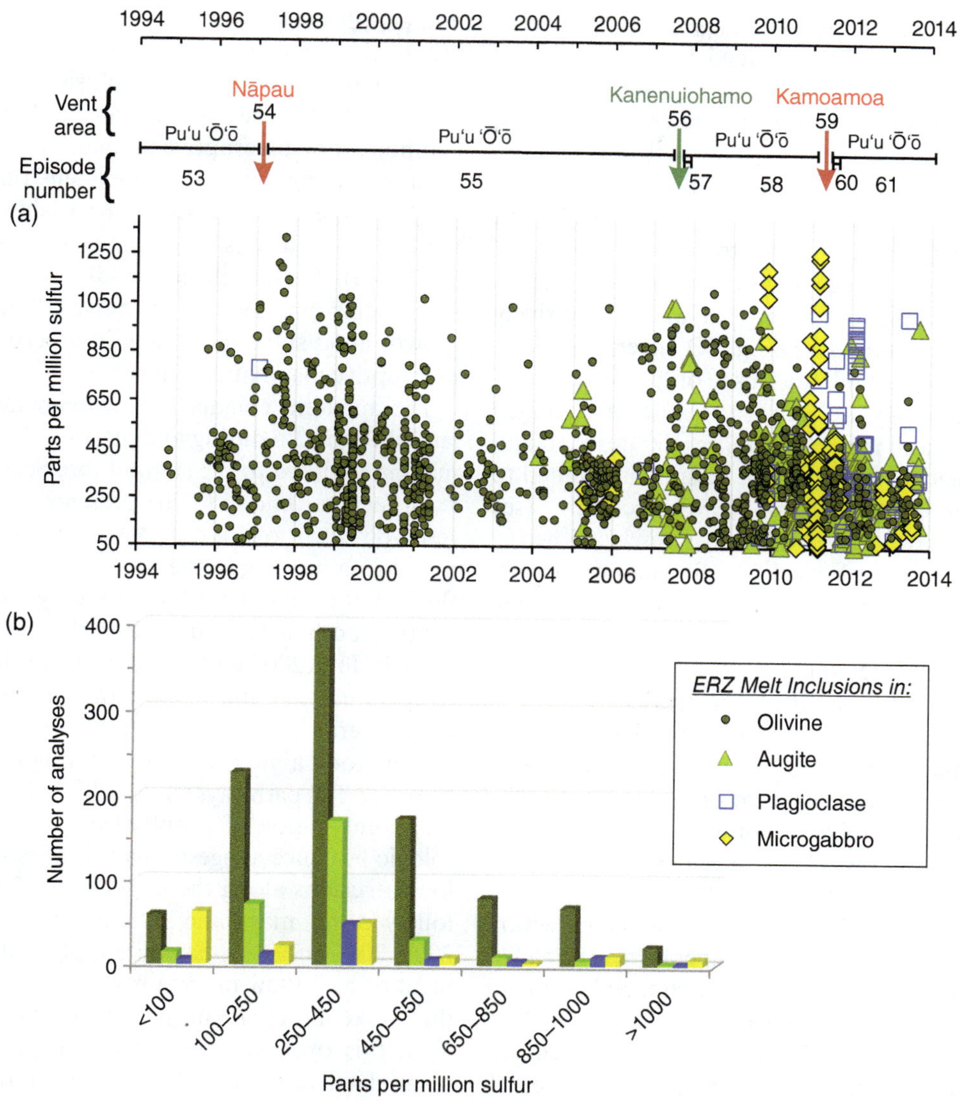

Figure 8.17 Sulfur in ERZ melt inclusions erupted during 1994–October 2013. (a) Temporal variation of elemental sulfur in melts included by olivine phenocrysts (dark green dots) and for melt inclusions in augite phenocrysts (bright green triangles) and plagioclase phenocrysts (empty blue squares) or trapped interstitially in glomerophyric clusters of augite ± plagioclase ± olivine (microgabbro, yellow diamonds). Symbols for inclusion hosts are as indicated in the legend. (b) Histogram showing number of analyses for incremental ranges of sulfur concentrations between 50 and 1400 ppm in ERZ melt inclusions hosted by olivine (dark green), augite (bright green), and plagioclase (blue) phenocrysts or trapped interstitially in glomerophyric clusters of clinopyroxene ± plagioclase ± olivine (microgabbro; yellow).

during ascent above 4 km depth beneath the summit of Kīlauea [*Gerlach*, 1986]. At lower pressures of ~30–0.1 MPa, as Kīlauea magma is transported from depths of ~1 km to the surface, nearly all of its initial ~1200–1500 ppm of elemental sulfur (S) is degassed as SO_2, and the majority of that SO_2 is generated within the upper few hundred meters [*Gerlach*, 1986; *Witham et al.*, 2012].

As established herein, the temperature of new magma delivered into the ERZ throughout the 1994–2013 interval (Figure 8.2) is defined by olivine-controlled MgO systematics of bulk lava and host glasses (Figure 8.9b). A strong correlation between running averages of olivine Fo content and inclusion sulfur (Figure 8.12d) attests to SO_2 degassing associated with cooling before and during transport along the ERZ during steady state eruption. The combination of olivine-liquid-sulfur relations in ERZ lava thus defines limits of temperature and depth regimes of olivine crystallization prior to eruption. Throughout the 1994–2001 periods of continuous effusion of olivine-spinel-phyric lava (episodes 53 and 55), a temperature/pressure array of ~1180–1160°C and 30–0.1 MPa (<1 km depth) is defined by Fo_{84}-$Fo_{81.5}$ olivine in equilibrium with bulk melts of 7.5–9 wt% MgO and melt inclusions with sulfur concentrations ranging up to 1300 ppm.

Synchronously declining running averages of ERZ lava MgO systematics and magmatic sulfur coupled with steady summit deflation from 1998 to 2001 (Figures 8.9b, c) suggest that ERZ magma was continuously drawn from a shallow summit reservoir that was progressively shrinking, cooling, and degassing SO_2 over this time. From 2001 to 2003, amid the onset of persistent multiphyric lava petrology, an upward flexure of averaged MgO systematics and magmatic sulfur (Figures 8.1b and 8.12d) was coincident with summit dilation attributed to magmatic backup along the ERZ conduit (see Section 8.4.1.). Petrology of ERZ lava described herein suggests that by 2001 a physical transformation was taking place within the shallow magmatic plumbing system that affected decreased efficiency of ERZ magma transport from the summit region.

From 2001 until October 2013, the array of Pu'u 'Ō'ō phenocryst and inclusion data attest to a dynamic environment of comagmatic mixing in an open-system reservoir of large enough dimension to effectively buffer the normal throughput of olivine-saturated, summit-derived magma at relatively low MgO, near-multitectic conditions. As a result, the post-2001 MgO variation of Pu'u 'Ō'ō bulk lava and glass (Figure 8.2) is a muted reflection of changing conditions in higher-temperature magma fed from the summit into the ERZ after being subjected to magmatic buffering during transit.

The systematic MgO overprint of summit-derived magma is inherently mimicked by running averages of ERZ olivine Fo content (Figure 8.9b) and shadowed by averaged sulfur levels trapped in olivine (Figure 8.12d). Running averages of temperature-sensitive parameters for Cr-spinel (Figure 8.12a) coincide with those of olivine. Also, the average Mg# variation of augite, formed in stasis during magma transit from summit to rift (Figure 8.12b), is likewise imparted by the changing condition of magma pulsing into the ERZ conduit from the summit. These running averages, however, mask observed mixtures of higher and lower Fo olivine phenocrysts that were grown in magma at disparate temperatures and with varying sulfur contents and then mixed together during summit-to-rift transit. Despite this bulk-magma MgO overprint, a record of kinetic disarray is frozen into the intra- and intersample compositional diversity of Pu'u 'Ō'ō olivine, augite, and plagioclase phenocrysts in most post-2001 ERZ lava. The textural and compositional character of simultaneously growing and dissolving phenocrysts and their spectrum of SO_2-degassed melt inclusions, as documented here, are a testament to dynamic interaction between cooler magma that is degassed of SO_2 and hotter, variably degassed, summit-derived magma. The persistent presence of inclusion-bearing higher-Fo olivine in ERZ lava documents a higher-temperature component of preeruptive mixing, but the conditions of shallow mixing are no more evident than in complexly zoned augite phenocrysts with variable Fe^{3+} and abundant melt inclusions with highly variable sulfur (up to 1000 ppm). These inclusion-riddled, emerald green, mm-sized gems provide an unequivocal glimpse of rapid growth amid sulfur degassing in a perpetually turbulent preeruptive magmatic environment.

The apparent enigma of low-temperature multiphyric lava at Pu'u 'Ō'ō during and after the 2003–2007 surge in magma supplied to the summit (see Sections 8.4.1–8.4.3) is, in fact, testimony to the presence of an open-system reservoir uprift of Pu'u 'Ō'ō. Cooler, degassed multiphyric magma residing in the rift reservoir was progressively flushed through the ERZ vents by hotter, sulfur-rich magma both before and after the culmination of the surge in June 2007 with an uprift conduit breach of high-temperature, summit-derived magma (episode 56). Like the proverbial "pig in a python," this glut of hot, fresh magma took a year to convey through the magmatically bloated ERZ pathways to Pu'u 'Ō'ō (Figure 8.9).

A combination of geophysical, geochemical, and petrologic evidence suggests that decades of repeated shallow intrusions along the upper part of the middle ERZ, followed by magmatic overpressurization during the 2003–2007 surge in magma supply, led to the development of a shallow magma reservoir along the ERZ conduit. Like an aneurysm in the well-worn artery to Pu'u 'Ō'ō, this open-system magma reservoir is perpetually recirculated and flushed by pulses of fresh, hot magma that sustain effusion of low-temperature, comagmatic hybrid lava from the vent. It is reasonable to speculate that one or more open-system reservoirs serve to buffer the throughput associated with recharge and eruption and could have developed anywhere along the conduit between the summit and Pu'u 'Ō'ō. Repeated down-rift dike propagation events from beneath Makaopuhi Crater (episodes 1–3, 54, and 59) [*Okamura et al.*, 1988; *Owen et al.*, 2000; *Lundgren et al.*, 2013] point to a plausible location for an open reservoir along the conduit.

8.7. 2008–2013 KĪLAUEA SUMMIT ERUPTION PETROLOGY

Premonitory seismic and fumarolic attributes of the current summit eruption became apparent in November 2007 [*Wilson et al.*, 2008], 25 five years after cessation of the 1982 summit eruptions that signaled the onset of the profuse eruptive era on the ERZ. On 19 March 2008, a newly formed fuming area on the southeast margin of

Halema'uma'u Crater collapsed to form a new eruptive vent, into which a circulating lava lake eventually rose [*Wooten et al.*, 2009; *Houghton et al.*, 2011; *Orr et al.*, 2013] (see also Chapter 20). With the exception of occasional explosive eruption of tephra instigated by rocks falls from the walls and rim of the vent [*Wooten et al.*, 2009, *Orr et al.*, 2013], eruptive activity from 2008 to 2013 was limited to nearly continuous lava lake convection within the slowly enlarging Halema'uma'u vent cavity.

Sampling lava directly within the vent has not been feasible, so petrologic monitoring is sustained by analyzing tephra routinely collected south of the vent (Figure 8.1). These collections usually yielded small amounts of Pele's hair, Pele's tears, tiny glassy spheres or broken rods, shards of bubble walls, and bits of frothy pumice (≤4 cm) that were wafted up in the outgassing plume. Our routine petrologic monitoring was, therefore, largely limited to electron microbeam analysis of hand-picked grain mounts comprising air-quenched glass particulates with occasional phenocrysts and melt inclusions. These data provide the means for tracking changes in eruption temperature and preeruptive conditions of magmatic crystallization.

The juvenile and predominant component of 2008–2013 summit tephra samples is sparsely olivine-(spinel) phyric with ~2 vol% of 100–300 μm, subhedral-to-euhedral olivine phenocrysts, typically with melt inclusions (Figures 8.18a, b, c, f). Olivine in summit glasses is consistently complemented by less than 0.05 vol% of chromian spinel microphenocrysts (Figures 8.18a, c, f), similar in appearance to spinel in ERZ lava described above. Finely contrasted scanning electron microscopy (SEM) images of summit glass particulates commonly reveal latest-stage magma mingling of hotter and cooler melts as micrometer-scale fluidal swirls around phenocrysts and spinning in tiny glass spherules (Figures 8.18b, e). The process responsible for these textures is easy to imagine when viewing webcam images of summit pond activity showing hot sputtering extrusions through a circulating cooler crust (http://hvo.wr.usgs.gov/multimedia/uploads/multimediaFile-430.mov). For this reason, the highest glass temperatures preserved in each summit tephra sample provide the clearest indication of temporal variations in eruption temperature of summit lava.

Aside from brief intervals of increased lava-pond activity that generated copious deposits of Pele's hair, the only substantive glassy pyroclastic samples in the 2008–2013 summit collection resulted from energetic ejections from the lava lake produced by rock falls from the vent walls and rim [*Orr et al.*, 2013]. Throughout this sampling interval, the diameter of the vent opening grew from an initial 35 m in March 2008 to 150 m by March 2011 [*Orr et al.*, 2013] and 160 m as of March 2013. Large glassy pyroclasts ejected during rock-fall-triggered events include vesicle-rich spatter and angular chunks of fresh or oxidized scoria- and glass-coated bombs and lapilli [*Wooten et al.*, 2009]. Based upon continuous observational and seismic records, it is apparent that, other than reticulite, any pyroclasts greater than 1–2 cm in the 2008–2013 Hawaiian Volcano Observatory (HVO) summit tephra sample archive are likely to have been ejected during a rock-fall-triggered event.

By virtue of their inherent association with rock falls into the vent, pyroclasts large enough for routine bulk analysis are often "salted" with anhydrite, inundated by fine lithic fragments, or cored by small chunks of holocrystalline basalt [*Wooten et al.*, 2009]. Bulk analysis of the glassy juvenile component in larger samples often entailed meticulous removal of both lithic and sulfate contaminants [*Thornber et al.*, 2014d]. Samples of several 2–5 cm size glassy lapilli erupted in March and April 2008 were cored by hyalocrystalline clots, as documented by *Wooten et al.* [2009]. These "crystal mush clots" all exhibit evidence of internal melting with remnants of resorbed and reacted low-temperature assemblages, including Fe-rich orthopyroxene and iron-titanium oxides [*Thornber et al.*, 2014b] in a heterogeneous array of low-temperature glasses. The clots are surrounded by an aureole of lower-temperature glasses intermingled with surrounding high-temperature, olivine-phyric host lava. An example of this clot texture is shown by *Rowe et al.* (Chapter 20), who present glass major- and trace-element compositions within and around mush clots.

8.7.1. Summit Tephra Bulk Geochemistry

Bulk sample major- and trace-element analyses were conducted on a wide range of tephra lithologies explosively erupted from the Halema'uma'u vent during March and April of 2008. These samples are described by *Wooten et al.* [2009] and include reddish-brown ash or rock-dust mixed with anhydrite; pumiceous scoria; glassy spatter variably inundated with rock-dust and anhydrite; and 2–5 cm glassy lapilli cored with dense melt-bearing fragments. Unsurprisingly, the trace-element data for these bulk tephra samples reveal a compositional spread that delineates a physical mixture of lithic and juvenile materials. The major- and trace- element array ranges from that which is equivalent to concurrent ERZ lava to that of older rocks composing the crater floor in the vent area [*Neal and Lockwood*, 2003], such as the 1919 lava (repeatedly analyzed as USGS basalt standard BHVO-1; Figures 8.19a–d).

Kīlauea summit lava throughout the last century is characteristically enriched in olivine-incompatible elements relative to nearly all Kīlauea ERZ lava erupted since 1983 [*Pietruszka and Garcia*, 1999a]. Bulk analyses of the ash erupted on 19 March 2008, yielded high SO_3 and CaO/MgO concentrations, indicative of significant

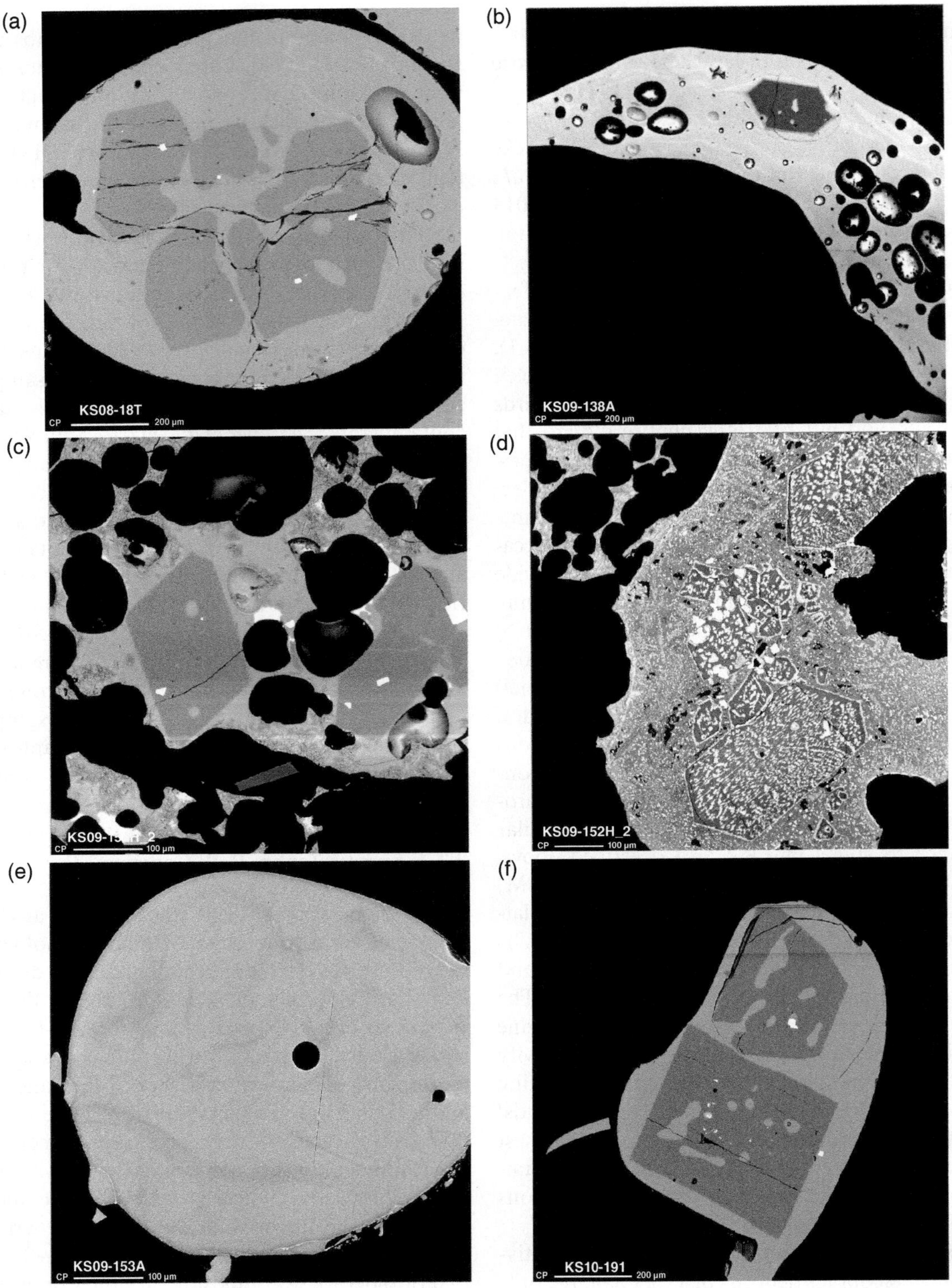

Figure 8.18 Backscattered electron images of juvenile (and recycled) 2008–2010 summit tephra. (a) KS08-18T (200 μm bar scale); $Fo_{82.6}$ olivine phenocrysts in a Pele's tear ejected with ash and pyroclastics on 27 March 2008 are close to equilibrium values of $Fo_{82.1}$ calculated for the host glass. Microphenocrysts of Cr-spinel, both free floating and included by olivine, appear as white flecks. The host glass has an MgO temperature of 1156°C with swirls of 1143°C glass around the phenocryst cluster. Sulfur concentrations of olivine melt inclusions and the glass interstitial to the phenocryst cluster are 700–800 ppm. The matrix glass has 200 ppm S and the lighter shaded, lower-temperature swirls have 100 ppm S. (b) KS09-138A (200 μm bar scale); Pele's tear deposited in a bucket within 24 h of collection on 8 March 2009. The glass is a mixture of two liquids with a ~10°C temperature difference. The higher-temperature glass at ~1160°C appears darker gray, and the more iron-rich glass at ~1150°C is lighter and contains microlites of plagioclase and pyroxene. Both liquids are degassed with less than 150 ppm sulfur.

anhydrite in basalt rock-dust with an enriched geochemical signature similar to 1919 summit lava. Physical mixtures of truly juvenile magma and lithic components are apparent in the bulk geochemistry of March and April 2008 pumice and scoria samples that were not hand picked for "clean glass" analysis. These samples have bulk compositions intermediate between incompatible-element-enriched 1919 lava and incompatible-element-depleted glassy tephra (Figures 8.19a,b, c). Three crystal mush clots, found as cores to March 2008 lapilli, were extracted for bulk analysis and also have relatively enriched incompatible-element signatures, consistent with a melted crater-crust derivation, as inferred from glass chemistry (Chapter 20), Pb isotope geochemistry [*Pietruszka et al.*, 2008], and textures indicating internal melting and crystal dissolution, thus supporting the conclusion that not all molten material ejected during Spring 2008 explosions was of a truly "juvenile" nature.

Since early April 2008, carefully prepared bulk aliquots of glassy summit pumice and spatter deposits and Pele's hair have major-element and trace-element compositions reflecting the same magmatic parentage as recent ERZ lava (Figures 8.19a–d). This observation, along with comparable data for melt inclusions (Chapter 20), provide geochemical confirmation of magmatic continuity between summit and ERZ eruption sites.

8.7.2. Summit Eruption Temperatures

The hottest lava vented at the summit, recorded by glass thermometry as 1169°C, was mingled with lithics during explosions in March 2008 (Figure 8.20a). This glass composition is directly comparable to 1168°C lava vented nearly a year earlier at Kanenuiohamo (episode 59) near the culmination of the 2003–2007 magma supply surge (Figure 8.2). The lowest-temperature summit glasses, also prevalent during late 2008, are found in the interstices of crystal-mush clots or in clasts of remelted scoria derived from recycled spatter and veneer on the walls in the vent cavity. Less-frequent bulk analyses of nearly 100% glassy samples Pele's hair and frothy pumice are closely matched by their matrix glasses, while some spatter samples have higher MgO due to variable crystal accumulation (Figure 8.20a).

Temperature steadily dropped by ~15°C during the first three years of the summit eruption, from near 1170°C down to ~1155°C in March 2011. Between that time and early 2012, eruption temperature rose to 1160–1165°C, where it remained through March 2013. This temporal variation of MgO in summit tephra mimics that of MgO systematics for coeval ERZ lava, overlapping MgO variation of contemporaneous bulk ERZ lava but erupting 20–25°C hotter than at Pu'u 'Ō'ō (Figure 8.20b).

8.7.3. Olivine-Spinel Relations in Summit Tephra

Changes in MgO of matrix glasses with time are consistent with changing forsterite content of olivine phenocrysts, the cores of which are usually constrained to compositions that are nearly in equilibrium with progressively cooler surrounding melts (Figure 8.21). The 2008–2009 summit olivine population is dominated by near-equilibrium $Fo_{82.5}$-$Fo_{81.5}$ compositions with progressively fewer $Fo_{83-83.5}$ phenocrysts observed after eruption onset. Two distinctly Mg-rich olivine phenocrysts (~Fo_{85}) were observed in explosively erupted glassy pyroclasts. The first is in an Mg-rich crystal-mush clot ejected on 23 March 2008. The $Fo_{85.4}$ olivine is resorbed, zoned, fragmented, and situated adjacent to low-temperature orthopyroxene (KS08-07S2C, Table 8.6s). The other is a free-floating, ragged-edged, and complexly cored crystal in a centimeter-sized lapillus ejected during a 2 September 2008 explosion.

By 2010, Mg-rich olivine phenocrysts were rarely observed as eruption temperatures declined. Typical olivine compositions fell to Fo_{81} by March 2011 before rising again to Fo_{82} in early 2012, in concert with the MgO

Figure 8.18 (Cont'd) The $Fo_{83.0}$ euhedral olivine has a thin olivine overgrowth, contains glass inclusions with up to 900 ppm sulfur, and was crystallized from 1184°C magma prior to sulfur degassing. (c) KS09-152H_2 (100 μm bar scale); pumiceous scoria among 17 September 2009 explosive ejecta. The olivine on the left ($Fo_{82.5}$) has glass inclusions bearing 608 ppm sulfur (top) and 562 ppm (bottom). The spinel-bearing olivine on the right is $Fo_{82.3}$. An equilibrium value of $Fo_{82.0}$ is calculated for the Mg-richest analysis of host glass (not proximal to quench-related micro-crystals of pyroxene, plagioclase, and olivine) erupted at 1163°C. (d) KS09-152H_2 (100 μm bar scale); same sample as (c), a recycled scoria or wall-veneer fragment incorporated by pumice. Annealing at high temperatures under oxidizing conditions is illustrated by olivine phenocrysts that are both exsolving and armored by magnetite (white), which is also peppered throughout the rapidly crystallized aphanitic matrix of the fragment. (e) KS09-153A (100 μm bar scale); an irregular shaped (~0.5 mm) droplet among hair and tephra formed sometime during 20–23 December 2009. Two glasses are apparent in the backscatter image, and the range in thermometry values obtained for this grain is 1157–1162°C. (f) KS10-191 (200 μm bar scale); among sparse tephra collected in early August 2010, this tear was erupted at 1156°C and contains olivine phenocrysts with $Fo_{81.3}$ and $Fo_{81.6}$ compositions, near to equilibrium values calculated for glass of $Fo_{81.4}$. Sulfur concentration in melt inclusions ranges from 280 to 430 ppm.

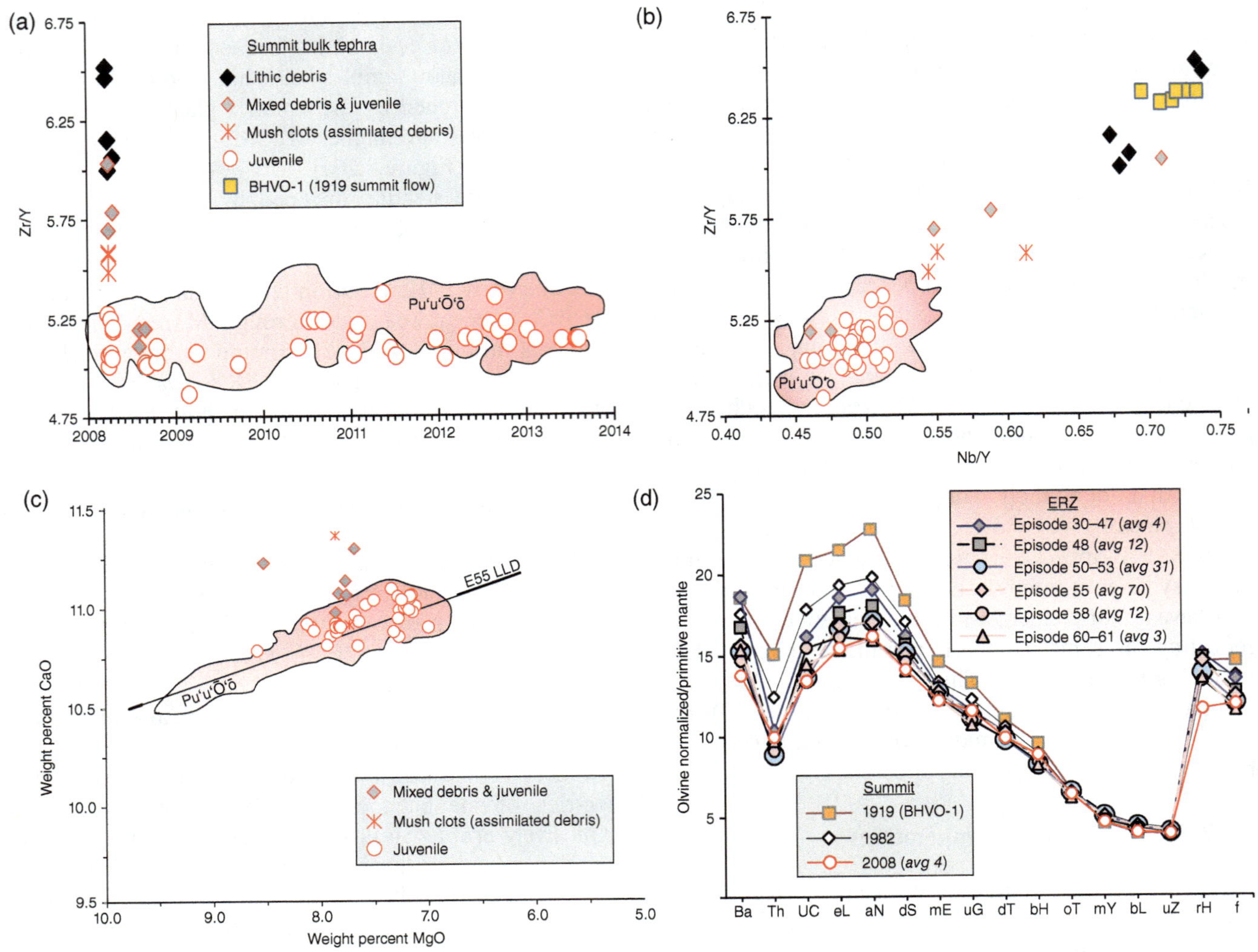

Figure 8.19 Bulk WDXRF analysis of 2008–August 2013 juvenile summit lava (open red circles) compared with explosive ash (rock-dust) and lithic debris (black filled diamonds), glassy pyroclasts containing lithic contaminants (gray filled diamonds), "crystal mush clots" (red asterisks), and nearby 1919 summit lava (orange squares; BHVO-1 standard, analyzed with these samples). (a) Temporal variation of Zr/Y in all bulk tephra samples analyzed since onset of the eruption. (b) Zr/Y versus Nb/Y for all bulk tephra samples. (c) CaO versus MgO for all glassy samples. The shaded area outlined in (a), (b), and (c) is that of all 1998–2013 steady state ERZ lava vented at or near Pu'u 'Ō'ō. The truly juvenile summit tephra compositions lie along the same incompatible-element-depleted liquid line of descent as the 1998–2013 ERZ lava (E55-LLD, as shown in Figure 8.3). The range of anomalous glassy pyroclasts are explained as a mixture of crater-floor materials, such as BHVO-1, with new summit magma. (d) Trace-element spidergram comparing the range of highly to moderately incompatible elements in juvenile 2008 summit tephra to that of 1919 (BHVO-1) and 1982 summit lava (white diamonds) and successive episodes of near–steady state ERZ eruption since 1983 (see Figure 8.2). All compositions are normalized to 7 wt% MgO by subtraction of olivine, and all data are plotted relative to primitive mantle compositions of *Sun and McDonough* [1989]. Summit data for September 1982 are from *Pietruszka and Garcia* [1999a]; sample number 1982-s14).

content of the host glasses. In a mid-June 2013 Pele's tear sample (KS13-263), $Fo_{84.2}$ olivine occurs in the small rounded core of a small euhedral phenocryst, outwardly zoned to an equilibrium composition of $Fo_{81.6}$. Overall, the temporal decrease in Fo content of summit olivine phenocrysts is synchronous with, but less dramatic than, that observed in Pu'u 'Ō'ō lava over the same interval (Figure 8.20).

Of the 27 chromian-spinel microphenocrysts in 2008–2013 summit tephra, 25 are close to equilibrium with coexisting olivine and liquid, as defined by "equipotential" lines from *Roeder et al.* [2003] that delineate spinel in equilibrium with Fo_{83} and Fo_{81} olivine over a 20°C temperature interval (Figure 8.22). Including two analyzed microphenocrysts with higher $Cr^{3+}/(Cr^{3+}+Al^{3+})$ in 2009 and 2013 tephra, all summit spinel compositions are within the range

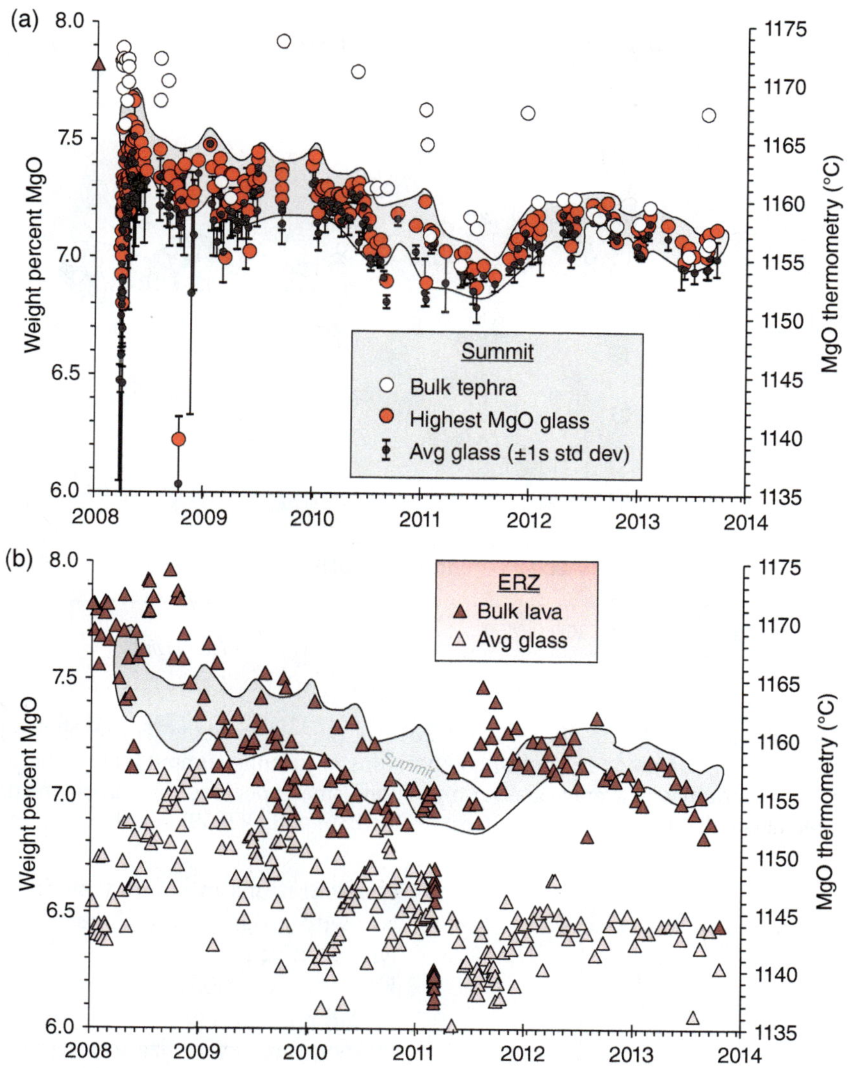

Figure 8.20 Temporal variation during 2008–October 2013 of MgO weight percent (left axis) and corresponding MgO thermometer values (right axis) for (a) Summit bulk-tephra samples (open red circles) and matrix glasses in all summit tephra samples, showing the highest (red filled circles), average (small dark circles), and ±1σ standard deviation for individual samples, and (b) ERZ bulk lava samples (dark-colored triangles) and average matrix glasses (light-colored triangles). The shaded area, outlining the temporal variation of highest-temperature summit glasses (a) is broadly comparable to the temporal variation ERZ bulk lava samples (b).

established for spinel in 1998–2013 ERZ lava (Figure 8.11). Compared to summit spinel, contemporaneous 2008–2013 Puʻu ʻŌʻō spinel compositions are richer in Fe and equilibrated at lower temperatures. This observation is consistent with the differences observed between ERZ and summit olivine-phenocryst compositions.

8.7.4. Tracking Magmatic Sulfur in Summit Melt Inclusions

The variation of sulfur in summit glasses provides important perspective on shallow preeruptive magmatic conditions. For example, the sulfur heterogeneity preserved in a single 1 mm Pele's tear generated by a rock fall on 27 March 2008 (KS08-18T, Figure 8.18a) is evidence for the depth of disruption caused by sudden and significant rock falls into the vent. That lava droplet is cored by an open-textured cluster of near-equilibrium olivine ($Fo_{82.6}$) containing 700–800 ppm sulfur and floating in a swirling mixture of melts with 100–200 ppm sulfur. Based on sulfur solubility relations as discussed above, this tiny Pele's tear records SO_2 degassing associated with sudden depressurization of magma at >100 m depth.

In juvenile summit tephra samples, the number of olivine melt inclusions typically exceeds the number of olivine phenocrysts. Routine microprobe analyses of these inclusions provide a means of tracking changes in magmatic sulfur concentrations with time. Maximum sulfur

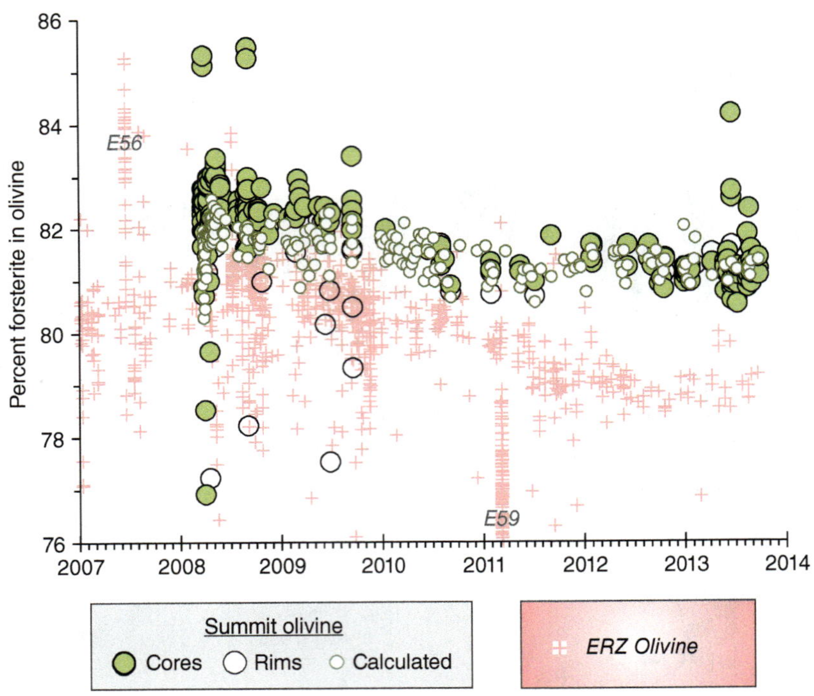

Figure 8.21 Temporal variation during 2008–August 2013 of percent forsterite (Fo) in olivine-phenocryst core (filled green circles) and rim (large open circles) compositions in juvenile summit tephra compared with values calculated to be in equilibrium with average glass compositions (small open circles). ERZ olivine data for the 2007–October 2013 interval (small crosses), including episodes 56 (E56) and 59 (E59), are shown for comparison.

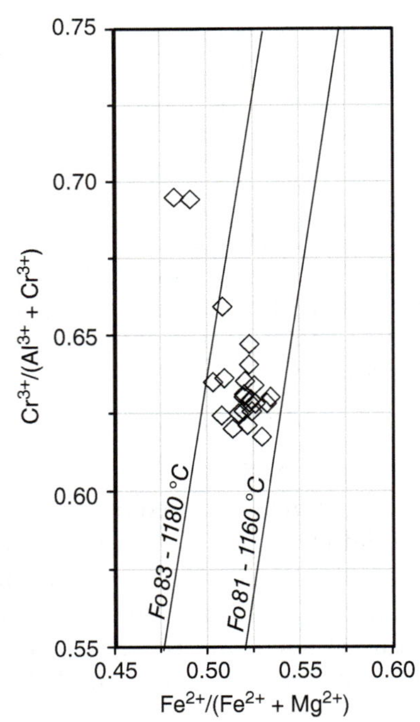

Figure 8.22 Spinel compositions in 27 samples of summit tephra from 2008–August 2013. $Cr^{3+}/(Cr^{3+}+Al^{3+})$ versus $Fe^{2+}/(Fe^{2+}+Mg^{2+})$ cations are calculated on the basis of four oxygens with Fe^{3+} and Fe^{2+} assigned according to stoichiometry and charge balance. "Equipotential" lines from *Roeder et al.* [2003] delineate spinel in equilibrium with Fo_{83} and Fo_{81} olivine at 1180°C and 1160°C.

concentrations in summit olivine melt inclusions correlate with progressively higher Fo content of host olivine, with relative maxima ranging from an initial 1100 ppm in Fo_{83} down to ~300 ppm in Fo_{81} (Figure 8.23). As deduced for ERZ melt inclusions, this relation suggests a pressure-temperature crystallization array from ~1180 to 1160°C (from MgO thermometry) at pressures less than ~30 MPa. *Rowe et al.* (Chapter 20) report sulfur concentrations in olivine melt inclusions from March-April and August 2008 pyroclasts ranging up to 1670 ppm, with a significant population greater than 1300 ppm. That study also compares melt inclusion data for a July 2008 olivine-phyric Puʻu ʻŌʻō spatter sample (KE58-2801S) with a significant population at or above 1300 ppm sulfur. These data are consistent with results presented here for inclusions in summit and rift samples over the 2008–2013 eruption interval.

In sync with the highest glass temperatures at the onset of eruption in 2008, the highest sulfur concentrations in olivine melt inclusions are also recorded (Figures 8.24a,b). As summit eruption temperature steadily declined from high values in March 2008 until March 2011, the maximum concentrations of magmatic sulfur preserved in progressively less forsteritic olivine phenocrysts dropped below 300 ppm. This pattern is mirrored by the decline in maximum sulfur contents of olivine melt inclusions in Puʻu ʻŌʻō lava during the same interval (Figure 8.24c). The subsequent increase in summit eruption temperature in 2012 is matched by an apparent steady increase to

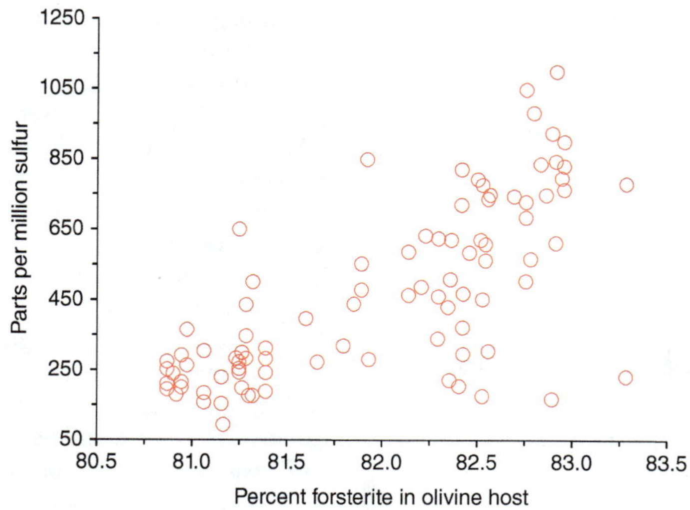

Figure 8.23 Sulfur concentrations in olivine melt inclusions compared with host olivine compositions in 2008–August 2013 juvenile summit tephra.

650 ppm of sulfur in melt inclusions. This is complemented by the appearance of high sulfur values (650–850) ppm at Pu'u 'Ō'ō in early 2012.

8.7.5. Magma Conditions Beneath the Halema'um'u Vent

Petrology of summit tephra throughout the 2008–2013 eruptive interval records changes in shallow magmatic conditions, probably within the upper few hundred meters of the Halema'uma'u vent conduit, where rising magma continually disgorges its sulfur and sinking lava is convectively assimilated into the underlying plumbing that connects to a deeper summit reservoir. In the first 5+ years of the summit eruption, the temperature and sulfur of magma reaching the surface at the summit appears to correlate with summit deformation (Figures 8.24a, b). This relation demonstrates the connection between magmatic pressurization of the shallow summit and the thermodynamic balance that is achieved between upwelling and draining magma within the crown of an artesian-like magma fountain in the summit vent. As with the steady state ERZ eruption, the temperature of summit lava appears to cycle in a manner regulated by conduit geometry and flux of magma into and out of the summit reservoir system.

8.8. KĪLAUEA SUMMIT-ERZ MAGMATIC CONTINUITY: IMPLICATIONS AND CONSEQUENCES

Routine petrologic monitoring of Kīlauea's ongoing ERZ and summit eruptions provides important new perspectives relevant to understanding shallow magmatic plumbing associated with prolific shield-building volcanism in Hawai'i. Throughout 30+ years of eruption, physical connectivity between a summit magma reservoir complex and ERZ vents has been inferred by correlations of deformation at the summit with ERZ eruption temperatures and activity levels [*Thornber*, 2003]. The geochemistry and petrology of lava erupted concurrently at the Halem'uma'u and Pu'u 'Ō'ō vents of Kīlauea between March 2008 and August 2013 provide an unequivocal testament to the magmatic connection between them via a shallow summit reservoir and a 20 km long rift conduit. As established herein and by Rowe et al. in Chapter 20 magma erupted at the summit has the same parental trace-element signature as contemporaneous ERZ lava that is characteristically "depleted" relative to the last century of Kīlauean magma [*Pietruszka and Garcia*, 1999a]. Furthermore, correlations of summit deformation with olivine-controlled MgO systematics and magmatic sulfur preserved in olivine melt inclusions in coeval summit tephra and ERZ lava are a clear indication that summit magma chamber pressurization during recharge is translated throughout interconnected magma pathways between Halema'uma'u and Pu'u 'Ō'ō.

Petrologic scrutiny of the ongoing burgeoning eruptive condition of Kīlauea Volcano, as constrained by geophysical and geologic observations at both ends of the eruptive plumbing system, has demonstrated that changes in eruption temperature and phenocryst content of lava vented at Pu'u 'Ō'ō are associated with physical maturation of magma pathways along the shallow ERZ conduit after three decades of magma recharge, repeated intrusions, and systematic overpressurization during the 2003–2007 surge in magma supply. The petrologic differences between summit tephra and ERZ lava are attributed, at least in part, to shallow open-system magma mixing in

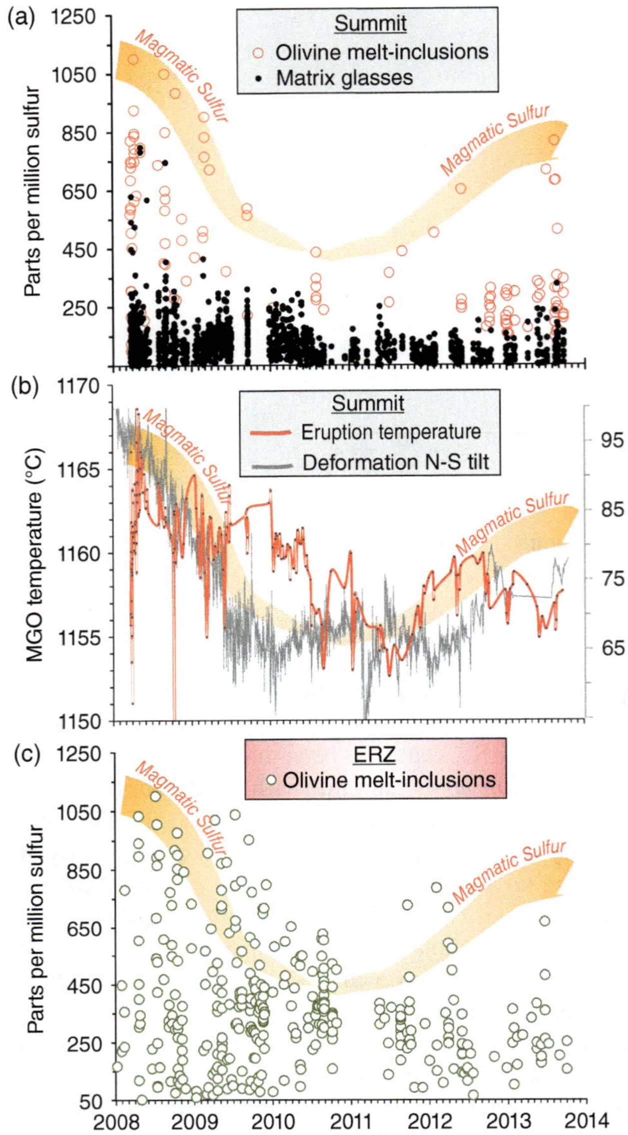

Figure 8.24 Temporal variation from 2008–August 2013 of: (a) sulfur concentration in olivine melt-inclusions (open red circles) and matrix glasses (black dots); (b) maximum MgO thermometry values of individual summit tephra samples (red line, left axis) with summit water tube tilt measurements (gray line, right axis), and (c) sulfur concentration in ERZ olivine melt inclusions. The lightly shaded swath, approximating temporal changes in the highest levels of magmatic sulfur observed in summit olivine melt inclusions (a), is broadly comparable to the temporal variations of inferred eruption temperature and summit deflation/inflation (b) and to coeval variations of the highest levels of magmatic sulfur observed in ERZ olivine melt inclusions (c).

one or more magma reservoirs that have developed along the ERZ and serve to buffer the magmatic throughput associated with ongoing recharge and eruption.

The demonstration of summit-to-rift magmatic continuity has significant implications pertaining to the overall magmatic condition of the Kīlauea edifice, as an example, raising questions about hypotheses that include the continued persistence of decades- or centuries-old magma in the summit reservoir [*Pietruszka and Garcia*, 1999b]. We believe that the balance of the evidence favors the theory posed by *Thornber* [2003]: that since 1982 and throughout the prolonged ERZ eruption, new magma derived from a uniform mantle source has completely flushed long-term resident magma from the shallow summit reservoir that is tapped by ongoing ERZ eruption. Furthermore, this petrologic testimony to continuous shallow magma mixing associated with prolonged recharge and eruption of Kīlauea Volcano should encourage advocates for mantle-sourced short-term changes in ERZ lava chemistry to consider the caveat of *Hoffman et al.* [1984] concerning such an interpretation for the 1969–1971 Mauna Ulu eruption: "[I]t is always possible to construct ad hoc models involving mixing of sources and or melts… that account for the same data. … For example, the monotonic decrease of incompatible–element concentrations observed during the course of the Mauna Ulu eruption might be explained by mixing of two end-member magmas." Likewise, interpreters of short-term Kīlauea trace-element and isotopic variations should consider evidence presented here in light of insight from *O'Hara* [1977]: "A realistic model of volcanic plumbing predicts that most of the established major, trace and isotopic chemical features of the common basalts could have been imposed during magma evolution in high-level magma chambers, for which process there is extensive independent field and phase equilibria evidence. These effects must be quantified before chemical features of lavas are ascribed to the nature of the underlying upper mantle."

SUPPLEMENTARY DIGITAL DATA

The following data items are available online at www.wiley.com/go/Carey/Hawaiian_Volcanoes

Table 8.1s. Bulk-lava major and trace element compositions of 2001–October 2013 Kīlauea ERZ lava.

Table 8.2s. Averaged microbeam analyses of matrix glasses of 2001–October 2013 Kīlauea ERZ lava.

Table 8.3s. Selected microbeam analyses of phenocrysts and melt-inclusions 1999–October 2013 Kīlauea ERZ lava.

Table 8.4s. Selected major and trace element compositions of 2008–2013 Kīlauea summit bulk-tephra.

Table 8.5s. Microbeam analyses of highest temperature matrix glasses in 2008–2013 Kīlauea summit tephra.

Table 8.6s. Representative microbeam analyses of phenocrysts and melt-inclusions 2008–2013 Kīlauea summit tephra.

ACKNOWLEDGMENTS

We gratefully acknowledge significant contributions to the post-2001 Kīlauea eruption sample archive made through intrepid collection efforts of Dave Sherrod, Don Swanson, Matt Patrick, Kelly Wooten, and Frank Trusdell, all at the U.S. Geological Survey (USGS) Hawaiian Volcano Observatory (HVO). Also, many vent samples would not have been collected without amazing helicopter support by David Okita (Volcano Helicopters). Glass, phenocryst, and melt inclusion data in post-2001 ERZ lava and summit tephra were achieved through careful grain-picking and polished grain-mount preparation at the USGS Cascades Volcano Observatory with the help of James Rea, Fiona Cooperthwaite, Evan Soderberg, and Mauri Fabio. More recently, sample triage and preparation for analysis have been done by Cheryl Ganzecki and Derek Gonzolas at the University of Hawai'i, Hilo. During 2001–2013, routine Kīlauea microprobe work was accomplished at USGS Denver and in collaboration with Heather Lowers, Greg Meeker, David Adams, and Celeste Mercer. Responsive and reliable analytical services directed by Rick Conrey and Laureen Wagoner at the Washington State University, Peter Hooper Geoanalytical Lab in Pullman, Washington, have enabled efficient geochemical monitoring of Kīlauea eruption products since 2006. James Budahn (USGS, Denver) continued to provide high-precision INAA trace-element analysis for 2001–2012 samples. Mike Poland and Asta Miklius (USGS, HVO) kindly provided the deformation data presented herein. We are very appreciative of the thoughtful and painstaking critique of this effort by Michelle Coombs (USGS), Heather Wright (USGS), and an anonymous reviewer, all of whom contributed to a significantly improved product. The senior author of this chapter is particularly grateful to associate editor Mike Poland for his helpful suggestions and patient encouragement along the road to its completion.

REFERENCES

Burgisser, A., and B. Scalliet (2007), Redox evolution of a degassing magma rising to the surface, *Nature*, *445*(1), 194–197, doi:10.1038/nature05509.

Cameron, M., and J. J. Papike (1980), Crystal chemistry of silicate pyroxenes, *Rev. Mineral. Geochem.*, *7*(1), 5–92.

Cervelli P., P. Segall, F. Amelung, H. Garbeil, C. Meertens, S. Owen, A. Miklius, and M. Lisowski (2002), The 12 September 1999 upper east rift intrusion at Kilauea Volcano Hawaii, *J. Geophys. Res.*, *107*(2150), doi:10.1029/2001JB000602

Cervelli, P. F., and A. Miklius (2003), The shallow magmatic system of Kilauea Volcano, in *The Pu'u 'Ō'ō-Kūpaianaha eruption of Kīlauea Volcano, Hawa'ii: The first 20 years*, Prof. Pap. 1676, edited by C. Heliker et al., pp. 149–163, U.S. Geol. Surv., Reston, Virginia.

Clague, D. A., and R. P. Denlinger (1994), Role of olivine cumulates in destabilizing the flanks of Hawaiian volcanoes, *Bull. Volcanol.*, *56*(6–7), 425–434, doi:10.1007/BF00302824.

Defant, M. J., and R. L. Nielson (1990), Interpretation of open system petrogenetic processes: Phase equilibria constraints on magma evolution, *Geochim. Cosmochim. Acta*, *54*, 87–102.

Dohmen, R., and S. Chakraborty (2007), Fe–Mg diffusion in olivine II: Point defect chemistry, change of diffusion mechanisms and a model for calculation of diffusion coefficients in natural olivine. *Phys. Chem. Min.*, *34*(6), 409–430, doi:10.1007/s00269-007-0157-7.

Droop, G. (1987), General equation for estimating Fe3+ concentrations in ferromagnesian silicates and oxides from microprobe analyses, using stoichiometric criteria, *Mineral. Mag.*, *51*, 431–435.

Evans, B. W., and J. G. Moore (1968), Mineralogy as a function of depth in the prehistoric Makaopuhi tholeiitic lava lake, Hawaii, *Contrib. Mineral. Petrol.*, *17*(2), 85–115, doi:10.1007/BF00373204.

Garcia, M. O., R. A. Ho, J. M. Rhodes, and E. W. Wolfe (1989), Petrologic constraints on rift-zone processes: Results from episode 1 of the Pu'u 'O'o eruption of Kilauea Volcano, Hawaii. *Bull. Volcanol.*, *52*, 81–96.

Garcia, M. O., J. M. Rhodes, E. W. Wolfe, G. E. Ulrich, and R. A. Ho (1992), Petrology of lavas from episodes 2–47 of the Puu Oo eruption of Kilauea Volcano, Hawaii: Evaluation of magmatic processes, *Bull. Volcanol.*, *55*(1–2), 1–16, doi:10.1007/BF00301115.

Garcia, M. O., J. M. Rhodes, F. A. Trusdell, and A. J. Pietruszka (1996), Petrology of lavas from the Puu Oo eruption of Kilauea Volcano: III. The Kupaianaha episode (1986–1992), *Bull. Volcanol.*, *58*(5), 359–379, doi:10.1007/s004450050145.

Garcia, M. O., A. J. Pietruszka, J. M. Rhodes, and K. Swanson (2000), Magmatic processes during the prolonged Pu'u 'O'o eruption of Kilauea Volcano, Hawaii, *J. Petrol.*, *41*(7), 967–990, doi:10.1093/petrology/41.7.967.

Garcia, M. O., A. J. Pietruszka, and J. M. Rhodes (2003), A petrologic perspective of Kīlauea volcano's summit magma reservoir, *J. Petrol.*, *44*(12), 2313–2339. doi:10.1093/petrology/egg079.

Gerlach, T. M. (1986), Exsolution of H2O, CO2, and S during eruptive episodes at Kilauea Volcano, Hawaii, *J. Geophys. Res.*, *91*(B12), 12,177–12,185, doi:10.1029/JB091iB12p12177.

Gerlach, T. M. (1993), Oxygen buffering of Kilauea volcanic gases and the oxygen fugacity of Kilauea basalt, *Geochim. Cosmochim. Acta*, *57*(4), 795–814, doi:10.1016/0016-7037(93)90169-W.

Gerlach, T. M. (2004), Comment on paper: "Morphology and compositions of spinel in Pu'u 'Ō'ō lava (1996–1998), Kilauea Volcano, Hawaii"—Enigmatic discrepancies between lava and gas-based fO_2 determinations of Pu'u 'Ō'ō lava, *J. Volcanol. Geotherm. Res.*, *134*(3), 241–244, doi:10.1016/j.jvolgeores.2004.02.004.

Gerlach, T. M., K. A. McGee, T. Elias, A. J. Sutton, and M. P. Doukas (2002), Carbon dioxide emission rate of Kilauea Volcano: Implications for primary magma and the summit reservoir, *J. Geophys. Res.*, *107*(2189), doi:2110.1029/2001JB000407.

Girard, G., M. K. Reagan, K. W. W. Sims, M. O. Garcia, A. J. Pietruszka, and C. R. Thornber (2012), From mantle to ash cloud: Quantifying magma ascent rates and degassing processes at Kilauea using short-lived U-series radionuclide disequilibria [abstract], Fall Meet. of the AGU, San Francisco, Dcember 3–7, Abstract V13D-288.

Greene, A. R., M. O. Garcia, A. J. Pietruszka, D. Weis, J. P. Marske, M. J. Vollinger, and J. Eiler (2014), Temporal geochemical variations in lavas from Kīlauea's Pu'u 'Ō'ō eruption (1983–2010): Cyclic variations from melting of source heterogeneities, *Geochem., Geophys., Geosyst.*, doi:10.1002/2013GC004719, in press.

Grove, T. L., and A. E. Bence (1977), Experimental study of pyroxene-liquid interaction in quartz-normative basalt 15597, *Proc. Lunar Sci. Conf.*, *8*, 1549–1579.

Heliker, C., and T. Mattox (2003), The first two decades of the Pu'u 'Ō'ō-Kūpaianaha eruption; chronology and selected bibliography, in *The Pu'u 'Ō'ō-Kūpaianaha eruption of Kīlauea Volcano, Hawai'i: The First 20 years*, Prof. Pap. 1676, edited by C. Heliker, D. A. Swanson, and T. J. Takahashi, pp. 1–27, U.S. Geol. Surv., Reston, Virginia.

Helz, R. T. (1980), Crystallization history of Kilauea Iki lava lake as seen in drill core recovered in 1967–1979, *Bull. Volcanol.*, *43*(4), 675–701, doi:10.1007/BF02600365.

Helz, R. T., and C. R. Thornber (1987), Geothermometry of Kilauea Iki lava lake, Hawaii, *Bull. Volcanol.*, *49*(5), 651–668, doi:10.1007/BF01080357.

Hofmann, A. W., M.D. Feigenson, and I. Raczek (1984), Case study on the origin of basalt: III Petrogenesis of the Mauna Ulu eruption, Kilauea, 1969–1971, *Contrib. Mineral. Petrol.*, *88*, 24–25

Houghton, B. F., D. A. Swanson, R. J. Carey, J. Rausch, and A. J. Sutton (2011), Pigeonholing pyroclasts; insights from the 19 March 2008 explosive eruption of Kīlauea Volcano, *Geology*, *39*(3), 263–266, doi:10.1130/G31509.1.

Huebner, J. S. (1980), Pyroxene phase equilibria at low pressure, *Rev. Mineral. Geochem.*, *7*(1), 213–288.

Huebner, J. S., and N. Turnock (1980), The melting relations at 1 bar of pyroxenes composed largely of Ca-, Mg-, and Fe-bearing components, *Am. Mineral.*, *65*, 225–271.

Lindsley, D. H. (1983), Pyroxene thermometry, *Am. Mineral.*, *68*, 477–493.

Lundgren, P., et al. (2013), Evolution of dike opening during the March 2011 Kamoamoa fissure eruption, Kīlauea Volcano, Hawai'i, *J. Geophys. Res.*, *118*(3), 897–914, doi:10.1002/jgrb.50108.

Marske, J. P., M. O. Garcia, A. J. Pietruszka, J. M. Rhodes, and M. D. Norman (2008), Geochemical variations during Kīlauea's Pu'u 'Ō'ō eruption reveal a fine-scale mixture of mantle heterogeneities within the Hawaiian plume, *J. Petrol.*, *49*(7), 1297–1318, doi:10.1093/petrology/egn025.

Martin, V. M., D. J. Morgan, D. A. Jerram, M. J. Caddick, D. J. Prior, and J. P. Davidson (2008), Bang! Month-scale eruption triggering at Santorini volcano, *Science*, *321*(5893), 1178–1178, doi:10.1126/science.1159584.

Matzen, A. K., M. B. Baker, J. R. Beckett, and E. M. Stolper (2011), Fe–Mg partitioning between olivine and high-magnesian melts and the nature of Hawaiian parental liquids, *J. Petrol.*, *52*(7–8), 1243–1263, doi:10.1093/petrology/egq089.

Miklius, A. (2005), Recent inflation of Kilauea Volcano during the ongoing eruption—Harbinger of change? *Eos Trans. AGU*, *86*(52), Fall Meet. Suppl., Abstract G53B-0883.

Montgomery-Brown, E. K., D. K. Sinnett, M. Poland, P. Segall, T. Orr, H. Zebker, and A. Miklius (2010), Geodetic evidence for en echelon dike emplacement and concurrent slow-slip during the June 2007 intrusion and eruption at Kīlauea Volcano, Hawaii, *J. Geophys. Res.*, *115*(B07405), doi:10.1029/2009JB006658.

Neal, C. A., and Lockwood, J. P., 2003, Geologic map of the summit region of Kīlauea Volcano, Hawaii, Invest. Ser. Map I-2759, scale 1:24,000, U.S. Geol. Surv. Geol., Reston, Virginia.

Norman, M., M. O. Garcia, and A. J. Pietruszka (2005), Trace-element distribution coefficients for pyroxenes, plagioclase, and olivine in evolved tholeiites from the 1955 eruption of Kilauea Volcano, Hawai'i, and petrogenesis of differentiated rift-zone lavas, *Am. Mineral.*, *90*(5–6), 888–899, doi:10.2138/am.2005.1780.

O'Hara, M. J. (1977), Geochemical evolution during fractional crystallization of a periodically refilled magma chamber, *Nature*, *266*, 503–507, doi:10.1038/266503a0.

O'Hara, M. J., and R. E. Mathews (1981), Geochemical evolution in an advancing, periodically replenished, periodically tapped and continuously fractionating magma chamber, *J. Geol. Soc. Lond.*, *38*, 237–277

Okamura, A. T., J. J. Dvorak, R. Y. Koyanagi, and W. J. L. Tanagawa (1988), Surface deformation during dike propagation, in *The Puu Oo Eruption of Kilauea Volcano, Hawaii: Episodes 1 Through 20, January 3, 1983, Through June 8, 1984*, edited by E. W. Wolfe, pp. 165–182, Prof. Paper, 1463, Reston, Virginia.

Orr, T., C. Heliker, and M. R. Patrick (2012), The ongoing Pu'u 'Ō'ō eruption of Kīlauea Volcano, Hawai'i—30 years of eruptive activity, Fact Sheet 2012-3127, U.S. Geol. Surv., Reston, Virginia.

Orr, T. R., W. A. Thelen, M. R. Patrick, D. A. Swanson, and D. C. Wilson (2013), Explosive eruptions triggered by rockfalls at Kīlauea Volcano, Hawaii, *Geology*, *41*(2), 207–210, doi:10.1130/G33564.1.

Owen, S., P. Segall, M. Lisowski, A. Miklius, M. Murray, M. Bevis, and J. Foster (2000), January 30, 1997 eruptive event on Kilauea Volcano, Hawaii, as monitored by continuous GPS, *Geophys. Res. Lett.*, *27*(17), 2757–2760, doi:10.1029/1999GL008454.

Pietruszka, A. J., and M. O. Garcia (1999a), A rapid fluctuation in the mantle source and melting history of Kilauea Volcano inferred from the geochemistry of its historical summit lavas (1790–1982), *J. Petrol.*, *40*(8), 1321–1342, doi:10.1093/petroj/40.8.1321.

Pietruszka, A. J., and M.O. Garcia (1999b), The size and shape of Kilauea Volcano's summit magma storage reservoir: A geochemical probe, *Earth Planet. Sci. Lett.*, *167*(3–4), 311–320, doi:10.1016/S0012-821X(99)00036-9.

Pietruszka, A. J., E. H. Hauri, R. W.Carlson, and M. O. Garcia (2006), Remelting of recently depleted mantle within the Hawaiian plume inferred from the 226Ra–230Th–238U disequilibria of Pu'u 'Ō'ō eruption lavas, *Earth Planet. Sci. Lett.*, *244* (1–2), 155–169, doi:10.1016/j.epsl.2006.01.039.

Pietruszka, A. J., M. O. Garcia, J. M. Rhodes, M. D. Norman, I. N. Bindeman, J. P. Marske, and D. H. Burns (2008), Rapid temporal geochemical and isotopic variations of tephra from the ongoing 2008 summit eruption of Kilauea Volcano, Hawaii, *Eos Trans. AGU*, *89*(53), Fall Meet. Suppl., Abstract V11B-2021.

Poland, M. P., A. Miklius, T. Orr, A. J. Sutton, C. R. Thornber, and D. Wilson (2008), New episodes of volcanism at Kilauea Volcano, Hawaii, *Eos Trans. AGU*, *89*(5), 37–38, doi:10.1029/2008EO050001.

Poland, M. P., A. J. Sutton, and T. M. Gerlach (2009), Magma degassing triggered by static decompression at Kīlauea Volcano, Hawai'i, *Geophys. Res. Lett.*, *36*(L16306), doi:10.1029/2009GL039214.

Poland, M. P., A. Miklius, A. J. Sutton, and C. R. Thornber (2012), A mantle-driven surge in magma supply to Kīlauea Volcano during 2003–2007, *Nature Geosci.*, *5*(4), 295–300, doi:10.1038/ngeo1426.

Poustovetov, A. A., and P. L. Roeder (2001), The distribution of Cr between basaltic melt and chromian spinel as an oxygen geobarometer, *Can. Mineral.*, *39*(2), 309–317, doi:10.2113/gscanmin.39.2.309.

Putirka, K. (1997), Magma transport at Hawaii: Inferences based on igneous thermobarometry, *Geology*, *25*(1), 69–72, doi:10.1130/0091-7613(1997)025<0069:MTAHIB>2.3.CO;2.

Putirka, K. D. (2008), Thermometers and barometers for volcanic systems, *Rev. Mineral. Geochem.*, *69*(1), 61–120, doi:10.2138/rmg.2008.69.3.

Rhodes, J. M. and S. R. Hart (1995), Episodic trace element and isotopic variations in historical Mauna Loa lavas: Implications for magma plume dynamics, in *Mauna Loa Revealed*, Geophys. Monogr. 92, edited by J. M. Rhodes and J. P. Lockwood, pp. 241–262, AGU, Washington, D. C.

Roeder, P. L., and R. Emslie (1970), Olivine-liquid equilibrium, *Contrib. Mineral. Petrol.*, *29*(4), 275–289, doi:10.1007/BF00371276.

Roeder, P. L., C. Thornber, A. Poustovetov, and A. Grant (2003), Morphology and composition of spinel in Pu'u 'O'o lava (1996–1998), Kilauea Volcano, Hawaii, *J. Volcanol. Geotherm. Res.*, *123*(3), 245–265, doi:10.1016/S0377-0273(02)00508-5.

Roeder, P. L., C. R. Thornber, and A. Grant (2004), Reply to comment on paper: "Morphology and composition of spinel in Pu'u 'O'o lava (1996–1998), Kilauea Volcano, Hawaii"—Enigmatic discrepancies between lava and gas-based fO_2 determinations of Pu'u 'O'o lava, *J. Volcanol. Geotherm. Res.*, *134*(3), 245–248, doi:10.1016/j.jvolgeores.2004.03.001.

Roeder, P., E. Gofton, and C. Thornber (2006), Cotectic proportions of olivine and spinel in olivine-tholeiitic basalt and evaluation of pre-eruptive processes, *J. Petrol.*, *47*(5), 883–900, doi:10.1093/petrology/egi099.

Ross, M., and J. S. Huebner (1979), Temperature-composition relationships between naturally occurring augite, pigeonite, and orthopyroxene at one bar pressure, *Am. Mineral.*, *64*, 1133–1155.

Scowen, P. A. H., P. L. Roeder, and R. T. Helz (1991), Reequilibration of chromite within Kilauea Iki lava lake, Hawaii, *Contrib. Mineral. Petrol.*, *107*(1), 8–20, doi:10.1007/BF00311181.

Sun, S. S., and W. F. McDonough (1989), Chemical and isotopic systematics of oceanic basalts: Implications for mantle composition and processes, in *Magmatism in the Ocean Basins*, Geol. Soc. Lon. Sp. Pub. 42, edited by A. D. Saunders and M. J. Norry, pp. 313–345, doi:10.1144/GSL.SP.1989.042.01.19.

Thompson, R. N., and C. E. Tilley (1969), Melting and crystallisation relations of Kilauean basalts of Hawaii. The lavas of the 1959–60 Kilauea eruption, *Earth Planet. Sci. Lett.*, *5*, 469–477, doi:10.1016/S0012-821X(68)80081-0.

Thornber, C. R. (1994), Ultramafic inclusions from Harrat Hutaymah: A record of mantle magmatism beneath north central Arabia, in *Kimberlites, Related Rocks and Mantle Xenoliths, 5th International Kimberlite Conference Proceedings Volume, Compania de Pasquisa de Recursos Minerais, Special Publication 94/A, Brazilia,* edited by H. O. A. Meyer and O. H. Leonardos, pp. 434–454. Compania de Pasquisa de Recursos Minerais, Special Publication 94/A, Brazilia, Brazil.

Thornber, C. R. (2001), Olivine-liquid relations of lava erupted by Kilauea Volcano from 1994 to 1998; implications for shallow magmatic processes associated with the ongoing east-rift-zone eruption, *Can. Mineral.*, *39*(2), 239–266, doi:10.2113/gscanmin.39.2.239.

Thornber, C. R. (2003), Magma-reservoir processes revealed by geochemistry of the Pu'u 'Ō'ō-Kūpaianaha eruption, in *The Pu'u 'Ō'ō-Kūpaianaha Eruption of Kīlauea Volcano, Hawaii: The First 20 Years*, Prof. Pap., 1676, edited by C. Heliker, D. A. Swanson, and T. J. Takahashi, pp. 121–136, U.S. Geol. Surv., Reston, Virginia.

Thornber, C. R., and J. S. Huebner (1985), Dissolution of olivine in basaltic liquids; experimental observations and applications, *Am. Mineral.*, *70*(9–10), 934–945.

Thornber, C. R., J. R. Budahn, I. W. Ridley, and D. M. Unruh (2003a), Trace element and Nd, Sr, Pb isotope geochemistry of Kilauea Volcano, Hawai'i, near-vent eruptive products: 1983–2001, Open-File Rep. 03-493, U.S. Geol. Surv., Reston, Virginia.

Thornber, C. R., C. Heliker, D. R. Sherrod, J. P. Kauahikaua, A. Miklius, P. G. Okubo, F. A. Trusdell, J. R. Budahn, W. I. Ridley, and G. P. Meeker (2003b), Kilauea east rift zone magmatism: An episode 54 perspective, *J. Petrol.*, *44*(9), 1525–1559, doi:10.1093/petrology/egg048.

Thornber, C. R., K. Hon, C. Heliker, and D. A. Sherrod (2003c), A compilation of whole-rock and glass major-element geochemistry of Kīlauea Volcano, Hawai'i, near-vent eruptive products: January 1983 through September 2001, Open-File Rep. 03-477, U.S. Geol. Surv., Reston, Virginia.

Thornber, C. R., L. Keszthelyi, C. Lewis-Kenedi, M. Cazeneuve, and D. Goehring (2004), Drilling an active pahoehoe lava flow, *Eos Trans. AGU*, *85*(47), Fall Meet. Suppl., Abstract V31D-09.

Thornber, C. R., T. Orr, H. Lowers, C. Heliker, and R. Hoblitt (2007), An Episode 56 perspective on post-2001 comagmatic mixing along Kilauea's east rift zone, *Eos Trans. AGU*, *88*(52), Fall Meet. Suppl., Abstract V51H-04.

Vinet, N., and M. D. Higgins (2010), Magma solidification processes beneath Kilauea Volcano, Hawaii: A quantitative textural and geochemical study of the 1969–1974 Mauna Ulu lavas, *J. Petrol.*, *51*(6), 1297–1332, doi:10.1093/petrology/egq020.

Wilson, D., T. Elias, T. Orr, M. Patrick, J. Sutton, and D. Swanson (2008), Small explosion from new vent at Kilauea's summit, *Eos Trans. AGU*, *89*(22), 203, doi:10.1029/2008EO220003.

Witham, F., J. Blundy, S. C. Kohn, P. Lesne, J. Dixon, S. V. Churakov, and R. Botcharnikov (2012), SolEx: A model for mixed COHSCl-volatile solubilities and exsolved gas compositions in basalt, *Comp. Geosci.*, *45*, 87–97, doi:10.1016/j.cageo.2011.09.021.

Wooten, K. M., C. R. Thornber, T. R. Orr, J. F. Ellis, and F. A. Trusdell (2009), Catalog of tephra samples from Kīlauea's summit eruption, March–December 2008, Open-File Rep. 2009–1134, U.S. Geol. Surv., Reston, Virginia.

9

Shallow Magma Storage at Piton de la Fournaise Volcano After 2007 Summit Caldera Collapse Tracked in Pele's Hairs

Andrea Di Muro[1], Thomas Staudacher[1], Valérie Ferrazzini[1], Nicole Métrich[2], Pascale Besson[2], Christine Garofalo[2], and Benoit Villemant[3]

ABSTRACT

Very fast cooled ejecta (Pele's hairs and tears) of the first cycle of eruptions (September, November, and December 2008) of Piton de la Fournaise (PdF) that occurred soon after the 2007 major caldera-forming eruption are studied in order to assess the influence of summit collapses on the functioning of a basaltic volcano plumbing system. The ejecta were collected during the initial and most intense phase of each of the three 2008 intracaldera eruptions. In spite of their fast cooling rates, pyroclasts contain widespread microphenocrysts and microlites of plagioclase, pyroxene, olivine, and spinel, with a broad range of textures and compositions. In September ashes, calcic (up to An_{79}) to intermediate plagioclase (An_{72-76}) is the main or only phase, a very uncommon occurrence in historical magmas of PdF. Increasing amounts of pyroxene ($En_{46}Fs_{12}Wo_{42}$) and low-Fo olivine (Fo_{77-81}) occur in November and December products, which contain more sodic plagioclase (An_{71-61}). Moderately high dissolved water contents (0.8–0.6 wt%) and storage depths (2.5–3.2 km below the volcano summit) are required to crystallize the calcic-intermediate plagioclase, while ascent-driven decompression and degassing of magma provoked the nucleation of sodic plagioclase. We here demonstrate that the shallow part of the PdF edifice (upper 3 km) hosts heterogeneous and on average evolved melts with variable An/Di ratios and olivine normative contents, whose evolution converges toward a common ternary eutectic minimum. Frequent summit collapses favor the formation of discontinuities for shallow temporary magma storage and possibly enhance the assimilation rate of host rocks. At least two distinct shallow pockets of relatively evolved melts (MgO: 6.2 wt%) were involved in the 2008 eruptions. Extrusion of these shallow and evolved magmas is triggered by small-volume inputs of deeper and hotter magnesian melts (MgO: up to 8.7 wt%), previously stored in the depth range 1–4 km below sea level. Their preeruptive injection is recorded by rare magnesian to intermediate Pele's hairs, calcic rims on plagioclases, and high-Fo (up to Fo_{85}) olivines. Regular monitoring of fast-quenched melts is an effective strategy for tracking magma evolution, storage, and dynamics at very active volcanoes such as Piton de la Fournaise and Kīlauea.

[1]*Observatoire Volcanologique du Piton de la Fournaise, Institut de Physique du Globe de Paris, La Plaine des Cafres, La Réunion, France*
[2]*Institut de Physique du Globe de Paris, UMR CNRS 7154, Sorbonne Paris Cité, Université Paris Diderot, Paris, France*
[3]*Institut de Sciences de la Terre de Paris, Université Pierre et Marie Curie, Paris, France*

9.1. INTRODUCTION

Caldera collapses occur in response to changes in the volume and shape of the underlying shallow reservoir or network of reservoirs (sills + dykes) subsequent to magma withdrawal by an eruption or a lateral intrusion. These catastrophic events induce drastic modification in a

Hawaiian Volcanoes: From Source to Surface, Geophysical Monograph 208. First Edition.
Edited by Rebecca Carey, Valérie Cayol, Michael Poland, and Dominique Weis.

Companion Website: www.wiley.com/go/Carey/Hawaiian_Volcanoes

volcano plumbing system and eruptive behaviors and can result in major and fast chemical evolution and spatial redistribution of the magma [i.e., *Marti and Gudmundsson*, 2000]. For instance, the Halema'uma'u collapse in 1924 ended a century-long phase of nearly continuous activity in the Kīlauea Caldera. In the new plumbing system formed after a caldera collapse, magma can evolve rapidly as a consequence of mixing, decompression, and degassing, increasing interaction with and assimilation of cold fractured rocks and hydrothermal fluids [i.e., *Wright and Fiske*, 1971; *Bindeman and Valley*, 2000]. To quantify the signature of these shallow processes in volcanic products turns out to be challenging because of late-stage phenomena, which occur after magma fragmentation or during magma extrusion and cooling at Earth's surface. Very fast quench rates of high-temperature melts are thus required to preserve the record of processes occurring in the shallow part of a plumbing system. Pele's hairs and tears are ideal candidates as they are the fastest naturally cooled glasses [*Potuzak et al.*, 2008; *Porritt et al.*, 2012].

At Piton de la Fournaise (PdF) volcano, the 2007 eruptive sequence was associated with the largest lava volume since at least three centuries, resulting in the collapse of a small (1 km wide; 340 m deep) summit caldera [*Staudacher et al.*, 2009; *Bachèlery et al.*, 2010; *Michon et al.*, 2009]. This event ended a nine-year-long period (March 1998–April 2007) of continuous edifice inflation and sustained eruptive activity (three eruptions per year on average) [*Staudacher et al.*, 2009; *Roult et al.*, 2012]. In contrast, the post-2007 activity is associated with a trend of continuous volcano deflation and consists of small-volume (<3 Mm^3) summit eruptions of moderate/low MgO magmas (MgO: 7.1 ± 0.6 wt%) and frequent shallow magma intrusions [*Roult et al.*, 2012]. The 2007 caldera collapse has been attributed either to partial emptying of about one-third of a single magma reservoir located at about 2–2.5 km below the volcano summit [*Staudacher et al.*, 2009; *Peltier et al.*, 2009; *Michon et al.*, 2009] or to a combination of several processes (emptying of shallow magma pockets, fluidization and drainage of dense cumulate bodies and lateral magma intrusion under the eastern volcano flank) [*Di Muro et al.*, 2014]. Geochemical and petrological studies have also demonstrated that the 2007 eruptive sequence was triggered by a new magma injection, whose eruption shortly preceded the summit caldera collapse [*Villemant et al.*, 2009; *Di Muro et al.*, 2014]. Unexpectedly and in spite of the large volume of magma erupted in 2007, volcano unrest and eruptive activity resumed quickly in 2008, soon after PdF summit collapse, and produced three closely spaced intracaldera eruptions and two shallow intrusions [*Staudacher*, 2010; *Roult et al.*, 2012]. Remarkably, the first 2008 eruption in September was preceded by an unusual event of noneruptive tremor, the first one recorded since the foundation of the Piton de la Fournaise Volcanological Observatory (OVPF) in 1979 [*Roult et al.*, 2012]. Noneruptive tremor and an increase in SO_2 emissions were related to magma intrusion at shallow depth (<2.0 km) below the volcano summit [*Garofalo et al.*, 2009].

The 2007–2008 phase of PdF activity represents an ideal case study to analyze the influence of caldera collapse on the evolution of the plumbing system at a persistently active basaltic volcano. Abundant Pele's hairs and tears were erupted at the beginning of each one of these eruptions and are the object of this study. In order to track magma storage and ascent, we compare new petrological and geochemical data for fast-quenched glasses with the composition of 2007 lavas, minerals, and melt inclusions [*Di Muro et al.*, 2014]. The new data integrate and complete the data set of *Villemant et al.* [2009] on fast-cooled 2007 and 2008 samples (Pele's hairs, coarse ash fragments produced by lava-seawater interaction, glassy crust of lavas, high-temperature lavas quenched in water, matrix glasses, melt inclusions).

Our new petrochemical data are consistent with the *Lenat and Bachèlery* [1990] geophysical model, which suggests that the shallow (subaerial) plumbing system of PdF is formed by a network of small-sized magma pockets (sills). We here explicitly link its formation and evolution with periodic magma recharges from deeper levels and repeated summit caldera collapses, which frequently affect the central cone of PdF.

9.1.1. Post-2007 Piton de la Fournaise Summit Eruptions

The 2007 eruptive sequence consisted of three closely spaced eruptions (in February, March, and April), the last one dominant in terms of intensity and volume. The 2007 eruptive sequence was exceptional in producing (i) the most voluminous lava field (at least 0.21 km^3) [*Bachèlery et al.*, 2010; *Roult et al.*, 2012], (ii) the most intense lava fountaining activity (>200 m high) [*Staudacher et al.*, 2009], (iii) the largest SO_2 plume (>230 kt) [*Tulet and Villeneuve*, 2011], (iv) the largest summit collapse (1 km wide × 0.34 km deep) [*Michon et al.*, 2009] and (v) the main flank slip event (up to 1.4 m eastward) ever documented at PdF [*Froger et al.*, 2007]. The bulk volume (~0.21 km^3; ~0.15 km^3 dens rock equivalent (DRE)) of lava extruded in 2007 is similar to that emitted during the entire 1998–2006 period (~0.25 km^3; ~0.16 km^3 (DRE) [*Roult et al.*, 2012]. Hence, the volume of lavas emitted during the 1998–2007 cycle is remarkably close to that estimated (~0.35 km^3) for the whole shallow plumbing system of PdF [*Albarède et al.*, 1997; *Sigmarsson et al.*, 2005; *Peltier et al.*, 2008]. Moreover, the 2007 eruptive sequence produced the complete range of textures (from aphyric basalt to olivine basalt to crystal-rich oceanite) typically found in PdF products [*Boivin and Bachelery*, 2009].

Before and after the caldera collapse of 2007, volcano-tectonic (VT) seismicity was concentrated at shallow levels (above sea level) below the summit cone of PdF (Figure 9.1).

Interestingly, hypocenter migration at deep levels was detected in two phases after the April 5, 2007 climactic eruption: (i) up to 7 km depth below sea level (bsl) on May–June 2007 and (ii) at much deeper levels (up to 30 km bsl) in the period August 2007–January 2008. The first group of deep hypocenters was centered below the PdF summit cone and clearly reveals the occurrence of an a-seismic zone in the range -1 to -5 km bsl (Figure 9.1a). This zone was already identified as a possible magma storage zone by *Prono et al.* [2009]. The second group was located about 20 km far from the volcano summit, along the northwest rift zone running between Piton des Neiges and PdF volcanoes. The origin of the eccentric and deep seismic activity is still unclear. Between January and July 2008, the global navigation satellite system (GNSS) recorded a continuous trend of deflation, associated with a very low number of VT seismic events. (Table 9.1).

Volcano unrest started on 28 July and culminated in a 10 h long event of noneruptive tremor and significant SO_2 emissions on 12 September, interpreted as a first intrusive event at shallow depth below the central cone. This first intrusion was followed by three small-volume intracaldera eruptions (21 September–2 October, 27–28 November, and 14 December–4 February, 2009). The 2008 activity has been described in detail by *Staudacher* [2010] and is summarized in Table 9.1. All VT events associated with the 2008 activity were localized at shallow depth, above sea level (Figure 9.1b).

9.2. SAMPLING AND ANALYTICAL PROCEDURES

Eruptive vents of the three 2008 eruptions were located on the intracaldera cliffs of Dolomieu and their ash have coated some sectors of the caldera rim (Figure 9.2). Ash samples have been collected on the ground, on solar panels of the OVPF monitoring stations, or on fabrics installed before the starting of the eruptions on the western and northern rims of the Dolomieu Caldera (Figure 9.2; Table 9.2).

Ash emissions have been most important at the beginning of each eruptive event (Table 9.2), when lava fountains and projections attained their maximum height. Actually, ash represents the only available samples of the 2008 eruptive products. The high instability of the steep caldera cliffs have prevented up to now the access to the 2008 lava flows emplaced on the floor of the caldera [*Staudacher*, 2010]. Moreover, subsequent intracaldera activity in 2009 and 2010 has almost completely buried the 2008 products (Figure. 9.2).

Ash comprises prevalently millimeter to centimeter long hairs, tears, rare scoria, lithics, and free crystals. In September samples, an increasing proportion of millimeter-sized tears is found from the first to the second day of eruption. The "DRAP" sample (Table 9.2) contains relatively abundant free olivine crystals lacking a matrix coating. These olivines were actually ejected at the vent, because the fabric has been folded in a way to avoid any contamination by material scattered on the ground. Small scoria and lithics are most abundant in November samples. The December eruption emitted the lowest amount of ash, which is made only of fine-grained hairs.

Hairs, tears, scoria, lithic fragments, and free crystals have been selected using a binocular microscope on the basis of textural criteria (shape, color, vesicularity) and subsequently doubly polished for Fourier transform infrared (FTIR) spectroscopic analyses or mounted on epoxy resin for electron probe microanalysis (EPMA) and micro-Raman analysis. Groups of hairs have been ground in a mortar and mounted on copper grids for transmission electron microscopy (TEM) observation. Depending on sample heterogeneity, about 12–24 hairs, 5–21 tears, and 4–7 scoria, plus 2 mm olivine crystals and a few lithic fragments were mounted for microanalysis (scanning electron microscopy (SEM), EPMA, micro-Raman, FTIR). As a whole, 140 hairs, 70, tears and 11 scoria have been mounted and carefully characterized in order to guarantee that the observations are representative of the samples.

Nanoscale analysis of ash with TEM (at 200 kV; IPGP, Paris) and Raman microspectrometry (RENISHAW; CEA, Saclay) were performed in order to check for possible alteration processes. FTIR mapping (VARIAN; Laboratoire Géosciences Réunion) and micro-Raman spot analyses were performed to quantify the amount and the spatial distribution of dissolved water and carbon. These analytical procedures are described in detail by *Mercier et al.* [2010]. For Raman microanalysis, samples were excited with a green diode laser during short (H_2O: 60 s; Silicate bands "TOT": 30 s) counting times and low (W: 50%) power on the sample in order to avoid oxidation or heating of the absorbing basaltic glasses. Dissolved water content was quantified by comparison with a set of basaltic standards (Etna basalt) spanning the H_2O range 0.4–1.5 wt%. In total, eight hairs from 2008, one glass inclusion in one olivine from 2008, and one hair from 2007 have been analyzed. In most cases, analyses have been repeated three times and averaged. Estimated relative error in water-poor glasses is 12% and decreases to about 7% at higher water contents. Major- and minor-element composition of glasses and minerals in glasses were measured using a SX50 Cameca electron microprobe

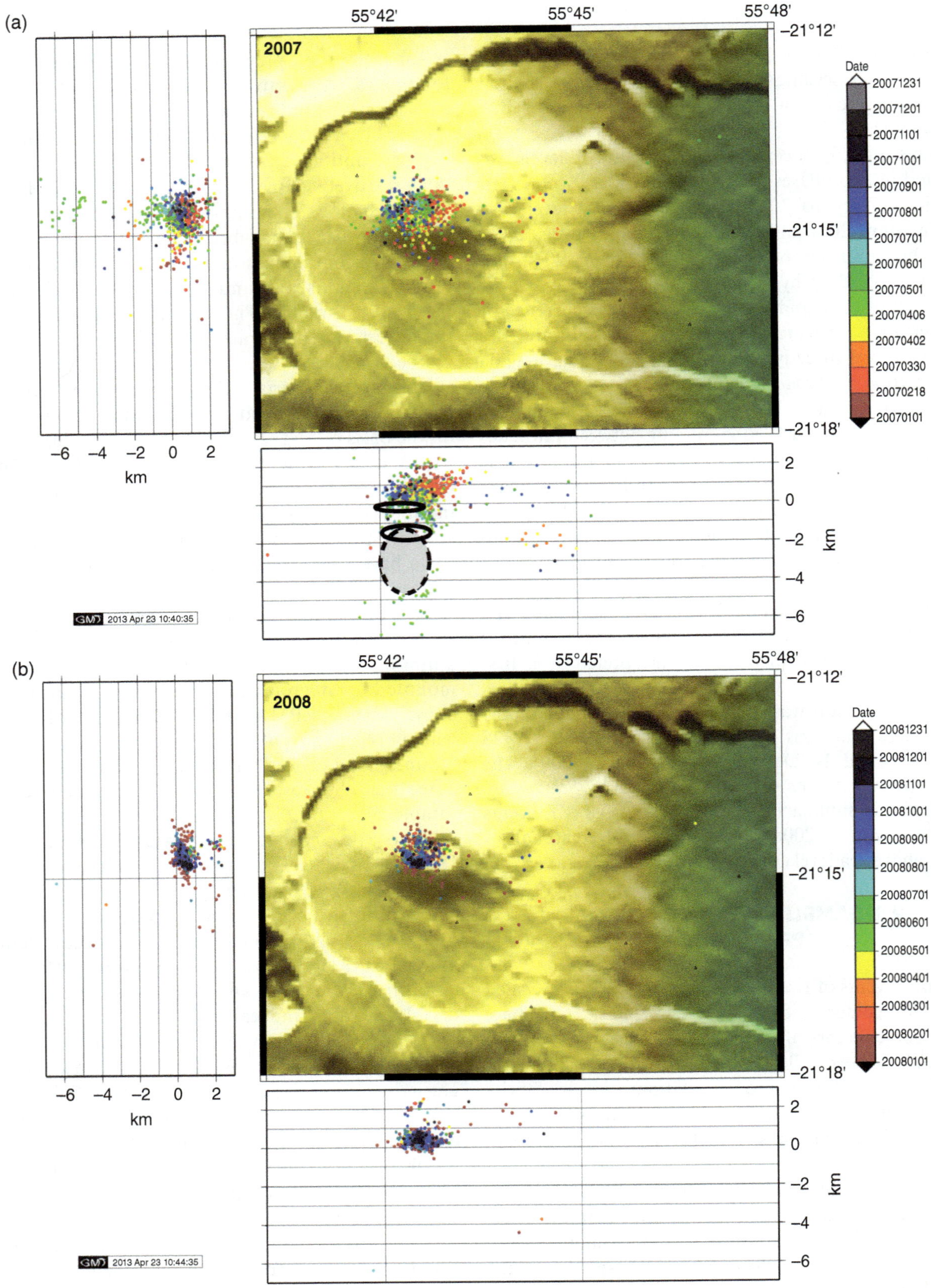

Figure 9.1 Spatial and time distribution of volcano-tectonic (VT) seismic events below PdF central cone in (a) 2007 and (b) 2008. In the depth scale, zero corresponds to sea level. PdF summit cone is 2632 m high. The base of Réunion Island is located at approximately 4 km below sea level. Ellipses indicate the possible magma storage zones near and below sea level evidenced by *Prono et al.* [2009]. Gray area contours a large aseismic zone clearly revealed by 2007 hypocenter distribution. High density of VT events above sea level is consistent with the occurrence of only small pockets of magma in the shallow and subaerial part of the edifice [*Lenat and Bachèlery*, 1990; *Prono et al.*, 2009; *Massin et al.*, 2011]. The scale of the color bar represents the monthly evolution of hypocenters.

Table 9.1 Eruption precursors and description of 2008 intrusive and eruptive events at PdF Volcano; VT: volcano-tectonic seismic events.

Date	Recorded Events	Interpretation
Precursory Phase of 2008 Eruptive Cycle		
1st May 2007	End of the tremor signal	End of the April 2007 caldera-forming eruption.
August 2007–January 2008	Deep and shallow VT swarms (hypocenters at -8 and -30 km bsl, about 20 km SW from the PdF edifice). In shallow VT swarms, most hypocenters are located below PdF between ±1 km bsl.	Posteruptive seismicity related to stress release. Possible migration of deep magma toward the shallow part of PdF plumbing system.
February–27 July 2008	Deflation of PdF volcanic edifice. Low level of VT seismic activity (<10/day). A few (1–4/month) deep VT events (>6.5 km below volcano summit).	Postcaldera deflation. No geophysical evidence of new injections in the main reservoir.
Shallow Magma Intrusion		
28 July 2008–September 2008	Inflation of summit cone recorded on GNSS network. High level of shallow VT seismic activity (5 seismic swarms). Hypocenters located at 0–1 km asl.	Pressurization of the shallow part of PdF plumbing system by deeper magma input. Opening of new magma ascent path at shallow level.
12 September 2008	10 h long weak tremor (04:40–14:20 UTC) and slight inflation increase detected by GNSS. Formation of new fumaroles on the SW wall of the caldera. SO_2 emission detected by portable DOAS	Magma ascent at shallow depth and possible formation of a shallow small-volume intrusion/sill.
Summit Eruption 1		
15 September 2008	VT seismic swarm. Hypocenters located at 0–1 km asl.	Formation of new magma ascent path toward the surface.
21 September–2 October 2008	Very small edifice inflation and short (0.5 h) VT swarm. Opening of 2 eruptive vents along a fracture on the SW summit caldera wall, at lower altitude with respect to the fumaroles. SO_2 emission detected by portable DOAS. Tremor peaks at the beginning of the eruption. Emission of Pele's hairs.	Eruption 1. Total emitted magma volume: 1×10^6 m^3.
Summit Eruption 2		
3 October–November 2008	High level of shallow VT activity: 4 seismic swarms. 31 October, VT seismic swarm detected by tiltmeters AM: 3.3 earthquake on 12 November. Most hypocenters located at 0.5–1 km asl.	Opening of new magma ascent path Possible magma propagation at shallow depth.
27–28 November 2008	Significant edifice inflation recorded on GNSS and tiltmeter networks Opening of 2 eruptive vents on the SW caldera wall near September vents. Tremor peaks at the beginning of the eruption. Emission of Pele's hairs.	Eruption 2. Total emitted magma volume: 0.1×10^6 m^3.
Summit Eruption 3		
29 November, 13 December 2008	High level of VT seismic activity. Inflation trend of the summit cone detected on tiltmeter network. Tremor peaks at the beginning of the eruption. Emission of Pele's hairs. Most hypocenters located at 0.5–1 km asl.	Opening of new magma ascent path. Possible magma propagation at shallow depth.
14 December 2008–1 January 2009	Significant Inflation detected on GNSS network. 19 h long VT seismic swarm. Opening of 2 eruptive vents on the northeast caldera wall. Progressive increase in intensity of eruptive tremor. On 1 January[t], tremor stops. Emission of a few Pele's hairs at the beginning of the eruption.	Eruption 3, first phase.
1 January, 4 February 2009	Long-lasting, low-intensity tremor. Magma eruption from the same vents on the northeast caldera wall. Most hypocenters located at 0–1 km asl.	Eruption 3, second phase. Total emitted magma volume (including phase 1 + 2): 1.3×10^6 m^3.

Note: VT refers to volcano-tectonic seismic events.

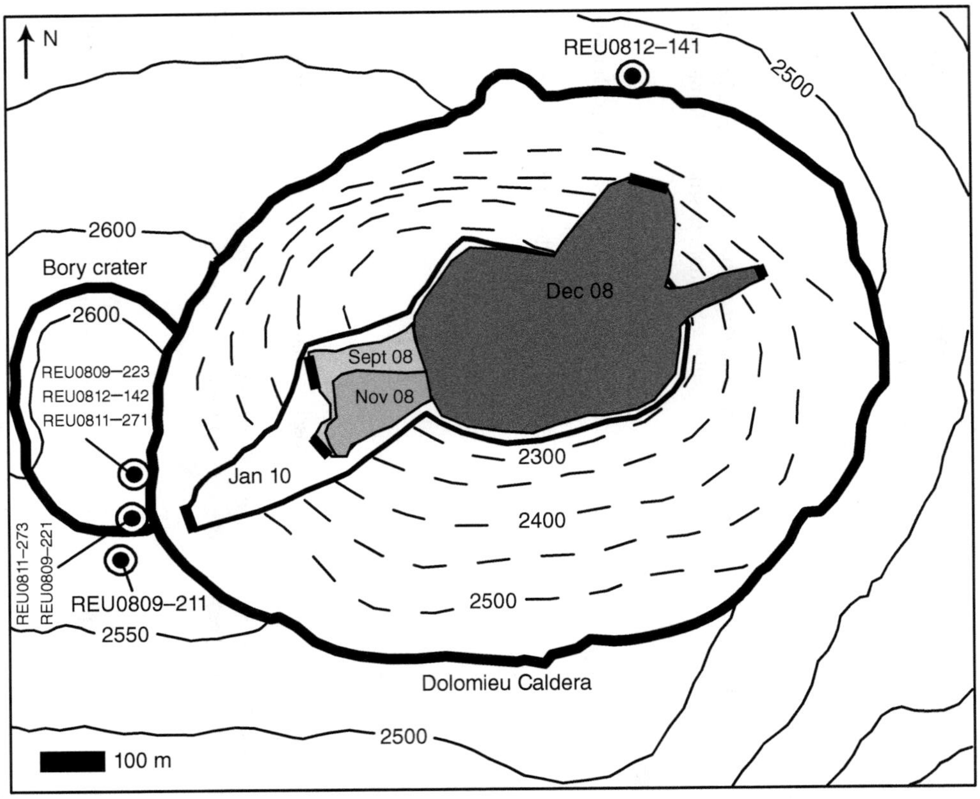

Figure 9.2 Summit topography of PdF central cone showing Bory crater (on the west) and Dolomieu Caldera (on the east). On the cliffs of Dolomieu Caldera, thick segments indicate eruptive fractures, which fed the three 2008 intracaldera eruptions. Color code of lavas is: light gray for September 2008, medium gray for November 2008, and dark gray for December 2008. Dots close to the western and northern caldera rims correspond to the sites of ash sampling. Labels (REUYYMM-DD) identify the ash samples collected during the 2008 activity (Table 9.2). Most of September, and November and part of December 2008 lavas have been covered by the lava field emplaced during the January 2010 eruption.

(Camparis, Paris), with the following conditions: 4 nA, 15 kV, 10 s dwell time for all elements except Na, for which a shorter time (5 s) ensures no significant Na loss under an electron beam. A 10 nA defocused beam and longer counting times on peak (120 s) were selected for accurate analysis of S, F, Cl, and P in glasses. Olivine compositions were determined with a focused beam of 20 nA.

For bulk analysis, Pele's hairs were dissolved after careful removal of crystals and lithics by hand picking under the microscope. Trace-element concentration was determined for bulk ash samples by inductively coupled mass spectrometry ICP-MS (University Paris VI). The analytical procedure, including reproducibility, is described in detail by *Salaün et al.* (2010).

9.3. TEXTURE OF ERUPTIVE PRODUCTS

Ash samples collected during the 2008 and 2007 eruptions contain (in order of abundance) Pele's hairs, Pele's tears, scoria, lithic fragments (black and reddish fine-grained lavas), and rare millimeter-sized olivine crystals. Juvenile fragments are glassy. Crystal-free clasts are rare and most fragments contain a variable amount of fine-grained microphenocrysts <500 μm long and microlites <100 μm long. Feldspar (plagioclase) is a widespread phase in all 2008 eruptive products (Figure 9.3).

Olivine, pyroxene, and oxide (chromite) are rare in the September clasts (Figures 9.3a,b,c). The abundance of these phases increases in November (Figure 9.3d) and December products (Figure 9.3e). We note that 2007 products contain olivine (±spinel) as the only phenocryst phase (Figure 9.3f) and olivine, plagioclase, and rare clinopyroxene in the groundmass.

Glasses: TEM data reveal that all observed samples are homogeneous and glassy, and no traces of microcrystallization or alteration are detected. This conclusion is further confirmed by rim-to-rim SEM analyses across Pele's hairs and tears that do not evidence any physical or chemical profile (e.g., coating of Si, alkali depletion; variation in closure of analyses) and demonstrate that glasses are fully homogeneous. Moreover,

Table 9.2 Samples of the 2008 and 2007 eruptive products analyzed in this research; up to three types of clasts (1, 2, 3) are identified on the basis of mineral paragenesis.

Sample	Eruption Phase	Sampling Location	Sampling Method	Sample Description	Mineral Paragenesis
September 2008					
REU0809-211	day 1 ~14 h (TU)	Southern rim of Bory Crater	On the ground	Hairs, rare tears	1. pl; 2. Rare pl + cpx; 3. Rare pl + cpx + ol
REU0809-221	day 2	Eastern margin of Bory Crater	On the ground	Hairs, rare tears	1. pl; 2. Rare pl + cpx; 3. Rare pl + cpx + ol
REU0809-223	day 2	Eastern margin of Bory Crater	On solar panel	Hairs, abundant tears	1. pl; 2. Rare pl + cpx; 3. Rare pl + cpx + ol
REU0809-223-DRAP	day 2	Eastern margin of Bory Crater	On the fabric	Hairs, abundant tears, mm-sized olivine cx	1. pl; 2. Rare pl + cpx; 3. Rare pl + cpx + ol; 4. Rare ol + sp
November 2008					
REU0811-2711	day 1	Eastern rim of Bory Crater	On solar panel	Hairs, tears, scorias, lithics (black and reddish lavas)	1. pl; 2. pl + cpx + ol
REU0811-273-1	day 1	Southeastern margin of Bory Crater	On the ground	Hairs, rare tears	1. pl; 2. pl + cpx + ol
REU0811-273-2	day 1	Southeastern margin of Bory Crater	On the ground	Fine-grained hairs	1. pl; 2. pl + cpx + ol
December 2008					
REU0812-141	day 1	Around tripod near Soufrière pit crater	On the ground	Hairs	1. pl + cpx + ol; 2. Rare pl; 3. Rare ol + sp
REU0812-142	day 1	Eastern margin of Bory Crater.	On the fabric	Hairs	1. pl + cpx + ol; 2. Rare pl;
April 2007					
REU0704-CH1	nd	St. Philippe	On a table	Fine-grained hairs	ol
REU0704-CH2	nd	Bras Panon	Floating on a swimming pool	Fine-grained hairs	ol
REU0704-051	day 4	Northern edge of lava flow	Sampled from a lava flow	Lava (quenched in water)	ol + (ol, pl, sp)
REU0704-061	day 5	Close to southern margin of enclos	Sampled from a lava flow	Microcrystalline scoria (not quenched)	ol + (ol, pl, sp, cpx)
REU0704-301	day 29	Near Tremblet ?	On the ground ?	Glassy scoria	ol + (ol, pl, sp)

Mineral phases: pl - plagioclase; cpx - clinopyroxene; ol - olivine; sp - spinel;

SEM imagery confirms that Pele's hairs do not contain a filling of fine-grained tears as reported at Masaya by *Moune et al.* [2007]. Most important, the occurrence of secondary hydration processes in the conduit or during the travel time in the gaseous plume, as suggested at Masaya [*Moune et al.*, 2007], can be discarded on the basis of FTIR mapping and Raman microanalyses. Water appears efficiently and homogeneously degassed from the matrix melt. Finally, Raman spectra further confirm the glassy state of all samples, irrespective of their composition, crystal content, vesicularity, or sampling date. Altogether, these observations exclude any syn- (inside the gas plume) or post- (depositional) eruptive modification of the studied glasses by acid attack or thermal processes (e.g., annealing; strong fluctuations in quench rate).

Vesicles: Most juvenile clasts (hairs and scoria) of September eruption are highly vesicular. Some low-vesicularity fragments are found only in the tear group. Pele's hairs contain bubbles that are stretched in both the filaments and the knots, consistently with the very fast quenching rate expected for these clasts [*Potuzak et al.*, 2008]. On the contrary, the bubbles of tears are systematically spherical or only moderately elongated, a likely consequence of their slower cooling rate permitting post-fragmentation relaxation [*Porritt et al.*, 2012]. In November ash, the amount of dense, microlite-rich tears increases with respect to September. The bulk

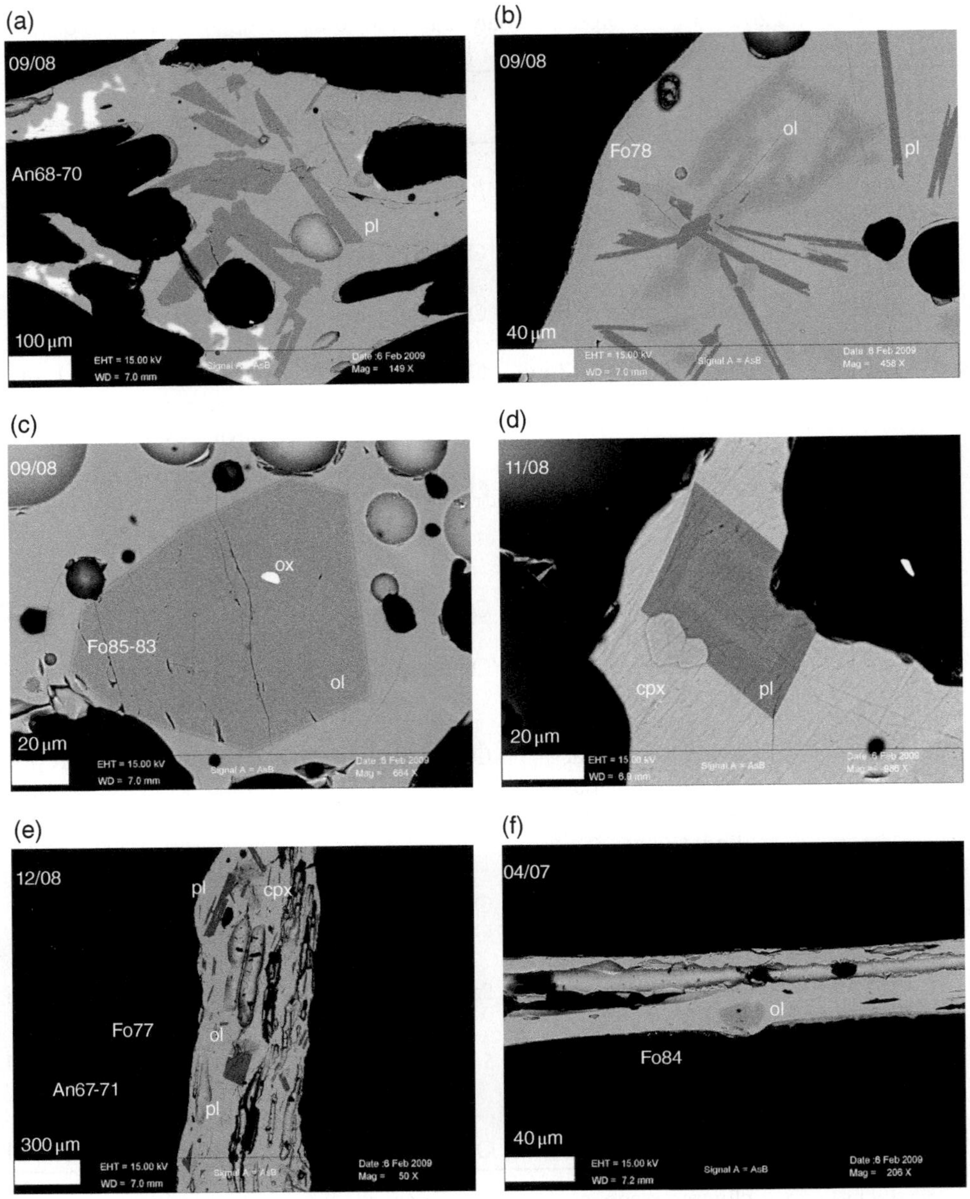

Figure 9.3 Textures representative of the Pele's hairs (a,d,e,f) and tears (b,c) emitted during the September, November, and December 2008 eruptions. In September hairs, the main and often only mineral phase is plagioclase, which often occurs concentrated in the hair knots (a). Dendritic microlites of femic phases (olivine; pyroxene) are best observed in tears (b), which have experienced a lower cooling rate. In November and December ashes, plagioclase is often associated with pyroxene (d) and pyroxene+low-Fo (Fo≤81) olivine (e). Complex and sometimes reverse zoning (up to An_{79}) can be observed on some plagioclase individuals (d). September and December ashes contain rare magnesian to intermediate melts and high-Fo (Fo_{83-85}) olivine with spinel inclusions (c). These mafic composition and mineral assemblages are relatively common in hairs emitted during the most intense phase of 2007 eruption (f).

vesicularity in hairs is also lower than in September samples. December fragments are those with the lowest vesicularity among the 2008 products. In those hairs, skeletal microlites are widespread.

Crystals: In 2008 Pele's hairs, microphenocrysts are often concentrated in their thickest portions ("knots", Figure 9.3a) while the thin glass filaments have lower crystal content. Pele's tears and scoria exhibit an almost homogeneous distribution of crystals and higher average content of microlites with respect to Pele's hairs, consistently with their lower cooling rate. On the basis of paragenesis and crystal size, we distinguish four types of pyroclasts (Table 9.2):

(i) Type 1: Tabular plagioclase microphenocrysts represent the most abundant phase, eventually associated with a secondary population of skeletal plagioclase microlites and rare pyroxene microlites (Figure 9.3a). Small swallowtail ends can occur on the corner of tabular microphenocrysts.

(ii) Type 2: Skeletal microlites of plagioclase, clinopyroxene, and olivine occur as isolated crystals or aggregates (Figure 9.3b).

(iii) Type 3: Plagioclase, clinopyroxene, and olivine occur as microphenocrysts (Figure 9.3e).

(iv) Type 4: Olivine and chromite are the only phase observed (Figure 9.3c).

In September ash, type 1 represents the main type of clasts. There, plagioclase microphenocrysts attain a maximum crystal length of 360 μm and have an average length of about 200 μm. A few (<5%) plagioclase microphenocrysts contain small (<20 μm long) melt inclusions (MI) in their core. Only rare pyroxene microlites (<20 μm) are found. Type 2 and type 3 clasts are rare and are recognized only in the tear group. In those clasts, maximum crystal size decreases from plagioclase (320 μm) to clinopyroxene (140 μm) to olivine (90 μm). Type 4 is extremely rare and contains euhedral olivine grains up to 120 μm that include small grains of anhedral chromite (Figure 9.3c).

In November ash, type 1 and type 2 are the main clast types. Maximum plagioclase size is the same as September (360 μm) in type 1 and attains a maximum length of 170 μm in type 2. Tabular plagioclase crystals with MI-rich cores are found. The amount of microlites in type 1 increases in November fragments with respect to September. In both September and November ashes, complex zoning is observed in plagioclase microphenocrysts and microlites (Figure 9.3d).

In December ash, the type 1 population becomes less abundant and contains larger tabular plagioclase microphenocrysts (up to 420 μm) and microlites (up to 150 μm) with respect to September and November (Figure 9.3e). Type 2 clasts are common and coarser grained than in November (plagioclase up to 230 μm). The dominant clast population belongs to type 3. These crystals are euhedral and plagioclase is coarser (140–330 μm) than pyroxene (130–290 μm) and olivine (110–150 μm).

As a whole, only rare clasts with an olivine-spinel paragenesis (type 4) are found in September and December products, while this paragenesis is characteristic of 2007 products (scoria, lavas, and hairs) (Figure 9.3f). In those rare hairs, maximum olivine diameter decreases from 130 μm in September to only 30 μm in December hairs.

In 2007 water-quenched lavas, large and abundant olivine phenocrysts (millimeter to centimeter sized; up to 60 wt%) are set in a fine-grained matrix (maximum crystal length <150 μm) composed of olivine (up to 150 μm), plagioclase (<60 μm), oxides, and rare pyroxene. Pyroxene is in most cases absent or very small (<30 μm) and included in larger plagioclase microlites.

Ash monitoring and analysis permit two important observations. First, the vast majority of the clasts emitted in the 2008 eruptions share a common crystallization sequence that is plagioclase first followed by clinopyroxene and then olivine. This sequence is profoundly distinct from that of 2007, where olivine is the first phase followed by plagioclase and rare pyroxene. Only a few 2008 hairs have this paragenesis. Second, all 2008 clasts exhibit an unusual abundance of plagioclase, a quite uncommon observation in PdF products [e.g., *Boivin and Bachelery*, 2009; *Welsch et al.*, 2009]. At PdF, plagioclase is usually considered as a late-stage phase, formed during lava emplacement [e.g., *Salaün et al.*, 2010]. Lavas containing only plagioclase, as of September 2008, have never been reported at PdF and plagioclase-phyric lavas are only known in its oldest phases of activity [*Albarède et al.*, 1997].

9.4. MAGMA CHEMISTRY AND MINERALOGY

9.4.1. Bulk Sample Composition

Major and trace-element composition: Bulk composition (major elements) has been measured only on one ash sample of September eruption (Table 9.3).

September ash has low MgO content (6.5 wt%), low CaO/Al_2O_3 ratio (0.77), and relatively high K_2O/Na_2O (0.28) and K_2O/TiO_2 (0.26) ratios. For comparison, the ranges measured in lavas of the 1998–2007 period are 5.9–30 wt% (MgO), 0.69–0.84 (CaO/Al_2O_3), 0.22–0.31 (K_2O/Na_2O), and 0.21–0.30 (K_2O/TiO_2). Low-MgO aphyric magmas represent a typical product of the summit activity at PdF [*Vlastélic et al.*, 2005; *Peltier et al.*, 2009; *Boivin and Bachelery*, 2009; *Roult et al.*, 2012].

Pele's hairs and tears of the three 2008 eruptions share similar concentrations and ratios of incompatible trace elements such as Th, Ta, Hf, rare earth elements (REE), and U (Figure 9.4; Table 9.4).

Incompatible-element concentrations of 2008 glasses fall in a narrow range, similar to those measured by LA-ICPMS in glassy matrices of 2007 eruptive products. Concentrations are slightly less variable than in 2007 products and slightly higher than values measured in 2007 bulk aphyric samples (Figure 9.4a).

Most important, ratios between incompatible elements are almost indistinguishable for 2007 and 2008 products. For instance, both groups of eruptions share similar La/Yb (2008: 9.0; 2007: 9.1) or U/Th (2008: 0.24; 2007: 0.28) ratios. In terms of compatible elements (Sc, Co) 2008 samples are very similar. The only difference between

Table 9.3 Average composition of glasses in 2008 tephra and 2007 lavas (values normalized to 100) compared with the bulk composition of September 2008 ash (XRF data; courtesy of V. Famin; Laboratoire Géoscience Réunion).

DATE	2007		2007		Sep. 2008		Sep. 2008		Sep. 2008	Nov. 2008		Dec. 2008		Dec 2008	
Sample	Matrix 301		Matrix 051		Hair		Hair		Bulk	Hair		Hair		Hair	
Notes	Glassy crust		Quenched lava		Average		Most mafic		Sample	Average		Average		Most mafic	
n	3		4		52		4			67		44		20	
SiO_2	49.81	(0.68)	50.39	(0.07)	50.06	(0.61)	49.40	(0.13)	50.20	50.44	(0.35)	50.36	(0.32)	49.37	(0.31)
TiO_2	2.95	(0.06)	2.94	(0.08)	3.23	(0.14)	3.05	(0.09)	2.88	3.14	(0.08)	3.26	(0.05)	2.84	(0.07)
Al_2O_3	14.36	(0.12)	14.35	(0.16)	14.39	(0.17)	14.40	(0.13)	14.50	14.24	(0.14)	14.01	(0.11)	13.34	(0.15)
FeOT	10.51	(0.03)	10.57	(0.25)	10.88	(0.26)	10.43	(0.33)	10.90	10.82	(0.18)	11.05	(0.14)	11.18	(0.26)
MnO	0.19	(0.05)	0.15		0.18	(0.03)	0.17	(0.05)	0.16	0.18	(0.03)	0.19	(0.03)	0.18	(0.05)
MgO	6.86	(0.04)	6.45	(0.19)	6.29	(0.11)	7.39	(0.10)	6.47	6.30	(0.07)	6.22	(0.05)	8.73	(0.13)
CaO	11.43	(0.02)	11.25	(0.03)	11.17	(0.24)	11.79	(0.10)	11.20	11.12	(0.20)	11.20	(0.10)	11.02	(0.09)
Na_2O	2.77	(0.14)	2.81	(0.04)	2.66	(0.10)	2.35	(0.14)	2.62	2.66	(0.12)	2.59	(0.04)	2.34	(0.05)
K_2O	0.74	(0.04)	0.70	(0.07)	0.80	(0.06)	0.74	(0.07)	0.74	0.75	(0.04)	0.77	(0.02)	0.58	(0.07)
P_2O_5	0.38	(0.03)	0.38	(0.05)	0.35	(0.05)	0.29	(0.06)	0.31	0.35	(0.02)	0.35	(0.02)	0.32	(0.04)
MgO/(MgO + FeOT)	0.40		0.38		0.37		0.41		0.37	0.37		0.36		0.44	
CaO/Al_2O_3	0.80		0.78		0.78		0.82		0.77	0.78		0.80		0.83	
K_2O/Na_2O	0.27		0.25		0.30		0.32		0.28	0.28		0.30		0.25	
K_2O/TiO_2	0.25		0.24		0.25		0.24		0.26	0.24		0.24		0.20	
P	nd		nd		0.169	(0.012)	0.176	(0.023)		0.172	(0.004)	0.176	(0.006)	0.149	(0.011)
S	nd		nd		0.022	(0.003)	0.013	(0.004)		0.024	(0.002)	0.022	(0.002)	0.019	(0.003)
Cl	nd		nd		0.022	(0.002)	0.022	(0.006)		0.024	(0.001)	0.024	(0.002)	0.020	(0.003)
F	nd		nd		0.050	(0.013)	0.042	(0.010)		0.056	(0.009)	0.058	(0.026)	0.062	(0.019)
TMgO (H&T87)	1152		1144		1140		1163			1141		1139		1190	
TCaO (H&T87)	1158		1155		1153		1164			1153		1154		1151	
Quartz (Q)			0.13		0.49					1.12		1.40			
Orthoclase (Or)	4.35		4.14		4.71		4.38			4.44		4.53		3.43	
Albite (Ab)	23.41		23.79		22.50		19.86			22.51		21.94		19.81	
Anorthite (An)	24.60		24.48		24.98		26.56			24.69		24.32		24.19	
Diopside (Di)	24.16		23.57		23.03		24.40			23.05		23.68		23.10	
Hypersthene (Hy)	11.67		15.12		14.97		14.18			15.04		14.69		17.17	
Olivine (Ol)	3.04						1.88							3.65	
Magnetite (Mt)	2.32		2.34		2.41		2.31			2.39		2.44		2.47	
Ilmenite (Il)	5.61		5.58		6.14		5.79			5.96		6.20		5.39	
Apatite (Ap)	0.88		0.89		0.80		0.67			0.82		0.81		0.73	

Note: Number in parentheses, are standard deviatieres.

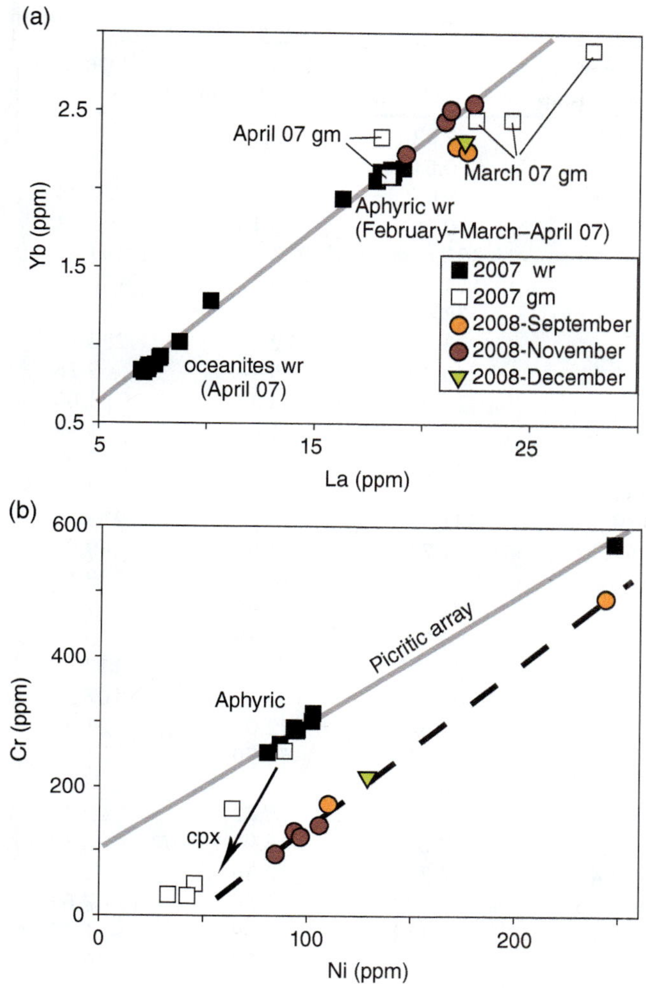

Figure 9.4 Concentration in incompatible (a) and compatible (b) trace elements in 2008 ashes. Composition of 2007 products (bulk lavas, groundmasses; from *Di Muro et al.* [2014]) is shown for comparison. Ratios in incompatible elements are indistinguishable for 2007 and 2008 magmas. The evolved 2008 melts have relatively high contents in incompatible trace elements, close to that of the matrices in March 2007 aphyric basalt (a). The apparent low Cr content of 2008 ashes (b) results from the variable amounts of free olivine crystals mixed with low Cr-Ni evolved glasses (hairs; tears).

2007 and 2008 products concerns the Cr and Ni concentrations, Cr being lower in 2008 samples with respect to 2007 samples for a same Ni content (Figure 9.4b). This is likely an artifact of 2008 sample composition in which low Ni-Cr glasses (Pele's hairs; tears), similar to those of March 2007 matrices, are mixed with some olivine crystals. The degree of sample contamination is lower in olivine-poor samples, e.g., those emitted in September 2008.

In conclusion, 2008 ashes belong to the same magmatic series emitted in 2007 and result to be slightly more evolved. In terms of trace-element contents, 2008 ashes and the matrix glasses of March 2007 aphyric lavas are very similar.

9.4.2. Olivine, Pyroxene, and Oxides

Olivine composition in the crystal-rich products (up to 60 wt% crystals in oceanite lavas) of the large April 2007 eruption is representative of the average composition observed in historical products of PdF [*Villemant et al.*, 2009; *Boivin and Bachelery*, 2009; *Di Muro et al.*, 2014].

Di Muro et al. [2014] report that, in 2007 products, olivine displays distinct compositional ranges in March (aphyric lavas) and April (from aphyric to olivine basalts to oceanite lavas) with an increasing content in CaO (Figure 9.5a) and decreasing content in NiO (Figure 9.5b) as the olivines become richer in FeO.

Most microcrysts of March 2007 lava span a chemical range between $Fo_{80.5}$ and Fo_{82} and are distinctly richer in Fe, richer in Ca, and poorer in Ni than olivines from April 2007 lavas (Fo_{82}-$Fo_{85.5}$) (Figure 9.5). Olivine phenocrysts in 2007 Pele's hairs and fast-quenched tephra documented by *Villemant et al.* [2009] fall in the range Fo_{78}-Fo_{84}. In 2007 and 2008 olivines, the asymptotic trend of CaO evolution with FeO can be attributed to late cotectic crystallization of clinopyroxene together with Fe-rich olivines ($Fo < 80$). We stress that clinopyroxene occurs only as rare microlites in the matrix of 2007 rocks.

In the hairs and tears of September and December 2008, two olivine populations can be identified (Figure 9.5). The main population is made of low-Fo olivine microphenocrysts (Fo_{77-80}) and is systematically associated with pyroxene and plagioclase. Skeletal olivine microlites can have even lower Mg content (e.g., Fo_{71} in December hairs). The second and minor population corresponds to magnesian olivines (Fo_{82-85}) found in type 4 clasts (Fo_{82}) and as free olivines (up to Fo_{85}). The composition of the most magnesian olivines is very close to that of April 2007 lavas, except for a higher Ca content (Figure 9.5). In November hairs, we identified only one population (Fo_{81}) of olivine microphenocrysts, whose composition is intermediate between the two main populations found in September and December products. This population has a composition close to that of March 2007 aphyric lavas (Figure 9.5).

Low-Fo olivines (Fo_{78}) are texturally and chemically in equilibrium with the 2008 evolved melt, on the basis of a (FeO/MgO)ol/(FeO/MgO) melt K_d value of 0.306 [*Fisk et al.*, 1988] (Table 9.5).

More magnesian olivines (Fo_{81} in November; Fo_{82-85} in September and December) are clearly not in equilibrium with the main melt composition found in 2008 ashes (Table 9.5). A further observation is that microlites often

Table 9.4 Trace-element concentration (ICP-MS) in 2008 Pele's hairs and tears; September data are from *Villemant et al.* (2009).

Eruption Sample	September REU0809-2203	September REU0809-211	September REU0809-2201	September REU0809-drap	November 0811-271-tears	November 0811-271-hairs	December 0812-19-homog
Th	2.36	2.48	2.52	2.59	2.43	2.53	2.60
U	0.56	0.59	0.61	0.61	0.61	0.62	0.63
Rb	18.50	20.90	20.80	22.90	20.45	20.78	20.24
Cs					0.16	0.16	0.13
Sr	338	376	377		387	383	393
Ba	144	160	160	172	154	156	156
Y					28.41	28.86	29.09
Zr					218	222	224
Hf	4.91	5.25	5.28	5.48	5.20	5.24	5.18
Nb					23.00	23.22	23.80
Ta	1.28	1.39	1.39	1.44	1.42	1.43	1.42
Sc	29.00	31.10	30.80		33.36	31.93	32.60
V	291	333	329	365	331	329	337
Cr	94	129	121	139	492	173	216
Co	40.50	46.10	46.70	51.90	51.73	47.62	49.93
Ni	85	94	97	106	243	110	129
Mo					34.01	1.92	1.87
W					0.37	0.18	0.16
Cu					125	114	117
Zn					104	103	107
La	19.16	20.97	21.22	22.29	21.49	22.01	21.89
Ce	44.20	48.34	48.98	51.53	49.51	50.59	51.58
Pr	5.92	6.44	6.54	6.87	6.56	6.71	6.88
Nd	25.80	28.17	28.48	29.78	28.72	28.72	29.81
Sm	6.26	6.91	6.91	7.29	7.02	6.95	7.30
Eu	2.08	2.26	2.30	2.42	2.39	2.36	2.43
Gd	6.45	7.12	7.26	7.56	7.33	7.44	7.64
Tb	0.990	1.080	1.100	1.130	1.112	1.099	1.149
Dy	5.61	6.14	6.24	6.43	6.35	6.38	6.59
Ho	1.09	1.18	1.22	1.24	1.22	1.24	1.28
Er	2.76	3.02	3.07	3.12	3.13	3.09	3.27
Tm	0.380	0.410	0.420	0.420	0.390	0.388	0.389
Yb	2.23	2.44	2.51	2.55	2.28	2.25	2.32
Lu	0.310	0.340	0.350	0.360	0.340	0.349	0.360
Pb					2.19	1.94	1.80

September data are from *Villemant et al.* (2009).

have higher CaO content than larger olivine phenocrysts and microphenocrysts (Figure 9.5a), which is independent of the type of hosting pyroclast. The slightly higher concentration in a slow-diffusing cation such as Ca has probably a kinetic origin.

The Mg-rich olivine microlites of September and December are associated with small *chromite* grains. In September samples, chromite has an anhedral shape and is included in olivine crystals (Figure 9.3c). On the contrary, chromite is euhedral and is embedded in matrix glass in December hairs. The assumption of equilibrium between spinel and olivine permits the calculation of the magma temperature [*Jianping et al.*, 1995] and oxidation state [*Ballhaus et al.*, 1991] (Table 9.5). Calculated redox state (NNO −0.4) for an apparent equilibrium temperature of 1130°C fits very well with the general trend of redox state defined for midocean ridge basalts (MORBs) by *Bezos and Humler* [2005]. However, as discussed in the following, such a low temperature is likely to indicate partial thermal reequilibration between high- and low-temperature melts before their final eruption.

Pyroxene in 2008 pyroclasts is a clinopyroxene whose composition plots in the field of aluminum augite [*Morimoto*, 1988] of the enstatite-ferrosilite-wollastonite diagram (Table 9.5). Pyroxene average composition is $En_{46}Fs_{12}Wo_{42}$ and Mg* ($Mg/Mg+Fe^{2+}$) spans the range 76–92 (average Mg* = 80). In December hairs and tears, only a single analysis of a microphenocryst core plots in

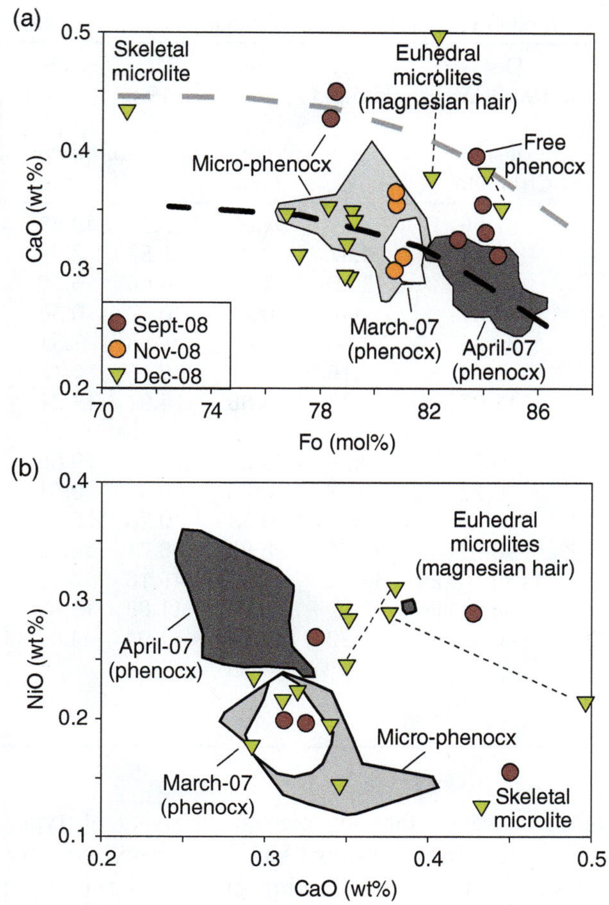

Figure 9.5 Forsterite vs. CaO (a) and NiO (b) contents in olivine microphenocrysts and microlites occurring in Pele's hairs, tears, and scoria emitted during the three 2008 eruptions of PdF. Low-Fo (~Fo_{78}) olivines are in equilibrium with the 2008 evolved melts. More magnesian olivines, like those found as millimeter-sized free olivine crystals in 2008 ash (Fo_{83-85}), in November hairs (Fo_{81}) are not in equilibrium with the evolved carrying melt. Olivines in lavas of April 2007 (dark gray area) and March 2007 eruptions (light gray area; white area contours the main composition of March 2007 phenocrysts) are shown for comparison. Rare magnesian hairs include high-Fo olivines like those of April 2007, but with higher CaO content. Fo content decreases from phenocrysts to microphenocrysts and microlites, while the content in slowly diffusing cations as Ca is highest in fast-grown microlites. Dotted lines connect core-rim analyses in a same crystal. Dashed lines: trends of asymptotic increase in CaO with decreasing Fo content that records the late crystallization of pyroxene.

the diopside field. In 2008 pyroxenes, Al_2O_3 and TiO_2 contents increase with decreasing Mg* from 2 to 6 wt% and from 0.8 to 2.9 wt%, respectively (Table 9.5). Aluminum and Ti are thus positively and linearly correlated and both increase from core to rim in larger crystals and between larger and smaller crystals or crystal micro-aggregates (Figure 9.6a).

The average correlation line has a slope of 3.88, a value very close to that (3.98) obtained by *Boivin and Bachèlery* [2009] for pyroxene microlites in PdF lavas for the whole 1972–1998 eruptive period. A global positive correlation between Al- and Ti is likely to stem from the increasing concentration of these elements in the melt as a consequence of the crystallization of an Al- and Ti-poor phase such as olivine.

A closer look at 2008 data reveals a clear evolution of pyroxene composition during the eruption sequence, the Al/Ti ratio of December pyroxenes being higher than that of September-November pyroxenes for a given Mg* ratio (Figure 9.6b). This difference in Al_T/Ti ratio in the pyroxenes can, at least in part, be ascribed to the higher proportion of plagioclase (with respect to femic minerals) in September and November than in December. Plagioclase crystallization induces melt depletion in alumina and enrichment in titanium. The observed mutual relationship between the minerals suggests that plagioclase nucleation preceded olivine and pyroxene crystallization in 2008 products. This is very uncommon in PdF magmas, where the first phase, at least at low pressure, is commonly olivine [e.g., *Famin et al.*, 2009]. Rapid nucleation of plagioclase microlites at shallow depth, such as that promoted by low-pressure magma degassing (see next paragraph for a discussion of plagioclase crystallization), caused a decrease in the Al/Ti ratio of December melts (Figure 9.6a).

9.4.3. Feldspars

In 2008 products, plagioclase composition ranges from An_{79} to An_{61} and spans most of the interval reported by *Boivin and Bachèlery* [2009] for recent PdF lavas (1977–1998 eruptive period; Figure 9.7). Very sodic plagioclase (An_{45-60}) is reported for recent PdF products but is not observed in 2008 ashes.

In 2008 ashes, the range of plagioclase composition is large in September and November and becomes much narrower in December (Figure 9.7). In September, plagioclase composition ranges from $An_{60.6}Ab_{38.3}Or_{1.1}$ to $An_{78.5}Ab_{21.0}Or_{0.5}$ and in November from $An_{63}Ab_{36}Or_{1.0}$ to $An_{77.9}Ab_{21.6}Or_{0.5}$. In December, the range is smaller ($An_{61.2}Ab_{37.3}Or_{1.5}$ to $An_{70.5}Ab_{28.6}Or_{0.9}$) because of the absence of calcic compositions (An > 70.5). On average, plagioclase becomes slightly more sodic from September (An_{72}) to November (An_{70}) to December (An_{67}). The sodic population (An_{61-71}) of September and November ashes becomes the only plagioclase type in December. The composition of this sodic population fits well with that of microlites scattered in the matrices of 2007 lavas (Figure 9.7) and of microphenocrysts and microlites in the matrix of 2005 oceanites [*Welsch et al.*, 2009].

Table 9.5 Representative composition of olivine, clinopyroxene, and chromite in 2008 Pele's hairs and tears.

Eruption	Sept.	Sept.	Sept.	Nov.	Dec.		Dec.		Sept.	Dec.	Dec.	Dec.
Label	220-biglia	22 drap ol	drap-pl1	271-tear2	19 10		19 10 AV		drap-pl3	19-x	19-x	19-3.
Paragenesis	pl-cpx-ol	ol	cpx-ol	pl-cpx-ol	ol-sp		ol-sp		pl-cpx	pl-cpx-ol	pl-cpx-ol	pl + (pl-cpx-ol)
Phase	Olivine	Olivine	Olivine	Olivine	Olivine		Chromite		cpx	cpx	cpx	cpx
SiO_2	38.80	40.03	39.91	39.22	39.87		2.96		52.25	52.03	49.59	48.85
TiO_2	bdl	bdl	bdl	bdl	0.07		2.61		0.87	0.81	1.52	2.04
Al_2O_3	bdl	bdl	bdl	bdl	0.12		14.36		2.29	2.87	4.66	6.00
Cr_2O_3	bdl	bdl	bdl	bdl	0.08		38.22		nd	0.77	0.77	0.58
FeOt	19.78	14.67	14.82	17.57	15.60		26.36		5.46	5.83	6.83	7.31
MnO	0.31	0.11	0.19	0.23	0.24		0.23		0.16	0.15	0.16	0.17
MgO	40.79	45.25	44.04	41.99	41.04		11.07		16.72	15.86	14.91	13.21
NiO	0.29	0.20	nd	nd	0.25		0.17		nd	nd	nd	nd
CaO	0.43	0.31	0.35	0.35	0.43		0.75		20.79	20.06	19.71	19.08
Na_2O	bdl	bdl	bdl	bdl	0.00		0.22		0.27	0.26	0.30	0.51
P_2O_5	bdl	bdl	bdl	bdl	0.16		bdl		bdl	0.33	0.28	0.29
Total	100.40	100.58	99.31	99.36	97.86		96.95		98.81	98.98	98.71	98.02
Fo	78.4	84.5	84.0	80.8	82.2	T (°C)	1130	En	48.01	47.16	45.18	42.45
Fa	21.3	15.4	15.8	19.0	17.5	Log fO2	−8.9	Fs	9.07	9.99	11.88	13.49
Tp	0.3	0.1	0.2	0.3	0.3	NNO	−0.4	Wo	42.92	42.85	42.94	44.07
K_d-MgO 6.3	0.311	0.208	0.216	0.269	0.244							
K_d-MgO 7.4	0.382	0.256	0.265	0.330	0.300							
K_d-MgO 8.7	0.421	0.282	0.292	0.364	0.330							

In September and November ashes, calcic plagioclase (An_{76-79}) forms the outer rim of some of the largest microphenocrysts (Figure 9.8a). Small microphenocrysts and microlites (10–70 µm long) lack calcic rims (<An_{76}) (Figure 9.8b). In November 2008 ashes, the distribution of plagioclase composition is bimodal, with a main mode at An_{63-72} and a second calcic mode at An_{74-78}. Intermediate plagioclase composition is very rare with a single analysis at An_{73}.

Intermediate plagioclase compositions (An_{72-76}) are the dominant type in September 2008 but are rare in November and absent in magmatic products of December 2008, 2007, and 2005 and more generally very rare in the whole 1977–1992 eruptive period [*Boivin and Bachèlery*, 2009] (Figure 9.8). These intermediate compositions of September 2008 fill a gap, which clearly exists for historical PdF magmas in the range An_{71}-An_{75}, between the rare calcic and the widespread sodic compositions.

A further important feature of plagioclase in 2008 products is the occurrence of individuals with contrasting zoning patterns coexisting side by side in the same pyroclast. Zoning patterns are here described in detail. Plagioclases of September 2008 exhibit several styles of zoning (Figure 9.8a). Moreover, some differences are observed between clasts containing only plagioclase (type 1 clasts) or the paragenesis plagioclase + clinopyroxene + olivine (type 2 and type 3 clasts), but crystal composition does not change significantly as a function of crystal size. Zoning styles of tabular microphenocrysts of type 1 clasts are: (a) normal zoning (An_{76-68}), (b) oscillatory zoning (An_{76-70}), (c) reverse zoning on crystal rim (up to An_{79}), and (d) reverse zoning on the whole crystal (up to An_{76}). In this last case, the crystal mantle embeds a resorbed sodic core (An_{61-64}) hosting many small (<20 µm) melt inclusions (Figure 9.8a). Plagioclase found in the pl-cpx-ol bearing tears (type 2 and type 3 clasts) is poorly zoned and slightly more sodic (An_{70-67}) than that found in type 1 clasts. On the contrary, in December and November products, the plagioclase found in type 1, type 2, and type 3 clasts shares very similar compositions.

In November ashes, four types of zoning patterns are found in the type 1 clasts, as in September ashes: (a) normally zoned sodic plagioclase (An_{70-66}), (b) oscillatory zoned plagioclase (An_{75-70}), (c) reversely zoned calcic plagioclase (An_{71-78}), (d) reversely zoned sodic plagioclase (An_{63-71}) with resorbed MI-bearing cores. Plagioclase in type 2 and 3 clasts is usually oscillatory zoned with a general trend of normal zoning (An_{77-63}).

Sodic plagioclase (An_{65-59}) with a low MgO content (MgO < 0.2 wt%) represents the only feldspar occurring in microcrystalline lithic fragments emitted during the November 2008 eruption (Figure 9.7b). Low MgO contents are also measured in the resorbed sodic cores hosting very small melt inclusions.

In December ashes, plagioclase zoning is much smaller than in September and November. Plagioclase ranges from An_{70-65} in the type 1 clasts to An_{71-61} in the dominant

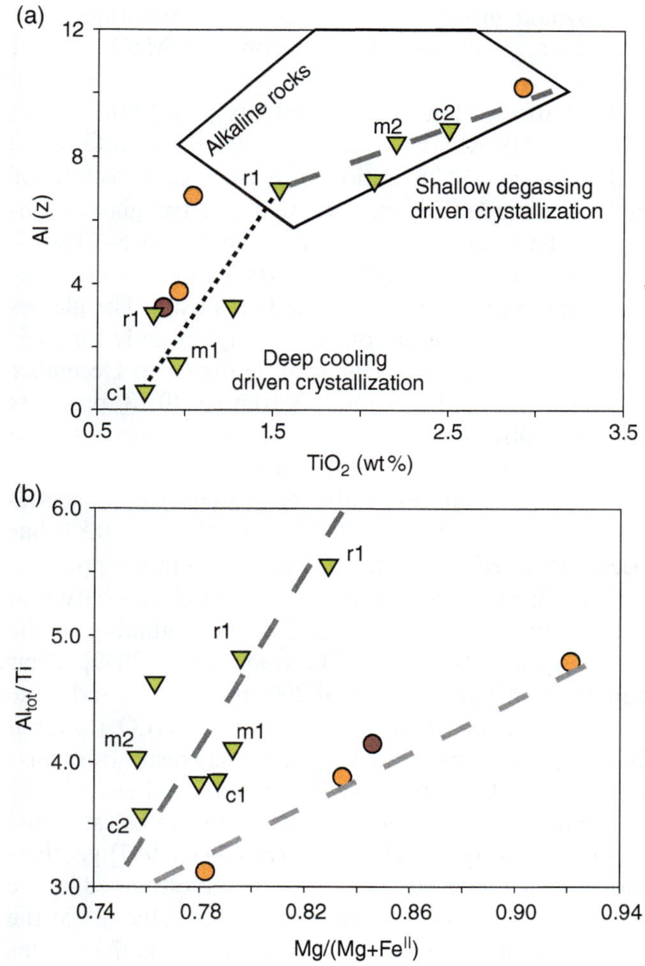

Figure 9.6 (a) Percentage of tetrahedral sites (Alz) occupied by Al vs. TiO_2 content (wt%) in 2008 clinopyroxenes. (b) Altot/Ti vs Mg* of the pyroxenes microphenocrysts and microlites in 2008 PdF pyroclasts. Plagioclase crystallization, enhanced in September and November evolved melts, increases the Ti content in the melt and decreases the Al/Ti ratios of the pyroxenes (c: core; m: mantle; r: rim; numbers indicate different crystals). The field of pyroxenes in alkaline rocks is from *Le Bas* [1962], pyroxene structural formulaes according to *Sturm* [2002].

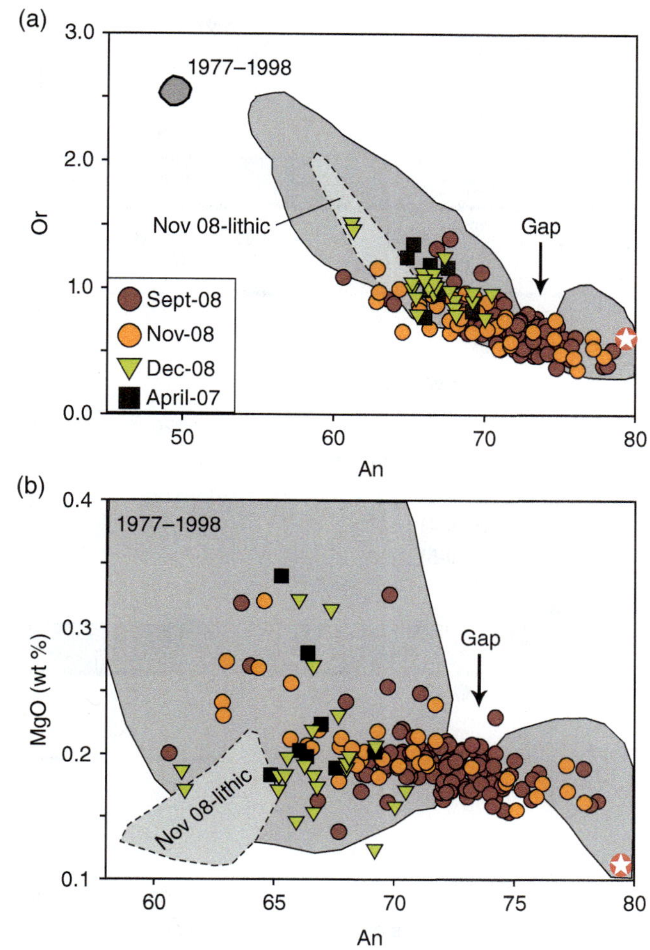

Figure 9.7 Plagioclase composition in eruptive products of PdF. In An vs. Or (a) and An vs. MgO (b) panels, the dark gray areas correspond to the range of plagioclase compositions (phenocrysts to microlites) measured in lavas erupted in the 1977–1998 period [*Boivin and Bachèlery*, 2009]. The star is the average composition of calcic plagioclase phenocrysts in old lavas since the beginning of PdF activity, 530,000 years ago [*Albarede et al.*, 1997]. The arrow points to a gap in plagioclase composition evidenced by the 1977–1998 data set. Most of September and part of November 2008 plagioclases fill this gap. December plagioclase has a more sodic composition; Or and MgO increase with decreasing An. In the three 2008 eruptions, the highest MgO content is found in sodic plagioclases coexisting with ol-cpx, consistently with an increase in Kd_{MgO} with decreasing An [*Bindeman et al.*, 1998]. The light gray area contours the composition of sodic plagioclases in a fine-grained lava clast (lithic) emitted in November 2008. Low-An resorbed cores in December and September plagioclases have low MgO and high Or contents, similar to the plagioclases in the lithic fragment.

type of clasts (types 2 and 3). Only a slight increase in Ca from core to rim and moderate increase in Ca from large to small plagioclases is detected. The most sodic compositions correspond to microaggregates of microlites and have a relatively high MgO content (Figure 9.7b).

9.4.4. Glasses

Matrix glass in Pele's hairs erupted during the September 2008 eruption has been already analyzed by *Villemant et al.* [2009]. The average of our new analyses (52) of glasses in hairs, tears, and some small scoria is very close to the previous average obtained on 43 analyses of *Villemant et al.* [2009]. Only TiO_2 and Na_2O are about 10% higher and lower, respectively, with respect to previous average (Electronic Appendix 1). In addition, we have here analyzed the matrix glasses of November and December eruptions.

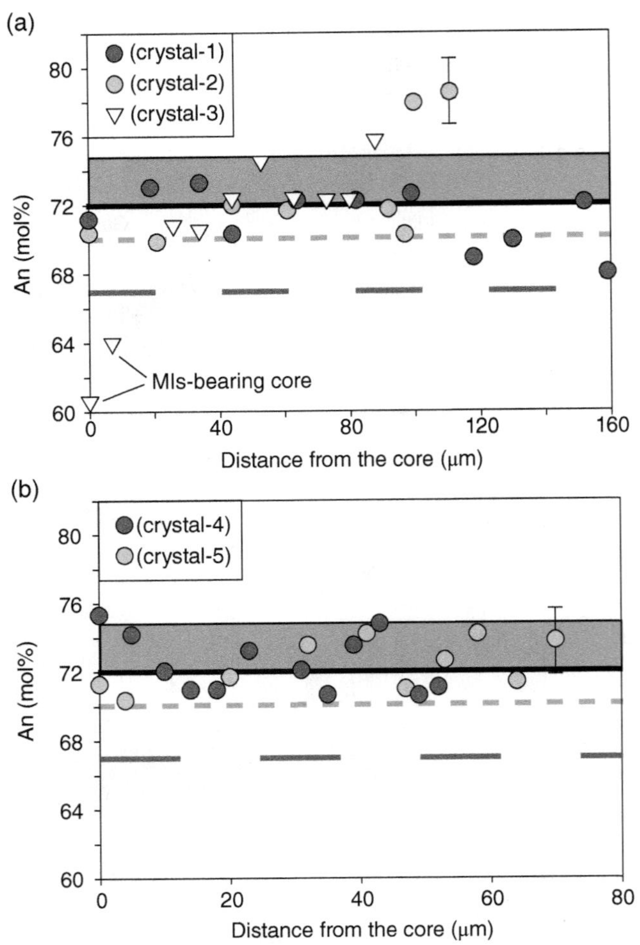

Figure 9.8 Zoning of plagioclase in September 2008 pyroclasts. Plagioclase microphenocrysts (a) are often concentrated in the knots of Pele's hairs, where contrasting types of zoning patterns are observed in crystals located side by side. On the contrary, small plagioclases (b) are only weakly zoned. Most of the largest microphenocrysts have a weak normal zoning (crystal 1; a) or are homogeneous. A secondary crystal population exhibits a sudden increase in An at the outer rim (crystal 2; a). This population is only found in September and November pyroclasts. In some of those crystals, a resorbed sodic core is found (crystal 3; a). This core is often partially resorbed and encloses many small melt inclusions. Those sodic and MgO-poor compositions are also found in fine-grained lithic lava fragments ejected during the November 2008 eruption (Figure 9.7). Average plagioclase composition becomes more sodic from September (continuous line) to November (dashed line) to December (long-dashed line). Only slight normal zoning is commonly observed in December plagioclase. Light gray area in panels (a) and (b) corresponds to the gap in plagioclase composition identified in the 1977–1998 data set [*Boivin and Bachèlery*, 2009].

Two main groups of glasses can thus be identified. The dominant group has low and constant MgO content (6.2 ± 0.1 wt%), moderate K_2O content (0.75–0.8 wt%), and relatively low CaO contents (11.2 ± 0.1 wt%) (Figure 9.9a). All glasses share similar K_2O/Na_2O (0.30) and K_2O/TiO_2 (0.25) ratios. The close composition of bulk ash sample (Figures 9.9a,b) and matrix glasses constrains the total crystal content to less than 5–10 wt%. Slightly higher K_2O (up to 0.97 wt%) and microlite content are typically found in some Pele's tears. The glasses of the three 2008 eruptions differ slightly only for their CaO/Al_2O_3, this ratio being slightly higher in December (0.80) than in September and November (0.78) products (Figure 9.9b). This is apparently in contrast with the higher pyroxene abundance in December ash.

A minor group of mafic and intermediate glasses (MgO: 7.4–8.7 wt%; K_2O: 0.7–0.6; CaO/Al_2O_3: 0.83) has been identified in September and December products (Table 9.3; Figure 9.9). Rare magnesian glasses have also been identified in hairs emitted at the beginning of the major April 2007 eruption [*Villemant et al.*, 2009]. Their composition is close to that of 2008 mafic hairs, and some minor differences are observed in the CaO/Al_2O_3 (lower in 2007). The rare intermediate glasses may result from mixing between the mafic and evolved end members.

The temperature of low-MgO glasses can be estimated at 1140°C using the *Helz and Thornber* [1987] geothermometer, a temperature identical to that estimated on the basis of olivine-spinel equilibrium. This value lies at the lower end of temperatures estimated for 2007 magmas (Figure 9.9c). A slight discrepancy can be observed between T_{MgO} and T_{CaO} (T_{CaO}: 7–20°C higher) that may imply that the 2008 glasses have not crystallized the equilibrium amount of plagioclase, even though some plagioclase is present.

The narrow range of estimated temperature, coupled with measured glass and feldspar compositions, permits an estimation of the water content dissolved in the magmatic melt during crystallization of plagioclase. Dissolved water content was calculated using the *Lange et al.* [2009] plagioclase-liquid hygrometer and can be compared with the micro-Raman analyses performed on 2007 and 2008 glasses (matrices and melt inclusions). Maximum initial dissolved water contents up to 0.8 wt% (September), 0.7 wt% (November), and 0.4 wt% (December) can thus be estimated. These values fit perfectly with those typically measured in melt inclusions hosted in olivines of shallow magmas at PdF [*Bureau et al.*, 1998; *Famin et al.*, 2009; *Di Muro et al.*, 2014]. Average H_2O content decreases from September to December and its distribution is bimodal in November products (Figure 9.10a).

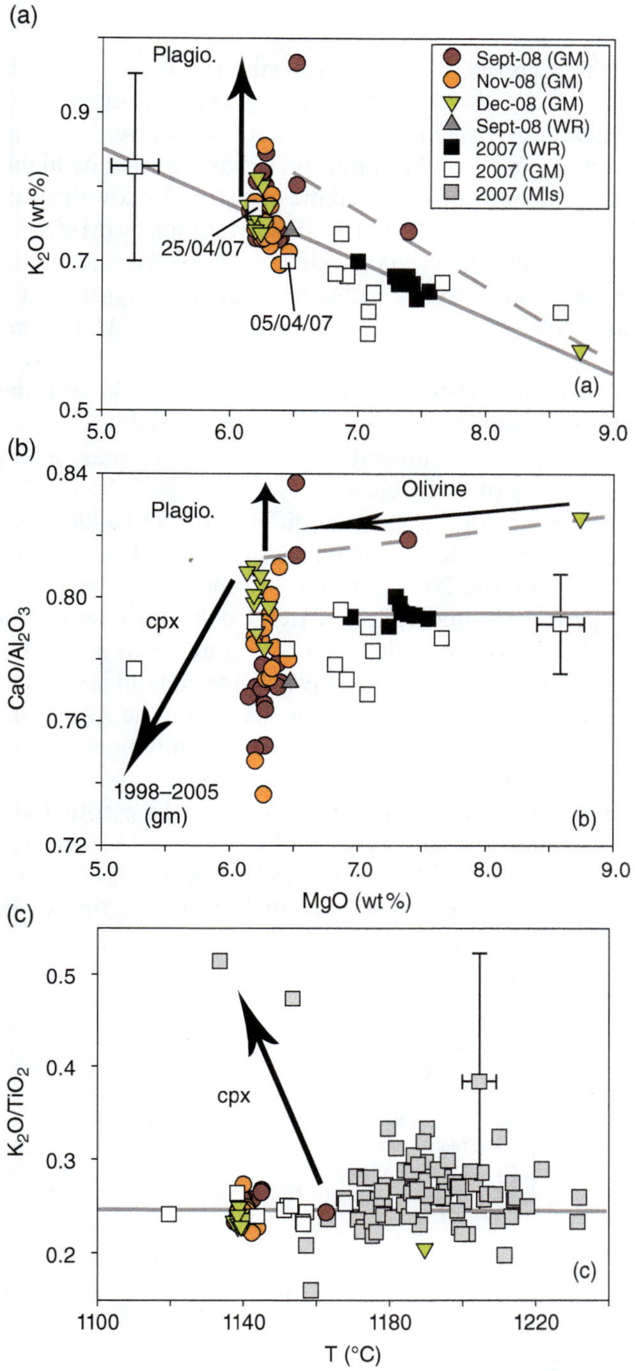

Figure 9.9 Composition of the glasses in 2008 hairs, tears, and scoria. Most 2008 glasses have a low MgO content (6.2 wt%) and differ slightly for their K_2O (a) and CaO/Al2O3 (b) ratios. In September and December ashes, a few intermediate and magnesian (MgO: 8.7 wt%) glasses are found, with a composition close to that of melts and lavas emitted by the 2007 eruptions. As a whole, the 2008 small-volume melts have lower *T* (1140°C) with respect to the large magma volume emitted in 2007 (c). Calculated melt temperatures [*Helz and Thornber*, 1987] are fully consistent with thermal imagery measurements performed during the eruptions (900–1150°C [*Staudacher*, 2010]. Melt inclusion data have been corrected for postentrapment crystallization.

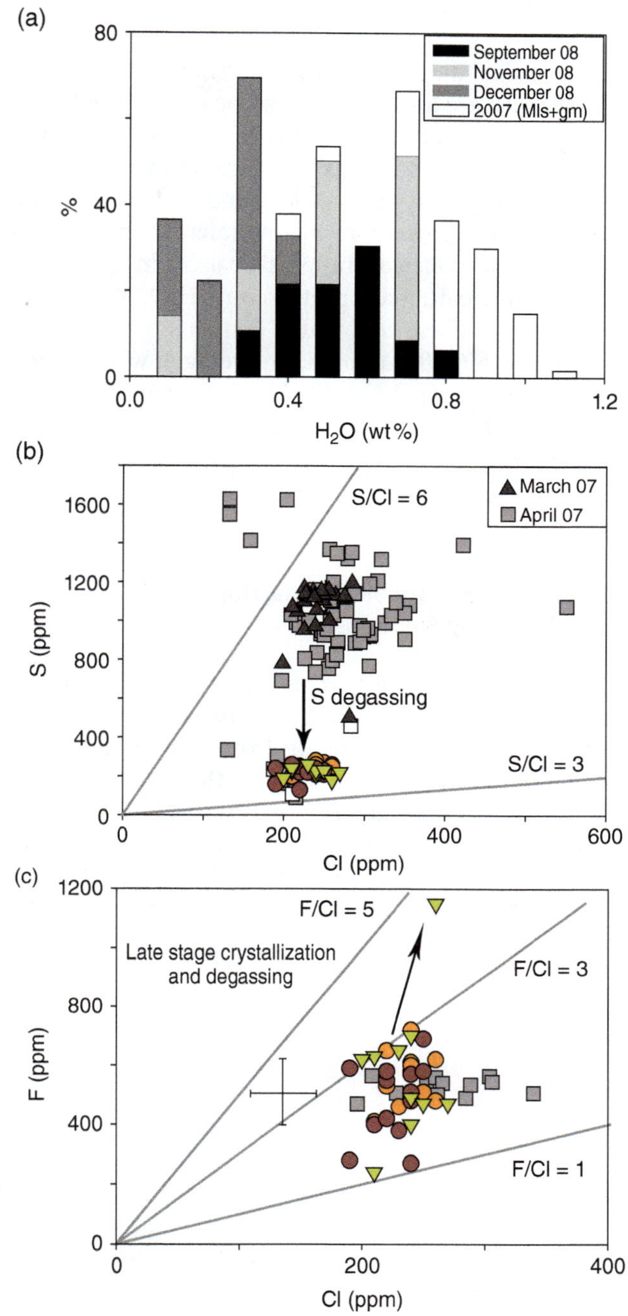

Figure 9.10 (a) Dissolved water contents in the melt crystallizing plagioclase at 1150°C and erupted during the September, November, and December 2008 eruptions. The dissolved water content is calculated on the basis of glass–feldspar equilibrium using the calibration of *Lange et al.* [2009]. Pressure effect is negligible at these shallow levels, where temperature mostly influences the calculated water content. The obtained values are fully comparable with the water contents measured in one glass inclusion in a skeletal olivine (0.6 wt%) and the average of 19 measurements in the groundmass glass by micro-Raman spectroscopy (0.2 wt%). Measured and calculated water contents are lower than those measured in 2007 melt inclusions and groundmasses [data from *Di Muro et al.*, 2014]. Progressively decreasing water contents explain the increasingly more sodic composition of plagioclase from September to December 2008. Dissolved water (a), S, Cl (b), and F (c) contents are fully compatible with Pele's hair recording equilibrium degassing in spite of their very fast quench rate. Shallow chlorine degassing and melt crystallization result in rare high F/Cl ratios in the melt (c).

On one side, the maximum and average water contents of 2008 melts are lower than those measured in 2007 melt inclusions (H_2O up to 1.1 wt%) (Figure 9.10a). On the other side, the low amounts compare well with the moderate content ($H_2O_T = 0.6$ wt%) found in the single analyzed glass inclusion in an olivine microphenocryst and with the low concentrations measured in the matrix glasses by micro-Raman spectroscopy ($H_2O_T = 0.22 \pm 0.04$ wt%; $n = 19$). These low water contents are consistent with the low S (220–240 ppm, on average), while F (500–580 ppm, on average) and Cl (220–240 ppm, on average) are not significantly degassed with respect to 2007 melt inclusions (Figures 9.10b,c; Table 9.3).

9.5. DISCUSSION

9.5.1. Heterogeneity of Magma Stored in PdF Plumbing System

The major 2007 eruptive sequence (bulk volume 0.21 km^3), extruded in a short time, spans a broad set of magma compositions, ranging from aphyric intermediate-MgO (7–7.5 wt% MgO) lavas at the beginning to olivine basalts and oceanites (up to 30 wt% MgO) at the end [*Staudacher et al.*, 2009; *Villemant et al.*, 2009; *Di Muro et al.*, 2014]. The most violent phase of the eruption (April) was also exceptional in emitting both a small volume of highly evolved trachyte melt mingled with the basalt [*Vlastélic et al.*, 2011] and high-MgO basaltic melts (up to MgO 8.6 wt% [*Villemant et al.*, 2009]. The very unusual trachytic melt has been interpreted as a pocket of highly differentiated and degassed magma stored at low pressure [*Vlastélic et al.*, 2011]. The high-MgO melts (MgO: 8.6 wt%) identified in 2007 Pele's hairs are circumstantial evidence for magma recharge being the trigger of the large April eruption [*Villemant et al.*, 2009; *Di Muro et al.*, 2014]. In spite of their very modest volume (2.6 Mm^3 [*Staudacher*, 2010; *Roult et al.*, 2012], 2008 eruptive products provide further record of the variability in melt composition and mineral paragenesis of magmas feeding the activity of PdF volcano.

However, some important differences can be identified between 2007 and 2008 magmas. Glasses (MIs and matrix glasses) of the 2007 eruption are mostly Ol normative (Figure 9.11) and follow a trend controlled by olivine crystallization (picritic trend in Figures 9.4b and 9.11). On the contrary, the normative olivine content is very low in the 2008 melts. Most 2008 melts fall on the An-Di join or converge toward the eutectic ternary minimum (plagioclase + pyroxene + olivine). Only rare magnesian and intermediate hairs found in 2008 follow the picritic trend and contain nonequilibrium olivine ± spinel as the only phase. *Boivin and Bachelery*, [2009] suggest that the normative bulk rock composition of PdF plot on the An-Fo

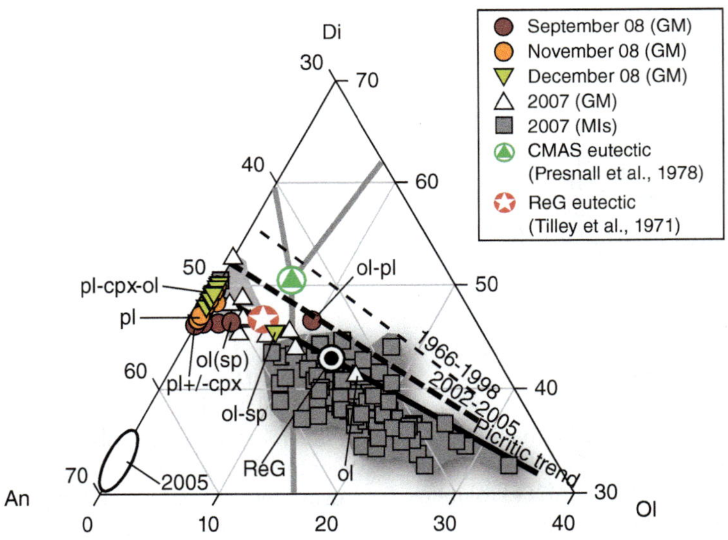

Figure 9.11 Normative diopside, anorthite, and olivine (recalculated to 100) contents in PdF glasses (Pele's hairs, matrices of lavas and scoria). The continuous gray lines and the green triangle correspond to the location of the liquidus boundary lines and of the ternary eutectic point at 1 bar in the $CaO–MgO–Al_2O_3–SiO_2$ (CMAS) synthetic system [*Presnall et al.*, 1978]. Pressure increase expands the pyroxene field at the expenses of plagioclase and shifts the cotectic lines toward the An pole. Actual position of liquidus lines in PdF melts is possibly shifted toward the An pole. Temperature and sequence of phase precipitation were determined at 1 bar by *Tilley et al.* [1971] on the ReG glass (1939 PdF eruption; red star). The normative composition of the ReG glass (7.2% MgO; black circle) is very close to that of the mafic glass erupted in September 2008 and April 2007. Black lines track the time evolution of possible mixing trends between ol-normative mafic melts and more evolved liquids at PdF (glass data from *Boivin and Bachèlery* [2009] and OVPF database).

cotectic line of the CMAS system. We believe that this is an artifact produced by the cumulative nature of most PdF magmas. In the CMAS system, the diopside stability field expands at expenses of olivine and anorthite with increasing pressure [*Presnall et al.*, 1978]. However, our analysis (Figure 9.11) suggests that the actual liquidus lines of PdF basalts differ from the liquidus lines in the synthetic CMAS system. The shift is not controlled by pressure but is related to the effect of alkalis and iron on the position of phase boundaries in natural melts. The observed shift in the An-Di cotectic line toward higher An contents would require very large (up to 7 kbars) pressure changes in the CMAS system. *Suzuki and Ohtani* [2003] show that mafic melts are more compressible with respect to ultramafic melts because of their higher amount of compressible linkages and the observed shift of the liquidus lines could occur in a lower range than reported for the CMAS system. However, a more likely explanation is that the CMAS system is not fully representative of natural PdF basalts. This is in agreement with the measured evolution of melts, which coexist in the PdF plumbing system (Figure 9.11). There, high- and low-Mg melts (Qz and Ol normative melts) never follow the CMAS cotectic lines. Part of the melt heterogeneity can be ascribed at mixing processes between Ol and Qz normative melts, which can have both variable An/Di normative ratio. The variability observed in evolved melts could, at least in part, arise from variable degrees of contamination, especially for extremely high An/Di ratios. Minor heterogeneity in Qz normative melts is likely to correspond to variable degrees of differentiation of distinct groups of evolved melts as they move toward the common cpx-plagioclase cotectic. Plot of Ol-rich normative bulk compositions along the ol-pl cotectic [*Boivin and Bachelery*, 2009] is thus only an artifact of olivine accumulation.

With the notable exception of 2008 melts, plagioclase becomes the second phase only at very low pressure (e.g., 2007 lavas) or plays a prominent role in low-MgO (MgO < 5 wt%) melts [*Tilley et al.*, 1971]. The rare (plagio-phyric) paragenesis of 2008 melts is consistent with the normative composition of these low-MgO melts (6.2 wt%). The normative melt composition is consistent with the observation that the main (in most cases the only) phase found in fast-quenched 2008 hairs is plagioclase, associated with clinopyroxene and Fo-poor olivine in November and December products. The average An/Di ratio decreases slightly from September-November (1.07–1.09) to December (1.03). We stress that the assumed iron redox state has only a minor effect on the calculated An/Di ratio. Increasing the $Fe^{3+}/\sum Fe_T$ molar ratio over a broad range (from 0 to 0.35) increases only slightly the An/Di ratio (from 50.5 to 51.2). On the other side, the iron redox state plays a major role in stabilizing magnetite (not observed in our samples) and strongly affects the Ol/Hy ratio. Hypersthene normative content increases with increasing iron oxidation at expenses of olivine and normative olivine disappears for $Fe^{3+}/\Sigma\ Fe_T > 0.35$.

Petrological analysis permits us thus to draw several conclusions. First, 2007 melts had a higher temperature and higher olivine normative content than the evolved melts emitted in 2008. Widespread plagioclase content in the products erupted in 2008 is consistent with a higher degree of evolution and lower temperature. Second, September 2008 and November 2008 melts cannot be derived by simple cooling and crystallization of 2007 melts, which have higher normative olivine and, sometimes, diopside contents. Third, November 2008 melts could result from September 2008 melts cooling toward the cotectic minimum. However, November 2008 melts contain nonequilibrium olivine (Fo_{81}) and bimodal plagioclase composition, thus revealing that their story is more complex. Fo_{81-82} olivine is in equilibrium with the rare intermediate and mafic melts found in September 2008 and December 2008. Fourth, December 2008 melts correspond to a true cotectic composition. A main and important conclusion is that "cotectic magmas" of PdF [sensu *Boivin and Bachèlery*, 2009] can be formed by cooling and crystallization of distinct melts evolving toward a common eutectic minimum. This is evident in both 2007 and 2008 products, which correspond to (at least) three distinct poles evolving in this same direction: (i) the "Ol-normative pole" (e.g., April 2007 melts), (ii) the "Di-normative pole" (e.g., March 2007 and December 2008), and (iii) the "An-normative pole" (September 2008 and November 2008). A consequence of this observation is that several groups of evolved liquids coexist below the summit of PdF. These melts have been fractionated by their magnesian magma parents after early olivine crystallization and evolve independently in distinct storage zones.

This is the first clear proof that several types of evolved melts occur stored in independent magma pockets in the shallow part of the PdF plumbing system. This conclusion is further validated when looking to longer time series. Highly variable An/Di ratios are found in the whole 1966–2008 period (Figure 9.11). The An/Di ratio is remarkably lower in the 1966–1998 period and increases after 1998. Rare high An/Di melts (65–70) are found in the 2005 oceanite [*Famin et al.*, 2009]. In high-temperature basaltic systems, magma contamination by interaction with plagioclase-bearing [*Pietruszka et al.*, 2009] or pyroxene-bearing [*Pichavant et al.*, 2009] cumulates has been invoked to explain anomalous melt compositions (e.g., An rich) or isotopic ratios (e.g., high $^{87}Sr/^{86}Sr$). In 2008 plagioclases, sodic cores record assimilation of surrounding rocks (older dykes?). Contamination by interaction with older lavas or gabbroic cumulates could explain the highest An/Di ratios, but it does not explain the significant shift in the An/Di ratio after 1998. In this last case, source processes are the most likely cause [*Pietruszka et al.*, 2009].

9.5.2. Feldspar Record of Magma Storage, Ascent, and Degassing

Modal fraction of plagioclase is usually small in historical lavas of PdF. A notable exception is represented by the low-MgO lavas of the long-lasting (six month) 1998 Kapor eruption [*Boivin and Bachelery*, 2009], which occurred after an unusually long (six year) phase of quiescence [*Roult et al.*, 2012]. Actually, most available data on plagioclase composition at PdF come from naturally or artificially (water) quenched lavas, while very few data exist regarding mineral content and composition in fast-quenched pyroclasts. When we consider the products emitted by PdF during its whole activity (530,000 years), plagioclase composition is clearly bimodal irrespective of the eruption age and rock type (pyroclast or lava). Calcic compositions (An_{75-80}) are rather uncommon and are mainly found in large phenocrysts of old lavas [*Albarède et al.*, 1997]. In recent times, calcic plagioclase was only detected in cores of crystals from lavas erupted in 1981, 1983, and primarily 1998 [*Boivin and Bachèlery*, 2009]. In most reported cases, plagioclases are described as small (<500 μm) sodic laths (<An_{71}) dispersed in the lava groundmass and often associated with late-formed clinopyroxene [*Boivin and Bachelery*, 2009; *Welsch et al.*, 2009; *Salaün et al.*, 2010].

In spite of their very fast quench rate and relatively homogeneous composition, all pyroclasts emitted in 2008 (hairs, tears, scoria) contain plagioclases. Moreover, 2008 plagioclases span an unusually broad range in texture, composition, and zoning pattern, even at the scale of a single Pele's hair. September and November pyroclasts are anomalous not only for the widespread occurrence of plagioclase (plagioclase is the only phase found in September hairs) but also for the in uncommon "intermediate" composition (An_{72-76}), which is very seldom observed in the recent eruptive history of the volcano. Plagioclase exhibiting normal to oscillatory zoning (An_{68-76}) likely records the internal convective dynamics of magma storage zones as suggested by *Welsch et al.* [2009] on the basis of pyroxene texture and composition. The unusual composition of the "intermediate" plagioclase is explained by the relatively high dissolved water content (0.8–0.6 wt%) required to crystallize it in the evolved 2008 melts. As a whole, 2008 plagioclases crystallized over a considerable depth range. Zoning patterns, texture, size distribution, and composition of September and November 2008 plagioclase are quite complex and exclude that their crystallization has occurred only at very shallow depth under the effect of magma ascent, decompression, and degassing. In most PdF lavas, sodic plagioclase forms near or at Earth's surface when melt water content drops below 0.4 wt%. This classical scenario is found in December products, where (i) plagioclase nucleation likely resulted from magma degassing during its ascent toward the surface and (ii) moderately variable crystal size and texture were likely produced by variable cooling rate in different parts of the dyke (e.g., center vs. border).

Ash from the 2008 eruptions bring us the final proof that evolved basaltic magmas reside and crystallize at relatively shallow depth below the PdF summit cone. As the initial CO_2 content of 2008 melts is unknown, only some crude estimation of the crystallization pressures (depths) can be performed. When looking at the OVPF database, evolved (MgO < 6.7 wt%) melt inclusions in olivines usually contain <0.4 wt% of H_2O and <100 ppm CO_2. Using *Newman and Lowestern* [2002] solubility model, these concentrations correspond to a lithostatic pressure of 230 bars (900–1200 m below Earth's surface, depending on assumed rock density distribution). Plagioclase erupted in December 2008 eruption and the more sodic population of September and November eruptions probably started to nucleate at these shallow depths. If we assume a higher dissolved CO_2 content of about 280 ppm as often measured in melts with 0.8 wt% H_2O, plagioclase crystallization in September and November would require higher pressures (about 660 bars), corresponding to an approximate maximum depth of 2500–3200 m. Interestingly, this depth range corresponds to the deepest hypocenters recorded before and during 2008 eruptions (Figure 9.1). Sodic plagioclase population likely tracks dyke ascent toward the surface.

In 2008 plagioclase, calcic compositions are not found at the cores of crystals (as in most PdF lavas) but form thin rims on some individuals. Abrupt Ca increases (up to An_{79}) likely result from the preeruptive injection of hotter magma from deeper levels. The rare magnesian and intermediate glasses together with the nonequilibrium high-Fo olivine crystals found in 2008 ash support this hypothesis. These magnesian and intermediate melts are very rare, suggesting that the new input had a quite small volume. The homogeneous low-MgO composition of most hairs suggests that calcic plagioclase rims record a recent event of magma heating. A further constraint comes from the detection of significant SO_2 emissions before the September 2008 eruption [*Garofalo et al.*, 2009]. As recently reviewed by *Di Muro et al.* [2014], the starting pressure of sulfur exsolution in PdF magmas is ~600 bars (3 km below volcano summit). Significant increase in SO_2 emission would thus require magma injection of new magma inside the subaerial part of the volcanic edifice (2600 m above sea level high).

Periodic injection of high-MgO, low-K_2O melts is consistent with the time evolution of K_2O recorded in glasses of PdF. Average K_2O is about 1.1 wt% in 1977 and decreases to 0.9–0.7 wt% in 2008. Low-K_2O glasses have been regularly reported in 1966 (0.53 wt%), 1977 (0.14 wt%) [*Clocchiatti et al.*, 1979], and 1998 (0.33 wt %) eruptions [*Boivin and Bachelery*, 2009]. Interestingly, *Boivin et al.* [2009] have identified a sharp and marked decrease in K_2O content of bulk lavas since 1988. The Mg number of glasses first decreases in the 1977–1998 period and then progressively increases. Since 1998, K_2O decrease is associated with an increase in the range of Mg_n (0.14–0.32 in 1998 and 0.36–0.44 in 2008). The mafic Pele's hairs of 2007 [*Villemant et al.*, 2009] and 2008 (this chapter) eruptions have the highest Mg number ever reported in glasses since at least 1940 (with the exception of MIs). This geochemical evidence supports the hypothesis that the 1998–2007 eruptive cycle corresponds to a phase of continuous injection of deeper melts into the shallower and subaerial part of the PdF edifice, hosting more evolved and crystallized magmas. The aseismic zone identified in Figure 9.1 at about 1–4 km below sea level could correspond to the storage depth range of these mafic magmas. Small mafic inputs from below sea level can trigger the eruption of evolved magma pockets stored in the shallower and subaerial part of the edifice. The scarcity of mafic melts in 2008 products implies that most of the erupted lava and pyroclast correspond to magma previously stored, cooling and crystallizing at relatively shallow depth, in the subaerial portion of the volcanic edifice. The rare occurrence of olivine and pyroxene in early erupted magmas (September, November) and the absence of calcic plagioclase in late erupted (December) evolved melts can point to either two distinct batches stored in distinct zones or to a single well zoned reservoir. The quite homogeneous melt composition of the ash from the three 2008 eruptions suggests that the hypothesis of several distinct magma pockets is more likely. We stress that December 2008 ash is quite similar in bulk and mineral composition to the small volume of aphyric magma emitted by the March 2007 proximal eruption. Those evolved magmas strongly differ in isotopic, major, trace-element, and mineral composition from the main lava volume emitted in April 2007 [*Di Muro et al.*, 2014]. The heterogeneities we here document in lavas emitted in a small time and space interval argue against the occurrence of a single large reservoir in the subaerial edifice proposed by *Peltier et al.* [2009]. Finally, the low-MgO content in partially assimilated sodic plagioclases at the core of larger crystals suggests that at least part of the plagioclase nucleated heterogeneously on assimilated fragments of older rocks (dykes, lavas). This assimilation possibly started with caldera collapse, shortly before eruption to avoid significant MgO reequilibration [*Costa et al.*, 2010].

9.6. CONCLUSIONS

Hawaiian calderas and associated plumbing systems are highly dynamic systems shaped by the alternation of frequent, often small to moderate volume collapses and subsequent refilling phases. Most important, caldera collapses mark major changes in eruptive behavior, as demonstrated by the dramatic decrease in volcanic activity following the 1924 Halema'uma'u collapse at Kīlauea and the 2007 Dolomieu collapse at PdF. The well-monitored 2007 PdF eruption represents a case study to unravel the link between caldera collapses and volcanic activity at basaltic volcanoes.

Geophysical studies reveal that the major 2007 caldera-forming event of PdF resulted in an enhanced tendency of the magma to experience multistep propagation toward the surface [*Roult et al.*, 2012]. Geophysical [*Staudacher*, 2010; *Peltier et al.*, 2010; *Taisne et al.*, 2011] and geochemical [*Garofalo et al.*, 2009] analysis of recent PdF eruptions have evidenced that magma propagates at variable velocity during its travel toward the surface and undergoes temporary stop and partial degassing at shallow depth. Occurrence of shallow storage zones below the PdF summit cone is consistent with our new data on glasses and minerals in 2008 fast-quenched eruptive products, together with the review of interstitial melt compositions for the whole 1966–2007 period. Mineral assemblages and composition together with glass normative compositions clearly indicate that the shallow plumbing system of PdF (upper 3 km, near or above sea level) hosts several pockets of heterogeneous and, on average, evolved melts. These shallow melts are physically separated from more mafic melts, which occur in a higher pressure range (below sea level; likely in the 1–4 km bsl depth range). Small volumes of these hot and magnesian melts periodically ascend, mix with, and reheat the shallower and more evolved magmas, which are at variable stages of differentiation toward a common eutectic minimum. Frequent summit collapses (calderas, pit craters) provide the needed structural conditions for (i) the creation of new storage zones and (ii) small inputs to provoke the eruption of shallow magma pockets [*Michon et al.*, 2013]. This interpretation is also consistent with (i) the progressive upward propagation of cavities detected during 2007 summit caldera collapse [*Peltier et al.*, 2009] and (ii) the shallow magma sources identified for the 2008 summit eruptions [*Peltier et al.*, 2010]. Our petrochemical analysis highlights that at least two shallow magma pockets fed the three 2008 eruptions. The first two eruptions (September and November) involved rare plagio-phyric magmas, whose degree of crystallization increased from September to November. On the contrary, the December eruption occurred on the opposite side of the summit caldera and involved a shallow magma source with a composition close to that of March 2007 aphyric lava. Heterogeneity of magmas extruded during both the 2007 and 2008 eruptive sequences demonstrates that shallow magmas of PdF are stored in several magma pockets. Together with cooling, degassing, and crystallization, assimilation of the surrounding rocks and periodic digestion of small inputs of hotter and more mafic melts determine the time evolution of the shallow and evolved magmas. Multidecadal evolution of their normative and chemical composition, on the other hand, is mostly controlled by source processes [*Pietruszka et al.*, 2009; *Di Muro et al.*, 2014].

Monitoring of fast-quenched glasses is a very effective strategy to track magma storage and dynamics in the plumbing system of very active volcanoes such as Piton de la Fournaise and Hawai'i.

ACKNOWLEDGMENTS

We thank Gilles Montagnac (ENS Lyon; micro-Raman lab) and Agnes Michel (IPGP, Paris; ICP-MS) for their highly skilled technical support in sample analysis. The OVPF team is acknowledged for its constant and continuous effort in volcano monitoring and in sampling of eruptive products. Constructive reviews of M. Edmonds and an anonymous reviewer improved the clarity of our arguments and data presentation. This is IPGP contribution number 3473.

REFERENCES

Albarède, F., B. Luais, G., Fitton, M. Semet, E. Kaminski, B. G. J. Upton, P. Bachèlery, and J.L. Cheminée (1997), The geochemical regimes of Piton de la Fournaise volcano (Réunion) during the last 530000 years, *J. Petrol.*, *38*, 171–201.

Bachèlery, P., F. Saint-Ange, N. Villeneuve, B. Savoye, A. Normand, E. Le Drezen, A. Barrère, J. P. Quod, and C. Deplus (2010), Huge lava flows into the sea and caldera collapse, April 2007, Piton de la Fournaise volcano, IAVCEI Workshop on Collapse Calderas, La Réunion, 2010, Abstract.

Ballhaus, C., R. F. Berry, and D. H. Green (1991), High pressure experimental calibration of the olivine-orthopyroxene-spinel oxygen geobarometer: Implications for the oxidation state of the upper mantle, *Contrib. Mineral. Petrol.*, *107*, 27–40.

Bezos, A., and E. Humler (2005), The Fe3+/ΣFe ratios of MORB glasses and their implications for mantle melting, *Geochim. Cosmochim. Acta*, *69*, 711–725.

Bindeman, I. N., and J. V. Valley (2000), Formation of low-δ18O rhyolites after caldera collapse at Yellowstone, Wyoming, USA, *Geology*, *28*, 719–722.

Bindeman, I. N., A. M. Davis, and M. J. Drake (1998), Ion microproble study of plagioclase-basalt partition experiments at natural concentration levels of trace elements, *Geochim. Cosmochim. Acta*, *62*, 1175–1193.

Boivin, P., and P. Bachèlery (2009), Petrology of 1977 to 1998 eruptions of Piton de la Fournaise, La Réunion Island, *J. Volcanol. Geotherm. Res.*, *184*, 109–125.

Bureau, H., F. Pineau, N. Métrich, P. M. Semet, and M. Javoy (1998), A melt and fluid inclusion study of the gas phase at Piton de la Fournaise volcano (Reunion Island). *Chem. Geol.*, *147*, 115–130.

Clocchiatti, R., A. Havette, and P. Nativel (1979), Relations pétrogénétiques entre les basaltes transitionnels et les océanites du Piton de la Fournaise (Ile de la Réunion, océan Indien) à partir e la composition chimique des inclusions vitreuses des olivines et des spineless, *Bul. Minéral. 102*, 511–525.

Costa, F., L. A. Coogan, and S. Chakraborty (2010), The time scales of magma mixing and mingling involving primitive melts and melt-mush interaction at mid-ocean ridges, *Contrib. Mineral. Petrol.*, *159*, 371–387.

Di Muro, A., N. Métrich, D. Vergani, M. Rosi, P. Armienti, T. Fougereox, E. Deloule, I. Arienzo, and L. Civetta (2014), The shallow plumbing system of Piton de la Fournaise volcano (La Réunion Island, Indian Ocean) revealed by the major 2007 caldera forming eruption, *J. Petrol.*, *55*, 1287–1315.

Famin, V., B. Welsch, S. Okumura, P. Bachèlery, and S. Nakashima (2009), Three differentiation stages of a single magma at Piton de la Fournaise (Réunion hotspot), *Geochem. Geophys. Geosyst.*, *10*(Q01007), doi:10.1029/2008GC002015.

Fisk, M. R., B. G. J. Upton, C. E. Ford, and W. M. White (1988), Geochemical and experimental study of the genesis of magmas of Reunion Island, Indian Ocean. *J. Geophys. Res., B. Solid Earth Planets 93*, 4933–4950.

Froger, J. L., Y. Fukushima, P. Tinard, V. Cayol, T. Souriot, O. Mora, T. Staudacher, P. Durand, B. Fruneau, and N. Villeneuve (2007), Multi sensors InSAR monitoring of volcanic activity: The February and April 2007 eruptions at Piton de la Fournaise, Reunion Island, imaged with Envisat-ASAR and ALOS-PALSAR data, *Proceedings of FRINGE*, ESA, 2007. Available at http://earth.esa.int/fringe07/programme.html

Garofalo, K., T. Staudacher, V. Ferrazzini, P. Kowalski, P. Boissier, A. Dupont, A. Peltier, B. Villemant, and G. Boudon (2009), Eruptive SO_2-plume measurements at Piton de la Fournaise (Ile de la Réunion) by stationary NOVAC scanning MAX-DOAS instruments, p. 12,113, 19–24 April, EGU General Assembly, Vienna.

Helz, R. T., and C. R. Thornber (1987), Geothermometry of Kīlauea Iki lava lake, Hawai'i, *Bull. Volcanol.*, *49*, 651–668.

Jianping, L., J. Kornprobst, D. Vielzeuf, and J. Fabries (1995), An improved experimental calibration of the olivine-spinel geothermometer, *Chin. J. Geochem.*, *14*, 68–77.

Lange, R. A., H. M. Frey, and J. Hector (2009), A thermodynamic model for the plagioclase-liquid hygrometer/thermometer, *Am. Mineral.*, *94*, 494–506.

Le Bas, M. J. (1962), The role of aluminium in igneous clinopyroxenes with relation to their parentage, *Am. J. Sci.*, *260*, 267–288.

Lenat, J. F., and P. Bachèlery (1990), Structure and dynamics of the central zone of Piton de la Fournaise volcano, Le *Volcanisme de la Réunion*, edited by J. R. Lenat, pp. 257–296. Clermont-Ferrand, France.

Marti, J., and A. Gudmundsson (2000), The Las Canadas Caldera (Tenerife, Canary Islands): An overlapping collapse caldera generated by magma-chamber migration, *J. Volcanol. Geotherm. Res.*, *103*, 161–173.

Massin, F., V. Ferrazini, P. Bachèlery, A. Nercessian, Z. Duputel, and T. Staudacher (2011), Structures and evolution of the plumbing system of Piton de la Fournaise volcano inferred from clustering of 2007 eruptive cycle seismicity, *J. Volcanol. Geotherm. Res.*, *202*, 96–106.

Mercier, M., A. Di Muro, N. Métrich, D. Giordano, O. Belhadj, and C. W. Mandeville, C (2010) Spectroscopic analysis (FTIR, Raman) of water in mafic and intermediate glasses and glass inclusions. *Geochim. Cosmochim. Acta*, 74, 5641–5656.

Michon, L., N. Villeneuve, T. Catry, and O. Merle (2009), How summit calderas collapse on basaltic volcanoes: New insights from the April 2007 caldera collapse of Piton de la Fournaise volcano, *J. Volcanol. Geotherm. Res.*, *184*, 138–151.

Michon, L., A. Di Muro, N. Villeneuve, P. Fadda, F. Manta, and C. Saint-Marc (2013), Explosive activity of the summit cone of Piton de la Fournaise volcano: A historical and geological review, *J. Volcanol. Geotherm. Res.*, *263*, 117–133.

Morimoto, N. (1988), Nomenclature of pyroxenes, *Mineral. Petrol.*, *39*, 55–76.

Moune, S., F. Faure, P. J. Gauthier, and K.W.W. Sims (2007), Pele's hairs and tears: Natural probe of volcanic plume, *J. Volcanol. Geotherm. Res.*, *164*, 244–253.

Newman, S., and B. Lowenstern (2002), VOLATILECALC: A silicate melt-H_2O-CO_2 solution model written in Visual Basic for Excel, *Comput. Geosci.*, *28*, 597–604.

Peltier, A., V. Famin, P. Bachelery, V. Cayol, Y. Fukushima, and T. Staudacher, (2008) Cyclic magma storages and transfers at Piton de La Fournaise volcano (La Réunion hotspot) inferred from deformation and geochemical data. *Earth and Planet. Sci. Lett.*, *270*, 180–188.

Peltier, A., P. Bachèlery, and T. Staudacher (2009), Magma transport and storage at Piton de la Fournaise (La Réunion) between 1972 and 2007: A review of geophysical and geochemical data, *J. Volcanol. Geotherm. Res.*, *184*, 93–108.

Peltier, A., T. Staudacher, and P. Bachèlery (2010), New behaviour of the Piton de la Fournaise volcano feeding system (La Réunion Island) deduced from GPS data: Influence of the 2007 Dolomieu caldera collapse. *J. Volcanol. Geotherm. Res.*, *192*, 48–56.

Pichavant, M., I. Di Carlo, Y. Le Gac, S. G. Rotolo, and B. Scaillet (2009), Experimental constraints on the deep magma feeding system at Stromboli volcano, Italy, *J. Petrol.* *50*, 601–624.

Pietruszka, A. J., E. H. Hauri, and J. Blichert-Toft (2009), Crustal contamination of mantle-derived magmas within Piton de la Fournaise, Réunion Island, *J. Petrol.*, *50*, 661–684.

Porritt, L. A., J. K. Russell, and S. L. Quane (2012), Pele's tears and spheres: Examples from Kīlauea Iki, *Earth Planet. Sci. Lett.*, *333-334*, 171–180.

Potuzak, M., A. R. L. Nichols, D. B. Dingwell, and D. A. Clague (2008), Hyperquenched volcanic glass from Loihi Seamount, Hawai'i, Earth Planet. *Sci. Lett.*, *270*, 54–62.

Presnall, D. C., S. A. Dixon, J. R. Dixon, T. H. O'Donnell, N. L. Brenner, R. L. Schrock, and D. W. Dycus (1978), Liquidus phase relations on the join diopside-forsterite-anorthite. From 1 atm to 20 kb: Their bearing on the generation and crystallisation of basaltic magma, *Contrib. Mineral. Petrol.*, *66*, 203–220.

Prono, E., J. Battaglia, V. Monteiller, J. L. Got, and V. Ferrazzini (2009), P-wave velocity structure of Piton de la Fournaise volcano deduced from seismic data recorded between 1996–1999, *J. Volcanol. Geotherm. Res.*, *184*, 49–62.

Roult, G., A. Peltier, T. Staudacher, V. Ferrazzini, B. Taisne, A. Di Muro, and the OVPF team (2012), A comprehensive classification of the Piton de la Fournaise eruptions (La Réunion Island) spanning the 1986–2010 period: Search for eruption precursors from the broad-band GEOSCOPE RER station analysis and interpretation in terms of volcanic processes, *J. Volcanol. Geotherm. Res.*, *241*, 78–104.

Salaün, A., B. Villemant, M. Semet, and T. Staudacher (2010), Cannibalism of olivine-rich cumulate xenoliths during the 1998 eruption of Piton de la Fournaise (La Réunion hotspot): Implications for the generation of magma diversity, *J. Volcanol. Geotherm. Res.*, *198*, 187–204.

Sigmarsson, O., M. Condomines, and P. Bachèlery (2005), Magma residence time beneath the Piton de la Fournaise volcano, Reunion Island, from U-series disequilibria, *Earth Planet. Sci. Lett.*, *234*, 223–234.

Staudacher, T. (2010), Field observations of the 2008 summit eruption at Piton de la Fournaise (Ile de La Réunion) and implications for the 2007 Dolomieu collapse, *J. Volcanol. Geotherm. Res.*, *191*, 60–68.

Staudacher, T., V. Ferrazzini, A. Peltier, P. Kowalski, P. Boissier, P. Catherine, F. Lauret, and F. Massin (2009), The April 2007 eruption and the Dolomieu crater collapse, two major events at Piton de la Fournaise (La Réunion Island, Indian Ocean), *J. Volcanol. Geotherm. Res.*, *184*, 126–137.

Sturm, R. (2002), PX-NORM—An interactive spreadsheet program for the computation of pyroxene analyses derived from the electron microprobe, *Comput. Geosci.*, *28*, 473–483.

Suzuki, A., and E. Ohtani (2003), Density of peridotite melts at high pressure, *Phys. Chem. Min.*, *30*, 449–456.

Taisne, B., F. Brenguier, N. M. Shapiro, and V. Ferrazini (2011), Imaging the dynamics of magma propagation using radiated seismic intensity, *Geophys. Res. Lett.*, *38*, L04304.

Tilley, C. E., R. N. Thompson, W. J. Wadsworth, and B. G. J. Upton (1971), Melting relations of some lavas of Réunion Island, Indian Ocean, *Mineral. Mag.*, *38*, 344–352.

Tulet, P., and N. Villeneuve (2011), Large scale modeling of the transport, the chemical transformation of the sulphur emitted during the April 2007 eruption of Piton de la Fournaise, *Atm. Chem. Phys.*, *11*, 4533–4546.

Villemant, B., A. Salaün, and T. Staudacher (2009), Evidence for a homogeneous primary magma at Piton de la Fournaise (La Réunion): A geochemical study of matrix glass, melt inclusions and Pélé's hairs of the 1998–2008 eruptive activity, *J. Volcanol. Geotherm. Res.*, *184*, 79–92.

Vlastélic, Y., T. Staudacher, and M. Semet (2005), Rapid change of lava composition from 1998 to 2002 at Piton de la Fournaise (Réunion) inferred from Pb isotopes and trace elements: Evidence for variable crustal contamination, *J. Petrol.*, *46*, 79–107.

Vlastélic, I., T. Staudacher, P. Bachèlery, P. Télouk, D. Neuville, and M. Benbakkar (2011), Lithium isotope fractionation during magma degassing: Constraints from silicic differentiates and natural gas condensates from Piton de la Fournaise volcano (Réunion Island), *Chem. Geol.*, *284*, 26–34.

Welsch, B., F. Faure, P. Bachèlery, and V. Famin (2009), Microcrysts record transient convection at Piton de la Fournaise Volcano (La Réunion Hotspot), *J. Petrol.*, *50*, 2287–2305.

Wright, T. L., and R. S. Fiske (1971), Origin of the differentiated and hybrid lavas of Kīlauea Volcano, Hawai'I, *J. Petrol.*, *12*, 1–65.

10

Analysis of Seismicity Rate Changes and Tilt During Early Episodic Fountaining Stage of Pu'u 'Ō'ō, Hawai'i, Eruption: Implications for Magma Storage and Transport

Harmony V. Colella[1,2,3] and James H. Dieterich[1]

ABSTRACT

During the early stages of the Pu'u 'Ō'ō eruption, beginning in 1983 and continuing into 1986, the summit region of Kīlauea Volcano experienced 47 inflation/deflation cycles associated with fountaining episodes at the eruption site. The periods of inflation were characterized by an increased rate of shallow tectonic earthquakes (or short-period events), which culminated in a fountaining episode at Pu'u 'Ō'ō, rapid deflation of the summit, and a spike in volcanic earthquakes (or long-period events). This process is understood as magma accumulation beneath the summit followed by downrift transport to the eruption site. We employ an earthquake rate formulation from rate-state-dependent friction to quantify Coulomb stress changes during deflation cycles at Kīlauea Volcano. Results illuminate two magma reservoirs—one beneath Kīlauea's summit near Halema'uma'u Crater and one near Keanakāko'i and Puhimau pit craters near the intersection between the summit caldera and East Rift Zone. An increase in deflation rates throughout the studied period also suggests an evolution of the East Rift Zone conduit from a blocked system to one where magma is able to move freely from the summit to Pu'u 'Ō'ō.

10.1. INTRODUCTION

Seismicity, magma movement, and deformation are intimately associated processes in active volcanic systems. Kīlauea Volcano, Hawai'i, which has erupted nearly continuously from vents along the East Rift Zone (ERZ) from 3 January, 1983 through date of acceptance of this chapter, provides an excellent natural laboratory to study these interactions (Figure 10.1). The first three-and-a-half years of the eruption were characterized by 47 quasi-periodic lava fountaining events that reached heights of 460 m [*Wolfe et al.*, 1987, 1988; *Okubo and Nakata*, 2003]. *Koyanagi et al.* [1988] recognized a pattern in seismicity and tilt in the summit region throughout the first 20 eruptive episodes (January 1983–June 1984). For 15–25 days prior to the fountaining episode, the summit experienced an increase in the number of short-period earthquakes (SP events) coincident with inflation of the caldera region. The peak of SP events and subsequent termination were followed by initiation of shallow long-period earthquakes (LP events) and rapid deflation of the summit, which followed the onset of high lava fountaining at Pu'u 'Ō'ō approximately 19 km from Kīlauea's summit. This pattern of fluctuating SP and LP events persisted throughout 1983 until mid-July 1986 [*Okubo and Nakata*, 2003] and is presumed to reflect magma accumulation in the summit followed by transport downrift to the Pu'u 'Ō'ō eruption site. The increased seismicity during inflation, as recorded by tilt records in the caldera region, represents stress accumulation in the summit from intrusion of magma. Similarly, the decreased seismicity during deflation reflects stress relaxation as magma is transport into the rift zone toward the eruption site. The high levels

1 Department of Earth Sciences, University of California at Riverside, Riverside, California, USA

2 Department of Geology, Miami University at Ohio Oxford, Ohio, USA

3 School of Earth and Space Exploration, Arizona State University, Tempe, Arizona, USA

Hawaiian Volcanoes: From Source to Surface, Geophysical Monograph 208. First Edition.
Edited by Rebecca Carey, Valérie Cayol, Michael Poland, and Dominique Weis.

Companion Website: www.wiley.com/go/Carey/Hawaiian_Volcanoes

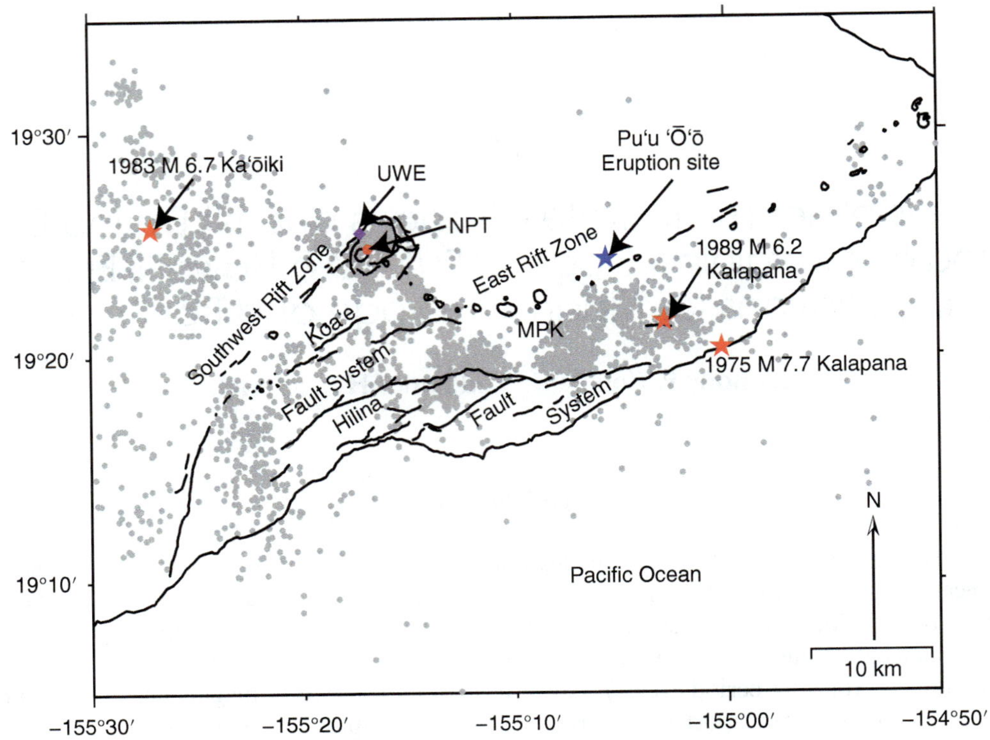

Figure 10.1 Map of Kīlauea Volcano that features the prominent structures. Gray dots represent locations of M_w ≥3.0 earthquakes since the start of the Pu'u 'Ō'ō eruption through December 2009, while red stars indicate earthquakes M_w ≥6. The blue star indicates the location of Pu'u 'Ō'ō, the site of fountaining events referred to in our study. The purple diamond indicates the location of the Uēkahuna vault. The red octagon indicates NPT seismic station used to record daily count data.

of LP events which accompany deflation of the summit, are indicative of magma movement [*Chouet*, 1981].

This study expands on previous studies of the episodic inflation/deflation sequences during the early stages of the Pu'u 'Ō'ō eruption. In particular, we focus on an integrated examination of tilt and seismicity for the entire period of episodic fountaining, May 1983–May 1986, which we explore with the rate-state-dependent friction formulation [*Dieterich*, 1994; *Dieterich et al.*, 2000] and invert earthquake rates to stressing rates. Results show a nonlinear correlation between the changes in tilt and seismicity and a linear relationship between tilt and stress. The results suggest evolution of the conduit that supplies magma from the summit to Pu'u 'Ō'ō.

10.2. GEOLOGIC BACKGROUND

Figure 10.1 illustrates prominent features of Kīlauea Volcano as well as seismicity recorded from January 1983 to December 2009. The structure of Kīlauea is dominated by the summit caldera, two rift zones (East Rift Zone and Southwest Rift Zone), a low-angle basal thrust fault that underlies the volcanic edifice seaward of the rift zones (not shown in Figure 10.1) [*Ando*, 1979; *Furumoto and Kovach*, 1979; *Thurber and Gripp*, 1988], and normal faults (Koa'e Fault System and Hilina Fault System). The faults mentioned above do not appear to play a role in the episodic inflation/deflation events in the study; therefore, they will not be discussed further. Surface deformation and shallow seismicity indicate that magma fed into the volcano accumulates in one or more shallow reservoirs in the vicinity of the summit caldera [*Eaton*, 1962; *Fiske and Kinoshita*, 1969; *Decker et al.*, 1983; *Klein et al.*, 1987]. The caldera region and flank of Kīlauea Volcano have high rates of seismicity. Seismicity rates in the rift zones are much lower and dominated by intrusion events, where seismicity migrates from the summit region along the rift zones as magma moves through the system [*Koyanagi et al.*, 1974; *Ryan et al.*, 1981; *Klein et al.*, 1987].

The geometry of magma storage at Kīlauea Volcano is complex. *Fiske and Kinoshita* [1969] were the first to document migration of the locus of maximum vertical deformation over time, which suggests the presence of multiple magma bodies beneath the summit caldera. They suggested a primary reservoir centered at a depth 2–3 km beneath the southern portion of the caldera based on a point source model. Alternatively, *Dieterich and Decker* [1975] employed a finite reservoir model, which suggested the south caldera reservoir is vertically elongated and extends close to surface (<1 km). Recent deformation studies of the volcano conclude there are a minimum of three reservoirs in the summit area: (1) located below the eastern margin of Halema'uma'u Crater (centered at 1–2 km depth), (2) near Keanakāko'i Crater (centered at ~3 km depth), and (3) the primary reservoir below the southern region of the caldera (centered at ~4–5 km depth) [*Baker and Amelung*, 2012; *Poland et al.*, 2014]. Magma migration to and from the summit is concentrated within the ERZ and Southwest Rift Zone (SWRZ) [e.g., *Holcomb*, 1987; *Fiske et al.*, 1993]. Other small magma reservoirs may exist along the rift zones as relicts of past intrusive events and fissure eruptions (e.g., near Nāpau and Makaopuhi craters) [*Swanson et al.*, 1976].

10.3. OBSERVATIONS AT KĪLAUEA'S SUMMIT

This study utilizes tilt and seismicity data collected by the staff of the Hawaiian Volcano Observatory (HVO) to analyze the relation between surface deformation, magma movement, and seismicity.

10.3.1. Tilt

Tilt data for the summit region for this study were recorded at the Uēkahuna Vault (UWE in Figure 10.1). The site consists of an Ideal Arrowsmith electronic tiltmeter and a water tube tiltmeter. The Ideal Arrowsmith is a mercury capacitance instrument that only records tilt in the east-west direction. The water tube tiltmeter records north-south and east-west tilt data on a daily (and occasionally on a subdaily) basis.

Figure 10.2 shows the average direction of deflation obtained from the water tube tiltmeter for Pu'u 'Ō'ō eruptive episodes between May 1983 and April 1986. The inset is a polar histogram that displays results for all deflation vectors during this time. The mean direction of deflation is calculated to be S75°E and varies between due east and S60°E. Consistency of the tilt direction indicates that a stable magma reservoir was activated during inflation/deflation cycles throughout the first three years of the Pu'u 'Ō'ō eruption. The stability of the source and nearly east-west cycling of the tilt vector further indicates that the continuously recorded data set from the Ideal Arrowsmith electronic tiltmeter can be used as a continuous quantitative measure of the summit inflation/deflation.

Tilt data collected from the electronic tiltmeter for the three-year period of May 1983 to May 1986 is shown in Figures 10.3a, 10.3d, and 10.3g, where gray lines indicate the onset of fountaining at Pu'u 'Ō'ō as published by *Heliker and Mattox* [2003]. Tilt gradually increases (summit inflation) between each eruptive episodes, then rapidly decreases (summit deflation) after the onset of fountaining. The average inflation period is approximately 24 days, and the average inflation magnitude is 11.97 μrads with a maximum of 18 μrads. The average deflation magnitude is 16.63 μrads, which indicates an overall deflation of the summit throughout this time period.

Individual inflation cycles show a roughly linear increase in tilt with time during the beginning of each

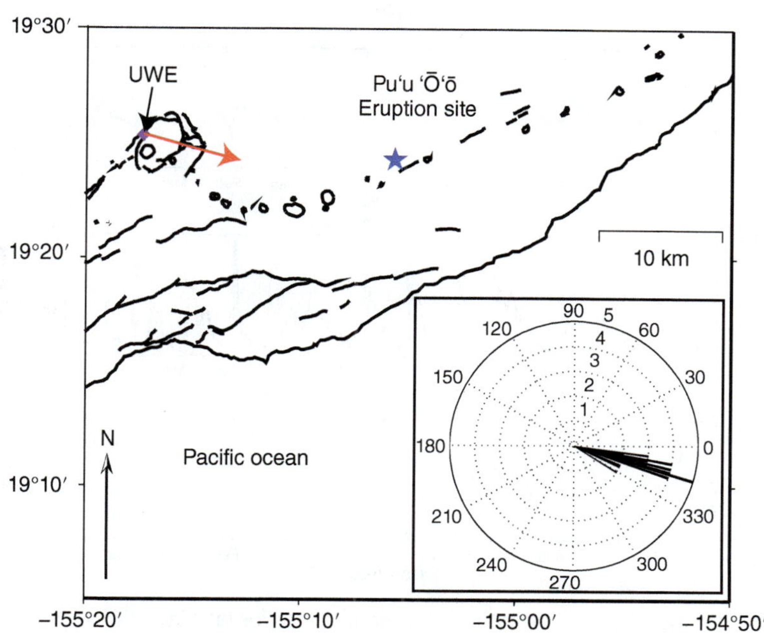

Figure 10.2 Map of Kīlauea Volcano where red arrow indicates mean direction of deflation from water tube tilt data from UWE, and the blue star indicates the eruption site. Inset is a polar histogram of deflation directions for all episodes between May 1983 and May 1986. A mean deflation direction of S75°E was obtained.

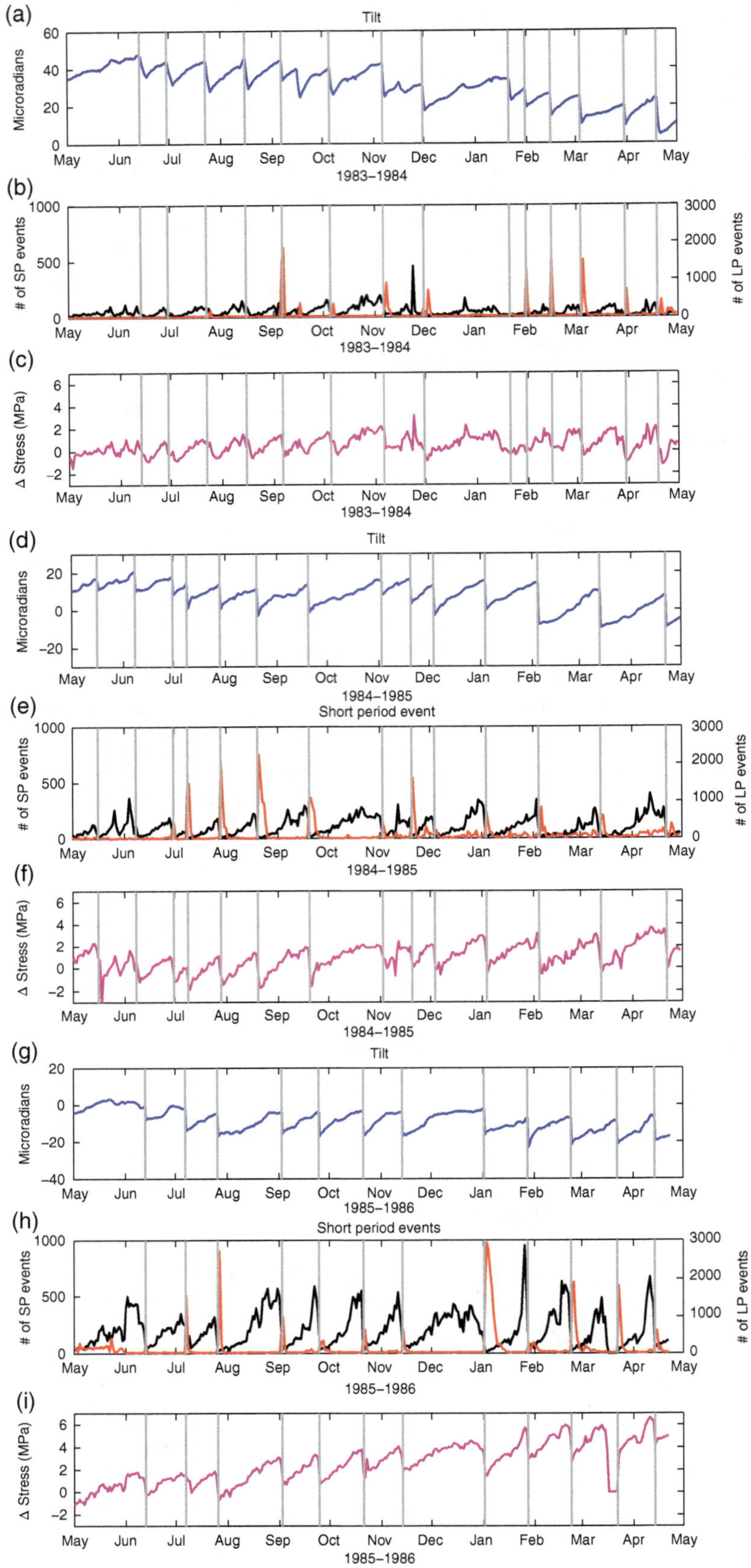

Figure 10.3 Tilt data from the Ideal Arrowsmith electronic tiltmeter for (a) May 1983–April 1984, (d) May 1984–April 1985, and (g) May 1985–April 1986. Note that there is no offset in the vertical axis between plots and the overall decrease in tilt level over the three year period indicates an overall deflation of the summit. Earthquake count data for (b) May 1983–April 1984, (e) May 1984–April 1985, and (h) May 1985–April 1986. Black curves show short-period events and red curves show long-period events. Results for temporal Coulomb stress solutions (magenta) for (c) May 1983–April 1984, (f) May 1984–April 1985, and (i) May 1985–April 1986. The gray lines indicate onset of fountaining at Pu'u 'Ō'ō [*Heliker and Mattox*, 2003]. Tilt and daily count data provided by HVO staff.

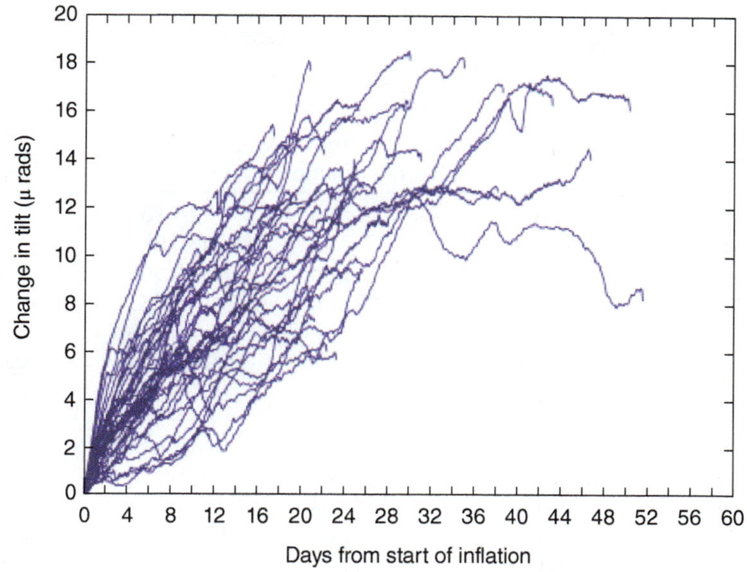

Figure 10.4 Change in tilt for each inflation event superimposed on one another. Data provided by HVO staff.

inflationary period; however, in the later stages the rate of inflation generally decreases and in some cases becomes zero or in some cases begins to slowly deflate prior to the rapid deflation events. Figure 10.4 shows superimposed tilt records for the inflation episodes, where time is measured from the onset of inflation to the time fountaining begins. The inflation periods vary from 8 to 51 days throughout the early stages of Pu‘u ‘Ō‘ō eruptions. Although there is considerable scatter in the curves, inflation rates begin to slow at 10–15 μrads and, in some cases, fluctuate between deflation and inflation prior to the main deflation event. This flattening suggests the summit magma reservoir may have a maximum volume capacity or critical pressure that limits inflation.

While the initial inflation rates remain relatively constant through fountaining cycles, the rates of summit deflation progressively increase. *Dvorak and Okamura* [1987] identified a progressive increase in deflation rates for the first six deflation episodes. An examination of the subsequent deflation episodes shows this trend continued to the end of the fountaining episodes (Table 10.1). This suggests that magma transport through the lateral conduit to the eruptive site became progressively easier with time due to evolution of the conduit system.

10.3.2. Earthquake Count Data

We make use of two types of earthquake data collected by staff at HVO: earthquake count data and standard catalog data. Earthquake count data consist of recorded events that are too small to be accurately located. The standard catalog data are part of the Advanced National Seismic System (ANSS) catalog with uniform reporting of earthquake magnitudes to $M_w = 1.5$.

The count data we employ were supplied by the HVO staff and previously described by *Koyanagi et al.* [1988] and *Okubo and Nakata* [2003]. Earthquakes recorded at selected seismic stations that are too small to locate are classified into regional categories based on arrival times and amplitude differences and reported as daily counts. Count data for the summit region of Kīlauea consists of events recorded by the NPT seismometer (see Figure 10.1). Within this region, the events are further classified as SP or LP events based on frequency and waveform characteristics. SP events have the characteristics of typical tectonic earthquakes and usually occur within 15 km of the surface. They are associated with ground deformation and formation of new cracks from intrusion of magma into the summit reservoir [*Koyanagi et al.*, 1988]. LP events are associated with active volcanism and are indicative of magma movement [*Chouet and Julian*, 1985; *Chouet*, 1996].

Figures 10.3b, 10.3e, and 10.3h shows the SP and LP count data together with electronic tilt data (Figures 10.3a, 10.3d, and 10.3g) for May 1983 through April 1986. As reported by *Koyanagi et al.* [1988] for the first 20 episodes, the earthquake count data, similar to the tilt measurements, show regular fluctuations throughout the 41 cycles shown by the black curves in Figures 10.3b, 10.3e, and 10.3h (the initial cycles are not shown and were not used in the analysis due to less cyclic behavior and the final few episodes are not shown due to a lack of earthquake count data). As tilt increases, the rate of SP events at the summit also increases. When tilt begins to rapidly decrease at the onset of deflation, the SP events essentially stop. We interpret the increase of SP seismicity to represent an increase in the state of stress in the vicinity of the summit reservoir and the decrease in seismicity to

Table 10.1 Start and end times of fountaining and deflation episodes.

Fountaining Episode Number	Start Date of Eruption	State Time of Eruption	Time to Rapid Deflation from Start of Eruption	End Date of Eruption	End Time of Eruption	Time to Minimum Tilt After Termination of Eruption (±30 min)
4	06/13/83	10:25 AM	+4 h 35 min	6/17/1983	1:13 PM	+4 h 47 min
5	06/29/83	12:51 PM	+8 h 49 min	7/3/1983	7:15 AM	+2 h 15 min
6	07/22/83	3:30 PM	+3 h 30 min	7/25/1983	4:30 PM	+2 h 30 min
7	08/15/83	7:41 AM	+2 h 19 min	8/17/1983	4:00 PM	+3 h
8	09/06/83	5:11 AM	0	9/7/1983	5:26 AM	+4 h 34 min
9	09/15/83	3:41 PM	0	9/17/1983	5:20 PM	+3 h 40 min
10	10/05/83	1:06 AM	+2 h 54 min	10/7/1983	4:50 PM	+10 min
11	11/05/83	11:50 PM	+1 h 10 min	11/7/1983	6:45 PM	+15 min
12	11/30/83	4:47 AM	+1 h 13 min	12/1/1983	3:45 PM	+2 h 15 min
13	01/20/84	5:24 PM	+3 h 36 min	1/22/1984	11:23 AM	+5 h 37 min
14	01/30/84	5:45 PM	+1 h 15 min	1/31/1984	1:18 PM	+5 h 42 min
15	02/14/84	7:40 PM	+20 min	2/15/1984	3:01 PM	+5 h 59 min
16	03/03/84	2:50 PM	+1 h 10 min	3/4/1984	10:31 PM	+2 h 31 min
17	03/30/84	4:48 AM	+2 h 12min	3/31/1984	3:24 AM	+4 h 36 min
18	04/18/84	6:00 PM	+1 h	4/21/1984	5:33 AM	+1 h 26 min
19	05/16/84	5:00 AM	Indeteminant	5/18/1984	12:50 AM	-16 h 50 min
20	06/07/84	9:04 PM	Indeteminant	6/8/1984	6:25 AM	+9 h 35 min
21	06/30/84	10:28 AM	Indeteminant	6/30/1984	6:27 PM	+9 h 33 min
22	07/08/84	7:30 PM	−6 h 30 min	7/9/1984	10:17 AM	+5 h 43 min
23	07/28/84	12:00 PM	0	7/29/1984	5:40 AM	+5 h 20 min
24	08/19/84	9:52 PM	+8 min	8/20/1984	5:25 PM	+3 h 35 min
25	09/19/84	4:04 PM	0	9/20/1984	5:32 AM	+4 h 28 min
26	11/02/84	11:40 AM	−40 min	11/2/1984	4:36 PM	+10 h 24 min
27	11/20/84	12:05 AM	0	11/20/1984	10:06 AM	+5 h 54 min
28	12/03/84	7:05 PM	0	12/4/1984	9:41 AM	+4 h 19 min
29	01/03/85	1:15 PM	−3 h 15 min	1/4/1985	5:04 AM	+2 h 56 min
30	02/04/85	5:46 AM	+14 min	2/5/1985	2:46 AM	+9 h 14 min
31	03/13/85	6:00 AM	+1 h	3/14/1985	4:55 AM	+2 h 5 min
32	04/21/85	3:16 PM	+2 h 44 min	4/22/1985	9:06 AM	+3 h 54 min
33	06/12/85	11:06 PM	0	6/13/1985	4:53 AM	+10 h 7 min
34	07/06/85	5:03 PM	+2 h 57 min	7/7/1985	8:50 AM	+5 h 10 min
35	07/26/85	2:52 AM	−52 min	7/26/1985	9:52 AM	+4 h 8 min
36	09/02/85	2:00 PM	−1 h	9/2/1985	11:35 PM	+3 h 25 min
37	09/24/85	6:08 PM	0	9/25/1985	6:19 AM	+2 h 41 min
38	10/21/85	3:00 AM	Indeteminant	10/21/1985	11:24 AM	+3 h 36 min
39	11/13/85	3:34 PM	0	11/14/1985	1:24 AM	+1 h 36 min
40	01/01/86	1:09 PM	−2 h 9 min	1/2/1986	2:38 AM	+3 h 22 min
41	01/27/86	8:35 PM	−2 h 35 min	1/28/1986	7:57 AM	+3 h 3 min
42	02/22/86	3:15 PM	+5 h 15 min	2/23/1986	4:20 AM	+2 h 40 min
43	03/22/86	4:50 AM	+5 h 50 min	3/22/1986	3:56 PM	+2 h 4 min
44	04/13/86	8:54 PM	+34 h 54 min	4/14/1986	7:56 AM	+2 h 4 min

Note: Eruption number and start and end times reflect those published by *Heliker and Mattox* [2003].

represent a relaxation of stresses around the reservoir. However, the earthquake count data are concave up while the tilt data are concave down, which indicates a nonlinear relationship between earthquake rates and tilt. The peaks in LP counts coincide with summit deflation and typically begin after the onset of fountaining at Puʻu ʻŌʻō (Table 10.1).

The cumulative number of earthquakes from daily count data at the summit for the studied period is shown in Figure 10.5 (black curve). Between January and May 1983 background earthquake count rates were 22 events/day. During the next year (May 1983–April 1984) earthquake rates were ~52 events/day. The rate doubles to ~105 events/day the following year (May 1984–April 1985) and

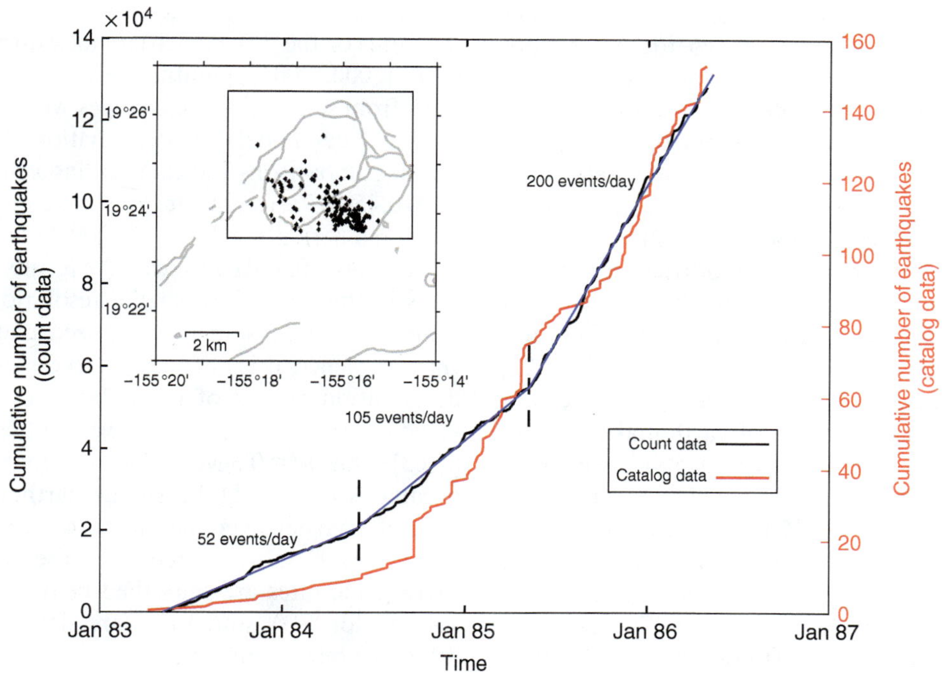

Figure 10.5 Cumulative number of earthquakes in the summit region. Solid black curves represent cumulative SP earthquake from daily count data. Red curves represent cumulative number of earthquakes from earthquake catalog data. Blue curve represents three different sub-intervals, with different average rates, separated by the black dotted lines. Inset is a map of the summit, including the location of catalog earthquakes from the ANSS catalog (black dots).

doubles yet again to ~200 events/day during the final year (May 1985–April 1986). The vertical, dotted, black lines in Figure 10.5 denote these changes in rate. Figure 10.5 also shows the cumulative number of earthquakes for catalog data reported for the summit region and upper ERZ (red, seismicity seen in the inset). Catalog data show a similar trend, although recorded events are sparse in comparison to earthquake count data.

10.4. ANALYSIS OF STRESS CHANGES FROM EARTHQUAKE RATE DATA

10.4.1. Method

We interpret the changes in SP seismicity rates during inflation-deflation cycles to be related to the stress changes in the summit region caused by magma reservoir inflation and deflation. The Coulomb stress function has been widely used to model the effect of stress changes on seismicity patterns [*King et al.*, 1994; *Jaumé and Sykes*, 1996; *Harris et al.*, 1995; *Stein*, 1999; *Cayol et al.*, 2000; *Troise*, 2001; *Feuillet et al.*, 2004; *Amelung et al.*, 2007]. Where there is an increase or decrease in the Coulomb failure stress function, the earthquake rate will increase or decrease, respectively. Changes in Coulomb failure stress are given as

$$\Delta S = |\Delta\tau| - \mu\Delta\sigma, \qquad (10.1)$$

where $\Delta\tau$ is the shear stress acting across the fault plane in the positive slip direction, $\Delta\sigma$ is the change in effective normal stress (normal stress less pore fluid pressure), and μ is the coefficient of fault friction.

The correlation between earthquake rates and stress changes is highly nonlinear and history dependent [*Dieterich*, 1994]. To analyze the relation between earthquake rates and tilt, we use the formulation of *Dieterich* [1994, 2007] to calculate stress changes from earthquake rates during inflation-deflation cycles. This formulation is based on laboratory observations of rate- and state-dependent fault strength. Earthquake rate is given by

$$R = \frac{r}{\gamma \dot{S}_r}, \qquad (10.2)$$

Where

$$d\gamma = \frac{1}{a\sigma}(dt - \gamma dS) \qquad (10.3)$$

and γ is a state variable, t is time, and S is the Coulomb stress function using a modified coefficient friction described below. Parameter a sets the rate dependence of friction on slip speed in the rate state friction formulation: It typically takes values of about 0.01 [*Dieterich*, 2007]. The constant r is a reference steady state earthquake rate in some volume for the reference stressing rate $\dot{S}_r$. The Coulomb stress in equations (10.2) and (10.3) is defined using the effective coefficient of friction $\mu' = \mu - \alpha$, where α is a dimensionless parameter $0 \leq \alpha \leq \mu$, which defines the evolution of frictional state in response to a change of normal stress [*Linker and Dieterich*, 1992]. Because the formulation is explicitly designed to account for the clustering characteristics of earthquakes, the full earthquake catalog is used at every stage of the analysis: no declustering of the catalog is done. For the inversions reported here, we use $\alpha = 0.25$ based on the laboratory measurements of *Linker and Dieterich* [1992]. In equation (10.3), the term $a\sigma$ is treated as a constant, i.e., changes in σ are negligible relative to total σ. See *Dieterich* [2007] for review of rate-state friction and applications to seismicity models, which include foreshocks and aftershocks.

Following *Dieterich et al.* [2000, 2003], we employ equations (10.2) and (10.3) to solve for stress changes based on observed changes in seismicity rate. Note from equation (10.2) that parameter γ is inversely proportional to the observed earthquake rate R. Hence, the seismicity rate as a function of time, $R(t)$, for a specified volume is used to directly calculate γ as a function of time (that is, $\gamma(t) = r / R(t)\dot{S}_r$). From $\gamma(t)$, a time-marching procedure is then used to integrate equation (10.3) to find the stress as a function of time, $S(t)$.

This method requires estimates of the constant formed by the ratio $r / \dot{S}_r$ and the term $a\sigma$. The amplitude, but not the form, of the solution for stress depends on the numerical value of $a\sigma$. Hence, in practice, the computations are done in terms of normalized Coulomb stress ($S/a\sigma$). This allows the normalized form of parameter $\dot{S}_r$ to be expressed in terms of parameter t_a, the aftershock duration. From *Dieterich* [1994]

$$t_a = a\sigma / \dot{S}_r \tag{10.4}$$

and is defined as the characteristic time for seismicity rates to return to the background rate r. If r and $\dot{S}_r$ are defined as the average values of earthquake rate and stressing rate, respectively, over the period of analysis, then t_a is directly estimated from observations of aftershocks. For the current study, we use t_a = 3 years, which was the value employed by *Dieterich et al.* [2000, 2003] from several aftershock sequences in this time interval. Test inversions show that the solutions are rather insensitive to variations of t_a by a factor of $\pm 0.3 t_a$, which is within the range of the observed aftershock durations. *Dieterich et al.* [2000, 2003] quantitatively compared stress solutions from seismicity rate changes with stresses obtained from elastic models of deformation data at Kīlauea. Those comparisons indicate $a\sigma$ has values in the range 0.25–0.45 MPa for the regions covered in the analyses presented here. *Dieterich et al.* [2003] adopted a preferred value of $a\sigma = 0.3$ MPa to scale the results. The stress solutions in this study lack spatial resolution and represent some average of a highly localized and variable stress change. Hence, it is not feasible to directly utilize elastic deformation models of the inflation events to scale the value of $a\sigma$. For this reason, we use the *Dieterich et al.* [2003] value $a\sigma = 0.3$ MPa. The reference earthquake rate parameter r is simply the average earthquake rate for the area of analysis over the period of time considered by the study. Finally, to reduce noise in the solutions, earthquake rates are smoothed before defining $\gamma(t)$ and solving for $S(t)$, and therefore determination of the reference rate parameter r.

Two somewhat different approaches are employed to analyze the summit area earthquake count and ANSS catalog data sets. First, the large number of events in the earthquake count data set provides good temporal resolution of stress changes for the individual inflation-deflation cycles; however, there is no spatial information from the the data set because the events are not located. Therefore, the stress inversions for the count data are presented as single time series (Figures 10.3c, 10.3f, and 10.3i, magenta curves). We regard the stresses determined from the count data to represent some average measure of the complex stress state in the summit region associated with magma injection withdrawal as recorded by the tilt changes. Because earthquake counts in the summit area changed over the period of this study (Figures 10.3b, 10.3e, and 10.3h, black curves), solutions are obtained for three subintervals, each with a different value for the reference rate r. The values for r were set to the average subinterval rates as indicated in Figures 10.3b, 10.3e, and 10.3h. Second, the ANSS earthquake catalog data are employed to determine earthquake rates as a function of time, $R(t)$, for overlapping cylindrical volumes centered at regular grid points within the summit region. This yields a solution for stress changes at each grid point as a function of time. From these time series, maps, and cross sections stress changes over some time interval can then be constructed (Figure 10.6). The earthquakes in the ANSS catalog are sparse for the period of this study. To better resolve possible spatial patterns of stress changes, the catalog events from individual inflation-deflation cycles are stacked relative to a common time in each cycle. The result is a composite catalog of events for several cycles. The event times in the stacked catalog are

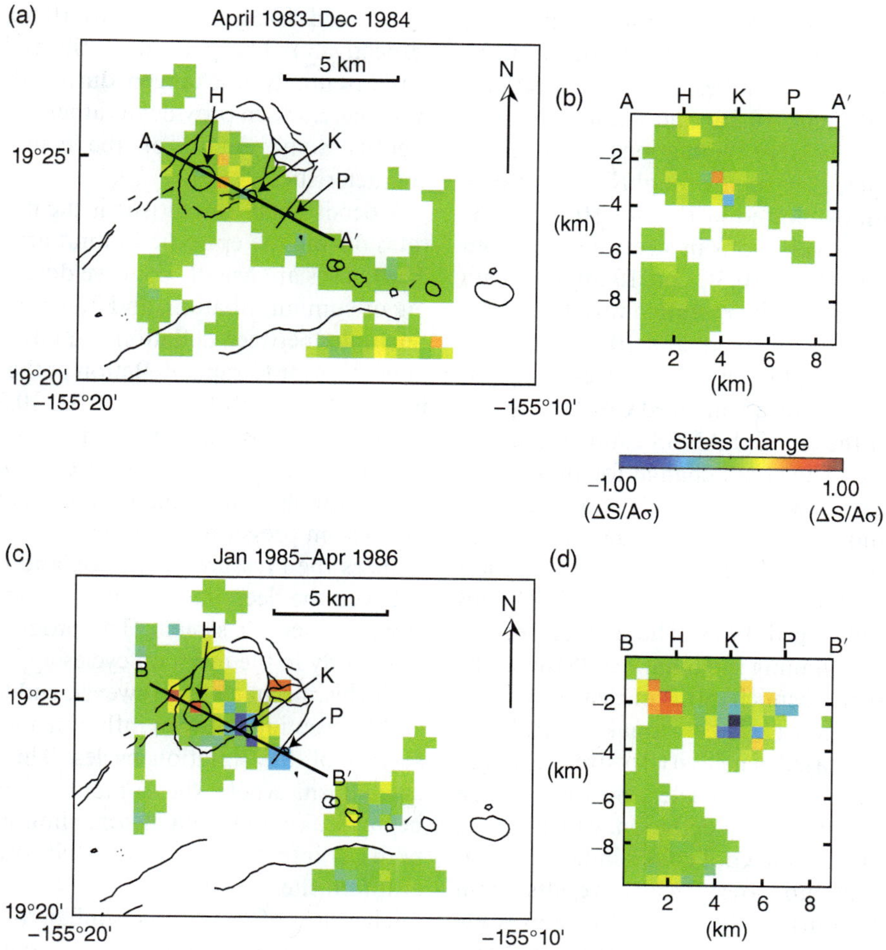

Figure 10.6 Spatial distribution of Coulomb stress solutions based on seismicity rate changes. (a) Map view solutions for April 1983–December 1984. (b) Cross-sectional view for solutions for April 1983–December 1984 for A–A′. (c) Map view solutions for January 1985–June 1986. (d) Cross-sectional view for solutions for January 1985–June 1986 for B–B′. Warm colors indicate an increase in stressing rate while cool colors indicate a decrease in stressing rate. H = Halema'uma'u Crater, K = Keanakāko'i Crater, P = Puhimau Crater.

referenced to the time of the highest daily count (peak) of the short-period events for each individual cycle, which typically marks the end of an inflation cycle. This stacking appears reasonable given the stability of the inflation-deflation cycles. To avoid signals unrelated to the inflation-deflation episodes, eight cycles were removed from the stacking process due to M_w >3.5 and their aftershocks, which occurred during the studied period (episodes 5–7, 12, 18, 24, 27, and 34). The average rate in the volume from the stacked catalog was used to set the parameter r for the solution. To generate maps of possible stress changes, a grid spacing of 0.5 km was used and the cylindrical volumes extended from the surface to depths of 20 km. The radii of the cylindrical volumes were incremented from 1 to 4 km to give a minimum average rate of eight earthquakes per year (M_w ≥1.5). If the minimum rate requirement was not met with a radius ≤4 km, a solution was not calculated for that grid point. Solutions for spatial stress changes for cross sections were obtained using a similar method, with a 1 km grid and horizontal cylindrical volumes extending ±1.5 km from the vertical plane of the cross section.

10.4.2. Results

Temporal solutions for Coulomb stress changes based on SP count data are shown in Figures 10.3c, 10.3f, and 10.3i. Although the stress solutions are quite similar to the observed seismicity rate data, some differences can be noted that arise from the nonlinear dependence of earthquake rates on stress changes. The SP count data tend to have a concave upward form (Figures 10.3b, 10.3e, and 10.3h, in black), which reflects an increase in seismicity rates, while the stress solutions tend to have a concave downward form, which indicates a decrease in stressing rates during inflation stages of the cycles. In this respect, the form of the stress solution curves more closely reflects the tilt data (Figures 10.3a, 10.3d, and 10.3g). In addition, SP count data show large fluctuations in seismicity rates from cycle to cycle (varying by roughly a factor of 4 in the 12 month periods plotted in Figures 10.3b, 10.3e, and 10.3h). In contrast, variations in stress change from cycle to cycle (Figures 10.3c, 10.3f, and 10.3i) are much smaller and, again, more closely resemble variations seen in tilt records (Figures 10.3a, 10.3d, and 10.3g), which is what one would expect if tilt is a proxy for stress changes.

Maps and cross sections of normalized Coulomb stress changes ($S/a\sigma$) from the stacked ANSS catalog data are shown in Figure 10.6. The stress change for these solutions is the difference between the seven day average stress prior to deflation and the seven day average stress after deflation. Results were obtained by stacking deflation events from April 1983 through December 1984 and January 1985 through April 1986. These time periods were chosen because beginning in January 1985 there was a significant increase in seismicity. It is unclear why there was such a significant increase in earthquake rates, but it certainly indicates increased sensitivity to stress changes during this time. Warm colors indicate a positive change in Coulomb stress during summit deflation, which results from increased seismicity, and cool colors indicate a negative change in Coulomb stress, which results from decreased seismicity during deflation. In map view (Figures 10.6a and 10.6c) and cross section (Figures 10.6b and 10.6d) two areas with persistent Coulomb stress changes during deflation emerge. These regions are most prominent in the January 1985–April 1986 interval (Figures 10.6b and 10.6d). The first is a decrease in Coulomb stress near Keanakāko‘i and Puhimau pit craters in the vicinity of the junction between the upper ERZ and Kīlauea Caldera. In cross section, this area of Coulomb stress decrease is located at ~3 km depth, again, near Keanakāko‘i and Puhimau craters. The second area is an increase in Coulomb stress below the eastern section of the caldera, in the northeastern part of Halema‘uma‘u Crater. In cross section, this area of Coulomb stress increase is located at ~1–2 km depth, which is thought to be the depth of a small reservoir perched above the main 3 km depth magma chamber beneath the south caldera [*Rowan and Clayton*, 1993; *Okubo et al.*, 1997; *Dawson et al.*, 1999].

10.5. DISCUSSION AND CONCLUSIONS

Between May 1983 and May 1986 Pu‘u ‘Ō‘ō experienced 41 episodic fountaining episodes associated with remarkably regular cyclic changes of tilt and seismicity in the area of Kīlauea's summit caldera (Figures 10.3a, 10.3d, and 10.3g and Figures 10.3b, 10.3e, and 10.3h, respectively). The pattern of tilt and seismicity change was essentially unchanged during the three years analyzed here, which provides a rather unique opportunity to identify some details of the causal mechanisms and characteristics.

A persistent characteristic is the nonlinearity of the tilt rates during the episodic summit area inflation episodes. The curves are usually concave down, indicative of slowing of summit inflation, and in many cases tilt flattens or fluctuates between deflation and inflation for an extended interval prior to rapid deflation at the time of a fountaining event downrift (Figures 10.3a, 10.3d, and 10.3g). Two mechanisms possibly operate to produce this slowing of inflation rates. First, as proposed by *Dvorak and Okamura* [1987], the slowing of inflation rates may be caused by an increase in pressure in the magma storage system, which reduces the pressure difference, and hence the flow rate, between the deep magma supply system and the shallow storage reservoir system. The progressive slowing of tilt rates early in the inflation cycles appears to be consistent with this mechanism. However, sudden flattening of tilt and fluctuation between inflation and deflation is seen in several of the inflation cycles. This suggests a second mechanism where the summit magma storage system begins leaking magma at some limiting magma pressure, perhaps into the lateral conduit system that feeds the eruption site.

The rates of SP events, similar to tilt, increase during the inflation stage of each cycle, but the SP rate curves show an exponential-like acceleration (Figures 10.3b, 10.3e, and 10.3h in black and Figure 10.5, black curves) that is opposite of the slowing of tilt rates (Figures 10.3a, 10.3d, and 10.3g). We infer this feature to be a direct consequence of the short-term exponential response of seismicity rates to stress change increases predicted by the rate-state seismicity formulation of equations 10.2 and 10.3. This exponential dependence is illustrated in Figures 10.7a–c, which summarizes the relationships between tilt data, SP rates, and temporal stress changes. The top panels (Figures 10.7a–c) show the stressing rate changes on the x-axis and summit area SP rates on the y-axis. The bottom panels (Figures 10.7d–f) show tilt on the x-axis and stressing rate changes on the y-axis. Notably the plots of stress against tilt give a linear relationship, which is consistent with the linear elastic deformation of the summit reservoir system due to injection of magma.

Spatial solutions for Coulomb stress changes show two persistent regions of change, which are more obvious between January 1985 and June 1986, related to deflation of Kīlauea's summit during fountaining at Pu‘u ‘Ō‘ō (Figures 10.6c and 10.6d): (1) a positive Coulomb stress change below the northeastern portion of Halema‘uma‘u Crater and (2) a negative Coulomb stress change near

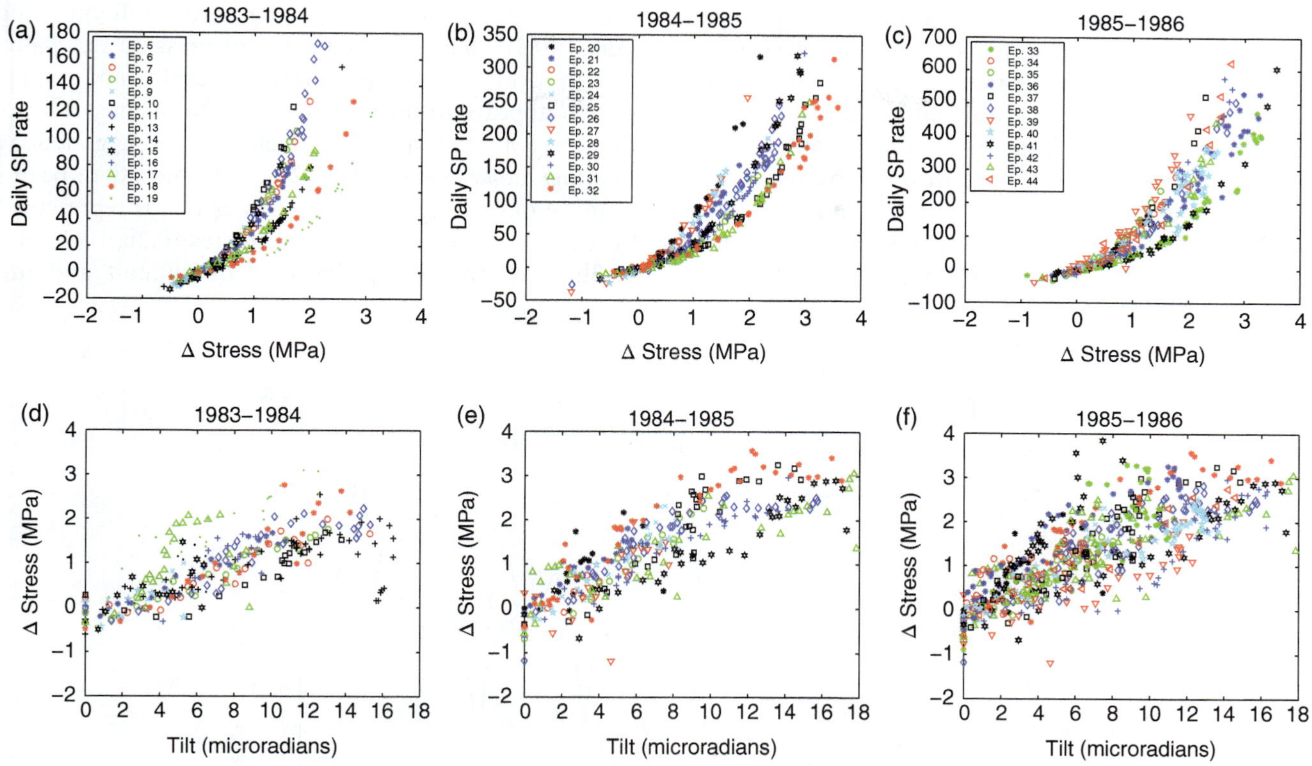

Figure 10.7 (a–c) Relation between Coulomb stressing rates and SP daily rates for May 1984–April 1984, May 1984–April 1985, and May 1985–April 1986, respectively. (d–f) Relation between tilt data and Coulomb stressing rates for May 1984–April 1984, May 1984–April 1985, and May 1985–April 1986, respectively. Different symbols in each plot are used to differentiate between Pu'u 'Ō'ō events.

Keanakāko'i and Puhimau craters near the start of the ERZ. We interpret the region of increased stress during deflation at ~1–2 km depth beneath the northeast portion of Halema'uma'u Crater to be related to a small magma reservoir located in that area. Our depth and location correlate well with seismic and deformation studies of Kīlauea Volcano and the location of the small magma chamber beneath Halema'uma'u Crater [*Rowan and Clayton*, 1993; *Okubo et al.*, 1997; *Dawson et al.*, 1999; *Battaglia et al.*, 2003; *Baker and Amelung*, 2012; *Poland et al.*, 2014]. It is less clear as to why this region undergoes a stress increase, since magma moving out of the reservoir should result in a stress decrease. Earthquake count data document an abundance of LP events (Figures 10.3b, 10.3e, and 10.3h, red curves) in the summit region during deflation, and one would expect a significant increase in larger locatable LP events during these intervals as well. The ANSS catalog should flag LP events; however, very few exist within the catalog during this time, while we know that thousands of LP event occurred. This suggests that the apparent positive Coulomb stress change near Halema'uma'u is an artifact of unflagged LP events during deflation rather than increased rate of SP events.

The negative Coulomb stress change is located between Keanakāko'i and Puhimau craters in the vicinity of the junction between the summit caldera and ERZ (Figure 10.6c) at approximately 3–4 km depth (Figure 10.6d). We interpret this region of decreased stress during deflation to reflect the decrease of pressure as magma is released into the lateral conduit that feeds the ERZ, which is ultimately erupted at Pu'u 'Ō'ō. The depth of the stress change is consistent with the ERZ conduit system [*Delaney et al.*, 1990; *Gillard et al.*, 1996]. A negative stress change in this region may also be indicative of an additional magma reservoir in the region of Puhimau Crater, which is also suggested by *Baker and Amelung* [2013] and *Poland et al.* [2014]. Several studies identify this area as a "hotspot" with high heat flow [*Dunn et al.*, 1985], soil enriched in mercury, carbon dioxide and sulfur [*Hinke et al.*, 1978], and high levels of CO_2 output [*McGee et al.*, 2006], all of which suggest a shallow magma body. Additionally, seismic studies identify low seismic velocities [*Dawson et al.*, 1999] and locate a cluster of LP events [*Battaglia et al.*, 2003] in this region, and *Koyanagi* [1988] identifies an intrusive event beneath Puhimau Crater in 1980.

No immediate precursors to the fountaining episodes have been identified. However, as noted by *Koyanagi et al.* [1988], tremor detected near Pu'u 'Ō'ō and Makaopuhi Crater (on the ERZ between the caldera and Pu'u 'Ō'ō) is coincident with the duration of the eruption, and it decreases in intensity with distance uprift from Pu'u 'Ō'ō. They also document that tremor at the summit is delayed with respect to fountaining at Pu'u 'Ō'ō. This is consistent with the delay (minutes to hours) of LP events and deflation at the summit after the onset of fountaining (Table 10.1). Hence, the reduction of pressure associated with the draining of magma from the conduit and storage system appears to take a short time to propagate from Pu'u 'Ō'ō to the summit, which is also suggested by *Montagna and Gonnermann* [2013]. Magma then begins to move out of the summit region, which is represented by the occurrence of LP events (Figures 10.3b, 10.3e, and 10.3h, red curves) and deflation of the summit, and into

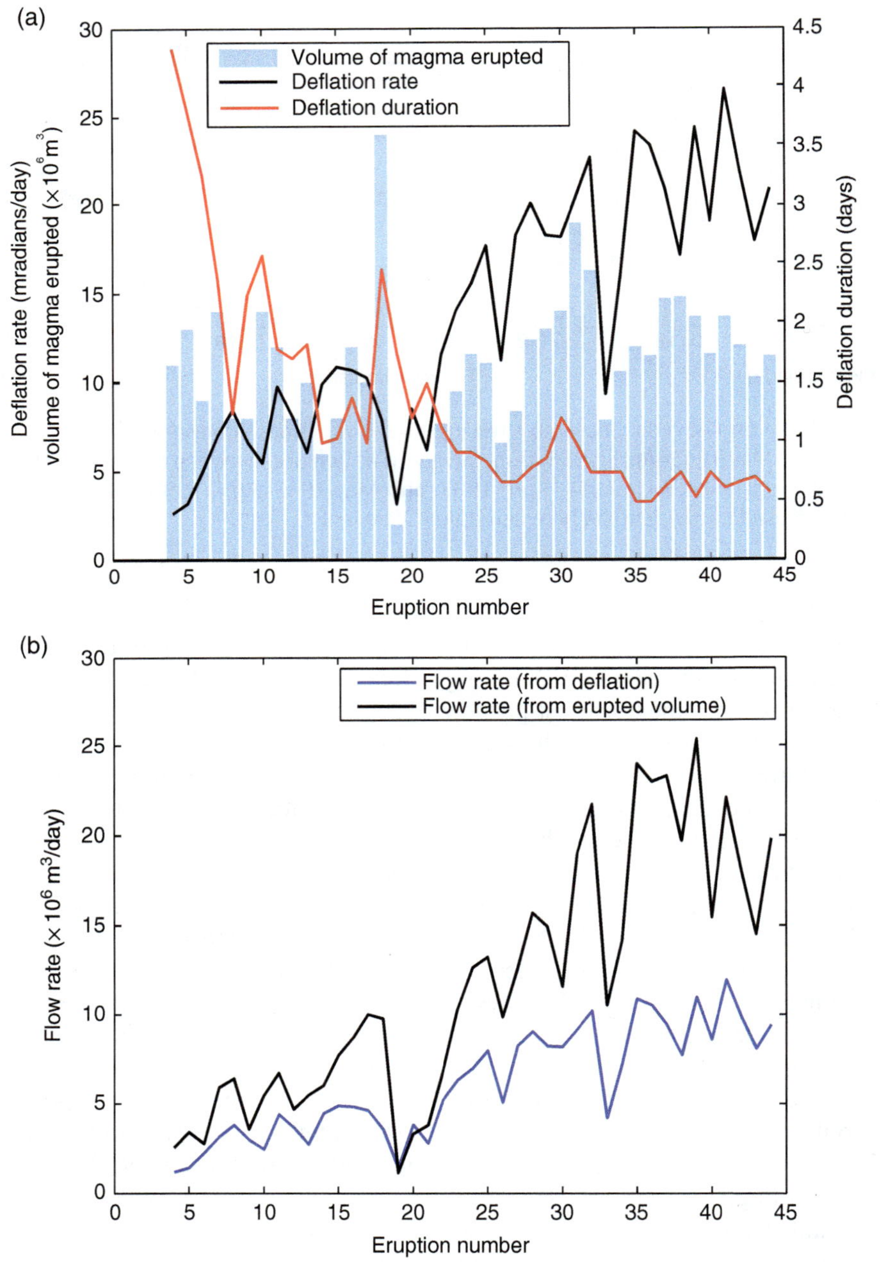

Figure 10.8 (a) Volume of erupted magma (cyan bars), deflation rates (black curve), and duration of deflation (red curve). (b) Flow rates based on *Dvorak and Okamura* [1987] (blue curve) and erupted volume divided by deflation duration (black curve). Data from Table 10.2.

Table 10.2 Tilt rates of Kīlauea summit and volume of magma erupted at Pu'u 'Ō'ō.

Eruption Number[a]	Volume of Magma (10^6 m^3)	Deflation Duration (days)	Deflation Amount (μrad)	Deflation rate (μrad/day)	Magma Flow Rate Based on rates from *Dvorak and Okamura*, [1985] (10^6 m^3/day)	Magma Flow Rate Based on Volume Erupted/Deflation Rates (10^6 m^3/day)
4	11	4.33	11.31	2.61	1.18	2.54
5	13	3.81	12.11	3.18	1.43	3.41
6	9	3.25	16.23	4.99	2.25	2.77
7	14	2.38	16.67	7.00	3.15	5.88
8	8	1.25	10.59	8.47	3.81	6.40
9	8	2.25	15.02	6.68	3.00	3.56
10	14	2.58	14.13	5.48	2.46	5.43
11	12	1.79	17.5	9.78	4.40	6.70
12	8	1.71	13.91	8.13	3.66	4.68
13	10	1.83	11.1	6.07	2.73	5.46
14	6	1	9.91	9.91	4.46	6.00
15	8	1.04	11.31	10.88	4.89	7.69
16	12	1.38	14.77	10.70	4.82	8.70
17	10	1	10.27	10.27	4.62	10.00
18	24	2.46	19.44	7.90	3.56	9.76
19	2	1.76	5.56	3.16	1.42	1.14
20	4	1.21	10.32	8.53	3.84	3.31
21	5.7	1.5	9.32	6.21	2.80	3.80
22	7.7	1.13	13.07	11.57	5.20	6.81
23	9.5	0.92	12.92	14.04	6.32	10.33
24	11.6	0.92	14.28	15.52	6.98	12.61
25	11.1	0.84	14.88	17.71	7.97	13.21
26	6.6	0.67	7.54	11.25	5.06	9.85
27	8.4	0.67	12.27	18.31	8.24	12.54
28	12.4	0.79	15.89	20.11	9.05	15.70
29	13	0.87	15.91	18.29	8.23	14.94
30	14	1.21	22.02	18.20	8.19	11.57
31	19	1	20.38	20.38	9.17	19.00
32	16.3	0.75	17.02	22.69	10.21	21.73
33	7.9	0.75	7.01	9.35	4.21	10.53
34	10.6	0.75	12.03	16.04	7.22	14.13
35	12	0.5	12.09	24.18	10.88	24.00
36	11.5	0.5	11.69	23.38	10.52	23.00
37	14.7	0.63	13.18	20.92	9.41	23.33
38	14.8	0.75	12.87	17.16	7.72	19.73
39	13.7	0.54	13.17	24.39	10.98	25.37
40	11.6	0.75	14.32	19.09	8.59	15.47
41	13.7	0.62	16.46	26.55	11.95	22.10
42	12.1	0.67	14.68	21.91	9.86	18.06
43	10.3	0.71	12.77	17.99	8.09	14.51
44	11.5	0.58	12.14	20.93	9.42	19.83
45	9.4	ND	ND	ND	ND	ND
46	9.8	ND	ND	ND	ND	ND
47	8.8	ND	ND	ND	ND	ND

Note: ND = no data.
[a] Eruption numbers reflect those published by *Heliker and Mattox* [2003].

the lateral conduit. Additionally, we observe that summit deflation typically continues for several hours after fountaining terminates at Pu'u 'Ō'ō, which extends the observation of *Koyanagi et al.* [1988] that LP events at the summit continue after termination of Pu'u 'Ō'ō fountaining and tremor at Makaopuhi beyond June 1984 and throughout the time of episodic inflation-deflation fountaining cycles. Together, these observations suggest that magma continued to flow from the summit into the conduit that feeds Puʻu ʻŌʻō even after fountaining stopped.

As discussed above, the tilt records and stress solutions appear to be indicative of a pressure limit for the storage system prior to fountaining. From this, we infer that the mechanism for initiation of the fountaining events involves the breakdown of a barrier, or barriers, within the conduit that transports magma from the summit region to Puʻu ʻŌʻō. It is unclear where this barrier is located; however, *Wolfe et al.* [1987] suggest that it may be located near the Puʻu ʻŌʻō vent. When the pressure limit is reached, volatile-rich magma is pushed into the shallow eruptive vent, which initiates rapid degassing and fountaining. The pressure drop that results within the vent due to fountaining then drives high flow rates from the summit to feed a self-sustaining eruption [*Montagna and Gonnermann*, 2013]. The relatively constant volume erupted in the fountaining episodes (Figure 10.8) suggests that eruptions end when the storage system, and hence the pressure difference that drives magma flow to the eruption site are sufficiently depleted.

The rates of summit deflation during the fountaining episodes progressively increase throughout the 3.5 year fountaining phase (Table 10.2), while the volume of magma erupted in the individual fountaining events essentially remains constant (Figure 10.8a). Flow rates were determined two ways. First, they are based on the calibration factor of *Dvorak and Okamura* [1987] that relates tilt at Uēkahuna to volume injected or withdrawn from the summit magma reservoir (Figure 10.8b, blue curve). This relation is based on an elastic point source model and was calibrated using tilt changes and level surveys between 1966 and 1983. Second, flow rates are based on the volume of erupted magma divided by the deflation duration (Figure 10.8b, black curve). These average flow rates at the eruption site both approximately double the estimates from tilt and show significant increased rates of flow with time. This indicates an evolution of the conduit system that allows magma to move more freely through the conduit system from the summit. *Parfitt and Wilson* [1994] also suggest a gradual evolution of the lateral conduit throughout this eruptive series due to progressive widening of the narrowest sections of the conduit. The transition from episodic fountaining to continuous eruption in July 1986 likely represents the final step in this evolution to a completely open conduit system.

ACKNOWLEDGMENTS

We thank Valérie Cayol, Michael Poland, and Rebecca Carey for the invitation to submit to this monograph. Associate Editor Michael Poland and two anonymous reviewers provided helpful comments that improved the manuscript. We thank the USGS and the Hawaiian Volcano Observatory for access to the data that are included in this chapter, with special thanks to Asta Milkius, Jennifer Nakata, and Paul Okubo. This study benefitted from in-depth discussions with Paul Okubo, Michael Poland, and Tom Wright.

REFERENCES

Amelung, F., S.-H. Yun, T. R. Walter, P. Segall, and S-W Kim (2007), Stress control of deep rift intrusion at Mauna Loa volcano, Hawaiʻi, *Science*, *316*, 1026–1029.

Ando, M. (1979), Hawaiʻi earthquake of November 29, 1975: Low dip angle faulting due to forceful injection of magma, *J. Geophys. Res.*, *84*, 7616–7626.

Baker. S., and F. Amelung (2012), Top-down inflation and deflation at the summit of Kīlauea Volcano, Hawaiʻi observed with InSAR, *J. Geophys. Res., 117*, B12406, doi:10.1029/2011JB009123.

Battaglia, J., J.-L. Got, and P. Okubo (2003), Location of long-period events below Kīlauea Volcano using seismic amplitudes and accurate relative relocations, *J. Geophys. Res., 108,* 2553, doi:10.1029/2003JB002517, B12.

Cayol, V., J. H. Dieterich, A. T. Okamura, and A. Miklius (2000), High magma storage rates before the 1983 eruption of Kīlauea, Hawaiʻi, *Science*, *30*(288), 2343–2346. doi:10.1126/science.288.5475.2343.

Chouet, B. A. (1981), Ground motion in the near field of a fluid-driven crack and its interpretation in the study of shallow volcanic tremor, *J. Geophys. Res.*, *86,* 5985–6016.

Chouet, B. A. (1996), Long period volcano seismicity: Its source and use in eruption forecasting, *Nature*, *380*, 309–316.

Chouet, B. A., and B. R. Julian (1985), Dynamics of an expanding fluid filled crack, *J. Geophys. Res.*, *90*, 11,187–11,198.

Dawson, P. B., B. A. Chouet, P. G. Okubo, A. Villaseñor, and H. M. Benz (1999), Three-dimensional velocity structure of the Kilauea Caldera, Hawaiʻi, *Geophys. Res. Letts., 26*(18), 2905–2808, doi:10.129/1999GL005379.

Decker, R. W., R. Y. Koyanagi, J. J. Dvorak, J. P. Lockwood, A. T Okamura, K. M. Yamashita, and W. R. Tanigawa (1983), Seismicity and surface deformation of Mauna Loa volcano, Hawaiʻi, *Eos Trans. AGU*, *64*, 545–547.

Delaney, P. T., R. S. Fiske, A. Miklius, A. T. Okamura, and M. K. Sako (1990), Deep magma body beneath the summit and rift zones and Kīlauea Volcano, Hawaiʻi, *Science*, *247*, 1311–1316.

Delaney, P. T., A. Miklius, T. Arnadottir, A. T. Okamura, and M. K. Sako (1993), Motion of Kīlauea Volcano during sustained eruption from the Puʻu ʻŌʻō and Kupaianaha vents, 1983–1991, *J. Geophys. Res.*, *98*(B10), 17,801–17,820.

Dieterich, J. H. (1994), A constitutive law from rate of earthquake production and its application to earthquake clustering, *J. Geophys. Res.*, *99*(B2), 2601–2618.

Dieterich, J. H. (2007), Applications of rate-and-state-dependent friction to models of fault slip and earthquake occurrence, in *Treatise of Geophysics*, vol. 4, edited by G. Schubert, Elsevier, Oxford.

Dieterich, J. H., and R. W. Decker (1975), Finite element modeling of surface deformation associated with volcanism, *J. Geophys. Res.*, *80*(29), 4094–4102, doi:10.1029/JB080i029p04094

Dieterich, J. H., V. Cayol, and P. G. Okubo (2000), The use of earthquake rate changes as stress meter at Kīlauea Volcano, *Nature*, *408*, 457–460.

Dieterich, J. H, V. Cayol, and P. G. Okubo (2003), Stress changes before and during the Pu'u 'Ō'ō -Kupaianaha eruption, Prof. Pap., 1676, pp. 187–201, U. S. Geol. Surv., Reston, Virginia.

Dunn, J. C., and H. C. Hardee (1985), Surface heat flow measurements at the Puhimau hot spot, *Geophysics. 50*(7), 1108–1112, doi: 10.1190/1.1441983.

Dvorak, J. J., and A. T. Okamura (1987), A hydraulic model to explain variations in summit tilt rate at Kīlauea and Mauna Loa volcanoes, Prof. Pap., 1350, pp. 1281–1296, U. S. Geol. Surv., Reston, Virginia.

Dzurisin, D., R. Y. Koyanagi, and T. T. English (1984), Magma supply and storage at Kīlauea Volcano, Hawai'i: 1956–1983, *J. Volcanol. Geotherm. Res.*, *21*, 177–206.

Eaton, J. P. (1962), Crustal structure and volcanism in Hawai'i, in The Crust of the Pacific Basin, Geophys. Monogr. *6*, edited by G. A. McDonald and H. Kuno, pp. 13–29, AGU, Washington, D. C.

Fiske, R. S., and W. T. Kinoshita (1969), Inflation of Kīlauea Volcano prior to its 1967–1968 eruption, *Science, 165*, 341–349, doi:10.1126/science.165.3891/341.

Fiske, R. S., D. A. Swanson, and T. L. Wright (1993), A model of Kīlauea Volcano's rift-zone magma system, V22F-5, Annual Meet. AGU, Washington, D. C.

Feuillet, N., C. Nostro, C. Chiarabba, and M. Cocco (2004), Coupling between earthquake swarms and volcanic unrest at the Alban Hills Volcano (central Italy) modeled through elastic stress transfer, *J. Geophys. Res., 109,* B02309, doi:10.1029/2003JB002419.

Furumoto, A. S., and R. L. Kovach (1979), The Kalapana earthquake of November 29, 1975; an intraplate earthquake and its relation to geothermal processes, *Phys. Earth Planet. Interiors*, *18,* 197–208.

Gillard, D., A. M. Rubin, and P. Okubo (1996), Highly concentrated seismicity caused by deformation of Kīlauea's deep magma system, *Nature, 384,* 343–346.

Harris, R. A., R. W. Simpson, and P. A. Reasenberg (1995), Influence of static stress changes on earthquake locations in southern California, *Nature, 375,* 221–224.

Heliker, C., and T. N. Mattox (2003), The first two decades of the Pu'u 'Ō'ō–Kupaianaha eruption: Chronology and selected bibliography, in *The Pu'u 'Ō'ō-Kupaianaha eruption of Kīlauea Volcano, Hawai'i: The First 20 Years*, edited by C. Heliker et al., pp. 1–27, U. S. Geol. Surv. Prof. Pap. 1676.

Hinkle, M. E. (1978). *Helium, mercury, sulfur compounds, and carbon dioxide in soil gases of the Puhimau Thermal Area,* Hawai'i Volcanoes National Park, Hawai'i. US Department of the Interior, Geological Survey.

Holcomb, R. T. (1987), Eruptive history and long-term behavior of Kīlauea Volcano, Prof. Pap. 1350, pp. 261–350, U. S. Geol. Surv., Reston, Virginia.

Jaumé, S. C., and L. R. Sykes (1995), Evolution of moderate seismicity in the San Francisco Bay region, 1850–1993; seismicity changes related to the occurrence of large and great earthquakes, *J. Geophys. Res.*, *101*, 765–789.

Jaumé, S. C., and L. R. Sykes (1996), Evolution of moderate seismicity in the San Francisco Bay region, 1850 to 1993: Seismicity changes related to the occurrence of large and great earthquakes, *J. Geophys. Res.*, *101*(B1), 765–789, doi:10.1029/95JB02393.

Johnston, M. J. S (1976), Testing the physical parameters of short baseline tiltmeters intended for earthquake prediction, Open-File Rep. 76–556:11, U. S. Geol. Surv., Reston, Virginia.

King, G. C. P., R. S. Stein, and J. Lin (1994), Static stress changes and the triggering of earthquakes, *Seism. Soc. Am.*, *84*, 935–953.

Klein, F. W., R. Y. Koyanagi, J. S. Nakata, and W. R. Tanigawa (1987), The seismicity of Kīlauea's magma system, Prof Pap. 1350, pp. 1019–1185, U. S. Geol. Surv., Reston, Virginia.

Koyanagi, R. Y., J. D. Unger, E. T. Endo, and A. T. Okamura (1974), Shallow earthquakes associated with inflation episodes at the summit of Kīlauea Volcano, Hawai'i, in *Symposium on Andean and Antarctic Volcanology Problems, Santiago, Chile, 1974, Proceedings*, edited by F. O. Gonzales, pp. 621–631, International Association of Volcanology and Chemistry of the Earth's Interior, Naples.

Koyanagi, R. Y., W. R. Tanigawa, and J. S. Nakata JS (1988), Seismicity associated with the eruption, Prof Pap., 1463, pp. 183–235, U. S. Geol. Surv., Reston, Virginia.

Kumagai, H., B. A. Chouet, and P. B Dawson (2005), Source process of a long-period event at Kīlauea Volcano, Hawai'i, *Geophys. J. Int.*, *161*, 243–254.

Linker, M. F., and J. H. Dieterich (1992), Effects of variable normal stress on rock friction: Observations and constitutive relationship, *J. Geophys. Res.*, *97*, 4923–4940.

McGee, K. A., A. J. Sutton, T. Elias, M. P. Doukas, and T. M. Gerlack (2006), Puhimau thermal area: A window into the upper East Rift Zone of Kilauea volcano, Hawai'i, *Pure. Appl. Geophys.*, *163*, 837–851, doi:10.1007/s00024-006-0036-z.

Montagna, C. P., and H. M. Gonnermann (2013), Magma flow between the summit and Pu'u 'O'o at Kīlauea Volcano, Hawai'i, *Geochem. Geophys. Geosyst.*, *14*, doi:10.1002/ggge.20145.

Moore, R. B. (1983), Distribution of differentiated tholeiitic basalts on the lower east rift zone of Kīlauea Volcano, Hawai'i: A possible guide to geothermal exploration, *Geology*, *3*, 136–140.

Okamura, A. T., J. J. Dvorak, R. Y. Koyanagi, and W. R Tanigawa (1988), Surface deformation during dike propagation, Prof. Pap. 1463, pp. 165–181, U. S. Geol. Surv., Reston, Virginia.

Okubo, P. G., H. M., Benz, and B. A. Chouet (1997), Imaging the crustal magma sources beneath Mauna Loa and Kilauea volcanoes, Hawai'i, *Geology*, *25*(10), 867–870, doi:10.1130/0091-7613(1997)025<0867:ITCMSB>2.3.CO;2.

Okubo, P. G., and J. S. Nakata (2003), Tectonic pulses during Kīlauea's current long-term eruption, Prof. Pap. 1676, pp. 173–186, U. S. Geol. Surv., Reston, Virginia.

Parfitt, E. A., and L. Wilson (1994), The 1983–86 Pu'u 'O'o eruption of Kīlauea Volcano, Hawai'i: A study of dike geometry and eruption mechanisms for a long-lived eruption, *J. Volcanol. Geotherm. Res.*, *59*, 179–205, doi:10.1016/0377-0273(94)90090-6.

Poland, M. P., A. Miklius, and E. K. Montgomery-Brown (2014), Magma supply, storage, and transport at shield-stage Hawaiian volcanoes, Chapter 5 of Poland, M. P., T. J. Takahashi, and C. M. Landowski, (eds.), *Characteristics of Hawaiian Volcanoes*, U.S. Geol. Surv. Prof. Pap. 1801, pp. 179–234, doi:10.3133/pp18015.

Rowan, L. R., and R. W. Clayton (1993), The three-dimensional structure of Kīlauea Volcano, Hawai'i, from travel time tomography, *J. Geophys. Res.*, *98*(B3), 4355–4375, doi:10.1029/92JB02531.

Ryan, M. P., R. Y. Koyanagi, and R. S. Fiske (1981), Modeling of the three-dimensional structure of macroscopic magma transport systems: Application to Kīlauea Volcano, Hawai'i, *J. Geophys. Res.*, *86*, 7111–7129.

Stein, R. A. (1999), The role of stress transfer in earthquake occurrence, *Nature*, *402*, 605–609.

Swanson, D. A., D. B. Jackson, R. Y. Koyanagi, and T. L Wright (1976), The February 1969 east rift eruption of Kīlauea Volcano, Hawai'i, Prof. Pap. 891, pp. 30, U. S. Geol. Surv., Reston, Virginia.

Thurber, C. H., and A. E. Gripp (1988), Flexure and seismicity beneath south flank of Kīlauea Volcano and tectonic implications, *J. Geophys. Res.*, *93*, 4271–4281.

Troise, C. (2001), Stress changes associated with volcanic sources: Constraints on Kīlauea rift dynamics, *J. Volcanol. Geotherm. Res.*, doi:10.1016/S0377-0273(00)00311-5.

Wolfe, E. W., M. O Garcia, D. B. Jackson, R. Y. Koyanagi, C. A. Neal, and A. T. Okamura (1987), The Pu'u' O'o eruption of Kīlauea Volcano, episodes 1–20, January 3, to June 8, 1984, Prof. Pap. 1350, pp. 471–508, U. S. Geol. Surv., Reston, Virginia.

Wolfe, E. W., C. A. Neal, N. G. Banks, and T. J. Duggan (1988), Geologic observations and chronology of eruptive events, Prof. Pap. 1463, pp. 1–97, U. S. Geol. Surv., Reston, Virginia.

Wright, T. L., W. T. Kinoshita, and D. L Peck (1968), March 1965 eruption of Kīlauea Volcano and the formation of Makaopuhi lave lake, *J. Geophys. Res.*, *73*, 3181–3205.

11

Episodic Deflation–Inflation Events at Kīlauea Volcano and Implications for the Shallow Magma System

Kyle R. Anderson[1], Michael P. Poland[1], Jessica H. Johnson[1,2,3,4], and Asta Miklius[1]

ABSTRACT

Episodic variations in magma pressures and flow rates at Kīlauea Volcano, defined by a characteristic temporal evolution and termed deflation-inflation (DI) events, have been observed since at least the 1990s. DI events consist of transient, days-long deflations and subsequent reinflations of the summit region, accompanied since 2008 by fluctuations in the surface height of Kīlauea's summit lava lake. After a delay of minutes to hours, these events also often appear along the volcano's East Rift Zone in ground deformation data and as temporary reductions in eruption rate (sometimes followed by brief surges). Notable pauses in DI activity have preceded many eruptive events at Kīlauea. We analyzed more than 500 DI events recorded by borehole tiltmeters at the summit during 2000–2013. Inverse modeling suggests that DI-related ground deformation at the summit is generated by pressure transients in a shallow magma reservoir located beneath the east margin of Halema'uma'u Crater and that this reservoir has remained remarkably stable for more than a decade. Utilizing tilt data and variation in the level of the summit lava lake during a large DI event, we estimate a reservoir volume of approximately 1 km^3 (0.2–5.5 km^3 at 95% confidence).

11.1. INTRODUCTION

Perturbations in magma pressures and flow rates at erupting volcanoes yield transient geophysical signals that can provide valuable insights into the properties of magma systems. Episodic or periodic perturbations, recorded in deformation, seismic, and other data sets [e.g., *Mori et al.*, 1996; *Voight et al.*, 1998; *Yamashina et al.*, 1999; *Bluth and Rose*, 2004; *Iverson et al.*, 2006; *Anderson et al.*, 2010; *Genco and Ripepe*, 2010], can be particularly useful for understanding volcanic behavior, as repeating activity provides more opportunities for observation, and much can be inferred from variations of episodic behavior with time and in association with other eruptive activity.

At Kīlauea Volcano, Hawai'i, episodic activity has been observed at various temporal and spatial scales for decades. Well-documented examples include the early part of the 1969–1974 Mauna Ulu eruption and the first few years of the 1983-to-present East Rift Zone (ERZ) eruption, both of which involved cycles of summit inflation punctuated by rapid deflation during lava fountaining [*Swanson et al.*, 1979; *Wolfe et al.*, 1987].

Beginning at least in the 1990s, a new type of episodic event began to be recognized at Kīlauea [*Heliker and Mattox*, 2003]. Superimposed on long-term trends of summit inflation and deflation, rift zone opening, slow slip events, intrusions, the formation of new eruptive vents, and other volcanic and tectonic changes, these deflation-inflation (DI) events have now become one of the most common geophysical signals associated with the ongoing ERZ eruption of Kīlauea.

DI events (not to be confused with earlier cyclic activity also involving summit inflation and deflation) occur over periods of hours to days, are recorded in interferometric

[1] *Hawaiian Volcano Observatory, U.S. Geological Survey, Hawai'i National Park, Hawaii, USA*

[2] *Center for the Study of Active Volcanoes, University of Hawai'i, Hilo, Hawaii, USA*

[3] *School of Earth Sciences, University of Bristol, Bristol, United Kingdom*

[4] *School of Environmental Sciences, University of East Anglia, Norwich, United Kingdom*

Hawaiian Volcanoes: From Source to Surface, Geophysical Monograph 208. First Edition.
Edited by Rebecca Carey, Valérie Cayol, Michael Poland, and Dominique Weis.

Companion Website: www.wiley.com/go/Carey/Hawaiian_Volcanoes

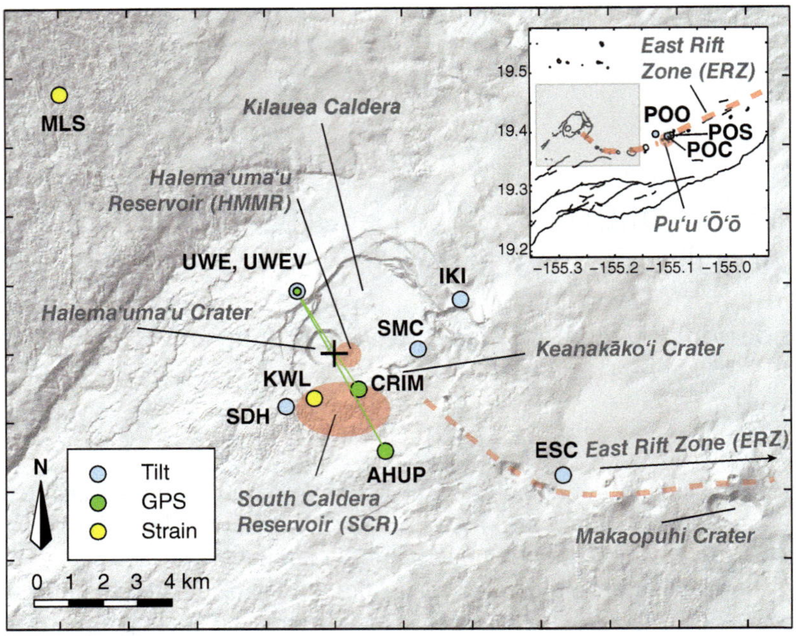

Figure 11.1 Locations of instruments and features discussed in this study. The center of the summit coordinate system, from which the radial component of summit tilt is calculated, is located at the summit eruptive vent and is shown by a +. Summit-crossing lengths between GPS stations, shown in some figures, are indicated by green lines. Locations and sizes of magma reservoirs are approximate. The inset shows the location of the map (shaded area) with relation to the ERZ.

synthetic aperture radar (InSAR), global positioning system (GPS), strain, and tilt data at the summit of Kīlauea (Figure 11.1), and are strongly correlated with variations in the level of the summit lava lake (Figure 11.2). DI events also often appear, after a short delay, as variations in tilt and effusion rate at the ERZ eruptive vent, ~20 km from the summit. The temporal evolution of DI events typically involves relatively gradual summit deflation followed by a recovery to roughly the level of pre-event tilt. Like the cyclic activity associated with lava fountaining, DI events reflect perturbations in magma flow and pressure conditions throughout much of Kīlauea's interconnected shallow magma plumbing system.

The process that produces DI events is not well understood, but may involve a temporary blockage in supply to Kīlauea's shallow magma reservoir [*Cervelli and Miklius*, 2003] and the ERZ or, alternatively, convective overturn in the summit magma reservoir, with degassed magma being flushed downward (deflation) before being replaced by gas-rich magma (inflation) [*Poland et al.*, 2009a].

DI events are particularly well recorded on the summit tilt network because of the high sensitivity and temporal resolution of the borehole tiltmeters. *Cervelli and Miklius* [2003] analyzed four large, self-similar DI events that occurred during 2000–2002 and located a source 0.5 km east of Halema'uma'u Crater at 500–700 m below ground level. Similarly, in a review paper on magma supply, storage, and transport in Hawai'i, *Poland et al.* [2014] inverted data from 152 DI events that occurred from 2000 to 2009 and found that sources were located slightly to the south of the *Cervelli and Miklius* [2003] location, at a poorly constrained depth of 1–2 km.

These results are consistent with a variety of observations over many years that suggest the presence of magma at a shallow depth just east of Halema'uma'u Crater. These include very-long-period (VLP) seismicity and tremor [*Ohminato et al.*, 1998; *Almendros et al.*, 2002; *Dawson et al.*, 2004; *Chouet et al.*, 2010], long-period seismicity [*Battaglia*, 2003], P- and S-wave tomography [*Dawson et al.*, 1999], and gravity [*Dzurisin et al.*, 1980; *Johnson et al.*, 2010]. Deformation studies [*Fiske and Kinoshita*, 1969; *Dvorak et al.*, 1983; *Johnson*, 1992], including those associated with rapid summit draining events that feed ERZ eruptions and intrusions [*Owen et al.*, 2000; *Poland et al.*, 2009b; *Montgomery-Brown et al.*, 2010; *Baker and Amelung*, 2012; *Lundgren et al.*, 2013], also suggest magma storage at shallow depths beneath the east Halema'uma'u region. Synthesizing these studies, *Poland et al.* [2014] suggested the long-term presence of magma at ~1–2 km below the east margin of Halema'uma'u Crater or somewhat to the east. This reservoir has been termed the Halema'uma'u magma reservoir (HMMR) [*Cervelli and Miklius*, 2003] and is believed to supply magma to Kīlauea's summit eruption (2008 to present).

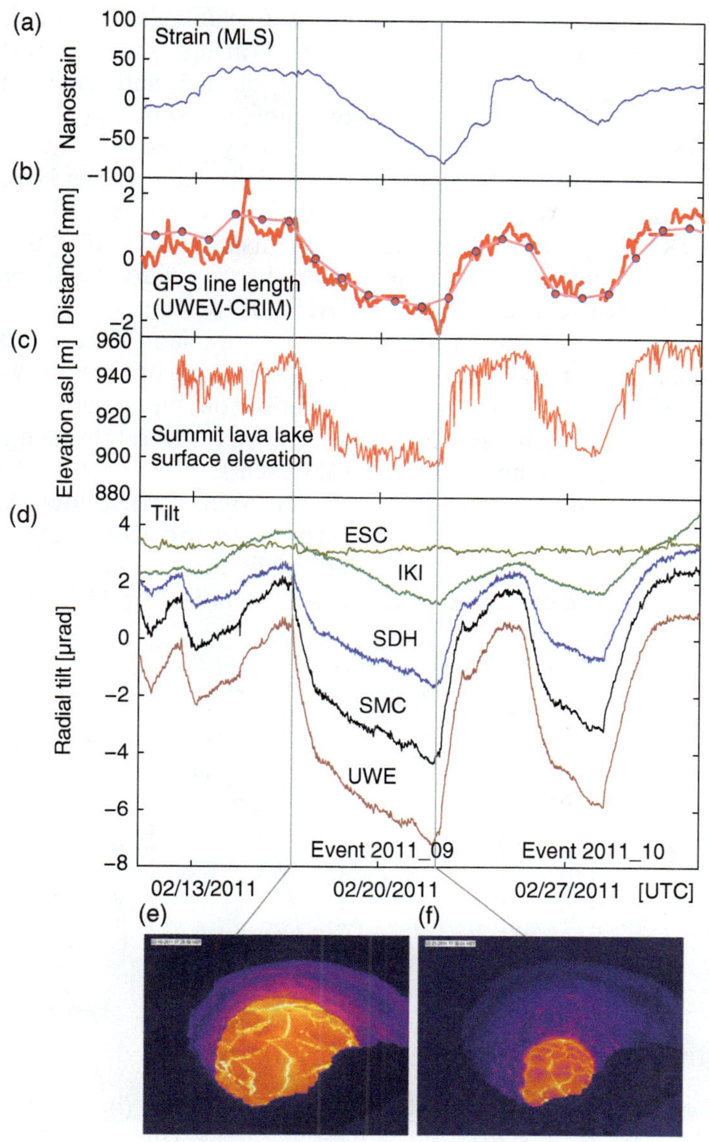

Figure 11.2 DI activity at Kīlauea's summit during large events just prior to the March 2011 Kamoamoa fissure eruption. (a) Strain at MLS (contractional strain positive). (b) Summit-crossing GPS line length (4min and daily solutions as red line and blue dots, respectively) (c) Height of the surface of the summit lava lake. (d) Correlated radial tilt recorded on five instruments (e, f) Thermal images of the lava lake in the Halema'uma'u vent at the onset of a DI event and the end of DI deflation.

Because DI events have occurred for so many years and have been so well recorded in a variety of data sets, they offer a remarkable opportunity to infer properties of Kīlauea's shallow magma plumbing system. Analysis of tilt and other data associated with DI events provides a valuable constraint on magmatic processes within the summit magma system, and the manifestation of DI events at Pu'u 'Ō'ō (Figure 11.1) can elucidate properties of the magma pathway linking the summit and the ERZ. In this work we summarize observations associated with DI events at Kīlauea, analyze 14 years of DI-related ground tilt, invert to estimate the location of the HMMR, describe temporal changes in DI activity and the relation with the ERZ, speculate about possible source mechanisms, and use the tilt data in conjunction with the level of the summit lava lake to place constraints on the total volume of the HMMR.

11.2. DATA AND PROCESSING

Although DI events sometimes appear in GPS, strain, and InSAR data (particularly the larger events), tiltmeters offer a combination of sensitivity and temporal resolution unmatched by other instruments at Kīlauea. We therefore analyze only tilt data recorded on the borehole tilt network, although we describe the appearance of DI events in a number of other data sets.

The U.S. Geological Survey (USGS) Hawaiian Volcano Observatory maintains a network of ~20 electronic tiltmeters on the Island of Hawai'i. DI events are recorded clearly at Kīlauea's summit at stations UWE, SMC, and SDH, less clearly at station IKI, and not at all at the more distant ESC (Figures 11.1 and 11.2). (Station KWL, located at the southern edge of Kīlauea Caldera, has also recorded DI events, but the instrument is considered unreliable and the data are not used here.) At Pu'u 'Ō'ō, DI-related ground deformation is recorded on station POC and sometimes weakly at POO and POS. All borehole tiltmeters used in this study sample at 1 min intervals and telemeter their data to the observatory in near-real time.

To isolate DI-related signals from other processes and sources of tiltmeter noise, we first remove spikes and offsets (including those associated with volcanic activity such as rift intrusions) from the data. We remove long-period, diurnal, and semidiurnal ground tilt (the latter two probably caused by thermoelastic strain in the near surface) by high-pass filtering the data and applying notch filters with daily and half-daily periods. The simple notch filter we use can reduce the recorded amplitude of DI events by roughly 10%, but we find that its use is necessary because diurnal and semidiurnal noise is strong at several of the stations.

The Hawaiian Volcano Observatory classifies DI events into one of three types based on their temporal evolution on the summit tilt network: "V-type" (375 events; 74% of total through the end of 2013), "U-type" (122 events; 24% of total), and "DID-type" (10 events; 2% of total). We follow this classification scheme throughout this work, although events do not always fall easily into one category or another. Event types are described in Section 11.3.

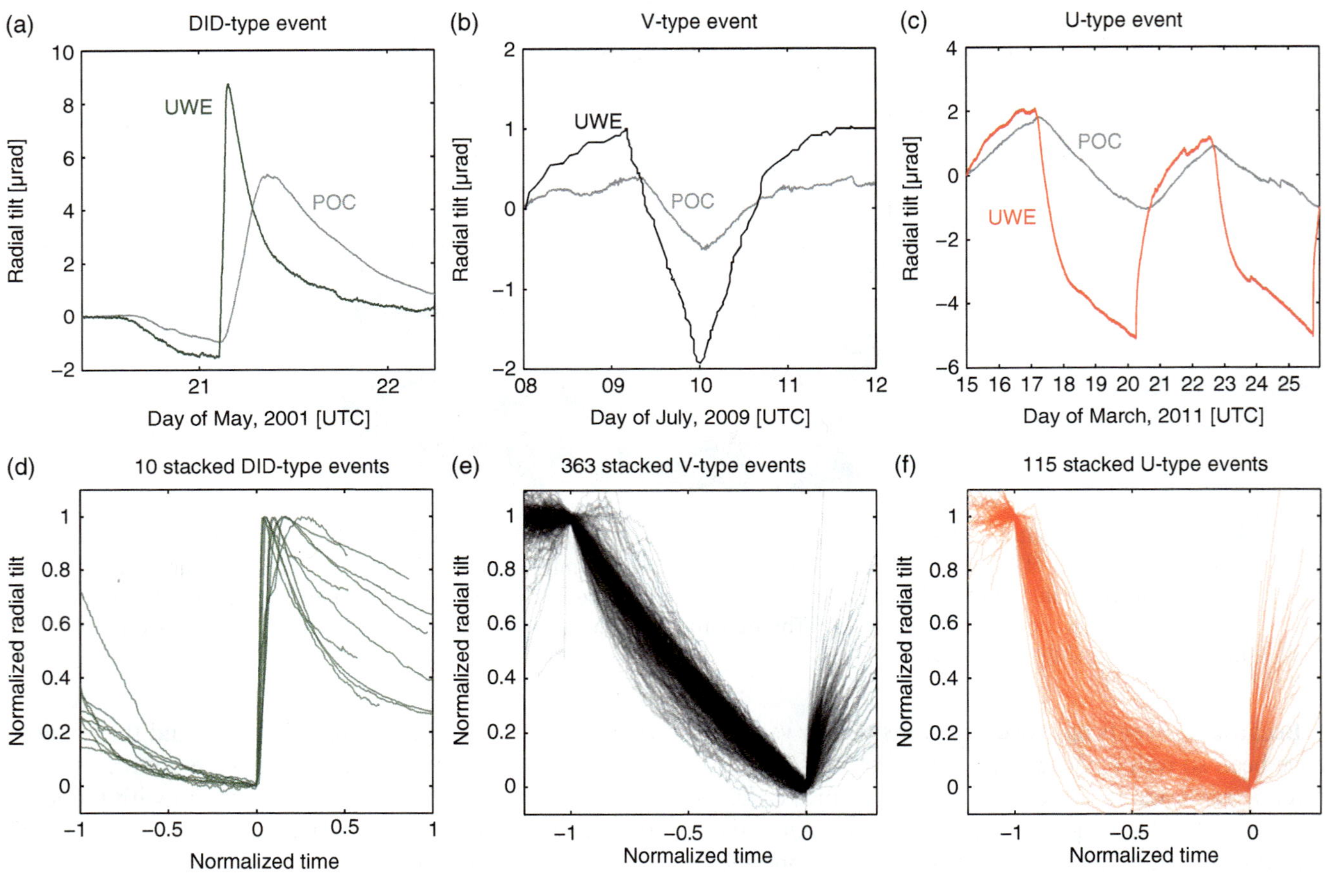

Figure 11.3 Classification of DI events and relation between summit tilt (low-pass filtered radial channel at UWE, calculated from center of coordinate system shown in Figure 11.1) and Pu'u 'Ō'ō tilt (POC, north channel). (a–c) Examples of DID-, V-, and U-type events, respectively, with delayed tilt at POC (note differing scales for the three different event types). (d–f) Stacks of deflations and rapid inflations recorded at UWE for all DID-type, V-type, and U-type events analyzed in this study. DID-type time series are normalized by setting onset of rapid inflation to (0,0) and setting the amplitude of all events to 1. U- and V-type time series are normalized by setting the onset of deflation to (−1,1) and the onset of rapid inflation to (0,0).

For the 14 year period from 1 January 2000 to 31 December 2013 we hand pick a total of 507 DI events based on their characteristic temporal evolution in data from multiple tiltmeters. For each DI event we identify the following times: (1) onset of deflation, (2) onset of inflation (3) transition to a lower rate of inflation (cessation of rapid inflation), and (4) the end of the event (DID-type events have a final deflationary phase and require a slightly different picking scheme). These picks yield three time intervals: deflation, rapid inflation, and slow inflation. We divide inflations into two time periods because inflations often occur at a decaying rate, making the cessation time difficult to pick; also, because of temporally correlated noise in the data, rapid inflations usually have a higher signal-to-noise ratio than total inflations.

We exclude from analysis those events during which the tiltmeters suffered data quality issues, caused, for instance, by heavy rainfall (which can produce transients unrelated to magmatic processes) or local earthquakes (unless the earthquakes induce only a simple offset to the tilt data, which we remove). Fewer than 10% of events were removed for these reasons. For each time segment and tiltmeter channel (east and north), we calculate the net tilt by differencing an average of 15 min of data around the start and end of the segment. Finally, we determine tilt magnitudes and azimuths for each time segment of each event; estimated uncertainties are described in Section 11.4.1.

11.3. CHARACTERISTICS OF DI EVENTS

11.3.1. Temporal Evolution

Deflation-inflation-deflation (DID) events consist of slow summit deflations lasting hours to days, very rapid inflations over minutes to hours that greatly exceed the preceding deflation, and quasi-exponential deflation to roughly the level of pre-event tilt (Figures 11.3a,d). Although the borehole tilt network was not installed until 1999, DID events were recognized by at least February 1990 and were referred to initially as "surge" events based on their association with surges in ERZ effusion (although a lack of tilt data from Pu'u 'Ō'ō and the poor resolution of summit tilt prior to 1999 preclude a detailed analysis) [*Heliker and Mattox*, 2003]. Since 2004, no DID events have been unambiguously recorded.

U- and V-type events lack the inflationary overshoot of DID events. V-type events (Figures 11.3b,e) exhibit a near-constant-rate deflation lasting ~0.4–1 day, while U-type events (Figures 11.3c,f) exhibit deflation at a decreasing (near-exponential or sometimes piecewise linear) rate over a period of ~1–3 days. Both types of events transition nearly instantaneously to rapid inflation (U-type events sometimes after "bottoming out" at a near-constant tilt for some time) that typically recovers to pre-event tilt. Both types of events are sometimes preceded by small inflationary "bumps" just prior to the onset of deflation. Periods of rapid inflation last about 0.25–3 h for V-type events and 0.7–8 h for U-type events. U- and V-type DI events probably also occurred throughout the 1990s, as suggested by surface tiltmeter measurements at UWE, but were not well characterized until the borehole tilt network was installed.

11.3.2. Summit Deformation

The pattern of deformation recorded in tilt data associated with DI events is generally radial from a region just east or southeast of Halema'uma'u Crater (Figures 11.4a,b and 11.5). This pattern is similar to that recorded during summit draining events associated with ERZ intrusions and eruptions (Figures 11.4c,d); the tilt data also include many small transients that are not associated with DI events or clearly associated with eruptive activity but share a similar spatial pattern.

Tilt azimuths, and their variances, are similar for all event types (Figure 11.6e), although there are few DID-type events available for analysis. (Estimated uncertainties in DID-type azimuths are much smaller and do not encompass the scatter in DID azimuths, suggesting that the noise model is inadequate or that the source position varied.) Source azimuths appear largely stable over time, although some change could be obscured by noise.

Deflationary tilt amplitudes at station UWE, 2 km from the summit eruptive vent, are typically on the order of 1–2.5 and 2–5 μrad for V- and U-type events, respectively (Figure 11.6d). DID-type deflations are usually less than 2 μrad, but subsequent inflations are very large, sometimes exceeding 10 μrad.

Large DI events are recorded on the summit continuous GPS network. Deflations are associated with inward motion of stations generally toward an area just east of Halema'uma'u Crater, and vice versa for inflations, consistent with the tilt observations. InSAR data spanning DI events also show a pattern of subsidence and inflation located in the central part of the caldera (P. Lundgren, personal communication, 2013).

The closest strainmeter that reliably records strain amplitude is located at MLS, 11 km NW of the summit eruptive vent (Figure 11.1). Larger DI deflations are recorded on this instrument as extensional strain (increase in volume) and inflations as contractional strain (reduction in volume) (Figure 11.2). Unfortunately, a strainmeter located in the Keller Well (KWL) [*Keller et al.*, 1979], less than 1.5 km from the vent, does not accurately record strain amplitude, although the sign of strain change is probably correct. This instrument also records extensional strain associated with deflation and vice versa.

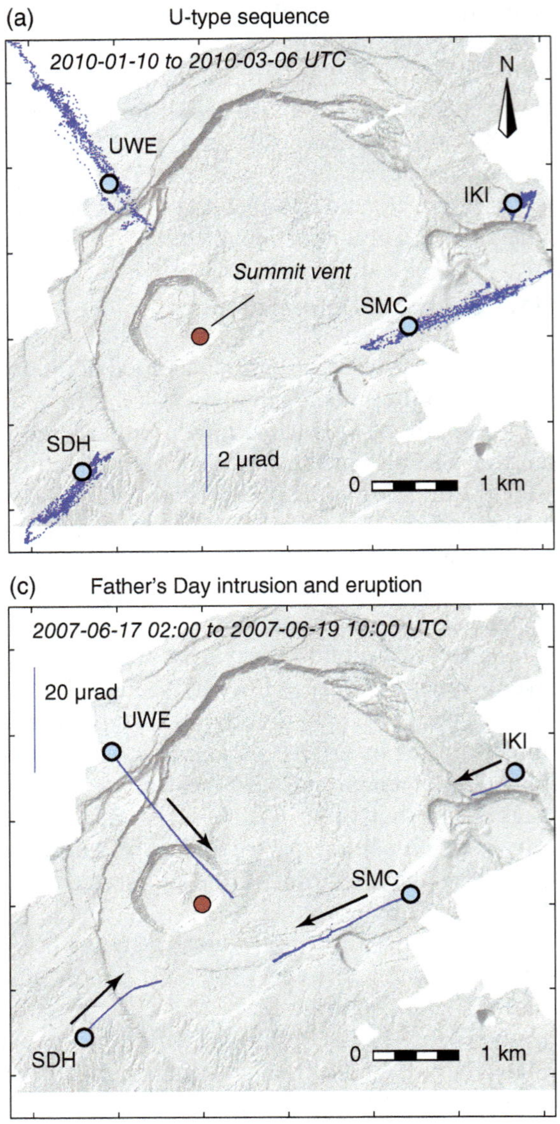

Figure 11.4 East vs. north tilt for DI events (top row; one point every hour) and summit draining events (bottom row; one point every 2 min). Background is shaded topography derived from lidar data acquired in 2009. (a) Roughly two months of back-to-back U-type events (~7 events; Figure 11.8, point 4). (b) Roughly one month of back-to-back V-type events (~16 events; Figure 11.8, point 6). (c) The June 2007 Father's Day intrusion and eruption (Figure 11.8, point 2). (d) The March 2011 Kamoamoa eruption (Figure 11.8, point 5).

11.3.3. Summit Lava Lake

An eruptive vent opened at Kīlauea's summit in March 2008 and an active lava lake was visible by 5 September; occasionally choked by collapse through 2009, the lake has been continuously visible since 2010 [*Orr et al.*, 2012; *Patrick et al.*, 2013]. DI deflations are closely associated with decreases in the surface height of the lava lake and vice versa (Figure 11.2c). Large DI events cause the lake surface to rise and fall by as much as 50 m [*Patrick and Orr*, 2013].

11.3.4. East Rift Zone Deformation

DI events are manifested along the ERZ and are best resolved on tiltmeter POC, located on the north flank of the Pu'u 'Ō'ō eruptive vent. This instrument often records DI transients (sequences of tilt toward and away from the Pu'u 'Ō'ō cone) that are similar to those recorded at the summit, but after a delay of anywhere from 30 min to 10 h (Figure 11.6c) and with a temporal evolution that is somewhat attenuated compared to that at the summit (particularly evident with DID events) (Figure 11.3).

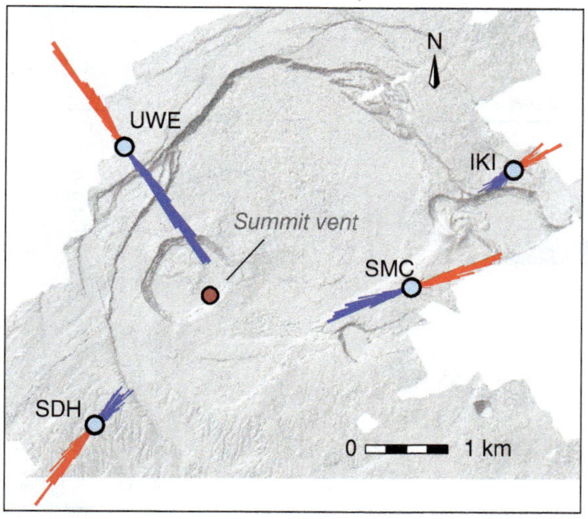

Figure 11.5 Azimuthal histograms of deflations (blue, pointing inward) and rapid inflations (red, pointing outward); one-quarter of events with lowest signal-to-noise ratio are not plotted. Station ESC does not record DI events and is not shown.

Delay times between the onset of inflation at the summit and Pu'u 'Ō'ō are variable (Figure 11.6c). Large DI events are sometimes also recorded weakly on tiltmeters POO and POS, located 1.5 km north of Pu'u 'Ō'ō and on the northeast flank of the cone, respectively, but the data are inadequate for constraining the location of the tilt source beneath Pu'u 'Ō'ō.

DI events are not always recorded at Pu'u 'Ō'ō. We note, in particular, two significant time periods during which summit tilt patterns (including DI events as well as other transient tilt events) were clearly manifested at Pu'u 'Ō'ō: (1) from the installation of the tiltmeter network in 1999 until late January 2005 (at about the time of a brief rapid summit inflation and south flank slow slip event [e.g., see Chapter 13, this volume, as well as Section 11.3.6]) and (2) from the 21 July 2007 ERZ fissure eruption until around the time of a M4.8 south flank earthquake on 23 January 2012 (from that date until the time of this writing, events are manifested at P'u 'Ō'ō but at a reduced magnitude compared with before January 2012). Figure 11.7 shows two transitions between times of greater or less DI activity at Pu'u 'Ō'ō.

Away from Pu'u 'Ō'ō on the ERZ, instrumentation is relatively sparse and DI events are not resolved in InSAR data; however, the line length between GPS stations spanning the ERZ near Makaopuhi Crater (about halfway between the summit and Pu'u 'Ō'ō; see Figure 11.1) sometimes reflects large DI events, with summit deflations corresponding to decreases in line length and vice versa. Unfortunately, the temporal resolution of the GPS data are inadequate to characterize the delay time from the summit, and it is also unknown if the tilt transients are generated by pressure changes in the presumed ERZ conduit or in a small reservoir possibly located beneath Makaopuhi [e.g., *Jackson et al.*, 1975; *Swanson et al.*, 1976; *Owen et al.*, 2000; *Thornber et al.*, 2003]. Regardless, the observations suggest that DI-related pressure transients may generate deformation along much or all of the ERZ. (DI events do not appear at station ESC; this instrument is apparently too far from the summit to record summit-related tilt and may be too close to the rift, since no tilt would be expected directly above it, to record the propagation of DI events down the rift.)

11.3.5. East Rift Zone Effusion

Short-term changes in effusion rate at Kīlauea are not usually well resolved by available observations, but it has been qualitatively observed that DI deflations are often associated with decreases in lava effusion or even eruptive pauses, and subsequent inflations are typically followed by a resumption of flow or even a short-lived surge in effusion rate. DI-related deflations have been responsible for the termination of long-lived ocean entries and the abandonment of lava tubes, while lava surges associated with inflations have resulted in flows breaking out of lava tubes [*Orr et al.*, in press], dormant vents becoming active, and effusion rates temporarily increasing by a factor of 10 or more above normal [*Heliker and Mattox*, 2003]. DI events can thus have an important impact on the ERZ eruption.

Similar to tilt at Pu'u 'Ō'ō, changes in effusion rate are delayed from tilt changes at the summit. *Heliker and Mattox* [2003] note 4–8 h delays between the peak of summit inflation and the resumption of effusion in the ERZ during the 1990s, consistent with tilt delays in this study (Figure 11.6c).

11.3.6. Occurrence Rate and Pauses in DI Activity

The rate of DI event occurrence increased significantly after the opening of the summit vent in 2008 (Figure 11.6b); summit gas emissions also increased dramatically at this time. Long trains of back-to-back events began to appear in 2009, and 2012 was characterized by near-constant DI activity (Figure 11.8).

Rates of summit extension, which are a proxy for summit inflation, appear to be somewhat correlated with DI activity (Figure 11.6a), but the relation is complex. The inflation of Kīlauea's summit from 2003 to 2007 in response to a surge in magma supply to the volcano [*Poland et al.*, 2012] correlated crudely with relatively few DI events and a poor connection to Pu'u 'Ō'ō (Section 11.3.4); yet, when summit extension resumed in 2010 (after deflation following the July 2007 fissure

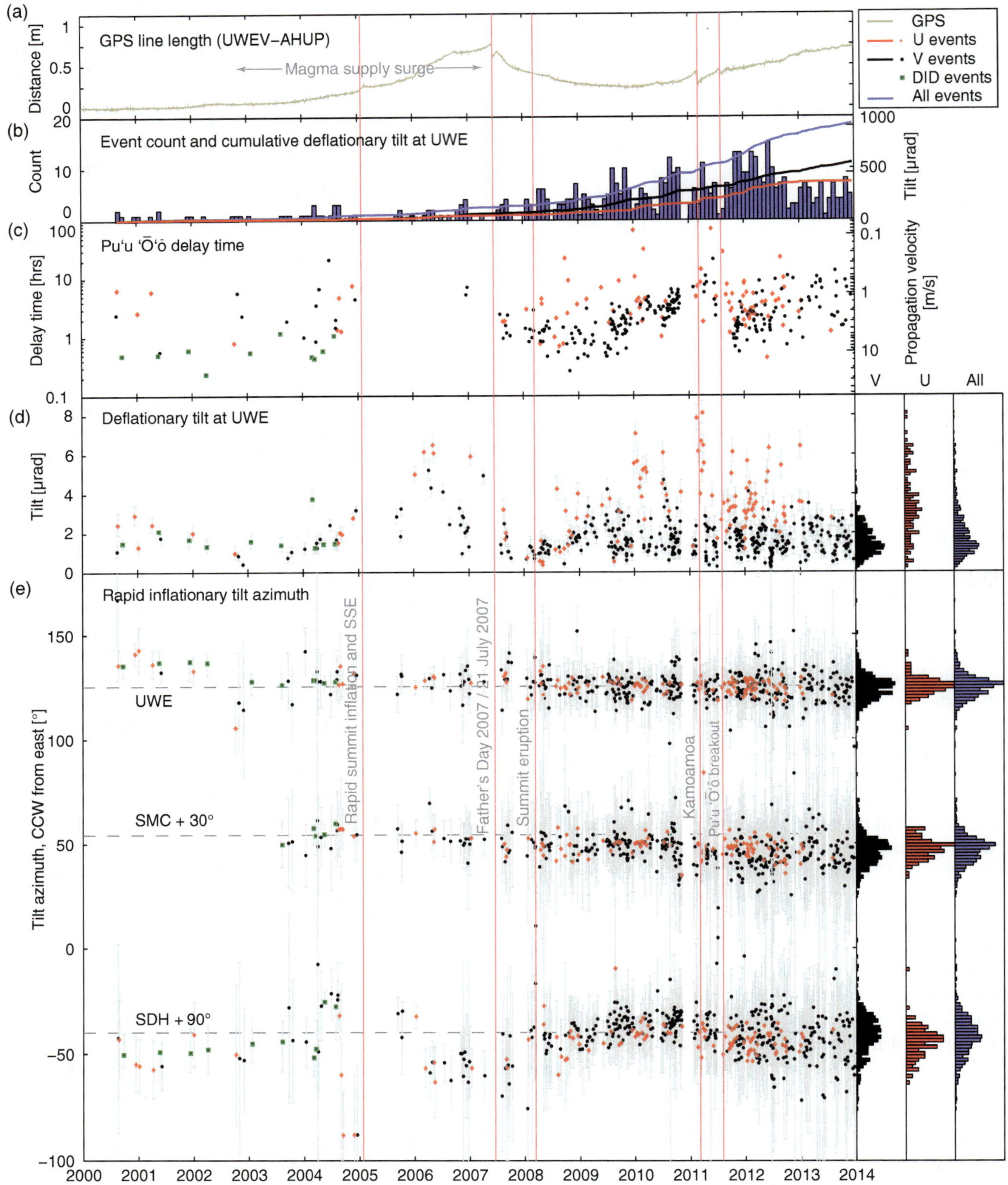

Figure 11.6 Time series of hand-picked DI events along with summit extension. Error bars are 1σ. Vertical red lines indicate significant volcanic events. (a) GPS line length across the summit caldera. (b) Cumulative deflationary DI tilt (solid lines) and binned event count. (c) Delay between onset of inflationary phase of DI event at the summit and at Pu'u 'Ō'ō and propagation velocity based on a 20 km distance. Note log scales. (d) Magnitude of deflationary tilt at UWE (inflations associated with DID events were much larger than the deflations shown here). (e) Azimuths for rapid inflationary tilts, averaged with back azimuths for corresponding deflations.

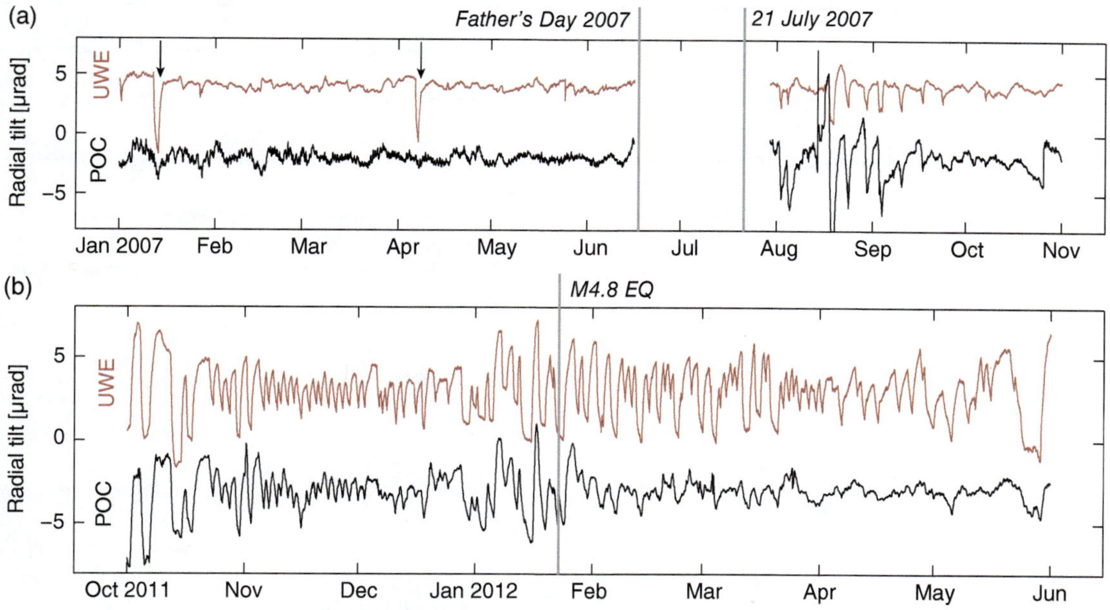

Figure 11.7 Changes in DI activity between the summit and Pu'u 'Ō'ō. Time series are in local radial coordinate systems (center of summit coordinate system shown in Figure 11.1; center of Pu'u 'Ō'ō coordinate system located at top of Pu'u 'Ō'ō cone). (a) Summit DI events were barely recorded, if at all, at Pu'u 'Ō'ō for more than two years before the 21 July 2007 fissure eruption (arrows point to prominent events not recorded well at Pu'u 'Ō'ō). (b) Gradual decay in amplitude of DI events at Pu'u 'Ō'ō following a M4.8 south flank earthquake in January 2012.

eruption), DI activity was common and the connection with the ERZ was very clear (Figure 11.6c).

Significant pauses in DI activity clearly preceded a number of eruptive events at Kīlauea, including the 2007 Father's Day intrusion/eruption and 21 July fissure eruption, the onset of the summit eruption in 2008, and the 2011 Kamoamoa fissure eruption (Figures 11.6 and 11.8), although in some cases DI events briefly reoccurred prior to the onset of these events (see the Kamoamoa eruption in Figure 11.8). Shorter pauses also appeared to precede or be associated with other events such as a short surge in ERZ lava effusion in late June to early July of 2008, the August 2011 Pu'u 'Ō'ō breakout, and a possible small intrusion in the south caldera region in October 2012 (Hawaiian Volcano Observatory, unpublished material). We also note that a rapid inflation on 25 January 2005, associated in time with a south flank slow slip event [*Segall et al.*, 2006; *Brooks et al.*, 2006] (see Chapter 13, this volume), heralded a brief surge in summit extension, a doubling in magma flux to the ERZ [*Poland et al.*, 2012], and the longest hiatus in DI activity yet detected by the borehole tilt network (Figure 11.8). Most of these pauses in DI activity are associated with increases in the rate of summit extension and inflation, or possibly with higher rates of magma flux through the summit, but not all rapid summit extensions lead to pauses.

11.3.7. Seismicity

Cervelli and Miklius [2003] described the evolution of volcanic tremor recorded near Halema'uma'u and associated with DID-type events, which included an increase in high-frequency tremor (≥3 Hz) associated with the inflationary tilt phase (Figures 11.9a,b). They attributed the seismicity changes to varying flow rates into the ERZ.

VLP energy associated with the inflation phase of the DID-type events has been located using the method of radial semblance, with the origin locations defining a region beneath the eastern edge of Halema'uma'u Crater from about 500 to 1200 m depth [*Dawson et al.*, 2004]. Similar source locations have been inferred for VLP seismicity associated with other tilt signals, such as a rapid inflation during February 1996 [*Ohminato et al.*, 1998].

Seismic tremor increased at Kīlauea's summit after the formation of the summit eruptive vent in 2008 [*Wilson et al.*, 2008]. Between late 2005 and 2012, broadband tremor (≤18 Hz), which was probably associated with gas release, was at times correlated with the tilt signal. Until mid-2010 (with occasional pauses in early 2008 and 2009), the deflationary phase of DI events was accompanied by an increase in tremor amplitude (Figures 11.9c,d). The character of the correlation changed briefly from the time of the August 2011 Pu'u 'Ō'ō breakout until early 2012 such that the deflationary phase of DI events was accompanied by a decrease in tremor amplitude that recovered

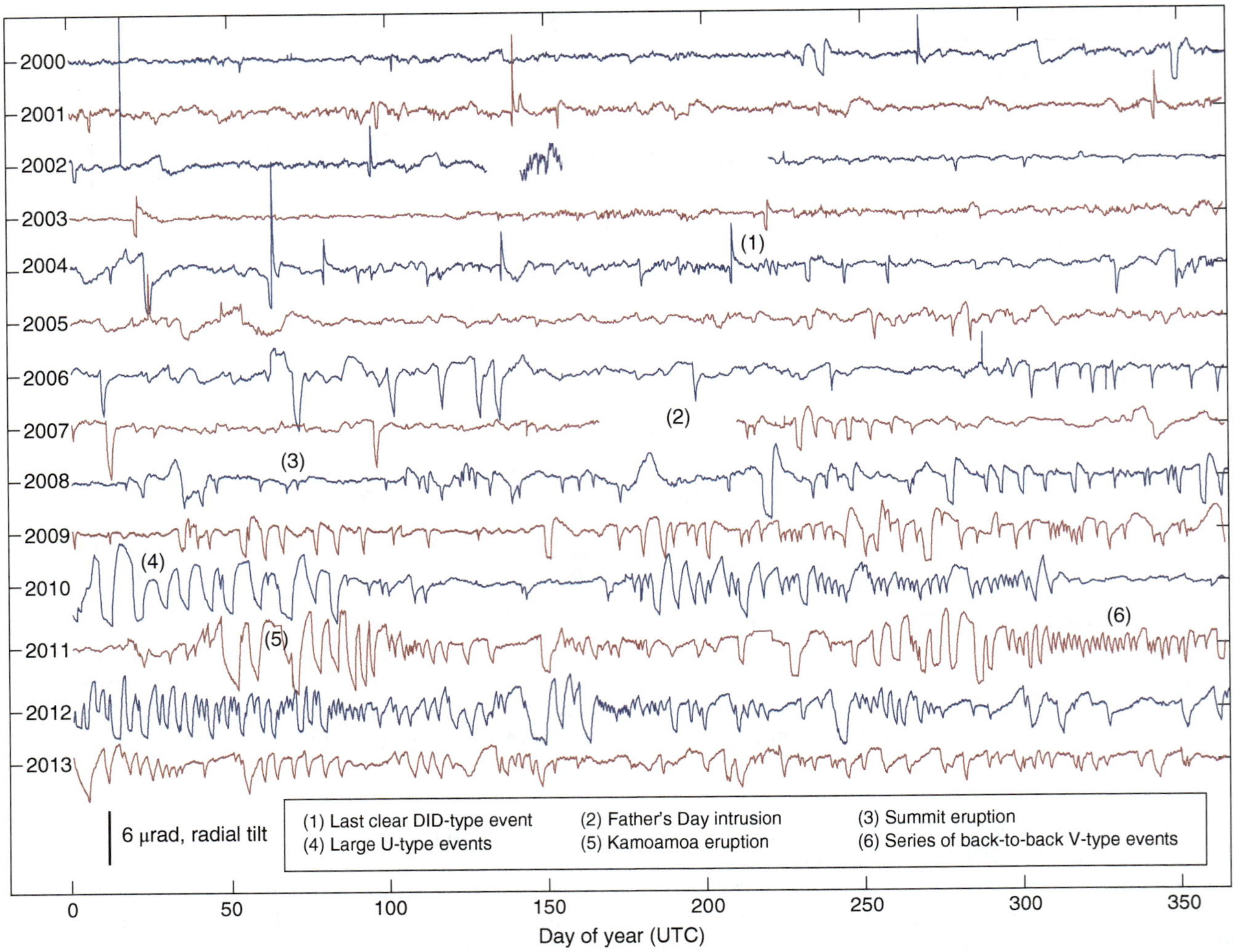

Figure 11.8 Fourteen years of radial tilt at UWE. Significant transients associated with volcanic activity have been removed and the data are high-pass filtered with a cutoff of 30 days. Most data gaps are from removal of large transients.

during inflation (e.g., Figures 11.9e,f). These correlations were observed for both U- and V-type events. In early 2012, roughly coincident with the M4.8 south flank earthquake discussed previously, the correlation between tilt and seismic tremor became gradually less clear, and no correlation is apparent as of the time of this writing.

11.4. SOURCE INVERSIONS

We use net ground tilts observed during DI events to estimate the location and volume change of the deformation source. A radially symmetric source model is consistent with the general symmetry of the deformation pattern recorded in tilt, GPS, and InSAR data. We argue that pressure changes in the subvertical conduit connecting the HMMR to the surface since 2008 do not play an important role in generating the observed tilt since there is no evidence for the existence of a shallow conduit beneath the summit from the installation of the borehole tilt network beginning in 1999 until the opening of the summit vent in 2008, and the opening of the vent was not associated with a change in the nature of tilt at the summit (except for an increase in event occurrence rate). For these reasons and because the existing tilt network is inadequate to constrain more complex source models, we follow *Cervelli and Miklius* [2003] and assume that observed ground deformation is generated by pressure changes in a spherical magma reservoir in a homogeneous elastic crust, although we acknowledge that the source is probably more complex. Radial ground tilt θ_r may be derived from expressions by *Mogi* [1958],

$$\theta_r = \frac{3(1-\nu)\Delta V}{\pi}\frac{rd}{(r^2+d^2)^{5/2}}, \qquad (11.1)$$

where ν is Poisson's ratio (here, $\nu = 0.25$), ΔV the volume change in the source, r the radial source-receiver distance

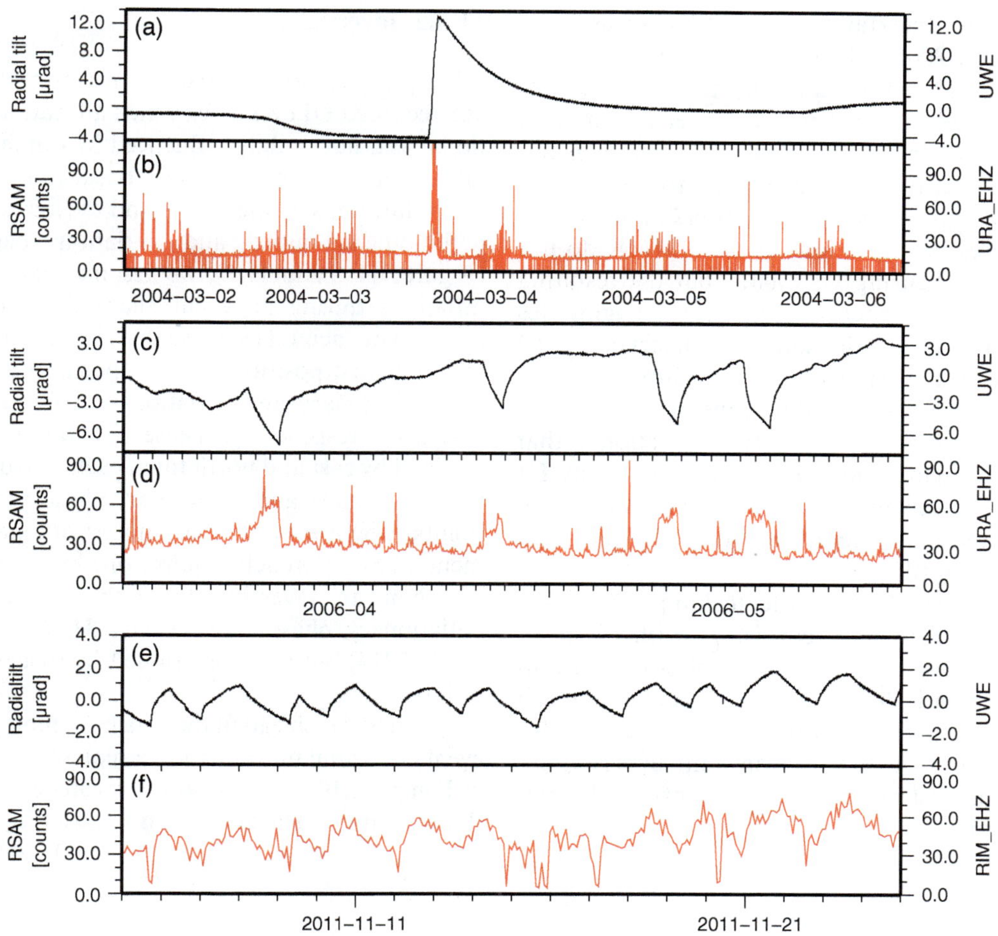

Figure 11.9 Radial tilt at UWE compared to RSAM. (a, b) Increase in RSAM associated with tremor during DID-type DI events. (c, d) Increase in RSAM associated with DI events. (e, f) Decrease in RSAM associated with DI events.

on the surface of the half-space, and d the source depth. We also include a simple correction to account for the source-receiver distance change caused by topography, but we do not fully account for topographic effects. The surface of the elastic half-space is set at the floor of Halema'uma'u Crater at roughly 1020 m elevation.

The inverse problem is formulated in a Bayesian sense, in which probability distributions are used to characterize data uncertainty, prior information on model parameters, and posterior probability density functions (posterior PDFs). We assume that data errors are Gaussian and formulate using an L2 norm. We set uniform prior bounds on all model parameters. During the Kamoamoa eruption the summit lava lake drained out of sight 200 m below the floor of Halema'uma'u Crater, so this is a reasonable minimum depth to the top of the reservoir; assuming the reservoir is larger than at least 0.03 km^3 (smaller than most existing estimates [e.g., *Poland et al.*, 2014] and corresponding to a reservoir radius of just ~200 m), we use a minimum centroid depth of 400 m. We assume a priori that volume changes ΔV_i associated with inflations are positive and volume changes ΔV_d associated with deflations are negative, and estimate the logarithm of ΔV_i and of $-\Delta V_d$.

We invert using data from stations UWE, SDH, SMC, IKI, and ESC (DI events are not detected at ESC but the data help to constrain source depth), only inverting the data from events in which all stations were operational. Posterior PDFs are sampled using a Markov chain Monte Carlo approach [e.g., *Mosegaard and Tarantola*, 1995].

11.4.1. Noise Model

White (temporally uncorrelated) noise in the tilt data may be reduced by time averaging, but temporally correlated noise due to temperature variations and many other causes is clearly evident and cannot be reduced in this manner. We find that the temporally correlated noise in the tilt data can be significant even over relatively short time periods and can be overwhelming at longer time periods.

To characterize uncertainty in net tilts we examine data from before 2006, when relatively few DI events were recorded, and remove all hand-picked DI events from the data. We calculate net tilts over one million randomly selected time intervals of fixed length τ and assume that the net tilts are due entirely to noise in the data (although this is certainly an oversimplification; for instance, pressure variations in the HMMR do not happen exclusively during DI events, but the resulting signal is treated as noise here). Using net tilt magnitudes on both channels of the tilt data, we construct a 2×2 covariance matrix Σ. We repeat this procedure for a wide range of τ encompassing the durations of all observed DI events and produce an empirical function f that allows us to calculate Σ for the duration τ of any DI event: $\Sigma = f(\tau)$. Uncertainties in derived quantities such as tilt magnitude and azimuth are then derived from uncertainties in net tilts.

We find that this Monte Carlo technique produces error estimates much more consistent with observed wander in the time series than white-noise-derived estimates, although it is limited by the assumption that noise is constant over the entire time series. We also assume there are no correlations in uncertainties between stations, and we neglect uncertainty in tiltmeter installation angle.

11.4.2. Inversions

We first perform a separate inversion for each of the 458 recorded DI events for which all stations were operational and not unduly influenced by rainfall, instrumental artifacts, or other effects. Because of the relatively short durations of the events and the agreement between deflation azimuths and inflation back azimuths (Figures 11.4 and 11.5), we assume that deflations and inflations during an event share the same unmoving source. For each DI event we therefore estimate the source location (east position x, north position y, and depth d) and inflationary and deflationary volume changes ΔV_i and ΔV_d, respectively. These five unknowns are constrained by east and north tilt at each instrument for both the deflationary and inflationary phase, yielding $4n_s$ data points, where $n_s = 5$ is the number of operating instruments. This approach requires one more model parameter than an inversion for only the inflationary or deflationary phase [*Cervelli and Miklius*, 2003; *Poland et al.*, 2014] but it is constrained by twice as many data points.

The model is able to fit the data from most U- and V-type events to within uncertainties; a representative fit is shown in Figure 11.10a. The DI source locates consistently under the east rim of Halemaʻumaʻu Crater (Figures 11.10a,b).

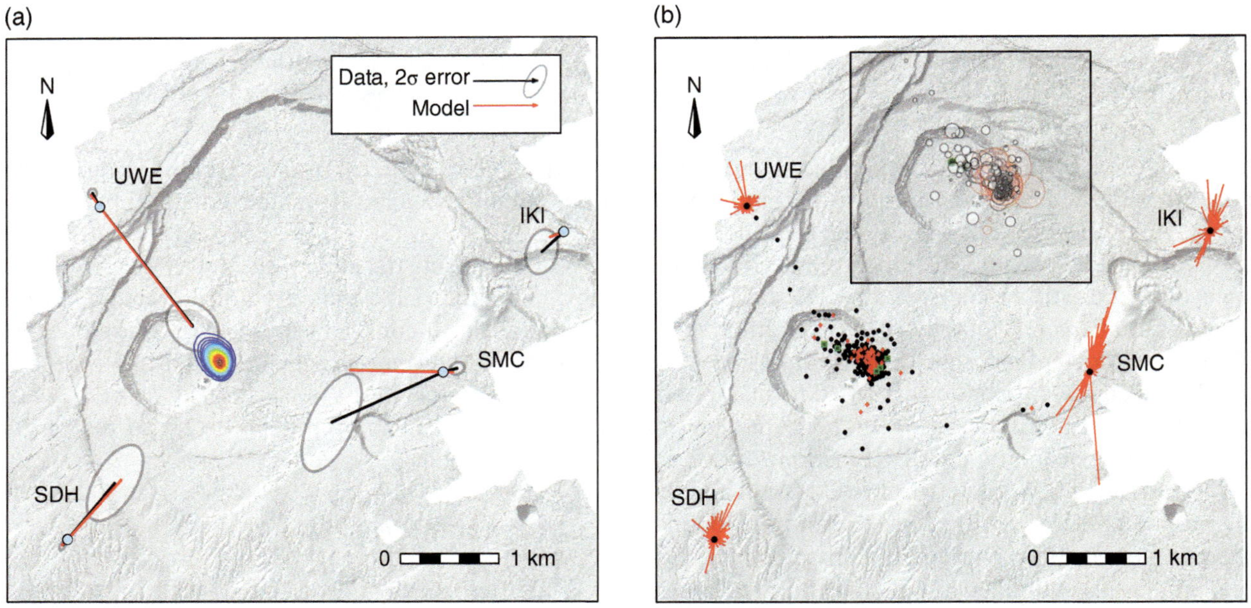

Figure 11.10 (a) Data (black vectors, with 95% error ellipses) and best-fitting model (red vectors) for event 2011_10 (see Figure 11.2). Contours indicate probabilistic location of the point source (warmer colors indicate more likely source locations). Note that the rapid inflationary part of this event happens to be very small; most are larger. (b) Residual between data and best-fitting model for all individual inversions (red lines), showing consistent misfit at SMC. The horizontal positions of best-fitting models are shown as dots with the same color convention as Figure 11.6. Inset shows best-fitting models with circle sizes scaled by the amplitude of tilt at UWE; larger events (usually with a higher signal-to-noise ratio) cluster at the east margin of Halemaʻumaʻu.

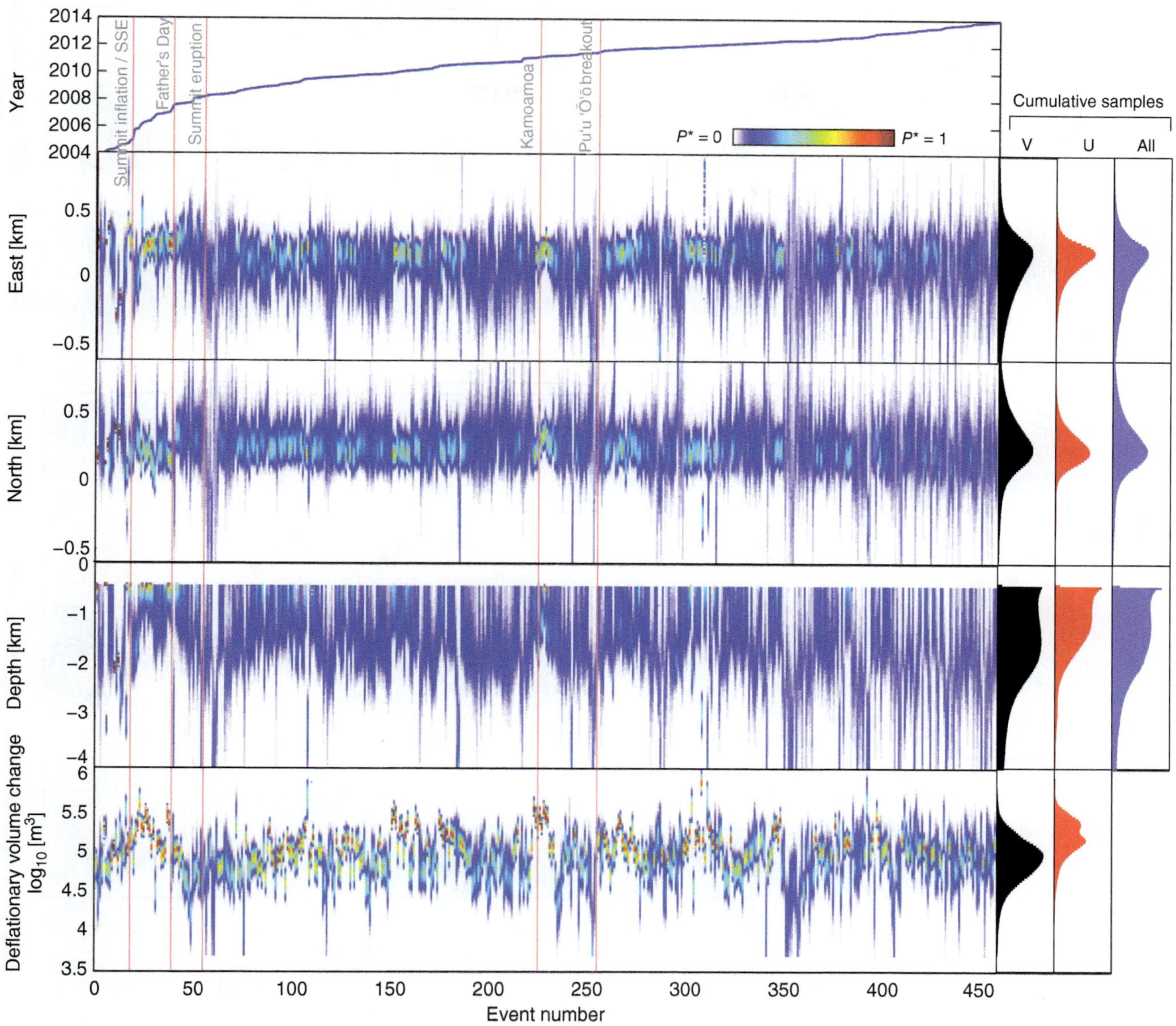

Figure 11.11 Time evolution of estimated model parameters. All posterior PDFs are plotted on the same color scale $0 \leq P^* \leq 1$. The total number of models accepted by the MCMC algorithm for each parameter value is shown using histograms. The time scale at the top of the figure relates event number to actual occurrence time. Volume change shown is for deflations. Large U-type events (such as those just before and after the Kamoamoa eruption) are better resolved in position because of their higher signal-to-noise ratio. Color scheme of histograms follows Figure 11.6.

This location lies northeast of the active lava lake and confirms, in part, our assumption that shallow conduit processes do not contribute significantly to the tilt signal. East and north positions are well constrained (Figure 11.11) but source depth is not, with probability distributions for most events extending from 400 m (the a priori minimum) to ~2 km. Deflationary volume changes are mostly on the order of 100,000 m^3, with larger events closer to 300,000 m^3. For comparison, modeled volume loss during the Kamoamoa fissure eruption was around $1.7 \times 10^6 m^3$ [*Lundgren et al.*, 2013], which is about 35,000 m^3 of volume loss per microrad of observed tilt at UWE. We estimate a similar ratio for DI events of roughly 50,000 m^3/μrad (25,000–100,000 m^3/μrad at 95% confidence).

Inversions reveal a consistent misfit between the model and data at station SMC (Figure 11.10). For most U- and V-type events the model is able to fit the data at SMC to within, or very nearly to within, uncertainties, but DID-type events (which have a much higher signal-to-noise ratio) are not adequately fit by the data. These issues are discussed later in this section.

No change in source position is clearly resolved outside the uncertainty in the inversions (Figure 11.11); although some apparent variations in position are evident over

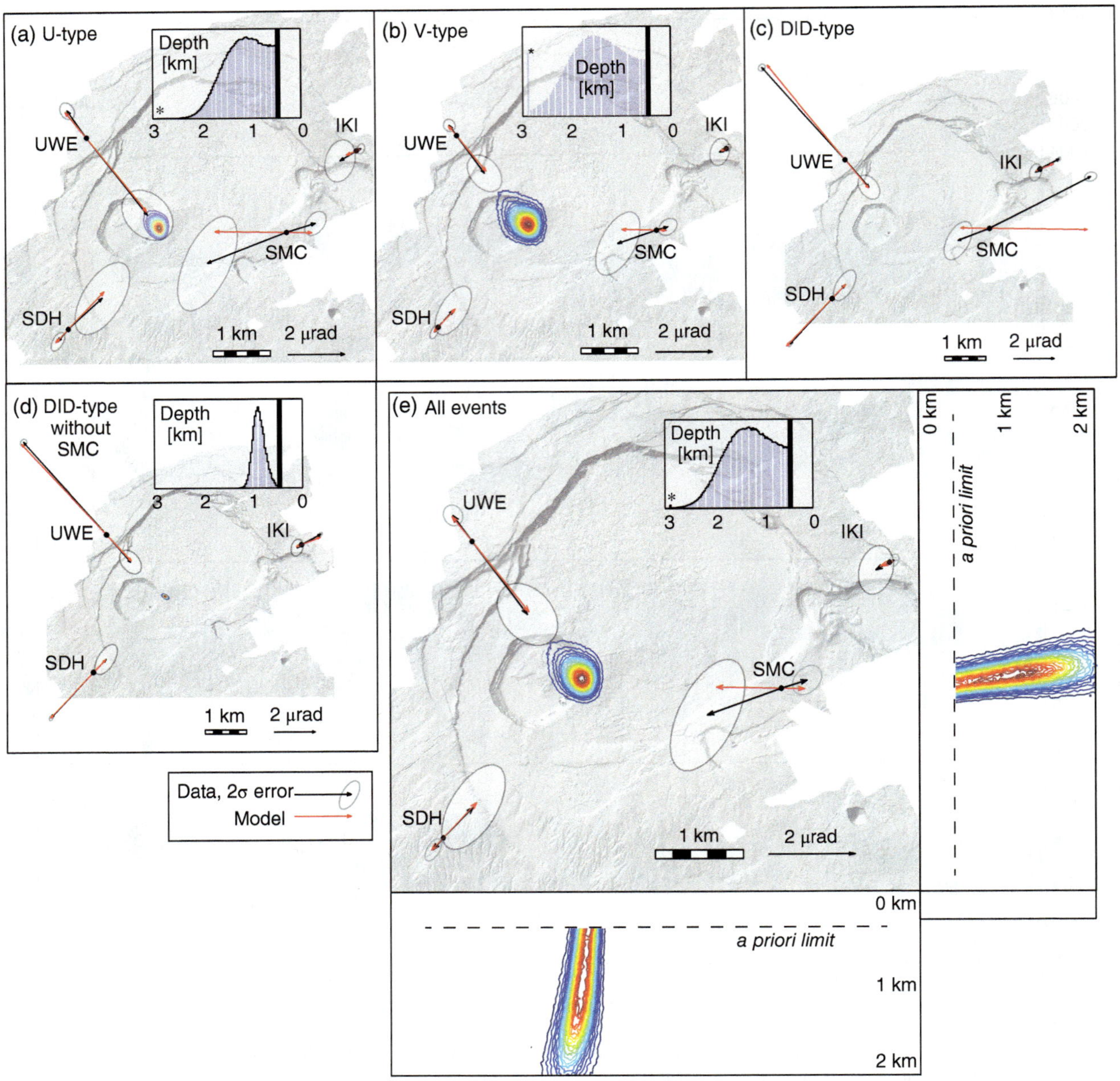

Figure 11.12 Inversions using averaged tilt data for (a) U-type events, (b) V-type events, (c) DID-type events, (d) DID-type events without station SMC, and (e) all event types. Inset histograms show estimated depth. No histograms are shown in (c) because the model does not fit the data. Bins beneath asterisks in histograms indicate the number of values below 3 km depth. Heavy vertical bars indicate the a priori minimum depth.

periods of months, they do not exceed uncertainty in the estimates. We therefore average events to reduce uncorrelated noise in the data and perform inversions for the averaged tilt vectors.

Results from inverting averaged data are consistent with those described above and indicate a source located beneath the east rim of Halema'uma'u Crater (Figure 11.12). The source location for U-type events is better resolved than for V-type events due to their higher signal-to-noise ratio. Inverting using an average of all DI types yields a best-fitting model about 300 m northeast of the summit eruptive vent; the east position is resolved to within about 500 m and the north position to within about 600 m, at 95% confidence (distributions are roughly Gaussian; Figure 11.13). Source depth is estimated in this inversion to be less than 2.3 km at 95% confidence, but the minimum is constrained by the a priori limit at 400 m rather than the data. Correlations between estimated volume change and source depth indicate that improved constraint on the latter would yield improved constraint on the former and vice versa (Figure 11.13).

The model fits data from averaged U- and V-type events to within uncertainties, but as noted earlier there is

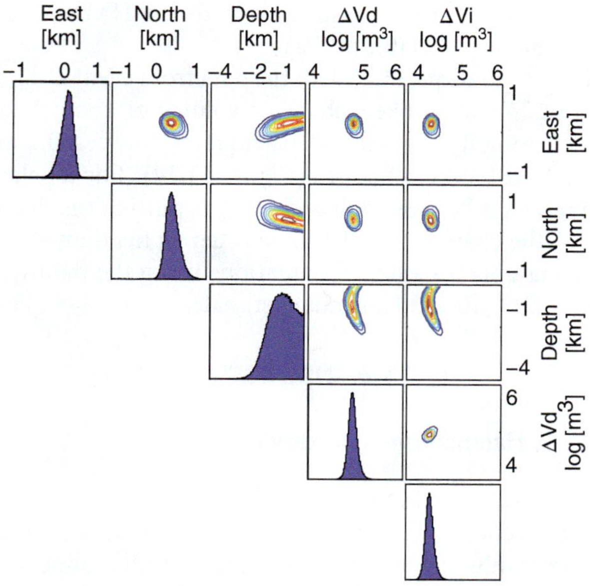

Figure 11.13 Estimated model parameters from inversion shown in Figure 11.12e. Marginal posterior PDFs for individual parameters are on the diagonal, and correlations between pairs of parameters are on the off-diagonals.

a consistent misfit at station SMC, with recorded tilt azimuths pointing significantly to the south of the modeled source. For DID-type events this misfit is much larger than the estimated uncertainty in the data. There are several possible explanations for this discrepancy. First, the instrument's installation azimuth could be incorrect, but SMC was pulled from its borehole twice in order to verify the azimuth. Second, the source model may be too simple. The intersection of summit and ERZ magma systems near Keanakāko'i Crater (Figure 11.1), along with possible transient magma storage in this area [e.g., *Poland et al.*, 2014], might produce a complex deformation signal to which SMC is particularly sensitive, and that is not accounted for in the simple Mogi model. Finally, it is possible that local material heterogeneity or topography [e.g., *Meertens and Wahr*, 1986] or poor coupling between the instrument and the borehole [e.g., *Kohl and Levine*, 1993] could alter the ground tilt at this station. Supporting this hypothesis, we note that during large summit deflations associated with intrusions and eruptions in 2007 and 2011, the SMC tilt azimuth was consistent with that recorded during DI events (Figure 11.4) whereas nearby continuous GPS stations moved in a more northerly direction toward the HMMR. Such factors are not accounted for in our formal error model, so estimated data uncertainties are minimums and may be too small.

To examine the influence of SMC on our results, we invert the averaged DID-type data without including observations from this station. The model fits the data, and the estimated source location is very tightly constrained (Figure 11.12d). The horizontal source position is located near the east margin of Halema'uma'u Crater, consistent with other results; the estimated source depth is approximately 850 m (500–1000 m at 95% confidence). We also invert individually for every DI event, without SMC data, and find that source locations are more poorly constrained but generally in the same location as inversions using all stations. We thus conclude that data from SMC do not significantly alter results, although the misfit at this station does indicate that our model is too simple and/or data uncertainties are too small.

11.5. SUMMIT RESERVOIR VOLUME

Simple geodetic inversions cannot in general be used to constrain the absolute volume of a magma reservoir. The geodetically derived volume change ΔV estimated in Section 11.4 is proportional to the product of the reservoir's volume V and its pressure change Δp; to estimate absolute reservoir volume independently thus requires constraint on pressure change.

At Kīlauea, changes in the surface heights of lava lakes often correlate directly with summit deformation [e.g., *Tilling*, 1987; *Johnson*, 1992; *Denlinger*, 1997; *Segall et al.*, 2001], suggesting that the lava lake can be used as a piezometer to estimate changes in reservoir pressure associated with geodetically estimated volume changes.

Because DI events take place slowly over a period of hours to days and changes in the lava lake level are not significantly (if at all) delayed from tiltmeter-measured changes in reservoir pressure, we assume that pressure p in the shallow reservoir is in magmastatic equilibrium with the surface of the lava lake. In this case $p=\rho gh$, where ρ is the average magma density and h is the height of the lava lake surface above the reservoir. Linearizing, $dp=\rho g\,dh+gd\rho$. Utilizing magma compressibility $\beta_m=(1/\rho)d\rho/dp$ and reservoir compressibility $\beta_{ch}\equiv(1/V)dV/dp$ and solving for V yield

$$V=\frac{(1-p\beta_m)}{\rho g\beta_{ch}}\frac{dV}{dh} \tag{11.2}$$

[e.g., *Segall et al.*, 2001]. For a spherical chamber, $\beta_{ch}\approx 3/4\mu$, where μ is the shear modulus [*McTigue*, 1987]. The $p\beta_m$ term accounts for compression and decompression of lava as the lake fills and drains, respectively. For a shallow chamber at 20 MPa, magma compressibility β_m must exceed $5\times10^{-9}\,\text{Pa}^{-1}$ for $p\beta_m$ to exceed just 0.1. Although the upper part of the lava lake is very vesicular [*Carbone et al.*, 2013] and must be highly compressible, vesicularity and compressibility in the reservoir must be much lower (e.g., $\beta_m=1\times10^{-10}\,\text{Pa}^{-1}$ [*Rivalta and Segall*, 2008]). We therefore neglect the $p\beta_m$ term and calculate volume using

$$V=\frac{4\mu}{3\rho g}\frac{dV}{dh}. \tag{11.3}$$

For ρ and μ we use a priori probability distributions (Figure 11.14) so that uncertainties in these estimates are propagated to the estimate for V. The distribution for ρ is based on the gravity modeling presented by *Carbone et al.* [2013] that suggests a density of just ~950 kg/m³ in the upper lava lake, together with the assumption that density must be significantly greater deeper in the conduit and chamber. Host rock shear modulus is not well known but is probably relatively low at Kīlauea [e.g., *Johnson*, 1992]; we use a distribution centered on ~3 GPa. We also constrain the depth to the top of the chamber to exceed 200 m (Section 11.4).

Lava lake surface height time series, derived from thermal camera imagery and corrected using occasional laser rangefinder measurements [*Patrick et al.*, 2013], yield height changes of $dh = 53 \pm 3$ m and $dh = 48 \pm 3$ during two large U-type DI events from February 2011, which for convenience we will refer to as events 2011_09 and 2011_10, respectively, as events 9 and 10 of 2011 (Figure 11.2). Comparing to radial tilt at UWE yields ratios of roughly 6–9 m of lake surface height change per microrad of tilt at UWE.

Using probability distributions for dV ($=\Delta V_d$) obtained from geodetic inversions in Section 11.4.2, we estimate $dV/dp \approx 0.29\,m^3/Pa$ (0.21–0.46 m³/Pa at 95% confidence); this is smaller than estimates of *Johnson* [1992] (0.6–2.3 m³/Pa for events apparently associated with the HMMR) and *Segall et al.* [2001] (0.8 m³/Pa).

Using equation (11.3) with the data from event 2011_09 we obtain a most likely chamber volume of about 1 km³, with a roughly log-normal distribution between 0.2 and 5.5 km³ at 95% confidence (Figure 11.14). For a sphere, the radius is between 400 and 1100 m, putting the diameter of the chamber on the same order as the diameter of Halema'uma'u Crater. Calculations using the data from event 2011_10 yield a similar estimate.

11.6. DISCUSSION

11.6.1. Halema'uma'u Reservoir

Tilt inversions in Section 11.4 clearly point to a source located beneath the east margin of Halema'uma'u Crater. These results agree with many other studies suggesting shallow magma storage in this area (Section 11.1), and we agree with *Cervelli and Miklius* [2003] and *Poland et al.* [2014] that DI-related ground deformation is most likely generated by pressure variations in the shallow Halema'uma'u reservoir. DI-related deformation can therefore be used to better constrain the location of this

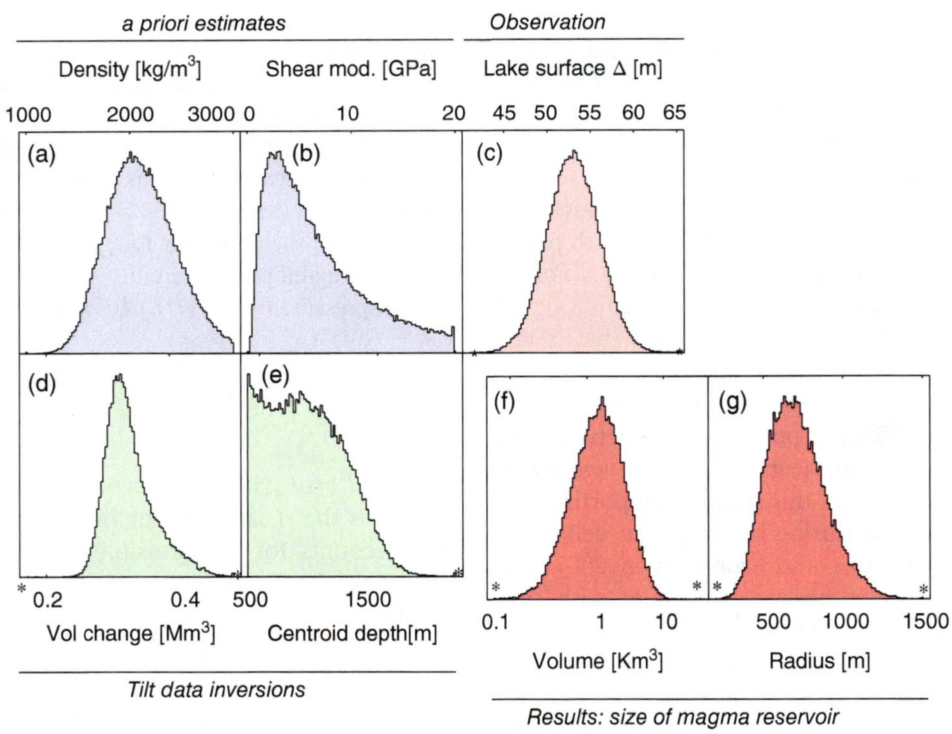

Figure 11.14 Reservoir volume is estimated from (a, b) a priori estimates of magma and host rock properties, (c) an observation of the change in lava lake surface height associated with event 2011_09 (Figure 11.2), and (d, e) estimated source volume change and centroid depth for this same DI event (centroid depth is used only to exclude too-shallow reservoirs). All inputs take the form of probability distributions, in which asterisks indicate that some values exceed the bounds of the plot. Estimated source volume and radius are shown in (f) and (g).

chamber. The horizontal location of the tilt source is constrained here to within a region of about 600 m diameter at the 95% confidence level.

Most inversions for individual DI events (Section 11.4.2) do not resolve the source depth between ~2 km and very shallow depths; however, we note that DI events with a higher signal-to-noise ratio (such as those bracketing the March 2011 Kamoamoa eruption), as well as the inversion using all averaged data (Figure 11.12), tend to favor a source depth of ~1–2 km. These results are consistent with other DI inversions suggesting depths of less than 2 km [*Poland et al.*, 2014] and with many other studies suggesting a reservoir depth somewhere between 0.5 and 5 km (but generally ~1–2 km) [e.g., *Ohminato et al.*, 1998; *Almendros et al.*, 2002; *Owen et al.*, 2000; *Dawson et al.*, 2004; *Poland et al.*, 2009b; *Montgomery-Brown et al.*, 2010; *Baker and Amelung*, 2012; *Lundgren et al.*, 2013]. We favor a deeper source depth than the 350 m estimated by *Cervelli and Miklius* [2003], although we acknowledge that the tilt data alone cannot rule out such a shallow reservoir.

The sign of strain changes observed during DI events might be used to further constrain source depth. Strainmeters KWL and MLS both record positive strain associated with DI inflations, establishing a maximum source depth of $r\sqrt{2}$, where r is the horizontal distance between KWL and the source [e.g., *Lisowski*, 2006]. This yields a maximum source depth of roughly 2 km, which does not improve the tiltmeter-derived constraint.

A striking result from the present study is the stability of the source location. Some minor variations in source location are shown in Figure 11.11, for instance around the time of the Kamoamoa eruption, but the variations are small (probably less than 100 m) and may not be real.

The volume we calculate for the HMMR is consistent with estimates of 1.6–6 and 0.2–1.2 km^3 from *Johnson* [1992] and *Poland et al.* [2009b], respectively (the latter using dV/dp calculated in the former). *Segall et al.* [2001] estimated $V = 20\,\text{km}^3$ using $\mu = 20\,\text{GPa}$; with $\mu = 2\,\text{GPa}$, the volume estimated in that study would be 2 km^3. Estimates by *Johnson* [1992] and *Segall et al.* [2001] were also based on a somewhat deeper reservoir than we estimate here, which yields a larger estimated volume.

The volume of the reservoir has implications for modeling. First, large spherical chambers cannot be very shallow without breaching the surface. For instance, a 1 km^3 chamber must be at least ~800 m deep, assuming a minimum top depth of 200 m (Section 11.4). Better constraints on volume may thus help to constrain reservoir depth and vice versa.

Second, for relatively large and shallow chambers a finite source model should be used. For the point source model used here [equation (11.1)], *McTigue* [1987] showed that the ratio of the source radius to depth should not much exceed 0.5. If the volume of the HMMR is 5 km^3 and its top depth is just 200 m below the surface, the radius/depth ratio is roughly 0.8; however, this is an upper bound and the reservoir is likely deeper and smaller. We compared results using equation (11.1) to results obtained using the finite element method for a 2 km^3 chamber with a centroid depth of 1 km (this yields a top depth of just over the minimum of 200 m). Numerical results differed from the analytical approximation by more than 20% only at surface distances of less than ~2 km from the source. Since all of the tiltmeters at Kīlauea exceed this horizontal distance from the inferred location of the HMMR, we conclude that equation (11.1) is adequate for the simple modeling we perform here. Future work considering observations closer to the source (i.e., GPS and InSAR) should however utilize finite source models.

11.6.2. Source Mechanism

It is generally accepted that magma supplied to Kīlauea rises into a reservoir in the south caldera (the south caldera reservoir, or SCR) at ~3–5 km depth [e.g., *Eaton*, 1962; *Dvorak et al.*, 1983; *Yang et al.*, 1992; *Baker and Amelung*, 2012; *Poland et al.*, 2014] before making its way to the HMMR. Precisely where magma departs the summit for the ERZ is unclear. *Cervelli and Miklius* [2003] favored a model in which the ERZ connects to the summit magma system relatively high in the HMMR, but a deeper ERZ conduit is more consistent with observed seismicity [*Klein et al.*, 1987]. The summit lava lake probably connects directly with the upper part of the HMMR.

We assume that deflations associated with U- and V-type DI events are caused by outflux q_o from the HMMR temporarily exceeding influx q_i (this simple conceptual model neglects changes in density and pressure due to outgassing). Increased outflux could be caused by increased effusion through Pu‘u ‘Ō‘ō, but this is contrary to observations showing reduced effusion during DI events. Alternatively, a transient diversion of magma to a sink somewhere in the ERZ between the summit and Pu‘u ‘Ō‘ō, such as a reservoir beneath Makaopuhi Crater, could explain reduced pressure at the summit and Pu‘u ‘Ō‘ō (the sink would need to be closer to the summit to explain the appearance of DI events there first), but there is no direct evidence for such a process, and indeed observations indicate that the ERZ contracts rather than inflates during DI events (Section 11.3.4).

It seems more likely that DI events begin with a reduction of influx q_i into the HMMR. A transient interruption of magma supply caused by a temporary blockage in the conduit leading from the SCR toward the HMMR, or else deeper in the system, could explain such a reduction. *Cervelli and Miklius* [2003] hypothesized that a blockage resulted in deflation due to continued withdrawal of magma into the ERZ, followed by rapid inflation upon removal of the blockage as accumulated magma rushed into the HMMR, and finally by deflation to near-pre-event levels as the system returned to a condition of steady magma flux. We hypothesize that a blockage either elsewhere in the system or of a different type could produce U- and V-type events as well.

If a blockage occurs in the conduit between the SCR and HMMR, blocked magma could either be diverted past the HMMR or stored in the shallow magma system. For reasons outlined previously we do not believe that the magma supply bypasses the HMMR into the ERZ. It is possible that the magma is diverted to deep storage beneath the summit or upper ERZ, but we find this scenario unlikely. The temporarily blocked magma would, therefore, most likely be stored somewhere in the upper part of Kīlauea's summit magma system.

At most, the volume of blocked magma could equal the magma supply rate times the duration of blockage (assuming that the supply rate does not decrease during the event). Using a maximum supply rate of 6 m^3/s (during the magma supply surge [*Poland et al.*, 2012]) and DI deflation of three days, the volume V_m of blocked magma could be as much as 1.5 Mm^3. Neglecting compressibility and assuming that magma is stored in the SCR at 4 km depth yield uplift of at most ~2 cm (using *Mogi* [1958]) and maximum ground tilt of about 5 µrad. Since most DI events are shorter and smaller, the magma supply rate is usually lower, this estimate neglects magma compressibility, and the supply rate into the shallow magma system might decrease during a DI event due to the blockage, it seems possible that DI-related SCR deformation could be small enough to go undetected.

The volume of magma lost from the HMMR system into the ERZ during events can also shed light on DI processes. During event 2011_09 (Figure 11.2), geodetically determined volume loss in the HMMR was of order 0.3–0.4 Mm^3, while volume loss in the lava lake was roughly 0.9 Mm^3 (assuming a conduit radius of 75 m). This yields a total loss from the shallow magma system of ~ 1.1–1.3 Mm^3, or an average flux out of the HMMR system of 2.3–2.7 m^3/s during the DI event (assuming that the geodetic estimate is insensitive to the conduit pressure change, which cannot be entirely true). Pre-event ERZ flux and hence magma supply rate must be higher since driving pressure and flow rate dropped during the event. The magma supply rate during this time period is poorly known, but the long-term average is ~3 m^3/s [*Poland et al.*, 2004]. Thus, although these estimates are crude, magma supply rate may be higher than ERZ flux during DI events, consistent with a simple blockage (or partial blockage) model.

What could cause a blockage in magma flow to the HMMR? One hypothesis involves small-scale convective overturns within Kīlauea's shallow summit magma storage area, with degassed magma being flushed downward and replaced by less dense gas-rich magma [*Poland et al.*, 2009a]. The downward motion of the denser, degassed magma would then act to impede flow to the ERZ. This model is consistent with the increasing number of DI events associated with increased gas emissions from Kīlauea's summit eruptive vent in March 2008; however, it does not explain the variation in occurrence of DI events over shorter time periods when gas emission rates did not measurably change.

The ultimate cause of DI events therefore remains ambiguous, although we have been able to place useful constraints on the likely mechanism. Modeling and further analysis are required to test these hypotheses.

11.6.3. Relation to the ERZ

The propagation of DI events between the summit and Pu'u 'Ō'ō, and the varying manifestation of the events at the latter, offer tantalizing clues into the nature of the ERZ. A careful analysis of these events may yield clues about the geometry of the molten core of the rift zone, changes in the rift zone with time and in association with earthquakes and eruptions, and about Pu'u 'Ō'ō's magma reservoir and its variability with time.

Observations of time-delayed pressure or flow transients migrating between the ERZ and summit have been observed for decades. In 1955 and 1960, Kīlauea's summit subsided seven and four days, respectively, after eruptions in lower Puna [*Eaton*, 1962; *Wright and Helz*, 1996]. For intrusions and eruptions near Pu'u 'Ō'ō in the middle ERZ, delay times to the summit have typically been on the order of only a few hours. Examples include the 2–3 January 1983 dike emplacement that led to the start of the Pu'u 'Ō'ō eruption (2 h delay), renewed motion of the 1983 dike downrift on January 6 (4 h [*Okamura et al.*, 1988]), and episodic eruptive activity over the next several years (3–12 h [*Dvorak and Okamura*, 1985; *Wolfe et al.*, 1987; *Heliker and Mattox*, 2003]). Taking the delay time between ERZ and summit pressure changes, inferred propagation velocities for 1955 and 1960 were of order 0.07 and 0.14 m/s, respectively; for activity near Pu'u 'Ō'ō, velocities have ranged from 1.4 to 2.7 m/s. DI events propagate from the summit to the ERZ, not vice versa as in these previous examples, but we calculate downrift propagation velocities of roughly 1–10 m/s (Figure 11.6c), which are not inconsistent with velocities inferred for earlier activity near Pu'u 'Ō'ō.

A model for the propagation of DI events through the ERZ should explain: (1) summit–ERZ delay times; (2) similarity of delay times during different types of activity, occurring as much as decades apart; (3) large variations in delay times over much shorter time periods and even different phases of the same inflation-deflation cycle; and (4) variability in the manifestation of DI events at Pu'u 'Ō'ō.

To date, two principal models have been proposed to explain the summit-ERZ delay times: one using a conduit geometry and the other a dike. *Cervelli and Miklius* [2003] argue that the propagation velocities inferred from DI events represent the flow rate of a pulse of magma in the conduit between the summit and Pu'u 'Ō'ō. Indeed, as noted by *Poland et al.* [2014], estimated propagation velocities are consistent with the velocities of steady state gravity-driven lava flows in tubes [e.g., *Kauahikaua et al.*, 1998], although invididual packets of magma might travel at a significantly different rate. *Cervelli and Miklius* [2003] also argue that the velocities are much too slow to represent the propagation of a pressure pulse, which they argue would be of order 1000 m/s (the speed of a P wave through magma).

Montagna and Gonnermann [2013] model delays during episodic activity at Pu'u 'Ō'ō (1983–1985) using a one-dimensional model of incompressible magma flow in an elastic dike, in which delay times are explained by the spatiotemporal evolution of magma pressure and flow rate in the dike. They hypothesize that longer-term changes in delay time were a consequence of changes in dike width (such as thermal erosion and solidification), while for a given episode, the inflation delay was slightly longer because lower pressure associated with the event reduced the dike width and increased the resistance to flow.

What might explain the temporary disappearances of DI events at Pu'u 'Ō'ō? Possibilities include (1) DI-related pressure pulses at the summit no longer enter the ERZ, (2) pulses enter the upper ERZ but do not propagate as far as Pu'u 'Ō'ō, or (3) pulses propagate to Pu'u 'Ō'ō but do not produce significant tilt there due to some change in the local magma system.

It seems unlikely that summit pressure changes would fail to enter the ERZ during times of lava effusion, but we cannot completely discount changes in the system that would buffer DI-related pressure changes before they reached the upper ERZ.

Changes in the degree to which DI events are manifested at Pu'u 'Ō'ō occurred apparently in association with a slow slip event (in 2005) and magma supply rate surge, an eruption, and a south flank earthquake. All of these events would introduce stress changes on the ERZ and/or pressure changes in the ERZ itself; however, it must be noted that many other similar events occurred that had no apparent effect on ERZ DI occurrence (slow slip events, for instance, occur roughly every two years). Interestingly, changes in Pu'u 'Ō'ō delay times tend to occur independently from how clearly DI events manifest themselves at Pu'u 'Ō'ō.

It is possible that DI events are sometimes not recorded at Pu'u 'Ō'ō because of local conditions near Pu'u 'Ō'ō itself. For instance, if the magma system at Pu'u 'Ō'ō is relatively "open" such that the magma supply can easily pass through or bypass Pu'u 'Ō'ō's magma reservoir, then DI-related pressure transients might not result in much deformation. Changes in DI activity associated with the Pu'u 'Ō'ō breakout in August 2011 might support such a mechanism because the event was clearly local, although it is not obvious how this mechanism might control delay times. We also note that the Pu'u 'Ō'ō breakout was associated with a change in the relation between tilt and tremor at the summit (Section 11.3.7), suggesting an impact on the larger magma system.

11.7. CONCLUSIONS

Episodic DI events at Kīlauea Volcano perturb much of the volcano's shallow magma system and produce transient geophysical signals that provide a remarkable window into the volcano's magma system dynamics and geometries, from the summit to the ERZ.

We inverted tilt data recorded during more than 450 DI events and conclude that the deformation is generated by pressure transients in Kīlauea's shallow magma reservoir (the Halema'uma'u reservoir). The data are well modeled by an inflating and deflating spherical pressure source except for a consistent misfit at station SMC, possibly caused by a local site effect or a more complex source than modeled here. The reservoir lies beneath the east margin of Halema'uma'u Crater; reservoir depth is poorly constrained by the tilt data but probably falls within a range of 1–2 km. We find that the reservoir has remained remarkably stable for more than a decade, spanning two major rift zone eruptions and the opening of an eruptive vent at Kīlauea's summit. DI events can be classified into several types based on their temporal evolution, but all types appear to be associated with pressure transients in this reservoir.

Total reservoir volume may be estimated by combining results from geodetic inversions with independent estimates of reservoir pressure change. We use pressure change estimated from variations in Kīlauea's summit lava lake during a large DI event and estimate a volume for the Halema'uma'u reservoir of around 1 km^3, or 0.2–5.5 km^3 at 95% confidence. Although this estimate varies by more than an order of magnitude, it includes uncertainty in magma density and host rock properties, which are neglected in previous studies.

The tilt data analyzed here are remarkable for their ability to record pressure transients that would otherwise go undetected, and they allow us to look for changes in the reservoir over more than a decade; they also clearly record the temporal evolution of these pressure transients. However, the geometry of the tilt network is not ideal for constraining the depth of the reservoir, leading also to higher uncertainty in estimated volume and volume change. Future work utilizing additional data sets (strain, GPS, InSAR, additional lava lake time series, gravity, and possibly tilt derived from broadband seismometers) and taking into account topography and finite source models should yield a significantly improved estimate of source properties.

The relation between summit DI events and their manifestation along the ERZ remains enigmatic. The appearance of DI events at Pu'u 'Ō'ō is delayed from their appearance at the summit by tens of minutes to hours; inferred propagation velocities along the ERZ are remarkably similar to those observed during historic activity in the ERZ, which were subsequently manifested after a short delay at the summit. Delay times, however, can vary widely and change systematically, and summit DI events are not always manifested at Pu'u 'Ō'ō. Changes in DI activity in the ERZ are associated with earthquakes and eruptive activity. These observations support the idea of a magmatic connection between the summit and ERZ that, in general, has remained remarkably stable for years or even decades but can change in response to variations in magma pressure and/or tectonic stress.

The source mechanism and processes controlling the temporal evolution of DI events are not understood. DI events may involve complex interactions between variable rates of influx and outflux into and out of the Halema'uma'u reservoir, degassing, convection, and elastically deforming host rock, as well as flow of magma in a compressible dike- or conduit-like structure into the ERZ. Simple models assuming a transient blockage of influx into the Halema'uma'u reservoir may explain the quasi-exponential decay of pressure variations, and be used to further constrain reservoir volume and conduit properties.

At the time of writing, DI activity continues at Kīlauea Volcano. This rich and evolving data set should inform many future studies of the volcano's magma system.

ACKNOWLEDGMENTS

The authors would like to thank Matt Patrick for providing the processed lava lake data, Adam Soule for the processed lidar data used in some figures, Tyler Huth for picking the times of many of the early DI events, and the former and present staff of the Hawaiian Volcano Observatory, especially Mike Lisowski, for the installation and maintenance of the deformation network. Early analysis of the DI events was also carried out by Alex Pesch. This work was supported in part by the USGS Mendenhall Research Fellowship Program. We appreciate valuable comments from Emily Montgomery-Brown, two anonymous reviewers, and the associate editor.

REFERENCES

Almendros, J., B. Chouet, P. Dawson, and T. Bond (2002), Identifying elements of the plumbing system beneath Kilauea Volcano, Hawaii, from the source locations of very-long-period signals, *Geophys. J. Int.*, *148*(2), 303–312, doi:10.1046/j.1365-246X.2002.01010.x.

Anderson, K., M. Lisowski, and P. Segall (2010), Cyclic ground tilt associated with the 2004–2008 eruption of Mount St. Helens, *J. Geophys. Res.*, *115*(B11), B11201, doi:10.1029/2009JB007102.

Baker, S., and F. Amelung (2012), Top-down inflation and deflation at the summit of Kīlauea Volcano, Hawaii observed with InSAR, *J. Geophys. Res.*, *117*(B12), B12406, doi:10.1029/2011JB009123.

Battaglia, J., J-L. Got, P. Okubo (2003), Location of long-period events below Kilauea Volcano using seismic amplitudes and accurate relative relocation, *J. Geophys. Res.*, *108*(B12), 2553, doi:10.1029/2003JB002517.

Bluth, G. J., and W. I. Rose (2004), Observations of eruptive activity at Santiaguito volcano, Guatemala, *J. Volcanol. Geotherm. Res.*, *136*(3–4), 297–302, doi:10.1016/j.jvolgeores.2004.06.001.

Brooks, B., J. Foster, M. Bevis, L. Frazer, C. Wolfe, and M. Behn (2006), Periodic slow earthquakes on the flank of Kīlauea volcano, Hawaii, *Earth Planet. Sci. Lett.*, *246*(3–4), 207–216, doi:10.1016/j.epsl.2006.03.035.

Carbone, D., M. P. Poland, M. R. Patrick, and T. R. Orr (2013), Continuous gravity measurements reveal a low-density lava lake at Kīlauea Volcano, Hawaii, *Earth Planet. Sci. Lett.*, *376*, 178–185, doi:10.1016/j.epsl.2013.06.024.

Cervelli, P., and A. Miklius (2003), The shallow magmatic system of Kilauea Volcano, in *The Pu'u 'O'o-Kupaianaha Eruption of Kilauea Volcano, Hawai'i: The First 20 Years,* edited by C. Heliker et al., pp. 149–164, Prof. Pap. 1676, U.S. Geol. Surv, Reston, Virginia.

Chouet, B. A., P. B. Dawson, M. R. James, and S. J. Lane (2010), Seismic source mechanism of degassing bursts at Kilauea Volcano, Hawaii: Results from waveform inversion in the 10–50 s band, *J. Geophys. Res.*, *115*(B9), B09311, doi:10.1029/2009JB006661.

Dawson, P., D. Whilldin, and B. Chouet (2004), Application of near real-time radial semblance to locate the shallow magmatic conduit at Kilauea Volcano, Hawaii, *Geophys. Res. Lett.*, *31*(21), L21606, doi:10.1029/2004GL021163.

Dawson, P. B., B. A. Chouet, P. G. Okubo, A. Villaseñor, and H. M. Benz (1999), Three-dimensional velocity structure of the Kilauea Caldera, Hawaii, *Geophys. Res. Lett.*, *26*(18), 2805–2808, doi:10.1029/1999GL005379.

Denlinger, R. P. (1997), A dynamic balance between magma supply and eruption rate at Kilauea volcano, Hawaii, *J. Geophys. Res.*, *102*(B8), 18091, doi:10.1029/97JB01071.

Dvorak, J., A. Okamura, and J. H. Dieterich (1983), Analysis of surface deformation data, Kilauea Volcano, Hawaii: October 1966 to September 1970, *J. Geophys. Res.*, *88*(B11), 9295, doi:10.1029/JB088iB11p09295.

Dvorak, J. J., and A. T. Okamura (1985), Variations in tilt rate and harmonic tremor amplitude during the January-August 1983 east rift eruptions of Kilauea Volcano, Hawaii,

J. Volcanol. Geotherm. Res., *25*(3–4), 249–258, doi:10.1016/0377-0273(85)90015-0.

Dzurisin, D., L. A. Anderson, G. P. Eaton, R. Y. Koyanagi, P. W. Lipman, J. P. Lockwood, R. T. Okamura, G. S. Puniwai, M. K. Sako, and K. M. Yamashita (1980), Geophysical observations of Kilauea Volcano, Hawaii, 2. Constraints on the magma supply during November 1975–September 1977, *J. Volcanol. Geotherm. Res.*, *7*(3–4), 241–269, doi:10.1016/0377-0273(80)90032-3.

Eaton, J. P. (1962), Crustal structure and volcanism in Hawaii, in *The Crust of the Pacific Basin*, vol. 6, edited by G. A. Macdonald and H. Kuno, pp. 13–29, Geophys. Monog. 6., AGU, Washington, D. C., doi:10.1029/GM006p0013.

Fiske, R. S., and W. T. Kinoshita (1969), Inflation of Kilauea Volcano prior to its 1967–1968 eruption, *Science*, *165*(3891), 341–349, doi:10.1126/science.165.3891.341.

Genco, R., and M. Ripepe (2010), Inflation-deflation cycles revealed by tilt and seismic records at Stromboli volcano, *Geophys. Res. Lett.*, *37*(12), doi:10.1029/2010GL042925.

Heliker, C., and T. Mattox (2003), The first two decades of the Pu'u 'O'o-Kupaianaha eruption: Chronology and selected bibliography, in *The Pu'u 'O'o-Kupaianaha Eruption of Kilauea Volcano, Hawai'i: The First 20 Years,* edited by C. Heliker et al., Prof. Pap. 1676, U.S. Geol. Surv., Reston, Va.

Iverson, R. M., et al. (2006), Dynamics of seismogenic volcanic extrusion at Mount St Helens in 2004–05, *Nature*, *444*(7118), 439–443, doi:10.1038/nature05322.

Jackson, D. B., D. A. Swanson, R. Y. Koyanagi, and T. L. Wright (1975), *The August and October 1968 East Rift Eruptions of Kilauea Volcano, Hawaii,* Prof. Pap. 890, U.S. Geol. Surv., Reston, VA.

Johnson, D. J. (1992), Dynamics of magma storage in the summit reservoir of Kilauea Volcano, Hawai'i, *J. Geophys. Res.*, *97*(B2), 1807, doi:10.1029/91JB02839.

Johnson, D. J., A. A. Eggers, M. Bagnardi, M. Battaglia, M. P. Poland, and A. Miklius (2010), Shallow magma accumulation at Kīlauea Volcano, Hawai'i, revealed by microgravity surveys, *Geology*, *38*(12), 1139–1142, doi:10.1130/G31323.1.

Kauahikaua, J., K. V. Cashman, T. N. Mattox, C. C. Heliker, K. A. Hon, M. T. Mangan, and C. R. Thornber (1998), Observations on basaltic lava streams in tubes from Kilauea Volcano, island of Hawai'i, *J. Geophys. Res.*, *103*(B11), 27303, doi:10.1029/97JB03576.

Keller, G. V., L. Grose, J. C. Murray, and C. K. Skokan (1979), Results of an experimental drill hole at the summit of Kilauea Volcano, Hawaii, *J. Volcanol. Geotherm. Res.*, *5*(3–4), 345–385, doi:10.1016/0377-0273(79)90024-6.

Klein, F. W., R. Y. Koyanagi, J. S. Nakata, and W. R. Tanigawa (1987), The seismicity of Kilauea's magma system, in *Volcanism in Hawaii*, vol. 2, edited by R. W. Decker et al., pp. 1019–1185, Prof. Pap. 1350, U.S. Geol. Surv., Reston, VA.

Kohl, M. L., and J. Levine (1993), Measuring low frequency tilts, *J. Res. Nat. Inst. Stand. Technol.*, *98*(2), 191–202.

Lisowski, M. (2006), Analytical volcano deformation source models, in *Volcano Deformation: Geodetic Monitoring Techniques*, edited by D. Dzurisin, pp. 279–304, Springer Praxis Books, Berlin.

Lundgren, P., M. Poland, A. Miklius, T. Orr, S.-H. Yun, E. Fielding, Z. Liu, A. Tanaka, W. Szeliga, S. Hensley, and S. Owen (2013), Evolution of dike opening during the March 2011 Kamoamoa fissure eruption, Kīlauea Volcano, Hawai'i, *J. Geophys. Res.*, *118*(March 2011), 1–18, doi:10.1002/jgrb.50108.

McTigue, D. F. (1987), Elastic stress and deformation near a finite spherical magma body: Resolution of the point source paradox, *J. Geophys. Res.*, *92*, 12,912–931,940, doi:10.1029/JB092iB12p12931.

Meertens, C. M., and J. M. Wahr (1986), Topographic effect on tilt, strain, and displacement measurements, *J. Geophys. Res.*, *91*(B14), 14,057, doi:10.1029/JB091iB14p14057.

Mogi, K. (1958), Relations between the eruptions of various volcanoes and the deformations of the ground surfaces around them, *Bull. Earthquake Res. Inst. Univ. Tokyo*, *36*, 99–134.

Montagna, C. P., and H. M. Gonnermann (2013), Magma flow between summit and Pu'u 'Ō'ō at Kilauea Volcano, Hawaii, *Geochem., Geophys., Geosyst.*, *14*(7), 2232–2246, doi:10.1002/ggge.20145.

Montgomery-Brown, E. K., P. Segall, and A. Miklius (2009), Kilauea slow slip events: Identification, source inversions, and relation to seismicity, *J. Geophys. Res.*, *114*, B00A03, doi:10.1029/2008JB006074.

Montgomery-Brown, E. K., D. K. Sinnett, M. Poland, P. Segall, T. Orr, H. Zebker, and A. Miklius (2010), Geodetic evidence for en echelon dike emplacement and concurrent slow slip during the June 2007 intrusion and eruption at Kīlauea volcano, Hawaii, *J. Geophys. Res.*, *115*(B7), B07,405, doi:10.1029/2009JB006658.

Mori, J., R. A. White, D. H. Harlow, P. Okubo, J. A. Power, R. P. Hoblitt, E. P. Laguerta, A. Lanuza, and B. C. Bautista (1996), Volcanic earthquakes following the 1991 climactic eruption of Mount Pinatubo: Strong seismicity during a waning eruption, in *Fire and Mud: Eruptions and Lahars of Mount Pinatubo, Philippines*, edited by C. G. Newhall and R. S. Punongbayan, pp. 339–350, Philippine Institute of Volcanology and Seismology and University of Washington Press, Seattle and London.

Mosegaard, K., and A. Tarantola (1995), Monte Carlo sampling of solutions to inverse problems, *J. Geophys. Res.*, *100*(B7), 12,431, doi:10.1029/94JB03097.

Ohminato, T., B. A. Chouet, P. Dawson, and S. Kedar (1998), Waveform inversion of very long period impulsive signals associated with magmatic injection beneath Kilauea volcano, Hawaii, *J. Geophys. Res.*, *103*(B10), 23,839, doi:10.1029/98JB01122.

Okamura, A. T., J. J. Dvorak, R. Y. Koyanagi, and W. R. Tanigawa (1988), Surface deformation during dike propagation, in *The Puu Oo Eruption of Kilauea Volcano, Hawaii; Episodes 1 through 20, January 3, 1983, through June 8, 1984*, pp. 165–181, Prof. Pap. 1463, U.S. Geol. Surv., Reston, Va.

Orr, T. R., W. A. Thelen, M. R. Patrick, D. A. Swanson, and D. C. Wilson (2012), Explosive eruptions triggered by rockfalls at Kīlauea volcano, Hawai'i, *Geology*, *41*(2), 207–210, doi:10.1130/G33564.1.

Orr, T. R., J. E. Bleacher, M. R. Patrick, and K. M. Wooten (in press), A sinuous tumulus over an active lava tube at Kīlauea Volcano: Evolution, analogs, and hazard forecasts, *J. Volcanol. Geotherm. Res.*, doi:10.1016/j.jvolgeores.2014.12.002.

Owen, S., P. Segall, M. Lisowski, A. Miklius, M. Murray, M. Bevis, and J. Foster (2000), January 30, 1997 eruptive event on Kilauea Volcano, Hawaii, as monitored by continuous GPS, *Geophys. Res. Lett.*, *27*(17), 2757–2760, doi:10.1029/1999GL008454.

Patrick, M. R., and T. R. Orr (2013), Controls on lava lake level at Halema'uma'u Crater, Kilauea Volcano, Fall Meet. 2013, Abstract V52C-04, AGU, Washington, D. C.

Patrick, M., T. Orr, A. J. Sutton, T. Elias, and D. Swanson (2013), The first five years of Kilauea's summit eruption in Halema'uma'u Crater, Fact Sheet 2013–3116, U.S. Geol. Surv., Reston, VA.

Poland, M. P., T. E. Huth, and A. Miklius (2009a), Source processes of short-term, transient tilt events at Kilauea Volcano, Hawaii, Fall Meet. 2009, Abstract V43G-2331, AGU, Washington, D. C.

Poland, M. P., A. J. Sutton, and T. M. Gerlach (2009b), Magma degassing triggered by static decompression at Kīlauea Volcano, Hawaii, *Geophys. Res. Lett.*, *36*(16), L16306, doi:10.1029/2009GL039214.

Poland, M. P., A. Miklius, A. Jeff Sutton, and C. R. Thornber (2012), A mantle-driven surge in magma supply to Kīlauea Volcano during 2003–2007, *Nature Geosci.*, *5*(4), 295–300, doi:10.1038/ngeo1426.

Poland, M. P., A. Miklius, E. K. Montgomery-Brown (2014), Magma supply, storage, and transport at shield-stage Hawaiian volcanoes, Chapter 5 of Poland, M. P., T. J. Takahashi, and C. M. Landowski, (eds.), *Characteristics of Hawaiian Volcanoes*, U.S. Geol. Surv. Prof. Pap. 1801, pp. 179–234, doi:10.3133/pp18015.

Rivalta, E., and P. Segall (2008), Magma compressibility and the missing source for some dike intrusions, *Geophys. Res. Lett.*, *35*(4), L04306, doi:10.1029/2007GL032521.

Segall, P., P. Cervelli, S. Owen, M. Lisowski, and A. Miklius (2001), Constraints on dike propagation from continuous GPS measurements, *J. Geophys. Res.*, *106*(B9), 19,301, doi:10.1029/2001JB000229.

Segall, P., E. K. Desmarais, D. Shelly, A. Miklius, and P. Cervelli (2006), Earthquakes triggered by silent slip events on Kīlauea volcano, Hawaii, *Nature*, *442*(7098), 71–74, doi:10.1038/nature04938.

Swanson, D. A., D. B. Jackson, R. Y. Koyanagi, and T. L. Wright (1976), *The February 1969 East Rift Eruption of Kilauea Volcano Hawaii*, Prof. Pap. 89, U.S. Geol. Surv, Reston, VA.

Swanson, D. A., W. A. Duffield, D. B. Jackson, and D. W. Peterson (1979), *Chronological Narrative of the 1969–71 Mauna Ulu Eruption of Kilauea Volcano, Hawaii*, Prof. Pap. 1056, U.S. Geol. Surv., Reston, VA.

Thornber, C. R., C. Heliker, D. R. Sherrod, J. P. Kauahikaua, A. Miklius, P. G. Okubo, F. A. Trusdell, J. R. Budahn, W. I. Ridley, and G. P. Meeker (2003), Kilauea East Rift Zone magmatism: An Episode 54 perspective, *J. Petrol.*, *44*(9), 1525–1559, doi:10.1093/petrology/egg048.

Tilling, R. I. (1987), Fluctuations in surface height of active lava lakes during 1972–1974 Mauna Ulu Eruption, Kilauea Volcano, Hawaii, *J. Geophys. Res.*, *92*(B13), 13,721, doi:10.1029/JB092iB13p13721.

Voight, B., R. P. Hoblitt, A. B. Clarke, A. B. Lockhart, A. D. Miller, L. Lynch, and J. McMahon (1998), Remarkable cyclic ground deformation monitored in real-time on Montserrat, and its use in eruption forecasting, *Geophys. Res. Lett.*, *25*(18), 3405–3408, doi:10.1029/98GL01160.

Wilson, D., T. Elias, T. Orr, M. Patrick, J. Sutton, and D. Swanson (2008), Small explosion from new vent at Kilauea's summit, *Eos Trans. AGU*, *89*(22), 203, doi:10.1029/2008EO220003.

Wolfe, E. W., M. O. Garcia, D. E. Jackson, R. Y. Koyanagi, C. A. Neal, and A. T. Okamura (1987), The Puu Oo eruption of Kilauea Volcano, episodes 1–20, January 3, 1983, to June 8, 1984, in *Volcanism in Hawaii*, vol. 1, pp. 471–508, Prof. Pap. 1350, U.S. Geol. Surv., Reston, VA.

Wright, T. L., and R. T. Helz (1996), Differentiation and magma mixing on Kilauea's east rift zone: A further look at the eruptions of 1955 and 1960. Part II. The 1960 lavas, *Bull. Volcanol.*, *57*(8), 602–630, doi:10.1007/s004450050115.

Yamashina, K., T. Matsushima, and S. Ohmi (1999), Volcanic deformation at Unzen, Japan, visualized by a time-differential stereoscopy, *J. Volcanol. Geotherm. Res.*, *89*(1–4), 73–80, doi:10.1016/S0377-0273(98)00124-3.

Yang, X., P. M. Davis, P. T. Delaney, and A. T. Okamura (1992), Geodetic analysis of dike intrusion and motion of the magma reservoir beneath the summit of Kilauea Volcano, Hawaii: 1970–1985, *J. Geophys. Res.*, *97*(B3), 3305, doi:10.1029/91JB02842.

12

Crustal Stress and Structure at Kīlauea Volcano Inferred from Seismic Anisotropy

Jessica H. Johnson[1,2,3,4], Donald A. Swanson[2], Diana C. Roman[5], Michael P. Poland[2], and Weston A. Thelen[2]

ABSTRACT

Seismic anisotropy, measured through shear wave splitting (SWS) analysis, can be indicative of the state of stress in Earth's crust. Changes in SWS at Kīlauea Volcano, Hawai'i, associated with the onset of summit eruptive activity in 2008 hint at the potential of the technique for tracking volcanic activity. To use SWS observations as a monitoring tool, however, it is important to understand the cause of seismic anisotropy at the volcano throughout the eruptive cycle. To address this need, we analyzed SWS results from across Kīlauea in combination with macroscopic surface structures (mapped fractures, faults, and fissures) and stress orientations inferred from fault plane solutions. Seismic anisotropy seems to be due to pervasive aligned structures in most regions of the volcano. The upper East and Southwest Rift Zones, however, show a bimodality in stress and SWS, suggesting a stress discontinuity with depth, perhaps related to magma conduits that trend obliquely to the dominant structure. Other areas in and around Kīlauea Caldera display principal stresses of similar magnitudes, indicating that small stress perturbations can rotate the maximum horizontal compressive stress direction by up to 90°. In these locations, static structures generally control SWS, but dynamic conditions due to magmatic activity can override the structural control. Monitoring of SWS may therefore provide important signs of impending volcanism.

12.1. INTRODUCTION

Seismic anisotropy is the variation of seismic wave speed with direction. In Earth's crust, seismic anisotropy can be caused by alignment of minerals, layering of materials, and aligned fractures or microcracks [*Crampin*, 1994]. In general, seismic wave speeds in the direction that crosses features such as layers, cracks and fractures will be slower than those in the direction parallel to the aligned features because the discontinuities and fluid spaces will affect the wave speed to a greater extent. An applied stress field may also create a seismically anisotropic medium by causing microcracks to preferentially open parallel to the maximum compressive stress. Stress-aligned microcracks are thought to be the only mechanism that allows seismic anisotropy to change on geophysically observable timescales; however, the dominant control of seismic anisotropy can switch between a static cause, such as aligned fractures, and a dynamic process, including stress-aligned microcracks, in areas where the stress field is highly variable over time, such as at active volcanoes [e.g., *Keats et al.*, 2011].

At Kīlauea Volcano, Hawai'i, the excellent geophysical, geological, and geochemical time series spanning frequent and accessible volcanic activity provides an outstanding opportunity to advance volcano monitoring beyond simple pattern recognition and toward a better understanding of the physical processes that are occurring. Such a

[1]Center for the Study of Active Volcanoes, University of Hawai'i, Hilo, Hawaii, USA
[2]Hawaiian Volcano Observatory, U.S. Geological Survey, Hawai'i National Park, Hawaii, USA
[3]School of Earth Sciences, University of Bristol, Bristol, United Kingdom
[4]School of Environmental Sciences, University of East Anglia, Norwich, United Kingdom
[5]Department of Terrestrial Magnetism, Carnegie Institution for Science, Washington, D.C., USA

Hawaiian Volcanoes: From Source to Surface, Geophysical Monograph 208. First Edition.
Edited by Rebecca Carey, Valérie Cayol, Michael Poland, and Dominique Weis.

Companion Website: www.wiley.com/go/Carey/Hawaiian_Volcanoes

setting is ideal for assessing the mechanism(s) for temporal change in seismic anisotropy, which has shown promise as a monitoring tool at a number of volcanoes worldwide [e.g., *Volti and Crampin,* 2003; *Gerst and Savage*, 2004; *Bianco and Zaccarelli* , 2009; *Savage et al.*, 2010; *Unglert et al.*, 2011; *Roman et al.*, 2011; *Johnson and Savage*, 2012].

Johnson and Poland [2013] examined the temporal evolution of seismic anisotropy at the summit of Kīlauea spanning the 2008 onset of the ongoing summit eruption. They noted changes in the fast direction of shear waves associated with the start of summit eruptive activity in 2008, from tangential to the center of the caldera to radial, which they interpreted as due to temporary dilation of cracks aligned radial to the caldera due to increased gas pressure associated with the start of the eruption. Greater understanding of these and any future changes associated with volcanic activity can be gained by determining the cause of seismic anisotropy in particular regions of the volcano throughout the eruptive cycle. Only with this knowledge can seismic anisotropy data be associated in real time with other monitoring data streams.

Here, we endeavor to investigate the broader picture of seismic anisotropy at Kīlauea, both to create a baseline for future research and monitoring and to elucidate the interaction between stress and structure at this dynamic volcano, thereby enabling a better understanding of its crustal evolution. In addition, by determining the regions in which crustal stress is actively influencing volcanic processes (and vice versa), we can identify seismic stations that are most likely to display variations in shear wave splitting (SWS) associated with eruptive activity at Kīlauea and that hence may be useful for monitoring future changes as they occur. We perform SWS analysis on local earthquakes that occurred at Kīlauea since 2007 to determine the state of anisotropy throughout the volcano. To determine the contributions of stress-induced and structurally controlled anisotropy to our SWS results, we calculated fault plane solutions (FPSs) from local earthquakes to infer stress conditions in the crust. We also developed a database of surface cracks, fractures, and fissures that indicate macroscopic structure that could influence seismic anisotropy and can therefore be compared to the SWS.

12.2. CONTROLS ON SEISMIC ANISOTROPY AT KĪLAUEA

Traditionally, crustal SWS in volcanic regions is assumed to be completely due to microcracks aligned with the maximum horizontal compressive stress due to magma pressure and movement. Under this assumption, the fast directions of anisotropy would be radial to overpressurized magma reservoirs and tangential to underpressurized magma reservoirs. In the case of Kīlauea, SWS fast directions would be expected to be radial to the magma reservoirs beneath the two current eruption sites at Halema'uma'u Crater (at the summit) and Pu'u 'Ō'o (on the East Rift Zone (ERZ)) and potentially perpendicular to the magma pathway between the two. Eruptive activity at Kīlauea, measured by effusion rate, lava lake height, deformation, and seismicity, tends to wax and wane, implying that the pressure in the magma system increases and decreases over time. The fact that eruptions continue through these cycles suggests that the system is still overpressurized and that radial anisotropy should dominate. At times when the pressure of the magmatic system is lower and the volcano is deflating, however, the maximum horizontal compressive stress may rotate to be tangential to the center of deformation, and other factors such as aligned fractures in fault zones may assume the dominant role in determining the anisotropy directions at Kīlauea.

Regional or tectonic stress can also affect alignment of microcracks. SWS studies often attribute anisotropy directions in volcanic regions at times of quiescence to these regional stresses. In Hawai'i, large-scale stress comes from the gravitational spreading of the volcanic pile on top of the oceanic crust. At Kīlauea, the buttressing of Mauna Loa volcano to the north coupled with southward motion of the south flank defines the regional stress regime. If seismic anisotropy is purely attributed to regional stress-aligned microcracks, then SWS fast directions should be perpendicular to the contact between Mauna Loa in the north [e.g., *Wyss et al.*, 1992a,b] and parallel to the coast in the south [e.g., *Gillard et al.*, 1996a].

Many studies have shown that crustal seismic anisotropy can be contolled predominantly by crustal structure, regardless of the local fluctuations in stress [e.g., *Balfour et al.*, 2005; *Boness and Zoback*, 2006; *Booth et al.*, 1992; *Cochran et al.*, 2003]. At Kīlauea, the only structures that are likely to have some control on seismic anisotropy are aligned fractures in fault zones and aligned dike complexes in rift zones, which, despite being composed of isotropic material, can create an anisotropic effect. Magmatic intrusions at Kīlauea generally tend to propagate along the rift systems that are evident at the surface from fissures, faults, and cracks [*Klein et al.*, 1987]. SWS fast directions should align parallel to the fractures in these cases. At Kīlauea, the trends of the structures seen at the surface are thought to continue to a depth of several kilometers. Evidence for this comes from deformation studies [*Delaney et al.*, 1990; *Cayol*, 2000], gravity surveys [*Hill and Zucca*, 1987; *Kauahikaua*, 1993], magnetotelluric surveys [*Hoversten et al.*, 2003], and seismic studies [Okubo et al., 1997; *Dawson et al.*, 1999]; therefore, anisotropy would be expected to be locally parallel to the rift zones, caldera walls, normal faults on the south flank, and the contact between Mauna Loa and Kīlauea.

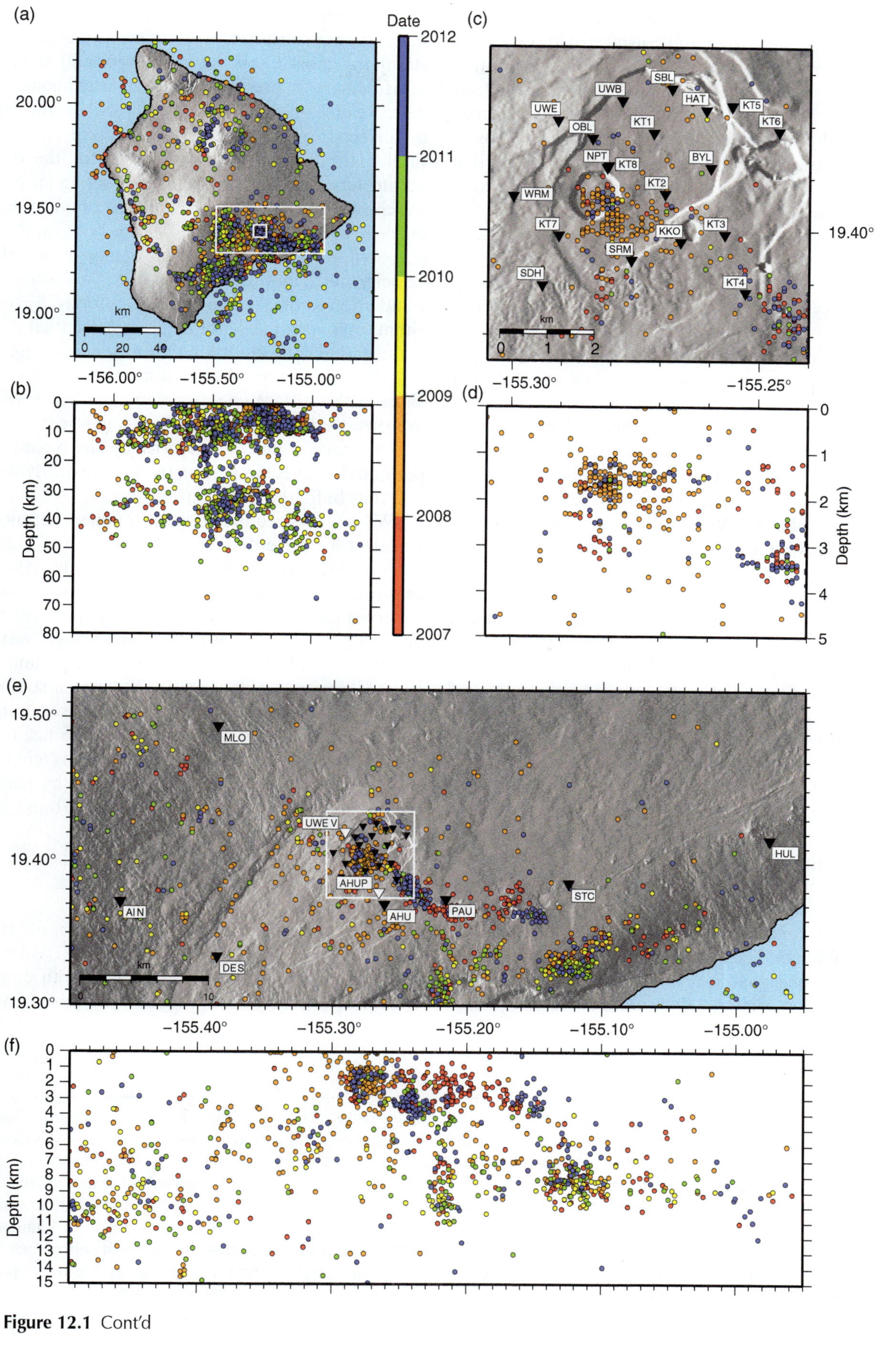

Figure 12.1 Cont'd

All three of the above-mentioned mechanisms (magmatic stress, regional stress, and crustal structure) have an effect on seismic anisotropy at Kīlauea. The contribution of each may change over time as local magmatic stress fluctuates across eruption cycles. This study is designed to identify the current control on seismic anisotropy in different regions of Kīlauea and to recognize areas that may be prone to changes in seismic anisotropy related to magmatic activity.

12.3. METHODS

12.3.1. Shear Wave Splitting

A shear wave in an anisotropic medium will be split into a fast and slow component with fast orientation ϕ and delay time δt. An automated program [*Savage et al.*, 2010] was used to perform all SWS measurements in this study to process large volumes of data and remove observer bias. The program is based on the algorithms of *Silver and Chan* [1991] and *Teanby et al.* [2004]. The algorithm for calculating SWS uses a grid search inversion over ϕ and δt for a given time window. It has been incorporated into a code that conducts cluster analysis over a range of time windows to find the most stable result. The automated method selects the most reliable SWS results from a range of frequency filters and time windows. The frequency filters applied to individual earthquake recordings are selected on the basis of signal-to-noise ratio, and the maximum and minimum time windows around the S-wave arrival are based on the dominant frequency of the first three seconds of the S waveform.

Results from the SWS analysis that gave ϕ to within 20° of the polarization of the incoming wave were considered null results. Nulls signify that no splitting was reliably detected, either because there was no anisotropy or because the initial polarization of the shear wave was already aligned with one of the fast or slow orientations, rendering the corresponding value of δt meaningless and ϕ a 90° ambiguity. The measurements were automatically graded based on signal-to-noise ratios and characteristics from the cluster analysis. Only nonnull results with a measurement grade of A or B and delay time smaller than 0.5 s were included. Measurements that differed substantially across filters were removed, and at most one measurement is presented for each earthquake-station pair. Refer to *Savage et al.* [2010] for details on these quality control steps. Uncertainties were calculated by finding the 95% confidence interval for the optimum parameters of anisotropy after conducting an *F* test for the chosen time window. The incidence angle of each ray at each station is calculated using the TauP Toolkit [*Crotwell et al.*, 1999] with the one-dimensional (1-D) velocity model for the Kīlauea region [*Okubo et al.*, 1997]. In addition to the selection criteria enforced by the algorithm, rays with incidence angles greater than 35° from vertical were not included in the analyses, as these lie outside the "shear wave window" (meaning that S-P conversions at the surface could contaminate the waveforms). We observe that the steep velocity gradient in the top 0.5 km ensures that the majority of the recorded local earthquake waves impinge on the surface nearly vertically and thus fulfill this criterion.

SWS analysis requires three-component seismometers to detect the three-dimensional motion of the incoming wave, and the measured SWS from a local earthquake is sensitive to the entire ray path. In general, however, the observed fast directions of anisotropy represent the last anisotropic layer that the ray passed through, which is in practice several kilometers below the recording station [e.g., *Silver and Savage*, 1994]. The seismic stations and local volcano-tectonic (VT) earthquakes used in the SWS analysis are displayed in Figure 12.1. SWS fast polarization (ϕ) results are plotted as rose diagrams (circular histograms) at the recording station, scaled by number of measurements (Figure 12.2), and mean directions are listed in Table 12.1.

12.3.2. Fault Plane Solutions

To independently determine the state of stress at Kīlauea, we calculated FPSs for local VT earthquakes with $M_L \geq 1$ and depths ≤8 km. This depth cutoff was chosen as it incorporates most of the seismicity in the

Figure 12.1 (Cont'd) Maps of the study area with earthquakes used in the SWS analysis. Earthquakes are color coded by year of occurrence. (a) Map of the Island of Hawai'i showing all earthquakes used in SWS analysis (see text for selection criteria) and study regions (white rectangles). (b) Depth section across the Island of Hawai'i west to east, showing depth distribution of earthquakes. (c) Map of Kīlauea Caldera showing a subset of earthquakes. Black triangles indicate three-component seismometer stations around the caldera used in SWS analysis. (d) Depth section across Kīlauea Caldera west to east, showing shallow details of depth distribution of Kīlauea earthquakes. (e) Map of Kīlauea Volcano showing a subset of earthquakes and three-component seismometer stations used in SWS analysis. White triangles indicate GPS stations used in temporal analysis. (f) Depth section across Kīlauea Volcano west to east, showing details of depth distribution of Kīlauea earthquakes.

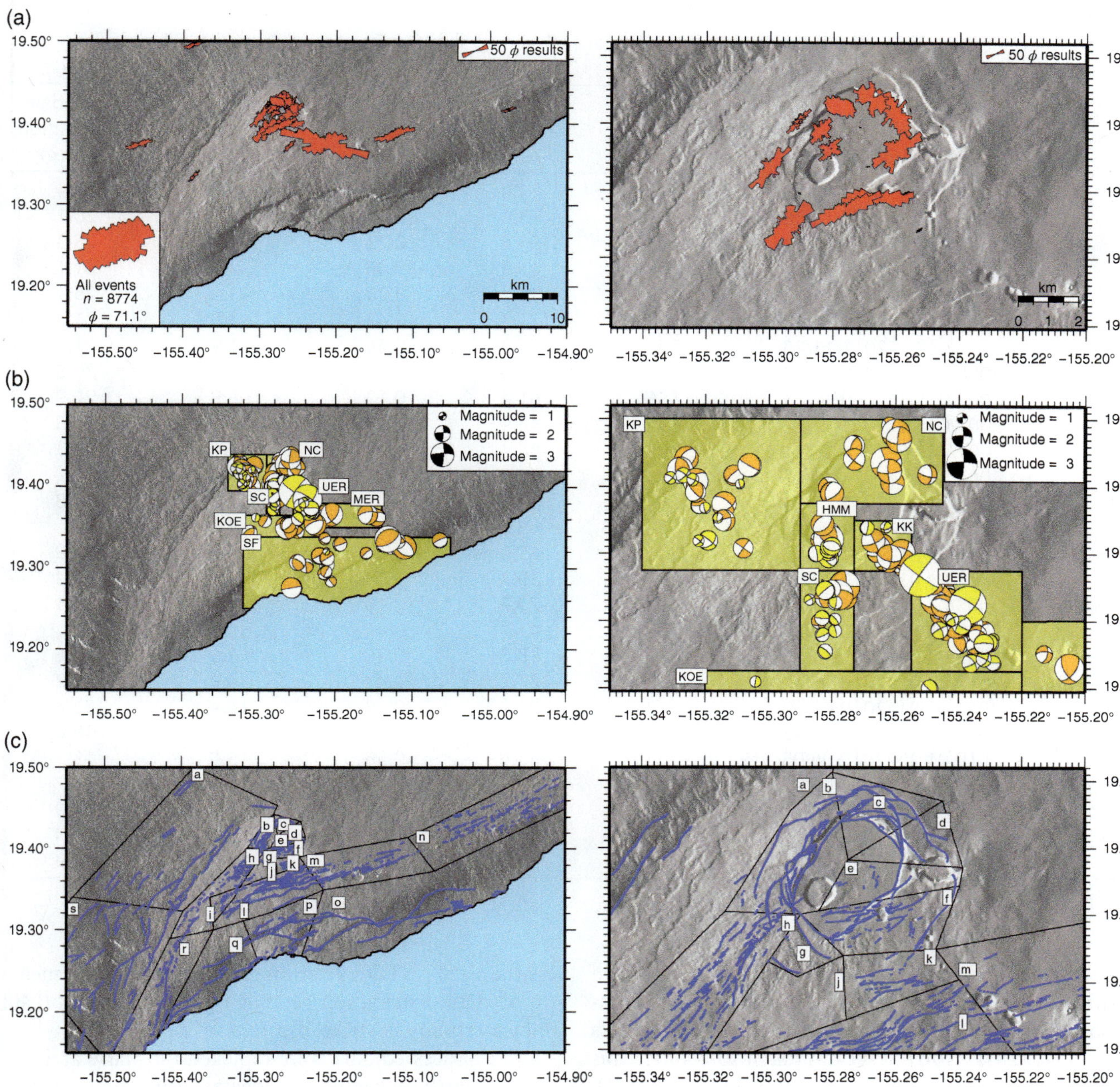

Figure 12.2 (a) Shear wave splitting fast direction of anisotropy (ϕ) results plotted as rose diagrams at the recording station and scaled by number of measurements. Right panels show details of Kīlauea summit. (b) Fault plane solutions for earthquakes at Kīlauea. Orange are results inverted from catalog data and yellow are inverted from reprocessed data. Shaded regions indicate sections of the volcano for further discussion: Ka'ōiki Pali (KP), northern caldera (NC), Halema'uma'u (HMM), Keanakāko'i (KK), southern caldera (SC), upper East Rift Zone (UER), middle East Rift Zone (MER), Koa'e Fault System (KOE), south flank (SF). (c) Structural features (ground cracks, faults, and eruptive fissures) compiled from *Wolfe and Morris* [1996], *Neal and Lockwood* [2002], and Hawaiian Volcano Observatory archives, plotted in blue. Black boxes delineate areas with consistent structural trends.

volcanic pile. Below this depth is the décollement, which is the surface between the oceanic plate and the overlying volcanoes. Significant seismicity occurs on the décollement and therefore some of the earthquakes included in this study are considered décollement earthquakes, but these only impact the south flank section of this study.

FPSs provide information about the orientation of the fault plane that slipped as well as the slip vector for each earthquake. From the FPS, the pressure and tension axes can be identified, which give information about the stress regime. The FPS *P*-axis azimuth, which represents the orientation of the principal compressional axis of the

Table 12.1 Statistical analysis results for shear wave splitting at each station showing the mean value $\bar{\theta}$, standard deviation, and standard error of the mean.

Station	Data Time Span	*n*	$\bar{\theta}$ (deg)	$\bar{R}$	Standard Error (deg)	Standard Deviation (deg)
AHU	2007/04/04–2010/12/30	243	45.9	0.39	3.2	22.4
AIN	2008/09/17–2010/12/26	362	73.9	0.51	1.9	20.1
BYL	2007/04/03–2011/11/23	616	50.5	0.22	3.7	25.3
DES	2008/08/12–2010/11/01	200	41.3	0.46	2.9	21.1
HAT	2007/04/02–2011/11/23	599	–36.4	0.23	3.6	25.1
HUL	2008/11/29–2010/12/29	168	71.4	0.42	3.5	21.8
KKO	2007/04/01–2011/11/28	522	88.7	0.44	1.9	21.4
MLO	2008/11/23–2010/12/30	282	73.6	0.49	2.3	20.5
NPT	2007/04/03–2010/12/15	356	61.4	0.12	8.9	26.9
OBL	2007/04/05–2011/11/24	390	44.9	0.14	7.3	26.6
PAU	2007/05/08–2011/11/29	1309	–85.4	0.34	1.6	23.3
SBL	2007/04/03–2011/11/23	498	–51.1	0.24	3.7	25
SDH	2007/04/04–2011/11/23	664	37.2	0.25	3.1	24.8
SRM	2007/04/01–2011/06/26	536	69.5	0.41	2	22
STC	2007/04/01–2011/11/26	586	65.6	0.36	2.2	22.9
URA	2007/04/03–2010/12/29	232	39.2	0.39	3.3	22.4
UWB	2007/04/08–2011/11/24	479	–74.7	0.16	5.7	26.3
UWE	2010/05/12–2010/12/29	132	41.7	0.41	4.1	22
WRM	2007/04/03–2011/11/24	457	38.7	0.26	3.6	24.6
KT1	2010/02/02–2010/06/20	11	–87.2	0.4	14.6	22.2
KT2	2010/02/01–2010/06/20	12	76.9	0.39	14.4	22.4
KT3	2010/01/31–2010/06/15	28	64.3	0.52	6.8	19.8
KT4	2010/01/26–2010/06/05	32	53.6	0.56	5.8	19
KT5	2010/02/01–2010/06/20	20	–19.9	0.48	8.8	20.7
KT6	2010/02/02–2010/05/30	17	–44.8	0.18	27.1	25.9
KT7	2010/02/02–2010/06/20	19	–40.4	0.22	20.9	25.3
KT8	2010/04/05–2010/05/30	4	34.3	0.12	84.1	26.9

Note: The $\bar{R}$ value is a statistical measure of dispersion for circular data sets, and *n* is the number of measurements.

moment tensor, may in reality differ significantly from the orientation of maximum compressive stress, depending on the orientation of preexisting planes of weakness (faults) [*McKenzie*, 1969]. Our approach to counter this inherent ambiguity follows that of *Roman et al.* [2006], which is to calculate large numbers of FPS *P*-axis azimuths and plot them as rose diagrams to highlight the major trends that most likely represent the actual orientations of maximum horizontal compressive stress.

We used FPSs from catalog earthquake data and confirmed them through comparison with reprocessed catalog data (Supplementary Data Table S1). The catalog data were located using HYPOINVERSE [*Klein*, 2002] and a 1D velocity model [*Okubo et al.*, 1997]. Double-couple FPSs based on first-motion polarities were calculated using the program FPFIT [*Reasenberg and Oppenheimer*, 1985] with catalog phase arrivals. The resulting FPSs were evaluated for quality and rejected if they did not meet all of the following criteria: at least 12 clear first-motion polarities; unique solutions; misfit of less than 15%; STDR (station distribution ratio, a measure of data distribution around the hypocenter) ≥0.40; and uncertainty in strike, dip, and rake of ≤25°.

The reprocessed data were repicked for phase arrivals, and the earthquakes were relocated using the program HYPOELLIPSE [*Lahr*, 1999]. Earthquakes were chosen for this analysis if they had a high-quality hypocenter location (defined as those with horizontal and vertical location errors of less than 1.5 km), an azimuthal gap of less than 180°, and a root-mean-square (RMS) of less than 0.20. For earthquakes with a high-quality hypocenter location, we calculated double-couple FPSs, also using FPFIT [*Reasenberg and Oppenheimer*, 1985]. The additional criterion of having polarities widely distributed and away from nodal planes on the focal sphere was applied to obtain high-quality FPS results.

Figure 12.2b displays in orange the focal mechanisms calculated from catalog data that passed the criteria and the reprocessed FPS in yellow. We divided the earthquakes into clusters based on the geographical

Table 12.2 Statistical analysis results for the *P*-axis azimuth data from FPSs divided by cluster, showing the mean value $\bar{\theta}$, standard deviation, and standard error of the mean.

Cluster	*n*	$\bar{\theta}$ (deg)	$\bar{R}$	Standard Error (deg)	Standard Deviation (deg)
HMM	16	68.1	0.19	26.4	25.8
KK	15	102.4	0.66	6.8	16.7
KOE	9	120.1	0.29	22.8	24.1
KP	26	91.2	0.53	6.9	19.6
MER	6	61.1	0.43	18.3	21.6
NC	11	127.3	0.44	13.1	21.4
SC	18	34.8	0.55	7.9	19.2
SF	24	162.8	0.23	17.7	25.1
UER	65	88.2	0.49	4.8	20.5

Note: The $\bar{R}$ value is a statistical measure of dispersion for circular data sets, and *n* is the number of measurements.

Table 12.3 Statistical analysis results for the structure data divided by box, showing the mean value $\bar{\theta}$, standard deviation, and standard error of the mean.

Box	*n*	$\bar{\theta}$ (deg)	$\bar{R}$	Standard Error (deg)	Standard Deviation (deg)
a	82	41:2	0.68	2.8	16.2
b	123	31:5	0.59	2.8	18.3
c	112	−74:9	0.4	4.6	22.2
d	26	−32:5	0.65	5.3	16.9
e	33	66:6	0.52	6.3	19.8
f	115	61	0.74	2.1	14.6
g	15	−43:9	0.73	5.8	14.9
h	217	41:4	0.91	0.8	8.6
i	46	50:8	0.94	1.5	7
j	32	72:1	0.97	1.2	5
k	36	70	0.9	2.2	9.1
l	404	68:3	0.85	0.8	11.1
m	235	68:2	0.94	0.7	7
n	162	64	0.98	0.5	4.1
o	157	75:4	0.78	1.6	13.4
p	116	59:9	0.69	2.3	16
q	95	53:1	0.73	2.3	14.9
r	120	21:3	0.75	2	14.3
s	134	33:9	0.69	2.1	16

Note: The $\bar{R}$ value is a statistical measure of dispersion for circular data sets, *n* is the number of measurements.

distribution of epicenters. The earthquakes were divided in this way because they all represent upper crustal earthquakes (between 0 and 8 km depth), and therefore we initially assume that they represent the same crustal volume. They were also not divided by origin time because our analysis is spatial rather than temporal. Other authors have observed changes in stress orientation or magnitude at Kīlauea [e.g., *Klein et al.*, 1987; *Gillard et al.*, 1996b; *Wauthier et al.*, 2013], and we investigate the possibility of a temporal evolution of stress in Section 12.5. However, as a first-order approximation to analyze the spatial patterns of stress, we initially assume no temporal variations. *P*-axis azimuths from each of the results in each cluster were combined to display an average maximum horizontal compressive stress for each region (Table 12.2).

12.3.3. Crack and Fault Orientations

Structural features (ground cracks, faults, and eruptive fissures) listed in digital versions of published geological maps by *Wolfe and Morris* [1996] and *Neal and Lockwood* [2002] were plotted in MapInfo and their trends calculated to the nearest degree using the MapInfo tool GeoMapVector [*Holcombe*, 2001]. Where each map showed slightly different locations for a feature, the location in *Neal and Lockwood* [2002] was used because of the larger map scale. The data are displayed in Figure 12.2c and provided in Supplementary Data Table S2. These data were subjectively grouped into regions based on general trends, labeled boxes a–s, and average orientations were calculated (Table 12.3). A straight line vector was used to approximate the orientation of each fracture. For sinuous faults, such as in the caldera (boxes b, c, d, e, f, and g, Figure 12.2c) and Koa‘e Fault System (box l, Figure 12.2c), short, straight-line segments approximating the curvatures were used for the purpose of trend calculation. In most cases, the transition between trends was clear and divisions could be made confidently; however, some areas display multiple dominant overlapping trends. In these cases, several divisions were tested to see if the trends could be isolated. Boxes g and j provide an example of areas that were originally combined but later divided so as to separate the caldera and ERZ structure. Other regions did not lend themselves to further division, and so the trends within those boxes are scattered or bimodal, for example, box s. A total of 2275 separate trends were calculated.

12.4. RESULTS

12.4.1. Shear Wave Splitting

We have performed SWS analyses at Kīlauea Volcano on 2815 local earthquakes with $M_L \geq 1.5$ recorded at 19 permanent three-component seismic stations from the beginning of 2007 until the end of 2011, as well as at 8 temporary three-component seismic stations that were deployed between January and June 2010 (Figure 12.1), achieving 8774 reliable measurements. The time period spans the onset of Kīlauea's currently ongoing summit eruption, which began on 19 March 2008 [*Wilson et al.*, 2008; *Houghton et al.*, 2011], and several other major

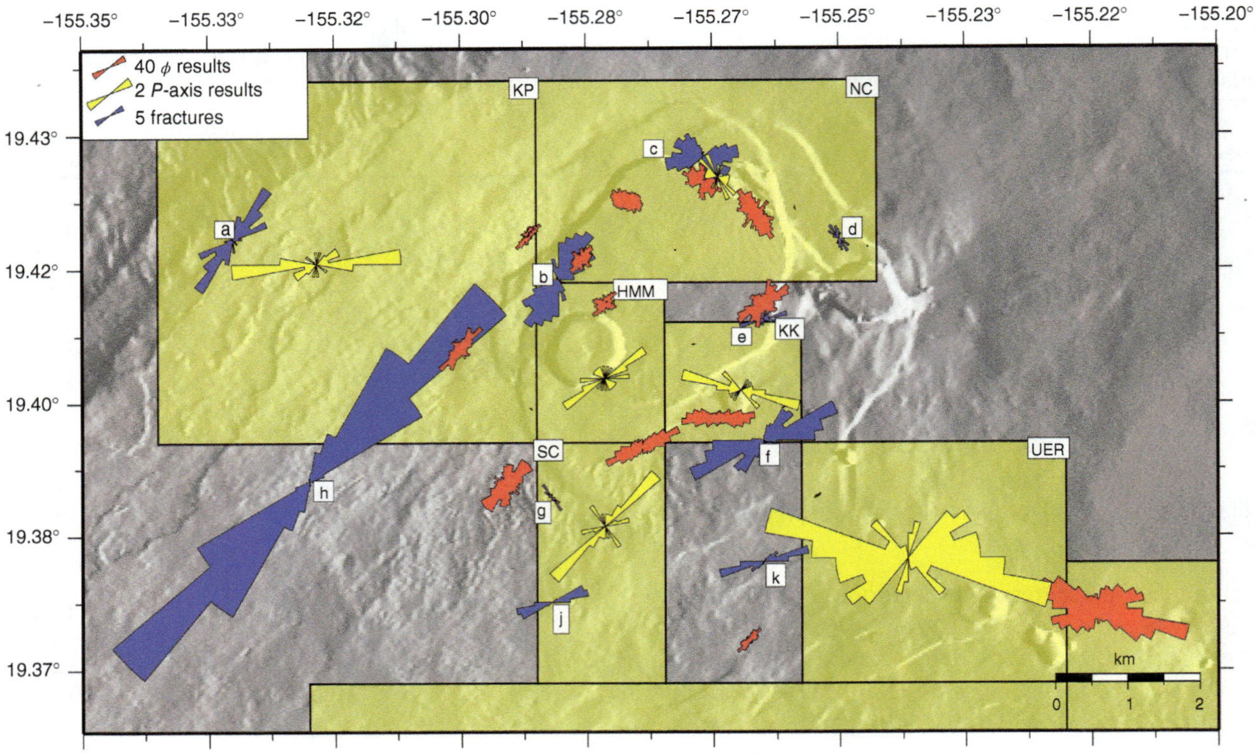

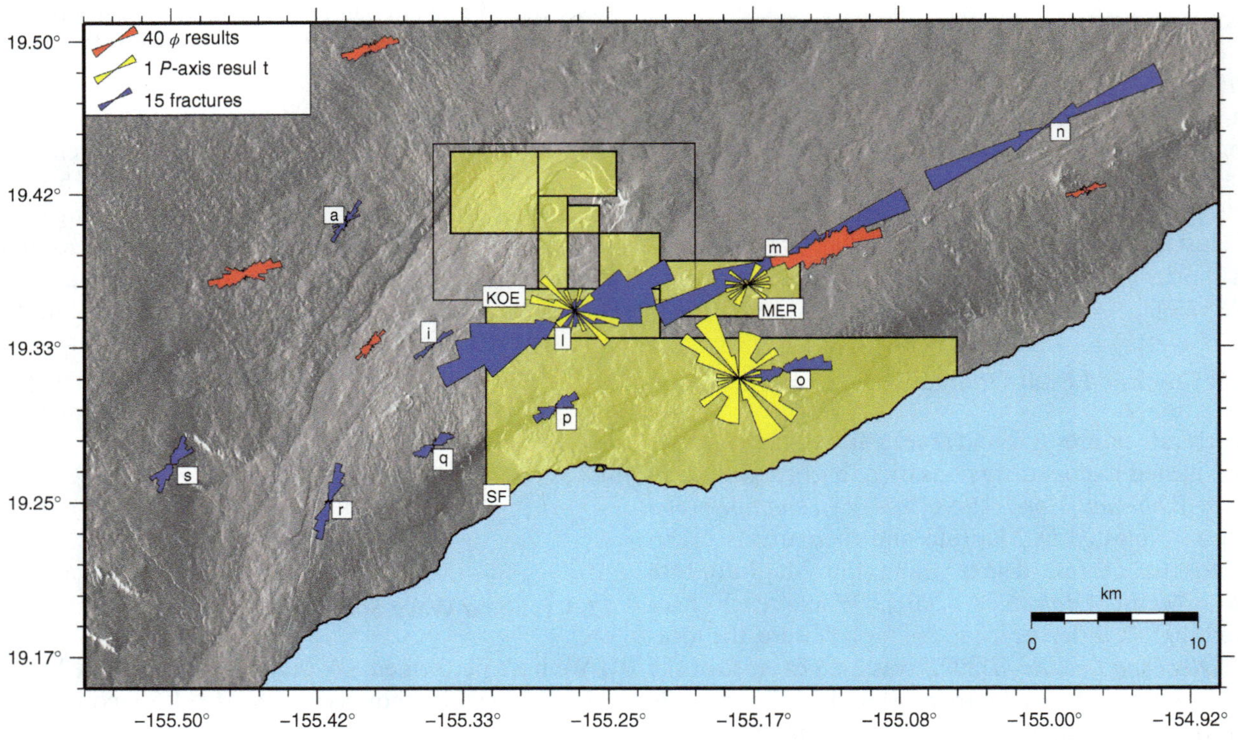

Figure 12.3 Rose diagrams (circular histograms) displaying SWS results, structure orientations, and maximum horizontal stress directions from fault plane solutions, scaled by number of measurements. The ϕ result: Fast direction of SWS anisotropy since 2007 in red, plotted at the seismic station at which the measurements were made. Fractures: Orientations of structural features for each box, plotted in blue at the center of each box. Lowercase letters refer to boxes defined in Figure 12.2c. *P*-axis result: *P*-axis azimuths of FPSs plotted in yellow at the center of the earthquake cluster. Shaded regions indicate sections of the volcano, labeled as in Figure 12.2b. Top plot shows detail around Kīlauea summit, bottom plot shows all of Kīlauea.

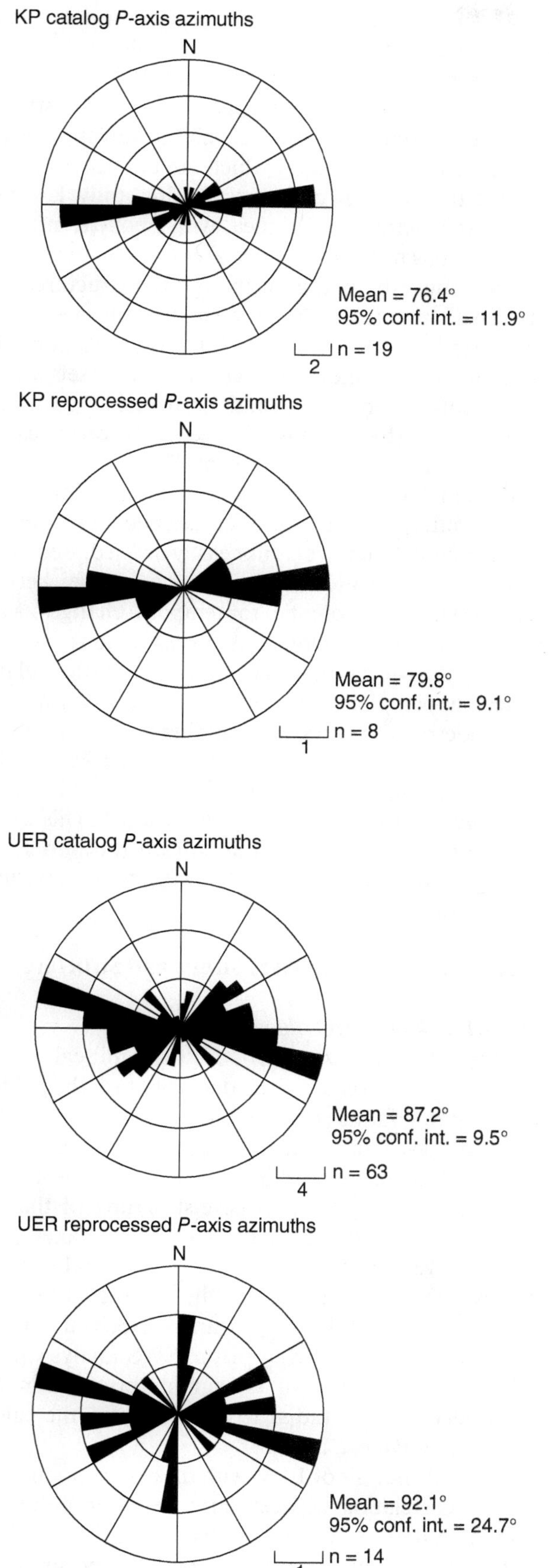

Figure 12.4 Cont'd

changes in the ongoing eruption of Kīlauea (for instance, dike intrusions, fissure eruptions, and changes in the location of the ERZ eruption site). Figure 12.2a displays the fast orientation of SWS (ϕ) for the entire time period as rose diagrams at the station where the recording was made, scaled by the number of measurements, as well as the average of all results combined (bottom left) to give a "background" anisotropy. The single-modal results, combined with the differences between closely spaced stations, suggest that the anisotropy recorded is acting directly beneath the station [*Zinke and Zoback*, 2000].

In general, anisotropy fast orientations display an east northeast to west southwest (ENE–WSW) direction (mean of 071.1 ± 2.2°; see Section 12.5.2 for a discussion). Stations within 1 km of significant faults (those that are large and shallow enough to have a surface expression) display structural anisotropy parallel to the faults. At the summit caldera, the background anisotropy is generally parallel to the caldera-bounding ring faults. On the ERZ, the anisotropy is parallel to the eruptive fissures and associated ground cracks. The exception is the bend in the ERZ (between the upper and the middle ERZ), where the fast direction of anisotropy appears to follow the bend even though there are no surface faults with this trend (Figures 12.2 and 12.3).

12.4.2. Fault Plane Solutions

We combined the *P*-axis azimuths from the FPS of 206 earthquakes around Kīlauea since 2007 in clusters based on different regions of the volcano. Rose diagrams of the *P*-axis azimuths for clusters of earthquakes are displayed in yellow in Figure 12.3, scaled by number of measurements, as a proxy for the maximum horizontal compressive stress direction (hereafter refered to as $S_{H,max}$). There is no single prominent direction of $S_{H,max}$ at Kīlauea, different regions display markedly different *P*-axis orientations.

Two regions yielded enough reprocessed events to enable a comparison between the catalog and reprocessed $S_{H,max}$ directions. Figure 12.4 displays the *P*-axis azimuths for the Ka'ōiki Pali (KP) and the upper East Rift Zone (UERZ). The mean directions of the *P*-axis azimuths (a proxy for the directions of $S_{H,max}$) are the same to within 95% confidence, calculated using circular statistics. We can therefore use the larger data set of catalog FPSs to infer $S_{H,max}$ directions in different regions around Kīlauea.

Figure 12.4 Rose diagrams (circular histograms) displaying FPS *P*-axis azimuths for catalog and reprocessed data in the Ka'ōiki Pali (KP) and upper East Rift Zone (UER) regions. Mean and 95% confidence intervals are calculated using circular statistics; *n* is the number of measurements.

12.4.3. Crack, Fault, and Fissure Orientations

With few exceptions, the mapped structures presented in this study are dilatational features, either gaping cracks and fissures or normal fault scarps. Figure 12.3 displays rose diagrams of crack, fault, and fissure strikes within each box defined in Figure 12.2c, plotted at the center of each box and scaled by the number of measurements. Two main trends are apparent at Kīlauea. The first, and most pervasive, is correlated with the strike of the middle and lower ERZ and trends approximately 70° from north. This trend appears in most of the regions east and south of Kīlauea's summit. Box 1 displays the ERZ trend but has a wider scatter than other boxes due to the inclusion of segments of sinuous faults in the Koa'e Fault System.

The second main trend is correlated with the strike of the Southwest Rift Zone (SWRZ), which strikes approximately 30° from north and dominates the western side of Kīlauea. The location where the ERZ and SWRZ trends meet (box i) appears to contain overlapping fractures of both trends (Figure 12.2c). The rose diagram of crack data in box i displays a mean direction of 50.8°, with a small standard deviation of 7.0°, standard error of 1.5°, and no obvious bimodality (Figure 12.3 and Table 12.2); therefore, even though there is influence from both the ERZ and the SWRZ in box i, the two trends do not appear to be distinct where they join.

Boxes b–g include the caldera-bounding ring faults at the summit of Kīlauea. Bimodality and scatter are evident due to the boxes containing arced segments of the faults. The anomaly is box e, which incorporates the center of the caldera, as well as Kīlauea Iki Crater. Even though box e contains some of the caldera-bounding ring faults, the trend is almost perpendicular to those faults and better approximates that of the ERZ.

12.5. DISCUSSION

To investigate the broad picture of seismic anisotropy and determine its controls, we compared the general trends of SWS, FPSs, and surface fractures in different regions. In general, seismic anisotropy in regions where SWS correlates with structure but not stress estimates can be interpreted as being structure controlled, and in regions where the SWS correlates with stress estimates but not structural trends the seismic anisotropy can be interpreted as stress controlled. The mapped surface structures can, however, provide independent evidence for the state of stress. As the mapped features all display dilatation, the opening occurs in the direction of tension, which is perpendicular to the maximum horizontal compressive stress. The strikes of these structures do not always align exactly with $S_{H,\max}$ [see *Duffield*, 1975], but the mean strike of the structures is a reasonable first-order approximation. It is therefore not surprising that the stress estimates from FPSs and the strikes of the fractures are aligned in many places on Kīlauea. In some of these regions, it is not practical to ask whether stress or structure is controlling the seismic anisotropy because they share the same trends, which are unlikely to change over the time periods of interest in this study. However, the dynamic nature of Kīlauea, with frequent inflations and deflations near the eruptive sites (see Chapter 11, this volume), suggests that, even though the structures indicate one dominant stress regime, small changes over short time periods may be manifested in SWS variations. The question then becomes "to what extent is the seismic anisotropy influenced by the static structures?" and hence "how large a stress change is needed before it can be observed in the SWS measurements?"

Additional information from past investigations can be used to interpret the cause of seismic anisotropy in regions where the control is not easily defined. Regions of particular interest are those with discrepancies between the data sets or the potential for temporal changes associated with volcanic activity. Past investigations also shed light on long-term (decadal) trends and the relation between upper crustal dynamics at Kīlauea Volcano and more widespread processes, including activity associated with the magmatic system of neighboring Mauna Loa volcano and slip along the décollement between the volcanic pile and the underlying oceanic crust. This discussion examines different regions in turn (Section 12.5.1) and then compares our results with previous studies (Section 12.5.2).

12.5.1. Comparison of SWS, Stress, and Structure

12.5.1.1. Northern Caldera (NC)

In the northern part of Kīlauea's summit caldera, the surface structure data reflect the strikes of the caldera-bounding ring faults (Figures 12.2c and 12.3, data sets b, c, and d). SWS fast orientations appear to follow the outlines of the caldera, implying that the anisotropy is mostly governed by the pervasive structures of the area. The *P*-axis azimuths from FPSs of earthquakes in the area, however, display a maximum horizontal compressive stress that is also parallel to these faults. The orientation of this inferred $S_{H,\max}$ is an unexpected outcome given that the inferred $S_{H,\max}$ from SWS analysis in this region during the start of the summit eruption in 2008 was observed to be radial to the center of the caldera [*Johnson and Poland*, 2013].

Temporal analysis of FPSs with the deformation record shows a correlation between large deflation events and earthquakes chosen for analysis in this area (Figure 12.5a), suggesting that, when these earthquakes occurred, the area was indeed under tensional stress radial to the center

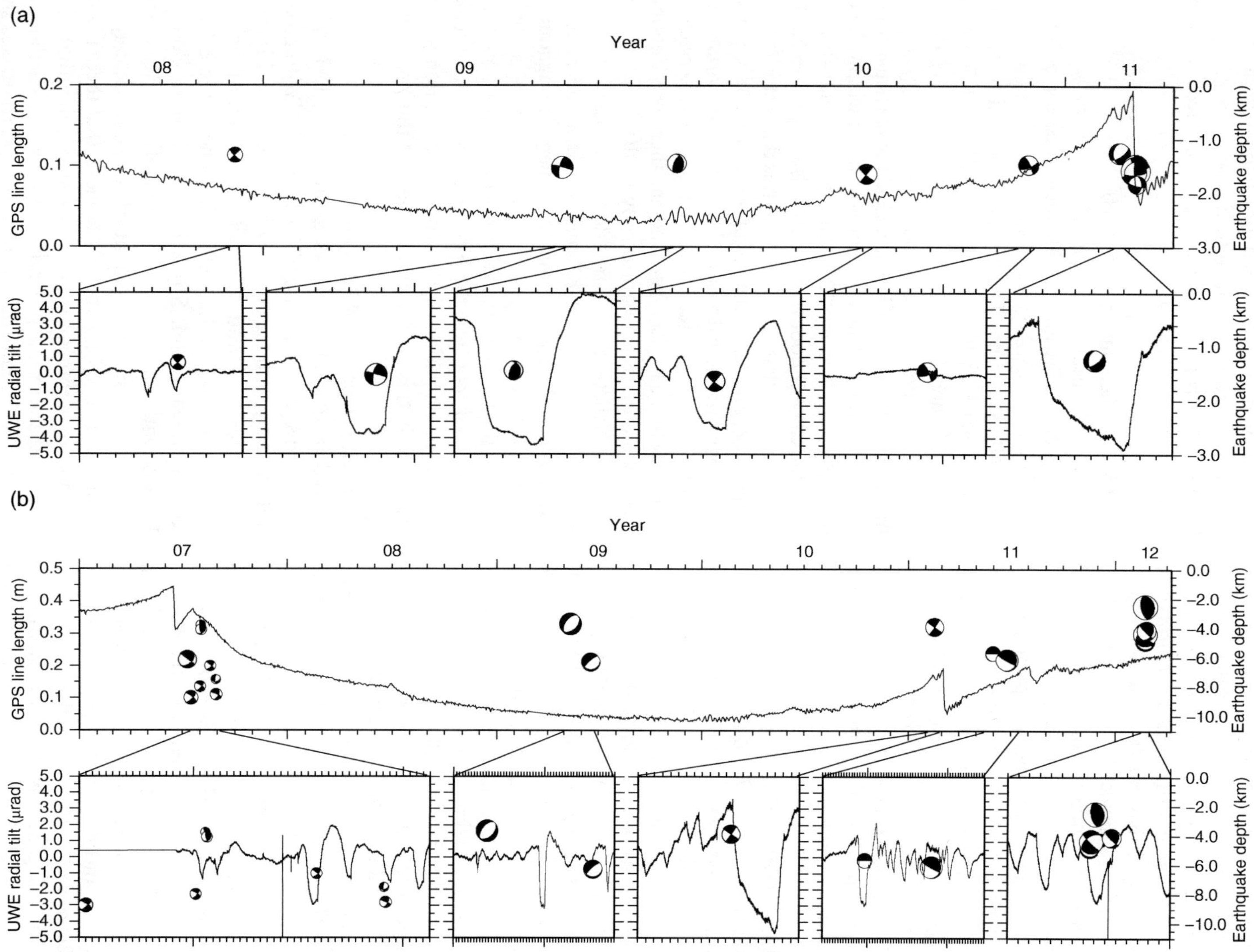

Figure 12.5 (a) GPS line length between stations UWEV and AHUP (Figure 12.1) from 18 July 2008 until 10 April 2011 and detrended radial (to the center of the caldera) tilt signal from station UWE for selected time periods. Focal mechanisms (beachballs) indicate the time, depth and FPS of earthquakes, scaled by magnitude, that occurred in the northern part of the caldera. (b) GPS line length between stations UWEV and AHUP from 1 January 2007 until 9 April 2012 and detrended radial (to the center of the caldera) tilt signal from station UWE for selected time periods. Focal mechanisms indicate the time, depth, and FPS of earthquakes, scaled by magnitude, that occurred in the Ka'ōiki Pali region.

of the caldera. Under such conditions, $S_{H,max}$ rotates to be tangential to the caldera walls. Not only do earthquakes in this area occur at times of deflation (Figure 12.5a), but those that occur during summit deflation are better recorded and therefore have been preferentially selected for this study. When this observation is combined with evidence from *Johnson and Poland* [2013] that the fast direction of SWS in the caldera changed in 2008 due to the prevalent cause switching between structural and stress-induced anisotropy, it is clear that the horizontal stress in the north part of Kīlauea Caldera is finely balanced and may change with inflation or deflation of the underlying magma reservoirs. These conditions could occur when the minimum and intermediate principal values of the stress tensor are similar in magnitude, allowing small changes in horizontal stress magnitudes to cause $S_{H,max}$ to rotate [e.g., *Villamor and Berryman*, 2006]. The lack of changes of SWS during the relatively small deflation-inflation (DI) events backs up the inference that structure is the dominant control of anisotropy in this region under most circumstances.

12.5.1.2. Ka'ōiki Pali (KP) Region

Surface structures in the Ka'ōiki Pali region (Figures 12.2c and 12.3, data set a) have a dominant NE-SW trend, and the $S_{H,max}$ direction from FPS *P*-axis azimuths displays one primary and one secondary mode. Both modes are oblique to the dominant surface structures, with the secondary mode approximately perpendicular to structure. This secondary mode of FPS *P*-axis azimuths correlates with estimates of $S_{H,max}$ by *Wyss et al.* [1992a,b], even though our study region is further SE and generally has shallower earthquakes. We do not see any distinction of FPS *P*-axis azimuth with depth or with sense of deformation (inflation or deflation) of Kīlauea. There are no seismometers in the Ka'ōiki Pali region; however, those nearby exhibit SWS fast directions parallel to the surface structure. The SWS does not seem to be related to stress in the Ka'ōiki Pali region and can be inferred to be controlled by structure.

The Ka'ōiki Pali region between Mauna Loa and Kīlauea volcanoes (also referred to as Nāmakani Paio in the literature) is a location of persistent earthquake activity. This region is not to be confused with the Ka'ōiki region of the east flank of Mauna Loa volcano, which also hosts significant seismicity such as the 16 November 1983 M6.6 earthquake and aftershocks [e.g., *Lockwood et al.*, 1987]. *Okubo and Nakata* [2003] showed that Ka'ōiki Pali earthquakes generally display a mixture of focal mechanisms, with the largest being consistent with normal faulting. They also demonstrated that Ka'ōiki Pali earthquakes generally occur at depths of 3–8 km, and many have a long-period component. Seismicity in this region generally occurs in swarms during both inflation and deflation of Kīlauea's summit (Figure 12.5b). It was noted by *Okubo and Nakata* [2003] that, between 1990 and 1993, swarms of earthquakes in this region preceded magmatic intrusions in Kīlauea's ERZ. The evidence that earthquakes occur more frequently at times of inflation or deflation of Kīlauea may suggest that the region is somewhat sensitive to volcanic strains and that large stress perturbations may be manifested as a variation in FPSs or SWS.

12.5.1.3. Halema'uma'u (HMM)

SWS results near Halema'uma'u Crater show scattered fast orientations, with a weak preference for a NE-SW trend. FPSs for earthquakes near Halema'uma'u also display a NE-SW trend. The nearest structural trend, other than the crater itself, represents the NW edge of Kīlauea Caldera (Figures 12.2c and 12.3, data set b), which displays the same direction. However, there are few mapped surface fractures on the floor of Kīlauea Caldera, particularly close to Halema'uma'u due to recent lava flows within the caldera. Figures 12.2c and 12.3, data set e, contain structures from the caldera floor that are still visible after recent lava flows. The fact that the trend from this data set is the same as the ERZ, even though it might be expected to be perpendicular due to the caldera faults, is quite important as it shows that the ERZ trend is transmitted into the caldera.

It is difficult to decipher whether there is a dominant mechanism of seismic anisotropy in the Halema'uma'u area. The structural trend that dominates most of Kīlauea could be present inside Kīlauea Caldera and affect the SWS. Alternatively, magma movement thoughout the period of this investigation may be causing sufficient stress in the surrounding rock to control the anisotropy, as it probably controls the FPS of the local earthquakes. Further investigation of the temporal evolution of SWS and FPSs in this area may shed light on this problem.

12.5.1.4. Upper East Rift Zone (UERZ)

SWS in the UERZ shows a prominent ESE–WNW trend and a lesser ERZ trend. The ESE-WNW trend does not correlate with the surface features used in this analysis (Figures 12.2c and 12.3, data sets f and k); however, the chain of craters that define the bend in the ERZ follows a similar trend. Stress is also bimodal along the UERZ, with one mode approximately 70° from north and the other, more dominant trend approximately 110° from north. The FPS analysis suggests that there is a distinction by depth of these two *P*-axis azimuthal directions, with the former generally resulting from earthquakes deeper than 3.5 km and the latter from earthquakes shallower than about 3.5 km (Figure 12.6). This discrepancy may be due to a stress discontinuity at about 3.5 km depth in this area, possibly due to local stress from the pressurized magma transport conduit and rifting [*Gillard et al.*,

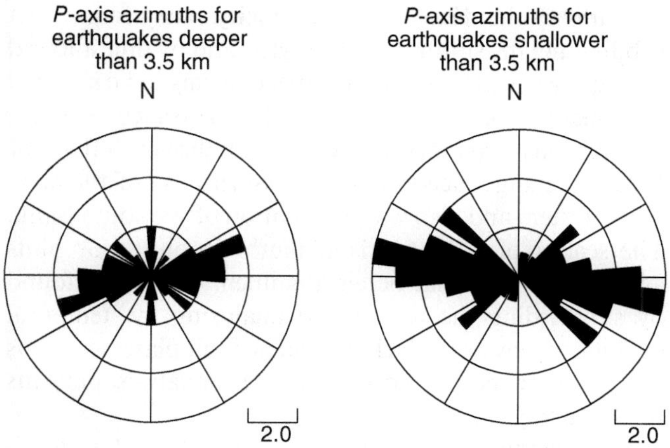

Figure 12.6 Rose diagrams of *P*-axis azimuths from FPS analysis separated by depth of earthquakes in the upper ERZ.

1996a] or due to a lithological boundary [*Moore and Fiske*, 1969].

Variable stress with depth could be caused by a pressurized pipe or dikelike conduit creating horizontal compressive stress perpendicular to the conduit walls. If the conduit is at approximately 3.5 km depth (which is consistent with seismic and deformation studies of magma transport through the ERZ, as suggested by, for example, *Klein et al.* [1987] and *Lundgren et al.* [2013]), $S_{H,\max}$ at this depth would be expected to reflect an orientation of approximately 60°−70°. Above the conduit $S_{H,\max}$ would not be greatly affected by the vertical stress. *Gillard et al.* [1996a] suggested that there is a stress concentration above the aseismic part of the UERZ (at about 3–3.5 km depth) due to the seaward displacement of Kīlauea's south flank.

A lithological boundary could also cause a stress discontinuity at depth. There is hyaloclastite layer (fragmental basaltic debris) that marks the transition from submarine eruptive activity to subaerial. This layer is known to be at about 1 km depth (500 m below sea level) in the lower ERZ [*Quane et al.*, 2000] and was not encountered in a 1.2 km deep borehole (which reached to about sea level) inside Kīlauea Caldera [*Keller et al.*, 1979]. The hyaloclastite layer is thought to dip to the north, with the same attitude as the decollement [*Moore and Fiske*, 1969]. It is possible that this layer could occur around 3−4 km depth in the UERZ. As SWS is more sensitive to the anisotropy directly beneath the station (i.e., shallow depths), the fact that the ESE-WNW orientation of $S_{H,\max}$ is the shallower of the two regimes could also mean that the anisotropy in the middle ERZ is predominantly controlled by the shallower stress regime.

12.5.1.5. Southern Caldera (SC) and Southwest Rift Zone

South of Kīlauea Caldera, the SWS fast direction and FPS-derived $S_{H,\max}$ direction correlate well, indicating that stress may be governing the anisotropy in this area. These SWS and $S_{H,\max}$ directions also correlate with the trend of the SWRZ discussed in Section 12.4.3 (Figures 12.2c and 12.3, data set h), although many structures in this region (directly south of the caldera) display the ERZ trend as well (Figures 12.2c and 12.3, data set j). As is the case in the UERZ, this discrepancy is suggestive of a stress discontinuity at depth in this area. In the SWRZ, few earthquakes during the study period and limited seismometer coverage have resulted in no FPS data for this region. However, SWS and structure orientations agree well, both displaying the SWRZ trend.

Other evidence for a stress discontinuity at depth comes from historic seismicity trends. *Klein et al.* [1987] analyzed seismicity associated with magmatic intrusions and eruptions in the SWRZ. The eruption in December 1974 and intrusion in August 1981, as well as other intrusive events throughout the 1960s–1980s, were accompanied by seismicity that propagated south from the caldera at approximately 3 km depth. The region of seismicity, known as the seismic SWRZ [e.g., *Wright and Klein*, 2006], bends west about 2 km south of the caldera and assumes a SWRZ trend. The 1974 eruptive fissures, however, individually have trends closer to that of the ERZ. An eruption in September 1971, located to the north of the 1974 eruption site, appears to have a shallower source, since accompanying earthquakes, which followed the trend of the SWRZ, occurred in a depth range of 0–2 km. The 1971 and 1974 eruptions probably occurred on different structural "strands" of the SWRZ, with the seismic SWRZ strand deeper than the more northerly strand [*Holcomb*, 1987. Structural trends [*Holcomb*, 1987], seismicity [*Klein et al.*, 1987], and eruptions and intrusions [e.g., *Lockwood et al.*, 1999] therefore support the existence of a stress discontinuity at depth. The difference in trend of the shallow and deeper events in the SWRZ suggest that magma movement in the different strands and at different depths is driven by different stress regimes.

12.5.1.6. Keanakākoʻi (KK)

The structure in the Keanakākoʻi region (Figures 12.2c and 12.3, data set f) displays a prominent trend consistent with fissure orientations in the middle and lower ERZ (Figures 12.2c and 12.3, data sets m and n), despite the NW–SE orientation of the ERZ in this area (as evidenced by aligned craters). Conversely, the FPSs display dominant *P*-axis azimuths in an ENE-WSW direction, parallel to the line of craters that makes up the UERZ. The SWS fast orientation at Keanakākoʻi is aligned between the two, in an E–W direction. The SWS could be a product of the two different directions of stress and structure, although two modes would normally be expected if that were the case. It is more likely that there is a complex interaction of stress and structure due to the intersection of the ERZ conduit with the summit magma system in the Keanakāko'i area. This complex interaction may change or rotate with depth, having a dominant E-W direction at the depths to which SWS is sensitive (1–2 km).

12.5.1.7. *Koa'e (KOE) Fault System*

The surface structure in the Koa'e Fault System (Figures 12.2c and 12.3, data set l) is consistent with the trend of the ERZ. There are no seismic stations inside the KOE box, but the SWS fast-orientation trend from nearby stations is also consistent with the ERZ trend. The FPS *P*-axis azimuths are variable with a general trend approximately perpendicular to the structural and anisotropic trends. The stress in this region may be governed by the interaction between the SWRZ and the ERZ; however, the SWS does not appear to be affected by the stress regime, as it consistently correlates with the structural trend. Changes in stress associated with eruptive or intrusive activity and due to tectonic activity of the rift zones and unstable south flank may be observed by monitoring the earthquake FPS but are unlikely to be manifested in SWS observations.

12.5.1.8. *Middle East Rift Zone (MERZ)*

As in many other regions of Kīlauea, anisotropic fast directions correlate well with surface structural features in the MERZ (Figures 12.2c and 12.3, data set m). There are few reliable FPSs in the region due to large azimuthal gaps in the network; however, there is a suggestion of a trend in FPS *P*-axis azimuths that also correlates with the structure and SWS directions. The MERZ was volcanically active for the duration of this study, and so temporal changes in stress would not be unexpected in this region. If there have been variations in stress, they have not been reflected in the SWS results. It is therefore likely that the strong alignments of cracks, fissures, and repeated dike intrusions in the MERZ are the main governing factor for SWS.

12.5.1.9. *Lower East Rift Zone and South Flank (SF)*

FPS data from the south flank of Kīlauea display $S_{H,\max}$ in a NNW-SSE orientation, perpendicular to both the surface structure data (Figures 12.2c and 12.3, data sets q, p, and o) and the SWS results. The majority of seismicity on the south flank is deeper than the cutoff used in this analysis. Those deeper earthquakes represent slip on the décollement [*Denlinger and Okubo*, 1995], which is distinct from the processes that occur closer to the surface. This NNW-SSE orientation of $S_{H,\max}$ correlates with estimates from FPS analysis carried out on the deeper décollement seismicity by *Crosson and Endo* [1982] and therefore probably reflects upper décollement slip.

Gillard et al. [1996b] also examined deeper earthquakes and found reverse and normal focal mechanisms in addition to the décollement seismicity. They found that subdividing the south flank by region and by time allowed earthquakes with similar focal mechanisms to be inverted for stress tensors with smaller misfits. The success of the subdivisions suggests that the stress on the south flank of Kīlauea changes according to pressurization of the magmatic system and also with the motion of the south flank. The scatter of FPS *P*-axis azimuths in our study hints at the possibility that better alignments would be found by subdividing the region and analyzing the temporal evolution; however, a lack of reliable fault plane solutions for the time period covered by our analysis prevents further analysis.

The structure at the surface of the south flank is dominated by normal faults that strike approximately parallel to both the coast and the ERZ and are a result of gravitational instability due to the seaward displacement of the south flank [*Swanson et al.*, 1976]. A lack of three-component seismic stations on the south flank of Kīlauea means that little can be said about the anisotropy in this region. To the east, where the south flank intersects the lower ERZ, SWS fast polarizations are parallel to the structure. Structure, rather than stress, therefore appears to govern the anisotropy of the lower ERZ.

12.5.2. Comparison of SWS with Past Results and Regional Trends

Past studies on crustal SWS at Kīlauea include *Savage et al.* [1989], *Booth et al.* [1992], *Munson et al.* [1995], and *Kenedi et al.* [2010]. The *Savage et al.* [1989] study used an array on and around the ERZ. Their results (Figure 12.7, green) show good agreement with our data in this region. Further along the lower ERZ (outside our study region), *Kenedi et al.* [2010] investigated SWS around the Puna geothermal field measured at eight borehole seismometers. They also found that the shear wave fast direction and inferred dominant fracture orientation correlated with the trend of the ERZ at that location.

Booth et al. [1992] carried out SWS analysis on data recorded at station AIN in the Ka'ōiki region. They found a fast polarization of $\sim 074 \pm 20°$ and attributed the anisotropy to stress-aligned microcracks, rather than pervasive structure, because the direction correlates with stress estimates from *Wyss et al.* [1992a] instead of surface faulting observed by *Jackson et al.* [1992]. Our results at station AIN are in accordance with the results of *Booth et al.* [1992].

The *Munson et al.* [1995] study used arrays around the ERZ and the SWRZ, one of which is close to our study region. Results from *Munson et al.* [1995] (Figure 12.7, yellow) also agree well with the SWS results that we present. The similarity between results from past studies and our own suggests that the "background" anisotropy is

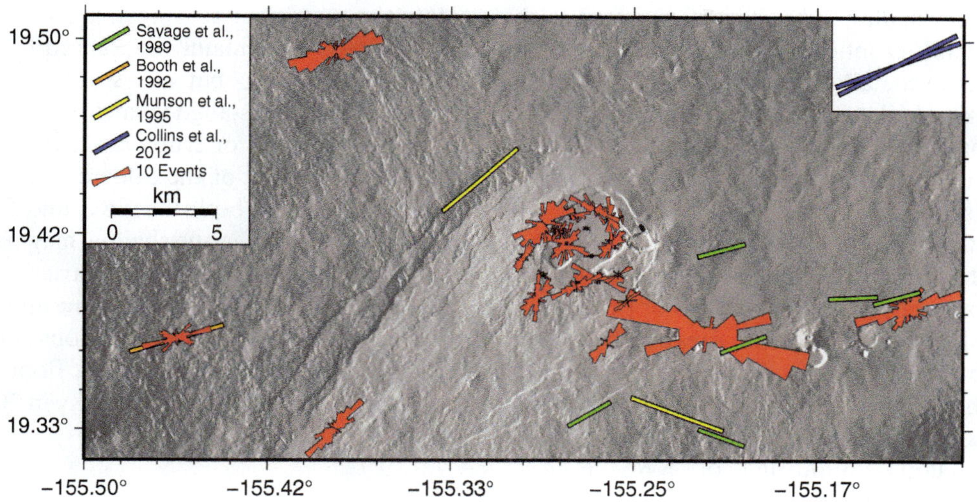

Figure 12.7 Rose diagrams displaying fast direction of anisotropy in 2010 as a representative year in red, plotted at the seismic station at which the measurements were made. Colored bars indicate anisotropy from past studies [*Savage et al.*, 1989; *Booth et al.*, 1992; *Munson et al.*, 1995]. The *Collins et al.* [2012] study used teleseisms for mantle anisotropy recorded at stations BIG2 and DLAH, outside of our study area, and so results are plotted separately (top right).

relatively stable and has not changed significantly for at least tens of years.

Collins et al. [2012] carried out SWS analysis on teleseisms recorded on the PLUME broadband land and ocean bottom seismometer deployment to characterize upper mantle anisotropy across the Hawaiian Swell. They found that SWS observations were predominantly 076 ± 19°, reflecting fossil lithospheric anisotropy from plate spreading. The two measurements closest to our study region are displayed in blue in Figure 12.7. The mantle SWS results are remarkably similar to our regional crustal result. This could be a coincidence, with the "regional" direction at Kīlauea and Mauna Loa reflecting the superposition of volcanoes or some other process. Alternatively, the correlation between mantle and local seismic anisotropy may be due to mantle processes affecting crustal stress and structure.

12.6. IMPLICATIONS FOR MONITORING SEISMIC ANISOTROPY

To monitor SWS in an authoritative and conscientious manner, a thorough understanding of the causes and controls of seismic anisotropy is necessary. The above analyses indicate that seismic anisotropy at Kīlauea Volcano is mostly controlled by crustal structure and is therefore unlikely to reflect small changes in stress conditions. One of the exceptions may be the UERZ, where SWS results correlate more strongly with $S_{H,\max}$ estimates from FPSs than with surface structure data. All of these areas, however, do show evidence of structure-controlled anisotropy, and so confidently identifying a temporal change in SWS may prove challenging.

Even though seismic anisotropy at Kīlauea is usually controlled by static crustal structure, large changes in magmatic and tectonic activity can affect the mechanism of anisotropy. The question is how large do the magmatic and tectonic changes have to be before they can be observed in SWS? We can go some of the way toward answering this question by examining events that have and have not altered the anisotropy. For example, in the NC, even though there is evidence that $S_{H,\max}$ may change during summit deflation, it is not until the conditions with the microcracks (i.e. pore content and pressure) change significantly that we see stress-induced anisotropy. *Johnson and Poland* [2013] showed that variations in seismic anisotropy in the northern part of the caldera in 2008 may reflect changes to pore and crack content caused by an increase in magmatic gas emissions associated with the start of the ongoing summit eruption. This interpretation for changing anisotropy was supported by corresponding decreases in the V_p/V_s ratio. The stations in the northern part of the caldera appear to be sensitive to changes in the shallow magmatic system at the summit and may therefore display variation in SWS associated with future changes in eruptive activity that can be monitored in near-real time. *Johnson and Poland* [2013] also identified a small but similar deviation from background in July 2007. While not a statistically significant change in ϕ or V_p/V_s, the presence of fluctuations in all of the time series and concurrence with the "Father's Day" intrusion and eruption [*Poland et al.*, 2009] suggest that the changes were driven by the same mechanism, namely increased gas content of the shallow subsurface. Compared to the start of the summit eruption in 2008, the Father's Day event was short-lived, allowing the system to quickly return to a background state and preventing strong SWS signals. Still, the data serve as a reminder that SWS may provide important supporting data when interpreting small VT events.

Data from other areas of the volcano, such as the KP, indicate that the stress field in those locations is finely balanced and so, even though the SWS appears to be controlled by structure, changes in stress with volcanic activity may be manifested as SWS variations. These observations show that, even though the general trends are relatively stable and mostly controlled by crustal structure, continuous monitoring of SWS, even in locations that seem unlikely to change, could detect unexpected volcanic signals.

12.7. CONCLUSIONS

Shear wave splitting has shown promise as a monitoring tool for detecting changes in activity at Kīlauea and other volcanoes around the world. To confidently interpret temporal variations in SWS as due to variations in seismic anisotropy, however, it is essential to understand the background controls of seismic anisotropy. We compiled a database of SWS results and fault plane solutions (which indicate stress conditions) from earthquakes that occurred during the late 2000s and early 2010s and combined those with structural trends deduced from mapped fractures and eruptive fissures across Kīlauea. Comparing these data sets provides a means of determining the dominant control of seismic anisotropy in specific areas of the volcano. Our results highlight a dominant SWS fast direction oriented ENE-WSW, roughly parallel to the trend of Kīlauea's ERZ away from the summit caldera. This direction is consistent with past results from Kīlauea, indicating stability of SWS on decadal timescales, as well as with regional SWS orientations from the lithosphere, which may be a coincidence or may reflect the influence of mantle processes on crustal structure.

The majority of SWS results correlate with stress orientations and surface structure, suggesting a structural control on most seismic anisotropy at Kīlauea. A few locations, however, are noteworthy in their departures from this pattern. The upper parts of both the ERZ and SWRZ show multiple modes of stress and anisotropy, only one of which correlates with surface structure. Such a result may indicate a stress discontinuity at depth (perhaps due to the presence of a magma conduit). In the Ka'ōiki Pali area and the northern part of Kīlauea Caldera, the minimum and intermediate principal values of the stress tensor appear similar in magnitude, so that small changes in horizontal stress magnitudes could rotate the maximum compressive stress. In these areas, SWS appears to be mainly influenced by the static structure of the volcano but can switch dynamically to be influenced by stress-aligned microcracks as a result of magmatic, tectonic, or eruptive activity. Seismic stations in the northern part of the caldera, the Ka'ōiki Pali, and the upper parts of both the ERZ and SWRZ could be sensitive to changes in the shallow magma plumbing system and may therefore display variations in SWS that could be monitored in near-real time and associated with volcanic or tectonic activity. Previous results documenting changes in SWS on the caldera floor associated with the onset of summit eruptive activity in 2008 support this conclusion.

SUPPLEMENTARY DIGITAL DATA

The following data items are available online at www.wiley.com/go/Carey/Hawaiian_Volcanoes

README. Explanation of supplementary tables.

Table 12.S1. Table of earthquakes used in the fault plane solution analysis.

Table 12.S2. Locations of mapped fault, fracture and fissure traces for Kilauea.

ACKNOWLEDGMENTS

We thank Paul Okubo, Kyle Anderson, Marco Bagnardi, and Ben McLeod for helpful discussions about the data, processing, and interpretations. This work was supported by a cooperative agreement between the University of Hawai'i at Hilo and the U.S. Geological Survey. We also thank Eleanora Rivalta, Paul Okubo, and an anonymous reviewer for helpful and constructive comments.

REFERENCES

Balfour, N. J., M. K. Savage, and J. Townend (2005), Stress and crustal anisotropy in Marlborough, New Zealand: Evidence for low fault strength and structure-controlled anisotropy, *Geophys. J. Int.*, *163*(3), 1073–1086, doi:10.1111/j.1365-246X.2005.02783.x.

Bianco, F., and L. Zaccarelli (2009), A reappraisal of shear wave splitting parameters from Italian active volcanic areas through a semiautomatic algorithm, *J. Seismol.*, *13*(2), 253–266, doi:10.1007/s10950-008-9125-z.

Boness, N. L., and M. D. Zoback (2006), Mapping stress and structurally controlled crustal shear velocity anisotropy in California, *Geology*, *34*(10), 825, doi:10.1130/G22309.1.

Booth, D. C., M. Wyss, and D. Gillard (1992), Shear-wave polarization alignments recorded above the Kaoiki fault zone, Hawaii, *Geophys. Res. Lett.*, *19*(11), 1141–1144, doi:10.1029/92GL01173.

Cayol, V. (2000), High magma storage rates before the 1983 eruption of Kilauea, Hawaii, *Science*, *288*(5475), 2343–2346, doi:10.1126/science.288.5475.2343.

Cochran, E. S., J. E. Vidale, and Y. G. Li (2003), Near-fault anisotropy following the Hector Mine earthquake, *J. Geophys. Res. Solid Earth*, *108*(B9), 2436, doi:10.1029/2002JB002352.

Collins, J. A., C. J. Wolfe, and G. Laske (2012), Shear wave splitting at the Hawaiian hot spot from the PLUME land and ocean bottom seismometer deployments, *Geochem. Geophys. Geosyst.*, *13*(2), Q02,007, doi:10.1029/2011GC003881.

Crampin, S. (1994), The fracture criticality of crustal rocks, *Geophys. J. Int.*, *118*(2), 428–438, doi:10.1111/j.1365-246X.1994.tb03974.x.

Crosson, R. S., and E. T. Endo (1982), Focal mechanisms and locations of earthquakes in the vicinity of the 1975 Kalapana earthquake aftershock zone 1970–1979: Implications for tectonics of the south flank of Kilauea Volcano, Island of Hawaii, *Tectonics*, *1*(6), 495–542.

Crotwell, H. P., T. J. Owens, and J. Ritsema (1999), The TauP Toolkit: Flexible seismic travel-time and ray-path utilities, *Seismol. Res. Lett.*, *70*(April), 154–160, doi:10.1785/gssrl.70.2.154.

Dawson, P. B., B. A. Chouet, P. G. Okubo, A. Villasenor, and H. M. Benz (1999), Three-dimensional velocity structure of the Kilauea Caldera, Hawaii, *Geophys. Res. Lett.*, *26*(18), 2805–2808, doi:10.1029/1999GL005379.

Delaney, P. T., R. S. Fiske, A. Miklius, R. T. Okamura, and M. K. Sako (1990), Deep magma body beneath the Summit and Rift Zones of Kilauea Volcano, Hawaii, *Science*, *247*(4948), 1311–1316, doi:10.1126/science.247.4948.1311.

Denlinger, R. P., and P. Okubo (1995), Structure of the mobile south flank of Kilauea Volcano, Hawaii, *J. Geophys. Res.*, *100*(B12), 24,499–24,507, doi:10.1029/95JB01479.

Duffield, W. A. (1975), Structure and origin, Koae fault system, Kilauea Volcano, Hawaii, Prof. Pap. 856, pp. 1–12, U.S. Geol. Sur., Reston, VA.

Gerst, A., and M. K. Savage (2004), Seismic anisotropy beneath Ruapehu Volcano: A possible eruption forecasting tool, *Science*, *306*(5701), 1543–1547, doi:10.1126/science.1103445.

Gillard, D., A. M. Rubin, and P. G. Okubo (1996a), Highly concentrated seismicity caused by deformation of Kilauea's deep magma system, *Nature*, *384*, 343–346, doi:10.1038/384343a0.

Gillard, D., M. Wyss, and P. Okubo (1996b), Type of faulting and orientation of stress and strain as a function of space and time in Kilauea's south flank, Hawaii, *J. Geophys. Res.*, *101*(B7), 16,025–16,042, doi:10.1029/96JB00651.

Hill, D. P., and J. J. Zucca (1987), Geophysical constraints on the structure of Kilauea and Mauna Loa volcanoes and some implications for seismomagmatic processes, in Volcanism in Hawaii, edited by R. Decker et al. pp. 903–918, Prof. Pap. 1350, U.S. Geol. Surv., Reston, VA.

Holcomb, R. T. (1987), Eruptive history and long-term behavior of Kilauea Volcano, in Volcanism in Hawaii, edited by R. W. Decker et al. pp. 261–350, Prof. Pap. 1350, U.S Geol. Surv., Reston, VA.

Holcombe, R. J. (2001), GeoMapVector. Available at http://www.holcombe.net.au/software/rodh_software_downloads.htm.

Houghton, B. F., D. A. Swanson, R. J. Carey, J. Rausch, and A. J. Sutton (2011), Pigeonholing pyroclasts: Insights from the 19 March 2008 explosive eruption of Kilauea Volcano, *Geology*, *39*(3), 263–266, doi:10.1130/G31509.1.

Hoversten, G. M., E. Gasperikova, G. A. Newman, J. Kauakihaua, and N. Cuevas (2003), Magnetotelluric imaging of the Kilauea Volcano, Hawaii, SEG Technical Program Expanded Abstracts 2012, pp. 2387–2390, doi:10.1190/1.1817851.

Jackson, M. D., E. T. Endo, P. T. Delaney, T. Arnadottir, and A. M. Rubin (1992), Ground ruptures of the 1974 and 1983 Kaoiki earthquakes, Mauna Loa volcano, Hawaii Pacific Ocean, *J. Geophys. Res.*, *97*(92), 8775–8796, doi:10.1029/92JB00649.

Johnson, J. H., and M. P. Poland (2013), Seismic detection of increased degassing before Kilauea's 2008 summit explosion, *Nature Commun.*, *4*, 1668, doi:10.1038/ncomms2703.

Johnson, J. H., and M. K. Savage (2012), Tracking volcanic and geothermal activity in the Tongariro Volcanic Centre, New Zealand, with shear wave splitting tomography, *J. Volcanol. Geotherm. Res.*, *223–224*, 1–10, doi:10.1016/j.jvolgeores.2012.01.017.

Kauahikaua, J. (1993), Geophysical characteristics of the hydrothermal systems of Kilauea Volcano, Hawai'i, *Geothermics*, *22*(4), 271–299, doi:10.1016/0375-6505(93)90004-7.

Keats, B. S., J. H. Johnson, and M. K. Savage (2011), The Erua earthquake cluster and seismic anisotropy in the Ruapehu region, New Zealand, *Geophys. Res. Lett.*, *38*(L16), L16315, doi:10.1029/2011GL049014.

Keller, G. V., L. T. Grose, J. C. Murray, and C. K. Skokan (1979), Results of an experimental drill hole at the summit of Kilauea volcano, Hawaii, *J. Volcanol. Geotherm. Res.*, *5*(3–4), 345–385, doi:10.1016/0377-0273(79)90024-6.

Kenedi, C. L., E. Shalev, A. Lucas, and P. Malin (2010), Microseismicity and 3-D mapping of an active geothermal field, Kilauea Lower East Rift Zone, Puna, Hawaii, in *Proceedings World Geothermal Congress*, April, pp. 1–6, Bali, Indonesia. Available at http://www.geothermal-energy.org/pdf/IGAstandard/WGC/2010/1352.pdf

Klein, F. W. (2002), User's Guide to HYPOINVERSE-2000, a Fortran Program to Solve for Earthquake Locations and Magnitudes 4/2002, Open File Rep. 02-171, U.S. Geol. Surv., Reston, VA.

Klein, F. W., R. Y. Koyanagi, J. S. Nakata, and W. R. Tanigawa (1987), The seismicity of Kilauea's magma system, in *Volcanism in Hawaii*, edited by R. W. Decker et al. pp. 1019–1185, Prof. Pap. 1350, U.S Geol. Surv., Reston, Virginia.

Lahr, J. C. (1999), HYPOELLIPSE: A computer program for determining local earthquake hypocentral parameters, magnitude, and first-motion pattern, U. S. Open-File Rep. 99–23, Geol. Surv., Reston, VA.

Lockwood, J. P., J. J. Dvorak, T. T. English, R. Y. Koyanagi, A. T. Okamura, M. L. Summers, and W. R. Tanigawa (1987), Mauna Loa 1974–1984: A decade of intrusive and extrusive activity, in *Volcanism in Hawaii*, edited by R. W. Decker et al., pp. 537–570, Prof. Pap. 1350, U.S Geol. Surv., Reston, Virginia.

Lockwood, J. P., R. I. Tilling, R. T. Holcomb, F. Klein, A. T. Okamura, and D. W. Peterson (1999), Magma Migration and Resupply During the 1974 Summit Eruptions of Kilauea Volcano, Hawaii, Prof. Pap. 1613, U.S. Geol. Surv., Reston, VA.

Lundgren, P., M. Poland, A. Miklius, T. Orr, S.-H. Yun, E. Fielding, Z. Liu, A. Tanaka, W. Szeliga, S. Hensley, and

S. Owen (2013), Evolution of dike opening during the March 2011 Kamoamoa fissure eruption, Kilauea Volcano, Hawai'i, *J. Geophys. Res. Solid Earth*, *118*(3), 897–914, doi:10.1002/jgrb.50108.

McKenzie, D. P. (1969), The relation between fault plane solutions for earthquakes and the directions of the principal stresses, *Bull. Seismol. Soci. Am.*, *59*(2), 591–601.

Moore, J. G., and R. S. Fiske (1969), Volcanic substructure inferred from dredge samples and ocean-bottom photographs, Hawaii, *Geol. Soc. Am. Bull.*, *80*(July), 1191–1202, doi:10.1130/0016-7606(1969)80[1191:VSIFDS]2.0.CO;2.

Munson, C. G., C. H. Thurber, Y. Li, and P. G. Okubo (1995), Crustal shear wave anisotropy in southern Hawaii: Spatial and temporal analysis, *J. Geophys. Res.*, *100*(B10), 20,367–20,377, doi:10.1029/95JB01288.

Neal, C. A., and J. P. Lockwood (2002), Geologic map of the summit region of Kilauea Volcano, Hawaii, Tech. Rep., Geologic Investigations Series I-2759, scale 1:24,000, U.S. Geol. Surv., Reston, VA.

Okubo, P. G., and J. S. Nakata (2003), Tectonic pulses during Kilauea's current long-term eruption, in *The Pu'u 'O'o–Kupaianaha Eruption of Kilauea Volcano, Hawai'i: The First 20 Years*, pp. 173–186, Prof. Pap. 1676, U.S Geol. Surv., Reston, Va.

Okubo, P. G., H. M. Benz, and B. A. Chouet (1997), Imaging the crustal magma sources beneath Mauna Loa and Kilauea volcanoes, Hawaii, *Geology*, *25*(10), 867–870, doi:10.1130/0091-7613(1997)025<0867:ITCMSB>2.3.CO;2.

Poland, M. P., A. J. Sutton, and T. M. Gerlach (2009), Magma degassing triggered by static decompression at Kilauea Volcano, Hawai'i, *Geophy. Res. Lett.*, *36*(L16), L16306, doi:10.1029/2009GL039214.

Quane, S. L., M. O. Garcia, H. Guillou, and T. P. Hulsebosch (2000), Magmatic history of the East Rift Zone of Kilauea Volcano, Hawaii based on drill core from SOH 1, *J. Volcanol. Geotherm. Res.*, *102*, 319–338, doi:10.1016/S0377-0273(00)00194-3.

Reasenberg, P., and D. Oppenheimer (1985), FPFIT, FPPLOT and FPPAGE: Fortran computer programs for calculating and displaying earthquake fault-plane solutions, Tech. Rep. 109, Open-File Rep. 85–739, U.S. Geol. Surv., Reston, Va.

Roman, D. C., J. Neuberg, and R. R. Luckett (2006), Assessing the likelihood of volcanic eruption through analysis of volcanotectonic earthquake fault plane solutions, *Earth Planet. Sci. Lett.*, *248*(1–2), 244–252, doi:10.1016/j.epsl.2006.05.029.

Roman, D. C., M. K. Savage, R. Arnold, J. L. Latchman, and S. De Angelis (2011), Analysis and forward modeling of seismic anisotropy during the ongoing eruption of the Soufriere Hills Volcano, Montserrat, 1996–2007, *J. Geophys. Res. Solid Earth*, *116*(B3), B03201, doi:10.1029/2010JB007667.

Savage, M. K., X. R. Shih, R. P. Meyer, and R. C. Aster (1989), Shear-wave anisotropy of active tectonic regions via automated S-wave polarization analysis, *Tectonophysics*, *165*(1–4), 279–292, doi:10.1016/0040-1951(89)90053-X.

Savage, M. K., T. Ohminato, Y. Aoki, H. Tsuji, and S. M. Greve (2010), Stress magnitude and its temporal variation at Mt. Asama Volcano, Japan, from seismic anisotropy and GPS, *Earth Planet. Sci. Lett.*, *290*(3–4), 403–414, doi:10.1016/j.epsl.2009.12.037.

Silver, P. G., and W. W. Chan (1991), Shear-wave splitting and subcontinental mantle deformation, *J. Geophys. Res. Solid Earth*, *96*(B10), 16,429–16,454, doi:10.1029/91JB00899.

Silver, P. G., and M. K. Savage (1994), The interpretation of shear-wave splitting parameters in the presence of two anisotropic layers, *Geophys. J. Int.*, *119*(3), 949–963, doi:10.1111/j.1365-246X.1994.tb04027.x.

Swanson, D. A., W. A. Duffield, and R. S. Fiske (1976), Displacement of the south flank of Kilauea Volcano: The result of forceful intrusion of magma into the rift zones, *Prof. Pap. 963*, p. 39, U.S. Geol. Surv., Reston, VA.

Teanby, N. A., J.-M. Kendall, and M. van der Baan (2004), Automation of shear-wave splitting measurements using cluster analysis, *Bull. Seismol. Soc. Am.*, *94*(2), 453–463, doi:10.1785/0120030123.

Unglert, K., M. K. Savage, N. Fournier, T. Ohkura, and Y. Abe (2011), Shear wave splitting, vP/vS, and GPS during a time of enhanced activity at Aso caldera, Kyushu, *J. Geophys. Res. Solid Earth*, *116*(B11), B11203, doi:10.1029/2011JB008520.

Villamor, P., and K. R. Berryman (2006), Late Quaternary geometry and kinematics of faults at the southern termination of the Taupo Volcanic Zone, New Zealand, *New Zealand J. Geol. Geophys.*, *49*(1), 1–21, doi:10.1080/00288306.2006.9515144.

Volti, T., and S. Crampin (2003), A four-year study of shear-wave splitting in Iceland: 2. Temporal changes before earthquakes and volcanic eruptions, *Geol. Soci. Lond. Spec. Publ.*, *212*(1), 135–149, doi:10.1144/GSL.SP.2003.212.01.09.

Wauthier, C., D. C. Roman, and M. P. Poland (2013), Moderate-magnitude earthquakes induced by magma reservoir inflation at Kilauea Volcano, Hawaii, *Geophys. Res. Lett.*, *40*(October), doi:10.1002/2013GL058082.

Wilson, D., T. Elias, T. Orr, M. Patrick, J. Sutton, and D. Swanson (2008), Small explosion from new vent at Kilaueas summit, *EOS Trans. AGU 89*(22), 203, doi:10.1029/2008EO220003.

Wolfe, E. W., and J. Morris (1996), Geologic map of the island of Hawaii, Tech. Rep., Misc. Invest. Ser. Map I-252-A, scale 1:100,000, U.S. Geol. Surv., Reston, VA.

Wright, T. L., and F. W. Klein (2006), Deep magma transport at Kilauea volcano, Hawaii, *Lithos*, *87*(1-2), 50–79, doi:10.1016/j.lithos.2005.05.004.

Wyss, M., B. Liang, W. R. Tanigawa, and X. Wu (1992a), Comparison of orientations of stress and strain tensors based on fault plane solutions in Kaoiki, Hawaii, *J. Geophys. Res.*, *97*(B4), 4769–4790, doi:10.1029/91JB02968.

Wyss, M., D. Gillard, and B. Liang (1992b), An estimate of the absolute stress tensor in Kaoiki, Hawaii, *J. Geophys. Res.*, *97*(B4), 4763–4768, doi:10.1029/91JB01455.

Zinke, J. C., and M. D. Zoback (2000), Structure-related and stress-induced shear-wave velocity anisotropy: Observations from microearthquakes near the Calaveras Fault in central California, *Bull. Seismol. Soc. Am.*, *90*(5), 1305–1312, doi:10.1785/0119990099.

13

Delicate Balance of Magmatic-Tectonic Interaction at Kīlauea Volcano, Hawai'i, Revealed from Slow Slip Events

Emily K. Montgomery-Brown[1], Michael P. Poland[2], and Asta Miklius[2]

ABSTRACT

Eleven slow slip events (SSEs) have occurred on the southern flank of Kīlauea Volcano, Hawai'i, since 1997 through 2014. We analyze this series of SSEs in the context of Kīlauea's magma system to assess whether or not there are interactions between these tectonic events and eruptive/intrusive activity. Over time, SSEs have increased in magnitude and become more regular, with interevent times averaging 2.44 ± 0.15 years since 2003. Two notable SSEs that impacted both the flank and the magmatic system occurred in 2007, when an intrusion and small eruption on the East Rift Zone were part of a feedback with a SSE, and 2012, when slow slip induced 2.5 cm of East Rift Zone opening (but without any change in eruptive activity). A summit inflation event and surge in East Rift Zone lava effusion was associated with a SSE in 2005, but the inferred triggering relation is not clear due to a poorly constrained slip onset time. Our results demonstrate that slow slip along Kīlauea's décollement has the potential to trigger and be triggered by activity within the volcano's magma system. Since only three of the SSEs have been associated with changes in magmatic activity within the summit and rift zones, both the décollement and magma system must be close to failure for triggering to occur.

13.1. INTRODUCTION

Magma-tectonic interactions occur at volcanoes worldwide and have been postulated since the earliest days of geology, when *Lyell* [1889] mapped relations between faults and volcanic vents [*La Femina et al.*, 2004]. Numerous authors have subsequently studied the relations between volcanoes and earthquakes [*Linde and Sacks*, 1998; *Hill et al.*, 2002; *Marzocchi*, 2002, and references within], producing evidence of tectonic earthquakes that are inferred to have triggered eruptions [e.g., *Walter*, 2007; *Walter and Amelung*, 2007] as well as intrusions that are inferred to have triggered tectonic earthquakes [e.g., *Mori et al.*, 1996; *Baer et al.*, 2008].

[1] *California Volcano Observatory, U.S. Geological Survey, Menlo Park, California, USA*

[2] *Hawaiian Volcano Observatory, U.S. Geological Survey, Hawai'i National Park, Hawaii, USA*

Basaltic shield volcanoes are particularly prone to magma-tectonic interactions owing to their inherent instability due to large mass and weak substratum that promotes spreading, and high magma flux [*Borgia*, 1994; *Borgia et al.*, 2000]. For example, interaction of the magmatic system with flank spreading at Mount Etna, Italy, is frequent and was especially well documented during 2002–2003 [*Acocella et al.*, 2003] when fault slip and magmatic intrusions/eruptions were inferred to trigger one another. *Neri et al.* [2004] discuss the role of the Pernicana Fault in the instability of Etna's eastern flank and suggest the existence of a deep décollement structure to accommodate the flank motion. By analyzing Coulomb stress computations from the 2002–2003 sequence, *Walter et al.* [2005] showed that the rift zone and décollement at Etna presented mutually favorable conditions for a positive-feedback cycle. A similar observation of inflation followed by accelerated flank motion during Etna's 2011 eruptive cycle is reported by *Patanè et al.* [2013].

Hawaiian Volcanoes: From Source to Surface, Geophysical Monograph 208. First Edition.
Edited by Rebecca Carey, Valérie Cayol, Michael Poland, and Dominique Weis.

Companion Website: www.wiley.com/go/Carey/Hawaiian_Volcanoes

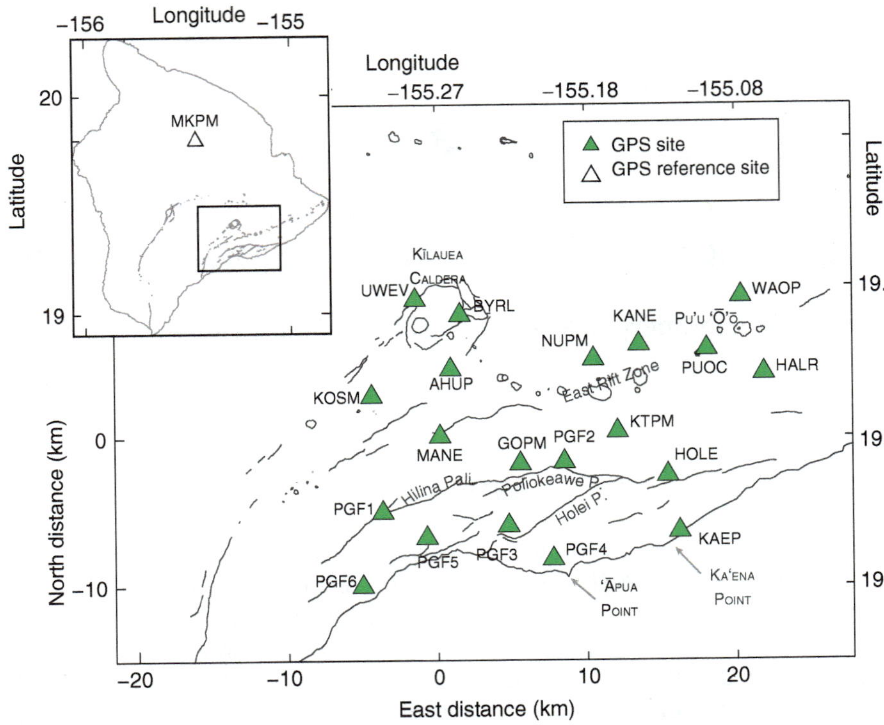

Figure 13.1 Reference map of Kīlauea showing locations of GPS stations (green triangles) and relevant geographic location names.

Like Etna, Kīlauea Volcano, Hawai'i (Figure 13.1), is characterized by rift zones and a décollement that form two boundaries of a mobile volcano flank. The rift zones and décollement interact with one another and promote instability of the volcano's south flank, which moves seaward via steady creep [*Delaney et al.*, 1990; *Owen et al.*, 2000a; *Cayol et al.*, 2000], strong earthquakes [*Lipman et al.*, 1985; *Ando*, 1979; *Árnadóttir et al.*, 1991; *Wyss*, 1988], and slow slip events (SSEs) [*Cervelli et al.*, 2002a; *Brooks et al.*, 2006; *Montgomery-Brown et al.*, 2009]. Improved understanding of the feedback between magmatic and tectonic processes at Kīlauea is critical to better forecasting eruptive and earthquake activity and can be elucidated by an analysis of ground deformation recorded by Kīlauea's continuous global positioning system (GPS) network, which has been in place since 1997. Here, we analyze 11 SSEs recorded by continuous GPS at Kīlauea to explore possible influence of the magmatic system on SSEs as well as the response of the magmatic system to stressing from SSEs.

13.2. KĪLAUEA VOLCANO

Kīlauea Volcano is characterized by magma storage beneath a central caldera, with rift zones radiating to the east and southwest (Figure 13.1). Eruptions are currently ongoing at the summit, in the form of a degassing vent that hosts a lava lake that opened on 19 March 2008 [*Wilson et al.*, 2008] and along the East Rift Zone from the long-lived Pu'u 'Ō'ō vent system, active since January 1983 [*Heliker and Mattox*, 2003]. The active conduit at 2–3 km depth transports magma from the summit reservoir out to the East Rift Zone eruption site [*Ryan*, 1988; *Poland et al.*, 2014]. The rift zone is not liquid throughout its entire depth down to the base of the volcano. The deep rift is likely a mix of cumulates and possibly some interstitial melt [*Ryan*, 1988; *Clague and Denlinger*, 1994; *Plattner et al.*, 2013]. Petrologic models of magma mixing, however, suggest that the rift can store magma for long periods of time in a closed system [*Thornber et al.*, 2003]. It is possible that repeated dike injections and the constant supply of magma out to Pu'u 'Ō'ō help keep Kīlauea's rift zone relatively hot (within a few hundred degrees of the melting temperature of basalt), increasing the solidification time of intruded dikes and in the meantime providing a structural weakness that is vulnerable to stress release.

Kīlauea's southern flank is a type locality for volcano instability, with GPS-measured seaward velocities at the coast of up to 7 cm/yr [*Miklius et al.*, 2005] (e.g., KAEP-MKPM baseline, Figure 13.2d). The most prominent surface featureson the south flank are the cliffs (pali in Hawaiian) that mark the headscarps of a large normal fault system (the Hilina Pali) [*Stearns and MacDonald*,

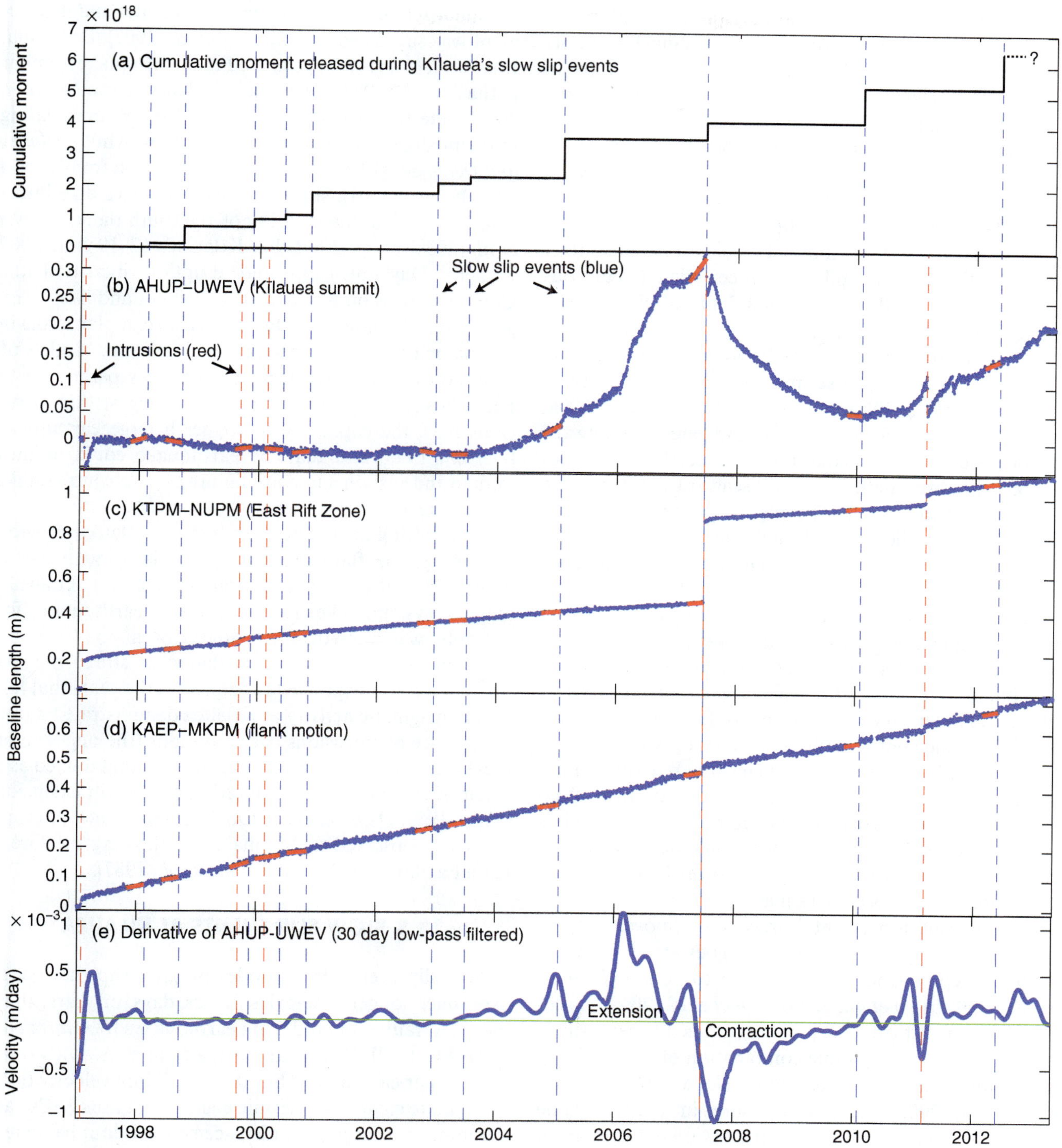

Figure 13.2 (a) Cumulative, geodetically derived moment released during Kīlauea's SSEs. (b–d) Long-term GPS line length time series (blue dots) starting in January 1997 with intrusions (red vertical dashed lines) and SSEs (blue vertical dashed lines) starting in January 1997. Small red line segments indicate the length of data prior to each SSE used to compute the velocities presented in Table 13.1. (e) Derivative of the KTPM-NUPM baseline after a 30 day period low-pass filter was applied. This representation of the relative velocities highlights the two main time spans of extension and contraction and how they relate temporally to the SSEs.

1946; *Moore et al.*, 1994]. The only documented slip at the surface across the largest of the Hilina Pali faults occurred during the 1975 Kalapana earthquake [*Lipman et al.*, 1985; *Denlinger and Okubo*, 1995]. Beneath the flank is a nearly horizontal décollement fault that is inferred to accommodate much of the observed seaward motion via nearly steady slip [*Nakamura*, 1980; *Dieterich*, 1988; *Thurber and Gripp*, 1988; *Delaney et al.*, 1993; *Owen et al.*, 2000a; *Cayol et al.*, 2000].

The décollement is thought to mark the contact between the volcanic pile and preexisting Cretaceous seafloor [*Hill*, 1969]. It occurs at a depth of about 8 km, extends across much of the southern flank, and was the most likely source of the 1975 Kalapana earthquake [*Ando*, 1979]. Previous seismic [*Zucca and Hill*, 1980; *Morgan et al.*, 2000] and geodetic [*Owen et al.*, 2000a; *Cayol et al.*, 2000] models have indicated a landward dipping décollement plane; relocated earthquake hypocenters, however, define a gently seaward dipping plane [*Syracuse et al.*, 2010]. The physical properties of the décollement, while relatively unknown, could have significant influence on the frictional behavior of the fault. The high-impedance contrast observed during a seismic reflection study across a reflective horizon, interpreted as the décollement, suggests a porous, weak layer of pelagic or clastic sediments [*Morgan et al.*, 2000]. As the volcanic pile builds, the overburden will compact these sediments and expel water, possibly allowing for high pore pressures on the décollement as well. Temperature is also important for understanding fault behavior, but other than insight from a few heat flow measurements from deep wells in the lower East Rift Zone [*Kauahikaua*, 1993], it is unknown whether the décollement could be warmer than expected from typical oceanic geotherm estimates because of its proximity to Kīlauea's magma system.

Décollement movement is related to movements of Kīlauea's rift system, which undergoes steady opening between depths from about 3–4 km to 10 km [*Delaney et al.*, 1990; *Owen et al.*, 2000a; *Cayol et al.*, 2000]. Given the effects of rift zone opening on the décollement, slip is likely to be driven by some combination of gravitational spreading due to topographic slope [*Morgan and McGovern*, 2005], magma pressure in the rift zone [*Swanson et al.*, 1976b], and gravitational forces from dense cumulates within the rift system [*Clague and Denlinger*, 1994; *Plattner et al.*, 2013]. The 1975 Kalapana earthquake was forecast on the basis of rift-zone-induced flank compression [*Swanson et al.*, 1976b].

Several authors have proposed and discussed detailed mechanical relationships between flank sliding and Kīlauea's magmatic system. For example, *Dieterich et al.* [2000] used earthquake rates on the south flank as a proxy for stress rates to show that rift expansion is coupled to aseismic décollement fault slip in an area close to the rift. A similar relationship between the rift zones and décollement was suggested by *Amelung et al.* [2007] for Mauna Loa. *Dieterich* [1988] speculated that earthquakes similar to the M_w=7.7 1975 Kalapana event [*Nettles and Ekström*, 2004] were the primary mechanism for accommodating frequent dike intrusion in the rift zones, while *Delaney and Denlinger* [1999], with further discussion from *Parfitt and Peacock* [2001], suggested that Kīlauea's dike intrusions result from the mobility of the south flank. *Brooks et al.* [2008] pointed out that Kīlauea's 17–19 June 2007 Father's Day intrusion and eruption increased the Coulomb stress on Kīlauea's décollement and likely triggered an SSE, indicating that SSEs also can play a role in the magma-tectonic feedback at the volcano. Models of dike-induced pore fluid pressure changes pertaining to Hawaiian-type shield volcanoes also suggest that intrusions have the capability to trigger a large earthquake [*Iverson*, 1995] and that the earthquake could in turn initiate and sustain a massive failure of a volcano's flanks [*Elsworth and Voight*, 1995].

While rifting at Kīlauea can influence flank behavior, conversely, the flank extension associated with earthquakes can also have a profound effect on Kīlauea's magma system. The 1975 Kalapana earthquake, for example, was immediately followed by a small summit eruption as well as summit deflation as stored magma drained into the expanded rift zones. In the years that followed, magmatic activity at Kīlauea demonstrated a predominance of intrusions over eruptions (the opposite of pre-1975 behavior) as magma filled areas that dilated as a result of the earthquake-induced flank and rift extension [*Klein*, 1982; *Dzurisin et al.*, 1984]. A similar lack of eruptions was observed at Mauna Loa following the Great Ka'ū Earthquake of 1868 [*Tilling et al.*, 1987].

13.3. SLOW SLIP EVENTS AT KĪLAUEA

Slow slip events are episodes of fault slip that occur over much longer timescales (hours, days, or years) than the rapid fault slip producing earthquakes [*Schwartz and Rokosky*, 2007]. During an SSE, a fault slips as it would in a regular earthquake, but the low sliding velocity does not radiate significant seismic energy, and thus SSEs can accommodate large fault displacement without being felt at the surface. Classic type-locations for observing SSEs are in subduction zones (e.g., *Southwest Japan and Cascadia* [*Obara*, 2002; *Dragert et al.*, 2001]). SSEs are also similar to creep events that have been observed on the San Andreas Fault for many years [*Louie et al.*, 1985; *Sieh and Williams*, 1990; *Linde et al.*, 1996; *Wei et al.*, 2009].

We identified 11 SSEs on Kīlauea's southern flank between 1997 and 2013 (Figure 13.2 and Table 13.1). Kīlauea's SSEs typically last about two days and each can

Table 13.1 Slow slip events at Kīlauea between 1997 and 2012.

						Interevent	Magma System State[a]	
ID	Year	Month	Day	Mag.	References	Time (yr)	AHUP-UWEV	KTPM-NUPM
A	1998	02	20	5.3	2,4	—	0.031	0.038
B	1998	09	19	5.7	1,2,4	0.57	−0.018	0.036
C	1999	11	21	5.5	2,4	1.17	0.014	0.117
D	2000	05	29	5.3	4	0.52	−0.007	0.040
E	2000	11	09	5.8	1,2,3,4	0.44	0.009	0.034
F	2002	12	17	5.6	2,4	2.10	−0.018	0.011
G	2003	07	03	5.4	1,2,4	0.54	0.002	0.017
H	2005	01	26	6.0	1,2,4	1.56	0.059	0.024
I	2007	06	18	5.7	4,5	2.39	0.148	0.025
J	2010	02	01	5.9	6	2.62	−0.006	0.029
K	2012	05	28	6.1	Here	2.32	0.044	0.034

Sources: (1) *Segall et al.* [2006]; (2) *Brooks et al.* [2006]; (3) *Cervelli et al.* [2002a]; (4) *Montgomery-Brown et al.* [2009]; (5) *Montgomery-Brown et al.* [2010]; (6) *Poland et al.* [2010].

[a] Magma system state is roughly represented by proxy measurements: The velocity of line length change between each of two pairs of GPS stations (AHUP-UWEV and KTPM-NUPM) for the 90 days prior to each SSE. Positive velocities correspond to extension (usually indicating inflation), negative velocities correspond to contraction (usually indicating deflation). Secular flank motion also contributes to extension of these baselines, but such motion is small in these 90 day windows. These time periods are noted by the red bars on the time series in Figure 13.2.

release an equivalent amount of moment [*Kanamori*, 1977] as a regular tectonic earthquake with magnitudes ranging from $M_w = 5.3$ to $M_w = 6.1 \pm 0.6$. Fault slip amplitudes are typically tens of centimeters and result in slip velocities of about 10^{-6} m/s. While recurrence intervals of Kīlauea's SSEs are highly variable early in the sequence, the most recent events have recurrence intervals of 2.44 ± 0.15 years (Figure 13.2).

Slow slip events at Kīlauea and in most subduction zones likely occur in the transition zone between a locked and a creeping portion of their respective faults (e.g., Cascadia [*Rogers and Dragert*, 2003; *Wech et al.*, 2009]). Although subduction zones are clearly a different tectonic regime and the SSEs differ by location and depth, there are many SSEs with similar durations, magnitudes, and recurrence intervals [*Ide et al.*, 2007; *Schwartz and Rokosky*, 2007] to which we can look to help understand the mechanisms of slow slip at Kīlauea.

For example, there were five "short-term" SSEs in the Bungo Channel, Japan [*Hirose and Obara*, 2005], in 2003–2004 with 0.79–2.64 cm of slip (M_w ranging from 5.8 to 6.2). Bungo Channel short-term SSEs are about the same size as Kīlauea's, but they have smaller slip amplitudes and occur more irregularly yet more frequently than Kīlauea's SSEs. In Cascadia, SSEs typically last longer than at Kīlauea (about two weeks) and occur relatively regularly with magnitudes ranging from $M_w = 6.1$ to $M_w = 6.7$ and maximum slip amplitudes of 1–5 cm [*Schmidt and Gao*, 2010]. These SSEs have recurrence intervals about 14 months in the north, 10 months in the mid-subduction zone, and 19 months in the southern end [*Brudzinski and Allen*, 2007]. Given the similarities between SSEs at Kīlauea and those in subduction zones, which also host large megathrust earthquakes, it is tempting to look to Kīlauea's SSEs for insight into the next large flank earthquake. Also, since Kīlauea's flank is much more accessible to measurement than the updip extent of subduction faults that are typically beneath the ocean, Kīlauea may provide some insights for shallow subduction zones.

Numerous numerical models have been able to produce SSEs in a variety of fault conditions. *Shibazaki and Iio* [2003] and *Shibazaki and Shimamoto* [2007] suggest a change in frictional behavior related to increasing slip speeds. *Liu and Rice* [2005, 2007] show that oscillatory behavior near neutral stability can also produce SSEs in numerical simulations, as can low effective normal stresses (likely resulting from high pore pressure) within slow slip zones. Continuing to explore the effects of friction, *Perfettini and Ampuero* [2008] show that external stress perturbations can induce slow slip in regions of steady state velocity-strengthening friction. Numerical simulations that include pore-pressure-dilatancy coupling have shown that a SSE can accelerate into a large earthquake [*Segall et al.*, 2010; *Segall and Bradley*, 2012], although the SSE immediately preceding the dynamic rupture is indistinguishable from other SSEs.

The $M_w = 9.0$ Tohoku earthquake provides a field observation of slow slip preceding a subduction zone earthquake. It was preceded by two slow fault slip events manifested as migrating swarms of small earthquakes [*Kato et al.*, 2012], the second of which, involving large slip rates, began about two days before the main shock. The large slip rates observed during this SSE may have caused substantial stressing near the future main shock epicenter and prompted the rupture [*Kato et al.*, 2012].

Assuming the mechanics of SSEs are related to fault physics and not tectonic regime, it is possible to speculate that at some future time an SSE may accelerate into a large earthquake at Kīlauea. Such events may also be associated with collapse of the volcano's south flank, as suggested by motion of both the Hilina faults (considered to be an example of active slumping [*Moore et al.*, 1994]) and the décollement during the 1975 Kalapana earthquake. SSEs could therefore provide insights that might lead to a better understanding of the potential for sector collapse at Kīlauea.

A majority of SSEs in subduction zones are correlated with swarms of tectonic tremor (also referred to as nonvolcanic tremor). Tectonic tremor is a persistent, low-amplitude seismic signal with dominant frequencies of 2–8 Hz [*Obara*, 2002]. Tectonic tremor episodes may occur at any time but are more frequent during episodes of slow slip [*Rogers and Dragert*, 2003]. The phenomenon of correlated episodic slow slip and tremor (ETS) was originally described in Cascadia and Japan [*Rogers and Dragert*, 2003; *Obara*, 2002; *Schwartz and Rokosky*, 2007]. Tremor has not yet been observed on the décollement surface during Kīlauea's SSEs [*Montgomery-Brown et al.*, 2013], although the SSEs are inferred to have triggered swarms of small earthquakes that follow a temporal pattern consistent with that of an aftershock sequence [*Segall et al.*, 2006].

13.4. MODELING KĪLAUEA'S SSES

SSEs at Kīlauea have been modeled in several studies [*Cervelli et al.*, 2002a; *Brooks et al.*, 2006; *Segall et al.*, 2006; *Montgomery-Brown et al.*, 2009]. Here, we compile the results from *Montgomery-Brown et al.* [2009] for events through 2005 and *Montgomery-Brown et al.* [2010] for the 2007 SSE and intrusion. We then apply the same methods to the events in 2010 and 2012. All models presented here are kinematic and use planar dislocations for fault slip and dike opening [*Okada*, 1985]. In the 2007 model, a summit reservoir is also included as a spherical point source [*Mogi*, 1958].

We invert for fault slip on a mesh of 2 km^2 dislocations [*Okada*, 1985] using a least squares algorithm that enforces positivity. The positivity constraint requires slip to be in the seward direction. Layered elastic structure is approximated using four parallel layers with varying shear moduli: (1) shallow volcanic crust (0–3 km, $\mu = 24$ GPa), (2) volcanic midcrust (3–9 km, $\mu = 40$ GPa), (3) Cretaceous sea floor (9–20 km, $\mu = 60$ GPa), and (4) upper mantle (>20 km, $\mu = 80$ GPa). Poisson's ratio is assumed to be 0.25 for all layers. The propagator matrix method provides the analytical solution for a horizontally layered elastic half-space [*Segall*, 2010]. To account for topographic effects, a first-order correction is computed from a uniform half-space and added to the layered elastic Green's function [*Williams and Wadge*, 2000]. Each model is smoothed with a Laplacian operator with a weight chosen by the L-curve criterion [*Hansen*, 1992].

Despite recent seismic evidence for a seaward dip of the décollement plane [*Syracuse et al.*, 2010], we follow previous geodetic and active seismic studies in using a slight (5°) landward dip [*Owen et al.*, 2000a; *Cayol et al.*, 2000; *Morgan et al.*, 2000; *Brooks et al.*, 2006; *Montgomery-Brown et al.*, 2009], because a small amount of dip either landward or seaward does not significantly influence our conclusions. During the 2010 and 2012 SSEs we also allow for the possibility of dike opening from near the surface to a depth of 8 km. Since there was no eruption observed during either event, the dike model is smoothly forced to zero at the surface.

Slip models for each event are presented in Figure 13.3. Moment magnitudes (M_w) in Table 13.1 are computed by integrating the slip on each fault patch over the total slip area. The magnitudes of the events increase with more recent events. In all cases, the cumulative seismic moment released by the triggered earthquakes is at least an order of magnitude less than the geodetically derived moment for the same slip event. Small and irregularly timed SSEs occurred early in the sequence of events (Figure 13.2a). Since 2005, however, Kīlauea's SSEs have been more regular (Figure 13.2a), and the smaller events have disappeared. This does not appear to be an effect of changing instrumentation, since the GPS network has remained largely the same throughout the sequence. A continuation of the match-filter search presented by *Montgomery-Brown et al.* [2009] and using the same set of GPS stations failed to find any small SSEs after 2005, which is a period characterized by well-defined events in the GPS time series (2007, 2010, and 2012).

13.4.1. Summary of 1998–2005 SSEs

Kīlauea's SSEs from 1998 to 2005 were variable in size, recurrence interval, and spatial extent (Figure 13.3). We define a characteristic slip area for Kilauea's SSEs based on the cleanest observations with the best signal-to-noise ratio. For example, the largest SSEs in September 1998 (event B), November 2000 (E), and January 2005 (H) slipped the décollement across the characteristic area. SSEs in November 1999 (C) and July 2003 (G) primarily slipped on the eastern half of the characteristic slip area, while one event in December 2002 (F) slipped the western half. These events were thus categorized by *Foster et al.* [2006, 2013] as eastern and western families of SSEs. The SSEs in December 2002 (F) and July 2003 (G) (Figure 13.3)

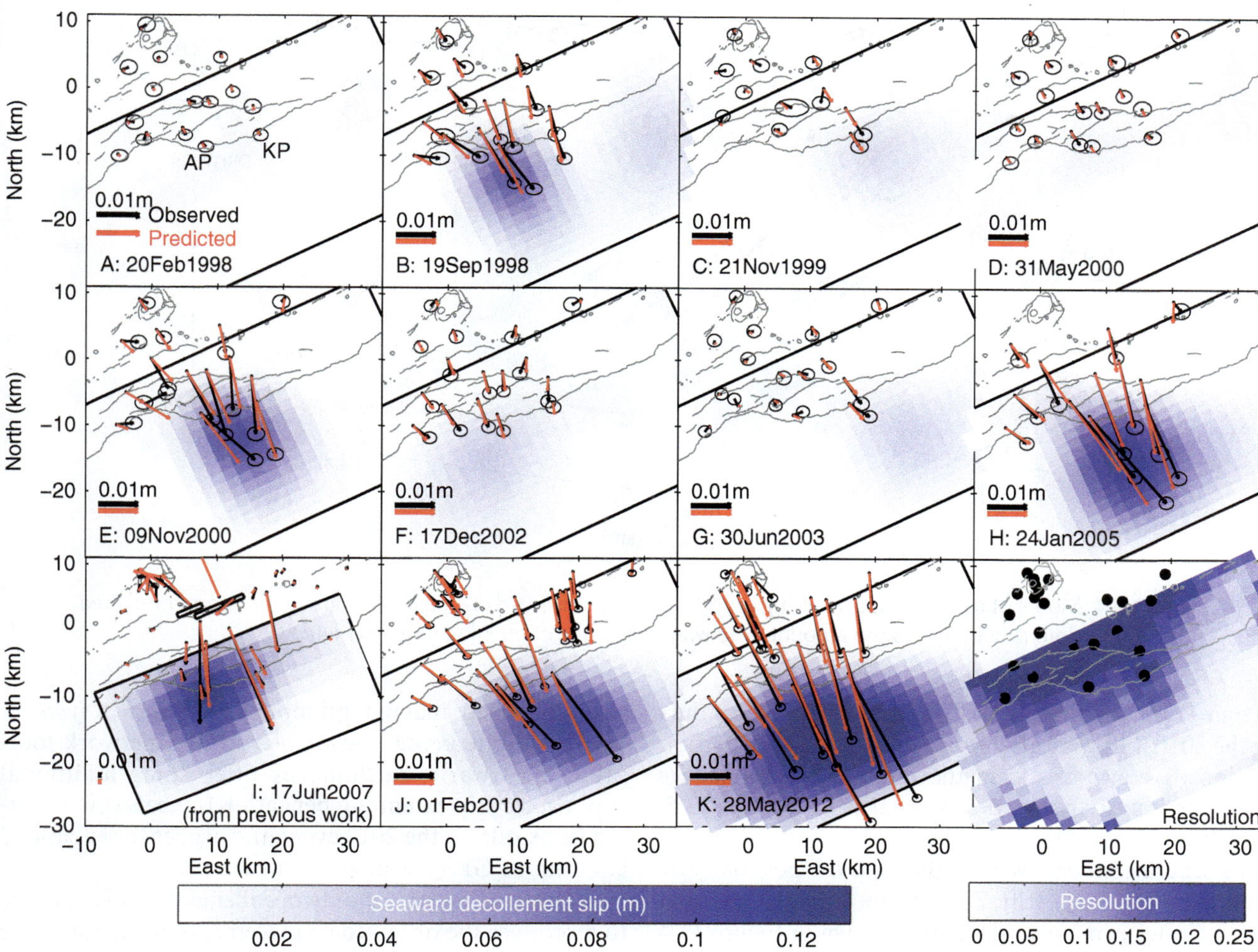

Figure 13.3 Slip models of all Kīlauea SSEs. Observed displacement vectors (black) with 2σ error ellipses and red displacement vectors predicted by the model are presented at the same scale for all events except 2007, which has been reproduced here from previous work [*Montgomery-Brown et al.*, 2010]. Color intensity shows the amount of seaward décollement fault slip (also note this diffierence for the 2007 event). The larger black rectangle shows the extent of décollement slip allowed in the inversion. Deeper color intensity in the model resolution indicates better resolved parts of the fault from the diagonal of the model resolution matrix [*Aster et al.*, 2012, Section 3.2].

formed a space-time pair (i.e., they were close together in time and slipped complimentary parts of the fault that summed roughly to a characteristic event). Although not as clearly complimentary, the November 1999 (C) and May2000 (D) events could also be a pair.

13.4.2. 2007 Intrusion/SSE Summary

The 2007 intrusion and concurrent SSE (event I) were reported by *Brooks et al.* [2008] and further analyzed by *Montgomery-Brown et al.* [2010, 2011]. The sequence of events involving the 2007 SSE began with an intrusion in Kīlauea's upper East Rift Zone on 17 June 2007 (the "Father's Day" holiday in the United States). The initial intrusion was followed about 6 h later by the onset of slow slip on the south flank. Despite the intrusion, the slow slip appeared to be of a typical size and duration with a maximum of about 10 cm of fault slip. Fault slip is, however, less well resolved during this event due to the nearly overwhelming intrusion displacements. The slow slip in turn facilitated additional rift opening and dike intrusion over the subsequent two days, resulting in a maximum of 2.4 m of opening between Makaopuhi and Nāpau Craters.

13.4.3. 2010 SSE Model Results

A model of the 2010 SSE (event J) was also presented briefly for context in a search for tectonic tremor accompanying Kīlauea's SSEs [*Montgomery-Brown et al.*, 2013]. The maximum fault slip that occurred during the 2010 event is typical for Kīlauea's SSEs with about 10±1.5 cm of seaward slip occurring about 10 km offshore, south of

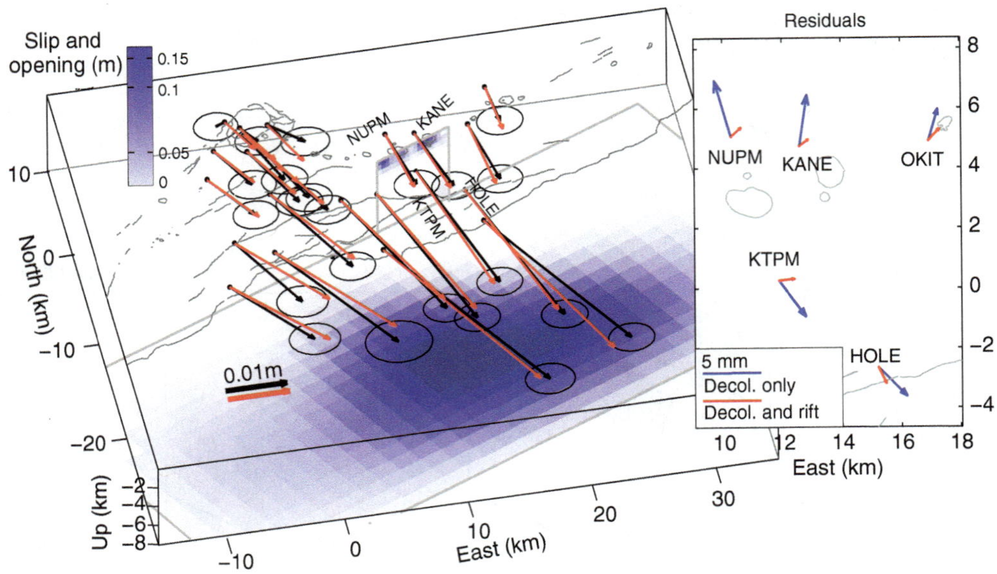

Figure 13.4 Model of the 2012 SSE allowing for rift opening. Observed displacements are in black and predicted in red. Color intensity indicates the amount of seaward slip and rift opening (both in meters).

Ka'ena Point (Figure 13.3). Triggered earthquakes during the 2010 SSE were also typical in size and distribution and mostly concentrated in the Poliokeawe cluster (see the discussion of triggered earthquakes in Section 13.5. and Figure 13.6).

In a separate model, we also allow for the possibility of opening in the East Rift Zone from Pauahi Crater to Nāpau Crater (similar to the model shown in Figure 13.4 for 2012). Such a model only permits about 3 mm of opening, suggesting that rift opening was not an important contribution to the deformation during the 2010 SSE.

13.4.4. 2012 SSE Model Results

For the 2012 SSE (the largest at the time of this writing, event K), the maximum amount of slip was 16±1.5 cm occurring about 10 km offshore and to the south of 'Āpua Point (Figure 13.3). As in previous events, the 2012 SSE triggered a small swarm of earthquakes in the Poliokeawe cluster. There is a suggestion in the geodetic model of the 2012 SSE (Figure 13.3) that slip may have extended farther west than in any previous event. While resolution from the continuous GPS network that far west is limited, and neither campaign GPS nor interferometric synthetic aperture radar (InSAR) has been able to resolve significant SSE displacements in 2012 (2010 SSE observations, however, were presented by *Chen et al.,* [2014]), a small swarm of shallow (0–5 km) earthquakes occurred in that region at the time of the 2012 SSE, a phenomenon not observed during any previous Kīlauea SSE. Previous earthquake swarms in this western area are usually deeper and possibly associated with the deep magma plumbing [*Aki and Koyanagi,* 1981]. Despite being outside of the GPS network footprint, the swarm earthquakes offer some additional support for slip extending beneath Kīlauea's southwestern flank during the 2012 event and argues for expanded continuous GPS coverage in this area.

A model allowing for only décollement slip is not able to reproduce the observed displacements at rift zone sites KTPM, NUPM, KANE, OKIT, and HOLE (e.g., 16 mm of extension (3.2 microstrain) between KTPM and NUPM, Figure 13.4, inset). Residual displacement vectors of about 5 mm northward at NUPM and KANE and 3.5 mm southward at KTPM suggest additional extension across the rift zone. In the continuum represented by the linear elastic model of *Okada* [1985], a small amount (~9 mm) of extension would be expected between KTPM and NUPM as a result of flank slip. After modeling flank slip, however, the residual horizontal line length change between KTPM and NUPM is still about 7 mm (1.3 microstrain), which is much larger than can be accounted for by observation errors (Figure 13.4). Thus, to account for the inability to fit the rift zone stations with décollement fault slip alone, we introduce a nearly vertical dike parallel to the strike of the rift zone extending laterally from Makaopuhi Crater to Nāpau Crater, a segment of the East Rift Zone that has been frequently and recently active [*Owen et al.*, 2000b; *Cervelli et al.*, 2002b; *Montgomery-Brown et al.*, 2010; *Lundgren et al.*, 2013]. Given the limited constraints on the geometry from the distribution of continuous GPS stations near this location, we could also have incorporated the possible magma reservoir beneath Makaopuhi Crater, suggested by *Swanson et al.* [1976a] and modeled by *Owen et al.* [2000b] among other studies, into this model. With the limited constraints on the geometry of that reservoir, however, results would have been similar to the planar model we use.

The change in distance between station pairs KTPM-NUPM, and KANE-HOLE during the 2012 SSE can be accommodated by about 2.6±1 cm of shallow opening within the rift zone (Figure 13.4), extending from near the surface to about 2 km deep. This corresponds to a total volume change of 0.13×10^6 m^3. In contrast, during the 2010 SSE, fewer than 3 mm of opening is allowable by the minimal extension between KTPM and NUPM. Because of the significant increase in the number of free model parameters, a quantitative comparison of the misfits could be misleading (more free parameters will always result in a better fit). Instead, we simply make a qualitative assessment that the rift zone GPS stations are much better fit during 2012 when a small amount of rift opening is allowed and the data are sufficiently dense to determine that, in some cases (e.g., 2010), rift opening likely did not occur during the SSE.

13.4.5. Observations from the Sequence

The recent SSEs at Kīlauea (i.e., since 2005) are among the most regular observed worldwide. Regular recurrence could suggest, for example, that the loading rate is constant. We can test the time-dependence of Kīlauea's SSEs by utilizing models for probabilistic earthquake forecasting. Two common models are the time-predictable [*Shimazaki and Nakata*, 1980] and slip-predictable [*Bufe et al.*, 1977] models. The time-predictable model suggests that the next earthquake will occur at the time when a fault recovers the stress that was relieved during the previous earthquake. Conversely, the slip-predictable model suggests that the amount of slip expected in the next earthquake will be equal to the amount of slip deficit accumulated in the elapsed time since the previous earthquake. Early in Kīlauea's GPS-observed SSE sequence (prior to 2005) the frequent, smaller SSEs make it difficult to assess whether the sequence conforms to either time- or slip-predictable occurrence. The mean interevent time for the sequence is 1.42±0.77 years (0.99±0.41 until 2003 and 2.44±0.02

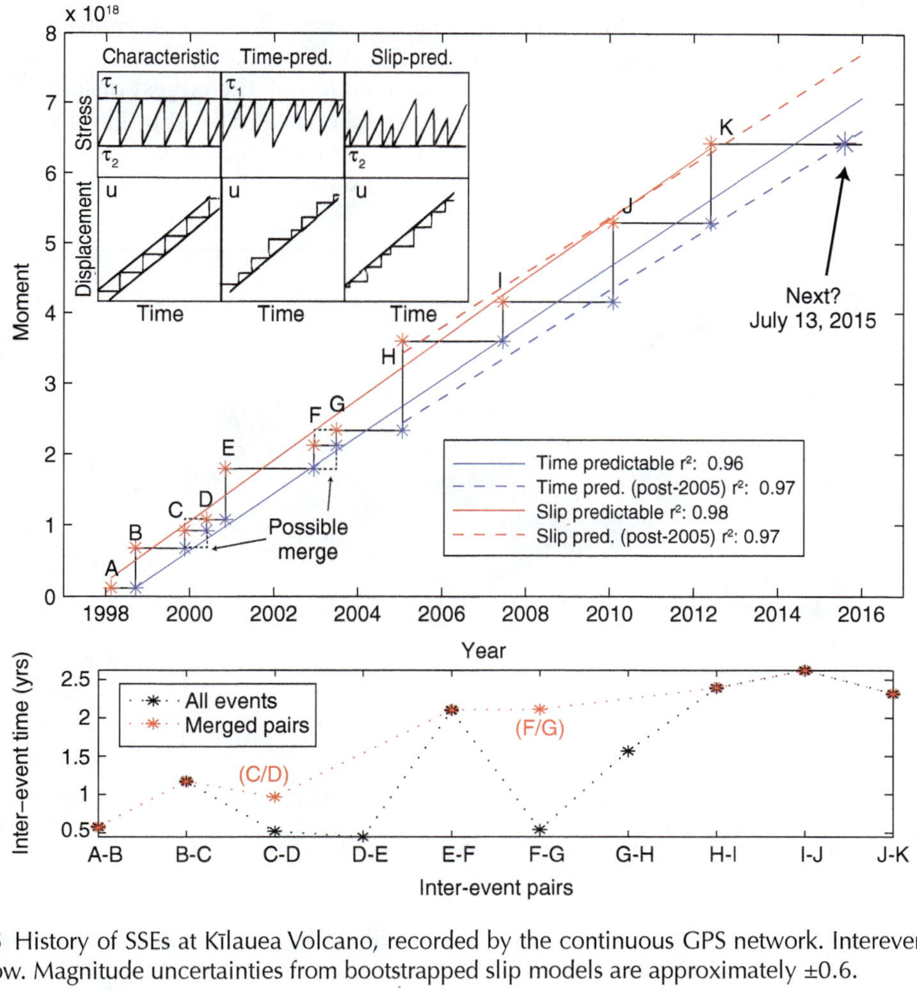

Figure 13.5 History of SSEs at Kīlauea Volcano, recorded by the continuous GPS network. Interevent times are shown below. Magnitude uncertainties from bootstrapped slip models are approximately ±0.6.

for 2005 and later). Since the smaller events disappeared after 2005, the size and interevent times have been much more regular.

The time- and slip-predictable models are not any more applicable when considering the eastern and western families [*Foster et al.*, 2006, 2013] of events separately. If, however, we merge the complimentary pairs of SSEs (F/G and C/D) we can again test the time- and slip-predictable models with the merged pairs acting as a single characteristic event (Figure 13.5). There are several ways to approach merging the pairs. We can decide that the start of the event is the important measurement and merge the slip from both events at the time of the first event in the pair. This approach favors a better fit to the slip-predictable model. Alternatively, we could decide that the finish of the event is a complete release of the built-up stress and merge the slip from both events at the time of the second event in the pair. This approach favors the time-predictable model. A third approach is to use an average time, but this has no physical basis. Considering the complimentary pairs of events as a merged event suggests that the sequence is more like a repeating characteristic event with minor deviations. Merging the complimentary pairs using the first SSE in the pair as the merged event time results in a longer interevent time with smaller variance (1.78±0.58 years). Using the second event in the pair as the event time results in the same interevent time (1.78±0.77 years) but a larger variance similar to that of the complete sequence.

13.5. SLOW SLIP AND EARTHQUAKES AT KĪLAUEA

Here we consider two ways in which SSEs at Kīlauea relate to earthquakes: (1) by triggering swarms of small earthquakes and (2) by contributing to the potential for large flank earthquakes. Since Kīlauea's SSEs are known to trigger small earthquakes on the décollement [*Brooks et al.*, 2006; *Segall et al.*, 2006; *Wolfe et al.*, 2007; *Syracuse et al.*, 2010], the potential exists for one of these triggered décollement earthquakes to also grow into a destructive earthquake.

Each of Kīlauea's SSEs are accompanied by swarms of 10–50 small flank earthquakes, with the largest magnitudes being about $M_w = 3.5$. *Segall et al.* [2006] showed that these earthquakes followed a temporal pattern consistent with an aftershock sequence, and *Wolfe et al.* [2007] demonstrated that they occur on streaks that are aligned with the SSE displacement direction. The SSE seismic swarms typically follow the onset of SSE deformation by about a day, based on SSE start times determined from slip models of the characteristic events in 1998, 2000, 2005, and 2010 derived from 30 s GPS data as well as tilt records collected since 2007 [*Montgomery-Brown et al.*, 2009, 2013]. A tiltmeter at Ka'ena Point was installed near the location of the largest on-land tilts predicted from a model of 2005 SSE slip. The 1 min recording interval of the station provides good temporal resolution of slip onset when filtered to remove diurnal temperature

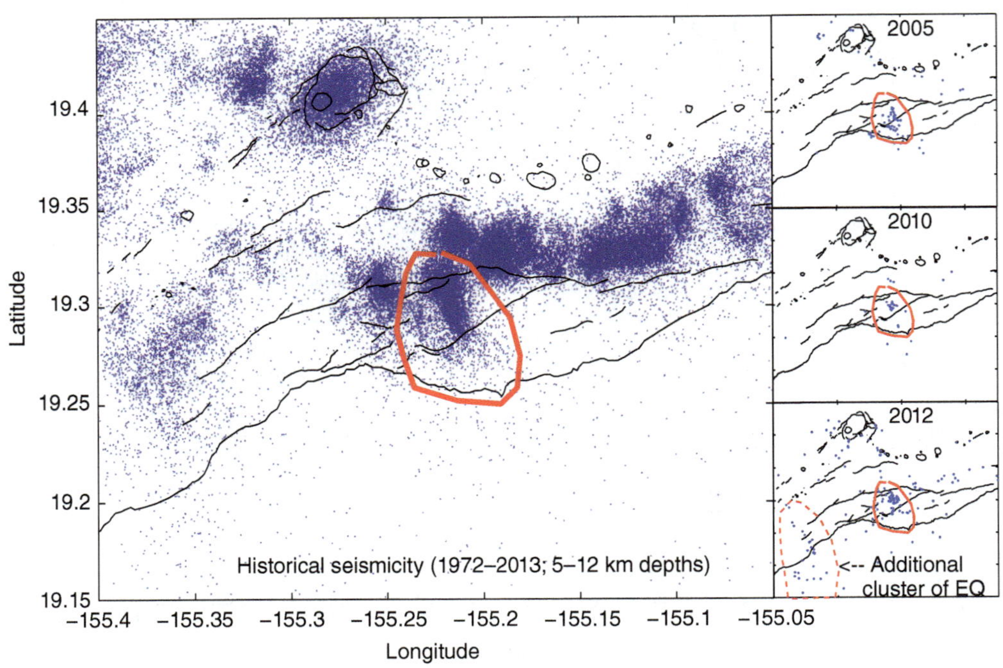

Figure 13.6 Map of 5–12 km deep earthquakes from 1972 to 2013. The area outlined in red (referred to here as the Poliokeawe cluster) is frequently activated during SSEs.

variations [*Montgomery-Brown et al.*, 2010]. For the 2012 SSE, triggered seismicity followed the onset of tilting by about one day. Given the consistent relation between SSE slip and triggered seismicity, it may be possible to use these swarms of flank earthquakes to search for SSEs farther back than the continuous GPS network allows.

The inferred triggering relations between Kīlauea's SSEs and swarms of flank microearthquakes was established by *Segall et al.* [2006] using earthquakes located in the box bounded by longitude −155.2551 and −155.0296 and latitudes 19.2264 and 19.3538 (Figure 13.6). We focus on a smaller area (referred to here as the Poliokeawe cluster) outlined in red in Figure 13.6, and we plot the earthquakes through time in Figure 13.7, south of Poliokeawe Pali and just north of 'Āpua Point. While active at other times, a swarm of small earthquakes occurs consistently in this area during Kīlauea's SSEs [*Montgomery-Brown et al.*, 2009]. While SSEs were not detected at Kīlauea prior to the installation of the continuous GPS network in 1997, *Wright and Klein* [2014] suggest that earthquake swarms occurred frequently in the past on the décollement in the area that is activated during SSEs. We therefore use this cluster of earthquakes to look for potential SSEs prior to the installation of Kīlauea's permanent continuous GPS network.

Typically, the Poliokeawe area as shown in Figure 13.6 sees about one earthquake every three days (0.35 earthquakes per day with a standard deviation of 0.9); thus, we flag days with more than four earthquakes per day (about four standard deviations) to be possibly associated with a SSE. For example, during the era of continuous GPS measurements when we have been able to identify SSEs geodetically, all of Kīlauea's large characteristic SSEs are associated with a Poliokeawe swarm ranging from a few earthquakes during the smaller SSEs to over 50 earthquakes during the 2005 SSE. Events that are very small (e.g., event A) or could be considered members of the "eastern" family (e.g., events C and G) do not show significant earthquake activity in the Poliokeawe area, although they do trigger earthquakes elsewhere that are less clustered and therefore harder to distinguish in space and time over this long of a time period.

During the same period of time as the continuous GPS coverage, there were nine additional days where the Poliokeawe cluster has had more than four earthquakes in a day but no slow earthquake occurred (Figure 13.7). The additional days where the Poliokeawe cluster was active did not coincide with nearby $M > 5$ earthquakes and thus do not likely represent aftershocks of regular earthquakes. Therefore, while earthquakes in the Poliokeawe area are not a perfect indicator of slow slip, since that area of the south flank appears to also respond to other unknown sources of stress, increased seismic activity in the Poliokeawe cluster is suggestive of an SSE.

Since the 1970s, Poliokeawe swarms have recurred frequently and with a similar regularity to the known SSEs, regardless of the background rate of seismicity (Figure 13.7). For example, the rate of earthquake occurrence in the area was quite high during 1982–1994, perhaps due to the emplacement and evolution of the dike feeding the currently ongoing East Rift Zone eruption [*Heliker and Mattox*, 2003]. The number of Poliokeawe swarms that meet our criteria during this time period, however, occur no more frequently or infrequently than at other times. The persistence of isolated swarms of earthquakes in the Poliokeawe cluster throughout time regardless of other magmatic or tectonic activity suggests that the swarms may warrant further consideration as times of potential past SSEs, but that would need to be corroborated with additional evidence.

13.6. SLOW SLIP AND KĪLAUEA'S MAGMA SYSTEM

Of the 11 SSEs recorded by GPS since 1997, at least 3 have been directly associated with magmatic activity at Kīlauea's summit or rift zones or both. The possibility of opening of the rift zone during an SSE is supported by an analysis of the stresses induced by slow slip, but magma system effects are not routinely seen during SSEs. This inconsistency suggests that the state of stress within the magma system is variable and may only be influenced by SSEs when it is already close to failure. During each SSE, the compressive stresses are reduced on nearly vertical planes near Kīlauea's rift zones (Figure 13.8). The magnitude of the stress change is small (0.1–0.3 MPa) but exceeds the tens of kilopascals suggested as a minimum limit needed for triggering earthquakes [*Stein*, 1999; *Hill*, 2010]. Analyzing the stresses with the source and receiver reversed, décollement slip is encouraged toward failure by rift opening. The resulting change in Coulomb failure stress ($\Delta CFS = \Delta\tau + \mu(\Delta\sigma_n + \Delta P)$ [*Stein*, 1999]) on the décollement is about 0.1–0.3 MPa for typical intrusions with about 2 m of opening in the upper few kilometers of the rift zone (e.g., as occurred in 2007).

13.6.1. January 2005: Summit Deformation and East Rift Zone Magma Effusion Associated With Slow Slip

Transient summit deformation and a surge in magma effusion may have been associated with a south flank SSE in January 2005. On January 25 at approximately 11 h

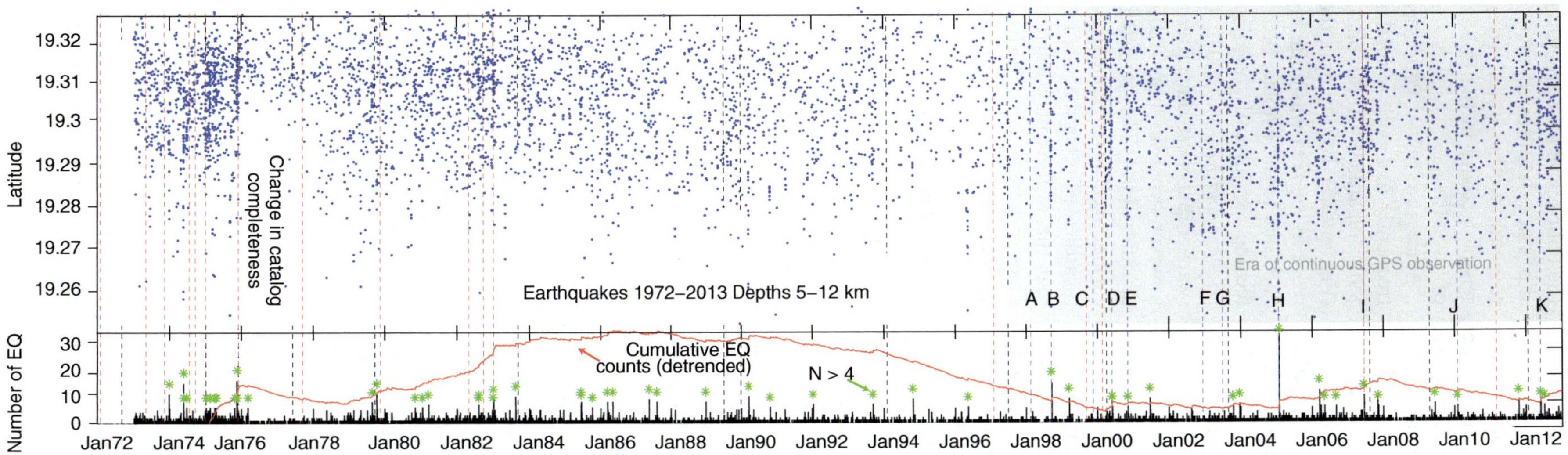

Figure 13.7 Time series and histogram of 5–12 km deep earthquakes from 1972 to 2013. Vertical dashed lines indicate eruptions/intrusions (red, list modified for all Kīlauea eruptions by HVO from *Macdonald et al.* [1986]), catalog *Mw* >5 earthquakes (black), and SSEs (blue). Gray shading indicates the era of continuous GPS observations and thus the time span through which we can identify SSEs geodetically. Green stars indicate days when there were more than four earthquakes in the Poliokeawe cluster. A steady rate of 120 earthquakes per year has been removed from the cumulative counts to highlight the deviations from background at this scale. Note that following the 1975 Kalapana earthquake, the completeness level of the catalog was temporarily raised to 2.6 because of the overwhelming aftershock activity (Fred Klein, personal communication, 6 May 2013). This accounts for the apparent lack of flank earthquakes for about two years after the Kalapana earthquake.

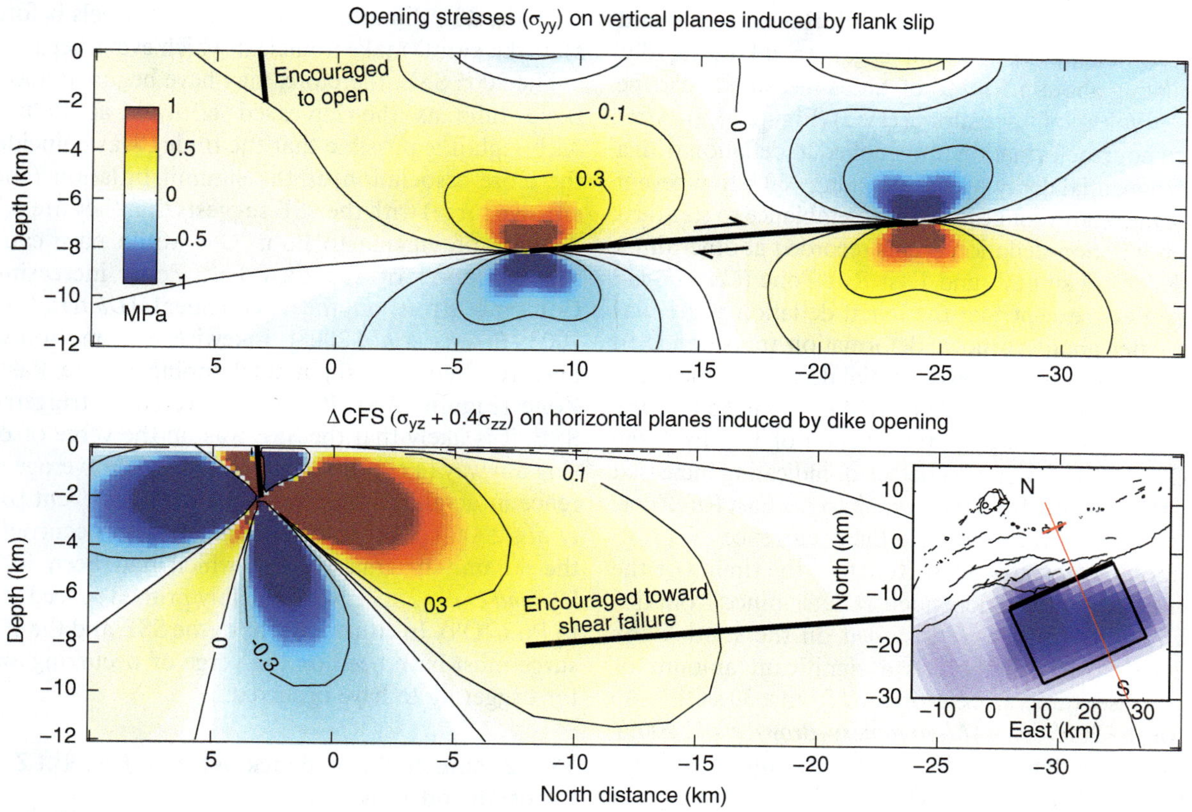

Figure 13.8 Top: Example opening stress change induced on nearly vertical planes by décollement slip similar to that which occurs during SSEs. Bottom: Coulomb stress change [*Stein*, 1999] on nearly horizontal planes induced by dike opening similar to that which occurred during the 2007 Father's Day intrusive events. Note that the sign convention on the top frame was flipped such that warm colors indicate an increased likelihood of failure in both frames.

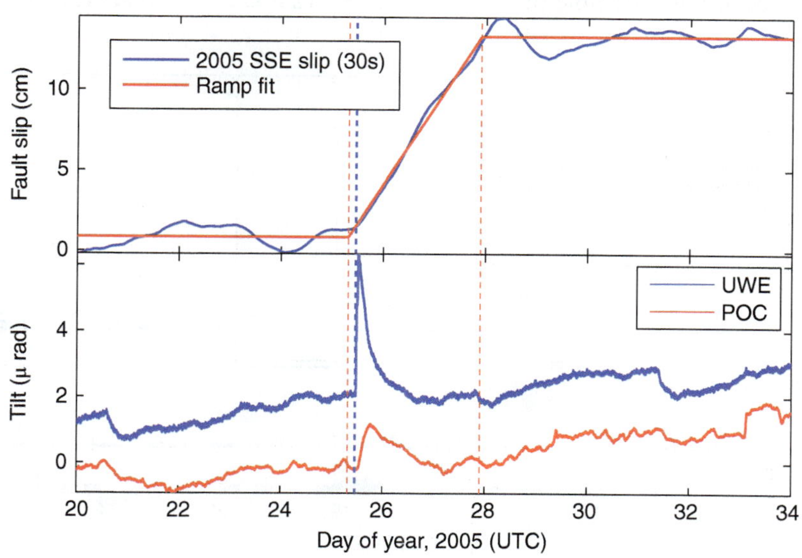

Figure 13.9 Slip model derived from 30 s GPS data (top) from Montgomery-Brown et al. [2009] and radial tilts (bottom) from Kīlauea's summit (UWE, 314°) and Pu'u 'Ō'ō vent (POC, 322°) demonstrating the coincidence in time of slow slip and summitination. Red dashed lines indicate the onset and termination of the SSE from the GPS model determined by the fit of a ramp function. Blue dashed lines indicate the onset of tilt at the summit tiltmeter UWE.

coordinated universal time (UTC) (01 h Hawaiian standard time (HST)), Kīlauea's summit began to inflate rapidly, accumulating about 5.5 μ rad of tilt as measured near the Hawaiian Volcano Observatory (HVO) (Figure 13.9). After about an hour, deformation transitioned to deflation with a quasi-exponential decay over the ensuing ~30 h to preevent levels of tilt. The event bears some resemblance to transient deflation-inflation-deflation events recorded at other times at the Kīlauea's summit and East Rift Zone [*Cervelli and Miklius*, 2003], except that the initial deflation phase was absent. After about an hour, deformation transitioned to deformation with a quasi-exponential decay over the ensuing 30 h to preevent levels of tilt. Within about 2.5 h of the summit inflation, a surge in the effusion of lava from the Pu'u 'Ō 'ō eruptive vent was detected, indicating increased transport of magma from the summit to the East Rift Zone.

Daily GPS positions indicate the occurrence of a SSE on or about 26 January. Unfortunately, the timing of the SSE onset cannot be independently determined from tiltmeters because none were installed on the south flank until 2007. Although there is a significant amount of noise, an onset time can be estimated from a 30 s GPS slip model of the 2005 event [*Montgomery-Brown et al.*, 2009], which suggests that the SSE began first on 25 January 2005 at about 07 h 30 min UTC (21 h 30 mins HST on 24 January). It is also instructive to examine seismicity in the area excited by SSEs at other times, as described above. A seismic swarm associated with the SSE began at about 12 h UTC (02 h HST) on 26 January, a little more than 24 h after the start of the summit tilt transient. Given that the East Rift Zone effusive surge and SSE occur very close in time, the large uncertainties in the start time of the SSE from the 30 s slip model (note the noise levels before and after the event) make it unclear which event began first.

The 2005 SSE, therefore, may have begun at about the same time as the increased effusion at Pu'u 'Ō'ō. Although it is possible that the timing was coincidental, the close association of the summit inflation (and the effusive surge) with the SSE suggests that they are related. A surge of magma to Pu'u 'Ō'ō would have caused a slight dilation of the East Rift Zone, increasing the Coulomb stress on the décollement [*Dieterich et al.*, 2003; *Brooks et al.*, 2008]. Likewise, SSE motion would have resulted in a slight unclamping of the East Rift Zone (Figure 13.8). If the effusive surge triggered the SSE, it is likely that the SSE was on the verge of occurring anyway (a condition suggested by the average recurrence interval and time since the previous event [*Brooks et al.*, 2006]). Conversely, if SSE motion occurred first, the summit magma system, which had been inflating [*Poland et al.*, 2012], was probably primed to feed magma to Pu'u 'Ō'ō. In either case, both the SSE and the effusive surge must have been on the verge of occurring anyway for triggering to have occurred.

13.6.2. June 2007: Feedback Between East Rift Zone Intrusion and Slow Slip

Interaction between the décollement and the rift zone was amply demonstrated in June 2007, when an East Rift Zone intrusion triggered an SSE on the décollement, which was followed by additional rift opening [*Montgomery-Brown et al.*, 2011]. The intrusion began on 17 June at about 12 h UTC (02 h HST, t_A in Figure 13.10). Geodetic

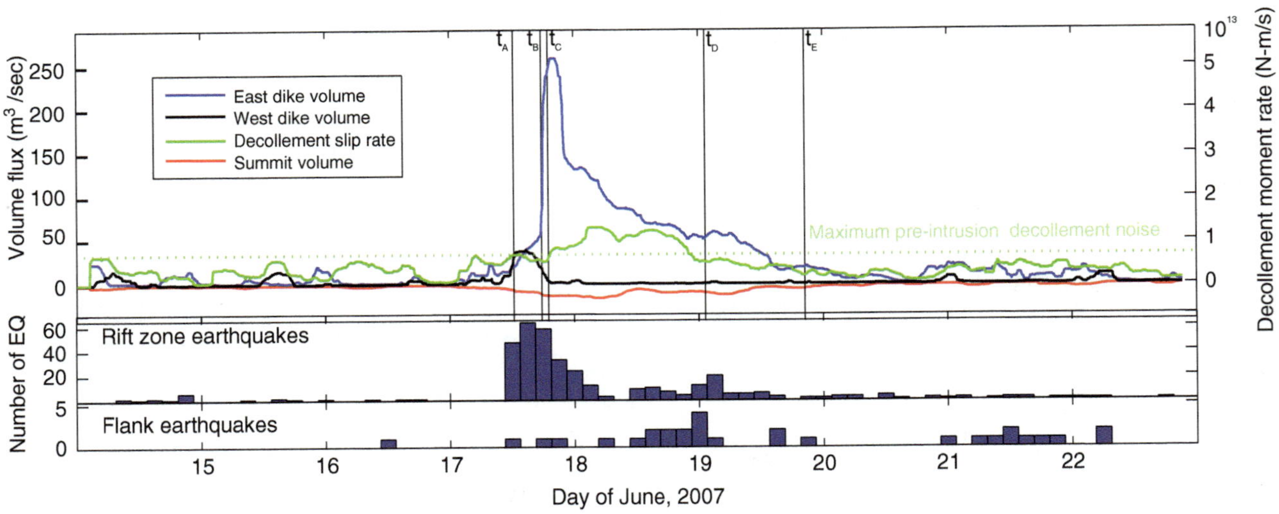

Figure 13.10 Time series of the 2007 intrusion and SSE. Lines (red, green, black, blue) show the magma accumulation rates in each of the dikes as well as the summit magma reservoir and décollement fault slip rate. Histograms below show earthquake counts for the rift zone and the flank showing the delay of triggered flank earthquakes following the onset of décollement slip. Black vertical lines (t_{A-E}) correspond to observed intrusion events discussed by *Poland et al.* [2008]. Figure reprinted from *Montgomery-Brown et al.* [2011].

models suggest emplacement of a small dike (less than 10^6 m^3 in volume) near Mauna Ulu on Kīlauea's East Rift Zone [*Montgomery-Brown et al.*, 2010, 2011]. Within a few hours, the rate of summit deflation due to drainage of magma from summit reservoirs to feed the intrusion began to decrease, suggesting that the end of the intrusion was near. At about 18 h UTC (08 h HST, t_C in Figure 13.10), however, tiltmeters and GPS stations on the south flank suggested the start of an SSE on the décollement. The rate of summit deflation increased at about the same time as another dike was intruded into the rift zone. This second dike segment was arranged en echelon and to the east with respect to the first but was much larger (modeled as 18×10^6 cm^3 [*Montgomery-Brown et al.*, 2010]), was active for 1.5 days, and resulted in a small eruption from its downrift extent [*Poland et al.*, 2008].

That the intrusions and SSE might have triggered one another is supported not only by their relative timing but also by stress modeling. Both *Dieterich et al.* [2003] and *Brooks et al.* [2008] demonstrated that opening of the upper few kilometers of the rift zone, as occurred during emplacement of the intrusions, increases the Coulomb stress on that part of the décollement fault that slips during SSEs. That slip, in turn, decreases the normal stress on the rift zone (Figure 13.8). It is therefore easy to understand how intrusion of the western dike might trigger an SSE, which in turn facilitated the emplacement of the eastern dike. As in 2005, however, it is likely that the triggering was only possible because both the magmatic system and décollement were highly stressed to begin with. An SSE had been forecast for March 2007 based on the recurrence of previous events [*Brooks et al.*, 2006]; therefore, the décollement was probably on the verge of slip in June 2007, and the intrusion of the western dike segment provided the impetus needed for the slip to begin.

Although summit inflation is sometimes related to a backup of magma in the rift zone (e.g., due to a blockage in the conduit), in 2007, Kīlauea's magma system was highly pressurized owing to an increase in magma supply to the volcano during the preceding years [*Poland et al.*, 2012]. Magma was readily available, therefore, to exploit the unclamping of the rift zone that was promoted by the SSE. If the SSE had not been on the verge of occurring anyway, the Father's Day intrusion might have ended with the intrusion of the western dike segment. Likewise, if the magma system had not been highly pressurized, the eastern segment might not have been triggered by the SSE.

13.6.3. May 2012: Rift Opening in Response to Slow Slip

The $M_w = 6.1$ SSE in 2012 induced a small amount of rift zone opening that was not an effect of magma overpressure (Figure 13.4). Expansion across the rift zone in excess of that expected solely from décollement slip was not observed during the 2010 SSE ($M_w = 5.9$), which was nearly as large as the 2012 event. It is possible that opening occurred during the 2012 SSE because Kīlauea's magma system was already inflating, whereas in 2010 the magma system was deflating. Evidence for such a relation, however, is contradictory, as Kīlauea was inflating in 2005 but no rift extension was observed during the 2005 SSE and rift opening was not required by models of the 2005 SSE.

Analysis of the stress associated with this SSE suggests that the rift opening was likely triggered by the décollement slip, and not vice versa. The 2.5 cm of rift opening modeled during the 2012 SSE would result in less than 0.002 MPa (2 kPa) of stress change on the décollement, which is about an order of magnitude below the suggested limit for static stress triggering of regular earthquakes. Assuming that a similar magnitude of static stress change would be required to trigger a slow slip, as was the case in June 2007, this small amount of rift opening likely would not have been enough to trigger the slow slip, and thus we favor the interpretation that the slow slip induced the rift opening.

Interestingly, there is no evidence that the 2012 SSE or associated rift zone opening had any impact on Kīlauea's magmatic system: There was no change in eruptive activity or summit deformation in May 2012. The section of the rift that was induced to open by the 2012 SSE had most recently been active during a major fissure eruption in March 2011 [*Lundgren et al.*, 2013]. The March 2011 intrusion occurred in an area of repeated recent intrusions [*Owen et al.*, 2000b; *Cervelli et al.*, 2002b; *Montgomery-Brown et al.*, 2010] and may have exploited or contributed (or both) to a persistent pocket of magma in the rift zone [*Thornber et al.*, 2003]. While a very simple cooling model [*Turcotte and Schubert*, 2002, p. 166] of a 2–3 m thick dike injected into ambient temperature host rock will solidify in a couple of weeks, the active conduit supplying the ongoing Pu'u 'Ō'ō eruption and the existence of persistent magma pockets suggested by *Thornber et al.* [2003] for a 1997 Nāpau Crater intrusion suggest that the rift is persistently warmer than the ambient temperatures used in these models. A warm rift zone allows for the possibility that the 2012 SSE could have induced passive rift opening with a much smaller amount of stress change than would be required for solid rock.

The 2012 SSE thus again demonstrates that a delicate balance exists between magmatic and tectonic activity at Kīlauea. The magmatic system must be favorably primed to be affected by SSEs, and likewise the décollement must be near to slip for SSEs to be triggered by magmatic activity.

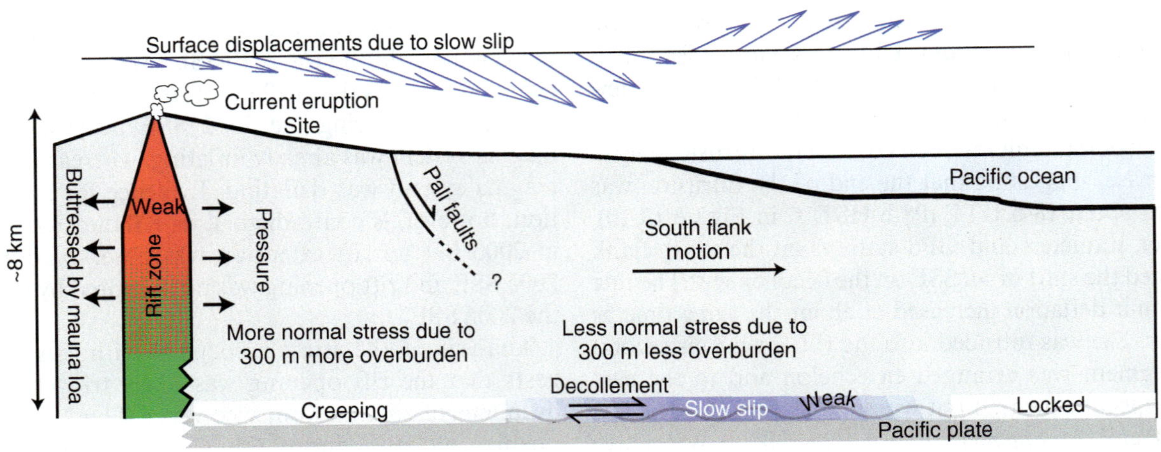

Figure 13.11 Schematic north-south cross section of Kīlauea (roughly through Pu'u 'Ō'ō vent) indicating the relationships between the magmatic system and décollement SSEs. The rift zone grades from red to green to represent the occurrence of pressurized magma in the active transport system likely with a base of cumulates, with possible interstitial melt. The décollement marks the contact between the volcanic pile and the Pacific Plate; it can be roughly separated into three zones based on the typical behaviors observed.

13.7. DISCUSSION AND CONCLUSIONS

Analysis of the 11 SSEs that have occurred at Kīlauea between 1997 and 2013 suggests that while décollement slip and rift zone magmatism have the potential to trigger one another, the tectonic and magmatic systems must be close to their respective critical states to be prone to triggering. Such a relation is supported by stress modeling of the impact of rift zone opening on the décollement and vice versa (Figure 13.8). Insights from stress modeling also have implications for Kīlauea's evolution. It is probably not a coincidence that the rift zone occurs near the location of maximum opening stresses on vertical planes due to flank slip and similarly that décollement slip (both SSEs and coseismic slip, such as the 1975 Kalapana earthquake) occurs where the ΔCFS due to rift intrusions is highest on horizontal planes. This observation lends support to the model that décollement slip is required to accommodate rift zone extension and over the long term is responsible for the persistence of Hawaiian rift zones [*Dieterich*, 1988].

The cause of the change in behavior of SSEs, with smaller, more frequent events prior to 2003 and larger, less frequent events thereafter, remains a conundrum. SSEs that occurred between 1997 and 2003 are widely variable in terms of their size and recurrence interval, with geodetic moments equivalent to earthquake magnitudes ranging from 5.3 to 5.6 and interevent times of 0.99±0.64 years. In contrast, SSEs in 2005 and later have been larger, with equivalent magnitudes ranging from 5.7 to 6.1 and more regular interevent times of 2.44±0.15 years. The transition from smaller and more frequent to larger and less frequent SSEs coincided with the onset of increased magma supply to Kīlauea [*Poland et al.*, 2012]. How the 2003–2007 supply surge might have changed the mode of SSE behavior is not understood. Indeed, the timing of the increased magma supply with respect to the change in SSE behavior may be coincidental. Alternatively, the variability induced by the increase in supply to Kīlauea's eruptive patterns (with many changes in eruptive vent locations and styles and the onset of summit eruptive activity) may reflect a modulation of SSEs by variations within Kīlauea's magmatic system.

Kīlauea's SSEs are similar to subduction zone events in that they occur at the transition between a locked zone and a creeping zone of the fault. Numerical models suggest that a SSE could accelerate into a large earthquake, which is also a possibility at Kīlauea. Regardless of SSEs, an earthquake similar to the 1975 Kalapana event is likely to occur again, but currently the mechanics of SSEs are not well-enough understood to use as a forecasting tool, and 11 events are still a small set to interpret.

Observations from Kīlauea's SSEs and their interactions with the magmatic system can provide some insight into the small amount of stress required to initiate fault slip on the décollement or opening of the rift zone if those systems are on the verge of slipping/opening (Figure 13.11). Kīlauea's décollement is weak due to various factors, including the likely presence of pelagic ocean sediments on the décollement interface, fluids available from the sediment compaction, and potentially high heat flow. Thus, if the décollement is on the verge of failure, very little magma pressure can induce fault slip, as suggested by the 2005 effusive surge and associated SSE and more conclusively demonstrated by the 2007 Father's Day intrusion and SSE. Similarly, Kīlauea's rift zone is weak due to existing cracks and the presence of magma conduits and storage areas, so a small amount of stress resulting from décollement fault slip can induce rift opening, as seen during the 2012 SSE. If the magma system is pressurized, extension of the rift zone may also initiate a rift intrusion, which apparently occurred during the 2007 Father's Day sequence. Overall, a delicate balance between inflation and deflation of Kīlauea's magma system and flank spreading is inferred from multiple observations of SSEs immediately preceding magmatic activity and vice versa.

ACKNOWLEDGMENTS

The authors would like to acknowledge comments from Dave Hill, Fred Klein, and Tom Wright, who provided helpful feedback on early drafts of this chapter, as well as internal USGS reviews from Benjamin Brooks and Roger Denlinger. The chapter also benefited from the reviews of the associate editor and two anonymous reviewers. This research was partially supported by a National Science Foundation (NSF) post-doctoral fellowship, No.EAR-0846959. The GPS network is operated in collaboration with the USGS, Stanford University, and Pacific GPS Facility at the University of Hawai'i. GPS data are archived at UNAVCO.

REFERENCES

Acocella, V., B. Behncke, M. Neri, and S. D'Amico (2003), Link between major flank slip and 2002–2003 eruption at Mt. Etna (Italy), *Geophys. Res. Lett.*, *30*(24), p. 2286.

Aki, K., and R. Koyanagi (1981), Deep volcanic tremor and magma ascent mechanism under Kilauea, Hawaii, *J. Geophys.Res.*, *86*(B8), 7095–7109, doi:10.1029/JB086iB08p07095.

Amelung, F., S.-H. Yun, T. R. Walter, P. Segall, and S.-W. Kim (2007), Stress control of deep rift intrusion at Mauna Loa volcano, Hawaii, *Science*, *316*(5827), 1026–1030.

Ando, M. (1979), The Hawaii earthquake of November 1975: Low dip angle faulting due to forceful injection of magma, *J. Geophys. Res.*, *84*, 7616–7626.

Árnadóttir, T., P. Segall, and M. Matthews (1991), A fault model for the 1989 Kilauea south flank earthquake from leveling and seismic data, *Geophys. Res. Lett.*, *18*(12), 2217–2220.

Aster, R. C., B. Borchers, and C. H. Thurber (2012), Parameter Estimation and Inverse Problems, 2nd ed., Academic Press, Waltham, Mass.

Baer, G., Y. Hamiel, G. Shamir, and R. Nof (2008), Evolution of a magma-driven earthquake swarm and triggering of the nearby Oldoinyo Lengai eruption, as resolved by InSAR, ground observations and elastic modeling, East African Rift, 2007, *Earth Planet. Sci. Lett.*, *272*(1), 339–352.

Borgia, A. (1994), Dynamic basis of volcanic spreading, *J. Geophys. Res.*, *99*(B9), 17,791–17,804.

Borgia, A., P. T. Delaney, and R. P. Denlinger (2000), Spreading volcanoes, *Annu. Rev. Earth Planet. Sci.*, *28*, 539–570.

Brooks, B., J. Foster, D. Sandwell, C. Wolfe, P. Okubo, M. Poland, and D. Myer (2008), Magmatically triggered slow slip at Kilauea Volcano, Hawaii, *Science*, *321*(5893), 1177, doi:10.1126/science.1159007.

Brooks, B. A., J. H. Foster, M. Bevis, L. N. Frazer, C. J. Wolfe, and M. Behn (2006), Periodic slow earthquakes on the flank of Kilauea Volcano, Hawai'i, *Earth Planet. Sci. Lett.*, *246*(3/4), 207–216, doi:10.1016/j.epsl.2006.03.035.

Brudzinski, M., and R. Allen (2007), Segmentation in episodic tremor and slip all along Cascadia, *Geology*, *35*(10), 907–910, doi:10.1130/G23740A.1.

Bufe, C., P. Harsh, and R. Burford (1977), Steady-state seismic slip: A precise recurrence model, *Geophys. Res. Lett*, *4*(2), pp. 91–94.

Cayol, V., J. Dieterich, A. Okamura, and A. Miklius (2000), High magma storage rates before the 1983 eruption of Kilauea, Hawaii, *Science*, *288*(5475), 2343–2346.

Cervelli, P., and A. Miklius (2003), The shallow magmatic system of Kilauea Volcano, Prof. Pap. 1676, pp. 149–163, U.S. Geol. Suru., Reston, Va.

Cervelli, P., P. Segall, K. Johnson, M. Lisowski, and A. Miklius (2002a), Sudden aseismic fault slip on the south flank of Kilauea Volcano, *Nature*, *415*(6875), 1014–1018, doi:10.1038/4151014a.

Cervelli, P., P. Segall, F. Amelung, H. Garbeil, C. Meertens, S. Owen, A. Miklius, and M. Lisowski (2002b), The 12 September 1999 Upper East Rift Zone dike intrusion at Kilauea Volcano, Hawaii, *J. Geophys. Res.*, *107*(B7), 2150–2150, doi:10.1029/2001JB000602.

Chen, J., H. A. Zebker, P. Segall, and A. Miklius (2014), The 2010 slow slip event and secular motion at Kīlauea, Hawai'i, inferred from TerraSAR-X InSAR data, *J. Geophys. Res.* Solid Earth, *119*, 6667–6683, doi:10.1002/2014JB011156.

Clague, D. A., and R. P. Denlinger (1994), Role of olivine cumulates in destabilizing the flanks of Hawaiian Volcanos, *Bull. Volcanol.*, *56*(6–7), 425–434.

Delaney, P., R. Fiske, A. Miklius, A. Okamura, and M. Sako (1990), Deep magma body beneath the summit and rift zones of Kilauea Volcano, Hawaii, *Science*, *247*(4948), 1311–1316.

Delaney, P., A. Miklius, T. Arnadottir, A. Okamura, and M. Sako (1993), Motion of Kilauea Volcano during sustained eruption from the Pu'u O'o and Kupianaha vents, 1983–1991, *J. Geophys. Res.*, *98*(B10), 17,801–17,820.

Delaney, P. T., and R. P. Denlinger (1999), Stabilization of volcanic flanks by dike intrusion: An example from Kilauea, *Bull. Volcanol.*, *61*(6), 356–362.

Denlinger, R. P., and P. Okubo (1995), Structure of the mobile south flank of Kilauea Volcano, Hawaii, *J. Geophys. Res.*, *100*(B12), 24,499–24,507.

Dieterich, J. (1988), Growth and persistence of Hawaiian volcanic rift zones, *J. Geophys. Res.*, *93*(B5), 4258–4270.

Dieterich, J., V. Cayol, and P. Okubo (2000), The use of earthquake rate changes as a stress meter at Kilauea Volcano, *Nature*, *408*(6811), 457–460.

Dieterich, J., V. Cayol, and P. Okubo (2003), Stress changes before and during the Pu'u 'Ō'ō-Kūpianaha eruption, Prof. Pap. 1676, pp. 187–202, U.S. Geol. Surv. Reston, Va.

Dragert, H., K. L. Wang, and T. S. James (2001), A silent slip event on the deeper Cascadia subduction interface, *Science*, *292*(5521), 1525–1528.

Dzurisin, D., R. Koyanagi, and T. English (1984), Magma supply and storage at Kilauea Volcano, Hawaii, 1956–1983, *J. Volcanol. Geotherm. Res.*, *21*(3), 177–206.

Elsworth, D., and B. Voight (1995), Dike intrusion as a trigger for large earthquakes and the failure of volcano flanks, *J. Geophys. Res.*, *100*(B4), 6005–6024, doi:10.1029/94JB02884.

Foster, J., C. Wolfe, B. A. Brooks, and C. Shacat (2006), Detailed study of slow earthquakes at Kilauea Volcano, Hawaii, EOS Trans. AGU, Fall Meet. Suppl. 87, Abstract T41A-1534.

Foster, J. H., A. R. Lowry, and B. A. Brooks (2013), Fault frictional parameters and material properties revealed by slow slip events at Kilauea Volcano, Hawai'i, *Geophys. Res. Lett.*, doi:10.1002/2013GL058234.

Hansen, P. (1992), Analysis of discrete ill-posed problems by means of the L-curve, *SIAM Rev.*, *34*(4), 561–580.

Heliker, C., and T. N. Mattox (2003), The first two decades of the Pu'u 'Ō'ō–Kupaianaha eruption: Chronology and selected bibliography, in *The Pu'u 'Ō'ō-Kupaianaha eruption of Kīlauea Volcano, Hawai'i: The First 20 Years*, edited by C. Heliker et al., pp. 1–27, U. S. Geol. Surv. Prof. Pap. 1676.

Hill, D. (2010), Surface-wave potential for triggering tectonic (nonvolcanic) tremor, *Bull. Seismol. Soc. Am.*, *100*(5A), 1859–1878, doi:10.1785/0120090362.

Hill, D. P. (1969), Crustal structure of the island of Hawaii from seismic-refraction measurements, *Bull. Seismol. Soc. Am.*, *59*(1), 101–130.

Hill, D. P., F. Pollitz, and C. Newhall (2002), Earthquake-volcano interactions, *Phys. Today*, pp. 41–47.

Hirose, H., and K. Obara (2005), Repeating short- and long-term slow slip events with deep tremor activity around the Bungo channel region, southwest Japan, *Earth Planets Space*, 55:11, 961–972.

Ide, S., D. R. Shelly, and G. C. Beroza (2007), Mechanism of deep low frequency earthquakes: Further evidence that deep non-volcanic tremor is generated by shear slip on the plate interface, *Geophys. Res. Lett.*, *34*(3), L03308, doi:10.1029/2006GL028890.

Iverson, R. M. (1995), Can magma-injection and groundwater forces cause massive landslides on Hawaiian Volcanos, *J. Volcanol. Geotherm. Res.*, *66*(1–4), 295–308.

Kanamori, H. (1977), The energy release in great earthquakes, *J. Geophys. Res.*, *82*(20), 2981–2987.

Kato, A., K. Obara, T. Igarashi, H. Tsuruoka, S. Nakagawa, and N. Hirata (2012), Propagation of slow slip leading up to the 2011 Mw 9.0 Tohoku-Oki earthquake, *Science*, *335*(6069), 705–708, doi:10.1126/science.1215141.

Kauahikaua, J. (1993), Geophysical characteristics of the hydrothermal systems of Kilauea Volcano, Hawaii, *Geothermics*, *22*(4), 271–299, doi:10.1016/0375-6505(93)90004-7.

Klein, F. (1982), Patterns of historical eruptions at Hawaiian volcanoes, *J. Volc. Geotherm. Res.*, *12*, 1–35.

La Femina, P. C., C. B. Connor, B. E. Hill, W. Strauchd, and J. A. Saballos (2004), Magmatectonic interactions in Nicaragua: the 1999 seismic swarm and eruption of Cerro Negro volcano, *Journal of Volcanology and Geothermal Research*, *137*, 187–199, doi:doi:10.1016/j.jvolgeores.2004.05.006.

Linde, A. T., and I. S. Sacks (1998), Triggering of volcanic eruptions, *Nature*, *395*(6705), 888–890, doi:10.1038/27650.

Linde, A. T., M. T. Gladwin, M. J. Johnston, R. L. Gwyther, and R. G. Bilham (1996), A slow earthquake sequence on the San Andreas fault, *Nature*, *383*(6595), 65–68.

Lipman, P. W., J. P. Lockwood, R. T. Okamura, D. A. Swanson, and K. M. Yamashita (1985), Ground deformation associated with the 1975 magnitude 7.2 earthquake and resulting changes in activity of Kilauea Volcano, Hawaii, Prof. Pap. 1276, pp. 1–45, U.S. Geol. Surv. Reston, Va.

Liu, Y. J., and J. R. Rice (2005), Aseismic slip transients emerge spontaneously in three-dimensional rate and state modeling of subduction earthquake sequences, *J. Geophys. Res.*, *110*(B8), B08307.

Liu, Y. J., and J. R. Rice (2007), Spontaneous and triggered aseismic deformation transients in a subduction fault model, *J. Geophys. Res.*, *112*(B9), B09404, doi:10.1029/2007JB004930.

Louie, J. W., C. R. Allen, D. C. Johnson, P. C. Haase, and S. N. Cohn (1985), Fault slip in southern California, *Bull. Seismol. Soc. Am.*, *75*(3), 811–833.

Lundgren, P., et al. (2013), Evolution of dike opening during the March 2011 Kamoamoa fissure eruption, Kilauea volcano, Hawai'i, *J. Geophys. Res.*, *118*, doi:10.1002/jgrb.50108.

Lyell, C. (1889), Principles of Geology, 11th ed., D. Appleton and Company, London.

Macdonald, G., A. Abbott, and F. Peterson (1986), Volcanoes in the Sea: The Geology of Hawaii, 2nd ed., pp. 80–81, University of Hawaii Press, Honolulu.

Marzocchi, W. (2002), Remote seismic influence on large explosive eruptions, *J. Geophys. Res.*, *107*(B1), 2018.

Miklius, A., P. Cervelli, M. Sako, M. Lisowski, S. Owen, P. Segal, J. Foster, K. Kamibayashi, and B. Brooks (2005), Global positioning system measurements on the Island of Hawai'i: 1997 through 2004, Open File Rep. 2005–1425, pp. 1–48, U.S. Geol. Surv., Reston, Va.

Mogi, K. (1958), Relations between the eruptions of various volcanoes and the deformations of the ground surfaces around them, *Bull. Earthquake Res. Inst. Univ. Tokyo*, *36*, 111–123.

Montgomery-Brown, E., P. Segall, A. Miklius, P. Cervelli, and D. Shelly (2009), Kilauea slow slip events: Identification, source inversions, and relation to seismicity, *J. Geophys. Res.*, *114*, B00A03, doi:10.1029/2008JB006074.

Montgomery-Brown, E., D. Sinnett, M. Poland, P. Segall, T. Orr, H. Zebker, and A. Miklius (2010), Geodetic evidence for en echelon dike emplacement and concurrent slow-slip during the June 2007 intrusion and eruption at Kilauea Volcano, Hawai'i, *J. Geophys. Res.*, *115*, B07405, doi:10.1029/2009JB006658.

Montgomery-Brown, E., K. Larson, M. Poland, D. Sinnett, and P. Segall (2011), Spatiotemporal evolution of dike opening and décollement slip at Kīlauea Volcano, Hawai'i, *J. Geophys.* Res., 116, B03401, doi:10.1029/2010JB007762..

Montgomery-Brown, E., C. Thurber, C. Wolfe, and P. Okubo (2013), Slow slip and tremor search at Kilauea Volcano, Hawaii, *Geochem. Geophys. Geosyst.*, *14*, 367, 384, doi:10.1002/ggge.20044.

Moore, J. G., W. R. Normark, and R. T. Holcomb (1994), Giant Hawiian landslides, *Annu. Rev. Earth Planet. Sci.*, *22*, 119–144.

Morgan, J., and P. McGovern (2005), Discrete element simulations of gravitational volcanic deformation: 1. Deformation structures and geometries, *J. Geophys. Res.*, *110*, B05402.

Morgan, J. K., G. F. Moore, D. J. Hills, and S. Leslie (2000), Overthrusting and sediment accretion along Kilauea's mobile south flank, Hawaii: Evidence for volcanic spreading from marine seismic reflection, *Geology*, *28*(7), 667–670, doi:10.1130/0091-7613.

Mori, J., R. A. White, D. H. Harlow, P. Okubo, J. A. Power, R. P. Hoblitt, E. P. Laguerta, A. Lanuza, and B. C. Bautista (1996), Volcanic earthquakes following the 1991 climactic eruption of Mount Pinatubo: Strong seismicity during a waning eruption, *Fire and Mud: Eruptions and Lahars of Mount Pinatubo, Philippines*, pp. 339–350. http://pubs.usgs.gov/pinatubo/

Nakamura, K. (1980), Why do long rift zones develop in Hawaiian volcanoes—A possible role of thick oceanic sediments, *Bull. Volcanol. Soc. Japan*, *25*, 255–267, in Japanese.

Neri, M., V. Acocella, and B. Behncke (2004), The role of the Pernicana Fault System in the spreading of Mt. Etna (Italy) during the 2002–2003 eruption, *Bull. Volcanol.*, *66*(5), 417–430.

Nettles, M., and G. Ekström (2004), Long-period source characteristics of the 1975 Kalapana, Hawaii, earthquake, *Bull. Seismol. Soc. Am.*, *94*(2), 422–429, doi:10.1785/0120030090.

Obara, K. (2002), Nonvolcanic deep tremor associated with subduction in southwest Japan, *Science*, *296*, 1679–1681, doi:10.1126/science.1070378.

Okada, Y. (1985), Surface deformation due to shear and tensile faults in a half-space, *Bull. Seismol. Soc. Am.*, *75*, 1135–1154.

Owen, S., P. Segall, M. Lisowski, A. Miklius, R. Denlinger, and M. Sako (2000a), Rapid deformation of Kilauea Volcano: Global positioning system measurements between 1990 and 1996, *J. Geophys. Res.*, *105*(B8), 18,983–18,998.

Owen, S., P. Segall, M. Lisowski, A. Miklius, M. Murray, M. Bevis, and J. Foster (2000b), January 30, 1997 eruptive event on Kilauea Volcano, Hawaii, as monitored by continuous GPS, *Geophys. Res. Lett.*, *27*(17), 2757–2760.

Parfitt, E., and D. Peacock (2001), Faulting in the south flank of Kilauea Volcano, Hawai'i, *J. Volcanol. Geotherm. Res.*, *106*(3), 265–284, doi:10.1016/S0377-0273(00)00247-X.

Patanè, D., et al. (2013), Insights into magma and fluid transfer at Mount Etna by a multi-parametric approach: A model of the events leading to the 2011 eruptive cycle, *J. Geophys. Res. Solid Earth*, 118, 3519–3539, doi:10.1002/jgrb.50248.

Perfettini, H., and J.-P. Ampuero (2008), Dynamics of a velocity strengthening fault region: Implications for slow earthquakes and postseismic slip, *J. Geophys. Res. Solid Earth (1978–2012)*, *113*(B9).

Plattner, C., F. Amelung, S. Baker, R. Govers, and M. Poland (2013), The role of viscous magma mush spreading in volcanic flank motion at Kilauea Volcano, Hawaii, *J. Geophys. Res. Solid Earth*, doi:10.1002/jgrb.50194.

Poland, M., A. Miklius, T. Orr, A. J. Sutton, C. Thornber, and D. Wilson (2008), New episodes of volcanism at Kilauea Volcano, Hawaii, *EOS Trans. AGU*, *89*(5), 37–48.

Poland, M., et al. (2010), Slow slip event at Kilauea Volcano, *EOS Trans.*, *91*, 118–119.

Poland, M. P., A. Miklius, and E. Montgomery-Brown (2014), Magma supply, storage, and transport at shield-stage Hawaiian volcanoes, Chapter 5 of Poland, M. P., T. J. Takahashi, and C. M. Landowski, (eds.), *Characteristics of Hawaiian Volcanoes*, U.S. Geol. Surv. Prof. Pap. 1801, pp. 179–234, doi:10.3133/pp18015.

Poland, M. P., A. Miklius, A. J. Sutton, and C. R. Thornber (2012), A mantle-driven surge in magma supply to Kilauea Volcano during 2003–2007, *Nature Geosci.*, *5*(4), 295–300, doi:10.1038/ngeo1426.

Rogers, G., and H. Dragert (2003), Episodic tremor and slip on the Cascadia subduction zone: The chatter of silent slip, *Science*, *300*(5627), 1942–1943.

Ryan, M. P. (1988), The mechanics and three-dimensional internal structure of active magmatic systems: Kilauea Volcano, Hawaii, *J. Geophys. Res.*, *93*(B5), 4213–4248.

Schmidt, D., and H. Gao (2010), Source parameters and time-dependent slip distributions of slow slip events on the Cascadia subduction zone from 1998 to 2008, *J. Geophys. Res.*, *115*, B00A18.

Schwartz, S., and J. Rokosky (2007), Slow slip events and seismic tremor at circum-Pacific subduction zones, *Rev. Geophys.*, *45*, RG3004, doi:10.1029/2006RG000208.

Segall, P. (2010), Earthquake and Volcano Deformation, Princeton Univ. Press, Princeton, N. J.

Segall, P., and A. M. Bradley (2012), Slow-slip evolves into megathrust earthquakes in 2d numerical simulations, *Geophys. Res. Lett.*, *39*(18), doi:10.1029/2012GL052811.

Segall, P., E. Desmarais, D. Shelly, A. Miklius, and P. Cervelli (2006), Earthquakes triggered by silent slip events on Kilauea Volcano, Hawaii, *Nature*, *442*, doi:10.1038/nature04938.

Segall, P., A. M. Rubin, A. M. Bradley, and J. R. Rice (2010), Dilatant strengthening as a mechanism for slow slip events, *J. Geophys. Res. Solid Earth (1978–2012)*, *115*(B12), doi:10.1029/2010JB007449.

Shibazaki, B., and Y. Iio (2003), On the physical mechanism of silent slip events along the deeper part of the seismogenic zone, *Geophys. Res. Lett.*, *30*(9), doi:10.1029/2003GL017047.

Shibazaki, B., and T. Shimamoto (2007), Modelling of short-interval silent slip events in deeper subduction interfaces considering the frictional properties at the unstable-stable transition regime, *Geophys. J. Int.*, *171*(1), 191–205, doi:10.1111/j.1365-246X.2007.03434.x.

Shimazaki, K., and T. Nakata (1980), Time-predictable recurrence model for large earthquakes, *Geophys. Res. Lett*, *7*(4), 279–282.

Sieh, K. E., and P. L. Williams (1990), Behavior of the southernmost San Andreas Fault during the past 300 years, *J. Geophys. Res.*, *95*(B5), 6629–6645.

Stearns, H., and G. MacDonald (1946), Geology and groundwater resources of the Island of Hawaii, *Bull. Hawaii Div. Hydrogr.*, *9*, p. 363.

Stein, R. S. (1999), The role of stress transfer in earthquake occurrence, *Nature*, *402*(6762), 605–609, doi:10.1038/45144.

Swanson, D., D. Jackson, R. Koyanagi, and T. Wright (1976a), The February 1969 East Rift Eruption of Kilauea Volcano, Hawaii, Prof. Pap. 891, U.S. Geol. Surv., Reston, Va.

Swanson, D. A., W. A. Duffield, and R. S. Fiske (1976b), Displacement of the south flank of Kilauea Volcano, Hawaii: The result of forceful intrusion of magma into the rift zones, Prof. Pap. 963, pp. 1–39, U.S. Geol. Surv., Reston, Va.

Syracuse, E. M., C. H. Thurber, C. J. Wolfe, P. G. Okubo, J. H. Foster, and B. A. Brooks (2010), High-resolution locations of triggered earthquakes and tomographic imaging of Kilauea Volcano's south flank, *J. Geophys. Res.*, *115*, B10,310.

Thornber, C. R., et al. (2003), Kilauea east rift zone magmatism: An Episode 54 perspective, *J. Petrol.*, *44*(9), 1525–1559, doi:10.1093/petrology/egg048.

Thurber, C., and A. Gripp (1988), Flexure and seismicity beneath the south flank of Kilauea Volcano and tectonic implications, *J. Geophys. Res.*, *93*(B5), 4271–4278.

Tilling, R., J. Rhodes, J. Sparks, J. Lockwood, and P. Lipman (1987), Disruption of the Mauna Loa magma system by the 1868 Hawaiian earthquake: Geochemical evidence, *Science*, *235*(4785), 196–199.

Turcotte, D. L., and G. Schubert (2002), *Geodynamics*, Cambridge University Press, New York.

Walter, T. R. (2007), How a tectonic earthquake may wake up volcanoes: Stress transfer during the 1996 earthquake-eruption sequence at the Karymsky Volcanic Group, Kamchatka, *Earth Planet. Sci. Lett.*, *264*(3), 347–359, doi:10.1016/j.epsl.2007.09.006.

Walter, T. R., and F. Amelung (2007), Volcanic eruptions following M>9 megathrust earthquakes: Implications for the Sumatra-Andaman volcanoes, *Geology*, *35*(6), 539–542.

Walter, T. R., V. Acocella, M. Neri, and F. Amelung (2005), Feedback processes between magmatic events and flank movement at Mount Etna (Italy) during the 2002–2003 eruption, *J. Geophys. Res.: Solid Earth (1978–2012)*, *110*(B10), B10205, doi:10.1029/2005JB003688.

Wech, A. G., K. C. Creager, and T. I. Melbourne (2009), Seismic and geodetic constraints on Cascadia slow slip, *J. Geophys. Res.*, *114*(B10), B10316, doi:10.1029/2008JB006090.

Wei, M., D. Sandwell, and Y. Fialko (2009), A silent Mw 4.7 slip event of October 2006 on the Superstition Hills fault, southern California, *J. Geophys. Res. Solid Earth (1978–2012)*, *114*(B7), B07402, doi:10.1029/2008JB006135.

Williams, C. A., and G. Wadge (2000), An accurate and efficient method for including the effects of topography in three-dimensional elastic models of ground deformation with applications to radar interferometry, *J. Geophys. Res.*, *105*(B4), 8103–8120, doi:10.1029/1999JB900307.

Wilson, D., T. Elias, T. Orr, M. Patrick, J. Sutton, and D. Swanson (2008), Small explosion from new vent at Kilauea's summit, *Eos, Trans. AGU*, *89*(22), 203, doi:10.1029/2008EO220003.

Wolfe, C., B. Brooks, J. Foster, and P. Okubo (2007), Microearthquake streaks and seismicity triggered by slow earthquakes on the mobile south flank of Kilauea Volcano, Hawai'i, *Geophys. Res. Lett*, *34*, doi:10.1029/2007GL031625.

Wright, T. L., and Klein, F. W., (2014), Two hundred years of magma transport and storage at Kīlauea Volcano, Hawai'i, 1790–2008: *U.S. Geological Survey Professional Paper* 1806, 240 p., 9 appendixes, http://dx.doi.org/10.3133/pp1806.

Wyss, M. (1988), A proposed source model for the great Kau, Hawaii, earthquake of 1868, *Bull. Seismol. Soc. Am.*, *78*, 1450–1462.

Zucca, J. J., and D. P. Hill (1980), Crustal structure of the southeast flank of Kilauea Volcano, Hawaii, from seismic refraction measurements, *Bull. Seismol. Soc. Am.*, *70*(4), 1149–1159.

14

From Reservoirs and Conduits to the Surface: Review of Role of Bubbles in Driving Basaltic Eruptions

Sylvie Vergniolle and Yves Gaudemer

ABSTRACT

In this review chapter, we discuss the key role played by the bubbles in driving basaltic eruptions. First, we describe the various processes at work for small bubbles during bubble nucleation, growth and coalescence and the differences between CO_2 bubbles and H_2O bubbles. Estimates of the gas volumes at the vent can be obtained from direct measurements (photoballistics, radar, Fourier transform infrared (FTIR) spectroscopy) or indirect measurements (seismic and acoustic records). The order-of-magnitude difference obtained between direct and indirect measurements may be both a consequence of the type of degassing measured, active (i.e., still overpressurized at the vent) for indirect methods and total (passive and active) for direct methods and to bias due to air entrainment.

Eruptive activity at open vents, including lava lake activity, gas pistons, and Strombolian explosions, is then discussed at Erta 'Ale, Nyiragongo, Kīlauea, and Stromboli. The current models explaining Strombolian and permanent activity, i.e., when the hot magma is present at the surface for extended periods, are reviewed. The most violent manifestations of basaltic eruptions, such as large basaltic eruptive columns, Vulcanian explosions during paroxysms at Stromboli, and high-level lava fountains, are finally discussed.

14.1. INTRODUCTION

Basaltic volcanoes can be found in various geodynamics settings and their eruptions are frequent globally. Classical basaltic eruptions include those displayed by hotspots such as at Kīlauea Volcano (Hawai'i) or Piton de la Fournaise (Réunion Island, France) as well as those existing in continental rift systems, such as at Erta 'Ale Volcano (Ethiopia) and Nyiragongo (Democratic Republic of Congo). Another type of classical basaltic eruption are those presenting permanent Strombolian activity, such as at Stromboli (Italy). Less frequent, but increasingly recognized, are the basaltic eruptions of subduction zone volcanoes, which produce very large basaltic eruptive columns, such as at Shishaldin (Alaska) or Lopevi (Vanuatu). High eruption columns, although very rare, have also been recognized at Kīlauea in the past [*Swanson et al.*, 2012].

Basaltic eruptions can be subaerial or submarine. At the bottom of the ocean, typical eruptions consist of gentle lava flows, sometimes lava lakes [*Fouquet et al.*, 1995; *Eissen et al.*, 2003], but can also produce Strombolian explosions [*Head and Wilson*, 2003; *Sohn et al.*, 2008; *Clague et al.*, 2009; *Helo et al.*, 2011; *Deardroff et al.*, 2011].

The strongest manifestation that an aerial basaltic eruption can take, although rarely, is one that produces a high column rising to heights of several kilometers in the troposphere (Figure 14.1d). The weakest activity is either the effusion of lava flows (Figure 14.1a) or when the magma remains in the conduit, with sporadic lava lake bubbling activity [e.g., *Richter et al.*, 1970; *Tilling et al.*, 1987; *Orr et al.*, 2013; *Edmonds etal.*, 2013].

All basaltic magmatic eruptions share the common feature of being driven by the gas phase, regardless of the geodynamical setting. Importantly the gas content differs with hotspot volcanoes being mostly driven by CO_2 and basaltic

Institut de Physique du Globe de Paris, UMR CNRS 7154, Sorbonne Paris Cité, Université Paris Diderot, Paris, France

Hawaiian Volcanoes: From Source to Surface, Geophysical Monograph 208. First Edition.
Edited by Rebecca Carey, Valérie Cayol, Michael Poland, and Dominique Weis.

Companion Website: www.wiley.com/go/Carey/Hawaiian_Volcanoes

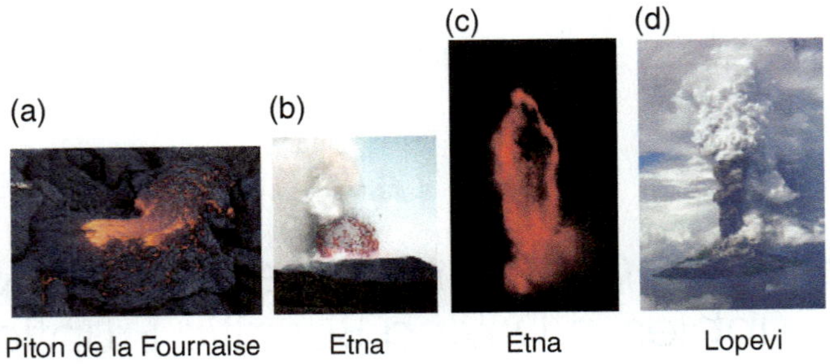

Figure 14.1 Main surface activity at a basaltic volcano. (a) Lava flow at Piton de la Fournaise (Réunion Island) on 27 December 2006 (photo taken by A. Dupont). (b) Strombolian explosion on 25 July 2001 at Etna volcano and clearly showing a very large bubble bursting with a diameter of several tens of meters (photo taken by T. Pfeiffer (//www.decadevolcano.net) and adapted from Figure 1 in *Vergniolle and Ripepe* [2008]). (c) High-level lava fountain at Etna on 26 September 1989 that reached 600 m height and clearly displayed the inner gas core, which drives the magma upward (adapted from Figure 3 for 26 September 1989 in *Bertagnini et al.* [1990] and Figure 2 in *Vergniolle and Ripepe* [2008]. (d) Basaltic magmatic eruptive column at Lopevi volcano (Vanuatu) on 8 June 2003 (courtesy of P. Leloup, Vanair). The height of the dark part within the column is ≈2 km.

volcanoes from subduction zones being mostly driven by H_2O. The process by which bubbles are formed is extremely complicated. Our current understanding relies on theory, laboratory experiments, unfortunately very rare for basaltic magma [*Lensky et al.*, 2006; *Bai et al.*, 2008; *Pichavant et al.*, 2013; *Hardiagon et al.*, 2013], and vesicularity measurements in eruptive products [e.g., *Polacci et al.*, 2006, 2009; *Houghton and Gonnermann*, 2008; *Sable et al.*, 2006a; *Mangan et al.*, 2014]. The new bubbles, once formed, can grow by diffusion, decompression, and coalescence. An extremely large range of bubble diameters is expected, from several micrometers at nucleation to several tens of meters at the vent. The new bubbles can be formed anywhere in the magma reservoir and the conduit or at the surface. Their formation leads to an increase in volume, which may produce an overpressure in a closed magma reservoir sufficient to trigger eruptions [*Tait et al.*, 1989]. The bubbles, significantly more buoyant than the magma, may also carry some viscous magma from depth to the surface and may enhance convection in open-vent systems. In summary, bubbles are fundamental to the eruption processes from the reservoir to the surface, although the physical mechanisms are very complicated to unravel. However our knowledge and insight into eruptive processes is increasing via multidisciplinary approaches to basaltic volcanoes, which are our laboratories for understanding how Earth works.

In this chapter, we explore the current understanding of the various eruptive regimes and review the constraints on eruption dynamics obtained by field measurements, laboratory experiments, and numerical calculations. We start by describing different styles of basaltic eruptions and associate the volcanic activities to various regimes driven by mixtures of bubbles and magma. In the next section, we compare the available techniques, which can constrain the gas volume at the vent. Then, we describe the physical processes at the origin of bubble formation and growth. Our current understanding of Strombolian explosions and the characteristics of open-vent volcanoes are then presented. Finally the classical non-permanent basaltic activity, such as leading to basaltic sub-Plinian/Plinian columns, paroxysms at Stromboli, and high-level lava fountains, is discussed.

14.2. DYNAMICS OF BASALTIC ERUPTIONS

14.2.1. Description of Basaltic Eruptions

The typical activity of a hotspot volcano, such as Kīlauea, consists of lava effusion and lava fountaining, spattering, sometimes coexistent with lava lakes and ponds [*Richter et al.*, 1970; *Swanson et al.*, 1979; *Wolfe et al.*, 1987]. Lava effusion (Figure 14.1a) at Kīlauea erupts magma with up to 20%–30% of small bubbles of diameters from millimeters to a few centimeters [e.g., *Wolfe et al.*, 1987], formed within the conduit. The most impressive activity observed at Kīlauea is that presented by high-level lava fountains, reaching heights of a few hundreds meters (Figure 14.1c). The vesicularity of the eruptive products can be very large, up to 98% as for the reticulite formed during very high lava fountains [e.g., *Richter et al.*, 1970; *Stovall et al.*, 2011, 2012]. Lava lakes are sometimes formed either above the conduit or in a pool separated from the deep plumbing system. Eruptions at Kīlauea can occur at its summit, such as during the Kīlauea Iki eruption (1959–1960), where 16 vigorous high-level lava fountains were produced [*Richter et al.*, 1970], or on the East Rift Zone on the flanks of the volcano, such as during the Mauna Ulu (1969–1974) and Pu'u 'Ō'ō (1983 to now) eruptions, with 12 and 47 episodes, respectively, of high-level lava fountains often higher than 50 m [*Swanson et al.*, 1979; *Wolfe et al.*, 1987]. Kīlauea eruptions usually start as an eruptive fissure often feeding relatively low-level fountains with the fissure focusing into a few single eruptive vents. The focusing of the long eruptive fissure in few single vents is explained by a thermal effect: The part of the fissure that has initially the smallest width undergoes more cooling, due to its largest contact with the surrounding cold rocks, than the largest portions of the fissure, leaving only active the initially largest spots [*Bruce and Huppert*, 1989].

Stromboli volcano is constantly erupting and is characterized by a regular series of mild explosions reaching heights at most 100–200 m above the vent and expelling black scoria at an average velocity of 20–30 m/s [e.g., *Chouet et al.*, 1974; *Ripepe et al.*, 1993; *Patrick et al.*, 2007; *Allard*, 2010]. Explosions are short, a few to several tens of seconds, and are separated by repose periods of 5–20 min [e.g., *Chouet et al.*, 1974; *Ripepe et al.*, 1993; *Patrick et al.*, 2007; *Allard*, 2010]. Stromboli also presents paroxysmal activity, such as in 1919, 1930, 1936, 1959, 2003, and 2007. Paroxysms usually produce a Vulcanian explosion, which only lasts several minutes but releases a strongly overpressurized gas that forms a small eruptive column a few kilometers high [*Barberi et al.*, 1993; *Rosi et al.*, 2006; *Bertagnini et al.*, 2008]. During the last two centuries 25 large paroxysms have been documented [*Barberi et al.*, 1993]. About twice a year, a major explosion also occurs [*Barberi et al.*, 1993; *Harris and Ripepe*, 2007b; *Bertagnini et al.*, 2008].

Basaltic volcanoes can produce kilometer-high eruptive columns, previously considered to be mostly associated with a phreato-magmatic fragmentation. However, the existence of purely magmatic eruptions and high columns, although rare [*Williams*, 1983], is now increasingly recognized. Examples include the 1999 eruption at Shishaldin (16 km) [*Nye et al.*, 2002; *Stelling et al.*, 2002], the 1992 eruption at Cerro Negro (Nicaragua) (7 km) [*Roggensack et al.*, 1997], the 3930 BP and the 122 BC eruptions of Etna (18–20 and 25 km, respectively) [*Coltelli et al.*, 1998; *Houghton et al.*, 2004; *Coltelli et al.*, 2005; *Sable et al.*, 2006a], or even at Kīlauea during the Keanakāko'i eruption ("unit 6") [*Swanson et al.*, 2012]. The height of the largest eruptive columns has been recently estimated for the Fontana lapilli (Masaya, Nicaragua) at 32 km [*Costantini et al.*, 2009, 2010]. Here we exclude from our review phreatic and phreatomagmatic eruptions.

The duration of these eruptive columns is relatively short, from 45 min to 4–6 hr, at Shishaldin and for the Fontana lapilli or the Plinian phase of the 1886 Tarawera eruption (New Zealand), respectively [*Thompson et al.*, 2002; *Caplan-Auerbach and McNutt*, 2003; *Vergniolle and Caplan-Auerbach*, 2006; *Sable et al.*, 2006a, 2006b; *Carey et al.*, 2007; *Costantini et al.*, 2009, 2010]. The ejecta velocity is estimated at between 70 m/s and 100 m/s at Shishaldin if similar to that of the gas [*Vergniolle and Caplan-Auerbach*, 2006] to 150–170 m/s at Masaya [*Williams*, 1983].

14.2.2. Two-Phase Flows in Volcanic Conduits

All types of basaltic activity can be understood and classified using the framework of two-phase flows, which describes the spatial organization of a mixture containing bubbles and liquid within a conduit [*Vergniolle and Jaupart*, 1986; *Vergniolle and Mangan*, 2000]. The two-phase flow framework implicitly assumes that crystals are absent or small and passively following the flow. In reality, this is not the case and many complexities arise when considering three-phase flows (gas, liquid, and solid). Adding crystals to a liquid can increase the apparent viscosity of the mixture for mild concentrations in solid fragments [*Kerr and Lister*, 1991; *Lejeune and Richet*, 1995]. At Stromboli, aphyric golden pumices produced by paroxysmal activity have a viscosity of 7–15 Pas whereas the crystal-rich black scoria, with 45–55% of crystals, correspond to a highly viscous mixture ($\approx 10^4$ Pa.s) [*Métrich et al.*, 2001]. When crystal abundances are large, the mixture may also have a non–Newtonian behavior. For simplicity, we consider here that the crystals are passively following the magma and gas flow, only changing the apparent viscosity of the mixture. In these conditions, the flow in the conduit can be described using a two-phase flow approach involving only gas and liquid.

In basaltic volcanism, the first two-phase flow regime is observed during the effusion of lava, in which the liquid phase (magma) is continuous and the bubbles small, less than a few centimeters (Figure 14.1a). The second regime is that displayed by Strombolian explosions (Figure 14.1b), driven by the bursting, at the surface, of large overpressurized bubbles whose diameter is only slightly smaller than that of the conduit. The third regime corresponds to high-level lava fountains whose height is large compared to its own diameter. During such an episode, the magma is expelled at a several tens to several hundreds of meters above the vent (Figure 14.1c). Although most of the time the gas inside the high-level lava fountain is not apparent, there are a few occurrences in which the inner gas core that drives the flow is very clear, such as during the high-level lava fountain (600 m) at Etna on 26 September 1989 (Figure 14.1c). The fourth regime (Figure 14.2d) is clearly displayed during the formation of high basaltic eruptive columns, such as those mostly associated with basaltic eruptions from volcanoes associated with subduction settings (Figure 14.1d). At the vent and in the uppermost part of the conduit, the gas phase is continuous and carries fragments of magma of various sizes upward.

This first volcanic regime, that of lava flows, is driven from the conduit by a mixture of magma containing small bubbles in suspension. This is called a bubbly flow in the two-phase flow literature (Figure 14.2a). The second regime, observed during Strombolian activity, corresponds to a vertical series of large bubbles rising in the conduit and is called a slug flow (Figure 14.2b). The third regime, displayed by high-level lava fountains typical of some eruptions at Etna (Figure 14.1c), is driven by an inner gas core surrounded by large fragments of magma (Figure 14.2c). This is typical of an annular flow, in which

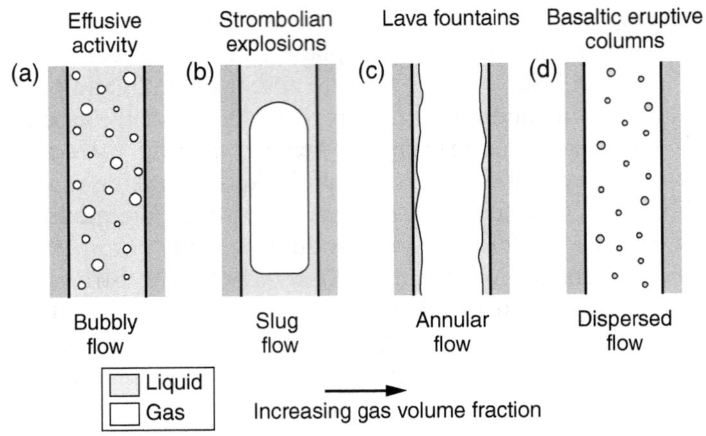

Figure 14.2 Two-phase flow (liquid and gas) in a vertical conduit (adapted from Figure 1a in *Vergniolle and Jaupart* [1986] and Figure 6 in *Vergniolle and Mangan* [2000]. (a) Bubbly flow, i.e., a suspension of small bubbles within a liquid, here the magma (shaded part). This regime corresponds to effusive activity (Figure 14.1a). (b) Slug flow, i.e., driven by a quasi-cylindrical bubble almost as large as the conduit. The flow in the conduit is made by a vertical series of slugs corresponding to Strombolian explosions (Figure 14.1b). (c) Annular flow, i.e., a central inner gas core surrounded by a liquid annulus. During a high-level lava fountain, such as at Etna (Figure 14.1c), the inner gas core drives the magma upward. (d) Dispersed flow, i.e., a suspension of liquid clots carried upward by a gas phase. This regime, which corresponds to a very high gas content, is displayed during basaltic eruptive columns (Figure 14.1d). The gas volume fraction α increases from bubbly ($\alpha \leq 10\%$) to slug ($10\% \leq \alpha \leq 80\%$) annular and dispersed flow [*Wallis*, 1969]. The above values, given for a vertical conduit, are only indicative because each transition depends on many parameters, such as gas volume fraction, gas flux, and magma viscosity.

the annulus of liquid around the gas is carried up by the central gas flow in the conduit (Figure 14.2c). In this regime, the shearing of the annular liquid film by the high-velocity central gas core can sometimes partly fragment the liquid into small droplets. These droplets are then carried upward by the gas jet [*Wallis*, 1969], potentially obscuring the view of the inner gas core, such as perhaps at Kīlauea. The fourth regime in the conduit, that producing high-level eruptive columns, occurs when the droplets of magma are carried upward by a continous gas phase (Figure 14.2d). This regime, called a dispersed flow, is opposite to that of the bubbly flow in which the continuous phase is the liquid phase (Figure 14.2a). The potential existence of liquid droplets inside the inner gas core during a strong annular flow makes the third and the fourth regimes (Figure 14.2c and 14.2d) representative of two end members. When the concentration in small liquid droplets is high, the inner central gas core driving the annular flow can be hidden, as proposed for the high-level lava fountains at Kīlauea [*Jaupart and Vergniolle*, 1988, 1989; *Vergniolle and Jaupart*, 1990]. The lack of visual observation of a central gas core at Kīlauea has led to an alternate model where high-level lava fountains are driven by a dispersed flow [*Parfitt and Wilson*, 1995, 1999, *Parfitt et al.*, 1995].

The gas volume fraction is small, a few to a few tens of percents solely for the bubbly flow, around 70% for slug flow, and greater for annular and dispersed flows [*Wallis*, 1969]. Although these classical values can change with magma viscosity and the angle of the conduit with the vertical axis, the fact remains that the gas volume fraction increases from bubbly flow to annular and dispersed flow. One way to make a transition from a bubbly flow to the slug flow, and then annular flow and dispersed flow, is by increasing the gas content in the conduit [*Vergniolle and Jaupart*, 1986]. It has been recently proposed, based on laboratory experiments performed in a long vertical tube (6.5 m), that the transition from bubbly flow toward annular flow (Figures 14.2a and 14.2c) is expected for gas volume flow rates of 10^3–10^4 m^3/s [*Pioli et al.*, 2012]. Other ways for the transition between regimes involving two-phase flows will be reviewed below. The characteristic of all these basaltic regimes is that they are driven by the gas phase, highlighting the critical need for measuring the gas volume during eruptions and understanding its formation at depth.

14.3. CONSTRAINING DEGASSING THROUGH MULTIDISCIPLINARY METHODS

The degassing at a vent can take various forms during Strombolian activity and Hawaiian activity. Different techniques are more or less suitable to different types of volcanic degassing. Four types of degassing may exist during classical Strombolian activity. The first type corresponds to the diffuse degassing from the flanks of the volcanoes and fumarolic activity [*Allard et al.*, 1994, 2008]. The second type is represented by small bubbles freely bursting at the top of the magma column, not detected on infrasonic records due to their relative small diameter [*Vergniolle and Brandeis*, 1996]. The third type is made by the very small pressure transients called puffs [*Ripepe et al.*, 1996, 2002; *Ripepe and Gordeev*, 1999; *Harris and Ripepe*, 2007a; *Patrick et al.*, 2007], which form at a very shallow depth, 11–150 m [*Burton et al.*, 2007a]. They are often very frequent, one every 2s, and represent most of the persistent background degassing. The fourth type is that of Strombolian explosions. In fact, small explosions may even need to be considered separately to typical explosions, due to their relatively short repose periods and shallower origin, 0.76–1.9 km instead of 2.6–3 km [*Burton et al.*, 2007a]. For simplicity, all Strombolian explosions here are considered as typical Strombolian explosions. The third and the fourth types are called active degassing, as they correspond to over-pressurized gas. This classification can be extended to Hawaiian eruptions, perhaps associating the puffs at Stromboli to the gas-piston activity at Kīlauea and the Strombolian explosions to the high-level lava fountains.

Our current understanding of eruptive behavior relies on constraints brought by measurements of the gas volume. This parameter can be estimated either quasi-directly from measuring gas velocity or gas composition or indirectly from seismic or acoustic records.

14.3.1. Quasi-Direct Measurements of Gas Volume

14.3.1.1. Gas Volume From Gas Velocity

Any technique that constrains the gas velocity at the vent, U_g, can also give an estimate of the gas volume V_g. Conservation of the gas flux at the vent leads to the total gas volume expelled by a high-level lava fountain, a Strombolian explosion or any other regime, equal to [*Vergniolle and Jaupart,* 1990]:

$$V_g = U_g S_{\text{vent}} t_g, \tag{14.1}$$

in which S_{vent} and t_g are the vent area and the duration of the event, respectively. The gas velocity can be estimated by using photos, classical videos [*Chouet et al.*, 1974; *Wilson*, 1980; *Ripepe et al.*, 1993], and high-speed videos [*Taddeucci et al.*, 2012a, 2012b]. Thermal video cameras can also follow the gas front propagating upward [*Calvari et al.*, 2006; *Harris and Ripepe*, 2007a; *Patrick et al.*, 2007; *Delle Donne and Ripepe*, 2012]. Videos have a better accuracy than a series of photos for following the temporal evolution of the gas velocity during the event, however in a reduced amount of spatial resolution. More sophisticated devices, such as radar and sodar measurements, are also used [*Weill et al.*, 1992; *Hort and Seyfried*, 1998; *Dubosclard et al.*, 1999; *Gouhier and Donnadieu*, 2011]. These instruments can provide both the gas and the ejecta velocity at a rate of about 1 s. All the techniques described above give an estimate of the gas velocity above the vent. The distance above the vent is sometimes significant for technical reasons related to the site of measurements, such as for radars. However, the release of hot gas with an upward velocity entrains air from the surrounding atmosphere with a horizontal velocity U_{horg} that depends on the vertical gas velocity U_{vertg} as [*Sparks et al.*, 1997]

$$U_{\text{horg}} = \alpha U_{\text{vertg}}, \tag{14.2}$$

in which α is the entrainement coefficient, varying between 0.12 for plumes, i.e., a sustained gas flow, and 0.28 for thermal, i.e., an instantaneous release of hot gas. Therefore, measurements of gas velocity above the vent may be significantly different from that at the vent.

14.3.1.2. Gas Volume From Gas Composition

The mass flux of CO_2 was estimated at the summit of Kīlauea, between 4900 and 8500 t/day in June–July 2003 and 1995–1999, respectively, from measuring the CO_2/SO_2 ratio during summit traverses and simultaneously measuring the SO_2 flux [*Gerlach et al.*, 2002; *Hager et al.*, 2008]. The apparent decrease of the mass flux of CO_2, with the very low value of 80 t/day in 2009 [*Edmonds et al.*, 2013], could mark the progressive depletion of the deep magmatic gas (CO_2). Similarly at Stromboli, the time-averaged CO_2 flux, estimated at 1.4×10^3 t/day during normal activity [*Allard et al.*, 2008], reaches peaks at $\approx 1.1 \times 10^4$ t/day during paroxysmal activity [*Aiuppa et al.*, 2010a]. The dramatic increase in CO_2 flux during paroxysms has been the basis of several models aimed at understanding paroxysmal activity at Stromboli (see Section 14.6.2.).Unfortunately, this technique does not constrain the mass flux of H_2O.

The gas composition can now be determined by using spectroscopy, such as an open-path Fourier transform infrared (FTIR) spectroscopy [*Oppenheimer et al.*, 2004; *Allard et al.*, 2005; *Burton et al.*, 2007a; *Edmonds and Gerlach*, 2007; *Sawyer et al.*, 2008]. This instrument, now light and portable, measures the ratio between diatomic gases such as H_2O/SO_2 and CO_2/SO_2 directly just above the vent. The amount of air, which is engulfed into volcanic gases and hence measured by FTIR, depends on both the gas velocity (equation 14.2) and the distance between the field of view and the vent. However, the correction for such an atmospheric contamination is often deduced by plotting all the data points of the time series with the concentration on a magmatic gas on the x axis (SO_2 or HCl) versus an atmospheric gas (CO_2 or H_2O) on the *y* axis. The *y* intercept represents the average atmospheric concentration in CO_2 and H_2O in the field of view for the whole measurement period.

Therefore high-energy explosions, associated with a high gas velocity, are corrected exactly as low-energy explosions, with low gas velocity, despite a potentially significant difference between air entrainment (equation 14.2) and atmospheric contamination. Using the average atmospheric contamination leads to overestimates of CO_2 and H_2O content for high-energy explosions and underestimates for inter-explosion periods and perhaps also for low-energy explosions. In some cases, the CO_2 and H_2O concentrations during explosions are obtained by substracting the gas composition measured prior the explosion to that during the explosion [*Burton et al.*, 2007a], but the difference in gas velocity prior and during an explosion is significant.

Once the FTIR measurements are corrected for atmospheric contamination, they are then combined to SO_2 measurements to estimate the gas volume, such as at Kīlauea [*Edmonds and Gerlach*, 2007], Erta 'Ale [*Oppenheimer et al.*, 2004; *Saywer et al.*, 2008] and Stromboli [*Burton et al.*, 2007a]. SO_2 measurements are now routinely performed on many volcanoes, such as at Kīlauea [e.g., *Sutton et al.*, 2001; *Koeppen et al.*, 2013], and Stromboli [e.g., *Allard et al.*, 1994; *Tamburello et al.*, 2012]. However, these measurements are sometimes difficult to interpret when analyzed in parallel with volcanic activity. For example, at Erta 'Ale, the SO_2 flux expelled close to the encrusted and apparently inactive lava lake within the northern crater, 0.6 kg/s, is almost as high as that released by the hot and active lava lake in the southern crater, 0.7 kg/s [*Oppenheimer et al.*, 2004]. This suggests that the value of 0.7 kg/s measured at the active lava lake within the southern crater results from diffuse and fumarollic degassing, with a SO_2 flux equal to 0.6 kg/s by analogy to the northern crater, and from the degassing associated with the active pool of hot magma, with a SO_2 flux only equal to 0.1 kg/s. Therefore, a value of 0.1 kg/s should be used for assessing the SO_2 flux expelled by the hot magma located in the active lava lake within the southern crater rather than 0.7 kg/s, as currently used. This interpretation decreases the total gas volume by a factor 7 when using the gas composition given by FTIR measurements and the SO_2 flux estimated at the southern hot lava lake.

14.3.2. Indirect Measurements of Gas Volume

Gas volume can also be estimated by a relatively new geophysical technique, using the infrasound or audible sound produced by volcanoes (Figure 14.3, redrawn from Figure 7 in the work of *Vergniolle and Ripepe* [2008]). This technique can only measure the overpressurized gas, the only one that produces a signal detectable by acoustic devices. The distance to the vent, which is variable depending on the danger or accessibility to the vent, can reduce the number of detectable signals. The results hence depend indirectly on the distance between the site of recordings and the vent. Acoustic measurements therefore only provide a minimum estimate of the active gas volume, i.e., the one corresponding to overpressurized gas.

Three methods currently exist for estimating the gas volume from acoustic measurements. The first one uses the acoustic power and a dimensionless analysis to estimate the gas velocity if the type of the acoustic source is assumed (monopole, dipole, quadrupole) [*Woulff and MacGetchin*, 1976; *Vergniolle et al.*, 2004; *Vergniolle and Caplan-Auerbach*, 2006]. This technique is very robust if the size of the source is known, e.g., when the conduit directly finishes at the surface without any large and deep pool of magma above the conduit in which the gas can expand. The second type is developed for the monopole source, a sudden variation of mass flux as for Strombolian explosions. In this case, a series of two successive integrations of acoustic pressure gives access to the gas volume [*Johnson et al.*, 2008a]. However, this method is very sensitive to the choice of the beginning and the end of the explosion and the likely existence of an offset in the acoustic recordings. But the size of the source is not required, in contrast to first method, and an estimate of the gas volume after decompression at atmospheric pressure can easily be obtained. The third method is more complicated to develop, as it requires a model for the source of the sound. It consists in calculating a synthetic acoustic waveform and by using a best fit to the measured acoustic waveform, from which the characteristics of the source, mostly its gas volume and overpressure, are determined. For example, the sound produced by a Strombolian explosion can be produced by the volume mode of an overpressurized bubble prior to its bursting (Figure 14.4a) [*Vergniolle and Brandeis*, 1996] as reproduced by laboratory experiments under specific conditions [*Kobayashi et al.*, 2010; *Lane et al.*, 2013] or by the formation of a small hole releasing the gas trapped within the remaining cavity of the bubble, a model called the Helmhotz resonator [*Vergniolle and Caplan-Auerbach*, 2004]. The interest in using a direct model and the calculation of synthetic waveforms (Figure 14.4b) is that it provides constraints on two independant parameters, the gas volume and its overpressure. The gas velocity can also be deduced and can potentially be compared to other geophysical techniques described above. The comparison between the gas velocity deduced from acoustic power and that obtained from assuming that the sound is produced by the volume mode of the bubble prior to its bursting gives a very good agreement for Strombolian explosions at Shishaldin (Figure 14.4c). This agreement, which exists because the source of Strombolian explosions

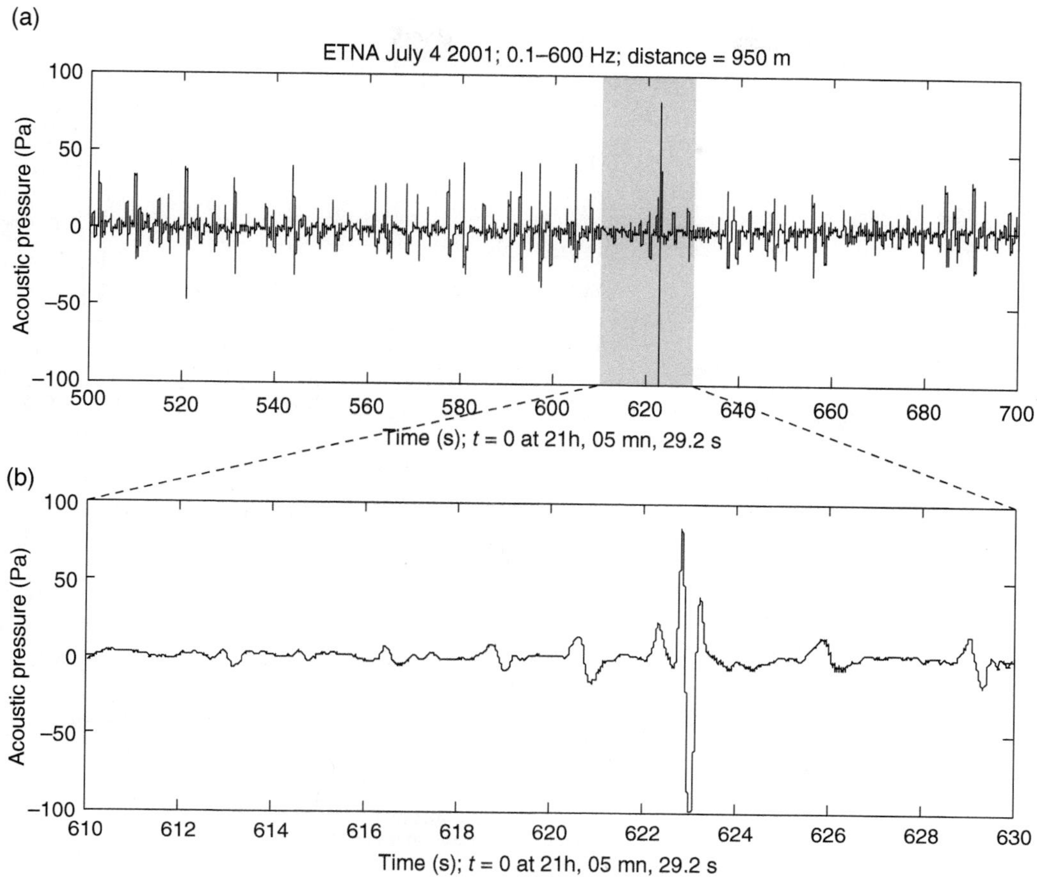

Figure 14.3 Acoustic pressure measured at 950 m from the southeast crater at Etna volcano on 4 July 2001 during a typical Strombolian activity (redrawn from Figure 7 in *Vergniolle and Ripepe* [2008]). Each main pulse in acoustic pressure, such as enlarged in (b), results from the breaking of a large bubble at the top of the magma column.

is relatively confined, reinforces the validity of the model in that case [*Vergniolle et al.*, 2004].

Inversions of seismic waveforms have also been performed for explosions at Stromboli. The best fit between synthetic and measured seismic waveforms gives also access to gas volume, ≈200 m^3, and to the force, 10^8 N, driving the seismic signal [*Chouet et al.*, 2003, 2008]. As for inversion of acoustic waveforms, this technique requires a fairly good knowledge of the physical processes at work and can solely estimate the overpressurized gas.

Infrasound measurements prior to a Strombolian explosion may contain some information on the rising nearby slug. The rise of an overpressurized slug far from the top of the magma column, i.e., at a distance below the surface exceeding 10 times the slug radius, oscillates in volume by changing its length (Figure 14.5) rather than by changing its radius, as when reaching the surface (Figure 14.4b) [*Vergniolle et al.*, 1996]. The existence of the slug length mode, as being the main mode excited during the rise of a slug in a conduit, has been shown using analog laboratory experiments [*James et al.*, 2004]. The slug length mode has also been observed on acoustic recordings at Stromboli [*Vergniolle et al.*, 1996]. However, it is difficult to estimate the slug length from the frequency of the main acoustic pulse with accuracy, because this requires knowledge of the slug depth below the surface at each time prior to the slug breaking. The location of the slug in the conduit can be roughly deduced if the slug diameter is known, hence its velocity. The slug length estimated from acoustic records, just prior to the explosion and decompressed at the vent pressure, is in good agreement with its value at the surface deduced from synthetic acoustic waveforms [*Vergniolle et al.*, 1996]. It has been suggested on qualitative grounds that this longitudinal slug mode could also act as a seismic source [*James et al.*, 2004].

The comparison between the gas volumes estimated by different techniques might be complicated by the use of different techniques, which may measure different types of degassing. Estimating the amount of overpressurized gas is crucial because it corresponds to gas coming from depth, and hence it may carry some information on the processes in the conduit or in the reservoir.

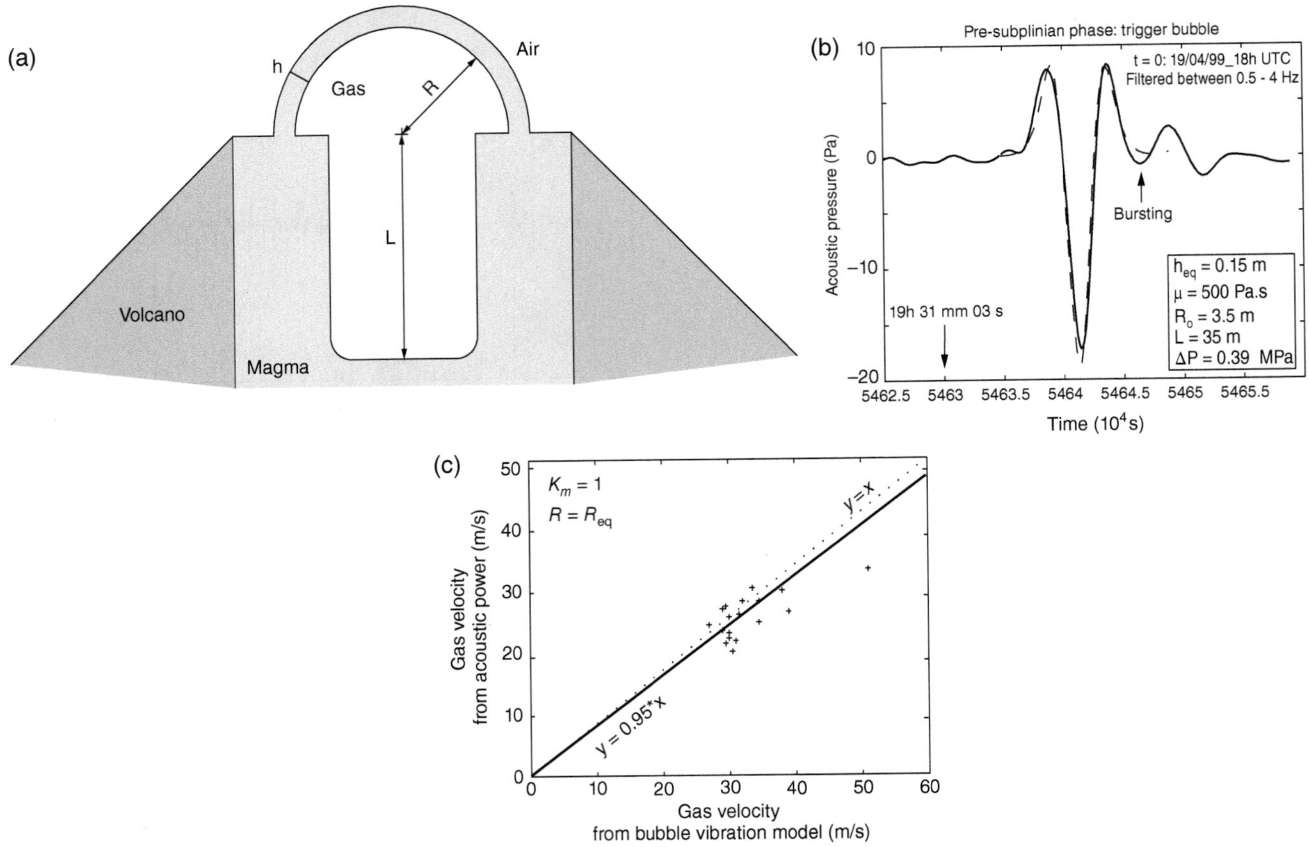

Figure 14.4 (a) Model used to explain how an overpressurized bubble can radiate acoustic pressure (Figure 3b in *Vergniolle and Brandeis* [1996]). (b) Best fit between synthetic (dashed line) and measured (plain line) acoustic pressure radiated by a Strombolian explosion during the 1999 eruption of Shishaldin (Figure 12 in *Vergniolle and Caplan-Auerbach* [2004]). (c) Comparison between the gas velocity estimated from the bubble vibration model and from acoustic power when assuming that the radius of the source is equal to the equilibrium radius deduced from the bubble vibration model (Figure 17b in *Vergniolle et al.* [2004]).

14.3.3. Comparison Between Methods Estimating the Gas Volume

At Stromboli, the use of synthetic acoustic waveforms constrains the gas volume of an explosion to 10–100 m^3 [*Vergniolle and Brandeis*, 1996], in agreement with synthetic seismic waveforms [*Chouet et al.*, 2003], while the puffs are slightly smaller, 25–35 m^3 [*Ripepe and Marchetti,* 2002]. The photoballistics study from photos taken in 1971 of two explosions leads to a gas volume of $\approx 10^3$ m^3 [*Chouet et al.*, 1974], which is similar to that estimated from more recent radar measurements at $\approx 1.4 \times 10^3 m^3$ [*Weill et al.*, 1992; *Hort et al.*, 2003]. FTIR measurements deduce similar values for the gas volume, 1.5–4 $\times 10^3 m^3$ [*Burton et al.*, 2007a; *Mori and Burton*, 2009]. The comparaison between all the direct techniques (photoballistics, radar, and FTIR) show an excellent agreement. Acoustic and seismic recordings give values lower to that obtained by the more direct methods by one order of magnitude. This difference could be due to different vents or activity being measured. But the main cause of discrepancy is related to the fact that acoustic and seismic measurements provide an estimate on the overpressurized gas and not on the total gas (overpressurized and non-overpressurized). Furthemore, all techniques, except acoustic and seismic measurements, measure the total gas volume after air entrainment and this could also lead to a significant overestimate, even for insufficiently corrected FTIR measurements, .

The same discrepancy between acoustic measurements and other techniques exists at Etna. The inversion of acoustic waveforms gives a gas volume during its Strombolian activity of $\approx 2 \times 10^3 m^3$ [*Vergniolle and Ripepe*, 2008; *Cannata et al.*, 2009], in agreement with video recordings. Radar and FTIR measurements give a gas volume at $\approx 7 \times 10^4$ m^3 [*Dubosclard et al.*, 1999] and at several 10^4 m^3 [*Allard et al.*, 2005], respectively. Such large bubbles rise in the conduit with a high Reynolds number, a dimensionless number that assesses the relative importance of inertia versus viscous forces [*Brennen*, 1995],

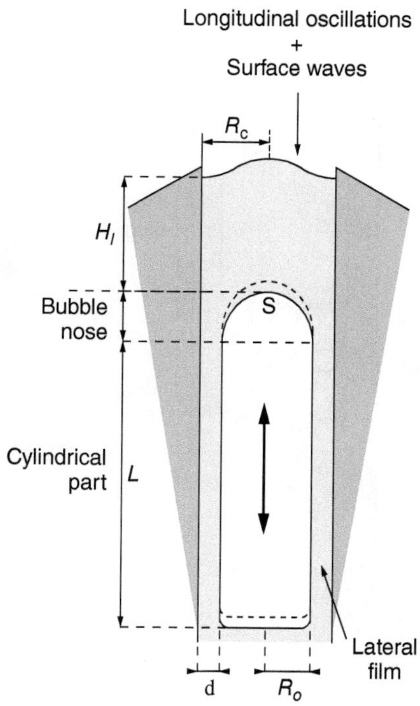

Figure 14.5 Sketch showing how an overpressurized slug rising in a conduit can contribute to the frequencies present at the surface of the magma column (Figure 8a in *Vergniolle et al.* [1996]). When the slug is not yet very close to the surface, the slug oscillates in volume by changing its length *L* (longitudinal mode). The periodic change in the slug volume pushes the magma column up and down with the same frequency and may induce surface waves at the top of the magma column. Both can be recorded as infrasonic acoustic pressure.

equal to 10^3–10^4 [*Vergniolle and Ripepe*, 2008]. In this condition, the flow at the bottom of the slug is turbulent and a bubbly wake develops (Figure 14.6). The bubbly wake, with a volume between 4 and 6 times that of the driving slug, is a source of non-overpressurized gas, untraceable on acoustic records [*Vergniolle and Ripepe*, 2008]. If the tendency of acoustic measurements is to underestimate the total gas volume when compared to radar or FTIR measurements, this can be explained at Etna with the existence of a bubbly wake. However, this interpretation is not valid for Strombolian explosions at Stromboli, whose Reynolds number, ≈80 for a magma viscosity equal to 300 Pa.s (as detected from acoustic measurements [*Vergniolle et al.*, 1996]), is too small to induce the formation of a large wake. Another alternative to explain the discrepancy at Stromboli could be related to the existence of a thin foam at the top of the magma column, which can be broken by the breaking slug. The existence of an uppermost foam overlying the magma at the surface or of a bubbly wake is important to consider when interpreting results on gas composition at the vent.

Passive degassing, i.e., that related to non-overpressurized gas and sometimes estimated from FTIR measurements, is overwhelmingly volumetrically dominant at Stromboli due to the short duration of explosions: It represents 92%–97% of the total degassing, which includes the active degassing related to overpressurized gas [*Allard et al.*, 1991, 1994, 2000; *Burton et al.*, 2003, 2007a]. At Erta 'Ale, FTIR measurements also show the major contribution of passive degassing on the total gas volume,with passive degassing representing 85% of the total degassing [*Oppenheimer et al.*, 2004]. As described above, various modes of degassing exist at relatively "simple" systems, such as at Stromboli or Erta 'Ale, which underpins the interest in performing acoustic and seismic measurements together with other techniques to assess the active gas volume whose origin is at depth.

14.4. BUBBLE FORMATION AND GROWTH

Bubbles can form in the magma reservoir, in the conduit, or at the surface. Their formation results from the depressurization of the magma in which the gas was initially dissolved or from the relative enrichment of volatile species, such as during fractional crystallization. In this section, we describe in detail the various processes at work during bubble nucleation, growth, and coalescence when the bubble remains relatively small, less than a few centimeters in diameter.

14.4.1. Bubble Formation

The first stages of a bubble's life involve many physical and chemical processes in the magma with complex feedback loops [e.g., *Sparks*, 1978; *Sparks et al.*, 1994; *Navon and Lyakvosky*, 1998], making the full problem extremely difficult to solve. The bubble has first to nucleate and the newly formed bubble should have a radius exceeding a critical value for the nuclei to be stable. The bubble nucleation requires that a significant amount of energy is available to form a new surface in order to overcome the interfacial forces governed by surface tension. The extra energy is given by the supersaturation existing in the liquid between the dissolved volatile species and that given by the equilibrium law, called Henry's law [*Mysen*, 1977; *Stolper and Holloway*, 1988; *Dixon and Stolper*, 1995; *Dixon et al.*, 1995; *Lesne et al.*, 2011a,b,c].

Homogeneous nucleation predicts that a very large amount of supersaturation, 10–30 MPa, is required for a new bubble to be formed [*Lensky et al.*, 2006]. Indeed, a bubble at its critical radius has a very large interfacial energy, due to its very small diameter, a few to several tens of micrometers [e.g., *Sparks et al.*, 1994: *Lensky et al.*, 2006; *Pichavant et al.*, 2013]. An ongoing study on basalts saturated in CO_2 at 2 GPa suggests that the

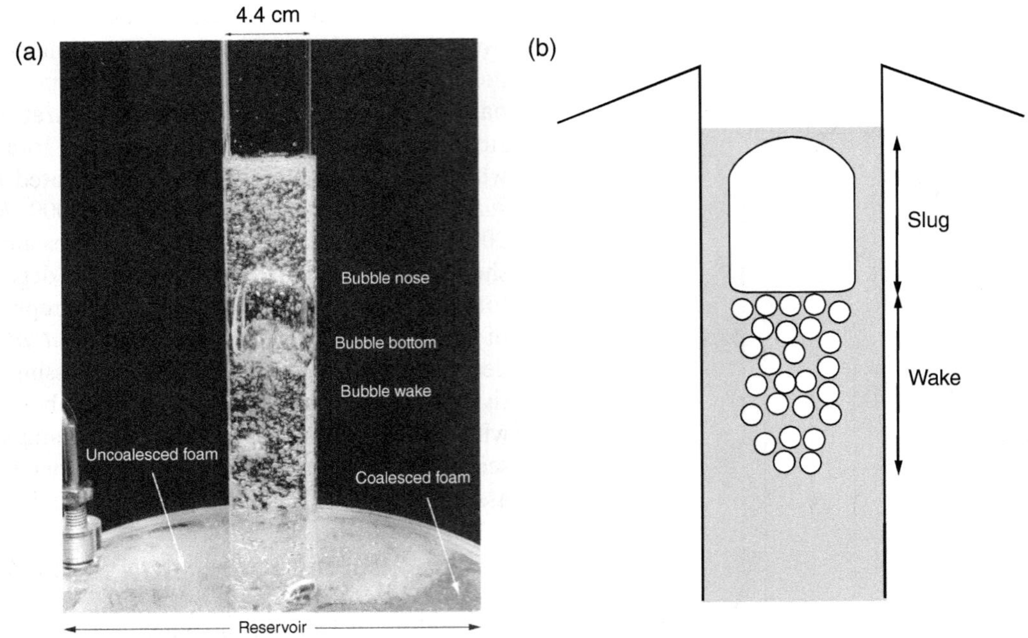

Figure 14.6 (a) Laboratory experiment of a slug flow rising in a tube of diameter 4.4 cm and produced from the coalescence of a foam trapped in the reservoir (Figure15 in *Vergniolle and Ripepe* [2008]). The slug bottom has the shape of an inward dome. The slug wake is well visualized by the trajectory of the small bubbles, which are solely in fast motion at the rear of the slug. The slug Reynolds number, ≈96, is much smaller than close to the surface on Etna, and the wake has a size between half and equal to the one of the slug. At Etna and Erta 'Ale, the volume of the wake is four to six times larger than the volume of the slug [*Vergniolle and Ripepe*, 2008; *Bouche et al.*, 2010]. The small bubbles in the slug wake correspond to passive degassing whereas the slug is associated to active degassing, i.e., corresponds to overpressurized gas. (b) Sketch showing the rise of a slug for relatively fluid basalts (Figure 17 in *Vergniolle and Ripepe* [2008]). When the Reynolds number of the bubble exceeds 30 [*Brennen*, 1995], a wake develops at the bottom of the slug that can engulf small bubbles, making it locally bubbles rich.

homogeneous nucleation may require a very large supersaturation (1 GPa) [*Hardiagon et al.*, 2013].

The nucleation of bubbles can also occur at preferential nucleation sites, such as given by crystal surfaces, as shown for silicic magma. In this case, heterogeneous nucleation of bubbles often is easier in terms of energy and requires less supersaturation than homogeneous nucleation for a silicic magma [*Hurwitz and Navon*, 1994; *Gardner and Denis*, 2004]. The supersaturation required for homogeneous nucleation in a rhyolitic magma is equal to 15–20 MPa, while the heterogeneous nucleation occurs at 1–5 MPa [e.g., *Mourtada-Bonnefoi and Laporte*, 1999, 2002, 2004; *Mangan and Sisson*, 2000; *Gardner and Denis*, 2004]. A similar difference between homogeneous and heterogeneous nucleation may be expected for basaltic magma. For heterogeneous nucleation of bubbles on crystals, it is highly dependent on the wetting angle between the crystal and the bubble, i.e., the angle between the crystal-liquid interface and the bubble-liquid interface seen through the liquid side [*Hurwitz and Navon*, 1994; *Gardner and Denis*, 2004; *Cluzel et al.*, 2008]. This parameter is not well known for basaltic magma due to the difficulty of measurements. Some microlites, e.g., Fe-Ti oxides, are efficient nucleation sites due to their high wetting angle between the crystal-liquid interface and the bubble-liquid interface seen through the liquid side (158° for magnetite) [*Hurwitz and Navon*, 1994; *Mourtada-Bonnefoi and Laporte*, 2002; *Cluzel et al.*, 2008]. In contrast, plagioclase and quartz are inefficient nucleation sites due to their low wetting angle, ≤20°, [*Hurwitz and Navon*, 1994]. Another important factor to consider is the number of crystals available for heterogeneous nucleation [*Hurwitz and Navon*, 1994; *Navon and Lyakvosky*, 1998; *Mangan and Sisson*, 2000; *Cluzel et al.*, 2008], and this value in situ is not very well known. In any case, once triggered, the bubble nucleation, quantified by the nucleation rate, is very fast [*Gardner*, 2012].

The critical bubble radius, above which a growing bubble is stable, depends mostly on surface tension [e.g. *Proussevitch et al.*, 1993b; *Hurwitz and Navon*, 1994; *Lensky et al.*, 2006; *Cluzel et al.*, 2008]. A small surface tension leads to a large critical radius and a relatively easy bubble nucleation compared to when the surface tension is large [*Hurwitz and Navon*, 1994]. The surface tension coefficient is large for a basalt depleted in water, 0.36–0.4 kg/s^2 [*Walker and Mullins*, 1981; *Proussevitch and Kutolin*, 1986], and decreases significantly when the water content increases [*Khitarov et al.*, 1979; *Mangan and Sisson*, 2005], reaching 0.1 kg/s^2 at Etna due to the 3.1 wt% water [*Métrich et al.*, 2004]. Hence the nucleation is more easily achieved in a magma rich in water than in one depleted in water, making H_2O bubbles larger and easy to form in comparison with CO_2 bubbles in a dry magma [e.g., *Proussevitch et al.*, 1993b; *Hurwitz and Navon*, 1994; *Lensky et al.*, 2006; *Cluzel et al.*, 2008]. The surface tension coefficient, independent of pressure for a dry basalt, decreases with pressure when the magma is enriched in water: The decrease exceeds a factor of 2 when increasing the ambient pressure from atmospheric pressure to that corresponding to a depth of 4 km [*Khitarov et al.*, 1979; *Proussevitch and Kutolin*, 1986]. So H_2O bubbles do not form at the same critical radius at the depth of the magma reservoir than at the top of the conduit.

14.4.2. Bubble Growth

After its formation, a single bubble can grow either by diffusion or by decompression [e.g. *Lensky et al.*, 2004]. The growth by diffusion is driven by a local supersaturation in volatiles in the magma around the bubble than within the bubble. Just after a nucleation event, the magma may be still very enriched in the dissolved gas species far from any newly formed bubbles, due to the significant amount of supersaturation required at nucleation. This enhances the initial diffusive growth of the bubble nucleus, a mechanism already very efficient for very small bubbles [*Navon and Lyakvosky*, 1998].

The diffusivities of H_2O and CO_2 in basalts increase with temperature and water content but decrease with pressure [*Zhang and Ni*, 2010]. The diffusivity of CO_2 in basalt is independent of CO_2 concentration and only very weakly dependent on the melt composition, a significant difference from H_2O. The diffusivity of CO_2 at 1200°C and a depth of 2 km varies between 6.5×10^{-12} and $3.7 \times 10^{-11} m^2/s$ for a water content between 0.15 and 3.1 wt%, such as at Erta 'Ale [*Field et al.*, 2012] and Etna [*Métrich et al.*, 2004; *Spillaert et al.*, 2006], respectively. The diffusivity of H_2O, between 6.7×10^{-11} and $1.4 \times 10^{-9} m^2/s$ at 1200°C in the same conditions [*Zhang and Ni*, 2010], is larger than that of CO_2, making the diffusive growth a much slower process for CO_2 bubbles than H_2O bubbles.

The bubble can also grow by decompression, following the perfect gas law for depths below several kilometers [*Papale*, 1999; *Papale et al.*, 2006]. The decrease of the ambient pressure around a bubble can result from its rise in the plumbing system or from following a decompression event, such as at the initiation of an eruption or when a significant magma volume is removed from the conduit at the vent. For example, the magma rises in the conduit during the emplacement of a lava flow but also during convective motions in the conduit, such as those currently existing in the open conduit at the summit of Kīlauea [*Carey et al.*, 2013]. The differential motion of the bubbles with respect to the magma is to be expected in the shallow magma reservoir: The bubble diameters, estimated during lava fountains at Kīlauea and Etna, during the lava lake activity at Erta 'Ale are significant, with values equal to 0.2–0.3 mm [*Vergniolle and Jaupart*, 1990; *Vergniolle*, 2008], 0.2–0.7 mm [*Vergniolle*, 2008; *Vergniolle and Gaudemer*, 2012], and 0.4–0.6 mm [*Vergniolle and Bouche*, submitted], respectively. Independently, the minimum bubble diameter is estimated to be 0.1–1 mm at a pressure of 10–100 MPa using the bubble diameter measured in basalts erupted at a pressure of 30 MPa along the mid-Atlantic ridge [*Sarda and Graham*, 1990] and decompression experiments using basalts with mixed H_2O-CO_2 fluids [*Mangan et al.*, 2014]. However the increase in the bubble diameter between the bottom and the top of the magma reservoir is never very large due to the large confining pressure at several kilometers depth. For example, the bubble diameter increases by a factor of 1.3, from 0.2 to 0.25 mm, in a reservoir located between 2 and 4 km, assumed for simplicity to be that of the summit reservoir of Kīlauea [*Yang et al.*, 1992; *Baker and Amelung*, 2012; *Poland et al.*, 2014]. Furthermore, the differential rise of a small isolated bubble, which is given by the Stokes equation [*Wallis*, 1969], occurs in magma reservoir located between 4 and 2 km depth on a very long time scale, equal to seven months at Kīlauea, for a bubble diameter and a magma viscosity equal to 0.2 mm and 10 Pa.s, respectively. The transition between a regime dominated by a bubble growth due to diffusion and a regime due to decompression depends on viscosity, initial supersaturation pressure, and decompression rate [*Toramaru*, 1995].

14.4.3. Bubble Coalescence

Another mechanism that can significantly increase bubble diameter is coalescence. Coalescence of bubbles in the conduit has been modeled as a binary process between two growing bubbles, ignoring the liquid flow between the two adjacent bubbles [*Lovejoy et al.*, 1994]. However, the different steps leading to coalescence involve first the approach of the two bubbles toward one another. In a dilute suspension, with a gas volume fraction less than 2%, the small bubbles are at a far distance from one another, making the occurrence of coalescence probably unlikely. Gas volume fraction at the top of shallow magma reservoirs is expected to be small at Kīlauea during the Mauna Ulu and Pu'u 'Ō'ō eruptions with 0.7%–1.3% [*Vergniolle and Jaupart*, 1990; *Vergniolle*, 2008], at

Etna with 0.005%–1% [*Vergniolle*, 2008; *Vergniolle and Gaudemer*, 2012], and at Erta 'Ale with 0.03%–0.12% [*Vergniolle and Bouche*, Submitted] but largely above 2% during the Kīlauea Iki eruption [*Vergniolle*, 2008] and the 1999 eruption at Shishaldin [*Vergniolle*, Submitted]. Therefore, the number of bubbles during the Kīlauea Iki eruption or the 1999 eruption at Shishaldin may have been sufficient to allow two bubbles to become in close contact in the magma reservoir. Furthermore, a vertical series of rising bubbles have more chance to interact and coalesce if their Reynolds number is below 1 as for small bubbles (≤3 and ≤14 cm for magma viscosity of 10 and 100 Pa·s, respectively), making the occurrence of interaction and coalescence between bubbles with higher Reynolds numbers (Re ≥ 10) unlikely [*Suckale et al.*, 2010].

Other important factors, described below, also have to be considered. When the two bubbles are in close contact, the thinning of the interstitial liquid separating the bubbles becomes a difficult process. It first occurs by drainage under gravity when the film is still relatively large and is then combined with the resisting effect of capillary forces when the film has become very thin [*Nguyen et al.*, 2013]. The thinning can be enhanced by bubble growth or deformation [*Castro et al.*, 2012]. The coalescence between two bubbles is also difficult when the bubbles are non-deformable, as for small bubbles, but becomes much easier for deformable bubbles, although they may later break up [*Manga and Stone*, 1993, 1994, 1995; *Suckale et al.*, 2010]. Coalescence is also difficult when the two bubbles have a very different diameter, with the large one pushing aside the small ones during its rise [*Wallis*, 1969; *Sahagian*, 1985; *Sahagian et al.*, 1989]. In this case, the coalescence, if it happens, occurs at the bottom of the large bubble.

14.4.4. Bubbles in Foams

A large number of bubbles can also be in close contact and form a foam, such as at the top of degassing magma reservoirs [*Jaupart and Vergniolle*, 1988; 1989; *Vergniolle and Jaupart*, 1990]. The gas volume fraction in a pneumatic foam, such as formed by the rise of small and non-deformable bubbles, is between 0.525 (cubic packing) and 0.62 for a random settling of uniform spheres and can reach higher values when the distribution in bubble diameters is wide [*Happel and Brenner*, 1973). The gas volume fraction is equal to 0.74 when the bubbles within the foam are spheres of the same diameter, touching one another (i.e. without interstitial film at contact), and organized in a rhombohedral packing. This value is often taken as the threshold for foam coalescence in fragmentation studies [e.g., *Sparks*, 1978; *Proussevitch et al.*, 1993a; *Mangan and Cashman*, 1996; *Cashman and Mangan*, 1996; *Navon and Lyakvosky*, 1998]. Larger values of the gas volume fraction are possible when the bubbles deform to become polyhedral. The largest value found in natural samples is 0.98, as for the reticulites, ejected from high-level lava fountains [*Mangan and Cashman*, 1996]. Above a gas volume fraction of 0.74, the foam is called dry whereas below 0.74 the foam is named wet, the difference in name reflecting differences in the dominant physical processes at work. The bubbles within a foam, regardless of being wet or dry, can coalesce by drainage of interstitial films, by enlarging large bubbles at the expense of small bubbles, by a diffusive process called Ostwald ripening [*Proussevitch et al.*, 1993a; *Mangan and Cashman*, 1996; *Navon and Lyakvosky*, 1998], by massive coalescence when the uppermost bubbles within the foam are highly deformed under a solid surface [*Jaupart and Vergniolle*, 1989], and under fast decompression [*Namiki and Manga*, 2005, 2006, 2008].

14.4.5. Role of Gas Composition

The chemistry of the gas phase plays a large role in the bubble formation and growth, the effect of CO_2 being significant at concentrations above 880 ppm in the case of basalts sampled at Stromboli [e.g., *Bai et al.*, 2008]. The most important parameters, such as equilibrium composition, given by the exsolution law, surface tension (see Section 14.4.1.), gas diffusivity (see Section 14.4.2.), and magma viscosity, are very different for CO_2 and H_2O.

The amount of gas that can be dissolved in a magma depends on the pressure, the dependence being more marked for CO_2 [*Stolper and Holloway*, 1988; *Dixon et al.*, 1995] than for H_2O [*Mysen*, 1977; *Dixon et al.*, 1995]. The depth at which a gas phase should nucleate if nucleation could occur at equilibrium depends also on the initial gas content. It is classically assumed that CO_2 bubbles are formed much below a depth of several kilometers, hence within or below the magma reservoir [e.g., *Bottinga and Javoy*, 1990a,b; 1991; *Vergniolle and Jaupart*, 1990; *Métrich et al.*, 2004; *Allard et al.*, 2005], while H_2O bubbles are mostly formed in the conduit [*Wilson and Head*, 1981; *Parfitt and Wilson*, 1995, 1999]. However the exsolution of H_2O at equilibrium is possible at a depth of several kilometers if the initial H_2O content exceeds a few weight percent, such as for Etna (at 5.2 km depth for a magma containing 3.4 wt% of water [*Métrich et al.*, 2004; *Vergniolle*, 2008]) or at Cold Bay, an old volcanic center close to Shishaldin where the water content is 4.1 wt% [*Sisson and Grove*, 1993] and the exsolution depth is between 6.0 and 6.6 km [*Mysen*, 1977; *Dixon et al.*, 1995; *Vergniolle*, submitted]. Once CO_2 bubbles are formed in a basalt exsolving CO_2 before water, any volatile phase such as H_2O can diffuse into the pre existing bubbles to reach the equilibrium state given by the exsolution law [*Holloway and Blank*, 1994]. This process, called CO_2 fluxing, which is increasingly recognized in shallow basaltic magma reservoirs, enhances water loss [*Spilliaert et al.*, 2006; *Johnson et al.*, 2008b; *Métrich and Wallace*, 2008; *Collins et al.*, 2009]. The H_2O concentration around a bubble is often likely to be at equilibrium with the ambient pressure at the bubble surface with a concentration profile around the bubble dominated by the diffusion time scale prior to equilibrium. However, this is not the case when the bubble is carried rapidly downward by a convective loop, such as currently existing in the shallow conduit at Kīlauea's summit [*Carey et al.*, 2013]. The fast growth of a crystal depleted in H_2O (anhydrous plagioclase or quartz) can create a boundary layer enriched in water that could locally saturate the melt in water [*Best*, 2002] and trigger bubble nucleation. Similarly, the melt layer around CO_2-free crystals (olivine, plagioclase amphibole, clinopyroxene) growing in basalts will be enriched by the CO_2 unable to enter the crystal structure. Hence, in the vicinity of a quasi-stagnant CO_2 bubble and in the absence of fast crystallization, no significant supersaturation in H_2O is to be expected. However, when the bubble nucleation leads to a low gas volume fraction and in the absence of pre-existing bubbles, a significant H_2O supersaturation or CO_2 supersaturation might be left in the magma after the nucleation far from the diffusive pumping of newly formed bubbles when a large supersaturation existed prior to nucleation [e.g., *Mourtada-Bonnefoi and Laporte*, 1999, 2002; *Mangan and Sisson*, 2000].

The magma viscosity, which plays a role in the differential motion of a small bubble and also in its deformability during its rise [*Wallis*, 1969], is also very sensitive to the amount of water: The viscosity decreases by two orders of magnitude between a dry basalt and that with 3 wt% of water, similar to that of Etna [*Giordano and Dingwell*, 2003, 2008; *Métrich et al.*, 2004].

We now leave the relatively small bubbles to turn directly to the large bubbles at the vent, postponing the discussion on how initially small bubbles can be transformed into larger ones.

14.5. DYNAMICS OF DEGASSING AT OPEN VENTS

14.5.1. Lava Lake Activity

Lava lakes can be both short and long timescale features, such as at Kīlauea [e.g., *Jaggar*, 1920, 1924; *Richter et al.*, 1970; *Swanson et al.*, 1979; *Tilling et al.*, 1987; *Flynn et al.*, 1993], Erta 'Ale [*LeGuern et al.*, 1979; *Oppenheimer and Francis*, 1997; *Sawyer*, 2008a] (see also J.-L. Cheminee, personal communication 2003), or Nyiragongo [*Tazieff*, 1975, 1984; *Sawyer*, 2008b]. The surface is quite often composed of a mosaic of colder, quasi-rigid plates separated by long incandescent zones, sometimes with zigzag patterns [*Karlstrom and Manga*, 2006], representing active spreading [e.g., *Swanson et al.*, 1979; *Flynn et al.*, 1993; *Oppenheimer and Yirgu*, 2002; *Harris et al.*, 2005; *Harris*, 2008].

At Erta 'Ale, the velocity of the cold plates, often at 0.01 m/s, can reach peaks of 0.4 m/s, associated with periods of intense activity at the surface of the lava lake and active overturning [*Harris et al.*, 2005; *Harris*, 2008]. Periods of active overturning, lasting approximately several hours, have been associated with periods of high magma convection in the conduit, whereas periods of relative quiescence correspond to reduced convection velocities [*Harris et al.*, 2005; *Harris*, 2008]. Alternatively, the existence of regular periods of active overturning has been explained by the sudden breaking of the crust induced by the massive coalescence of a foam accumulated below the surface [*Vergniolle and Bouche*, submitted].

At Erta 'Ale, the degassing at the surface of a lava lake occurs at fixed positions, likely to be above the underlying conduit, at the boundary between cold plates, or as low-level lava fountain, i.e., a vertical and roughly cylindrical incandescent jet of gas and hot magma with a height and duration less than 20 m and 5 min, respectively (Figure 14.7; from Figure 3 in *Bouche et al.*, [2010]), not to be confused with the Hawaiian high-level lava fountains. Large bubbles rising from the conduit have a volume of 36–700 m^3 at Erta 'Ale and are only weakly overpressurized, $4 \times 10^3 - 1.8 \times 10^3$ Pa [*Bouche et al.*, 2010]. The degassing of these large bubbles, about one per hour, likely corresponds to mild Strombolian activity, whereby they rise in a conduit and a bubbly wake forms at the bubble bottom and periodically detaches (Figure 14.8; from Figure 8 in *Bouche et al.* [2010]). Detached wakes could be at the origin of the small and short lava fountains, previously recognised as a zone rich in gas [*Palma et al.*, 2008].

Studies on Nyiragongo lava lake are even rarer than that at Erta 'Ale, due to the difficult access and political unrest in the area. However, the description of the lava lake, much larger than that of Erta 'Ale, resembles that of Erta 'Ale, with a variable yet higher crust velocity due to the low magma viscosity [*Giordano et al.*, 2007], low-level lava fountains, small Strombolian explosions with visual observations of large bubbles with diameter of 10 m, and continuous degassing [*Tazieff*, 1975, 1984, 1996; *Sawyer*, 2008b].

The behavior of lava lakes at Kīlauea is similar to Erta 'Ale and Nyiragongo, where the lava lake is connected to an underlying conduit, although perhaps with different typical gas volumes and return times, i.e. delay between two successive slugs, than those of Erta 'Ale and Nyiragongo. Periodic active overturning episodes and low-level lava fountaining (mostly ≤30 m) were described between July 1986 and August 1990 at the Kupaianaha lava lake [*Flynn et al.*, 1993] as well as during 1972–1974 at Mauna Ulu [*Tilling*, 1987].

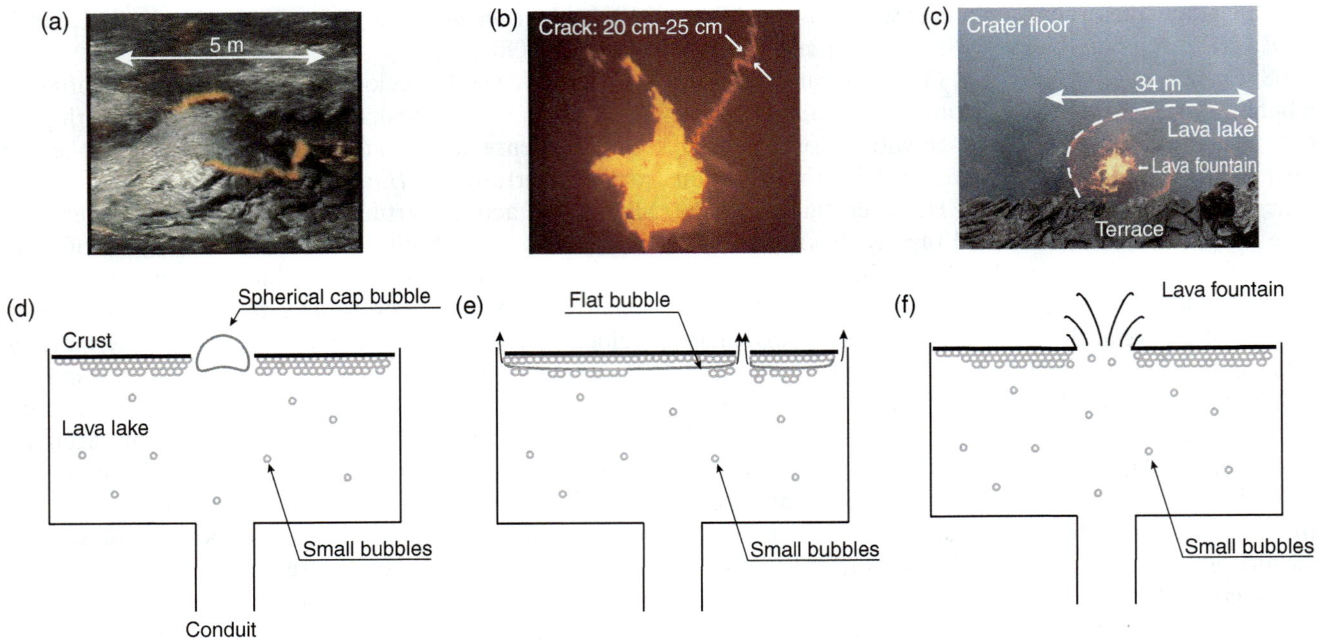

Figure 14.7 Different types of degassing at Erta 'Ale lava lake from photos (top figures) and from the interpretative sketchs of the lava lake (bottom figures) (from Fig. 3 in *Bouche et al.* [2010]): (a, d) large bubbles, coming from the conduit and breaking at fixed positions at the surface of the lava lake; (b, e) flat bubbles, resulting from the accumulation and collapse of a thin foam below the cooling crust; (c, f) lava fountains, resulting from a clustering of small bubbles.

Lava lake activity has restarted at the summit of Kīlauea in March 2008 but was not observed until September 2008, with a diameter of ≈35 m, which has enlarged to ≥160 m [*Orr et al.*, 2013]. Current activity at the Halema'uma'u lava lake consists of frequent weakly overpressurized degassing bursts [*Fee et al.*, 2010] accompanied by very long period seismic data, interpreted by the expansion and bursts of large slugs [*Chouet et al.*, 2010]. This view has now been revisited in the light of the results brought by a combined study of video and seismic measurements [*Orr et al.*, 2013]. This multidisciplinary study suggests that these small pressure transients are triggered by rockfalls [*Orr et al.*, 2013]. The summit lava lake at Kīlauea also degasses via passive bubble bursting at the surface, which has diameters between centimeters and meters. Current activity at the Halema'uma'u lava lake also includes small-level lava fountains (≤20 m) and crust motion at the surface, with a velocity of 0.1–0.2 m/s,in agreement with a model of the velocity of the convective loop in the underlying conduit [*Carey et al.*, 2013]. Occasionally the summit lava lake partially drains to feed a dike intrusion and a fissure eruption on the East Rift Zone, as on 5 March 2011 [*Carbone et al.*, 2013]. With continuous gravity measurements, when combined with visual and thermal images, this study suggests that the magma draining during that episode was of low density, 950 ± 300 kgm^3, indicating that the lava lake was rich in gas [*Carbone et al.*, 2013].

Lava lakes can also occur due to ponded perched lavas [*Patrick et al.*, 2011], but these are not discussed here.

14.5.2. Gas Piston Activity

Two types of degassing at Kīlauea have been called gas piston activity. The first type, also the first to be recognized, is characterized by a cyclic rise and fall of the lava column associated with a series of large bubbles breaking at the vent [*Swanson et al.*, 1979]. The second type, only recently characterized, is isolated from the magma plumbing system and is associated with gas accumulation beneath the viscoelastic component of the surface crust [*Patrick et al.*, 2011]. Gas pistons from type 2 are not discussed here. Gas pistons from type 1 exist during the whole eruption, such as during the Mauna Ulu and Pu'u 'Ō'ō eruptions, although their number increases prior to a high-level lava fountaining [*Swanson et al.*, 1979; *Wolfe et al.*, 1987, 1988]. The regular occurrence of gas pistons [*Chouet and Shaw*, 1991; *Ferrazzini et al.*, 1991] can be particularly clear on thermal records, ≈3 min in June–July 2002 at Pu'u 'Ō'ō Crater [*Johnson et al.*, 2005]. The gas piston activity, which originates at a depth of 400–900 m below the surface at Kīlauea [*Edmonds and Gerlach*, 2007], is hence that of a slug flow (Figure 14.2) and a

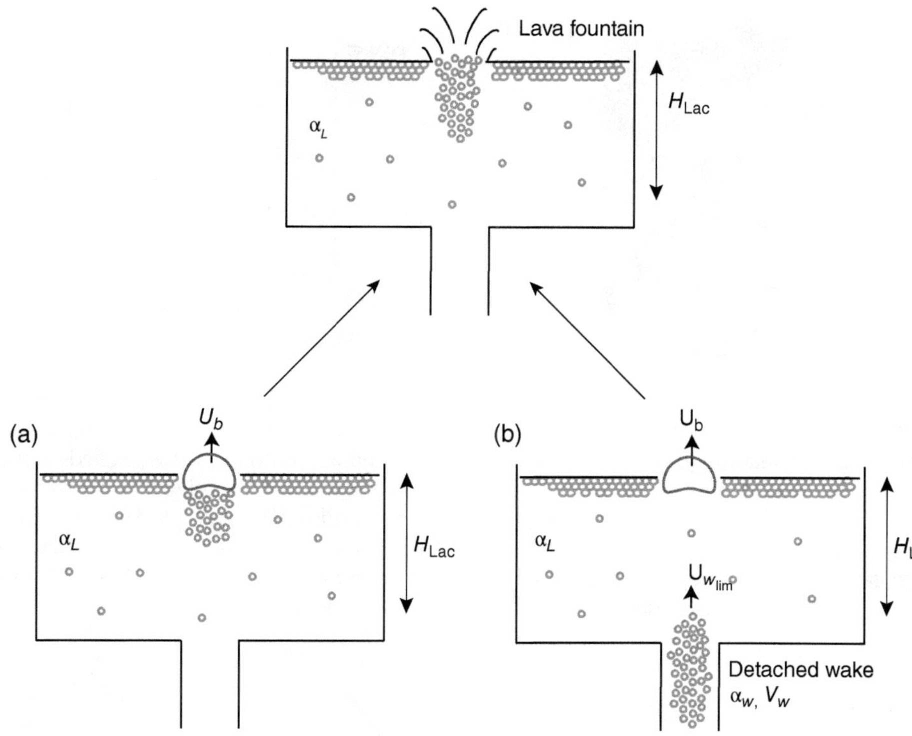

Figure 14.8 Sketch of the formation of small lava fountains at Erta 'Ale lava lake (from Figure 8 in *Bouche et al.* [2010]). They can result from either (a) the breaking of the large bubble with its bubbly wake still attached to it or (b) a former bubbly wake previously attached from its driving large bubble, as in (a).

milder version of Strombolian explosions. However, gas-piston activity at Kīlauea differs from Strombolian-type activity at Stromboli in the gas composition, the crystal content of the ejecta, and the importance of dike and fissure eruptions at Kīlauea [*Mangan et al.*, 2014].

14.5.3. Strombolian Explosions

14.5.3.1. Origin of Strombolian Explosions

Laboratory experiments have proposed that the accumulation of a foam at the top of the large reservoir is unstable and can produce a regular series of large gas pockets rising in the conduit and breaking at the surface as Strombolian explosions (Figure 14.9) [*Jaupart and Vergniolle*, 1988, 1989]. This view was later confirmed at Stromboli using FTIR measurements that constrained the depth of the slug formation, and hence the constriction, at ≈3 km [*Burton et al.*, 2007a]. The foam coalescence at the top of a reservoir has also been proposed as a potential source of seismic and acoustic signals [*Ripepe et al.*, 2001]. However, the shallow magma reservoir is not observed on the inversion of very long period seismic events, and the foam coalescence may be a too small source of pressure to be significant [*Chouet et al.*, 2003, 2008]. But the formation of a foam at the top of the reservoir does not require a large trapping area; hence the magma reservoir could be undetected by seismic techniques. It is also possible that a fast massive foam coalescence event, by releasing quasi-simultaneously numerous bubbles and their capillary pressure due to surface tension, could be a source of pressure at depth [*Vergniolle and Brandeis*, 1996].

For the purpose of modeling Strombolian explosions, it has been assumed that the conduit is vertical; however, the uppermost part of the conduit is fairly inclined at Stromboli, as shown by seismic studies [*Chouet et al.*, 2003, 2008]. Laboratory experiments on inclined conduits (Figure 14.10) show that the gas flux neccessary to make the transition between a bubbly and slug flow is decreased by an order of magnitude compared to that neccessary for a vertical conduit [*Lane et al.*, 2001; *Chouet et al.*, 2003]. This geometry has been called for to explain Strombolian explosions at Stromboli. The third model to explain Strombolian explosions is related to a slow and progressive bubble coalescence in the conduit [*Parfitt and Wilson*, 1995; *Parfitt et al.*, 1995; *Parfitt*, 2004]. In this case, the formation of the slug is relatively shallow and without any obvious source of overpressure within the gas.

Alternatively, large gas pockets can be formed in the conduit from the development of a concentration wave [*Manga*, 1996]. This wave develops from an initial

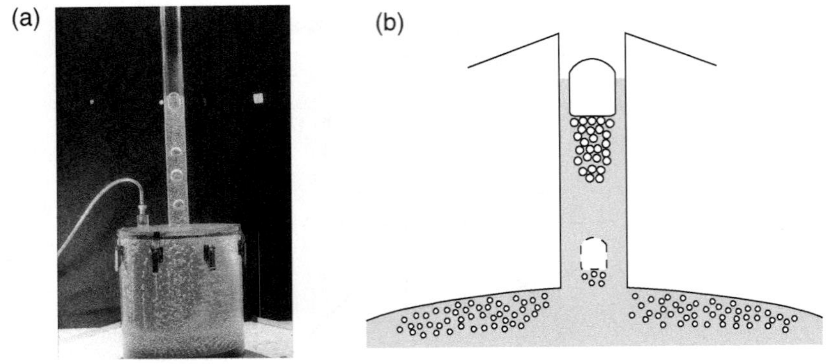

Figure 14.9 (a) Laboratory experiments mimicking the formation of Strombolian activity (Figure 3c in *Jaupart and Vergniolle* [1988]). The bubbles are produced at the base of the reservoir, an analog to the magma reservoir, rise and accumulate at its top, leading to the formation of a foam. When the foam thickness exceeds a critical thickness, the foam undergoes a massive coalescence, here limited by the magma viscosity. This coalescence induces the formation of a series of large gas pockets rising in the vertical conduit, an analog to Strombolian activity. (b) Sketch of a magma reservoir and a conduit showing the accumulation of a foam at the roof of the reservoir and a series of large gas pockets inflating while rising (Figure 17 in *Vergniolle and Ripepe* [2008]). A bubbly wake is shown as for a slug whose Reynolds number exceeds 30 [*Brennen*, 1995].

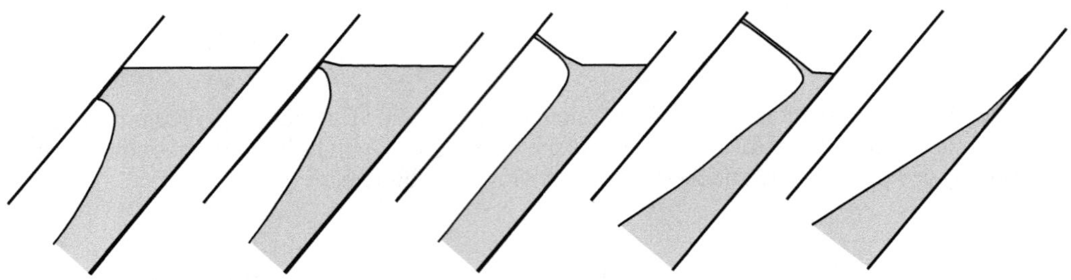

Figure 14.10 Sketch of deformations and upward motion of a slug rising in an inclined conduit (adapted from Figure 6c in *James et al.* [2004]). Inclined conduits, such as observed at Stromboli [*Chouet et al.*, 2003, 2008], correspond to a reduced friction on the sides of the slug when compared with vertical conduits. The rise velocity of bubbles in an inclined conduit is enhanced by a value between 10% [*Wallis*, 1969] and 40% for an angle of 30° [*James et al.*, 2004] when compared with a vertical conduit of the same radius [*James et al.*, 2004]. The conditions for forming a slug from a underlying bubbly flow are more easily reached in an inclined conduit than for a vertical conduit [*Lane et al.*, 2001; *James et al.*, 2004].

homogeneous bubbly liquid if the bubble velocity is much larger than the magma velocity and if the bubble velocity depends on the gas volume fraction (Figure 14.11). Therefore this wave does not exist for small gas volume fractions, less than 2% (Figure 14.11a). For gas volume fractions above 2%, the bubbles interact together, and this effect increases with the gas volume fraction. In this case, the local bubble velocity decreases with increasing gas volume fraction. A bubble rising faster than its surrounding bubbles increases locally the gas volume fraction above, and hence decreases the overall velocity of nearby bubbles. This favors the development of bubble-rich layers, as bubbles rising from below are slowed down by the locally higher gas volume fraction. As this process develops, the initial bubbly flow becomes organized into a series of horizontal layers, alternatively rich and poor in small bubbles (Figures 14.11b, c). Bubble coalescence is enhanced within the bubbly layers and can lead to a vertical series of large gas pockets (Figure 14.11d). This model has been called for to explain the frequent degassing associated with puff events described at Stromboli [*Ripepe et al.*, 2002].

A bubbly flow in which the gas volume fraction exceeds 2% displays both an upward and downward bubbly flow at the same height (Figure 14.12). This configuration is favorable to the formation of bubble clusters within the bubbly rising zones, independently of the formation of a concentration wave, usually developing for much larger gas volume fraction.

Bubble clusters may also coalesce during their rise. This mechanism, which is independent of that based on a concentration wave [*Manga*, 1996], could also lead to the formation of large gas pockets, such as presented by puffs or Strombolian explosions. However, these large bubbles

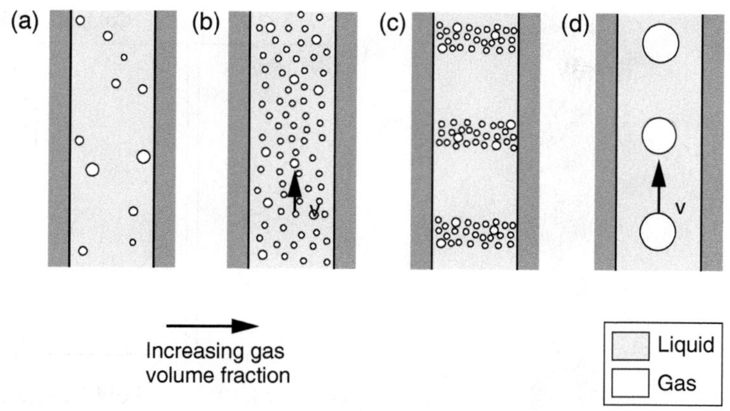

Figure 14.11 Formation of large gas pockets by the development of a concentration wave within a bubbly liquid [*Manga*, 1996]. These waves develop when the gas flux depends on the local concentration in bubbles. This is the case when bubbles start to interact together. (a) Homogeneous bubbly liquid with a very small gas volume fraction, i.e., in the dilute regime ($\alpha \geq 2\%$), for which the concentration wave does not develop. (b) Homogeneous bubbly liquid with a mild gas volume fraction, hence in favorable conditions for initiation of a concentration wave. A bubble rising faster than the average population will be slowed during its rise by a locally higher gas volume fraction. The local gas volume fraction increases, decreasing even more the bubble rise velocity. This enhances the formation of zones with a local high gas volume fraction. (c) Hypothetical series of horizontal layers presenting a regular alternance of zones enriched and depleted in bubbles. (d) The bubbly layers are quickly transformed into bubble clusters, which can coalesce on themselves when their size is large enough. Bubble clusters can then be transformed into large gas pockets and may lead to Strombolian activity [*Manga*, 1996]. This scenario has been proposed to explain the puffing activity at Stromboli [*Ripepe et al.*, 2002].

form at shallow depths without any obvious source of gas overpressure. Hence they are more likely to contribute to the puffing activity rather than to be responsible for Strombolian explosions.

14.5.3.2. Stability of Large Bubbles

A recent numerical study has suggested that large bubbles may not be stable in shape during their rise in a conduit [*Suckale et al.*, 2010]. They should break into much smaller bubbles as a result of instabilities at the bubble surface (Figure 14.13). This happens when the Reynolds number, based on the bubble radius, exceeds 50, leading to catastrophic break-up. The comparison with only a few existing laboratory experiments, the sole use of the Reynolds number as a criterion for bubble fission, and the initial starting condition, chosen to be spherical although spherical is not the equilibrium shape for large bubbles distorted by their rise, have been questioned by *James et al.*, [2011], then replied by *Suckale et al.* [2011]. Clearly, more dimensionless numbers than the sole Reynolds number (see details by *Suckale et al.* [2010]) should be considered to ensure scale invariance between laboratory experiments, numerical modeling, and volcanic systems.

A bubble whose radius exceeds a few meters should not be stable in shape in a basaltic magma far from the surface and should undergo a fission process [*Suckale et al.*, 2010]. However, bubbles much larger than a few meters have existed at the vent, such as at Etna (Figure 14.1b), Kīlauea [*Richter et al.*, 1970], Marum (Ambrym, Vanuatu) [Figure 2 in *Rouland et al.*, 2009; *Seach*, 1999], and Nyiragongo [*Tazieff*, 1975, 1996; Figure 7 in *Burgi*, 2006]. Three main processes can be called for to explain observations at the vent and results from this numerical study, which does not include the presence of an additonal free interface. First, the large bubble may be breaking and reforming all the time in the conduit, as the smaller bubbles resulting from the fission process are very close to one another. While this is certainly an important process, *Suckale et al.* [2010] suggest that it may probably not lead to slug-sized gas pockets, because bubbles with diameter equivalent to that of the conduit should be unstable. Second, the large change in bubble radius due to its inflation during rise could prevent the development of the instability because the wavelength, based on the bubble radius, is changing all the time. Third, the existence of an additional free interface, particularly when coupled with a temperature drop and potentially a rheological transition, will very likely affect bubble dynamics sensitively. In some cases, non-Newtonian effects could be dominant in a magma close to the surface, due to a significant amount of crystals and tiny bubbles in suspension.

The existence of multiple peaks in the ejecta velocity displayed by the long explosions (≥10 s) of Stromboli and Yasur [*Taddeucci et al.*, 2012a, 2012b], although previously associated with a resonance in the conduit [*Chouet*

Figure 14.12 Bubble clusters rising in a conduit filled with a silicone oil (Rhodorsil 47V100, 0.1 Pa.s, zoom from Figure 14.1b [*Jaupart and Vergniolle*, 1988]). These can be formed when the gas volume fraction corresponding to the small bubbles exceeds 2%. When the size of the bubble cluster is sufficient, it can coalesce on itself and be transformed into a large gas pocket, which could be responsible for the puffing activity at Stromboli [*Ripepe et al.*, 2002].

et al., 1974], could also result from breaking a close series of slugs [*Ripepe et al.*, 1993; *Ripepe and Marchetti*, 2002], as potentially formed from an instability of the bubble surface [*Suckale et al.*, 2010].

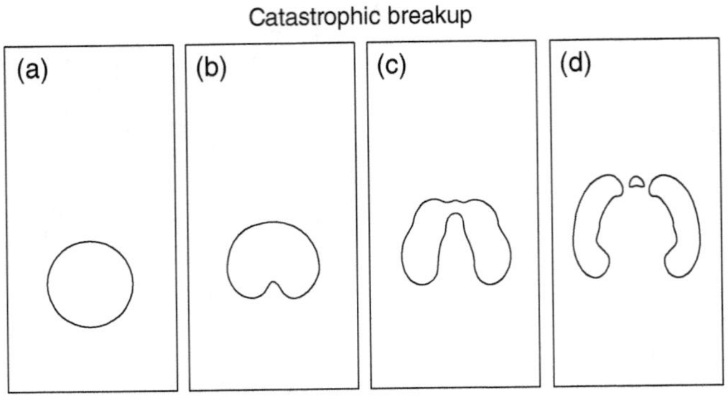

Figure 14.13 Large bubbles whose Reynolds number exceeds 50 (based on the bubble radius) undergo a Rayleigh-Taylor instability while rising, leading to their catastrophic break-up (adapted from Figure 4c in *Suckale et al.* [2010]): (a) initial spherical bubble rising in a conduit; (b) deformations of the bubble, starting at its nose, are advected at its bottom by the bubble rise; (c) bubble deformations becoming stronger at its bottom than at its nose; (d) break-up of the bubble into two large fragments and a small one.

14.5.3.3. Origin of Overpressure Within Strombolian Explosions

In this section, we only review the causes of gas overpressure in large bubbles (spherical cap bubbles or slugs). We do not discuss the dynamical overpressure of small undeformable bubbles, which is fairly well understood [*Proussevitch and Sahagian*, 1996, 1998].

We have described several interpretations of the origin of Strombolian explosions, driven by the breaking of large bubbles, either spherical cap bubbles or slugs. However, they are difficult to test and more than one type of bubbles, i.e., sharing the same mechanism of formation regardless of their diameter and shape, may exist at the same time. Bubbles can also coalesce during ascent, erasing their history of formation at depth. One way to gain more information on the physical processes at work between bubble formation and bubble breaking at the vent is to constrain the gas overpressure within the slugs prior to its breaking. The order of magnitude on the gas overpressure can be obtained from ejecta velocity using a very simplified ballistic approximation [*Wilson*, 1980]. However, a better estimate of the gas overpressure requires synthetic acoustic or seismic sources and hence a good knowledge of the source of acoustic pressure.

The existence of gas overpressure when the slug breaks is clear from direct measurements of the gas velocity [*Wilson*, 1980] but also from infrasonic records (Figure 14.3). If the acoustic pressure results from the formation of a hole during bursting, as for a only slightled overpressurized soap bubble, the acoustic pressure is unrelated to the initial gas overpressure prior to bursting [*Vidal et al.*, 2010]. Alternatively, the main positive peak in acoustic pressure may result from the propagation of a hole, first accelerating and then decelerating [*Kobayashi et al.*, 2010]. In this case, the velocity at which the hole is formed and enlarged must be related to the initial bubble overpressure. Furthermore, if the main pulse in acoustic pressure is due to the bubble volume mode or to the growth of instabilities [*Vergniolle and Brandeis*, 1996; *Vergniolle and Caplan-Auerbach*, 2004], its modeling, although difficult to perform, could give additional constraints on the bubble overpressure prior to bursting. The source of gas overpressure is likely to be related to the depth of slug formation or magma viscosity, and hence knowing its value is crucial.

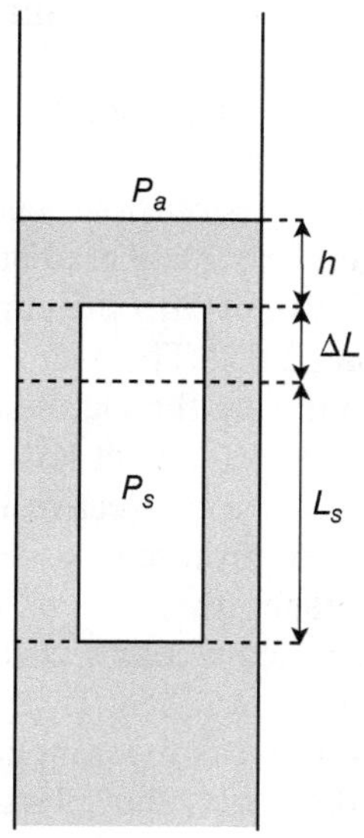

Figure 14.14 Sketch of the simplified geometry, i.e., a cylinder, used for a slug approaching the top of the magma column (adapted from Figure 2b in *Del Bello et al.* [2012]). When the bubble length L exceeds a critical value, it becomes unstable and the bubble breaks. The perturbation theory, based on the bubble length and assuming that the gas pressure corresponds to the magma weight, gives constraints on the bubble length and its depth at bursting. This model [*Del Bello et al.*, 2012] predicts a good transition between passive degassing and explosive bursting at Stromboli. There, explosive bursting occurs for gas volume exceeding 24–230 m^3.

The stability of a slug assumed to be a cylinder and stopped near the surface during its rise in the conduit has been calculated using perturbation methods. In this mathematical technique, the slug length is forced to increase further than equilibrium (Figure 14.14), and the equations for the slug behavior predict either a further increase in the slug length, leading to the slug instability and breaking, or a decrease, restoring the equilibrium, as for a stable slug. In this model, the slug does not have the pointy nose or the potential inward saucer-shaped bottom, both due to its rise (Figure 14.6a). This simplified model, which also ignores the slug rise during breaking, and hence the thinning of the overlying magma layer and the rounded shape of the slug, leads to estimates on the slug length and overpressure during breaking [*Del Bello et al.*, 2012], in very good agreement with the results obtained from inversion of acoustic waveforms performed for Stromboli [*Vergniolle and Brandeis*, 1996]. Furthermore, this study predicts a relationship between gas volume and overpressure prior to slug breaking, which can be tested when a valid source model for the inversion of acoustic or seismic waveforms exists. *Del Bello et al.* [2012] also propose that the transition between passive and explosive bursting occurs for s slug volume of 24–230 m^3 at Stromboli.

Laboratory experiments have shown that a dynamic pressure could exist during the ascent of individual slugs in inclined or vertical conduits [*James et al.*, 2004]. This dynamic pressure could act as a seismic source. The deformations of a slug at a flare have also been investigated by laboratory experiments [*James et al.*, 2006]. The slug, passing through a flare, is strongly deformed because its nose is rising at a larger velocity than its bottom, due to the

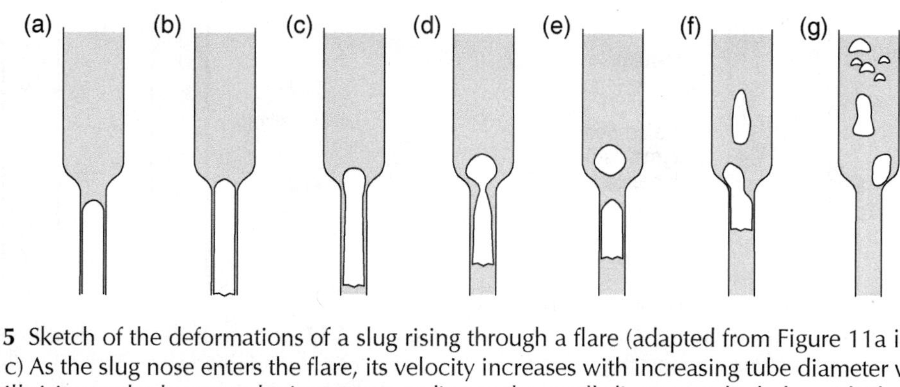

Figure 14.15 Sketch of the deformations of a slug rising through a flare (adapted from Figure 11a in *James et al.* [2006]). (a, c) As the slug nose enters the flare, its velocity increases with increasing tube diameter while the slug bottom is still rising at the lowest velocity corresponding to the small-diameter tube below. (d) The bubble nose is rising so fast that a neck is starting to form between the two parts of the slug. (e) The bubble nose detaches and forms an isolated bubble of a much smaller volume than the initial bubble. (f) The process repeats itself. The deformations can be strong and may induce a loss of axisymetry. (g) The slug has been dismantled into smaller fragments.

enlargement of the section (Figure 14.15). The slug can also break for a severe change in the tube diameter. The slug deformations, when passing the flare, can trigger the slug length mode [*James et al.*, 2006], the main volume mode for a cylindrical bubble far from the surface [*Vergniolle et al.*, 1996]. When the slug is separated into two smaller bubbles (Figure 14.15e), the closure of the newly formed slug can trigger some oscillations in the slug length. An overpressure of 1–10 MPa is predicted at Stromboli, in agreement with results from very long period events [*James et al.*, 2006] and acoustic signals. Another source of overpressure, although not yet quantified, could be related to the massive foam coalescence at magma reservoir depths.

Although several models have been proposed to explain the overpressure within the slug prior to its breaking, these models are difficult to test because they require a valid inversion of acoustic or seismic signals. However, it is crucial to estimate the overpressure independently of the gas volume as it provides an independent constraint. This is particularly crucial when two populations of large bubbles are present, such as at Stromboli or Shishaldin [*Ripepe et al.*, 2002; *Vergniolle et al.*, 2004; *Vergniolle and Caplan-Auerbach*, 2004].

14.5.4. Enigma of Permanent Activity at Open-Vent Volcanoes

The question of how open vents can be permanently active without significant magma cooling, such as during the current Pu'u 'Ō'ō eruption at Kīlauea, remains an enigma, perhaps related to the conduit flow, now reviewed.

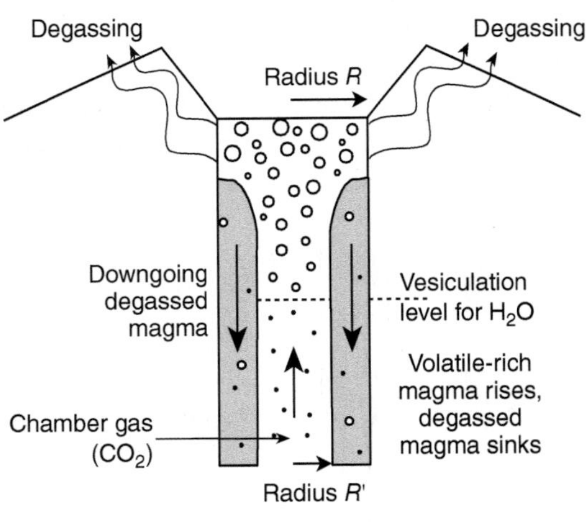

Figure 14.16 Processes developing in the upper conduit of open-vent volcanoes (redrawn from Figure 1b in *Stevenson and Blake* [1998]). The fresh magma coming from depth, hot and potentially bubbly, rises in the center of the conduit while the magma, having degassed at the surface, is denser than the fresh magma and is falling down in an annulus at the conduit walls. Note that when the downward flow is sufficient, small bubbles can be carried downward, against gravity.

Long-lasting eruptive activity from an open vent requires a permanent gas flow related to conduit flow or the rise of a volatile-rich hot magma to prevent cooling. Several explanations have been proposed to explain the conduit dynamics of open-vent volcanoes and the existence of a permanent activity. The spatial evolution in height of an initially bubbly flow rising in a long vertical tube shows that the gas flow promotes an effective convection of the liquid, which favors magma homogeneity and a stable condition [*Pioli et al.*, 2012]. A bimodal bubble population is observed in the tube, with the small one being recirculated with the liquid and contributing to less than 20% of the total gas outflow. The gas flow circulation may counteract the cooling gradient and may explain how the magma stays hot into the conduit and the vent still open [*Pioli et al.*, 2012]. The microtextures of juvenile pyroclasts expelled during the October 2008 "explosive" event from the active lava lake at Kīlauea have probably partially recorded such a convective loop [*Carey et al.*, 2013]. Other causes for promoting an efficient gas flow in the conduit, and hence the lack of cooling, could be related to the massive foam coalescence limited by viscosity [*Jaupart and Vergniolle*, 1988, 1989], the flow of a stable foam in the case of weak Strombolian activity, such as observed at lava lakes [*Stix*, 2007; *Vergniolle and Bouche*, in revision], or the existence of long and inclined conduits [*Lane et al.*, 2001; *James et al.*, 2004].

Open-vent volcanoes have a curious characteristic of expelling much more gas than magma [*Francis et al.*, 1993; *Allard*, 1997]. This paradox has been explained by endogeneous growth and by the existence of a convective loop between a volatile-rich magma rising up and the

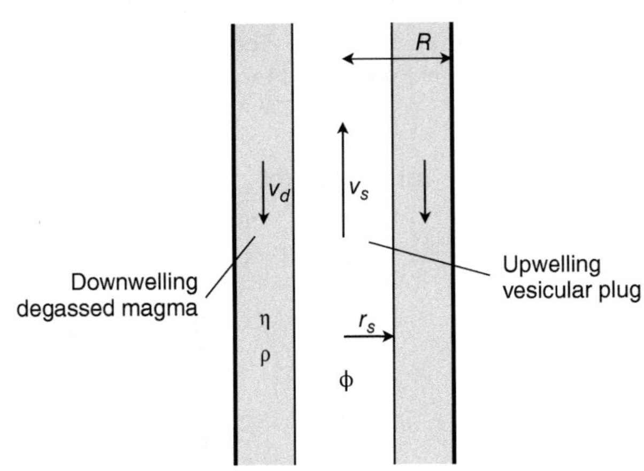

Figure 14.17 Sketch of the model used to understand open-vent volcanoes (adapted from Figure 3 in *Burton et al.* [2007b]). The light vesicular magma rises in the center while the degassed, hence dense, magma falls in an annulus at the conduit wall. In this model [*Burton et al.*, 2007b], the radius and the velocity of the upward core flow and that of the outer annulus flow are calculated.

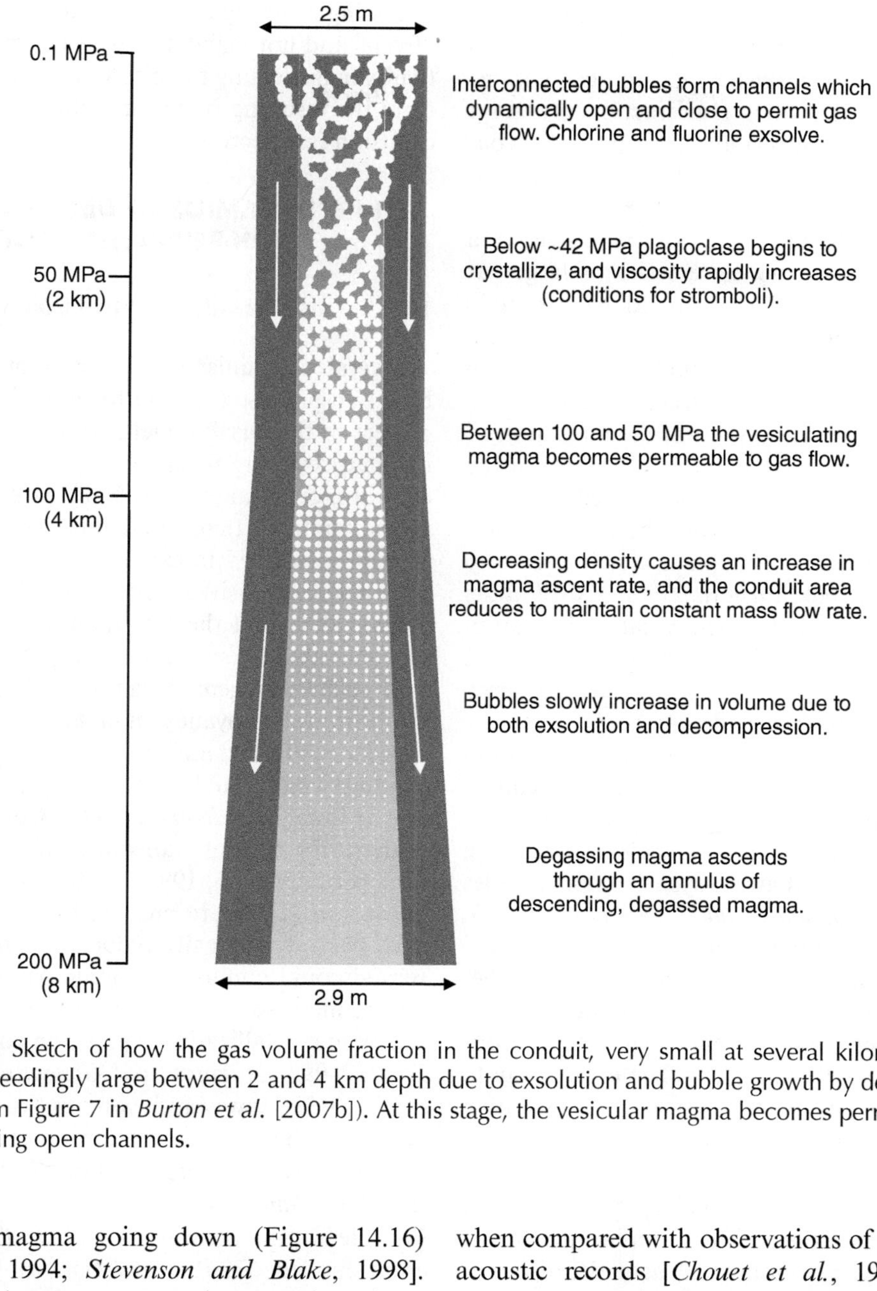

Figure 14.18 Sketch of how the gas volume fraction in the conduit, very small at several kilometers depth, becomes exceedingly large between 2 and 4 km depth due to exsolution and bubble growth by decompression (adapted from Figure 7 in *Burton et al.* [2007b]). At this stage, the vesicular magma becomes permeable to gas flow by forming open channels.

mostly degassed magma going down (Figure 14.16) [*Kazahaya et al.*, 1994; *Stevenson and Blake*, 1998]. However, the endogeneous growth, resulting from emplacing the degassed magma at depth, is a questionable process at some open-vent volcanoes due to the lack of deformation, such as at Erebus [*Otway et al.*, 1994; *Kelly et al.*, 2008]. The degassed magma may simply be mixed and recycled within the plumbing system at depth.

Alternatively, the bubbly flow could rise on one side of the conduit while the degassed magma is falling on the other side [*Burton et al.*, 2007b]. In this biflow model, the radius of the rising inner core is calculated (Figure 14.17) [*Burton et al.*, 2007b]. This model, in order to explain a magma flow rate of 575 kg/s neccessary to maintain the quiescent SO_2 flux of 200 t/day, predicts a conduit diameter at Stromboli of 2.5–2.9 m, a very reasonable value when compared with observations of vent diameters and acoustic records [*Chouet et al.*, 1974; *Vergniolle and Brandeis*, 1996]. However, injecting fresh and undegassed magma in the central core and degassing all the volatiles at equilibrium, with the exsolution law, lead to the formation of a foam ≥60% in gas volume fraction at a depth of 3.6–1.8 km at Stromboli (Figure 14.18) [*Burton et al.*, 2007b]. At this depth and toward the surface, the gas within each bubble of the foam can be connected to the rest of the bubbles above via degassing channels, forming a percolation zone. Hence, a transition from a closed system, below the foam, to an open system, in the percolation zone, occurs. The final gas emission is hence a superposition of gases from below and above the percolation transition.

This model, which assumes that the differential motion of the bubbles with respect to the magma is negligible, can only be applied when the magma velocity is large compared to that of the bubbles. Furthermore, it ignores the bubble interaction that exists when the gas volume fraction exceeds 2%. This bubble interaction promotes the formation of bubble clusters and their potential coalescence (Figure 14.12). This configuration is not favorable to the formation of a foam, which could be disturbed by the rise of large coalesced bubble clusters. Instead of a foam, the formation of a very close series of large gas pockets is likely. This could correspond to the puffs observed at Stromboli but the formation of puffs is very shallow, ≤12–160 m [*Burton et al.*, 2007a], in contrast to the depth at which the gas volume fraction reaches 2%. One way to adapt the model to prevent the formation of a deep foam is to consider that the vesiculation in the conduit is out of equilibrium and postponed until the shallow depth. This is compatible with the existence of a large supersaturation prior to bubble nucleation [*Lensky et al.*, 2006; *Bai et al.*, 2008; *Pichavant et al.*, 2013] in the absence of preexisting bubbles. The volatile disequilibrium could also also be generated when gas diffusion is too slow compared to bubble rise. In this case, the buoyancy of the rising magma is provided by heat.

Furthermore, building a foam in a conduit is complicated for Newtonian liquids. Even if a foam is built, it acts as a trap for the bubbles rising from below because the bubbles in the foam are stagnant. Having a long foam in a conduit also requires that the system present cycles of foam collapse [*Jaupart and Vergniolle*, 1988; *Proussevitch et al.*, 1993a] in which a dispersed flow (Figure 14.2d) is produced at the vent. If the magma rheology is non-Newtonian, percolation channels can be maintained between open and closed periods (Figure 14.18) [*Burton et al.*, 2007b]. The easiest way to transform a Newtonian magma into a non-Newtonian one is to severely increase the number of crystals (≥40%) or bubbles (≥70%) [*Kerr and Lister*, 1991; *Lejeune et al.*, 1999]. Having a magma with a non-Newtonian rheology, as proposed by *Burton et al.* [2007b], implies that many crystals or bubbles are accumulated at a depth below 1.8–3.6 km by an unknown mechanism.

Another point to consider with the biflow model (Figures 14.17 and 14.18) [*Burton et al.*, 2007b] is the possibilty that large bubbles may be produced at depth in the conduit, at a depth of ≈3 km at Stromboli [*Burton et al.*, 2007a], and below the point where a bi-flow and a foam may develop. In this condition, the bi-flow will be disturbed and potentially erased by the rising large bubbles. This only leaves the short delay between two successive large bubbles, only ≈15 min at Stromboli, to restore the bi-flow before being interrupted by the next large bubble. For all the above reasons, the current biflow model may need to be adapted. One way to prevent the formation of a deep foam [*Burton et al.*, 2007b] is to only have a partial degassing resulting from the solubility law at each rise of "fresh" and hot magma. This is compatible with the difficulty in nucleating bubbles and in the complexity arising from the resulting bubble growth by decompression and diffusion (see Section 14.4.).

14.6. DYNAMICS OF DEGASSING DURING NON-PERMANENT ACTIVITY

14.6.1. Large Basaltic Eruptive Columns

Several mechanisms have been proposed to explain how the gas, usually able to move in the conduit with respect to the basaltic melt, could be trapped within the magma, resulting in large eruptive columns. The existence of a deep caldera, probably with a floor at or below the water table that could interact with the magmatic system, is invoked to explain the two prolonged periods of episodic explosive eruptions at Kīlauea, leading to the Keanakāko'i and the Uēkahuna tephra deposits [*Fiske et al.*, 2009; *Swanson et al.*, 2012]. A sudden decompression, such as triggered by a flank collapse associated with the powerful buoyancy of a ≥1km^3 magma body, may have induced a fast and extensive bubble nucleation in the conduit, which could be at the origin of the 122 BC eruption at Etna [*Coltelli et al.*, 1998]. Conversely, the lack of evidence for any simultaneous and large tectonic events at Etna [*Coltelli et al.*, 1998] led *Houghton et al.* [2004] and *Sable et al.* [2006a] to propose that the vent radius could have been dynamically reduced by the formation of a very viscous annulus of magma. Another alternative is a drastic increase in viscosity, such as associated with the microlite crystallization within the conduit [*Sable et al.*, 2001, 2009; *Houghton and Gonnermann*, 2008; *Moitra et al.*, 2013]. However, the amount of microlites is insufficient during the formation of the Fontana lapilli [*Goepfert and Gardner*, 2010] or the 122 BC eruption at Etna [*Giordano and Dingwell*, 2003]. The lack of anomalous preeruptive conditions, with 3 wt% of H_2O, for both eruptions led *Goepfert and Gardner* [2010] to conclude that the highly intense basaltic Plinian activity must result from a very high magma ascent rate, which leads to an intense nucleation, thereby trapping the newly formed bubbles within the magma.

Basaltic magma can be water-rich in subduction zones [e.g., *Wallace*, 2005], with an average value of 4 wt% [*Plank et al.*, 2013] and peaking up to ≥6 wt% for Chikurachki and Kliuchevskoy (Kamchatka) [*Gurenko et al.*, 2005; *Auer et al.*, 2009] and for Augustine (Aleutians) [*Zimmer et al.*, 2010]. These values, similar to that of silicic volcanoes, can possibly explain their explosivity [*Gurenko et al.*, 2005; *Plank et al.*, 2013]. A high volatile content of 4 wt% in water leads to bubble exsolution at a depth of 6.0–6.6 km when water is the sole volatile phase [*Dixon et al.*, 1995; *Lesne et al.*, 2011a; *Mysen*, 1977, respectively], possibly below the top of the shallow magma reservoir. Furthermore, the maximum volatile content, recorded by fluid inclusions, might still be underestimated [e.g., *Plank et al.*, 2013], especially for CO_2 due to its very low solubility [*Papale*, 2005; *Wallace*, 2005; *Wade et al.*, 2006; *Houghton and Gonnermann*, 2008; *Métrich and Wallace*, 2008; *Barsanti et al.*, 2009]. At Shishaldin, 2.4 wt% of water estimated from Holocene scoria [*Plank et al.*, 2013] is compatible with exsolution at a depth of 3.0–3.2 km when H_2O is the sole volatile phase [Figure 4 in *Dixon et al.* [1995]; *Lesne et al.*, 2011a; *Mysen*, 1977; *Vergniolle*, submitted] and possibly leads to a bubble exsolution just within the shallow magma reservoir located at 3.9 km below the summit [*Vergniolle and Caplan-Auerbach*, 2006]. The small bubbles, when formed within the magma reservoir, can accumulate at its top and produce a foam during basaltic sub-Plinian eruptions as for classical basaltic eruptions. However, the foam in the reservoir leading to the sub-Plinian column at Shishaldin is stable due to tiny bubbles (0.020–0.025 mm in diameter) [*Vergniolle*, submitted] and not unstable as for Strombolian activity or high-level lava fountains [*Jaupart and Vergniolle*, 1988, 1989]. The existence of a stable foam at the top of the reservoir combined with a large gas flux could produce a long foam in the conduit, 1–2 km, whose rupture may have induced the sub-Plinian activity [*Vergniolle and Caplan-Auerbach*, submitted; *Vergniolle*, submitted].

The existence of several models for explaining large eruptive columns wordwilde is still partially unsatisfactory due to the lack of constraints. For example, continuous acoustic measurements could be used to estimate the active gas volume and the velocity of the fragmentation front [*Vergniolle and Caplan-Auerbach*, submitted], while FTIR measurements could provide an estimate of the total gas volume and the depth of the gas formation, perhaps following the deepening of the fragmentation zone in time (see Section 14.3.). In the future, using multiple streams of data could be decisive in supporting the best one of the proposed model.

14.6.2. Paroxysmal Activity at Stromboli

Activity at Stromboli can exhibit major explosions and paroxysms [e.g., *Harris and Ripepe*, 2007b]. Paroxysms, which mostly display Vulcanian-type explosions, are often associated with lava flows at a low elevation on the volcano flank. Typically several days prior to paroxysms at Stromboli, the CO_2 flux dramatically increases [*Aiuppa et al.*, 2010b]. This feature has been explained by the fast rise of a gas-rich primitive high-potassium basalt from a depth of 7–10 km [*Bertagnini et al.*, 2003, 2011; *Métrich et al.*, 2010] or by an abnormal voluminous slug [*Del Bello et al.*, 2012]. Alternatively, paroxysms could be a consequence of a sudden depressurization of the magma within the plumbing system leading to an intense bubble nucleation event triggered by lava effusion at low elevations on the slope of the volcano when it exists [*Aiuppa et al.*, 2010b; *Calvari et al.*, 2011]. The increase in the CO_2 flux could also result from the accumulation of a foam growing at the roof of the deep magma reservoir (≥6 km) while spreading toward the deep conduit and the surface [*Allard et al.*, 2008; *Allard*, 2010]. Finally, ascending parcels of magma, very oversaturated in CO_2 due to inhibited bubble nucleation or a limited bubble growth, could during rise (≤50 MPa) undergo a rapid bubble nucleation event at the origin of paroxysmal activity at Stromboli [*Pichavant et al.*, 2013; *Rivalta et al.*, 2013].

14.6.3. High-Level Lava Fountains

Two main models have been proposed to explain the formation of high-level lava fountains whose height is large compared to its own diameter (Figure 14.1c). These models also apply to episodes at Etna called quasi-lava fountains, which consist in a phase of Strombolian activity not always leading to a high-level lava fountain during its paroxysm. In the first one, a high-level lava fountain results from the relatively fast rise of the magma, which vesiculates during its upward motion, in contrast to a relatively slow magma velocity that produces progressive bubble coalescence as for Strombolian activity [*Parfitt and Wilson*, 1995, 1999; *Parfitt et al.*, 1995; *Parfitt*, 2004]. In this case, a high-level lava fountain is driven by a dispersed flow in the conduit (Figure 14.2d).

In the second model, a high-level lava fountain is produced when foam accumulated at the top of the reservoir (Figure 14.19a) becomes unstable, coalesces quasi-instantaneously, and releases the voluminous gas pocket (Figure 14.19b) [*Jaupart and Vergniolle*, 1988, 1989; *Vergniolle and Jaupart*, 1990]. These laboratory experiments also show that massive foam coalescence is regular. The periodic behavior of the foam can explain the monthly cyclicity observed in the regular series of high-level lava fountains, ≈3 weeks during the Mauna Ulu and Pu'u 'Ō'ō eruptions at Kīlauea [*Swanson et al.*, 1979; *Wolfe et al.*, 1987, 1988]. This cyclicity is not explained by the first model [*Parfitt and Wilson*, 1995, 1999; *Parfitt et al.*, 1995; *Parfitt*, 2004]. Recent FTIR measurements suggest that the high-level lava fountains at Etna result from a foam accumulated at ≈2 km depth [*Allard et al.*, 2005; *Spilliaert et al.*, 2006], thereby validating the foam model [*Jaupart and Vergniolle*, 1988, 1989; *Vergniolle and Jaupart*, 1990]. This view is in very good agreement with several independent measurements of seismo-acoustic waves, eruptive patterns, and coupled textural and compositional characteristics of basaltic

Figure 14.19 Laboratory experiments proposed as an analog to Hawaiian activity (from Figure 1b in *Jaupart and Vergniolle* [1988] and Figure 12 in *Vergniolle and Mangan* [2000]) [*Jaupart and Vergniolle*, 1988, 1989; *Vergniolle and Jaupart*, 1990]. Small bubbles are produced at the base of the reservoir, rise, and accumulate at the top to produce a foam. When the foam reaches a critical thickness, which depends on gas flux, liquid viscosity, and bubble diameter, the foam becomes unstable and coalesces. This massive foam coalescence leads to the formation of a very large gas pocket, which rises in the conduit and is associated with the high-level lava fountain activity (Figure 14.1c). The fluid is a silicone oil Rhodorsil 47V1000 (1 Pa.s) and the upper tube has a diameter of 4.4 cm. (a) The thickening of the foam pushes the liquid in the conduit upward. (b) Analog to the high-level lava fountain with a central gas core and an annulus of liquid.

scoria at Etna [e.g., *Gresta et al.*, 2004; *Behncke et al.*, 2006; *Polacci et al.*, 2006].

14.7. CONCLUSION

14.7.1. Summary

In this review chapter, we have shown how bubbles in magma can play a key role in driving basaltic eruptions. The spatial organization of the bubbles in the conduit, i.e., a bubbly flow, a slug flow, an annular flow, or a dispersed flow, is representative of effusive activity, Strombolian explosions, high-level lava fountain, and large eruptive columns, respectively, although high-level lava fountains are also sometimes associated with a dispersed flow.

The description of various techniques available to measure the gas volume at the vent has highlighted the various forms that degassing can take at the vent. If all the direct techniques (photoballistics, radar, FTIR) are in good agreement at Stromboli and Etna, indirect techniques (seismic and acoustic records) find values that are one order of magnitude less. Direct techniques may overestimate the gas volume at the vent due to air entrainment within the volcanic plume. Furthermore, direct methods estimate the total gas volume (active, i.e., still overpressurized at the vent, and passive degassing) while indirect methods solely give access to the active degassing, whose origin is at depth.

The observations of eruptive activity at open vents, such as Erta 'Ale, Nyiragongo, Kīlauea, and Stromboli, have enabled us to continually improve our understanding of lava lake activity, gas pistoning events, and Strombolian explosions. Strombolian explosions result from massive coalescence of a foam trapped in the reservoir (in agreement with FTIR measurements), progressive coalescence in a long and potentially inclined conduit, or a series of bubble-rich layers in the conduit (as for puffs). The permanent activity in open-vent volcanoes is thought to be maintained by a biflow between a hot and vesiculated magma rising up in the conduit and a cold and mostly degassed magma sinking. The period of intense activity at the surface of a lava lake can result from either a high convective velocity for the hot magma or the sudden breaking of the crust induced by the massive foam coalescence beneath.

Non-permanent eruptive activity also consists of rare but impressive large basaltic eruptive columns and Vulcanian explosions during paroxysmal activity at Stromboli. Large basaltic eruptive columns can be formed as a consequence of a sudden decompression of the magma in the plumbing system, a dramatic increase in the magma viscosity and closed-system degassing, an abnomalously high magma ascent rate, and the existence of a stable foam accumulated at the top of the magma reservoir and fed by a large underlying gas flux. The large eruptive columns observed during the paroxysmal phases at Stromboli volcano have been explained by the fast rise of a gas-rich primitive basalt, an abnormally voluminous slug or the sudden decompression of the plumbing system during the lava flows at low elevation. Paroxysms could also be triggered by the release of a deep slug (7–10 km) or by the deep release of a parcel of magma highly oversaturated in CO_2. High-level lava fountains are intrepreted as resulting from either the massive coalescence of the foam accumulated at the top of the magma reservoir (in agreement with FTIR measurements at Etna) or a progressive bubble coalescence in the conduit.

14.7.2. Perspectives

Although our understanding of physical processes at work during basaltic eruptions has significantly increased in the past 10 years, several questions remain to be answered. For example, laboratory experiments on bubble nucleation in basalts are few, due to the complexities arising when working with very hot magma and the potential coupled effect of both CO_2 bubbles and H_2O bubbles. The value of the critical supersaturation in these cases is hardly known, although bubble nucleation at depth plays a key role to initiate eruptions or to provide initial buoyancy for the magma. Combining laboratory experiments, the study of vesicularity in eruptive products, and perhaps geophysical measurements could significantly increase our knowledge of the dynamics of basaltic eruptions.

Our current interpretation of fluid dynamical mechanisms present during basaltic eruptions [*Gonnermann and Manga*, 2007], based on geology, geochemistry, geophysics, and multidisciplinary data sets, has allowed us to define questions associated with basaltic eruptions to be solved in the future. Advances in the field, either with laboratory experiments or with computational numerical modeling, are very promising, as marked by the increasing number of studies in this field. In the future, field observations and measurements should continue to inform fluid dynamic models and results tested with variable field data.

Open-vent volcanoes, which respond relatively rapidly at the surface to physical processes in the shallow magma reservoir, could be extremely useful to refine our understanding of mechanisms at depth. Etna, Stromboli, and Kīlauea are all highly accessible, well monitored, and highly active at the time of writing. They are available to test ideas and hypotheses for the dynamics of basaltic volcanism. They provide us, by having a permanent activity, with a window to observe deep eruptive phenomena in continuous fashion. The active and accessible lava lake at Halema'uma'u is fairly easy to observe and measure continuosly with a multidisciplinary monitoring approach, in contrast to the remote Erta 'Ale and Nyiragongo lava lakes, both in areas of political unrest.

The existence of permanent open-vent lava lake activity and the very reduced amount of cooling for several decades are so far only qualitatively explained by a convective loop involving the rise of a hot and gas-rich magma and the sinking of the cold and degassed magma. In the future, it is very important to model the thermal budget and quantify the conditions allowing a conduit with actively degassing magma to stay open.

The advantages of studying basaltic volcanoes in a multidisciplinary fashion are also becoming very clear now. In the future, it is important to increase coupled petrological, geochemical, and geophysical studies as results from individual studies can conflict and also one technique may give important constraints that others need. For example, the existence of magma reservoirs, required by petrologists to explain the crystallization sequence, are not always observed by seismologists with geophysical data. Conduit and magma storage zones and geometries need better resolution. The exsolution, degassing, and vesiculation processes that occur in magma reservoirs are still poorly understood. In this respect, key information to determine in the future is how bubbles can be produced in a magma reservoir in phases separated by quiescent periods without bubble formation and growth, such as during inter-eruptive periods. It is also crucial to revisit the idea that an eruption has to be triggered by the injection of fresh magma from below.

ACKNOWLEDGMENTS

We thank R. Carey, J. Suckale, and two anonymous referees for their careful reviews and D. Laporte for discussions. We thank T. Pfeiffer for the use of his photo at Etna (//www.decadevolcano.net). This is an IPGP contribution number 3472. We also thank our funding bodies, i.e., CNRS-INSU, Université Paris Diderot 7 (BQR), IPGP (BQR and PPF), and ANR (Undervolc, Arc-Vanuatu, Risk-Volc-An, Domoscan), for their support.

REFERENCES

Aiuppa, A., A. Bertagnini, N. Métrich, R. Moretti, A. Di Muro, M. Liuzzo and G. Tamburello (2010a), A model of degassing for Stromboli volcano, *Earth Planet. Sci. Lett.*, *295*, 195–204.

Aiuppa, A., M. Burton, T. Caltabiano, G. Giudice, S. Guerrieri, M. Liuzzo, F. Murè, and G. Salerno (2010b), Unusually large magmatic CO_2 gas emissions prior to a basaltic paroxysm, *Geophys. Res. Lett.*, *37*(L17303), 1–5.

Allard, P. (1997), Endogeneous magma degassing and storage at Mount Etna, *Geophys. Res. Lett.*, *24*, 2219–2222.

Allard, P. (2010), A CO_2-rich gas trigger of explosive paroxysms at Stromboli basaltic volcano, Italy, *J. Volcanol. Geotherm. Res.*, *189*, 363–374.

Allard, P., J. Carbonelle, D. Dajlevic, J. Le Bronec, P. Morel, J. M. Maurenas, M.C. Robe, R. Faivre-Pierret, J.C. Sabroux, and P. Zettwoog (1991), Eruptive and diffuse of CO_2 emissions from Mount Etna, *Nature*, *351*, 387–391.

Allard, P., J. Carbonelle, N. Métrich, and H. Loyer (1994), Sulphur output and magma degassing budget of Stromboli volcano, *Nature*, *368*, 326–330.

Allard, P., A. Aiuppa, H. Loyer, F. Carrot, A. Gaudry, G. Pinte, A. Michel, and G. Dongarra (2000), Acid gas and metal emissions rates during long-lived basalt degassing at Stromboli volcano, *Geophys. Res. Lett.*, *27*(8), 1207–1210.

Allard, P., M. Burton, and F. Filippo (2005), Spectroscopic evidence for a lava fountain driven by previously accumulated magmatic gas, *Nature*, *433*, 407–410.

Allard, P., A. Aiuppa, M. Burton, T. Caltabiano, C. Federico, G. Salerno, and A. La Spina (2008), Crater gas emissions and the magma feeding system of Stromboli volcano, in *The Stromboli Volcano: An Integrated Study of the 2002–03 Eruptive Crisis*, Geophys. Monogr. 182, edited by S. Calvari, et al., pp. 65–80, AGU, Washington, D.C., doi:10.1029/182GM07.

Auer, S., I. Bindeman, P. J. Wallace, V. Ponomareva, and M. Portnyagin (2009), The origin of hydrous, high-$\delta^{18}O$

voluminous volcanism: Diverse oxygen isotope values and high magmatic water contents within the volcanic record of Klyuchevskoy volcano, Kamchatka, Russia, *Contrib. Mineral. Petrol.*, *157*, 209–230.

Bai, L., D. R. Baker, and M. Rivers (2008), Experimental study of bubble growth in Stromboli basalt melts at 1 atm, *Earth Planet. Sci. Lett.*, *267*, 533–547.

Baker, S., and F. Amelung (2012), Top-down inflation and deflation at the summit of Kilauea Volcano, Hawai'i, observed with InSAR, *J. Geophys Res.*, *117*(B12406), doi:10.1029/2011JB009123.

Barberi, F., M. Rosi, and A. Sodi (1993), Volcanic hazard assessment at Stromboli based on review of historical data, *Acta Vulcanol.*, *3*, 173–187.

Barsanti, M., P. Papale, D. Barbato, R. Moretti, E. Boschi, E. Hauri, and A. Longo (2009), Heterogeneous large total CO_2 abundance in the shallow magmatic system of Kīlauea Volcano, Hawai'i, *J. Geophys Res.*, *114*(B12201), doi:10.1029/2008JB006187.

Behncke, B., M. Neri, E. Pecora, and V. Zanon (2006), The exceptional activity and growth of the Southeast crater, Mt Etna (Italy) between 1996 and 2001, *Bull. Volcanol.*, *69*, 149–173.

Bertagnini, A., S. Calvari,M. Coltelli, P. Landi, and M. Pompilio (1990), The 1989 eruptive sequence, in *Mt Etna: the 1989 Eruption*, edited by F. Barberi et al., pp. 10–22, Gruppo Nationale Vulcanologia, Italy.

Bertagnini, A., N. Métrich N.P. Landi, and M. Rossi (2003), Stromboli volcano (Aeolian archipelago, Italy): An open window on the deep-feeding system of a steady state basaltic volcano, *J. Geophys. Res.*, *108*(B7), 2336, ECV4-1–15, doi:10.1029/2002JB002146.

Bertagnini, A., N. Métrich, L. Francalanci, P. Landi, S. Tommasini, and S. Conticelli (2008), Volcanology and magma geochemistry of the present-day activity: Constraints on the feeding system, in *The Stromboli Volcano: An Integrated Study of the 2002–2003 Eruption*, Geophys. Monogr. 182, edited by S. Calvari et al., pp. 19–37, AGU, Washington, D. C.

Bertagnini, A., A. Di Roberto, and M. Pompilio (2011), Paroxysmal activity at Stromboli: Lessons from the past, *Bull. Volcanol.*, *73*, 1229–1243.

Best, M. G. (2002), *Igneous and Metamorphic Petrology*, 2nd ed., Wiley, New York.

Bottinga, Y., and M. Javoy (1990a), MORB degassing: Bubble growth and ascent, *Chem. Geol.*, *81*, 255–270.

Bottinga, Y., and M. Javoy (1990b), Mid-ocean ridge basalt degassing: Bubble nucleation, *J. Geophys Res.*, *95*, 5125–5131.

Bottinga, Y., and M. Javoy (1991), The degassing of Hawaiian tholeiite, *Bull. Volcanol.*, *53*, 73–85.

Bouche, E., S. Vergniolle, T. Staudacher, A. Nercessian, J. C. Delmont, M. Frogneux, F. Cartault, and A. Le Pichon (2010), The role of large bubbles detected from acoustic measurements on the dynamics of Erta 'Ale lava lake (Ethiopia), *Earth Planet. Sci.Lett.*, *295*, 37–48.

Brennen, C. E. (1995), *Cavitation and Bubble Dynamics*, Oxford Eng. Sci. Seri. *44*, Oxford University Press, Oxford, U.K.

Bruce, P. M., and H. E. Huppert (1989), Thermal control of basaltic fissure eruptions, *Nature*, *342*(6250), 665–667.

Burgi, P.Y. (2006), Nyiragongo, *Soc. Volcanol. Geneve, 58*, March.

Burton, M., P. Allard, F. Mure, and C. Oppenheimer (2003), FTIR remote sensing of fractional magma degassing on Mt. Etna volcano, Sicily., in *Volcanic Degassing*, Geol. Soc. London Spec. Pub. 213, pp. 281–293.

Burton, M., P. Allard, F. Muré, and A. La Spina (2007a), Magmatic gas composition reveals the source depth of slug-driven strombolian explosive activity, *Science*, *317*(5835), 227–230.

Burton, M. R., H. M. Mader, and M. Polacci (2007b), The role of gas percolation in quiescent degassing of persistently active basaltic volcanoes, *Earth Planet. Sci. Lett.*, *264*, 46–60.

Calvari, S., L. Spampinato, and L. Lodato (2006), The 5 April 2003 vulcanian paroxysmal explosion at Stromboli volcano (Italy) from field observations and thermal data, *J. Volcanol. Geotherm. Res.*, *149*, 160–175.

Calvari, S., L. Spampinato, A. Bonaccorso, C. Oppenheimer, E. Rivalta, and E. Boschi (2011), Lava effusion—A slow fuse for paroxysms at Stromboli volcano? *Earth Planet. Sci. Lett.*, *301*, 317–323.

Cannata, A., P. Montalto, E. Privitera, and G. Russo (2009), Characterization and location of infrasonic sources in active volcanoes: Mt. Etna, September–November 2007, *J. Geophys. Res.*, *114*, B08308, doi:10.1029/2008JB006007.

Caplan-Auerbach, J., and S. R. McNutt (2003), New insights into the 1999 eruption of Shishaldin volcano based on acoustic data, *Bull. Volcanol.*, *65*, 405–417.

Carbone, D., M. P. Poland, M. R. Patrick, and T.R. Orr (2013), Continuous gravity measurements reveal a low-density lava lake at Kīlauea Volcano, Hawai'i, *Earth Planet. Sci. Lett.*, *376*, 178–185.

Carey, R. J., B. F. Houghton, J. E. Sable, and C. J. N. Wilson (2007), Contrasting grain size and componentry in complex proximal deposits of the 1886 Tarawera basaltic plinian eruption, *Bull. Volcanol.*, *69*, 903–926.

Carey, R., M. Manga, W. Degruyter, H. Gonnermann, D. A. Swanson, B. F. Houghton, T. Orr, and M. Patrick (2013), Convection in a volcanic conduit recorded by bubbles, *Geology*, *41*(4), 395–398.

Castro, J. M., A. Burgisser, C. I. Schipper, and S. Mancini (2012), Mechanisms of bubble coalescence in silicic magmas, *Bull. Volcanol.*, *74*, 2339–2352.

Chouet, B., and B. Shaw (1991), Fractal properties of tremor and gas piston events observed at Kilauea Volcano, Hawaii, *J. Geophys. Res.*, *96*(B6), 10,177–10,189.

Chouet, B., N. Hamisevicz, and T. R. McGetchin (1974), Photoballistics of volcanic jet activity at Stromboli, Italy, *J. Geophys. Res.*, *79*, 4961–4975.

Chouet, B. A., P. A. Dawson, T. Ohminato, M.Martini, G. Saccorotti, F. Giudicepietro, G.De Luca, G.Milana and R.Scarpa (2003), Source mechanisms of explosions at Stromboli volcano, Italy, determined from moment-tensor inversions of very-long-period data, *J. Geophys. Res.*, *108*(B1), 2019, ESE 7, 1–25.

Chouet, B., P. Dawson, and M. Martini (2008), Shallow-conduit dynamics at Stromboli volcano, Italy, imaged from waveforms inversions, in *Fluid Motion in Volcanic Conduits: A Source of Seismic and Acoustic Signals*, Geol. Soc. London Spec. Pub. *307*, pp. 57–84.

Chouet, B. A., P. B. Dawson, M. R. James, and S. J. Lane (2010), Seismic source mechanism of degassing bursts at Kīlauea Volcano, Hawai'i: Results from waveform inversion in the 10–50 s band, *J. Geophys Res.*, *115*, B09311, doi:10.1029/2009JB006661.

Clague, D. A., J. B. Paduan, and A. S. Davis (2009), Widespread strombolian eruptions of mid-ocean ridge basalt, *J. Volcanol. Geotherm. Res.*, *180*, 171–188.

Cluzel, N., D. Laporte, A. Provost, and I. Kannewischer (2008), Kinetics of heterogeneous bubble nucleation in rhyolitic melts: Implications for the number density of bubbles in volcanic conduits and for pumice textures, *Contrib. Mineral. Petrol.*, *156*, 745–763.

Collins, S. J., D. M. Pyle, and J. Maclennan (2009), Melt inclusions track pre-eruption storage and dehydration of magmas at Etna, *Geology*, *37*, 571–574.

Coltelli, M., M. Pompilio, P. Del Carlo, S. Calvari, S. Pannucci, and V. Scribano (1998), Etna, 1. Eruptive activity, *Acta. Vulcanol.*, *26*(10), 1, 141–148.

Coltelli, M., P. Del Carlo, M. Pompilio, and L. Vezzoli (2005), Explosive eruption of a picrite: The 3930 BP subplinian eruption of Etna volcano (Italy), *Geophys. Res. Lett.*, *32*, L23307, doi:10.1029/2005GL024271.

Costantini, L., C. Bonadonna, B. F. Houghton, and H. Wehrmann (2009), New physical characterization of the Fontana Lapilli basaltic Plinian eruption, Nicaragua, *Bull. Volcanol.*, *71*, 337–355.

Costantini, L., B. F. Houghton, and C. Bonadonna (2010), Constraints on eruption dynamics of basaltic explosive activity derived from chemical and microtextural study; the example of the Fontana Lapilli Plinian eruption, Nicaragua, *J. Volcanol. Geotherm. Res.*, *189*(3–4), 207–224.

Deardorff, N. D., K. V. Cashman, and W. W. Chadwick, Jr. (2011), Observations of eruptive plume dynamics and pyroclastic deposits from submarine explosive eruptions at NW Rota-1, Mariana arc, *Volcanol. Geotherm. Res.*, *202*(1), 47–59.

Del Bello, E., E. W. Llewellin, J. Taddeucci, P. Scarlato, and S. Lane (2012), An analytical model for gas overpressure in slug-driven explosions: Insights into Strombolian volcanic eruptions, *J. Geophys. Res.*, *117*, B02206, doi:10.1029/2011JB008747.

Delle Donne, D., and M. Ripepe (2012), High-frame rate thermal imagery of Strombolian explosions: Implications for explosive and infrasonic source dynamics, *J. Geophys. Res.*, *117*, B09206.

Dixon, J. E., and E. M. Stolper (1995), An experimental study of water and carbon dioxide solubilities in mid-ocean ridge basaltic liquids. Part II: Applications to degassing, *J. Petrol.*, *36*, 1633–1646.

Dixon J. E., E. M. Stolper, and J. R. Holloway (1995), An experimental study of water and carbon dioxide solubilities in mid-ocean ridge basaltic liquids. Part I: Calibration and solubility models, *J. Petrol.*, *36*, 1607–1631.

Dubosclard, G., R. Cordesses, P. Allard, C. Hervier, M. Coltelli, and J. Kornprobst (1999), First testing of a volcano Doppler radar (Voldorad) at Mount Etna, Italy, *Geophys. Res. Lett.*, *26*, 3389–3392.

Edmonds, M., and T. M. Gerlach (2007), Vapor segregation and loss in basaltic melts, *Geology*, *35*(8) 751–754.

Edmonds, M., et al. (2013), Magma storage, transport and degassing during the 2008–10 summit eruption at KilaueaVolcano, Hawai'i, *Geochim. Cosmochim. Acta*, *123*, 284–301.

Eissen, J. P., Y. Fouquet, D. Hardy, and H. Ondréas (2003), Recent MORB volcaniclastic explosive deposits formed between 500 and 1750 m.b.s.l. on the axis of the Mid-Atlantic Ridge, south of the Azores, in *Explosive Subaqueous Volcanism*, Geophys. Monogr. 140, edited by J. D. L. White, et al. pp. 143–166, AGU, Washington, D.C. 10.1029/140GM09.

Fee, D., M. Garces, M. Patrick, B. Chouet, P. Dawson, and D. Swanson (2010), Infrasonic tremor and degassing bursts from Halema'uma'u Crater, Kīlauea Volcano, Hawai'i, *J. Geophys. Res.*, *115*, B11316, doi:10.1029 /2010JB007642.

Ferrazzini, V., K. Aki, and B. Chouet (1991), Characteristics of seismic waves composing Hawaiian volcanic tremor and gas-piston events observed by a near-source array, *J. Geophys. Res.*, *96*, 6199–6209.

Field, L., T. Barnie, J. Blundy, R. A. Brooker, D. Keir, and E. Lewi (2012), Integrated field, satellite and petrological observations of the November 2010 eruption of Erta 'Ale, *Bull. Volcanol.*, *74*, 2251–2271.

Fiske, R. S., T. R. Rose, D. A. Swanson, D.E. Champion, and J.P. McGeehin (2009), Kulanaokuaiki Tephra (ca. A.D. 400–1000); newly recognized evidence for highly explosive eruptions at Kīlauea Volcano, Hawai'i, *Geolo. Soc. Am. Bull.*, *121*(5–6), 712–728, doi:10.1130/B26327.1.

Flynn, L. P., P. J. Mougginis-Mark, J. C. Gradie, and P. G. Lucey (1993), Radiative temperature at Kupaianaha lava lake, Kilauea Volcano, Hawaii, *J. Geophys. Res.*, *98*(B4), 6461–6476.

Francis, P., C. Oppenheimer, and D. Stevenson (1993), Endogenous growth of persistently active volcanoes, *Nature*, *366*, 554–557.

Fouquet, Y., J-L Charlou, J. P. Donval, J. Radford-Knoery, I. Costa, N. Lourenço, and M. K. Tivey (1995), Atlantic lava lakes and hot vents, *Nature*, *377*(6546), 201–201.

Gardner, J. E. (2012), Surface tension and bubble nucleation in phonolite magmas, *Geochim. Cosmochim. Acta*, *76*, 93–102.

Gardner, J. E., and M. H. Denis (2004), Heterogeneous bubble nucleation on Fe–Ti oxide crystals in high-silica rhyolitic melts, *Geochim. Cosmochim. Acta*, *68*, 3587–3597, doi:10.1016/j.gca.2004.02.021.

Gerlach, T. M., K. A. McGee, T. Elias, A. J. Sutton, and M. P. Doukas (2002), Carbon dioxide emission rate of Kilauea Volcano: Implications for primary magma and the summit reservoir, *J. Geophys. Res.*, *107*(B9), 2189, ESV3, 1–15.

Giordano, D., and D. B. Dingwell (2003), Viscosity of hydrous Etna basalt: Implications for Plinian-style basaltic eruptions, *Bull. Volcanol.*, *65*, 8–14.

Giordano, D., M. Polacci, A. Longo, P. Papale, D. B. Dingwell, E. Boschi, and M. Kasereka (2007), Thermo-rheological magma control on the impact of highly fluid lava flows at Mt Nyiragongo, *Geophys. Res. Lett.*, *34*, L06301, doi.10.1029/2006GL028459.

Giordano, D., J. K. Russell, and D. B. Dingwell (2008), Viscosity of magmatic liquids: A model, *Earth Planet. Sci. Lett.*, *271*, 123–134.

Goepfert, K., and J. E. Gardner (2010), Influence of pre-eruptive storage conditions and volatile contents on explosive Plinian style eruptions of basic magma, *Bull. Volcanol.*, *72*, 511–521, 4758, doi:10.1007/s00445-010-0343-1.

Gonnermann, H. M., and M. Manga (2007), The fluid mechanics inside a volcano, *Annu. Rev. Fluid Mech.*, *39*, 321–356.

Gouhier, M., and F. Donnadieu (2011), Systematic retrieval of ejecta velocities and gas fluxes at Etna volcano using L-band Doppler radar, *Bull. Volc.*, *73*, 1139–1145.

Gresta, S., M. Ripepe, E. Marchetti, S. D'Amico, M. Coltelli, A. J. L. Harris, and E. Privitera (2004), Seismo-acoustic measurements during July-August 2001 eruption of Mt Etna volcano, Italy, *J. Volcanol. Geotherm. Res.*, *137*, 219–230.

Gurenko, A. A., A. B. Belousov, R. B. Trumbull, and A. V. Sobolev (2005), Explosive basaltic volcanism of the Chikurachki Volcano (Kurile arc, Russia): Insights on pre-eruptive magmatic conditions and volatile budget revealed from phenocryst-hosted melt inclusions and groundmass glasses, *J. Volcanol. Geotherm. Res.*, *147*, 203–232.

Hager, S. A., T. M. Gerlach, and P. J. Wallace (2008), Summit CO_2 emission rates by the CO_2/SO_2 ratio method at Kīlauea Volcano, Hawai'i, during a period of sustained inflation, *J. Volcanol. Geotherm. Res.*, *177*, 875–882.

Happel, J., and H. Brenner (1973), *Mechanics of Fluids and Transport Processes: Low Reynolds Number Hydrodynamics*, Kluwer Academic Publishers, Dordrecht, Netherlands.

Hardiagon, M., D. Laporte, Y. Morizet, and A. Provost (2013), CO_2 degassing in a haplo-basaltic magma: An experimental approach, Goldschmit Conference, Firenze, Italy, *Mineral. Mag.*, *77*(5), 1258.

Harris, A. J. L, (2008), Modelling lava heat loss, rheology and convection, *Geophys. Res. Lett.*, *35*(L07), 303.

Harris, A. J. L and M. Ripepe (2007a), Temperature and dynamics of degassing at Stromboli, *J. Geophys. Res.*, *112*, B3, doi:10.1029/2006JB004393.

Harris, A. J. L., and M. Ripepe (2007b), Synergy of multiple geophysical approaches to unravel explosive eruption conduit and source dynamics: A case study from Stromboli, *Chem. Erde*, *112,* 67, 1–37.

Harris, A. J. L., R. Carniel, and J. Jones (2005), Identification of variable convective regime at Erta 'Ale lava lake, *J. Volcanol. Geotherm. Res.*, *142*, 207–223.

Head, J. W., and L. Wilson (2003), Deep submarine pyroclastic eruptions: Theory and predicted landforms and deposits, *J. Volcanol. Geotherm. Res.*, *121*, 155–193.

Helo, C., M.A. Longpré, N. Shimizu, D.A. Clague, and J. Stix (2011), Explosive eruptions at mid-ocean ridges driven by CO_2-rich magmas, *Nature Geosci.*, *4*, 260–263.

Holloway, J. R., and J. G. Blank (1994), Application of experimental results to C-O-H species in natural melts, in Volatiles in Magmas, Rev. Mineral., *30*, edited by M.R. Caroll and J. R. Holloway, Mineral. Soc. of America, pp. 187–230, Washington, D. C.

Hort, M., and R. Seyfried (1998), Volcanic eruption velocities measured with a micro radar, *Geophys. Res. Lett.*, *154*, 515–532.

Hort, M., R. Seyfried, and M. Vöge (2003), Radar Doppler velocimetry of volcanic eruptions: Theoretical considerations and quantitative documentation of changes in eruptive behaviour at Stromboli volcano, Italy, *Geophys. J. Int.*, *25*(1), 113–116.

Houghton, B. F., and H. M. Gonnermann (2008), Basaltic explosive volcanism: Constraints from deposits and models, *Chemie der Erde*, *68*, 117–140.

Houghton, B. F., C. J. N. Wilson, P. Del Carlo, M. Coltelli, J. E. Sable, and R. Carey (2004), The influence of conduit processes on changes in style of basaltic Plinian eruptions: Tarawera 1986 and Etna 122 BC, *J. Volcanol. Geotherm. Res.*, *137*, 1–14.

Hurwitz, S., and O. Navon (1994), Bubble nucleation in rhyolitic melts: Experiments at high-pressure, temperature, and water-content, *Earth Planet. Sci. Lett.*, *122*, 267–280.

Jaggar, T. A., R. H. Finch, and O. H. Emerson (1924), The lava tide, seasonal tide and the volcanic cycle, *Monthly Weather Rev.*, 142–145.

James, M. R., S. L. Lane, B. Chouet, and J. S. Gilbert (2004), Pressure changes associated with the ascent and bursting of gas slugs in liquid-filled vertical and inclined conduits, *J. Volcanol. Geotherm. Res.*, *129*, 61–82.

James, M. R., S. L. Lane, and B. Chouet (2006), Gas slug ascent through changes in conduit diameter: Laboratory insights into a volcano-seismic source process in low-viscosity magmas, *J. Geophys. Res.*, *111*, B05201, doi:10.1029/2005JB003718.

James, M. R., E. W. Llewellin, and S. L. Lane (2011), Comment on "It takes three to tango: 2. Bubble dynamics in basaltic volcanoes and ramifications for modeling normal Strombolian activity" by J. Suckale, B. H. Hager, L. T. Elkins-Tanton and J-C Nave, *J. Geophys. Res.*, *116*, B06207, doi:10.1029/2011JB008167, 1–16.

Jaupart, C., and S. Vergniolle (1988), Laboratory models of Hawaiian and Strombolian eruptions, *Nature*, *331*, 6151, 58–60.

Jaupart, C., and S. Vergniolle (1989), The generation and collapse of a foam layer at the roof of a basaltic magma chamber, *J. Fluid Mech.*, *203*, 347–380.

Johnson, J. B., A. J. L. Harris, and R. P. Hoblitt (2005), Thermal observations of gas pistoning at Kīlauea volcano, Hawai'i, *J. Geophys. Res.*, *110*, B11201, doi:10.1029/2005JB003944, 1–12.

Johnson, J. B., R. C. Aster, P. R. Kyle, and W. McIntosh (2008a), Acoustic source characterization of impulsive Strombolian eruptions from the Mount Erebus lava lake, *J. Volcanol. Geotherm. Res.*, *117*, 673–686.

Johnson, E. J., P. J. Wallace, K. Cashman, V. Delgado Granados, and A. Kent (2008b), Magmatic volatile contents and degassing-induced crystallization at Volcan Jorullo, Mexico: Implications for melt evolution and the plumbing systems of monogenetic volcanoes, *Earth Planet. Sci. Lett.*, *269*, 477–486.

Karlstrom, L., and M. Manga (2006), Origin and implications of zigzag rift patterns on lava lakes, *J. Volcanol. Geotherm. Res.*, *154*, 317–324.

Kazahaya, K., H. Shinohara, and G. Sato (1994), Excessive degassing of Izu-Oshima volcano: magma convection in a conduit, *Bull Volcanol.*, *56*, 207–216.

Kelly, P.J., P.R. Kyle, N.W. Dunbar, and K.W. Sims (2008), Geochemistry and mineralogy of the phonolite lava lake, Erebus volcano, Antartica: 1972–2004 and comparison with older lavas, *J. Volcanol. Geotherm. Res.*, *177*, 589–605.

Kerr, R. C., and J. R. Lister (1991), The effect of shape on crystal settling and on the rheology of magmas, *J. Geol.*, *99*, 457-467.

Khitarov, N. I., Y. B. Lebedev, A. M. Dorfman, and N. S. Bagdasarov (1979), Effects of temperature, pressure, and volatiles on the surface tension of molten basalt, *Geochem. Int.*, *16*(5), 78–86.

Kobayashi, T., A. Namiki, and I. Sumita (2010), Excitation of airwaves caused by bubble bursting in a cylindrical conduit: Experiments and a model, *J. Geophys. Res.*, *115*, B10201, doi:10.1029/2009JB006828.

Koeppen, W. C., M. Patrick, T. Orr, A. J. Sutton, D. Dow, and R. Wright (2013), Constraints on the partitioning of Kīlauea's lavas between surface and tube flows, estimated from infrared satellite data, sulfur dioxide emission rates, and field observations, *Bull. Volcanol.*, *75*, 716.

Lane, S. J., B. A. Chouet, J. C. Phillips, P. Dawson, G. A. Ryan, and E. Hurst (2001), Experimental observations of pressure oscillations and flow regimes in an analogue volcanic system, *J. Geophys. Res.*, *106*(B4), 6461–6476.

Lane, S. J., M. R. James, and S. B. Corder (2013), Volcano infrasonic signals and magma degassing: First-order experimental insights and application to Stromboli, *Earth Planet. Sci. Lett.*, *377 –378*, 169–179.

LeGuern F., J. Carbonelle, and H. Tazieff (1979), Erta 'Ale lava lake: Heat and gas transfer to the atmosphere, *J. Volcanol. Geotherm. Res.*, *6*, 27–48.

Lejeune, A. M., and P. Richet (1995), Rheology of crystal-bearing silicate melts: An experimental study at high viscosities, *J. Geophys. Res.*, *100*, 4215–4229.

Lejeune, A.M., Y. Bottinga, T.W. Trull and P. Richet (1999), Rheology of bubble-bearing magmas, *Earth Planet. Sci. Lett.*, *166*, 71–84.

Lensky, N. G., O. Navon, and V. Lyakhovsky (2004), Bubble growth during decompression of magma: Experimental and theoretical investigation, *J. Volcanol. Geotherm. Res.*, *129*, 7–22.

Lensky, N. G., R. W. Niebo, J. R. Holloway, V. Lyakhovsky, and O. Navon (2006), Bubble nucleation as a trigger for xenolith entrapment in mantle melts, *Earth and Planet. Sci. Lett.*, *245*, 278–288.

Lesne, P., B. Scaillet, M. Pichavant, G. Iacono-Marziano, and J.-M. Beny (2011a), The H_2O solubility of alkali basaltic melts: An experimental study, *Contrib. Mineral. Petrol.*, *162*, 133–151.

Lesne, P., B. Scaillet, M. Pichavant, and J.-M. Beny (2011b), The carbon dioxide solubility in alkali basalts: An experimental study, *Contrib. Mineral. Petrol.*, *162*, 153–168.

Lesne, P., S. C. Kohn, J. Blundy, F. Witham, R. E. Botcharnikow, and H. Behrens (2011c), Experimental simulation of closed-system degassing in the system basalt-H_2O−CO_2-S-Cl, *J. Petrol.*, *52*(9), 1737–1762.

Lovejoy, S., H. Gaonac'h, and D. Schertzer (2004), Bubble distributions and dynamics: The expansion-coalescence equation, *J. Geophys Res.*, *109*, B11203, doi:10.1029/2003JB002823.

Manga, M. (1996), Waves of bubbles in basaltic magmas and lavas, *J. Geophys. Res.*, *101*, 17,457–17,465.

Manga, M., and H. A. Stone (1993), Buoyancy-driven interactions between two deformable viscous drops, *J. Fluid Mech.*, *256*, 647–683.

Manga, M., and H. A. Stone (1994), Interactions between bubbles in magmas and lavas: Effects of bubble deformations, *J. Volcanol. Geotherm. Res.*, *63*, 267–279.

Manga, M., and H. A. Stone (1995), Collective hydrodynamics of deformable drops and bubbles in dilute low Reynolds number suspensions, *J. Fluid Mech.*, *300*, 231–263.

Mangan, M. T., and K. V. Cashman (1996), The structure of basaltic scoria and reticulite and inferences for vesiculation, foam formation, and fragmentation in lava fountains, *J. Volcanol. Geotherm. Res.*, *73*, 1–18.

Mangan, M., and T. Sisson (2000), Delayed, disequilibrium degassing in rhyolite magma: Decompression experiments and implications for explosive volcanism, *Earth Planet. Sci. Lett.*, *183*, 441–455.

Mangan, M. T., K. V. Cashman, and D. A. Swanson (2014), The dynamics of Hawaiian-style eruptions: A century of study, Chapter 8 of Poland, M. P., T. J. Takahashi, and C. M. Landowski, (eds.), *Characteristics of Hawaiian Volcanoes*, U.S. Geol. Surv. Prof. Pap. 1801, pp. 323–354, doi:10.3133/pp18018.

Mangan, M. T. and T. Sisson (2005), Evolution of melt-vapor surface tension in silicic volcanic systems: Experiments with hydrous melts, *J. Geophys. Res*, *110*, B01202,1–9.

Métrich, N., and P. J. Wallace (2008), Volatile abundances in basaltic magmas and their degassing paths tracked by melt inclusions, *Rev. Mineral. Geochem.*, *69*, 363–402.

Métrich, N., A. Bertagnini, P. Landi, and M. Rosi (2001), Crystallisation driven by decompression and water loss at Stromboli volcano (Aeolian Islands). *J. Petrol.*, *42*(8), 1471–1490.

Métrich, N., P. Allard, N. Spilliaert, D. Andronico, and M. Burton (2004), 2001 flank eruption of the alkali- and volatile-rich primitive basalt responsible for Mount Etna's evolution in the last three decades, *Earth Planet. Sci. Lett.*, *228*, 1–17.

Métrich, N., A. Bertagnini, and A. Di Muro (2010), Conditions of magma storage, degassing and ascent at Stromboli: New insights into the volcanic plumbing system with inferences on the eruptive dynamics, *J. Petrol.*, *51*, 603–626.

Moitra, P., H. M. Gonnermann, B. F. Houghton, and T. Giachetti. (2013), Relating vesicle shapes in pyroclast to eruption styles, *Bull. Volcanol.*, *75*, 691, 1–14, doi:10.1007/s00445-013-0691-8.

Mori, T., and M. Burton (2009), Quantification of the gas mass emitted during single explosions on Stromboli with the SO_2 imaging camera, *J. Volcanol. Geotherm. Res.*, *188*, 395–400.

Mourtada-Bonnefoi, C. C., and D. Laporte (1999), Experimental study of homogeneous bubble nucleation in rhyolitic magmas, *Geophys. Res. Lett.*, *26*(23), 3505–3508.

Mourtada-Bonnefoi, C. C., and D. Laporte (2002), Homogeneous bubble nucleation in rhyolitic magmas: An experimental study of the effect of H_2O and CO_2, *J. Geophys Res.*, *107*(B4), 2066, 1–19, doi:10.1029/2001JB000290.

Mourtada-Bonnefoi, C. C., and D. Laporte (2004), Kinetics of bubble nucleation in a rhyolitic melt: An experimental study of the effect of ascent rate, *Earth Planet. Sci. Lett.*, *218*, 521–537.

Mysen, B. O. (1977), The solubility of H_2O and CO_2 under predicted magma genesis conditions and some petrological and geophysical implications, *Rev. Geophys. Space Phys.*, *15*, 351–361.

Namiki, A., and M. Manga (2005), Response of a bubble bearing viscoelastic fluid to rapid decompression: Implications for explosive volcanic eruptions, *Earth Planet. Sci. Lett.*, *236*, 269–284.

Namiki, A., and M. Manga (2006), Influence of decompression rate on the expansion velocity and expansion style of bubbly fluids, *J. Geophys Res.*, *111*, B11208, doi:10.1029/2005JB004132.

Namiki, A., and M. Manga (2008), Transition between fragmentation and permeable outgassing of low viscosity magmas, *J. Volcanol. Geotherm. Res.*, *169*, 48–60.

Navon, O., and V. Lyakhovsky (1998), Vesiculation processes in silicic magmas, in The Physics of Explosive Volcanic Eruptions, Spec. Pub. 145, edited by J. S. Gilbert and R. S. J. Sparks, pp. 27–50, Geol. Soc. London, Uk.

Nguyen, C. T., H. M. Gonnermann, Y. Chen, C. Huber, and A. A. Maiorano (2013), Film drainage and the lifetime of bubbles, *Geochem., Geophys., Geosyst.*, *14*(9), 3616–3631, doi:10.1002/ggge.20198.

Nye, C. J., T. Keith, J. C. Eichelberger, T. P. Miller, S. R. McNutt, S. C. Moran, D. J. Schneider, J. Dehn, and J. R. Schaefer, (2002), The 1999 eruption of Shishaldin volcano, Alaska: Monitoring a distant eruption, *Bull. Volcanol.*, *64*, 507–519.

Oppenheimer, C., and P. Francis (1997), Remote sensing of heat, lava and fumarole emissions from Erta 'Ale vocano, Ethiopia, *Int. J. Remote Sens.*, *18*(8), 1661–1692.

Oppenheimer, C., and G. Yirgu (2002), Thermal imaging of an active lava lake: Erta 'Ale volcano, Ethiopia, *Int. J. Remote Sens.*, *23*, 4777–4782.

Oppenheimer, C., A. J. S. Mcgonigle, P. Allard, M. J. Wooster, and V. Tsanev (2004), Sulfur, heat and magma budget of Erta 'Ale lava lake, Ethiopia, *Geology*, *32*(6), 509–512.

Orr, T. R., W. A. Thelen, M R. Patrick, D.A. Swanson, and D. C. Wilson (2013), Explosive eruptions triggered by rockfalls at Kīlauea volcano, Hawaii, *Geology*, *41*(2), 207–210.

Otway, P. M, G.H. Blick, and B. J. Scott (1994), Volcanic deformations on Mount Erebus: Methods and results of geodetic surveys, 1980–1985, in Volcanological and Environmental Studies of Mount Erebus, Antartic Res. Ser. 66, edited by P. R. Kyle, pp. 57–68, AGU, Washington, D. C.

Palma, J. L., E. S. Calder, D. Basualto, S. Blake, and D. A. Rothery (2008), Correlations between SO_2 flux, seismicity and outgassing activity at the open vent of Villarica volcano, Chile, *J. Geophys. Res.*, *113*, B8,B10201, doi:10.1029/2008JB005577.

Papale, P. (1999), Modeling of the solubility of a two-component H_2O+CO_2 fluid in silicate liquid, *Am. Mineral.*, *84*, 477–492.

Papale, P. (2005), Determination of total H_2O and CO_2 budgets in evolving magmas from melt inclusion data, *J. Geophys Res.*, *110*, B03208, 1–13.

Papale, P., R. Moretti, and D. Barbato (2006), The compositional dependence of the saturation surface of H_2O+CO_2 fluids in silicate melts, *Chem. Geol.*, *229*, 78–95.

Parfitt, E. A. (2004), A discussion of the mechanisms of explosive basaltic eruptions, *J. Volcanol. Geotherm. Res.*, *134*, 77–107.

Parfitt, E. A., and L. Wilson (1995), Explosive volcanic eruptions: IX The transition between Hawaiian-style lava fountaining and Strombolian explosive activity, *Geophys. J. Int.*, *121*, 226–232.

Parfitt, E. A., and L. Wilson (1999), A plinian treatment of fallout from Hawaiian lava fountains, *J. Volcanol. Geotherm. Res.*, *88*, 67–75.

Parfitt, E. A., L. Wilson, and C. A. Neal (1995), Factors influencing the height of Hawaiian lava fountains: Implications for the use of fountain height as an indicator of magma content, *Bull. Volcanol.*, *57*, 440–450.

Patrick, M. R., A. J. L. Harris, M. Ripepe, J. Dehn, D. A. Rothery, and S. Calvari (2007), Strombolian explosive styles and source conditions: Insights from thermal (FLIR) video, *Bull. Volcanol.*, *69*, 769–784.

Patrick, M. R., T. Orr, D. Wilson, D. Dow, and R. Freeman (2011), Cyclic spattering, seismic tremor, and surface fluctuation within a perched lava channel, Kīlauea Volcano, *Bull. Volcanol.*, *73*, 639–653, doi:10.1007/s00445-010-0431-2.

Pichavant, M., I. Di Carlo, S. G. Rotolo , B. Scaillet, A. Burgisser, N. Le Gall, and C. Martel (2013), Generation of CO_2-rich melts during basalt magma ascent and degassing, *Contrib. Mineral. Petrol.*, *166*, 545–561, doi:10.1007/s00410-013-0890-5.

Pioli, L., C. Bonadonna, B. J. Azzopardi, J.C. Phillips, and M. Ripepe, M. (2012), Experimental constraints on the outgassing dynamics of basaltic magmas, *J. Geophys. Res.*, *117*, B03204, 1–17, doi:10.1029/2011JB008392.

Plank, T., K. A. Kelley, M. M. Zimmer, E. H. Hauri, and P. J. Wallace (2013), Why do mafic arc magmas contain ≈4 wt% water on average? *Earth Planet. Sci. Lett.*, *364*, 168–179.

Polacci, M., R.A. Corsaro, and D. Andronico (2006), Coupled textural and compositional characterization of basaltic scoria: Insights into the transition from Strombolian to fire fountain activity at Mount Etna, *Geology*, *34*, 201–204.

Polacci, M., D. R. Baker, L. Mancini, S. Favretto, and R. J. Hill (2009), Vesiculation in magmas from Stromboli and implications for normal Strombolian activity and paroxysmal explosions in basaltic systems, *J. Geophys. Res.*, *114*, B01206, doi:10.1029/2008JB005672.

Poland, M. P., A. Miklius, and E. K. Montgomery-Brown (2014), Magma supply, storage, and transport at shield-stage Hawaiian volcanoes, Chapter 5 of Poland, M. P., T. J. Takahashi, and C. M. Landowski, (eds.), *Characteristics of Hawaiian Volcanoes*, U.S. Geol. Surv. Prof. Pap. 1801, pp. 179–234, doi:10.3133/pp18015.

Proussevitch, A. A., and V. A. Kutolin (1986), Surface tension of magmatic melts (in russian), *Geol. Geophys.*, *9*, 58–67.

Proussevitch, A. A. and D. L. Sahagian (1996), Dynamics of coupled diffusive and decompressive bubble growth in magmatic systems, *J. Geophys. Res.*, *101*, B8, 17 447–17455.

Proussevitch, A. A., and D. L. Sahagian (1998), Dynamics and energetics of bubble growth in magmas: Analytical formulation and numerical modeling, *J. Geophys. Res.*, *103*(B8), 18,223–18,251.

Proussevitch, A. A., D. L. Sahagian, and V. A. Kutolin (1993a), Stability of foams in silicate melts, *J. Volcanol. Geotherm. Res.*, *59*, 161–178.

Proussevitch, A. A., D.L. Sahagian, and A.T. Anderson (1993b), Dynamics of diffusive bubble growth in magmas: Isothermal case, *J. Geophys. Res.*, *98*(B8), 22,283–22,308.

Richter, D. H., J. P. Eaton, K. J. Murata, W. U. Ault, and H. L. Krivoy (1970), Chronological narrative of the 1959–60 eruption of Kilauea volcano, Hawaii, Prof. Pap. 537, pp. 1–70, U.S. Geol. Surv., Reston, Va.

Ripepe, M., and E. Marchetti (2002), Array tracking of infrasonic sources at Stromboli volcano, *J. Geophys. Res.*, *29*(22), 33-1–33-4.

Ripepe, M., M. Rossi, and G. Saccorotti (1993), Image processing of explosive activity at Stromboli, *J. Volcanol. Geotherm. Res.*, *54*, 335–351.

Ripepe, M., P. Poggi, T. Braun and E. Gordeev (1996), Infrasonic waves and volcanic tremor at Stromboli, *Geophys. Res. Lett.*, *23*, 2, 181–184.

Ripepe, M. and E. Gordeev (1999), Gas bubble dynamics model for shallow volcanic tremor at Stromboli, *J. Geophys. Res.*, *104*, B5, 10,639–10,654.

Ripepe, M., S. Ciliberto, and M. D. Schiava (2001), Time constraints for modeling source dynamics of volcanic eruptions at Stromboli, *J. Geophys. Res.*, *106*, 8713–8727.

Ripepe, M., A. J. L. Harris, and R. Carniel (2002), Thermal, seismic and infrasonic evidence of variable degassing rates at Stromboli volcano, *J. Volcanol. Geotherm. Res.*, *118*, 285–297.

Rivalta, E., K. Pascal, J. Phillips, and A. Bonaccorso (2013), Explosive expansion of a slowly decompressed magma analogue: Evidence for delayed bubble nucleation, *Geochem. Geophys. Geosyst.*, *14*(8), 3067–3084, doi:10.1002/ggge.20183.

Roggensack, K., R. L. Hervig, S. B. McKnight, and S. N. Williams (1997), Explosive basaltic volcanism from Cerro Negro Volcano: Influence of volatiles on eruptive style, *Science*, *277*(5332), 1639–1642.

Rosi, M., A. Bertagnini, A. J. L. Harris, L. Pioli, M. Pistolesi, and M. Ripepe (2006), A case history of paroxysmal explosion at Stromboli: Timing and dynamics of the April 5, 2003 event, *Earth Planet. Sci. Lett.*, *243*, 594–606.

Rouland, D., D. Legrand, M. Zhizhin, and S. Vergniolle (2009), Automatic detection-discrimination of volcanic tremors and tectonic earthquakes: An application to Ambrym volcano, Vanuatu, *J. Volcanol. Geotherm. Res.*, *181*, 196–206.

Sable, J. E., B. F. Houghton, P. del Carlo, and M. Coltelli (2006a), Changing conditions of magma ascent and fragmentation during the Etna 122 BC basaltic eruption: Evidence from clast microtextures, *J. Volcanol. Geotherm. Res.*, *158*, 333–354.

Sable, J. E., B. F. Houghton, C. J. N. Wilson, and R. J. Carey (2006b), Complex proximal sedimentation from Plinian plumes: The example of Tarawera 1886, *Bull. Volcanol.*, *69*, 89–103.

Sable, J. E., B. F. Houghton, C. J. N. Wilson, and R. J. Carey (2009), Eruption mechanisms during the climax of the Tarawera 1886 basaltic Plinian eruption inferred from microtextural characteristics of the deposits, in Studies in Volcanology: The Legacy of George Walker, Geol. Soc. London Spec. Pub. IAVCEI, vol. 2, edited by T. Thordason et al., pp. 129–154.

Sahagian, D. (1985), Bubble migration and coalescence during the solidification of basaltic lava flows, *J. Geol.*, *93*(2), 205–211, doi:10.1086/628942.

Sahagian, D. L., A. T. Anderson, and B. Ward (1989), Bubble coalescence in basalt flows: Comparison of a numerical model with natural examples, *Bull. Volcanol.*, *52*, 49–56.

Sarda, P., and D. Graham (1990), Mid-ocean ridge popping rocks: Implications for degassing at ridge crests, *Earth Planet. Sci. Lett.*, *97*, 268–289.

Sawyer, G. M., C. Oppenheimer, V. I. Tsanev, and M. Burton (2008a), Magma degassing at Erta 'Ale volcano, Ethiopia, *J. Volcanol. Geotherm. Res.*, *178*, 837–846.

Sawyer, G. M., S. A. Carn, V. I. Tsanev, C. Oppenheimer, and M. Burton (2008b), Investigation into magma degassing at Nyiragongo volcano, Democratic Republic of the Congo, *Geochem. Geophys. Geosyst.*, *9*(2), 1–17.

Seach, J. (1999), Ambrym volcano, *Bulletin of Global Volcanism Network, Monthly Reports, 324:*08, (August).

Sisson, T. W., and T. L. Grove (1993), Temperatures and H_2O contents of low-MgO high-aluminia basalts, *Contrib. Mineral. Petrol.*, *113*, 167–184.

Sohn, R. A., et al. (2008), Explosive volcanism on the ultraslow-spreading Gakkel ridge, *Arctic Ocean, Nature*, *453*, 1236–1238.

Sparks, R. S. J. (1978), The dynamics of bubble formation and growth in magmas: A review and analysis, *J. Volcanol. Geotherm. Res.*, *3*, 1–37.

Sparks, R. S. J., M. I. Bursik, S. N. Carey, J. S. Gilbert, L. S. Glaze, H. Sigurdsson, and A. W. Woods (1997), *Volcanic Plumes*, Wiley, Chichester.

Sparks, R. S. J., J. Barclay, C. Jaupart, H. M. Mader, and J. C. Phillips (1994), Physical aspects on magma degassing, I. Experimental and theoretical constraints on vesiculation, *Rev. Mineral.*, *30*, 413–445.

Spilliaert, N., N. Métrich, and P. Allard (2006), S-Cl-F degassing pattern of water-rich alkali basalt: Modelling and relationship with eruption styles on Mount Etna volcano, *Earth Planet. Sci. Lett.*, *248*(3–4), 772–786.

Stelling, P., J. Beget, C. Nye, J. Gardner, J. D. Devine, and R. M. M. George (2002), Geology and petrology of ejecta from the 1999 eruption of Shishaldin Volcano, Alaska, *Bull. Volcanol.*, *64*, 548–561.

Stevenson, D. S., and S. Blake (1998), Modelling the dynamics and thermodynamics of volcanic degassing, *Bull. Volcanol.*, *60*, 307–317.

Stix, J. (2007), Stability and instability of quiescently active volcanoes: The case of Massaya, *Nicaragua, Geology*, *35*, 535–538.

Stolper, E., and J. R. Holloway (1988), Experimental determination of the solubility of carbon dioxide in molten basalt at low pressure, *Earth Planet. Sci. Lett.*, *87*, 397–408.

Stovall, W. K., B. F. Houghton, H. Gonnermann, S. A. Fagents, and D. A. Swanson (2011), Eruption dynamics of Hawaiian-style fountains: The case study of episode 1 of the Kīlauea Iki 1959 eruption, *Bull. Volcanol*, *73*, 511–529, doi:10.1007/s00445-010-0426-z.

Stovall, W. K., B. F. Houghton, J. E. Hammer, S. A. Fagents, and D. A. Swanson (2012), Vesiculation of high fountaining Hawaiian eruptions: Episodes 15 and 16 of 1959 Kīlauea Iki, *Bull. Volcanol.*, *74*, 441–455, doi :10.1007/s00445-011-0531-7.

Suckale, J., B. H. Hager, L. T. Elkins-Tanton, and J-C Nave (2010), It takes three to tango: 2. Bubble dynamics in basaltic volcanoes and ramifications for modeling normal Strombolian activity, *J. Geophys. Res.*, *115*, B07410, doi:10.1029/2009JB006916, 1-16.

Suckale, J., B.H. Hager, L. T. Elkins-Tanton, and J-C Nave (2011), Reply to the comment by M. R. James et al. on "It takes three to tango: 2. Bubble dynamics in basaltic volcanoes and ramifications for modeling normal Strombolian activity," *J. Geophys. Res.*, *116*, B06208, doi:10.1029/2011JB008351, 1–16.

Sutton, A. J., T. Elias, T. M. Gerlach, and J. B. Stokes (2001), Implications for eruptive processes as indicated by sulfur dioxide emissions from Kīlauea Volcano, Hawai'i, 1979–1997, *J. Volcanol. Geotherm. Res.*, *108*, 283–302.

Swanson, D. A., D. A. Duffield, D. B. Jackson, and D. W. Peterson (1979), Chronological narrative of the 1969–1971 Mauna-Ulu eruption of Kīlauea volcano, Hawaii, Prof. Pap. 1056, U.S. Geol. Surv., Denver, CO.

Swanson, D. A., T. R. Rose, R. S. Fiske, and J. P. McGeehin (2012), Keanakako'i Tephra produced by 300 years of explosive eruptions following collapse of Kīlauea's caldera in about 1500 CE, *J. Volcanol. Geotherm. Res.*, *215–216*, 8–25.

Taddeucci, J., P. Scarlato, A. Capponi, E. Del Bello, C. Cimarelli, D. M. Palladino, and U. Kueppers (2012a), High-speed imaging of Strombolian explosions: The ejection velocity of pyroclasts, *Geophys. Res. Lett.*, *39*, L02301, doi:10.1029/2011GL050404.

Taddeucci, J., M. A. Alatorre-Ibargengoitia, M. Moroni, L. Tornetta, A. Capponi, P. Scarlato, D. B. Dingwell, and D. De Rita (2012b), Physical parameterization of Strombolian eruptions via experimentally-validated modeling of high-speed observations, *Geophys. Res. Lett.*, *39*, L16306, doi:10.1029/2012GL052772.

Tait, S. R., C. Jaupart, and S. Vergniolle (1989), Pressure, gas content and eruption periodicity of a shallow crystallising magma chamber, *Earth Planet. Sci. Lett.*, *92*, 107–123.

Tamburello, G., A. Aiuppa, E. P. Kantzas, A. J. S. McGonigle, and M. Ripepe (2012), Passive vs. active degassing modes at an open-vent volcano (Stromboli, Italy), *Earth Planet. Sci. Lett.*, *359–360*, 106–116.

Tazieff, H. (1975), *Niragongo ou le Volcan Interdit*, Flamarion, Germany.

Tazieff, H. (1984), Niragongo: Renewed activity at the lava lake, *J. Volcanol. Geotherm. Res.*, *20*, 267–280.

Tazieff, H. (1996), *Volcans*, Editions Bordas, France.

Thompson, G., S. R. McNutt, and G. Tytgat, (2002), Three distinct regimes of volcanic tremor associated with the eruption of Shishaldin volcano, Alaska, April 1999, *Bull. Volcanol.*, *64*, 535–547.

Tilling, R. I. (1987), Fluctuations in surface height of active lava lakes during 1972–1974 Mauna Ulu eruption, Kīlauea volcano, Hawaii, *J. Geophys. Res.*, *92*(B13), 13,721–13,730.

Tilling, R. I., C. Heliker and T. L. Wright (1987), Eruptions of hawaiian Volcanoes: Past, Present, Future, edited by department of the Interior, U.S. Geol. Surv., Denver, CO.

Toramaru, A. (1995), Numerical study of nucleation and growth of bubbles in viscous magmas, *J. Geophys. Res.*, *100* (B2), 1913–1931.

Vergniolle, S. (2008), From sound waves to bubbling within a magma reservoir: Comparison between eruptions at Etna (2001, Italy) and Kilauea (Hawai'i), in Fluid Motion in Volcanic Conduits: A Source of Seismic and Acoustic Signals, Geol. Soc. London Spec. Pub. 307, edited by S. J. Lane and J. S. Gilbert, pp. 125–146, doi:10.1144/SP307.8.

Vergniolle, S. (submitted), Insights into magma chamber behaviour at Shishaldin volcano from the chronology of the 1999 eruption, *J. Geophys. Res.*, submitted.

Vergniolle, S., and E. Bouche (submitted), Massive foam coalescence detected from acoustic measurements at the surface of Erta 'Ale lava lake (Ethiopia): Its role in the recycling of the crust, *Bull Volcanal*, Submitted.

Vergniolle, S., and G. Brandeis (1996), Strombolian explosions: A large bubble breaking at the surface of a lava column as a source of sound, *J. Geophys. Res.*, *101*(B9), 20,433–20,448.

Vergniolle, S., and J. Caplan-Auerbach (2004), Acoustic measurements of the 1999 basaltic eruption of Shishaldin volcano, Alaska: 2) Precursor to the Subplinian activity, *J. Volcanol. Geotherm. Res.*, *137*, 135–151.

Vergniolle, S., and J. Caplan-Auerbach (2006), Basaltic thermals and Subplinian plumes: Constraints from acoustic measurements at Shishaldin volcano, Alaska, *Bull. Volcanol.*, *68*(7–8), 611–630.

Vergniolle, S., and J. Caplan-Auerbach (submitted), Foam disruption as the origin of the Subplinian basaltic plume at Shishaldin volcano (Alaska), *J. Geophys. Res.*, submitted.

Vergniolle, S., and Y. Gaudemer (2012), Decadal evolution of a degassing magma reservoir unravelled from fire fountains produced at Etna between 1989 and 2001, *Bull. Volcanal.*, *74*, 725–742.

Vergniolle, S., and C. Jaupart (1986), Separated two-phase flow and basaltic eruptions, *J. Geophys. Res.*, *91*(B12), 12,842–12,860.

Vergniolle, S., and C. Jaupart (1990), The dynamics of degassing at Kilauea Volcano, Hawaii, *J. Geophys. Res.*, *95*(B3), 2793–2809.

Vergniolle, S., and G. Brandeis (1994), Origin of the sound generated by strombolian explosions, *Geophys. Res. Lett.*, *21*, 1959–1962.

Vergniolle, S., and M. Mangan (2000), Hawaiian and Strombolian eruptions, *Encyclopedia of Volcanoes*, pp. 447–461, Academic Press, San Diego, CA.

Vergniolle, S., and M. Ripepe (2008), From Strombolian explosions to fire fountains at Etna Volcano (Italy): What do we learn by acoustic measurement? in *Fluid Motion in Volcanic Conduits: A Source of Seismic and Acoustic Signals, Geol. Soc. London Spec. Pub., 307*, edited by S.J. Lane and J.S. Gilbert, pp. 103–124, doi:10.1144/SP307.7.

Vergniolle, S., G. Brandeis, and J. C. Mareschal (1996), Strombolian explosions: Eruption dynamics determined from acoustic measurements, *J. Geophys. Res.*, *101*(B9), 20,449–20,466.

Vergniolle, S., M. Boichu, and J. Caplan-Auerbach (2004), Acoustic measurements of the 1999 basaltic eruption of Shishaldin volcano, Alaska: 1) Origin of Strombolian activity, *J. Volcanol. Geotherm. Res.*, *137*, 109–134.

Vidal, V., M. Ripepe, T. Divoux, D. Legrand, J-C Géminard, and F. Melo (2010), Dynamics of soap bubble bursting and its implications to volcano acoustics, *Geophys. Res. Lett.*, *37*, L07302.

Wade, J. A., T. Plank, W. G. Melson, G. J. Soto, and E. H. Hauri (2006), The volatile content of magmas from Arenal volcano, Costa Rica, *J. Volcanol. Geotherm. Res.*, *157*, 94–120.

Walker, D., and O. Mullins, Jr. (1981), Surface tension of natural silicates from 1200–1500°C and implications for melt structure, *Contrib. Mineral. Petrol.*, *76*, 455–462.

Wallace, P. (2005), Volatiles in subduction zone magmas: Concentrations and fluxes based on melt inclusion and volcanic gas data, *J. Volcanol. Geotherm. Res.*, *140*, 217–240.

Wallis, G. B. (1969), *One Dimensional Two-Phase Flows*, McGraw-Hill, New York.

Weill, A., G. Brandeis, S. Vergniolle, F. Baudin, J. Bilbille, J.-F. Fevre, B. Piron, and X. Hill (1992), Acoustic sounder measurements of the vertical velocity of volcanic jets at Stromboli volcano, *Geophys. Res. Lett.*, *19*, 2357–2360.

Williams, S. N. (1983), Plinian airfall deposits of basaltic composition, *Geology*, *11*, 211–214.

Wilson, L. (1980), Relationships between pressure, volatile content and ejecta velocity, *J. Volcanol. Geotherm. Res.*, *8*, 297–313.

Wilson, L., and J. W. Head (1981), Ascent and eruption of basaltic magma on the Earth and Moon, *J. Geophys. Res.*, *86*, 2971–3001.

Wolfe, E. W., M. O. Garcia, D. B. Jackson, R. Y. Koyanagi, C. A. Neal, and A. T. Okamura (1987), The Puu O'o eruption of Kilauea Volcano, episodes 1–20, January 3, 1983 to June 8, 1984, in *Volcanism in Hawaii*, Prof. Pap., 1350, edited by R. W. Decker et al. 471–508, U.S. Geol. Surv., Reston, Va.

Wolfe, E. W., C. A. Neal, N. G. Banks, and T. J. Duggan (1988), Geologic observations and chronology of eruptive events, in *The Puu O'o Eruption of Kilauea Volcano, Episodes 1–20, January 3, 1983 to June 8, 1984*, Prof. Pap. 1463, edited by E Wolfe, pp. 471–508, U.S. Geol. Surv., Reston, Va.

Woulff, G., and T. R. McGetchin (1976), Acoustic noise from volcanoes: Theory and experiments, *Geophys. J. R. Astron. Soc.*, *45*, 601–616.

Yang, X., P. M. Davis, P. T. Delaney, and A. T. Okamura (1992), Geodetic analysis of dike intrusion and motion of the magma reservoir beneath the summit of Kilauea Volcano, Hawaii, *J. Geophys. Res.*, *97*, 3305–3324.

Zhang, Y., and H. Ni (2010), Diffusion of H, C, and O components in silicate melts, *Rev. Mineral. Geochem.*, *72*, 171–225.

Zimmer, M. M., T. Planck, E. H. Hauri, G. M. Yogodzinski, P. Stelling, J. Larsen, B. Singer, B. Jicha, C. Mandeville, and C.J. Nye (2010), The role of water in generating the calc-alkaline trend: New volatile data for Aleutian magmas and a new tholeiitic index, *J. Petrol.*, *51*(12), 2411–2444.

15

Insights Into Mixing, Fractionation, and Degassing of Primitive Melts at Kīlauea Volcano, Hawai'i

Marie Edmonds, Isobel Sides, and John Maclennan

ABSTRACT

The nature of primitive magmas and the deep plumbing system at ocean island volcanoes is enigmatic and poorly understood. Melt inclusions provide windows into the magmatic history of primitive melts, allowing insights into melt heterogeneity, mixing, fractionation, and degassing. We present the major, trace, and volatile element composition of melt inclusions from tephra and lava samples erupted during 25 prehistoric and historic eruptions of Kīlauea Volcano over the past 600 years. Once corrected for post-entrapment crystallization, the most primitive melts have 14.7 wt% MgO, making them close to primary. The melts form a compositional array broadly controlled by olivine fractionation, but with significant heterogeneity superimposed. This heterogeneity exceeds that observed in previous studies of whole-rock compositions and glasses of historical Kīlauea magmas and is caused by mixing of compositionally distinct melts during fractionation. Primitive melt inclusions show a range in H_2O contents up to ~0.7 wt% and H_2O/Ce up to ~290, suggesting that some primary melts may be inherently enriched in H_2O, inherited from the mantle melting region. Temporal trends in melt compositions are similar to those observed in whole rocks, suggesting that there are significant changes in the degree of mantle melting and mantle source composition at Kīlauea Volcano over 50–100 year timescales.

15.1. INTRODUCTION

Petrographic analysis of erupted tephras and lavas from Kīlauea Volcano provides a powerful tool to interrogate the deep plumbing. A vast body of work has been carried out to attempt to reconstruct the compositions of primary melts and the subsequent processes of melt differentiation [*Garcia et al.*, 1989, 2003; *Dixon et al.*, 1991; *Clague et al.*, 1991, 1995; *Wallace*, 1998]. We focus here on quantifying the abundances of volatiles in melts and understanding the controls on their concentrations in primitive melts. Volatiles have an importance that belies their abundance in magmas. The presence of volatiles, as exsolved or dissolved components, influences magma-phase equilibria, transport, storage, and eruption. Volatile exsolution and degassing underpins an important component of volcano monitoring and an understanding of how volatiles affect magma properties and eruption rates and styles is critical for hazard assessment. The volatile contents of Kīlauea melts have been quantified from a range of perspectives. The most direct methods to study volatiles and degassing look at the composition and fluxes of volcanic gases, which represent the integrated sum of degassing over the magma column, variably modified during its transport though rocks, fluids, and atmosphere [*Edmonds and Gerlach*, 2007; *Edmonds et al.*, 2013a; *Gerlach*, 1986; *Gerlach and Graeber*, 1985; *Gerlach et al.*, 2002; *Greenland et al.*, 1985; *Hager et al.*, 2008; *Sutton et al.*, 2001]. A complementary line of enquiry seeks to reconstruct magma degassing by analyzing volatiles (and other elements) in melt inclusions and submarine glasses, which preserve the volatile concentrations that existed in the melt at high pressures.

Department of Earth Sciences, University of Cambridge, Cambridge, United Kingdom

Hawaiian Volcanoes: From Source to Surface, Geophysical Monograph 208. First Edition.
Edited by Rebecca Carey, Valérie Cayol, Michael Poland, and Dominique Weis.

Companion Website: www.wiley.com/go/Carey/Hawaiian_Volcanoes

Submarine glasses are erupted beneath a column of seawater and therefore degassing of the more soluble volatiles are partially inhibited. The compositions of submarine glasses from dredge hauls have been characterized for four of the five volcanoes on the Island of Hawai'i (Kīlauea, Mauna Loa, Mauna Kea, and Hualālai) and for the submarine volcanic center Lō'ihi [*Coombs et al.*, 2006; *Dixon and Clague*, 2001; *Dixon et al.*, 1991; *Garcia et al.*, 1989; *Kent et al.*, 1999; *Seaman et al.*, 2004; *Clague et al.*, 1991, 1995]. A suite of melt inclusions allows, in theory, reconstruction of magma transport, differentiation, and degassing through the crust [*Kamenetsky and Clocchiatti*, 1996; *Kent*, 2008; *Lowenstern*, 1995; *Métrich and Wallace*, 2008; *Sobolev*, 1996]. The compositions of melt inclusions from Kīlauea and Lō'ihi have been characterized in terms of their major, trace, and volatile elements [*Anderson and Brown*, 1993; *Edmonds et al.*, 2013b; *Hart and Wallace*, 2000; *Kent et al.*, 1999; *Wallace and Anderson*, 1998]. There are other petrologic indicators of degassing, for example, bubble textures, including parameters such as bubble size and number densities, which may illustrate processes of volatile supersaturation and bubble nucleation, growth, and coalescence [e.g., *Mangan et al.*, 1993; *Stovall et al.*, 2012; *Parcheta et al.*, 2012; *Carey et al.*, 2012]. Degassing may be observed indirectly through records of degassing phenomena, for example, recent images of gas pistoning at Halema'uma'u [*Patrick et al.*, 2011]. Degassing processes may be inferred from geophysical data, such as the occurrence of shallow long-period seismicity that might be linked to bubble growth [*Chouet et al.*, 2010] and acoustic measurements, linked to bubble bursting at the surface [*Fee et al.*, 2010]. There is clearly also potential for ground deformation measurements to quantify changes in the proportions of melt and gas in subsurface reservoirs based on changes in magma bulk modulus [*Rivalta and Segall*, 2008]. These latter methods to understand degassing, however, are all focused on the shallow plumbing system; understanding volatile systematics in the deeper plumbing, in more primitive melts, is more enigmatic. Below we review what is understood about the abundance of volatiles in primary and primitive melts and during differentiation and degassing for Kīlauea tholeiites, before presenting a new dataset of major, trace, and volatile data for melt inclusions from historical eruptions of Kīlauea Volcano.

15.1.1. Primary Melts

Primary tholeiitic melts (with ~16.5 wt% MgO [*Clague et al.*, 1991, 1995]) are generated at pressures of 2–35 GPa (to stabilize garnet in the residue [*Clague and Frey*, 1982]) and temperatures of 1450–1550°C beneath Kīlauea [*Eggins*, 1992; *Maaløe*, 2004], where the plume head impacts the base of the lithosphere [*Ribe and Christensen*, 1999]. Primary melts contain dissolved volatile species water (H_2O), carbon dioxide (CO_2), sulfur, and halogens [*Gerlach and Graeber*, 1985; *Dixon et al.*, 1991]. Melts become saturated with a CO_2-rich vapor at high pressures [*Dixon*, 1997; *Pan et al.*, 1991; *Spera and Bergman*, 1980; *Witham et al.*, 2012], likely supraliquidus between 10 and 20 kbar if the primary melt contains between 0.5 and 1 wt% CO_2 [*Spera*, 1981; *Eggins*, 1992; *Gerlach et al.*, 2002]. Bubble nucleation occurs at an unknown degree of supersaturation. Calculations and experiment suggest that supersaturation during ascent and eruption might be substantial [*Bottinga and Javoy*, 1991; *Pichavant et al.*, 2013], but this has not been borne out by observation. Whereas many mid-ocean ridge basalt liquids are supersaturated (primarily with CO_2) with respect to their recovery pressures by as much as 60 MPa [*Dixon and Stolper*, 1995; *Dixon et al.*, 1988; *Fine and Stolper*, 1986; *Soule et al.*, 2012; *Stolper and Holloway*, 1988], saturation pressures for tholeiitic glasses from the Puna Ridge are no more than 10 MPa greater than the recovery pressures [*Coombs et al.*, 2006; *Dixon and Stolper*, 1995; *Dixon et al.*, 1991]. However, supersaturation deeper in the lithosphere might be commonplace. From an early stage in their evolution, the melts coexist with a substantial lower density CO_2-rich supercritical fluid phase, which contributes to magma buoyancy [*Anderson*, 1995].

Primary melts with 16.5 wt% MgO have been proposed to contain 0.35 wt% ± 0.05 wt% H_2O on the basis of extrapolation back along a line of olivine control for the high-MgO glasses from the Puna Ridge [*Clague et al.*, 1991,1995]. It has been proposed that melt H_2O contents can only reach 0.8 wt% H_2O in extensively fractionated melts (with <7 wt% MgO) [*Dixon et al.*, 1991]. Estimation of primary melt CO_2 content is not straightforward, as most melts (submarine quenched and inclusions) are substantially degassed of CO_2. Parental melts have been proposed to contain 0.7 wt% CO_2 on the basis of combining fluxes of CO_2 gases measured at the summit of Kīlauea [*Gerlach et al.*, 2002] with independent estimates of magma supply rate to the shallow summit and rift zones [*Cayol et al.*, 2000]. The fluxes of CO_2 from the summit have recently, however, been shown to be variable and not clearly related to summit inflation or deflation. Fluxes of >8000 t/day were measured in 1998 during a period of summit deflation [*Gerlach et al.*, 2002], and fluxes of almost 5000 t/day in 2003 during a period of inflation [*Hager et al.*, 2008]. More recently, fluxes of >18000 t/day were measured in 2004 and >20,000 t/day in 2006 during a period of more intense summit inflation which has been interpreted to reflect an increase in magma supply rate [*Poland et al.*, 2012]. The uncertainty inherent in modeled estimates of magma supply are large [*Poland et al.*, 2012]. Other recent studies have shown that whole-rock ratios involving trace elements with similar partitioning behavior to the volatiles exhibit large changes over 10^1–10^2 year timescales at Kīlauea [*Pietruszka and Garcia*, 1999], suggesting that the relative abundances of the volatiles in primary melts might also be variable.

15.1.2. Fractionation, Degassing, and Mixing in Summit and Rift Zones

Primary melts undergo substantial modification through fractionation and mixing. Primary melts undergo extensive olivine fractionation (>20 wt%, of which around 17 wt% is fractionated into cumulates) involving olivines with forsterite contents of 78.9–90.7 mol% [*Clague et al.*, 1995], over a temperature range of 1346–1170 °C [*Helz and Thornber*, 1987]. The cumulate-forming olivines are frequently remobilized, in the form of small clusters or individual crystals, in later eruptions [*Clague et al.*, 1995]. At 7.8 wt% MgO, augite and then plagioclase join the liquidus [*Helz and Thornber*, 1987]. Magma mixing has been proposed to be important in controlling the compositions of lavas erupted subaerially [*Helz and Wright*, 1992; *Wright*, 1973; *Wright and Fiske*, 1971] and along the submarine portion of the East Rift Zone (ERZ) [*Dixon et al.*, 1991]. Evidence for mixing is in the form of resorbed phenocrysts, reaction rims, and reverse-zoned crystals [*Clague et al.*, 1995; *Helz and Wright*, 1992; *Wright*, 1973]. Additional evidence for magma mixing comes from the greater variability in melt inclusion compositions over the composition of glassy pillow rinds, indicating that the inclusions are compositionally more variable than the carrier liquids, which have been homogenized by mixing [*Clague et al.*, 1995], consistent with melt inclusion studies in many settings [*Kent and Elliott*, 2002; *Sobolev*, 1996; *Maclennan*, 2008].

The intermediate values of H_2O and S in fractionated glassy rinds from the Puna Ridge have been interpreted as the products of mixing of two liquids (one degassed and one undegassed) and subsequent fractionation [*Dixon et al.*, 1991]. The degassed end-member magma is suggested to form in the summit magma reservoir as a result of either prolonged gas flushing by CO_2-rich gases or long-lived lava lake activity and convection involving magma ascending to the shallow magma reservoir beneath Halema'uma'u and then sinking back into the main reservoir at 2–5 km (Figure 15.1) [*Dixon et al.*, 1991]. A recent study of volatile degassing during the 2008–present summit eruption has confirmed that magma convection and mixing take place, generating degassed carrier liquids with 7.4–7.6 wt% MgO that are transported back to depth [*Edmonds et al.*, 2013b], testifying as to the plausibility of this mechanism. High-fountaining eruptions with lava drain-back also provide a mechanism for degassed lavas to be recycled [*Wallace and Anderson*, 1998]. The mixed magma is then transported laterally out to the rift zones, where it may erupt.

Volcanic gas geochemistry is in broad agreement with the model of the magma "plumbing system" at Kīlauea that has been reconstructed from detailed analysis of seismic and ground deformation data over decades [*Ryan*, 1988], which highlights the near-vertical ascent of magma beneath the summit, followed by lateral injection of magma into the rift zones at a shallow level (Figure 15.1). The primary measurement of volcanic gases over the past few decades at Kīlauea Volcano is the SO_2 flux from both the summit and rift eruption sites [*Sutton et al.*, 2001; *Sutton et al.*, 2003; *Elias and Sutton*, 2007; *Elias and Sutton*, 2012; *Elias et al.*, 1998]. The SO_2 flux generally ranges from 500 to 3000 t/day at the East Rift Zone eruption site and is used as a proxy for lava extrusion rate [*Sutton et al.*, 2003], based on evidence from glass geochemistry that suggests that sulfur exsolves at very low pressures pre- and syn-eruption [*Harris and Anderson*, 1983; *Gerlach and Graeber*, 1985; *Dixon et al.*, 1991]. During magma ascent to the surface, H_2O and sulfur begin to partition strongly into vapor, becoming significant only at pressures of <400 bars [*Gerlach and Graeber*, 1985; *Gerlach*, 1986; *Dixon*, 1997; *Dixon et al.*, 1991]. CO_2 fluxes at the ERZ are consistently low [*Gerlach et al.*, 1998], consistent with the extensive fractionation of CO_2 gases from the magma beneath the summit (Figure 15.1) [*Gerlach and Graeber*, 1985].

The earliest volcanic gas collections at Kīlauea Volcano were obtained in Halema'uma'u crater in 1912 [*Day and Shepherd*, 1914], 1917 [*Shepherd*, 1921], and 1918–1919 [*Shepherd*, 1921; *Jaggar*, 1940]. In 1918, Thomas Jaggar sampled volcanic gases (the J series) directly from the Halema'uma'u lava lake at the summit of Kīlauea. The restored gas analyses illustrate a broad continuum of compositions related by progressive CO_2 degassing [*Gerlach*, 1980]. These compositions were used, along with samples collected during the 1983 fountaining activity at the start of the Pu'u 'Ō'ō-Kupianaha eruption, and melt inclusion data [*Harris and Anderson*, 1983] to formulate a model for volatile degassing at Kīlauea Volcano, illustrating the two "stages" of degassing, involving removal of segregated CO_2 at the summit and then the more soluble gases at the eruption site [Figure 15.1; *Gerlach and Graeber*, 1985].

15.1.3. Geochemical Heterogeneity From Mantle

Variability in the abundance of minor, trace, and volatile elements, as well as isotopic composition, at a fixed MgO has long been recognized for Hawaiian lavas [*Clague and Frey*, 1982; *Garcia et al.*, 1989; *Wagner and Grove*, 1998] and has been linked to the degree of partial

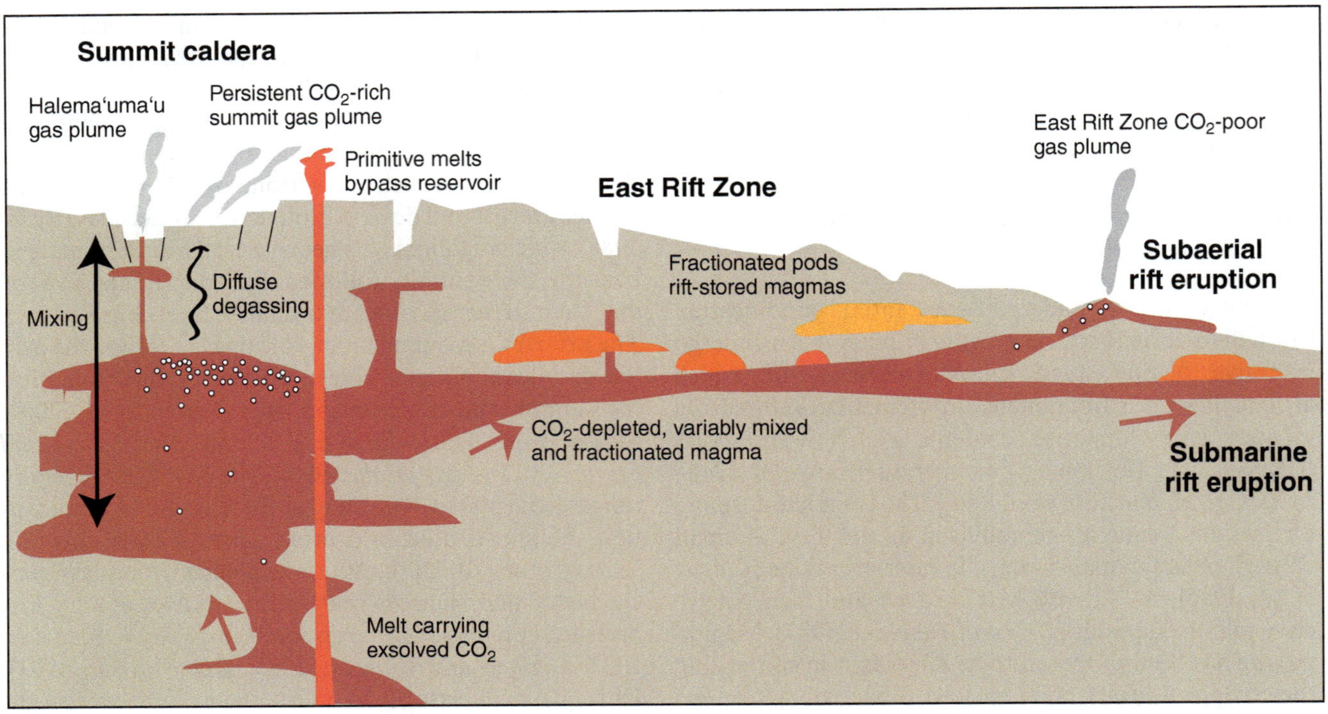

Figure 15.1 Schematic diagram of the shallow plumbing system of Kīlauea Volcano as inferred from geophysics [*Ryan*, 1988], volcanic gas geochemistry [*Gerlach and Graeber*, 1985], and petrology [e.g., *Dixon et al.*, 1991, *Sides et al.*, 2014]. The locations of the persistent outgassing plumes are indicated.

melting, progressive source depletion, and/or mantle source heterogeneity. Long-timescale intra- and inter-volcano geochemical evolution [*Blichert-Toft et al.*, 2003; *Kurz et al.*, 1996; *Lassiter and Hauri*, 1998] and the isotopic heterogeneities along two trends "Kea" and "Loa" have been linked to a geochemically zoned or streaked plume [*Dixon and Clague*, 2001; *Frey and Rhodes*, 1993; *Hauri et al.*, 1996; *Jackson et al.*, 1972; *Lassiter and Hauri*, 1998; *Weis et al.*, 2011]. The water content of the mantle plume has been quantified by study of a range of Hawaiian submarine glasses: Lō'ihi melts are relatively depleted in H_2O (at a fixed Ce content) relative to tholteiites from Kīlauea derived from the plume core, and melts upstream (South Arch) are enriched, metasomatized by hydrous partial melts [*Dixon and Clague*, 2001]. Within the plume core, two mantle source end members can be recognized: a relatively dry "KOO" component, with higher Sr isotope ratios, perhaps representing dehydrated oceanic crust and sediments, and a more H_2O-rich "KEA," with higher $^{206}Pb/^{204}Pb$ ratios, perhaps representative of more H_2O-rich recycled lithosphere. On a smaller scale, isotopic heterogeneities have been recognized for individual volcanoes over timescales of 10^1–10^2 years [*Frey and Rhodes*, 1993; *Marske et al.*, 2008; *Pietruszka and Garcia*, 1999; *Pietruszka et al.*, 2006, 2012; *Rhodes and Hart*, 1995].

At Kīlauea, alongside Pb isotopic variability are changes in the ratios of incompatible elements by almost a factor of 2 between 1790 and the present day [*Pietruszka and Garcia*, 1999]. These data support the interpretations of major-element studies of historical summit and rift eruption lavas, which show that each eruption is characterized by a "unique" geochemical fingerprint that is independent of olivine control, suggesting a degree of source heterogeneity [*Wright*, 1973; *Wright and Fiske*, 1971]. Modeling the isotopic and trace-element composition of whole rocks erupted during historical Kīlauea summit eruptions suggests that the lavas with higher proportions of the highly incompatible elements were the result of a smaller degrees of melting of a more depleted source, and their eruption correlates with periods of low eruption rate [*Pietruszka and Garcia*, 1999], suggesting a fundamental link between mantle geochemistry and volcanic processes.

15.1.4. Aim of This Chapter

In this chapter, we present geochemical data for olivine, melt inclusions, and matrix glasses contained within tephra and lavas erupted during 25 eruptions at or very close to the summit of Kīlauea Volcano over the past 600 years. This work is the first melt inclusion study for Kīlauea to encompass major, trace, and volatile elements together and to incorporate a careful correction for postentrapment processes. Our melt inclusions reach 14.7 wt% MgO (after correction for post-entrapment crystallization), which is close to the composition of the most primitive glasses from the Puna Ridge [*Clague et al.*, 1991], and the data set yields important new insights into near-primary melts from Kīlauea Volcano. A subset of the melts has avoided the extensive fractionation and mixing processes that characterize melts erupted further downrift (Figure 15.1). The combination of trace and volatile elements is a powerful tool for examining variability in primitive melt compositions and processes of magma mixing, fractionation, and degassing as well as postentrapment modification. We examine the temporal trends in melt composition related to primary melt heterogeneity and quantify the range in melt compositions that have existed in the deep plumbing of Kīlauea Volcano over the past 600 years.

15.2. ERUPTIONS SAMPLED AND MICROANALYTICAL METHODS

15.2.1. Summary of Stratigraphy

Samples were collected from the deposits of 25 historical eruptions over a 600 year period (Table 15.1, with sample localities shown in Figure 15.2). In addition to the subaerial samples from Kīlauea, a submarine tephra deposit from Lō'ihi volcano was also analyzed in this study (supplied by D. Clague). The sample (deposit D, P5-401-Sc3) [*Clague et al.*, 2003] was a tholeiitic olivine-rich clastic sand comprized of angular glass fragments produced during submarine strombolian explosions, from which olivines were picked for melt inclusion analysis. The material was collected from a spatter cone on the northeast rim of Lō'ihi by the Hawai'i Undersea Research Laboratory's Pisces-V submersible.

The oldest samples from Kīlauea Volcano analyzed in this study were erupted prior to formation of the current caldera. These include the products of a fissure eruption in what is now the Kīlauea Military Camp on the northern margin of the caldera, thought to have erupted during the summit overflow period during growth of the Observatory shield, or possibly earlier (ca. AD1400; D. Swanson, personal communication, May 2014). Another precaldera sample was a pristine glassy upper surface of one of the 'Ailā'au lava flows AD ~1445, preserved beneath thick forest soils to the east of Kīlauea Iki Crater [*Clague et al.*, 1999].

Other prehistoric samples collected were from units within the Keanakāko'i Tephra Member, the deposits produced from largely explosive eruptions during the 300 year period following caldera collapse [*McPhie et al.*, 1990; *Swanson et al.*, 2006, 2012; *Swanson and Christiansen*, 1973; *Weaver et al.*, 2010]. These deposits were widespread around the caldera rim, with the thickest and best-preserved sections exposed within 1–2 m wide rifted gullies within a 1 km distance of the southern caldera boundary fault (Figure 15.2). Seven samples were collected from within the Keanakāko'i, including two from the basal reticulite produced during high Hawaiian fountains immediately after the summit collapse; lapilli and ash falls produced during the phreatomagmatic events; dense scoria and ash from the layer 6 unit [*McPhie et al.*, 1990]; spatter from circumferential fissures on the southwest (SW) margin of the caldera and dense, nonvesicular juvenile lava blocks and lava-coated bombs from block falls attributed to transients subplinian explosions. Also sampled was the 1–3 m thick golden pumice deposit [*Sharp et al.*, 1987], which directly overlies the Keanakāko'i Tephra and was produced during vigorous high fountaining within the caldera.

Historical eruptions between 1823 and 1924 were dominated by lava lake activity, which produced numerous pahoehoe flows across the caldera floor [*Brigham*, 1909]. Repeated eruptions buried many earlier flows and thus limited the sampling potential of this period. However, spatter from the 1832 fissure eruption on Byron Ledge and from a fissure on the east caldera wall in 1877 was collected. In addition, two samples from an overflow from the 1882–1885 Halema'uma'u lava lake were collected from the caldera floor, along with glassy selvages from the 1919–1920 Mauna Iki eruption in the Southwest Rift Zone (SWRZ) which was fed by summit lavas [*Rowland and Munro*, 1993].

No juvenile material was produced during the phreatic summit explosions in 1924, and the products from the seven brief eruptions that followed the event are no longer exposed within the caldera. Vesicular, scoracious spatter produced during the summit fissure eruptions in 1954, 1971, 1974, and 1982 were sampled from ramparts on the caldera floor (Figure 15.2; Table 15.1). Most of these ramparts lie above the location of the summit magma reservoir, east of Halema'uma'u Crater, and as a result they are sites of well-developed gas fumaroles that have led to significant mineral coating of the spatter and scoria clasts. Care was therefore taken to select samples that were as pristine as possible. Tephras produced during the 1959 Kīlauea Iki eruption were sampled extensively from multiple pits. Lapilli-sized pumice clasts from a smaller point source fountain within Halema'uma'u Crater in 1961 were also collected from a surface-exposed deposit within the caldera.

The most recent Kīlauea summit samples analyzed were vesicular scoria and scoracious "cow-pat" bombs produced during the 2008–current eruption from the open vent within Halema'uma'u Crater. The bombs, though juvenile, were coated in lithic ash and contained many wall-rock fragments. They were ejected during four

Table 15.1 Summary of the eruption deposits, with dates, description, and references, sampled in this study.

Date	Sample	Eruption Style	Locality	Deposit Type	References[a]
~1400	1400-K	Fissure	Summit	Pumiceous spatter	1
~1445	1445-A	Effusive	Summit	Pahoehoe flow top	2
~1500	1500-1	Hawaiian fountain	Summit	Pure reticulite	3
~1500	1500-BR	Hawaiian fountain	Summit	Pure reticulite	3
~1550	1550	Phreatomagmatic	Summit	Scoria, lithics, ash, and accretionary lapilli	3
~1650	1600-6	Transient explosive	Summit	Scoria, lapilli	4, 5
1700	1700	Fissure	SWRZ	Dense spatter	3
1790	1790-1	Transient explosive	Summit	Dense, pasty bomb crust	3
1790	1790-w	Transient explosive	Summit	Dense, rounded juvenile bombs	6
1820–1823	1823	Hawaiian fountain	Summit	Golden pumice	7
Jan. 1832	1832	Fissure	Summit	Vesicular spatter	8
5 May 1877	1877	Fissure	Summit	Vesicular spatter	8
1882–1885	1885-A	Effusive	Summit	Tumulus flow top	8
1882–1885	1885-B	Effusive	Summit	Pahoehoe flow top	8
1919–1920	1920	Effusive	SWRZ	Pahoehoe flow top	9
31 May 1954	1954	Fissure	Summit	Vesicular spatter	10
14 Nov. 1959	(1959)-E1	Hawaiian fountain	Summit	Pumice lapilli, achneliths, and Pele's hair	11
26 Nov. 1959	(1959)-E2	Hawaiian fountain	Summit	Pumice and Pele's hair	
28 Nov. 1959	(1959)-E3	Hawaiian fountain	Summit	Pumice	
6 Dec. 1959	(1959)-E5	Hawaiian fountain	Summit	Fluidal pumice lapilli	
7 Dec. 1959	(1959)-E6	Hawaiian fountain	Summit	Fluidal pumice lapilli	
8 Dec. 1959	(1959)-E7	Hawaiian fountain	Summit	Fluidal pumice lapilli	
10 Dec. 1959	(1959)-E8	Hawaiian fountain	Summit	Fluidal pumice lapilli	
14 Dec. 1959	(1959)-E10	Hawaiian fountain	Summit	Pumice lapilli and achneliths	
17 Dec. 1959	(1959)-E15	Hawaiian fountain	Summit	Golden pumice	
19 Dec. 1959	(1959)-E16	Hawaiian fountain	Summit	Golden pumice	
17 Jan. 1960	1960-KO	Hawaiian fountain	ERZ	Golden pumice	11
Feb. 1961	1961	Hawaiian fountain	Summit	Pumice lapilli	12
24 May 1969	1966-mu	Fissure	ERZ	Vesicular spatter	13
6 Sep. 1969	1969-7	Hawaiian fountain	ERZ	Golden pumice/reticulite	13
30 Dec. 1969	1969-8	Hawaiian fountain	ERZ	Pumice lapilli and achneliths	13
14 Aug. 1971	1971	Fissure	Summit	Vesicular spatter	14
10 Nov. 1973	1973	Fissure	ERZ	Pumiceous spatter	15
19 July 1974	1974-J1	Fissure	ERZ	Vesicular spatter	16
19 July 1974	1974-J2	Fissure	Summit	Vesicular spatter	16
19 Sept. 1974	1974-S	Fissure	Summit	Vesicular spatter	16
30 Apr. 1982	1982-1	Fissure	Summit	Vesicular spatter	17
Sept. 1982	1982-2	Fissure	Summit	Vesicular spatter	17
9 Apr. 2008	2008-S1	Transient explosive	Summit	Pumiceous bombs/lapilli and lithics	18
27 Aug. 2008	2008-S2	Transient explosive	Summit	Pumiceous bombs/lapilli and lithics	
2 Sept. 2008	2008-18	Transient explosive	Summit	Pumiceous bombs/lapilli and lithics	
26 Apr. 2010	2010-S3	Transient explosive	Summit	Pumiceous bombs/lapilli and lithics	

Note: See Fig. 2 for sampling locations.
[a]References 1: *Neal and Lockwood*, 2003; 2: *Clague et al.*, 1999; 3: *Swanson et al.*, 2012; 4: *McPhie et al.*, 1990; 5: *Swanson et al.*, 2006; 6: *Weaver et al.*, 2010; 7: *Sharp et al.*, 1987; 8: *Brigham*, 1909; 9: *Rowland and Munro*, 1993; 10: *Macdonald and Eaton*, 1957; 11: *Richter et al.*, 1970; 12: *Richter*, 1964; 13: *Swanson et al.*, 1979; 14: *Duffield et al.*, 1982; 15: *Tilling et al.*, 1987; 16: *Lockwood et al.*, 1999; 17: *Neal and Lockwood*, 2003; 18: *Houghton et al.*, 2011.

transient explosive events on 9 April 2008, 27 August 2008, 2 September 2008, and 26 April 2010. The September 2008 cow-pat bombs were scattered across a 500 m^2 area south of the eruption vent. All other scoria clasts were collected by D. Swanson as part of ongoing U.S. Geological Survey (USGS) monitoring of the current eruption using "tear catcher" sampling boxes and buckets placed at specific localities downwind of the vent [*Houghton et al.*, 2011; *Orr et al.*, 2013; *Swanson et al.*, 2009; *Swavely et al.*, 2010].

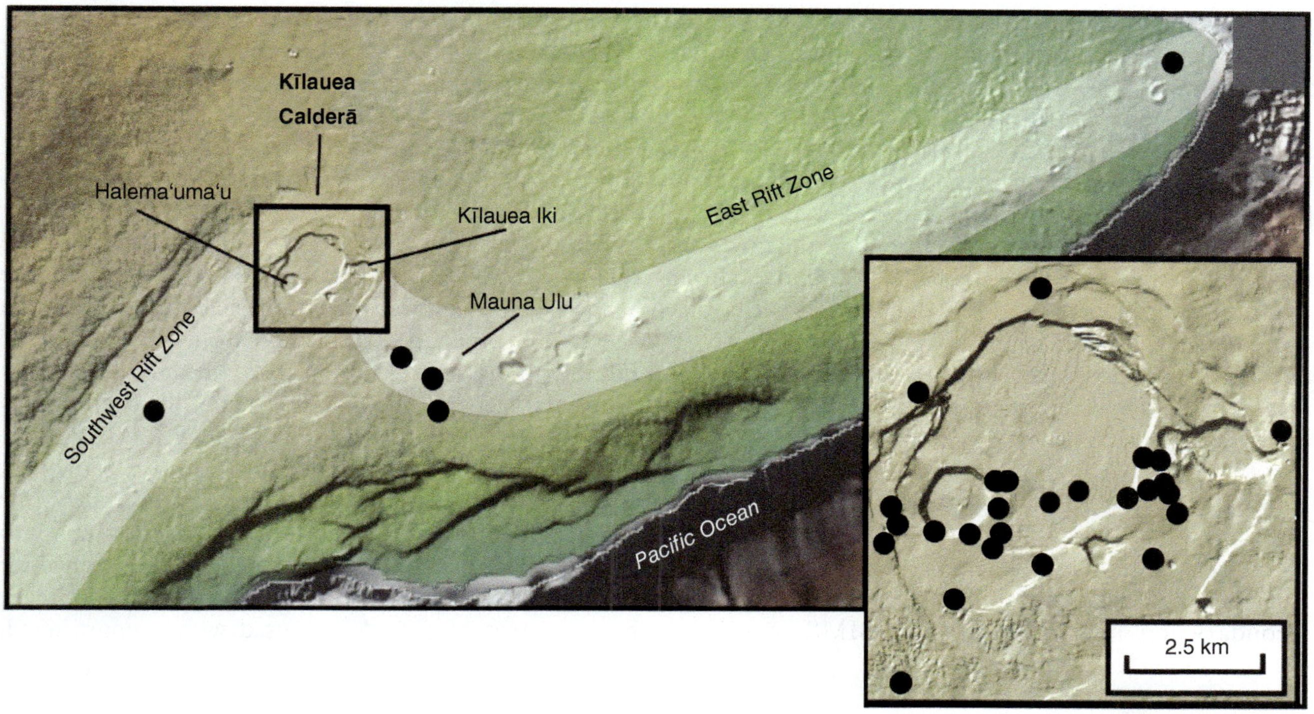

Figure 15.2 Map to show the locations of samples collected in this study. Details of each eruption and sample descriptions are given in Table 15.1.

Rift zone sampling was largely restricted to the upper rift zone portions (Figure 15.2) and included material from the 1969–1974 Mauna Ulu eruption. Spatter produced both during the initial fissure stage and reticulite samples from the later high fountains were collected. The majority of the other rift zone samples were produced during fissure activity, and these included the 1973 fissure near Pauahi Crater and the 1974 fissure in the ERZ.

15.2.2. Deposit Types

A primary objective was to sample rapidly quenched, glassy eruption products from a broad range of prehistoric and historic eruptions in order to cover the greatest variation in eruption styles and temporal range. In all, 79 samples were collected but only 43 were selected for analysis, which encompassed material from 25 eruptions over a 600 year period (Table 15.1). The sampled deposits arose from a wide range of eruption styles. Fissure or weak fountain deposits, e.g., from the 1832, 1877, 1971, 1973, 1974, and 1982 eruptions, covered a spectrum of highly vesicular to dense scoria clasts from spatter ramparts. Eruptions characterized by point source fountains (e.g., 1500, 1823, 1959, 1960, and 1969 eruptions; Table 15.1) typically produced deposits with pumice clasts with fluidal surfaces and highly vesicular interiors as well as ragged, highly vesicular golden to light brown pumice and reticulite with up to 98–99% vesicularity (produced in fountains >300 m). The higher fountains produced large (up to 25 cm diameter) spatter bombs. High-fountaining deposits were usually sampled at localities downwind of the eruption vent in beds up to several meters thick. Clasts were collected from sections exposed in existing collapse pit structures and from 1–2 m deep pits dug by hand.

The 1550 and 1790 eruptions were phreatomagmatic [*Swanson et al.*, 2012; *Weaver et al.*, 2010], and their deposits were finer grained than the pyroclasts produced from fissures and high fountains. These "wet" pyroclastic deposits were well-bedded as a result of their deposition by fall or displayed low-angle cross stratification as a result of base surge formation [*McPhie et al.*, 1990]. The deposits were a mixture of juvenile (targeted for sampling) and lithic components with grain sizes ranging from well-sorted fine ash, loose crystals, and lapilli to scattered large blocks up to 1 m and with clear "bomb-sags" in the underlying stratigraphy. Unlike fissure and high-fountaining products, fluidal clasts were absent from the phreatomagmatic ash deposits.

The samples collected from the 2008–2009 summit eruption at Halema'uma'u and also the Keanakāko'i eruption deposits included Pele's hair and tears and juvenile bombs up to 20 cm in diameter. The bombs were discrete globules of vesicle-rich spatter and pumice, angular chunks of fresh and oxidized scoria, or glass-coated bombs and lapilli. The material from the 2008–2010 summit activity was scattered downwind of the vent as discrete clasts, and despite the broad range of clast morphology, the volumes of ejecta are too small to form well-sorted beds. Samples were also collected from effusive eruptions such as the 1445, 1882, and 1919 eruptions (Table 15.1), from summit overflows of pahoehoe lavas [*Clague et al.*, 1999], where the upper 1–1.5 cm of the glass-rich rope-texture on small sections of the flow surfaces was sampled. While glass could be obtained from these effusive products, it is likely that the melt inclusions hosted by the olivines in such samples are likely to be affected by water loss by diffusive reequilibration [*Gaetani et al.*, 2012]. However, samples were pristine and the upper selvages were typically vesicular.

15.2.3. Microanalysis Procedures

Glass chips and olivine phenocrysts were hand picked and mounted with epoxy to be analyzed first by secondary ion mass spectrometry (SIMS) and then by electron microprobe and laser ablation inductively coupled mass spectrometry (LA ICP-MS). All of the inclusions analyzed were naturally quenched, 40–200 μm in size, and not necked or breached by cracks (Figure 15.3). There were no visible daughter crystals in most of the inclusions, although some contained rare, small sulfides.

The concentrations of H_2O, CO_2, Li, and B were obtained by SIMS on a Cameca IMF 4f ion microprobe at the North American Electric Reliability Corporation (NERC) microanalytical facility at the University of Edinburgh using a 15 kV primary beam of O^- ions using the general methodology published previously [*Hauri et al.*, 2002b; *Blundy and Cashman*, 2008]. Positive secondary ions were accelerated to 4500 eV, with an offset of −75 eV (for ^{1}H and trace elements) and −50 eV (for ^{12}C) (±20 eV) to reduce transfer of molecular ions into the secondary column. A 50 μm raster was performed for 3 min prior to the start of each analysis, and a primary beam current of 5–6 nA was used with a nonrastered,

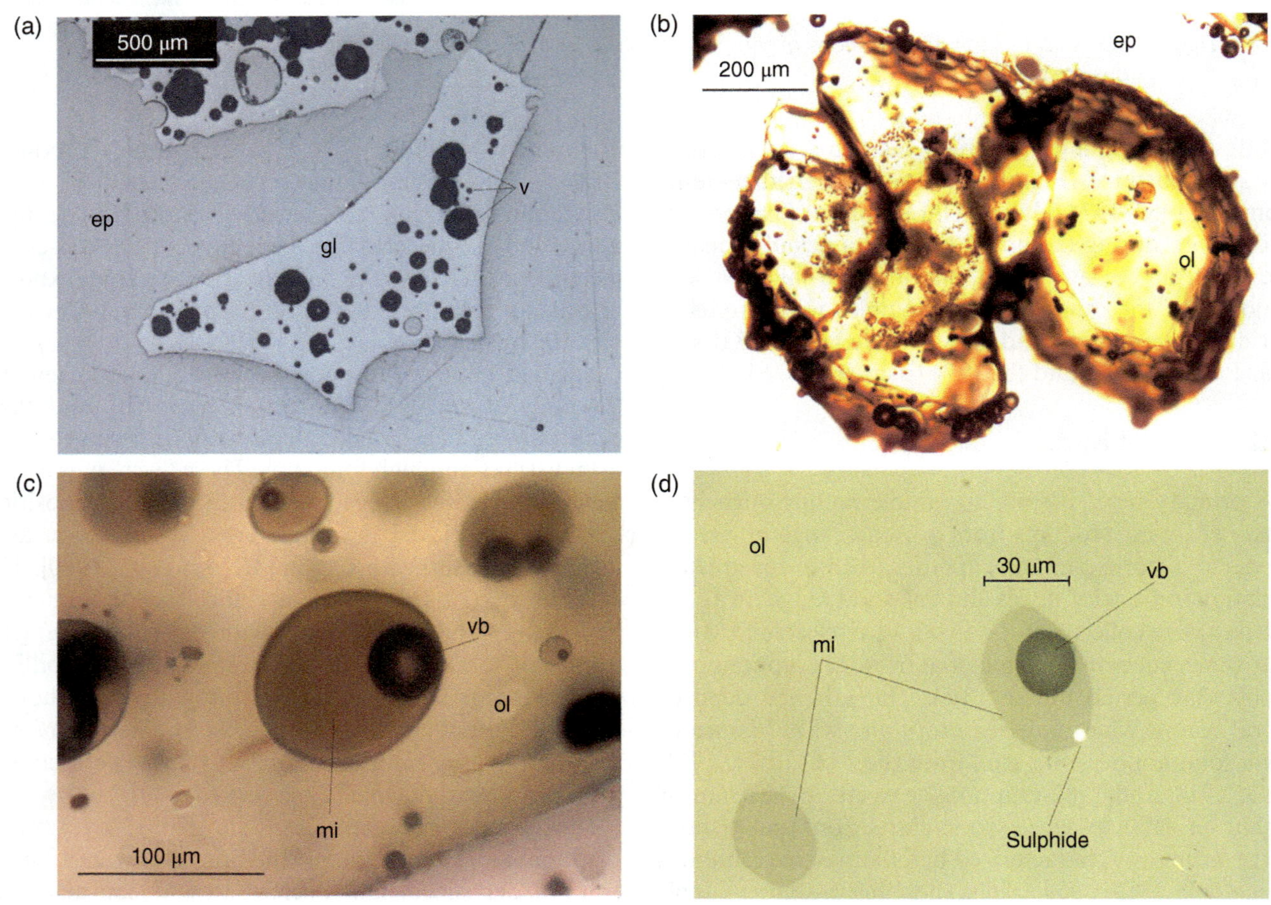

Figure 15.3 Images of glass chips, olivines, and melt inclusions analyzed in this study: (a) reflected light image of a vesicular glass chip (ep: epoxy resin, gl: glass); (b) transmitted light image of a cluster of olivine (ol) crystals, with melt inclusions visible in their interiors; (c) transmitted light image of melt inclusions (mi) contained within olivine, showing their clear brown pristine nature, and the included shrinkage bubbles (vb); (d) reflected light image of melt inclusions, showing a rare sulfide inclusion.

oval-shaped beam covering a 15–20 μm area on single spots within the boundaries of the melt inclusions. Peak positions were verified before each analysis. The following elements were analyzed by counting for 3 s in each of a 10 cycle run: ^{1}H, ^{7}Li, ^{11}B, ^{25}Mg, ^{30}Si. These counts were then normalized to ^{30}Si and converted to concentrations using a calibration curve populated by glass standards. The relative ion yield for H correlates with SiO_2 content such that plotting $^{1}H/^{30}Si$ versus H_2O yields a single working curve for glasses of variable SiO_2 content [*Blundy and Cashman*, 2008]. CO_2 concentrations, however, require a correction for SiO_2 content. The full list of glass standards used is shown in Table 15.2. Accuracy (given by $100[(x_{obs} - x_{ref})/(x_{ref})]$, where x_{obs} is the mean of the observed concentrations and x_{ref} is the reference concentration in percent) and precision (given by $100(\sigma / \overline{x_{obs}})$, where σ is the standard deviation on repeat measurements of concentration in percent) were monitored throughout the sessions by repeat analysis of the standards as unknowns: for H_2O analyses these were >9% and <6%, respectively, and for CO_2, they were <11% and <8%, respectively. The average CO_2 and H_2O backgrounds over seven sessions were 56 ppm and 0.03 wt%, respectively.

Carbon was measured independently of ^{1}H using the same beam conditions but with a 50 μm image field to improve transmission at moderate mass resolution, which was sufficient to resolve $^{24}Mg^{2+}$ at the ^{12}C peak position for background olivine measurements and inclusion analyzes. Carbon-12 was analyzed for 3 s in each of 20 cycle runs in which $^{24}Mg^{2+}$, $^{28}Si^{2+}$, and ^{30}Si were also measured. During data processing, the first five cycles of the ^{1}H analyzes and the first 10 cycles of the ^{12}C data were discarded to avoid the effects of surface contamination on the samples that may have survived the cleaning process. Instrumental backgrounds were minimized by allowing samples held in epoxy to outgas in a separate vacuum for at least 10 h prior to use in the SIMS instrument.

The major-element and volatile (S, Cl, and F) compositions of the glasses, inclusions, and host olivines were determined using the Cameca SX100 electron microprobe at the University of Cambridge. Quantitative determinations of elements were made using the wavelength-dispersive system. A range of metal, oxide, and silicate (e.g., jadeite, wollastonite) standards were used for calibration of the spectrometers. All analyzes used an accelerating voltage of 15 kV. For olivine, a spot size of 4 μm and a 100 nA beam current were used. For glasses, a 10 μm spot was used with a beam current of 60 nA for Cl, F, S, P, Cr, and Ni and 4 nA for all other elements, with counting times of 50–200 s per analysis. During glass measurements, Na peaks were counted first to avoid significant migration during the run. In addition to calibration of each X-ray line, a series of secondary reference standards (olivines, pyroxenes, feldspars, and glasses) were measured daily to check accuracy, precision, and totals. Accuracy was in general better than 5% for most elements, based on 77 repeat analyzes of EMPA secondary standard 2390-5 and by comparison with reference concentrations for the standard, with the exception of TiO_2, K_2O, P_2O_5, and Cl, which were better than 20%–35%. Detection limits for S, Cl, and F were 40, 38, and 170 ppm, respectively, and precision was typically <5% for all oxides, with the exception of MnO, P_2O_5, and F, which were better than 20%.

LA ICP-MS analysis of trace elements in melt inclusions was performed at the Department of Earth Sciences, University of Cambridge, using a Perkin-Elmer Elan DRC II ICP-MS. This was coupled to a UP213 Nd:YAG laser operating at 266 nm. A laser repetition rate of 10 Hz and laser power of 1 mJ (10 J/cm) were used with a 60 μm diameter beam for most analyses and 40 μm beam for smaller inclusions. Smaller beam sizes were avoided due to the reduction of signal intensity with size. The standard NIST 610 (trace elements in a glass matrix 3 mm wafer, National Institute of Standards and Technology, Gaithersburg, Maryland) was used for calibration of element sensitivity [*Pearce et al.*, 1997]. The ^{44}Ca concentration in each glass and inclusion sample was used for internal standard normalization of the trace-element signals and ^{12}Mg was used for olivine measurements. Calibration accuracy was verified by repeat analysis of the USGS standards BCR-2G, BHVO-2G, and KL2-G (MPI-DING, Mainz) during each analytical run. Drift during the analytical session was less than 10% each day and was compensated for by the internal standard calculations. For most elements, accuracy and precision, based on 70 repeat analyzes of glass KL2-G, were generally better than 4% and 10%, respectively. Detection limits were in the sub-ppm range for many of the elements.

15.2.4. Post-entrapment Crystallization Corrections

Once the melt inclusion has become trapped and the temperature of the system decreases, crystallization of the host phase will begin to take place on the inclusion walls [*Kent*, 2008; *Sobolev*, 1996]. In the absence of reequilibration between the host olivine and the newly crystallized olivine on the inclusion wall, the new crystallization will be fractional. The olivine will be more magnesian than the trapped melt and will follow established Fe/Mg partitioning relationships [*Roeder and Emslie*, 1970], where

$$K_D = \left({}^{X_{Fe}}\!\big/_{X_{Mg}} \right)_{olivine} \Big/ \left({}^{X_{Fe}}\!\big/_{X_{Mg}} \right)_{melt} = 0.30 \pm 0.04.$$

Table 15.2 Standards used for secondary ion mass spectrometry in this study on a Cameca 4f ion probe at the NERC facility, Edinburgh.

	Composition	SiO_2	H_2O	CO_2
TPF	Basalt	46.1	6.21	
ST-1	Basalt	48.0	2.96	629
ST-2	Basalt	48.1	2.84	1,261
ST-6	Basalt	48.7	1.58	914
CFD	Basalt	48.0	1.391	
ST-3	Basalt	48.8	1.33	
2390-5	Basalt	48.6	0.68	176
TPA	Basalt	48.2	0.403	
2678-6	Basalt	50.0	0.21	229
Rb480	Rhyolite	73.6		9,717
Rb497	Rhyolite	80.4		10,055
Siss51	Rhyolite	75.6		424

This process depletes the trapped melt in Mg relative to Fe^{2+} and the next increment of olivine will therefore be slightly less MgO rich. The overall effects of postentrapment crystallization (PEC) are to produce a Fe-rich rim of olivine immediately next to the melt inclusion, deplete the melt inclusion of all olivine-compatible elements, and increase slightly the concentrations of incompatible elements. The melt inclusion compositions in this study were corrected for variable amounts of PEC by calculation (using Petrolog3 [*Danyushevsky and Plechov*, 2011]). Petrolog3 applies an iterative correction scheme [*Danyushevsky et al.*, 2000a], requiring specification of the measured major- and minor-element compositions of each inclusion, the forsterite content of the host olivine, the H_2O content of the inclusion, and the oxidation state of the melt (the magnetite-wüstite buffer [*Rhodes and Vollinger*, 2005]). The elements that are most affected by the PEC process (and therefore the correction) are the olivine-compatible elements, and therefore these are associated with the greatest degree of uncertainty. Variation of just ±1 Fo unit in the host olivine is equivalent to a range of corrected MgO contents of up to ±2.0 wt% [*Kent*, 2008], compared to just 15% change in the concentration of an incompatible element.

A complicating factor in the PEC correction process is Fe loss caused by diffusive reequilibration between the melt inclusion, the olivine host, and the carrier liquid [*Gaetani et al.*, 2012; *Qin et al.*, 1992]. Fe loss occurs at slow cooling rates, whereby the Fe-rich olivine rim that crystallizes in the inclusion equilibrates with the more magnesian host olivine. As Fe diffuses from the rim to the host, so the inclusion also begins to equilibrate with the rim by "adding" Fe to maintain the Fe/Mg equilibrium. In order to correct for Fe loss, the initial FeO content of the inclusion at the time of entrapment must be known. Primitive MgO contents can be difficult to estimate, but the FeO content of primitive melts is little affected by olivine crystallization [*Putirka*, 2005] and is nearly constant along the olivine control line. The input FeO value specified for the model calculations was 11.33 wt%, which was the average of the matrix glass FeO concentrations and agrees well with the estimate of primitive melt FeO for Hawai'i [*Putirka*, 2005] (11.4 wt% ± 0.03). The assumption of a constant FeO value may carry with it significant error if the olivines analyzed are mantle-derived xenocrysts and not the result of low-pressure crystallization or if the melt inclusions have variable compositions (which they do; see below). PEC corrections typically required 0–20% olivine addition, although 5% of the inclusions required 20%–35% (see online data supplement).

15.3. PETROGRAPHY OF OLIVINE HOSTS, MELT INCLUSIONS, AND MATRIX GLASSES

All tephras and lavas sampled were olivine phyric, with the abundance and size of the phenocryst population appearing to increase with increasing eruption rate such that many of the intracaldera lava flows and fissures were notably phenocryst poor (<5 vol%), in stark contrast with all high-fountaining products and the 1973 Pauahi fissure eruption in the ERZ, which contained abundant olivine (10–20 vol% in some samples), often with a diverse range of morphologies. Olivine is by far the most abundant phenocryst present in all samples and usually occurs as equant, euhedral grains 0.5–2.0 mm in size, although skeletal, elongate, rounded, and aggregate forms up to 8 mm were also observed in some fountain tephras. Clinopyroxene and plagioclase phenocrysts were rare or absent in the tephras and glassy selvages collected, although some rare microcrysts of plagioclase were observed within several glass samples from the selvages of lava flows or relatively degassed clasts erupted from lava lakes.

15.3.1. Olivine Petrography

Olivine cores range in composition from $Fo_{89.2}$ to $Fo_{77.5}$, very close to the range observed in the Puna Ridge samples [*Clague et al.*, 1995] and encompassing the ranges observed in lavas from the subaerial rift zone [*Helz and Wright*, 1992; *Wright and Fiske*, 1971], from Kīlauea Iki [*Helz*, 1987], and from tholeiites from Mauna Loa and Mauna Kea [*Davis et al.*, 2003; *Garcia et al.*, 1989; *Norman and Garcia*, 1999; *Seaman et al.*, 2004]. There are two peaks in olivine composition (Figure 15.4a; online data supplement), corresponding to Fo_{83-84} and Fo_{87-89}, which correspond to the two peaks recognized by *Clague et al.* [1995] for lavas from the Puna Ridge (both peaks), turbidite grains (the higher peak only), and xenolith olivines (both peaks). The two peaks were ascribed to the large euhedral olivines having high-Fo cores and smaller anhedral or subhedral crytals, often partly resorbed, forming the lower-Fo peak [*Clague et al.*, 1995]. The more forsteritic olivines, which are often deformed, have been ascribed to the disaggregation of crystal mushes [*Clague et al.*, 1995], although more recent work on the texture and composition of similar phenocrysts from Réunion suggests that the full range of textures and compositions might be derived by rapid growth in suspension and may not be derived from cumulates [*Welsch et al.*, 2013]. Minor elements in olivine vary systematically so that, as Fo content decreases from 89.2 to 77.5 mol%, MnO increases from ~0.13 to ~0.30 wt%, CaO increases weakly from ~0.22 to ~0.33 wt%, and NiO decreases from ~0.50 to ~0.20 wt% (Figures 15.4b–d).

The olivine cores are in general not in equilibrium with the carrier liquids (Figure 15.5), similar to the Puna Ridge lavas [*Clague et al.*, 1991]. However, a subset of 79 olivine crystal rims (outer 10–20 μm) were analyzed in addition to their cores. In accordance with previous work [*Garcia et al.*, 2003; *Clague et al.*, 1995], the majority of Kīlauea's

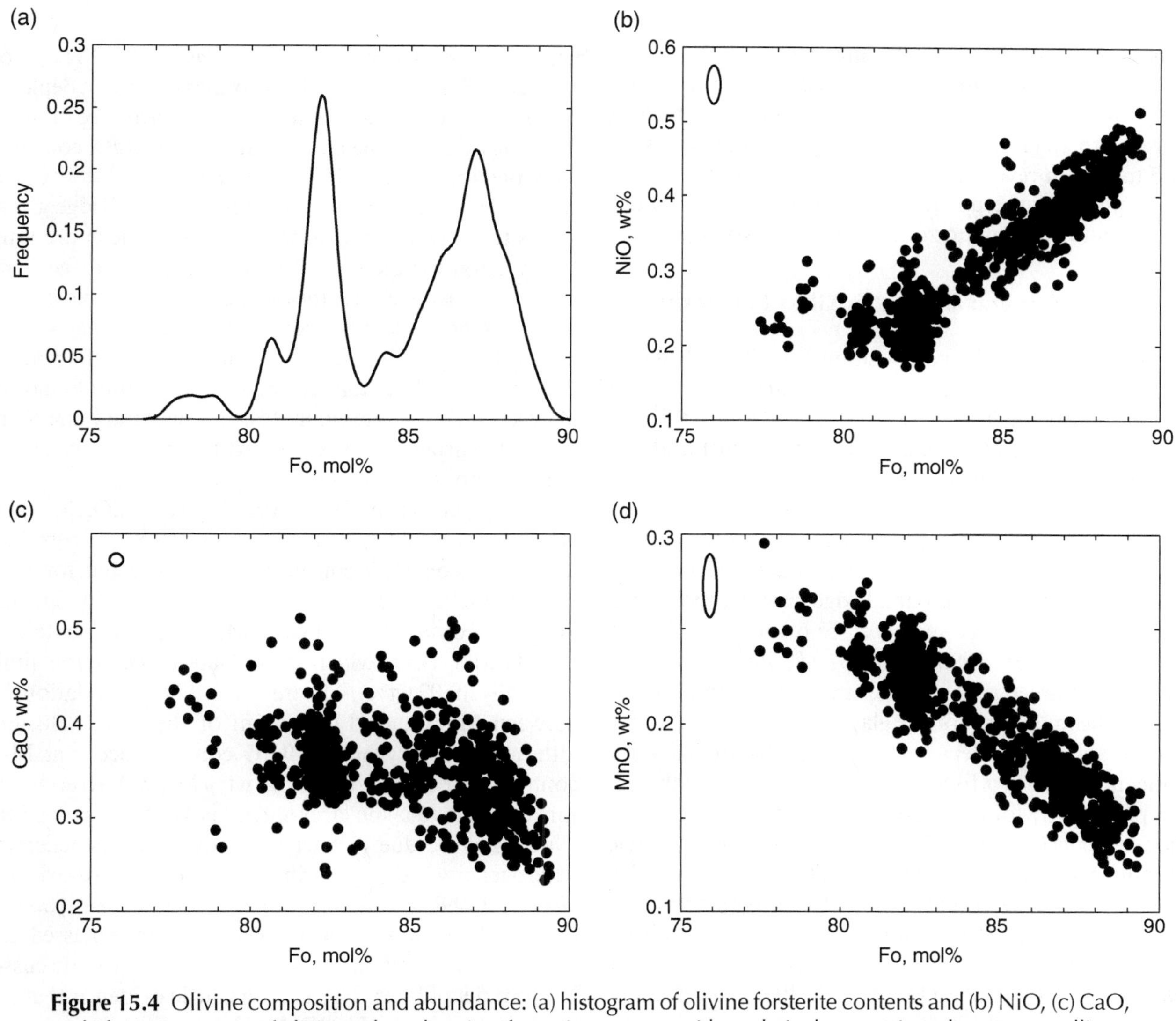

Figure 15.4 Olivine composition and abundance: (a) histogram of olivine forsterite contents and (b) NiO, (c) CaO, and (d) MnO content of olivines plotted against forsterite content, with analytical uncertainty shown as an ellipse.

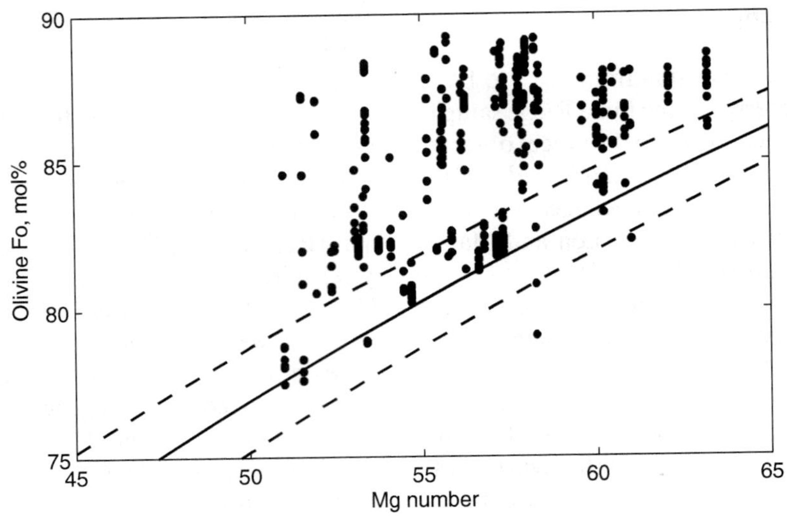

Figure 15.5 Rhodes plot of the Mg# of the postentrapment-corrected melt inclusions against the forsterite content of olivine. Solid and dashed lines mark where $\left(X_{Fe}/X_{Mg}\right)_{olivine} \Big/ \left(X_{Fe}/X_{Mg}\right)_{melt} = 0.30 \pm 0.04$ [*Roeder and Emslie*, 1970].

olivines are normally zoned or unzoned, with the maximum normal zoning observed in one crystal from a core of $Fo_{88.6}$ to a rim of $Fo_{88.2}$. The rims are generally in equilibrium with their carrier melts. Only two olivines were found to exhibit reversely zoned rims thought to be indicative of pre- or syn-eruptive magma mixing [*Garcia et al.*, 2003], with the maximum zoning only $Fo_{86.7}$ to $Fo_{87.8}$.

15.3.2. Melt Inclusion and Matrix Glass Compositions

We have analyzed the compositions of 374 melt inclusions and 165 matrix glasses. The PEC-corrected MgO content of the melt inclusions ranges from 14.73 to 6.71 wt% (Figure 15.6; online data supplement) and the MgO content of the matrix glasses ranges from 10.28 to 4.37 wt%. The melt inclusion data are plotted in terms of MgO content, instead of host forsterite content (which is preferable for inclusions that have undergone PEC) to allow comparison with the Puna Ridge glass compositions, which are marked in red in Figure 15.6 and other figures [*Clague et al.*, 1991, 1995; *Dixon et al.*, 1991]. The melt compositions are tholeiitic in terms of their total alkali content. The main array of the glass compositions in terms of major and minor elements are olivine controlled, with significant scatter at a fixed MgO that is outside analytical error, particularly for the alkalis, Ca, and the incompatible minor elements (Figure 15.6). At a fixed MgO of 10.0 wt%, for example, K_2O ranges from 3.0 to 6.8 wt%, TiO_2 from 2.1 to 2.8 wt%, and P_2O_5 from 0.23 to 0.4 wt%, with an outlier at 0.62 wt% (Figure 15.6). Similar heterogeneity has been recognized in melt inclusion compositions for other Kīlauea [*Clague et al.*, 1995], Lō‘ihi [*Kent et al.*, 1999], and Mauna Loa lavas [*Davis et al.*, 2003] and ascribed to melts from a heterogeneous source [*Sobolev et al.*, 2000]. The more evolved melt inclusions and matrix glasses show a depletion in CaO and Al_2O_3 consistent with the appearance of augite and plagioclase on the liquidus at <7 wt% MgO, consistent with previous studies [*Helz and Thornber*, 1987; *Clague et al.*, 1995]. The comparison with the Puna Ridge glasses shows that the melt inclusion array spans the entire range of evolution for the diverse parental melts and the upper range approaches closely the composition of the most primitive glass grain composition presented by *Clague et al.* [1991] (15 wt%). The melt inclusion data set, with its higher number of analyses and coverage of the entire liquid line of descent, shows additional features over the Puna Ridge data set, particularly with regard to the heterogeneity in liquid compositions at a fixed MgO content.

The concentrations of the volatiles H_2O, CO_2, S, and Cl are shown plotted against MgO and K_2O in Figure 15.7. Melt inclusion H_2O concentrations, corrected for PEC, range from below detection up to 0.69 wt%, with one melt inclusion containing 0.91 wt%. Matrix glasses contain up to 0.17 wt% H_2O, which is well above detection limits (typically 0.03 wt%). There is no clear relationship between MgO and H_2O content of the melt inclusions, although the very highest H_2O contents occur at MgO contents of between 12 and 14 wt% MgO. A striking feature of the data is the large spread in H_2O concentrations, which may be due in part to fundamental geochemical heterogeneity of the primary melts [*Pietruszka and Garcia*, 1999], degassing, diffusive loss of H^+ [*Gaetani et al.*, 2012], or magma mixing between degassed and undegassed melt [*Dixon et al.*, 1991]. This is discussed in more detail later. There is very little evidence that primary melts with 16.5 wt% MgO have a fixed H_2O content

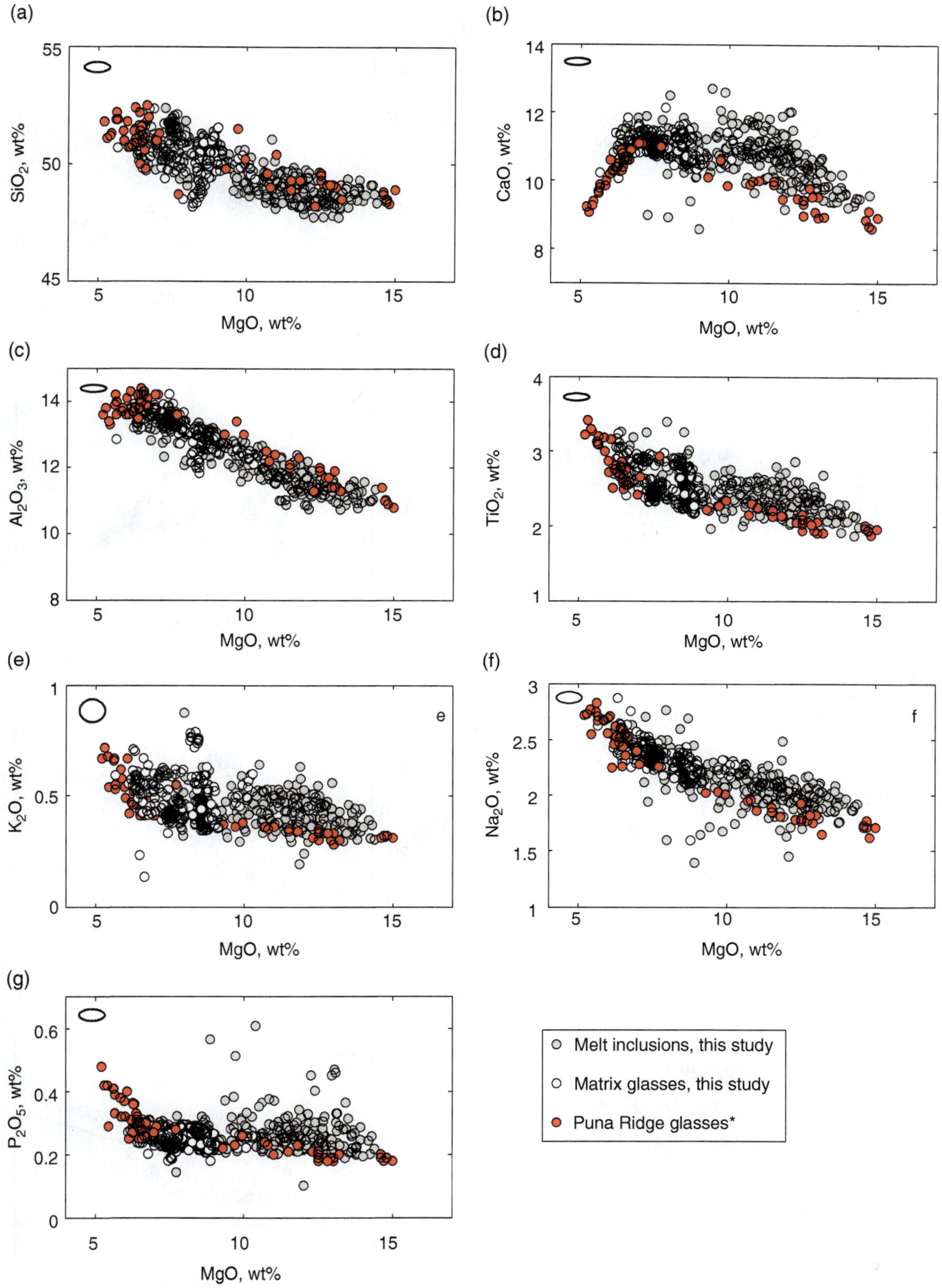

Figure 15.6 Major-element concentrations of the melt inclusions (after correction for PEC) plotted against the MgO content: (a) SiO_2; (b) CaO; (c) Al_2O_3; (d) TiO_2 (wt%); (e) K_2O (wt%); (f) Na_2O; (g) P_2O_5 against MgO (all wt%). Gray dots, Kīlauea melt inclusions; open circles, Kīlauea matrix glasses; red dots, Puna Ridge glasses [*Dixon et al.*, 1991; *Clague et al.*, 1991]. Maximum analytical uncertainty (assuming no interdependence of errors) shown as an ellipse.

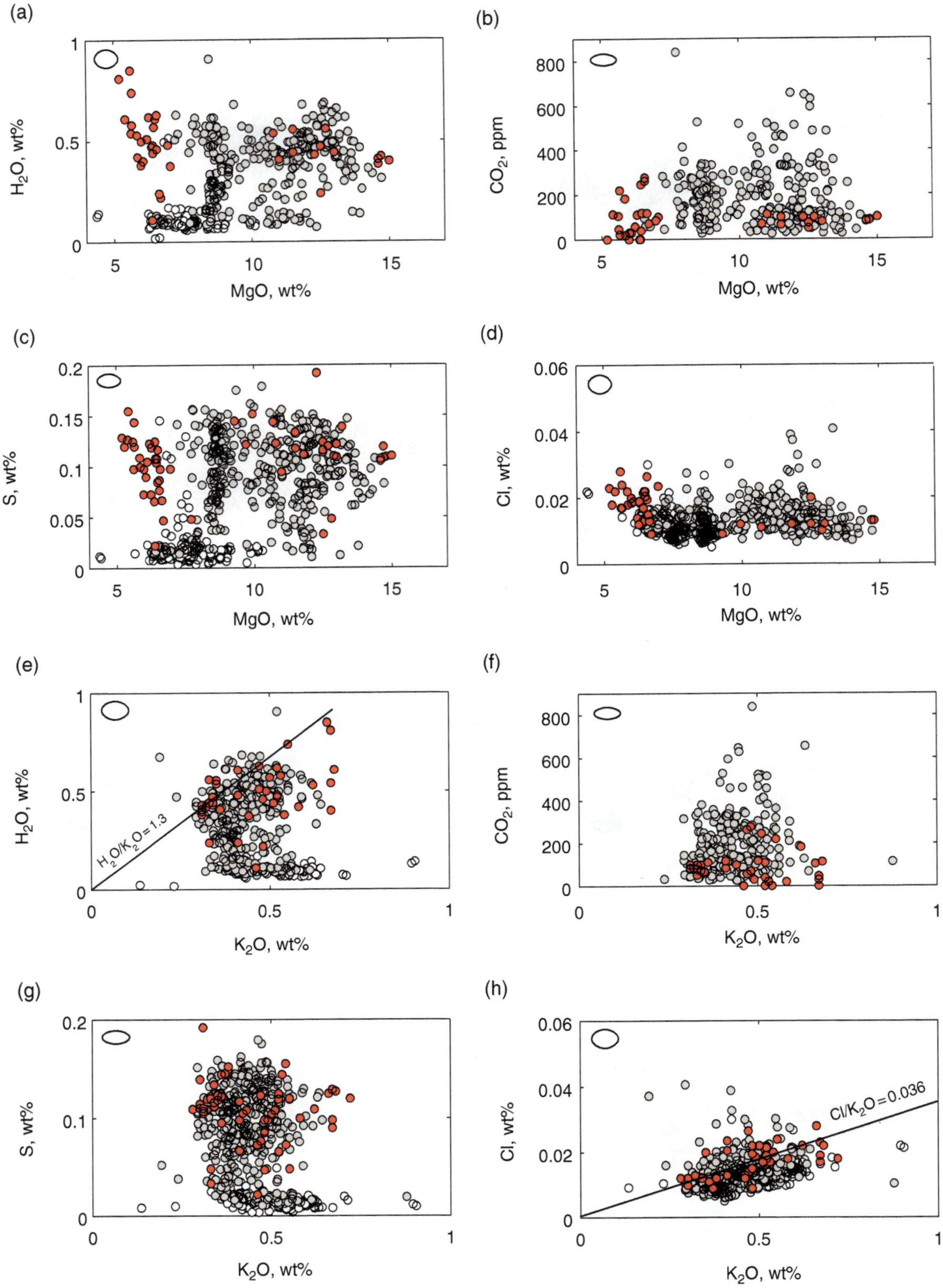

Figure 15.7 Volatile abundances plotted against MgO and K_2O: (a) H_2O; (b) CO_2; (c) S; and (d) Cl plotted against MgO (all wt% except CO_2, which is in ppm). (e–h) The same volatiles plotted against K_2O (wt%). Gray dots, Kīlauea melt inclusions; open circles, Kīlauea matrix glasses; red dots, Puna Ridge glasses [*Dixon et al.*, 1991; *Clague et al.*, 1991]. Maximum analytical uncertainty (assuming no interdependence of errors) shown as an ellipse. The H_2O/K_2O ratio of *Wallace* [1998] and the Cl/K_2O ratio of 0.036 of *Kent et al.* [1999] are shown for reference.

of 0.35 wt% ±0.05 H_2O, which then increases to 0.7 wt% at MgO contents of 7.0 wt% MgO and below, as has been previously proposed [*Dixon et al.*, 1991]. Instead, there is considerable heterogeneity in melt H_2O concentrations up to 14.7 wt% MgO, with the maximum melt H_2O concentration of 0.69 wt% occurring in a melt inclusion with 12.64 wt% MgO (Figure 15.7a); this particular melt inclusion suffered only very little (<7%) PEC. The more evolved, degassed matrix glass compositions are similar to the end member proposed to be involved in magma mixing to produce the more evolved Puna Ridge glasses [*Dixon et al.*, 1991]. The plot of H_2O versus K_2O again shows considerable scatter, but only two data points sit above (outside the error) the H_2O/K_2O line of 1.3 proposed by *Wallace* [1998] as being representative of primary melts, similar to the distribution of melts from Puna Ridge and from Mauna Loa [*Davis et al.*, 2003]. The very low H_2O/K_2O ratios are due to degassing, but the intermediate values may be due to mixing between degassed and undegassed melts or to their being derived from primary melts that are inherently depleted in H_2O, as described by *Dixon and Clague* [2001] for Lō'ihi melts.

Carbon dioxide (CO_2) concentrations are <640 ppm, apart from one melt inclusion with 849 ppm CO_2 (Figure 15.7b). Unlike the Puna Ridge glasses, which were substantially degassed of CO_2 on the seafloor, these melt inclusions retain a larger, albeit still small, fraction of their original CO_2, the rest having been lost by degassing. The highest CO_2 concentrations are in the most primitive melts, suggesting that these melts last equilibrated at higher pressures and are therefore sampling deeper parts of the Kīlauea plumbing system [*Dixon*, 1997]. The fraction of the CO_2 sequestered into the shrinkage bubbles is unknown, so the H_2O and CO_2 systematics cannot reliably be used to estimate the pressure of equilibration.

Sulfur concentrations reach 0.18 wt% but are more generally <0.15 wt% (Figures 15.7c,g), consistent with previous studies [*Davis et al.*, 2003; *Dixon et al.*, 1991; *Garcia et al.*, 1989; *Wallace and Anderson*, 1998]. Water and sulfur clearly have common controls, judging by their similar relationship with MgO and K_2O, and this has been observed previously in glasses from Kīlauea [*Dixon et al.*, 1991] and from Mauna Loa and Mauna Kea [*Davis et al.*, 2003; *Seaman et al.*, 2004]. The positive correlation

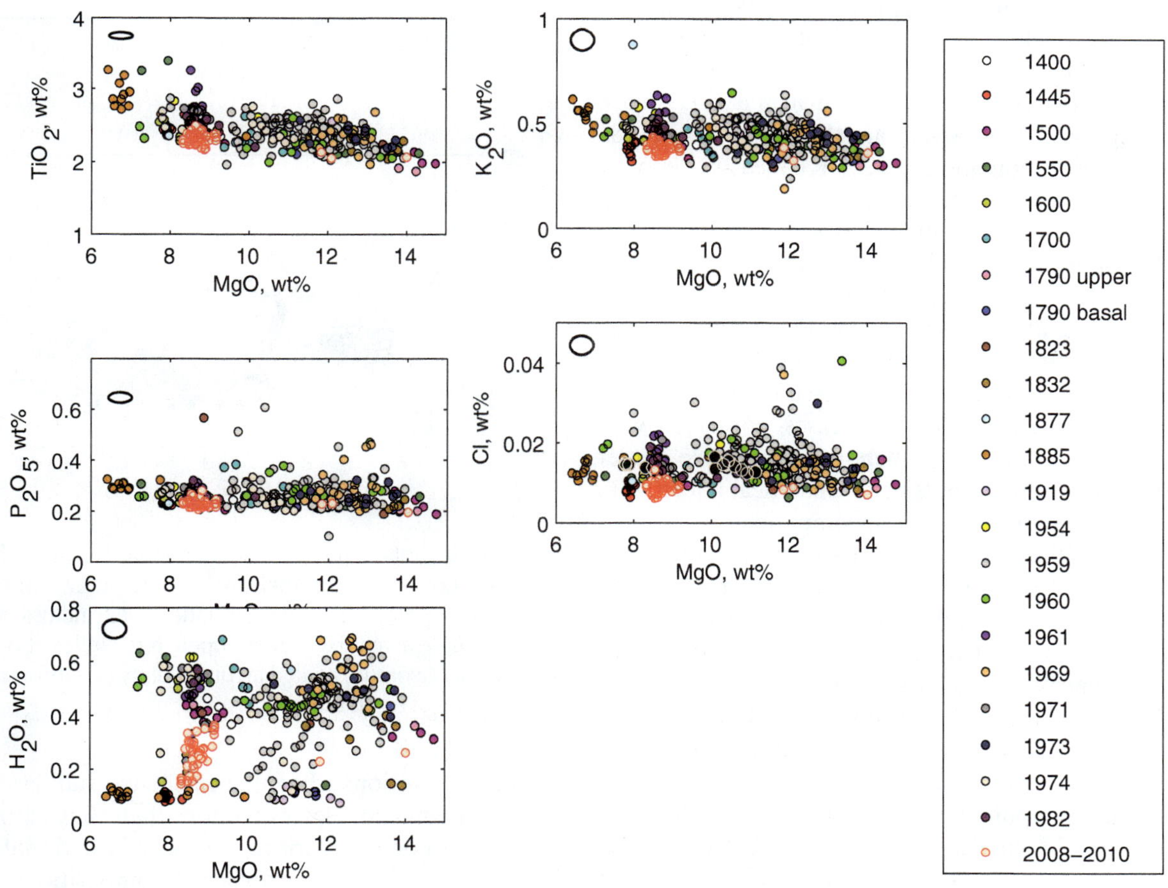

Figure 15.8 Plots of incompatible minor and volatile elements against MgO color coded for eruption: (a) TiO_2; (b) K_2O, (c) P_2O_5; (d) Cl; (e) H_2O, all in wt%. For details of the eruptions, see text and Table 15.1. 1790 basal is the basal reticulite and 1790 upper is "layer 6" of the Keanakāko'i Tephra.

between H_2O and S is consistent with most of the spread in the data being caused by degassing, as these volatiles are coupled in the degassing process at Kīlauea [*Gerlach*, 1986]. Chlorine concentrations in the melt inclusions range from 0.005 to 0.023 wt%, with some outliers reaching as high as 0.044 wt%. Matrix glasses exhibit a similar range of chlorine concentrations, consistent with only limited degassing during fractionation, which is borne out by volcanic gas analyzes [*Gerlach and Graeber*, 1985; *Mather et al.*, 2012]. The Cl/K_2O ratio is similar to the ratio for uncontaminated Lōʻihi melt inclusions (0.036; Figure 15.7 h), indicating there is little or no assimilation of brine [*Kent et al.*, 1999].

The data are shown with the different eruptions distinguished in Figure 15.8. The eruptions associated with the most primitive melts and lowest concentrations of incompatible elements are the ~1500 fissure eruption and the 1973, 1959, 1960, and the 1832 eruptions. The most evolved melt inclusions are those from the 1885, 1960, 1550, and 1974 eruptions. The eruptions with the largest degree of scatter in incompatible, nonvolatile elements are the 1959 (Kīlauea Iki), 1969 (Mauna Ulu), and 1960 (Kapoho) eruptions, all of which were high-fountaining eruptions. The H_2O-MgO plot shows that the 1969 and 1959 eruptions produced melt inclusions with the highest H_2O contents (>6.5 wt% at MgO contents of 12–14 wt%). The only two eruptions that display a continuous range in H_2O contents, from undegassed to degassed, are those from the 2008–2010 summit eruption and the 1959 Kīlauea Iki deposits. These eruptions are/were characterized by the draining back of degassed lava to depth [*Wallace and Anderson*, 1998; *Edmonds et al.*, 2013], making it likely that these trends are caused by mixing between degassed and undegassed magmas, as proposed for the Puna Ridge glasses [*Dixon et al.*, 1991], discussed in more detail later.

Trace-element concentrations increase as the MgO content of the melt inclusion decreases, consistent with fractionation (Figure 15.9). There is considerable scatter at a fixed MgO content, outside of analytical uncertainty (Figure 15.9), similar to the incompatible major and minor compositions (Figure 15.6) and to previous work [*Clague et al.*, 1995]. Plots of La against Yb and Nb against Y show considerable scatter (independent of MgO content, which is affected by the PEC calculations). For La/Yb, the matrix glasses have an overall standard deviation of 1.36 and the melt inclusions 1.19. For Nb/Y, the matrix glasses have an overall standard deviation of 0.13 and the melt inclusions 0.14. It is not clear that the melt inclusions are sampling melts that are substantially more variable than the carrier liquids over ~1000 year timescales. The variation within individual eruptions, however, is very marked. The standard deviation of melt inclusion La/Yb, for example, in individual eruptions, is on average 5 times greater than that of the matrix glasses, with only 3 out of 38 eruptions where the calculated standard deviation of the matrix glasses is more than that of the melt inclusions. For comparison, the standard deviations in La/Yb and Nb/Y of the whole-rock compositions for similar eruption deposits (Figure 15.10) are 0.95 and 0.09, respectively [*Pietruszka and Garcia*, 1999]. Lōʻihi melt inclusions range to considerably higher Nb/Y ratios.

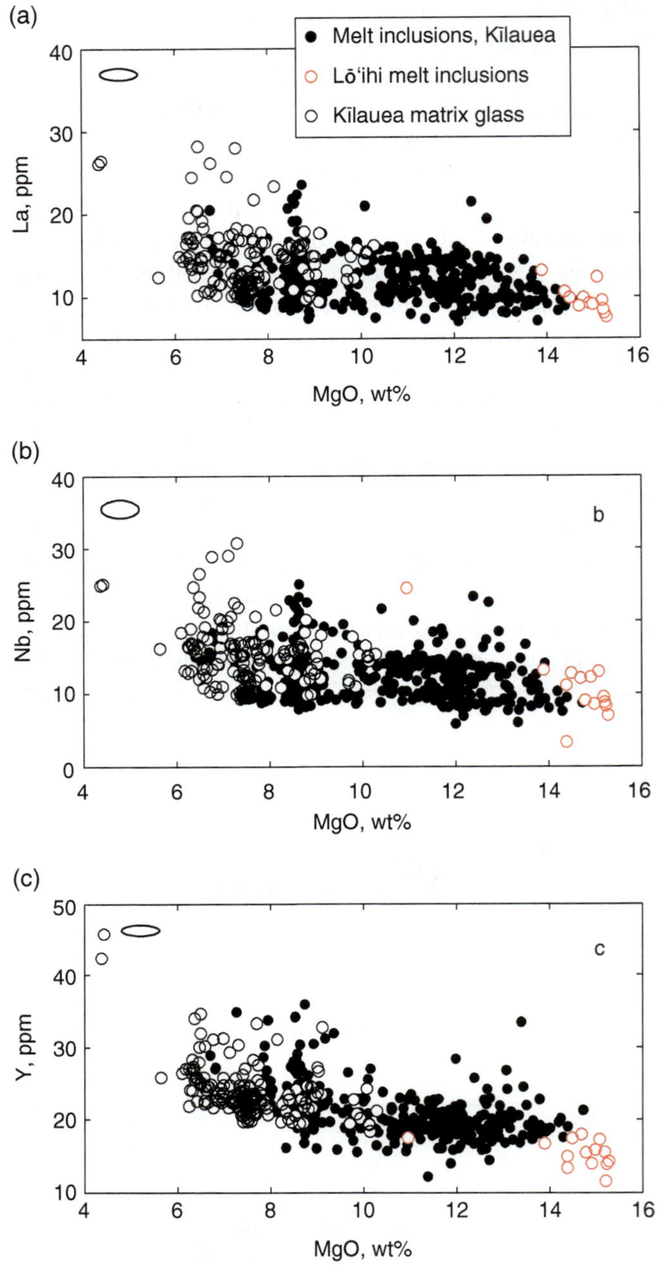

Figure 15.9 Plots of trace element abundances plotted against the MgO content of the melt inclusions: (a) La; (b) Nb; and (c) Y (all in ppm). Black dots, Kīlauea melt inclusions; open circles, Kīlauea matrix glasses; open red circles, Lōʻihi melt inclusions. Maximum analytical uncertainty (assuming no interdependence of errors) shown as an ellipse.

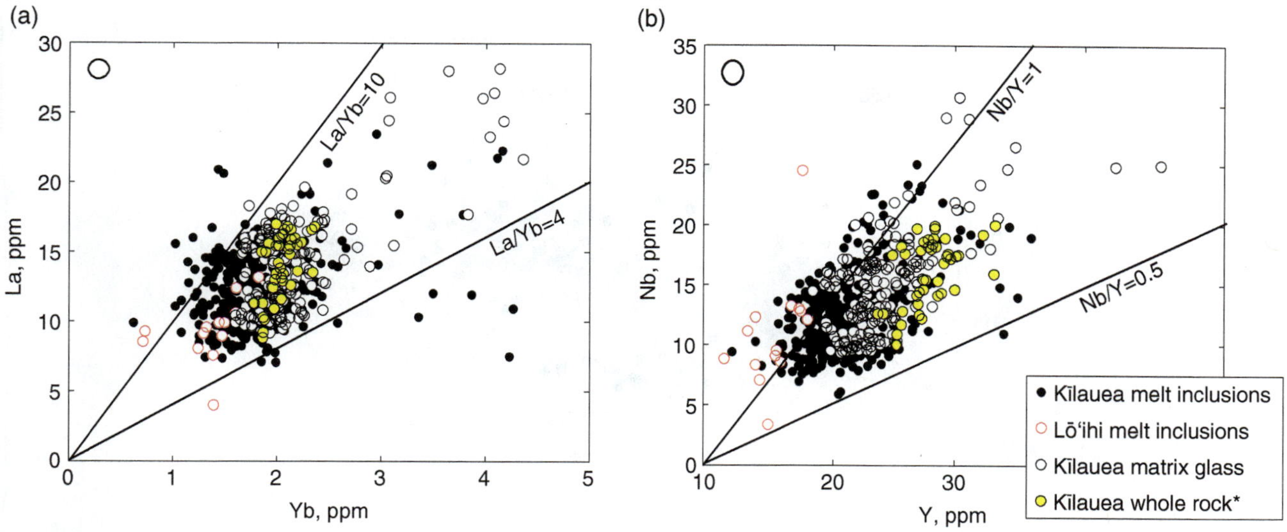

Figure 15.10 Plots of (a) La versus Yb and (b) Nb versus Y (all in ppm). Black dots, Kīlauea melt inclusions; open circles, Kīlauea matrix glass; open red circles, Lō'ihi melt inclusions; yellow dots, whole rocks from historical Kīlauea eruptions [*Pietruszka and Garcia*, 1999]. Maximum analytical uncertainty (assuming no interdependence of errors) shown as an ellipse.

The elements La and Ce have a similar incompatibility to H_2O [*Danyushevsky et al.*, 2000b; *Dixon et al.*, 1988; *Dixon and Clague*, 2001] and therefore H_2O/Ce is a useful indicator of enrichment relative to other incompatible trace elements. The ratio H_2O/Ce varies from 18 up to 315, with the highest values corresponding to the most primitive melts (Figure 15.11a). The Kīlauea melt inclusion H_2O/Ce ratios reach higher values than a grain of primitive Puna Ridge glass, which has a H_2O/Ce of 182 [*Dixon and Clague* 2001], but not such high values as the Lō'ihi melt inclusions, which record H_2O/Ce of up to 400 (Figure 15.11), although these melts may have assimilated some H_2O from brine [*Kent et al.*, 1999]. Submarine-quenched Lō'ihi glasses record much lower H_2O/Ce of up to 237 but more generally between 153 and 196 [*Dixon and Clague*, 2001] (in red, Figure 15.11). The Kīlauea matrix glasses have H_2O/Ce of between 7 and 65, indicating that they are degassed (open black circles, Figure 15.11). The melt inclusion array also extends to these low values, indicating that they have degassed prior to entrapment, have lost H^+ after entrapment [e.g., *Gaetani et al.*, 2012], and/or represent mixed melts [*Dixon et al.*, 1991]. Analogously, CO_2 and Nb may behave similarly during melting, and it might be expected, based on measurements of undersaturated mid-ocean ridge basalt (MORB) and ocean island basalt (OIB), and measurements of "bulk" CO_2 (dissolved plus exsolved fractions), that the CO_2/Nb might be between 200 and 400 for undegassed melts [*Saal et al.*, 2002; *Cartigny et al.*, 2008; *Hauri et al.*, 2002a]. Kīlauea melt inclusions have a CO_2/Nb ratio of <80 (Figure 15.11b), a similar range to the Lō'ihi submarine glasses [*Dixon and Clague* 2001], indicating they are substantially degassed.

Similar to the findings of previous whole-rock studies at Kīlauea [e.g., *Pietruszka and Garcia*, 1999; *Garcia et al.*, 2003; *Marske et al.*, 2008], the matrix glass and inclusion compositions vary significantly with time, shown in Figure 15.12. The melt inclusion compositions show wide variability, with a peak in K_2O concentrations and Nb/Y ratio for the eruptive products of the 1959 Kīlauea Iki eruption, but the melt inclusions from this eruption span the entire range of compositions observed for Kīlauea. The matrix glasses are less variable, consistent with their origin from the mixing of heterogeneous liquids [*Sobolev et al.*, 2000; *Kent*, 2008], and show a clearer trend with time, with a slow decrease in K_2O concentrations and Nb/Y from AD ~1500 to around 1790, followed by a steeper increase through the 18th and 19th centuries, reaching a peak in 1959 before decreasing sharply to the present day. These trends are mirrored in other compatible- and incompatible-element abundances and ratios (e.g., the trend for SiO_2 is inverse to that for K_2O, TiO_2, CaO, and La; La/Yb shows a similar trend to Nb/Y).

15.4. MIXING, FRACTIONATION, AND DEGASSING

The dominant trends in composition in the melt inclusion suite in both major and incompatible trace elements (Figures 15.6–15.9) are caused by olivine fractionation between 1346 and 1160 °C [*Helz and Thornber*, 1987], consistent with previous interpretations [e.g., *Helz*, 1987; *Garcia et al.*, 1989; *Clague et al.*, 1995]. There is significant variation in minor and trace elements in the melt inclusions at a constant MgO content, with scatter

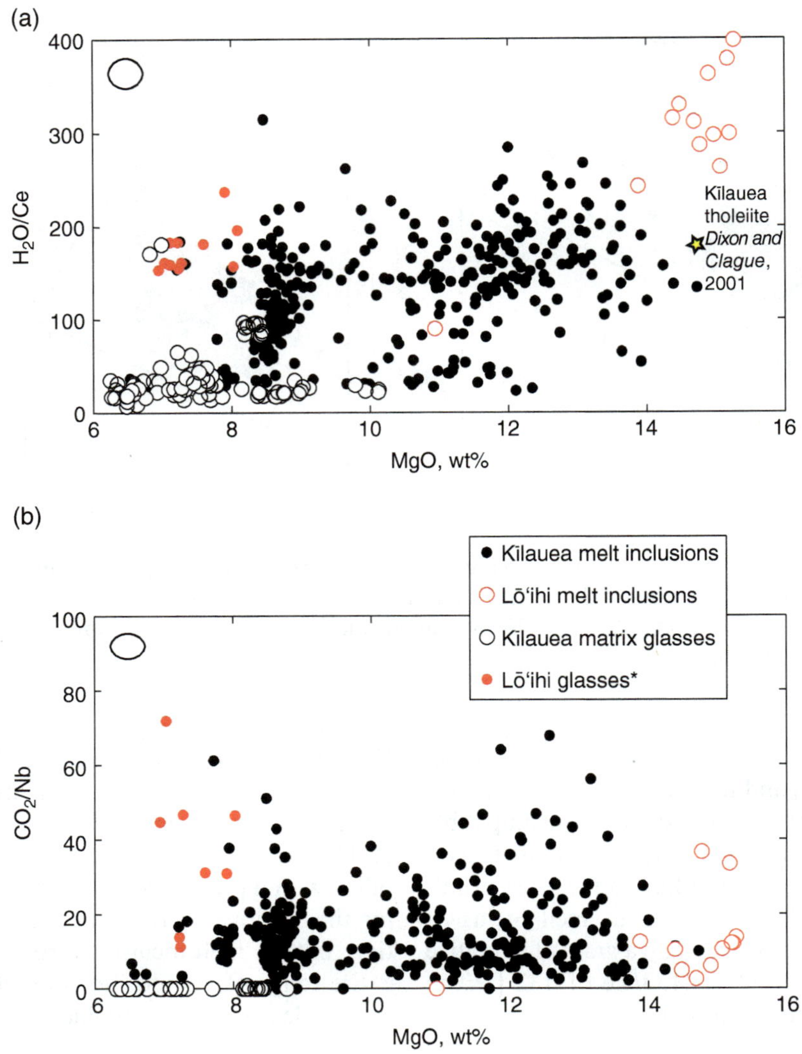

Figure 15.11 Plots of (a) H_2O/Ce and (b) CO_2/Nb against MgO content of the melt inclusions. Black dots, Kīlauea melt inclusions; open circles, Kīlauea matrix glasses; open red circles, Lōʻihi melt inclusions; red dots, Lōʻihi submarine quenched glasses [from *Dixon and Clague* 2001]. Maximum analytical uncertainty (assuming no interdependence of errors) shown as an ellipse.

outside the uncertainty associated with either analytical error or the PEC corrections (Figures 15.9 and 15.10). This variability is not consistent with batch fractionation of parental melts with a uniform composition. The variability in P_2O_5/K_2O and TiO_2/K_2O, for example, mirrors that found in glass inclusions in Puna Ridge lavas [*Clague et al.*, 1995]. The variability may be explained if melts entering the reservoir from depth are chemically heterogeneous as a result of variable source composition or changes in the degree of partial melting.

The relationship between the matrix glass geochemistry and the olivine host composition (Figure 15.5) is a clue to olivine provenance, and whether the melt inclusion geochemistry can be related in any meaningful way to the whole rocks, and to the eruptions themselves. The poor correlation between the host olivines and their average carrier melt compositions indicates a high degree of crystal-melt disequilibrium for the majority of the samples, perhaps due to the olivines being antecrysts (crystals associated with the system but not in chemical equilibrium with their carrier melts at the time of eruption [e.g., *Davidson et al.*, 2007]). The low Ca contents of the olivines rule out a mantle origin [*Garcia et al.*, 2003]. We concur with previous workers [*Murata and Richter*, 1966; *Clague et al.*, 1995] that the crystal habits, zoning, and compositions of the olivines all suggest that petrogenesis is complex, with fractionation and accumulation of olivines followed by reentrainment during eruption. Disequilibrium between the olivine and the carrier liquid is likely the result of olivine reentrainment on timescales

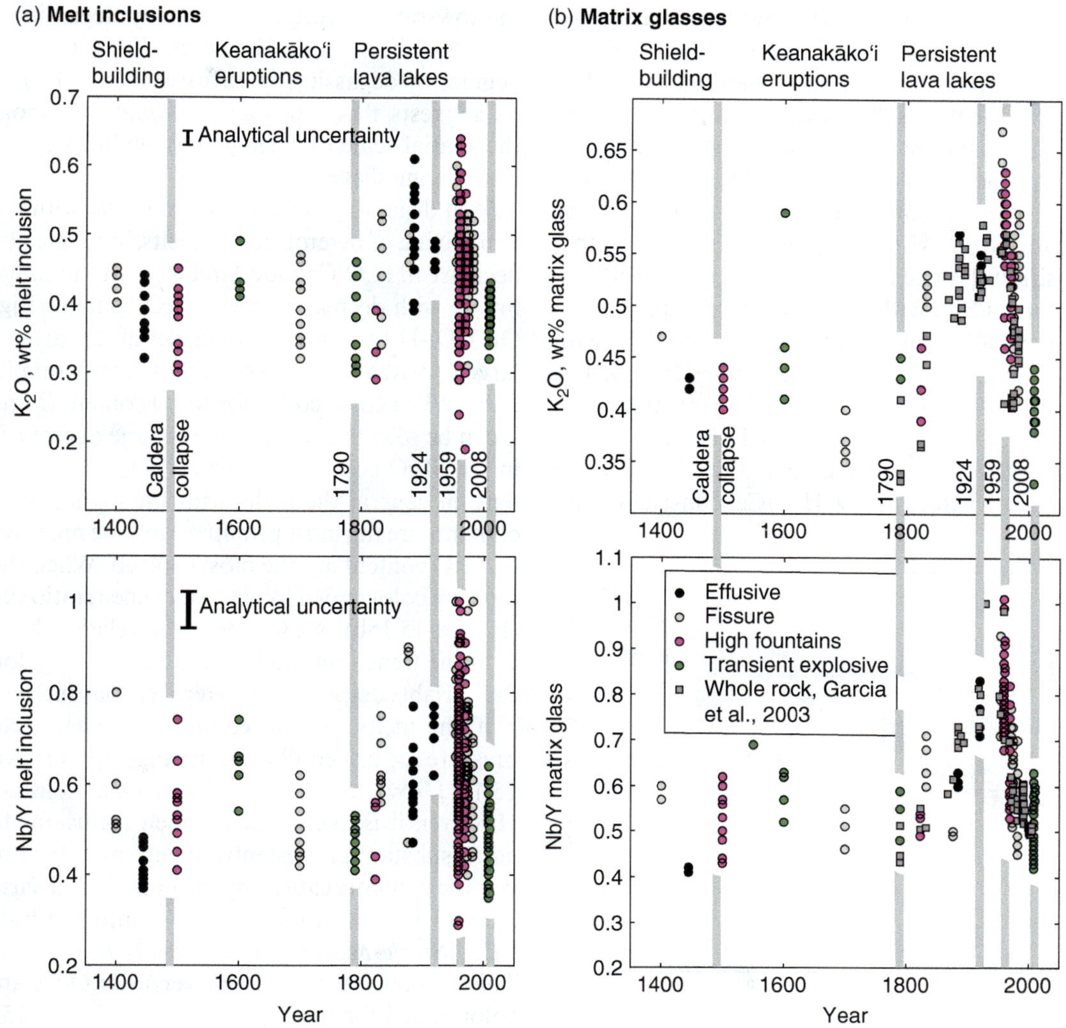

Figure 15.12 (a) Melt inclusion and (b) Matrix glass geochemistry with time in the samples analyzed in this study from Kīlauea Volcano color coded by eruption style. Top: K_2O (wt%) concentrations with time and bottom: Nb/Y ratio with time. Also shown are whole-rock data from *Garcia et al.* [1989] and some important eruptions and periods of activity. Analytical uncertainty is marked with the vertical bar.

short compared to the diffusive equilibration of Mg-Fe throughout the olivine crystal.

The question then arises how can the melt inclusion compositions be related to the carrier liquids given that the host olivines are not in equilibrium with the carrier liquids? In agreement with previous studies in Iceland [*Maclennan*, 2008] and a similar study for Lō'ihi [*Kent et al.*, 1999], the mean composition of the inclusions for each eruption is approximately equal to the matrix glass composition (Figure 15.13), with only a few eruptions plotting away (outside the error) from the 1:1 line relating the two parameters. A simple least-squares fit of the data gives an r^2 value of 0.51 and shows that the correlation is significant at the 99.9% level. If the fit is weighted using the errors associated with each point, then the r^2 value rises to 0.68. The single point (corresponding to the 1877 eruption) with low matrix glass and high melt inclusion values has large errors because this is based on a sample size of 4. Removal of this outlier of the data set would further improve the significance of the correlation. The melt inclusion compositions show that a much wider range of liquids are present in the plumbing system that actually erupt, consistent with many other melt inclusion studies from a range of settings [*Sobolev*, 1996; *Sobolev et al.*, 2000; *Kent and Elliott*, 2002; *Kent*, 2008; *Maclennan*, 2008]. This observation indicates that although diverse parental melts rise, crystallize olivine, and mix, there is still a broad association between the olivines, the melt inclusion population, and the carrier liquid.

Inclusions within the same olivine are variable in composition, but in general they do not exhibit the full range of variability seen in the entire data set. In 55 cases, two melt inclusions from the same olivine were analyzed and in 7 cases three to four melt inclusions were analyzed. The compositions of the two melt inclusions, or the inclusions with the minimum and maximum Nb/Y value, are shown plotted against one another in Figure 15.14. In general, the pairs of inclusion compositions are within 25% of one another for Nb/Y, with three outliers (compared to a total compositional range for the full data set of ~300%), within 40% for La/Sm, with five outliers (compared to a total range of ~400%), and within 40% for H_2O/Ce, with five outliers (compared to a total range of ~1500%). The maximum levels of analytical uncertainty on the ratios are 13%, 7%, and 9%, respectively. For H_2O/Ce there is a higher degree of variability at high H_2O/Ce, with more consistent pairs of values at low H_2O/Ce, consistent with

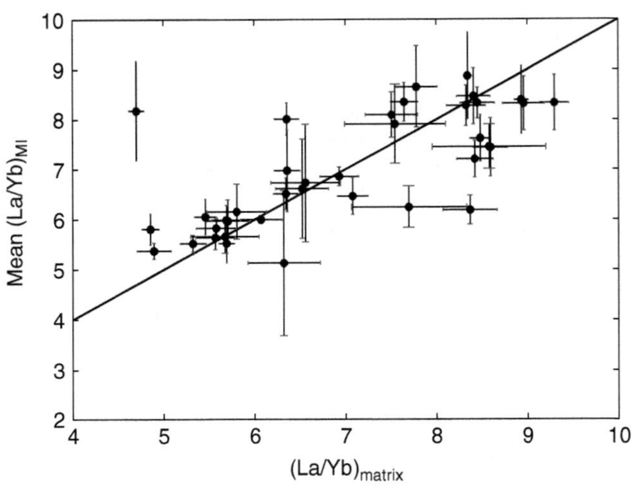

Figure 15.13 Mean melt inclusion La/Yb for each eruption plotted against the matrix glass La/Yb. Error bars are the 1σ standard error of estimate of the means.

the low ratios being controlled by degassing (the inclusions within a single olivine are likely to exhibit similar degrees of degassing or diffusive loss of H^+). This analysis suggests that melts are reasonably homogeneous on the spatial scale of a single olivine but heterogeneous at interolivine scales.

Melt degassing occurs synchronously with fractionation and mixing of heterogeneous melts, illustrated by the broad decrease in H_2O/Ce ratio (and CO_2/Nb) in range and magnitude with decreasing melt MgO content (Figure 15.11). On CO_2-H_2O plots, the data define an array that has a large spread and does not conform to a simple degassing path. When color coded for MgO content (Figure 15.15a), it can be seen that while there is large degree of variability in the MgO content of melts at a fixed H_2O or CO_2 content, in general the melts with the highest H_2O and CO_2 contents are the most primitive, and the melts with the lowest H_2O content are the most evolved. When the H_2O-CO_2 data are color coded for a trace-element ratio such as Nb/Y (Figure 15.15b), we see no clear relationship, consistent with the trends on the CO_2-H_2O plot being dominated by the variable degassing of heterogeneous melts.

If the melts were in equilibrium when quenched, the pressures of last equilibration range up to 1.5 kbar (Figure 15.15) [*Dixon et al.*, 1997; *Newman and Lowenstern*, 2002]. However, it is likely that the melt inclusions have undergone substantial postentrapment modification of their volatile contents caused by the growth of a shrinkage bubble and the partitioning of CO_2 into the bubble during PEC [*Steele-Macinnis et al.*, 2011; *Bucholz et al.*, 2013]. This is shown in plots of Fo versus H_2O/Ce and CO_2/Nb, color coded for the degree of PEC (Figure 15.16), which shows that most PEC occurs in melt inclusions hosted by high-Fo olivines (which is to be expected, given that they are crystallized from and trapped the hottest liquids), and furthermore, there is no correlation between the H_2O

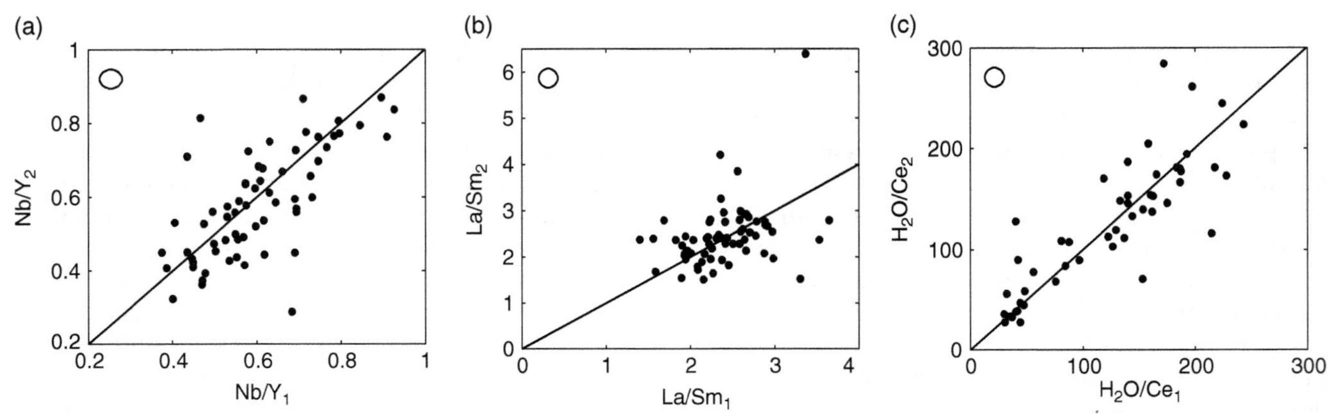

Figure 15.14 Melt inclusion compositional heterogeneity on a single olivine scale. Plots of the compositions of pairs of inclusions contained within the same olivine in terms of (a) Nb/Y, (b) La/Sm, and (c) H_2O/Ce. The solid lines mark compositional equivalence. Maximum analytical uncertainty (assuming no interdependence of errors) shown as an ellipse.

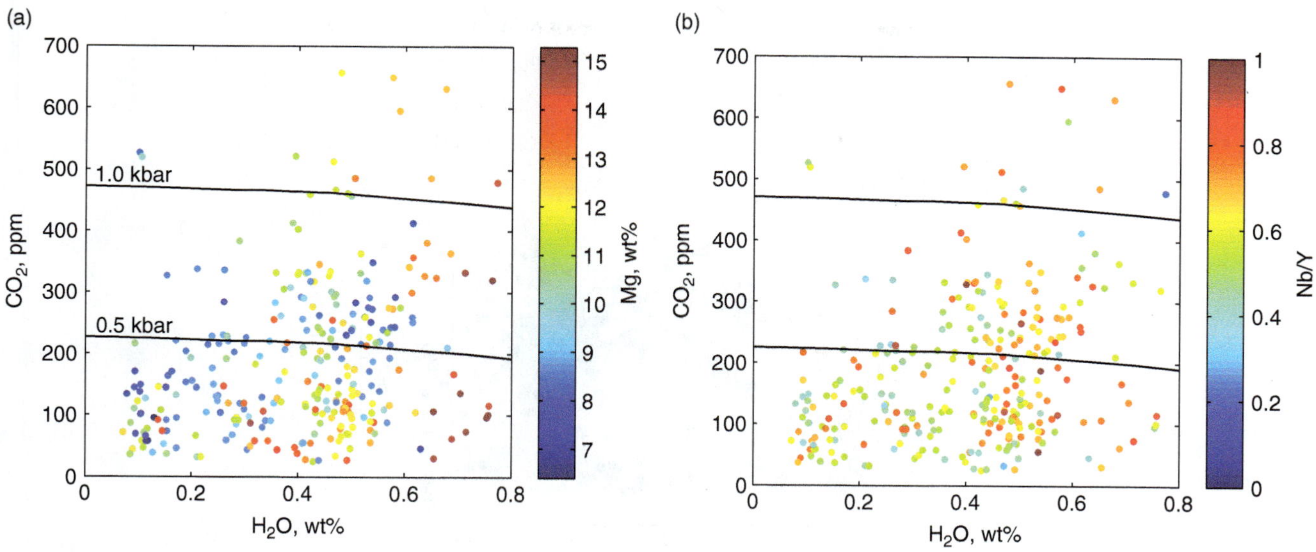

Figure 15.15 Volatile systematics in the melt inclusions. Plots of CO_2 (ppm) against H_2O color coded for (a) MgO content (wt%) and (b) Nb/Y. Isobars are from *Newman and Lowenstern* [2002].

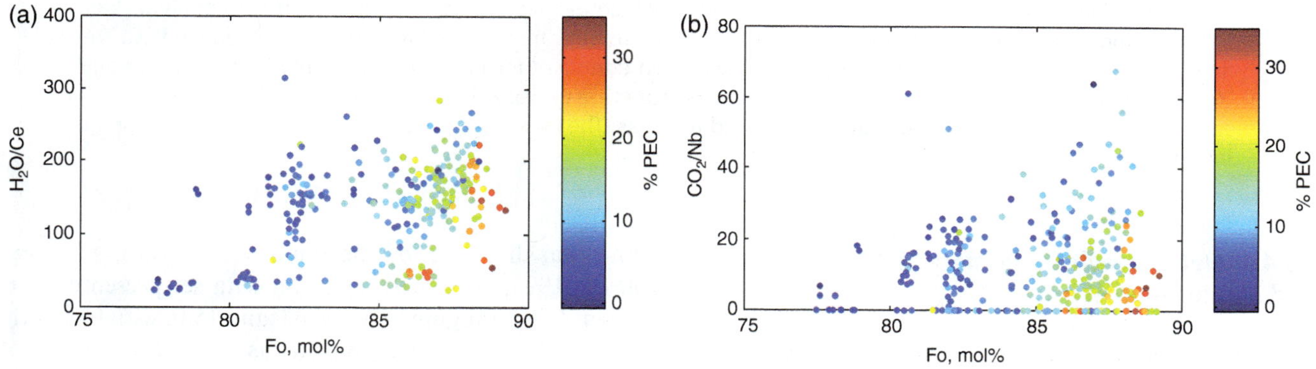

Figure 15.16 Effects of post-entrapment crystallization (PEC) on the volatile contents of melt inclusions. Plots of (a) H_2O/Ce and (b) CO_2/Nb against olivine forsterite content (mol%), color coded for %PEC.

content and the amount of PEC. In contrast, inclusions that have undergone the highest amounts of PEC have the lowest CO_2 concentrations (Figure 15.16b), consistent with a large fraction of the CO_2 being sequestered into a growing shrinkage bubble. These results imply that the pressures of equilibration derived from the CO_2-H_2O data are minimum values. In addition, it is likely that a large fraction of the melt inclusions may be vapor undersaturated if they are the products of reservoir-depth mixing of surface degassed melts with undegassed melts, similar to the low-Mg Puna Ridge glasses and melt inclusions associated with the 2008–current eruption of Halema'uma'u [*Dixon et al.*, 1991; *Edmonds et al.*, 2013].

In summary, the melt inclusion data set shows that there is significant heterogeneity in melt composition and likely in primary melt composition in terms of trace and volatile species (Figure 15.17). A large fraction of the melt inclusions are substantially degassed or mixed with degassed end-member magmas (Figure 15.17), but there is a significant fraction with high H_2O/Ce content at high MgO and olivine forsterite contents (Figures 15.11 and 15.16) that are likely to be undegassed, and they exhibit a far greater range than previously observed for Kīlauea, indicating that primary melts may vary in their volatile content (as well as in their abundances of other incompatible trace elements). In comparison to previous studies, the melt inclusion compositions broadly define a liquid line of descent similar to the high-Mg Puna Ridge glasses, albeit with significantly greater heterogeneity. Melt inclusions from Kīlauea Iki [*Anderson and Brown*, 1993] were not corrected for PEC, resulting in their high H_2O at low MgO (Figure 15.17).

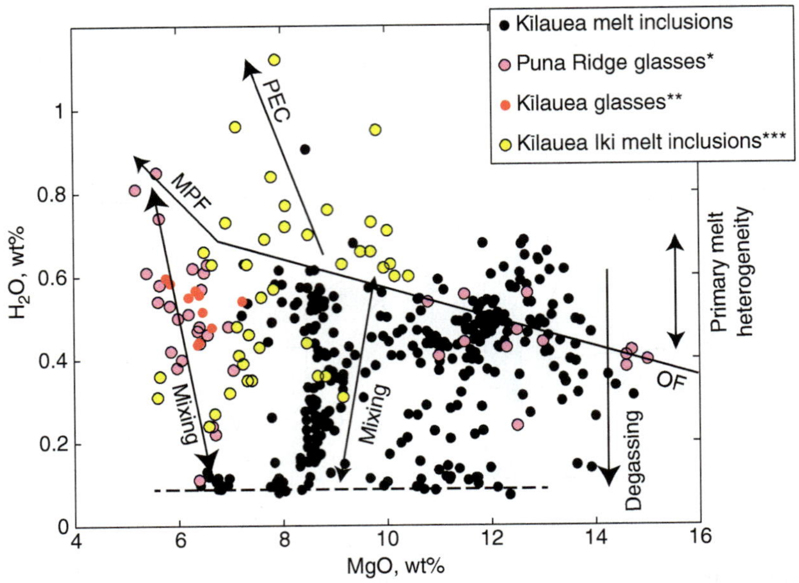

Figure 15.17 Summary of the processes acting to modify and control volatile abundances in melt inclusions from Kīlauea. Black dots, Kīlauea melt inclusions; yellow dots, Kīlauea Iki melt inclusions from *Anderson and Brown* [1993], which were not corrected for PEC; pink dots, Puna Ridge submarine quenched glasses from *Dixon et al* [1991] and *Clague et al.* [1991]. Primary melts are likely heterogeneous with respect to their H_2O content. Melt H_2O may increase by fractionation (OF, olivine fractionation; MPF, multiphase fractionation) and due to PEC. The melt inclusions in the data set presented here have been corrected for PEC; those presented by *Anderson and Brown* [1993] were not. H_2O concentrations may decrease due to degassing, diffusive loss of H^+ from the inclusion, or mixing between degassed and undegassed magmas. The latter process explains the origin of most of the intermediate H_2O/Ce glasses.

15.4.1. Heterogeneity in Primary Melts and Mantle Source

Considerable variability in the composition of Kīlauea lavas, which cannot be explained by fractionation (either olivine only or multiphase), has long been recognized [*Hofmann et al.*, 1984; *Rhodes et al.*, 1989; *Clague et al.*, 1995; *Pietruszka and Garcia*; 1999; *Marske et al.*, 2008]. The compositional variability has been ascribed to changes in the degree of partial melting [*Hofmann et al.*, 1984; *Clague et al.*, 1995], invasion of Kīlauea's magma reservoir by melts from Mauna Loa [*Rhodes et al.*, 1989], and mantle source heterogeneity coupled with changes in the degree of melting [*Pietruszka and Garcia*, 1999; [*Garcia et al.*, 2003; *Marske et al.*, 2008; *Pietruszka et al.*, 2006]. The composition of primary melts has been estimated to vary between 13.4% and 18.4% MgO, caused by differing degrees of partial melting [*Clague et al.*, 1995], with a corresponding variability in trace-element composition, depending on their partitioning behavior. The melt inclusion data set presented here shows variability far exceeding that documented by *Rhodes et al.* [1989] for historic and prehistoric Kīlauea whole-rock compositions (which have, for example, a Zr/Nb of 9–15, with a mean of 11.5 and a standard deviation of 1.5, $N \sim 150$) and that displayed by the Puna Ridge samples [*Clague et al.*, 1995]. The melt inclusion data set presented here has a Zr/Nb ranging between 6.0 and 15.0, with five outliers (reaching as high as 18.0 and as low as 1.0), a mean of 10.7, and a standard deviation of 2.36 (with $N = 373$).

Recent studies of isotopic and trace-element ratio changes in whole-rock compositions suggest the source region for melting at Kīlauea is isotopically and chemically heterogeneous over historic and prehistoric timescales and that mantle source composition is coupled strongly to the degree of partial melting [*Marske et al.*, 2008; *Pietruszka and Garcia*, 1999]. It has been proposed that the mantle source tapped during the late 19th and early 20th centuries at Kīlauea corresponded to a time of low degrees of melting of a relatively depleted source, both chemically and isotopically (with low $^{87}Sr/^{86}Sr$ and high ε_{Nd} and $^{206}Pb/^{204}Pb$ [*Pietruszka and Garcia*, 1999]), which is thought to represent a long-term depleted component within the Hawaiian plume. Our melt inclusion data set shows a similar pattern, with this period corresponding to low abundances of incompatible trace elements and low Nb/Y ratios and lower H_2O/Ce. In contrast, lavas erupted during the later 20th century, which included many of the fissure and high-fountaining eruptions between 1959 and 1982 (Table 15.1), were formed by the highest melt fractions of a less depleted source, perhaps corresponding to recycled, hydrated oceanic crust [*Kingsley and Schilling*, 1995; *Pietruszka et al.*, 2012], reflected in the higher abundances of the more incompatible trace elements, and higher H_2O contents and H_2O/Ce ratios of the melt inclusions. There is also no evidence, contrary to the findings of *Dixon and Clague* [2001], that Lō'ihi has a particularly "dry" source. In contrast, our analysis suggests H_2O/Ce extending to higher values than for Kīlauea (Figure 15.11). It seems clear that there are fundamental links between mantle source and primary melt composition and surface activity, caused by variability in magma supply rate, primitive melt volatile content and buoyancy, and magma transport paths.

15.5. CONCLUSIONS

• A large data set of melt inclusions from tephras and lava samples erupted during historical eruption of Kīlauea Volcano, once corrected for PEC, forms a detailed record of fractionation, mixing, and degassing in the deep plumbing of Kīlauea Volcano. The most primitive melts have 14.7 wt% MgO, making them close to being primary liquids.

• The melts form a compositional array that is broadly controlled by olivine fractionation, but with significant heterogeneity superimposed, particularly in terms of minor, trace, and volatile elements. This heterogeneity exceeds that observed in previous studies of whole-rock compositions and glasses of historical magmas at Kīlauea Volcano.

• The heterogeneity is caused by the mixing of compositionally distinct primary melts and variable amounts of fractionation in the deep plumbing of Kīlauea Volcano. Mixing also occurs in the shallow plumbing system between degassed melts that sink beneath the summit caldera after degassing at or near to the surface.

• The melt inclusion data set shows a large degree of heterogeneity as a whole, but a much smaller degree of heterogeneity on the spatial scale of a single olivine, within which melt inclusions vary in composition by <40%.

• While the H_2O/K_2O ratios of the melt inclusions are similar to previous studies, the H_2O/Ce exceeds previous measurements, and the H_2O contents are higher in the more primitive melts, suggesting that some primary melts may be inherently enriched in H_2O. The range in primary melt H_2O concentrations is larger than previously proposed.

• The melt inclusion CO_2 content in some of the more primitive inclusions has been significantly altered by PEC and the sequestration of CO_2 into a shrinkage bubble, making them of little use for estimates of equilibration pressure.

• There are temporal trends in melt inclusion and matrix glass compositions that are similar to those observed in whole-rock compositions, suggesting that there are significant changes in the degree of mantle melting and mantle source composition over 50–100 year timescales. There is a fundamental link between the composition and volume flux of primary melts and eruption style at the surface.

ACKNOWLEDGMENTS

We are indebted to Don Swanson (USGS) and Bruce Houghton (University of Hawai'i), whose help in the field made this work possible, through identifying the deposits and carefully guiding sampling. The Hawaiian Volcano Observatory and National Park Service granted us permission to access the caldera and other areas in the Park. Reviews from Andrea DiMuro and Jacqueline Dixon were immensely helpful and improved the chapter greatly. The work was funded by the Natural Environment Research Council (NERC), UK: IRS acknowledges a NERC studentship and ME acknowledges a NERC urgency grant NE/G001537/1 and a NERC ion probe facility grant IMF376/0509. IRS acknowledges a USGS Jack Kleinman Grant for Volcano Research. In Edinburgh, Richard Hinton and Cees-Jan de Hoog assisted with the ion probe measurements. In Cambridge, Jason Day, Chiara Petrone, Martin Walker, and Iris Buisman assisted with sample preparation and analysis.

REFERENCES

Anderson, A. T. (1995), CO2 and the eruptibility of picrite and komatiite, *Lithos*, *34*(1–3), 19–25.

Anderson, A. T., and G. G. Brown (1993), CO2 contents and formation pressures of some Kilauean melt inclusions, *Am. Mineral.*, *78*(7–8), 794–803.

Blichert-Toft, J., D. Weis, C. Maerschalk, A. Agranier, and F. Albarède (2003), Hawaiian hot spot dynamics as inferred from the Hf and Pb isotope evolution of Mauna Kea volcano, *Geochem., Geophys., Geosyst.*, *4*(2), 8704, doi: 10.1029/2002GC000340.

Blundy, J., and K. Cashman (2008), Petrologic reconstruction of magmatic system variables and processes, *Rev. Mineral. and Geochem.*, *69*(1), 179–239.

Bottinga, Y., and M. Javoy (1991), The degassing of Hawaiian tholeiite, *Bull. Volcanol.*, *53*(2), 73–85.

Brigham, W. T. (1909), *The Volcanoes of Kīlauea and Mauna Loa on the Island of Hawaii*, Bishop Museum Press.

Bucholz, C. E., G. A. Gaetani, M. D. Behn, and N. Shimizu (2013), Post-entrapment modification of volatiles and oxygen fugacity in olivine-hosted melt inclusions, *Earth Planet. Sci. Lett.*, *374*, 145–155.

Carey, R. J., M. Manga, W. Degruyter, D. Swanson, B. Houghton, T. Orr, and M. Patrick (2012), Externally triggered renewed bubble nucleation in basaltic magma: The 12 October 2008 eruption at Halema'uma'u Overlook vent, Kīlauea, Hawai'i, USA, *J. Geophys. Res. Solid Earth*, *117*(B11), B11202.

Cartigny, P., F. Pineau, C. Aubaud, and M. Javoy (2008), Towards a consistent mantle carbon flux estimate: Insights from volatile systematics (H_2O/Ce, δD, CO_2/Nb) in the North

Atlantic mantle (14° N and 34° N), *Earth Planet. Sci. Lett.*, *265*(3–4), 672–685.

Cayol, V., J. H. Dieterich, A. T. Okamura, and A. Miklius (2000), High magma storage rates before the 1983 eruption of Kilauea, Hawaii, *Science*, *288*(5475), 2343–2346.

Chouet, B. A., P. B. Dawson, M. R. James, and S. J. Lane (2010), Seismic source mechanism of degassing bursts at Kīlauea Volcano, Hawaii: Results from waveform inversion in the 10–50 s band, *J. Geophys. Res. Solid Earth*, *115*(B9), B09311.

Clague, D., and F. A. Frey (1982), Petrology and trace element geochemistry of the Honolulu Volcanics, Oahu: Implications for the oceanic mantle below Hawaii, *J. Petrol.*, *23*(3), 447–504.

Clague, D. A., W. S. Weber, and J. E. Dixon (1991), Picritic glasses from Hawaii, *Nature*, *353*(6344), 553–556.

Clague, D. A., J. G. Moore, J. E. Dixon, and W. B. Friesen (1995), Petrology of submarine lavas from Kilauea's Puna Ridge, Hawaii, *J. Petrol.*, *36*(2), 299–349.

Clague, D. A., J. T. Hagstrum, D. E. Champion, and M. H. Beeson (1999), Kīlauea summit overflows: Their ages and distribution in the Puna District, Hawai'i, *Bull. Volcanol.*, *61*(6), 363–381.

Clague, D. A., R. Batiza, J. W. Head, and A. S. Davis (2003), Pyroclastic and hydroclastic Deposits on Loihi Seamount, Hawaii, in *Explosive Subaqueous Volcanism*, edited by J. D.L. White, J. L. Smellie and D. A. Clague, AGU, Washington, D. C. doi: 10.1029/140GM05.

Coombs, M. L., T. W. Sisson, and P. W. Lipman (2006), Growth history of Kilauea inferred from volatile concentrations in submarine-collected basalts, *J. Volcanol. Geothermal Res.*, *151*(1), 19–49.

Danyushevsky, L. V., and P. Plechov (2011), Petrolog3: Integrated software for modeling crystallization processes, *Geochem., Geophys., Geosyst.*, *12*(7), Q07021.

Danyushevsky, L. V., F. N. Della-Pasqua, and S. Sokolov (2000a), Re-equilibration of melt inclusions trapped by magnesian olivine phenocrysts from subduction-related magmas: Petrological implications, *Contr. Mineral. Petrol.*, *138*(1), 68–83.

Danyushevsky, L. V., S. M. Eggins, T. J. Falloon, and D. M. Christie (2000b), H_2O abundance in depleted to moderately enriched mid-ocean ridge magmas; Part I: Incompatible behaviour, implications for mantle storage, and origin of regional variations, *J. Petrol.*, *41*(8), 1329–1364.

Davidson, J. P., D. J. Morgan, B. L. A. Charlier, R. Harlou, and J. M. Hora (2007), Microsampling and isotopic analysis of igneous rocks: Implications for the study of magmatic systems, *Ann. Rev. Earth Planet. Sci.*, *35*(1), 273–311.

Davis, M., M. Garcia, and P. Wallace (2003), Volatiles in glasses from Mauna Loa Volcano, Hawai'i: Implications for magma degassing and contamination, and growth of Hawaiian volcanoes, *Contr. Mineral. Petrol.*, *144*(5), 570–591.

Day, A. L., and E. S. Shepherd (1913), Water and volcanic activity, *Bull. Geol. Soc. Am.*, *24*, 573–606.

Dixon, J. E. (1997), Degassing of alkalic basalts, *Am. Mineral.*, *82*, 368–378.

Dixon, J. E., and D. A. Clague (2001), Volatiles in basaltic glasses from Loihi Seamount, Hawaii: Evidence for a relatively dry plume component, *J. Petrol.*, *42*(3), 627–654.

Dixon, J. E., and E. M. Stolper (1995), An experimental study of water and carbon dioxide solubilities in mid-ocean ridge basaltic liquids. Part II: Applications to degassing, *J. Petrol.*, *36*(6), 1633–1646.

Dixon, J. E., E. Stolper, and J. R. Delaney (1988), Infrared spectroscopic measurements of CO_2 and H_2O in Juan de Fuca Ridge basaltic glasses, *Earth Planet. Sci. Lett.*, *90*(1), 87–104.

Dixon, J. E., D. A. Clague, and E. M. Stolper (1991), Degassing history of water, sulfur, and carbon in submarine lavas from Kīlauea Volcano, Hawaii, *J. Geol.*, *99*, 371–394.

Dixon, J. E., D. A. Clague, P. Wallace, and R. Poreda (1997), Volatiles in alkalic basalts form the North Arch Volcanic Field, Hawaii: Extensive degassing of deep submarine-erupted alkalic series lavas, *J. Petrol.*, *38*(7), 911–939.

Duffield, W. A., R. L. Christiansen, R. Y. Koyanagi, and D. W. Peterson (1982), Storage, migration, and eruption of magma at Kīlauea volcano, Hawaii, 1971–1972, *J. Volcanol. Geothermal Res.*, *13*(3), 273–307.

Edmonds, M., and T. M. Gerlach (2007), Vapor segregation and loss in basaltic melts, *Geology*, *35*(8), 751–754.

Edmonds, M., I. Sides, D. Swanson, C. Werner, R. Martin, T. Mather, R. Herd, R. Jones, M. Mead, and G. Sawyer (2013), Magma storage, transport and degassing during the 2008–10 summit eruption at Kilauea Volcano, Hawaii, *Geochim. Cosmochim. Acta*, *123*, 284–301.

Eggins, S. M. (1992), Petrogenesis of Hawaiian tholeiites: 1, phase equilibria constraints, *Contr. Mineral. Petrol.*, *110* (2–3), 387–397.

Elias, T., and A. J. Sutton (2007), Sulfur dioxide emission rates from Kīlauea Volcano, Hawaii, an update: 2002–2006, Open File Rep. U. S. Geol. Sur., 37.

Elias, T., and A. Sutton (2012), Sulfur dioxide emission rates from Kīlauea Volcano, Hawaii 2007–2010, Open File Rep., *U.S. Geol. Surv.*, *1107*, 25.

Elias, T., A. Sutton, J. Stokes, and T. Casadevall (1998), *Sulfur Dioxide Emission Rates of Kīlauea Volcano, Hawaii, 1979–1997*, U.S. Dep. of the Int., U.S. Geol. Surv., Washington, D.C.

Fee, D., M. Garcés, M. Patrick, B. Chouet, P. Dawson, and D. Swanson (2010), Infrasonic harmonic tremor and degassing bursts from Halema'uma'u Crater, Kīlauea Volcano, Hawaii, *J. Geophys. Res.: Solid Earth*, *115*(B11), B11316.

Fine, G., and E. Stolper (1986), Dissolved carbon dioxide in basaltic glasses: Concentrations and speciation, *Earth Planet. Sci. Lett.*, *76*(3), 263–278.

Frey, F., and J. Rhodes (1993), Intershield geochemical differences among Hawaiian volcanoes: Implications for source compositions, melting process and magma ascent paths, *Philos. Trans. R. Soc. Lond. Ser. A Phys. Eng. Sci.*, *342*(1663), 121–136.

Gaetani, G. A., J. A. O'Leary, N. Shimizu, C. E. Bucholz, and M. Newville (2012), Rapid reequilibration of H_2O and oxygen fugacity in olivine-hosted melt inclusions, *Geology*, *40*(10), 915–918.

Garcia, M. O., D. W. Muenow, K. E. Aggrey, and J. R. O'Neil (1989), Major element, volatile, and stable isotope geochemistry of Hawaiian submarine tholeiitic glasses, *J. Geophys. Res. Solid Earth*, *94*(B8), 10525–10538.

Garcia, M. O., A. J. Pietruszka, and J. M. Rhodes (2003), A petrologic perspective of Kīlauea Volcano's summit magma reservoir, *J. Petrol.*, *44*(12), 2313–2339.

Gerlach, T. M. (1980), Evaluation of volcanic gas analyses from Kilauea volcano, *J. Volcanol and Geothermal Res.*, *7*(3), 295–317.

Gerlach, T. M. (1986), Exsolution of H_2O, CO_2, and S during eruptive episodes at Kīlauea Volcano, Hawaii, *J. Geophys. Res. Solid Earth*, *91*(B12), 12,177–12,185.

Gerlach, T. M., and E. J. Graeber (1985), Volatile budget of Kīlauea volcano, *Nature*, *313*(6000), 273–277.

Gerlach, T. M., K. A. McGee, A. J. Sutton, and T. Elias (1998), Rates of volcanic CO_2 degassing from airborne determinations of SO2 Emission rates and plume CO_2/SO_2: Test study at Pu'u 'O'o Cone, Kīlauea Volcano, Hawaii, *Geophys. Res. Lett.*, *25*(14), 2675–2678.

Gerlach, T. M., K. A. McGee, T. Elias, A. J. Sutton, and M. P. Doukas (2002), Carbon dioxide emission rate of Kīlauea Volcano: Implications for primary magma and the summit reservoir, *J. Geophys. Res. Solid Earth*, *107*(B9), 2189.

Greenland, L., W. Rose, and J. Stokes (1985), An estimate of gas emissions and magmatic gas content from Kilauea volcano, *Geochim. Cosmochim. Acta*, *49*(1), 125–129.

Hager, S. A., T. M. Gerlach, and P. J. Wallace (2008), Summit CO_2 emission rates by the CO_2/SO_2 ratio method at Kīlauea Volcano, Hawai'i, during a period of sustained inflation, *J. Volcanol. Geothermal Res.*, *177*(4), 875–882.

Harris, D. M., and A. T. Anderson Jr. (1983), Concentrations, sources, and losses of H_2O, CO_2, and S in Kilauean basalt, *Geochimica et Cosmochimica Acta*, *47*(6), 1139–1150.

Hart, D., and P. J. Wallace (2000), Volatiles in melt inclusions from the Keanakakoi ash member, Kīlauea volcano: Implications for explosive Hawaiian eruptions, *Eos*, *81*, 1291.

Hauri, E. H., J. C. Lassiter, and D. J. DePaolo (1996), Osmium isotope systematics of drilled lavas from Mauna Loa, Hawaii, *J. Geophys. Res. Solid Earth (1978–2012)*, *101*(B5), 11,793–11,806.

Hauri, E., K. Gronvold, N. Oskarsson, and D. McKenzie (2002a), Abundance of carbon in the Icelandic mantle: Constraints from melt inclusions, paper presented at AGU Spring Meet. Abstracts.

Hauri, E., J. Wang, J. E. Dixon, P. L. King, C. Mandeville, and S. Newman (2002b), SIMS analysis of volatiles in silicate glasses: 1. Calibration, matrix effects and comparisons with FTIR, *Chem. Geol.*, *183*(1–4), 99–114.

Helz, R. T. (1987), Differentiation behavior of Kilauea Iki lava lake, Kilauea Volcano, Hawaii: an overview of past and current work. *Magmatic processes: physicochemical principles*, *1*, 241–258.

Helz, R., and C. Thornber (1987), Geothermometry of Kīlauea Iki lava lake, Hawaii, *Bull. Volcanol.*, *49*(5), 651–668.

Helz, R. T., and T. L. Wright (1992), Differentiation and magma mixing on Kilauea's east rift zone, *Bull. Volcanol.*, *54*(5), 361–384.

Hofmann, A. W., M. D. Feigenson, and I. Raczek (1984), Case studies on the origin of basalt: III. Petrogenesis of the Mauna Ulu eruption, Kilauea, 1969–1971, *Contr. Mineral. Petrol.*, *88*(1–2), 24–35.

Houghton, B. F., D. A. Swanson, R. J. Carey, J. Rausch, and A. J. Sutton (2011), Pigeonholing pyroclasts: Insights from the 19 March 2008 explosive eruption of Kīlauea volcano, *Geology*, *39*(3), 263–266.

Jackson, D. E., E. A. Silver, and G. B. Dalrymple (1972), Hawaiian-Emperor chain and its relation to Cenozoic circumpacific tectonics, *Geol. Soc. Am. Bull.*, *83*(3), 601–618.

Jaggar, T. A. (1940), Magmatic gases, *Am. J. Sci.*, *238*(5), 313–353.

Kamenetsky, V., and R. Clocchiatti (1996), Primitive magmatism of Mt. Etna: Insights from mineralogy and melt inclusions, *Earth Planet. Sci. Lett.*, *142*(3), 553–572.

Kent, A. J., and T. R. Elliott (2002), Melt inclusions from Marianas arc lavas: Implications for the composition and formation of island arc magmas, *Chem. Geol.*, *183*(1), 263–286.

Kent, A. J. R. (2008), Melt inclusions in basaltic and related volcanic rocks, *Rev. Mineral. Geochem.*, *69*(1), 273–331.

Kent, A. J. R., D. A. Clague, M. Honda, E. M. Stolper, I. D. Hutcheon, and M. D. Norman (1999), Widespread assimilation of a seawater-derived component at Loihi Seamount, Hawaii, *Geochim. Cosmochim. Acta*, *63*(18), 2749–2761.

Kingsley, R. H., and J.-G. Schilling (1995), Carbon in Mid-Atlantic Ridge basalt glasses from 28 N to 63 N: Evidence for a carbon-enriched Azores mantle plume, *Earth Planet. Sci. Lett.*, *129*(1), 31–53.

Kurz, M. D., T. C. Kenna, J. Lassiter, and D. DePaolo (1996), Helium isotopic evolution of Mauna Kea Volcano: First results from the 1-km drill core, *J. Geophys. Res. Solid Earth (1978–2012)*, *101*(B5), 11,781–11,791.

Lassiter, J., and E. Hauri (1998), Osmium-isotope variations in Hawaiian lavas: Evidence for recycled oceanic lithosphere in the Hawaiian plume, *Earth Planet. Sci. Lett.*, *164*(3), 483–496.

Lockwood, J. P., R. I. Tilling, R. T. Holcomb, F. Klein, A. Okamura, and D. Peterson (1999), *Magma Migration and Resupply During the 1974 Summit Eruptions of Kīlauea Volcano, Hawaii*, U.S. Gov. Print. Office, Washington, D.C.

Lowenstern, J. B. (1995), Applications of silicate-melt inclusions to the study of magmatic volatiles, *Magmas, Fluids, and Ore Deposits*, *23*, 71–99.

Maaløe, S. (2004), The PT-phase relations of an MgO-rich Hawaiian tholeiite: The compositions of primary Hawaiian tholeiites, *Contr. Mineral. Petrol.*, *148*(2), 236–246.

Macdonald, G. A., and J. P. Eaton (1957), *Hawaiian Volcanoes During 1954*, U.S. Gov. Print. Office, Washington, D.C.

Maclennan, J. (2008), Concurrent mixing and cooling of melts under Iceland, *J. Petrol.*, *49*(11), 1931–1953.

Mangan, M. T., K. V. Cashman, and S. Newman (1993), Vesiculation of basaltic magma during eruption, *Geology*, *21*(2), 157–160.

Marske, J. P., M. O. Garcia, A. J. Pietruszka, J. M. Rhodes, and M. D. Norman (2008), Geochemical variations during Kīlauea's Pu'u 'Ō'ō eruption reveal a fine-scale mixture of mantle heterogeneities within the Hawaiian Plume, *J. Petrol.*, *49*(7), 1297–1318.

Mather, T. A., et al. (2012), Halogens and trace metal emissions from the ongoing 2008 summit eruption of Kīlauea volcano, Hawai'i, *Geochim. Cosmochim. Acta*, *83*(0), 292–323.

McPhie, J., G. Walker, and R. Christiansen (1990), Phreatomagmatic and phreatic fall and surge deposits from explosions at Kīlauea volcano, Hawaii, 1790 a.d.: Keanakakoi Ash Member, *Bull. Volcanol.*, *52*(5), 334–354.

Métrich, N., and P. J. Wallace (2008), Volatile abundances in basaltic magmas and their degassing paths tracked by melt inclusions, *Rev. Mineral. Geochem.*, *69*(1), 363–402.

Murata, K. J., and D. H. Richter (1966), The settling of olivine in Kilauean magma as shown by lavas of the 1959 eruption, *Am. J. Sci.*, *264*(3), 194–203.

Neal, C. A., and J. P. Lockwood (2003), Geologic map of the summit region of Kīlauea Volcano, Hawaii, Map I-2759, *U.S. Geol. Surv.*, *1*(24,000).

Newman, S., and J. B. Lowenstern (2002), VolatileCalc: A silicate melt–H_2O–CO_2 solution model written in Visual Basic for Excel, *Comput. Geosci.*, *28*(5), 597–604.

Norman, M. D., and M. O. Garcia (1999), Primitive magmas and source characteristics of the Hawaiian plume: Petrology and geochemistry of shield picrites, *Earth Planet. Sci. Lett.*, *168*(1–2), 27–44.

Orr, T. R., W. A. Thelen, M. R. Patrick, D. A. Swanson, and D. C. Wilson (2013), Explosive eruptions triggered by rockfalls at Kīlauea volcano, Hawai'i, *Geology*, *41*(2), 207–210.

Pan, V., J. R. Holloway, and R. L. Hervig (1991), The pressure and temperature dependence of carbon dioxide solubility in tholeiitic basalt melts, *Geochim. Cosmochim. Acta*, *55*(6), 1587–1595.

Parcheta, C. E., B. F. Houghton, and D. A. Swanson (2012), Hawaiian fissure fountains 1: Decoding deposits—Episode 1 of the 1969–1974 Mauna Ulu eruption, *Bull. Volcanol.*, *74*(7), 1729–1743.

Patrick, M., D. Wilson, D. Fee, T. Orr, and D. Swanson (2011), Shallow degassing events as a trigger for very-long-period seismicity at Kīlauea Volcano, Hawai'i, *Bull. Volcanol.*, *73*(9), 1179–1186.

Pearce, N. J. G., W. T. Perkins, J. A. Westgate, M. P. Gorton, S. E. Jackson, C. R. Neal, and S. P. Chenery (1997), A compilation of new and published major and trace element data for NIST SRM 610 and NIST SRM 612 glass reference materials, *Geostand. Newslett.*, *21*(1), 115–144.

Pichavant, M., I. Di Carlo, S. G. Rotolo, B. Scaillet, A. Burgisser, N. Le Gall, and C. Martel (2013), Generation of CO_2-rich melts during basalt magma ascent and degassing, *Contr. Mineral. Petrol.*, *166*, 545–561.

Pietruszka, A. J., and M. O. Garcia (1999), A rapid fluctuation in the mantle source and melting history of Kīlauea Volcano inferred from the geochemistry of its historical summit lavas (1790–1982), *J. Petrol.*, *40*(8), 1321–1342.

Pietruszka, A. J., E. H. Hauri, R. W. Carlson, and M. O. Garcia (2006), Remelting of recently depleted mantle within the Hawaiian plume inferred from the 226Ra-230Th-238U disequilibria of Pu'u 'Ō'ō eruption lavas, *Earth Planet. Sci. Lett.*, *244*(1–2), 155–169.

Pietruszka, A. J., M. D. Norman, M. O. Garcia, J. P. Marske, and D. H. Burns (2012), Chemical heterogeneity in the Hawaiian mantle plume from the alteration and dehydration of recycled oceanic crust, *Earth Planet. Sci. Lett*, *361*, 298–309.

Poland, M. P., A. Miklius, A. Jeff Sutton, and C. R. Thornber (2012), A mantle-driven surge in magma supply to Kīlauea Volcano during 2003–2007, *Nature Geosci*, *5*(4), 295–300.

Putirka, K. D. (2005), Mantle potential temperatures at Hawaii, Iceland, and the mid-ocean ridge system, as inferred from olivine phenocrysts: Evidence for thermally driven mantle plumes, *Geochem., Geophys., Geosyst.*, *6*(5), Q05L08.

Rhodes, J., K. Wenz, C. Neal, J. Sparks, and J. Lockwood (1989), Geochemical evidence for invasion of Kilauea's plumbing system by Mauna Loa magma, *Nature*, *337*(6204), 257–260.

Rhodes, J. M., and S. R. Hart (1995), Episodic trace element and isotopic variations in historical Mauna Loa Lavas: Implications for magma and plume dynamics, in *Mauna Loa Revealed: Structure, Composition, History, and Hazards*, edited by J. M. Rhodes and J. P. Lockwood, pp. 263–288, AGU, Washington, D.C.

Rhodes, J. M., and M. J. Vollinger (2005), Ferric/ferrous ratios in 1984 Mauna Loa lavas: A contribution to understanding the oxidation state of Hawaiian magmas, *Contr. Mineral. Petrol.*, *149*(6), 666–674.

Ribe, N. M., and U. R. Christensen (1999), The dynamical origin of Hawaiian volcanism, *Earth Planet. Sci. Lett.*, *171*(4), 517–531.

Richter, D. H. (1964), *The 1961 Eruption of Kīlauea Volcano, Hawaii*, U.S. Gov. Print. Office, U.S. Geol. Surv., Washington, D.C.

Richter, D. H., J. Eaton, K. Murata, W. Ault, and H. Krivoy (1970), *Chronological Narrative of the 1959–60 Eruption of Kīlauea Volcano, Hawaii*, U.S. Govt. Print. Office, Washington, D.C.

Rivalta, E., and P. Segall (2008), Magma compressibility and the missing source for some dike intrusions, *Geophys. Res. Lett.*, *35*(4), L04306.

Roeder, P. L., and R. F. Emslie (1970), Olivine-liquid equilibrium, *Contr. Mineral. Petrol.*, *29*(4), 275–289.

Rowland, S., and D. Munro (1993), The 1919–1920 eruption of Mauna Iki, Kīlauea: Chronology, geologic mapping, and magma transport mechanisms, *Bull. Volcanol.*, *55*(3), 190–203.

Ryan, M. P. (1988), The mechanics and three-dimensional internal structure of active magmatic systems: Kīlauea Volcano, Hawaii, *J. Geophys. Res. Solid Earth*, *93*(B5), 4213–4248.

Saal, A. E., E. H. Hauri, C. H. Langmuir, and M. R. Perfit (2002), Vapour undersaturation in primitive mid-ocean-ridge basalt and the volatile content of Earth's upper mantle, *Nature*, *419*(6906), 451–455.

Seaman, C., S. B. Sherman, M. O. Garcia, M. B. Baker, B. Balta, and E. Stolper (2004), Volatiles in glasses from the HSDP2 drill core, *Geochem., Geophys., Geosyst.*, *5*(9), Q09G16, doi:10.1029/2003GC000596.

Sharp, R. P., D. Dzurisin, and M. C. Malin (1987), Volcanism in Hawaii, *Prof. Pap., U.S. Geol. Surv.*, *1*(1350), 395.

Shepherd, E. (1921), Kīlauea gases, 1919, *Hawaiian Volcanol. Obs. Bull.*, *9*, 83–88.

Sides, I. R., M. Edmonds, J. Maclennan, D. A. Swanson, and B. F. Houghton(2014), Eruption style at Kilauea Volcano in Hawai'i linked to primary melt composition, *Nature Geosci.*, *7*(6), 464–469.

Sobolev, A. (1996), Melt inclusions in minerals as a source of principal petrological information, *Petrology*, *4*(3), 209–220.

Sobolev, A. V., A. W. Hofmann, and I. K. Nikogosian (2000), Recycled oceanic crust observed in "ghost plagioclase" within the source of Mauna Loa lavas, *Nature*, *404*(6781), 986–990.

Soule, S. A., D. S. Nakata, D. J. Fornari, A. T. Fundis, M. R. Perfit, and M. D. Kurz (2012), CO_2 variability in mid-ocean ridge basalts from syn-emplacement degassing: Constraints on eruption dynamics, *Earth Planet. Sci. Lett.*, *327–328*(0), 39–49.

Spera, F. (1981), Carbon dioxide in igneous petrogenesis: II. Fluid dynamics of mantle metasomatism, *Contr. Mineral. Petrol.*, *77*(1), 56–65.

Spera, F., and S. Bergman (1980), Carbon dioxide in igneous petrogenesis: I, *Contr. Mineral. Petrol.*, *74*(1), 55–66.

Steele-Macinnis, M., R. Esposito, and R. J. Bodnar (2011), Thermodynamic model for the effect of post-entrapment crystallization on the H_2O–CO_2 systematics of vapor-saturated, silicate melt inclusions, *J. Petrol.*, *52*(12), 2461–2482.

Stolper, E., and J. R. Holloway (1988), Experimental determination of the solubility of carbon dioxide in molten basalt at low pressure, *Earth Planet. Sci. Lett.*, *87*(4), 397–408.

Stovall, W., B. Houghton, J. Hammer, S. Fagents, and D. Swanson (2012), Vesiculation of high fountaining Hawaiian eruptions: Episodes 15 and 16 of 1959 Kīlauea Iki, *Bull. Volcanol.*, *74*(2), 441–455.

Sutton, A. J., T. Elias, T. M. Gerlach, and J. B. Stokes (2001), Implications for eruptive processes as indicated by sulfur dioxide emissions from Kīlauea Volcano, Hawai'i, 1979–1997, *J. Volcanol. Geothermal Res.*, *108*(1–4), 283–302.

Sutton, A. J., T. Elias, and J. Kauahikaua (2003), Lava-effusion rates for the Pu'u' O'o–Kupaianaha eruption derived from SO_2 emissions and very low frequency (VLF) measurements, Prof. Pap, *U.S. Geol. Surv.*, *1676*, 137–148.

Swanson, D., W. Duffield, D. Jackson, and D. Peterson (1979), Chronolo ical narrative of the 1969–71 Mauna Ulu eruption of Kīlauea volcano. Hawaii, Prof. Pap, *U.S. Geol, Surv. 1056*, 55.

Swanson, D., T. Rose, and R. Fiske (2006), The layer 6 scoria fall about 400 years ago at Kilauea, paper presented at AGU Fall Meet. Abstracts.

Swanson, D., K. Wooten, and T. Orr (2009), Buckets of ash track tephra flux from Halema'uma'u Crater, Hawai'i, *Eos, Trans. AGU*, *90*(46), 427–427.

Swanson, D. A., and R. L. Christiansen (1973), Tragic base surge in 1790 at Kīlauea Volcano, *Geology*, *1*(2), 83–86.

Swanson, D. A., T. R. Rose, R. S. Fiske, and J. P. McGeehin (2012), Keanakāko'i tephra produced by 300 years of explosive eruptions following collapse of Kīlauea's caldera in about 1500CE, *J. Volcanol. Geothermal Res.*, *215*, 8–25.

Swavely, L., D. Swanson, B. Houghton, and R. Carey (2010), Changing componentry for eruptions of Halemaumau Crater, Kīlauea in 2008: Implications for ascent and eruption dynamics, paper presented at AGU Fall Meet. Abstracts.

Tilling, R. I., R. L. Christiansen, W. A. Duffield, E. T. Endo, and R. T. Holcomb (1987), Volcanism in Hawaii, Prof. Pap, *U.S. Geol. Surv.*, *1*(1350), 405.

Wagner, T., and T. Grove (1998), Melt/harzburgite reaction in the petrogenesis of tholeiitic magma from Kilauea volcano, Hawaii, *Contr. Mineral. Petrol.*, *131*(1), 1–12.

Wallace, P. J. (1998), Water and partial melting in mantle plumes: Inferences from the dissolved H_2O concentrations of Hawaiian basaltic magmas, *Geophys. Res. Lett.*, *25*(19), 3639–3642.

Wallace, P. J., and A. T. Anderson (1998), Effects of eruption and lava drainback on the H_2O contents of basaltic magmas at Kīlauea Volcano, *Bull. Volcanol.*, *59*(5), 327–344.

Weaver, S., B. Houghton, and D. Swanson (2010), Kilauea's explosive past: Understanding violent explosions at Hawai'i's most active volcano, paper presented at AGU Fall Meet. Abstracts.

Weis, D., M. O. Garcia, J. M. Rhodes, M. Jellinek, and J. S. Scoates (2011), Role of the deep mantle in generating the compositional asymmetry of the Hawaiian mantle plume, *Nature Geosci.*, *4*(12), 831–838.

Welsch, B., F. Faure, V. Famin, A. Baronnet, and P. Bachèlery (2013), Dendritic crystallization: A single process for all the textures of olivine in basalts? *J. Petrol.*, *54*(3), 539–574.

Witham, F., J. Blundy, S. C. Kohn, P. Lesne, J. Dixon, S. V. Churakov, and R. Botcharnikov (2012), SolEx: A model for mixed COHSCl-volatile solubilities and exsolved gas compositions in basalt, *Comput. Geosci.*, *45*, 87–97.

Wright, T. L. (1973), Magma mixing as illustrated by the 1959 eruption, Kīlauea Volcano, Hawaii, *Geol. Soc. Am. Bull.*, *84*(3), 849–858.

Wright, T. L., and R. S. Fiske (1971), Origin of the differentiated and hybrid lavas of Kilauea volcano, Hawaii, *J. Petrol.*, *12*(1), 1–65.

16

Reticulite-Producing Fountains From Ring Fractures in Kīlauea Caldera ca. 1500 CE

Michael May[1], Rebecca J. Carey[2], Donald A. Swanson[3], and Bruce F. Houghton[4]

ABSTRACT

A widely dispersed reticulite bed occurs close to the base of the Keanakāko'i Tephra at Kīlauea Volcano. It can be divided into six subunits in the northern sector of the volcano; the reticulite also occurs in the southern sector, but outcrops are sparse owing to penecontemporaneous erosion and burial. Multilobate isopachs for each subunit and the total deposit suggest that multiple fountaining vents were distributed in the northern half of the caldera, possibly along ring fractures for the newly formed caldera. Isopach maps also show a sharp decline in thickness along the dispersal axis of each lobe, which could be explained by remobilization of tephra and/or inclined fountains. Despite such isopach characteristics, thinning rates calculated from the isopach data indicate that the fountains were among the most intense and powerful of all studied Kīlauea fountains. Density analyses of the pyroclasts suggest that fountaining was high (>600 m) yet complex, possibly due to lava ponding and reentrainment. The calculated volume of the reticulite deposited around the caldera rim is approximately 0.2 km^3, more voluminous than the deposits of the 1959 Kīlauea Iki eruption; this volume is a minimum, however, as the low-density tephra is easily remobilized, and 600 m high caldera walls probably trapped tephra within the caldera, which is deeply buried today and not accounted for in the volume calculations. The duration of this eruption was most likely at least a few days to weeks, based on the calculated volume and estimated discharge rates as seen during the Kīlauea Iki 1959 eruption.

16.1. INTRODUCTION

Kīlauea Volcano, the southernmost shield volcano on the Island of Hawai'i, is one of the most studied and best understood basaltic volcanoes in the world (Figure 16.1). Kīlauea's eruptive past since about 1470 CE is complex, with major changes in summit architecture that likely enabled highly variable explosive magmatic and phreatomagmatic behavior [*Swanson et al.*, 2012]. The collapse of Kīlauea Caldera probably occurred sometime in the interval 1470–1510 CE and initiated ca. 300 years of explosive eruptions that produced the Keanakāko'i Tephra. *Swanson et al.* [2012] assembled evidence from a variety of historical reports, publications, and field data to suggest that the caldera formed before the Keanakāko'i eruptions began and was probably greater than 600 m deep. The duration and timing of caldera development are uncertain, though the geologic relations agree with Hawaiian oral tradition and suggest at least one large collapse event [*Swanson*, 2008, 2010]. High caldera walls restricted the dispersal of erupted products on the summit and flanks of the volcano for ca. 300 years following the large-collapse event.

Just after the caldera formed or during its final stage of collapse, high fountains produced a deposit of reticulite that encircles the caldera. The reticulite drapes the

[1]*Department of Geology, University of Otago, Dunedin, New Zealand*

[2]*School of Physical Sciences and Centre of Excellence in Ore Deposits, University of Tasmania, Hobart, Tasmania, Australia*

[3]*Hawaiian Volcano Observatory, U.S. Geological Survey, Hawai'i National Park, Hawaii, USA*

[4]*Department of Geology and Geophysics, University of Hawai'i at Mānoa, Honolulu, Hawaii, USA*

Hawaiian Volcanoes: From Source to Surface, Geophysical Monograph 208. First Edition.
Edited by Rebecca Carey, Valérie Cayol, Michael Poland, and Dominique Weis.

Companion Website: www.wiley.com/go/Carey/Hawaiian_Volcanoes

Figure 16.1 (a) Location map of Kīlauea Caldera on the Island of Hawai'i (yellow square); (b) IKONOS image of various features around the caldera; (c–e) oblique angle views (widths indicated in (b)) of present-day caldera morphology.

caldera and burned vegetation, producing charcoal used to date the reticulite and constrain the age of caldera collapse. Part of the dispersal of this unit was originally mapped by *Stone* [1926], and his data have been incorporated together with data from our field work and mapping. From these combined data, we calculate fountaining heights and thinning rates, and we interpret these data in terms of the dynamics of the fountains, the eruption in general, and its relation to the caldera and its formation.

In a pioneering study of the Keanakāko'i Tephra, *McPhie et al.* [1990] considered that the reticulite may be "a composite record of many summit fire-fountain eruptions." Those authors chose to exclude the reticulite from the Keanakāko'i on the grounds that the overlying tephra deposits are "genetically related to a coherent, though punctuated, eruption episode" separate from that of the reticulite. *Swanson et al.* [2012] showed that there were many distinct eruptive episodes after the reticulite, rather than just one, and that there was no good reason to exclude the reticulite from the Keanakāko'i. We follow this usage in this chapter.

The following interpretation of the reticulite eruption and dispersal profits from comparison with reticulite-producing eruptions during the past several decades at Kīlauea, in particular, the high fountains at Mauna Ulu in 1969 and Pu'u 'Ō'ō in 1983–1986, both on Kīlauea's East Rift Zone. The production of reticulite and frothy pumice during the Mauna Ulu eruption was associated with fountain heights >300 m above the vent, with a maximum height of 540 m [*Swanson et al.*, 1979]. At Pu'u 'Ō'ō, similar fountain heights of 250–450 m were recorded, with a maximum height of 510 m, and these were associated with the production of reticulite and less vesicular pumice [*Wolfe et al.*, 1987, 1988]. At Kīlauea Iki in 1959, fountain heights were up to 580 m, but the conduit and eruption dynamics were complicated by the partially degassed lava lake draining into the vent, leading to the production of a wide variety of clast types but, importantly, far more pumice than reticulite.

16.2. DEPOSITS

16.2.1. Deposit Characteristics and Juvenile Clast Density

The Keanakāko'i reticulite overlies a millimetric to centimetric gray-pink lithic ash [*Stone*, 1926; *Swanson et al.*, 2012] and is clearly a fallout deposit from lava fountains. The reticulite unit was originally distributed radially around the caldera but now is poorly preserved in the southern portion of the caldera and beyond, where the reticulite fell on unvegetated pāhoehoe and was rapidly stripped away by strong wind that often sweeps across this area, as well as by heavy rainfall, before later tephra was deposited [*Swanson et al.*, 2012]. We concentrated our investigation along the western, northern, and eastern sides of the caldera, where the reticulite was mostly protected from wind erosion by vegetation. The reticulite unit was located and mapped in proximal exposures along the topographic rim of the caldera and on the flank of the volcano; the most distal site is in Volcano Village, approximately 5.3 km north of the caldera. Juvenile components were identified (Figure 16.2, Table 16.1), and clast density was calculated and converted to clast vesicularity following the methods of *Houghton and Wilson* [1989]. To convert clast density to vesicularity, a dense rock equivalent (DRE) of 2950 kg/m^3 was used. Juvenile components within the deposits were identified, but their grain size and percentage were impossible to measure due to their fragility. The componentry was therefore done in a qualitative way at every outcrop, and grain size measurements using sieves were not conducted. The components are described according to the terminology of *Walker and Croasdale* [1972] and include the following:

(i) *Reticulite* Typically golden brown in color, in places oxidized pink or red, with clast diameters from subcentimeter to 15–20 cm (Figure 16.2a). The reticulite clasts display nearly perfect, open-cell polyhedral structure as described by *Mangan and Cashman* [1996]. Reticulite clasts have vesicularities of 95–98 vol%.

(ii) *Skinned Reticulite* Reticulite clasts with discontinuous black glassy outer margins, or "skin." The skins range in thickness but are dominantly <1 mm (Figure 16.2b). The internal structure of a clast appears similar whether or not a skin is present. Rarely, a skin is abnormally thick (2–3 mm) and dull black. The skinned reticulite clasts have vesicularities between 85 and 95 vol% and are typically smaller than the nonskinned clasts, rarely reaching 15 cm in diameter.

(iii) *Denser Clasts* Denser clasts are dull to glassy black and have ragged to slightly fluidal morphologies (Figure 16.2c). These clasts are less vesicular than reticulite but still moderately to highly vesicular (69–89 vol%). Denser clasts rarely exceed 5 cm in diameter and are most commonly in the 2–3 cm range.

(iv) *Achneliths* Small (1–10 mm), dense, glassy pyroclasts with fluidal morphologies including spheres, dumbbells, rods, and irregular shapes (Figure 16.2d). These clasts have smooth outer surfaces with low vesicularity. Larger achneliths are moderately vesicular, with measured values between 65 and 74 vol%.

(v) *Pele's Tears and Hair* Pele's hair is a thin strand of glass, while Pele's tears are shaped in the form of a tear or raindrop (Figures 16.2e,f).

(vi) *Ash* Medium to coarse-grained ash composed of reticulite fragments.

The reticulite deposit is thickest and best preserved along the northwestern sector of the caldera rim, near the

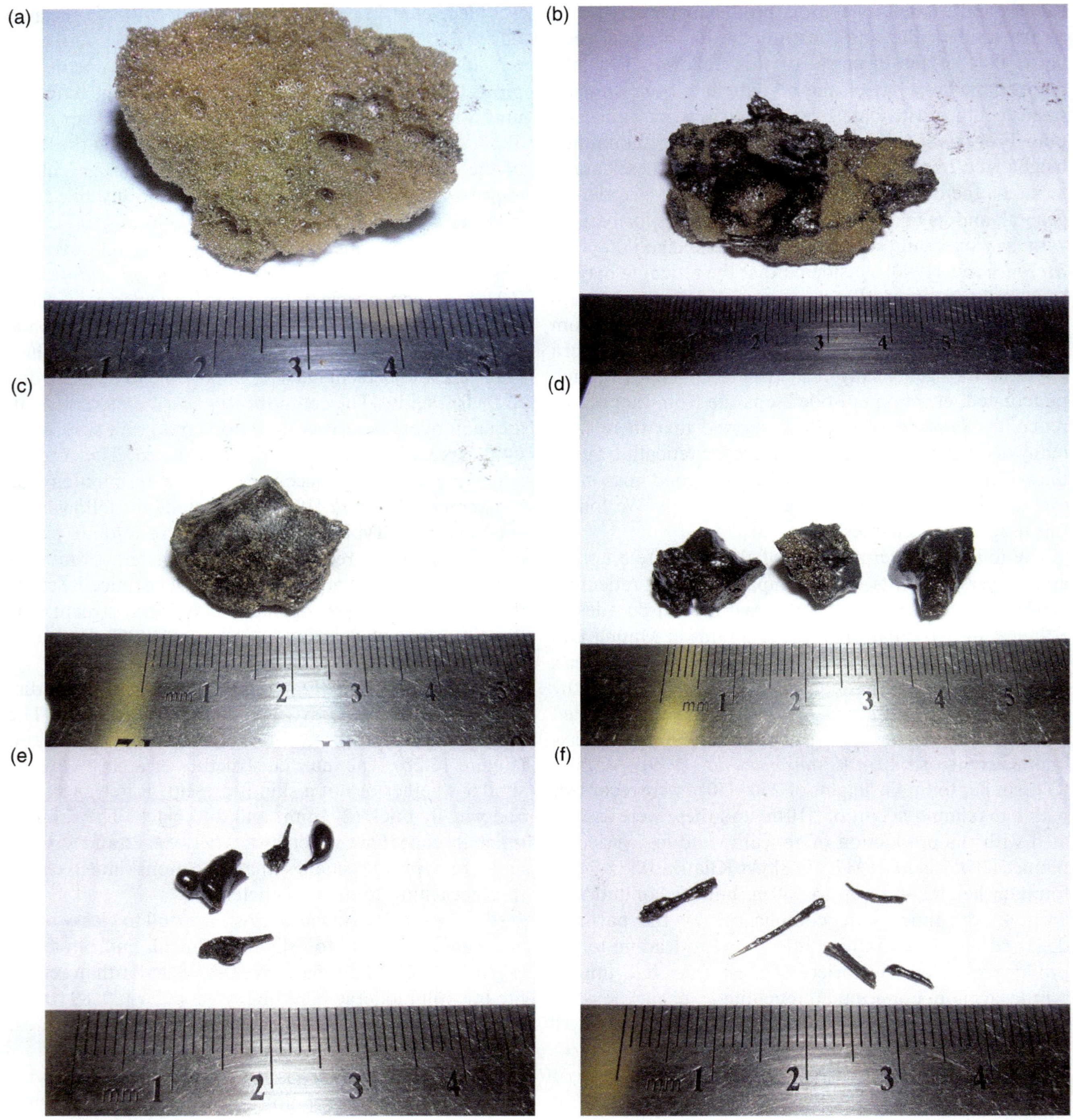

Figure 16.2 Components present in the Keanakāko'i reticulite deposit: (a) reticulite; (b) skinned reticulite; (c) denser clasts; (d) achneliths; (e) Pele's tears; (f) Pele's hair.

Hawaiian Volcano Observatory (HVO; Figures 16.3 and 16.4). Here, five subunits can be identified (NW1–NW5; Figures 16.3 and 16.5) defined by the relative abundance of components, sorting, grading, oxidation, and abundance of ash. In the northeast sector, the stratigraphy is cryptic, and the deposit could not be subdivided. Its distribution, however, makes it clear that it is a subunit (E) distinct from NW1–NW5 with no obvious stratigraphic changes throughout the sections (Figure 16.6).

Samples of 100 clasts were collected from two sites for density analysis, KR08-01 in the northwestern sector and KR08-14 in the eastern sector (Figure 16.4). At location KR08-01, 100 clasts were collected from each subunit, whereas at location KR08-14, five samples of 100 clasts

Table 16.1 Volume percentages of components across all size fractions as estimated in the field.

Subunit	Reticulite %	Skinned Reticulite %	Dense %	Achnelith %	Haris and Tears %	Ash %
NW5	33	19	8	11	3	26
NW4	30	33	7	13	6	11
NW3	14	19	29	21	9	8
NW2	33	30	9	13	6	9
NW1	13	27	10	21	13	16
Ev	30	30	15	10	10	5
Eiv	60	15	10	5	5	5
Eiii	25	30	25	10	8	2
Eii	25	30	20	15	8	2
Ei	45	15	15	10	10	5

Figure 16.3 View of reticulite deposit at location KR08-01 near the Hawaiian Volcano Observatory. Subunit boundaries are indicated. Scraper is 15 cm long.

each were collected throughout the thickness of the deposit. The density histograms obtained for all Keanakāko'i subunits have a sharp unimodal peak, with a small skewed tail of denser clasts (Figure 16.7). Regardless of the subunit, the modal densities are about 100 kg/m^3 (97 vol%). The first material erupted in both the northwestern and eastern sectors has a density tail that extends to 900 kg/m^3 (69 vol%), and both are followed by a more uniform population of low-density clasts. The northwestern subunits alternate between

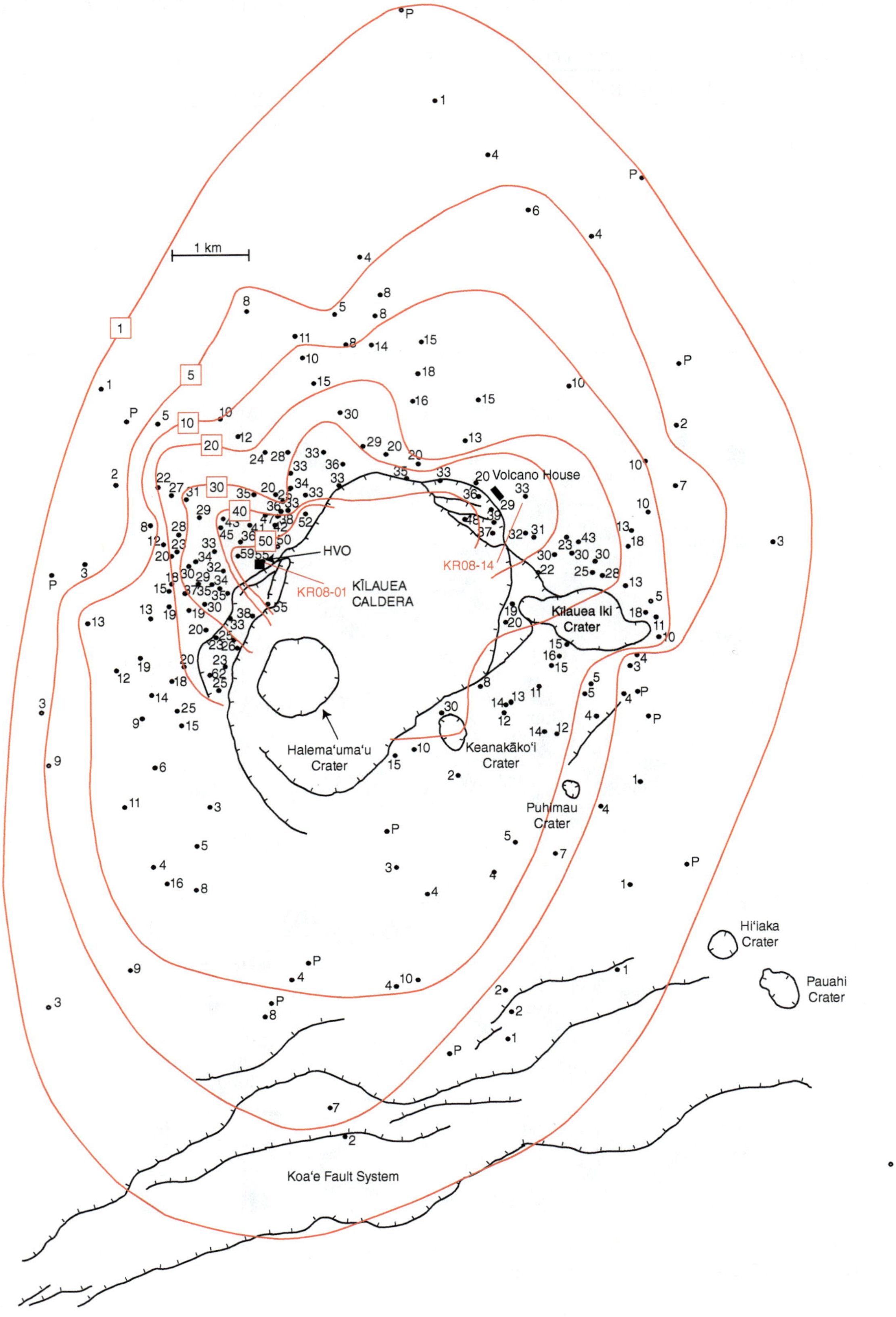

Figure 16.4 Map of the present-day caldera showing locations and whole deposit thicknesses (cm) where primary reticulite was located and described by either *Stone* [1926] or us. Isopachs of total deposit thickness show distribution around the caldera. Locations where juvenile clast density were taken (KR08-01 and KR08-14) are shown in red. Dots located in the south of the caldera are locations where in situ reticulite was recorded. "P" indicates that the deposit is present but a thickness measurement could not be made.

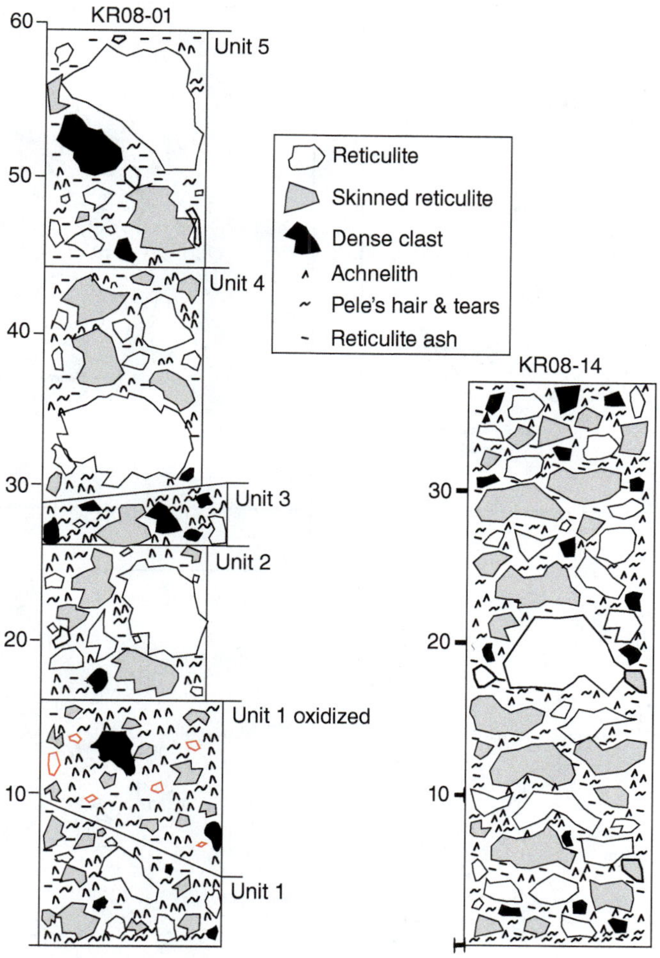

Figure 16.5 Stratigraphic logs of the Keanakākoʻi reticulite from the NW (at left) and E (at right) sectors. The locations KR08-01 and KR08-14 were chosen as representative localities. Vertical scale in centimeters. Subunits NW1 to NW5 could be defined around the caldera. Stratigraphy within subunit E was cryptic, and therefore the deposits were mapped as one subunit. Ei to Ev denote locations within the stratigraphy where density samples were collected.

dominantly low density reticulite (NW2, NW4) and a mixture of low and moderately dense clasts (NW3, NW5). In comparison with juvenile clast density distributions of the Kīlauea Iki 1959 eruption (specifically episodes 1, 15, 16), the Keanakākoʻi reticulite clasts are more highly unimodal and skewed to much lower (200–1000 kg/m^3) modal densities (*Stovall et al.*, 2011, 2012). Clast densities from the Keanakākoʻi reticulite eruptions rarely exceed 600 kg/m^3. The Keanakākoʻi reticulite deposits have similar clast density distributions to that of the high-intensity fountaining episode (episode 9) of the 1969 Mauna Ulu eruption [*Parcheta et al.*, 2013].

16.2.2. Stratigraphy

16.2.2.1. Northwestern Sector

The stratigraphy of the deposit is distinguished based upon the dominance of different components. Subunits 1 and 3, which contain relatively high abundances of dense juvenile achneliths, Pele's tears and hair, sandwich the reticulite-dominant subunit 2. Subunits 4 and 5 are subdivided based on the abundance of skinned reticulite, which is noticeably higher in subunit 4 than in subunit 5. However, this subdivision is cryptic if deposits are thinner than ~15 cm. The subdivision is likely critical for changes of eruption or fountaining dynamics with time, but for the purposes of deposit dispersal we grouped subunits 4 and 5.

1. NW Subunit 1 (NW1) NW1 can commonly, but not always, be divided into two parts, dominated by denser clasts, Pele's hair and tears at the base, and larger 1–3 cm reticulite and skinned reticulite clasts (commonly oxidized) dominating the top (Figures 16.3 and 16.5). Generally, denser clasts, achneliths, and Pele's hair and tears represent 40–50 vol% of the subunit (Table 16.1), reticulite ash is ~15 vol%, and reticulite clasts represent ~35–45 vol% of the subunit.

2. NW Subunit 2 (NW2) NW2 mostly contains tightly packed reticulite clasts with 5–25 cm diameters

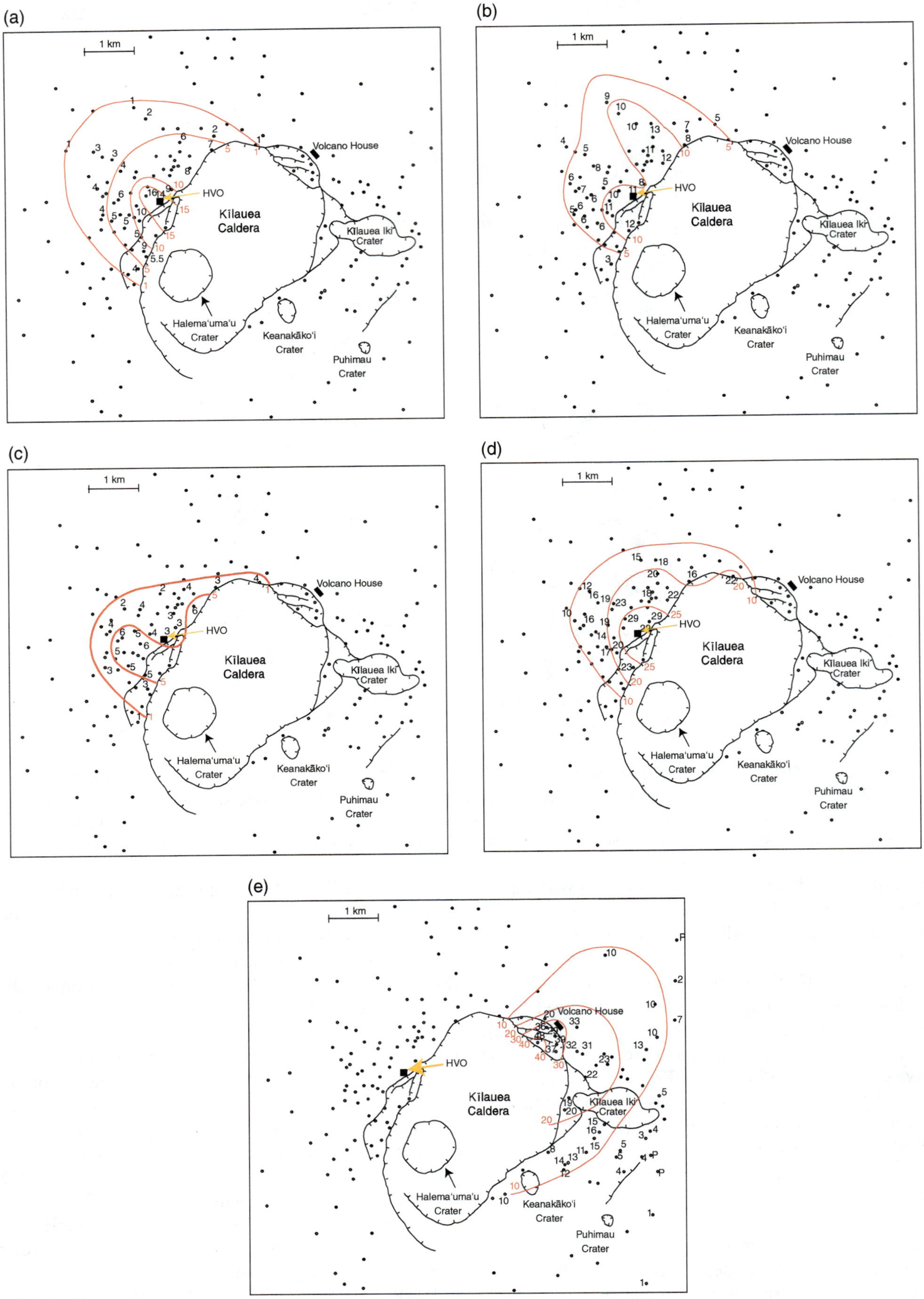

Figure 16.6 Isopach maps of all subunits identified during mapping: (a) subunit NW1; (b) subunit NW2; (c) subunit NW3; (d) subunit NW4/5; (e) subunit E. Isopach lines are in centimeters.

(Figures 16.3 and 16.5). Reticulite and skinned reticulite each form ~30 vol% of NW2 (Table 16.1). Denser clasts, achneliths, and Pele's hair and tears together make up, on average, 30 vol% of the subunit, and ash derived from reticulite comprises 10 vol%. In some localities there is a hint of reverse grading.

3. NW Subunit 3 (NW3) NW3 is rich in denser clasts and achneliths (~30 and 20 vol%, respectively), and skinned reticulite (~30 vol%) is more abundant than reticulite (~20 vol%) (Table 16.1). Both clast varieties of reticulite are smaller in size than in NW1–NW2. The top of the subunit is oxidized at some localities (Figures 16.3 and 16.5). NW3 is typically thin, with variable thickness within an outcrop.

4. NW Subunit 4 (NW4) NW4 is rich in reticulite and skinned reticulite, with sizes comparable to that of NW2 (Figures 16.3 and 16.5). Skinned reticulite is slightly more abundant than reticulite, and achneliths are more abundant than denser clasts, Pele's hair and tears, and reticulite ash.

5. NW Subunit 5 (NW5) NW5 is similar to NW4, except the skinned reticulite component decreases to

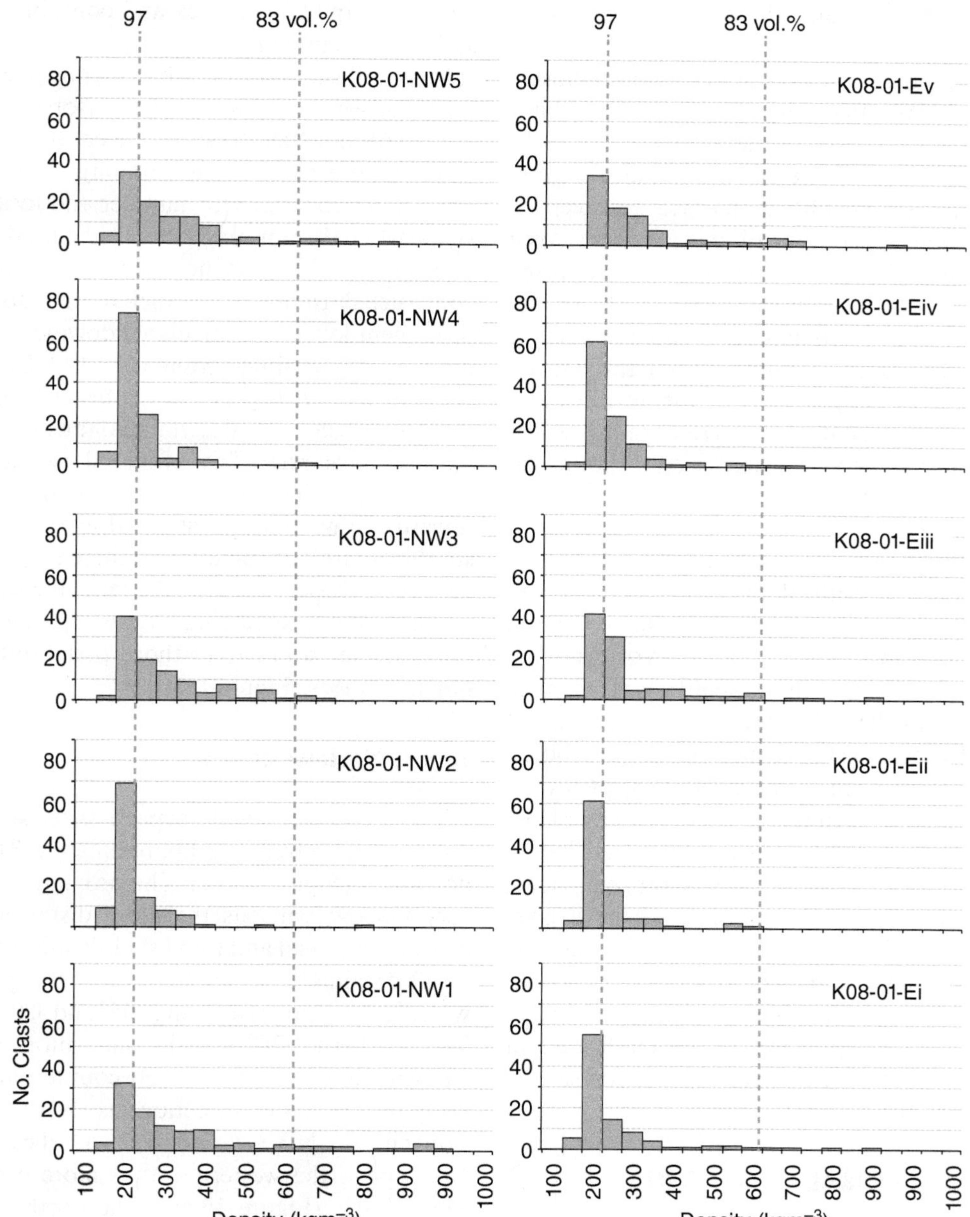

Figure 16.7 Density histograms for subunit samples NW1 to NW5 at location KR08-01. At location KR08-14, samples were collected at five different stratigraphic positions. See Figure 16.5 for more stratigraphic detail.

~20 vol%, and NW5 has a higher ash content than NW4. Denser clasts, achneliths, and Pele's hair and tears have similar vol% to those in NW4. It is often difficult to distinguish between NW4 and NW5, due to similarities in grain size and components, particularly given the limited outcrop available for detailed stratigraphic and component analysis. For this reason they have been grouped for the purposes of isopach mapping.

16.2.2.2. Eastern Sector

We could not distinguish distinct subunits in the field in the eastern sector because the deposits have cryptic stratigraphy as expressed by changes of the juvenile clast density distribution data throughout the deposit thickness.

6. Subunit E The basal portion of the subunit closely resembles that of NW1 from the western and northern sectors. It is rich in Pele's hair, contains Pele's tears, achneliths, and denser clasts (Table 16.1), and has a slightly higher vol% of skinned reticulite than reticulite, except in the layers where density samples Ei and Eiv were taken (Figures 16.5 and 16.7; Table 16.1). The two reticulite components are dominant, whereas denser clasts, achneliths, ash, and Pele's tears and hair are focused in pockets (interstices between larger reticulite clasts) throughout the deposit. In general, the grain size of the reticulite and skinned reticulite components appears finer in this sector compared to that in the northwestern sector, though no measurements were made.

16.2.2.3. Southern Sector

The reticulite occurs south of the caldera and within the Koa'e Fault System (Figure 16.4). It is poorly exposed proximally, however, owing to several meters of overlying Keanakāko'i deposits. Also, the reticulite was, in most places, stripped from the pāhoehoe surface on which it fell by wind and water. Distally, the reticulite is exposed only in protected areas, such as in cracks on the pāhoehoe surface. It has probably been redeposited in many places owing to eolian processes, though this occurred before subsequent Keanakāko'i Tephra deposits were emplaced. Given these unfavorable circumstances, we have been unable to do the detailed stratigraphic analysis that was possible in the northern and eastern sectors. The scattered presence of the reticulite, however, and its clast size (commonly several centimeters in diameter) indicate that the reticulite was once an extensive deposit for several kilometers south of the caldera.

16.3. DISPERSAL OF DEPOSITS

The six subunits of reticulite are not evenly distributed around the whole caldera but instead were deposited in distinct overlapping lobes. The stratigraphic relations between the northwestern and eastern lobes cannot be defined due to their reduced cross-wind thicknesses; in addition, proximal sites in the northern caldera area are lacking. We considered aeolian remobilization of low-density reticulite while mapping and paid particular attention to deposit stratigraphy to ensure measurement of insitu deposits. The most distal location where the subunits can be mapped is approximately 5.7 km north of the caldera, where the deposit is <1 cm thick. No sites were logged in detail in the southern portion of the caldera due to the poor quality of exposure. Possible in situ reticulite was mapped up to 6 km south of Halema'uma'u in the Koa'e Fault System, where the deposit is up to 7 cm thick and contains lapilli as much as 3 cm in diameter.

Maps of subunit and whole-deposit thickness of the reticulite were used to construct isopachs (Figures 16.4 and 16.6), intended to show the contrasting dispersal of subunits and to calculate thinning rates. Isopach maps are used to constrain the number and location of source vents, while the thinning rates can be used to estimate the intensity (or fountain height) at each source vent. We constructed transects parallel to the dispersal axis of prominent lobes and calculated downwind thinning half-distance along single transects ($T_{1/2}$) [*Houghton et al.*, 2004] instead of b_t [Pyle, 1989], because source locations cannot be accurately determined, thus rendering b_t calculations inaccurate. The proximal isopach maps show three main lobes, two toward the northwest and one toward the northeast. Dispersal axes toward the south-southwest are not apparent on these maps for the reasons described above. However, the reticulite was found >5 km south of the caldera rim, suggesting one or more fountains farther south than those that produced the three northern lobes (Figure 16.4).

16.3.1. Northwestern Sector

The NW1 proximal isopach map shows one main dispersal axis trending approximately 315° toward the northwest (Figure 16.6a). The proximal 15 cm isopach is very narrow, whereas the more distal isopachs have a broader dispersal and trend slightly more northward. The NW2 proximal isopach map shows two prominent lobes with distinct sources trending 305° and 320° (Figure 16.6b). The northern, 320° lobe is much more prominent and elongate than its more westerly counterpart, which trends roughly the same direction as NW1. The proximal isopachs of NW3 delineate two lobes, approximately 1 km apart, the western with a more obvious dispersal axis at 305° (Figure 16.6c). The northerly lobe lacks a well-defined dispersal axis, but both lobes appear coincident with those of NW2 (Figure 16.6b). Thickness values for NW4 and NW5 are combined and indicate a dispersal axis trending about 325° (Figs. 16.6d). Dispersal directions indicated by proximal subunit deposit isopachs do not correlate with the polylobate dispersal pattern of the outer whole-deposit isopachs in the northern region (Figures 16.4 and 16.6). The northwest lobe shifts from a unilobate proximal 30 cm isopach to an increasingly wider 20 cm isopach, followed by bilobate shape for 10 and 5 cm isopachs.

16.3.2. Eastern Sector

Proximal isopachs for subunit E outline its dispersal to the northeast and east, with a single well-defined lobe trending approximately 50° (Figure 16.6e). The lobe is quite asymmetric; contours are much farther apart south of the dispersal axis than north. The deposit rapidly thins 3–4 km from the present-day caldera wall, similar to the thinning for the NW subunits.

16.4. INTERPRETATION AND DISCUSSION

16.4.1. Source Vents

The location of the source vents was estimated by closely examining the isopach maps for the whole eruption as well as for each subunit. The proximal isopachs on the total-thickness map shows three main dispersal axes, two to the northwest and one to the northeast (Figure 16.8). The position and trends of the three lobes of reticulite cannot be fit to a single point source. Each of the dispersal axes can be extended into the caldera to determine potential sites for vent locations. We suggest that a minimum of three vents erupted the reticulite in this sector of the volcano. Two vents were likely located near the western wall of the caldera, one southeast of HVO and the other northwest of HVO (Figure 16.8). These two vents, however, cannot be responsible for the dispersal toward the northeast, even with varying wind directions; therefore, we propose that a third active vent was located along the northeastern rim of the caldera, southwest of the Volcano House hotel (Figure 16.8). The data south and southeast of the caldera show thickness accumulations trending back to two broad locations east of Halema'uma'u, but we are not able to define vent locations adequately. The presence of a further southeasterly vent is also suggested by the asymmetry of the southern limb of the subunit E isopachs, although this pattern could also be explained by tradewind conditions dispersing tephra crosswind to the southeast. The presence and locations of several fountaining vents are best explained by eruption from a ring fracture related to caldera collapse, not to a linear fissure as are most Hawaiian fountains. The timing of the eruption, determined by stratigraphic study to have been just after or possibly during the closing stage of caldera collapse, is consistent with our hypothesis that several vents along an arcuate fissure, i.e., a ring fracture, supplied the eruption products.

16.4.2. Eruption Parameters

16.4.2.1. Eruption Intensity

Qualitative estimates of eruption intensity can be determined from the geometry of fall deposits by plotting thickness versus the square root of the area enclosed by each isopach used in the calculation [e.g., *Fierstein and Nathenson*, 1992] (Figure 16.9). The complicated dispersal of the reticulite subunit geometry limits the degree to which these plots are useful for the interpretation of eruption intensity. The whole deposit data show an exponential thinning trend with three segments (KR_WD in Figure 16.9a), whereas individual subunit trends in the northwestern sector show proximal exponential behavior that steepens abruptly for distal data (except for NW2 and NW3, each of which has only two points) (Figure 16.9b). Two possibilities could be driving this dispersal behavior: aeolian erosion and inclined fountains. These two possibilities would reduce the measured thicknesses in distal areas, leading to calculations of volume and mass that are lower than actual erupted values.

Aeolian removal of reticulite soon after deposition is a reasonable scenario due to its low density, especially in the northwestern area, which was probably vegetated with grass and shrubs but not forest. Such wind erosion probably caused even more dramatic erosion south of the caldera, an area virtually devoid of vegetation and subject today to stronger winds than any other part of Kīlauea's summit.

The other possibility for the abrupt drop-off in exponential thinning is that inclined fountains directed tephra outward from steeply dipping ring fractures within the caldera. Brief periods of inclined fountaining have been observed at Kīlauea during episodes 3 and 8 of the 1959 Kīlauea Iki eruption [*Richter et al.*, 1970] and episode 11 of the 1969 Mauna Ulu eruption [*Swanson et al.*, 1979]. During Mauna Ulu episode 11, vent blockage and inclined fountaining dispersed hot pumice 2.5 km upwind from the vent, igniting forest fires. Ring fractures dipping 70°–85° toward the center of the caldera are plausible structures related to caldera collapse and may be structurally responsible for inclined fountains. A significant proportion of the fallout from such inclined fountains would have been trapped within the caldera bounded by 500–600 m high walls, and outside the caldera proximal deposits would have accumulated far faster than distal deposits.

As a separate method to infer eruption intensity, we draw transects for each dispersal axis identified on isopach

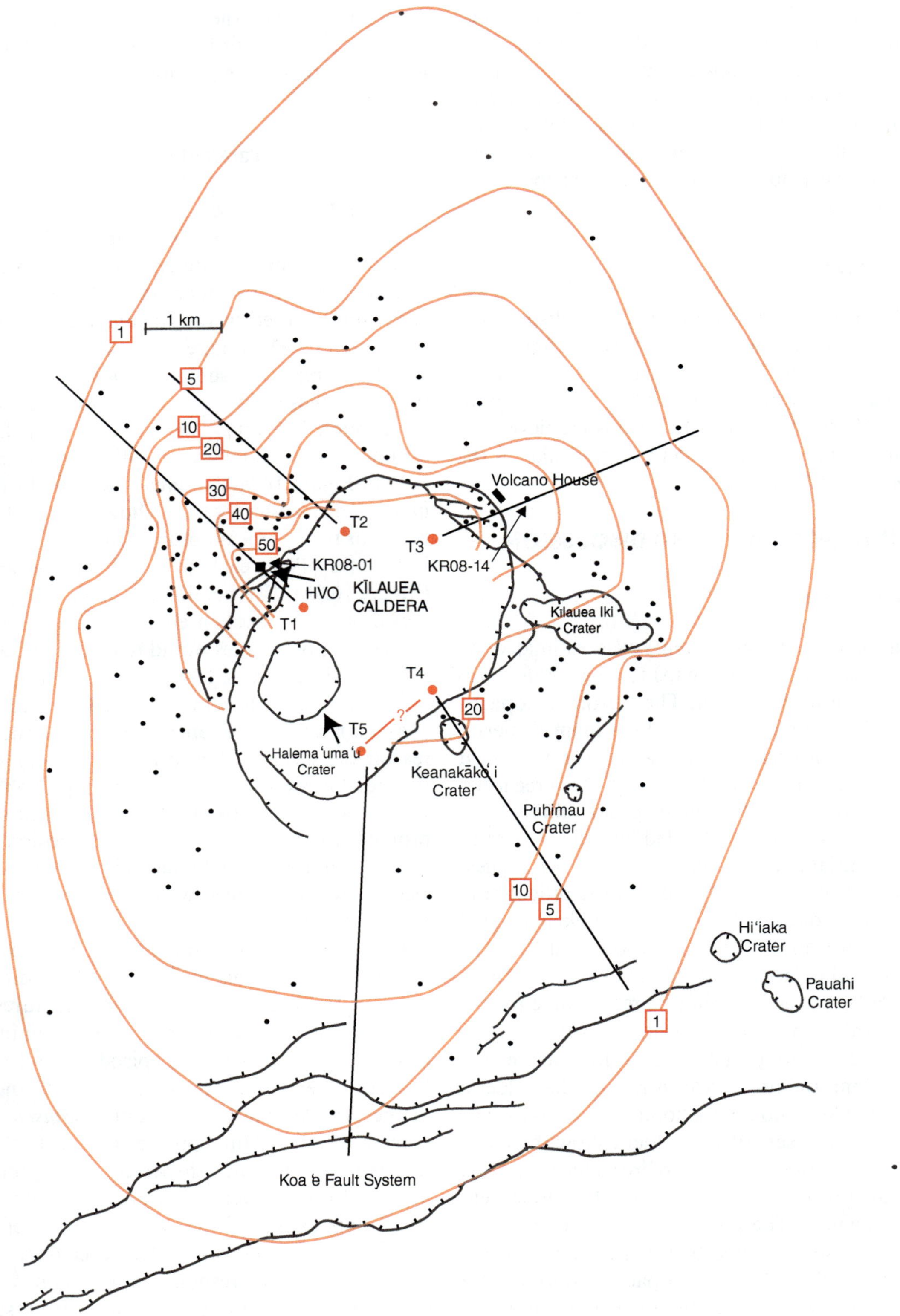

Figure 16.8 Inferred locations of vents (red dots) and locations of traverses identified from isopach maps that were used to define $T_{1/2}$ thinning rates (T1 to T5 in black) are illustrated. Proximal whole-deposit isopachs are listed in centimeters.

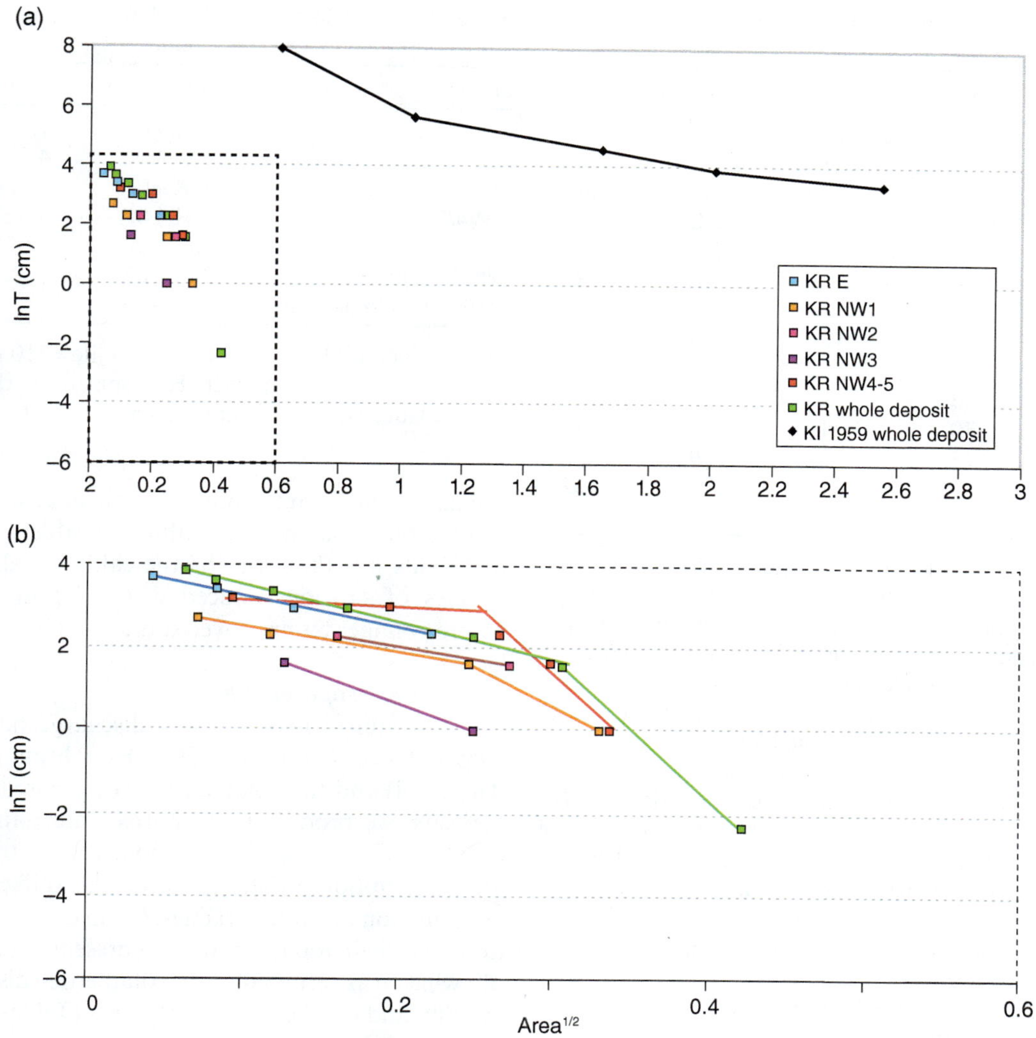

Figure 16.9 (a) Ln thickness versus area$^{1/2}$ for the whole reticulite deposit, its subunits, and whole deposit of the Kīlauea Iki 1959 eruption: KR, Kīlauea reticulite; KI, Kīlauea Iki. Dashed box: area illustrated in Figure 16.9b. (b) Ln thickness versus area$^{1/2}$ for reticulite subunits at increased proximal resolution.

maps and along two other directions south of the caldera and calculate thinning half-distances, or $T_{1/2}$ (Figure 16.8; Table 16.2). We calculate $T_{1/2}$ for each subunit to estimate the relative intensity and fountain heights from each vent [after *Houghton et al.*, 2004]. We use this parameter rather than thickness half-distance from the vent (b_t, *Pyle* 1989) because the vent locations are unknown and there is the potential that multiple vents contribute to the thickness of individual subunits. Analogous to b_t, a small $T_{1/2}$ value indicates local dispersal, while a large value reflects wider dispersal and, by inference, greater fountain height if the fountain were vertical.

These data show that subunits NW2 and NW4/5 have the greatest values of $T_{1/2}$ and hence are more widely dispersed and likely the products of more intense or steeper fountaining than subunits NW1, NW3, and E, with lower $T_{1/2}$'s. Thinning rates in the southeastern sector of the caldera have comparable $T_{1/2}$ values to those of the NW2 and NW4/5 subunits. The calculated thinning half-distances for the reticulite subunits range from 0.2 to 4 km (Table 16.2). For comparison, fountains during episodes 15 and 16 of the 1959 Kīlauea Iki eruption reached maximum heights of 580 and 460 m, respectively, and the thinning half-distance for these combined deposits is 0.22 km (B. F. Houghton, unpublished data 2014). This comparison, coupled with vent locations in the deep caldera, is evidence for very high fountaining and (or) directed fountaining.

Independent of thinning behavior, mass and volumetric discharge rates can be approximately correlated with fountain height when using data from previously observed Kīlauea fountains. For example, lava volumetric discharge rates in excess of 400 m^3/s were responsible for 580 m high fountains during the Kīlauea Iki 1959 eruption [*Richter et al.*, 1970]. Our data suggest that the reticulite fountain heights were in significantly in excess of 600 m and therefore we're probably erupted at volumetric discharge rates

Table 16.2 Transect thinning half-distances for different subunits of the reticulite deposit.

Transect and Subunit	Proximatl *T* (cm)	Distal *T* (cm)	Distance (km)	$T_{1/2}$ (km)
T1 (NW1)	14	1	2.2	**0.6**
T1 (NW2)	11	4	2.3	**1.4**
T2 (NW1)	8	1	2.8	**0.9**
T2 (NW2)	12	9	2	**4.0**
T1 (NW3)	4	1	1.6	**0.8**
T2 (NW3)	6	1	0.5	**0.2**
T1 (NW4/5)	29	12	1.9	**1.5**
T2 (NW4/5)	22	15	1	**1.4**
T2 (E1)	48	2	3	**0.7**
T4 (?SE)	30	1	4.3	**1.1**
T5 (S)	15	2	4.9	**1.7**
KI1959 Ep 15 & 16				**0.22**

Note:Transect locations identified in Figure 16.8. Proximal *T* and distal *T* refer to the two locations used for the thinning calculation, and the distance between these two locations is shown in the adjacent column. For each subunit, thinning half-distances have been calculated on traverses where subunit thicknesses are greatest. Thinning half-distances on two traverses have therefore been calculated for the northwestern subunits, whereas only one thinning half-distance has been calculated for eastern, southeastern, and southern deposits. For comparison, the $T_{1/2}$ value for combined episodes 15 and 16 of the 1959 Kīlauea Iki eruption is shown. Example calculations:
T2 (NW1): The deposit thins 3 times over a distance of 2.8 km, so $T_{1/2}$ = 2.8/3 = 0.93, rounded to 0.9.
T1 (NW3): The deposit thins twice over 1.6 km, so $T_{1/2}$ = 1.6/2 = 0.8 km.
T1 (NW1): The deposit thins ~3.8 times in 2.2 km, so $T_{1/2}$ = 2.2/3.8 = 0.58 km, rounded to 0.6 km.

of ~ 500 m³/s, which equates to mass discharge rates of ~106 kg/s, assuming a lava density of 2950 kg/m³.

16.4.2.2. Eruptive Volumes

The whole-deposit thickness data (Figure 16.4) can approximate the minimum total volume for the reticulite in the northern sector of the volcano, and isopachs of subunits (Figure 16.6) allow their volumes to be calculated [e.g., *Bonadonna and Houghton*, 2005]. A thickness-area$^{1/2}$ plot was used to calculate each volume. One complication is that the proximal (intracaldera) and distal data are missing, and the locations of the source vents are only interpretative though fairly well constrained (Figure 16.8). The results are given in Table 16.3. Our best estimate for the minimum volume of this eruption is 2×10^7 m³ (approximate mass of 4.2×10^9 kg). For comparison, the calculated volume of the 1959 Kīlauea Iki eruption is 1.8×10^6 m³, as determined from the isopach map of *Richter et al.* [1970] (Table 16.3). The volumes

Table 16.3 Calculated volume and mass for the reticulite whole deposit and for each subunit.

Subunit	Volume (m³)	Mass (kg)[a]	VEI
NW1	8.1E+05	1.7E+08	1
NW2	1.9E+06	4.0E+08	2
NW3	4.9E+05	1.0E+08	1
NW4/5	1.1E+07	2.3E+09	3
E	2.8E+06	5.9E+08	2
Whole deposit	2.0E+07	4.2E+09	3
KI1959 whole deposit	1.8E+06		3

[a] Average deposit bulk density for reticulite = 210 kgm^{-3}
Note: VEIs are shown for each. For comparison, data from the 1959 Kīlauea Iki eruption are shown.

calculated for the reticulite are minima, as it is likely that much tephra was trapped within the caldera (i.e., ejected to heights less than that of the caldera wall), and high degrees of erosion reduced distal deposit thicknesses before younger tephra covered the reticulite.

16.4.2.3. Eruption Duration

Using equivalent volumetric discharge rates from the 1959 Kīlauea Iki eruption (B. F. Houghton, unpublished data 2014) and the volume calculated from the reticulite deposits, we produced an approximate estimate of the duration of the reticulite eruption. To constrain an envelope of eruption duration, we calculated likely durations as a function of three variables: (1) an estimation of eruption volume if today's deposit represents only a fraction of what was erupted, (2) volumetric discharge rate (VDR), and (3) the number of vents (Table 16.4). Given our mapping data, measurements, and calculations, three to five vents were responsible for the deposit, and these vents were likely erupting at a VDR > 400 m³/s (and more probably 500 m³/s) to disperse tephra over the top of the ~600 m high caldera wall. The minimum duration determined from present-day deposits is based upon a VDR of 500 m³/s at a single vent and equates to an approximate duration of 11 h, given the estimated volume of 0.02 km³; an eruption rate of 500 m³/s for four vents would yield a duration of 3 h. We suspect that our calculated eruption volumes are minima, and consequently our best estimations of eruption durations are one to two days. These calculations are not well constrained, but nonetheless they are helpful in considering the minimum duration of fountaining during this eruption.

16.4.3. Eruption Setting

The deposits of this eruption are mostly reticulite with the exception of NW1 and NW3 and two parts of the E stratigraphy (Ei and Eiii). As presented by *Swanson et al.* [2012], the fountains likely took place in a deep caldera,

Table 16.4 Calculations for estimated envelope of eruption duration

	Volume (m^3)	Mass (kg)	VDR (m^3/s)	Seconds	Hours	Days	Two Vents (days)	Four Vents (days)	Eight Vents (days)
Measured	2.0E+07	4.2E+09	100	200000	55.6	2.3	1.2	0.6	0.3
			400	50000	13.9	0.6	0.3	0.1	0.1
			500	40000	11.1	0.5	0.2	0.1	0.1
50%[a]	4.0E+07	8.4E+09	100	400000	111.1	4.6	2.3	1.2	0.6
			400	100000	27.8	1.1	0.6	0.3	0.1
			500	80000	22.2	0.9	0.5	0.2	0.1
10%[b]	2.0E+08	4.2E+10	100	2000000	555.6	23.2	11.6	5.8	2.9
			400	500000	138.9	5.8	2.9	1.4	0.7
			500	400000	111.1	4.6	2.3	1.2	0.6

[a] If we assume that the volume of reticulite deposited around the summit of Kīlauea is ~50% of the total erupted volume
[b] If we assume that the volume of reticulite deposited around the summit of Kīlauea is ~10% of the total erupted volume
Note: Duration was calculated as a function of three variables: (1) an estimation of eruption volume if today's deposit represents only a fraction (10% or 50%) of what was erupted; (2) volumetric discharge rate (VDR; 100–500 m^3/s); and (3) the number of vents (2–8).

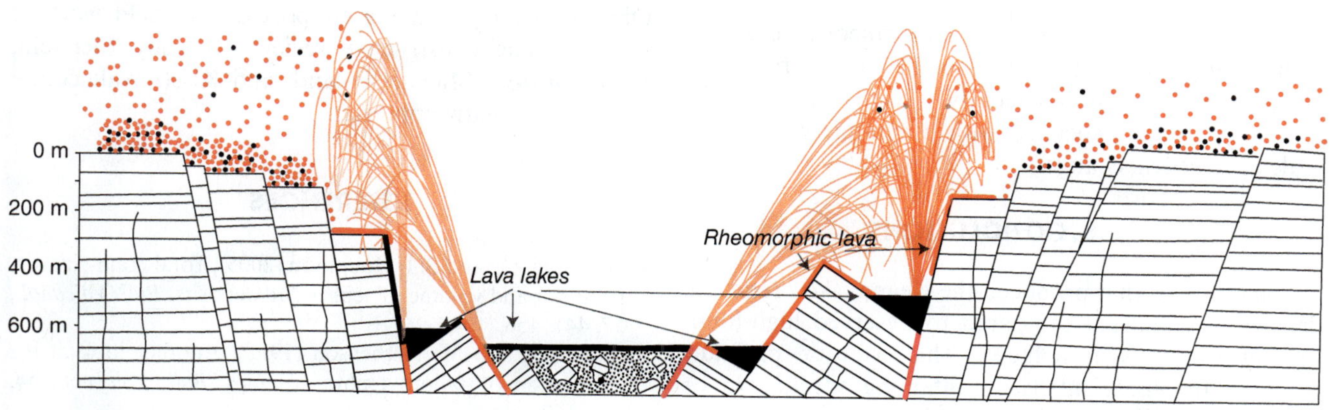

Figure 16.10 Conceptual model to show a scenario in which inclined and/or vertical fountains may have been active from some, or all, vents during the eruption. Both styles of fountaining would have plastered reticulite on the caldera walls and potentially fed rheomorphic lava flowing back into the caldera. Multiple vents at different locations within the caldera were present, and lava ponds may have developed during early phases of the eruption. Fountaining through such lava lakes would result in higher production of denser juvenile clasts and achnelith components. Red dots are reticulite clasts; black dots are denser clasts.

so that only small amounts of the denser components could rise above the rim and thus be transported beyond the caldera. This can explain the relative purity of the reticulite deposits. The production of a denser component could be related to quenching on the margin of an inclined fountain or to eruption through a fountain-fed pond or rheomorphic lava around the source vents. The morphology and vesicularity of the denser lapilli-sized components appear similar to those from the 1959 eruption of Kīlauea Iki; however, limited proximal exposure of the Keanakāko'i reticulite limits the interpretation of near-vent eruption dynamics and deposit geometry during this event.

16.4.4. Timing of Keanakāko'i Reticulite in Context of Caldera Formation

Figure 16.10 presents a conceptual model of the fountaining within the caldera. For the caldera to collapse, there must have been a void in the summit reservoir system that developed over a timescale undeterminable from current information [*Swanson et al.*, 2012]. The caldera collapsed in piecemeal fashion, presumably along a series of ring fractures. The inferred vent locations from our deposit mapping are near the locations of step faults and benches along the circumference of the caldera (Figure 16.8). The reticulite was erupted just after or during the latest stage of the collapse of the caldera [*Swanson et al.*, 2012]. Estimates for the initial minimal depth are at least 600 m deep based on historical records of the depth in the 1820s [*Mastin*, 1997] and on oral tradition that the collapse might have intersected the water table, today at a depth of ~615 m below the high point on the caldera rim [*Swanson*, 2010].

The lines of evidence that point to caldera formation as a principal factor driving the eruption are the following:

(i) All of the available data point to an unprecedented high intensity of fountaining. Fountain heights without caldera collapse observed at Kīlauea are limited to less than 600 m (Kīlauea Iki in 1959 [*Richter et al.*, 1970]). Two well-documented caldera or pit-crater collapses of basaltic volcanoes (Miyakejima, Japan, and Piton de la Fournaise, Réunion Island) have each shown a transition to higher magma mass discharge rates as a response to caldera formation [*Geshi et al.*, 2002; *Michon, et al.*, 2009]. In each of these cases, collapse occurred over hours.

(ii) Vigorous (and perhaps simultaneous) eruption from multiple vents, probably on the caldera ring-fault system. For such unusual activity, it seems likely that the fountaining was associated with a summit wide large-scale geometrical change.

16.5. CONCLUSIONS

The eruption that produced the Keanakākoʻi reticulite deposit was unusual compared to observed high-fountaining eruptions at Kīlauea. This eruption probably included the most intense fountaining yet known or inferred from Kīlauea. The eruption produced at least three fountains with heights likely greater than 600 m, which were probably distributed along the northern ring fractures of the summit caldera. Although our data cannot pinpoint other fountaining locations in the south, we infer that additional vents were likely active in those regions. The volume of the erupted material that we can constrain with our mapping data is an order of magnitude larger than that produced by historical fountains at Kīlauea, and this volume is likely a minimum, as we can constrain neither the proximal deposit in the south nor the volume of denser tephra that did not escape the caldera.

The multilobate northerly, northwesterly, and easterly dispersals of proximal subunit deposits are not entirely consistent with the present-day wind patterns of Kīlauea, even during periods of southerly "Kona winds" as proposed by *Stone* [1926]. The observation that the lobes are approximately normal to the caldera margin is consistent with inclined fountains from one or more steeply dipping ring fractures. The broadening of the 320° proximal isopach and transition to a bilobate distal pattern of the outer whole-deposit isopachs is an indication of advected fallout depositing north of the caldera during southerly winds and contemporaneous with inclined fountaining. The calculated minimum volume of the deposit and the estimated discharge rates provide a likely envelope of eruption duration, probably one to two days, but this estimate does not take into account potential pauses in the eruption. We suggest that the collapse and formation of the caldera along ring faults both provided pathways for the magma to ascend synchronously to the surface at multiple locations. This large-scale event influenced the shallow storage region such that magma was erupted at very high discharge rates.

ACKNOWLEDGMENTS

This research was supported by NSF grants EAR-1049662 and EAR-0810332 and USGS grant SV-ARRA-0004. We thank the staff of the Hawaiian Volcano Observatory for ongoing support during field work at Kīlauea. The constructive reviews of Judy Fierstein, Lucia Gurioli, Michael Poland, and Kristi Wallace are valued and greatly appreciated.

REFERENCES

Bonadonna, C., and B. F Houghton (2005), Total grain-size distribution and volume of tephra-fall deposits, *Bull. Volcanol.*, *67*, 441–456.

Fierstein, J., and M. Nathenson (1992), Another look at the calculation of fallout tephra volumes, *Bull. Volcanol.*, *54*, 156–167.

Geshi, N., T. Shimano, T. Chiba, and S. Nakada (2002), Caldera collapse during the 2000 eruption of Miyakejima Volcano, Japan, *Bull. Volcanol.*, *64*, 55–68.

Houghton, B. F., and C. J. N Wilson (1989), A vesicularity index for pyroclastic deposits, *Bull. Volcanol.*, *51*, 451–462.

Houghton, B. F., C. J. N Wilson, J. Fierstein, and W. Hildreth (2004), Complex proximal deposition during the Plinian eruptions of 1912 at Novarupta, Alaska, *Bull. Volcanol.*, *66*, 95–133.

Mangan, M. T. and K. V. Cashman (1996), The structure of basaltic scoria and reticulite and inferences for vesiculation, foam formation, and fragmentation in lava fountains, *J. Volcanol. Geotherm. Res.*, *73*, 1–18.

Mastin, L. G. (1997), Evidence for water influx from a caldera lake during the explosive eruption of 1790, Kīlauea Volcano, Hawaiʻi, *J. Geophys. Res.*, *102*, 20,093–20,109.

McPhie, J., G. P. L. Walker, and R. L. Christiansen (1990), Phreatomagmatic and phreatic fall and surge deposits from explosions at Kilauea volcano, Hawaii, 1790 A.D.: Keanakakoi Ash Member, *Bull. Volcanol.*, *52*, 334–354.

Michon, L., N. Villeneuve, T. Catry, and O. Merle (2009), How summit caldera collapse on basaltic volcanoes: New insights from the April 2007 caldera collapse of Piton de la Fournaise volcano, *J. Volcanol. Geotherm. Res.*, *184*, 138–151.

Parcheta, C. E., B. F. Houghton, and D. A. Swanson (2013), Contrasting patterns of vesiculation in low, intermediate, and high Hawaiian fountains: A case study of the 1969 Mauna Ulu eruption, *J. Volcanol. Geotherm. Res.*, *255*, 79–89.

Pyle, D. M (1989), The thickness, volume and grain-size of tephra fall deposits, *Bull. Volcanol.*, *51*, 1–15.

Richter D. H., J. P. Eaton, K. J. Murata, W. U. Ault, and H. L Krivoy (1970), Chronological narrative of the 1959–60 eruption of Kilauea volcano, Hawai'i. in *The 1959–60 Eruption of Kilauea Volcano, Hawai'i*, U.S. Geol. Surv. Prof. Pap. 537-E: E1-E73.

Stone, J. B. (1926), The products and structure of Kīlauea: Bernice P, *Bishop Museum Bulletin*, *33*, p.59, Honolulu.

Stovall, W. K., B. F. Houghton, H. Gonnermann, S. A. Fagents, and D. A. Swanson (2011), Eruption dynamics of Hawaiian-style fountains: The case study of episode 1 of the Kīlauea Iki 1959 eruption, *Bull. Volcanol.*, *73*, 511–529.

Stovall, W. K., B. F. Houghton, J. E. Hammer, S. A. Fagents, and D. A. Swanson (2012), Vesiculation of high fountaining Hawaiian eruptions: Episodes 15 and 16 of 1959 Kīlauea Iki, *Bull. Volcanol.*, *74*, 441–455.

Swanson, D. A. (2008), Hawaiian oral tradition describes 400 years of volcanic activity at Kīlauea, *J. Volcanol. Geotherm. Res.*, *176*, 427–431.

Swanson, D. A. (2010), Hawaiian oral tradition clarifies 400 years of volcanic activity at Kīlauea: Hūlili, *Multidisciplinary Research on Hawaiian Well-Being*, *6*, 25–33.

Swanson, D. A., W. A. Duffield, D. B. Jackson, and D. W Peterson (1979), Chronological narrative of the 1969-71 Mauna Ulu eruption of Kīlauea Volcano, Hawai'i, U.S. *Geol. Surv. Prof. Pap. 1056*, 1–55.

Swanson, D. A., T. R. Rose, R. S. Fiske, and J. P. McGeehin (2012), Keanakāko'i Tephra produced by 300 years of explosive eruptions following collapse of Kīlauea's caldera in about 1500 CE, *J. Volcanol. Geotherm. Res.*, *215–216*, 8–25.

Walker, G.P.L. and R. Croasdale (1972), Characteristics of some basaltic pyroclastics, *Bull. Volcanol.*, 35, 303–317, doi: 10.1007/BF02596957.

Wolfe, E. W., M. O. Garcia, D. B. Jackson, R. Y. Koyanagi, C. A. Neal, and A. T. Okamura (1987), The Pu'u 'Ō'ō eruption of Kīlauea volcano, episodes 1–20, January 3, 1983, to June 8, 1984, U.S. *Geol. Surv. Prof. Pap. 1350*, 471–508.

Wolfe, E. W., C. A. Neal, N. G. Banks, and T. J. Duggan (1988), Geological observations and chronology of eruptive events, U.S. *Geol. Surv. Prof. Pap. 1463*, 1–97.

17

Hawaiian Fissure Fountains: Quantifying Vent and Shallow Conduit Geometry, Episode 1 of the 1969–1974 Mauna Ulu Eruption

Carolyn Parcheta[1,2], Sarah Fagents[3], Donald A. Swanson[4], Bruce F. Houghton[1], and Todd Ericksen[3]

ABSTRACT

Geometries of shallow magmatic pathways feeding volcanic eruptions are poorly constrained, yet many key interpretations about eruption dynamics depend on knowledge of these geometries. Direct quantification is difficult because vents typically become blocked with lava at the end of eruptions. Indirect geophysical techniques have shed light on some volcanic conduit geometries, but the scales are too coarse to resolve narrow fissures (widths typically 1 m). Kīlauea's Mauna Ulu eruption, which started with <50 m high Hawaiian fountains along a 4.5 km fissure on 24 May 1969, provides a unique opportunity to measure the detailed geometry of a shallow magmatic pathway, as the western vents remain unobstructed to depths >30 m. Direct measurements at the ground surface were augmented by tripod-mounted lidar measurements to quantify the shallow conduit geometry for three vents at a resolution <4 cm. We define the form of the fissure in terms of aspect ratio, flaring ratio, irregularity, sinuosity, and segmentation and discuss the factors influencing these parameters. In the past, simplified first-order fissure geometries have been used in computational modeling. Our data can provide more accurate conduit shapes for better understanding of shallow fissure fluid dynamics and how it controls eruptive behavior, especially if incorporated into computer models.

17.1. INTRODUCTION

Our understanding of basaltic fissure eruptions, the most common type of terrestrial eruptions [*Sigurdsson*, 2000], is poorly constrained due to a lack of empirical and observational data (vent and conduit geometry, temporal and spatial variations in mass and gas flux, etc.). The typically agglutinated nature of Hawaiian pyroclastic fall deposits and the common burial of vents make quantification and constraint of eruption dynamics a challenging task. This knowledge is of the upmost importance, though, as it significantly impacts calculations of mass flux and our understanding of eruption fluid dynamics, including magma fragmentation and syn-eruptive nucleation and growth of crystals and bubbles.

The preserved fissure vents of 24 May 1969, which initiated the five-year-long Mauna Ulu eruption of Kīlauea, provide an excellent opportunity to study fissure system geometries and derive input geometry parameters for models of basaltic fountain dynamics. The westernmost 880 m of the original 4.5 km fissure system remains accessible today to depths that are, in places, greater than 30 m. Lava drained back into the deeper system, leaving an open cavity that allows quantification of the geometry of much of the vent and shallow conduit system. Here we use detailed field measurements, including light detection and ranging (lidar), to precisely characterize and quantify the fissure

[1] *Department of Geology and Geophysics, University of Hawai'i at Mānoa, Honolulu, Hawaii, USA*

[2] *Jet Propulsion Laboratory, Pasadena, California, USA*

[3] *Hawai'i Institute of Geophysics and Planetology, University of Hawai'i at Mānoa, Honolulu, Hawaii, USA*

[4] *Hawaiian Volcano Observatory, U.S. Geological Survey Hawai'i National Park, Hawaii, USA*

Hawaiian Volcanoes: From Source to Surface, Geophysical Monograph 208. First Edition.
Edited by Rebecca Carey, Valérie Cayol, Michael Poland, and Dominique Weis.

Companion Website: www.wiley.com/go/Carey/Hawaiian_Volcanoes

and vent geometries using the parameters of aspect ratio, flaring ratio, irregularity, and sinuosity.

17.2. BACKGROUND

17.2.1. Propagation of Dikes and Fractures

When magmatic dikes propagate to the surface, they move upward either by intruding a self-created and self-propagating crack or by exploiting preexisting fractures and joints oriented in a direction preferential to propagation [*Delaney et al.*, 1986; *Valentine and Krogh*, 2006]. Both dike propagation direction and opening behavior react to the local and regional stress fields. A single fracture at depth may splay into shallow en echelon segments if the principal stresses rotate around the propagation axis [*Sommer*, 1969], with the number of segments proportional to the stress rotation angle [*Pollard et al.*, 1982]. Lengths of en echelon segments are always smaller than the parent fracture [*Pollard et al.*, 1982]. Fractures dilate in the direction of least compressive stress (i.e., greatest tensile stress σ_t) and propagate in the direction of greatest compressive stress (σ_c). Therefore, if the principal stresses rotate toward the ground surface, the dike will realign to the new stress field to conserve energy and continue propagation [*Pollard et al.*, 1982]. A dike may exploit existing local cracks and fractures in the country rock. A cross-sectional dike shape, akin to an inverted teardrop, is obtained when the dike driving pressure is constant while the regional pressure is gradational along the dike length [*Pollard*, 1987]. Jagged, subvertical scarps along the fracture faces are relicts of cooling joints, tumuli clefts, monclinal edges, or sutures in older lava flows forming the conduit walls [*Pollard et al.*, 1982, *Walker*, 1991; *Self et al.*, 1996]. Joint growth in the older lava flows likely occurred long ago as those flows were emplaced and cooled, with cracks nucleating and propagating until reaching the brittle-to-ductile transition in the lava crust [*DeGraff and Aydin*, 1987].

17.2.2. Vent Geometry and Eruption Dynamics

Theoretical work in the early 1980s [*Wilson*, 1980; *Wilson and Head*, 1981] provided a solid first-order foundation for understanding the dynamics of Hawaiian-style eruptions. Results of these models cover the range of fountain behavior given various factors, including gravity, atmospheric density, magma water content, vent geometry, ascent velocity, fountain spread angle, and fountain height suppression when erupted through ponded lava. Minimum vent widths of 0.2–0.6 m were derived for basaltic fissures erupting 30–500 kg/s per meter of fissure from conduit lengths of 0.5–20 km [*Wilson and Head*, 1981]. *Giberti and Wilson* [1990] developed this model further to allow for a prescribed vent geometry, analyzing the effects of geometry on dike flow to the surface from 1000 m depth. In the previous models, which required pressure within the ascending magma to equal lithostatic pressure, the vent always flares near the surface to accommodate expansion of the gas phase as atmospheric pressure is approached. *Giberti and Wilson* [1990] constrained the fissure to be (1) straight walled, (2) convergent toward the surface, or (3) of a variable width (lenticular shape). In all cases, and for steady state eruption dynamics, exit pressures and mass discharge are dependent on magmatic source driving conditions (e.g., mass flux, gas content, viscosity). Effective exit velocities in their simulations, however, are broadly similar to the lithostatic case because, after the overpressured pyroclastic mixture exits the vent, it quickly expands and decompresses to ambient atmospheric pressure. The choice of fissure geometry (and therefore lithostatic versus nonlithostatic pressure condition in the ascending magma-gas mixture) is not likely to produce significant or observable differences in fountain structure. *Giberti and Wilson* [1990] conclude that fissure widths less than 1 m wide are most likely to reflect nonlithostatic pressure gradients in the ascending mixture.

Mitchell [2005] further investigated shallow conduit shape and how it changes through time during an eruption. In two scenarios, surface-choked conditions (i.e., exit velocity less than or equal to Mach 1) and balanced lithostatic pressure conditions, vent widening through time allowed the choking point of the flow to descend lower into the conduit. For basaltic fissures, *Mitchell* [2005] found that wall stresses generally are not large enough to disrupt the conduit and erode walls, even at the surface where they are greatest. He inferred that original dike shape is likely preserved in the early part of the eruption or for eruptions of short duration. Eruptions of longer durations can evolve toward complex vent shapes via mechanical erosion on a time scale of hours.

17.2.3. Previous Documentation of Fissures

In Kīlauea's recent history (1950 to present), 37 fissures have erupted on the rift zones and within the summit caldera (Table 17.1). Documentation of these eruptive episodes is constantly improving, with the 2011 Kamoamoa eruption constituting the most detailed documentation of a Hawaiian fissure fountain to date (Chapter 18, this volume). Measurements of vent geometry, however, are unconstrained from all documentation. There may be several reasons for this, including: (a) many fissures may be buried by contemporaneous or subsequent lava flows (including drain-back) or may become clogged with talus or pyroclastic debris; (b) fissures

Table 17.1 Summary of eruptive fissures at Kīlauea Volcano since 1950.

Eruption Location	Duration (h)	Fissure Opening	Max Fountain Height (m)	Final Fissure Geometry	Initial Total Vent Length (m)	Mean Vent Width (m)	Vent Variability: Sinuosity	Vent Variability Irregularity	Vent Variability: Roughness	Total Erupted Volume (m³)	Ensuring Activity (Duration)	Source
1952, sc	3264 (136 d)	? (ne, sw), ?	240	C	700	—	—	—	—	4.90E+07	Const H&E (5 mo)	1
1954 sc	72	u (ne), ?	200	O	1150	—	—	—	—	6.50E+06	1954 sc	1, 2
1955, sc	144 (6 d)	b (ne, sw), sp	125	L	~6000	—	—	—	—	9.20E+07	Intermit H&E (3 mo)	1, 3
1959, sc *1	167 (7 d), sc *1	b (e, w), sp	360	C	800	—	—	—	—	5.00E+07	15 H (1 mo)	4
1960, erz *^18	24	u (e), sp	100	C	1100	—	—	—	—	1.13E+08	Const focused H&E (1 mo)	4
1961, sc	24	—(ne, n)—	30	C	~400	—	—	—	—	2.45E+05	N	5
1961, sc	528 (22 d)	? (?), ?	30	C	<50	—	—	—	—	2.68E+05	N	5
1961, sc	167 (7d)	—(ne),—	200	C	~450	—	—	—	—	1.32E+07	N	5
1961, erz	72	—	~120	L	700	—	—	—	—	1.01E+07	N	5
1962, erz	48	b (ne, sw), rp	85	L	~3300	—	—	—	—	3.26E+05	N	6
1963, erz	48	b? (ne?, sw?), ?	50?	C	~400?	—	—	—	—	4.00E+04	N	7
1963, erz	24	? (ne, sw),?	45	L	~6400	—	—	—	—	6.90E+07	N	8
1965, erz	240 (10 d)	b (ne, sw) rp	< 100	C	8500?	—	—	—	—	1.80E+07	N	9
1965, erz	24	—(ne, sw)—	—	L	~2400	13	—	—	—	9.00E+05	N	10
1967, sc	6024 (251 d)	? (n, s), ?	<50?	C	<1000?	—	—	—	—	8.40E+07	30 H (5 mo)	11
1968, erz	120 (5 d)	u (ne), sp	20	L	20000	—	—	—	—	9.90E+04	N	12
1968, erz	360 (15 d)	u (ne), sp	100	L	6500	—	—	—	—	7.00E+06	N	12
1969, erz	144 (6 d)	b (ne, sw), sp	100	L	10750	1	—	—	—	1.70E+07	N	13
1969, **1	18	b (e, w), sp	50	fL	5000	1.1	This chapter	This chapter	This chapter	2.25E+06	11 H (6 mo), E (54 mo)	This study, 14
1971, sc	10	u (w), ?	75	L	2000	1	—	—	—	9.60E+06	N	15
1971, sc	121 (5 d)	u (w), ?	50	L	12000	2	—	—	—	8.00E+06	N	15
1972, erz **^16	<24?	—	—	—	—	—	—	—	—	8.00E+03	2 H&E (1 mo)	14
1972, erz **^17	144 (6 d)	—	20	L	200	—	—	—	—	1.25E+08	E (36 mo)	14
1973, erz **^18	24	u (w), sp	50	L	1800	—	—	—	—	1.00E+06	E (6 mo)	16.17
1973, erz **20	720 (30 d)	u (e), sp	70	C	3100	—	—	—	—	3.00E+06	12 H&E (7 mo)	16, 17
1974, sc/erz	96	u (w), sp	55	L	2700	—	—	—	—	6.50E+06	N	18
1974, sc	8,	u (sw), sp	100	L	1300	—	—	—	—	1.09E+07	N	18
1974, swz	6	b (e, w), sp	100	L	4400	0.2	—	—	—	5.9E+06	N	18
1975, sc	24	u (w), rp	100	L	500	—	—	—	—	2.54E+05	N	19
1977, erz	480 (20 d)	b (e, w), rp	150	C	7000	—	—	—	—	3,50E+07	N	20
1979, erz	24	u (w), sp	<50?	L	<550?	—	—	—	—	5.00E+05	N	21

(Continued)

Table 17.1 Cont'd

Eruption Location	Duration (h)	Fissure Opening	Max Fountain Height (m)	Final Fissure Geometry	Initial Total Vent Length (m)	Mean Vent Width (m)	Vent Variability: Sinuosity	Vent Variability Irregularity	Vent Variability: Roughness	Total Erupted Volume (m³)	Ensuring Activity (Duration)	Source
1980, erz	24	? (?), ?	?	?	—	—	—	—	—	3.00E+00	N	21
1982, sc	19	u (w), rp	50	L	1000	—	—	—	—	5.00E+05	N	22
1982, sc	14	b (?), ?	70	L	1000	—	—	—	—	3.00E+06	N	22
1983, erz ***1	99 (4 d)	u (e) rp	100	L	6500	—	—	—	—	1.40E+07	46H (36 mo) E (376 mo)	23, 24, 25
1991, erz ***49	432 (18 d)	u (e), sp	30	L	2000	—	—	—	—	—	—	24
2011, erz ***59	102 (5 d)	b (ne, sw) rp	50	L	2000	1	—	—	—	—	E (29 mo)	26, 27

Note: For eruption location: sc=summit caldera, erz=East Rift Zone, swrz=Southwest Rift Zone, *#=a specific episode from the Kīlauea Iki-Kapoho eruption, **#=a specific episode from the Mauna Ulu eruption, and ***#=a specific episode from the Pu'u 'Ō'ō eruption. The symbol ^ indicates an "episode" of the eruption where the episode number, if not previously assigned, was counted onward from the last episode previously determined by original authors. For the fissure opening: u=unidirectional, b=bidirectional, n=north, e=east, s=south, w=west, rp=random progression of vent opening, sp=sequential progression of vent opening. For vent geometry, L=linear/fissure, C=circular source, f=minor flaring, F=major flaring. For ensuing activity: #H (# m) are the number of ensuing Hawaiian fountain events (nonfissure) and in parentheses are the number of months over which those fountains occurred, E (#m) is ensuing effusion of lava with no fountain activity, and the duration in months, K=lava lake N=no further activity, and A=currently active/ongoing activity.

Sources: 1: *MacDonald,* 1959; 2: *MacDonald and Eaton,* 1957; 3: *MacDonald and Eaton,* 1964; 4: *Richter et al.*, 1970; 5: *Richter et al.*, 1964; 6: *Moore and Krivoy,* 1964; 7: *Peck and Kinoshita*, 1976; 8: *Moore and Koyanagi*, 1969; 9: *Wright et al.*, 1968; 10: *Fiske and Koyanagi,* 1968; 11: *Kinoshita et al.*, 1969; 12: *Jackson et al.*, 1975; 13: *Swanson et al.*, 1976; 14: *Swanson et al.*, 1979; 15: *Duffield et al.*, 1982; 16: *Peteron et al.*, 1976; 17: *Tilling et al.*, 1987; 18: *Lockwood et al.*, 1999; 19: *Tilling et al.*, 1976; 20: *Moore et al.*, 1980; 21: *Banks et al.*, 1981; 22: *Banks et al.*, 1983; 23: *Wolfe et al.*, 1985; 24: *Heliker and Mattox*, 2003; 25: Hawalian Volcano Observatory internal monthly reports 26: Chapter 18, this volume 27: *Parcheta et al.*, in preparation.

remain too hot to approach for several months after eruptive activity ends and researchers may not return for such measurements at a later date; and (c) given all the other eruption response and documentation tasks taking place simultaneously, rough estimates were deemed to suffice for vent width. For the few fissures with documented widths, the values may be for specific locations or are representative ranges of widths, without mention of how these values were obtained (estimate, tape measure, laser-based method, etc.). Kīlauea's Mauna Ulu eruption started as a fissure, the western kilometer of which remains unburied by lava flows and largely unclogged with talus or vegetation, providing a unique opportunity to measure the detailed geometry of fissure vents and a shallow magmatic pathway.

17.3. FIELD AREA AND METHODS

17.3.1. The 24 May 1969 Eruption (Mauna Ulu Episode 1)

An 18h episode of fissure fountains on 24 May 1969 [*Swanson et al.*, 1979] initiated the five year long Mauna Ulu eruption of Kīlauea (Figure 17.1) [*Tilling et al.*, 1987]. The eruption took place on the upper East Rift Zone (ERZ) of Kīlauea, where it crosses the periphery of the east-northeast trending fractures of the Koa'e Fault System. Five eruptions had occurred in the Mauna Ulu area between January 1952 and May 1969. The 1969 fissure system crosses eruptive fissures from December 1965 and lies less than 100m south of the December 1962 eruption

Figure 17.1 (a) Map of Kīlauea's summit and the upper to middle portions of the ERZ (gridded area). The light gray line from top to bottom is the old Chain of Craters Road (C.c. Rd), and branching from it is the old 'Āinahou Rd (A. Rd). The original length of the 24 May 1969 Mauna Ulu fissure system is indicated in red, along with the outlines of Aloi and Alae craters, which were buried during the eruption. The yellow box outlines the Mauna Ulu episode 1 field area, which is enlarged in part b. (b) Present-day exposure of the fissure system, indicated by yellow box in (a), is 880 m long. The fissure system has five en echelon segments, labeled I–V. Spatter rampart trace is shown in light gray, vents in black. The purple boxes designate the locations of Figure 17.5 (LiDAR 1), Figure 17.6 (LiDAR 2), and Figure 17.7 (LiDAR 3) from west to east, respectively.

location [*Swanson et al.*, 1979]. Activity on 24 May 1969 was the only fountaining episode, out of 12 total fountaining episodes in 1969, that was associated with a fissure source [*Swanson et al.*, 1979].

The eruption began at 04:45 Hawaiian standard time (HST), preceded by a propagating earthquake swarm that woke local residents [*Swanson et al.*, 1979]. The fissure system opened rapidly, extending 2.5 km in 15 min (by 05:00 HST) to 200 m north of Alae Crater in the east and cutting Chain of Craters Road to the west. The fissure system crossed Aloi Crater and partially filled it with 25 m of lava. The entire ponded volume drained back into the fissure by the time aerial photographs were taken at 09:40 HST. By 08:30 HST, the fissure system reached its full extent, crossing 'Āinahou Road (1 km west of Chain of Craters Road). During fissure opening, vents along the whole length erupted low fountains ≤50 m in height. The fissure evolved into two fountaining zones by 09:40 HST [*Parcheta et al.*, 2012]: an eastern zone (EZ) between Aloi and Alae craters and a western zone (WZ, our field area) between 'Āinahou and Chain of Craters Roads. The ≤50 m high fountains showed small variations in height temporally and spatially in the few photographs available from 24 May 1969. The fissure was flooded to heights of 2–3 m above the current ground surface at peak discharge [*Parcheta et al.*, 2012]. *Swanson et al.* [1979] calculated that a total magma volume of 4×10^6 m^3 was erupted on 24 May; no separate estimates were made for EZ or WZ. Both zones produced lava flows that were impounded 1.2 km downslope by a north facing fault scarp in the Koa'e Fault System [*Swanson et al.*, 1979].

17.3.2. Measurements of Vent/Fissure Geometry

The field study area measures 880 m long by 60 m wide on an 80° orientation and encompasses the WZ fissure [*Parcheta et al.*, 2012] between the old Chain of Craters Road (now Mauna Ulu pullout) and the old 'Āinahou Ranch Road (now the modern Chain of Craters Road). This area is the westernmost kilometer of an originally 4.5 km long fissure system (Figure 17.1a, red line) and is the western of the two loci of eruptive activity in 1969 (the eastern locus was where the Mauna Ulu shield is now located) [*Swanson et al.*, 1979]. The fissure system consists of five segments, labeled I–V (Figure 17.1b), and each segment contains several vents (Table 17.2). Vents are

Table 17.2 Number (*n*) of width measurements per vent.

Seg. I	*n* Width			Seg. II	*n* Width			Seg. II	*n* Width			Seg IV	*n* Width			Seg. V	*n* Width		
Vents	2009	2011	TOT	Vents	2009	2011	TOT	Vents	2009	2011	TOT	Vents	2009	2011	TOT	Vents	2009	2011	TOT
1	1	0	1	9	7	4	11	15	2	2	4	34	1	0	1	41	1	0	1
2	1	1	2	10	1	0	1	16	1	2	3	35	5	8	13	42	1	0	1
3	1	0	2	11	2	2	4	17	3	0	3	36	3	5	18	43	1	0	1
4	1	0	1	12	4	3	7	18	1	0	1	37	1	0	1	44	4	2	6
5	2	1	3	13	1	0	1	19	1	0	1	38	4	0	4	45	1	0	1
6	1	1	2	14	1	1	2	20	1	1	2	39	1	0	1	46	4	0	4
7	5	4	9					21	1	0	1	40	1	0	1	47	4	0	4
8	2	1	3					22	1	3	4					48	1	0	1
								23	1	4	5					49	1	0	1
								24	1	0	1					50	1	0	1
								25	1	0	1					51	5	3	8
								26	1	0	1					52	14	8	22
								27	1	0	1					53	13	9	22
								28	1	0	1					54	23	15	38
								29	1	8	9								
								30	1	0	1								
								31	6	5	11								
								32	1	0	1								
								33	1	0	1								

Note: Segments I–V are in consecutive groups of columns; Seg = segment. Two field sessions measured vent width. The first, in 2009, was exclusive and semisystematic; i.e., we measured at every noticeable change for vents with aspect ratios $L > 2W$. Vents that had aspect ratios of $L < 2W$ had only one measurement made across the width as the width was essentially uniform. The second field campaign, in 2011, consisted of tape measurements of width correlated to measurements of flaring, irregularity wavelength, and irregularity amplitude of the vent walls. Vent widths were measured at locations of flare and from the amplitude of each irregularity on both the north and south walls to the opposing wall. For our average vent widths, we used the cumulative number (TOT) of width measurements (and therefore the spacing of all measurements considered is not regular or uniform).

numbered consecutively eastward from vent 1 to vent 54. Lava is preserved west of vent 1, on the western side of the new Chain of Craters Road, but no source vents or ramparts are observed. We use vent 1 as the point from which distances are reported in this chapter. In the aerial photographs, four plumes of gas and minor ash can be seen emanating from linear vents immediately east of the old Chain of Craters Road. There is fresh lava in the photograph, but the plumes obscure most of the fountains, with only the easternmost of WZ visible. These vents, if preserved today, would constitute an additional segment VI but were buried by 1971 pāhoehoe and 1974 ʻaʻā lava flows. Only the crest of one spatter rampart remains visible.

Vent surface dimensions (length and width) along the exposed fissure system were determined using tape measures (July 2009; horizontal error 10 cm), tape and Bruntons (Sepember 2011; horizontal error 10 cm), and kinematic global positioning system (GPS) (November 2011; horizontal error 25 cm, vertical error 50 cm relative to the continuous GPS receivers). The 2009 campaign was exclusively aimed at determining vent widths, and we made one measurement for every substantial change in width where vents were at least twice as long as their width. When vent lengths were shorter than twice their width, we recorded only one measurement, as the width was generally uniform. The September 2011 measurements were aimed at measuring the wavelength and amplitude of the vent wall irregularities, as well as vent flaring, where widths were subsequently measured from each irregularity and at locations where vents were flared. The kinematic GPS campaign mapped out the exact vent surface shape in detail that was too complex for simpler methods. To do the mapping, the antenna was held above the edge of the vents on an extended pole and was walked around the edge of the vent at a slow pace (~1 m/s). The end points of each vent were recorded in the GPS after 10 s of stationary delay. Average vent widths quoted in this chapter use the cumulative number of width measurements made from all the campaigns.

Vent geometry is defined as the combination of width, length, and depth (Figures 17.2a,b). We utilize these parameters as an aspect ratio in x-y space (map view) and a flaring ratio in y-z space (cross section), with x and y measured parallel and perpendicular, respectively, to the strike of the fissure and z denoting the depth coordinate. The mean width of each linear vent was calculated from multiple measurements, as described in the preceding paragraph (Figure 17.2c, Table 17.2). Vent orientation was also measured if the long axis was significantly greater than the short axis (Figure 17.2d). Aspect ratios (ARs) were calculated by dividing the short axis by the long axis. Flaring ratios (FRs) were calculated by dividing the vent width at the inflection point between sloping (vent) walls and subvertical (conduit) walls by the vent width at the ground surface. The AR and FR both have values ≤1. Trigonometry was used to calculate the flaring ratio (Figure 17.2b). The two inclined walls were measured for inclination and length (distance to the change in wall angle of the vent/upper conduit system), and we calculated the horizontal extent corresponding to that length using

$$w_c = w_v - (f_N \cos\theta_N) - (f_S \cos\theta_S),$$

where w_c is the conduit width, w_v is the flared vent width, f_N is the length of the northern vent wall, θ_N is the angle of the northern vent wall relative to horizontal, f_S is the length of the southern flared vent wall, and θ_S is the angle of the southern vent wall relative to horizontal. Each horizontal extent due to flaring was subtracted from the flared surface vent width to calculate the conduit width.

Depths along the entire exposed fissure system were measured using a plumb bob. Depths may only be minima if the plumb bob landed on a ledge, overhang, or false roof, none of which could be distinguished from the "true bottom" of the fissure system.

17.3.3. Light Detection and Ranging Measurements

Three vents (Figure 17.1b), encompassing the range of observed geometries, were chosen to image with a tripod-mounted lidar device (February 2010) (as described by *Morris et al.*, 2008) in order to quantify the subterranean geometry and identify any distinct geometric change between the vent and the shallow conduit. Twenty-two tripod-mounted scans centered at 34 m (vent 3), 243 m (vent 17), and 689 m (vent 47) were acquired with 4 cm image resolution at a distance of 30 m from the tripod for sweeping radial scans encompassing the ramparts and 1 cm resolution at 10 m from the tripod to image the vents and conduits. The first lidar study used one scan directly above the vent on the west side, looking at 70° below the horizontal to image a small ellipsoidal vent. The second study combined three scans to image the widest linear vent in the field area (on the NE, E, and SE edges of the vent looking at 50°–70° below the horizontal allowing the deepest view into the fissure), together with five radial sweep scans of the associated northern ramparts and isolated areas of spatter south of the fissure. The third study combined six scans to image a flared circular vent overlying a straight-walled conduit (on the NE, E, SE, SW, W, and NW edges of the vent looking at 50°–70° below the horizontal allowing the deepest view into the fissure), together with six consisting of radial sweep scans of the associated northern and southern ramparts. Within each set of lidar data, cross sections were positioned to cut the fissure in

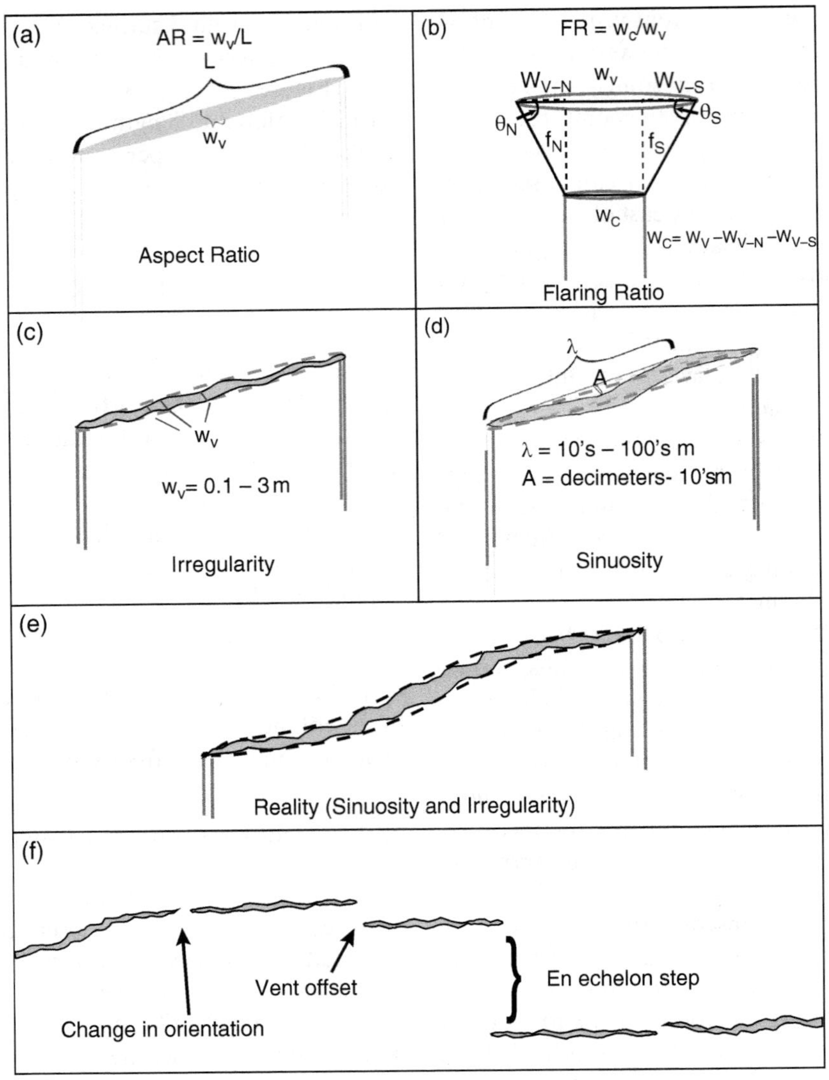

Figure 17.2 Measurements and parameters discussed throughout chapter. (a) Aspect ratio (AR, vent width divided by vent length) is the surface area geometry of the vent. (b) Flaring ratio (FR, calculated conduit width divided by measured vent width). Both measurements are taken in the same vertical plane. Conduit width is calculated by using trigonometry; measuring the length of the flared walls and the angle of the flare allows the calculation of the horizontal widths w_{vN} and w_{vS}, and subtracting those values from the vent width (from both sides of the fissure) yields the conduit width. (c) Irregularities in vent width, which we define as fluctuations in vent width *without a change in vent strike*. Measurements show that the irregularities and protrusions are all oriented along the length of the fissure, with a range of wavelengths from 1 to 10 m, corresponding amplitudes of 0.1–3 m, and a variation in fissure widths from 0.1 to 3.4 m with a mean of 0.6 m. (d) Sinuosity in vent geometry, which we define as a change in vent strike *without a change in vent width*. The line of strike is indicated by a dashed gray line. The wavelengths of sinuosity are several tens to hundreds of meters with corresponding amplitudes of 0.1–10 m, respectively. (e) Complexity of a vent observed in the field. Irregularities are the most common perturbation of geometry from an ideal circle, ellipse, or linear fissure, but they are superimposed on sinuosity when sinuosity is present. Sinuosity is typically difficult to discern, because the wavelengths are large and the amplitudes small, making the waveform subtle. (f) Complexity of sequential vent geometries observed in the field. Vents can be colinear or offset (upper left, displacement or orientation angle) from each other within each segment or separated by en echelon steps (lower right).

two manners: (1) parallel to the ground surface at 1 m depth intervals and (2) perpendicular to the strike of the vent at 1 m intervals to show the true width. Where depth and cross-sectional contours intersect the data point cloud, a width measurement was calculated, creating a 1 m × 1 m grid of measurements within the shallow conduit.

17.4. RESULTS

17.4.1. Pre- and Posteruption Topography

Digitized ground elevation derived from a 1965 topographic map, with the addition of lava flows from eruptions between 1965 and 1969 and fault scarps from 1965 aerial photographs [*Parcheta et al.*, 2012], provides a baseline map to identify the topographic features with which the fissure system interacted as it broke the surface. The elevation of the fissure is approximately constant, with no more than 6 m of vertical relief along its length; the lowest point along the fissure is between segments IV and V and the high point is at segment II [*Parcheta et al.*, 2012]. The fissure is aligned with the ERZ and preexisting Koa'e faults that also trend 80°, and *Swanson et al.* [1979] noted that the westernmost vents (i.e., those west of our vent 1) were located along a north facing Koa'e fault scarp. The WZ fissure spans and connects two visible fault scarps seen on the 1965 aerial photographs. It is not clear from the aerial photographs whether there was a fault trace connecting the two scarps prior to the eruption.

Lava flow thickness along the fissure is estimated as 0.5–3 m by subtracting the preeruption topographic map (10 m vertical resolution, digitized from 1965 U.S. Geological Survey (USGS) 1:24,000 Makaopuhi topographic quadrangle) from the current digital elevation model (10 m vertical resolution, obtained from the State of Hawaii GIS archive digitized from 2003 USGS 1:24,000 Makaopuhi topographic quadrangle) [*Parcheta et al.*, 2012]. The vertical transition from flared vents to steep-walled conduit may coincide with the base of the 1969 lava flow, but drain-back coats vent walls in 50 of the 54 vents, concealing the contact between the 1969 and older lava flows. There are four places where vent flare extends well below the estimated thickness of 1969 lava. In these vents, a surface veneer of juvenile lava has spalled off the vent wall to reveal older flows; this relation is evidence that flare cuts into wall rock.

17.4.2. Fissure System Orientation, Dimensions, and Segmentation

The Mauna Ulu fissure system strikes 80° along the ERZ at an elevation of 1000 m and is exposed for 470 m over a linear span of 880 m. Its five segments consist of alignments of closely spaced vents separated by four vent-free regions at distances from vent 1 of 82, 150, 336, and 475 m, respectively. Segments define an en echelon pattern; three vent-free regions are marked by meter-scale right-stepping en echelon steps, and the fourth by a meter-scale left-stepping step. Vent configuration of segments is complex; segments differ in the type, number, and relative positioning of vents. Within segments, smaller lateral shifts in vent alignment are observed. There are two typical planform geometries, circular vents and linear vents, both of which may have either flared or nonflared vertical profiles.

Segment I (Figure 17.3) (vents 1–8) contains two lateral vent offsets. Vents include three circular and five linear geometries. Segment II (vents 9–14) also has two minor offsets in vent alignment. Vents 9–14 all display linear geometries. Segment III (vents 15–33) has six vent offsets. Segment III consists of 2 circular vents, 14 linear vents, and 3 flared circular vents. A cluster of 6 vents (vents 18–23, Figure 17.3) forms two closely spaced and parallel alignments of 1 circular vent, 2 flared circular vents, and 3 linear vents. Segment IV (vents 34–40) has four vent displacements. Vents 34–40 contain 1 circular and 6 linear vent geometries. Segment V (vents 41–54) has four separate alignments of linear vents and one cluster containing 5 closely spaced vents that do not lie along a single trend. Vents 41–54 comprise a series of nine linear and five flared-circular vents that make up the vent cluster. In summary, individual fissure segments contain between 6 and 19 vents; all have some linear vents, but some also contain circular vents, flared linear vents, and/or flared circular vents. In two instances, vents are clustered in groups instead of being aligned along the fissure.

17.4.3. Vent Geometries: Sinuosity, Irregularity, and Roughness

On decameter or smaller scales, none of the vents are well approximated by a simple geometry (i.e., a rectangle (linear vent) or ellipse (circular vent)). All vents show centimeter-scale relief along the vent walls related to drain-back, which we refer to as roughness. Linear (elongate) vents also display meter-scale irregularity and decameter-scale sinuosity. Vent irregularity (Figure 17.2c) takes the form of pinching and swelling of the fissure and causes a deviation in vent width along the fissure length from an idealized geometry (Figure 17.2b). Irregularity was characterized by measuring variations in vent width. We define vent sinuosity (Figure 17.2d) as a longer wavelength deviation of vent geometry from a straight line, unrelated to variation in vent width, giving a curved or wavy appearance. Sinuosity was characterized by measuring trough-to-trough wavelength orientations and amplitudes in the geographic information system program ArcGIS. Even though the spacing of all measurements considered is not precisely regular, our vent width measurements and kinematic GPS trace should meet the

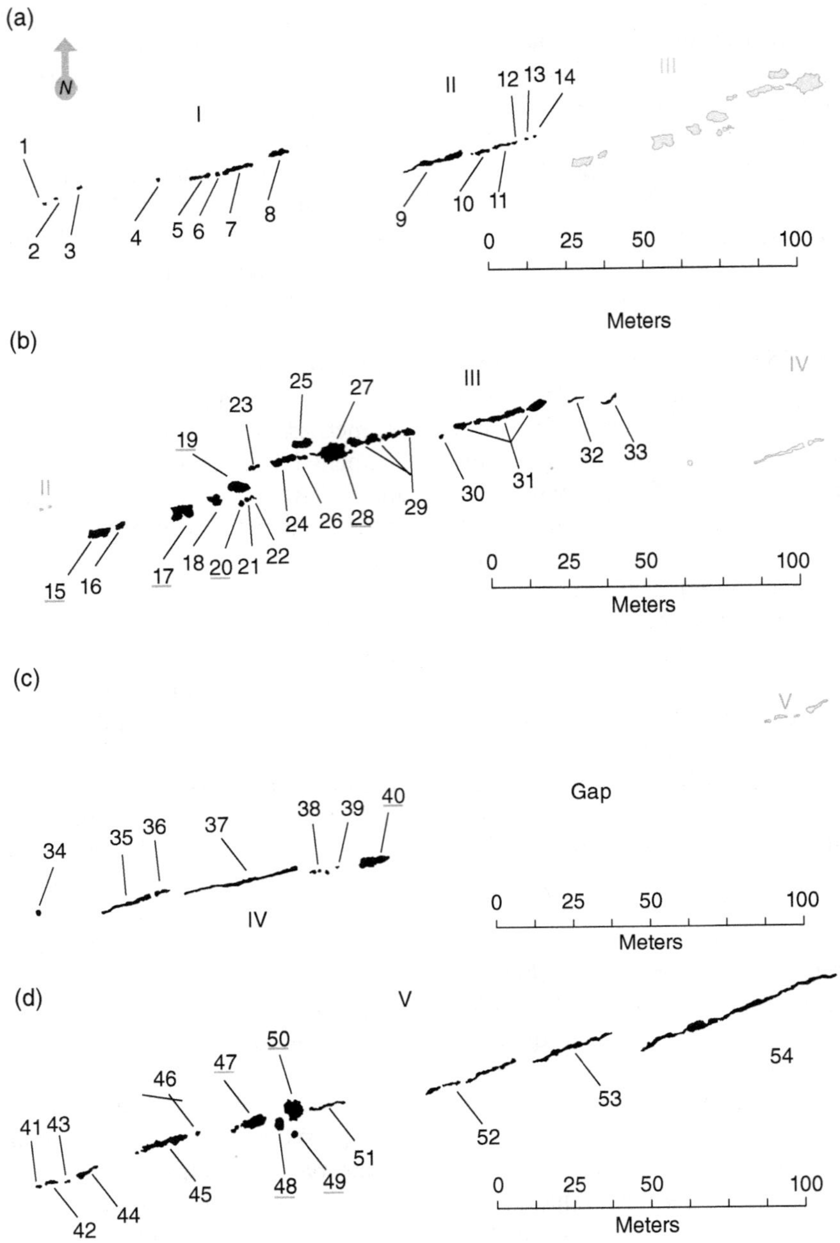

Figure 17.3 Kinematic GPS trace of 54 vents, showing their position, geometry, and size. Vents in light gray are shown in the next (or previous) panel: (a) vents 1–14 on fissure segments I and II; (b) vents 15–33 on fissure segment III; (c) vents 34–40 on fissure segment IV and rampart/vent gap; (d) vents 41–54 on fissure segment V.

Nyquist theorem [*Nyquist*, 1928] criterion for sampling of the irregularities and sinuosity, as there are at least two measurements per wavelength of these features.

Sinuosity occurs over wavelengths of tens to rarely hundreds of meters. Shorter waveforms are confined to single linear vents, specifically vents 29, 35, 37, 52, 53, and 54. Four multivent waveforms are found within segments III and IV. Wavelengths for these four waveforms are 78, 168, 206, and 454 m. Trough-to-peak amplitudes of these four waveforms are 11, 12, 21, and 39 m, respectively. The six vents displaying internal sinuosity are the six longest linear vents in the field area. They are long enough to incorporate the entire waveform, whereas other vents are too short. Measurements from vents 29, 35, 37, 52, 53, and 54 show a typical wavelength of 11 ± 6 m and amplitude of 63 ± 27 cm. This sinuosity and any associated irregularity also extend over the entire depth range that is visible within the fissure.

Irregularity of the fissure (Figure 17.2c) occurs on meter scales. The north wall irregularities can be characterized by a spacing of 3 ± 1.2 m and an amplitude of 80 ± 60 cm, oriented along an average strike of 77°. Irregularities on the south wall of the fissure are slightly more variable, with an average spacing of 4 ± 2 m and amplitude of 70 ± 40 cm, oriented along an average strike of 72°. Opposing fissure walls can touch each other if two large irregularities are aligned in a narrow fissure. Although the irregularities on each side of the fissure have different spacings and rarely match one to one, they are spaced within the stated standard deviations of each other. Irregularity is seen to continue with depth into the fissure and, while not measured for safety reasons, appears similar to the irregularity measured at the vent surface.

Late-stage drain-back of lava into the vents produced a smaller-scale undulation on the fissure walls over the entire length of the fissure. There are two mean spacings measured at this scale: (1) trough-to-trough spacing of 65 ± 10 cm with an amplitude of 30 ± 5 cm and (2) trough-to-trough spacing of 150 ± 25 cm with an amplitude of 65 ± 10 cm. As with irregularity and intravent sinuosity, the drain-back texture continues with depth into the fissure. In contrast, though, it does not always extend to the lowest depths visible [*Parcheta et al.*, 2012]. A lava veneer resulting from drain-back exists on many of the vent walls. In general, its thickness could not be measured because the veneer is a continuous coating, but where spalling exposes the thickness, it is <5 cm.

17.4.4. Vent Distribution, Dimensions, and End-Member Types

The two typical planform geometries—circular vents and linear vents—may have either flared or nonflared vertical profiles. Figure 17.4 plots the aspect ratio versus flaring ratio; the left side of the graph corresponds to linear vents and the right side to circular vents; the top of the graph corresponds to nonflared vents and the bottom to flared vents. The aspect ratio division between circular

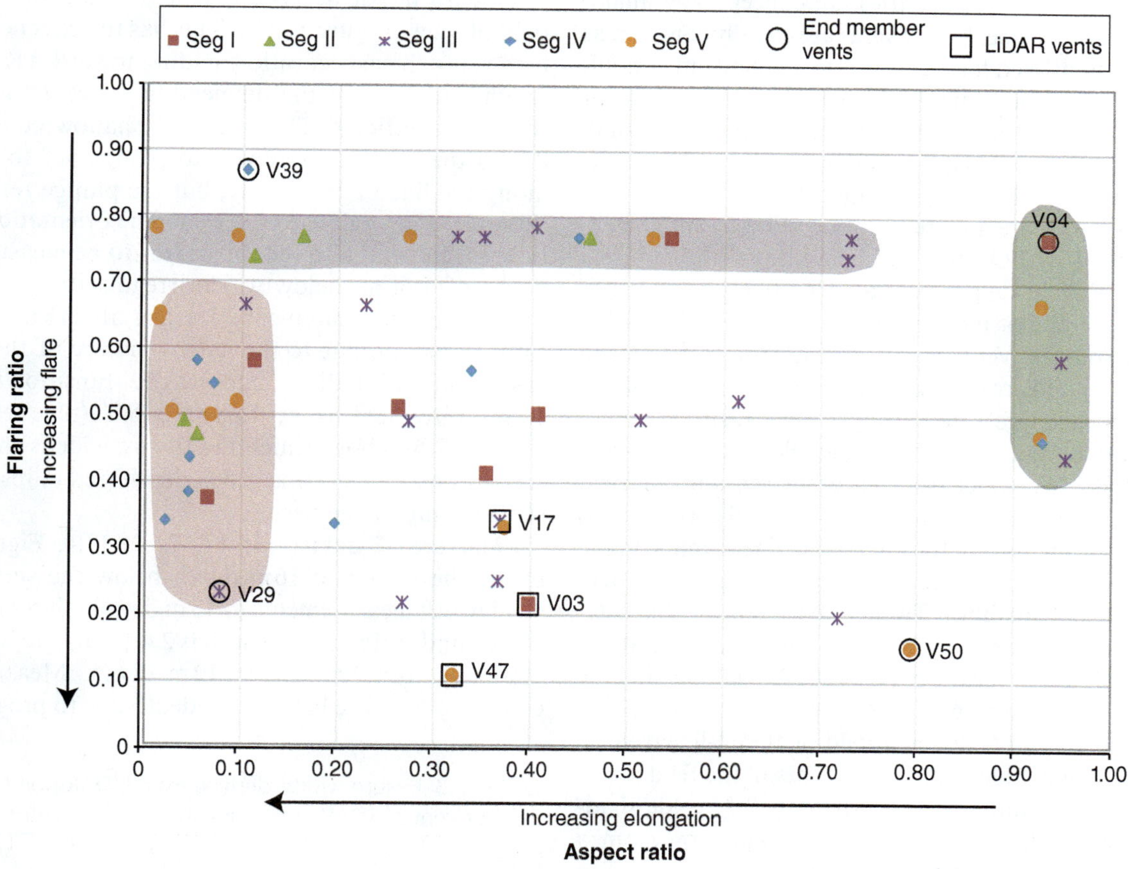

Figure 17.4 Plot of aspect ratio (vent width divided by vent length) versus flaring ratio (calculated conduit width divided by measured vent width). Symbols represent fissure segments: segment I = red squares, segment II = green triangles, segment III = purple asterisks, segment IV = blue diamonds, and segment V = orange circles. End-member-type vents are grouped by shading: circular (unflared to flared) = light green, unflared (circular to fissure) = light blue, and fissure (unflared to flared) = light red. Vent 4 (circled) is the most circular vent in the field area with the least flaring. Vent 39 (circled) represents the most linear vent with the least amount of flaring (compared with other linear vents). Vent 29 (circled) is the linear vent with the most flaring. Vent 50 (circled) is the most circular of highly flared vents. Vent 47 (boxed; Figure 17.7) is the vent with the most flaring in the entire field area. Vent 17 (boxed; Figure 17.6) has the clearest and widest view with depth of any vent in the field area, especially those of fissure geometry. Vent 03 (boxed; Figure 17.5) was our chosen "average" unflared circular vent with drainback.

and linear sources is taken to be AR = 0.5, where the vent length is twice the vent width. Smaller aspect ratios correspond to more elongate surface geometries. The division between flared and nonflared vents is taken to be FR = 0.7, where the vent width is ≥50% wider than the conduit width. This is larger than the scale of irregularities, and therefore the documented flaring is genuine and not an artifact of the irregularities. All vents show some degree of flaring, and flaring ratios vary continuously across the vents, with the exception of nearly circular vents. Transition points for flaring could not be documented everywhere due to the unsafe nature of several vents and the lack of an obvious transition point in other vents. However, most of the measured flaring ratios had a slope inflection point less than 2 m down the inclined vent wall.

Linear vents predominate. Circular vents, where present, are typically found at the ends of a segment. Flared circular vents are the least common form. Nonflared circular vent/shallow conduit geometries are conceptually analogous to a vertical pipe. Linear vents are equally common as flared and nonflared forms. Nonflared linear vent/conduit geometries are conceptually similar to a knife blade or polyhedron, where the longest axis extends into the ground, and the median and short axis dimensions visible at the ground surface yield an aspect ratio <0.5.

Several trends are apparent on the plot of vent geometry (Figure 17.4). On the far right side of the plot, six vertically aligned data points at AR = 0.93 (green shading) show a trend of increasing vent flare for nearly circular point sources. Vents 4, 20, 21, 34, 48, and 49 plot in this category and are typically 1.3 ± 0.2 m long by 1 ± 0.3 m wide, with the largest measuring 1.5 m × 1.4 m. We identify these data points as outlining a field of circular vents in Figure 17.4, distinct from all other vents in the field area. A similar trend, albeit slightly less well correlated, is shown in the increase in flaring among linear vents on the left side of the plot, indicated by data points where AR < 0.12 (red shading). Twenty vents plot in this category and are between 3 and 33 m long and 0.8 ± 0.4 m wide. The longest linear vent is vent 54, lying outside the standard deviation at 67.5 m long and 1 m mean width. These 20 data points define a field of flared-linear vents in Figure 17.4 distinct from other vents in the field area.

Along an FR value of approximately 0.77 (blue shading), 15 vents define a continuous progression from a nonflared circular vent to a nonflared linear vent. The remaining 17 vents plot in a wide region between AR = 0.2–0.8 and FR = 0.1–0.67. These vents are visibly more complex than others in the field area.

Segments III and V, the two longest segments, have vents that plot in every field of Figure 17.4. Segment I contains three of the four vent types, missing only flared circular vents. Of the two shortest segments, segment II resides almost exclusively in the nonflared linear vent field while segment IV, except for one vent, straddles the flared and nonflared linear vent fields. It is important to note that photographic evidence [*Swanson et al.*, 1979; *Parcheta et al.*, 2012] indicates long-lived fountain sources were associated with distinct positions along the linear vents.

17.4.5. Shallow Conduit Geometries

Lidar area 1 data (vent 03, Table 17.3, Figure 17.5) extend 4.5 m down into the circular vent, but the data become too sparse for analysis below 3.5 m depth. No vertical cross sections were made for this point cloud, but four horizontal cross sections could be measured. Across the fissure, the vent width narrows abruptly from the northern side, whereas the deeper conduit pinches more gradually (Figure 17.5c). Viewed in the plane of the fissure, the apparent vent length changes more gradually with depth because it is a function of the lidar field of view (Figure 17.5b). This has the effect of falsely increasing the aspect ratio trending toward AR = 1 with depth and making the delineation of vent to conduit transition difficult. The vent and shallow conduit dip 85° to the south and appear to plunge 80° to the east along the line of the fissure, but the plunge reflects the lidar viewing angle into the vent. The orientation of the lidar instrument allowed for widths to be measured to a depth of 1.68 m, below which irregularities cause data gaps. These measurements are not all oriented in the same direction, due to the narrow nature of the subcircular vent and shallow conduit. The transition from the conduit to vent is at 1.68 m depth (blue ellipse in Figure 17.5), above which the fissure widens sharply. A second break in slope at 0.8 m depth (red ellipse) represents flaring of the fissure.

Lidar area 2 data (vent 17, Table 17.4, Figure 17.6) image the fissure to 16 m depth below the surface and yield width measurements to 13 m deep in the conduit. At the ground surface, vent width is 2.4 ± 1 m, and it narrows gradually to 1.2 ± 0.4 m at 12 m depth. Measurements could not be made below 13 m depth due to progressively

Table 17.3 Fissure widths derived from LiDAR point cloud data for vent 3/LiDAR area 1 (Figures 17.5b,c, video 1).

Horizontal Transect	Depth (m)	Width (m)
A ●	0.83	0.34
B ●	1.38	0.46
C ●	1.68	0.78
D ●	3.58	0.21

Note: Measurements of the maximum width were taken at four depth intervals, as vent 3 was our "type vent" for the circular geometry. Colored circles correspond to ellipses in Figures 17.5b,c.

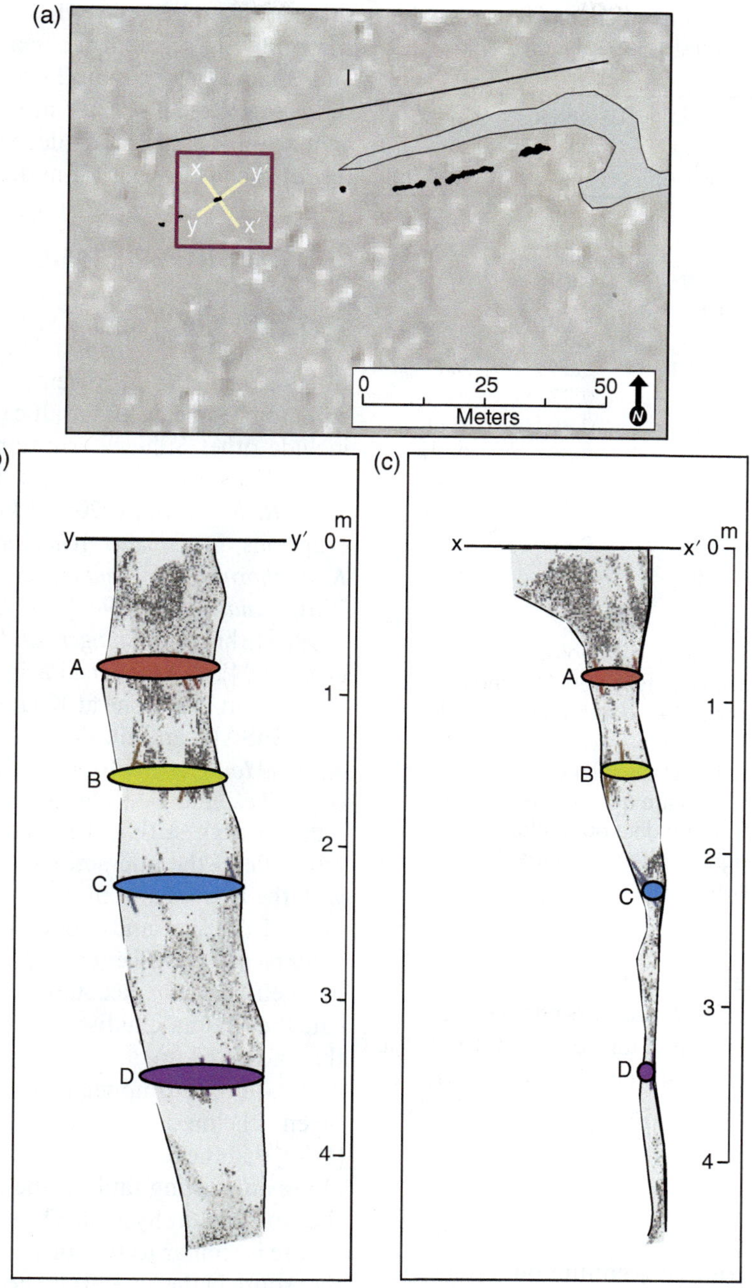

Figure 17.5 (a) Map showing segment I; location of LiDAR area 1 is marked with the purple box. Yellow lines indicate cross sections *y-y'* in (b) and *x-x'* in (c). (b) View of LiDAR point cloud perpendicular to vent/conduit strike. View is to the north-northwest. Black lines denote boundaries of the LiDAR cloud. Colored ellipses indicate locations of vent width and length measurements. *A, B, C,* and *D* correspond to the transects listed in Table 17.3. (c) View of LiDAR point cloud along strike. View angle is to the east. *A, B, C,* and *D* correspond to the transects listed in Table 17.3.

larger gaps in the data. The vent and shallow conduit system in the plane of the fissure dip 72° to the north. Viewing angles of the lidar cause the system to appear to plunge 53° to the west along the fissure strike. The transition from shallow conduit to vent is at 3 m depth.

Lidar area 3 data (vent 47, Table 17.5, Figure 17.7) record a flared vent and fissure conduit combination to a depth of 11 m below the surface. In the first 4 m, the vent is wide, with a mean width of 2.9 ± 1.3 m. Widths dramatically decrease below 4 m from the ground surface, narrowing to 0.8 ± 0.7 m. Below this depth, while parts of the system were imaged, there was no depth where data were recovered on both walls to allow width to be calculated. The transition from shallow conduit to

Table 17.4 Fissure widths and lengths derived from LiDAR point cloud data for vent 17/LiDAR area 2 (Figures 17.6b,c, video 2).

Horizontal Transect	Depth (m)	Mean Width (m)	1 Standard Deviation (m)	Linear Distance Covered (m)
A ○	0	2.44	1.05	8.96
B ○	1	2.19	0.81	8.19
C ○	2	1.79	0.85	5.15
D ○	3	1.82	0.97	3.99
E ○	4	1.63	0.32	5.18
F ○	5	1.48	0.31	6.44
G ○	6	1.46	0.27	5.11
H ○	7	1.38	0.59	4.71
I ○	8	1.25	0.56	2.78
J ○	9	1.39	0.02	1.68
K ○	10	1.33	0.31	1.94
L ○	11	0.96	0.16	3.45
M ○	12	0.65	NA	1.86
N ○	13	0.80	NA	1.73

Note: Colored circles correspond to data points in Figures 17.6b,c. This vent was our "type vent" for the fissure geometry. The point cloud could be quantified to 13 m depth. Each horizontal transect, labeled A–N, has multiple width measurements along its length; the mean and standard deviation of those measurements are quoted. Linear distance covered is the length visible in the data point cloud. Transects M and N did not have enough width measurements to calculate a standard deviation.

vent is at 4 m depth, where the upper 4 m comprises the flared vent and the lower 7 m represents shallow conduit (Figure 17.7). Of the sparse data returned from the lower 7 m, the fissure dips 84° north and appears to plunge 53° to the east, but the plunge may result from the viewing angle.

17.4.6. Vent Depths

We assume the feeder conduit is continuous along each fissure segment deeper than can be directly observed (i.e., greater than ~10 m). At even greater depth, which we have not been able to measure, the conduit must be continuous along the entire fissure system, as the vents were all fed by the same dike. The depths measured for the fissure system through the entire field area (Table 17.6) range from 0.4 m (vents filled in by talus or drain-back) to 41 m. The mean recorded depth is 11 m. In most vents, one could typically feel the descending plumb bob ricocheting off irregularities on the vent or conduit wall, indicating that irregularities persist with depth in the system.

The middle of segment II is consistently 20–39 m deep. On the west end of segment II, the plumb bob bounced off of irregularities during parts of the descent, but on the east side of segment II, especially vent 12, it dropped smoothly. On segment III, vents 17 (lidar area 2), 27, 31, and 32 are greater than 25 m deep. Vents 35 and 37 on segment IV reached 20+ m depth. The western end of vent 45 on segment V is deeper and separated from the rest of the linear vent (~5 m deep) by a spatter saddle.

17.5. INTERPRETATION AND DISCUSSION

17.5.1. Fissure System Geometry

An en echelon arrangement of fissure segments is a common geometry for basaltic volcanic systems. Examples include other Kīlauea eruptions [*Macdonald and Eaton*, 1957; *Fiske and Koyanagi*, 1968; *Moore and Koyanagi*, 1969; *Richter et al.*, 1970; *Lockwood et al.*, 1999] as well as eruptions in Iceland [*Gudmundsson*, 2003; *Grant and Kattenhorn*, 2004; *Jenness and Clifton*, 2009], Réunion [*Cayol and Cornet*, 1997], the Canary Islands [*Day et al.*, 1999], Ethiopia [*Ebinger and Casey*, 2001; *Wolfenden et al.*, 2004], and the East Pacific Rise [*Fornari et al.*, 2004]. Very recent eruptions at Kīlauea have been studied using GPS, InSAR, and tilt data to interpret en echelon geometries [*Montgomery-Brown et al.*, 2010]. The consistency of strikes and prevalence of en echelon segments at Kīlauea suggest that the overall stress field near Mauna Ulu reflects the interactions of the ERZ at shallow level with the Koaʻe Fault System, producing a combination of mode I (tensile) and mode III (shear cracking) opening [*Montgomery-Brown et al.*, 2010]. The overall fissure geometry and the account by *Swanson et al.* [1979] indicate the Mauna Ulu fissure was a rectilinear dike at depth that adjusted to a slightly rotated, near-surface stress orientation, causing segmentation into the observed series of en echelon steps. It was postulated by *Swanson et al.* [1979] that the portion of the fissure still preserved erupted along an existing fault in the extreme eastern portion of the Koaʻe Fault System. The geometry of the Mauna Ulu fissure is similar to that of noneruptive cracks (100 m farther south in the Koaʻe) and noneruptive fissures elsewhere on Kīlauea, supporting this interpretation. Exploitation of a preexisting fault may explain why the dike took 3 h to propagate the final 1 km westward as compared to 15 min for bilateral extension of the initial, but now buried, 4 km of fissure east of the field area.

17.5.2. Vent Geometry

17.5.2.1. Correlation to Rampart Geometry

There is no obvious relationship between the sparse data for eruption rates (or fountain heights) inferred from the heights of spatter ramparts [documented by *Parcheta et al.*, 2012] and flared or nonflared vent geometries. The

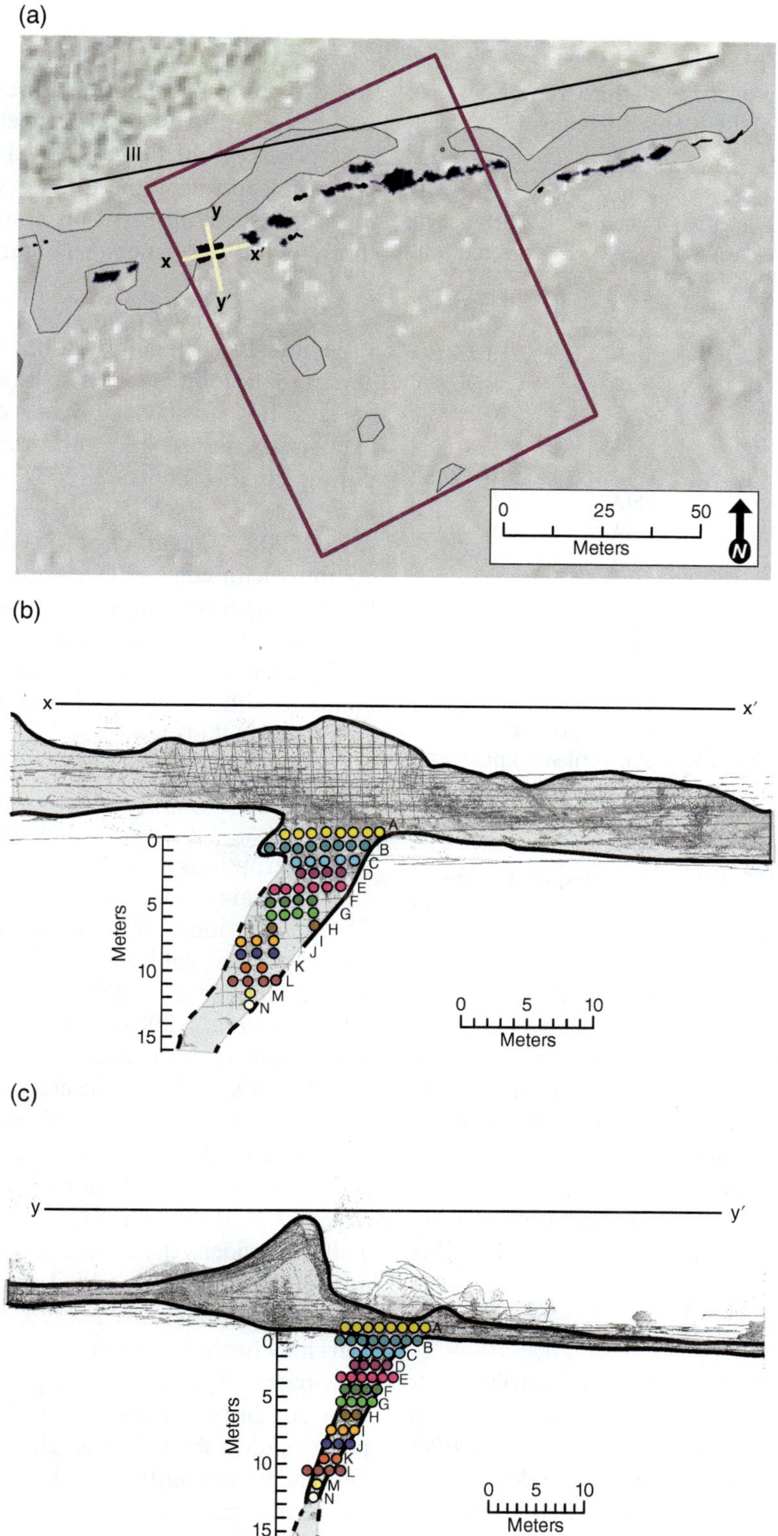

Figure 17.6 (a) Map showing all of segment III; location of LiDAR area 2 is marked with the purple box. Yellow lines indicate cross sections *x-x′* in (b) and *y-y′* in (c). (b) View of LiDAR point cloud perpendicular to vent/conduit strike. View angle is to the northwest. Black line is dashed where the fissure wall is projected and LiDAR data are too scarce to know with certainty. Colored circles at grid point intersections indicate locations of vent width measurements. *A–N* correspond to the transects listed in Table 17.4. (c) View of LiDAR point cloud along strike, looking east. Black lines denote ground surface in the closest parts of the LiDAR cloud, but ramparts east of the vent can be seen above this line. Colored circles extending across the vent/conduit correspond to the depths of width measurements and correlate to the same colors/measurements in (b). *A–N* correspond to transects listed in Table 17.4.

Table 17.5 Fissure widths and lengths derived from LiDAR point cloud data for vent 47/LiDAR area 3 (Figures 17.7b,c, video 3).

Horizontal Transect	Depth (m)	Mean Width (m)	1 Standard Deviation (m)	Linear Distance Covered (m)
A ●	0	3.97	1.68	11.33
B ●	1	2.59	1.42	10.10
C ●	2	2.42	0.88	8.75
D ●	3	2.52	1.07	9.90
E ●	4	0.80	0.67	10.25
F	5	NA	NA	0
G	6	NA	NA	0
H	7	NA	NA	0
I	8	NA	NA	0
J	9	NA	NA	0
K	10	NA	NA	0
L	11	NA	NA	0
M	12	NA	NA	0

Note: Colored circles correspond to data points in Figures 17.7b,c. This vent was our "type vent" for a general flared geometry and is a flared subcircular vent superposed on an unflared linear conduit. The point cloud could be measured to 4 m depth. Each horizontal transect, labeled A–N, has multiple width measurements along its length; the mean and standard deviation of those measurements are quoted. Linear distance covered is the transect length visible in the data point cloud. No part of the shallow conduit below 4 m depth has data on both fissure walls in any given location to allow width measurements in the point cloud (transects F–N).

highest spatter accumulation rates instead occur at the minor offsets and larger en echelon steps in the fissure, where ejecta accumulated from the combined output of two adjacent noncollinear vents. Walls of flared vents always possess flow textures indicative of lava drain-back, accompanied by low spattering that produced the isolated dense spatter seen around these vents (described by *Parcheta et al.* [2012].

17.5.2.2. Sinuosity

Dike and fault asymmetries have been attributed to gradients in the regional stress field, or, for a dike, an internal fluid pressure gradient [*Pollard and Muller*, 1976]. Fracture length influences the magnitude of asymmetry. Dikes divided into systematic en echelon segments through rotation of the principal stress field may interact with each other, may bend, and create a sinuosity within each segment [*Pollard*, 1987]. Whereas *Pollard et al.* [1982] showed that overlapping en echelon segments can have sinuosity in an elastic model, recent work by *Daniels et al.* [2012] demonstrates that dike shapes are not consistent with a purely elastic model of the host rock and also include some inelastic behavior and host-rock property variations. It is unclear whether segment sinuosity in the field area reflects primary features of the fault plane that was exploited by the dike or whether these features were superposed on the fault in 1969 by the propagating dike. Sinuous features of the longest wavelength span minor vent offsets and are most pronounced in segment III but subtly visible in segment V (Figure 17.3).

17.5.2.3. Irregularity

We interpret irregularities in fissure walls to be inherited from the joint patterns in the wall rock, which consists of lava flows ~400–600 years old. Fault propagation likely utilized these joints and weaknesses, which formed during cooling contraction [*Ryan and Sammis*, 1978; *Pollard et al.*, 1982; *Pollard and Aydin*, 1988; *Goehring et al.*, 2006; *Holland et al.*, 2006]. Cooling joints define columns with edge or face widths averaging 26 cm wide but ranging from millimeters to tens of meters [*Ryan and Sammis*, 1978; *Aydin and DeGraff*, 1988; *Dance et al.*, 2001; *Goehring and Morris*, 2008]; these measurements coincide with the range of documented irregularities of the Mauna Ulu fissure.

17.5.2.4. Flaring

In general, it is thought that vents can become enlarged during eruptions through thermal erosion, mechanical erosion, drain-back, or a combination of these processes. Thermal erosion of the cold wall rock is impossible on the time scale available here [*Bruce and Huppert*, 1989; *Kauahikaua et al.*, 1998; *Fagents and Greeley*, 2001; *Williams et al.*, 2004; *Petcovic and Dufek*, 2005], but mechanical removal (i.e., plucking) of blocks [*Peterson and Swanson*, 1974] by the eruptive gas-pyroclast mixture expanding to achieve a lithostatic (and subsequently atmospheric) pressure can occur over the span of hours [*Mitchell*, 2005] if the wall rock is prebrecciated or jointed [*Delaney and Pollard*, 1981; *Wylie et al.*, 1999]. Removal of wall rock widens the conduit, reduces flow resistance, and allows the eruptive material to focus in these preferred pathways [*Delaney and Pollard*, 1981; *Wylie et al.*, 1999]. This may further enhance mechanical erosion along those flow paths [*Wylie et al.*, 1999], especially during vigorous flow conditions [*Williams et al.*, 2004]. However, computer models show that conduit sheer stresses in basaltic fissures are not sufficiently high to promote syn-eruptive fracturing and removal of large pieces of wall rock [*Mitchell*, 2005]. Mitchell proposed that steady and continuous mechanical abrasion is the most feasible mechanism for widening a basaltic fissure, although this model does not exclude removal of prefractured rock. Since Mauna Ulu's episode 1 fissure has drain-back textures coating all but four of the vents, it is unclear if flaring takes place exclusively in the 24 May 1969 flow or in the conduit walls. We propose that flaring results either from fluid dynamical processes during drain-back (constrained

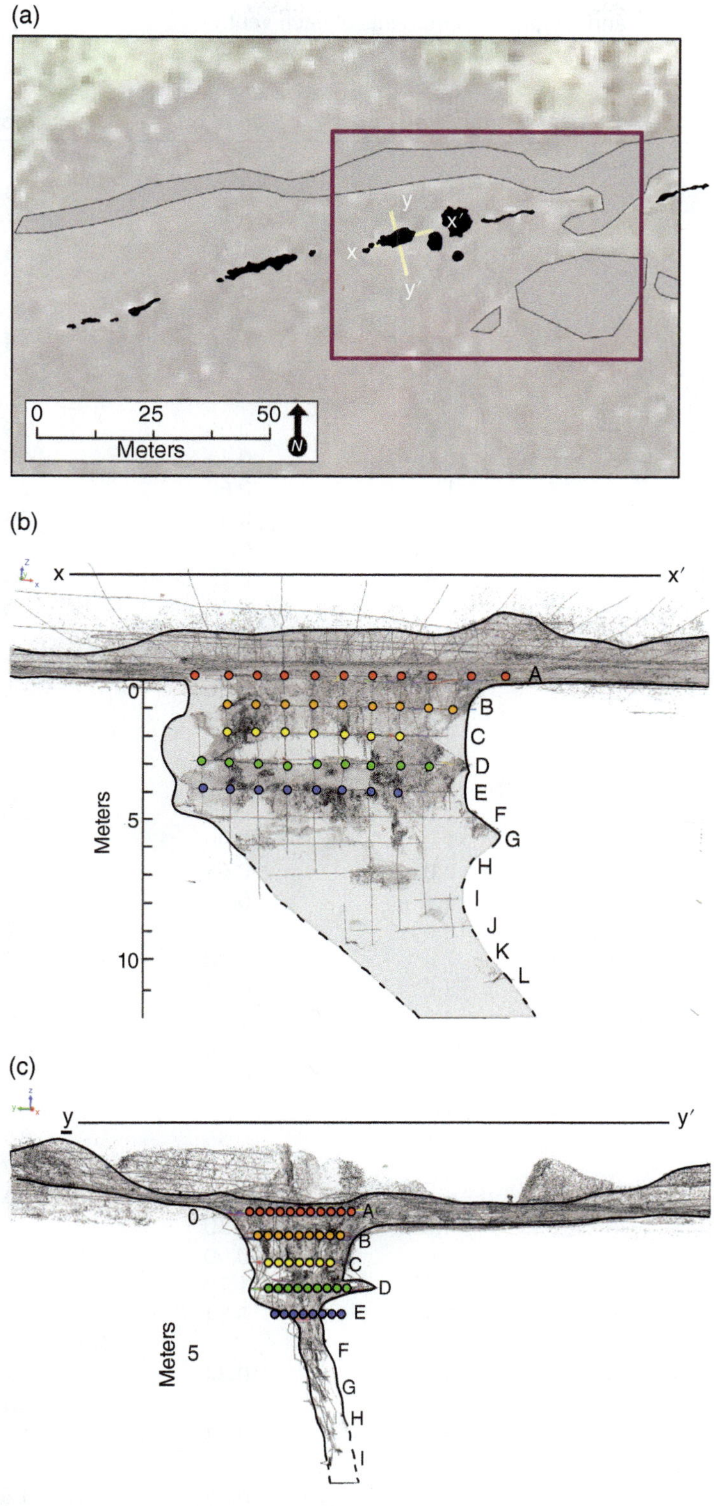

Figure 17.7 (a) Map showing part of segment V; location of LiDAR area 3 is marked with the purple box. Yellow lines indicate cross sections *x-x′* in (b) and *y-y′* in (c). (b) View of LiDAR point cloud perpendicular to vent/conduit strike. View angle is to the north. Black lines denote boundaries of the LiDAR cloud. Black line is dashed where the fissure wall is presumed to be, but LiDAR data are too scarce or have too many holes to know with certainty. Colored circles at grid point intersections indicate locations of vent width measurements. *A–L* correspond to the transects listed in Table 17.5. (c) View of LiDAR point cloud down strike to the east. Black lines denote ground surface in the closest parts of the LiDAR cloud, but ramparts east of the vent can be seen above this line. Colored circles extending across the vent/conduit correspond to the depths of width measurements and correlate to the same colors/measurements in (b). *A–I* correspond to transects listed in Table 17.5.

Table 17.6 Table of orientation, depth, and length measurements of each vent.

Seg. Number	Vent Number	Vent Strike	Vent Length (m)	Mean Vent Width (m)	Vent Depth (m)	Depth Location Along Strike (from west end) (m)
I	1	—	1.1	0.60	2.3	—
	2	—	1.7	0.41	3.3	—
	3	—	1.5	0.55	15	—
	4	—	1.5	1.40	2.3	—
	5	—	5.8	0.66	1.7	5
					1.5	1
	6	—	1.1	0.66	4.4	—
	7	77	9.6	0.51	4.8	4.9
					13	9.1
	8	78	2.6	0.96	6.9	4.3
II	9	75	21	0.44	4.6	19
	10	—	1.3	0.60	23	—
	11	—	4.3	0.52	39	4.0
	12	76	8	0.21	26	5.5
	13	—	1.5	0.25	30	—
	14	—	0.9	0.27	6.5	—
III	15	83	9.1	3.23	2.0	1.0
					3.6	9.0`
	16	—	3.2	1.83	5.2	3.0
	17	—	9.0	3.33	27	4.0
	18	—	4.4	3.20	2.9	2.0
	19	94	6.8	3.50	5.0	6.0
	20	—	1.9	2.21	1.9	1.0
	21	—	1.0	0.90	1.6	—
	22	—	2.0	0.65	na	na
	23	69	1.9	0.56	1.9	1.0
	24	73	8.0	1.83	1.6	7.0
	25	87	7.4	2.00	1.5	3.0
	26	92	2.8	0.30	2.5	1.4
	27	—	11	3.08	36	2.0
	28	—	1.3	0.80	4.1	—
	29	84	22	0.96	17	9.1
	30	—	1.3	0.95	0.40	—
	31	75	25	0.55	30	1.0
					4.7	3.7
					30	23
	32	—	4.0	0.30	41	1.5
	33	—	2.0	0.90	13	1.5
IV	34	—	1.4	1.30	1.4	—
	35	70	20	0.54	11	11
					20	20
	36	74	4.5	0.44	16	14
	37	—	25	0.63	30	27
	38	—	3.0	0.64	4.4	1.0
					4.1	2.5
	39	—	5.0	0.65	3.8	2.0
	40	—	11	3.40	4.8	2.0
V	41	—	3.8	0.38	1.2	1.0
	42	—	1.5	0.40	3.0	1.0
	43	—	6.5	0.47	1.5	1.4
	44	—	17	1.51	7.3	1.2
	45	—	10	0.60	4.2	8.0
	46	—	2.6	0.38	0.6	—

Table 17.6 Cont'd

Seg. Number	Vent Number	Vent Strike	Vent Length (m)	Mean Vent Width (m)	Vent Depth (m)	Depth Location Along Strike (from west end) (m)
	47	—	13	4.24	22	3.0
	48	—		3.80	1.5	2.0
	49	—	2.7	2.50	2.4	1.5
	50	—	6.2	7.80	25	2.0
	51	82	22	0.44	6.5	7.3
	52	72	35	0.53	4.2	5.5
					4.0	26
	53	73	27	0.77	30	8.4
					17	17
	54	75	68	0.58	13	8.5
					4.5	30
					30	56

Note: Strike, when the fissure vent was long enough to have an orientation, was measured with a Brunton compass. One strike measurement was taken for the whole vent, from the western tip to the eastern tip. Fissure depth was measured in each vent by a plumb bob attached to a 100 m tape measure. Depth was recorded at the transition between when the plumb bob would freely drop and when the tape became slack. For some vents, the plumb bob did not reach the bottom of the fissure and the maximum length of the tape is recorded as the measurement. The distance along the vent length of each depth measurement (as well as total vent length), from the *vent's western tip*, is listed when the vent is a fissure geometry. Vent 22 is filled with plant material, making a depth measurement impossible and is listed as "na."

by rheology, thermal gradients, and fissure width) if the flare is contained entirely within 1969 flow or from mechanical erosion of the vent walls if the flare cuts into to preexisting flows. Due to the 10 m resolution of the inferred lava thickness data by *Parcheta et al.* [2012] and the fact that 95% of vents are coated with drain-back, we cannot ascertain which explanation is more likely for the Mauna Ulu vents. Further work is needed to determine how much of the flaring is in 1969 lava versus wall rock.

17.5.2.5. Vent Localization

Flow focusing (i.e., vent localization) can be dynamically induced by fingering instabilities resulting when small temperature gradients cause large viscosity increases, which in turn cause the unfragmented magma to move sluggishly or to stagnate [*Wylie et al.*, 1999]. Fingering instabilities occur at low discharge rates [*Wylie et al.*, 1999] and would favor a tendency toward a few fountains (or a single fountain) along a fissure rather than many fountains. Focusing of activity to a few loci does not necessarily imply that the source driving pressure is dropping [*Wylie et al.*, 1999]; eruptive activity along a fissure can localize to a main vent(s) if the magma pressure supplying the eruption is weak but steady. The 24 May fissure provides conditions favorable for thermoviscous fingering instabilities in the magma prior to fragmentation as the eruption rate was low ($1.3 \times 10^5\,m^3/h$ [*Swanson et al.*, 1979]) and steady and the geometry broadly rectilinear. Viscosity-induced flow stagnation provides a powerful mechanism for flow localization and was found to be more significant than quenching for flow localization [*Wylie et al.*, 1999]. Flow localization also sets the premise that fountains did not shift position along the fissure but remained fixed in space during temporal evolution of the vents. *Delaney and Pollard* [1982] note that dike thickness affects neither the amount nor the rate of heat transferred to the wall rock during the eruption and that those magmatic thermal properties affect dike geometry only through their influence on internal magma pressure distribution. They conclude that rising magma will cool and quench within a day, when flow is laminar through a narrow dike. For Hawaiian eruptions to last longer than a day, localization to wider central vents is necessary [*Delaney and Pollard*, 1982].

As the Mauna Ulu fissure fountains waned, any lava ponded on the surface was able to drain back into the system [*Parcheta et al.*, 2012]. It seems likely that the same localized pathways used for magma ascent were also used for magma drain-back and descent. Furthermore, drain-back lava that descended in these pathways had a sufficiently low viscosity (i.e., remained hot enough) that vents and the shallow conduit were unobstructed for several meters to 41 m depth. The dike may currently be held open in the manner described by *Daniels et al.* [2012], where lava cooling at the tips of segments restricts the center from elastically closing after magma drains away.

The fault the dike intruded may have opened in a mode I manner [*Holland et al.*, 2006] during eruption and never relaxed back after the eruption was over. Geodetic data show minimal ground breakage over the last several decades in the field area (Hawaiian Volcano Observatory, unpublished data); we assume, therefore, that the widths we have measured are the same as they were in 1969 and that the fissure system has not structurally widened or narrowed in the 43 years since the eruption. Narrow fissures with no flaring at Mauna Ulu likely represent the original shape of the dike or exploited fault plane.

17.5.2.6. Vent Depths

Chilling of a dike's crack tips may act to prevent the dike from closing fully as the eruption wanes and magma overpressures drop, prolonging the duration over which an eruption may continue [*Daniels et al.*, 2012]. This could have aided the drain-back at Mauna Ulu and might help explain the depths of vents measured in the field (mean of 30 m).

17.5.3. Constraints on Fissure Width for Models of Weak Basaltic Fountains

Vent or fissure width is an important input parameter for models of Hawaiian eruption dynamics. Most studies do not explicitly state a source for this input but have used a range of values between 3 and 14 m for circular vents [*Wilson and Head*, 1981; *Head and Wilson*, 1989; *Parfitt and Wilson*, 1995] and 0.2–10 m for linear vents [*Wilson and Head*, 1981; *Hardee*, 1987; *Head and Wilson*, 1989; *Parfitt and Wilson*, 1994]. Our data suggest that a width value of ≤1 m is appropriate for terrestrial Hawaiian conduits and nonflared linear vents, with a surface width value of ~1.5 m for flared linear vents.

Oversimplification of the fissure geometry in computational and conceptual models, while providing a fruitful first step forward, can inhibit a complete understanding of eruption dynamics. When fissures are only 1–2 m thick, with irregularities and perturbations of the fissure wall averaging 50 cm (25%–50% of the dike thickness), simplification of the system as linear with parallel-sided walls removes the effects of important fluid-dynamic processes from model results and thus from our conceptualization and understanding of volcanic processes. The next step forward in our understanding of volcanic eruptions is to utilize a more realistic (if still simplified) conduit geometry in our models, especially over depth ranges for which important processes, such as bubble and crystal nucleation, occur. Future modeling of low-intensity fissure fountains should, therefore, adopt width values in the lower end of the range previously used by researchers and concentrate on 1–2 m widths with some irregularity included. Depending on the length of the fissure in 3D models, sinuosity may also need to be considered in conduit shape; for sufficiently short lengths (i.e., less than 10 m wavelengths), however, it can be disregarded in computational eruption models.

17.6. CONCLUSIONS

This contribution describes the detailed near-surface geometry of the west end of the 24 May 1969 Mauna Ulu eruptive fissure at Kīlauea. It is characterized by en echelon geometry, subtle sinuosity, noticeable irregularity of fissure width, and development of at least 54 vents along 880 m of fissure. We interpret the en echelon pattern as an early feature derived from a single 80° trending dike that rose with minimal lateral propagation to a shallow level from Kīlauea's ERZ before segmenting and subsequently extending westward 1 km through interaction with the ENE trending Koa'e Fault System. Sinuosity likely developed from the interactions of the stress field produced by the interaction of the ERZ and the eastern Koa'e Fault System and could be a primary feature of the fault plane exploited by the dike or could have been superimposed on the fault by the intruding dike. Fissure irregularity is most likely the result of preexisting cooling joints in the wall rock that were opened in a mode I fashion during dike emplacement. Vent localization was the inevitable consequence of cooling-induced increase in viscosity leading to flow stagnation toward the margins and ends of fissure segments. The distribution of flared versus nonflared vents shows no obvious relationship to fountain height (inferred from rampart geometry). Instead, we suggest that flaring, where present, is a result of either (1) syn-eruptive mechanical erosion or (2) late-stage dynamic response of lava ponded above the pre-1969 surface draining back into the system. Data on the contact of 1969 lava and pre-1969 lava are needed to differentiate between the two possibilities. Where vents are not flared and have narrow parallel-sided conduit walls, we interpret that the original geometry of the magmatic pathway is preserved, in some locations to depths of approximately 30 m.

ACKNOWLEDGMENTS

We thank Matt Patrick, Steve Self, and Nadine Friese for helpful and thoughtful reviews. We thank Mike Poland for use of the kinematic GPS and subsequent data processing as well as the rest of HVO for logistical support during field work and valuable discussions about the data. This work was funded by the NSF through grants EAR0810332 and EAR1145159 and a one-year Hawaiian Volcano Observatory Graduate Fellowship.

REFERENCES

Aydin, A., and J. M. DeGraff (1988), Evolution of polygonal fracture patterns in lava flows, *Science*, *239*, 471–4476.

Banks, N. G., F. W. Klein, R. B. Moore, D. B. Jackson, and A. T. Okamura (1981), Kilauea Volcano November 1979 eruptions, *Eos Trans. AGU*, *62*(45), 1072, Abstract.

Banks, N. G., E. W. Wolfe, T. A. Duggan, A. T. Okamura, R. Y. Koyanagi, L. P. Greenland, and D. B. Jackson (1983), Magmatic events at Kilauea Volcano, Hawaii, 1982, *Eos Trans. AGU*, *64*(45), 901–902, Abstract.

Bruce, P. M., and H. E. Huppert (1989), Solidification and melting along dykes by the laminar flow of basaltic magma, in Magma Transport and Storage, edited by M. P. Ryan, pp. 87–101, Wiley, New York.

Cayol, V., and F. H. Cornet (1997,) 3D Mixed boundary element for elastostatic deformation field analysis, *Int. J. Rock. Mech. Min. Sci.*, *34*(2), 275–287.

Dance, M., P. L., Hancock, R. S. J., Sparks, and A. Wallman (2001), Fracture and surface crust development in a Holocene pahoehoe lava flow on the island of Tenerife, *Canaries J. Struct. Geol.*, *23*, 165–182.

Daniels, K. A., J. L. Kavanagh, T. Menand, and S. Sparks (2012), The shapes of dikes: Evidence for the influence of cooling and inelastic deformation, *Geol. Soc. Am. Bull.*, *124*(7–8), 1102–1112, doi:10.1130/B30537.1.

Day, S. J., C. Carracedo, H. Guillou, and P. Gravestock (1999), Recent structural evolution of the Cumbre Vieja Volcano, La Palma, Canary Islands: Volcanic rift zone reconfiguration as a precursor to volcano flank instability? *J. Volcanol. Geotherm. Res.*, *94*, 135–167.

DeGraff, J. M., and A. Aydin (1987), Surface morphology of columnar joints and its significance to mechanics and direction of joint growth, *Geol. Soc. Am.*, *99*, 605–617.

Delaney, P. T., and D. D. Pollard (1981), Deformation of host rocks and flow of magma during growth of minnette dikes and breccias-bearing intrusions near Ship Rock, New Mexico, U.S. Geol. Surv. Prof. Pap. 1202:1–61.

Delaney, P. T., and D. D. Pollard (1982), Solidification of basaltic magma during flow in a dike, *Am. J. Sci.*, *282*, 85–885.

Delaney, P. T., D. D. Pollard, J. I. Ziony, and E. H. McKee (1986), Field relations between dikes and joints: Emplacement processes and paleo stress analysis, *J. Geophys. Res.*, *91*(B5), 4920–4938.

Duffield, W. A., R. L.,Christiansen, R. Y., Koyanagi, and D. W., Peterson (1982), Storage, migration, and eruption of magma at Kilauea Volcano, Hawaii 1971–1972, *J. Volcanol. Geotherm. Res.*, *13*, 273–307.

Ebinger, C. J., and M. Casey (2001), Continental Breakup in magmatic provinces: An Ethiopian example, *Geology*, *29*, 527–530.

Fagents, S. A., and R. Greeley (2001), Factors influencing lava-substrate heat transfer and implications for thermomechanical erosion, *Bull. Volcanol.*, *62*, 519–532, doi:10.1007/s004450000113.

Fiske, R. S., and R. Y. Koyanagi (1968), December 1965 eruption of Kilauea volcano, Hawaii, U.S. Geol. Surv. Prof. Pap. 607.

Fornari, D., et al. (2004), *Submarine Lava Flow Emplacement at the East Pacific Rise 9° 50′ N: Implications for Uppermost Ocean Crust Stratigraphy and Hydrothermal Fluid Circulation*, Geophys. Monogr. 148, AGU, Washington, D.C., doi:10.1029/148GM08.

Giberti, G., and L. Wilson (1990), The influence of geometry on the ascent of magma in open fissures, *Bull. Volcanol.*, *52*, 515–521.

Goehring, L., and S. W. Morris (2008), Scaling of columnar joints in basalt, *J. Geophys. Res.*, *113*, B10203, doi:10.1029/2007JB005018.

Goehring, L., S. W. Morris, and X. Lin (2006), Experimental investigation of the scaling of columnar joints, *Phys. Rev. E.*, *74*(3), 036115–036127, doi:10.1103/PhysRevE.74.036115.

Grant, J. V., and S. A. Kattenhorn (2004), Evolution of vertical faults at an extensional plate boundary, southwest Iceland, *J. Struct. Geol.*, *26*, 537–557.

Gudmundsson, A. (2003), Surface stresses associated with arrested dykes in rift zones, *Bull. Volcanol.*, *65*, 606–619.

Hardee, H. C. (1987), Replenishment rates of crustal magma and their bearing on potential sources of thermal energy, *J. Volcanol. Geotherm. Res.*, *28*(3–4), 275–296.

Head, J., and L. Wilson (1989), Basaltic pyroclastic eruptions: Influence of gas-release patterns and volume fluxes on fountain structure, and the formation of cinder cones, spatter cones, rootless flows, lava ponds, and lava flows, *J. Volcanol. Geotherm. Res.*, *37*(3–4), 261–271, doi:10.1016/0377-0273(89)90083-8.

Heliker, C., and T. N. Mattox (2003), The first two decades of the Pu'u 'Ō'ō–Kupaianaha eruption: Chronology and selected bibliography, in *The Pu'u 'Ō'ō-Kupaianaha eruption of Kīlauea Volcano, Hawai'i: The First 20 Years*, edited by C. Heliker et al., pp. 1–27, U. S. Geol. Surv. Prof. Pap. 1676.

Holland, M., J. L. Urai, and S. J. Martel (2006), The internal structure of fault zones in basaltic sequences, *Earth Planet. Sci. Lett.*, *246*(1–2), 301–315.

Jackson, J. B., D. A. Swanson, R. Y. Koyanagi, and T. L. Wright (1975), The August and October 1968 east rift eruptions of Kilauea Volcano, Hawaii, U.S. Geol. Surv. Prof. Pap. 890:1–33.

Jenness, M. H., and A. E. Clifton (2009), Controls on the geometry of a Holocene crater row: A field study from southwest Iceland, *Bull. Volcanol.*, *71*, 715–728.

Kauahikaua, J., K. V. Cashman, T. N. Mattox, C. C. Heliker, K. A. Hon, M. T. Mangan, and C. R. Thornber (1998), Observations on basaltic lava streams in tubes from Kilauea volcano, Hawaii, *J. Geophys. Res.*, *103*, 27,303–27,324.

Kinoshita, W. T., R. Y. Koyanagi, T.L Wright, and R. S. Fiske (1969), Kilauea Volcano; the 1967–1968 summit eruption, *Science*, *166*, 459–468.

Lockwood, J. P., R. I. Tilling, R. T. Holcomb, F. Klein, A. T. Okamura, and D. W. Peterson (1999), Magma migration and resupply during the 1974 summit eruptions of Kilauea Volcano, Hawai'i, U.S. *Geol. Surv. Prof. Pap. 1613*:1–43.

MacDonald, G. A. (1959), The activity of Hawaiian volcanoes during the years 1951–1956, *Bull. Volcanol.*, *2*(15), 120–179.

Macdonald, G. A., and J. P. Eaton (1957), Hawaiian volcanoes during 1954, *U.S. Geol. Surv. Bull.*, *1061*(B), 17–72.

Macdonald, G. A., and J. P., Eaton (1964), Hawaiian volcanoes during 1955, *U.S. Geol. Surv. Bull.*, *1171*, 170.

Mitchell, K. L. (2005), Coupled conduit flow and shape in explosive volcanic eruptions, *J. Volcanol. Geotherm. Res.*, *143*(1–3), 187–203 doi:10.1016/j.volgeores.2004.09.017.

Montgomery-Brown, E. K., D. K. Sinnett, M. Poland, P. Segall, T. Orr, H. Zebker, and A. Miklius (2010), Geodetic evidence for en echelon dike emplacement and concurrent slow slip during the June 2007 intrusion and eruption of Kilauea volcano, Hawaii, *J. Geophys. Res.*, *115*, B07405, doi:10.1029/2009JN006658.

Moore, J. G., and R. Y. Koyanagi (1969), The October 1963 eruption of Kilauea volcano, Hawaii, U.S. Geol. Surv. Prof. Pap, 614-c.

Moore, J. G., and H. L. Krivoy (1964), The 1962 flank eruption of Kilauea Volcano and structure of the east rift zone, *J. Geophys. Res.*, *69*, 2033–2045.

Moore, R. B., R. T. Helz, D. Dzurisin, G. P. Eaton, R. Y. Koyanagi, P. W. Lipman, J. P. Lockwood, and G. S., Puniwai (1980), The1977 eruption of Kilauea Volcano Hawaii, *J. Volcanol. Geotherm. Res.*, *7*, 189–210.

Morris, A. R., F. S. Anderson, P. J. Moughinis-Mark, A. F. C. Haldermann, B. A. Brooks, and J. Foster (2008), Roughness of Hawaiian volcanic terrains, *J. Geophys. Res.*, *113*, E12007, doi:10.1029/2008JE003079.

Nyquist, H. (1928), Certain topics in telegraph transmission theory, *Trans. Am. Inst. Electr. Eng.*, *47*, 617–644.

Parcheta, C., B. F. Houghton, and D. A. Swanson (2012), Hawaiian Fissure Fountains 1: Decoding deposits — Episode 1 of the 1969–1974 Mauna Ulu eruption, *Bull. Volcananal.*, *74*(7), 1729–1743, doi:10.1007/s00446-012-0621-1.

Parfitt, E. A., and L. Wilson (1994), The 1983–86 Pu'u 'O'o eruption of Kilauea Volcano, Hawaii: A study of dike geometry and eruption mechanisms for a long-lived eruption, *J. Volcanol. Geotherm. Res.*, *59*(3), 179–205.

Parfitt, E. A., and L. Wilson (1995), Explosive volcanic eruptions—IX. The transition between Hawaiian-style lava fountaining and Strombolian explosive activity, *Geophys. J. Int.*, *121*(1), 226–232.

Peck, D. L., and W. T. Kinoshita (1976), the eruption of August 1963 and the formation of Alae lava lake, Hawaii, U.S. Geol. Surv. Prof. Pap. 935(A):1–33.

Petcovic, H. L., and J. D. Dufek (2005), Modeling magma flow and cooling in dikes: Implications for emplacement of Columbia River flood basalts, *J. Geohpys. Res.*, *110*, B10201, doi:10.1029/2004JB003432.

Peterson, D. W. and D. A. Swanson (1974), Observed formation of lava tubes during 1970–1971, Kīlauea volcano, Hawaii, *Stud. Speleol.*, *2*, 209–222.

Peterson, D. W., R. L. Christiansen, W. A. Duffield, R. T. Holcomb, and R. I. Tililing (1976), Recent activity of Kilauea Volcano, Hawaii, in *Symposium on Andean and Antarctic Volcanology Problems,* vol. 75 edited by O. Gonzales-Ferran, 1974, pp. 271–293, Int. Assoc. Volcanol. Chem. Earth Int., Santiago, Chile.

Pollard, D. D. (1987), Elementary fracture mechanics applied to the structural interpretation of dykes, *Mafic Dike Swarms*, edited by H. C. Halls and W. F. Fahrig, pp. 5–24, Special Pap. 34, Geol. Assoc., Ontario, Canada.

Pollard, D. D., and A. Aydin (1988), Progress in understanding jointing over the past century, *Geol. Soc. Am. Bull.*, *100*(8), 1181–1204, doi:10.1130/0016-7606(1988)100<1181:PIUJOT>2.3.CO;2.

Pollard, D. D., and O. H. Muller (1976), The effect of gradients in regional stress and magma pressure on the form of sheet intrusions in cross section, *J. Geophys. Res.*, *81*, 975–984.

Pollard, D. D., P. Segall, and P. T. Delaney (1982), Formation and interpretation of dilatants echelon cracks, *Geol. Soc. Am. Bull.*, *93*(12), 1291–1303, doi:10.1130/0016-7606(1982)93<1291:FAIODE>2.0.CO;2.

Ryan, M. P., and C. G. Sammis (1978), Cyclic fracture mechanisms in cooling basalt, *Geol. Soc. Am. Bull.*, *89*(9), 1295–1308, doi:10.1130/0016-76-6(1978)89<1295:CFMICB>2.0.CO;2.

Richter, D. H., W. U. Ault, J. P. Eaton, and J. G. Moore (1964), The 1961 eruption of Kilauea Volcano, Hawaii, U.S. Geol. Surv. Prof. Pap. 474(D):1–34.

Richter, D. H., J. P. Eaton, K. J. Murata, W. U. Ault, and H. L. Krivoy (1970), Chronological narrative of the 1959–60 eruption of Kilauea Volcano, Hawaii, U.S. Geol. Surv. Prof. Pap. 537–E:E1–E73.

Self, S., T. Thordarson, L. Keszthelyi, G. P. L. Walker, K. Hon, M. T. Murphy, P. Long, and S. Finnermore (1996), A new model for the emplacement of the Columbia River basalts as large, inflated pahoehoe lava flow fields, *Geophys. Res. Lett.*, *23*(19), 2689–2692.

Sigurdsson, H. (2000), Introduction, in edited by H. Sigurdsson et al., Encyclopedia of Volcanoes, pp. 1–13, Academic Press, San Diego.

Sommer, E. (1969), Formation of fracture "lances" in glass, *Eng. Fract. Mech.*, *1*, 539–546.

Swanson, D. A., D. B. Jackson, R. Y. Koyanagi, and T. L. Wright (1976), The February 1969 east rift eruption of Kilauea Volcano, Hawaii, U.S. Geol. Surv. Prof. Pap. 891:1–30.

Swanson, D. A., D. A. Duffield, D. B. Jackson, and D. W. Peterson (1979), Chronological narrative of the 1969–1971 Mauna Ulu Eruption of Kilauea volcano, Hawaii, U.S. Geol. Surv. Prof. Pap. 1056.

Tilling, R. I., R. Y. Koyanagi, P. W. Lipman, J. P. Lockwood, J. G. Moore, and D. A. Swanson (1976), Earthquake and related catastrophic events, Island of Hawaii, November 29,1975: A preliminary report, U.S. Geol. Surv. Circ. 740:1–33.

Tilling, R. I., R. L. Christiansen, W. A. Duffield, E. T. Endo, R. T. Holcomb, R. Y. Koyanagi, D. W. Peterson, and J. D. Unger (1987), The 1972–1974 Mauna Ulu eruption, Kilauea Volcano: An example of quasi steady-state magma transfer, U.S. Geol. Surv. Prof. Pap. 1350:405–469.

Valentine, G. A. and K. E. C. Krogh (2006), Emplacement of shallow dikes and sills beneath a small basaltic volcanic center–the role of pre-existing structure (Paiute Ridge, southern Nevada, USA), *Earth Planet. Sci. Lett.*, *246*, 217–230.

Walker, G. P. L. (1991), Structure, and origin by injection of lava under surface crust, of tumuili, "lava rises," "lava-rise pits," and "lava inflation clefts" in Hawai'i, *Bull. Volcanol.*, *53*, 546–558.

Williams, D. A., S. D. Kadel, R. Greeley, C. M. Lesher, and M. A. Clynne (2004), Erosion by flowing lava: Geochemical evidence in the Cave Basalt, Mount St. Helens, Washington, *Bull. Volcanol.*, *66*:168–181, doi:10.1007/s00445-003-0301-2.

Wilson, L. (1980), Relationships between pressure, volatile contents, and ejecta velocity in three types of volcanic explosions, *J. Volcanol. Geotherm. Res.*, *8*(2–4), 297–313.

Wilson, L., and J. Head (1981), Ascent and eruption of basaltic magma on the Earth and Moon, *J. Geophys. Res.*, *86*(B4), 2871–3001.

Wolfe, E. W., C. A. Neal, N. G. Banks, and T. J. Dugagn (1987), The Pu'u 'O'o eruption of Kilauea Volcano, Hawaii, episodes 1 through 2, January 3, 1984 through June 8, 1984, U.S. Geol. Surv. Prof. Pap. 1463:1–97.

Wolfenden, E., C. Ebinger, G. Yirgu, A. Deino, and D. Ayalew (2004), Evolution of the northern Main Ethiopian rift: Birth of a triple junction, *Earth Planet. Sci. Lett.*, *224*, 213–228.

Wright, T. L., K. T., Kinoshita, and D. L. Peck (1968), March 1965 eruption of Kilauea Volcano and the formation of the Makaopuhi lava lake, *J. Geophys. Res.*, *73*, 3181–3205.

Wylie, J. J., K. R. Helfrich, B. Dade, J. R. Lister, and J. F. Salzig (1999), Flow localization in fissure eruptions, *Bull. Volcanol.*, *60*(6), 432–440.

18

Kīlauea's 5–9 March 2011 Kamoamoa Fissure Eruption and Its Relation to 30+ Years of Activity From Pu'u 'Ō'ō

Tim R. Orr[1,2], Michael P. Poland[1], Matthew R. Patrick[1], Weston A. Thelen[1], A. Jeff Sutton[1], Tamar Elias[1], Carl R. Thornber[3], Carolyn Parcheta[2,5], and Kelly M. Wooten[1,4]

ABSTRACT

Lava output from Kīlauea's long-lived East Rift Zone eruption, ongoing since 1983, began waning in 2010 and was coupled with uplift, increased seismicity, and rising lava levels at the volcano's summit and Pu'u 'Ō'ō vent. These changes culminated in the four-day-long Kamoamoa fissure eruption on the East Rift Zone starting on 5 March 2011. About $2.7 \times 10^6 m^3$ of lava erupted, accompanied by ~15cm of summit subsidence, draining of Kīlauea's summit lava lake, a 113m drop of Pu'u 'Ō'ō's crater floor, ~3m of East Rift Zone widening, and eruptive SO_2 emissions averaging 8500 tonnes/day. Lava effusion resumed at Pu'u 'Ō'ō shortly after the Kamoamoa eruption ended, marking the onset of a new period of East Rift Zone activity. Multiparameter monitoring before and during the Kamoamoa eruption suggests that it was driven by an imbalance between magma supplied to and erupted from Kīlauea's East Rift Zone and that eruptive output is affected by changes in the geometry of the rift zone plumbing system. These results imply that intrusions and eruptive changes during ongoing activity at Kīlauea may be anticipated from the geophysical, geological, and geochemical manifestations of magma supply and magma plumbing system geometry.

18.1. INTRODUCTION

The East Rift Zone of Kīlauea (Figure 18.1) has hosted most of the volcano's eruptions since 1955, including the ongoing Pu'u 'Ō'ō eruption, which began in 1983. The 30+ years of eruptive activity that ensued have provided an outstanding opportunity to study the evolution of a prolonged eruption, the first multidecadal flank eruption at Kīlauea in over 500 years, with an ever-improving set of geological, geophysical, and geochemical monitoring tools. The 2008–ongoing eruption at Kīlauea's summit (Figure 18.1), the first recorded instance of simultaneous rift zone and summit eruption at Kīlauea lasting more than a year, provides new insights into the connection between Kīlauea's summit and rift zone magmatic systems.

Over the course of more than 30 years, eruptive episodes included episodic high lava fountaining from Pu'u 'Ō'ō (1983–1986); lava effusion from the Kupaianaha vent, 3km east of Pu'u 'Ō'ō (1986–1992); lava effusion from Pu'u 'Ō'ō with occasional uprift intrusions (1992–2007); and lava effusion from a vent 2km east of Pu'u 'Ō'ō (2007–2011). Repeated intrusions and fissure eruptions have resulted in eruptive pauses, reorganizations of lava flow and tube systems, and even the formation of new eruptive vents (Table 18.1). The early 1990s, in particular, included several intrusions in the upper East Rift Zone that caused eruptive pauses lasting hours to days and some minor collapses at Pu'u 'Ō'ō (Table 18.2). More significant intrusions and eruptions (Table 18.2) occurred in 1997 [*Owen et al.*, 2000; *Thornber et al.*, 2003], 1999 [*Cervelli et al.*,

[1] *Hawaiian Volcano Observatory, U.S. Geological Survey, Hawai'i National Park, Hawaii, USA*

[2] *Department of Geology and Geophysics, University of Hawai'i at Mānoa, Honolulu, Hawaii, USA*

[3] *Cascades Volcano Observatory, U.S. Geological Survey, Vancouver, Washington, USA*

[4] *Department of Geological and Mining Engineering and Sciences, Michigan Technological University, Houghton, Michigan, USA*

[5] *Jet Propulsion Laboratory, Pasadena, California, USA*

Hawaiian Volcanoes: From Source to Surface, Geophysical Monograph 208. First Edition.
Edited by Rebecca Carey, Valérie Cayol, Michael Poland, and Dominique Weis.

Companion Website: www.wiley.com/go/Carey/Hawaiian_Volcanoes

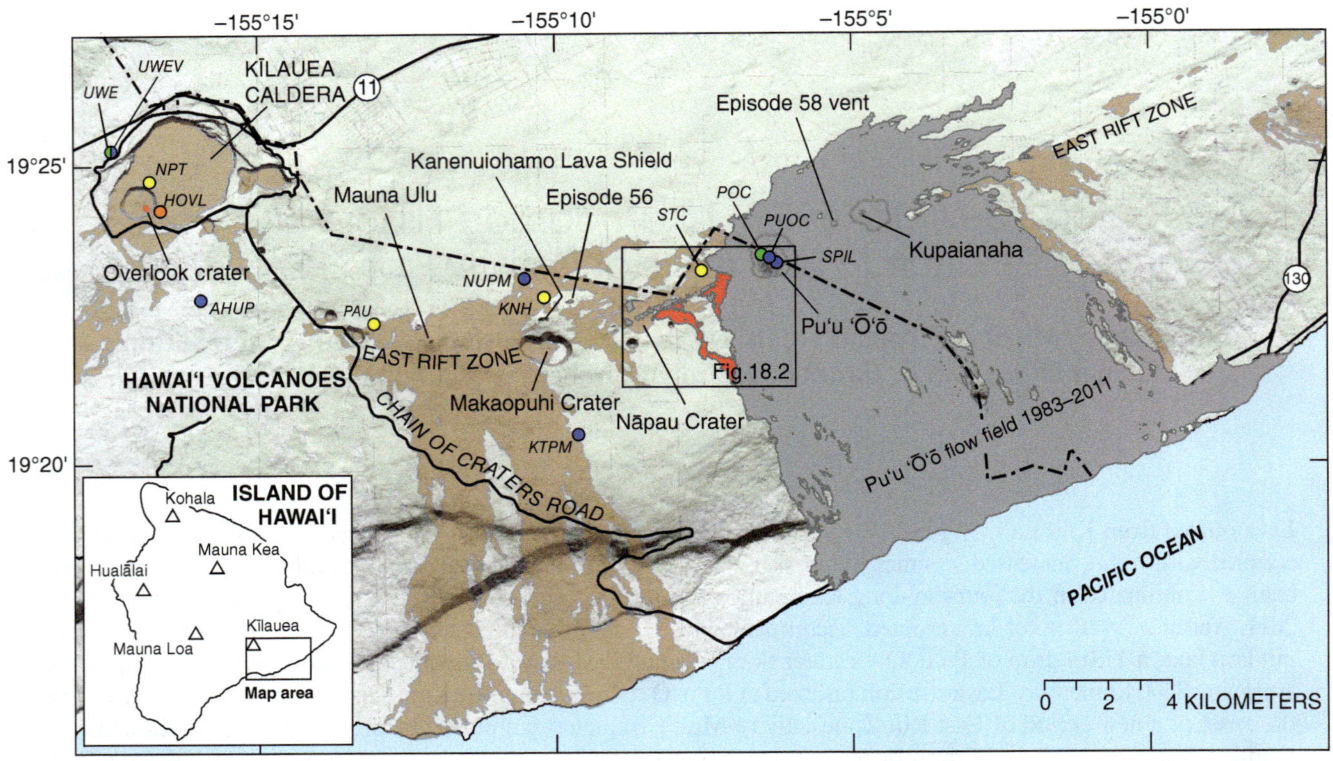

Figure 18.1 Map of Kīlauea's summit and East Rift Zone showing extent of flows erupted during 5–9 March 2011 Kamoamoa eruption (red) in relation to 1983–2011 Pu'u 'Ō'ō flow field (pre-Kamoamoa; gray) and historical flows (brown). Area depicted in Figure 18.2 outlined. Blue circles, continuously recording GPS stations spanning summit (UWEV–AHUP), east rift near Makaopuhi Crater (NUPM–KTPM), and Pu'u 'Ō'ō crater (PUOC–SPIL); orange circle, gravimeter (HOVL) near Overlook crater; yellow circles, seismic stations (NPT, PAU, KNH, STC); green circles, tiltmeters (UWE, POC). (Inset) Island of Hawai'i and its five volcanoes, showing map area.

2002], and 2007 [*Poland et al.*, 2008; *Montgomery-Brown et al.*, 2010], the latter of which was accompanied by a shift in the locus of eruption away from Pu'u 'Ō'ō to a new vent 2 km farther east (episode 58 vent; Figure 18.1).

During 5–9 March 2011, the Kamoamoa fissure eruption (Figures 18.1 and 18.2), a few kilometers west (uprift) of Pu'u 'Ō'ō, marked the end of the eruptive vent established east (downrift) of Pu'u 'Ō'ō in 2007 and presaged a return of activity to Pu'u 'Ō'ō that continued through 2013. The Kamoamoa eruption was exceptionally well monitored by an array of geophysical and geological observations, supplemented by gas measurements and lava sampling. We use these data to examine the dynamics of the eruption at a level of fine spatial and temporal detail, which provides insight into how such episodes occur and their effect on the long-lived Pu'u 'Ō'ō eruption.

We first review the history of the ongoing Pu'u 'Ō'ō eruption, which provides a context for our interpretation of the causes and consequences of the Kamoamoa fissure eruption. This review extends the work published by *Heliker and Mattox* [2003] and *Heliker et al.* [2003] by updating the eruption chronology in a generalized fashion through 2013. Eruptive activity is formalized as discrete episodes based on eruption location and continuity, as shown in Table 18.1, by summarizing the episodes discussed by *Heliker and Mattox* [2003] and defining new episodes to highlight changes in the ongoing eruption.

18.2. THE PU'U 'Ō'Ō ERUPTION

Eruptive activity at Kīlauea since 1983 has been dominated by discharge from East Rift Zone vents as part of a single prolonged sequence, the Pu'u 'Ō'ō eruption (Figure 18.3) [*e.g.*, *Heliker and Mattox*, 2003]. This eruptive activity is fed by basaltic magma thought to rise from a zone of partial melting in the upper mantle [e.g., *Eaton and Murata*, 1960; *Wright*, 1984; *Wyllie*, 1988; *Sen and Jones*, 1990; *DePaolo and Stolper*, 1996] and then to collect within storage reservoirs ~1–5 km beneath Kīlauea's summit [e.g., *Fiske and Kinoshita*, 1969; *Ryan*, 1987; *Baker and Amelung*, 2012]. From the summit reservoir, magma rises into Kīlauea's summit lava lake or is transported ~20 km through a conduit within the volcano's East Rift Zone (Figure 18.1) at a depth of ~3 km [e.g., *Klein et al.*, 1987; *Montgomery-Brown et al.*, 2010; *Lundgren et al.*, 2013] to the eruption site. Figure 18.4, a cross section of Kīlauea's summit and part of its East Rift Zone, conceptualizes the basic geometry of Kīlauea's magmatic system at the time of the Kamoamoa eruption.

Table 18.1 Eruption statistics for episodes 1–61 of Pu'u 'Ō'ō eruption as of November 2013.

Episode	Repose Period Before Episode (days)	Episode Start Date (mm/dd/yy)	Episode Start Time (HST)	Episode End Date (mm/dd/yy)	Episode End Time (HST)	Duration (days)	Subaerial Lava Coverage (km²)	DRE Volume (×10⁶ m³)
1–47		01/03/83	0031	06/26/86	1635	1,207.7	39.8	387
48–49		07/18/86	1205	02/07/92	(0000)	2,030	47.8	505
50–53		02/17/92	~1930	01/29/97	1852	1,808.0	32.7	618
54 (Nāpau eruption)		01/30/97	0240	01/31/97	0033	0.9	0.23	0.3
55	24.3	02/24/97	~0700	06/21/07	(0000)	3,768	62.1	1660
Crater filling		02/24/97	~0700	03/28/97	~1430	32.3	—	—
Ep 55 flow		03/28/97	~1430	08/17/02	(0000)	1,967	38.3	—
Mother's Day flow		05/12/02	~0835	09/23/04	(0000)	865	22.8	—
MLK flow		01/18/04	0554	01/24/04	(0000)	5.8	0.4	—
PKK flow		03/20/04	2112	06/21/07	(0000)	1,188	30.9	—
56 (Father's Day eruption)		06/19/07	0015	06/19/07	0045	0.02	.0022	[0.001]
57–58	12.7	07/01/07	~1800	03/07/11	(0000)	1,344	40.6	630
Crater filling		07/01/07	~1800	07/20/07	2355	19.2	0.07	[1.05]
Perched channel flow		07/21/07	0006	02/10/08	(0000)	204	10.3	—
TEB flow		11/21/07	0604	03/07/11	(0000)	1,202	30.6	—
59 (Kamoamoa eruption)		03/05/11	1709	03/09/11	~2230	4.2	1.39	[2.7]
60	16.5	03/26/11	1009	08/15/11	(1200)	142	2.31	40
Crater filling		03/26/11	1009	08/03/11	1418	130.2	.025	—
August breakout		08/03/11	1418	8/15/11	(1200)	12	2.28	—
61	5.2	08/20/11	1715	ongoing	—	—	22.09	209
Crater filling		08/20/11	1715	09/21/11	0220	31.4	0.048	—
Peace Day flow		09/21/11	0220	11/07/13	(0000)	778	16.03	—
Kahauale'a 1 flow		01/19/13	1123	04/18/13	(1200)	89	2.63	—
Kahauale'a 2 flow		05/06/13	1719	ongoing	—	—	4.27	—
Total							126.47	4052

Note: Episodes 1–54, shown in detail by *Heliker and Mattox* (2003), grouped here into appropriate categories based on eruption style and (or) location. Episodes 55–61 subdivided to highlight significant flows and eruptive periods. Episodes with imprecise end times shown in parentheses to indicate activity stopped overnight (0000) or during the day (1200). In these cases, duration rounded to nearest day. Determination of lava volumes for episodes 1–54 as described by *Heliker and Mattox* (2003). Determination of lava volumes for subsequent episodes derived from SO_2 emissions as described by *Sutton et al.* (2003) and *Elias and Sutton* (2012), except for volumes in brackets, which were calculated from mapped surface area and estimated thickness, assuming 25% vesicularity.

Table 18.2 Kīlauea East Rift Zone (ERZ) intrusions, with volume statistics where known, during ongoing Pu'u 'Ō'ō eruption July 1986–present.

Intrusion Date	DRE Volume ($\times10^6$ m^3)	Comment
[18–20 July 1986]		Episode 48 fissure eruption; intrusion from dike beneath Pu'u 'Ō'ō; precursory summit inflation
4 December 1990		Upper ERZ intrusion; eruptive activity increase at Pu'u 'Ō'ō; precursory summit inflation and upper ERZ seismicity
26 March 1991		Upper ERZ intrusion; precursory summit inflation and upper ERZ seismicity
21 August 1991		Upper ERZ intrusion; precursory summit inflation and upper ERZ seismicity
[8 November 1991]		Episode 49 fissure eruption; intrusion from Pu'u 'Ō'ō storage; precursory summit inflation
Pu'u 'Ō'ō crater collapse	(3.1)[3]	
[17 February 1992]		Episode 50 fissure eruption; intrusion from Pu'u 'Ō'ō storage; precursory summit inflation
Pu'u 'Ō'ō lava lake draining	(~0.2)[9]	
3 March 1992		Upper ERZ intrusion; precursory summit inflation; episode 50 terminated
Pu'u 'Ō'ō lava lake draining	(~0.1)[9]	
7–8 February 1993		Upper ERZ intrusion; no obvious precursors; episode 51 terminated
Pu'u 'Ō'ō crater collapse	(3.1)[3]	
29–30 January 1997		Middle ERZ intrusion; Nāpau fissure eruption (episode 54); no obvious precursors; passive rift opening[1]; episode 53 terminated
Pu'u 'Ō'ō crater collapse	(13)[3]	
Halema'uma'u reservoir loss	(1.5)[1]	
Makaopuhi reservoir loss	(0.12)[1]	
Episode 54 erupted volume	0.3[4]	
Dike volume	23[1]	
12 September 1999		Upper ERZ intrusion; no obvious precursors; passive rift opening[2]
Pu'u 'Ō'ō crater collapse	(<1)[3]	
Dike volume	3.3[2]	
23 February 2000		Upper ERZ intrusion; no obvious precursors
[12 May 2002]		Intrusion from Pu'u 'Ō'ō storage; Mother's Day breakout; precursory summit inflation, upper ERZ seismicity, and Pu'u 'Ō'ō filling
[18 January 2004]		Intrusion from Pu'u 'Ō'ō storage; MLK breakout; precursory summit inflation, upper ERZ seismicity, and Pu'u 'Ō'ō filling
17–19 June 2007		Upper and middle ERZ intrusion; Father's Day eruption (episode 56); precursory summit inflation and upper ERZ seismicity; episode 55 terminated
Pu'u 'Ō'ō crater collapse	(4.2)[8]	
Halema'uma'u reservoir loss	(1.8)[6]	
Pu'u 'Ō'ō reservoir loss	(0.02)[6]	
Episode 56 erupted volume	0.001[5]	
Dike volume	16.6[6]	
[21 July 2007]		Intrusion from Pu'u 'Ō'ō storage; episode 58 fissure eruption; precursory summit inflation, upper ERZ seismicity, and Pu'u 'Ō'ō filling
Pu'u 'Ō'ō crater collapse	(1.3)[10]	
March 2011		Middle ERZ intrusion; Kamoamoa fissure; precursory summit inflation, upper ERZ seismicity, and Pu'u 'Ō'ō and Halema'uma'u filling

Table 18.2 Cont'd

Intrusion Date	DRE Volume ($\times 10^6\, m^3$)	Comment
Pu'u 'Ō'ō crater collapse	(5.5)[10]	
Summit lava lake draining	(0.8)[10]	
Halema'uma'u reservoir loss	(1.7)[7]	
Episode 59 erupted volume	2.7[10]	
Outgassed volume	7.6[10]	
Dike volume	15.6[7]	
[3 August 2011]		Intrusion from Pu'u 'Ō'ō storage; Pu'u 'Ō'ō flank breakout; precursory summit inflation, upper ERZ seismicity, and Pu'u 'Ō'ō and Halema'uma'u filling
Pu'u 'Ō'ō crater collapse	(2.5)[10]	
[21 September 2011]		Intrusion from Pu'u 'Ō'ō storage; Peace Day fissure eruption; precursory summit inflation, upper ERZ seismicity, and Pu'u 'Ō'ō and Halema'uma'u filling
Pu'u 'Ō'ō crater sagging	(1.0)[10]	
October 2012		Upper ERZ intrusion; precursory summit inflation and upper ERZ seismicity
Summit lava lake draining	(0.2)[9]	

Note: Intrusions in brackets fed from Pu'u 'Ō'ō magma storage; intrusions not in brackets fed from summit or ERZ storage uprift from Pu'u 'Ō'ō. Volume losses shown in parentheses.
Source: [1]*Owen et al.*, 2000; [2]*Cervelli et al.*, 2002; [3]*Heliker et al.*, 2003; [4]*Thornber et al.*, 2003; [5]*Poland et al.*, 2008; [6]*Montgomery-Brown et al.*, 2010; [7]*Lundgren et al.*, 2013; [8]*Orr*, 2014; [9]Hawaiian Volcano Observatory data; [10]This study.

18.2.1. 1983–1986: High Fountaining at Pu'u 'Ō'ō

The Pu'u 'Ō'ō eruption began on 3 January 1983 following 24 h of seismicity that propagated down the East Rift Zone to the eventual eruption site, an 8 km long fissure system that first reached the surface about 15 km downrift of the summit. During the first few episodes of the eruption, activity jumped from one fissure to another (episodes 1–3) [*Wolfe et al.*, 1988] but eventually localized at a single vent, Pu'u 'Ō'ō, in June 1983. Thereafter, a series of high lava fountains produced short-lived 'a'ā flows that buried much of the surrounding area (episodes 4–47; Figure 18.3a; Table 18.1) [*Wolfe et al.*, 1988; *Heliker et al.*, 2003]. Tephra fallout and lava overflows from this vent built the Pu'u 'Ō'ō cone, which by 1986 had grown to a height of 255 m above the preeruption landscape of 719 m elevation [*Wolfe et al.*, 1988; *Heliker et al.*, 2003].

18.2.2. 1986–1992: Continuous Effusion from Kupaianaha

In July 1986, the conduit beneath Pu'u 'Ō'ō ruptured and fed fissures both uprift and downrift of Pu'u 'Ō'ō (Table 18.2). Within days, activity had focused 3 km downrift and eventually formed the Kupaianaha lava shield (Figure 18.3b). This marked a change in eruptive style from episodic lava fountaining to nearly continuous effusion (episode 48; Table 18.1). A broad, tube-fed pāhoehoe flow field developed over the next 5.5 years, extending southeast from the vent to the ocean and eventually engulfing the community of Kalapana in 1990 [*Heliker and Wright*, 1991; *Mattox et al.*, 1993; *Hon et al.*, 1994]. During effusion from Kupaianaha, the near-surface conduit at Pu'u 'Ō'ō widened progressively and formed a crater that often contained a small lava lake [*Heliker et al.*, 1998, 2003].

Lava effusion from Kupaianaha began to decline in mid-1990 [*Kauahikaua et al.*, 1996], and activity at Pu'u 'Ō'ō waxed commensurately, presumably as the conduit between the two vent systems became increasingly blocked [*Mangan et al.*, 1995]. The lava lake within Pu'u 'Ō'ō rose 65 m during this period and flooded the surrounding crater floor, and the summit of Kīlauea began to inflate [*Kauahikaua et al.*, 1996]. These changes were accompanied by several small intrusions into Kīlauea's upper East Rift Zone (Table 18.2). In November 1991, the lava lake withdrew and the crater floor fell 75 m as a fissure opened between Pu'u 'Ō'ō and Kupaianaha and erupted for three weeks (episode 49; Figure 18.3b; Table 18.2) [*Mangan et al.*, 1995; *Heliker and Mattox*, 2003]. Lava reappeared in Pu'u 'Ō'ō a few days after the end of the fissure eruption and refilled the crater to its previous level within weeks. The eruptive output from Kupaianaha, which dropped dramatically during episode 49, continued to fall afterward and finally ceased altogether on 7 February 1992 [*Kauahikaua et al.*, 1996].

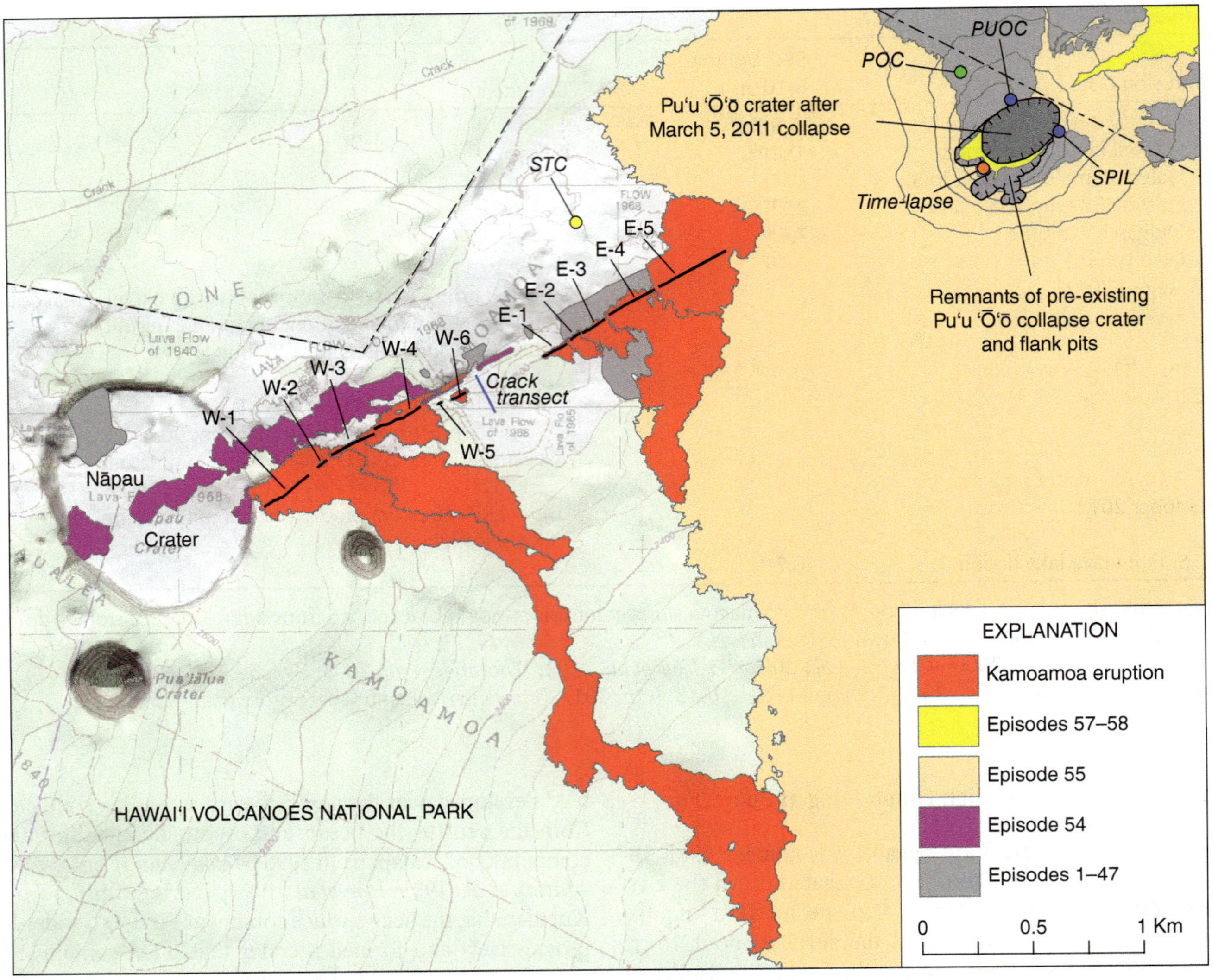

Figure 18.2 Map showing Kamoamoa lava flows (red) and eruptive fissures (heavy black lines) in relation to Pu'u 'Ō'ō, Nāpau Crater, and older Pu'u 'Ō'ō flows erupted 1983–2011. Eastern fissure system comprises fissures E-1 to E-5; western fissure system comprises fissures W-1 to W-6. Yellow circle, seismometer (STC); orange circle, time-lapse camera; green circle, tiltmeter (POC); blue circles, continuous GPS stations (PUOC, SPIL); blue line, crack measurement transect.

18.2.3. 1992–1997: Flank Vents and Shield Building at Pu'u 'Ō'ō

Ten days after Kupaianaha stopped erupting, new eruptive vents formed on the west flank of the Pu'u 'Ō'ō cone. This new epoch in the eruption, comprising episodes 50–53 (Figure 18.3c; Table 18.1), saw the progressive burial of the west and south flanks of Pu'u 'Ō'ō beneath a lava shield and the nearly continuous eruption of lava to widen the existing tube-fed pāhoehoe flow field south of the rift zone [*Heliker et al.*, 1998; *Heliker and Mattox*, 2003]. The lava level within Pu'u 'Ō'ō varied widely over this time period, with subsidence and collapse events accompanying intrusive activity along the upper East Rift Zone (Table 18.2) [*Clague and Heliker*, 1993; *Okubo et al.*, 1996; *Heliker et al.*, 1998, 2003].

Episode 53 ended on January 29 with an intrusion uprift of Pu'u 'Ō'ō in and adjacent to Nāpau Crater (Nāpau eruption; Table 18.2) [*Owen et al.*, 2000; *Thornber et al.*, 2003; *Heliker and Mattox*, 2003]. In response, Pu'u 'Ō'ō's crater floor which had been 60 m below the northeast rim, dropped 150 m, and part of the western wall of the Pu'u 'Ō'ō cone collapsed. The intrusion culminated in a brief eruption, episode 54 (Figure 18.3c; Table 18.1), which lasted for less than a day and was followed by a 24 day hiatus in eruptive activity [*Heliker and Mattox*, 2003].

18.2.4. 1997–2007: Resumption of Flank Vents and Shield Building

Lava reappeared within Pu'u 'Ō'ō's crater in late February 1997, marking the onset of episode 55 (Figure 18.3d; Table 18.1), and rose to within 50 m of the crater's

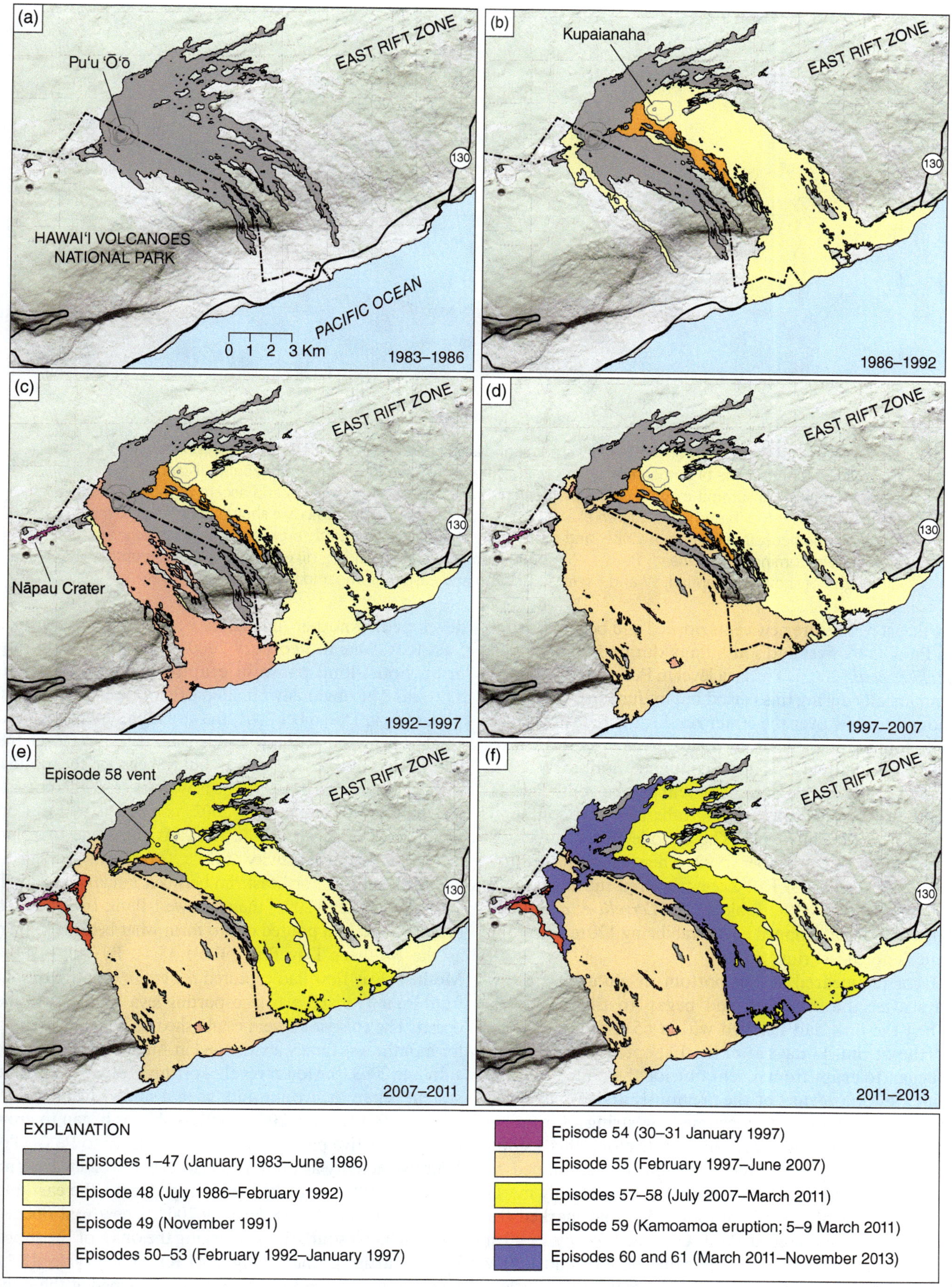

Figure 18.3 Maps of Kīlauea's middle East Rift Zone showing flows erupted 1983–2013, subdivided by flows emplaced during (a) 1983–1986, (b) 1986–1992, (c) 1992–1997, (d) 1997–2007, (e) 2007–2011, and (f) 2011–2013.

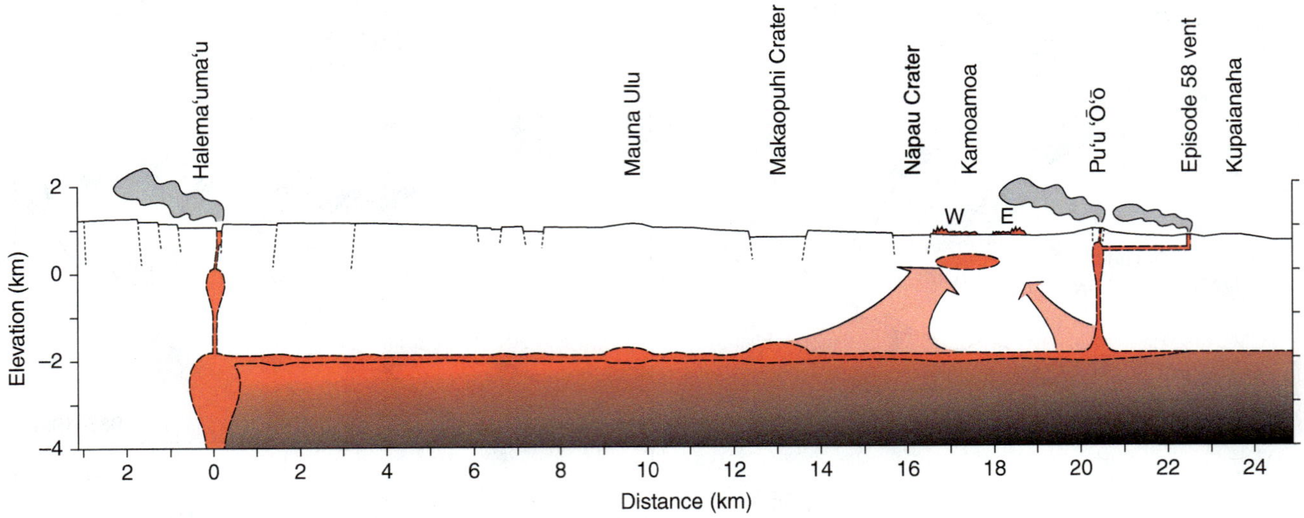

Figure 18.4 Cross section through Kīlauea's summit and along part of East Rift Zone showing conceptual model of magmatic system at time of Kamoamoa eruption. Elevation is relative to sea level; horizontal distance is flow path distance relative to summit magma chamber. Idealized pink arrows show sources and ascent of magma feeding Kamoamoa eruption (based on *Lundgren et al.* [2013]). Main part of Kamoamoa dike shown connected to magma storage zone beneath Makaopuhi Crater and intersecting shallow magma body en route to surface; other part of dike shown coming from beneath Pu'u 'Ō'ō. Connection between ascending dike and Kamoamoa eruption site, although not shown, is implied. W and E refer to western and eastern Kamoamoa fissures, respectively.

northeast rim before new vents opened and began to erupt on Pu'u 'Ō'ō's west and south flank during March to July [*Heliker et al.*, 2003]. The lava level in Pu'u 'Ō'ō fluctuated dramatically during this period but by June had risen high enough to flow over the crater rim for the first time since high fountaining ended in 1986 [*Heliker and Mattox*, 2003]. Intermittent overflows continued until early 1998, even though a single dominant vent on the south side of the Pu'u 'Ō'ō cone was sending tube-fed flows southeast to the ocean by July 1997. The episode 55 lava tube was active until 12 September 1999, when an intrusion into Kīlauea's upper East Rift Zone triggered subsidence and collapse at Pu'u 'Ō'ō (Table 18.2) [*Cervelli et al.*, 2002]. The crater floor dropped 65 m, stabilizing 120 m below the crater's northeast rim.

Lava reappeared at the bottom of Pu'u 'Ō'ō a few days after the intrusion and began to fill the crater. When the lava had reached within 65 m of the crater's northeast rim, 11 days after the 12 September intrusion, it began to erupt from a vent outside the crater, and the upper few kilometers of the previously active episode 55 lava tube system was reoccupied. A string of rootless shields formed above breakout points along this section of the tube during its reoccupation and continued to grow during the months that followed, eventually merging to form a broad, ~2 km long ridge that marked the tube trace. The crater at Pu'u 'Ō'ō filled to within 40 m of the northeast rim over the same period, and a new tube system extending downslope from the rootless shields to the coastline southeast of Pu'u 'Ō'ō developed. Another intrusion occurred in the upper East Rift Zone in February 2000 (Table 18.2), resulting in an hours-long pause in eruptive activity, but Pu'u 'Ō'ō was not obviously affected otherwise.

Starting in December 2001, the lava tube system began to break down, and rootless shields started to grow over the tube in the same general area as, but just east of, the rootless shields that formed in 1999. The new rootless shields continued to grow through April 2002, coalescing to form an ~3 km long ridge above the tube. At the same time, an increase in eruptive activity at Pu'u 'Ō'ō filled the crater to within 12 m of its northeast rim between January and April 2002. On 12 May 2002, the southwest flank of Pu'u 'Ō'ō ruptured and lava poured out to form what became known as the Mother's Day flow (Table 18.1). By August, the Mother's Day flow had captured the entire eruptive output from Pu'u 'Ō'ō and was transporting lava southeast to the ocean. The episode 55 vent established in 1997 and waning for months was finally abandoned in August 2002.

In late 2003, the lower reaches of the Mother's Day tube system began to atrophy and, as in a similar fashion during late 2001 to early 2002, rootless shields began to grow along the active part of the lava tube closer to Pu'u 'Ō'ō. Eruptive activity at Pu'u 'Ō'ō also began to increase, and the crater filled until lava overtopped both the east and west crater rims. On 18 January 2004, a new vent opened on Pu'u 'Ō'ō's south flank, marking the onset of the short-lived Martin Luther King (MLK) flow (Table 18.1). Eruptive activity in Pu'u 'Ō'ō's crater stopped at the same time, and the crater floor sagged by about 5 m.

The cessation of activity within Pu'u 'Ō'ō did not last long, however, and lava was once again erupting on the crater floor by early February 2004. This was followed by a brief resurgence of activity from the MLK vent in mid-February. In March, the Prince Kūhiō Kalaniana'ole (PKK) flow (Table 18.1) broke out from the lower south flank of Pu'u 'Ō'ō, finally relieving the pressure beneath the cone and marking the end of rootless shield growth along the Mother's Day tube system, which had remained active throughout this period. Lava fed from the Mother's Day rootless shields traveled downslope and entered the ocean from May until September, when the Mother's Day vent died. This activity was replaced by the PKK flow, which increased in vigor as the Mother's Day flow waned. The PKK flow reached the ocean southeast of Pu'u 'Ō'ō in November 2004, setting up a tube system that remained relatively stable for the next 2.5 years, until episode 55 ended in June 2007.

18.2.5. 2007–2011: Downrift Migration to Episode 58 Vent

An intrusion into Kīlauea's East Rift Zone near Mauna Ulu during 17–19 June 2007 (Table 18.2) resulted in the collapse of Pu'u 'Ō'ō's crater floor, which dropped about 80 m, and culminated in a brief eruption on the northeast flank of Kanenuiohamo Lava Shield (Father's Day eruption, episode 56; Figure 18.1; Table 18.1) [*Poland et al.*, 2008; *Fee et al.*, 2011]. A 12 day hiatus in Kīlauea's eruptive activity followed, ending with the return of lava to Pu'u 'Ō'ō on 1 July (episode 57; Table 18.1) [*Orr*, 2014]. The subsequent refilling of Pu'u 'Ō'ō crater was accompanied by several meters of uplift of the crater floor, evidence of pressure building beneath the cone. This pressure was relieved with the opening of a fissure system between Pu'u 'Ō'ō and Kupaianaha on 21 July 2007 (episode 58; Figure 18.3e; Tables 1 and 2) [*Poland et al.*, 2008]. Activity quickly localized at the eastern end of the fissure system, the main episode 58 vent, and built a lava shield and an ~2 km long "perched" lava channel, elevated up to ~50 m above the preexisting surface [*Patrick et al.*, 2011]. The channel facilitated flow advancement toward the northeast, sparking lava inundation concerns in populated areas downslope along the East Rift Zone [*Kauahikaua*, 2007]. In November 2007, the southern flank of the shield over the episode 58 vent ruptured, and flows northeast of the vent stagnated as lava was diverted southeastward. At first, a series of rootless shields grew over the developing lava tube [*Patrick and Orr*, 2012], but this stopped after the flow reached the ocean near Kalapana in March 2008. Thereafter, the lava tube system transported lava to the ocean until early 2011. It was during this period, in March 2008, that an eruptive vent opened in Halema'uma'u Crater at Kīlauea's summit (Figure 18.1) [*Wilson et al.*, 2008; *Patrick et al.*, 2013]. This marked the start of the first recorded instance of simultaneous summit and rift eruptions at Kīlauea lasting more than one year.

In late May and early June 2010, after nearly three years with negligible eruptive activity within Pu'u 'Ō'ō, lava erupted within and partly filled Pu'u 'Ō'ō's crater, and another effusive surge occurred there in October 2010 (Figure 18.5). This followed months of low discharge from the episode 58 vent [*Orr et al.*, in press] and repeated cycles of deflation and inflation (DI events) at Kīlauea's summit (Chapter 11, this volume) that resulted in pronounced oscillations in the summit lava lake level (Figure 18.5). The DI events were associated with decreases and increases in effusion through the episode 58 tube system, evidence that the cyclic deformation was associated with varying flux between the summit and Pu'u 'Ō'ō [*Orr et al.*, in press]. Starting on 7 November 2010, the DI events stopped occurring and relatively steady inflation ensued until mid-January 2011. This period was accompanied by an increase in the abundance of active surface flows fed by breakouts from the lava tube [*Orr et al.*, in press] as well as by crater filling at Pu'u 'Ō'ō (Figure 18.5) and a rise in lava lake level in the Overlook crater at Kīlauea's summit (Figure 18.5). DI events and associated fluctuations in tube effusion resumed in late January, while crater filling and rising lava lake level continued.

Pu'u 'Ō'ō's 2010–2011 filling was tracked by a number of geological and geophysical signals. At the summit, deformation measurements indicated inflation (Figure 18.5) as magma accumulated within a shallow reservoir beneath the east margin of Halema'uma'u Crater. Seismicity also increased beneath Kīlauea's summit and upper East and upper Southwest Rift Zones, especially after late January 2011 (Figures 18.5c and 18.6a). This pattern in seismicity is well correlated with times of high summit pressurization [*Klein et al.*, 1987]. Seismicity consisted mostly of long-period events in the summit and volcano-tectonic events in the rift zones. Across the East Rift Zone near Makaopuhi Crater, extension accompanied a modest increase in seismicity (Figure 18.5). At Pu'u 'Ō'ō (Figure 18.5), there was almost no increase in located seismicity and no obvious change in tremor, although extension across the crater suggested magma accumulation beneath the cone.

Beginning in February 2011, summit DI events became more distinct, with a series of three relatively large (>5 μrad) events immediately preceding the Kamoamoa eruption. The DI events were mirrored by large swings in lava lake level (Figure 18.5). Tilt and seismicity rates in the summit and upper East Rift Zone were also highly correlated, with seismicity rates approaching zero on the upper East Rift Zone during the deflation phases. Most of the seismicity on the upper East Rift Zone consisted of

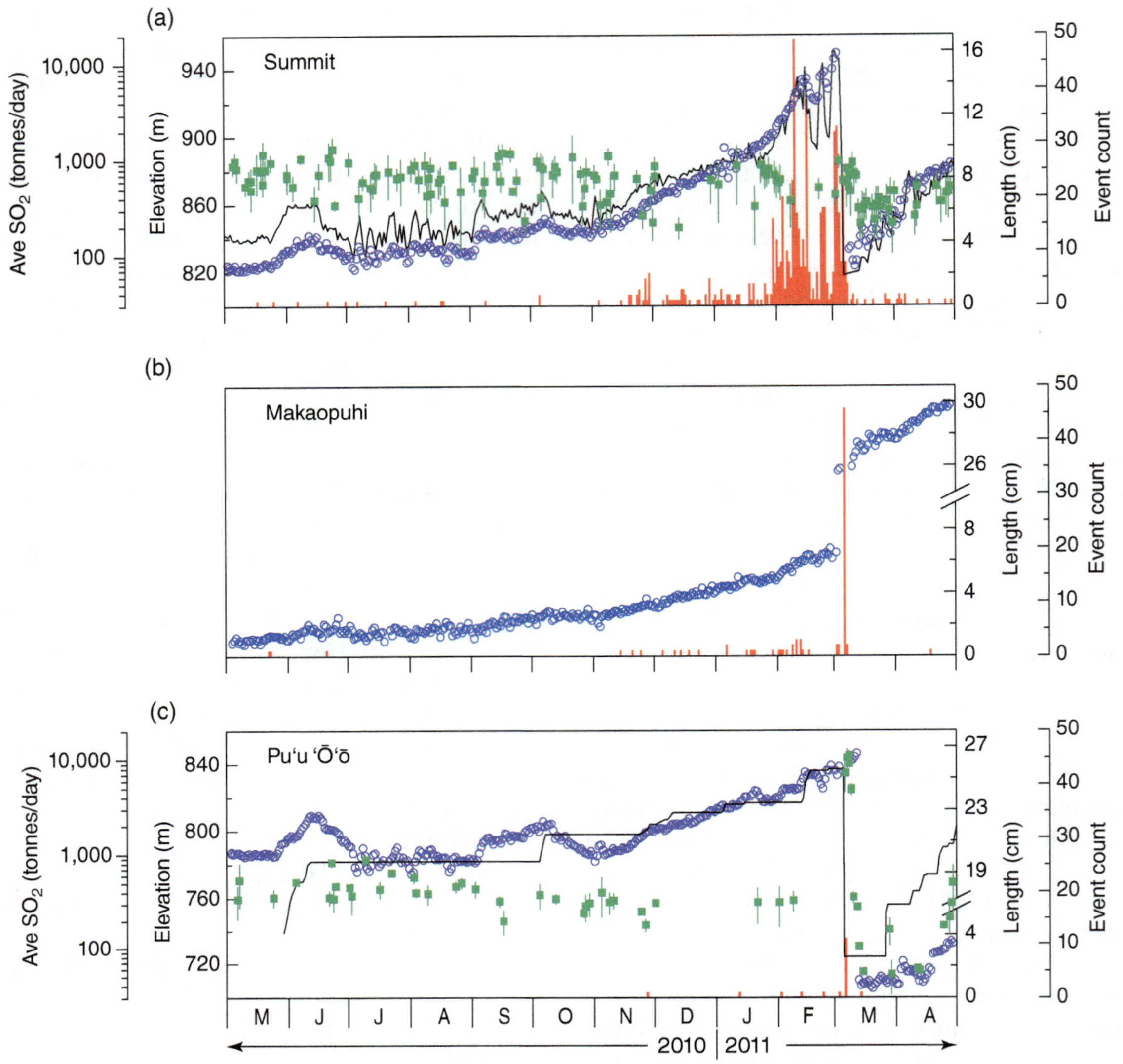

Figure 18.5 Time series of geological, geophysical, and geochemical data from Kīlauea during 2010–2011. (a) Graph showing average SO_2 emissions at summit (green squares; lines show one standard deviation), summit lava lake elevation (black line), summit and upper East Rift Zone earthquake count (red bars), and GPS line length between stations AHUP and UWEV (blue circles; see Figure 18.1 for location) during 1 May 2010–30 April 2011. Positive GPS line length change interpreted as inflation (although a few centimeters per year of line length increase also occurs from seaward south flank motion); March 2011 GPS offset ~12 cm contraction. Pronounced lava-level oscillations in response to summit DI events (see text). (b) Graph showing GPS line length variation between stations NUPM and KTPM and earthquake count in region around Makaopuhi and Nāpau craters. Note GPS scale broken to accommodate data. March 2011 GPS offset ~18 cm extension. (c) Graph showing average SO_2 emissions from East Rift Zone, Pu'u 'Ō'ō crater floor elevation, GPS line length between stations PUOC and SPIL, and Pu'u 'Ō'ō area earthquake count. Note GPS scale broken to accommodate data. March 2011 GPS offset ~25 cm contraction. SO_2 release at Kīlauea is sensitive to near-surface eruptive processes. Consequently, observed SO_2 emission rate variability incorporates influence of shallow changes in magmatic system, such as convective lava lake circulation and free-surface activity level. Another factor, radiative transfer, affects absolute uncertainty associated with measured emission rates at Kīlauea since 2008 [*Elias and Sutton*, 2012].

an event family (multiplet), whose occurrence was modulated by the large DI events occurring at the summit.

By the beginning of March 2011, small but widespread surface flows were scattered about the flow field, and Pu'u 'Ō'ō had filled to within ~20 m of the crater's northeast rim (Figure 18.7a). At the same time, the summit lava lake within the Overlook crater had risen to within ~62 m of the Halemau'uma'u crater floor (Figure 18.7b) [*Orr et al.*, 2013], the highest level recorded to that time. Starting on 1 March, seismicity intensified between Makaopuhi and

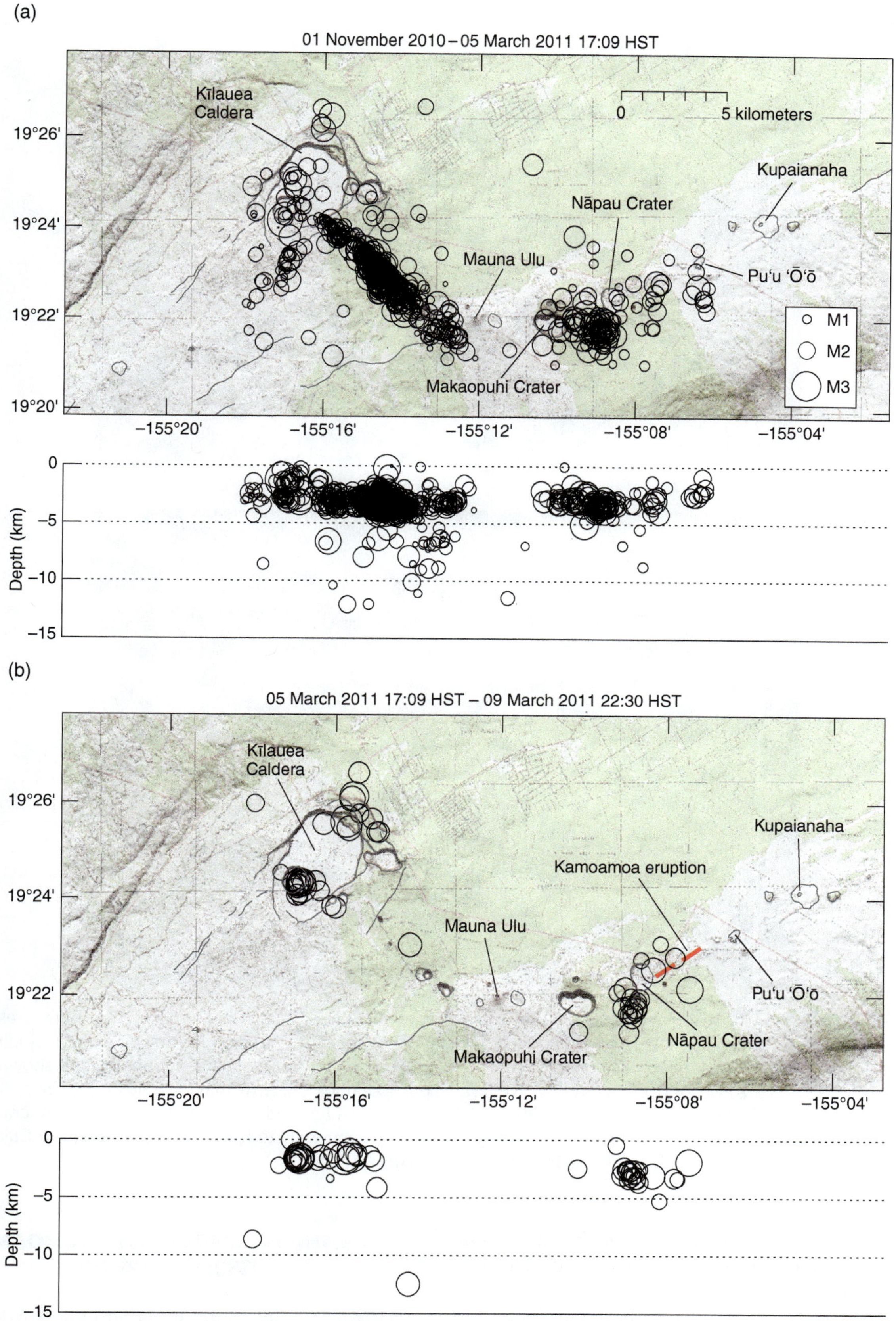

Figure 18.6 Locations (in map view and depth) of earthquakes at Kīlauea's summit and East Rift Zone during (a) 1 November 2010 until Kamoamoa dike reached surface on 5 March 2011 at 1709 HST and (b) effusion of lava from Kamoamoa fissure from 5 March 2011 at 1709 HST until 9 March 2011 at 2230 HST. Refer to Figure 18.5 for event rates in different areas. Red lines mark western and eastern Kamoamoa fissures.

Figure 18.7 (a) Oblique aerial photograph showing mostly filled Pu'u 'Ō'ō crater on 2 March 2011. Active vents and time-lapse camera are labeled; light-colored flows in crater are active or very recent. Crater (not including flank pits) ~430 × 270 m, major axis oriented upper right to lower left. (b) Oblique aerial photograph showing monitoring cameras, visitor overlook, and active lava lake in Overlook crater (within Halema'uma'u Crater) on 2 March 2011. Dashed white lines mark rim of Halema'uma'u Crater and Overlook crater. Overlook crater ~190 × 150 m, lava lake ~160 × 140 m (major axes oriented from upper right to lower left). Halema'uma'u Crater floor ~85 m below cameras; lava lake ~65 m below Halema'uma'u floor.

Nāpau craters (Figure 18.6a), a pattern similar to those observed prior to some previous eruptions on the East Rift Zone, such as those of 1974 and 1983 [*Klein et al.*, 1987]. Unlike the earthquakes on the upper East Rift Zone, most of the Makaopuhi-Nāpau earthquakes did not belong to any earthquake family recognized at the Hawaiian Volcano Observatory, and the few families that did occur were only active for a short time.

18.3. THE 5–9 MARCH 2011 KAMOAMOA FISSURE ERUPTION

The Kamoamoa fissure eruption of 5–9 March 2011 was the culmination of several months of precursory activity (Figures 18.5 and 18.6), which included (1) summit and Pu'u 'Ō'ō inflation, (2) accelerated East Rift Zone spreading, (3) increased upper East Rift Zone seismicity, and (4) rising

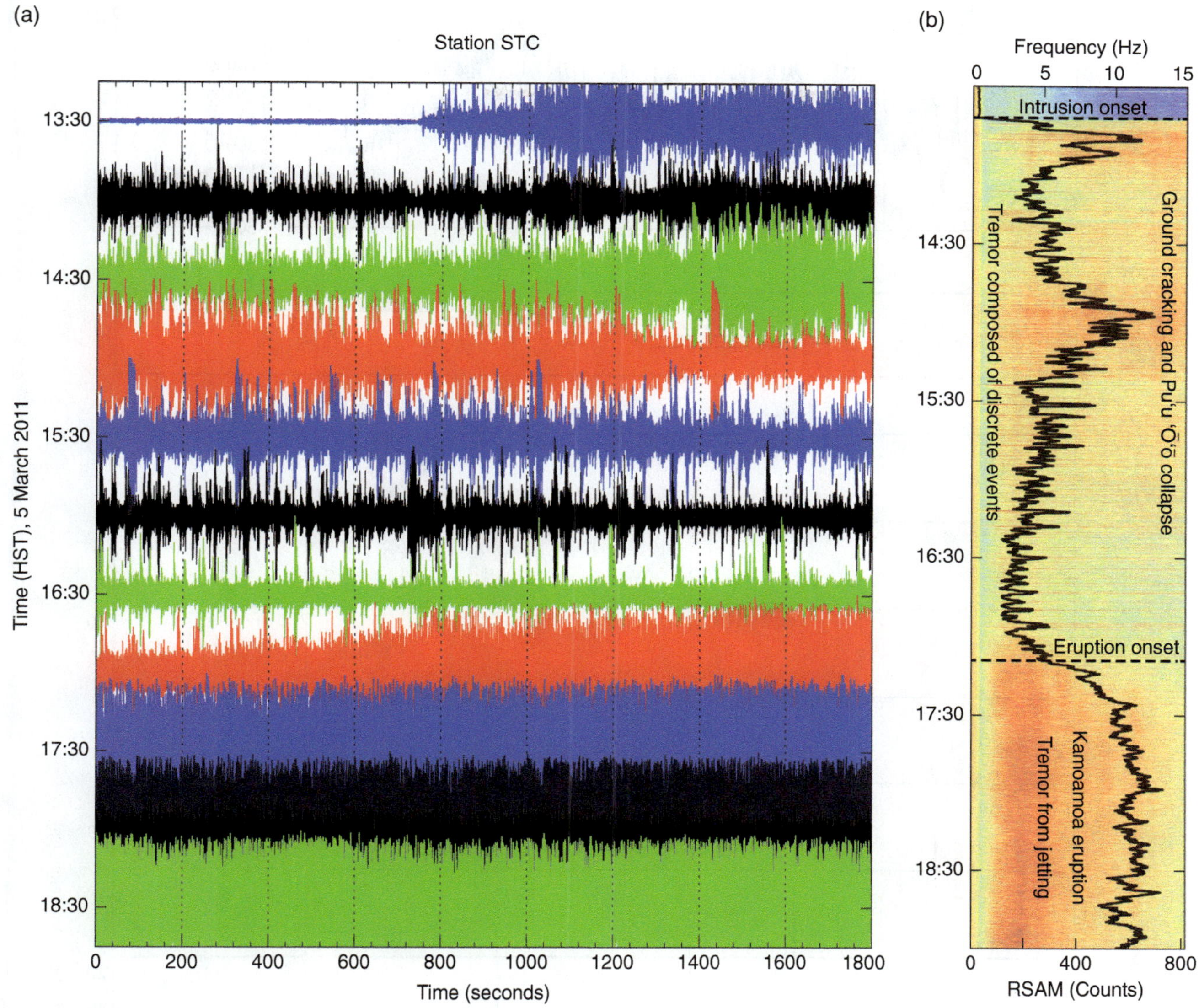

Figure 18.8 (a) Helicorder view and (b) spectrogram and 1 min averaged seismic amplitude (black line) showing seismicity for period 1330–1900 on 5 March 2011, recorded on STC seismometer. Period spans start of intrusion, collapse of Pu'u 'Ō'ō, and onset of Kamoamoa eruption.

lava levels within the summit eruptive vent and at Pu'u 'Ō'ō. We describe next the characteristics of the Kamoamoa eruption, including information from geological observations, geophysical data, and geochemical measurements. A detailed chronology of the Kamoamoa eruption, which we designate as episode 59 of the Pu'u 'Ō'ō eruption (Table 18.1), is provided as Appendix A.

18.3.1. Eruption Onset

The sudden onset of seismic tremor and elevated earthquake activity along Kīlauea's East Rift Zone at 1342 HST (Hawaiian standard time; coordinated universal time (UTC) – 10h) on 5 March 2011 (Figure 18.8) signaled the start of an intrusion a few kilometers uprift from Pu'u 'Ō'ō. Rapid deflation started at Pu'u 'Ō'ō at almost the same time and at Kīlauea's summit about 30 min later (Figure 18.9a). Active lava flows on the floor of Pu'u 'Ō'ō crater stagnated, and at ~1420 HST most of the crater floor began to undergo wholesale subsidence (Figures 18.10 and 18.A1; Movie 18.1).

Pu'u 'Ō'ō subsidence was most rapid during the first 1.5 h, when the crater floor dropped about 80 m, revealing both solid and molten (or partly molten) layers. As the crater deepened, small rockfalls from the preexisting crater walls ejected plumes of reddish lithic ash (Figure 18.A1) that dusted the southwest (downwind) flank of Pu'u 'Ō'ō. Rockfalls continued for the next few hours, even as the pace of subsidence slowed. Subsidence was mostly finished by 1800 HST, with the

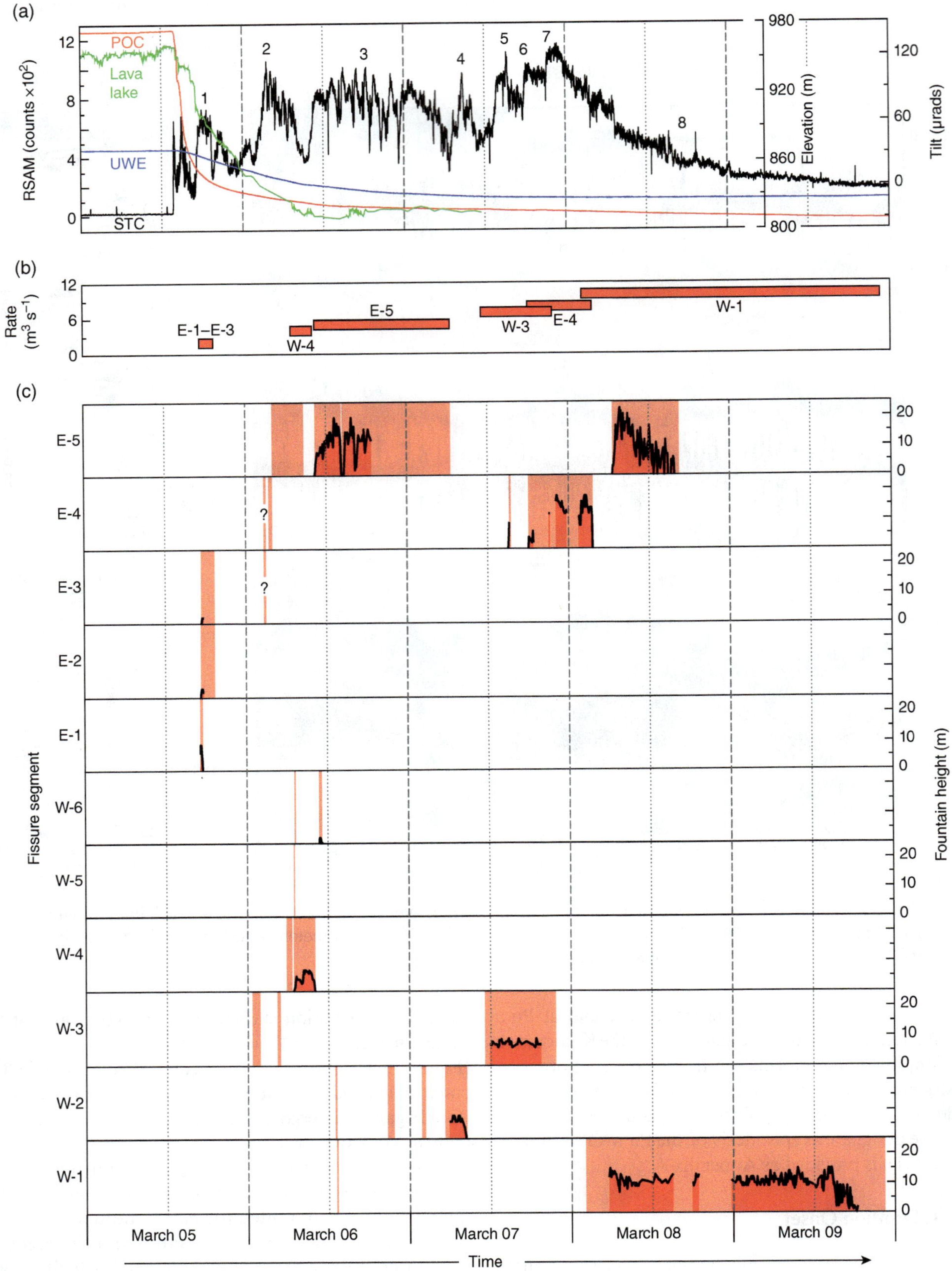

Figure 18.9 (a) Radial tilt at Pu'u 'Ō'ō (POC, 308° azimuth; red line) and summit (UWE, 327° azimuth; blue line), summit lava lake surface elevation (green line), and mean seismic amplitude (RSAM) near eruption site (STC; black line) during 5–9 May 2011. Numbers denote specific periods in RSAM record discussed in text (see Section 18.4.7). (b) Time-averaged discharge rate for significant eruptive periods. (c) Periods of activity for Kamoamoa eruption fissures (pink bars; bar width illustrates duration) showing fountain height (red areas; based on optically dense part of fountains) where quantifiable.

Figure 18.10 Time-lapse camera photos showing collapse of Pu'u 'Ō'ō crater floor on 5 March 2011. Distance from camera to opposite (northeast) rim of crater is ~430 m; camera location shown in Figures 18.2 and 18.7. (a) 1345 HST, immediately after onset of tremor but before onset of collapse. (b) 1440 HST, ~20 min after collapse onset. Central part of crater sagging to form elongate trough aligned with East Rift Zone orientation; ring fractures beginning to form. (c) 1520 HST, ~1 h after collapse onset. Crater floor has dropped along ring fractures, exposing incandescent lava fill accumulated over preceding several months. (d) 1240 HST on 13 March, eight days after collapse. Deepest part of crater ~133 m below northeast rim.

crater floor having dropped 113 m (Figures 18.5a and 18.10). For several hours afterward, slices of the preexisting crater floor occasionally collapsed and sent avalanches of incandescent debris to the bottom of the new depression.

At Kīlauea's summit, the surface of the lava lake within the Overlook crater began to drop shortly after 1400 HST on 5 March, reaching a relatively stable level ~205 m below the floor of Halema'uma'u (a drop of ~143 m) by about 0800 the following morning (Figures 18.9a and 18.11; Movie 18.2). During this draining, slabs of lava veneer coating the walls of the vent, as well as portions of the vent walls, fell away and plunged into the lava column. Some collapses were large enough to trigger very-long-period (VLP) seismicity recorded on summit seismometers (e.g., NPT, see Figure 18.1) in the fashion described by *Orr et al.* [2013].

18.3.2. Fissure Eruption

The Kamoamoa eruption began at 1709 HST on 5 March, when lava broke the surface ~2 km southwest of Pu'u 'Ō'ō and formed an eruptive fissure that propagated northeastward midway between Pu'u 'Ō'ō and Nāpau Crater (Figures 18.A1 and 18.A2; Movie 18.3). This "eastern" fissure system grew to a total length of 950 m and was composed of five distinct fissures (E-1 to E-5, labeled from southwest to northeast; Figures 18.2 and 18.12). A "western" fissure system (Figures 18.A2 and 18.A3; Movie 18.4) opened early on 6 March about 1 km

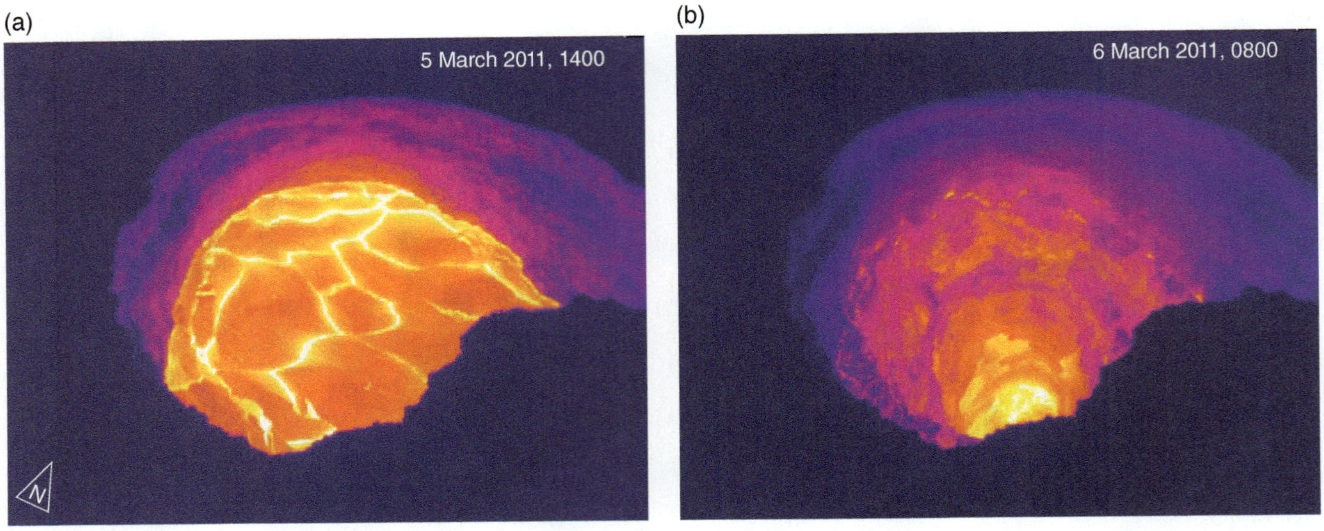

Figure 18.11 Infrared camera images of lava lake in Overlook crater at southeast edge of Halema'uma'u Crater at Kīlauea's summit. (a) 1400 HST on 5 March 2011, immediately before onset of draining. (b) 1400 HST on 6 March 2011, after lava lake level had dropped ~143 m and stabilized.

farther southwest and propagated both northeast and southwest. This fissure system reached a final length of 1050 m and comprises six individual fissures (W-1 to W-6, labeled from southwest to northeast; Figures 18.2 and 18.12). The eastern and western fissure systems were separated by 360 m of heavily cracked ground. Eruption along both fissure systems was initially sporadic, and activity migrated from one spot to another in semirandom fashion, repeatedly starting and stopping (Figure 18.9c). Early on 8 March, the eruption became focused near the opposing ends of the two fissure systems (Figure 18.12).

To the northeast, low fountaining from an ~75 m long line of discrete points along the southwest end of fissure E-5 (Movies 18.5 and 18.6) constructed an elongate (120 × 65 m) pyroclastic cone complex that eventually reached a maximum height of ~13 m (Figures 18.12 and 18.A3). The optically dense part of the fountains was up to ~25 m high, and small particles reached at least twice that height during discrete bursts. The instantaneous discharge rate was relatively low, with eruptive output slowly building the pyroclastic cone and feeding a small, slow-moving lava flow exiting from the northeast end of the cone. The gas output, however, was high, as suggested by loud pulsatory outgassing and relatively wide dispersal of tephra. At times, large pulses of incandescent (in daylight) fume broke through the fountains like a jet and produced a lava "spray" laced with sparse small clots of lava (Movie 18.6). Eruptive activity at E-5 waxed and waned constantly; at times, the fountains died down to almost nothing, then quickly rebuilt to their previous level (Figure 18.9c).

At the W-1 fissure (Movie 18.7), an ~200 m long line of fountains with the optically dense part up to ~15 m high (and small clasts reaching occasionally to ~50 m) constructed an elongate (115 × 55 m) pyroclastic cone complex with a maximum height of ~14 m (Figsures 18.12 and 18.A3). The instantaneous discharge rate at W-1 appeared to be far greater than at E-5, and W-1 fed a channelized 'a'ā flow that advanced southeastward (Figure 18.A3). A perched lava pond grew atop the northeastern half of the W-1 fissure at the head of the lava channel (Figure 18.A3). No pyroclastic edifice was constructed over the fountains on the northeastern half of W-1, because flowing lava constantly removed spatter from the site of the eruption [e.g., *Parcheta et al.*, 2012]. The smaller of the W-1 fountains were episodically drowned by the lava pond [e.g., *Wilson and Head*, 1981], sometimes disappearing completely for a few seconds before bursting back through the pond surface moments later (Movie 18.8).

Fissure E-5 shut down on the afternoon of 8 March (Figure 18.9c), although distant webcam views indicated occasional spatter bursts overnight. However, W-1 remained active through the next day. Late in the afternoon of 9 March, W-1 began to wane, and the lava level in the perched pond over the northeast end of the fissure began to drop. The fissure shut down at about 2230 HST on 9 March, marking the end of the Kamoamoa eruption (Figure 18.9c). The channelized 'a'ā flow stalled soon afterward, having reached a final length of 3.3 km (Figure 18.2).

18.3.3. Posteruption Recovery and Subsequent Activity

Kīlauea's East Rift Zone was quiet for several days following the end of the Kamoamoa eruption. During midmorning on 26 March, lava began erupting on Pu'u 'Ō'ō's collapsed crater floor, ending an East Rift Zone

Figure 18.12 Photographs of Kamoamoa eruption. (a) Eastern and western fissure systems at 1226 HST on 8 March 2011, after activity focused at fissures E-5 and W-1. Nāpau Crater at upper right; yellow circle, STC seismometer. (b) Fountaining and pyroclastic cone at fissure E-5 at 0809 HST on 8 March 2011. (c) Fountaining and pyroclastic cone at fissure W-1 at 0953 on 9 March 2011. Lava accumulated in perched pond (left side of photo with small fountain) before feeding channelized flow.

eruption hiatus of 16.5 days (Table 18.1). The crater continued to refill over the following several months. At Kīlauea's summit, accumulation of debris at the bottom of the summit eruptive vent from frequent rock falls (during and after the Kamoamoa eruption) buried the top of the lava column beneath rubble, extinguishing nighttime glow by 11 March. This did not last long, however, as lava burrowed its way through the overlying carapace of rubble and reappeared after dark on 13 March. As at Pu'u 'Ō'ō, the lava lake rose slowly thereafter, gradually filling the vent.

By late July 2011, Pu'u 'Ō'ō's crater had mostly refilled and begun to overflow. On 3 August, lava broke from the lower west flank of Pu'u 'Ō'ō (episode 60; Figure 18.3f; Tables 18.1 and 18.2), causing Pu'u 'Ō'ō's crater to collapse again and the summit lava lake to drop slightly. By 15 August, when the flow stopped, lava had buried most of the flows produced by fissures E-4 and E-5 of the Kamoamoa eruption. After an eruptive hiatus of five days on Kīlauea's East Rift Zone, lava reappeared in Pu'u 'Ō'ō on 20 August and began to fill the crater once again, this time comparatively quickly. Lava began overflowing the crater onto Pu'u 'Ō'ō's flanks on 10 September, and on 21 September a small fissure opened high on the northeastern side of the cone, feeding the Peace Day flow (episode 61; Figure 18.3f; Tables 18.1 and 18.2) and causing the crater floor to sag slightly. This breakout carried lava southeastward across older Pu'u 'Ō'ō flows and established a tube system that sent flows briefly into the ocean during December 2011.

Kīlauea's East Rift Zone eruption was relatively stable for the next 16 months, with activity focused through the episode 61 lava tube system, although a persistent ocean entry was not established until November 2012. By January 2013, the episode 61 vent was no longer able to erupt all of the magma reaching Pu'u 'Ō'ō. Lava, fed directly from a spatter cone on Pu'u 'Ō'ō's crater floor, exited the crater and moved northeastward, where it fed flows that encroached on the adjacent forest (Figure 18.3f). These northeastward-directed flows (Kahauale'a 1 and Kahauale'a 2; Table 18.1) captured all lava erupting at Pu'u 'Ō'ō in November 2013, when the Peace Day flow was abandoned. This transition in flow direction marked an important change in risk: The area far northeast of the cone, toward which flows are currently directed (as of November 2013), has a greater population density [e.g., *Kauahikaua*, 2007] than the area south of Pu'u 'Ō'ō, where most flows have gone over the past 30+ years.

18.4. RESULTS

18.4.1. Pu'u 'Ō'ō Crater Collapse on 5 March

Prior to collapse, Pu'u 'Ō'ō had filled to within about 20 m of the northeast rim of the cone (Figure 18.7); the crater floor was then at an elevation of about 840 m (the northeast rim elevation was 860 m). The crater floor, held up by the magma body beneath Pu'u 'Ō'ō, dropped as the magma drained away. After collapse, the crater floor was measured by a laser rangefinder as 133 m below the northeast rim, or at an elevation of 727 m. The result was a 380 × 245 m concave depression (Figure 18.10d) with a depth of 113 m (compared to the precollapse crater floor height; Figure 18.5). The shape of the Pu'u 'Ō'ō crater floor after collapse can be approximated by a half ellipsoid. A collapse volume (V_c) can therefore be calculated as

$$V_c = \frac{\pi l w h}{6}, \tag{18.1}$$

where l is the crater length, w the crater width, and h the collapse depth. Using the dimensions above, the volume of the crater collapse is calculated to be $5.5 \times 10^6\,m^3$ (Table 18.2). Assuming the duration of the collapse to be 3.7 h (based on camera observations), this equates to an average volume loss rate of ~400 m³/s. The magma body that drained from beneath Pu'u 'Ō'ō was likely a mix of (1) magma that had not yet circulated to the surface and was too deep to have experienced significant vesiculation and (2) magma that had mostly outgassed to atmospheric pressure and in which its remaining vesicles may have collapsed upon being carried back to the depth of the storage chamber; therefore, the calculated volume and volume loss rate could approach dense-rock equivalent (DRE) values.

In comparison with volume loss approximations for two previous crater collapses at Pu'u 'Ō'ō (Table 18.2), which were also calculated using the formula for a half ellipsoid (equation 18.1), the volume we derive for the 2011 collapse falls between the $13 \times 10^6\,m^3$ for the 1997 subsidence associated with episode 54 [*Heliker et al.*, 2003] and the $4.2 \times 10^6\,m^3$ for the 2007 subsidence associated with episode 56 [*Orr*, 2014]. Though collapse of Pu'u 'Ō'ō's crater floor was not fully brittle, involving both brittle and ductile layers, the subsequent crater fill was probably, in part, a column of brecciated material that occupied a volume greater than that of the same material preceding the collapse [*Roche et al.*, 2001]. Thus, the crater collapse volumes calculated in each of these cases may represent minimum values for the actual volume lost from the magma storage body beneath Pu'u 'Ō'ō.

Montgomery-Brown et al. [2010] calculated an approximate volume loss from Pu'u 'Ō'ō during the episode 56 eruption in 2007 by multiplying the time elapsed from the end of the eruption until lava reappeared in Pu'u 'Ō'ō by the long-term supply rate to Kīlauea (0.12 km³/yr [*Heliker and Mattox*, 2003]). Following this example, the 16.5 day hiatus between the end of the Kamoamoa eruption and the reappearance of lava in Pu'u 'Ō'ō corresponds to a volume of $5.4 \times 10^6\,m^3$, strikingly similar to the value calculated from the volume of crater collapse. As noted earlier, however, the discharge rate starting at least by mid-2010 was lower than the long-term discharge rate, perhaps by as much as 25%–50% [*Orr et al.*, in press], which could reflect a decrease in the rate of supply to Pu'u 'Ō'ō. While the actual supply rate at the time of the Kamoamoa eruption remains uncertain, the similarity between these two methods of approximating volume loss could be interpreted to indicate that there was no significant change in supply rate.

18.4.2. Summit Lava Lake Draining

Before the Kamoamoa eruption started, the level of the summit lava lake in the Overlook crater was ~62 m below the floor of Halema'uma'u Crater (Figure 18.7), an altitude of ~961 m. The lava lake had a northwest-southeast-oriented major axis of ~160 m and a northeast-southwest-oriented minor axis of ~140 m. By about 0800 HST on 6 March (the morning following the onset of the Kamoamoa eruption), the lake's surface had dropped to ~205 m below the vent rim to an altitude of ~818 m (a drop of 143 m; Figure 18.9a). The walls of the evacuated crater appeared to be roughly vertical, although they may have been slightly inclined, suggesting that the bottom of the crater was narrower than the top. Because we cannot resolve this subtle geometry, we treat the volume of lava drained from the lake (V_d) by modeling the summit vent as an elliptical cylinder:

$$V_d = \frac{\pi l w h}{4}, \tag{18.2}$$

where l is the lake length, w the lake width, and h the amount of lava lake drop. Using the vent dimensions stated above and a lava level drop of 143 m, the volume of lava drained from the lake is ~2.5×10^6 m^3. This suggests an average bulk-volume loss rate of ~39 m^3/s during the first 18 h of draining, although lava level measurements indicate that the volume loss rate was not constant, instead slowing during those 18 h (Figure 18.8). Using gravity data recorded on the southeast rim of Halema'uma'u Crater (station HOVL; Figure 18.1), *Carbone et al.* [2013] calculated an average lava lake density of ~950 kg/m^3, or about a third of the DRE density of ~2800 kg/m^3. This corresponds to an average DRE volume loss rate of ~13 m^3/s during draining and a total DRE volume loss of ~0.8×10^6 m^3 (Table 18.2). This estimate, added to the modeled (from interferometric synthetic aperture data (In SAR) and global positioning system (GPS) data) volume loss of 1.7×10^6 m^3 from the shallow summit magma reservoir [Table 18.1; *Lundgren et al.*, 2013], gives a nominal summit volume loss of ~2.5×10^6 m^3.

18.4.3. Kamoamoa Fissure Eruption Rates

The Kamoamoa eruption was active for a total duration of 4.2 days. The detailed observations made during the eruption, coupled with field mapping and estimates of flow thickness, allow calculation of the mean output rate for the entire eruption as well as for distinct periods during the eruption. Table 18.3 provides a summary of the duration, area covered, volume erupted, and time-averaged discharge rate for individual fissures over the course of the eruption. DRE volumes are based on a void space correction factor of 25%, the value used by *Wolfe et al.* [1988] for Pu'u 'Ō'ō 'a'ā flows. Durations are based on observed start and stop times of eruptive activity, which, when not precisely known, are defined as the temporal midpoint between bracketing observations. Inaccuracy in our estimates of flow thickness is the largest source of uncertainty in calculating volume, which could be in error by as much as ±50%.

During the eruption, lava flows covered 1.39 km^2 with a total volume of ~2.7×10^6 m^3 DRE (Table 18.3). This estimate is considerably lower than that calculated from gas emissions (a minimum of 7.6×10^6 m^3 DRE; Table 18.2; see gas geochemistry results in Section 18.4.5). The gas-based lava discharge estimate, however, reflects the volume of magma that outgassed even though not all of this magma necessarily erupted. In fact, the outgassed volume is no more than a factor of 2 lower than the dike volume modeled from deformation data (15.6×10^6 m^3 DRE [*Lundgren et al.*, 2013]; Table 18.2; see deformation results in Section 18.4.6). The outgassed volume may therefore be a better reflection of not only the magma that erupted but also that which intruded shallowly but did not erupt, either freezing in place or draining to deeper levels.

Using the mapped flow volume and duration, the time-averaged discharge rate of the Kamoamoa eruption was approximately 8 m^3/s. More than half of the volume was erupted from the W-1 fissure, with fissures W-3, E-4, and E-5 composing most of the remainder. The other fissures contributed very little to the overall volume of the flow field, although the amount erupted from fissure W-4 is unknown, since a significant portion of its volume poured into a deep ground crack in a graben north of the fissure between 0736 and 1003 HST on 6 March. If we disregard the mapped surface area of that part of the flow in the graben, we can calculate a new volume and time-averaged discharge rate for lava from fissure W-4 that was extruded outside the graben prior to 0736; these assumptions yield a time-averaged discharge rate that is 53% higher. Assuming this higher rate for the entire duration of W-4 activity, a revised volume of lava erupted from fissure W-4 can be calculated (Table 18.3). This is probably more representative of W-4 output than the initial value calculated from surface area and average thickness alone, but it does not change the mean output rate calculated for the entire eruption.

For much of the Kamoamoa eruption, discharge occurred simultaneously from multiple fissures, and discharge from a particular fissure was sometimes episodic and interspersed with activity from other fissures (Figure 18.9c). In general, however, the time-averaged discharge rate increased over the course of the eruption (Figure 18.9b), and near the end it had reached a level nearly three times higher (~10 m^3/s) than the long-term average rate (~4 m^3/s) over the course of the Pu'u 'Ō'ō eruption [*Sutton et al.*, 2003].

As with other fissure eruptions and dike intrusions during Kīlauea's ongoing East Rift Zone activity [e.g., *Owen et al.*, 2000; *Rivalta and Segall*, 2008; *Montgomery-Brown*

Table 18.3 Area, DRE volume (assuming 25% void space), duration, approximate time-averaged discharge, and distance to STC seismometer for Kamoamoa eruptive fissures.

Fissure	Area (ha)	DRE Volume (m^3)	Duration (h)	Discharge Rate (m^3/s)	Distance to STC (m)
E-1, E-2, and E-3[a]	4.12	0.015×10^6	2.2	~2	580; 510; 475
E-4[b]	23.82	0.27×10^6	9.5	~8	440
E-5[c]	19.0	0.33×10^6	17.7	~5	500
W-1[d]	68.61	1.68×10^6	44.4	~10	1765
W-2[e]	4.0	0.03×10^6	3.3	~3	1580
W3[f]	22.43	0.26×10^6	10.6	~7	1500
W-4[g]	5.70	0.043×10^6	4.1	~3	1120
W-4 w/no graben[g]	5.25	0.039×10^6	2.5	~4	
W-4 w/graben[g]	na	0.065×10^6	4.1	~4	
W-5 and W-6[h]	0.35	0.0013×10^6	1.0	~0.4	1020; 940
Total	139.06	2.7×10^6	89.5	~8	–

Notes: Three different sets of values shown for fissure W-4 (see text for discussion). Total volume uses "W-4 w/graben" value. Total duration is actual time spent in eruption (i.e., not counting inactive periods).
[a]Average flow thickness for fissures E-1, E-2, and E-3 estimated at 0.5 m.
[b]Average flow thickness for fissure E-4 estimated at 1.5 m. Duration of E-4 does not include initial E-4 activity on 6 March or ~20 min of E-4 activity during afternoon of 7 March. Both were periods of very low discharge.
[c]Flows from fissure E-5 subdivided into areas with average thicknesses of 1–3 m. Small volume of E-5 cone not included. Duration of E-5 considers main effusive period from this fissure.
[d]Flows from fissure W-1 subdivided into areas with average thicknesses of 1–6 m. Duration of W-1 considers only main effusive period from this fissure. Small volume of W-1 cone not included.
[e]W-2 lava buried by W-1 and W-3 flows before mapped. Area and thickness (1.5 m) approximated from oblique aerial and ground photographs.
[f]Flows from fissure W-3 subdivided into areas with average thicknesses of 1–3 m. Duration of W-3 considers main effusive period from this fissure.
[g]Average thickness of W-4 flow outside graben estimated at 1 m (based on thickness of lava exposed in tree molds).
[h]Fissures W-5 and W-6 represent two different eruptive periods on 6 March, but lumped because of small volume erupted. Average flow thickness estimated at 0.5 m.

et al., 2010], there is a discrepancy between the volume of magma withdrawal and that associated with dike intrusion and eruption. DRE volume loss at the summit and Pu'u 'Ō'ō combine to total $8.0 \times 10^6 m^3$, compared to the DRE erupted volume (from field mapping) plus modeled dike volume of $18.3 \times 10^6 m^3$ (Table 18.2). The erupted/intruded volume is therefore more than double that lost from beneath the summit and Pu'u 'Ō'ō [*Lundgren et al.*, 2013], nearly the same as the ratio between erupted/intruded and lost volumes calculated for the June 2007 intrusion and eruption [*Montgomery-Brown et al.*, 2010]. As pointed out by *Lundgren et al.* [2013], this falls within the range of expected volume ratios for compressible magma [*Rivalta and Segall*, 2008]; therefore, no additional sources of magma are required to explain the volume discrepancy.

18.4.4. Petrology

During 2000–2001, the petrologic character of East Rift Zone lava shifted to that of a hybrid comprising two components: (1) a volatile-rich, high-MgO, high-temperature magma fed to Kīlauea from the mantle and (2) a partially outgassed, lower-MgO, lower-temperature, partially crystallized resident magma stored at shallow levels beneath the summit and within the rift zones [*Thornber*, 2003] (see also Chapter 8, this volume). Through the early 2000s, lava composition showed a pattern of decreasing magnesia (MgO) and eruption temperatures, but this trend reversed in 2006 in tandem with a surge in magma supply to Kīlauea that caused an influx of hotter recharge magma to the eruption site [*Poland et al.*, 2012]. By 2009 (well into episode 58), however, whole-rock concentrations of MgO had begun declining again, reaching values near 7.0 wt% by early 2011. Corresponding magmatic temperatures, based on MgO glass thermometry [*Helz and Thornber*, 1987], were about 1140 °C. Mineral phenocrysts in erupted lavas were an assemblage of clinopyroxene, plagioclase, and lower-magnesium olivine. These observations suggest a progressive increase in the proportion of cooler resident magma (relative to hotter recharge magma) being erupted from the episode 58 vent after 2009. Chapter 8 gives a thorough overview of the petrologic trends and petrographic character of erupted products throughout this period as well as subsequent periods discussed below.

Lavas discharged during the Kamoamoa fissure eruption have significantly lower whole-rock magnesia concentrations than those erupted during episode 58, ranging from 6.1 to 6.7 wt% and corresponding to cooler magmatic temperatures (1130–1145°C). The mineral assemblage remains a mix of clinopyroxene, plagioclase, and olivine; however, these phenocrysts display an array of disequilibrium textures indicative of a complex heating and cooling history (Chapter 8). As the eruption progressed, the last products from fissure W-1 had chemical compositions that trended toward values typical of those observed during episode 58. It seems, then, that the dike that fed the Kamoamoa eruption either originated from or encountered (and mingled with) a pod of cooler magma that had been stored within the East Rift Zone (Figure 18.4). The East Rift Zone probably contains numerous such isolated magma storage zones, and the area beneath Makaopuhi Crater, where the dike originated (see deformation results in Section 18.4.6), has long been thought to be a region of magma storage (Figure 18.4) [e.g., *Swanson et al.*, 1976]. The cooler component was gradually flushed over the course of the eruption, and, by 9 March, lavas began to resemble the compositions erupted from the East Rift Zone prior to the Kamoamoa activity. In contrast, gases released from fissure W-1 showed an evolved composition even when last measured a few hours before the eruption shutdown (USGS, unpublished data).

After the Kamoamoa eruption ended, whole-rock magnesia concentrations that erupted within Pu'u 'Ō'ō crater varied erratically between 6.5 and 7.5 wt%, but the overall trend was a return to values seen during episode 58 (although magmatic temperatures were slightly cooler, with values around 1140°C or less) (Chapter 8). At the end of 2013, Mineral assemblages continue to include clinopyroxene, plagioclase, and lower-magnesium olivine, and phenocryst textures still suggest disequilibrium conditions, a testament to the disruption in the East Rift Zone plumbing system wrought by the Kamoamoa eruption.

18.4.5. Gas Geochemistry

The Kamoamoa eruptive episode produced the highest SO_2 emission rates recorded at Kīlauea since Pu'u 'Ō'ō's high-fountaining episodes during 1983–1986. For five sets of measurements spanning 6–9 March, emissions averaged 8500 ± 2500 t/d (tonnes/day), with a peak daily value of 11,000 ± 1500 t/d on 8 March (Figure 18.5). These reported emission rates represent a minimum constraint because recent studies indicate that dense, SO_2-rich plumes, like that of the Kamoamoa eruption, result in an underestimation of column SO_2 and, therefore, calculated emission rates [*Elias and Sutton*, 2012; *Kern et al.*, 2012].

Neither the East Rift Zone SO_2 emissions nor those of the summit showed any clear premonitory changes during the year leading up to the eruption; in fact, both sites displayed subtly declining SO_2 emissions (Figure 18.5). Furthermore, summit CO_2 emissions, which are closely tied to the magma supply rate to the volcano [*Gerlach et al.*, 2002; *Poland et al.*, 2012], showed relatively steady or even declining emissions for the year running up to the eruption, suggesting no significant changes in magma supply prior to the Kamoamoa eruption.

Post-Kamoamoa SO_2 emissions from the East Rift Zone declined rapidly, hovering at or below the lower limit of quantification (approximately 100 t/d) for six weeks before finally recovering and surpassing the preeruption levels (Figure 18.5). Summit SO_2 emissions after the Kamoamoa eruption decreased as well, although in a more muted fashion, values never dropping below 200 t/d (Figure 18.5).

Taken together, the posteruptive SO_2 extinction from Kīlauea's East Rift Zone and the sharp depression of summit emissions through and beyond the Kamoamoa eruption support the idea that the dike feeding the Kamoamoa fissure system tapped magma sources beneath both the summit and Pu'u 'Ō'ō. A rough gas-based estimate of lava output for the Kamoamoa eruption, using the method of *Sutton et al.* [2001], indicates that $7.6 \times 10^6 m^3$ of DRE magma was outgassed to atmospheric pressure during 6–9 March (Table 18.2), although only a fraction of this may have actually erupted. The gas data suggest that this value is a minimum magma-outgassed constraint, as it is based on minimum-constraint SO_2 emission rates. Actual values could be up to a factor of 3 higher.

18.4.6. Deformation

The Kamoamoa eruption was accompanied by deflation at the summit and Pu'u 'Ō'ō as well as by widening of the East Rift Zone (Figures 18.9a and 18.13). Subsidence at the summit occurred at a steadily decreasing rate over the course of the eruption and switched to inflation at about 1700 HST on 9 March (shortly before the end of activity at fissure W-1). GPS results indicate a maximum subsidence during 5–9 March of 15 cm along the east margin of Halema'uma'u Crater. Summit deformation modeled from InSAR and GPS data suggest a volume loss of $1.7 \times 10^6 m^3$ (Table 18.2) from a magma reservoir at 1.7 km depth [*Lundgren et al.*, 2013], which is a known area of magma storage that has deflated during previous East Rift Zone dike intrusions and eruptions [*Owen et al.*, 2000; *Montgomery-Brown et al.*, 2010]. At Pu'u 'Ō'ō, deflation also waned over time but continued beyond the end of the eruption until the afternoon of 13 March.

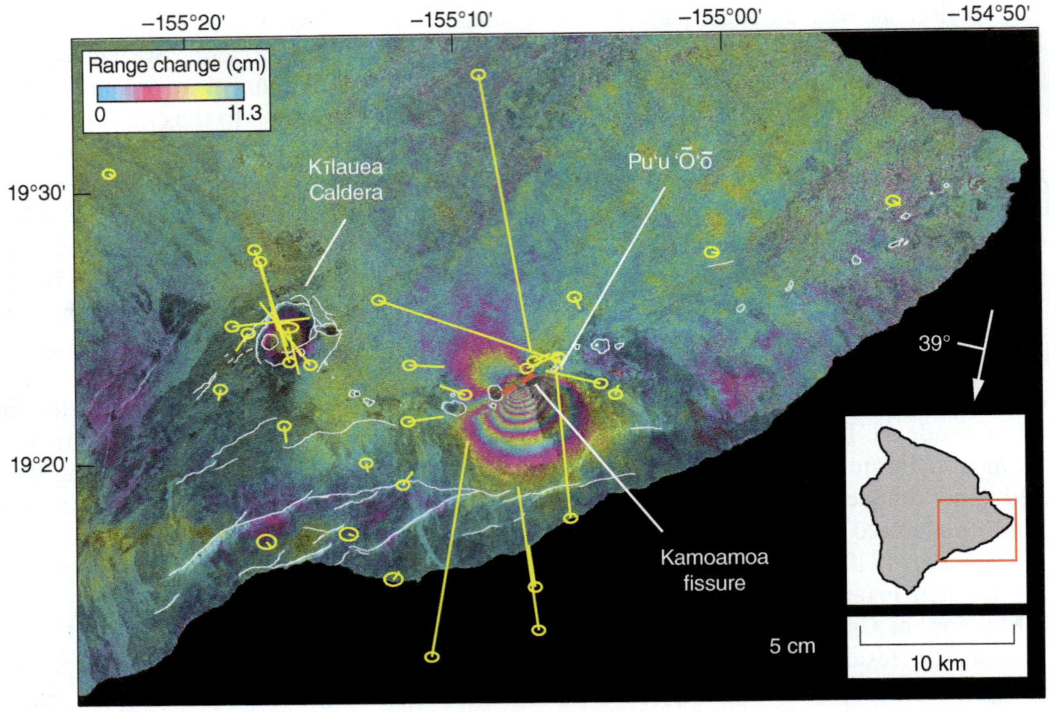

Figure 18.13 Interferogram from PALSAR instrument on ALOS satellite spanning 24 January–11 March 2011 (orbital path 601). White arrow and degree value at right give flight direction and look angle (in degrees from vertical), respectively. Yellow arrows are horizontal displacements from GPS with estimated error ellipses; length of yellow arrow scale in lower right shows 5 cm displacement. Red lines in area of greatest deformation denote western and eastern Kamoamoa eruptive fissures. (Inset) Island of Hawai'i, showing map area.

Deformation at Pu'u 'Ō'ō was highly localized, suggesting a source perhaps only a few hundred meters deep (consistent with previous estimates of Pu'u 'Ō'ō magma storage [*Heliker et al.*, 2003]).

Deformation modeling based on InSAR and GPS data from the East Rift Zone indicates that the magma initially feeding the Kamoamoa eruption originated at ~3 km depth near the eastern edge of Makaopuhi Crater [*Lundgren et al.*, 2013] and ascended along an inclined (~45°) trajectory. The eruption began when the magma arrived at the surface ~3.5 h later, after traveling an ascent path distance of ~4.25 km. This suggests that the dike tip advanced at an average velocity of ~1.2 km/h (0.3 m s^{-1}), similar to average dike propagation rates for past Kīlauea intrusions (~1 km/h [*Klein et al.*, 1987; *Okamura et al.*, 1988; *Mangan et al.*, 1995; *Owen et al.*, 2000; *Cervelli et al.*, 2002]). The horizontal component of this vector equates to ~0.8 km/h (0.2 m/s). Field observations of the opening of fissures E-1 to E-3 show that the fissures ruptured over a distance of 330 m in about 25 min, or 0.8 km/h, the same as the horizontal propagation rate of the dike as it ascended toward the surface.

The ERZ spread apart by several meters at the site of the Kamoamoa eruption, with progressive opening over time as indicated by both fracture measurements and InSAR data [*Lundgren et al.*, 2013]. The ground between the eastern and western fissure systems is covered by tephra from the 1983–1986 high fountains of the Pu'u 'Ō'ō eruption. As a result, cracks that formed due to emplacement of the Kamoamoa dike were very well exposed (Figure 18.14). A transect perpendicular to the rift zone on 6 March measured the opening amounts of cracks in the area, finding a total of 1.5 m of cumulative crack opening over a distance of 200 m (Figure 18.14). The total opening was 2.75 m when the transect was repeated on 9 March (Figure 18.14). These amounts are minima, since the transect's northern end terminated against a preexisting graben that experienced an unknown amount of opening. InSAR data from different days of the eruption corroborate these results, showing increasing displacement across the dike over time. The total dike volume at the end of the eruption, based on models of InSAR and GPS displacements, was $15.6 \times 10^6 m^3$ (Table 18.2) [*Lundgren et al.*, 2013].

18.4.7. Seismicity

At approximately the same time as the onset of subsidence and collapse at Pu'u 'Ō'ō, seismic amplitudes

Figure 18.14 (a) Oblique aerial photograph from 7 March 2011 showing fractured, tephra-covered ground between eastern (background) and western (foreground) fissure systems. Fissure W-6 at bottom. Dashed white line shows transect (walked south to north) along which crack opening was measured. (Inset) Photograph showing close-up view of ground cracks on 6 March 2011. (b) Graph showing cumulative crack width along south-to-north transect shown in part (a). Blue line, 6 March 2011 (~1130 HST); red line, 9 March (~1430 HST). Strike of western fissure intersects transect between 100 and 160 m.

(expressed in terms of real-time seismic amplitude measurement (RSAM) [*Endo and Murray*, 1991]) increased dramatically (Figures 18.8 and 18.9) on stations located along the East Rift Zone (PAU, STC, and KNH). The seismic signal was initially composed of discrete events that were often so close together temporally that they combined to create a tremorlike signal (Figure 18.8a). Most of these events did not record on enough stations to be located; however, it is clear that the seismicity was associated both with dike propagation between Makaopuhi Crater and Nāpau Crater and failure of Pu'u 'Ō'ō's crater floor. Long-period and volcano-tectonic events were simultaneously present. Once fountaining along the Kamoamoa fissure started, the character of tremor changed at the station closest to the fissure site (STC), becoming more energetic and also having a broader and higher-frequency spectral signature than previous activity (Figure 18.8b).

For the remainder of the eruption, tremor recorded at station STC (using RSAM as a proxy) had frequencies between 1.5 and 7 Hz, and tremor amplitudes generally matched changes in fountaining (Figure 18.9). The amplitude of tremor (and RSAM value) recorded on STC depended on two factors: (1) the intensity of the source and (2) the distance from the source to the seismometer. The distance between STC and the fountaining source centroid on each fissure ranged from 440 m (E-4) to 1765 m (W-1) (Table 18.3). Clearly, when two fountaining sources occurred simultaneously, the closer source was the dominant signal, assuming both sources had similar intensities.

Over the course of the eruption, large changes in RSAM corresponded to variations in fountain behavior. A significant bump in RSAM during ~1700–1930 HST on 5 March corresponded to the initiation of fountaining from fissures E-1, E-2, and E-3 (see 1, Figure 18.9a). Even though the fountain heights and rate of extrusion were low, STC was only ~500 m away, and thus the response in RSAM was significant. Activity resumed shortly after midnight on 6 March with fountaining at fissure W-3 (1500 m distance from STC) occurring during a small, temporary bump in RSAM. Another small rise in RSAM started at ~0140 HST on 6 March but began to climb more rapidly at ~0240, when sporadic activity started at fissures E-3 and (or) E-4 and propagated to E-5 (~500 m from STC; see 2, Figure 18.9a). Fissure W-4 activated during the decline of this RSAM anomaly, but W-4's distance from STC resulted in no significant contribution to the RSAM compared to the decline in activity at the eastern fissures. Another increase in RSAM started at ~0945 HST on 6 March. As the RSAM ramped up, eruptive activity resumed at fissure E-5 (~500 m from STC). Modulations in RSAM for the subsequent 12 h may be explained by changes in observed fountain height at E-5 (see 3, Figure 18.9a).

On 7 March at ~0700 HST, RSAM values started to increase at the same time as the cessation of eruptive activity from fissure E-5 and shortly after the onset of eruptive activity from fissure W-2 (see 4, Figure 18.9a). The RSAM increase peaks at about the same time as the cessation of W-2 activity. Since RSAM did not correlate with earlier changes in fountaining from western fissures and in this case ramped up and peaked as fissure W-2 was shutting down, it seems unlikely that this RSAM increase was related to W-2 activity. On the other hand, no eruptive activity from any eastern fissure matches the RSAM increase either. While the RSAM increase could correlate with loud outgassing from eastern fissures without eruption, no ground observations can verify if that was, indeed, the case.

A slight increase in RSAM began at ~1110 HST on 7 March, at about the same time as the onset of fountaining at fissure W-3. At ~1315 HST, RSAM began to increase rapidly. This increase does not correlate with obvious eruptive activity from any eastern fissure, but it does spike during a brief fountaining episode from fissure E-4 (440 m from STC) at ~1520 HST (see 5, Figure 18.9a). A step in RSAM at ~1800 HST can be attributed to another episode of fountaining from E-4 (see 6, Figure 18.9a), and a second step at ~2115 occurs in tandem with an increase in E-4 fountain height (see 7, Figure 18.9a). The subsequent long decline in RSAM that extended to the end of the eruption (see 8, Figure 18.9a) reflects a shift in activity from the eastern fissures to W-1, the fissure farthest from STC (~1765 m). In the analysis above, the link between amplitudes and eruption activity is very strong. This suggests that the source of the tremor with this spectral character can be attributed in most instances to near-surface processes associated with fountaining, which includes not only outgassing and the movement of magma but possibly also the impact of spatter on the ground surface.

Tremor with peak frequencies between 0.1 and 0.5 Hz was also recorded on near-summit stations during the Kamaomoa eruption. The tremor was recorded at distances >12 km but was strongest in the summit area. Previous studies of VLP seismic sources with nearly identical frequency content have been attributed to magma movement through the shallow conduit system at the summit of Kīlauea [*Dawson et al.*, 2010]. We believe that a source invoking magma flow out of the shallow summit conduit system is consistent with seismic observations, deflation, and falling lava lake level.

18.5. DISCUSSION

The exceptional suite of geological, geophysical, and geochemical observations collected before and during the Kamoamoa eruption provides an excellent window into the mechanism for eruptive activity that can be applied to future episodes of eruptive change at Kīlauea (and similar volcanoes). Based on these data, we conclude that the Kamoamoa eruption was caused by an imbalance between the rates of magma supply and eruption. Similar imbalances, regardless of whether they are controlled by an increase in supply or a decrease in output, have caused other changes in activity during the Puʻu ʻŌʻō eruption, such as at the June 2007 eruption at Kanenuiohamo Lava Shield (Table 18.1). This is in contrast to eruptive changes that occur in response to the "passive" opening of the rift (e.g., January 1997 and September 1999 intrusions [*Owen et al.*, 2000; *Cervelli et al.*, 2002]), which is related to seaward motion of Kīlauea's south flank.

The time-averaged discharge rate of the episode 58 vent in 2007 was approximately 6 m^3/s (0.18 km^3/yr [*Poland et al.*, 2012]). By mid-2010, however, the time-averaged discharge rate from that vent had dropped precipitously by several times (to ~1 m^3/s [*Orr et al.*, in press]). The decline in discharge rate was reflected in a waning of lava flow activity on the flow field between the East Rift Zone and the ocean as well as by a decrease in the SO_2 emission rate from the East Rift Zone. Unfortunately, the SO_2 emission rates cannot be used to precisely quantify the discharge rate as was possible prior to 2008 [*Sutton et al.*, 2001, 2003] because lava erupting from the East Rift Zone after 2008 is thought to have already partly outgassed as it circulated through the summit lava lake [*Elias and Sutton*, 2012]. Nevertheless, the decrease in SO_2 emission rates from the East Rift Zone during 2009–2010 (which, at the end of 2010, were the lowest recorded since vehicle-based measurements began in 1992 [*Elias and Sutton*, 2012]) supports direct geological observations of a decline in discharge rate.

The influence of a decrease in magma throughput on the shallow, subhorizontal conduit connecting Pu'u 'Ō'ō to the episode 58 vent, its top at a depth of no more than a few hundred meters (Figure 18.4) [*Poland et al.*, 2008], probably induced cooling and shrinking of the conduit. This would have the effect of reducing the conduit's cross-sectional area and, hence, its carrying capacity. The reduction in carrying capacity was probably exacerbated by interruptions in magma transport from the summit to the East Rift Zone caused by repeated DI events. When the DI events stopped occurring for several months starting in November 2010, the conduit was not able to accommodate the supply of magma being fed constantly from the summit, and pressure built quickly within Kīlauea's magma plumbing system (beneath the summit and within the East Rift Zone). The increase in pressure was manifested as uplift at the summit and Pu'u 'Ō'ō and as extension across the East Rift Zone (Figure 18.5) and also as an increase in the number of earthquakes beneath the summit and upper East Rift Zone (Figures 18.5 and 18.6). Eventually, the system reached its breaking point, and the weakest link, in this case, the East Rift Zone near Nāpau Crater, failed. This is not surprising given the history of the area, which has been the site of frequent eruptions over the past several decades, including in 1961, 1963, 1965, 1968, 1983 (episode 1 of the Pu'u 'Ō'ō eruption), and 1997 (episode 54 of the Pu'u 'Ō'ō eruption). These and previous eruptions might have left residual pods of melt in the subsurface that could be incorporated into subsequent eruptions, as indicated by petrologic evidence from early Pu'u 'Ō'ō lavas [*Garcia et al.*, 1992] and the 1997 Nāpau Crater eruption [*Thornber et al.*, 2003], as well as the Kamoamoa eruption.

We therefore interpret the pressurization of Kīlauea's magma plumbing system, as well as periods of filling of Pu'u 'Ō'ō crater during 2010 (in particular, the sustained filling that began in November of that year), as being caused by an imbalance between the amount of magma arriving at Pu'u 'Ō'ō and the amount being erupted from the episode 58 vent. We also cannot rule out a small increase in the rate of magma supply to Kīlauea, especially since supply has been shown to fluctuate on time scales of years [*Poland et al.*, 2012]. Such an increase in supply could not have been large, given the lack of any substantive change in the rate of CO_2 emission from the summit, and may be part of a natural waxing and waning of magma supply that is generally too small to be recognized. Regardless, the volume of magma reaching Pu'u 'Ō'ō exceeded that which could be erupted from the episode 58 vent, resulting in a backup that caused the volcano's summit and East Rift Zone magma systems to pressurize and eventually rupture, leading to the Kamoamoa fissure eruption.

Although fed initially by magma from both Pu'u 'Ō'ō and the summit, the summit probably contributed the bulk of the magma being erupted after the first few hours. Deflation at Kīlauea's summit decayed rapidly and smoothly following eruption onset (Figure 18.8), and there were no geophysical changes that matched the intermittency of the Kamoamoa eruption as it jumped between fissures during the initial three days of the eruption. If the transport of magma from the summit to the Kamoamoa eruption site occurred in a steady state fashion, as suggested by the smoothness of the tilt record, then variations in eruptive location and output must have been caused by more localized factors, such as changes in magma viscosity, the near-surface stress field, or the shape of the erupting fissures over time. As with most basaltic fissure eruptions, activity eventually focused on a single fissure (W-1), which was the dominant source of lava discharge over the last two days of the eruption (Figure 18.9c; Table 18.3).

Once the excess pressure that had accumulated during 2010 and early 2011 had dissipated, the pressure gradient between the summit and the Kamoamoa fissure was not sufficient to sustain the eruption. This allowed activity to cease, and the Kamoamoa dike presumably cooled thereafter due to the large thermal gradient between the dike and the surrounding country rock [*Wilson and Head*, 1981]. After the eruption ended, pressure began to climb within Kīlauea's magma storage areas and transport pathways in response to the continuous supply of magma to the volcano. This pressure led to the reappearance of the summit lava lake in the Overlook crater and was eventually directed toward restarting eruptive activity at Pu'u 'Ō'ō, probably because that well-worn conduit remained hot and provided an easy way for magma to reach the surface along the East Rift Zone.

The 2010–2011 gradual abandonment of the episode 58 vent and onset of the episode 59 Kamoamoa fissure eruption bears striking resemblance to the 1991–1992 slowdown of the episode 48 Kupaianaha vent and onset of the episode 49 fissure. In the latter case, output from Kupaianaha waned over several months, while the summit inflated and eruptive activity increased at Pu'u 'Ō'ō [*Kauahikaua et al.*, 1996]. Also during 1990–1991, a series of seismic swarms in the upper East Rift Zone, interpreted as intrusions, interrupted magma flow to the eruption site [*Kauahikaua et al.*, 1996; *Heliker and Mattox*, 2003]. *Kauahikaua et al.* [1996] interpreted these events as an indication that cooling and crystallization had caused the magma reservoir beneath Kupaianaha to become increasingly isolated from the rest of the magmatic system, which led to a decline in the magma pressure driving the eruption. The uprift buildup in pressure in response to the Kupaianaha slowdown probably led to the episode 49 fissure eruption between Pu'u 'Ō'ō and Kupaianaha in November 1991 (Table 18.2) [*Mangan et al.*, 1995], although Kupaianaha continued to be weakly active until February 1992 [*Kauahikaua et al.*, 1996].

Based on the examples from the Kupaianaha decline and episode 49 fissure eruption and from the episode 58 decline and Kamoamoa fissure eruption, it should be possible to project the likelihood of future changes in eruptive activity several months in advance, provided the activity follows a similar pattern. Such changes are most likely to occur when eruption is occurring at a vent away from Pu'u 'Ō'ō but still connected to Pu'u 'Ō'ō via a shallow conduit. Our model is not, however, suitable for forecasting fissure eruptions like that which occurred in 1997 from Nāpau Crater [*Owen et al.*, 2000; *Thornber et al.*, 2003]. As discussed above, that eruption, as well as a dike intrusion in 1999 [*Cervelli et al.*, 2002], was inferred to have resulted from passive opening of the East Rift Zone due to seaward motion of the south flank, and neither event was preceded by summit inflation or increased East Rift Zone spreading. The transitions of eruptive activity away from both the episode 48 (in 1991–1992) and episode 58 (2011) vents and the ensuing fissure eruptions were, in contrast, associated with waning lava effusion and pressurization of the magmatic system.

The magma transport pathway between the summit and Pu'u 'Ō'ō is robust after decades of activity and continues to be used, even when lava effusion is focused away from Pu'u 'Ō'ō. In contrast, conduits associated with vents that are satellitic to Pu'u 'Ō'ō tend to be shallow and more susceptible to choking by minor changes in magma supplied to the vent. Repeated interruptions in magma flow from the summit to the East Rift Zone, due to uprift intrusions, DI events, or other processes, could therefore cause the satellite vent system to cool and become isolated. The discharge rate would wane as a result, causing pressurization of Kīlauea's magma plumbing system manifested as inflation of the summit (along with inflation at the eruption site and spreading of the East Rift Zone), increases in seismicity along the upper East Rift Zone, and rising lava level within the summit eruptive vent. Eruptive activity at the satellite vent would decline, while lava effusion at Pu'u 'Ō'ō would increase. Once a critical threshold is reached, the conduit system will rupture in response to the building pressure.

18.6. SUMMARY AND CONCLUSIONS

The 5–9 March 2011 Kamoamoa fissure eruption occurred as part of a shift in vent location of the long-lived East Rift Zone eruption at Kīlauea, which has been ongoing since 1983. On the afternoon of 5 March, a dike began propagating from the East Rift Zone conduit between Makaopuhi and Nāpau Craters (a site of repeated fissure eruptions since the 1960s), originating from and (or) intersecting and incorporating older, cooler, more evolved magma. Magma from Pu'u 'Ō'ō was also diverted into the growing dike, causing the cone's crater floor to collapse. The lava lake in the Overlook crater began to drain as magma likewise withdrew from beneath Kīlauea's summit. The dike reached the surface about 3.5 h after its initiation and fed an eruption that extended over more than 2 km along two fissure system (each of which was composed of five to six fissures). Eruptive activity jumped between the fissures during the first three days before finally settling on the westernmost fissure for the final day of the eruption. The total volume erupted was approximately $2.7 \times 10^6 m^3$, and the peak eruption rate was about 3 times higher than the long-term average eruption rate from Pu'u 'Ō'ō.

The Kamoamoa eruption was probably a result of the gradual shutdown of the episode 58 vent, which was located about 2 km downrift of Pu'u 'Ō'ō. The vent had been active since 2007 but experienced a slow decline in eruption rate that became evident in 2010. Repeated interruptions in magma flow from the summit to the East Rift Zone due to cycles of summit deflation and inflation may have led to the gradual cooling and isolation of the episode 58 conduit, which in turn caused pressurization within Kīlauea's magma system. The pressurization was manifested by inflation of the summit and Pu'u 'Ō'ō, spreading of the East Rift Zone, increased seismicity along the upper East Rift Zone, and rising lava levels within both the summit eruptive vent and Pu'u 'Ō'ō's crater. The activity culminated in the Kamoamoa fissure eruption, after which the locus of long-term lava effusion moved back to Pu'u 'Ō'ō, and the episode 58 vent ceased erupting altogether. The 2010–2011 pattern of activity strongly resembled that of 1990–1992, when waning of the Kupaianaha eruptive vent culminated in an East Rift Zone fissure eruption and subsequent shift in eruptive activity to Pu'u 'Ō'ō. Based on these patterns and the excellent geological, geophysical, and geochemical monitoring of Kīlauea, it should be possible to recognize similar progressions months in advance of any shifts in eruption style and/or location.

SUPPLEMENTARY DIGITAL DATA

The following data items are available online at www.wiley.com/go/Carey/Hawaiian_Volcanoes

Appendix 18A. Detailed chronology of Kamoamoa fissure eruption.

Movie 18.1. Time-lapse movie showing collapse of Pu'u 'Ō'ō crater floor on 5 March 2011.

Movie 18.2. Time-lapse movie showing draining of summit lava lake on 5–6 March.

Movie 18.3. Movie clips showing fissure E-3 on 5 March 2011.

Movie 18.4. Movie clips showing fissure W-4 on 6 March 2011.

Movie 18.5. Movie clips showing fissure E-5 on 6 March 2011.

Movie 18.6. Movie clips showing fissure E-5 on 8 March 2011.

Movie 18.7. Movie clip showing fissure W-1 on 8 March 2011.

Movie 18.8. Movie clip showing fissure W-1 on 9 March 2011.

ACKNOWLEDGMENTS

We are grateful to the staff and volunteers of the Hawaiian Volcano Observatory for installing and maintaining the instruments upon which we based our conclusions and who also collected numerous observations that contributed to a better understanding of the eruption. Special thanks to David Okita for his expert piloting, which has facilitated unparalleled access to eruptive activity at Kīlauea for several decades. Don Swanson, Thomas Staudacher, and an anonymous reviewer provided helpful reviews that improved the manuscript.

REFERENCES

Baker, S., and F. Amelung (2012), Top-down inflation and deflation at the summit of Kīlauea Volcano, Hawaii observed with InSAR, *J. Geophys. Res.*, *117*(B12406), doi:10.1029/2011JB009123.

Carbone, D., M. P. Poland, M. R. Patrick, and T. R. Orr (2013), Continuous gravity measurements reveal a low-density lava lake at Kīlauea Volcano, Hawai'i, *Earth Planet. Sci. Lett.*, *376*(0): 178–185, doi:10.1016/j.epsl.2013.06.024.

Cervelli, P., P. Segall, F. Amelung, H. Garbeil, C. Meertens, S. Owen, A. Miklius, and M. Lisowski (2002), The 12 September 1999 Upper East Rift Zone dike intrusion at Kilauea Volcano, Hawaii, *J. Geophys. Res.*, *107*(2150), doi:10.1029/2001JB000602.

Clague, D. A., and C. Heliker (1993), The ten-year eruption of Kilauea Volcano, *U. S. Geol. Surv. Earthquakes and Volcanoes*, *23*(6), 244–254.

Dawson, P., M. Benítez, B. Chouet, and D. Wilson (2010), Monitoring very-long-period seismicity at Kilauea Volcano, Hawaii, *Geophys. Res. Lett.*, *37*(L18306), 1–5, doi:10.1029/2010GL044418.

DePaolo, D. J., and E. M. Stolper (1996), Models of Hawaiian volcano growth and plume structure: Implications of results from the Hawaii Scientific Drilling Project, *J. Geophys. Res.*, *101*(B5), 11,643–11,654, doi:10.1029/96JB00070.

Eaton, J. P., and K. J. Murata (1960), How volcanoes grow, *Science*, *132*(3432), 925–938, doi:10.1126/science.132.3432.925.

Elias, T., and A. J. Sutton (2012), Sulfur dioxide emission rates from Kīlauea Volcano, Hawai'i, 2007–2010, *U. S. Geol. Surv. Open-File Rep. 2012–1107*.

Endo, E. T., and T. Murray (1991), Real-time seismic amplitude measurement (RSAM): A volcano monitoring and prediction tool, *Bull. Volcanol. 53*(7), 533–545, doi:10.1007/BF00298154.

Fee, D., M. Garces, T. Orr, and M. Poland (2011), Infrasound from the 2007 fissure eruptions of Kīlauea Volcano, Hawai'i, *Geophys. Res. Lett.*, *38*(L06309), doi:10.1029/2010GL046422.

Fiske, R. S., and W. T. Kinoshita (1969), Inflation of Kilauea Volcano prior to its 1967–1968 eruption, *Science*, *165*(3891), 341–349, doi:10.1126/science.165.3891.341.

Garcia, M. O., J. M. Rhodes, E. W. Wolfe, G. E. Ulrich, and R. A. Ho (1992), Petrology of lavas from episodes 2–47 of the Puu Oo eruption of Kilauea Volcano, Hawaii: Evaluation of magmatic processes, *Bull. Volcanol.*, *55*(1–2), 1–16.

Gerlach, T. M., K. A. McGee, T. Elias, A. J. Sutton, and M. P. Doukas (2002), Carbon dioxide emission rate of Kilauea Volcano: Implications for primary magma and the summit reservoir, *J. Geophys. Res.*, *107*(2189), doi: 2110.1029/2001JB000407.

Heliker, C., and T. N. Mattox (2003), The first two decades of the Pu'u 'Ō'ō–Kupaianaha eruption: Chronology and selected bibliography, in *The Pu'u 'Ō'ō-Kupaianaha eruption of Kīlauea Volcano, Hawai'i: The First 20 Years*, edited by C. Heliker et al., pp. 1–27, U. S. Geol. Surv. Prof. Pap. 1676.

Heliker, C., and T. L. Wright (1991), Lava-flow hazards from Kilauea's current eruption, *Geotimes*, *36*(5), 16–18.

Heliker, C. C., M. T. Mangan, T. N. Mattox, J. P. Kauahikaua, and R. T. Helz (1998), The character of long-term eruptions: Inferences from episodes 50–53 of the Pu'u 'Ō'ō–Kūpaianaha eruption of Kīlauea Volcano, *Bull. Volcanol.*, *59*(6), 381–393, doi:10.1007/s004450050198.

Heliker, C., J. Kauahikaua, D. R. Sherrod, M. Lisowski, and P. Cervelli (2003), The rise and fall of Pu'u 'Ō'ō cone, 1983–2002, in *The Pu'u 'Ō'ō-Kupaianaha eruption of Kīlauea Volcano, Hawai'i: The First 20 Years*, edited by C. Heliker et al., pp. 29–51, U. S. Geol. Surv. Prof. Pap. 1676.

Helz, R. T., and C. R. Thornber (1987), Geothermometry of Kilauea Iki lava lake, Hawaii, *Bull. Volcanol.*, *49*(5), 651–668.

Hon, K., J. Kauahikaua, R. Denlinger, and K. Mackay (1994), Emplacement and inflation of pahoehoe sheet flows: Observations and measurements of active lava flows on Kilauea Volcano, Hawaii, *Geol. Soc. Am. Bull. 106*(3), 351–370.

Kauahikaua, J. (2007), Lava flow hazard assessment, as of August 2007, for Kīlauea east rift zone eruptions, Hawai'i Island, *U. S. Geol. Surv. Open-File Rep. 2007-1264*.

Kauahikaua, J., M. Mangan, C. Heliker, and T. Mattox (1996), A quantitative look at the demise of a basaltic vent: The death of Kupaianaha, Kilauea Volcano, Hawai'i, *Bull. Volcanol.*, *57*(8), 641–648, doi:10.1007/s004450050117.

Kern, C., T. Deutschmann, C. Werner, A. J. Sutton, T. Elias, and P. J. Kelly (2012), Improving the accuracy of SO2 column densities and emission rates obtained from upward-looking UV-spectroscopic measurements of volcanic plumes by taking realistic radiative transfer into account, *J. Geophys. Res. Atmos.*, *117*(D20), D20302, doi:10.1029/2012JD017936.

Klein, F. W., R. Y. Koyanagi, J. S. Nakata, and W. R. Tanigawa (1987), The seismicity of Kilauea's magma system, in Volcanism in Hawaii, edited by R. W. Decker et al., pp. 1019–1185, U. S. Geol. Surv. Prof. Pap. 1350.

Lundgren, P., M. Poland, A. Miklius, T. Orr, S.-H. Yun, E. Fielding, A. Liu, A. Tanaka, W. Szeliga, S. Hensley, and S. Owen (2013), Evolution of dike opening during the March 2011 Kamoamoa fissure eruption, Kīlauea Volcano, Hawai'i, *J. Geophys. Res.*, *118*(3), 897–914, doi:10.1002/jgrb.50108.

Mangan, M.T., C. C. Heliker, T. N. Mattox, J. P. Kauahikaua, and R. T. Helz (1995), Episode 49 of the Pu'u 'O'o–Kupaianaha eruption of Kilauea volcano — Breakdown of a steady-state eruptive era, *Bull. Volcanol.*, *57*(2), 127–135.

Mattox, T. N., C. Heliker, J. Kauahikaua, and K. Hon (1993), Development of the 1990 Kalapana flow field, Kilauea Volcano, Hawaii, *Bull. Volcanol.*, *55*(6), 407–413.

Montgomery-Brown, E. K., D. K. Sinnett, M. Poland, P. Segall, T. Orr, H. Zebker, and A. Miklius (2010), Geodetic evidence for en echelon dike emplacement and concurrent slow slip during the June 2007 intrusion and eruption at Kīlauea volcano, Hawaii, *J. Geophys. Res.*, *115*(B07405), doi:10.1029/2009JB006658.

Okamura, A. T., J. J. Dvorak, R. Y. Koyanagi, and W. R. Tanigawa (1988), Surface deformation during dike propagation, in The Puu Oo Eruption of Kilauea Volcano, Hawaii; Episodes 1 Through 20, January 3, 1983, Through June 8, 1984, edited by E. W. Wolfe, pp. 165–182, U. S. Geol. Surv. Prof. Pap. 1463.

Okubo, P., J. Nakata, B. Chouet, and P. Dawson (1996), The February 1, 1996, Kilauea summit earthquake swarm, *EOS Trans. AGU*, *77*, F798.

Orr, T. R. (2014), The June–July 2007 collapse and refilling of Pu'u 'Ō'ō Crater, Kīlauea Volcano, Hawai'i, *U. S. Geol. Surv. Sci. Invest. Rep.*, 2014–5124, 15 pp.

Orr, T. R., W. A. Thelen, M. R. Patrick, D. A. Swanson, and D. C. Wilson (2013), Explosive eruptions triggered by rockfalls at Kīlauea volcano, Hawai'i, *Geology*, *41*(2), 207–210.

Orr, T. R., J. E. Bleacher, M. R. Patrick, and K. M. Wooten (in press), A sinuous tumulus over an active lava tube at Kīlauea Volcano: Evolution, analogs, and hazard forecasts, *J. Volcanol. Geotherm. Res.*, doi:10.1016/j.jvolgeores.2014.12.002.

Owen, S., P. Segall, M. Lisowski, A. Miklius, M. Murray, M. Bevis, and J. Foster (2000), January 30, 1997 eruptive event on Kilauea Volcano, Hawaii, as monitored by continuous GPS, *Geophys. Res. Lett.*, *27*(17), 2757–2760.

Parcheta, C. E., B. F. Houghton, and D. A. Swanson (2012), Hawaiian fissure fountains 1: Decoding deposits—Episode 1 of the 1969–1974 Mauna Ulu eruption, *Bull. Volcanol.*, *74*(7), 1729–1743.

Patrick, M., T. Orr, A. J. Sutton, T. Elias, and D. Swanson (2013), The first five years of Kīlauea's summit eruption in Halema'uma'u, 2008–2013, *U. S. Geol. Surv. Fact Sheet 2013–3116*, 4 p.

Patrick, M. R., and T. Orr (2012), Rootless shield and perched lava pond collapses at Kīlauea Volcano, Hawai'i, Bull. *Volcanol.*, *74*(1), 67–78.

Patrick, M. R., T. Orr, D. Wilson, D. Dow, and R. Freeman (2011), Cyclic spattering, seismic tremor, and surface fluctuation within a perched lava channel, Kīlauea Volcano, *Bull. Volcanol.*, *73*(6), 639–653.

Poland, M. P., A. Miklius, T. Orr, A. J. Sutton, C. Thornber, and D. Wilson (2008), New episodes of volcanism at Kīlauea Volcano, Hawai'i, EOS Trans. *AGU*, *89*(5), 37–38.

Poland, M. P., A. Miklius, A. J. Sutton, and C. R. Thornber (2012), A mantle-driven surge in magma supply to Kilauea Volcano during 2003–2007, *Nature Geosci.*, *5*(4), 295–300.

Rivalta, E., and P. Segall (2008), Magma compressibility and the missing source for some dike intrusions, *Geophys. Res. Lett.*, *34*(L04306), doi:10.1029/2007GL032521.

Roche, O., B. van Wyk de Vries, and T. H. Druitt (2001), Subsurface structures and collapse mechanisms of summit pit craters, *J. Volcanol. Geotherm. Res. 105*(1–2), 1–18.

Ryan, M. P. (1987), Neutral buoyancy and the mechanical evolution of magmatic systems, in *Magmatic Processes: Physiochemical Principles*, edited by B. O. Mysen, pp. 259–287.

Sen, G., and R. E. Jones (1990), Cumulate xenolith in Oahu, Hawaii: Implications for deep magma chambers and Hawaiian volcanism, *Science*, *249*(4973), 1154–1157.

Sutton, A. J., T. Elias, T. M. Gerlach, and J. B. Stokes (2001), Implications for eruptive processes as indicated by sulfur dioxide emissions from Kilauea Volcano, Hawai`i, 1979–1997, *J. Volcanol. Geotherm. Res.*, *108*(1–4), 283–302.

Sutton, A. J., T. Elias, and J. Kauahikaua (2003), Lava-effusion rates for the Pu'u 'Ō'ō-Kupaianaha eruption derived from SO_2 emissions and very low frequency (VLF) measurements, in *The Pu'u 'Ō'ō-Kupaianaha Eruption of Kīlauea Volcano, Hawai'i: The First 20 Years*, edited by C. Heliker et al., pp. 137–148, U. S. Geol. Surv. Prof. Pap. 1676.

Swanson, D. A., D. B. Jackson, R. Y. Koyanagi, and T. L. Wright (1976), The February 1969 east rift eruption of Kilauea Volcano, Hawaii, U.S. Geol. Surv. Prof. Pap. 891.

Thornber, C. R. (2003), Magma-reservoir processes revealed by geochemistry of the Pu'u Ō'ō -Kupaianaha eruption, in *The Pu'u 'Ō'ō-Kupaianaha Eruption of Kīlauea Volcano, Hawai'i: The First 20 Years*, edited by C. Heliker et al., pp. 121–136, U. S. Geol. Surv. Prof. Pap., 1676.

Thornber, C. R., C. Heliker, D. R. Sherrod, J. P. Kauahikaua, A. Miklius, P. G. Okubo, F. A. Trusdell, J. R. Budahn, W. I. Ridley, and G. P. Meeker (2003), Kilauea east rift zone magmatism; an episode 54 perspective, *J. Petrol.*, *44*(9), 1525–1559.

Wilson, D., T. Elias, T. Orr, M. Patrick, A. J. Sutton, and D. Swanson (2008), Small explosion from new vent at Kilauea's summit, EOS Trans. *AGU*, *89*(22), 203.

Wilson, L., and J. W. Head III (1981), Ascent and eruption of basaltic magma on the Earth and Moon, *J. Geophys. Res.*, *86*(B4), 2971–3001.

Wolfe, E. W., C. A. Neal, N. G. Banks, and T. J. Duggan (1988), Geologic observations and chronology of eruptive events, in *The Puu Oo eruption of Kilauea Volcano, Hawaii;* Episodes 1 Through 20, January 3, 1983, Through June 8, 1984, edited by E. W. Wolfe, pp. 1–97, U. S. Geol. Surv. Prof. Pap., 1463.

Wright, T. L. (1984), Origin of Hawaiian tholeiite: A metasomatic model, *J. Geophys. Res.*, *89*(B5), 3233–3252.

Wyllie, P. J. (1988), Solidus curves, mantle plumes, and magma generation beneath Hawaii, *J. Geophys. Res.*, *93*(B5), 4171–4181.

19

Onset of a Basaltic Explosive Eruption From Kīlauea's Summit in 2008

Rebecca J. Carey[1], Lauren Swavely[2], Donald A. Swanson[3], Bruce F. Houghton[2], Tim R. Orr[3], Tamar Elias[3], and A. Jeff Sutton[3]

ABSTRACT

The onset of a basaltic eruption at the summit of Kīlauea Volcano in 2008 is recorded in the products generated during the first three weeks of the eruption and suggests an evolution of both the physical properties of the magma and also lava lake levels and vent wall stability. Ash componentry and the microtextures of the early erupted lapilli products reveal that the magma was largely outgassed, perhaps in the preceding weeks to months. An increase in the juvenile-lithic ratio and size of ash collected from 23 March to 3 April records an increasing level of the magma within the conduit. After 3 April until the explosive eruption of 9 April, a trend of decreasing juvenile-lithic ratio suggests that vent wall collapses were more frequent, possibly because lava level increased and destabilized the overhanging wall [*Orr et al.*, 2013]. Despite increasing lake height, the microtextural characteristics of the lapilli suggest that the outgassed end member was still being tapped between 26 March and 8 April. The 9 April rockfall triggered an explosive eruption that produced a new component in the eruption deposits not seen in the preceding weeks, microvesicular juvenile lapilli, the first evidence of an actively vesiculating magma. Two additional dense end-member pyroclast types were also erupted during the 9 April explosion, likely related to outgassed magma with longer residence times than the microvesicular magma. We link these pyroclasts to a stagnant viscous crust at the top of the magma column or to convecting, downwelling magma. Our study of ash componentry and the textures of juvenile lapilli suggests that the 9 April explosive event effectively cleared the conduit of largely outgassed magma. The degassing processes during this eruption are complex and varied: In the period of persistent degassing during 26 March – 8 April small resident bubbles at shallow levels in the lava lake were coupled to the magma whereas large bubbles ascended, expanded, and fragmented. During the rockfall-triggered explosion of 9 April, all bubbles were coupled to the host magma on the time scale of decompression, but additional exsolution, decompression, and expansion of deeper, more gas-rich resident magma likely occurred [cf. *Carey et al.*, 2012]. Where external conditions play a significant role in eruption dynamics, e.g., by triggering eruptions, vesiculation and degassing dynamics can be expected to be complex.

19.1. INTRODUCTION

The time scales of magma transport and degassing are complex and relatively difficult to quantify without the integration of well-constrained multidisciplinary data sets, such as those from Kīlauea [e.g., *Poland et al.*, 2009; *Patrick et al.*, 2011], Stromboli [e.g., *Rosi et al.*, 2006], and

[1] *School of Physical Sciences and Centre of Excellence in Ore Deposits, University of Tasmania, Hobart, Tasmania, Australia*

[2] *Department of Geology and Geophysics, University of Hawai'i at Mānoa, Honolulu, Hawaii, USA*

[3] *Hawaiian Volcano Observatory, U.S. Geological Survey, Hawai'i National Park, Hawaii, USA*

Hawaiian Volcanoes: From Source to Surface, Geophysical Monograph 208. First Edition.
Edited by Rebecca Carey, Valérie Cayol, Michael Poland, and Dominique Weis.

Companion Website: www.wiley.com/go/Carey/Hawaiian_Volcanoes

Etna [e.g., *Andronico et al.*, 2008; *Polacci et al.*, 2009]. The onset of unrest and eruption in Halema'uma'u Crater (Kīlauea) in 2008 was well captured and documented by the Hawaiian Volcano Observatory (HVO) in terms of geophysical, visual, gas flux, and field data sets [*Wilson et al.*, 2008; *Patrick et al.*, 2011; *Houghton et al.*, 2011; *Elias and Sutton*, 2012]. The first month of activity from the new summit vent produced tephra from three different eruption processes:

1. Mechanical decoupling of gas and melt phases, blockage of the magmatic volatile flux creating overpressure, and subsequent release and expansion, producing a purely lithic ballistic and windblown fall deposit [*Wilson et al.*, 2008; *Houghton et al.*, 2011].

2. Persistent continuous degassing, ash emission, and tephra transport in a convective plume [*Swanson et al.*, 2009; *Patrick et al.*, 2011].

3. Transient explosive events on time scales of seconds to minutes [*Orr et al.*, 2008, 2013; *Patrick et al.*, 2011].

Here we integrate the published field, gas, and geophysical data from a period within the first month of eruptive activity from Kīlauea's new summit vent (the "overlook vent") and combine them with new observations of ash componentry and lapilli microtextures. Ash and lapilli produced throughout the early stages of the 2008 summit eruption provide a direct way to understand the dynamics of the onset and continued activity of this new vent during a period when the lava lake was not visible. In addition, this understanding can be coupled with other geophysical observations to further elucidate degassing dynamics and storage conditions of the shallow conduit system beneath Kīlauea's summit.

We address the following questions: How did factors such as vent-wall stability and rising lava level influence the degassing and eruption dynamics of the magma? What were the time scales of transport and residence for melt and gas phases during different styles of basaltic eruptive activity in early 2008?

19.1.1. Kīlauea's Summit Explosive Activity

Halema'uma'u Crater has been the site of ongoing tephra emission since March 2008 (Figure 19.1), punctuated by larger explosive events. This is the first eruption at Kīlauea's summit since September 1982 and the first explosive eruption within the caldera since May 1924. The close proximity of the activity to the HVO facilitated ease of sampling, acquisition of a multitude of geophysical, geochemical, gas geochemistry, and physical volcanology data, and almost continuous recorded visual observations. Unlike the May 1924 summit activity, in which water interacted with hot wallrock to produce 17 days of violent eruptions [*Jaggar and Finch*, 1924; *Stearns*, 1925], the March–April 2008 activity involved the explosive formation of the summit overlook vent, persistent background magmatic outgassing, and nearly continuous low-level ash emission, punctuated by two short-lived explosive events.

19.1.2. Chronology: June 2007–April 2008

In June 2007, the summit of Kīlauea deflated rapidly as magma drained from the subsurface to feed an east rift zone intrusion and eruption 7–13 km from the summit,

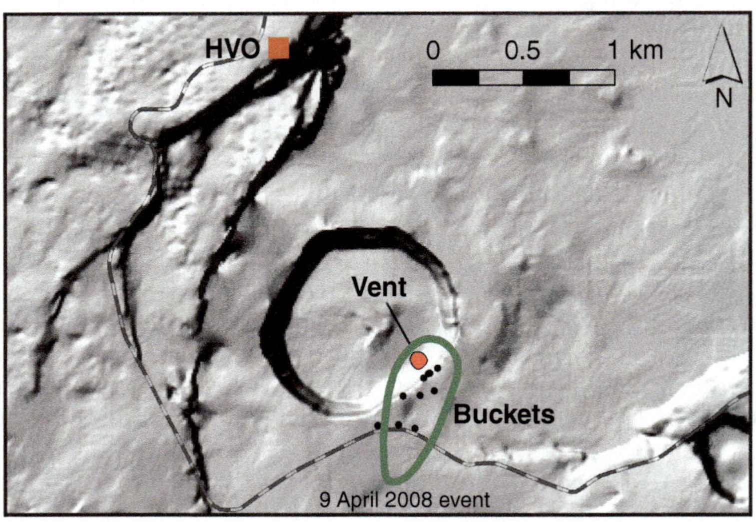

Figure 19.1 Map view of Kīlauea Caldera and Halema'uma'u Crater with location of the active vent shown as an orange area. The bucket network (black dots) is located downwind from the active vent parallel to the typical trade wind directions (winds from the northwest). The outline (green color) shows the area of tephra dispersal from the 9 April 2008 explosion.

Table 19.1 Multidisciplinary observations from the period of vent opening (19 March) to the 9 April explosion. Vent diameter data from *Orr et al.* [2008].

Date	Event	Ash samples	Visual/Audible Observation	Gas Production (SO_2)
19 March	Explosive: juvenile free	No juvenile		1500 tonnes/day
March 20–23	Passive degassing	No juvenile	While steam plume	1200–1300 tonnes/day
23, 27 March (lapilli sample March 27)	Passive degassing	Sample collections: 25 March	Gray ash-laden plume night incandescence at the base of the plume venting sounds	1200 decreasing steady to 950 tonnes/day
28 March–31 March	Passive degassing	Sample collections: 28, 29, 30, 31 March	Pulsatory white and gray ash-laden plumes	850 decreasing to 650 tonnes/day reduced emissions coincident with white plume
1 April–8 April (lapilli sample 1–8 April)	Passive degassing	Sample collections: 1–8 April	Pulsatory white and gray ash-laden plumes	500–800 tonnes/day
9 April (lapilli sample 9 April)	Explosive eruption juvenile rich	Sample collection 10 April		No measurement

the "Father's Day eruption" [*Poland et al.*, 2009]. This briefly resulted in a rapid increase of the SO_2 emission rate from the summit (~4 times background rates), followed by an exponential-style decay, to a somewhat elevated baseline within about one month. At least as early as December 2007, Kīlauea's summit region began a period of escalating SO_2 emission and seismicity, superimposed on an established long-lived trend of slow deflation. In early 2008, SO_2 emission rates were elevated (>400 tonnes/day) with respect to normal rates and continued to escalate. In mid-March, incandescent areas appeared in a talus cone at the base of the southeast wall of Halema'uma'u Crater. SO_2 emission rates were measured at 1630 tonnes/day on 18 March [*Elias and Sutton*, 2012]. At 02:58 Hawaiian standard time (HST) on 19 March, the incandescent talus fell into the previously roofed and hidden vent, and a weak explosive eruption took place, depositing purely lithic material over more than 77 ha [*Wilson et al.*, 2008; *Houghton et al.*, 2011]. The eruption opened a new vent with a diameter of approximately 35 m (830 m^2) and left a crater that was a minimum of 30 m deep, but likely much deeper [*Patrick et al.*, 2011]. SO_2 emission rates over the following week remained high (700–2000 tonnes/day).

Juvenile ejecta were first observed (via time-lapse camera and confirmed by HVO staff) on 22 March and consisted predominantly of ash to fine lapilli. On 23 March, recovered samples included juvenile morphologies comprised Pele's tears and hair, fluidal glassy achneliths, and coarse ash [*Swanson et al.*, 2009; *Wooten et al.*, 2009; *Swavely*, 2010]. Eruption of such tephra continued through the end of March and beginning of April, while SO_2 emission rates were 600–1600 tonnes/day. Minor vent-wall failures and crater widening occurred sporadically until 9 April, when a large wall failure increased the vent area by 200 m^2, triggering an explosive event that punctuated the ongoing eruption (Table 19.1). The volume of the collapse, estimated from time-lapse camera images and field photographs, was ~1800 m^3. At the time of the 9 April event, the SO_2 emission rate was around 700 tonnes/day. On 16 April, another relatively large collapse occurred, with an estimated volume of ~2000 m^3 occurring. A small explosive eruption was likewise triggered, but no tephra larger than 4 mm was deposited above the rim of Halema'uma'u (we do not report an analysis of that event here).

19.1.3. Analysis of Juvenile and Lithic Ejecta

Almost daily collections of ash and lapilli were made to obtain a high-resolution time series of textural and compositional variations and of tephra accumulation rate. The tephra was collected in 10 containers in down- and cross-wind locations for typical trade wind directions [Figure 19.1; *Swanson et al.*, 2009]. Accumulated tephra was collected soon after dawn and reflects the previous 24 h of deposition. Thus a collection on 29 March, for example, will mostly contain tephra erupted on 28 March. In this chapter we use the collection date to identify each sample unless otherwise specified. In addition, more than 100 lapilli in the 8–16 mm size fraction were collected from the periods of 26–27 March and 1–8 April. One complication of sample collection is that strong winds >5–5.5 m/s typically cause eolian reworking and increased amounts of ash in the collecting vessels. The periods of 28–31 March, 1–3 April, and 7 April were preceded by a 24 h

Figure 19.2 Images of juvenile and lithic ash components accumulated in the downwind collection network at Kīlauea. Juvenile components J1 to J9 and lithic components L1 to L6 were defined by *Swavely* [2010] for multiple samples collected throughout 2008. Components J7 and J8 are not present in the March–9April tephra.

average wind speed in excess of 5 m/s. Shallow collecting pans were replaced by deeper buckets after 31 March, and the quality of the collections improved accordingly but was still affected by strong winds. A third lapilli sample was collected on 10 April (9 April eruption) for microtextural analysis, representing a snapshot in time of the shallow conduit magma during fragmentation.

We have conducted ash abundance and componentry measurements on tephra collections where possible as these methods have proven useful as indicators of degassing dynamics within a shallow conduit, including any perturbations of the conduit system, e.g., Etna [*Taddeucci et al.*, 2002; *Lautze et al.*, 2012] and Stromboli [*D'Oriano et al.*, 2011a, 2011b]. Furthermore, we have conducted a quantitative microtextural study of the lapilli size fraction of the erupted products. Clast microtextures of volcanic products are the only preserved remnant of vesiculation history and hence the degassing behavior of magmas. Microtextural analysis of basaltic ejecta, in particular the number, size distribution, and shape of vesicles and crystals, are commonly used to understand the degassing and outgassing behavior of the magma in addition to its rheology [e.g., *Cashman and Mangan*, 1994; *Blower et al.*, 2001, 2002; *Lautze and Houghton*, 2005; *Polacci et al.*, 2006; *Sable et al.*, 2006, 2009; *Gurioli et al.*, 2008; *Degruyter et al.*, 2010]. Such microtextural studies have been conducted on products of different eruptive intensity, but importantly studies of explosive basaltic products from Kīlauea include the high-fountaining episodes of Kīlauea Iki during 1959 [*Stovall et al.*,

2011a, 2011b], Mauna Ulu during 1969 [*Parcheta et al.*, 2013], and Pu'u 'Ō'ō during 1983–1986 [*Mangan et al.*, 1993]. These studies show that, regardless of eruption duration or style, the dynamics of such eruptions is critically driven by shallow processes, for example, the late-stage nonequilibrium degassing of volatiles, rheological evolution of shallow resident magma, and degree and mechanism of volatile coupling to the magma.

19.2. DATA

19.2.1. Ash Componentry and Abundance

The largest sample from the bucket collections (by mass) was dried and sieved at either half phi or phi scale. With a binocular microscope at least 300 grains (where possible) were examined and separated into componentry classes. The juvenile components of Halema'uma'u tephra were split into nine categories by *Swavely* [2010] based on morphological, textural, and compositional criteria, but only seven are present in the March–April tephra collections: fine Pele's hair (J1), coarse Pele's hair (J2), spheres (J3), fluidal clasts (J4), angular dense clasts (J5), ragged scoria (J6), and vesicular fluidal clasts (J9) (Figure 19.2; Table 19.2). The wall-rock lithic components were split into six different classes: anhydrite rich (L1), phenocryst poor (L2), red oxidized (L3), microcrystalline (L4), altered wall (L5), and phenocryst-rich (L6) wall rock.

The componentry was determined for 14 ash samples collected over the period of 23 March (onset of juvenile ash production) to 10 April; the latter collection samples the first major juvenile-rich explosive event of the eruption on 9 April. We later compare this componentry to the microtextural observations of the three lapilli collected in that period.

One consideration in the interpretation of ash componentry data at Halema'uma'u is the role of gas flux in efficiently transporting juvenile, lithic, and recycled components out of the vent. With less gas delivered to the magma free surface, there is less opportunity for both lithic and juvenile components to be transported out of the vent, and with time there will be significant depletion of transportable size fractions. We tested this hypothesis with the March–April data and found that a positive correlation between ash flux and SO_2 flux with time does exist throughout late March and early April (Figure 19.3), with the caveats that the late March data are questionable owing to eolian contamination and SO_2 data throughout this period can show relatively large errors. Therefore, data interpretation require the consideration that lithic ash abundance and total grain size distribution are not a function of explosive efficiency but rather are inherited from vent-wall collapse either during the 19 March opening event or subsequent collapses and semicontinuous raveling [*e.g., Carey and Houghton,* 2010]. Lithic abundance and total grain size distribution in the collections from 24 March to 10 April are therefore a function of what is available in terms of size and mass and are a direct indication of vent-wall stability rather than explosive efficiency.

Ash samples collected in March show an increasing trend of relative juvenile abundance and an increase in the dominance of Pele's tears. On 31 March, however, the dominant erupted component was Pele's hair (Figure 19.4). Prior to 29 March, the dominant lithic components were microcrystalline and altered lithic clasts. Thereafter, the diversity of lithic components increased and became dominated by phenocrystpoor lithic clasts. These trends were superimposed on a gradual increase in the juvenile-lithic

Table 19.2 Tephra components present within the bucket network over the period 23 March–9 April (J, juvenile, L, lithic).

Class	Component	Description
J1	Fine Pele's hair	Fine strands of Pele's hair with diameters < 1 mm
J2	Coarse Pele's hair	Coarse strands of Pele's hair with diameters <2 mm
J3	Spherules	Spherical achnelith hollow balls <3 mm often with altered but smooth outer surfaces
J4	Pele's tears	Achnelith fluidal clasts often droplet shaped with smooth outer surfaces
J5	Dense clasts	Components with poor vesicularity (<30 vol %) and angular shapes
J6	Scoria	Ragged scoraceous clasts with coarse vesicularities (<60 vol %)
J9	Vesicular fluidal clasts	Fluidal shaped clasts with coarse vesicularity (<70 vol %); often with a broken angular side
LI	Anhydrite rich	Anhydrite-rich basaltic components
L2	Phenocryst poor	phenocryst poor low vesicularity (<20 vol %) components
L3	Red oxidized	Oxidized low to moderate (<50 vol %) components
L4	Macrocrystalline	Microcrystalline low; vesicularity (<20 vol %) components
L5	Altered	Altered components of variable vescfiWarity (<80 vol %)
L6	Phenocryst rich	Phenocryst-rich low vesicularity (<20 vol %) components
R	Recycled	Components with overall alteration and often adhered ash particles

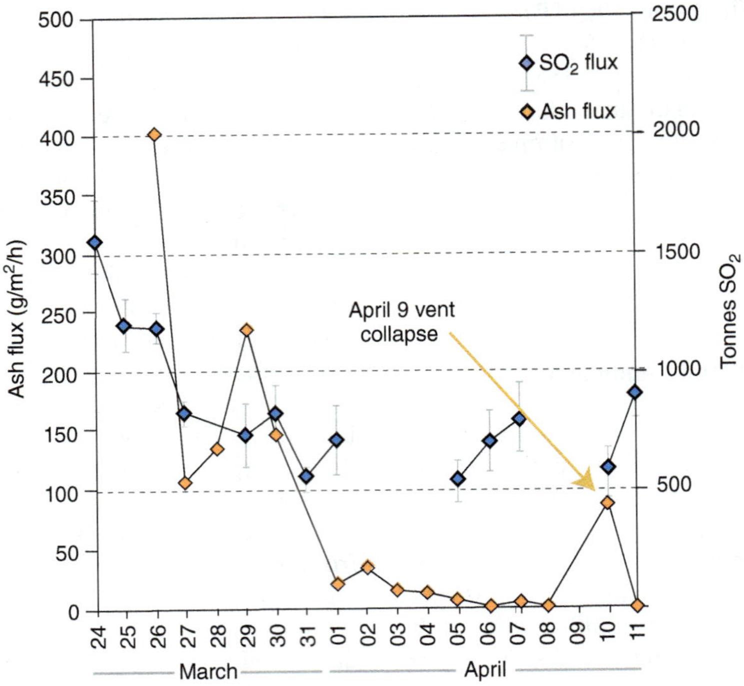

Figure 19.3 SO_2 flux (blue diamonds) in tonnes of SO_2 per day and nominal ash accumulation rate (orange diamonds) calculated using the method of *Swanson et al.* [2009]. This is an average rate for the time since the preceding collection; the instantaneous rate for the 9 April explosive event was much higher, estimated as 9×10^3 kg/s by *Houghton et al.* [2013]. Note that these data points are gas collection dates, and the tephra accumulation data have been altered to reflect eruption date rather than collection date.

ratio and a general dominance of the Pele's tear component (Figure 19.4). A twofold increase of the ash flux in the 29 March collection rapidly lessened to a low flux on 1 April (Figure 19.3), which we attribute, at least partly, to changing wind speed during this period.

Throughout 1–8 April there was a continuous trend of decreasing ash flux for both juvenile and lithic ash. The 1 April and 2 April collections contained a diverse population of both juvenile and lithic components. The first appearance of vesicular fluidal clasts was within the 1 April collection, and Pele's tears continued to dominate the juvenile components until the 10 April collection (Figure 19.4). Collections from 3 April to 6 April, except on 4 April, showed an increase in grain size; the juvenile-lithic ratio continued to remain very high but showed a slow decline and a corresponding increase in the variety of lithic clast components. Tephra in the period from 5 April until the 9 April explosion showed a continuation of the decreasing juvenile-lithic ratio, and the juvenile component continued to be dominated by Pele's tears. The lithic components in this period were dominated by microcrystalline, anhydrite-rich, and altered types, which are also major components collected on 10 April after the explosion. The 10 April collection contained mostly lithic (67 wt%), recycled (23 wt%), and juvenile clasts (10 wt%). The juvenile components also showed a heightened variability in type as described in the following section.

19.2.2. Juvenile Lapilli

19.2.2.1. Juvenile Sample Morphology

Juvenile lapilli from 22 March and 26–27 March have homogenous morphologies: fluidal and dense with glassy outer surfaces similar to achneliths or Pele's tears (Figures 19.5a,b,c). Rare vesicles observed within clasts are millimeter to centimeter size and spherical. The juvenile lapilli erupted during 1–8 April are slightly more diverse than those of 27 March, with millimeter-scale vesicles (Figures 19.5d,e,f). Macroscopically, however, these clasts are glassy and dense to moderately dense achneliths and Pele's tears.

Lapilli from the 9 April explosion are very different microtexturally and morphologically from those erupted previously. There are three end members: (1) high vesicularity, microvesicular pumice with complexly shaped vesicles up to 4 mm across (Figure 19.5g); (2) ragged dense spiny clasts with a vesicle-poor (<10 vol%) outer margin up to 1 cm in thickness and a more coarsely vesicular interior, which is glassy and contains predominantly millimeter size spherical vesicles (Figure 19.5h); and (3) fluidal fluted dense clasts

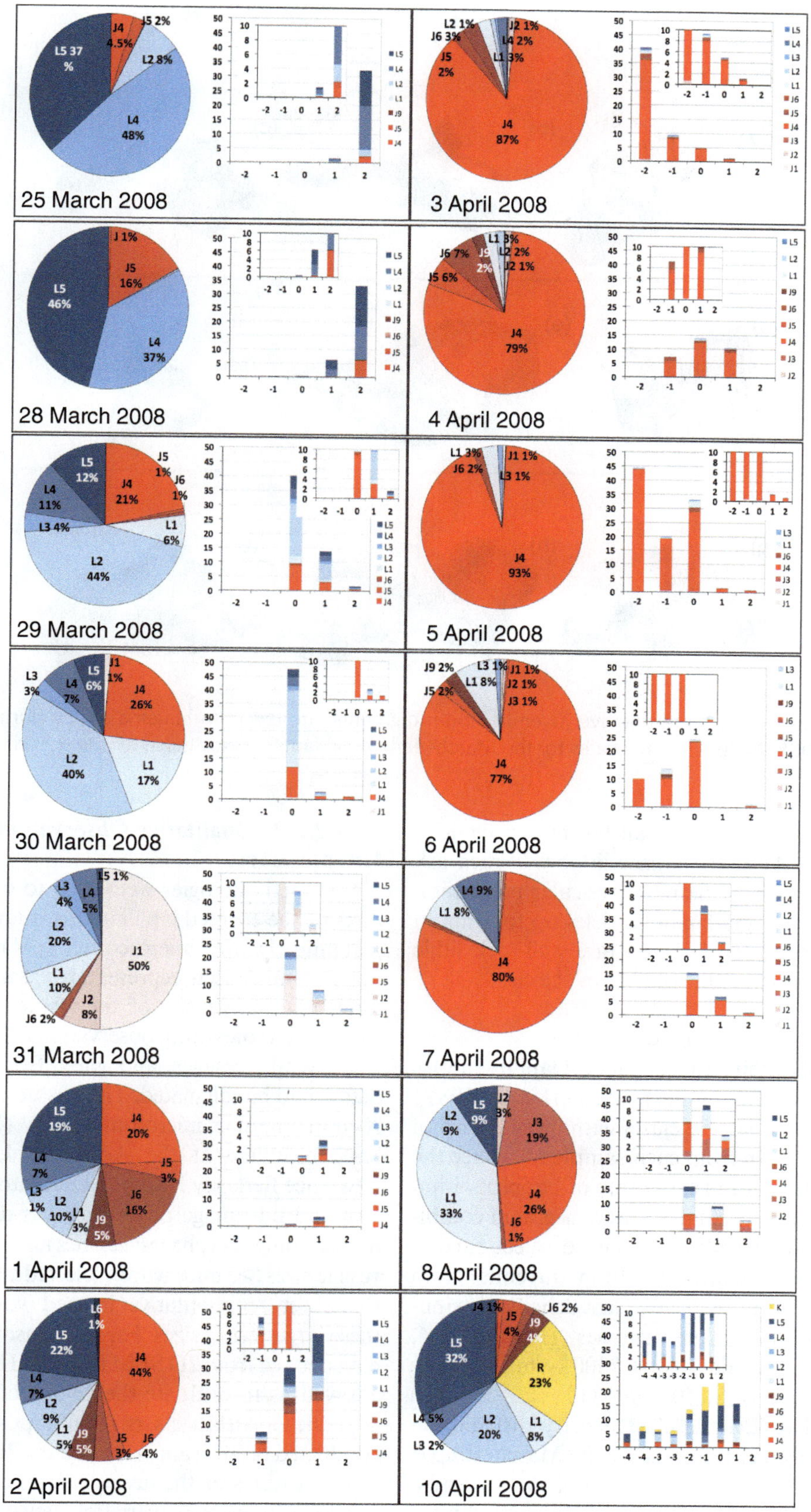

Figure 19.4 Grain size (wt% per phi unit in bar graphs) and componentry (wt% in pie charts) for ash components of tephra bucket collections throughout March and early April 2008. Red colors show juvenile components, blue colors lithic components. Components <1% of the collection are not labeled. Lithic and juvenile types are listed with relative abundances in the size fractions between −2 and 2 φ, except for the 10 April collection (−4 to 2 φ). In grain size distributions, larger-scale plots are shown at similar grain size scales together with insets of identical distributions but at higher *Y*-axis resolution; *X* axis in φ units. Dates indicate day of collection, generally between 0630 and 0730.

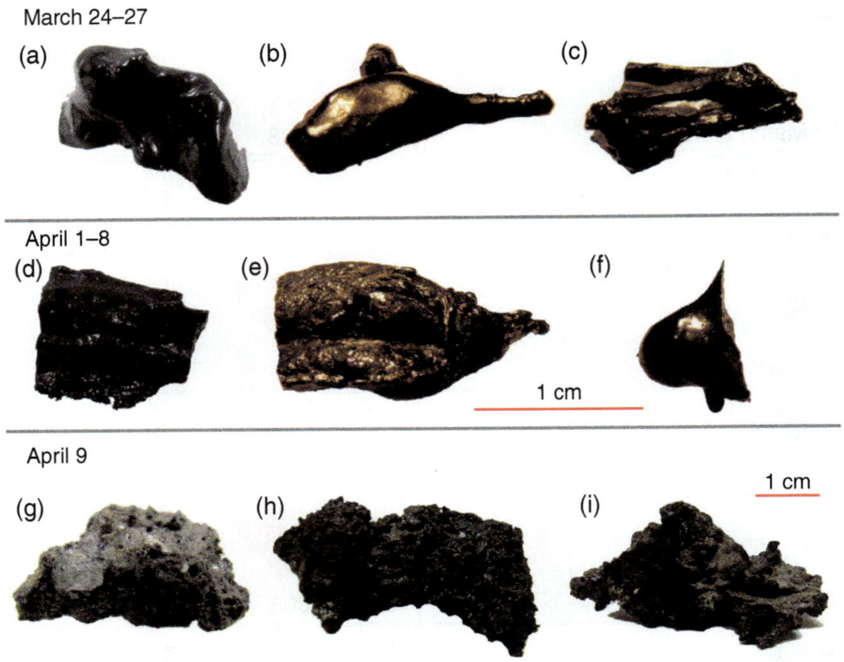

Figure 19.5 Photos of representative juvenile lapilli fragments erupted from Halema'uma'u during March and early April 2008. Note different scales for the March–early April lapilli versus those for the 9 April event.

whose dense outer margins have small 1–4 mm size vesicles (Figure 19.5i). All 9 April clasts have lithic wall-rock ash adhering to them, but only the more vesicular population has lithic wall-rock ash and lapilli particles up to 3 mm in diameter within them (<3 vol%) and rare wall-rock lithic clasts up to 1 cm adhering to the outside of the clast.

19.2.2.2. Juvenile Clast Density

The juvenile clast density analyses for all lapilli samples were conducted following *Houghton and Wilson* [1989]. When possible, the 100 largest clasts within the 8–16 mm size fraction were chosen from each sample to reduce the chance of postfragmentation expansion of bubbles within the clasts. Juvenile density analyses were useful in combination with macroscopic observations to select further clasts for microtextural characterization. Juvenile lapilli erupted during 26–27 March have a clast density distribution skewed toward higher densities (Figure 19.6, Table 19.3). The modal density is 2300 kg/m^3 (25 vol% vesicles) with a range of 2650 kg/m^3 (10 vol%) to 1710 kg/m^3 (42 vol%). The juvenile clast density distribution from 1–8 April is broader than that of the March sample, with a range from 2370 to 540 kg/m^3 (20–82 vol%) and a modal density of 1800 kg/m^3 (39 vol%). The 9 April juvenile clast density distribution shows obvious bimodality with two peaks, one at 850 kg/m^3 (71 vol%) and the other at 1750 kg/m^3 (41 vol%); the majority of the population is skewed toward lower densities (Figure 19.6, Table 19.3). The total range is 440–1940 kg/m^3 (34–85 vol%).

19.2.2.3. Qualitative Observations of Clast Microtextures

Two methodologies were used to observe the microtextures of the erupted lapilli in this study. The first employed scanning electron microscopy (SEM) images of thin sections from chosen representative samples that span the range of clast density of the whole sample. This method yielded microtextural observations of all phases present, i.e., crystals, vesicles, and glass, from which quantitative data could be calculated. The second methodology was 3D X-ray computed microtomography (XRCMT), using the 8.3.2 beamline at the Advanced Light Source at the Lawrence Berkeley National Laboratory. The image collection and postimage processing was designed for observations of only two phases: vesicles and glass. Due to the large vesicle sizes (>5 mm) within selected clasts, we were not able to collect representative imaged volumes that included these large vesicle sizes. For that reason we use quantitative data derived from 2D SEM analysis. However, the XRCMT allowed us to evaluate the degree of interconnectedness of vesicles within clasts. Where possible, XRCMT was conducted on the same clasts as the 2D SEM analysis.

The textures of the juvenile clasts of the March–9 April eruptive phase exhibit variation between and within samples with respect to all components present: crystals, vesicles, and wall-rock inclusions. Clast microtextures are shown as SEM images (Figure 19.7), where vesicles are black, groundmass glass is white, phenocrysts and microlites are light gray, and wall-rock lithic clasts are dark gray. The clasts do not

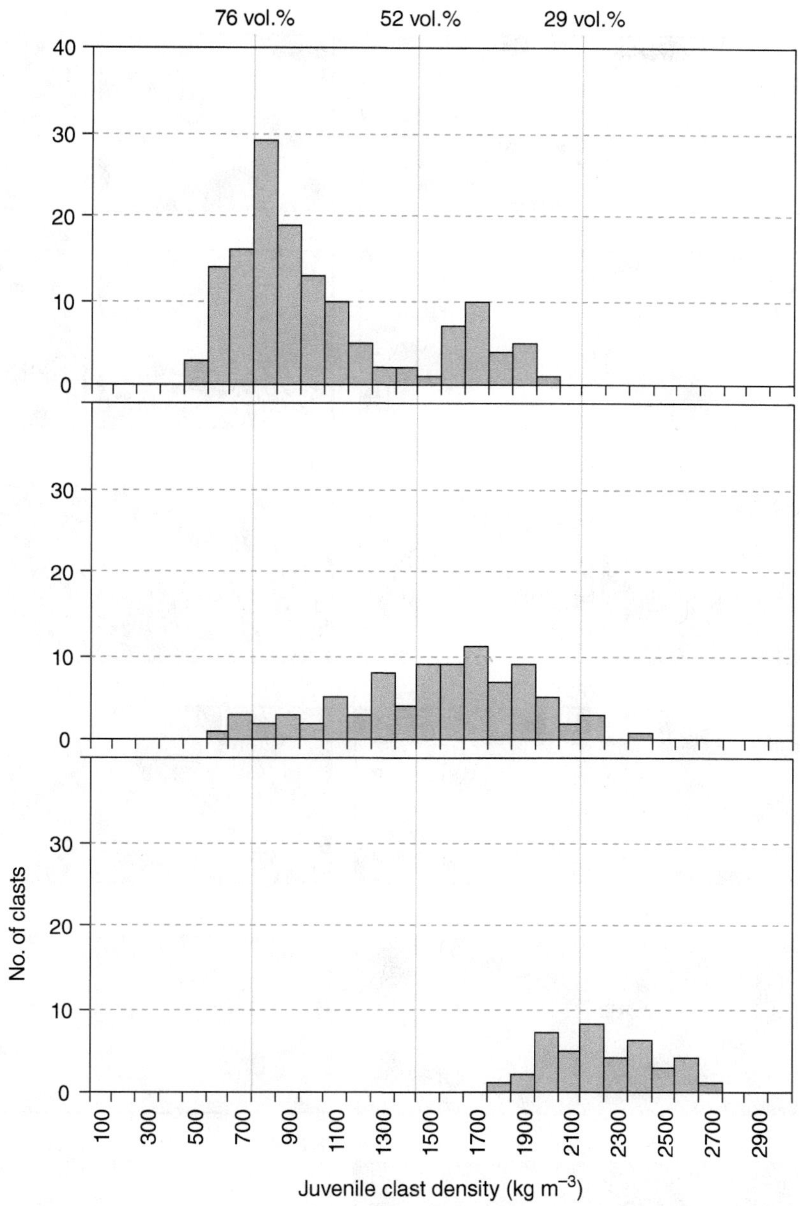

Figure 19.6 Juvenile density of the lapilli size fraction erupted from Halema'uma'u over the study period. Density histograms March 24-27 (bottom), April 1-8 (middle), April 9 (top) show that the juvenile lapilli become increasingly diverse and more vesicular with time. The density bimodality of the lapilli from 9 April shows a dominance of the microvesicular end member, yet continued eruption of a dense end member of two contrasting morphologies: ragged and platy. Arrows denote the density of bins from which clasts were chosen for microtextural analysis.

Table 19.3 Summary of juvenile lapilli density data from the three lapilli samples. Modal, range, and averages of the 3 minimum and 3 maximum density clasts are shown.

Sample	Modal density	Range density	3 min.	3 max.
26–27 March	1: 2300 kg/m^3 (25 vol%)	1710–2650 kg/m^3 (10–42 vol%)	1810 kg/m^3 (39 vol%)	2600 kg/m^3 (10 vol%)
1–8 April	1: 1800 kg/m^3 (39 vol%)	540–2370 kg/m^3 (20–82) vol%	610 kg/m^3 (78 vol%)	2230 kg/m^3 (17 vol%)
9 April	2: 800 (73 vol%) and 1800 kg/m^3 (39 vol%)	440–1940 kg/m^3 (34–85 vol%)	460 kg/m^3 (83 vol%)	1880 kg/m^3 (30 vol%)

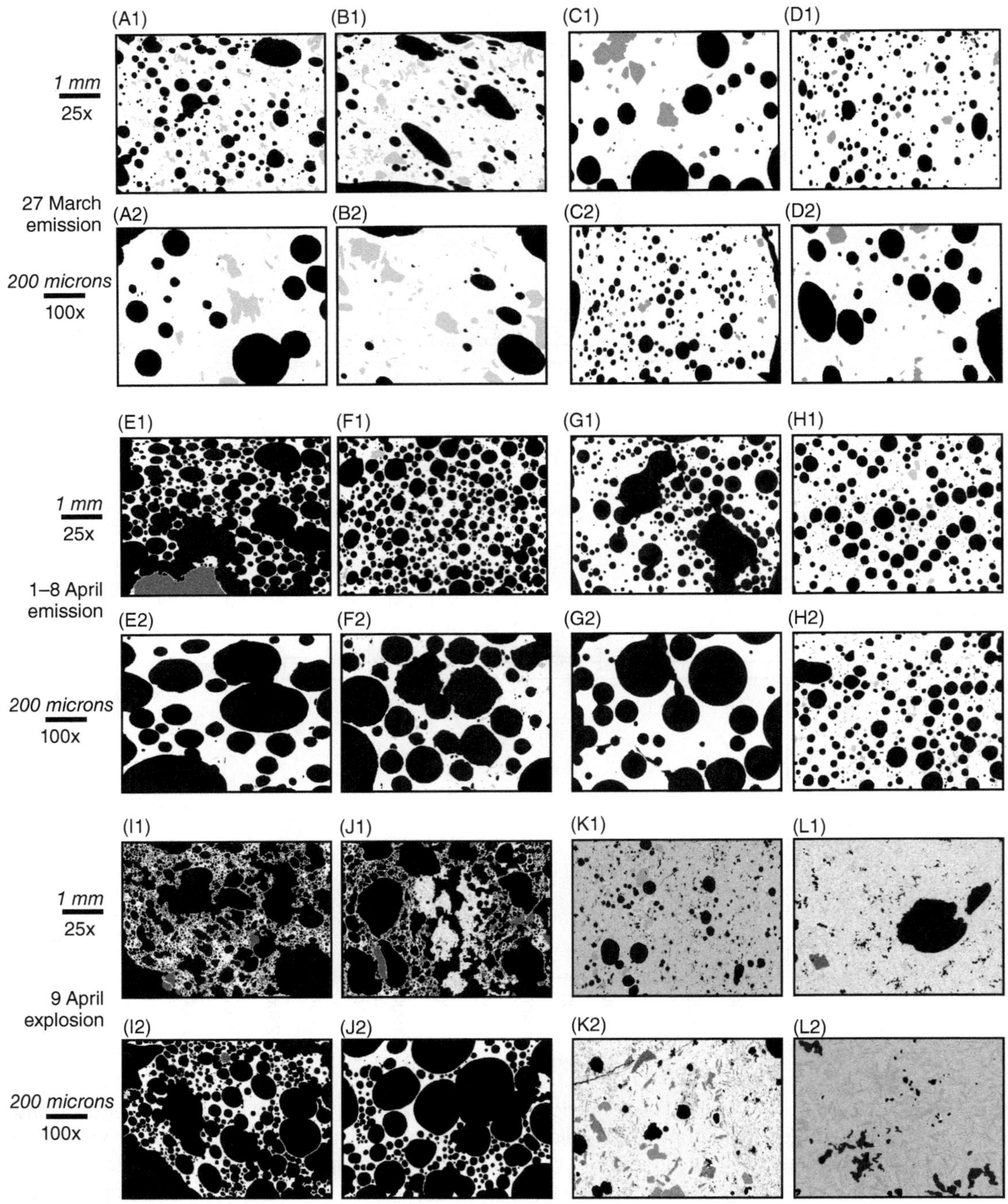

Figure 19.7 Scanning electron microscope images of representative clasts at two different magnifications. Upper images for each sample show a horizontal field of view of ~6 mm, lower images 1.3 mm. Images K1 and K2 represent the platy end member, and L1 and L2 show the ragged end member erupted during the 9 April explosion. Only two clasts were analyzed for the 27 March sample, and two 25x and 100x images are shown for each clast (A&B and C&D, respectively). Image J1 shows a crystalline wall-rock lithic clast within the highly vesicular end-member juvenile lapilli.

represent the heterogeneity of the magma within the shallow conduit either temporally or spatially, as the lapilli were erupted over extended periods rather than during one explosive event. In contrast, the material ejected during the 9 April explosive event represents the magma within the shallow conduit that was fragmented during a single explosion.

26–27 March: The lapilli textures are homogeneous between modal and end-member clasts and are overwhelmingly "mature"; there are relatively few vesicles, and those that are present are spheroidal within a narrow size distribution from 30 μm to 2 mm (Figures 19.7a–d and 19.8a,b). The vesicles are rather widely dispersed, and few are connected. These clasts have populations of both microlites and phenocrysts (Figure 19.7). The phenocrysts make up 5%–17% of the sample by area and include olivine, plagioclase, clinopyroxene, and iron-titanium oxides. They can be rounded, euhedral, or angular and display complex zonation patterns. Within the March products, plagioclase, clinopyroxene, and oxide microlites are present, but the dominant phase is plagioclase, often found as glomerophyric aggregates. Within these clasts, the most abundant plagioclase microlites and microphenocrysts range in length from 30 to 200 μm and are randomly distributed within the glassy groundmass (~<5% by area). Microlites are often associated with basaltic tephras, abundant in sub-plinian and Plinian basaltic tephras [e.g., *Sable et al.*, 2006, 2009; *Costantini et al.*, 2010] and less abundant in Hawaiian-Strombolian tephras [*Polacci et al.*, 2006; *Lautze and Houghton*, 2005]. The clasts representing the higher densities (2500–2900 kg/m^3) often contain mostly crystalline xenoliths coated by basaltic magma.

1–8 April: Lapilli collected during the 1–8 April period are very similar in morphology to those collected during March, except that the higher vesicularity samples more commonly display coalescence textures among the larger (100–200 μm diameter) vesicles (Figures 19.7e–h and 19.8c,d). In SEM images, the modal density clast (Figures 19.7g1, g2) has vesicles in the millimeter to 30 μm size range, similar to the lower density clast (Figures 19.7e1, e2). The higher density clast (Figures 19.7h1, h2) has a smaller size range of vesicles, 30–500 μm. The phenocryst phases are identical to those within the March sample, with a similarly diverse range of morphologies, but are much less abundant (<7% by area). Plagioclase microlites are common but less abundant than in the March sample and have a much smaller range of sizes (typically <50 μm).

9 April: The 9 April sample is unique in that it is the only collection of lapilli erupted during a single event. Based upon the bimodality of the density histogram, two pyroclasts were selected for microtextural analysis from the lower density population and two from the higher density population (Figure 19.6).

Within the lower density lapilli, we see the first evidence of mingling between two microtexturally distinct magmas (Figures 19.7i,j). The major contrast with the earlier samples is the appearance of a much less mature microtexture identified by a high number of relatively small vesicles (1–100 μm) (Figures 19.7i,j), a widening of the range of the vesicle sizes (to 6 mm), complexity of vesicle shapes, and paucity of microlites and phenocrysts in the groundmass (<5 area %). The two higher density lapilli also contrast with the earlier pyroclasts (Figures 19.7 and 19.8). The highest density clast shown (Figure 19.7l) belongs to an end member that has a ragged morphology, with very few vesicles, which are bimodal in size and either spheroidal or amoeba shaped. The juvenile end member, with textures

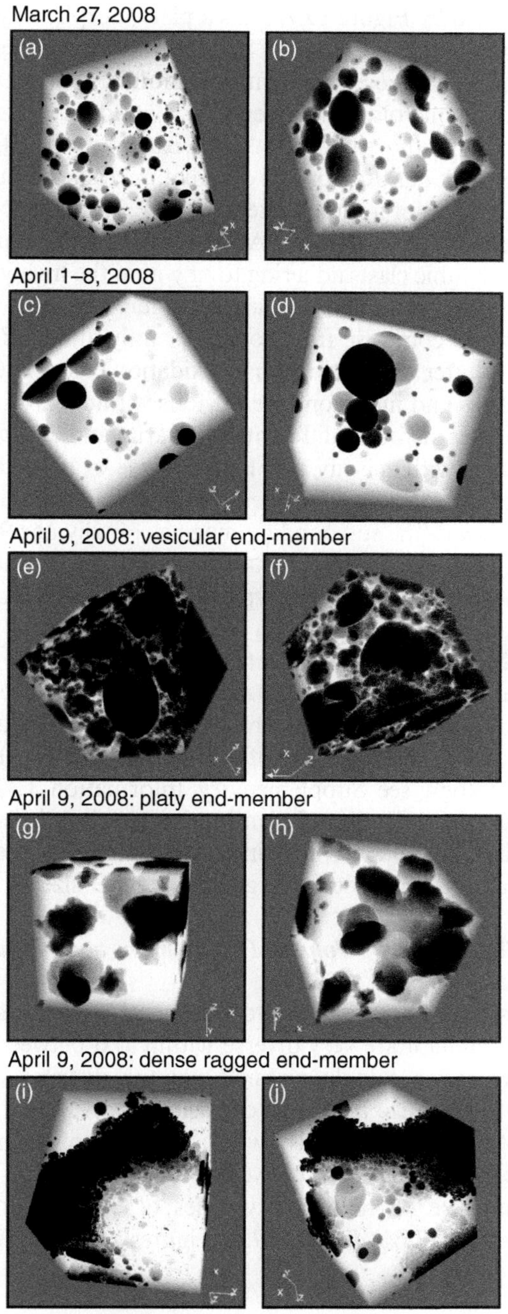

Figure 19.8 X-ray microtomographic renderings of segmented images from juvenile lapilli from each of the three sample suites 27 March, 1–8 April, and 9 April. The three end-member textures within the 9 April sample are also shown. In all images, the length of the shortest side is 0.9 mm. *X-Y-Z* orientation of imaged volume is shown by colored axes in corners of images. At energies of 20 keV, radiographs with 200 ms exposure time were taken with 1400 rotation steps between angles of 0° and 180°. Images were cropped to capture inside of volume. Image segmentation was conducted using 3dma-rock [*Lindquist and Venkatarangan*, 1999] as applied to volcanic particles by *Degruyter et al.* [2010]. Image visualization was conducted with open-source software ParaView (http://www.paraview.org).

represented by Figure 19.7k, has a platy morphology, with a bimodal size distribution of vesicles that are rounder than those in the ragged end member. Another contrast is the diversity in microlites and phenocryst abundances, size, and habit within the same clast. For example, the clast in Figure 19.7k has domains of different microlite abundance within an area of approximately 1.5 mm². Another contrast between the samples from 9 April is the high abundance of wall-rock lithic clasts adhering to or within the highly vesicular end-member juvenile lapilli (Figures 19.7j,i). These lithic clasts comprise fresh to significantly altered lithologies, the latter with a red-orange oxidation color. The wall-rock clasts and inclusions are obvious within SEM images due to their distinctly different vesicularity and crystallinity compared with the juvenile lapilli. Close examination of the lithic inclusions indicated no rounding, melting, or reaction with the surrounding magma. Finally, the 9 April clasts have significantly reduced crystallinities, and both phenocrysts and microlites typically total <2 area %.

19.2.2.4. Quantitative Data of Vesicularity

Quantitative vesicle data calculated from measurements taken from SEM images are presented for all clasts in Table 19.4 in order to quantify their textural diversity. For methodology, see Supplementary Information 1. Three-dimensional images sufficiently characterize the shapes and sizes of the small and intermediate size vesicles, but they cannot represent the largest (millimeter size) vesicles.

19.2.2.5. Vesicle Volume Distributions and Number Densities

As shown above, the microtextures of pyroclasts from Halema'uma'u are very diverse within and between samples. Here we calculate vesicle volume distributions in order to visualize the range of vesicle sizes (Figure 19.9). The March vesicle data show a 1–1.5 order-of-magnitude variability in the size of vesicles, which are of intermediate size (50–800 µm); vesicles of 150–500 µm size dominate (>75%) (Table 19.4; Figures 19.9a,b). Vesicle number densities (N_v^m) are within the 10^2–10^3 cm^{-3} range (Table 19.4). The early April clasts have a broader range in vesicle size than those from March and include vesicles in the 1–8 mm size range and have lower abundances of small vesicles (<100 µm) (Figures 19.9c,d,e,f). The vesicle number densities (N_v) from these clasts are similar to those before 1 April and contain vesicles dominantly (>75%) within the size range 500 µm to 8 mm (Table 19.4).

The 9 April eruption produced three populations of clasts. We discuss each in terms of quantitative vesicularity data. The vesicle sizes have extreme variability, on the same scale as those commonly seen in silicic magmas (10 µm–6 mm). The more vesicular subpopulation identified here (Figures 19.9i,j) has distinct polymodal distributions with multiple modes. Vesicles of millimeter size form one of the modes, and at least two other modes are present at different sizes (Table 19.4). The denser representative clasts (Figures 19.9k,l) have a similar range of vesicle sizes but are volumetrically dominated by the larger millimeter size vesicles, with an order-of-magnitude fewer small vesicles (Table 19.3). While these two populations have similar vesicle size ranges, the clasts with higher vesicularity have more abundant smaller (<100 µm) and coarser (millimeter size) vesicles and a higher vesicle number density (Figure 19.7, Table 19.4).

19.3. INTERPRETATION

Our approach to understanding the onset and early evolution of the eruption considers the degassing dynamics in the shallow conduit using ash morphology and microtextures of the erupted juvenile lapilli in March and April 2008. The high-resolution time series of ash morphology and flux can be linked to the eruption style and dynamics in the vent and at the magma free surface. Two multiple-day lapilli samples are considered to provide an understanding of how the characteristics of the magma and its degassing and outgassing history changed with time. We then contrast the erupted products from periods

Table 19.4 Vesiculation and microtextural quantitative data for juvenile lapilli clasts from the three sample suites (N_v^m = melt-corrected vesicle number density).

Eruption	Clast No.	Density kg/m³	Vesicularity	N_v Total (mm³)	N_v^m Total (mm³)	Size range (µm)	Xenolith (area %)
26–27 March	H08-64	2,300	23 vol%	197	256	80–1,600	0
	H08-60	2,400	19 vol%	1,174	1525	8–1,000	0
1–8 April	H08-02	1,090	63 vol%	224	295	20–2500	0
	H08-05	1,380	53 vol%	541	703	25–1000	0
	H08-06	2,000	32 vol%	305	396	8–80	0
	H08-08	1,740	41 vol%	59	77	25–3,200	0
9 April	H08-21	530	82 vol%	19,115	24,825	4–3,978	0.9
	H08-19	830	72 vol%	17,953	23,316	6–2,500	1.2–2.1
	H08-20	1,670	43 vol%	1,442	1,873	25–3,978	0
	H08-23	1,940	34 vol%	2,650	3,442	8–3,978	0

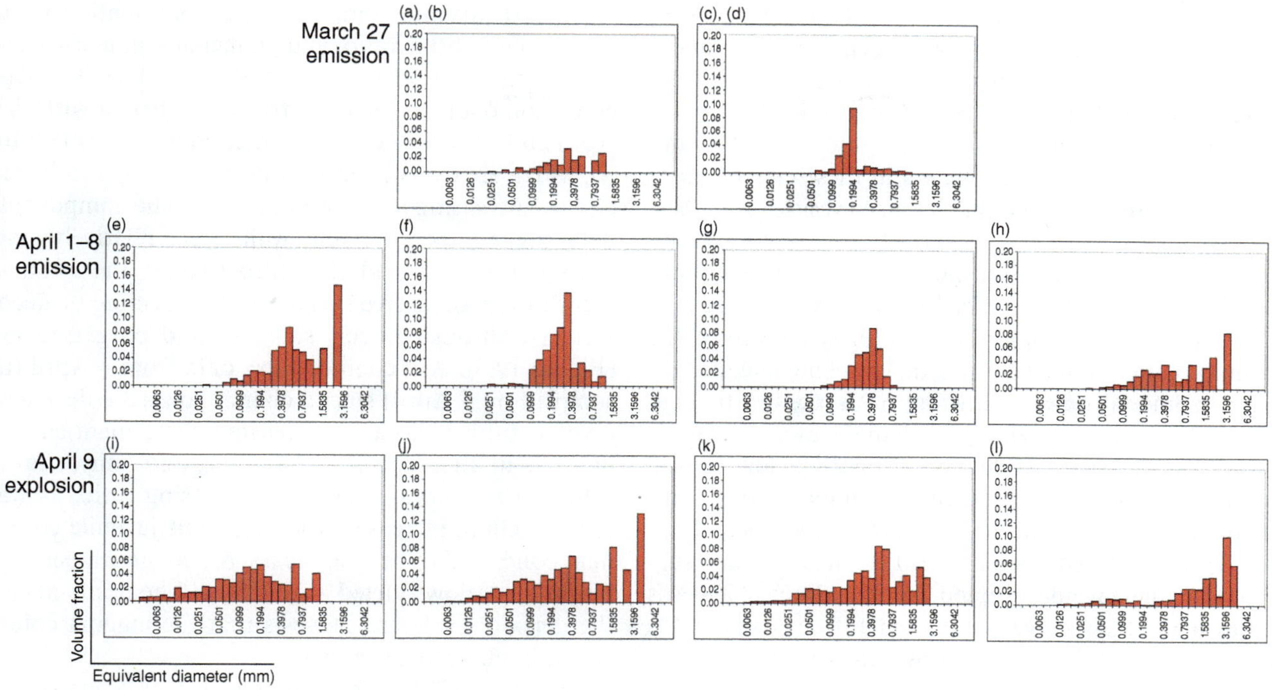

Figure 19.9 Plots of vesicle size distribution for representative juvenile lapilli from the three sample suites. The volume distributions for the March and 1–8 April samples are relatively simple with approximately unimodal distributions, whereas the 9 April samples show a broadening of vesicle size, greater tendency for polymodality and higher abundance of small vesicles (with exception of *k*).

of stable degassing with those from a larger discrete explosive event. We couple the data presented here with data and observations collected by other researchers to arrive at interpretations involving the early evolution of the eruption.

19.3.1. Time Series

19.3.1.1. First Appearance of Juvenile Tephra (24–31 March)

The period of ash emission from 25 – 30 March erupted dominantly crystalline, partly altered lithic clasts (Figure 19.4) that were almost certainly derived from the vent walls and associated talus rather than from the magma body itself. The walls were still unstable following the opening of the vent on 19 March, and some of the debris falling from the vent walls into the conduit was ejected by the exiting magmatic gas. The first juvenile ash sample collected is dominated by vesicle-poor Pele's tears, suggesting that the shallowest magma was significantly outgassed. The juvenile-lithic ratio increased with time despite the fallen wall debris, perhaps because the magma column was rising, and its free surface was nearer to the collection sites (the surface area of the unstable vent walls would also be reduced by the rising magma column). The plume was often brown during this time, and collections showed the presence of very small amounts of fine lithic ash. White or light gray plumes contained much less ash and were less common than the brown plumes during the first few days of the eruption. Nighttime observations showed an increasing abundance of incandescent particles in the plume (Table 19.1). Visual observations from late 2008 to the present indicate that brown plumes are related to rockfalls and vent-wall collapses of all sizes, consistent with the unstable nature of the wall throughout the early phase of the eruption.

On 27 March, juvenile bombs less than 10 cm in diameter, together with smaller juvenile pyroclasts, were landing on the rim of Halema'uma'u Crater, 85 m above the vent, yet ash flux had decreased by a factor of 3 (with some caveats as described above) (Figure 19.3). Most of the bombs had a lithic core coated with a juvenile rind. The presence of the small bombs suggests that the lava free surface was less than ~250–300 m below Halema'uma'u Crater's rim (the distance of possible observation), although no lava was observed despite numerous attempts. The presence of juvenile bombs at the crater rim and the trend of increasing relative juvenile abundance despite a period of crater-wall instability suggest that the magma free surface was rising without many wall collapses.

The first juvenile components erupted during the 23–31 March period were predominantly dense, glassy, fluidal ash particles and fine to medium lapilli with relatively few vesicles (~10s within an area of 1200 mm^3) that range over a single order of magnitude in size (Figures 19.7, 19.8, and 19.9). Rarely are vesicle diameters in these clasts greater than ~1 mm, and all of the characterized vesicles are spherical and isolated (Figures 19.7 and 19.8). These vesicles are in a groundmass with moderate abundances of microlites and phenocrysts (<20 area %). They suggest extended residence times, perhaps in a viscous layer at the top of the magma column or in convecting downwelling magma.

The unimodal density distribution is skewed to high (2200 kg/m^3) values, much higher than the average magma density (950 k/gm^3 +/– 300 kg/m^3) proposed for the top ~120 m of magma in the shallow conduit during 2011 based on gravity and lava-level data [*Carbone et al.*, 2013]. The density data combined with the microtextures suggests that the magma erupted in late March 2008 was mature and largely outgassed, potentially degassing in the months before 19 March when SO_2 emissions were high. The quantitative vesiculation data of the March samples support this interpretation. The very low number densities and limited sizes of vesicles are consistent with escape of the larger bubbles and suggest that the ejecta developed either from a thick shell around a much larger bubble that burst at the magma free surface or from melt adjacent to the expanding bubble. This requires that small bubbles are present at or just below the magma free surface, a likely situation since bubbles <10 cm in diameter are unable to decouple from magma with the convective flow velocities of 0.1–0.3 m/s observed at Halema'uma'u [*Carey et al.*, 2013]. During this phase, the free surface was likely at depths considerably less than 250–300 m, perhaps only a few tens of meters, from the base of Halema'uma'u Crater's floor; otherwise the bursting of bubbles would have been unlikely to deposit lapilli size material into the collection area on the rim of Halema'uma'u, even if the bursting was caused by rockfalls hitting the lava surface. We suggest that tephra was ejected into a hot jet of exsolved gas, which streamlined the particle shapes

19.3.1.2. Continued Production of Dominantly Juvenile Tephra (31 March–8 April)

This period is associated with decreasing ash and SO_2 flux until the 9 April explosion. On 31 March and 1 April, brown ashy plumes were observed, and the componentry of ash collected on 31 March and 1 and 2 April shows a diversity of lithic type similar to that of pre-31 March samples, but with an increase in the juvenile-lithic ratio (Figure 19.4, Table 19.1). The ratio probably increased because of rising lava level, and the lithic content of the brown plumes was small. The 2 April collection shows a similar degree of variability in the lithic and juvenile components, a continuation of the high juvenile-lithic ratio, and an increase in grain size of both lithic and juvenile components. The 3–5 April collection data continue the trend that began with the 1 April collection; it contains larger components (−2 to 0 φ), which are overwhelmingly dominated by Pele's tears, and an increasing juvenile-lithic ratio. The componentry data suggest that the vent walls were relatively stable during this time, and they also suggest the establishment of a magma free surface with streaming of decoupled gas through largely stagnant and outgassed melt (Figure 19.4). Ash componentry data from 4 April (i.e., sampled on 5 April) start a change toward a decreasing juvenile-lithic ratio and decreasing component size. An increase in altered, microcrystalline, and anhydrite-rich lithic clasts accompanied the decreasing ratio, as wall-rock ravelling increased; the dominant juvenile component continued to be Pele's tears. A two month period of bright vent glow started at this time. The dominance of the dense Pele's tears suggests that the magma column was still relatively outgassed.

Orr et al. (2013) show that wall-rock collapses are broadly correlated with high lava level, as indicated by the decreased juvenile-lithic ratio of the ash componentry data preceding the 9 April explosion. A heightened lava level would also permit greater likelihood of direct impact of large-volume collapses of wall rock overhanging the lava surface, shown throughout the eruption to drive heightened degassing of the lava lake and predicted to cause more violent explosive fragmentation during transient explosions [*Carey et al.*, 2012].

The glassy and fluidal clast morphologies and vesicle size distributions coupled with low vesicle number densities in lapilli between 24 March and 8 April confirm that the magma was largely outgassed and erupted during rise and bursting of bubbles from greater depth.

19.3.1.3. Evacuation of Dense Magma and Appearance of Vesicular Magma

The wall collapse on 9 April was relatively large (~200 m^2; ~1800 m^3), and ash components are mostly lithic clasts (67 wt%) and recycled juvenile clasts (23 wt%). The small amount of juvenile ejecta (10 wt%) showed greater textural variability (Figures 19.4 and 19.5) than was present in previous ash collections. This eruption occurred as a single explosion with a mass discharge rate of 8.9×10^3 kg/s and a relatively low volume of 100 m^3 [*Houghton et al.*, 2013]. The explosive eruption produced the most diverse population of juvenile lapilli of any event in 2008 in terms of clast density, morphology, and microtextures (Figures 19.7 and 19.8) and had a bimodal density distribution (Figures 19.6, 19.7 and 19.8).

The fragmentation mechanism for this eruption was significantly different from that acting before 9 April.

The rockfall apparently served as an external trigger for relatively powerful decompression, expansion, and fragmentation of magma residing in the shallow conduit [cf. *Carey et al.*, 2012]. The three textural magma types are all inferred to be primary juvenile pyroclasts, present and mingling in the shallow conduit. The two "mature" dense end members suggest extended residence times in the conduit under conditions permitting extended nucleation and microlite growth as well as considerable outgassing. This melt is similar to that erupted in March and early April based on crystal and vesicle populations and abundance. Situations favorable for cooling and extended residence are located at the conduit wall or magma free surface or within downwelling magma. The third type of magma erupted is microvesicular and, in some clasts, is intimately mingled with the two dense end-member magmas. Its vesicularity is dominated by vesicles with <600 μm diameters and complex shapes, suggesting that the magma was moderately to highly vesicular and that bubbles were actively coalescing prior to initiation of the explosion. High numbers of vesicles <50 μm in diameter with spherical shapes sandwiched between large and complexly shaped vesicles also suggest renewed nucleation just prior to quenching. The quantitative vesicle data show a 100-fold increase in the vesicle number density between the dense and vesicular end-member magmas, reflecting this renewed nucleation (Table 19.4).

19.3.2. Degassing Dynamics: Residence Times of Gas and Melt

The populations of pyroclasts produced during the first 22 days of the eruption vary within and between samples, with every phase present (crystals, vesicles, glass, wallrock), trends that are reflected in the clast densities.

19.3.2.1. March to Early April

We suggest that the first erupted magma had largely degassed in the days to weeks preceding its appearance at the surface. Petrological and geochemical evidence suggests that this magma is not related to the last of the 1982 magma in summit reservoirs (Chapter 20, this volume). The melt during this opening phase of the 2008 summit eruption was sufficiently degassed so that rising bubbles were only coupled to the melt for a short time during expansion and fragmentation of bubbles at the free surface. The vesicle size data, high crystallinity, and inferred raised viscosity favor a model in which the preserved vesicles (0.01–0.8 mm) are more coupled to the host magma than the other larger bubbles that nucleate at depth and arrive at the magma free surface as buoyantly rising, actively expanding bubbles. The small vesicles within the clasts therefore contributed little to the fragmentation process itself. Instead, they are a product of the bubble envelopes that surround larger expanding bubbles derived from two sources: (i) magma at depths of perhaps hundreds of meters, where SO_2 and H_2O gas is being exsolved, or (ii) CO_2 bubbles, which nucleate at or below the shallow storage system.

19.3.2.2. 9 April

The clasts erupted during the 9 April explosive event represent three different textural classes of magma. The vesicular texture replicates those examined by *Carey et al.* [2012] produced by the 12 October 2008 explosion, which was also triggered by vent wall collapse [*Orr et al.*, 2013]. Importantly, the 12 October sample lacks the dense end-member textures observed in 9 April clasts. We suggest that the shallow conduit magma erupted on 9 April was a complex mixture, reflecting mingling of melt remaining from the stagnant outgassing period in late March to early April with more juvenile vesiculating magma. The three different magma types could be related to temporal or spatial variations of the magma column. The denser end-member textures could be actively downwelling magma [*Carey et al.*, 2013] or cooler denser magma at the free surface that was underlain by more volatile-rich magma (Figures 19.7 and 19.8).

19.4. CONCLUSIONS

Ash componentry and microtextures of the juvenile products during the first 22 days of a new summit eruption at Kīlauea capture the rapid evolution of the physical properties of the magma and its degassing dynamics. Such information is not available from other techniques, for example, visual observations or geophysical measurements. The fluidal morphologies of the earliest erupted clasts suggest that the shallow magma was neither gas rich nor actively vesiculating nor rapidly ascending.

We favor a model in which, during this time period, centimeter to meter size bubbles were rising, expanding, and bursting at the free surface. Tephra could thus be related to the passive disruption of melt adjacent to the bubble bursts or in the shells around these large expanding bubbles. These passively rising bubbles were probably not nucleated from the shallow outgassed magma but rather by exsolution from deeper magma. The homogeneity of the erupted material in terms of vesicularity and crystallinity suggests that the shallow magma was uniform, at least at the free surface. We suggest that a largely outgassed magma was present at the top of the magma free surface; perhaps it had been degassing in the weeks to months before the onset of the 9 April explosive event.

Lithic ash abundances were higher than that of juvenile ash between 25 March and 30 March, and relative lithic ash abundances increased between 5 April and 8 April. We suggest that ravelling and small collapses of the vent wall were common during this time. Large collapses were rare, however, and probably seldom directly impacted and penetrated the magma free surface prior to 9 April; such events therefore did not generate pressure amplitudes sufficient to trigger decompression, expansion, and fragmentation of the shallow magma. Instead, we suggest that the magma underwent magmatic fragmentation associated with passive degassing.

The 9 April rockfall and subsequent explosive eruption produced a new component in the eruption deposits, i.e., microvesicular juvenile lapilli, the first evidence during the eruption for an actively vesiculating magma within the vent. As also suggested by petrological, geochemical, and gas data, this event was, in effect, a "vent- and shallow-conduit-clearing" process that erupted a large volume of magma with three different degassing histories [*Elias and Sutton*, 2012] (see also Chapter 20, this volume). The low-density end member was derived from a shallow actively vesiculating magma that underwent renewed nucleation, expansion of existing bubbles, and fragmentation. The two high-density end members have mature textures suggesting intermediate to long residence times, due to either a stagnant viscous crust at the top of the magma free surface or convecting, downwelling magma.

This study shows that smaller resident bubbles will be coupled to the host magma on the time scale of decompression both for the impact of a rockfall that triggers expansion of the gas phase (9 April) and in the melt film of a bubble that expands at the fragmentation surface (24 March to 8 April). Degassing dynamics of basaltic melts are complex, especially in scenarios where external conditions can play a significant role. Understandably, attempts to classify eruptions based on coupled or decoupled degassing continue to be a challenge.

ACKNOWLEDGMENTS

This research was supported by National Science Foundation grants EAR-104662, EAR-0810332, and EAR-1145187 and U.S. Geological Survey grant SV-ARRA-04. We would like to thank Michael Poland, Kimberly Genereau, and an anonymous reviewer for insightful comments. The XRT analyses were conducted at the Advanced Light Source (ALS) at the Lawrence Berkeley National Laboratory in Berkeley, California. We wish to sincerely thank the 8.3.2 ALS Beamline Scientists Alastair McDowell and Dula Parkinson for their time, helpful advice, and technical assistance.

REFERENCES

Andronico, D., S. Scollo., S. Caruso, and A. Cristaldi (2008), The 2002–03 Etna explosive activity: Tephra dispersal and features of the deposits, *J. Geophys. Res.*, *113*, B04209, doi:10.1029/2007JB005126.

Blower, J. D., J. P. Keating, H. M. Mader, and J. C. Phillips (2001), Inferring volcanic degassing processes from vesicle size distributions, *Geophys. Res. Lett.*, *28*, 347–350.

Blower, J. D., J. P. Keating, H. M. Mader, and J. C. Phillips (2002), The evolution of bubble size distributions in volcanic eruptions, *J. Volcanol. Geotherm. Res.*, *120*, 1–23.

Carbone, D., M. Poland, M. Patrick, and T. Orr (2013), Continuous gravity measurements reveal a low-density lava lake at Kilauea Volcano, Hawai'i, *Earth Planet. Sci. Lett.*, *376*, 178–185.

Carey, R. J., and B. F. Houghton (2010), "Inheritance": An influence on the particle size of pyroclastic deposits, *Geology*, *38*, 347–350.

Carey, R. J., M. Manga, W. Degruyter, D. A. Swanson, B. Houghton, T. Orr, and M.R Patrick (2012), Externally triggered renewed bubble nucleation in basaltic magma: The 12 October 2008 eruption at Halema'uma'u Overlook vent, Kīlauea, Hawai'i, USA, *J. Geophys. Res.*, *117*, doi:10.1029/2012JB009496.

Carey, R. J., M. Manga, W. Degruyter, H. Gonnermann, D. Swanson, B. F. Houghton, T. Orr, and M. R. Patrick (2013), Convection in a volcanic conduit recorded by bubbles, *Geology*, *41*, 395–398.

Cashman, K. V., and M. T. Mangan (1994), Physical aspects of magmatic degassing. 2. Constraints on vesiculation processes from textural studies of eruptive products, volatiles in magmas, *Rev. Mineral. Soc. Am.*, 447–478.

Costantini, L., B. F. Houghton, and C. Bonadonna (2010), Constraints on eruption dynamics of basaltic explosive activity derived from chemical and microtextural study: The example of Fontana Lapilli Plinian eruption, Nicaragua, *J. Volcanol. Geotherm. Res.*, *189*, 207–224.

Degruyter, W., O. Bachmann, and A. Burgisser (2010), Controls on magma permeability in the volcanic conduit during the climactic phase of the Kos Plateau Tuff eruption (Aegean Arc), *Bull. Volcanol.*, *72*, 63–74.

D'Oriano, C., A. Bertagnini, and M. Pompilio (2011a), Ash erupted during normal activity at Stromboli (Aeolian Islands, Italy) raises questions on how the feeding system works, *Bull. Volcanol.*, *73*, 471–477.

D'Oriano, C., R. Cioni, A. Bertagnini, D. Andronico, and P. D. Cole (2011b), Dynamics of ash-dominated eruptions at Vesuvius: The post-512 AD AS1a event, *Bull. Volcanol.*, *73*, 699–715.

Elias, T., and A. J. Sutton (2012), Sulfur dioxide emission rates from Kīlauea Volcano, Hawaii, 2007–2010, *U.S. Geol. Surv. Open-File Rep. 2012–1107*, 29 p.

Gurioli, L., A. J. L. Harris, B. F. Houghton, M. Polacci, and M. Ripepe (2008), Textural and geophysical characterization of explosive basaltic activity at Villarrica volcano, *J. Geophys. Res. Solid Earth*, *113*, B8, doi:10.1029/2007JB005328.

Houghton, B. F., and C. J. N. Wilson (1989), A vesicularity index for pyroclastic deposits, *Bull. Volcanol.*, *51*, 451–461.

Houghton, B. F., D. A. Swanson, R. J. Carey, J. Rausch, and A. J. Sutton (2011), Pigeonholing pyroclasts: Insights from the

19 March 2008 explosive eruption of Kılauea volcano, *Geology*, *39*, 263–266.

Houghton, B. F., D. A. Swanson, J. Rausch, R. J. Carey, and S. Fagents (2013), Pushing the Volcanic Explosivity Index to its limit and beyond: Constraints from exceptionally weak explosive eruptions at Kīlauea in 2008, *Geology*, *41*, 627–630.

Jaggar, T. A., and R. H. Finch (1924), The explosive eruption of Kīlauea in Hawaii, 1924, *Am. J. Sci.*, *8*, 353–374.

Lautze, N. C., and B. F. Houghton (2005), Physical mingling of magma and complex eruption dynamics in the shallow conduit at Stromboli volcano, Italy, *Geology*, *33*, 425–428.

Lautze, N. C., J. Taddeucci, D. Andronico, C. Cannata, L. Tornetta, P. Scarlato, and M. D. LoCastro (2012), SEM-based methods for the analysis of basaltic ash from weak explosive activity at Etna in 2006 and the 2007 eruptive crisis at Stromboli, *Phys. Chem. Earth*, *45*, 113–127.

Lindquist, W. B., and A. Venkatarangan (1999), Investigating 3D geometry of porous media from high resolution images, *Phys. Chem. Earth*, *24*, 593–599.

Mangan, M. T., K. V. Cashman, and S. Newman (1993), Vesiculation of basaltic magma during eruption, *Geology*, *21*, 157–160.

Orr, T. R., M. R. Patrick, K. M. Wooten, D. A. Swanson, T. Elias, A. J. Sutton, and M. P. Poland (2008), Explosions, tephra, and lava: A chronology of the 2008 summit eruption of Kīlauea volcano, Hawai'i, *AGU Fall Meeting Abstracts*, vol. *1*, p. 2018.

Orr, T. R., W. A. Thelen, M. R. Patrick, D. A. Swanson, and D. C. Wilson (2013), Explosive eruptions triggered by rockfalls at Kīlauea volcano, Hawai'i, *Geology*, *41*, 207–210.

Parcheta, C. E., B. F. Houghton, and D. A. Swanson (2013), Contrasting patterns of vesiculation in low, intermediate, and high Hawaiian fountains: A case study of the 1969 Mauna Ulu eruption, *J. Volcan. Geotherm. Res.*, *255*, 79–89.

Patrick, M. R., D. Wilson, D. Fee, T. Orr, and D. A. Swanson (2011), Shallow degassing events as a trigger for very-long-period seismicity at Kīlauea Volcano, Hawai'i, *Bull. Volcanol.*, *73*, 1179–1186.

Polacci, M., R. A. Corsaro, and D. Andronico (2006), Coupled textural and compositional characterization of basaltic scoria: Insights into the transition from Strombolian to fire fountain activity at Mount Etna, Italy, *Geology*, *34*, 201–204.

Polacci, M., M. R. Burton, A. La Spina, F. Mure, S. Favretto, and F. Zanini (2009), The role of syn-eruptive vesiculation on explosive basaltic activity at Mt. Etna, Italy, *J. Volcanol. Geotherm. Res.*, *179*, 265–269.

Poland, M. P., A. J. Sutton, and T. M. Gerlach (2009), Magma degassing triggered by static decompression at Kīlauea volcano, Hawai'i, *Geophys. Res. Lett.*, *36*(16), L16306, doi:10.1029/2009GL039214.

Rosi, M., A. Bertagnini, A. J. L. Harris, L. Pioli, M. Pistolesi, and M. Ripepe (2006), A case history of paroxysmal explosion at Stromboli: Timing and dynamics of the April 5, 2003 event, *Geophys. Res. Lett.*, *243*, 594–606.

Sable, J. E., B. F. Houghton, P. Del Carlo, and M. Coltelli (2006), Changing conditions of magma ascent and fragmentation during the Etna 122 BC basaltic Plinian eruption: Evidence from clast microtextures, *J. Volcanol. Geotherm. Res.*, *158*, 333–354.

Sable, J. E., B. F. Houghton, C. J. N. Wilson, and R. J. Carey (2009), Eruption mechanisms during the climax of the Tarawera 1886 basaltic Plinian eruption inferred from microtextural characteristics of the deposits, in *Studies in Volcanology: The Legacy of George Walker,* Spec. Publ. IAVCEI 2, pp. 129–154.

Sahagian, D. L., and A. A. Proussevitch (1998), 3D particle size distributions from 2D observations: Stereology for natural applications, *J. Volcanol. Geotherm. Res.*, *84*, 173–196.

Stearns, H. T. (1925), The explosive phase of Kīlauea Volcano, Hawaii, in 1924, *Bull. Volcanol.*, *2*, 193–208.

Stovall, W. K., B. F. Houghton, H. Gonnermann, S. A. Fagents, and D. A. Swanson (2011a), Eruption dynamics of Hawaiian-style fountains: the case study of episode 1 of the Kīlauea Iki 1959 eruption, *Bull. Volcanol.*, *73*, 511–529.

Stovall, W. K., B. F. Houghton, J. Hammer, S. A. Fagents, and D. A. Swanson (2011b), Vesiculation of high fountaining Hawaiian eruptions: Episodes 15 and 16 of 1959 Kīlauea Iki, *Bull. Volcanol.*, *74*, 441–455.

Swanson, D. A., K. Wooten, and T. Orr (2009), Mass flux of tephra sampled frequently during the ongoing Helema'uma'u eruption, *AGU Fall Meet, San Francisco*, Abstract V52B-01.

Swavley, L. (2010), Changing componentry for eruptions of Halema'uma'u Crater, Kīlauea, in 2008 and implications for ascent and eruption dynamics, thesis, University of Hawai'i at Mānoa.

Taddeucci, J., M. Pompilio, and P. Scarlato (2002), Monitoring the explosive activity of the July–August 2001 eruption of Mt. Etna (Italy) by ash characterization, *Geophys. Res. Lett.*, *29*(8), 1230, doi:10.1029/2001GL014372.

Wilson, D., T. Elias, T. Orr, M. P. Patrick, A. J. Sutton, and D. A. Swanson (2008), Small explosion from new vent at Kīlauea's summit, *Eos Trans. AGU*, *89*, 203.

Wooten, K. M., C. R. Thornber, T. R. Orr, J. F. Ellis, and F. A. Trusdell (2009), Catalog of tephra samples from Kilauea's summit eruption, March-December 2008, *U.S. Geol. Surv. Open-File Rep. 2009–1134*, 26 p.

20

Primitive Components, Crustal Assimilation, and Magmatic Degassing During the Early 2008 Kīlauea Summit Eruptive Activity

Michael C. Rowe[1], Carl R. Thornber[2], and Tim R. Orr[3]

ABSTRACT

Simultaneous summit and rift zone eruptions at Kīlauea starting in 2008 reflect a shallow eruptive plumbing system inundated by a bourgeoning supply of new magma from depth. Olivine-hosted melt inclusions, host glass, and bulk lava compositions of magma that erupted at both the summit and East Rift Zone demonstrate chemical continuity at both ends of a well-worn summit-to-rift pipeline. Analysis of glass within dense-cored lapilli that erupted from the summit in March–August 2008 show these are not samplings of compositionally distinct magmas stored in the shallow summit magma reservoir, but instead result from remelting and assimilation of fragments from conduit wall and vent blocks. Summit pyroclasts show the predominant and most primitive component erupted to be a homogenous, relatively trace-element-depleted melt that is compositionally indistinguishable from East Rift Zone lava. Based on a "top-down" model for the geochemical variation in East Rift Zone lava over the past 30 years, we suggest that the apparent absence of a 1982 enriched component in melt inclusions, as well as the proposed summit–rift zone connectivity based on sulfur and mineral chemistry, indicate that the last of the pre-1983 magma has been flushed out of the summit reservoir during the surge of mantle-derived magma from 2003 to 2007.

20.1. INTRODUCTION

Magma that had forged its way into Kīlauea's summit during a surge in magma supply from 2003 to 2007 [*Poland et al.*, 2012] made its debut with a small explosive eruption in Halema'uma'u Crater on 19 March, 2008, beginning an era of tandem Kīlauea eruptions [*Wilson et al.*, 2008]. The first glassy material erupted from the new 35 m diameter vent was observed on 23 March, 2008 (Figure 20.1) [*Wooten et al.*, 2009; *Wilson et al.*, 2008, *Houghton et al.*, 2011; *Mather et al.*, 2012]. Since its onset, the eruption has been characterized by persistent lava lake activity within a progressively enlarging collapse pit, punctuated by small explosions triggered by rock falls from the pit walls [*Orr et al.*, 2013].

Significant fluctuations in trace-element and isotopic compositions are recorded by historical eruptions at Kīlauea's summit (from 1790 to 1982; Figure 20.2) [*Pietruszka and Garcia*, 1999; *Garcia et al.*, 2003]. In particular, explosive summit eruptions (1790 and 1924) also record rapid changes in magma composition. The 1790 tephra has distinctive trace-element and isotope compositions compared to both 1790 lava and later erupted lava. Similarly, the 1924 explosive summit eruption was associated with a significant isotopic shift to relatively lower $^{206}Pb/^{204}Pb$ and εNd and higher $^{87}Sr/^{86}Sr$ [*Pietruszka and Garcia*, 1999]. In addition, longer-term (decades to centuries) trends in trace-element and isotopic variability in summit magmas are bracketed by explosive activity. Trace-element ratios Nb/Y and La/Yb record systematic increases between 1790 and 1924, followed by subsequent decreases

[1]*School of Environment, University of Auckland, Auckland, New Zealand*

[2]*Cascades Volcano Observatory, U.S. Geological Survey, Vancouver, Washington, USA*

[3]*Hawaiian Volcano Observatory, U.S. Geological Survey, Hawai'i National Park, Hawaii, USA*

Hawaiian Volcanoes: From Source to Surface, Geophysical Monograph 208. First Edition.
Edited by Rebecca Carey, Valérie Cayol, Michael Poland, and Dominique Weis.

Companion Website: www.wiley.com/go/Carey/Hawaiian_Volcanoes

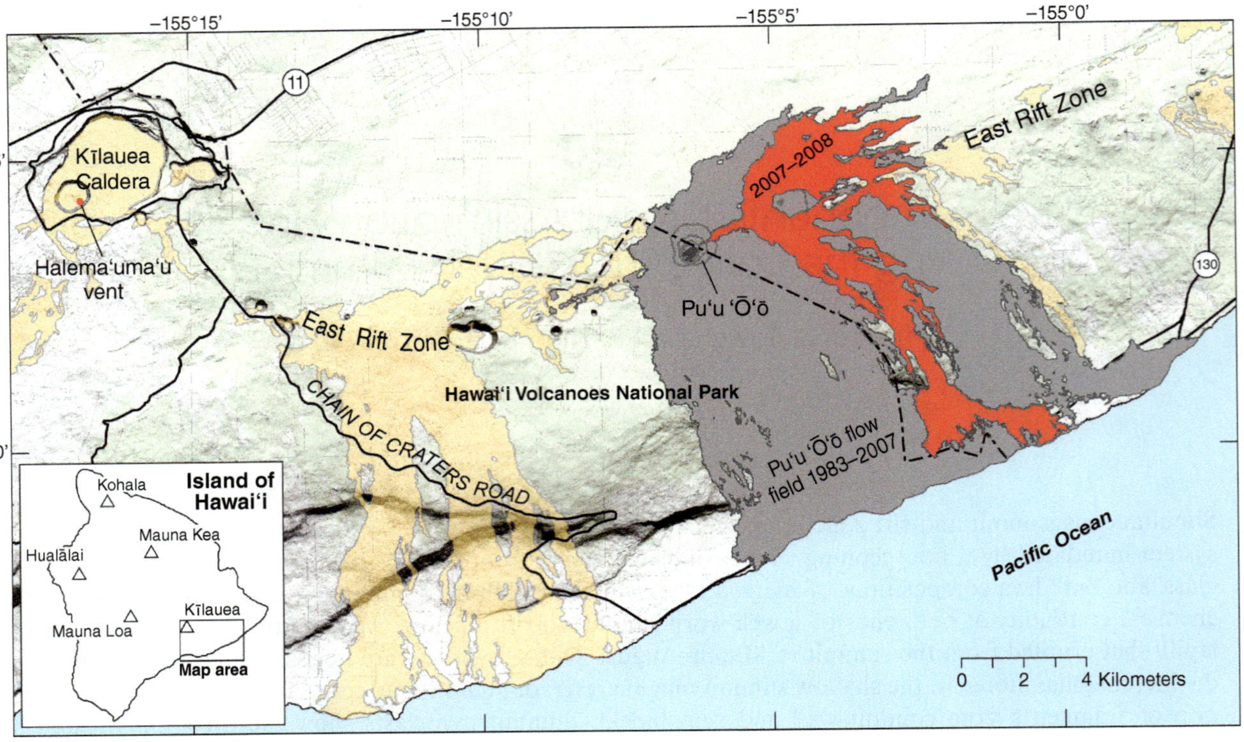

Figure 20.1 Location map of Halema'uma'u Crater (summit vent site) and Pu'u 'Ō'ō crater (East Rift Zone vent site) modified from *Orr et al.* [2012]. The new summit vent site is indicated by the dot. Red shading denotes the extent of ERZ lavas erupted from Pu'u 'Ō'ō in 2007 and 2008.

in highly incompatible to moderately incompatible element ratios until the cessation of summit eruptions in 1982, trending toward depleted Pu'u 'Ō'ō rift zone lava (Figure 20.2) [*Pietruszka and Garcia*, 1999; *Garcia et al.*, 2003]. Overall variations in Nb/Y in historic summit eruptions range from a minimum of 0.44 during the 1790 explosive eruption to a maximum of 0.70 during the 1924 explosive summit eruption [*Pietruszka and Garcia*, 1999].

Concurrent with the renewed activity at the summit, East Rift Zone (ERZ) lava has continued to outpour since 1983. The majority of lava generated on the upper-middle East Rift Zone during the first two decades of eruption was olivine (spinel) phyric containing 3–4 vol% of primarily near-equilibrium olivine phenocrysts (Fo_{81-83}) [*Thornber*, 2001, 2003]. Prior models have indicated olivine formed within two days in magma tapped from a summit reservoir circulating at 1180°C–1200°C (using the MgO glass geothermometer of *Helz and Thornber* [1987]), as it cooled another 30°C while flowing through the East Rift Zone conduit [*Thornber*, 2001].

Similar to prior summit eruptions, Pu'u 'Ō'ō lava displays a long-term trend of decreasing ratios of highly incompatible to moderately incompatible elements [*Thornber*, 2003]. The decrease in highly incompatible element ratios (e.g., La/Yb) coincides with decreases in MgO-normalized CaO and K_2O/TiO_2 contents of lava. La/Yb ratios, for example, vary from a maximum of ~7.5 in 1983 to a minimum of 5 in 2001 (Figure 20.2). Superimposed on these long-term trends, ERZ lava MgO contents display cyclic variation predominantly between 7 and 9 wt% [*Thornber*, 2003].

Systematic changes in ERZ whole-rock geochemistry since 1983 have led to a "top-down" versus "bottom-up" debate on the origin of variations. The bottom-up approach to geochemical variation explains the change in whole-rock ERZ compositions as a result of either variable mixing of melts derived from a heterogeneous mantle source or progressive depletion during decompression of a homogeneous mantle source [*Garcia et al.*, 1996, 2000]. The top-down model proposed by *Thornber* [2003] suggests instead that the progressive decrease in highly to moderately incompatible elements can instead result from efficient mixing of 1982 summit magma with a uniform composition recharge melt, whereby the decreasing ratios result from either an increase in the relative proportion of recharge magma or a greater diversion of recharge magma directly to the ERZ. From whole-rock geochemistry, both models have previously been deemed viable.

During the interval from 2003 to 2007, Kīlauea experienced a doubling of magma supply rates from the mantle, evidenced from increasing CO_2 emissions, seismicity, as well

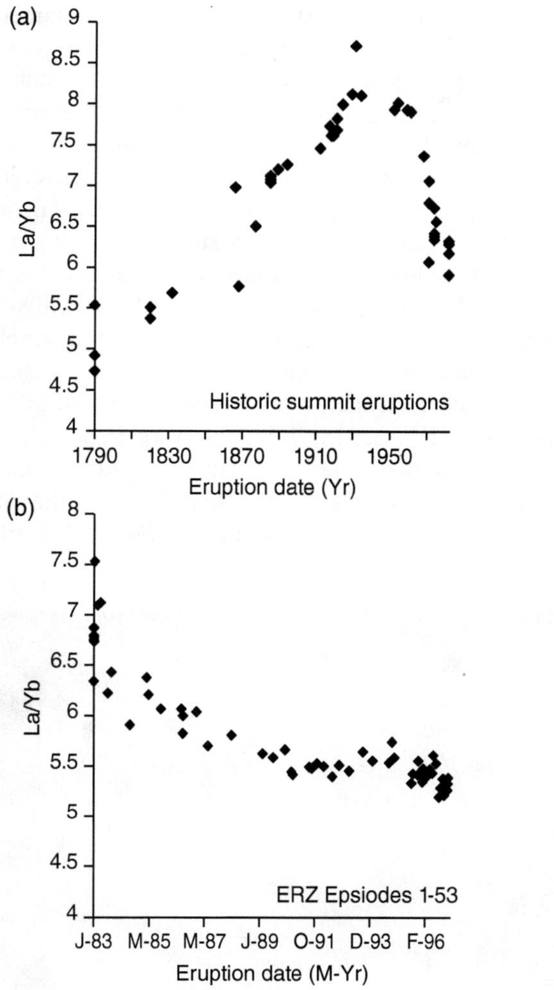

Figure 20.2 Compiled whole-rock geochemistry for (a) historic (1790–1982) summit eruptive products [*Pietruszka and Garcia*, 1999] and (b) ERZ episode 1–53 lavas (1983–1997) from *Thornber* [2003].

as summit inflation [*Poland et al.*, 2012]. During the period of edifice inflation associated with "the surge" [*Poland et al.*, 2012], most ERZ lava was erupted at relatively low temperatures (1140–1145°C) with a mixed multiphyric assemblage (olivine-spinel ± clinopyroxene ± plagioclase). In spring 2008, rift eruption temperatures began to increase, peaking at 1155°C in August 2008, after which time olivine-only lava erupted for nearly a year while eruption temperature gradually cooled to ~1145°C and multiphyric lava returned by July 2009. This interval of hotter eruption products reflects a delayed translation of surge-related magma through the overpressurized rift plumbing system [*Thornber et al.*, 2010].

In this chapter we present a succinct geochemical and petrologic study of concurrent ERZ and Kīlauea summit eruption products in 2008. In order to address key petrologic questions related to magmatic degassing and parental melt compositions, this study focuses largely on the analysis of silicate melt inclusions. Because melt inclusions are essentially isolated from the bulk magma by the trapping mineral phases, they provide a potential means of establishing the most primitive component within an eruptive plumbing system [e.g., *Roedder*, 1979; *Sobolev and Shimizu*, 1993; *McNeill and Danyushevsky*, 1996; *Kent*, 2008]. Analysis of melt inclusions is also one of the few direct means to measure magmatic volatile contents prior to eruption, thus providing an important opportunity to relate dissolved magmatic volatiles to measured gas ratios collected from volcanic plumes to better understand degassing processes [e.g., *Wallace*, 2003; *Wade et al.*, 2006; *Metrich and Wallace*, 2008]. For about three decades scientists have used Kīlauea melt inclusions to try to understand volcanic degassing processes for a range of geologic problems, including eruption forecasting and monitoring, fluxing of volcanic gases to Earth's atmosphere, and understanding the evolution of magma stored beneath Kīlauea's edifice [e.g., *Harris and Anderson*, 1983; *Wallace and Anderson*, 1998; *Wallace*, 2003; *Edmonds et al.*, 2009]. In addition, large numbers of analyzed melt inclusions per sample present an opportunity to study populations of inclusions that record processes lost in a bulk analysis [e.g., *Thornber*, 2001; *Kent et al.*, 2002].

In addition to melt inclusion analyses, we incorporate petrographic and geochemical analyses of partially molten crystal aggregates or "crystal mush clots" found within cores of lapilli erupted from the Kīlauea summit vent. Crystal clots as dense cores in lapilli ranging up to ~5 cm were erupted during March and April 2008 activity. Early trace-element and isotopic analysis of the glassy lapilli resulted in ambiguous results leading to different interpretations of magmatic processes at work beneath Kīlauea [*Pietruszka et al.*, 2008; *Thornber et al.*, 2010]. Analysis and interpretation of crystal clots in conjunction with melt inclusion analyses allow for unambiguous distinction between juvenile magma and lithic material.

Ultimately, by integrating petrography with the geochemical analysis of whole-rock samples, silicate melt inclusions, groundmass glass, and crystal clot glass, this study will identify the juvenile magmatic component in the summit eruptive material while addressing the issue of magmatic continuity between the summit and ERZ and lend further insight into the top-down versus bottom-up debate. Volatile contents, in particular sulfur, will be used to examine the degassing behavior at both the summit and ERZ and to examine the degassing history to better understand magma storage and the continuity between the ERZ and summit. As historic summit eruptions at Kīlauea tend to herald in changes in magmatic compositions, this study will also address magmatic variability at the summit and in comparison to the ERZ.

Understanding attributes of shallow magmatism is clearly a necessity if we are to try to distinguish and interpret subtle geochemical variations related to mantle processes and magma generation [*O'Hara*, 1977].

20.2. SAMPLES AND METHODOLOGY

This study utilizes seven samples of glassy pyroclasts (bombs, lapilli, tears, spatter, and scoria) for melt inclusion analysis. Six of the samples are from the summit of Kīlauea, and one spatter sample is from Kīlauea's ERZ. All samples utilized for melt inclusion analysis were erupted between 27 March 2008 and 1 August 2008. In addition, chemical data and petrographic interpretations are included from a representative "crystal mush clot" (KS08-02S-T1), erupted 23 March 2008 [see *Wooten et al.*, 2009, Figure 9]. Sample eruption dates and descriptions from *Wooten et al.* [2009] are provided in Table 20.1. Representative whole-rock or bulk sample analyses are provided in the online supplemental data with bulk rock data presented in detail in Chapter 8. Bulk rock analyses presented here were analyzed by X-ray fluorescence (XRF) and inductively coupled plasma mass spectrometry (ICP-MS) for major and trace elements, respectively, at the Washington State University GeoAnalytical Lab [*Knaack et al.*, 1994; *Johnson et al.*, 1999].

For melt inclusion work, glassy pyroclasts were lightly crushed and olivine grains were handpicked from the material. Following separation, grains were individually mounted and polished to expose silicate melt inclusions (Figure 20.3). Most olivine grains contain abundant, large glassy melt inclusions (Figure 20.3). Melt inclusion diameters range in size from <10 to 150 μm (online supplement Table 20.1) with generally circular to ovoid forms. Strongly elongated or irregular shaped inclusions may represent enclosed or partially enclosed melt channels. Clinopyroxene is also a significant phase in the ERZ lava. Clinopyroxene-hosted melt inclusions were analyzed for comparison to olivine-hosted inclusions of the ERZ and summit. All but one clinopyroxene-hosted inclusion have diameters less than 10 μm. In order to avoid melt inclusion geochemical variations correlating to subtle changes in bulk magma chemistry with time during the Pu'u 'Ō'ō eruption [e.g., *Garcia et al.*, 2003; *Thornber*, 2003] and to better assess geochemical populations of inclusions and host grains for both the summit and ERZ, we instead focused on a limited number of samples. From six summit samples, a total of 168 melt inclusions and adjacent host grains plus an additional 74 olivine core-rim pairs (127 from KS08-09S and KS08-108E; Table 20.2) were analyzed for major elements. Thirty olivine-hosted melt inclusions and 13 clinopyroxene-hosted melt inclusions were analyzed from the ERZ sample (KE58-2801S; Table 20.2) for major elements. A limited number of representative groundmass major and trace-element analyses were also collected at the same time as melt inclusions. From samples KS08-09S, KS08-109E,

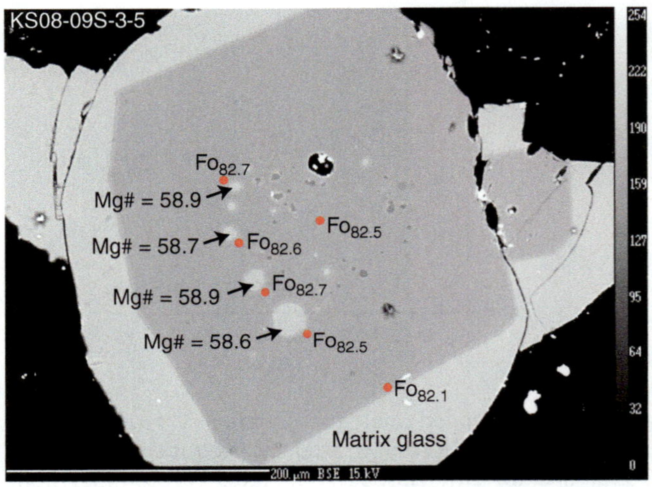

Figure 20.3 Backscatter electron (BSE) image of Kīlauea summit olivine-hosted melt inclusions (KS08-09S-3-5). Host olivine Fo number and melt inclusion Mg number, calculated as molar $MgO/(MgO+FeO_T)$, are shown for four of the larger melt inclusions plus the core and rim of the olivine host.

Table 20.1 Samples utilized for glass and melt inclusion analysis.

Sample ID	Eruption Date	Description
KS08-09S	27 March 2008	Black-vesicular-glass coated orbicular bomb, cored with juvenile-looking black-vesicular-microcrystalline scoria
KS08-39T	02 April 2008	Glassy, lithic-bearing, Pele's tears, Pele's hair, and shards
KS08-45T	04 April 2008	Glassy, lithic-bearing, Pele's tears, Pele's hair, and shards
KS08-54T	06 April 2008	Glassy, lithic-bearing, Pele's tears, Pele's hair, and shards
KS08-56T	07 April 2008	Glassy, lithic-bearing, Pele's tears, Pele's hair, and shards
KS08-108E	01 August 2008	Gray pumice and pumice agglomerate
KE58-2801S	09 September 2008	Spatter deposited on Pu'u 'Ō'ō crater rim
KS08-02S-T1	23 March 2008	Partially molten crystal cluster within glassy bomb

*Sample eruption dates and descriptions from *Wooten et al.* [2009], Chapter 8 (this volume), and HVO unpublished data. KS samples are erupted from the summit, KE from the ERZ.

Table 20.2 Average measured and corrected melt inclusion major element compositions.

	KS08-09S		KE58-2801S		KS08-39T		KS08-54T		KS08-09S		KE58-2801S		KS08-39T		KS08-54T	
		KS08-108E		KE55-2801S		KS08-45T		KS08-56T		KS08-108E		KE55-2801S		KS08-45T		KS08-56T
N=	*N*=(34)	*N*=1σ	*N*=(93)	*N*=1σ	*N*= (30 ol)	*N*=1σ	*N*=(13 px)	*N*=1σ	*N*=(6)	*N*=1σ	*N*=(6)	*N*=1σ	*N*=(12)	*N*=1σ	*N*=(17)	*N*=1σ
Measured Major Elements (wt%)																
SiO_2	51.33	0.43	51.16	0.50	51.37	0.38	51.61	0.52	51.75	0.38	52.13	0.68	51.56	0.70	51.60	0.43
TiO_2	2.47	0.15	2.47	0.21	2.51	0.18	2.58	0.15	2.38	0.01	2.60	0.37	2.41	0.11	2.33	0.03
Al_2O_3	13.92	0.29	13.89	0.34	14.01	0.18	14.22	0.43	13.99	0.09	13.71	0.68	13.87	0.20	13.82	0.11
FeO	11.21	0.21	11.38	0.43	11.34	0.27	12.36	0.50	10.60	0.34	10.65	0.57	10.88	0.33	10.88	0.17
MnO	0.17	0.02	0.16	0.03	0.17	0.02	0.18	0.02	0.17	0.02	0.15	0.02	0.18	0.02	0.16	0.03
MgO	6.32	0.64	6.56	0.76	6.27	0.64	5.79	0.72	6.78	0.08	6.68	0.45	6.70	0.47	6.92	0.13
CaO	11.70	0.39	11.58	0.46	11.39	0.59	10.31	0.43	11.22	0.15	11.33	0.25	11.32	0.24	11.13	0.11
Na_2O	2.13	0.07	2.16	0.12	2.19	0.13	2.26	0.13	2.24	0.07	2.16	0.19	2.18	0.12	2.21	0.05
K_2O	0.38	0.03	0.38	0.04	0.40	0.06	0.39	0.03	0.41	0.03	0.34	0.07	0.38	0.02	0.39	0.02
P_2O_5	0.27	0.10	0.25	0.09	0.24	0.09	0.25	0.09	0.27	0.03	0.28	0.07	0.25	0.02	0.25	0.01
S	1187	152	1028	317	794	376	690	134	1089	218	1187	331	1208	85	946	268
Cl	125	28	125	37	144	40	117	26	113	20	145	46	136	41	112	26
Total	100.04	0.57	100.13	0.52	100.00	0.41	100.06	0.53	100.03	0.60	100.24	0.45	99.95	0.46	99.88	0.65
Fo#m	82.57	0.27	82.55	0.48	81.19	0.53	79.9		82.50	0.15	82.42	0.20	82.45	0.23	82.37	0.23
Fo#eq	78.67	1.87	79.02	2.13	78.35	2.09			80.86	0.61	80.52	1.40	80.20	1.13	80.75	0.42
X Ol	5.95	2.28	5.67	2.79	4.13	2.40			2.52	1.01	2.80	1.83	3.37	1.34	2.53	0.80
Normalized Inclusions Corrected for Post-Entrapment Crystallization (wt%)																
SiO_2	50.57	0.27	50.40	0.51	50.83	0.44			51.38	0.41	51.72	0.75	51.09	0.58	51.23	0.40
TiO_2	2.32	0.14	2.33	0.18	2.40	0.16			2.32	0.02	2.53	0.36	2.32	0.10	2.27	0.03
Al_2O_3	13.10	0.27	13.10	0.31	13.43	0.23			13.63	0.09	13.32	0.71	13.40	0.27	13.46	0.15
FE_2O_3	1.17	0.04	1.19	0.05	1.21	0.03			1.19	0.04	1.20	0.07	1.23	0.04	1.23	0.02
FeO	10.54	0.27	10.64	0.45	10.54	0.37			9.67	0.35	9.74	0.58	9.98	0.31	9.92	0.19
MnO	0.16	0.02	0.15	0.02	0.16	0.02			0.17	0.02	0.15	0.02	0.17	0.02	0.15	0.03
MgO	8.41	0.22	8.54	0.38	7.71	0.32			7.68	0.30	7.68	0.41	7.89	0.27	7.82	0.24
CaO	11.00	0.17	10.91	0.23	10.91	0.44			10.93	0.13	11.01	0.26	10.93	0.13	10.84	0.09
Na_2O	2.00	0.09	2.03	0.12	2.10	0.14			2.18	0.06	2.10	0.19	2.11	0.14	2.15	0.05
K_2O	0.36	0.03	0.35	0.04	0.38	0.06			0.40	0.03	0.33	0.07	0.36	0.02	0.38	0.02
P_2O_5	0.25	0.09	0.23	0.08	0.23	0.08			0.26	0.03	0.27	0.07	0.24	0.01	0.24	0.01
S	1117	143	967	288	759	355			1059	207	1151	315	1166	81	922	262
Cl	118	27	117	34	139	38			110	20	141	44	131	41	109	26

Note: Inclusion correction and equilibrium olivine calculation assumes $Fe^{3+}/Fe_T=0.1$. Fo#m is the measured host Fo# (or Mg# for the cpx-hosted inclusions). Fo#eq is the calculated equilibrium forsterite number. X Ol is wt% olivine added for equilibrium calculations.

Table 20.3 Average trace-element analysis of melt inclusions.

Sample	KS08-09S		KS08-108E		KE58-2801S		BHVO-2G		BHVO-2G	
Analysis Method	SIMS	LA	SIMS	LA	SIMS	LA	SIMS		LA	
N=	*N*=(5)	*N*=(10)	*N*=(5)	*N*=(17)	*N*=(5)	*N*=(11)	*N*=(7)	*N*=1σ	*N*=(8)	*N*=1σ
Ti	14828	14629	14775	14357	15107	14912	16981	216	16448	238
Sr	328.3	334.3	326.6	323.2	330.3	340.5	395.6	10.5	401.4	9.8
Y	27.2	24.6	26.6	24.0	26.4	25.6	25.6	1.2	25.6	1.3
Zr	143.8	128.8	140.2	126.5	144.0	136.8	172.4	4.1	168.7	6.7
Nb	12.1	12.8	12.5	12.1	12.6	12.2	18.3	0.5	18.2	0.6
Ba	98.5	102.0	100.1	100.2	102.2	106.9	129.8	4.2	132.7	3.5
La	11.2	10.2	11.3	10.4	11.7	11.3	16.3	0.6	15.0	0.6
Ce	29.8	29.3	29.8	28.6	29.8	29.7	39.8	1.7	38.1	1.0
Nd		18.6		18.8		20.4			25.1	1.3
Sm		5.13		4.44		5.02			5.8	0.6
Dy		4.79		4.60		5.13			5.3	0.2
Er		2.41		2.42		2.54			2.5	0.2
Yb		1.89		2.07		1.83			2.0	0.3
Hf		3.25		3.22		3.31			4.3	0.2

Note: Analyses by SIMS and laser ablation (LA) ICP-MS.

and KE58-2801S, five olivine-hosted melt inclusions per sample were analyzed via SIMS and a total of 38 inclusions were analyzed by laser ablation ICP-MS (LA-ICP-MS) for trace elements (Table 20.3).

Melt inclusion major elements (including S and Cl) were determined by electron microprobe analysis on a Cameca SX100 at Oregon State University over two analytical sessions (Table 20.2). Host olivine grains were analyzed adjacent to the melt inclusions in addition to getting core and rim analyses (Figures 20.3 and 20.4). Clinopyroxene-hosted melt inclusions were also analyzed from ERZ spatter (KE58-2801S). Trace elements (Ti, Sr, Y, Zr, Nb, Ba, La, and Ce) were analyzed by secondary ion mass spectrometry (SIMS) at Arizona State University on a Cameca 6f ion microprobe from samples KS08-09S, KS08-108E, and KE58-2801S following procedures described by *Rowe et al.* [2011a]. In addition, inclusions and groundmass glass from these three samples were analyzed via LA-ICP-MS at the University of Iowa using a Thermo X-series II ICP-MS + collision cell with a NewWave UP213 nm laser ablating under He gas. Inclusions were analyzed using a 30 μm laser spot (output energy of 29 J/cm^2). Data were collected over 30 s dwell time (5 Hz pulse frequency) with a 45 s washout at the start and end of the analysis. Trace-element concentrations were determined relative to calibration standard BCR-2G with ^{43}Ca as an internal standard. Averaged results from both SIMS and ICP-MS are reported in Table 20.3, with all results available in the online supplement. BHVO-2G was utilized as a secondary standard in both instances (Table 20.3). For SIMS and LA-ICP-MS trace-element measurements, reproducibility based on repeated analysis of BHVO-2G is better than 5% for Ti, Sr, Y, Zr, Nb, Ba, La, and Ce. For heavier masses, LA-ICP-MS reproducibility is better than 5% for Dy and Hf, 10% for Nd, Sm, and Er, and 17% for Yb (Table 20.3). Repeat analyses via SIMS and LA-ICP-MS indicates BHVO-2G results are within 1σ between the two analytical techniques.

Major-element analysis of interstitial glass from a partially molten crystal clot was performed at the U.S. Geological Survey (USGS) electron microprobe lab in Denver, Colorad. Trace-element compositions (Ti, Sr, Y, Zr, Nb, Ba, La, Ce, Nd, Sm, Eu, Gd, Dy, Er, Yb, and Hf), determined on the same spots as microprobe analysis, were measured by SIMS analysis at Arizona State University during a separate analytical session from the melt inclusions following methodology from *Rowe et al.* [2011a]. Crystal clot glass analyses are presented in Table 20.4.

20.3. RESULTS

20.3.1. Mineral Chemistry

Olivine compositions at both the summit and ERZ are relatively homogenous over the examined time interval. Summit sample core and rim olivine compositions are overlapping and restricted from $Fo_{82.6\pm0.5}$ to $Fo_{82.4\pm0.3}$, respectively, with most core-rim pairs of analyses within 1σ error of each other (Figure 20.4). With a similar absence of compositional variability, ERZ olivine grains have an average core composition of $Fo_{81.2\pm0.7}$ and rim composition of $Fo_{81.1\pm0.4}$. Four (out of 22) olivine grains

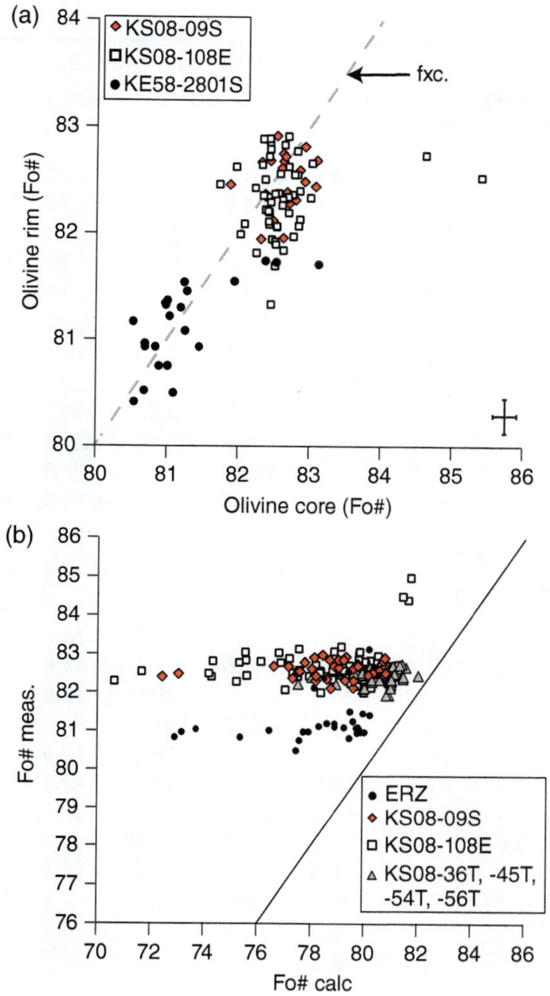

Figure 20.4 (a) Olivine rim versus core Fo number. Grains plotting significantly to the right of the dashed 1:1 line indicate xenocrystic grains or normally zoned phenocrysts (see text for details). Most grains plot within 1σ error of the 1:1 line. (b) Olivine analyses adjacent to melt inclusions with measured Fo and calculated Fo number based on the procedure described in the text.

from the ERZ have core compositions that overlie the compositional range of olivine cores recovered from the summit eruptive material with compositions of $Fo_{81.9}$ to $Fo_{83.1}$ (Figure 20.4a). Rim compositions of these four olivine grains are intermediate in composition between rims on summit and other ERZ olivine with a range of $Fo_{81.5}$ to $Fo_{81.7}$. This observation suggests that two populations of olivine are recorded in the ERZ lava. Several grains within the summit samples record a distinctly higher forsterite core value, up to $Fo_{85.4}$, which is comparable to the type B olivine of *Thornber* [2001]. A representative pyroxene composition with low Al (3.19 wt%) and high Mg number (79.9) is consistent with prior ERZ clinopyroxenes. Prior experimental and petrologic studies have indicated that clinopyroxenes are low pressure fractionates (<0.3 GPa) [*Helz and Thornber*, 1987; *Garcia et al.*, 2000]. All mineral chemistry is available as an online supplement.

20.3.2. Inclusion, Groundmass, and Whole-Rock Geochemistry

Major-element compositions of melt inclusions within a given sample are relatively homogenous. The greatest standard deviation was observed in inclusion MgO content (ranging from 10% to 12% relative variation) and may reflect minor postentrapment crystallization (Figure 20.4). Equilibrium olivine compositions are calculated for each inclusion assuming an Fe^{3+}/Fe_T of 0.1 after *Roeder et al.* [2003] and a K_D for Fe-Mg olivine-liquid exchange of 0.3 [*Roedder and Emslie*, 1970] and indicate, on average, ~5.4 wt% olivine crystallization following inclusion entrapment. Inclusion compositions are corrected to be in equilibrium with their host olivine compositions following procedures described by *Rowe et al.* [2011a, 2011b]. Average measured and corrected melt inclusion compositions are reported in Table 20.2. Average olivine-hosted melt inclusion compositions are offset to higher MgO (and lower SiO_2, Al_2O_3, and CaO) compared to whole-rock geochemical trends (Chapter 8, this volume). There is a large potential for error in melt inclusion MgO content associated with melt recalculations because of the leveraging capabilities of MgO in the host olivine. Errors in the recalculation process, particularly in regard to MgO, may result from the assumed K_D and Fe^{3+}/Fe_T. Calculated K_D from measured glass-olivine pairs compared to inclusion size indicates smaller inclusions typically have lower calculated Mg numbers and are further out of equilibrium with their host olivine, thus requiring more significant corrections (online supplement). This is most likely the source of most of the variation in major-element melt inclusion compositions and the elevation of melt inclusion MgO contents from whole-rock trends. In this study, we primarily use olivine-incompatible minor and trace-elements and trace element ratios. Minor- and trace-element concentrations are only slightly affected by the correction routines (dilution effect), and this amount is often within analytical uncertainty. Trace-element ratios are not affected by this correction process.

Groundmass glass Mg numbers range from 0.47 to 0.57 (average of 0.55 ± 0.02) with calculated average equilibrium olivine of $Fo_{80.4 \pm 1.9}$ slightly below that of measured rim compositions. Melt inclusions that are least modified by postentrapment crystallization have comparable Mg numbers ranging from 0.51 to 0.56. Glass temperatures are calculated using the MgO thermometer of *Helz and Thornber* [1987] (1014°C + 20.1(MgO wt%)). Calculated temperatures of tephra/spatter samples from both the summit and ERZ (based on groundmass glass analyses) range from 1136 to 1165°C but are dominated

Table 20.4 Crystal mush clot glass geochemistry from sample KS08-02S-T1.

Analysis no.	1	2	9	10	3	4	5	6	7	8	11	12	13
BSE (l/d)	dBSE	dBSE	dBSE	dBSE	lBSE	lBSE	lBSE	lBSE	lBSE	lBSE	lBSE	lBSE	lBSE
Major Elements (Oxide wt%)													
SiO_2	50.45	50.45	50.18	50.18	48.51	48.50	47.79	48.65	48.49	47.42	49.00	48.73	48.56
TiO_2	2.42	2.42	2.30	2.30	3.80	3.72	3.67	3.79	4.12	4.13	3.51	3.76	4.00
Al_2O_3	13.35	13.35	13.25	13.25	12.73	12.77	12.45	12.60	12.18	12.54	12.86	12.77	12.87
FeO	10.87	10.87	11.00	11.00	12.09	12.27	12.58	13.08	12.76	12.24	11.96	12.20	12.33
MnO	0.15	0.15	0.17	0.17	0.19	0.18	0.21	0.20	0.17	0.16	0.16	0.19	0.17
MgO	7.00	7.00	7.08	7.08	6.28	6.16	6.26	6.15	6.22	6.32	6.27	6.34	6.16
CaO	10.56	10.56	10.71	10.71	10.36	10.67	10.57	10.30	10.28	10.46	10.50	10.71	10.51
Na_2O	2.40	2.40	2.33	2.33	2.45	2.41	2.33	2.46	2.41	2.43	2.43	2.37	2.37
K_2O	0.42	0.42	0.41	0.41	0.57	0.52	0.52	0.54	0.54	0.55	0.53	0.52	0.57
P_2O_5	0.247	0.247	0.22	0.22	0.32	0.294	0.28	0.37	0.39	0.303	0.30	0.311	0.331
SO_3	0.016	0.016	0	0	0.015	0.027	0	0	0	0.02	0	0	0
Total	97.93	97.93	97.70	97.70	97.32	97.5	96.68	98.19	97.59	96.58	97.54	97.95	97.89
Trace Elements (ppm)													
Ti	14304	15683	15889	14850	22911	22326	22862	24633	24684	22995	21839	24159	24657
Sr	307.4	318.1	323.3	318.8	356.1	378.2	379.6	397.7	391.0	378.5	387.1	399.8	398.0
Y	21.3	23.0	25.8	25.4	22.3	24.2	23.0	28.8	30.4	27.4	27.1	28.0	27.8
Zr	127.4	141.9	144.8	138.4	168.9	165.7	157.9	204.8	211.8	179.6	175.6	183.8	184.4
Nb	12.2	13.3	13.6	12.1	19.8	19.7	19.5	27.0	28.0	22.5	20.9	22.8	23.1
Ba	91.6	101.4	102.2	99.6	142.2	144.1	141.4	174.7	172.4	147.7	151.4	160.0	155.4
La	10.1	11.6	11.6	10.9	15.5	15.7	14.4	20.1	20.1	16.8	16.5	17.4	17.3
Ce	25.8	29.8	29.4	28.6	37.7	38.4	36.9	48.4	50.2	41.8	40.9	43.7	43.1
Nd	18.5	20.9	20.6	20.7	23.7	24.1	23.3	29.8	30.9	26.7	26.0	27.4	27.5
Sm	4.6	4.7	5.4	5.5	5.1	5.3	5.6	6.6	7.2	6.2	5.9	6.3	6.2
Eu	1.4	1.4	1.7	1.6	1.5	1.8	1.7	1.8	1.9	1.8	1.7	1.9	1.9
Gd	4.1	4.4	5.1	4.8	5.0	5.2	5.1	6.6	6.7	6.0	6.0	6.1	5.8
Dy	4.4	4.6	5.0	4.5	4.6	4.9	5.0	5.8	6.2	5.3	5.6	5.7	5.7
Er	2.0	2.3	2.3	2.4	2.4	2.2	2.2	2.8	2.9	2.5	2.7	2.7	2.6
Yb	2.0	1.9	2.0	2.2	2.0	1.9	1.8	2.4	2.5	2.2	1.9	2.2	2.4
Hf	3.0	3.9	3.7	3.9	4.1	4.0	3.5	5.0	5.2	4.4	4.4	4.4	4.6

Note: BSE (l/d) indicates relative shading on backscatter imaging. Trace elements measured via SIMS. Points 1/2 and 9/10 are different SIMS trace-2 element analyses around the same microprobe analysis.

by the more abundant, higher temperature glasses (overall average 1155 ± 8°C). Corrected inclusion temperatures are only slightly higher in summit inclusions (1182 ± 9°C) compared to ERZ olivine-hosted melt inclusions (1169 ± 6°C). Given prior discussion around elevation of inclusion MgO contents, these melt inclusion estimates are likely high; however, direct comparisons between populations should remain valid. From these observations, there is no statistical difference in temperature between melt inclusion populations from the ERZ and summit, and inclusions are only slightly offset to higher temperatures compared to groundmass glass.

Volatile concentrations are of particular interest in this study for evaluating magmatic degassing, primarily sulfur, which preferentially degasses relative to chlorine [see review by *Metrich and Wallace*, 2008]. Sulfur concentrations in summit inclusions range from ~1670 to 200 ppm (Figure 20.5). A slightly more restricted range is observed in the ERZ, with sulfur contents from 1300 to 300 ppm. Groundmass sulfur contents are primarily low, predominantly from 200 to 100 ppm S with two outliers from ~400 to 500 ppm S. The primary difference in inclusion sulfur content between the summit and ERZ is not the overall range in concentrations but the skewed distribution toward lower concentrations in the ERZ sample. In the case of the summit samples, 57% of the inclusions range from 1300 to 1000 ppm S, increasing to 66% between 1300 and 900 ppm S. In contrast, ERZ inclusions from 1300 to 900 ppm S only account for 36% of the inclusions, with an equal split between high forsterite (Fo_{82-83}) and lower forsterite ($Fo_{80.5-81.5}$; Figure 20.5). Analysis of chlorine in inclusions and groundmass shows no appreciable difference between summit and ERZ samples, indicating only minor or no loss of chlorine during degassing. Concentrations are evenly distributed, ranging from the detection limit (~80 ppm) to ~200 ppm Cl.

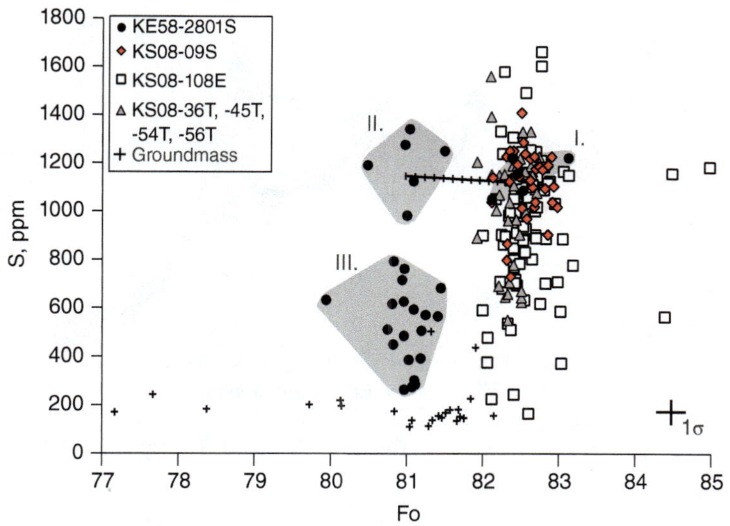

Figure 20.5 Host olivine Fo number versus melt inclusion S content for summit and ERZ melt inclusions and groundmass. Solid line indicates a 3% olivine fractional crystallization trend from an average summit melt inclusion and host olivine composition (increments are 0.2 wt% olivine). Type I, II, and II inclusions are indicated as gray fields (see text for discussion).

Olivine-incompatible trace-element compositions of melt inclusions, groundmass glass, and whole rocks are normalized to be in equilibrium with Fo_{84} olivine, allowing for direct comparison between samples without the influence of fractional crystallization. Trace-element concentrations of melt inclusions from both the ERZ and summit are relatively uniform (Figure 20.6; Table 20.3). The majority of trace elements within the summit and ERZ all have less than 10% variance in melt compositions. Higher variance in the middle rare-earth-element (MREE) and heavy rare-earth-element (HREE) are comparable to the increased analytical uncertainty in the laser ablation analyses and likely does not indicate a "real" increased variance. In addition, for all trace elements analyzed, average inclusion compositions from the summit and ERZ are within 9% of each other. The sum of the measured relative standard deviation of trace elements measured by SIMS and LA-ICP-MS for the summit and ERZ melt inclusions exceeds the difference between average summit and ERZ melt inclusion compositions, and as such there is no statistical difference between the two inclusion populations.

In contrast to the homogeneity of melt inclusions, representative whole-rock data presented in Figure 20.6 for spatter and tephra from the recent summit activity between 19 March and 5 April 2008 record a wide compositional range that varies from more depleted, ERZ-like trace-element abundances to more enriched values, akin to historic summit lava from 1919 (e.g., BHVO-1) [*Pietruszka and Garcia*, 1999]. Clean, handpicked, summit glass samples that erupted 23 and 27 March and 3 and 5 April have relatively depleted trace-element abundances that match ERZ lava (erupted on 21 March and 3 April) trace-element compositions (Figure 20.6). In contrast, bulk samples of explosive debris (referring here to erupted materials predominantly, if not exclusively, derived from lithic material from a range of older summit lava flows) deposited 19 and 27 March 2008 have distinctly enriched olivine-incompatible trace-element concentrations that more closely match BHVO-1 [*Pietruszka and Garcia*, 1999]. The term lithic is used here simply to refer to fragments of preexisting crystalline material, in contrast to the juvenile material (where juvenile hereafter refers specifically to previously unerupted melt with depleted trace-element characteristics of the clean, handpicked summit glass or its immediate crystal fractionates). Bulk analyses of spatter samples containing lithics, including partially molten crystal clots in dense cores of lapilli, that erupted 27 March have trace-element abundances that are intermediate between the clean, juvenile glass and explosive debris (Figure 20.6) (See Chapter 8, this volume). Nb/Y and La/Yb values range from 0.71 to 0.44 and 7.9 to 5.6, respectively, in bulk eruptive products from March 2008. The large range observed in La/Yb spans most of the compositional variation observed in historic summit eruptions (Figure 20.2). All melt inclusion and groundmass major- and trace-element compositions and representative whole-rock analyses used in later discussions are available as an online supplement. Whole-rock geochemistry is reported by Thornber et al. in Chapter 8.

20.3.3. Crystal Clots

Within individual spatter clasts/lapilli from the summit is a mixture of material including internally melted, micro-gabbroic fragments and glomerophyric clusters of

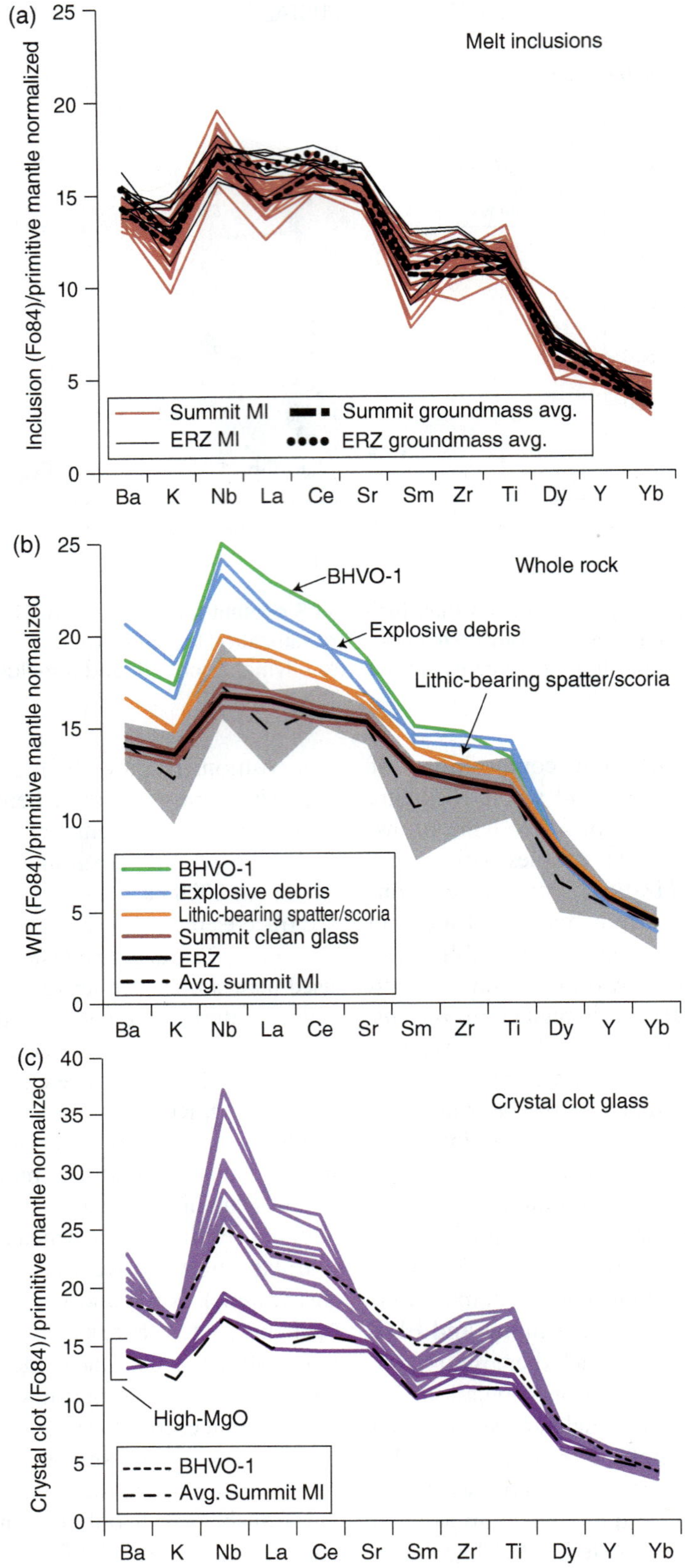

Figure 20.6 Spider diagrams of trace-element variation in (a) melt inclusions and average groundmass glass, (b) whole rocks, and (c) crystal mush clot glass. Melt inclusion and crystal clot glass analyzed via LA-ICP-MS. Trace-element abundances are corrected to a constant Fo_{84} host composition to eliminate compositional offsets resulting from differences in crystal fractionation. Whole-rock data from Chapter 8 (this volume) is available in the online supplement. Average melt inclusion compositions are indicated in both (b) and (c). The range of summit melt inclusion compositions is indicated by the gray field in (b). For comparison, BHVO-1 is also included in (c). BHVO-1 is the 1919 summit lava. Corrected trace elements are primitive-mantle normalized [*Sun and McDonough*, 1989].

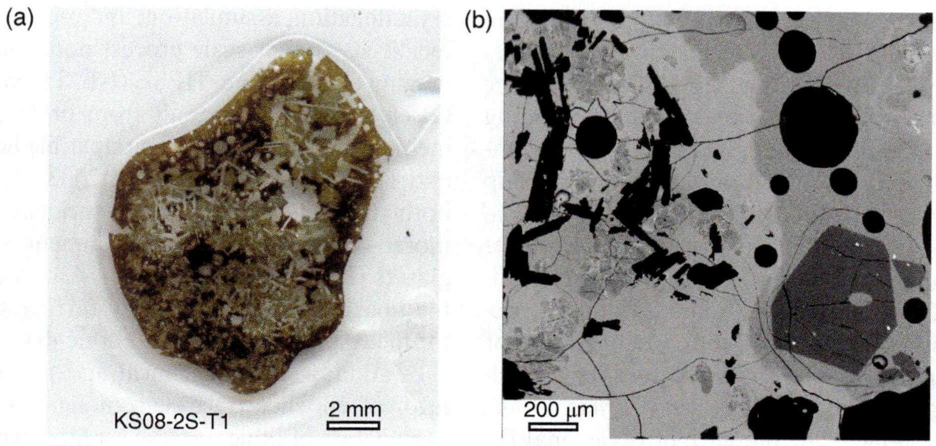

Figure 20.7 (a) Transmitted light image of crystal clot KS08-02S-T1 with "light" and "dark" glass. (b) Backscatter electron image of partially molten crystal clot KS08-02S-T1 displaying mingled zones of glass.

resorbed clinopyroxene + plagioclase ± olivine ± orthopyroxene ± Fe-Ti oxides, interladen with melt (referred to above as partially molten crystal clots in dense-cored lapilli; Figure 20.7). Through both petrographic and backscatter electron (BSE) imaging, two distinct glasses are evident in and around crystal clots: a light-colored glass (dark BSE) and a dark-colored glass (light BSE, predominantly surrounding crystal clots; Figure 20.7). The light BSE glass has distinctive major- and trace-element concentrations. It is noticeably enriched in FeO, TiO_2, and P_2O_5 and has lower MgO (6.2–6.3 wt% MgO) relative to the dark BSE glass (7.0–7.1 wt% MgO; Table 20.4). The light BSE glass (dark colored) and dark BSE glass (light colored) are hereafter referred to as the low-MgO and high-MgO glasses, respectively. Low-MgO glass is also degassed with regard to sulfur concentrations, with only a small fraction of the analyses having detectable S (Table 20.4). Major-element components of the high-MgO glass compositionally resemble juvenile groundmass glass (online supplemental data), and given the predominance of high-MgO glasses near the margin of the crystal clots, it seems reasonable to conclude that these are both representative of the juvenile material.

Trace-element concentrations are also significantly different between the two clot glasses, indicating that they are not petrogenically related to one another. Low-MgO glass is enriched in most olivine-incompatible trace elements (excluding Y and Yb) with concentrations equivalent to or greater than those observed in the 1919 Kīlauea summit eruption (BHVO-1; Figures 20.6 and 20.8; Table 20.4). In particular, high-field-strength elements (HFSEs), Ti and Nb, within crystal clot glasses all have significantly higher concentrations compared to BHVO-1 (Figure 20.6) and Kīlauea tholeiite in general [*Garcia et al.*, 2000]. As with whole-rock data, crystal clot glass is normalized to be in equilibrium with Fo_{84}

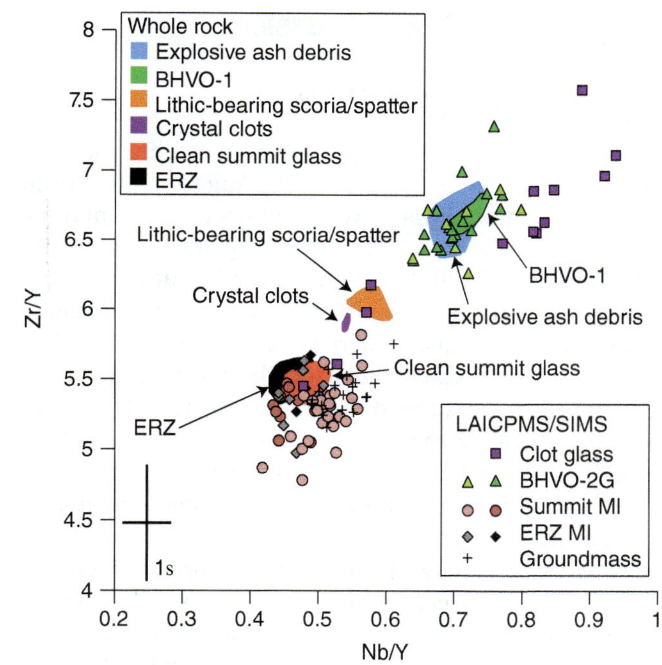

Figure 20.8 Zr/Y versus Nb/Y for representative whole-rock analyses (See Chapter 8, this volume), melt inclusions, and crystal clot glasses. Fields represent whole-rock data with individual symbols representing SIMS (dark symbols) and LA-ICP-MS (light symbols) analyses of inclusions and clot glass. BHVO-1 and BHVO-2G are both from the 1919 summit lava.

olivine for comparative purposes (Figure 20.6). From a comparison of trace-element abundances, as with major-element compositions, it is clear that the high-MgO glass is equivalent to the melt inclusion and groundmass glass compositions (Figure 20.5). Backscatter imaging indicating mingling (but incomplete mixing) of glasses may reflect a relatively short time during which the two glasses were coexisting (Figure 20.7).

Within an individual clot coexisting with surrounding groundmass glass of the lapilli, calculated temperatures of the low-MgO glass within the crystal clot are cooler (1140°C) compared to the high-MgO glass surrounding the clot (1156°C). This observation with regard to temperature is purely a function of the linear relationship between temperature and MgO content [*Helz and Thornber*, 1987]. However, both calculated temperatures overlap the overall range observed in the juvenile groundmass glass. The Fe-Ti oxide geothermometry from a second crystal clot indicate an equilibration temperature of 1148°C, within analytical uncertainty of juvenile groundmass glass compositions [*Ghiorso and Evans*, 2008] (Chapter 8, this volume). The presence of occasional Fe-rich orthopyroxene (a low-temperature, late-crystallization phase [*Helz and Thornber*, 1987]) supports lower temperatures of crystal clots (Chapter 8, this volume).

20.4. DISCUSSION

20.4.1. Primitive Component(s)

Many of the samples of the early tephra samples deposited between March and August 2008 from Kīlauea's summit eruptive vent contained abundant lithic fragments, including crystal clots and gabbroic fragments [*Wooten et al.*, 2009]. Wind-blown particulates, unrelated to juvenile magma, were also observed in many of these 2008 samples. Lithic materials are texturally variable, ranging from dense, coherent rock fragments to partially disaggregated mixtures of crystals and partial melt. In the absence of petrographic observations, bulk analysis of individual lapilli would lead to false conclusions about the degree of incompatible trace-element enrichment and compositional variability. While large, intact xenoliths [*Wooten et al.*, 2009] are obvious examples of crustal assimilation, more cryptic contamination is also observed in the mingling of highly incompatible enriched crystal clot glasses with the high-MgO relatively depleted "bulk" glass (Figure 20.7). Although we cannot definitively rule out some initial variation in juvenile basalt geochemistry, this more cryptic contamination may be reflected in the highly variable but distinctive trace-element and isotopic compositions reported for March 2008 explosive summit material [*Pietruszka et al.*, 2008] (see also Chapter 8). Despite the general variability inherent to melt inclusions, none of the inclusions analyzed from the summit have the highly incompatible trace-element contents observed in clot glasses or bulk samples. Additionally, analysis of handpicked glass where lithics (including crystal clots) were absent resulted in trace-element abundances consistent with melt inclusion results (Figure 20.8). This suggests that the crystal clots, and lithics in general, represent (1) late-stage, post-olivine crystallization, assimilation, and mixing and/or (2) a very localized, smaller-scale process not reflected by the bulk magma composition. HFSE (Nb, Ti) enrichment in crystal clot glass could result from preferential melting of mesostasis and ilmenite, a compatible host phase for Nb, and Ti [e.g., *Jang and Naslund*, 2003]. This would explain both the broad trace-element enrichment as well as the more significant HSFE enrichment observed in clot glasses (Figures 20.6 and 20.8). A partial melt origin for the unusual clot geochemistry further supports a model for melting of relatively cold lithic material.

In addition to geochemical and petrologic evidence in favor of enriched trace-element signatures resulting from assimilation of lithic material (or lithic partial melts), visual observations at the vent also support this idea. Since the start of the eruption in 2008, video documentation from webcams has provided unprecedented records of explosive activity. The video demonstrates how the initial eruptive vent within Halema'uma'u Crater evolved from a diameter of 35 m in 2008 [*Wilson et al.*, 2008] to 150 m by March 2011 [*Orr et al.*, 2013]. *Orr et al.* [2013] estimate that the crater within Halema'uma'u has a volume of $\sim 4 \times 10^6 m^3$, whereas estimated amounts of lithic debris that erupted from the vent total only $1.1 \times 10^3 m^3$ (by March 2011), or <0.1% of the total volume. Microgravity surveys of Kīlauea suggest the presence of void space beneath Halema'uma'u Crater prior to the 2008 eruption [*Johnson et al.*, 2010]. While most of the void space is hypothesized to have filled prior to the 2008 explosions and crater formation [*Johnson et al.*, 2010], the presence of void space may account for some of the volume discrepancy. Despite this, these estimates clearly indicate that a significant proportion of the wall rock incorporated into the juvenile magma must be assimilated and/or transported into the deeper magma system [*Swanson et al.*, 2009; *Orr et al.*, 2013].

Thus, melt inclusion and "clean juvenile" glass bulk analysis indicate that the sole primitive, juvenile magma component erupting at the summit, at least from 23 March 2008 to 1 August 2008, is relatively homogenous and compositionally indistinguishable from trace-element-depleted ERZ lava flows (Figure 20.6). This observation, while relatively straightforward, has implications for models focusing on the long-term temporal evolution of Kīlauea (see below). In addition, these results indicate that, unlike prior summit explosive eruptions in 1790 and 1924, there does not appear to be either a major shift in trace-element geochemistry or a high amount of diversity in the erupted juvenile component in early 2008.

20.4.2. Summit and ERZ Degassing

As discussed above, average sulfur concentrations of summit melt inclusions (1020 ± 260 ppm) are significantly elevated relative to summit groundmass glass (190 ± 90 ppm;

Figure 20.5). Since melt inclusions record a range of S content from ~1670 to ~200 ppm, this would suggest that crystallization and melt inclusion formation was concurrent with degassing from the summit reservoir. However, there is also no clear change in melt composition correlating to changing sulfur content of the melt inclusions. In contrast, Cl contents of melt inclusions (average 120 ± 30 ppm) are essentially identical to those of coexisting groundmass glass (110 ± 40 ppm), indicating no appreciable halogen degassing from the summit reservoir. These observations are consistent with plume gas measurements taken during this same time interval in 2008 that indicate very low molar HCl/SO_2 ratios (average 0.038 ± 0.03) [*Mather et al.*, 2012]. Overall, melt inclusion sulfur contents (primarily below ~1300 ppm) are consistent with prior studies of Kīlauea magmas [e.g., *Wallace and Anderson*, 1998].

The same overall range of S and Cl contents seen in summit inclusions is also observed in ERZ olivine-hosted melt inclusions. However, as previously mentioned, the distribution of sulfur concentrations is noticeably different in ERZ olivine-hosted inclusions. Based on sulfur concentrations and olivine Fo contents, ERZ inclusions can be subdivided into three populations (Figure 20.5). Type I inclusions are characterized by high sulfur contents (>1100 ppm S) and are hosted in high-Fo olivine (>Fo_{82}). These inclusions and their host olivine phenocrysts have compositions matching those of the summit region. Type II and III inclusions are hosted within olivine ranging from $Fo_{80.5}$ to $Fo_{81.5}$. Type II inclusions have >900 ppm S while type III inclusions contain <900 ppm S (Figure 20.5). Type III inclusions are the dominant population and average ~500 ppm S. The distinction between type II and type III inclusions is arbitrarily defined and may prove to be a single population with increased sampling. The prevalence of type III inclusions and the sulfur range observed in clinopyroxene-hosted inclusions may suggest that ERZ magmas typically have a maximum sulfur content of ~800 ppm and the higher S contents observed in smaller proportion may reflect an earlier stage of crystallization and degassing, perhaps during transport from the summit reservoir.

Mather et al. [2012] proposed that the similarity observed between gas compositions at both the ERZ and summit suggests that both vents are accessing magmas with similar degassing behavior and initial volatile contents. This hypothesis differs from previous models of Kīlauea degassing that instead call on a two-stage model, whereby initial degassing is dominated by C and S at the summit and subsequent degassing in the ERZ results in H, S, and Cl loss [*Gerlach*, 1993; *Gerlach and Graeber*, 1985]. Prior models required summit magmas to be stored at intermediate pressures, which allowed S degassing but inhibited loss of halogens, leading *Mather et al.* [2012] to propose a shallower magma storage area compared to previous summit eruptions, with sulfur contents in the plume a result of very shallow (<3 m) degassing processes. Our results of melt inclusion S and Cl contents are less clear. The Cl/S ratios vary significantly in summit melt inclusions from ~0.08 to 0.17. In summit inclusions, this ratio increases only slightly with decreasing sulfur contents down to 900 ppm S (Figure 20.9). Variation in Cl/S at constant sulfur contents likely results from analytical uncertainty of Cl concentrations close to detection limits (~80 ppm). The important observation here is, despite the scatter, average Cl contents are no different between summit, ERZ, and groundmass analyses, indicating little appreciable Cl degassing up into the point of groundmass glass quenching at the ERZ vent site. ERZ melt inclusions are offset to slightly higher Cl/S ratios for a given Cl content, indicating generally lower sulfur content, which is consistent with our observations on the relative distribution of sulfur in melt inclusions from the ERZ (Figure 20.5). Assuming the highest sulfur content (~1670 ppm) represents an undegassed end member, then up to 88% magmatic sulfur degassing is recorded between inclusions and groundmass glass. If high concentrations are anomalous and 1300 ppm S is more realistic of the magma within the summit reservoir based on sampling density, then, locally, 85% sulfur loss is recorded by summit melt inclusion and groundmass glass geochemistry (Figure 20.5). These values, while greater than those proposed by *Mather et al.* [2012], are consistent with estimates that less than 30% original S is retained in Kīlauea

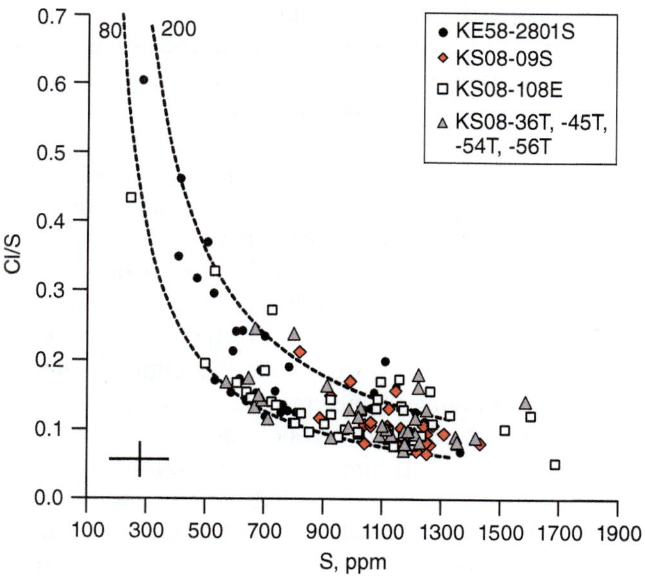

Figure 20.9 Cl/S versus S content in olivine- and clinopyroxene-hosted melt inclusions. Variation in Cl/S at constant sulfur contents likely reflects increased analytical uncertainty of Cl near detection limits. Dashed lines represent decreasing sulfur content with constant Cl values of 80 and 200 ppm. Error bars represent 1σ.

melts. Similarly, *Thordarson and Self* [1996], in a study of volatile degassing of Columbia River basalts, indicate that up to 70% S degassing had occurred at the time of eruption. The slightly higher calculated sulfur degassing recorded by rapidly quenched groundmass glass and melt inclusions compared to other low-water subaerial basaltic eruptions may reflect more efficient degassing within the shallow magma plumbing of Kīlauea.

20.4.3. Comparison of Rift Zone and Summit Melt Inclusions

Geophysical, remote sensing, and geochemical studies have previously discussed the potential connectivity of Kīlauea's summit magma storage region to the ERZ. These studies focus on the correlation between summit inflation and deflation and magma output rates, magma MgO content (and calculated melt geothermometry), and modal mineralogy [*Helz and Thornber*, 1987; *Thornber*, 2001, 2003; *Poland et al.*, 2008]. As previously discussed, ERZ melt inclusions and 2008 summit melt inclusions have normalized trace-element compositions indistinguishable from one another. While summit melt inclusion trace-element variations in Nb and La may appear distinct compared to the ERZ inclusions, given the error associated with analysis of both elements by either SIMS or LA-ICP-MS, this distinction is not resolvable. However, major-element analysis of ERZ olivine-hosted melt inclusions reveals that rift lava melt inclusions are predominantly more evolved compared to olivine-hosted melt inclusions in summit tephra.

Despite the predominance of more evolved melt and olivine compositions in ERZ lava samples, a significant component of melt inclusion sulfur contents (5 out of 30, or 16.6%) and olivine forsterite contents (4 out of 22, or 18.1%) matches that of the summit material (Figure 20.4, 20.5). Assuming an average summit melt inclusion composition, crystal fractionation of ~3 wt% olivine can account for the offset in host olivine compositions from the average of $Fo_{82.5}$ observed in summit ejecta compared to the ERZ with an average Fo_{81} olivine. With the exception of ~3 olivine grains from Kīlauea's summit material, none of the summit olivine grains have rim compositions that are as evolved as the cores of ERZ olivine, implying that the magma resulting in crystallization of olivine phenocrysts observed in the ERZ is unique to the ERZ (Figure 20.4). However, all four of the high-Fo ERZ olivine (hosting type I inclusions) have rim compositions overlapping the rim compositions on low-Fo ERZ olivine (hosting type II and III inclusions). This would be expected if these "summit" olivine grains shared a common late-stage growth history during the ~30°C cooling predicted by *Thornber* [2001] during magma transit along the ERZ.

Combining melt inclusion and crystal chemistry populations, we can propose a petrogenic model to further elucidate the connectivity between the summit and ERZ magma systems. In this scenario, the presence of the type I melt inclusions and their host olivine (which match summit compositions) are potential evidence for direct connectivity between summit and ERZ reservoirs. The striking bimodality of olivine compositions transitioning from the summit to the ERZ (Figure 20.5) and appearance of significant clinopyroxene in ERZ lava likely imply that ERZ magma is primarily evolving outside of the summit reservoir that is tapped by the summit eruption (with a degree of differentiation but a similar range of S degassing). This is in agreement with the model proposed by *Thornber* [2001] wherein "type A" near-equilibrium olivine grew during transport from the summit to the rift zone. Interestingly, the observation that the more evolved olivine-hosted melt inclusions from the ERZ contain sulfur contents up to ~1300 ppm despite evidence for significant sulfur degassing from the summit reservoir may imply a partially isolated reservoir for ERZ magma capable of preserving pre-summit-degassing sulfur contents while still allowing for potential influx from the summit magma reservoir. That, however, does not negate the fact that normalized melt inclusion trace-element compositions indicate the tapping of a compositionally indistinguishable source for both the summit and ERZ eruptions.

As presented earlier, the top-down model of *Thornber* [2003] explains geochemical diversity in Pu'u 'Ō'ō ERZ lava flows by mixing a homogenous, incompatible trace-element-depleted recharge magma with residual 1982 magma stored in the summit reservoir and ultimately flushing of this material through the ERZ. The overall decrease in incompatible elements with time in ERZ lava flows therefore reflects a decreasing proportion of 1982 magma stored in the summit over the past 30 years (Figure 20.2) [*Garcia et al.*, 2000; *Thornber*, 2003]. In a top-down framework, the apparent absence of a 1982 component in both the 2008 summit and ERZ melt inclusions, as well as the proposed connectivity between these two eruption sites, suggests the last of the pre-1983 magma has been flushed out of the summit reservoir during the "surge" from 2003 to 2007 [*Poland et al.*, 2012]. Continued volcanism with simultaneous summit and ERZ eruptions of a uniformly incompatible trace-element-depleted magma indicates the homogeneity of the magma composition currently feeding both eruptive sites.

20.5. CONCLUSIONS

Melt inclusion geochemistry, when combined with other microanalytical and petrographic techniques, provides insight into a number of petrologic problems surrounding the Kīlauea summit eruption beginning in 2008. Trace elements from melt inclusions from both erupted summit material and the ERZ indicate indistinguishable, highly incompatible element-depleted juvenile magmatic compositions. This observation is in contrast to bulk rock measurements, which are significantly influenced by the incorporation of lithic materials, as well as more cryptic assimilation of crystal clot partial melts, resulting in enriched trace-element concentrations similar to prior summit eruptions. Petrographic and geochemical investigations of the crystal clots indicate that these partially molten clots formed at lower temperatures than surrounding glass and are nearly completely degassed with regard to sulfur, supporting models that these clots and lithic fragments likely represent late-stage wall rock assimilation.

Melt inclusions crystallized at Kīlauea's summit record efficient sulfur degassing during magmatic decompression in the shallow magmatic system, with up to 85% S degassing recorded during olivine crystallization. Chlorine concentrations remain relatively uniform and do not indicate significant Cl loss to degassing, even in quenched groundmass glass, and as such, Cl/S ratios increase from the summit to ERZ due to progressive sulfur loss. The presence of a small proportion of high sulfur contents in ERZ samples implies that at least a portion of the magma feeding the ERZ is independent of degassing affecting the summit magma reservoir. This is further supported by the bimodality of olivine compositions in the ERZ, with the lower-Fo olivine population likely growing either in an independent storage region or during transit from the summit.

The proposed connectivity between the summit and ERZ based on inclusion sulfur contents and variations in host olivine composition, in conjunction with the ubiquitous presence of a highly incompatible element-depleted magma at both the summit and ERZ, indicate that any remaining 1982 magma stored within the summit reservoir was flushed during the 2003–2007 "surge" [*Poland et al.*, 2012]. Overall, the approach taken here for a population analysis of melt inclusions from a limited number of samples compliments prior studies of shallow magmatic processes based primarily on whole-rock data, providing insight into petrologic processes recorded during reactivation of Kīlauea's summit as an eruptive center.

SUPPLEMENTARY DIGITAL DATA

The following data items are available online at www.wiley.com/go/Carey/Hawaiian_Volcanoes

Table 20.1. Major element compositions of melt inclusions and host grains.

Table 20.2. Major and trace element concentrations of olivine-hosted melt inclusions.

Table 20.3. Bulk samples of summit tephra and ERZ lavas between March and April 2008.

ACKNOWLEDGMENTS

Samples used in this study were collected by T. Orr. We would like to thank R. Hervig and L. Williams at the Arizona State University SIMS facility (partially supported by a grant from the Instrumentation and Facilities Program, NSF) and F. Tepley at the Oregon State University Electron Microprobe Lab for their assistance. Thanks also to D. Peate and J. Thompson for access to and assistance with LA-ICP-MS data collection at the University of Iowa. Reviews by A. Pietruszka, T. Plank, and N. Metrich as well as additional comments by M. Poland and R. Carey greatly improved this manuscript. This work was partly supported by an NSF award (EAR0809309) to Rowe.

REFERENCES

Edmonds, M., T. M. Gerlach, and R. A. Herd (2009), Halogen degassing during ascent and eruption of water-poor basaltic magma, *Chem. Geol.*, *263*, 122–130.

Garcia, M. O., J. M. Rhodes, F. A. Trusdell, and A. J. Pietruszka (1996), Petrology of lavas from the Pu'u 'Ō'ō eruption of Kīlauea Volcano: III. The Kupaianaha episode (1986–1992), *Bull. Volcanol.*, *58*, 359–379.

Garcia, M. O., A. J. Pietruszka, J. M. Rhodes, and K. Swanson (2000), Magmatic processes during the prolonged Pu'u 'Ō'ō eruption of Kīlauea Volcano, Hawaii, *J. Petrol.*, *41*, 967–990.

Garcia, M. O., A. J. Pietruszka, and J. M. Rhodes (2003), A petrologic perspective of Kīlauea volcano's summit magma reservoir, *J. Petrol.*, *44*, 2313–2339.

Gerlach, T. M. (1993), Oxygen buffering of Kīlauea volcanic gases and the oxygen fugacity of Kīlauea basalt, *Geochim. Cosmochim. Acta*, *57*, 795–814.

Gerlach, T. M., and E. J. Graeber (1985), Volatile budget of Kīlauea volcano, *Nature*, *313*, 273–277.

Ghiorso, M. S., and B. W. Evans (2008), Thermodynamics of rhombohedral oxide solid solutions and a revision of the Fe-Ti two-oxide geothermometer and oxygen-barometer, *Am. J. Sci.*, *308*, 957–1039.

Harris, D. M., and A. T. Anderson, Jr. (1983), Concentrations, sources, and losses of H_2O, CO_2, and S in Kīlauean basalt, *Geochim. Cosmochim. Acta*, *47*, 1139–1150.

Helz, R. T., and C. R. Thornber (1987), Geothermometry of Kīlauea Iki lava lake, Hawaii, *Bull. Volcanol.*, *49*, 651–668.

Houghton, B. F., D. A. Swanson, R. J. Carey, J. Rausch, and A. J. Sutton (2011), Pigeonholing pyroclasts: Insights from the 19 March 2008 explosive eruption of Kīlauea volcano, *Geology*, *39*, 263–266.

Jang, Y. D., and H. R. Naslund (2003), Major and trace element variation in ilmenite in the Skaergaard Intrusion: Petrologic implications, *Chem. Geol.*, *193*, 109–125.

Johnson, D. J., A. A. Eggers, M. Bagnardi, M. Battaglia, M. P. Poland, and A. Miklius (2010), Shallow magma accumulation at Kīlauea Volcano, Hawai'i, revealed by microgravity surveys, *Geology*, *38*, 1139–1142.

Johnson, D. M., P. R. Hooper, and R. M. Conrey (1999), XRF analysis of rocks and minerals for major and trace elements on a single low dilution Li-tetraborate fused bead, *Adv. X-ray Anal,*, *41*, 843–867.

Kent, A. J. R. (2008), Melt inclusions in basaltic and related volcanic rocks., *Mineral. Soc. Am. Geochem. Soc. Rev. Mineral. Petrol. Minerals*, *69*, 273–331.

Kent, A. J. R., J. A. Baker, and M. Wiedenbeck (2002), Contamination and melt aggregation processes in continental flood basalts: Constraints from melt inclusions in Oligocene basalts from Yemen, *Earth Planet. Sci. Lett.*, *202*, 577–594.

Knaack, C., S. B. Cornelius, and P. R. Hooper (1994), Trace element analyses of rocks and minerals by ICP-MS, Technical Notes, GeoAnalytical Lab, Washington State University.

Mather, T. A., et al. (2012), Halogens and trace metal emission from the ongoing 2008 summit eruption of Kīlauea volcano, Hawaii, *Geochim. Cosmochim. Acta*, *83*, 292–323.

McNeill, A. W., and L. V. Danyushevsky (1996), Composition and crystallization temperatures of primary melts from hole 896A basalts: Evidence from melt inclusion studies, *Proc. Ocean Drilling Program, Scientific Results*, *148*, 21–35.

Metrich, N., and P. J. Wallace (2008), Volatile abundances in basaltic magmas and their degassing paths tracked by melt inclusions, *Mineral. Soc. Am. Geochem. Soc., Rev. Mineral. Petrol. Minerals. 69*, 363–402.

O'Hara, M. J. (1977), Geochemical evolution during fractional crystallization of a periodically refilled magma chamber, *Nature*, *266*, 503–507.

Orr, T. R., C. Heliker, and M. Patrick (2012), The ongoing Pu'u 'Ō'ō eruption of Kīlauea Volcano, Hawai'i—30 years of eruptive activity, U.S. Geol. Surv. Fact Sheet 2012-3127.

Orr, T. R., W. A. Thelen, M. R. Patrick, D. A. Swanson, and D. C. Wilson (2013), Explosive eruptions triggered by rockfalls at Kīlauea volcano, Hawaii, *Geology*, *41*, 207–210.

Pietruszka, A. J., and M. O. Garcia (1999), A rapid fluctuation in the mantle source and melting history of Kīlauea Volcano inferred from the geochemistry of its historical summit lavas (1790–1982), *J. Petrol.*, *40*(8), 1321–1342.

Pietruszka, A. J., M. O. Garcia, J. M. Rhodes, M. D. Norman, I. N. Bindeman, J. P. Marske, and D. H. Burns (2008), Rapid temporal geochemical and isotopic variations of tephra from the ongoing 2008 summit eruption of Kīlauea Volcano, Hawaii. *Eos Trans. AGU*, *89*, 53, Suppl., AGU 2008 Fall Meet., V11B-2021.

Poland, M., A. Miklius, T. Orr, J. Sutton, C. Thornber, and D. Wilson (2008), New episodes of volcanism at Kīlauea Volcano, Hawaii, *EOS Trans. AGU*, *89*(5), 37–38.

Poland, M. P., A. Miklius, A. J. Sutton, and C. R. Thornber (2012), A surge in magma supply to Kīlauea Volcano during 2003–2007, *Nature Geosci.*, *5*, 295–300, doi:10.10.38/NGEO1426.

Roedder, E. (1979), Origin and significance of magmatic inclusions, *Bull. Mineral.*, *102*, 487–510.

Roedder, E., and R. Emslie (1970), Olivine-liquid equilibrium, *Contrib. Mineral. Petrol.*, *29*, 275–289.

Roeder, P. L., C. Thornber, A. Poustovetov, and A. Grant (2003), Morphology and composition of spinel in Pu'u 'Ō'ō lava (1996–1998), Kīlauea Volcano, Hawaii, *J. Volcanol. Geotherm. Res.*, *123*, 245–265.

Rowe, M. C., D. W. Peate, and A. Newbrough (2011a), Compositional and thermal evolution of olivine-hosted melt inclusions in small-volume basaltic eruptions: A "simple" example from Dotsero Volcano, NW Colorado, *Contrib. Mineral. Petrol.*, *161*, 197–211.

Rowe, M. C., D. W. Peate, and I. Ukstins Peate (2011b), An investigation into the nature of the magmatic plumbing system at Paricutin Volcano, Mexico, *J. Petrol.*, *52* (11), 2187–2220.

Sobolev, A. V., and N. Shimizu (1993), Ultra-depleted primary melt included in an olivine from the Mid-Atlantic Ridge, *Nature*, *363*, 151–154.

Sun, S.-S., and W. F. McDonough (1989), Chemical and isotopic systematic of oceanic basalts: Implications for mantle composition and processes, in *Magmatism in the Ocean Basins*, Spec. Pub. 42, edited by A. D. Saunders, and M. J. Norry, pp. 313–345, Geol. Soc. of Lon., London.

Swanson, D., K. Wooten, and T. Orr (2009), Buckets of ash track tephra flux from Halema`uma`u Creater, Hawai'i, *Eos Trans., AGU*, *90*, 427.

Thordarson, Th., and S. Self (1996), Sulfur, chlorine, and fluorine degassing and atmospheric loading by the Roza eruption, Columbia River Basalt Group, Washington, USA, *J. Volcanol. Geotherm. Res.*, *74*, 49–73.

Thornber, C. R. (2001), Olivine-liquid relations of lava erupted by Kīlauea Volcano from 1994–1998: Implications for shallow magmatic processes associated with the ongoing east rift zone eruption, *Can. Mineral.*, *39*, 323–350.

Thornber, C. R. (2003), Magma reservoir processes revealed by geochemistry of the ongoing Pu'u O'o – Kupaianaha eruption, in *The Pu'u O'o – Kupaianaha Eruption of Kīlauea Volcano, Hawaii, the First Twenty Years*, edited by C. C. Heliker, et al. pp. 121–136. U.S. Geol. Surv. Prof. Pap. 1676, Reston, VA.

Thornber, C. R., M. C. Rowe, D. T. Adams, J. C. Rea, and T. R. Orr (2010), Application of microbeam techniques to identifying and assessing comagmatic mixing between summit and rift eruptions at Kīlauea Volcano, *Eos Trans. AGU*, V43F-1 Invited.

Wade, J., T. Plank, W. G. Melson, G. J. Soto, and E. H. Hauri (2006), The volatile content of magmas from Arenal volcano, Costa Rica, *J. Volcanol. Geotherm. Res.*, *157*, 94–120.

Wallace, P. J. (2003), From mantle to atmosphere: Magma degassing, explosive eruptions, and volcanic volatile budgets, in *Melt Inclusions in Volcanic Systems: Methods, Applications and Problems: Developments in Volcanology*, vol. 5, edited by B. De Vivo and R. J. Bodnar, pp. 105–127, Elsevier, Amsterdam.

Wallace, P. J., and A. T. Anderson, Jr. (1998), Effects of eruption and lava drainback on the H_2O contents of basaltic magmas at Kīlauea Volcano, *Bull. Volcanol.*, *59*, 327–344.

Wilson, D., T. Elias, T. Orr, M. Patrick, J., Sutton, and D. Swanson (2008), Small explosion from new vent at Kīlauea's Summit, *Eos Trans. AGU*, *89*(22), 203.

Wooten, K. M., C. R. Thornber, T. R. Orr, J. F. Ellis, and F. A. Trusdell (2009), Catalog of tephra samples from Kīlauea's summit eruption, March–December 2008, U. S. Geological Survey Open-File Rep. 26, OF 2009-1134, and online database: http://pubs.usgs.gov/of/2009/1134/.

21

FLOWGO 2012: An Updated Framework for Thermorheological Simulations of Channel-Contained Lava

Andrew J. L. Harris[1] **and Scott K. Rowland**[2]

ABSTRACT

A new version of the thermo-rheological lava flow model FLOWGO allows (i) incorporation of new rheological relations, (ii) use of stochastic lava flow path models, and (iii) greater flexibility to allow best-fitting of model output with natural data. We apply the model to four channel-fed lava flows for which good field data exist. These include Mauna Loa's 1859 and 1984 flows plus Kīlauea's December 1974 flow. Best fits between model output and field data require (i) selection of an appropriate viscosity and crystallization model, (ii) a thermal model that moves from poorly insulated across proximal channel reaches to well insulated in medial-to-distal reaches, and (iii) a lava flow emplacement event that is cooling limited, not volume limited. For Mauna Loa's 1859 flow, if we use a thermal regime that changes from poorly insulated to well insulated at 10 km, we can re-create a transition between rapid and slow cooling witnessed at this point. For Kīlauea's 1974 flow, we model an appropriate thermal regime (i.e., cooling rates of 1.2°C/km). However, when fed at the actual effusion rate of at least 270 m^3/s, the model gives a flow run out that is too long. A reasonable explanation is that this flow was volume limited.

21.1. INTRODUCTION

Lava flow emplacement modeling serves two purposes. First, models allow better understanding of the physical processes governing lava flow dynamics and emplacement as well as the complex feedbacks between cooling, crystallization, and rheology that govern flow dynamics [e.g., *Daneš* 1972; *Park and Iverson*, 1984; *Dragoni*, 1989; *Heslop et al.*, 1989; *Crisp and Baloga*, 1994; *Keszthelyi and Self*, 1998]. Second, models can be used to assess the likely lava flow path, run-out and inundation area for hazard assessment [e.g., *Young and Wadge*, 1990; *Ishihara et al.*, 1990; *Crisci et al.*, 2004; *Hidaka et al.*, 2005; *Favalli et al.*, 2005; *Vicari et al.*, 2007]. In addition, such simulations now have the potential to be run in real time using satellite and/or field-derived lava discharge rates as input [*Wright et al.*, 2008; *Herault et al.*, 2009; *Vicari et al.*, 2009, 2011; *Ganci et al.*, 2012].

The FLOWGO model of *Harris and Rowland* [2001] was originally intended for the first purpose, as framework within which we could organize the complex links and feedbacks governing the thermal and rheological evolution of a lava control volume moving down a channel. It was designed to be highly flexible, so that thermal and rheological relationships could be modified and updated as our understanding and constraint of such relations evolved. It was also designed to be adaptive, so that input parameters could be set on a case-by-case basis so as to be tailored to the compositional, rheological, or thermal conditions appropriate to the flow in question [e.g., *Rowland et al.*, 2004; *Mueller*, 2005; *Harris et al.*, 2007; *Riker et al.*, 2009]. However, we found that the cooling-limited distance that the control volume could reach before freezing provided a reasonable approximation for actual channel length [*Harris and*

[1] *Laboratoire Magmas et Volcans, Université Blaise Pascal, Clermont Ferrand, France*

[2] *Department of Geology and Geophysics, University of Hawai'i at Mānoa, Honolulu, Hawaii, USA*

Hawaiian Volcanoes: From Source to Surface, Geophysical Monograph 208. First Edition.
Edited by Rebecca Carey, Valérie Cayol, Michael Poland, and Dominique Weis.

Companion Website: www.wiley.com/go/Carey/Hawaiian_Volcanoes

Rowland, 2001]. Thus, FLOWGO has also been used to assess the likely distance a channel-fed flow can extend at a given rate of volumetric lava supply (i.e., effusion rate) and to assess the likely inundation hazard posed by emplacement of channel-fed lava flows [*Rowland et al.*, 2005; *Harris et al.*, 2011].

Since the initial presentation of FLOWGO by *Harris and Rowland* [2001], the code and its application have evolved significantly. We here present the latest version of FLOWGO (FLOWGO^S) that allows input and storage of a digital elevation model (DEM) and geolocated satellite-derived land cover map. This is used for projection of flow paths over the DEM, with a lava flow run-out model allowing estimation of the length down each flow path that a lava will likely extend. The reporting system then outputs results in map and table form, with simulations being projected over the satellite-derived base map [*Harris et al.*, 2011]. To achieve this, we have updated, combined, and tested two existing lava flow emplacement models: FLOWGO and the DOWN-FLOW model of *Favalli et al.* [2005]. DOWN-FLOW is used to yield flow paths, and FLOWGO estimates how far down each path a control volume will extend under given thermo-rheological and effusion rate conditions. Together, the two models provide a framework that allows us to explore our current understanding of thermo-rheological relations for active lavas.

21.2. PHYSICAL BASIS OF FLOWGO PROGRAM

The FLOWGO model tracks a control volume of lava as it moves down a channel of known starting dimensions. As the control volume is advanced down-channel, all heat losses and gains are calculated to determine the internal temperature and crystallinity of the control volume. These parameters, in turn, are used to calculate the viscosity and yield strength of the flow. These thermal, textural and rheological conditions are used to assess whether the lava is still capable of forward motion. The three stopping conditions can be summarized by the three following questions: Is the calculated velocity greater than zero? Is the temperature at or below the solidus? Is the yield strength at the base of the channel greater than the downhill stress? If none of these conditions are met, then advance of the control volume down-channel is continued. If any of the conditions is met, then the loop ends (Figure 21.1).

Velocity is therefore fundamental to the model as it determines whether the control volume is in motion or not. Mean velocity (v_{mean}) for a Newtonian fluid flowing in a channel can be calculated following *Jeffreys* [1925],

$$v_{mean} = \left(\frac{r^2 \rho g \sin(\theta)}{n\eta}\right), \tag{21.1a}$$

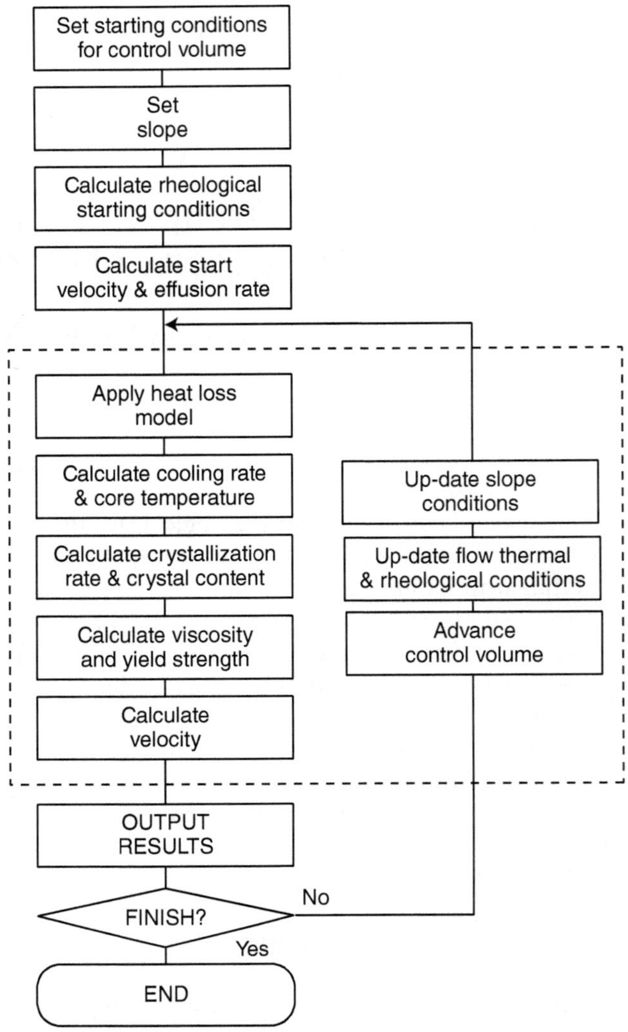

Figure 21.1 Flow chart giving the main steps executed by FLOWGO.

in which r is channel radius, ρ is lava density, g is acceleration due to gravity, θ is the underlying slope, n is a constant that depends on channel shape, and η is lava viscosity. However, the equation needs to be modified for a Bingham fluid so that, for a semicircular channel, we have [*Moore*, 1987]

$$v_{mean} = \left(\frac{r^2 \rho g \sin(\theta)}{8\eta}\right)\left[1 - \frac{4}{3}\frac{\tau_0}{\tau} + \frac{1}{3}\left(\frac{\tau_0}{\tau}\right)^4\right] \tag{21.1b}$$

and for a channel that is wider than it is deep,

$$v_{mean} = \left(\frac{d^2 \rho g \sin(\theta)}{3\eta}\right)\left[1 - \frac{3}{2}\frac{\tau_0}{\tau} + \frac{1}{2}\left(\frac{\tau_0}{\tau}\right)^3\right], \tag{21.1c}$$

where d is channel depth. The right-hand term takes into account Bingham conditions using the ratio of the lava yield strength (τ_0) to the basal shear stress (τ). Note that when $\tau_0 = 0$ (i.e., the flow is Newtonian), the entire right-hand term reduces to 1. Basal shear stress is given by [*Hulme*, 1974]

$$\tau = d\rho g \sin(\theta), \tag{21.2}$$

in which *d* is lava thickness, which in our case equates to the depth of the lava in the channel.

Appropriate starting conditions (i.e., at-vent parameters for all values in equation (21.1)) are given for Mauna Loa's 1984 channel in Table 21.1 and for a channel active on Mount Etna during September 2004 in Table 21.2. To set up these starting conditions, density has been calculated from lava chemistry following the method of *Bottinga and Weill* [1970]. This dense rock value (ρ_{DRE}) needs to be corrected for vesicularity (ϕ_b) to derive a bulk density using

$$\rho = (1-\phi_b)\rho_{DRE}, \tag{21.3}$$

ρ being the bulk density of the lava flowing in the channel.

Next, at-vent viscosity needs to be calculated. This can be obtained by applying models for fluid viscosity (η_f) for lava of given compositions and temperatures following the method of *Shaw* [1972] for Mauna Loa and *Giordano and Dingwell* [2003] for Etna. Bulk viscosity (η) has then, traditionally, been calculated using the fluid viscosity and lava crystal content from the Einstein-Roscoe relationship [e.g., *Crisp et al.*, 1994]. Likewise, yield strength can be calculated on the basis of temperature and crystal content, as shown in Tables 21.1 and 21.2.

The remaining at-vent variables in equation (21.1) are slope and channel radius or depth (*d*). Given an estimate

Table 21.1 Mean starting conditions for lava in Mauna Loa's 1984 channel and at-vent velocity and effusion rate that these give.

Parameter	Term	Value	Units	Source
Channel depth	r	3	m	Typical channel depth at 2850 m vent outlet given in Table 57.3 of *Lipman and Banks* [1987]
Channel width	w	21	m	Typical channel width at 2850 m vent outlet given in Table 57.3 of *Lipman and Banks* [1987]
Dense rock density	ρ_{DRE}	2720	kg/m	Density calculated from mean composition data given for Mauna Loa's 1984 lava by *Rhodes* [1988] using the method of *Bottinga and Weill* [1970]
Vesicularity	ϕ_b	32	%	Mean vesicularity calculated from bulk density data for channel samples given in Table 57.1 of *Lipman and Banks* [1987]
Bulk density	ρ	1850	kg/m^{-3}	$\rho = (1-\phi_b)\,\rho_{DRE}$
Slope	θ	2.7	deg	At-vent slope from DEM
Viscosity Model				
Fluid viscosity	$\eta_f(T)$	86	Pa·s	Viscosity calculated using the method of *Shaw* [1972] using the mean whole-rock composition for Mauna Loa's 1984 lava as given in Table 4 of *Rhodes* [1988] and an eruption temperature of 1140°C
Crystal content	ϕ	16.5	%	Median phenocryst content during 26 March to 6 April from data by *Lipman and Banks* [1987]
R	—	1.51	Dimensionless	Inverse of maximum crystal content
Bulk viscosity	η	177	Pa·s	$\eta(\phi) = \eta_f\,(1-R\phi)^{-2.5}$
Yield Strength Model				
Eruption temperature	T_{erupt}	1140	°C	Mean from *Lipman and Banks* [1987]
Liquidus temperature	T_0	1120	°C	
Constant 1	B	0.01	Pa	From *Dragoni* [1989]
Constant 2	C	0.08	K^{-1}	From *Dragoni* [1989]
Yield strength	τ_0	30	Pa	$\tau_0(T,\phi) = B\left[\exp^{C(T_0 - T_{core})} - 1\right] + \left[6500\,\phi^{2.85}\right]$
Basal shear stress	τ	4700	Pa	$\tau = d\rho g \sin(\theta)$
At-Vent (Starting) Velocity				
Mean velocity	V_{mean}	14.2	m/s	Equation (21.1c)
Effusion Rate				
Bulk effusion rate	E_r	893	m^3/s	$E_r = dwv_{mean}$

Table 21.2 Mean starting conditions for lava in Etna's 2004 channel and at-vent velocity and effusion rate that these give.

Parameter		Value			Units	Source
Channel depth	r	1	1	1	m	Mean channel depth over first 100 m from Lidar data of *Mazzarini et al.* [2007]
Channel width	w	6	6	6	m	Mean channel width over first 100 m from Lidar data of *Mazzarini et al.* [2007]
Dense rock density	ρ_{DRE}	2,720	2,720	2,720	kg/m	Density calculated from composition data given for Etna's 1991–93 lavas by *Tonarini et al.* [1995] using the method of *Bottinga and Weill* [1970]
Vesicularity	ϕ_b	22	22	22	%	Mean vesicularity from Etna literature values collated by *Harris et al.* [2005]
Bulk density	ρ	2,120	2,120	2,120	kg/m	$\rho=(1-\phi_b)\,\rho_{DRE}$
Slope	θ	29	29	29	deg	At-vent slope from *LIDAR* data of *Mazzarini et al.* [2007]
Viscosity Model						
Fluid Viscosity	$\eta_f(T)$	670	670	670	Pa·s	Method of *Giordano and Dingwell* [2003] applied using $H_2O=0.1$ wt % and $T_{erupt}=1065°C$; H_2O content selected for Etna summit eruptions following *Harris and Allen* [2008]
Crystal content	ϕ	27	34	41	%	Mean phenocryst ($\pm 1\sigma$) content from all literature values for Etna collated by *Harris et al.* [2005]
R	—	1.51	1.51	1.51	Dimensionless	Inverse of maximum crystal content
Bulk viscosity	η	2,480	4,054	7,477	Pa·s	$\eta(\phi)=\eta_f\,(1-R\phi)^{-2.5}$
Yield Strength Model						
Eruption temperature	T_{erupt}	1,065	1,065	1,065	°C	At-vent internal temperature for lava channel active on Etna in 2001 [*Bailey et al.*, 2006]
Liquidus temperature	T_0	1,160	1,160	1,160	°C	
Constant 1	B	0.01	0.01	0.01	Pa	From *Dragoni* [1989]
Constant 2	C	0.08	0.08	0.08	K^{-1}	From *Dragoni* [1989]
Yield strength	τ_0	163	308	519	Pa	$\tau_0(T,\phi)=B\left[\exp^{C(T_0-T_{core})}-1\right]+\left[6500\phi^{2.85}\right]$
Basal shear stress	τ	10,080	10,080	10,080	Pa	$\tau=d\rho g\sin(\theta)$
At-Vent (Starting) Velocity						
Mean velocity	V_{mean}	1.3	0.8	0.4	m/s	Equation (21.1c)
Effusion Rate						
Bulk effusion rate	E_r	8.0	4.7	2.2	m^3/s	$E_r=dwv_{mean}$

of at-vent channel depth and underlying slope, equation (21.1) can be solved to yield a mean velocity for lava at the head of the channel, as done in Tables 21.1 and 21.2. This now allows at-vent effusion rate (E_r) to be calculated from

$$E_r = dwv_{mean} \tag{21.4a}$$

which reduces to

$$E_r = d^2 v_{mean} \tag{21.4b}$$

for a square channel and

$$E_r = \pi d^2 v_{mean} \tag{21.4c}$$

for a semicircular channel.

In practice, a user may also have a measurement for E_r. Thus, a desired E_r can be input to the program and FLOWGO iterates equant (depth=width) at-vent channel dimensions until they, in combination with the underlying slope of the first 10 DEM pixels, yield the desired E_r. Beyond the first 10 DEM pixels, depth is held constant but width is allowed to vary so as to conserve volume (see below).

21.2.1. Down-flow Variation in Rheological Conditions

The main variables now controlling variations in flow velocity down the channel are slope, viscosity, and yield strength. These, in turn, can be linked to changes in lava internal temperature (T_{int}) and crystallinity (ϕ) through application of appropriate cooling and rheological models.

Viscosity can be defined for a mixture of liquid and solids (crystals) through the Einstein-Roscoe relationship whereby bulk viscosity can be defined as a function of crystallinity following [*Einstein*, 1906; *Roscoe*, 1952]

$$\eta(\phi)=\eta_f(1-R\phi)^{-2.5} \tag{21.5}$$

in which η_f is the dynamic viscosity of the fluid and $R=1/\phi_{max}$, ϕ_{max} being the maximum crystal content that the lava can attain before flow is rheologically impossible. In the original version of FLOWGO, η_f was calculated as a function of temperature (T) following [*Dragoni*, 1989]

$$\eta_f(T)=\eta_0 \exp^{A(T_0-T_{int})}, \tag{21.6}$$

η_0 being dynamic viscosity at the liquidus temperature (T_0) and A is a constant. By inserting equation (21.6) into (21.5), bulk viscosity can be written as a function of temperature and crystallinity:

$$\eta(T,\phi)=\eta_f(T)(1-R\phi)^{-2.5}, \tag{21.7a}$$

so that

$$\eta(T,\phi)=\eta_0 \exp^{A(T_0-T_{int})}(1-R\phi)^{-2.5} \tag{21.7b}$$

Likewise, yield strength can be written as a function of temperature and crystallinity following [*Dragoni*, 1989; *Pinkerton and Stevenson*, 1992]

$$\tau_0(T,\phi)=B\left[\exp^{C(T_0-T_{int})}-1\right]+\left[6500\phi^{2.85}\right]. \tag{21.8}$$

Now, given down-flow cooling and crystallization, viscosity and yield strength can be calculated as a function of flow internal temperature and crystal content.

These treatments ignore the effect of bubbles forming and deforming in the flowing lava. The bubble content may be more significant than the phenocryst content. For example, for the Mauna Loa 1984 lavas the at-vent phenocryst content was between 0 and 7% on the first day of the eruption, 25 March, increasing to 8%–15% during 26–28 March [*Lipman and Banks*, 1987]. In contrast, *Lipman and Banks* [1987] obtained maximum bubble contents of 85% and described near-vent lava as "fluffy near–reticulite" with "the consistency of egg white". The presence of bubbles in the lava mixture can increase the bulk viscosity if the bubbles are non-deformed and spherical or decrease the bulk viscosity if the bubbles are deformed and sheared [*Manga et al.*, 1998]. Hence, for mixtures of fluid and bubbles, *Pal* [2003] gives two forms of equation (21.5):

$$\eta(\phi_b)=\eta_f\left[1-\frac{\phi_b}{\phi_{b,max}}\right]^{-\phi_{b,max}} \tag{21.9a}$$

for regimes in which the bubbles are spherical and

$$\eta(\phi_b)=\eta_f\left[1-\frac{\phi_b}{\phi_{b,max}}\right]^{\frac{5\phi_{b,max}}{3}} \tag{21.9b}$$

in cases where the bubbles are sheared. Here, $\eta(\phi_b)$ is the viscosity of the fluid-bubble mixture, ϕ_b is the vesicularity, and $\phi_{b,max}$ is the maximum bubble content the mixture can attain before flow is rheologically impossible. If we set $\phi_{b,max}$ to unity, equations (21.9) and (21.10) reduce to [*Llewellin and Manga*, 2005]:

$$\eta(\phi_b)=\eta_f[1-\phi_b]^{-1} \tag{21.9c}$$

and

$$\eta(\phi_b)=\eta_f[1-\phi_b]^{5/3} \tag{21.9d}$$

However, a lava will be a mixture of fluid, bubbles, and crystals. Hence a three-phase treatment is likely more appropriate. *Phan-Thien and Pham* [1997] introduced a treatment for the viscosity of a three-phase mixture comprising a suspension of rigid spheres and bubbles. They give three cases:

Case 1 Crystals smaller than bubbles:

$$\eta(\phi,\phi_b)=\eta_f\left[1-\frac{\phi}{1-\phi_b}\right]^{-5/2}(1-\phi_b)^{-1}. \tag{21.10a}$$

Case 2 Crystals and bubbles of the same size range:

$$\eta(\phi,\phi_b)=\eta_f[1-\phi-\phi_b]^{\frac{(5\phi-2\phi_b)}{(2\phi-\phi_b)}}. \tag{21.10b}$$

Case 3 Crystals larger than bubbles:

$$\eta(\phi,\phi_b)=\eta_f\left[1-\frac{\phi_b}{1-\phi}\right]^{-1}(1-\phi)^{-5/2}. \tag{21.10c}$$

When applied using chemical, temperature, crystallinity, and vesicularity data from Mauna Loa's 1984 lavas, results that are in excellent agreement with the velocity-based viscosity estimates of *Moore* [1987] are obtained [*Harris and Allen*, 2008].

21.2.2. Down-flow Variation in Thermal Conditions

To apply these rheological relationships, we need to estimate the down-flow evolution of the interior temperature and crystallinity of the lava flowing in the channel. Both rely on an estimate of down-flow cooling, which can be obtained by applying a heat budget model [e.g., *Danes*, 1972; *Park and Iversen*, 1984; *Crisp and Baloga*, 1994; *Keszthelyi and Self*, 1998]. The heat budget applied by FLOWGO incorporates heat lost by radiation, convection, conduction, and rain with heat gained due to heat of crystallization and viscous dissipation. These heat losses and gains, plus the budget that is rearranged next, is given in Table 21.3, with input constants and variables being defined in Table 21.4. The heat budget given in Table 21.3 can be rearranged to estimate cooling per unit distance ($\delta T/\delta x$) from

$$\frac{\delta T}{\delta x}=\frac{-Q_{\text{rad}}-Q_{\text{conv}}-Q_{\text{rain}}-Q_{\text{cond}}+Q_{\text{visc}}}{E_r\rho L(\delta\phi/\delta T)} \quad (21.11)$$

Heat loss due to entrainment of chilled surface crust can also be included within the model, as well as surface crust thickness, as detailed in Appendix A of *Harris and Rowland* [2001].

Cooling per unit distance is used to calculate the new internal temperature after the lava advances by the step distance. It can also be used to estimate the rate of crystallization per unit distance ($\delta\phi/\delta x$) from:

$$\frac{\delta\phi}{\delta x}=\frac{\delta T}{\delta x}\frac{\delta\phi}{\delta T}. \quad (21.12)$$

Crystallization rate is used to determine the new crystal content of the control volume once it has advanced by one step.

21.2.3. Volume Conservation and Stopping Conditions

FLOWGO conserves volume down-channel. That is, no volume is lost to the levees. Thus, the starting (vent-leaving) effusion rate is maintained down-channel. Given down-channel changes in velocity, the effusion rate must be maintained at each channel station by varying channel depth and/or width. We take a simple approach whereby flow depth is held constant at down-channel point x so that width varies as a function of velocity:

$$w(x)=E_r\,/\,dv_{\text{mean}}(x), \quad (21.13)$$

where $v_{\text{mean}}(x)$ is the mean velocity calculated for channel station x and $w(x)$ is the channel width at the same station. Thus, as velocity declines, modeled channel width will increase, and vice versa.

Cooling and crystallization will eventually cause the viscosity and yield strength of the control volume to increase to a point where v_{mean} approaches zero. At this point, forward motion of the control volume stops. This point defines the cooling-limited distance the control volume will attain before freezing, that is, the distance a lava flow will move down a channel before cooling to a point at which additional motion is no longer possible. This point defines one of two potential stopping conditions of the FLOWGO loop. A second potential stopping condition occurs if the temperature of the control volume reaches the solidus of the particular lava composition that is input to the model. The value of the solidus temperature is independent of the FLOWGO model and must be input manually before running the program.

Table 21.3 Heat budget terms used in FLOWGO as modified from Table 3 of *Harris and Rowland* [2001].

Term	Definition	Derivation of flux in W/m
Q_{rad}	Radiative heat loss	$Q_{\text{rad}}=\sigma\varepsilon[fT^4_{\text{crust}}+(1-f)T^4_{\text{hot}}]^{0.25}\ w$
Q_{conv}	Convective heat loss (free or forced)	$Q_{\text{conv}}=h_c[fT^{1.3}_{\text{crust}}+(1-f)T^{1.3}_{\text{hot}}]^{0.75}\ w$
Q_{rain}	Heat loss due to rain vaporization	$Q_{\text{rain}}=\delta R/\delta t\ \rho_{H_2O}\ L_{H_2O}w$
Q_{cond}	Heat loss due to conduction through the flow base and levees	$Q_{\text{cond}}=\kappa_{\text{lava}}\ (\Delta T/\Delta h)\ w$
Q_{visc}	Heat from viscous dissipation	For a semi-circular channel: $Q_{\text{visc}}=\eta\ (v_{\text{mean}}/w)^2\ d$ Where $w>d$ $Q_{\text{visc}}=\eta\ (v_{\text{mean}}/d)^2\ w$
Q_{cryst}	Heat from crystallization	$Q_{\text{cryst}}=(\delta T/\delta x)\ E_r\rho L\ (\delta\phi/\delta T)$
$Q_{\text{in}}=Q_{\text{out}}$	Heat balance	$Q_{\text{rad}}+Q_{\text{conv}}+Q_{\text{rain}}+Q_{\text{cond}}=Q_{\text{visc}}+Q_{\text{cryst}}$ or $Q_{\text{rad}}+Q_{\text{conv}}+Q_{\text{rain}}+Q_{\text{cond}}=Q_{\text{visc}}+(\delta T/\delta x)\ E_r\rho L\ (\delta\phi/\delta i)$

Note: See Table 21.4 for definition of input parameters.

Table 21.4 Terms used for heat budget equations of Table 21.3.

Symbol	Definition	Units
d	Channel depth	m
f	Fractional crust cover	fraction
E_r	Effusion rate	m^3/s
h_c	Convective heat transfer coefficient	W/mK
L	Latent heat of crystallization	J/kg
L_{H_2O}	Latent heat of vaporization plus heat required to heat water to 100°C	J/kg
T_{crust}	Crust (surface) temperature	K
T_{hot}	Hot crack temperature	K
V_{mean}	Mean velocity	m/s
w	Channel width	m
Δh	Height of thermal boundry layer at flow base	m
ΔT	Temperature difference across thermal boundary layer.	K
$\delta R/\delta t$	Rain fall rate	m/s
$\delta T/\delta x$	Cooling with distance	K/m
$\delta \phi/\delta T$	Crystallization per degree cooling	K^{-1}
ε	Emissivity	
η	Lava viscosity	Pa·s
κ_{lava}	Thermal conductivity of lava	W/m·K
ρ	Lava bulk density	kg/m
ρ_{H_2O}	Water density	kg/m
σ	Stefan-Boltzmann constant	W/m·K

21.3. THERMAL AND VISCOSITY MODELS

FLOWGO provides a flexible framework within which thermal and rheological equations can be adjusted to assess the applicability of different thermo-rheological scenarios. The most effective way to achieve this is through adjustment of the thermal structure of the lava surface and/or the input viscosity to produce a best fit with available data.

21.3.1. Thermal Surface Model

Parameters that define the thermal surface from which heat is lost include (i) temperature of the surface crust (the cool component), (ii) temperature of lava exposed at cracks and across shear zones (the hot component), and (iii) fractional percentages of the cool and hot components (see Table 21.3). The better insulated (cooler and more heavily crusted) the surface, the lower the heat losses. Lower cooling rates mean that the control volume will extend farther before freezing, with increases in crystallization, viscosity, and yield strength being slower down-channel than in a high-cooling-rate (poorly insulated) case [*Keszthelyi*, 1995; *Keszthelyi and Self*, 1998]. Thus, crust cover percentage and crust and crack temperature are important model variables that control the distance the control volume will extend and the thermo-rheological changes that it will experience.

FLOWGO uses a two-component thermal surface whereby fraction f of the surface comprises a chilled crust at temperature T_{crust}, with the remainder $(1–f)$ of the surface comprising high-temperature zones at T_{hot}. The integrated temperature of that surface can be described by the effective radiation temperature (T_e) [*Crisp and Baloga*, 1994]:

$$T_e = \left[f T_{crust}^4 + (1-f) T_{hot}^4 \right]^{1/4}, \tag{21.14}$$

Based on analysis of air photographs of the active 1984 Mauna Loa flow where velocity had been measured [*Lipman and Banks*, 1987], we developed an emperical relationship between crust growth and velocity:

$$f(V) = z \exp(y v_{mean}). \tag{21.15}$$

Harris and Rowland [2001] give $z = 0.9$ and $y = -0.16$ for a poorly insulated flow and $z = 1.0$ and $y = -0.00756$ for a more heavily crusted flow. Alternatively, crust cover can be set to be constant down-channel at a user-defined value between zero (crust free, poorly insulated) and unity (complete crust coverage, well insulated). Note that complete crustal coverage is not equivalent to flow in a lava tube [*Rowland et al.*, 2005].

Similarly, crust temperature can be set to be constant or can be varied down-flow using a function whereby crust temperature is related to time since it formed (t) following [*Hon et al.*, 1994]

$$T_{\text{crust}}(t) = m\log(t) + n. \tag{21.16}$$

Appropriate values for m and n for Hawaiian pāhoehoe are given by *Hon et al.* [1994] as –140 and 303, with time being in units of hours and output temperature in degrees Celsius. Within the FLOWGO model, the time that a stable surface crust forming in the vent has had to cool can be calculated from the velocity of and distance traveled by the control volume. To achieve this, we calculate the time taken for the control volume to move across each step:

$$t_{\text{step}} = D_{\text{step}} / v_{\text{step}}, \tag{21.17}$$

where D_{step} is the step distance and v_{step} the mean velocity for that step. Summing these values for each step between the vent and down-flow point x gives the travel time for the control volume between the vent and position x (i.e., t for use in equation (21.16)). Note that this assumes that the crust forms at the vent and survives down-flow to wherever the calculation takes place.

21.3.2. Viscosity Model

Fluid viscosity will vary from case to case depending on composition, water content, and temperature [e.g., *Bottinga and Weil*, 1972; *Shaw*, 1972]. The original FLOWGO model of *Harris and Rowland* [2001] used the Arrhenian model of equation (21.6) to allow the viscosity for the fluid phase of the lava mixture to be estimated as a function of temperature. This requires a value for the constant A that will vary depending on composition. For Hawaiian tholeiitic samples *Shaw* [1969] used a value of 0.02 K^{-1} for A, whereas *Dragoni* [1989] used a value of 0.04 K^{-1} for Etna. Solution of equation (21.6) also requires

Table 21.5 Average compositions used for FLOWGO models applied here with density calculated using Bottinga and Weill [1970].

Oxide	Generic[a]	Etna 91–93[b]	ML 1984[c]	ML 1859[d]	Kīlauea[e]	Units
SiO_2	51.63	47.93	52.32	51.6	51.43	wt%
Al_2O_3	13.12	17.87	13.82	13.37	13.52	wt%
Fe_2O_3	2.58	3.30	0.00	0.00	0.00	wt%
FeO	8.48	6.96	10.99	10.95	10.96	wt%
MgO	8.53	5.40	6.81	8.49	6.76	wt%
CaO	9.97	9.98	10.63	10.07	11	wt%
TiO_2	2.63	1.74	2.11	2.06	2.44	wt%
Na_2O	2.28	3.82	2.36	2.26	2.37	wt%
K_2O	0.49	1.87	0.39	0.41	0.43	wt%
H_2O	0	0.1	0.1	0	0	wt%
Density	2774	2723	2724	2739	2744	kg/m
T_0	1200	1160	1200	1240	1200	°C
T_{erupt}	1140	1065	1140	1216	1154	°C
Slope	2.22	2.27	2.36	2.22	2.2	
Viscosity at T_0	20	42	44	14	19	Pa·s
Viscosity at T_{erupt}	39	129	86	18	30	Pa·s

Note: Also given are the liquidus (T_0) and eruption temperatures (T_{erupt}) plus the slope of the temperature – fluid viscosity relation that these compositions give (using the method of *Shaw* [1972]). Finally the fluid viscosities at T_0 and T_{erupt} calculated using the appropriate slope values in equation (21.18) are given.

[a]Tholeiitic composition for Mauna Loa given in Table 9.6 of *Wilson* [1989] using data taken, in turn, from the *Basaltic Volcanism Study Project* [1981].

[b]Etna 91–93: Mean whole rock composition for Etna's 1991–1993 lava given in Table 3 of *Tonarini et al.* [1995]

[c]ML 1984: Mean whole-rock composition for Mauna Loa's 1984 lava given in Table 4 of *Rhodes* [1988].

[d]ML 1859: Glass composition for Mauna Loa's 1859 lava (mean from near-vent spatter samples) given in Appendix 2 of *Riker et al.* [2009].

[e]Kīlauea: Mean glass composition for upper elevation tube samples obtained for Kīlauea's episode 53 (1994–1997) eruption and given in Appendix 1A of *Thornber* [2001].

a value of the fluid viscosity at its liquidus temperature (η_0). Viscosity at liquidus can be calculated using a range of methods, as reviewed for Hawaiian and Etna lavas by *Harris and Allen* [2008]. One method is that of *Shaw* [1972], whereby fluid viscosity can be calculated as a function of composition and temperature following

$$\ln(\eta_f) = s(10^4/T) - sc_T + c_\eta, \qquad (21.18)$$

in which s is the characteristic slope for the viscosity-temperature relationship of a given multicomponent mixture and c_T are c_η are temperature and viscosity-dependent constants with values of 1.5 and −6.4, respectively. This gives η_0 of 14–44 Pa·s for the Hawaiian samples considered here and 42 Pa·s for the Etna samples (Table 21.5). The resulting $\eta_f(T)$ relationships defined by all three approaches considered here are plotted, for the Etna and Hawai'i cases of Table 21.5, in Figure 21.2.

However, the relationship of *Shaw* [1972] assumes an Arrhenian temperature dependence for viscosity, which may not be valid. As a result, in the updated version of FLOWGO presented here, we have adopted the Vogel-Tammann-Fulcher (VTF) equation as an empirical means of providing a best fit to available temperature and viscosity data for a given melt. This relationship is given by *Mano and Pereira* [2004] as

$$\log\eta(T) = D + \frac{E}{T(K) - F}, \qquad (21.19)$$

D, E and F being adjustable (best-fit) parameters that depend on melt composition [e.g., *Hess and Dingwell*, 1996; *Whittington et al.*, 2001; *Zhang et al.*, 2003; *Giordano et al.*, 2006]. *Giordano and Dingwell* [2003] provide one such relationship for Etna lavas whereby

$$\log_{10}(\eta_f) = -4.643 + \frac{5812.44 - 427.04H_2O}{T_{int}(K) - 499.31 + 28.74\ln(H_2O)}, \qquad (21.20)$$

in which H_2O is the water content in wt% and $T(K)$ is lava temperature in Kelvin. As shown in Figure 21.2a, this provides a somewhat different temperature-dependent viscosity relationship for Etna lavas than that provided by *Dragoni* [1989] or *Shaw* [1972]. The cooling relation can now be used to up-date the lava temperature, estimate the crystal content based on the Table 21.6 look-up table, and calculate the fluid-crystal mixture viscosity by inserting any of the preceding equations as η_f into equation (21.5). These differing models can be folded into FLOWGO as described next.

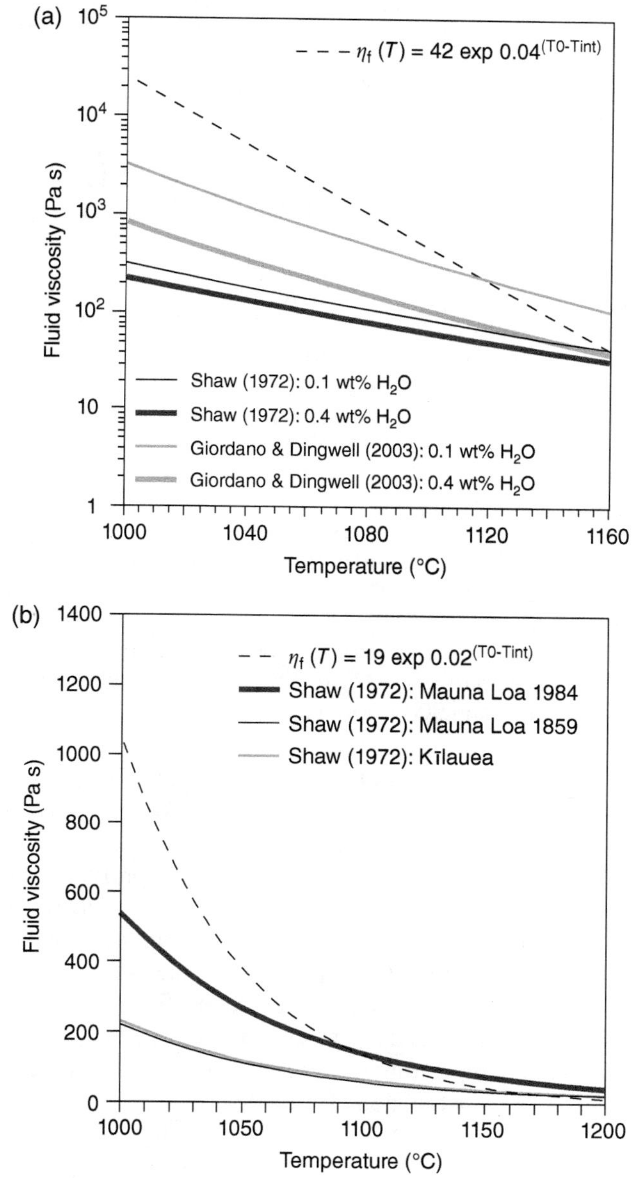

Figure 21.2 Temperature-dependent fluid viscosity relations used for (a) Etna and (b) Mauna Loa and Kīlauea.

21.3.3. Best-Fit Scenarios

The intent of FLOWGO was that the framework of input equations could be adjusted to test the applicability of different thermo-rheological models to different flow emplacement scenarios. For a given set of models, it would also allow thermo-rheological input parameters to be set and tested on a case by case basis (e.g., Tables 21.1 and 21.2). Field data for down-flow variations in channel width, velocity, internal temperature, and/or viscosity allow models and parameters to be adjusted to provide best-fit scenarios that likely describe the thermo-rheological emplacement conditions and their variation down-channel. We illustrate this using an example from Etna.

Table 21.6 Generic crystallization file used by FLOWGO.

Melt Temperature (°C)	Liquid Mass (g)	New Solids (g)	Solid Mass (g)	Total Mass (g)	Microlites Grown ($\delta\phi$)	Crystallization Rate ($\delta\phi/\delta T$)
1140	54.87		44.2	99.07		
1120	44.47	10.41	54.61	99.08	0.1051	5.25×10^{-3}
1100	32.55	11.95	66.56	99.11	0.1206	6.03×10^{-3}
1080	24.75	7.83	74.39	99.14	0.0790	3.95×10^{-3}
1060	20.85	3.92	78.31	99.16	0.0395	1.98×10^{-3}
1040	18.4	2.46	80.77	99.17	0.0248	1.24×10^{-3}
1020	16.63	1.77	82.54	99.17	0.0178	8.92×10^{-4}
1000	15.27	1.37	83.91	99.18	0.0138	6.91×10^{-4}
980	14.21	1.07	84.98	99.19	0.0108	5.39×10^{-4}
960	13.37	0.85	85.83	99.2	0.0086	4.28×10^{-4}
940	12.69	0.69	86.52	99.21	0.0070	3.48×10^{-4}
920	12.13	0.57	87.09	99.22	0.0057	2.87×10^{-4}
900	11.56	0.48	87.57	99.13	0.0048	2.42×10^{-4}
Total		43.37			0.4375	1.82×10^{-3}

Note: File is based on liquid and new solid masses calculated using the generic composition of Table 21.5 in MELTS and running the model in 10°C cooling steps between 1140 and 900°C. The mass fraction of microlites grown over each cooling step is obtained by dividing the new solids grown over that step by the total mass. Crytallization rate is then obtained by dividing the amount of microlites grown over each step (new solids) by the step range (10°C). Over the total cooling range of 240°C, a total mass of 43.37 g of new solids (microlites) are grown to give a volume fraction of crystallization of 0.4375 between 1140 and 900°C for this composition.

The starting conditions for a 1 m deep and 6 m wide channel active on Etna during September 2004 are given in Table 21.2. Excellent constraints on the dimensions of this channel are available, where cross-channel profiles were taken every 10 m down-channel from Lidar data by *Mazzarini et al.* [2007]. *Harris et al.* [2007] used the starting conditions of Table 21.2 to fit model output effusion rate, velocity, and width variation to these Lidar data. A best fit was achieved using a poorly insulated thermal surface over the first 350 m, with a crust temperature varying as a function of time and crust fraction as a function of velocity, following

$$T_{crust}(t) = -140\log(t) + 303 \qquad (21.21a)$$

and

$$f(V) = 1\exp(-0.00756\, v_{mean}). \qquad (21.21b)$$

Beyond 350 m, a well-insulated model was applied in which crust coverage was total (i.e., $f=1$) and crust temperature was constant at 150°C. The resulting fit between model output and Lidar-derived channel width is given in Figure 21.3a.

To obtain this fit, at-vent crystal contents of 25%–30% were required (Figure 21.3a), and these values are lower than the typical phenocryst contents for Etna lavas of 34%±7% [e.g., *Tanguy*, 1973; *Armienti et al.*, 1984; 1994]. Using the upper bounds of the typical range in our model for the 2004 flow causes the fit to break down (Figure 21.3b).

The original best-fit model [*Harris et al.*, 2007] used the viscosity model of equation (21.6), i.e., the relation of *Dragoni* [1989]. As seen in Figure 21.2a, this model may overestimate the fluid viscosity–temperature relation. If instead we use the empirical model derived for Etna lavas by *Giordano and Dingwell* [2003], i.e., the viscosity model of equation (21.20), a fit that can be applied over the full range of expected at-vent crystallinitiy conditions is obtained (Figure 21.3c). We can also cross-check at-vent model conditions, as given in Table 21.2, with available field measurements. Field measurements revealed maximum flow velocities of 0.7 m/s and effusion rates of 2–4 m^3/s at a measurement point ~5 m from the vent (www.ct.ingv.it/Etna2004/Default.htm). The Lidar-derived flow volume of $1.1\times10^6\pm0.4\ m^3$ and an emplacement duration of ~6 days [*Mazzarini et al.*, 2007] also indicated a time-averaged discharge rate of 2.2±0.8 m^3/s. These flow velocities and effusion rates are consistent with our model output values when run with a starting crystal content of 34%–41%, as given in Table 21.2.

Thus, a fit can be obtained if we use a variable thermal surface with appropriate viscosity and crystallinity inputs, with the resulting model-based flow run-outs of 1.6–2.9 km being in line with the mapped channel length of 1.6 km and total (channel-fed) flow unit length of 1.625 km [*Mazzarini et al.*, 2007]. The maximum bound for our model-based estimate may suggest that the flow had not reached its full, cooling-limited potential at the time of Lidar data acquisition. Indeed, the empirical relation

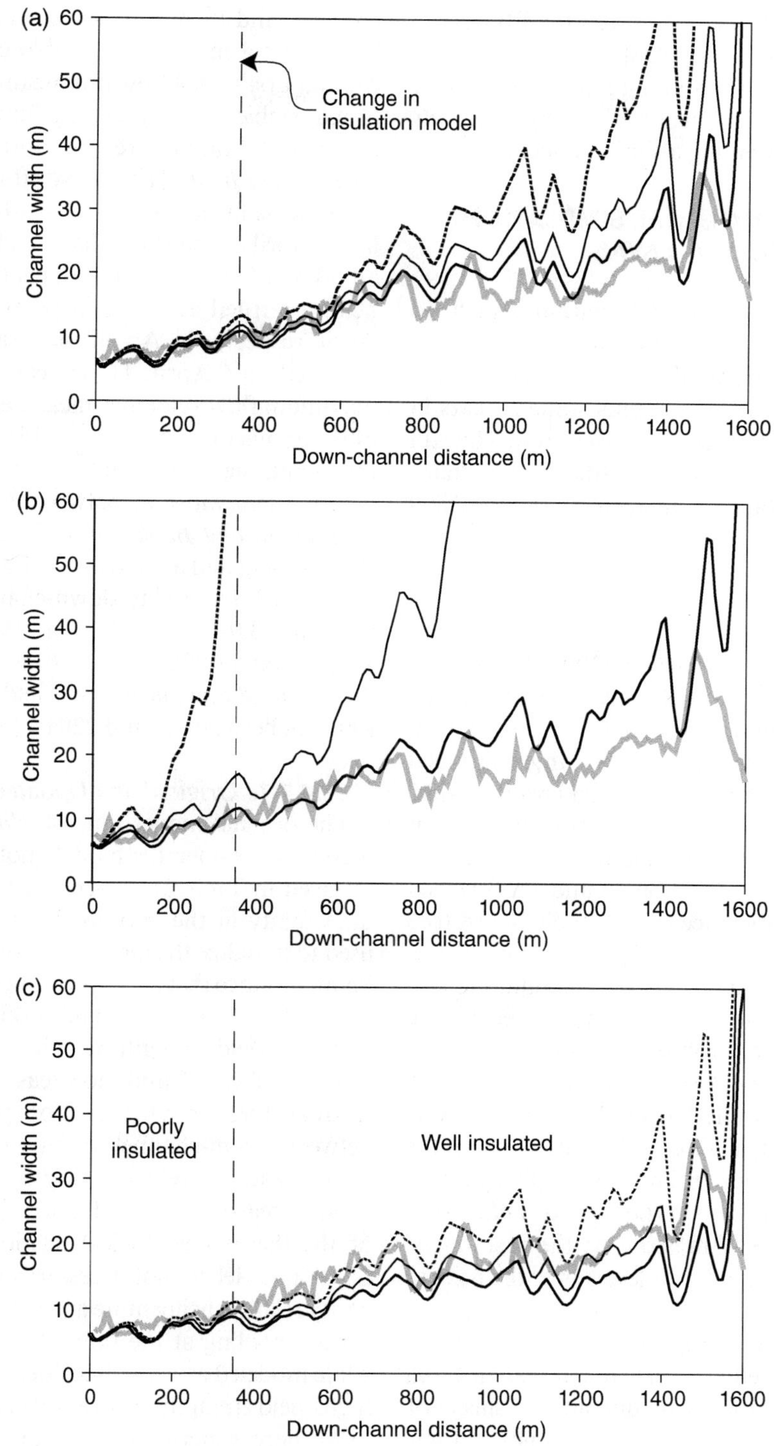

Figure 21.3 FLOWGO-estimated channel widths for Etna's September 2004 channel (black lines) and widths measured from the *Lidar* data of *Mazzarini et al.* [2007] (gray line). (a) Original-best fit of *Harris et al.* (2007) using the starting conditions of Table 21.2, the viscosity model of equation (21.6), and starting crystal (phenocryst) contents of 30% (dashed black line), 27.5% (solid black line, thin), and 25% (solid black line, thick). (b) Fit using the starting conditions of Table 21.2, the viscosity model of equation (21.6), and starting crystal (phenocryst) contents of 41% (dashed black line), 34% (solid black line, thin), and 27% (solid black line, thick). (c) Fit using the starting conditions of Table 21.2, the viscosity model of equation (21.20), and starting crystal (phenocryst) contents of 37% (dashed black line), 34% (solid black line, thin) and 27% (solid black line, thick).

between effusion rate and flow length (L) for Etna lavas given by *Calvari and Pinkerton* [1998], i.e., $L=10^{3.11}E_r^{0.47}$ ($R^2=0.86$) indicates that this flow had the potential to extend 1.8 km at 2 m^3/s and 2.5 km at 4 m^3/s, supporting our run-out estimates and prompting a similar conclusion.

21.4. THERMO-RHEOLOGICAL SIMULATIONS OF HAWAIIAN LAVAS

FLOWGO provides a framework within which to generate thermorheological histories for channel-contained lava control volumes. We demonstrate the best fits that can obtained from FLOWGO plus some caveats in applying the model using three test cases from Hawai'i for which excellent control data exist, these being channel-fed flows active on Mauna Loa during 1984 and 1859 as well as on Kīlauea in 1974.

21.4.1. Mauna Loa: 1984

Initial tests by *Harris and Rowland* [2001] focused on fitting model output values to field measurements made during Mauna Loa's 1984 eruption. This eruption began during the afternoon of 25 March 1984 and continued until 14 April 1984, building a channel-fed lava flow field whose most distant unit extended 26.5 km from the vent [*Lipman and Banks*, 1987]. The two main channel-fed flow units active during the eruption (flows 1 and 1A) reached 90% of their maximum distance during their first 4 to 5 days of activity [*Rowland et al.*, 2005]. Flow 1 was active between 25 and 29 March, reaching its maximum length (~25 km) on 29 March, at which point a levee failure ~13 km behind the flow front robbed flow 1 of supply and began supply to flow 1A [*Lockwood et al.*, 1985]. Flow 1A began to extend rapidly parallel to flow 1 but did not overtake flow 1 until 4 April, reaching its maximum extent the following day, when it too was robbed of supply by a new levee collapse higher in the system [*Lockwood et al.*, 1985]. Effusion rates peaked at ~800 m^3/s during the first day of the eruption, before falling to lower levels over the next 12 days [*Lipman and Banks*, 1987]. *Rowland et al.* [2005] pointed out that early in the eruption, when effusion rate was high, there was no well-established channel and lava lost considerable energy flowing as unconfined sheets of dispersed flow over rough ground. With time, a stable channel developed [*Lipman and Banks*, 1987] to allow more efficient delivery of lava to the flow front. However, by this time the effusion rate had decreased, meaning that the highest effusion rate never coincided with the greatest length of efficient, stable channel.

Because FLOWGO models the thermo-rheological evolution of a control volume moving down an established channel, and not advance of a flow front. FLOWGO is thus initialized with the effusion rates that fed the channel, and then outputs velocities and travel times for lava flowing in the same stable channel. In the case of Mauna Loa's 1984 flow this means that data for the established, stable channel feeding flows 1 and 1A on or after day 4 (29 March) are appropriate as input. Data by *Lipman and Banks* [1987] give effusion rates measured for these flows in the range 100–440 m^3/s between 29 March and 5 April (when flow 1 had reached its maximum extent and flow 1A was active), with peaks of ~1140 and 945 m^3/s being recorded at the main vent outlet on 2 April and 4 April, respectively. An effusion rate of 900 m^3/s was also recorded on 6 April. These peak values corresponded to maximum flow velocities measured in the at-vent channel of 17.65 m/s on 2 April and ~15 m/s on 4 and 6 April. At this point, the channel at the main vent outlet had a depth of 3 m and a width of between 20 and 21.5 m [Table 57.3 of *Lipman and Banks*, 1987]. At the same time, *Moore* [1987] calculated a viscosity of 140 Pa·s for lava at the vent on 2 April, increasing down-channel to 1.0–1.3×10^3 Pa·s at 3 km, 3.0–9.5×10^3 Pa·s at 9 km, 0.9–1.6×10^5 Pa·s at 15 km, and 5.6×10^6 Pa·s at 26 km (at the toe of flow 1A). At the same time, near-vent yield strength was calculated as being between 66 and 220 Pa [*Moore*, 1987].

21.4.1.1. Original and Updated Model Fits

The original model runs of *Harris and Rowland* [2001] used two end -member models, hot and cool. These models, as given in Table 21.7, were set to cover the full range in uncertainty in the various thermo-rheological conditions used to initialize the model. The two models yielded a maximum distance that lava flowing in the channel could extend of 23.7 km (hot model) and 30 km (cool model). At-vent viscosity, yield strength, velocity, and effusion rate are also given in Table 21.7 and show reasonable agreement with the field data for near-vent conditions given above. Comparisons between down-channel variations in model-derived and field-measured velocity, width, and temperature also showed reasonable fits (Figures 21.4a–c). However, some of the thermo-rheological relations used for the original best–fit model are not necessarily appropriate, with the viscosity relation being more appropriate for Etna lava and the surface cooling at the same rate as the core. In addition, while modeled at-vent viscosities and velocities were a little high, yield strengths were a little low (Table 21.7).

We here experiment with an updated version of the model which allows us to maintain these fits while using more appropriate input parameters and thermo-rheological models, as given in Table 21.8. This uses a viscosity relation set for the composition of Mauna Loa's 1984 lava, as given in Table 21.5, and a surface crust that cools with time and distance from the vent. The model output now spans a wider range of distances; 17.5 km (hot model) and 28 km (cool model). However, fits between modeled at-vent viscosity, yield strength,

Table 21.7 Cool and hot models applied to provide best fits to the 1984 Mauna Loa channel conditions by *Harris and Rowland* [2001].

Parameter (term, units)	Model Value Cool	Hot	Source
Channel width (w, m)	5.5	5.5	Mean channel width from all values given in Table 57.3 of *Lipman and Banks* [1987] for channel stations below the 1900 m elevation level
Channel depth (d, m)	5.5	5.5	Square channel ($w=d$)
Air temperature (T_{air}, °C)	10	30	
Eruption temperature (T_{erupt}, °C)	1137	1143	Mean (±1σ) eruption temperature given by *Lipman and Banks* [1987]
Solidus temperature (T_{solid}, °C)	970	990	
Phenocryst content (ϕ_{phen})	0.15	0.00	Range given by *Lipman and Banks* [1987] for the period 25–28 March 1984
Posteruption crystallization (ϕ_{post})	0.45	0.45	Posteruption crystallization calculated using MELTS
Crystallization rate ($\delta\phi/\delta T$)	—	—	Value for linear crystallization model [*Harris and Rowland*, 2001]: $\delta\phi/\delta T=\phi_{post}/(T_{erupt}-T_{solid})$
Crust temperature (T_{crust}, °C)	425	675	Approximate crust temperature range obtained at the margins of an active channel on Kīlauea by *Flynn and Mouginis-Mark* [1994]
Crack temperature (T_{hot}, °C)	A	B	Model A: $T_{hot}=T_{core}-140$; model B: $T_{hot}=T_{core}-0$
Basal contact temperature (T_{base}, °C)	500	900	Based on a value of 700±200°C from *Keszthelyi* [1995] and *Wooster et al.* [1997]
Basal crust thickness (H_b, %)	19	1	
Vesicularity (%)	1	15	Vesicularity range calculated from all bulk density data of *Lipman and Banks* [1987]
Crust cooling model	Fixed at vent: cools down flow as a function of core cooling rate		
Crust coverage model	$f(V)=x\exp(yV_{mean})$, $x=0.9023$, $y=-0.1601$ [*Harris and Rowland*, 2001]		
Rheological model	Arrhenian model (Equation 21.6) with $\eta_0=1000$ Pa·s, $T_0=T_{erupt}$, $A=0.04\,K^{-1}$		
Output			
At-vent viscosity (η, Pa s)	1900	1000	Viscosity for mixture of fluid and crystals
At-vent yield strength (τ_0, Pa)	29	0	Yield strength for mixture of fluid and crystals
At-vent velocity (V_{mean}, m/s)	13.3	21.7	
Effusion rate (E_r, m³/s)	402	656	
Distance (km)	28.77	23.69	

velocity, and effusion rate and field data are improved over the original runs (Table 21.8). These new fits between down-channel variations in model-derived and field-measured velocity, width, and temperature are given in Figures 21.4d–f.

21.4.1.2. The Viscosity Problem

For both the original and updated models, we note a problem when trying to fit down-flow variation in viscosity to the 2 April 1984 field data of *Moore* [1987]. The problem being that we underestimate the expected viscosity at all down-flow points after the vent (Figure 21.5a). The difficulty in matching the viscosities obtained by *Moore* [1987] when using a fluid viscosity calculated using the work of *Shaw* [1972] and the flow crystal content in a mixture viscosity calculation based on the Einstein-Roscoe relation (as given in equation (21.5)) was noted by *Crisp et al.* [1994]. *Crisp et al.* [1994] suggested that this difficulty was likely due to the failure to take into account the influence of bubbles on the mixture viscosity. *Harris and Allen* [2008] thus were able to match all of *Moore*'s [1987] viscosity regimes using appropriate crystallinity and vesicularity values in the three-phase viscosity treatment of *Phan-Thien and Pham* [1997].

Following *Harris and Allen* [2008], we obtain a fit with at-vent viscosity if we use the three-phase model of viscosity as given in equation (21.10) and the starting conditions of the FLOWGO hot model set up for Mauna Loa's 1984 channel (Figure 21.5b). To obtain this at-vent fit, we have to apply a case whereby the vesicles have no effect on the mixture viscosity and use a phenocryst content of 15%, this being the minimum phenocryst content during 2–6 April [*Lipman and Banks*, 1987]. However, hot model viscosity predictions for all

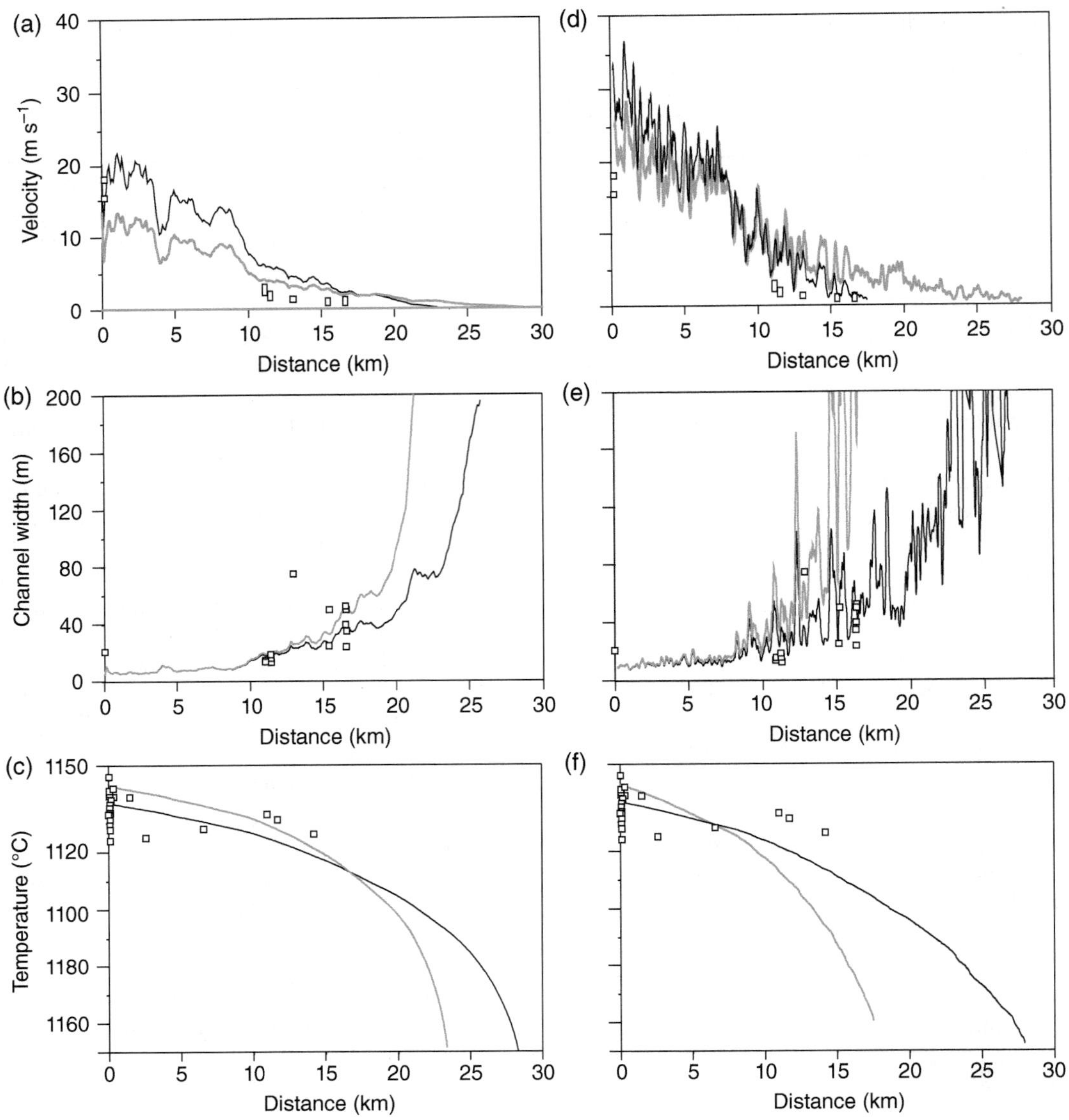

Figure 21.4 Original (a–c) and updated (d–f) FLOWGO runs for Mauna Loa's 1984 channel. Results for velocity, channel width, and interior temperature are given. For both cases results of two end member models are given as calculated using a "hot" model (gray line) and "cold" model (black line). Down-channel field measurements for velocity, channel width, and lava interior temperatures are from *Lipman and Banks* [1987] and are given using open squares. The original model is run using the initialization parameters given in Table 21.7, with an Arrhenian viscosity relation set for Etna's lava, as used by *Harris and Rowland* [2001] for initial model tests on this channel. The updated model is run using the initialization parameters given in Table 21.8, with an Arrhenian viscosity relation set using *Shaw* [1972] and a chemistry appropriate for this lava, as given in Table 21.5.

other points down-channel remain far too low and the flow run out distance is too far. To obtain a fit with viscosity between 3 and 9 km requires application of the cold model and the three-phase viscosity treatment applied using a bubble content of 43% and phenocryst content of 29% (Figure 21.5b). Both are values that are within the bounds of those measured at this channel, with 30% being the maximum phenocryst content given by *Lipman and Banks* [1987] and 43% being the maximum vesicularity calculated from the density measurements made on channel samples by *Lipman and Banks* [1987]. However, we can only achieve the fit if we assume a Newtonian flow regime. That is, we have to assume that the lava has no yield strength and drop the second term of equations (21.1b,c), thus reverting to equation (21.1a). After 10 km we can only obtain a fit if we also alter the thermal surface model to one of well-insulated flow conditions whereby the flow surface has a total coverage of crust at 125°C after this point (Figure 21.5c).

Table 21.8 Updated FLOWGO model run with Mauna Loa 1984 vent channel dimension data and parameter ranges for 2–6 April when flow 1A was active.

Parameter (term, units)	Model Value Cool	Hot	Source
Channel width (w, m)	21	21	Typical channel width and depth at 2850 m vent outlet given in Table 57.3 of *Lipman and Banks* [1987] for the period 2–6 April 1984 (days 8–12)
Channel depth (d, m)	3	3	
Air temperature (T_{air}, °C)	10	30	
Eruption temperature (T_{erupt}, °C)	1137	1143	Mean (±1σ) eruption temperature given by *Lipman and Banks* [1987]
Solidus temperature (T_{solid}, °C)	970	990	
Phenocryst content (ϕ_{phen})	0.14	0.25	Range given by *Lipman and Banks* [1987] for the period 2–6 April 1984
Posteruption crystallization (ϕ_{post})	0.45	0.45	Posteruption crystallization calculated using MELTS
Crystallization rate ($\delta\phi/\delta T$)	0.0027	0.0029	Value for linear crystallization model [*Harris and Rowland*, 2001]: $\delta\phi/\delta T = \phi_{post} / (T_{erupt} - T_{solid})$
Crust temperature (T_{crust}, °C)	Variable	Variable	Calculated as a function of time equation (21.13)
Crack temperature (T_{hot}, °C)	A	B	Model A: $T_{hot} = T_{core} - 140$; Model B: $T_{hot} = T_{core} - 0$
Basal contact temperature (T_{base}, °C)	500	900	Based on a value of 700 ± 200°C from *Keszthelyi* [1995] and *Wooster et al.* [1997]
Basal crust thickness (H_b, %)	19	1	
Vesicularity (%)	14	43	Vesicularity range calculated from bulk density data for channel samples given in Table 57.1 of *Lipman and Banks* [1987]
Crust cooling model	Crust cools as a function of distance and time from vent Equations (21.16) and 21.17		
Crust coverage model	$f(V) = x\ \exp(yV_{mean})$, $x = 0.9$, $y = -0.16$ [*Harris and Rowland*, 2001]		
Rheological model	Model of *Shaw* [1972], i.e. equation (21.18) with the slope calculated using the ML1984 chemistry of Table 21.5		
Output			
At-vent mixture viscosity (η, Pa·s)	292	151	Viscosity for mixture of fluid and crystals
At-vent yield strength (τ_0, Pa)	126	24	Yield strength for mixture of fluid and crystals
At-vent velocity (V_{mean}, m/s)	10.3	13.8	
Effusion rate (E_r, m³/s)	647	869	
Distance (km)	27.95	17.46	

This suggests a complex thermo-rheological regime, which requires application of a hybrid model that evolves down-channel if we are to obtain the best possible fit with all available field data. We suggest that at least three flow thermo-rheological regimes existed on 2 April: (1) a poorly insulated regime over the first 3 km across which vesicularities were high but vesicles were sheared so that they had little effect on the mixture viscosity; (2) a poorly insulated regime between 3 and 10 km across which vesicularities where high but vesicles were less sheared so that that they served to increase the mixture viscosity as the crystal content also increased; and (3) an insulated flow regime over the remaining channel length with both vesicles and crystals contributing to the viscosity gain. The need for such a down-channel-evolving thermo-rheological regime was suggested by *Moore* [1987], who stated that "it is entirely possible that the Mauna Loa lava obeyed different flow laws at different places and times the lava may have been a Newtonian fluid at the vents, a Bingham fluid at station 8, and a pseudoplastic fluid or some other kind of fluid in other reaches" [*Moore*, 1987, pp. 1555–1586]. In addition, to account for a change in bulk volume flux between the vent and 15 km, the lava bulk density had to have increased by a factor near 12 [*Moore*, 1987], pointing to a down-channel decrease in vesicularity alongside a cooling-induced increase in crystallinity, with maximum downstream crystallinity (of 39%, comprising 25% microlites and 14% microphenocrysts) being found in a quenched sample 14 km from the vent by *Crisp et al.*

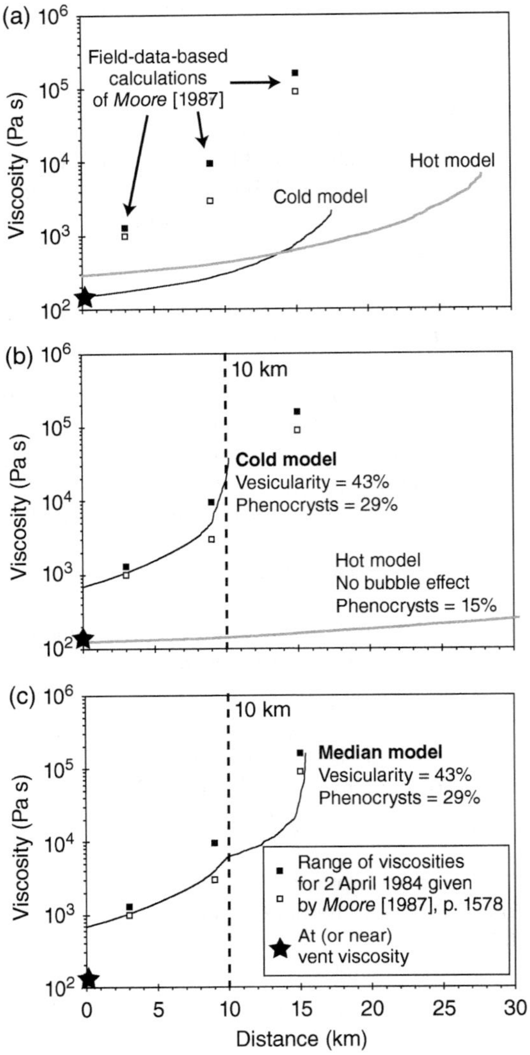

Figure 21.5 FLOWGO viscosity best-fit simulations for Mauna Loa's 1984 channel. (a) *Shaw* [1972] in the Einstein-Roscoe equation with a 15% phenocryst content and the time-dependent crust cooling (and hence heat loss) model of equation (21.16). (b) *Shaw* [1972] in the three-phase viscosity treatment of *Phan-Thien and Pham* [1997] with the time-dependent crust cooling (and hence heat loss) model of equation (21.16). (c) *Shaw* [1972] in the three-phase viscosity treatment of *Phan-Thien and Pham* [1997] with the time-dependent crust cooling (and hence heat loss) model until 10 km and an insulated model of complete crust coverage at 125°C thereafter. Models in (b) and (c) are Newtonian.

[1994]. This compares with a total crystallinity of 28%–31% obtained using our hot model at a distance of 14–15 km (14%–17% microlites, 14% microphenocrysts), with 25% microlites being grown by 17 km. Hence, the fluid-crystal-bubble mixture viscosity would have evolved as a function of not just vesicularity but also crystallinity.

Finally, at ~15 km *Moore* [1987] described a flow surface that was "a hummocky mass of slowly moving debris, rubble and blocks" compared with a more poorly crusted channel with incandescent zones nearer the vent. This indicates a change in the thermal surface character from poorly to well insulated between the proximal and medial-distal reaches. Clearly, the thermal and rheolgical conditions, as well as the associated flow regimes, changed significantly down-channel, necessitating application of a highly flexible thermo-rheological model, as we have had to do here.

21.4.2. Mauna Loa: 1859

Mauna Loa's 1859 eruption produced a 51 km long channel-fed flow field fed at effusion rates of ~391 m^3/s during the period of channel lengthening [*Riker et al.*, 2009]. This case can be used to demonstrate the best-fit approach in applying FLOWGO.

Initial fits between expected down-channel variation in lava temperature and crystallinity and those obtained using FLOWGO by *Riker et al.* [2009] were poor (Figure 21.6). Given that these fits were achieved using an unrealistic crystallization model as well as an inappropriate thermal and rheological model, the failure is not too surprising. As discussed by *Riker et al.* [2009], the default MELTS-based crystallization model used by FLOWGO is only appropriate up to temperatures of 1140°C (see Table 21.6), where eruption temperatures were as high as 1216°C [*Riker et al.*, 2009]. The MELTS-based crystallization model of FLOWGO also considers a tholeiitic composition for Mauna Loa as given in Table 9.6 of *Wilson* [1989] using data taken, in turn, from the *Basaltic Volcanism Study Project* [1981]. In addition, while the viscosity model used in the model runs of *Riker et al.* [2009] was appropriate to an Etna lava, the thermal (and crystallization) regime appeared to change down-channel. Interior cooling, for example, appeared to show a rather steep declining trend until 10 km and a trend of ~0.5°C/km thereafter (Figure 21.6). We therefore prefer to produce a best fit to this flow using appropriate input data and a hybrid model.

It is apparent from the data of *Riker et al.* [2009] that there were two cooling and crystallization regimes: (1) rapid cooling and crystallization between the vent and 10 km and (2) slower cooling and crystallization thereafter. Over the first 10 km, temperatures dropped from 1216 to 1160°C at a rate of 5.6°C/km, thereafter showing a steady, slower decline to 1146°C by 36 km at a rate of 0.5°C/km [*Riker et al.*, 2009]. Thus, we set up a hybrid model that applies one set of (poorly insulated) thermal and crystallization conditions above 10 km and a second (well-insulated) set of conditions below 10 km. Also, we need to adapt the viscosity model to be appropriate for the temperatures and composition relevant to this flow. To do this, we use the method of *Shaw* [1972], as given

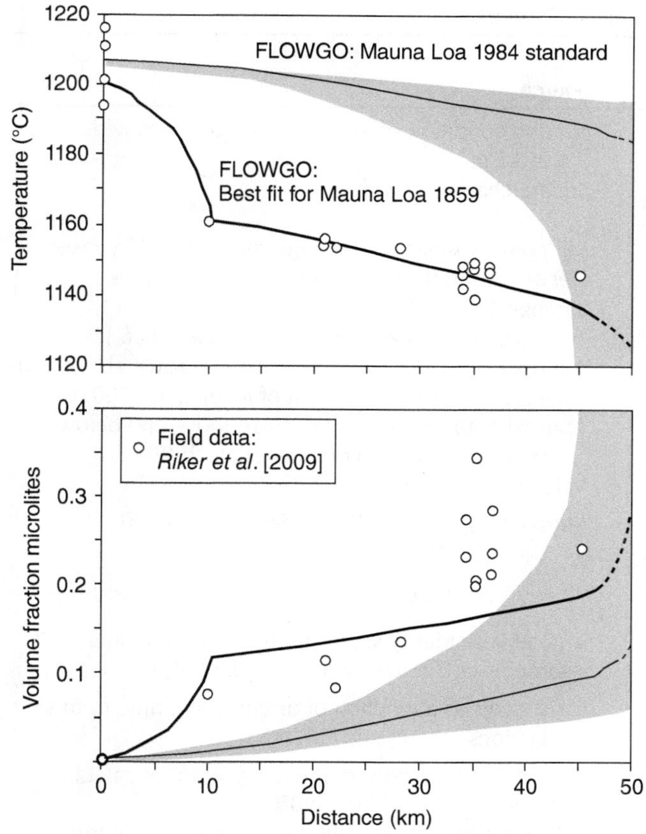

Figure 21.6 Comparison of FLOWGO model results and field data from the channel-fed phase of Mauna Loa's 1859 eruption as given by *Riker et al.* [2009] in their Figure 14. Input parameters are the standards set for Mauna Loa's 1984 eruption by *Harris and Rowland* [2001], and shaded regions show range of model results obtained by *Riker et al.*[2009] when input viscosity, vesicularity, and starting channel dimensions were varied within reasonable limits (10–1000 Pa·s, 8–40%, and 4.0–4.5 m, respectively). Comparison is given with a new hybrid FLOWGO run set up to provide a best fit to the field data of *Riker et al.* [2009], where (a) is a fit with down-flow interior temperature versus distance traveled and (b) fits the model to the volume fraction microlites grown versus distance.

here in equation (21.18), with a slope calculated for the composition of Mauna Loa's 1859 lavas as given in Table 21.5. Our revised FLOWGO input parameters for Mauna Loa's 1859 channel are given in Table 21.9. If we apply this hybrid model, we achieve excellent fits with the expected down-channel variation in crystallinity and temperature (Figure 21.6) as well as viscosity (Figure 21.7), producing a run out of 45.6 km at an effusion rate of 399 m^3/s and interior temperatures at 10 and 36 km of 1159 and 1145°C, respectively. In achieving this fit, as with our Etna 2006 and Mauna Loa 1984 fits, the most crucial adjustments were to input (i) an appropriate crystallization model, (ii) an appropriate viscosity model, and (iii) a thermal surface whose character varied down-channel between a poorly insulated proximal reach and a well-insulated medial-distal reach.

21.5. FLOW PATH AND RUN-OUT PROJECTIONS OF HAWAIIAN LAVAS

The initial FLOWGO model used the simple line of steepest descent, or deterministic eight-neighbor (D8) model [*Mark*, 1984; *O'Callaghan and Mark*, 1984], to set a flow line. This approach is not dissimilar to standard watershed models used in GIS to determine the channels water will take to exit a drainage basin via the highest order stream. In FLOWGO, the D8 model is corrected to allow flow lines to escape pits, basins and data drop-out holes in the DEM as well as to extend across flat zones [*Fairfield and Leymarie*, 1991; *Turcotte et al.*, 2001; *Jones*, 2002] so that a line of steepest descent is projected from the selected lava source (vent) to the edge of the DEM. FLOWGO then uses this line to extract a slope profile and uses this to estimate each thermo-rheological parameter at each point along the line.

Later, the flow path model was upgraded through combination with the DOWN-FLOW model of *Favalli et al.* [2005]. The DOWN-FLOW model finds the line of steepest descent, adds random noise *to* the DEM (which is equal to the DEM vertical resolution), and then finds the new path. This new path will be slightly different from the first due to the slight alteration of the DEM by addition of noise. The process is repeated until a field of flow paths is generated. This, in effect, provides the range in uncertainty of the flow line projection given that we are working with a quantized version of the terrain, with limits set (and potential topographic error being induced) by both the pixel size and DEM vertical resolution. We then run FLOWGO down each potential line of steepest descent.

We next consider output from the combined FLOWGO-DOWN-FLOW model for three cases for which we have reasonable control on effusion rates, down-flow thermo-rheological parameters, and flow length channel-fed lava flow units: Mauna Loa's 1984 flow units plus the 1859 flow and as Kīlauea's 1974 flow. We note that we are using DEMs in which these flow fields already exist as topographic highs. They will thus influence the flow line, which will not follow the same path as the original flow. Thus, our objective is not to re-create these flows but to assess the validity of the combined thermorheologcal and flow line projections. We are thus looking at fictional flows that have similar properties to the existing flows and are flowing over roughly similar terrain (albeit with the 1859, 1974, and 1984 flows in place). These new flow paths may differ from the historical paths but will be influenced by similar topography (e.g., regional slopes, valleys, hills, cones, preexisting local flow margins, levees, and fronts).

Table 21.9 FLOWGO input parameters for Mauna Loa's 1859 channel simulation.

Parameter (term, units)	Model Value <10 km	>10 km	Source
Channel width (w, m)	4	—	Starting condition of *Riker et al.* [2009] uses width of 4.0–4.5 m
Channel depth (d, m)	2	—	Square channel ($w=d$)
Air temperature (T_{air}, °C)	20	—	
Eruption temperature (T_{erupt}, °C)	1205	—	Midpoint of range of temperatures obtained by *Riker et al.* [2009] from lower vent spatter samples (range 1194–1216°C)
Phenocryst content (ϕ_{phen})	0.02	0.02	Typical values from Appendix 3 of *Riker et al.* [2009]
Posteruption crystallization (ϕ_{post})	0.05	0.3	Approximate crystal volume fractions grown above and below 10 km [see Figure 6a of *Riker et al.*, 2009]
Temperature range (δT, °C)	60	20	Approximate temperature decline above and below 10 km [see Figure 6b of *Riker et al.*, 2009]
Crystallization rate ($\delta\phi/\delta T$)	0.00083	0.015	Value for linear crystallization model: $\delta\phi/\delta T = \phi_{post}/\delta T$
Crack temperature (T_{hot}, °C)	B	B	Model A: $T_{hot} = T_{core} - 140$; model B: $T_{hot} = T_{core}$ 0
Basal contact temperature (T_{base}, °C)	500	500	
Basal crust thickness (H_b, %)	19	19	
Vesicularity (%)	40	25	Typical vesicularity values given for 'a'a channel samples in Figure 9 of *Riker et al.* [2009]
Crust cooling model	PI	—	Crust cools as a function of distance and time from vent Equations (21.16) and (21.17)
	—	WI	Stable (constant) crust temperature at value calculated for the 10 km position (480°C)
Crust coverage model	PI	—	$f(V) = x \exp(yV_{mean})$, $x = 0.9023$, $y = -0.04778$: poorly crusted
	—	WI	$f(V) = x \exp(yV_{mean})$, $x = 0.9023$, $y = -0.03652$: more heavily crusted
Rheological model	Model of *Shaw* [1972], i.e. Equation (21.18) with the slope calculated using the ML1859 chemistry of Table 21.5		

Note: Models: PI = poorly Insulated, WI = well Insulated.

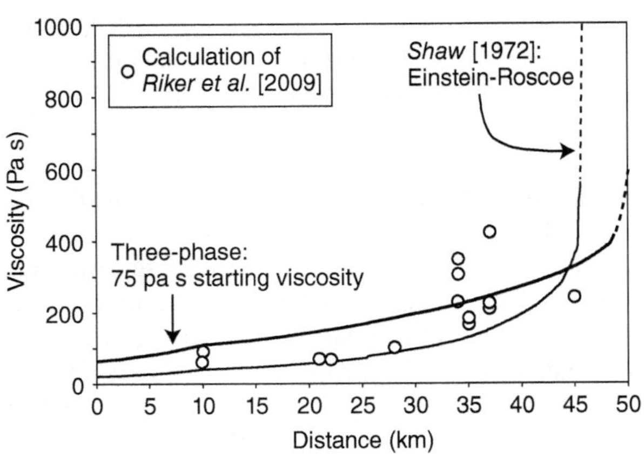

Figure 21.7 FLOWGO viscosity best-fit simulations for Mauna Loa's 1859 channel. Model is run using (i) a two-phase (melt and crystal) viscosity model and (ii) a three phase (melt, crystal, and bubble) viscosity model. Output is compared with the viscosities expected by *Riker et al.* [2009].

21.5.1. Mauna Loa: 1984

If we pick a vent location at the lower end of the Mauna Loa 1984 fissure system, we obtain a flow line projection that follows flow 1 for about a third of its total length (~8 km). At this point, the flow line falls off of flow 1 and takes a path parallel to the northern edge of the 1984 lava flow field, with the control volume advancing 26.7 km down that path to stop just short of the flow 1A toe (Figure 21.8a). To obtain this length, the model requires an effusion rate of 500 m^3/s and gives at-vent velocity of 16.7 m/s.

The problem is, our DEM is from 2000 and therefore contains the 1984 flow field, which, locally, creates a topographic high. In such a case, on reaching the edge of the lava pile, the flow line projection will drop off of that pile and then follow the base of its marginal levee. Thus, we cannot re-create the path of the 1984 flow using this DEM because the flow itself is already there. However, we do project the likely path that a new flow erupted from a similar vent position and subject to the topographic control exerted by the presence of the 1984 flow field will now follow.

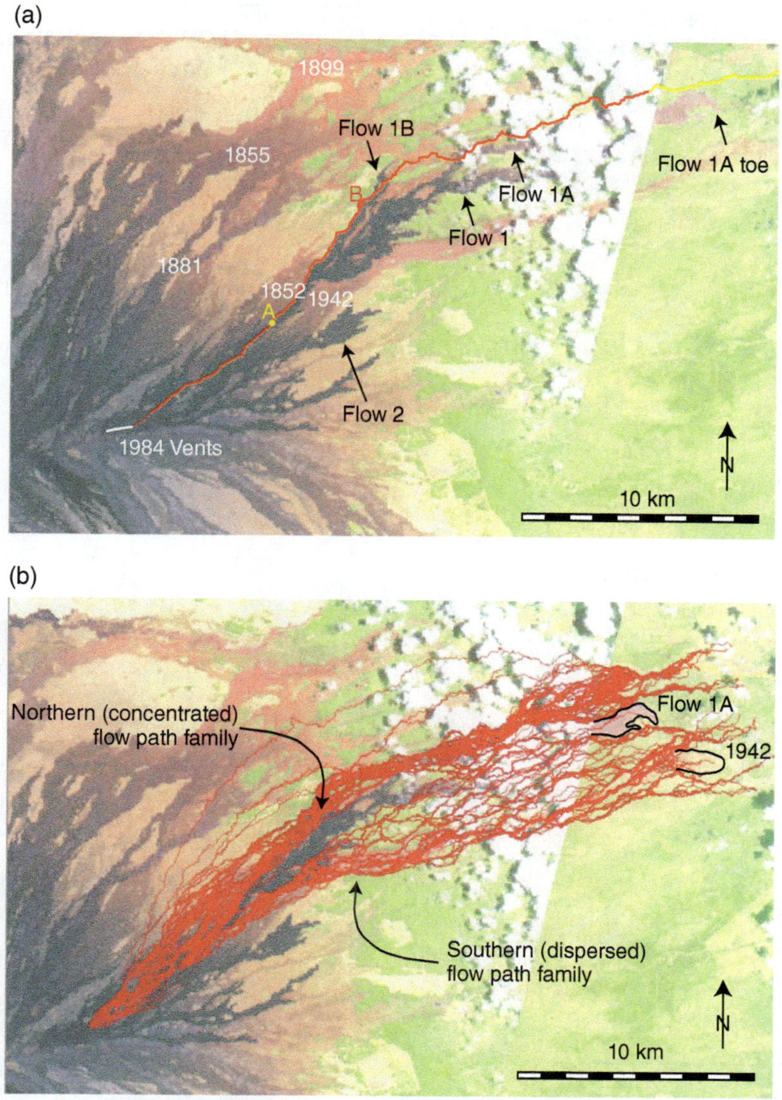

Figure 21.8 (a) FLOWGO single-path and (b) FLOWGO-DOWN-FLOW projections for a control volume advancing from the lower portion of Mauna Loa's 1984 eruptive fissure with the starting conditions of Table 21.1. (a) Yellow line projects the line of steepest descent between the designated vent position to the edge of the DEM. The red line gives the distance down the projection a FLOWGO control volume will advance. At point *A*, the projected path falls off of the local topographic high presented by the 1984 flow field and at *B* the projection encounters small basin, which is filled with the flow line leaving from the lowest elevation point around the basin rim. (b) Path and runout results for a 1000 iteration run at an effusion rate of 300 m^3/s, with DEM noise of 2 m and a channel aspect ratio of 1. In both cases, base image is a 30 m spatial resolution Thematic Mapper color composite linked to the underlying DEM, and flow units are located following Figure 57.1b of *Lipman and Banks* [1987].

We next run the flow path model iteratively (1000 iterations, DEM noise 2 m, square channel) from a vent location toward the lower end of the 1984 eruptive fissure and in the vicinity of the vent that fed the 1942 flow. At an effusion rate of 300 m^3/s, a rate typical of that feeding Mauna Loa's 1984 channel [*Rowland et al.*, 2005], we find that the flow run-outs approximate the distances obtained by both the 1984 and 1942 flows, to give an average run-out of 27.4 km (Figure 21.8b). We note that the family of paths that extend in the direction of the 1984 flow field divide to flow down either side of that flow field, so that two families of flow paths develop: to the north and south of the 1984 flow field, respectively (Figure 21.8b). This is a result of the 1984 flow field now being a local topographic high, around which a fluid will flow. To the south, the family of flow lines is somewhat more dispersed and encompasses the path of the 1942 flow. To the north, the flow line distribution is somewhat more concentrated, defining a dense zone of flow path concentration down the northern margin of the 1984 flow field. The presence of the 1984 flow field also creates a series of shadow zones, including a zone just downslope of the flow front (Figure 21.8b).

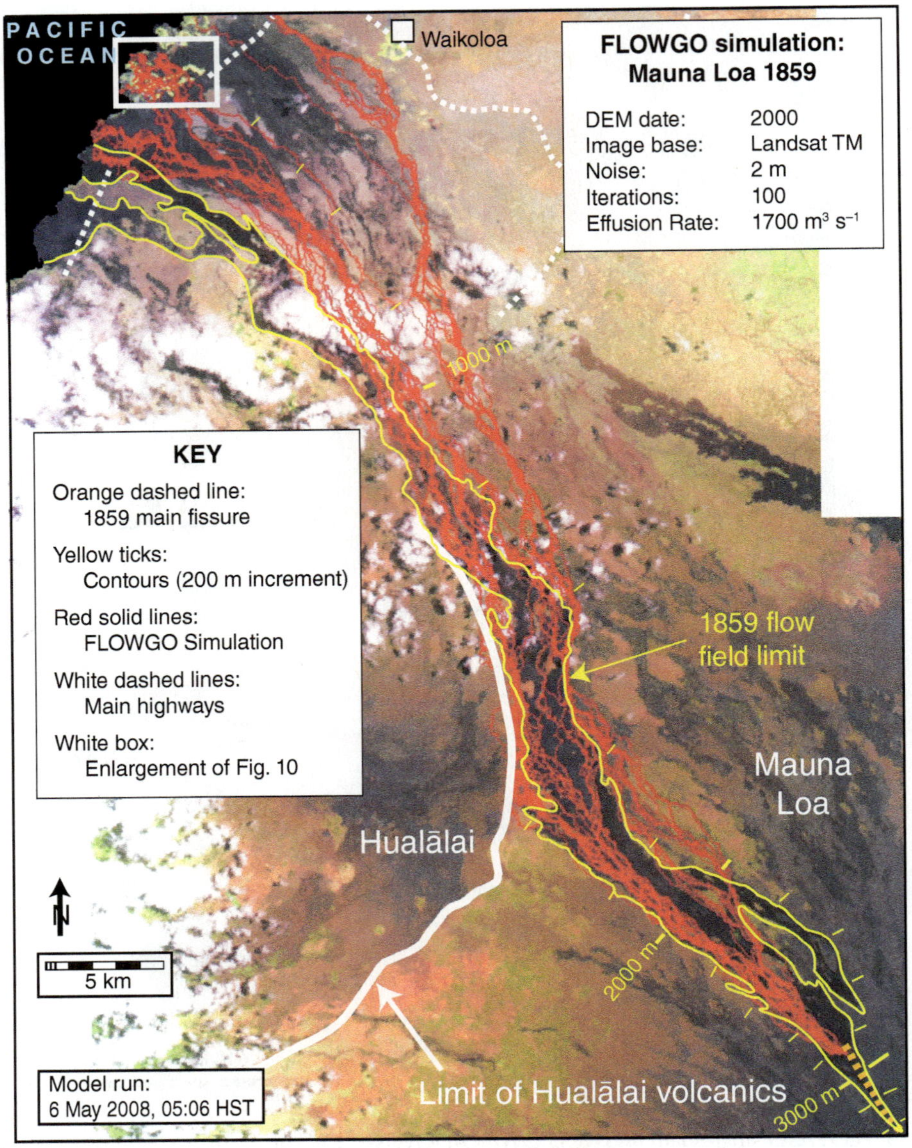

Figure 21.9 FLOWGO-DOWN-FLOW simulation of Mauna Loa's 1859 flow. Base image is a 30 m spatial resolution Thematic Mapper color composite linked to the underlying DEM and approximate flow field perimeter is located following Figure 3 of *Riker et al.* [2009]. White box locates the coastal section of the simulation as enlarged in Figure 21.10.

21.5.2. Mauna Loa: 1859

Running the DOWN-FLOW path model iteratively (100 iterations, DEM noise 2 m, square channel) from a vent location within Mauna Loa's 1859 eruptive fissure, we obtain a family of flow paths that matches the area of the actual flow field quite well (Figure 21.9). One family of flow lines strays to the north of the actual flow. Again, this is a result of the flow field being present in the DEM and thus influencing the post-1859 path projections, deflecting them north once they have fallen off the flow itself.

Across the coastal zone (after a down-flow distance of ~45 km and within ~5 km of the coast), slopes decline to less than 1° (average 0.4°). An effusion rate of 1000 m^3/s ensures that the control volume arrives at the back of the coastal zone, giving run-outs that stop just short of Queen Ka'ahumanu highway. An increase to 1700 m^3/s is required if the control volume is to advance the final 2–3 km, over the flat coastal zone, and enter the ocean. Across the coastal zone the family of flow lines also begins to spread, as the actual flow field did, to cut a broad swath of the main highway serving Hawai'i's Kona and Kohala coasts and covering land now

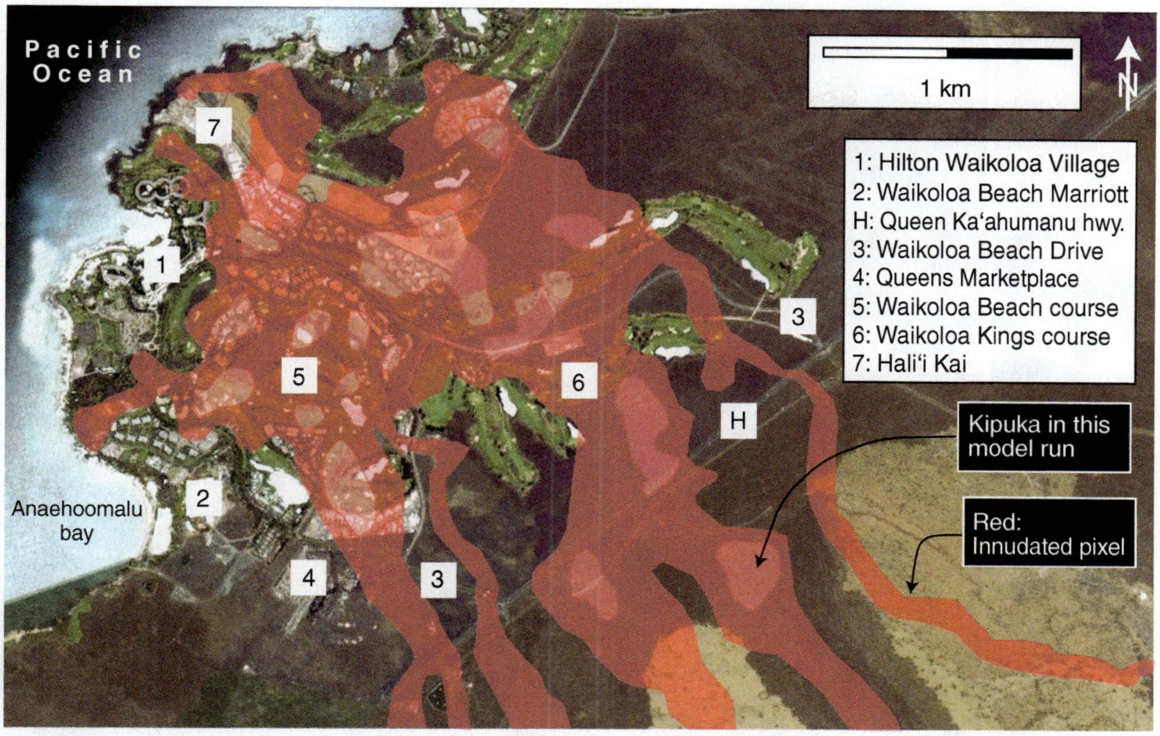

Figure 21.10 Enlargement of the coastal section of the FLOWGO-DOWN-FLOW simulation of Mauna Loa's 1859 flow given in Figure 21.9. Simulation is overlain on the Google Digital Globe [©2010] image of the Waikoloa resort.

occupied by Kona resorts and their associated golf courses (Figure 21.10).

The greens on the golf courses actually cause the flow lines to congregate or be deflected. While some lines follow the fairway edges (where they are raised), others are drawn down the line of the course, ponding within the greens or jumping from green to green. This is a result of the man-modified terrain that comprises topographic highs and lows in the surrounding lava that was locally cleared, heaped, and/or graded to construct the course. The two courses (Beach and Kings) were opened in 1981 and were thus not present during the emplacement of the 1859 flow. As a result, such modern man-made topography would not have had an influence on the 1859 flow field emplacement, the change in slope at the coast being the primary influence on the flow field form to cause distal spreading. However, the golf courses could influence flow paths should a similar effusive event occur today, with the courses possibly guiding the flows between and/or around the main resort structures, which are located just to the north and west (Hilton Waikoloa Village) as well as to south (Waikoloa Beach Marriott Resort & Spa) of the two courses. Certainly, all flow paths skirt north of the Waikoloa Beach Marriott Resort & Spa following the line of the Beach course, which also keeps the same flow paths just south of the Hilton Waikoloa Village.

A clever but unintentional piece of resort design to guard against lava flow inundation in a hazardous location? Of course, there is plenty of infrastructure and real estate within the two courses themselves (especially the Beach course), but could well placed and well-designed golf courses and tourist infrastructure be used as subtle defense structures in zones prone to lava inundation?

21.5.3. Kīlauea: 1974

Kīlauea's December 1974 eruption began from a ~700 m long fissure that opened in the upper East Rift Zone. First fountains were sighted at 02:56 Hawaiian standard time (HST). These fed "extremely fluid" lava that built a flow field with a total volume of $5.9\times10^6\,m^3$ [*Lockwood et al.*, 1999]. The eruption ended around 08:50 HST the same day to give a mean output rate over the ~6 hour-long event of $270\,m^3/s$; peak effusion rates were probably higher. The 12.4 km long flow contained a 10 km long channelized segment, down which cooling rates were 1°C/km across medial to distal portions [*Soule et al.*, 2004].

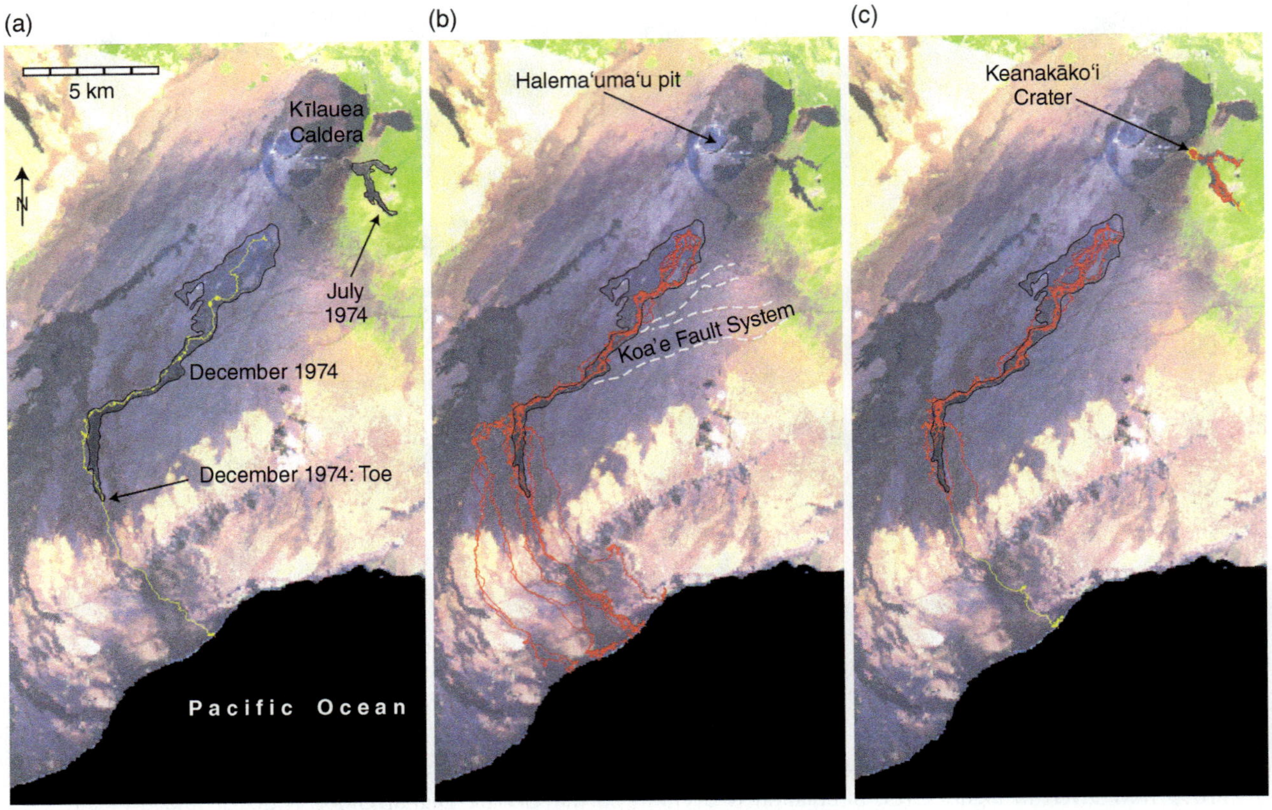

Figure 21.11 FLOWGO-DOWN-FLOW simulation of Kīlauea's July and December 1974 flows whose approximate flow field perimeters are outlined in black, as apparent on the Thematic Mapper color composite underlay. (a) Line of steepest descent from the December 1974 vent zone to the coast (line in yellow). (b) FLOWGO-DOWN-FLOW simulations of the December 1974 flow (red lines, iterations 20, noise 3 m, effusion rate 500 m^3/s), with main fault scarps (down throw to NNE) as apparent from shadows in the image marked using dashed white lines. (c) FLOWGO-DOWN-FLOW simulation of the July and December 1974 flows (red lines, iterations 20, noise 3 m, effusion rate 50 m^3/s). Modified from *Harris* [2013].

The line of steepest descent for this flow (Figure 21.11a) follows the flow field center line, without falling off of the topographic "high" caused by the flow field being present in the DEM, as was the case for our Mauna Loa 1984 flow path. This is probably due to the very thin character of this flow field, lava flows being typically less than 1 m thick [*Lockwood et al.*, 1999]. This thickness is less than the 3 m vertical resolution of the DEM, so that the flow is too thin to be properly registered and is thus not apparent in our DEM,

Run iteratively at 500 m^3/s, we obtain an array of flow lines all of which reach the ocean, attaining run-outs with an average length of 21.7 km (Figure 21.11b). Flow lines are well concentrated within the actual flow field area over the first ~8 km of flow, following the north – northeast facing fault scarps of the Koa'e Fault System in a southeast direction. Thereafter, flow lines turn south and fan out as they extend directly downslope to the ocean. Over the first 10 km slopes are rather shallow being, on average, 2.15°, thereafter having an average of 5.9°.

Reducing the effusion rate used in the model to 270 m^3/s still yields flow lines that attain the coast. However, over the first 10 km of the model run average velocity is 3.3 m/s (5.9 m/s at the vent), and we have a flow interior that cools by 12°C, to give a typical cooling rate of 1.2°C/km. This cooling rate is the same as that calculated by *Soule et al.* [2004] for this flow, and velocities are in line with the 1–6 m/s range estimated for this flow by *Lockwood et al.* [1999] and *Soule et al.* [2004]. To obtain the correct length, we have to reduce the effusion rate used in the model to 50 m^3/s (Figure 21.11c). This now gives a channel length of ~10 km and suggests that this flow was volume limited. That is, supply was terminated before the flow had attained its full, potential, cooling-limited extent. This seems consistent with (1) the short duration of the eruption, (2) the fact that the lava interior had only cooled by 10°C after 10 km, and (3) the flow field having only a length of 12 km when effusion rates were at least 270 m^3/s. FLOWGO modeling indicates that, had supply to this flow not been cut after 6 h, the channel-fed flow would have been capable of extending up to 22 km, thereby reaching the coast.

21.6. CONCLUSION

FLOWGO provides a framework within which to review and link thermo-rheological lava properties and relations and produce thermodynamic histories for channel-contained lava control volumes. This aim is best achieved through a best fit with all available field and sample data, and the current version of FLOWGO allows a high degree of manipulation and iteration to allow thermo-rheological and dimensional fitting, as we have done here for some Hawaiian lavas. This model is for an idealized lava flow and uses relatively simple models for cooling, rheology, and flow dynamics. The approach is thus intended to provide useful insights into these phenomena. FLOWGO is not meant to accurately re-create a natural flow. Simulations that allow lava to spread, thicken, and evolve require application of cellular automata models, such as SCIARA (*Crisci et al.*, 2004), MAGFLOW (*Del Negro et al.*, 2008), or LavaSIM (*Hidaka et al.*, 2005). In these models, the thermal and rheological conditions of each pixel across the flow field evolve with time as well as distance from the source (*Harris*, 2013).

Clearly, the thermal and rheological conditions as well as the associated flow regimes change significantly down a lava channel [e.g., *Soule et al.*, 2004; *Woodcock and Harris*, 2006; *Riker et al.*, 2009]. This points to a complex flow regime that requires application of an extremely sophisticated and highly flexible model if we are to obtain perfect fits between model output and natural data. To obtain a best fit with all dimensional, dynamic, thermal, and rheological variables, the thermo-rheological conditions as well as the relations themselves may have to be changed down-channel as flow conditions and regimes change. In addition, we need to generate relations that are currently not available. For FLOWGO, the most obvious improvement would be addition of a down-flow vesiculation model that allows the size, shape, and number of vesicles to be calculated at each step as well as a rheological treatment that varies depending on whether Newtonian, Bingham, or viscoelastic conditions are appropriate. For now, the approximations provided by FLOWGO indicate that the relations and input parameters remain reasonable starting points on which we can build.

Our ability to model is limited by our ability to parameterize and link the physics and dynamics behind the phenomena in question. Lava is a challenging fluid to model, and FLOWGO is an attempt to link available knowledge and provide a framework within which we can identify the holes in our understanding.

REFERENCES

Armienti, P., F. Barberi, F. Innocenti, M. Pompilio, R. Romano, and L. Villari (1984), Compositional variation in the 1983 and other recent Etnean lavas: Insights on the shallow feeding system, *Bull. Volcanol.*, *47*(2), 995–1007.

Armienti, P., R. Clocchiatti, M. D'Orazio, F. Innocenti, R. Petrini, M. Pompilio, S. Tonarini, and L. Villari (1994), The long-standing 1991–1993 Mount Etna eruption: Petrography and geochemistry of lavas, *Acta Vulcanol.*, *4*, 15–28.

Bailey, J. E., A. J. L. Harris, J. Dehn, S. Calvari, and S. K. Rowland (2006), The changing morphology of an open lava channel on Mt. Etna, *Bull. Volcanol.*, *68*, 497–515.

Basaltic Volcanism Study Project (1981) Basaltic Volcanism on the Terrestrial Planets, Pergamon Press, New York.

Bottinga, Y., and D. F. Weill (1970), Densities of liquid silicate systems calculated from partial molar volumes of oxide components, *Am. J. Sci.*, *269*, 169–182.

Bottinga, Y., and D. F. Weill (1972), The viscosity of magmatic silicate liquids: A model for calculation, *Am. J. Sci.*, *272*, 438–475.

Calvari, S., and H. Pinkerton (1998), Formation of lava tubes and extensive flow field during the 1991–1993 eruption of Mount Etna, *J. Geophys. Res.*, *103*(B11), 27,291–27,301.

Crisci, G. M., S. Di Gregorio, R. Rongo, and W. Spataro (2004), The simulation model SCIARA: The 1991 and 2001 at Mount Etna, *J. Volcanol. Geotherm. Res.*, *132*, 253–267.

Crisp, J., and S. Baloga (1994), Influence of crystallization and entrainment of cooler material on the emplacement of basaltic aa lava flows, *J. Geophys. Res.*, *99*(B6), 11,819–11,831.

Crisp, J., K.V. Cashman, J. A. Bonini, S.B. Hougen, and D. C. Pieri (1994), Crystallization history of the 1984 Mauna Loa lava flow, *J. Geophys. Res.*, *99*(B4), 7177–7198.

Daneš, Z. (1972), Dynamics of lava flows, *J. Geophys. Res.*, *77*, 1430–1432.

Del Negro, C., L. Fortuna, A. Herault, and A. Vicari, (2008), Simulations of the 2004 lava flow at Etna volcano by the magflow cellular automata model, *Bull. Volcanol.*, doi:10.1007/s00445-007-0168-8.

Dragoni, M. (1989), A dynamical model of lava flows cooling by radiation, *Bull. Volcanol.*, *51*, 88–95.

Einstein, A. (1906), Eine neue bestimmung der molekuldimensionen, *Ann. Phys.*, *19*, 289–306.

Fairfield, J., and P. Leymarie (1991), Drainage networks from grid digital elevation models, *Water Resourc. Res.*, *27*(5), 709–717.

Favalli, M., M. Pareschi, A. Neri, and I. Isola (2005), Forecasting lava flow paths by a stochastic approach, *Geophys. Res. Lett.*, *32*(L03305), 1–17.

Flynn, L. P. and P. J. Mouginis-Mark (1994), Temperature measurements of an active lava channel from spectral measurements, Kilauea Volcano, Hawai'i. *Bull. Volcanol.*, *56*, 297–301.

Ganci, G., A. Vicari, A. Cappello, and C. Del Negro (2012), An emergent strategy for volcano hazard assessment: From thermal satellite monitoring to lava flow modeling, *Remote Sens. Environ.*, *119*, 197–207.

Giordano, D., and D. B. Dingwell (2003), Viscosity of hydrous Etna basalt: Implications for Plinian-style basaltic eruptions, *Bull. Volcanol.*, *65*, 8–14.

Giordano, D., A. Mangiacapra, M. Potuzak, J. K. Russell, C. Romano, D. B. Dingwell, and A. Di Muro (2006), An expanded non-Arrhenian model for silicate melt viscosity: A treatment for metaluminous, peraluminous and peralkaline liquids, *Chem. Geol.*, *229*, 42–56.

Harris, A., J. Bailey, S. Calvari, and J. Dehn (2005), Heat loss measured at a lava channel and its implications for down-channel cooling and rheology, *Geol. Soc. Am. Spec. Pap.*, *396*, pp. 125–146.

Harris, A., M. Favalli, F. Mazzarini, and M. T. Pareschi (2007), Best-fit results from application of a thermo-rheological model for channelized lava flow to high spatial resolution morphological data, *Geophys. Res. Lett.*, *34*(L01301), doi:10.1029/2006GL028126.

Harris, A. J. L. (2013), Lava flows, in Modeling Volcanic Processes: The Physics and Mathematics of Volcanism, Chapter 5, edited by S. A. Fagents, T. K. P. Gregg, and R. M. C. Lopes, Cambridge University Press, pp. 85–106.

Harris, A. J. L., and J. S. Allen, III (2008), One-, two- and three-phase viscosity treatments for basaltic lava flows, *J. Geophys. Res.*, *113*, B09212, doi:10.1029/2007JB005035.

Harris, A. J. L., and S. K. Rowland (2001), FLOWGO: A kinematic thermo-rheological model for lava flowing in a channel, *Bull. Volcanol.*, *63*(1), 20–24.

Herault, A., A. Vicari, A. Ciraudo, and C. Del Negro (2009), Forecasting lava flow hazards during the 2006 Etna eruption: using the MAGFLOW cellular automata model. *Comp. Geosci.*, *35*, 1050–1060, doi:10.1016/j.cageo.2007.10.008.

Heslop, S. E., L. Wilson, H. Pinkerton, and J. W. Head III (1989), Dynamics of a confined lava flow on Kilauea volcano, Hawaii. *Bull. Volcanol.*, *51*, 415–432.

Hess, K-U., and D. B. Dingwell (1996), Viscosities of hydrous leucogranitic melts: A non-Arrhenian model, *Am. Minerol.*, *81*, 1297–1300.

Hidaka, M., A. Goto, S. Umino, and E. Fujita (2005), VTFS project: Development of the lava flow simulation code LavaSIM with a model for three-dimensional convection, spreading, and solidification, *Geochem. Geophys. Geosyst.*, *6*, Q07008, doi:10.1029/2004GC000869.

Hon, K., J. Kauahikaua, R. Denlinger, and K. Mackay (1994), Emplacement and inflation of pahoehoe sheet flows: Observations and measurements of active lava flows on Kilauea Volcano, Hawaii, *Geol. Soc. Am. Bull.*, *106*, 351–370.

Hulme G. (1974), The interpretation of lava flow morphology, *Geophys. J. R. Astron. Soc.*, *39*, 361–383.

Ishihara, K., M. Iguchi, and K. Kamo (1990), Numerical simulation of lava flows on some volcanoes in Japan. In *Lava Flows and Domes*, edited by J. Fink, Berlin: Springer, pp. 184–207.

Jeffreys, H. (1925), The flow of water in an inclined channel of rectangular section, *Philos. Mag.*, *49*, 793–807.

Jones, R. (2002), Algorithms for using a DEM for mapping catchment areas of stream sediment samples, *Comput. Geosci.*, *28*, 1051–1060.

Keszthelyi, L. (1995), A preliminary thermal budget for lava tubes on the Earth and planets, *J. Geophys. Res.*, *100*, No.(B10), 20,411–20,420

Keszthelyi, L., and S. Self (1998), Some physical requirements for the emplacement of long basaltic lava flows, *J. Geophys. Res.*, *103*(B11), 27,447–27,464.

Lipman, P. W., and N. G. Banks (1987), Aa flow dynamics, Mauna Loa 1984, U.S. Geol. Surv. Prof. Pap 1350, pp. 1527–1567.

Llewellin, E. W., and M. Manga (2005), Bubble suspension rheology and implications for conduit flow, *J. Volcanol. Geotherm. Res.*, *143*, 205–217.

Lockwood, J. P., N. G. Banks, T. T. English, L. P. Greenland, D. B. Jackson, D. J. Johnson, R.Y. Koyanagi, K.A. McGee, A. T. Okamura, and J. M. Rhodes (1985), The 1984 eruption of Mauna Loa volcano, Hawaii, *Eos Trans. AGU*, *66*, 169–171.

Lockwood, J. P., R. I. Tilling, R. T. Holcomb, F. Klein, A. T. Okamura, and D.W. Peterson (1999), Mama migration and resupply during the 1974 summit eruptions of Kilauea volcano, Hawaii, U.S. Geol. Surv. Prof. Pap., 1613, 37 pp.

Manga, M., J. Castro, K. V. Cashman, and M. Loewenberg (1998), Rheology of bubble-bearing magmas, *J. Volcanol. Geotherm. Res.*, *87*, 15–28.

Mano, J. F., and E. Pereira (2004), Data analysis with the Vogel-Fulcher-Tammann-Hesse equation, *J. Phys. Chem.*, *108*, 10,824–10,833.

Mark, D.M. (1984), Automated detection of drainage networks from digital elevation models, *Cartographica*, *21*, 168–178.

Mazzarini, F., M. T. Pareschi, M. Favalli, I. Isola, S. Tarquini, and E. Boschi (2007), Lava flow identification and aging by means of LiDAR intensity: The Mt. Etna case, *J. Geophys. Res*, *112*, B02201, doi:10.1029/2005JB004166.

Moore, H. J. (1987), Preliminary estimates of the rheological properties of 1984 Mauna Loa Lava, U.S. Geol. Surv. Prof. Pap 1350, pp. 1569–1588.

O'Callaghan, J.F. and D.M. Mark (1984), The extraction of drainage networks from digital elevation data, *Computer Vision, Graphics and Image Processing*, *28*, 323–344.

Pal, R. (2003), Rheological behavior of bubble-bearing magmas, *Earth Planet. Sci. Lett.*, *207*, 165–179.

Park, S., and J. D. Iversen (1984), Dynamics of lava flow: Thickness growth characteristics of steady two-dimensional flow, *Geophys. Res. Lett.*, *11*(7), 641–644.

Phan-Thien, N., and D. C. Pham (1997), Differential multiphase models for polydispersed suspensions and particulate solids, *J. Non-Newtonian Fluid Mech.*, *72*, 305–318.

Pinkerton, H., and R. J. Stevenson (1992), Methods of determining the rheological properties of magmas at sub-liquidus temperatures, *J. Volcanol. Geotherm. Res.*, *53*, 47–66.

Rhodes, J. M. (1988), Geochemistry of the 1984 Mauna Loa eruption: Implications for magma storage and supply, *J. Geophys. Res.*, *93*(B5), 4453–4466.

Riker, J. M., K.V. Cashman, J. P. Kauahikaua, and C. M. Montierth (2009), The length of channelized lava flows: Insight from the 1859 eruption of Mauna Loa volcano, Hawai'i, *J. Volcanol. Geotherm. Res.*, *183*, 139–156.

Roscoe, R. (1952), The viscosity of suspensions of rigid spheres, *Br. J. Appl. Phys.*, *3*, 267–269.

Rowland, S. K., A. J. L. Harris, and H. Garbeil (2004), Effects of Martian conditions on numerically modeled, cooling-limited, channelized lava flows, *J. Geophys. Res.*, *109*, E10010, doi: 10.1029/2004JE002288, 1–16.

Rowland, S., H. Garbeil, and A. Harris (2005), Lengths and hazards from channel-fed lava flows on Mauna Loa, Hawai'i, determined from thermal and downslope modeling with FLOWGO, *Bull. Volcanol.*, *67*, 634–647.

Shaw, H. R. (1969), Rheology of basalt in the melting range, *J. Petrol.*, *10*(3), 510–535.

Shaw, H. R. (1972), Viscosities of magmatic silicate liquids: an empirical method of prediction, *Am. J. Sci.*, *272*, 870–893.

Soule, S. A., K. V. Cashman, and J. P. Kauahikaua (2004), Examining flow emplacement through the surface morphology of three rapidly emplaced, solidified lava flows, Kilauea volcano, Hawai'i, *Bull. Volcanol.*, *66*, 1–14.

Tanguy, J. C. (1973), The 1971 Etna eruption: Petrology of the lavas, *Phios. Trans. R. Soc. Lond.*, *274*, 45–53.

Thornber, C. R. (2001), Olivine-liquid relations of lava erupted by Kilauea volcano from 1984 to 1998: Implications for shallow magmatic processes associated with the ongoing east-rift-zone eruption, *Can. Minerol.*, *39*, 239–266.

Tonarini, S., P. Armienti, M. D'Orazio, F. Innocenti, M. Pompilio, and R. Petrini (1995), Geochemical and isotopic monitoring of Mt. Etna 19891993 eruptive activity: Bearing on the shallow feeding system, *J. Volcanol. Geotherm. Res.*, *64*, 95–115.

Turcotte, R., Fortin, J.–P., Rousseau, A. N., Massicotte, S. and J.–P. Villeneuve (2001), Determination of the drainage structure of a watershed using a digital elevation model and a digital river and lake network, *J. Hydrol.*, *240*, 225–242.

Vicari, A., A. Herault, C. del Negro, M. Coltelli, M. Marsella, and C. Proietti (2007), Modeling of the 2001 lava flow at Etna volcano by a Cellular Automata approach, *Environmental Modelling and Software*, *22*, 1464–1471.

Vicari, A., A. Ciraudo, C. del Negro, A. Herault, and L. Fortuna (2009), Lava flow simultations using discharge rates from thermal infrared satellite imagery during the 2006 Etna eruption, *Natural Hazards*, *50*, 539–550.

Vicari, A., G. Ganci, B. Behncke, A. Cappello, M. Neri, and C. del Negro (2011), Near-real-time forecasting of lava flow hazards during the 12–13 January 2011 Etna eruption, *Geophys. Res. Lett.*, *38*, L13317, doi: 10.1029/2011GL047545.

Whittington, A., P. Richet, Y. Linard, and F. Holtz (2001), The viscosity of hydrous phonolites and trachytes, *Chem. Geol.*, *174*, 209–223.

Wilson, M. (1989), Igneous Petrology: A Global Tectonic Approach, Unwin Hyman, London, 466 pp.

Woodcock, D., and A. Harris (2006), The dynamics of a channel-fed lava flow on Pico Partido volcano, Lanzarote, *Bull. Volcanol.*, *69*, 207–215.

Wooster, M., R. Wright, S. Blake, and D. Rothery (1997), Cooling mechanisms and an approximate thermal budget for the 1991–1993 Mount Etna eruption, *Geophys. Res. Lett.*, *24*(24), 3277–3280.

Wright, R., H. Garbeil, and A. J. L. Harris (2008), Using infrared satellite data to drive a thermo-rheological/stochastic lava flow emplacement model: A method for near-real-time volcanic hazard assessment, *Geophys. Res. Lett.*, *35*, L19307, doi:10.1029/2008GL035228.

Young, P. and G. Wadge (1990), FLOWFRONT: Simulation of a lava flow, *Comp. Geosci.*, *16*, 1171–1191.

Zhang, Y., Z. Xu, and Y. Liu (2003), Viscosity if hydrous rhyolitic melts inferred from kinetic experiments, and a new viscosity model, *Am. Minerol.*, *88*, 1741–1752.

22

Lava Flows in 3D: Using Airborne Lidar and Preeruptive Topography To Evaluate Lava Flow Surface Morphology and Thickness in Hawai'i

Hannah R. Dietterich[1,2], S. Adam Soule[3], Katharine V. Cashman[4], and Benjamin H. Mackey[5]

ABSTRACT

Until recently, lava flow maps have been restricted to two dimensions and, largely, flow outlines rather than internal features. Airborne laser swath mapping (ALSM, or lidar), with its high spatial resolution and intensity data, offers new ways to map lava flows, as demonstrated by recent work in Italy. We extend this approach to Hawai'i and combine ALSM with preeruptive photogrammetry of the well-observed Mauna Loa 1984 and Kīlauea December 1974 lava flows to map both surface features and along-flow distributions of lava volume. Lidar intensity records progressive biological colonization of flows as a function of both elevation and flow age. Surface roughness can be used to map along-flow variations in flow type (e.g., the pāhoehoe to 'a'ā transition) and internal features such as channels and levees, or to develop automated classification schemes. The spatial and volumetric distributions of lava reflect the effusion rate and interactions with topography. Key observations are that the main channel serves to transport, rather than store, lava and that preexisting topography exerts a primary control on the (3D) spatial distribution of individual lava flows. In summary, ALSM data provide new tools for mapping lava flows and new constraints on lava flow emplacement.

22.1. INTRODUCTION

Mafic lava flows represent a persistent and widespread form of volcanic activity that affects communities around the world. Central to lava flow hazard assessment and mitigation are the construction of probabilistic flow hazard maps [e.g., *Favalli et al.*, 2005, 2009a,b, 2012; *Tarquini and Favalli*, 2010] and development of tools for real-time prediction of flow paths, flow advance rates, and final flow lengths. Both hazard assessment and predictive modeling require high-resolution digital topography as well as a fundamental understanding of the ways in which advancing lava flows interact with Earth's surface. For this reason, volcanologists have embraced new methods of imaging the landscape, focusing particularly on airborne lidar because of its combination of high spatial resolution and extensive spatial coverage [e.g., *Mazzarini et al.*, 2005; *Pyle and Elliot*, 2006; *Coltelli et al.*, 2007; *Ventura and Vilardo*, 2008; *Marsella et al.*, 2009; *Favalli et al.*, 2010; *Deardorff and Cashman*, 2012; *Tarquini et al.*, 2012]. New multitemporal imaging of active lava flows is now providing remarkable spatial and temporal coverage of the flow emplacement process [e.g., *Marsella et al.*, 2009; *Favalli et al.*, 2010]. Lidar data also provide important information on lava flow surfaces in the form of both intensity data and surface roughness. The combination of high-resolution topography and quantifiable surface characteristics has the potential to transform our understanding of lava flow emplacement. That this potential has not yet been reached attests to the need for both new analytical tools and

[1]*Department of Geological Sciences, University of Oregon, Eugene, Oregon, USA*

[2]*Volcano Science Center, U.S. Geological Survey, Menlo Park, California, USA*

[3]*Geology and Geophysics Department, Woods Hole Oceanographic Institution, Woods Hole, Massachusetts, USA*

[4]*School of Earth Sciences, University of Bristol, Bristol, United Kingdom*

[5]*Department of Geological Sciences, University of Canterbury, Christchurch, New Zealand*

Hawaiian Volcanoes: From Source to Surface, Geophysical Monograph 208. First Edition.
Edited by Rebecca Carey, Valérie Cayol, Michael Poland, and Dominique Weis.

Companion Website: www.wiley.com/go/Carey/Hawaiian_Volcanoes

improved models of lava flow behavior. We begin with an overview of previous work on the description of lava flow surfaces and the use of lidar to investigate flow morphology before introducing the work presented here.

22.1.1. Characterizing Lava Flow Surfaces

Lava flow surface characteristics have been examined primarily in the context of discriminating flows of either different ages or flow type [e.g., *Mazzarini et al.*, 2007; *Morris et al.*, 2008; *Bretar et al.*, 2013]. One measure of flow surface properties is derived from the intensity of the lidar return signal. A study of lava flows at Mt. Etna [*Mazzarini et al.*, 2007] shows that the lidar intensity is controlled primarily by surface reflectance and by the distance between the receiver and the imaged surface. When normalized to a standard elevation to correct for receiver distance (plane height in the case of airborne lidar), lidar data show intensity variations that reflect both flow surface roughness and flow surface characteristics such as surface weathering and vegetation coverage. For this reason, lidar intensity maps are difficult to interpret without additional information (such as flow age, vegetation cover, or an additional measure of surface roughness).

Studies of lava flow surface roughness have been developed primarily by the planetary community for mapping the surfaces of the terrestrial planets. The most straightforward measure of surface roughness is the root-mean-square (RMS) height within a cell [usually taken as a square matrix of pixels; e.g., *Shepard et al.*, 1995; *Campbell and Shepard*, 1996; *Morris et al.*, 2008; *Bretar et al.*, 2013]. The RMS height is strongly dependent on cell size and the scale of surface features and measures only variations in vertical elevation. It is also affected by slope and thus requires detrending of elevations prior to calculation [e.g., *Campbell and Shepard*, 1996]. Horizontal roughness can be assessed by determining the relationship between RMS height deviation and a correlation length scale [*Morris et al.*, 2008; *Bretar et al.*, 2013]. Periodic flow surface features, such as folds, can be measured by fast Fourier transform (FFT) of linear profiles perpendicular to the feature of interest [*Pyle and Elliott*, 2006]. Surface analysis has also been used to highlight abrupt topographic offsets [*Morris et al.*, 2008] and to discriminate among different flow surface morphologies (e.g., tephra, smooth pāhoehoe, slabby pāhoehoe and ʻaʻā) [*Bretar et al.*, 2013].

22.1.2. Characterizing Lava Flow Morphology

Volcanological applications of lidar data have focused primarily on developing high-resolution digital elevation models (DEMs) for probabilistic lava inundation models [e.g., *Favalli et al.*, 2005; 2009a,b, 2012] and for morphometric analysis of (mostly recent) lava flows [e.g., *Mazzarini et al.*, 2005; *Pyle and Elliot*, 2006; *Coltelli et al.*, 2007; *Neri et al.*, 2008; *Ventura and Vilardo*, 2008; *Marsella et al.*, 2009; *Favalli et al.*, 2010; *Deardorff and Cashman*, 2012; *Tarquini et al.*, 2012]. Morphometric analysis includes assessment of along-flow changes in flow characteristics, planform analysis of flow margins, and detailed studies of specific flow features. Many of these studies have had as a goal increasing the accuracy of flow volume estimates and effusion rates and, where detailed flow chronologies exist, changes in effusion rate with time. Another goal has been to identify topographic controls on both flow morphology and flow emplacement conditions.

These studies have been supplemented by photogrammetric studies [*Applegarth et al.*, 2010; *Marsella et al.*, 2012] and field measurements of flow cross sections [*Branca et al.*, 2013].

Key to the analysis of topographic controls on lava flow emplacement is an accurate preeruption DEM or other information about the landscape traversed by the flow. Preeruption lidar data are rare, although current lidar mapping campaigns, particularly at volcanoes such as Stromboli and Etna in Italy, will provide reference topography for future lava flows. For this reason, the resolution of thickness maps constructed using posteruption lidar is constrained by the resolution of the preeruptive DEM. Resulting maps have been used to measure total flow volumes and average effusion rates [e.g., *Coltelli et al.*, 2007; *Marsella et al.*, 2009] and to evaluate the effect of topography on the development of individual lava channels [e.g., *Favalli et al.*, 2010; *Tarquini et al.*, 2012].

22.1.3. Brief Review of Target Lava Flows

The studies reviewed above have developed methodologies for lidar processing and demonstrated the potential of lidar for investigations of lava flow emplacement. None of these studies, however, has used flow thickness data to examine mass distribution and dynamics of multilobed flows, nor have morphometric and surface analysis characteristics been combined to address along-flow changes in both flux and flow characteristics. Here we do both, with our focus on two recent (December 1974 and March–April 1984) and well-characterized lava flows from Kīlauea and Mauna Loa volcanoes, Hawaiʻi (Figure 22.1).

The December 1974 lava flow erupted along a series of right-stepping northeast-to-southwest (NE-SW) fissures southwest of Kīlauea Caldera (Figure 22.1b). Lava fountaining migrated from the NE to the SW end of the fissure system over the course of the ~6 h eruption. The average volumetric effusion rate was ~270 m^3/s and the total erupted volume was ~5.9×10^6 m^3 [*Lockwood et al.*, 1999]. Abutment against the Koaʻe Fault System forced the lava into a single channel along the last 9 km of the total 12 km flow length. At the same time, the flow surface changes from pāhoehoe at the vent to ʻaʻā at the distal

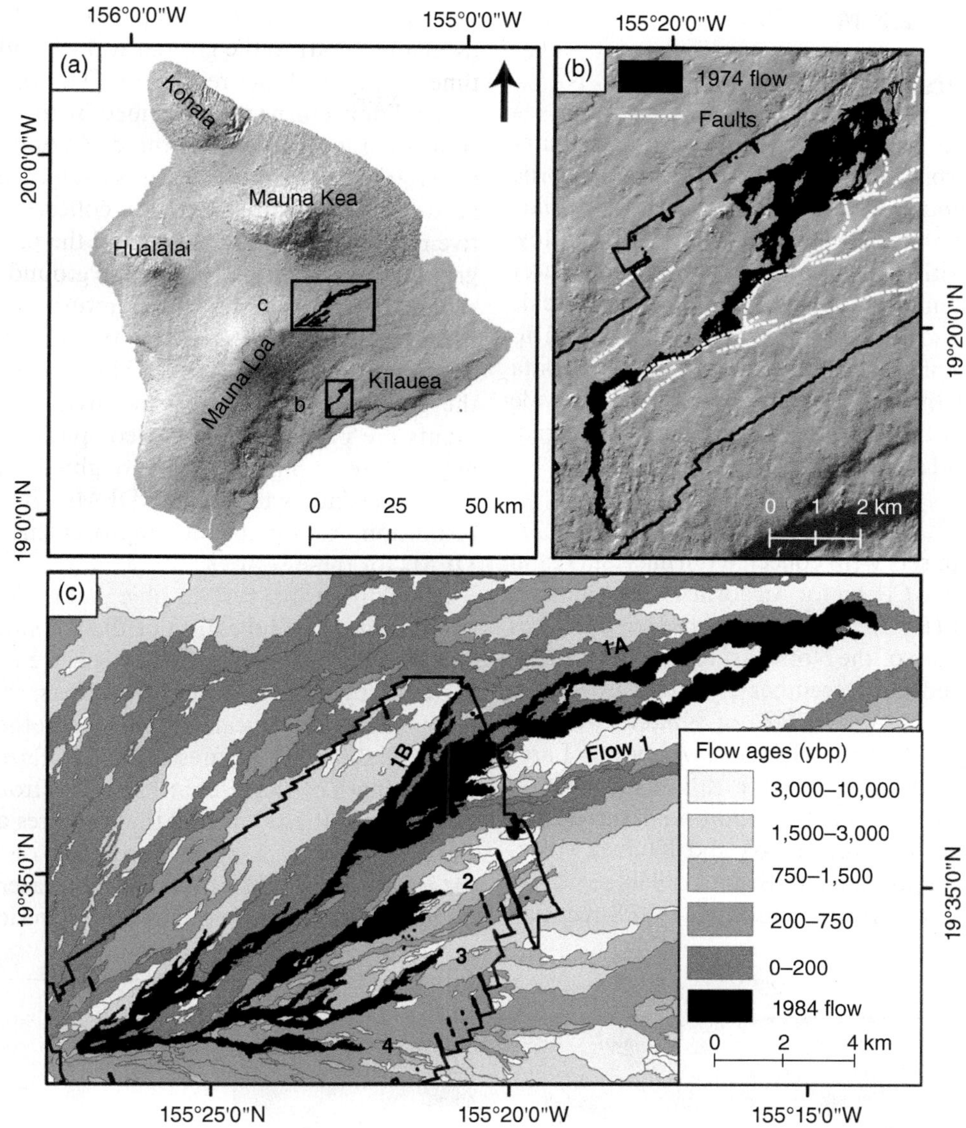

Figure 22.1 Location maps of the study area. (a) Hillshade map of the Island of Hawai'i with the extents of the lava flows included in this study shown. (b) Map of Kīlauea December 1974 flow in solid black, with the extent of lidar in the black outline and Koa'e Fault System marked by white dashed lines. (c) Map of Mauna Loa 1984 flow in solid black overlain on the *Wolfe and Morris* [1996] geologic map with the ages of surrounding flows. The lidar extent is shown in the black outline and the major branches of flow labeled after *Lipman and Banks* [1987] (flows 1–4 along with flows 1A and 1B).

end, a consequence of lava cooling and crystallization within the active channel [*Soule et al.*, 2004].

The March–April 1984 eruption of Mauna Loa volcano occurred on the Northeast Rift Zone (Figure 22.1c). The eruption lasted three weeks and produced a large (~$1.98 \times 10^8 m^3$) lava flow. The main fissures that opened on March 26 initially fed four main flow lobes (flows 1–4) with an effusion rate of ~$275 m^3/s$. Flows 2–4 stagnated within two days. Flow 1 advanced 27 km from the vents within the first few days of the eruption. A blockage of the flow 1 channel and subsequent overflow on March 29 created flow 1A, which had traveled nearly as far by April 5 (Q~$100 m^3/s$). Reduced flux and further blockages produced successive shorter overflows throughout the remainder of the eruption, including flow 1B on April 5. The velocity, temperature, and channel dimension measurements along the length of this flow were closely monitored because of concern that it could reach the city of Hilo. The intense monitoring effort produced an unprecedented set of observations along a large active 'a'ā channel that greatly improved understanding of channel construction and flow emplacement [*Lipman and Banks*, 1987].

22.2. METHODS

22.2.1. Data Sources

We use two types of airborne remote sensing to quantify lava flow morphology for both lava flows. Airborne laser swath mapping (i.e., ALSM or airborne lidar) provides topographic data of the present-day flow surfaces along with intensities of backscattered laser pulses. Photogrammetry allows us to view grayscale imagery from orthorectified aerial photographs and to construct topography of the preeruptive surfaces. Data products derived from these two data sources provide detailed views of the extents, properties, and internal structures of both lava flows.

22.2.1.1. Lidar

The ALSM data sets were collected during a survey in 2009 by the National Center for Airborne Laser Mapping (www.ncalm.org) (Figure 22.1). This chapter describes surveys over a portion of the Northeast Rift Zone (NERZ) of Mauna Loa and the December 1974 lava flow from the Southwest Rift Zone (SWRZ) of Kīlauea Volcano conducted with an Optech GEMINI Airborne Laser Terrain Mapper (specifications for this instrument at www.optech.ca). These and two additional surveys of the summit regions of Mauna Loa and Kīlauea volcanoes, as well as operational details, are available at www.opentopography.org. ALSM creates a GPS-rectified *x-y-z* point cloud of topography by sending laser pulses from an aircraft to the ground surface and recording the time of flight of the reflections. The accuracy of individual lidar returns is determined by the size of the laser spot when it reaches the ground. At our survey altitude of ~1800 m, vertical accuracy is 5–30 cm and horizontal accuracy is ~30 cm. The survey collected approximately five returns per square meter, and the point clouds were gridded to produce a DEM of the ground surface topography. We did not filter laser returns from vegetation before gridding, because of the minimal vegetation cover. For this reason, the DEM does not fully represent the ground surface in the few areas where pioneering plants are present. We removed spurious points manually before using a linear weighted mean within a 3 m × 3 m window to generate DEMs with 1 m horizontal resolution, a significant improvement over existing DEMs for these regions.

In addition to topographic data, ALSM provides surface imagery in the form of the amplitude of the laser returns. The near-infrared laser is more effectively backscattered by some surfaces than others, so that a map of lidar "intensity" reveals differences in the near-infrared color of different features on the surface (Figure 22.2). The strength of the laser reflection is strongly dependent on the aircraft altitude and thus requires altitude corrections to produce a normalized map of lidar intensity [*Mazzarini et al.*, 2007]. In addition, laser intensity may be affected by differences in atmospheric transmission,

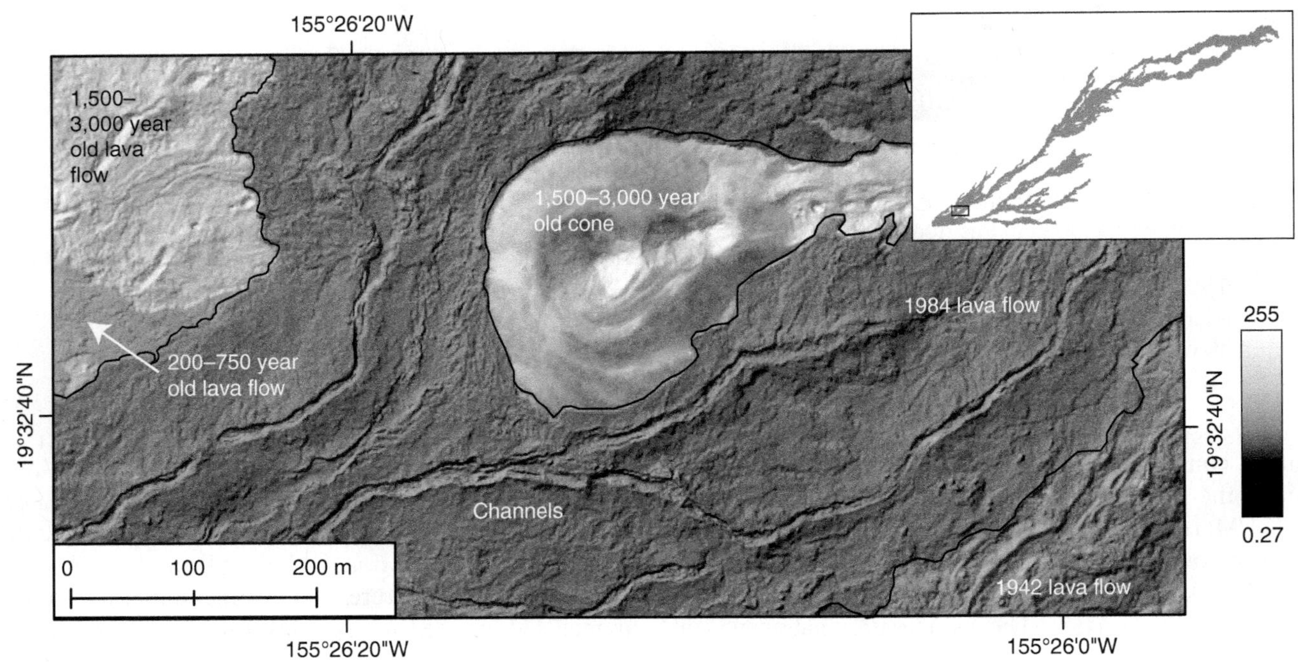

Figure 22.2 Annotated zoom of the lidar intensity map with lidar hillshade overlay to demonstrate the intensity differences between surfaces of different ages, cones, flows, and channels. Ages from *Wolfe and Morris* [1996]. The morphometric properties of this region are shown in Figure 22.5 and classified in Figures 22.9 and 22.10.

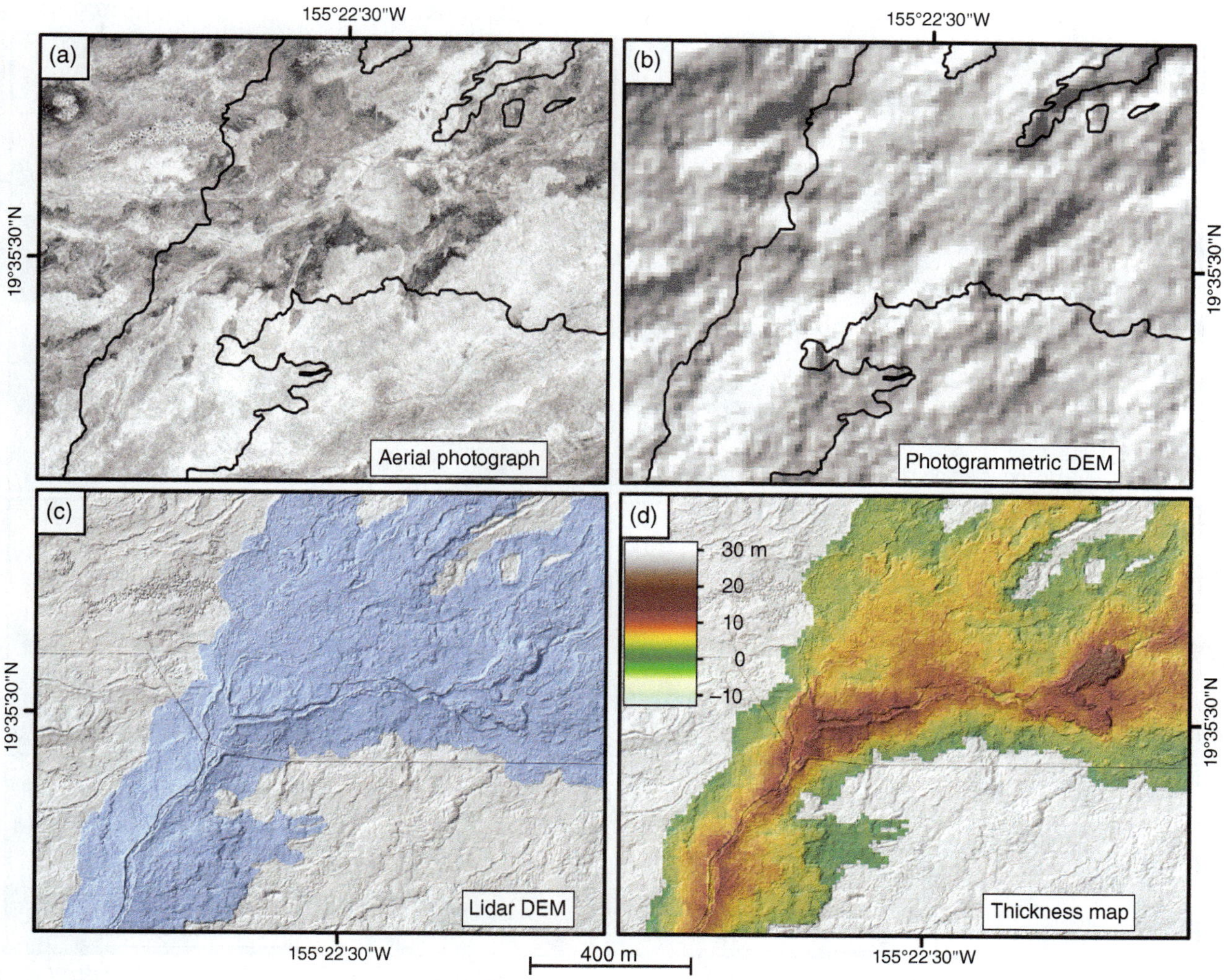

Figure 22.3 Local view of the assembly of the preeruptive DEM and thickness map for the Mauna Loa 1984 flow. (a) Orthorectified aerial photograph from a USGS survey in 1977 with an outline of the 1984 lava flow. (b) Hillshade of the 10 m resolution photogrammetric DEM with the 1984 lava flow outline. (c) Hillshade of the 1 m resolution lidar DEM with the 1984 lava flow extent shown in blue. (d) Thickness map produced by DEM differencing (c) minus (b) within the extent of the lava flow.

e.g., due to humidity or presence of volcanic gases. This was the case for the Mauna Loa 1984 flow where there were persistent clouds due to the orographic effect at lower elevations. We corrected for atmospheric transmission effects by shifting normalized intensities on parallel flight lines so that features covered by both flight lines had consistent intensity values [*Favalli et al.*, 2009a].

22.2.1.2. Photogrammetry

To visualize topography prior to flow emplacement, we used aerial photographs taken in 1977 by the U.S. Geological Survey (USGS) to construct a DEM of the preeruptive surface of the Mauna Loa 1984 flow (Figure 22.3). The aerial photographs were high-resolution (14 μm scan, ~1 m ground resolution) scanned photographs acquired from the USGS; the accompanying camera calibration report allowed us to constrain the internal orientation of the camera. A set of seven images provided stereo-pair coverage across the whole flow path (Figure 22.3a).

Both the DEM and accompanying orthophotos were generated using Leica Photogrammetry Suite (LPS) software. We identified 50 suitable ground control points (GCPs) on unfiltered shaded-relief lidar maps and used the 1984 flow outline mapped from newly collected lidar elevation and intensity data to ensure that the GCPs were outside the area covered by the 1984 flow. GCPs were typically distinctive small-scale patterns of vegetation (shrubs) or structural features in the older (pre-1984) lava flows. The lidar DEM was used for GCP elevation control. We then identified the corresponding points in the aerial photographs using the LPS graphical interface.

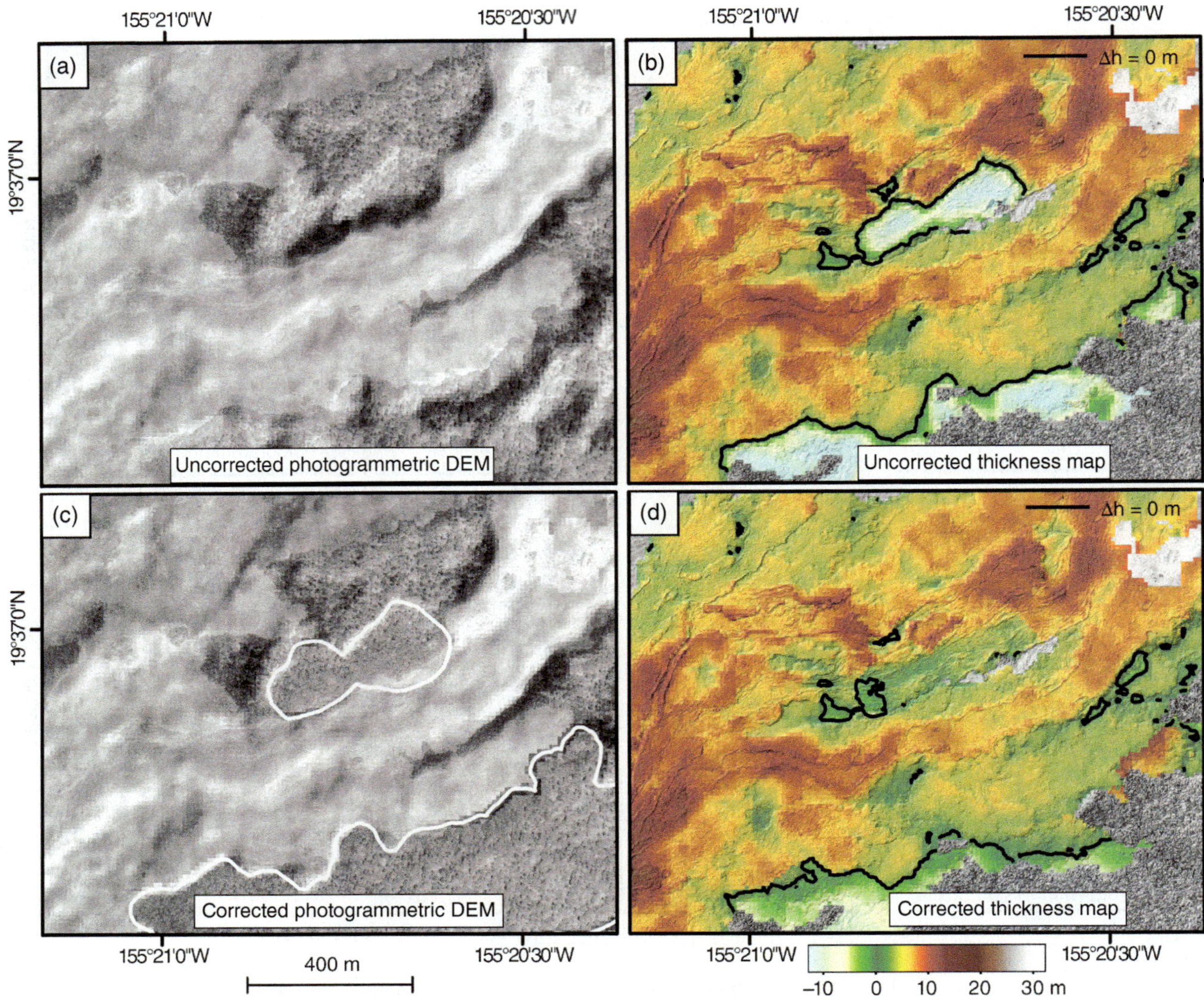

Figure 22.4 Demonstration of errors and corrections of the preeruptive photogrammetric DEM and resulting thickness map due to dense pockets of preeruptive vegetation. (a) Uncorrected photogrammetric DEM overlain on a 1977 orthophoto. Artificial "hills" are visible within the heavily vegetated areas. (b) Thickness (DEM difference) map produced from subtracting the DEM in (a) from the lidar DEM. Where the flow buried vegetation, the "hill" artifacts produce strongly negative thicknesses outlined by a 0 m thickness contour line. (c) Photogrammetric DEM that has been corrected by outlining the problem regions and fitting a plane to the pixels surrounding them in order to interpolate a new surface within them. (d) Thickness map resulting from the corrected photogrammetric DEM removes much of the error, while some negative regions remain, outlined by a 0 m thickness contour line. Reoccupation of a preeruptive channel by the 1984 flow is also apparent in the comparison between the preeruptive imagery and the posteruptive lidar.

We identified points common to multiple photographs, with at least three control points between overlapping photos, and augmented this set with the LPS automated tie point generation feature. The resulting triangulation produced a 10 m resolution DEM with horizontal and vertical RMS error (RMSE) of <2 m in all directions (Figure 22.3b).

After the photogrammetric DEM construction, comparison to the lidar DEM revealed that the entire data set was tilted. This is likely a product of minor misregistration across the whole DEM, which we corrected by fitting a planar ramp to the offset and subtracting it from the DEM [*Hollingsworth et al.*, 2013]. The photogrammetric DEM captures the upper surface (canopy) of vegetation, so where the forest has been cleared for development or buried by new lava, the difference between the lidar and the photogrammetric DEM is strongly negative (Figures 22.4a,b). We corrected for this by identifying a band of well-fit pixels around the negative excursions and interpolating a planar surface between them, replacing the artificially high pixels in the photogrammetric DEM

(Figure 22.4c). This does not fully remove the problem but greatly reduces the offsets (Figure 22.4d). Other positive excursions from pixels resulted from clouds in the lidar and other errors. We corrected these and produced a final DEM by removing any pixels with values outside $\pm 3\sigma$. We then used the difference between the lidar DEM and the photogrammetric DEM outside of the Mauna Loa 1984 lava flow to estimate the photogrammetric DEM error. The RMSE (vertical) of these areas is 2.36 m. In an unvegetated region near the Mauna Loa 1984 vent, the RMSE is 1.03 m, while a region within a cleared part of the forest has an RMSE of 4.46 m.

22.2.2. Derived Data Sets

22.2.2.1. Lidar Surface Properties

Lidar elevation data form the basis of the morphological analysis presented here. Additionally, we convert DEMs into maps of individual surface properties to highlight small-scale flow features and characteristics. For this purpose we use derived products of the lidar DEM that utilize neighborhoods of pixels (3 × 3 pixel region with the target pixel in the center) to calculate properties of the surface, such as slope, curvature, or roughness (Figure 22.5) to assist in feature identification.

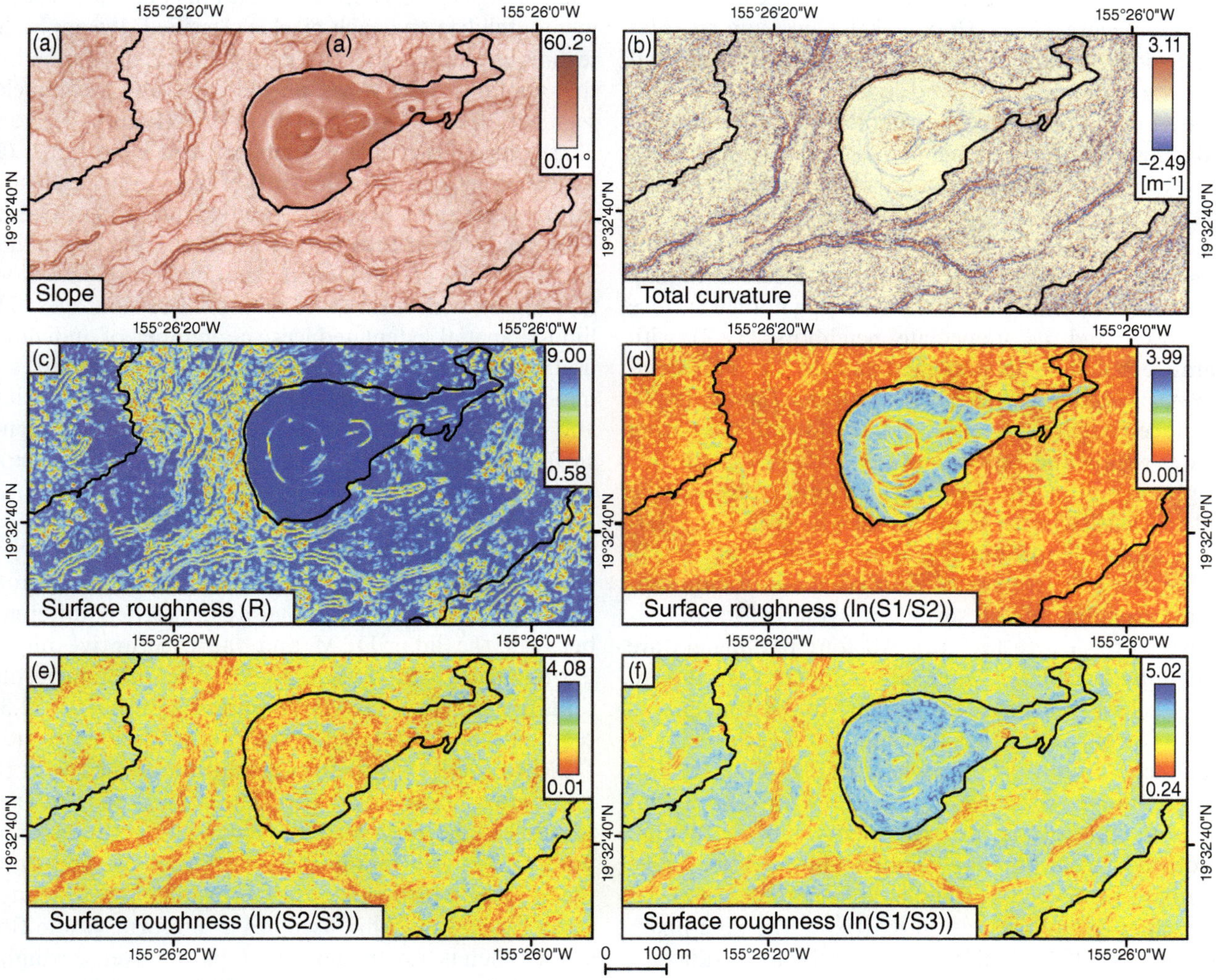

Figure 22.5 Example lidar surface properties maps within the extent of the classification analysis: (a) slope map in degrees, (b) total (surface) curvature, (c) surface roughness *R* parameter from *McKean and Roering* [2004] where high values represent smooth terrain, (d) natural log of the ratio of the first and second eigenvalues (ln(S1/S2)) showing clustering of vectors normal to the ground surface (larger values are more clustered and therefore smoother), (e) natural log of the ratio of the second and third eigenvalues (ln(S2/S3)) showing linearity of vectors normal to the ground surface (larger values are more linear or girdled), and (f) natural log of the ratio of the first and third eigenvalues (ln(S1/S3)) showing combined clustering and lineation. In roughness maps (c)–(f), blue is smooth and red is rough. All measures are produced from a resampled 2 m lidar DEM to match the 2 m intensity data set.

We discuss a variety of methods to process lidar data that allow extraction of quantitative morphometric information from within, between, and along lava flows.

Slope and curvature (the first and second spatial derivatives) are frequent metrics used to investigate surface morphology. Slope (the maximum slope at a pixel in units of degrees) reveals steeper areas within a lava flow, such as channel or flow margins, in comparison to the shallow slopes along a channel or flow lobe [*Ventura and Vilardo*, 2008]. A slope map of the preeruptive surface beneath the lava flow reveals the path of steepest descent, which is commonly used to predict flow paths [e.g., *Favalli et al.*, 2005]. We calculate a slope map by measuring the local gradient of a surface fit to a moving 3×3 neighborhood of pixels (Figure 22.5a) [*Horn*, 1981]. Curvature provides information on inflection points within a surface and whether a region is concave (positive curvature) or convex (negative curvature). In geomorphology, curvature is used to distinguish river channels (concave) from hillslopes (convex) and to calculate amounts of erosion and accumulation [e.g., *Passalacqua et al.*, 2010]. Curvature of a lava flow surface can highlight features such as levees [*Tarquini et al.*, 2012]. We create a second derivative map of curvature by fitting a fourth-order polynomial surface to a pixel and its immediate neighbors (3×3) with Lagrange polynomials and then calculating the second derivative (Figure 22.5b) [*Zevenbergen and Thorne*, 1987].

Surface roughness is also valuable for differentiating features within and between lava flows; when quantified using orientation surfaces, surface roughness can be used to identify linear features, such as levees and surface ridges in landslides [*McKean and Roering*, 2004]. We test the effectiveness of four of these orientation-based measurements of surface roughness for analysis of lava flow surface features. All four parameters are defined by calculating unit vectors perpendicular to each pixel within a moving window (a 3×3 neighborhood). The parameter R is the length of a vector sum of all of the unit vectors in the window (Figure 22.5c). For example, the maximum value of R is 9 for a 3×3 window with nine pixels; this represents a perfectly flat, smooth surface with nine parallel unit vectors. Three other parameters can be calculated from the normalized eigenvalues of an orientation matrix of the nine pixels in the window (S1, S2, S3). The natural logs of ratios of these eigenvalues reflect clustering and alignment of the vectors [Figure 5 in *McKean and Roering*, 2004]. High values of ln(S1/S2) designate highly clustered vectors and a smooth surface (Figure 22.5d), while high ln(S2/S3) represents highly lineated or girdled vectors (Figure 22.5e). Finally ln(S1/S3) is a combination of both clustering and lineation (Figure 22.5f). Together these parameters can be used to identify channel levees, flow margins, and surface roughness.

22.2.2.2. Surface Properties Classification

Surface property measurements can be used individually to highlight flow surface features or in concert to map elements of an entire flow surface. We use an unsupervised classification of all surface layers (intensity, slope, curvature, and roughness) to extract flow margins and internal features with the ERDAS Imagine software. Unsupervised classification differentiates classes statistically by finding pixels with similar values across different layers. We specify the number of classes and initialize the class clusters along the principal axis at the start of each classification. The ERDAS software assigns pixels to the nearest cluster and then calculates a new cluster mean and reassigns pixels iteratively until the pixel class assignments stabilize and each pixel is classified. We apply this technique to a small section of the NERZ of Mauna Loa that contains older flows, an old cone, and a channeled segment of the Mauna Loa 1984 flow (Figures 22.2 and 22.5) and use two different sets of these input layers (all of the layers and only the surface roughness eigenvalue ratio layers) and a range of specified numbers of classes. This small target area allows us to test the capabilities and shortfalls of automated classification of lava flows in a region with a range of features and flow ages. The limited spatial extent reduces the number of individual units, and therefore the number of classes required, and minimizes the effects of varying elevation (vegetation) and along-flow rheological changes that can influence surface characteristics within individual units across larger spatial extents.

22.2.2.3. Flow Thickness

We have generated a thickness map for the Mauna Loa 1984 lava flow by calculating the vertical difference between the lidar DEM and the preeruptive photogrammetric DEM (Figure 22.3d). The different resolutions of the two DEMs (Figure 22.3b versus 22.3c) required us to resample the 1 m (horizontal) lidar DEM to a DEM with 10 m resolution aligned with the photogrammetric DEM grid. The error in the thickness values can be estimated by propagating the error of both DEMs and taking the square root of the sum of the squares, which yields an approximate flow thickness error of 2.38 m. This assumes a uniform whole-flow error, which is not an ideal assumption, given the higher error of the photogrammetric DEM in vegetated areas. We therefore estimate a maximum error of 4.47 m using the photogrammetric DEM RMSE calculated among partially cleared vegetation, which encapsulates the most negative thickness values within the forest. A minimum error from the RMSE of unvegetated regions is 1.07 m. The thickness shows large variations across the flow that correspond to vents, levees, overflows, and the channel network.

22.3. RESULTS AND DISCUSSION

Our surface intensity, roughness, classification, and flow thickness results illustrate the types of data that can be derived from lidar and photogrammetric DEMs. The remarkable complexity and detail visible in the flows illustrate the value of using high-resolution topographic data to relate flow morphology to the dynamics of flow emplacement. These data, individually and in concert, also provide insight into postemplacement biological colonization of flows [e.g., *Deligne et al.*, 2012] and assist in mapping flows and flow features.

22.3.1. Intensity Analysis

Intensity of lidar reflections has been used to map the relative (and perhaps absolute) ages of volcanic surfaces [e.g., *Mazzarini et al.*, 2007]. On the NERZ of Mauna Loa, lidar intensities (Figure 22.2) are most strongly controlled by the abundance of low-lying vegetation (e.g., grasses, lichens, ferns). As vegetation on young lava flows is strongly controlled by elevation [*Raich et al.*, 2000], lidar intensity can discriminate flows by age only when corrected for elevation.

The NERZ of Mauna Loa is an ideal location to evaluate lidar intensity-age relationships because the lava flow boundaries and ages are well defined [*Wolfe and Morris*, 1996], as are the local climatic conditions that control revegetation rates on the flow surfaces [*Aplet and Vitousek*, 1994; *Aplet et al.*, 1998; *Raich et al.*, 1997]. The intensity map we have generated includes lava flows with ages between 30 and >5000 years [*Wolfe and Morris*, 1996] and elevations between 3100 and 2000 m (Figure 22.6). Variations in lidar intensities are clear and closely match flow boundaries mapped by aerial photographs and field studies. Intensities also vary within flows as a function of elevation (Figure 22.7); all flows have higher intensities at lower elevations. These data match measurements of changing biological productivity during primary succession on flow surfaces [e.g., *Aplet and Vitousek*, 1994], which mirrors patterns of decreasing temperature and precipitation with increasing elevation along the NERZ (Figure 22.7 inset). Primary succession on new lava surfaces initiates with the lichen *Stereocaulan vulcani*, which can become established on newly erupted lava flows within a few years [e.g., *Jackson*, 1971]. The productivity of this and other early colonizers is amplified by increasing temperature and to a lesser extent rainfall [*Anderson-Teixeira et al.*, 2008; *Raich et al.*, 2000].

We calculate an intensity value for individual flows and flow units (including older flows preserved in kīpuka within younger flows) by determining the mode of the

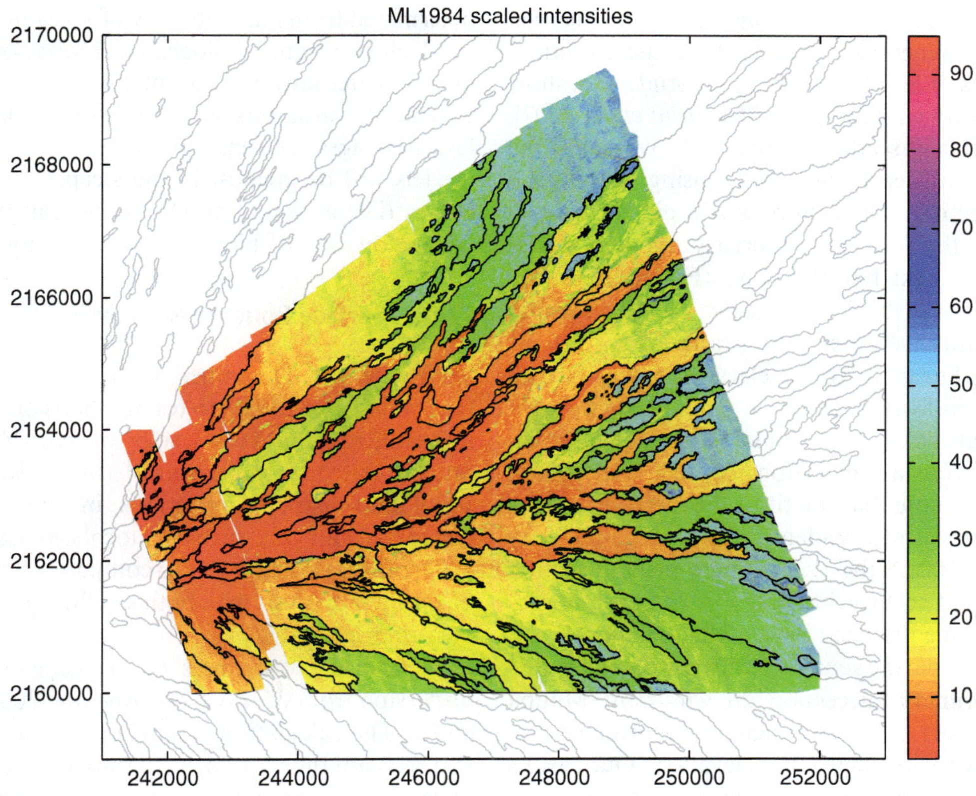

Figure 22.6 Map of normalized lidar intensities on Mauna Loa NERZ with mapped geologic units outlined [*Wolfe and Morris*, 1996]. This can be compared to Figure 22.1c for flow ages.

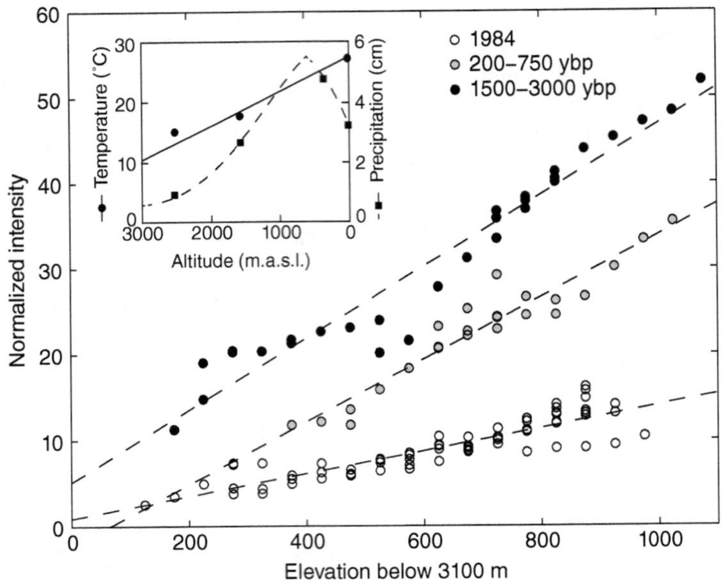

Figure 22.7 Intensity increases with decreasing elevation within individual flows (i.e., of constant age). This increase is consistent with increasing biological productivity due to climatic gradients along the NERZ of Mauna Loa including air temperature and precipitation which are shown in the inset [after *Aplet and Vitousek*, 1994]. The rate of intensity change is greater for older flows, perhaps due to changes in plant types that are present at different stages of succession. Linear fits to the data suggest that the intensity difference between flows of different ages will be small or negligible at high elevations.

distribution of intensity values from within the mapped flow boundaries. When viewed as a whole, the age-intensity relationships of lava flows within the study area show increasing intensity with age [e.g., *Mazzarini et al.*, 2007], albeit with significant scatter (Figure 22.8). We can fit the mean intensities of each age group using a (natural) logarithmic function, which implies a rapid increase in intensity within the first few hundred years, and slower increase over the next few thousand (Figure 22.8). We attribute the scatter in the data to the fact that flows within any age group traverse an elevation range along the rift zone. Using the linear fits to the elevation-intensity data determined for individual flows (Figure 22.7), we construct additional curves that would reflect age-intensity relationships at constant elevations of 2000 and 3000 m (Figure 22.8). We note that the fit to the mean intensities of each age group is nearly identical to the age-intensity relationship at 2500 m, the approximate mean elevation within the data. The logarithmic form of age-intensity relationship and the flattened form of that relationship at high elevations mimic the simulated form of plant productivity during primary succession on windward Mauna Loa slopes [*Raich et al.*, 2000], where initial primary productivity declines with increasing elevation. Our results support the hypothesis that biological colonization is the primary driver of age-intensity variations [*Mazzarini et al.*, 2007] and suggest that absolute ages could be determined from lidar intensity of volcanic surfaces where local climatic and biological colonization patterns are known. Alternatively, lidar intensity could be used to map vegetation variations with elevation in locations where lava flow ages are well constrained. The fidelity of these models will be greatest in the steeper regions of the fit curves (i.e., at young ages), the age range of which will depend on rates of biological productivity (elevation).

22.3.2. Surface Roughness Analysis

We use the December 1974 lava flow of Kīlauea [*Lockwood et al.*, 1999] to test methods of quantifying the fine-scale roughness resolved by the ALSM data. Previous satellite-based measurement of roughness [*Gaddis et al.*, 1990] and mapped variations in the proportions of pāhoehoe and ʻaʻā surface morphologies [*Soule et al.*, 2004] of this flow allow us to compare, directly, the results of the lidar analysis with both satellite- and ground-based measurements.

We measure roughness using the method of vector dispersion from *McKean and Roering* [2004] as described above. The calculations are made over a 3 × 3 m window and thus are tuned to identify roughness on 1–2 m length scales. We determined roughness along the ~11.5 km flow using ~1 km (along-flow) bins in order to resolve changes in morphology that occur over length scales of hundreds

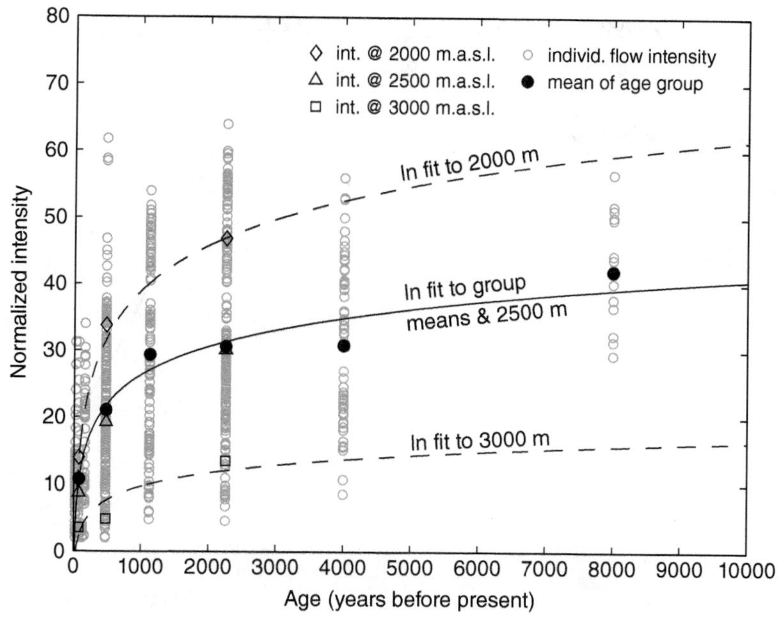

Figure 22.8 Age-intensity relationships for lava flows on the NERZ of Mauna Loa show an increase in intensity with age. Lava flow ages are from *Wolfe and Morris* [1996]. The data can be fit by a logarithmic function defining an initial rapid increase in intensity over a few hundred years that slows over the next several thousand years. The solid line is a fit to the mean of each age group (flows <200 years are grouped with a mean age of 78 years). Significant scatter in the data is due to the fact that intensities are calculated for flows that span a large range in elevation, over which climatic and thus biological productivity during primary succession will be highly variable. Additional fits to the data shown for altitudes of 2000 and 3000 m by dashed lines, constrained by fits to elevation-intensity relationships within single flows that span large elevation ranges (Figure 22.7), explain much of the variability within each age group.

of meters. This bin spacing is the same as used in field-based interpretations of flow morphology [e.g., *Soule et al.*, 2004], which facilitates comparison between the two methods. Bin boundaries are dictated by spacing along the presumed flow pathway and orientations that are normal to the dominant flow direction, although flow directions are more variable in the near-vent region (Figure 22.9a). Given the non-Gaussian distribution, the roughness within individual bins is determined by the mode of the ln(S2/S3) measurements within each bin. Of the four roughness parameters, we find that ln(S2/S3), which identifies lineations, and ln(S1/S3), which combines linear and clustered features, provide the most useful results.

Overall, ln(S2/S3) highlights a downflow sigmoidal shift from low roughness in the area mapped as entirely pāhoehoe to high roughness in the area mapped as entirely ʻaʻā (Figure 22.9b), consistent in form with that obtained from field and aerial photographs [*Soule et al.*, 2004]. Relative to that study, however, the inflection points marking the onset and conclusion of the pāhoehoe-to-ʻaʻā transition occur ~1 km closer to the fissure vents. This offset might reflect variations in the vertical scale of roughness along the flow from changes in the ʻaʻā clast size [*Gaddis et al.*, 1990; *Soule et al.*, 2004], or subtle changes in pāhoehoe surface textures that were not captured in the binary pāhoehoe-ʻaʻā analysis of *Soule et al.* [2004]. Despite these discrepancies, the common sigmoidal form of the mapped pāhoehoe-to-ʻaʻā transition indicates that the ALSM data and roughness metrics can be used to evaluate flow-scale surface morphology as well as the characteristic roughness of end-member morphologies.

22.3.3. Classification of Lidar Surfaces

As demonstrated above, lidar elevation and intensity data can be combined to discriminate flows of different ages and surface textures. Here we use all of these input layers to test their collective ability to map a section of the Mauna Loa 1984 flow and surrounding area using simple automated classification (Figure 22.2). We use two different sets of input layers: all layers and only the surface roughness eigenvalue ratio layers (Figure 22.5). Of the specified layers, we find that intensity and slope have the greatest variability between different features in area and so are most useful for classification.

Both the number of layers and the number of assigned classes determine the degree of detail represented by the classification. When we use all of the layers, an eight-class

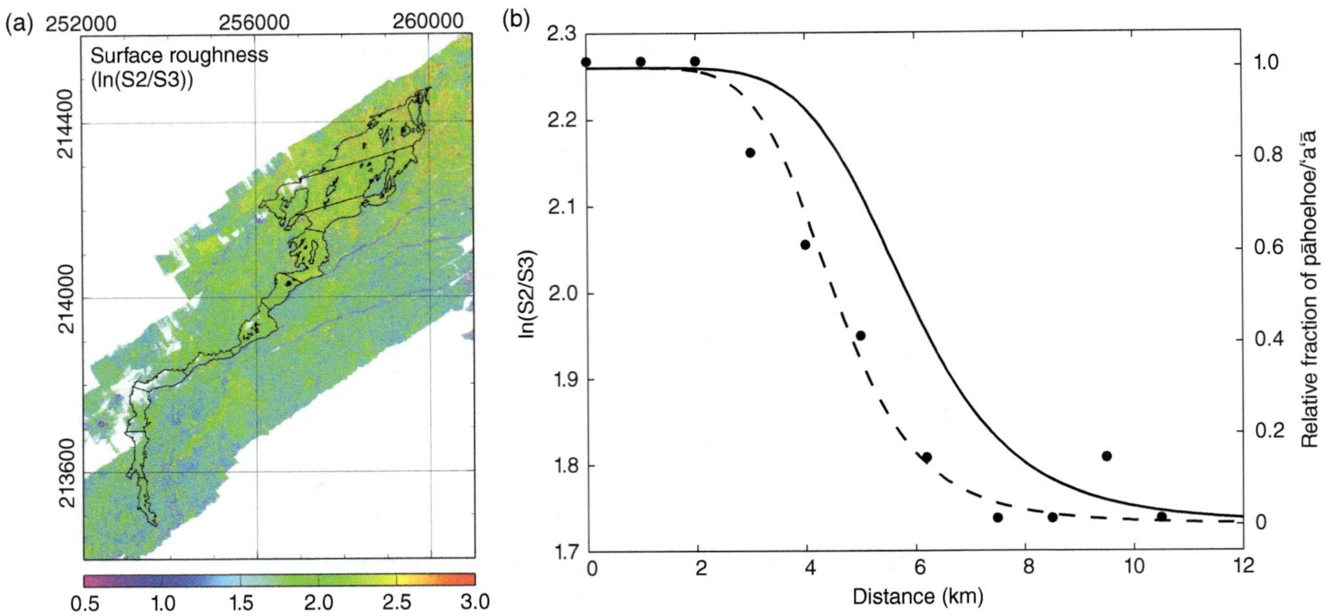

Figure 22.9 Derived roughness map and downflow roughness for the December 1974 lava flow on the SWRZ of Kīlauea. (a) Map of roughness and bins. (b) Plot of vector dispersion roughness metric, ln(S2/S3), for bins along the December 1974 flow of Kīlauea. A fit to the relative proportions of pāhoehoe and 'a'ā to observations from air-photo and field-based mapping of the flow is shown by the solid line. A modification to the fit line to better fit the roughness measurements is shown by the dashed line.

result captures most of the different units and features in the image (Figure 22.10a; Table 22.1). This particular classification can be used to identify differences in flow age as well as the channels within individual flows. Importantly, half of the classes (1, 3, 6, and 8) are distinct (unmixed) in the features that they include, while the other half range from somewhat mixed to random. Reclassifying the data set into only four classes yields simpler results and still captures differences in surface age but retains some mixed classes (Figure 22.10b; Table 22.1). As a result, some of the noise is reduced but the channels are not mapped as continuous features because the levees are partially grouped with the rest of the flow. In the four-class analysis, classes 3 and 4 each make up half the flow surface and could easily be combined, and yet they are not merged when a three-class analysis is run. Finally, requiring only two classes creates a map of the prehistoric surface (class 1) relative to the historic lava flows (class 2; Figure 22.10c).

Running the unsupervised classification with only the eigenvalue ratios as input layers produces a classification that discriminates among only the flow or cone features and not the flow ages. Allowing multiple classes increases the scatter within flow segments outside of the main channels (Figure 22.11a). Removal of all classes except the channels (class 1) and the cone (class 8), however, can generate a map that discriminates features from different textures (Figure 22.11b).

The results presented above demonstrate that unsupervised (automated) classification of lidar-derived surface metrics can be used to map distinctive features such as lava flows of different ages, cones, and lava channels. Here we have focused on a small part of the lidar image. Although the same statistical descriptions of each class (the class signature) produced could be used in a supervised classification of a larger area, several of the parameters are sensitive to elevation or distance from the eruptive vent. This sensitivity complicates development of automated classification protocols for the entire flow. Addition of input layers, such as high-resolution color, radar, or infrared imagery, could improve classification results. Radar amplitude, in particular, measures centimeter-scale surface roughness that could facilitate differentiation of flow ages and fine-scale changes in texture [e.g., *Gaddis et al.*, 1990]. Remote sensing of lava flows at higher spatial, spectral, and temporal resolutions will continue to improve our quantification of flow morphology and understanding of flow emplacement.

22.3.4. Flow Thickness Analysis

Our thickness map of the Mauna Loa 1984 flow provides important new insight into both the overall development of the flow field and the distribution of lava within the flow (Figure 22.12). Analysis of these data is enhanced by detailed observations, made both during and

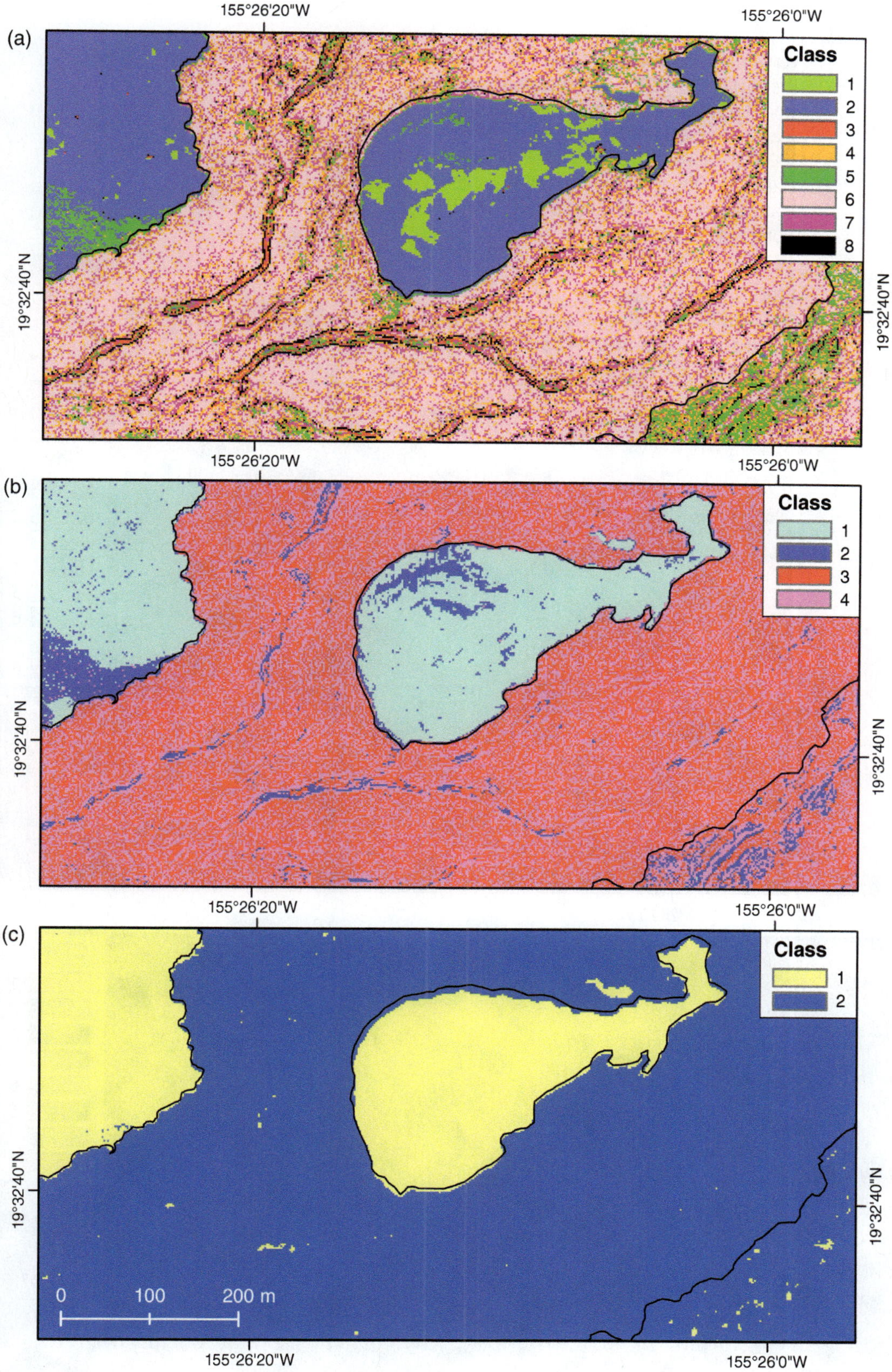

Figure 22.10 All-layer classification results for (a) eight classes, (b) four classes, and (c) two classes.

Table 22.1 All-layer unsupervised classification interpretation.

Class	Description
Eight Classes	
1	Very bright areas with the 1500–3000 year old units
2	The rest of the 1500–3000 year old units
3	Inner levee slopes
4	Scattered pixels within 1942 and 1984 lava flows
5	200–750 year old flow, 1942 channel, and overflows within the 1984 flow
6	Majority of the nonchannel 1984 and 1942 flow surface
7	Scattered pixels within 1942 and 1984 lava flows, some concentration in the channels
8	Outer side of levees
Four Classes	
1	1500–3000 year old units
2	200–750 year old flow, the 1942 channel, and the rest of the 1500–3000 year old units
3	Half the surface of the 1942 and 1984 lava flows
4	Half the surface of the 1942 and 1984 lava flows

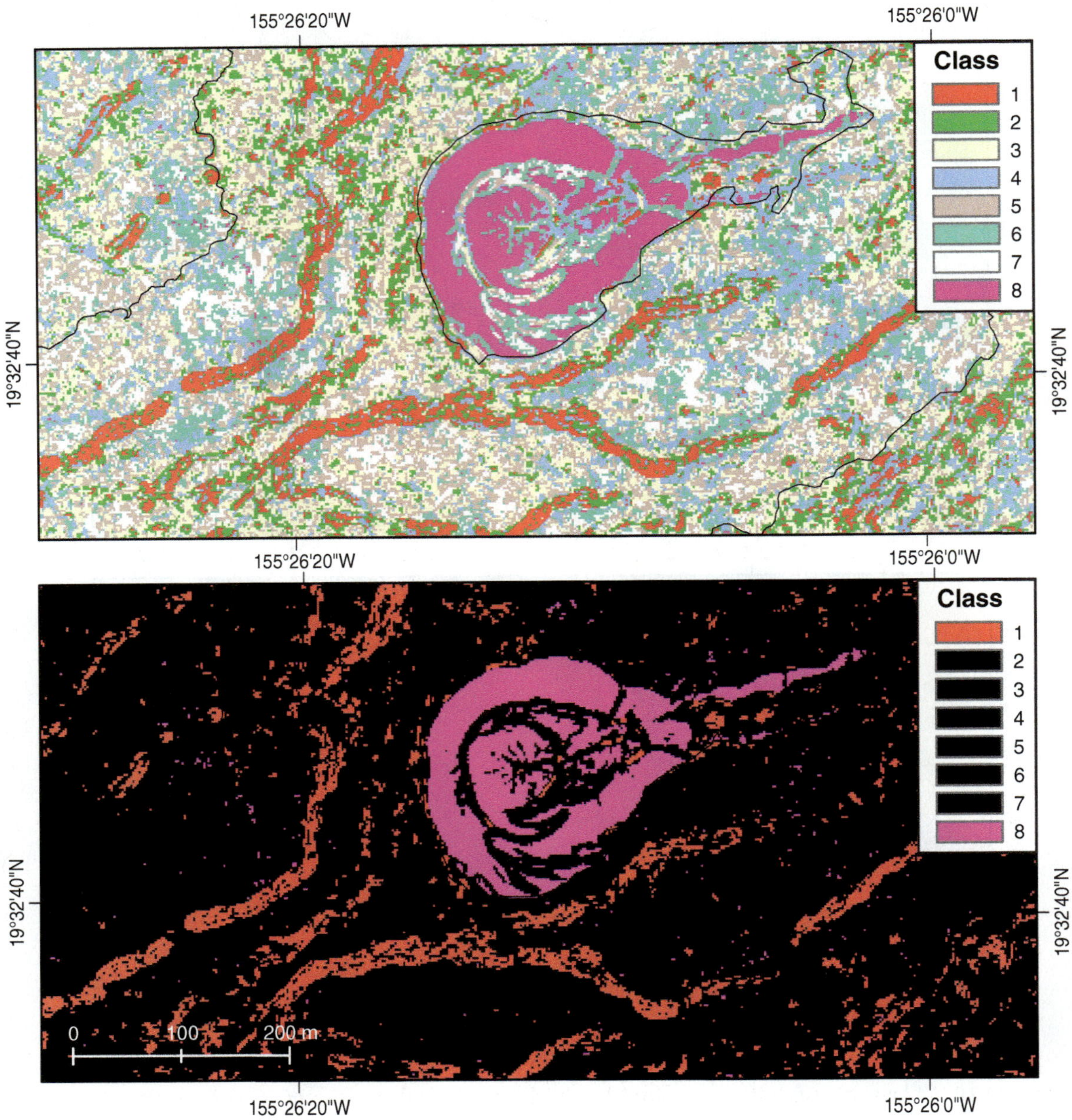

Figure 22.11 Surface roughness ratio classification results for eight classes showing (a) all of the classes or (b) just the two classes that show channels and the cone.

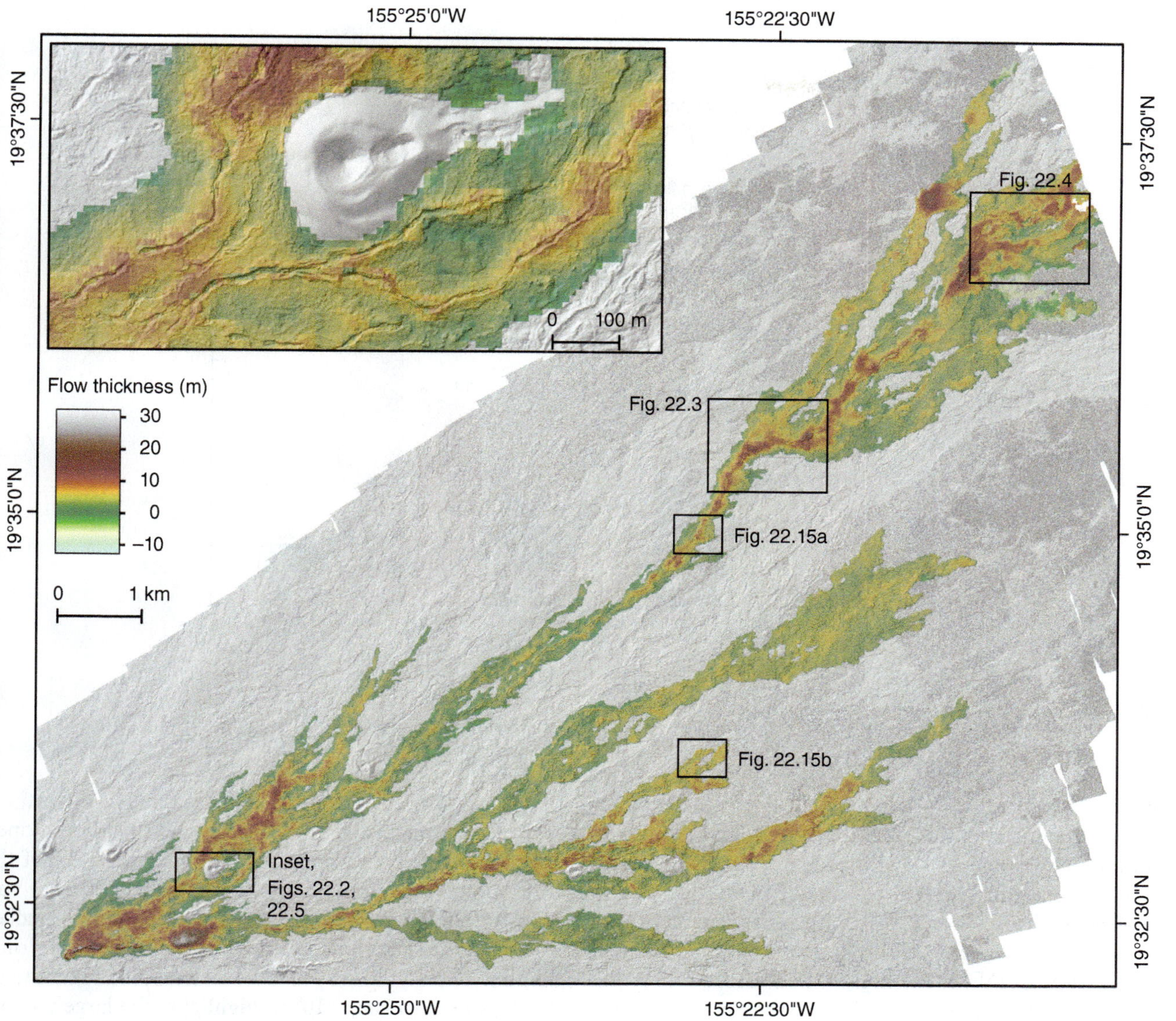

Figure 22.12 Flow thickness map with lidar hillshade of the Mauna Loa 1984 flow within the extent of the lidar data. Boxes indicate the extents of other figures.

after the 1984 eruption, related to the evolution of lava channels and spatial and temporal changes in lava rheology [e.g., *Lipman and Banks*, 1987; *Crisp et al.*, 1994].

22.3.4.1. Flow Volume

Although our lidar data set is limited to the upper portion of the Mauna Loa 1984 flow field, it captures much of the main flow lobe (flow 1 of *Lipman and Banks* [1987]) and all of the southern lobes that were active for the first 48 h of the eruption (flows 2–4 of *Lipman and Banks* [1987]; Figure 22.1). We calculate a total volume of the lidar-imaged 1984 lava flow by summing the flow thickness values of each pixel within the flow and multiplying by pixel area ($100\,m^2$). A maximum error can be estimated by multiplying the thickness error by the flow area. A minimum estimate assumes uncorrelated errors and is calculated by dividing the maximum error by the square root of the number of pixels to calculate an error range [*Favalli et al.*, 2010]. The resulting volume of 0.894 ($\pm$0.001–0.489) $\times 10^8\,m^3$ accounts for approximately 45% of the ~$1.98 \times 10^8\,m^3$ total volume of the flow estimated by *Lockwood et al.* [1985]. The area of the flow within the lidar extent is ~67% of the total flow area, which requires the more distal section of the flow to be significantly thicker, on average, than the upper two-thirds. In fact, the volume estimate of *Lockwood et al.* [1985] would require the distal section to have an average flow thickness of almost 11 m. If, instead, we apply our average thickness of 4.35 ± 0.01(standard error) m to the final third of the flow, we obtain a new total volume of 1.34

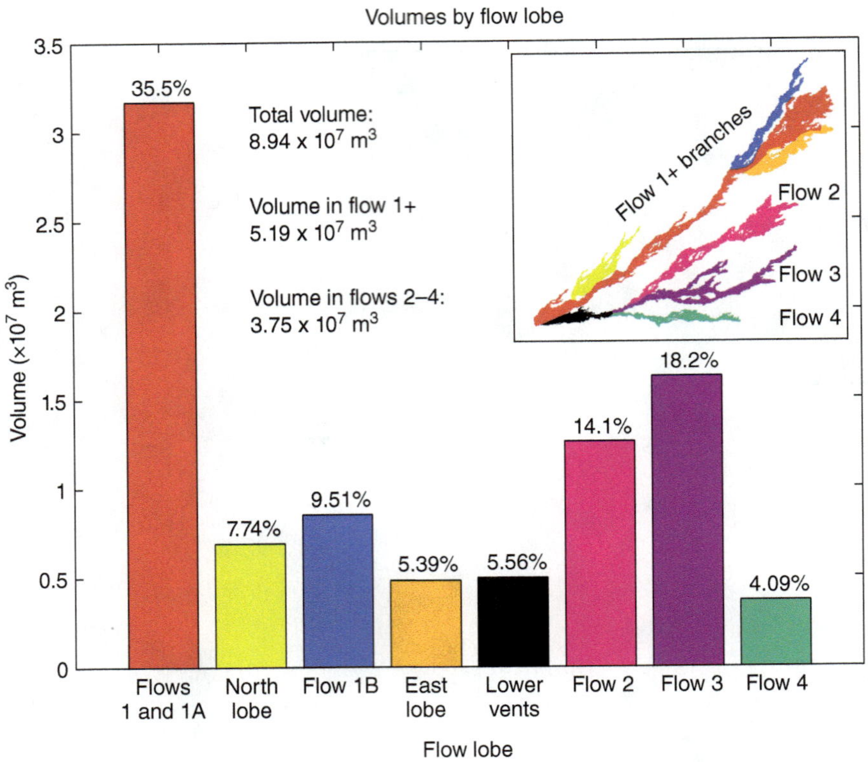

Figure 22.13 Volume distribution among flow lobes. Colors reference the inset map of the different major flows.

(±0.001–0.733) × 10^8 m^3. This value is 33% smaller than the estimate made at the time of eruption and represents a minimum volume, as flow thickness typically increases with distance from the vent because of increasing viscosity [*Moore*, 1987; *Lister*, 1992]. Although the uncertainty in these estimates could be large because of the error in the preeruptive DEM, the high spatial resolution of the DEM differencing improves volume estimates by accounting for local thickness variations. We note that part of the discrepancy in volume estimates may be accounted for by volume reduction due to postemplacement contraction and settling [e.g., *Stevens et al.*, 2001; *Lu et al.*, 2005; *Dietterich et al.*, 2012].

The thickness map illustrates the inhomogeneous distribution of lava volume within the flow; it also demonstrates that the flow is thickest in regions of major flow branching. To a first order, flow lobes of longer duration, which are subjected to repeated channel overflows and levee development, are also thicker [*Lipman and Banks*, 1987; *Coltelli et al.*, 2007; *Branca et al.*, 2013]. From this perspective, the Mauna Loa 1984 lava flow can be divided into two parts. The first is the portion of the channel that transported lava for the entire three weeks of the eruption, including flows 1 and 1A and their subsidiary branches, flow 1B, and what we term flow 1 north lobe and flow 1 east lobe (flow 1 + branches; Figure 22.13). This part of the flow contains 0.519 (±0.001–0.281) × 10^8 m^3, or ~58% of the total volume within the lidar image. The other three main flow branches, flows 2–4, plus the shield developed over the lower vents account for the remaining 42% of the flow volume within the lidar extent (individual lobe volumes are shown in Figure 22.13). These southern branches started at the same time as flow 1 but were active for only the first 48 h of the eruption. Their combined volume of 0.375 (±0.001–0.208) × 10^8 m^3 highlights the large volume that went into these branches to the south, instead of along the dominant channels of flow 1, during the early (high-effusion-rate) phase of the eruption.

The four primary flow lobes are separated by topography, with each occupying a different lavashed (equivalent to a watershed) delineated by drainage divides [*Kauahikaua et al.*, 2003]. Confinement of individual flow lobes within single lavasheds confirms the use of lavasheds as a simple hazard-mapping tool. We can also use the known flow lobe durations (48 h) to calculate time-averaged discharge rates for each flow component: the respective average fluxes for flows 2, 3, and 4 are 73, 94, and 21 m^3/s. These make up 68% of the ~275 m^3/s instantaneous effusion rate at the vents measured during the emplacement of these lobes, which requires that flow 1 account for the remaining 32%. No detailed observations of these southern flow lobes were made at the time of the eruption, but our estimated time-averaged discharge rates are similar to flow rates measured at distal sites of flows

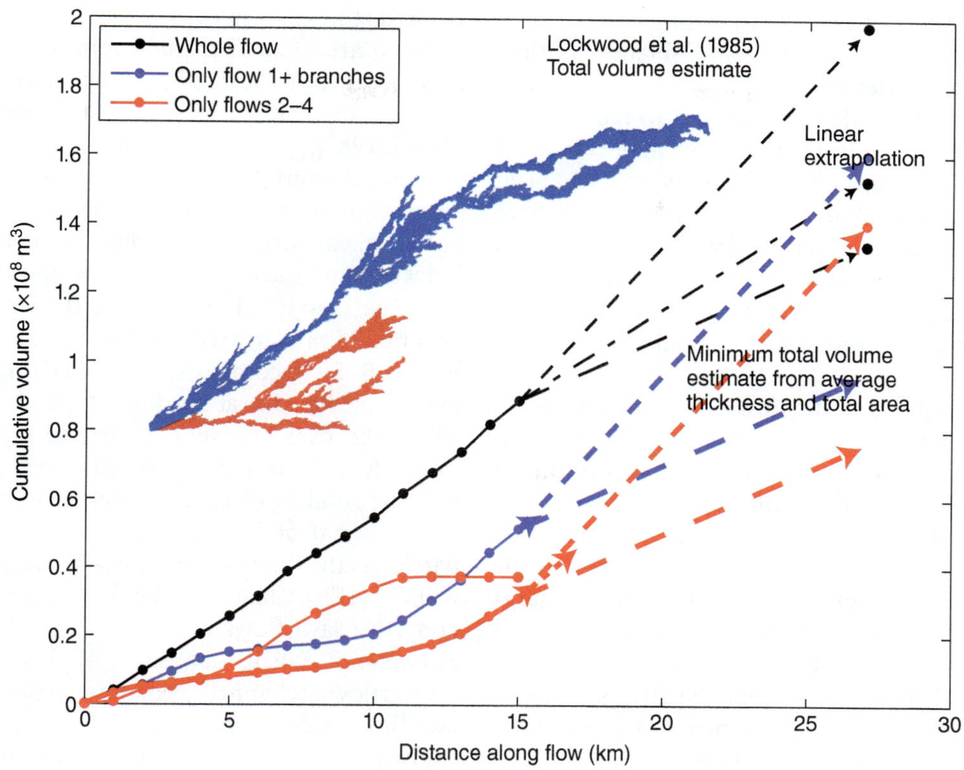

Figure 22.14 Distribution of volume with distance along the flow. Different lines show the volume distribution for the whole flow (black), flow 1 and all of its branches (blue), and flows 2–4 (red). Dashed lines extend to the total volume reported by *Lockwood et al.* [1985] (short dashes), our linear fit extrapolation of total volume (dash-dot), and our estimate of total volume calculated by extending the average flow thickness in the first 15 km of the flow to the total flow area (long dashes). The blue line could also be extrapolated to the maximum flow length with a set of dashed lines parallel the whole-flow ones.

1 and 1A during the eruption [*Lipman and Banks*, 1987]. The threefold variation in flux most likely reflects the initial flux from the fissure segment feeding each lobe, the partitioning of lava between different flow branches, and the actual duration of each flow branch (we have assumed a constant value of 48 h for all three). Understanding the differences in flux between branches is crucial for determining both rates of flow advance and final flow length; for this reason, collecting syn-eruptive multitemporal topographic data could provide important input data for future eruptions. Branching lobes with different fluxes have also been observed in other locations such as the 1669 flow field at Mt. Etna [*Branca et al.*, 2013].

Another way to view volume distribution is to measure the cumulative volume in the downflow direction, which shows where lava is being stored within the flow field (Figure 22.14). We have done this by measuring the cumulative volume upslope of generalized flow cross sections spaced every kilometer along the flow path. We plot separately curves for the entire flow (black), flow 1 and its associated branches (blue), and flows 2–4 (shown in red). The constant slope of the whole-flow line shows that the volume within each 1 km bin is approximately constant, and the flow (all branches) therefore has a consistent total cross-sectional area. In contrast, the curve showing flow 1 (blue) documents lava accumulation both near the vent (within the first few kilometers) and where the flow branches extensively and widens at a distance of 12 km from the vent. The red curve emphasizes the importance of flows 2–4 as major (early) components of the flow volume. Dashed lines show extrapolation of the whole-flow volume distribution curve to the total volume of the flow estimated by *Lockwood et al.* [1985] and our total calculated from the average flow thickness and total inundated area. A final estimate of the total volume of the flow can be made from our observation that the cumulative volume curve for the whole flow has a constant slope. If this slope is extrapolated with a linear fit, we find a total flow volume of $\sim 1.53 \times 10^8\,m^3$, which falls between the previous estimates. This value represents an average flow thickness of ~6.3 m over the final third of the flow, which is within the range of distal flow thicknesses measured by *Lipman and Banks* [1987].

22.3.4.2. Thickness of Lava Flow Features

Within the Mauna Loa 1984 flow, differences in flow thickness record the details of flow emplacement. The thickest portions of the flow correspond to the eruptive vents (≤32 m), flow levees (≤17 m), and major channel overflows and junctions where the flow splits (≤24 m; Figure 22.12). Flow thickness also varies with the degree of channel development, described by *Lipman and Banks* [1987] as the stabilized channel, transitional and dispersed flow zones, and the flow toe. In total, these patterns of flow thickness reveal much about the evolution of the flow field.

Locally thick regions of the flow show both the development and failure of lava channels. The main (continuously active) channel of flow 1, which makes up much of the narrow, central portion of the lidar-imaged flow, is characterized by distinct levees that form the margins of a confined channel that is only 10–20 m wide (Figure 22.15a). This section of the channel represents the stabilized channel zone, where flow was restricted to the central channel conduit [*Lipman and Banks*, 1987]. Here the channel is so narrow that the 10 m resolution of the thickness map is barely enough to capture the thin channel between the thick levees. A profile line across this section of the channel (A–A′) shows pre- and posteruptive elevations as well as the flow thickness from the thickness map. At 15 m height, the levees in this segment are some of the tallest in the whole flow, probably because the longevity of this part of the channel allowed prolonged levee construction [*Lipman and Banks*, 1987].

Channel overflows that caused local flow thickening were common throughout the Mauna Loa 1984 eruption and formed both at locations where the channel was blocked and during times of temporary increases in lava flux (surges). For example, the overflow shown in Figure 22.15a increased the flow thickness by 1–3 m (see profile B–B′ in comparison to A–A′). Here the thickest part of the flow lies at the downslope edge of the junction where the overflow splits from the main channel and probably reflects the combined effect of the overflow volume, ponding of lava in the channel prior to overflow, and material added by run-up of lava on the central barrier as the flow divided. Locally increased thicknesses from overflows have also been documented in the 2001 and 2006 lava flows at Mt. Etna [*Coltelli et al.*, 2007; *Favalli et al.*, 2010].

Overflows (channel breakouts) redirect much of the lava flowing down the channel. For example, field observations made along the flow 1/1A channel during a 24 h period show a fourfold decrease in the volumetric flux in the channel with distance from the vent (Figure 22.16)

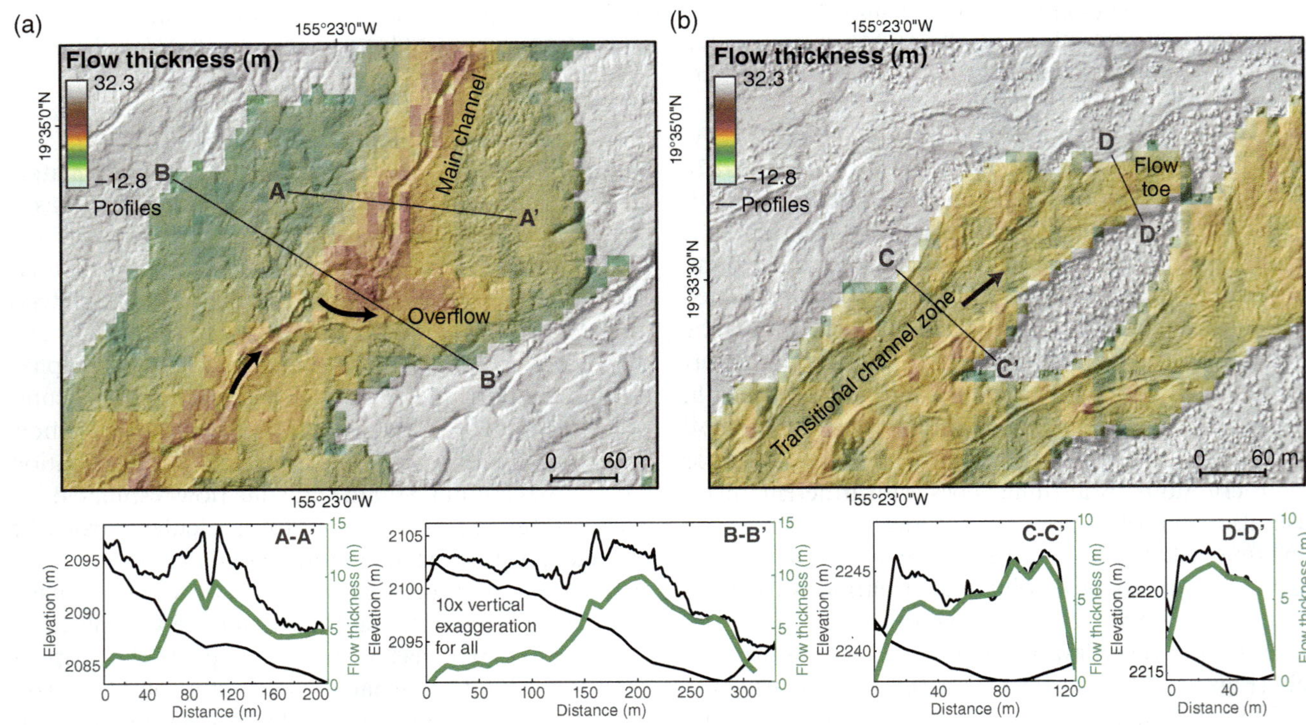

Figure 22.15 Flow thickness maps with lidar hillshade overlay of (a) the stable channel of flow 1 with a large overflow and (b) the distal portion of flow 2, showing the broad channel of the transitional channel zone and the flow toe. Profile lines of the main channel, overflow, transitional channel zone, and flow toe are also shown, with the pre- and posteruptive topography in black and the flow thickness in green.

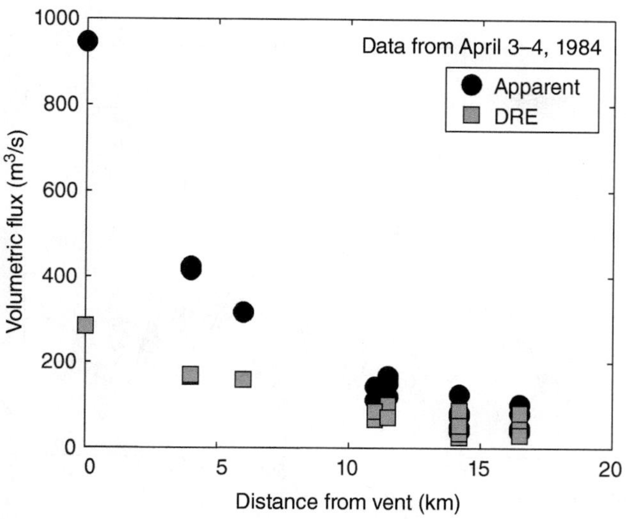

Figure 22.16 Decrease in flux down the main channel of flow 1/1A with distance from the vent as recorded on 3–4 April, 1984. Data from *Lipman and Banks* [1987].

[*Lipman and Banks*, 1987]. The thickness data show that much of this missing flux was diverted to levees, overflows, and branches. From this perspective, thickness maps provide important information about changes in lava flux along individual flow lobes, information that is key to accurate models of lava flow emplacement (and is not fully accounted for in existing flow models).

Flow thickness data can also inform our understanding of the more distal portions of a lava flow. Downflow of the stabilized channel zone is a broad transitional channel zone that leads to a zone of dispersed flow and flow toe with no central channel. These features, originally defined by observations of advancing flows during this eruption, are nicely preserved in flows 2–4, whose full lengths are captured by the lidar. A map and set of profiles from flow 2 shows the wide channel and marginal lenses and ridges of the transitional channel zone and the convex shape of the flow toe (Figure 22.15b). Also evident are multiple small lobes at the flow toe, a feature that is common in lava flows at Mt. Etna [*Applegarth et al.*, 2010; *Favalli et al.*, 2010]. The lobe directly downslope from the central channel traveled the furthest, as also seen at Mt. Etna.

22.3.4.3. Influence of Preeruptive Topography on Flow Thickness

Reconstruction of the preeruptive surface also allows us to investigate the influence of preeruptive topography on flow thickness. Cross-slope profiles (Figure 22.15) show that local topography affects both the path of the flow and asymmetry in flow thickness. All flow profiles show a dip in the cross-slope preeruptive topography, which shows that individual flow lobes followed preexisting depressions. Interestingly, in many places the flow appears to have traveled at an angle to the slope. This appears to have affected the flow thickness in that the downslope side of the flow is thicker than the upslope side (e.g., A–A′). The overflow in Figure 22.15a also occurs on the downslope side. This suggests that channels traveling across the slope may thicken preferentially, and overtop, in the downslope direction.

On a more fundamental level, viscous flow theory predicts that the slope of the preexisting surface is a major control on flow thickness such that flows thicken on low slopes and thin on high slopes [e.g., *Lister*, 1992; *Griffiths*, 2000]. As described above, when flows traverse irregular topography, the effect of slope may be overprinted by levee formation and overflows. To investigate the influence of slope on flow thickness, we track the thickness of the flow in a channel within a single, short-lived lobe (flow 2). By restricting our analysis to a short-lived flow lobe, we remove complications related to levee development and variations in lava flux. We calculate the local slope by measuring the maximum slope of a local surface approximated by a neighborhood of pixels around the pixel of interest [*Horn*, 1981] (Figure 22.17). Although this measure of preeruptive slope is noisy on the pixel scale (Figure 22.17a), the length scale of the slope that controls flow thickness is likely to be larger than the pixel size. To evaluate this hypothesis, we use the Soil Land Inference Model (SoLIM) software (University of Wisconsin-Madison) to change the neighborhood size used for the slope calculation [*Zhu et al.*, 2008]. Correlations between slope and flow thickness are weak (Figure 22.17b) but peak at a neighborhood size of 70–90 m (Figure 22.17c). This corresponds to approximately twice the flow width and suggests an upper limit to the DEM resolution required for accurate flow modeling. One portion of the channel is omitted from the analysis due to a pile-up of lava in the channel that did not drain. This is probably the origin of the overall general scatter in the plot, as different channel segments clearly drained different amounts.

Two other parameters important for determining the flow thickness are flux and viscosity. In most eruptions, volumetric flux from the vent decreases with time. Additionally, the complexity of the channel network within the flow causes downflow variations in flux within and between individual flow lobes. Viscosity also increases with distance because of increasing crystallinity and decreasing vesicularity [*Moore*, 1987; *Riker et al.*, 2009], although in high-effusion-rate flows from Mauna Loa, substantial viscosity increases occur only after flow transport distances of several kilometers.

Our analysis of the flow thickness map highlights variations in flux and volume distribution within flow lobes and between flow branches, variations that will control

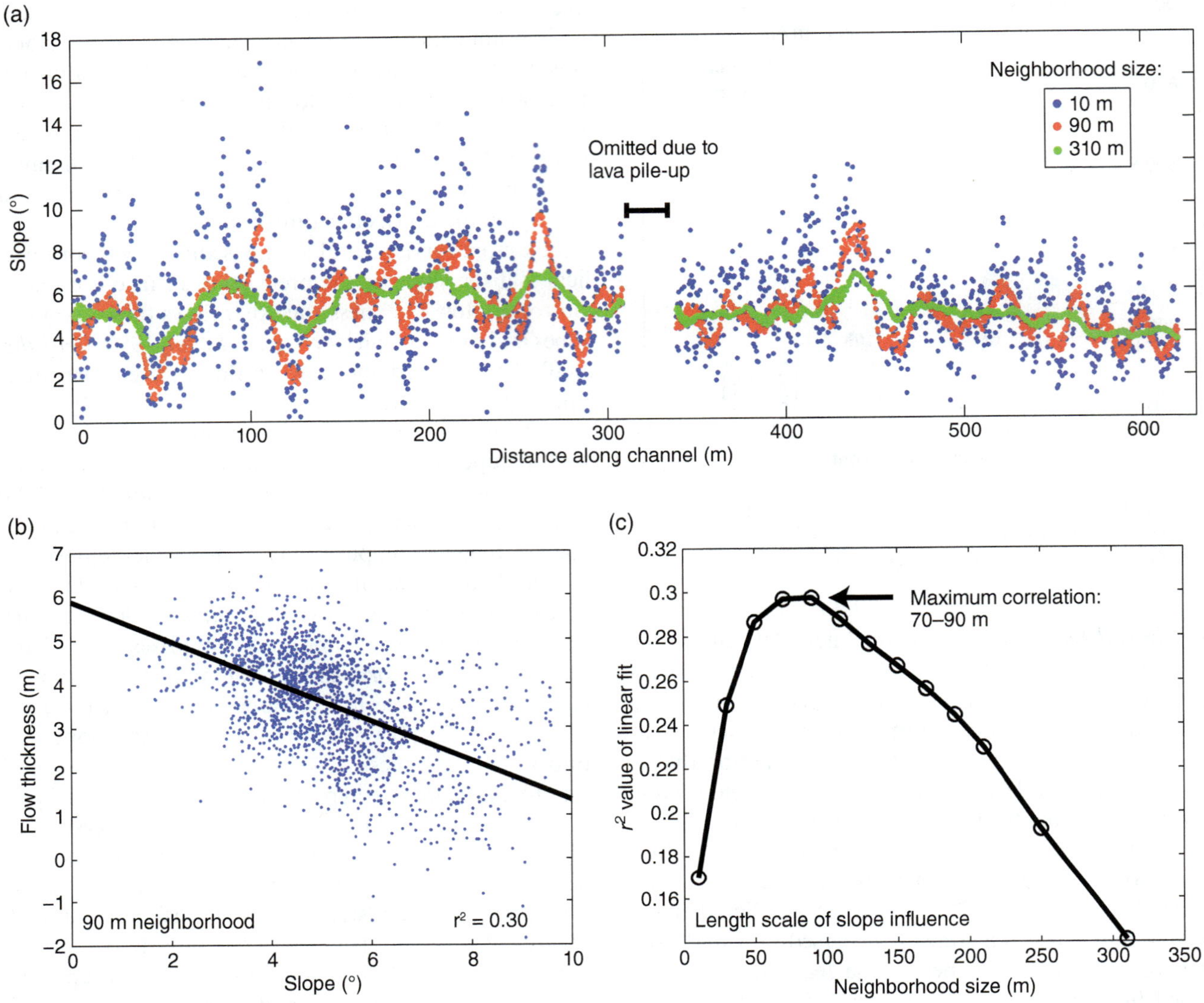

Figure 22.17 Comparison of lava flow thickness, represented along a channel in flow 2, to underlying slope at a variety of length scales. (a) Plot of slope along the channel showing the effects of using a different neighborhood size. (b) Example linear regression of slope and flow thickness for a 90 m neighborhood slope calculation. (c) Correlation, measured by the r^2 of linear fits, as in (b), for a range of neighborhood sizes.

flow advance rates and final flow lengths [*Kauahikaua et al.*, 2003]. Our data also reveal information about flow routing and conditions under which flows traverse topographic lavasheds. We show that flows travel down depressions, even when they cut across slopes, which can lead to overflows and thickening along the downslope side of the channel. The length scale over which slope influences flow thickness also reveals the necessary length scale, and therefore resolution, for modeling flow thickness changes and resulting flow behavior. We anticipate that increased availability of high-resolution DEMs will make routine the creation of high-quality flow thickness data that will permit more detailed comparisons of theoretical and actual conditions of lava flow emplacement.

22.4. CONCLUSIONS

Our use of lidar and photogrammetric topographic data to investigate the emplacement history of Hawaiian lava flows suggests new methods to quantify flow morphology and surface properties and to interpret the evolution and style of flow emplacement from 3D flow data. Importantly, high-resolution lidar topographic and intensity data allow unbiased analysis of surface characteristics that can be used to track the biological colonization of lava flows and to date flow surfaces. Surface roughness can also be used to detect morphological changes within lava flows, particularly the transition from ʻaʻā to pāhoehoe that is a characteristic indicator of downflow rheological changes in Hawaiian lava flows.

Together, these surface measures, combined with those of slope and curvature, offer an unbiased and unsupervised means of identifying lava flow features. The results of our classification provide statistics on the surface properties of flows and cones of separate ages and channel features that suggest potential for automated flow mapping.

When integrated with preeruptive imagery to build a photogrammetric DEM, the lidar data constrain measurements of flow thickness, adding a third dimension to our view of lava flow morphology. Analyses of volume distribution and local flow thickening reveal where lava was stored within the flow, emphasize the importance of side branches, and explain measured flux decreases along the Mauna Loa 1984 channel. Examination of the pre-eruptive surface with respect to the flow route and flow thickness completes our analysis, exposing some of the influence of preeruptive topography on flow emplacement and morphology. Taken together, these data illustrate the power of integrating pre- and postemplacement DEMs with field observations of emplacement chronology, effusion rates, and viscosities; our analysis demonstrates that such integration can provide significant insight into the dynamics of lava flow emplacement.

This work expands on new techniques and analyses that have been developed for Italian volcanoes by combining both surface and 3D analysis to address questions related to the emplacement of two Hawaiian lava flows. Detailed observations of Hawaiian lava flows, including work on flow surface morphologies and revegetation studies, provide necessary baseline data for our analysis. Our work also extends previous analyses by pioneering the 3D study of lava flow morphology in Hawai'i and at the same time provides a parallel and well-constrained data set to compare flow emplacement styles at Mt. Etna and Hawai'i. An important result of our work is demonstrating the effect of preexisting topography on Hawaiian lava, with associated implications for predictive models of future flow paths. Taken together, our combined analysis of lava flow surface characteristics and volume distribution shows the potential of such data for improving our understanding of lava flow emplacement and informing efforts to mitigate the hazards posed by effusive eruptions.

ACKNOWLEDGMENTS

We thank Natalia Deligne, Einat Lev, and Sam Nakata for help in the field and the Hawaiian Volcano Observatory staff for access to archive photos and information on the Mauna Loa 1984 eruption. Natalia Deligne also supplied the code to implement the McKean and Roering surface analysis. Lidar data were acquired by the National Center for Airborne Laser Mapping (NCALM) and aerial photography was collected by the U.S. Geological Survey. Reviews by Rebecca Carey, Francesco Mazzarini, and an anonymous reviewer greatly improved this manuscript. This work was funded by the National Science Foundation via a Graduate Research Fellowship under grant DGE-0829517 to HRD, NSF EAR 0738894 to KVC and SAS, and NSF EAR 1250554 to KVC. BHM was supported by a Rutherford Foundation Postdoctoral Fellowship.

REFERENCES

Anderson-Teixeira, K. J., P. M. Vitousek, and J. H. Brown (2008), Amplified temperature dependence in ecosystems developing on the lava flows of Mauna Loa, Hawai'i, *Proc. Natl. Acad. Sci.*, *105*, 228–233, doi:10.1073/pnas.0710214104.

Aplet, G. H., and P. M. Vitousek (1994), An age–altitude matrix analysis of Hawaiian rain-forest succession, *J. Ecol.*, *82*, 137–147.

Aplet, G. H., R. F. Hughes, and P. M. Vitousek (1998), Ecosystem development on Hawaiian lava flows: Biomass and species composition, *J. Veg. Sci.*, *9*, 17–26, doi:10.2307/3237219.

Applegarth, L. J., H. Pinkerton, M. R. James, and S. Calvari (2010), Morphological complexities and hazards during the emplacement of channel-fed 'a'ā lava flow fields: A study of the 2001 lower flow field on Etna, *Bull. Volcanol.*, *72*, 641–656, doi:10.1007/s00445-010-0351-1.

Branca, S., E. De Beni, and C. Proietti (2013), The large and destructive 1669 AD eruption at Etna volcano: Reconstruction of the lava flow field evolution and effusion rate trend, *Bull. Volcanol.*, *75*, 694, doi:10.1007/s00445-013-0694-5.

Bretar, F., M. Arab-Sedze, J. Champion, M. Pierrot-Deseilligny, E. Heggy, and S. Jacquemoud (2013), An advanced photogrammetric method to measure surface roughness: Application to volcanic terrains in the Piton de la Fournaise, Reunion Island, *Remote Sens. Environ.*, *135*, 1–11, doi:10.1016/j.rse.2013.03.026.

Campbell, B. A., and M. K. Shepard (1996), Lava flow surface roughness and depolarized radar scattering, *J. Geophys. Res.*, *101*(E8), 18,941–18,951, doi:10.1029/95JE01804.

Coltelli, M., C. Proietti, S., Branca, M. Marsella, D. Andronico, and L. Lodato (2007), Analysis of the 2001 lava flow eruption of Mt. Etna from three-dimensional mapping, *J. Geophys. Res.*, *112*, F02029, doi:10.1029/2006JF000598.

Crisp, J., K. V. Cashman, J. A. Bonini, S. B. Hougen, and D. C. Pieri (1994), Crystallization history of the 1984 Mauna Loa lava flow, *J. Geophys. Res.*, *99*(B4), 7177–7198, doi:10.1029/93JB02973.

Deardorff, N. D., and K. V. Cashman (2012), Emplacement conditions of the c. 1,600-year BP Collier Cone lava flow, Oregon: A LiDAR investigation, *Bull. Volcanol.*, *74*, 2051–2066, doi:10.1007/s00445-012-0650-9.

Deligne, N. I., K. V. Cashman, and J. J. Roering (2012), After the lava flow: The importance of external soil sources for plant colonization of recent lava flows in the central Oregon Cascades, *USA, Geomorphology*, *202*, 15–32, doi:10.1016/j.geomorph.2012.12.009.

Dietterich, H. R., M. P. Poland, D. A. Schmidt, K. V. Cashman, D. R. Sherrod, and A. T. Espinosa (2012), Tracking lava flow emplacement on the east rift zone of Kīlauea, Hawai'i, with synthetic aperture radar coherence, *Geochem. Geophys. Geosyst.*, *13*, Q05001, doi:10.1029/2011GC004016.

Favalli, M., M. T. Pareschi, A. Neri, and I. Isola (2005), Forecasting lava flow paths by a stochastic approach, *Geophys. Res. Lett.*, *32*, L03305, doi:10.1029/2004GL021718.

Favalli, M., A. Fornaciai, and M. T. Pareschi (2009a), LIDAR strip adjustment: Application to volcanic areas, *Geomorphology*, *111*, 123–135, doi:10.1016/j.geomorph.2009.04.010.

Favalli, M., F. Mazzarini, M. T. Pareschi, and E. Boschi (2009b), Topographic control on lava flow paths at Mount Etna, Italy: Implications for hazard assessment, *J. Geophys. Res.*, *114*, F01019, doi:10.1029/2007JF000918.

Favalli, M., A. J. L. Harris, A. Fornaciai, M. T. Pareschi, and F. Mazzarini (2010), The distal segment of Etna's 2001 basaltic lava flow, *Bull. Volcanol.*, *72*, 119–127, doi:10.1007/s00445-009-0300-z.

Favalli, M., S. Tarquini, P. Papale, A. Fornaciai, and E. Boschi (2012), Lava flow hazard and risk at Mt. Cameroon volcano, *Bull. Volcanol.*, *74*, 423–439, doi:10.1007/s00445-011-0540-6.

Gaddis, L., P. Mouginis-Mark, and J. A. Hayashi (1990), Lava flow surface textures—SIR-B radar image texture, field observations, and terrain measurements, *Photogramm. Eng. Remote Sens.*, *56*, 211–224.

Griffiths, R. W. (2000), The dynamics of lava flows, *Annu. Rev. Fluid Mech.*, *32*, 477–518, doi:10.1146/annurev.fluid.32.1.477.

Hollingsworth, J., S. Leprince, F. Ayoub, and J.-P. Avouac (2013), New constraints on dike injection and fault slip during the 1975–1984 Krafla rift crisis, NE Iceland, *J. Geophys. Res. Solid Earth*, *118*, 3707–3727, doi:10.1002/jgrb.50223.

Horn, B. K. P. (1981), Hill shading and the reflectance map, *Proc. IEEE*, *69*(1), 14–47, doi:10.1109/PROC.1981.11918.

Jackson, T. A. (1971), A study of the ecology of pioneer lichens, mosses, and algae on recent Hawaiian lava flows, *Pacific Sci.*, *25*, 22–32.

Kauahikaua, J., D. R. Sherrod, K. V. Cashman, C. Heliker, K. Hon, T. N. Mattox, and J. A. Johnson (2003), Hawaiian lava-flow dynamics during the Pu'u Ō'ō-Kupianaha eruption: A tale of two decades, in *The Pu'u 'Ō'ō-Kupaianaha Eruption of Kīlauea Volcano, Hawai'i: The First 20 Years*, edited by C. Heliker et al., U.S. Geol. Surv. Prof. Pap. 1676, pp. 63–87.

Lipman, P. W., and N. G. Banks (1987), Aa flow dynamics, Mauna Loa 1984, in *Volcanism in Hawaii*, edited by R. W. Decker et al., U.S. Geol. Surv. Prof. Pap. 1350, pp. 1527–1567.

Lister, J. R. (1992), Viscous flows down an inclined plane from point and line sources, *J. Fluid Mech.*, *242*, 631–653, doi:10.1017/S0022112092002520.

Lockwood, J. P., N. G. Banks, T. T. English, L. P. Greenland, D. B. Jackson, D. J. Johnson, R. Y. Koyanagi, K. A. McGee, A. T. Okamura, and J. M. Rhodes (1985), The 1984 eruption of Mauna Loa Volcano, Hawaii, Eos Trans. *AGU*, *66*(16), 169–171, doi:10.1029/EO066i016p00169-01.

Lockwood, J. P., R. I. Tilling, R. T. Holcomb, F. Klein, A. Okamura, and D. Peterson (1999), Magma migration and resupply during the 1974 summit eruptions of Kilauea volcano, Hawaii, U.S. Geol. Surv. Prof. Pap. 1613.

Lu, Z., T. Masterlark, and D. Dzurisin (2005), Interferometric synthetic aperture radar study of Okmok volcano, Alaska, 1992–2003: Magma supply dynamics and postemplacement lava flow deformation, *J. Geophys. Res.*, *110*, B02403, doi:10.1029/2004JB003148.

Marsella, M., C. Proietti, A. Sonnessa, M. Coltelli, P. Tommasi, and E. Bernardo (2009), The evolution of the Sciara del Fuoco subaerial slope during the 2007 Stromboli eruption: Relation between deformation processes and effusive activity, *J. Volcanol. Geotherm. Res.*, *182*, 201–213, doi:10.1016/j.jvolgeores.2009.02.002.

Marsella, M., P. Baldi, M. Coltelli, and M. Fabris (2012), The morphological evolution of the Sciara del Fuoco since 1868: Reconstructing the effusive activity at Stromboli volcano, *Bull. Volcanol.*, *74*, 231–248, doi:10.1007/s00445-011-0516-6.

Mazzarini, F., M. T. Pareschi, M. Favalli, I. Isola, S. Tarquini, and E. Boschi (2005), Morphology of basaltic lava channels during the Mt. Etna September 2004 eruption from airborne laser altimeter data, *Geophys. Res. Lett.*, *32*, L04305, doi:10.1029/2004GL021815.

Mazzarini, F., M. T. Pareschi, M. Favalli, I. Isola, S. Tarquini, and E. Boschi (2007), Lava flow identification and aging by means of lidar intensity: Mount Etna case, *J. Geophys. Res.*, *112*, B2, doi:10.1029/2005JB004166.

McKean, J., and J. Roering (2004), Objective landslide detection and surface morphology mapping using high-resolution airborne laser altimetry, *Geomorphology*, *57*(3–4), 331–351, doi:10.1016/S0169-555X(03)00164-8.

Moore, H. J. (1987), Preliminary estimates of the rheological properties of the 1984 Mauna Loa lava, in *Volcanism in Hawaii*, edited by R. W. Decker et al., U.S. Geol. Surv. Prof. Pap. 1350, pp. 1569–1588.

Morris, A. R., F. S. Anderson, P. J. Mouginis-Mark, A. F. C. Haldemann, B. A. Brooks, and J. Foster (2008), Roughness of Hawaiian volcanic terrains, *J. Geophys. Res.*, *113*, E12007, doi:10.1029/2008JE003079.

Neri, M., F. Mazzarini, S. Tarquini, M. Bisson, I. Isola, B. Behncke, and M. T. Pareschi (2008), The changing face of Mount Etna's summit area documented with Lidar technology, *Geophys. Res. Lett.*, *35*, L09305, doi:10.1029/2008GL033740.

Passalacqua, P., T. D. Trung, E. Foufoula-Georgiou, G. Sapiro, and W. E. Dietrich (2010), A geometric framework for channel network extraction from lidar: Nonlinear diffusion and geodesic paths, *J. Geophys. Res.*, *115*, F01002, doi:10.1029/2009JF001254.

Pyle, D. M., and J. R. Elliot (2006), Quantitative morphology, recent evolution, and future activity of the Kameni Islands volcano, Santorini, Greece, *Geosphere*, *2*, 253–268, doi:10.1130/GES00028.1.

Raich, J. W., A. E. Russell, and P. M. Vitousek (1997), Primary productivity and ecosystem development along an elevational gradient on Mauna Loa, Hawai'i, *Ecology*, *78*, 707–721.

Raich, J. W., W. J. Parton, A. E. Russell, R. L. Sanford, Jr., and P. M. Vitousek (2000), Analysis of factors regulating ecosystem development on Mauna Loa using the Century model, *Biogeochemistry*, *51*, 161–191, doi: 10.1023/A:1006495408992.

Riker, J. M., K. V. Cashman, J. P. Kauahikaua, and C. M. Montierth (2009), The length of channelized lava flows: Insight from the 1859 eruption of Mauna Loa Volcano, Hawai'i, *J. Volcanol. Geotherm. Res.*, *183*, 139–156, doi:10.1016/j.jvolgeores.2009.03.002.

Shepard, M. K., R. A. Brackett, and R. E. Arvidson (1995), Self-affine (fractal) topography: Surface parameterization and radar scattering, *J. Geophys. Res.*, *100*(E6), 11,709–11,718, doi:10.1029/95JE00664.

Soule, S. A., K. V. Cashman, and J. P. Kauahikaua (2004), Examining flow emplacement through the surface morphology of three rapidly emplaced, solidified lava flows, Kilauea Volcano, Hawai'i, *Bull. Volcanol.*, *66*, 1–14, doi:10.1007/s00445-003-0291-0.

Stevens, N. F., G. Wadge, C. A. Williams, J. G. Morley, J.-P. Muller, J. B. Murray, and M. Upton (2001), Surface movements of emplaced lava flows measured by synthetic aperture radar interferometry, *J. Geophys. Res.*, *106*(B6), 11,293–11,313, doi:10.1029/2000JB900425.

Tarquini, S., and M. Favalli (2010), Changes of the susceptibility to lava flow invasion induced by morphological modifications of an active volcano: The case of Mount Etna, Italy, Nat. *Hazards*, *54*, 537–546, doi:10.1007/s11069-009-9484-y.

Tarquini, S., M. Favalli, F. Mazzarini, I. Isola, and A. Fornaciai (2012), Morphometric analysis of lava flow units: Case study over LIDAR-derived topography at Mount Etna, Italy, *J. Volcanol. Geotherm. Res.*, *235–236*, 11–22, doi:10.1016/j.jvolgeores.2012.04.026.

Ventura, G., and G. Vilardo (2008), Emplacement mechanism of gravity flows inferred from high resolution Lidar data: The 1944 Somma-Vesuvius lava flow (Italy), *Geomorphology*, *95*, 223–235, doi:10.1016/j.geomorph.2007.06.005.

Wolfe, E. W., and J. Morris (1996), Geologic map of the Island of Hawaii, Misc. Invest. Ser. I–2524-A, U.S. Geol. Surv., Washington, D. C.

Zevenbergen, L. W., and C. R. Thorne (1987), Quantitative analysis of land surface topography, Earth Surf. *Proc. Land.*, *12*, 47–56, doi:10.1002/esp.3290120107.

Zhu, A., J. E. Burt, M. Smith, R. Wang, and J. Gao (2008), The impact of neighbourhood size on terrain derivatives and digital soil mapping, in Advances in Digital Terrain Analysis, edited by Q. Zhou et al., Springer, Berlin, doi:10.1007/978-3-540-77800-4_18.

23

Are Piton de la Fournaise (La Réunion) and Kīlauea (Hawai'i) Really "Analog Volcanoes"?

Aline Peltier[1], Michael P. Poland[2], and Thomas Staudacher[1]

ABSTRACT

The basaltic ocean island volcanoes of Kīlauea (Island of Hawai'i) and Piton de la Fournaise (La Réunion Island) are remarkable natural laboratories for volcanology. Both are near the active ends of long hotspot chains and host frequent eruptive activity (both effusive and explosive). Investigations of the geophysical, geochemical, and geologic manifestations of volcanism at the two locales provide insights into a variety of processes, from the supply of magma to its eruption at the surface. The oft-stated claim that Kīlauea and Piton de la Fournaise are "analog volcanoes" appears well founded, at least in part. The two volcanoes share similar characteristics with regard to magma plumbing, occurrence of flank instability, structural and compositional evolution, and eruptive activity. Noteworthy differences include the size of the edifices (Kīlauea is about an order-of-magnitude larger in volume), their life spans (Piton de la Fournaise has been active for twice as long), and flank instability mechanisms. These differences are attributable to Hawai'i's higher magma supply from and faster plate velocity relative to the hotspot compared to Piton de la Fournaise. Understanding these similarities and differences can be used to leverage studies of processes at one volcano by exploiting their occurrence or absence at the other.

23.1. INTRODUCTION

Basaltic hotspot volcanoes are common across Earth, particularly in the world's oceans, and arise from similar processes: an upwelling plume of material that generates melt in the upper mantle that can then erupt at the surface. The volcanic manifestations of oceanic hotspots, however, are often distinct from one edifice to the other. Kīlauea, on the Island of Hawai'i, and Piton de la Fournaise, on La Réunion Island, are two of the best-known basaltic hotspot volcanoes and are often cited as analogs for one another. Both are at the active ends of age-progressive hotspot tracks and host frequent eruptive activity that is accompanied by a variety of geophysical, geochemical, and geologic phenomena. But how similar are these two systems beyond these broad commonalities? The past several decades have seen an explosion in the monitoring systems and scientific investigations at both volcanoes as well as abundant eruptive activity with which to develop and test models. The time is therefore ripe to assess the similarities and differences between Kīlauea and Piton de la Fournaise, especially since, where processes are comparable, studies at one volcano may help to inform research at the other. By the same logic, understanding how the volcanoes differ will guard against the assumptions that accompany the "analog volcano" label, which may lead to unsupported lines of investigation.

We assess the similarities and differences between Kīlauea and Piton de la Fournaise across a number of parameters, including magma supply and tectonic setting, volcano evolution, flank instability, magma storage and transport, and eruption style. Our analysis reveals

[1]*Observatoire Volcanologique du Piton de la Fournaise, Institut de Physique du Globe de Paris, La Plaine des Cafres, La Réunion, France*

[2]*Hawaiian Volcano Observatory, U.S. Geological Survey, Hawai'i National Park, Hawaii, USA*

Hawaiian Volcanoes: From Source to Surface, Geophysical Monograph 208. First Edition.
Edited by Rebecca Carey, Valérie Cayol, Michael Poland, and Dominique Weis.

Companion Website: www.wiley.com/go/Carey/Hawaiian_Volcanoes

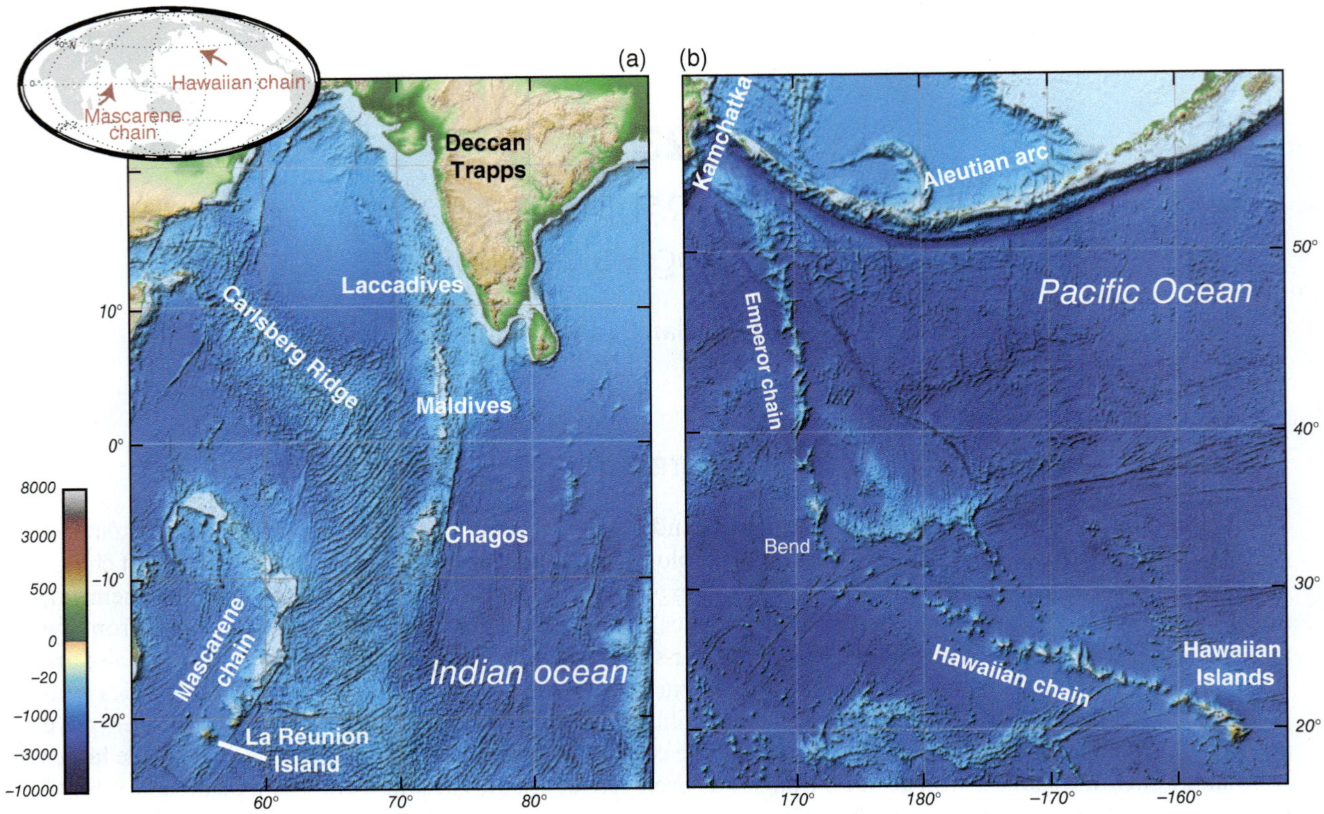

Figure 23.1 Topographic/bathymetric maps of (a) the Mascarene-Chagos-Maldives-Laccadives-Deccan Trapps and (b) the Hawaiian-Emperor chains of islands and seamounts. Data are from NOAA (ETOPO1).

that the biggest difference between the two volcanoes may be the timescale of their evolution (with Kīlauea erupting more material over a shorter time span). In many other ways, including magma plumbing, compositional and structural evolution, and eruptive activity, the two volcanoes are broadly similar.

23.2. PITON DE LA FOURNAISE VOLCANO

23.2.1. Geologic Setting

Piton de la Fournaise (21°14'37"S, 55°42'50"E) is the youngest volcano of La Réunion Island (Indian Ocean), which is the active expression of the 5000 km long hotspot chain that began with the Deccan Trap flood basalt 65 million years ago [*Duncan*, 1981; *Morgan*, 1981; *Courtillot et al.*, 1986] and led to the construction of Laccadives, Maldives, Chagos, and Mascarene islands (Figure 23.1a). A study of seismic anisotropy in the upper mantle beneath the Indian Ocean suggests that the Réunion mantle plume is located 100–200 km north of La Réunion Island and plume spreading reaches as far south as the island to feed the volcanism of Piton de la Fournaise [*Barruol and Fontaine*, 2013]. This is supported by the contemporaneous volcanism at La Réunion and Mauritius between 450 and 70 ka [*McDougall and Chamalaun*, 1969; *Gillot and Nativel*, 1982], suggesting that the plume lies somewhere in between the two islands and feeds the volcanoes of both.

Piton de la Fournaise first erupted about 500 ka [*McDougall*, 1971; *Gillot and Nativel*, 1989] on the eastern flank of the Piton des Neiges massif (5 million years old; e.g., *Gillot et al.* [1994]) (Figure 23.2a). Because of the uncertainty on the depth of Piton de la Fournaise's base, its volume cannot be determined with certainty, but it probably represents less than 3% of the whole volume of La Réunion Island, which is estimated at ~50,100 km^3 [*Villeneuve et al.*, 2014].

Piton de la Fournaise hosts three main calderas; the first formed ~150 ka, the second ~24 ka [*Gillot and Nativel*, 1989; *Staudacher and Allègre*, 1993], and the third, the Enclos Fouqué caldera (~13 × 9 km in size), ~4.5 ka. An active volcanic cone is located inside of the Enclos Fouqué caldera. Two craters cut this summit cone: the Bory crater (to the west) and the Dolomieu crater (to the east). The Dolomieu crater collapsed in April 2007 and formed a small kilometer-sized caldera [*Michon et al.*, 2007a; *Staudacher et al.*, 2009]. Three rift zones, elongated along N10–25 (northeast rift zone), N160–180 (southeast rift zone), and N120, radiate from the summit (Figure 23.2a).

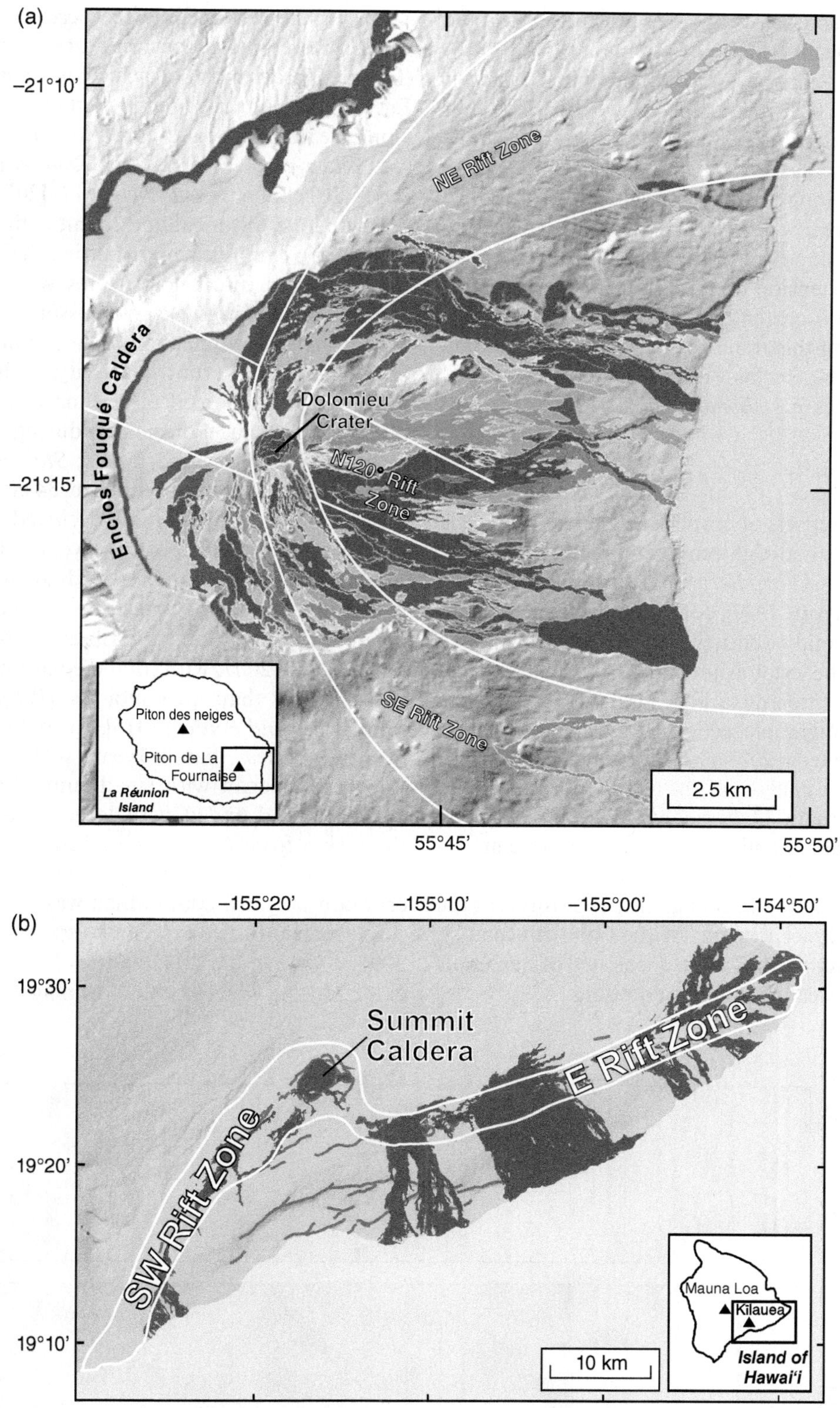

Figure 23.2 Shaded-relief maps of (a) Piton de la Fournaise and (b) Kīlauea volcanoes. (a) On the Piton de la Fournaise map, locations of the Dolomieu crater, the Enclos Fouqué caldera, the 1950–2011 lava flows (shaded), and the rift zones are shown. (b) On the Kīlauea map, the summit caldera and the rift zones are shown. The insets show the location of each volcano on their respective island.

The northeast and southeast rift zones are arcuate and fan shaped, and they extend outside of the Enclos Fouqué caldera to a few kilometers beyond the coastline [*Bachèlery*, 1981; *Bonali et al.*, 2011; *Michon et al.*, 2007b; *Saint-Ange et al.*, 2013]. The extension of the N120 rift zone from the summit cone to the east, highlighted by recent eruptive activity occurring along this direction in 2000, 2001, 2002, and 2007 [*Michon et al.*, 2007b], is limited and vanishes on the lower eastern slopes of the volcano (Figure 23.2a). Only the eastern extension of the N120 rift zone (east of the summit crater) is currently active, although eruptions have occurred along this trend to the west of the summit cone (outside of the Enclos Fouqué caldera) in historic and prehistoric times [e.g., *Bonali et al.*, 2011].

23.2.2. Recent Activity

The first written reports of eruptive activity at Piton de la Fournaise date to the mid-eighteenth century [e.g., *Héguerty*, 1754; *De Crémont*, 1770; *Bory de Saint Vincent*, 1804; *Hugoulin*, 1860; *Lacroix*, 1939], but the first detailed descriptions start in the mid-twentieth century and have been systematic since the establishment of the Observatoire Volcanologique du Piton de la Fournaise (OVPF) in 1980. The 340 m high stratigraphic section exposed by the collapse of the Dolomieu crater in 2007 allows for detailed lithostratigraphic and geologic analysis of the recent eruptive activity of Piton de la Fournaise [*Peltier et al.*, 2012; *Michon et al.*, 2013]. The eighteenth and nineteenth centuries were mostly characterized by long-lasting effusive eruptions and frequent phreatic and phreatomagmatic explosions at the volcano's summit. Since the start of the twentieth century, however, eruptive activity has mostly consisted of numerous discrete effusive or weakly explosive eruptions.

During the well-documented period from 1950 to 2010, 93 effusive eruptions occurred at Piton de la Fournaise (Figure 23.3a), most of which were located inside the Enclos Fouqué caldera (either within the summit craters or along the rift zones); a total of 970 Mm^3 of lava was emitted (see *Peltier et al.* [2009a, 2012], *Roult et al.* [2012], *Michon et al.* [2013] and *Staudacher et al.* [2015] for a complete review of the historical and recent activity of Piton de la Fournaise). Eruptive activity during that time consisted of the opening of en-echelon fissures, with activity eventually focusing on the lowest elevation fissure and forming a small cone from the accumulation of lava-fountaining products.

Two periods of eruptive activity, each preceded by five to six years of quiescence, have marked the recent history of Piton de la Fournaise: during 1972–1992 and 1998–2010 [*Peltier et al.*, 2009a; *Staudacher et al.*, 2015]. During the most recent sequence, 34 eruptions, each lasting for few days to months, occurred with a maximum inter-eruption repose of 1.3 years. Quasi-continuous summit inflation, a sign of a gradual build-up of pressure within the shallow magma reservoir due to long-term magma accumulation, was observed between eruptions during 2000–2007, with total displacement rates of up to 5 cm in less than two months [*Peltier et al.*, 2009a; *Staudacher and Peltier*, 2015]. Eruptive activity occurred in cycles of less than 1.5 years within this time period. Each cycle began with small summit (or near-summit) eruptions and ended with large distal eruptions more than 4 km from the summit [*Peltier et al.*, 2008, 2009a]. A change in this cyclicity occurred after the major eruption of April 2007, which was associated with large displacement of the east flank (up to 1.4 m [*Augier et al.*, 2008; *Clarke et al.*, 2013]) and collapse of the Dolomieu crater, during which the shallow magma plumbing system

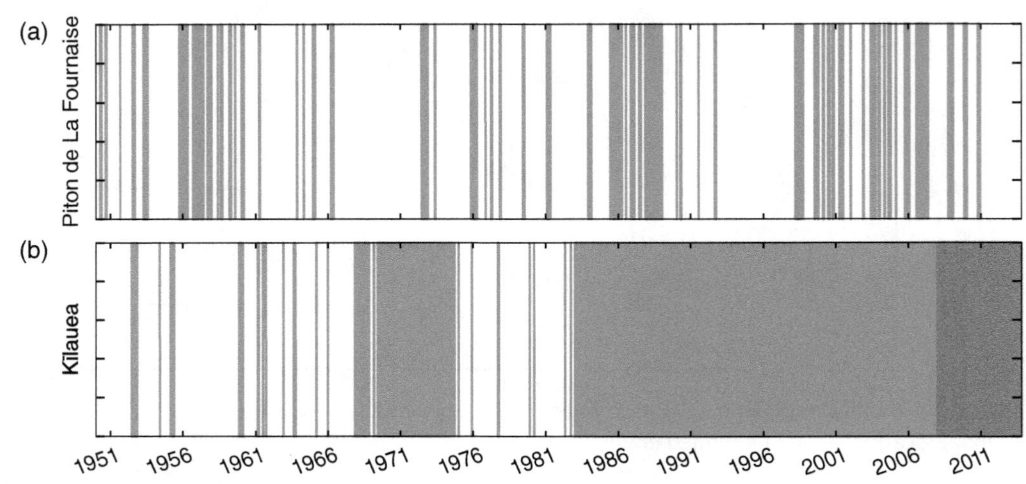

Figure 23.3 Eruptive activity over time between 1950 and 2013 at (a) Piton de la Fournaise and (b) Kīlauea. The dark shading during 2008–2013 for Kīlauea indicates two concurrent eruptions: one at the summit and the other on the East Rift Zone.

was largely drained [*Michon et al.*, 2007a; *Peltier et al.*, 2009b; *Staudacher et al.*, 2009]. The activity that followed (2008–2010) consisted of 8 low-volume eruptions and 18 non-eruptive intrusions [*Roult et al.*, 2012; *Staudacher et al.*, 2015]. This activity was not associated with detectable long-term magma accumulation (indicated by summit inflation), as was observed during 2000–2007; instead, surface deformation has been dominated by postcollapse relaxation. Stress perturbations inside the edifice and changes in the shape and refilling of the shallow plumbing system after the 2007 events may be the origin of the low level of eruptive activity observed in 2008–2010 [*Peltier et al.*, 2010; *Roult et al.*, 2012].

Since the end of 2010 through 2013, no eruption has occurred at Piton de la Fournaise; only slight summit deflation and ongoing seaward motion of the volcano's east flank have been recorded by the permanent global navigation satellite system (GNSS) (Figure 23.4a). Indeed, since the establishment of six permanent GNSS stations on the volcano's east flank in 2009 – 2010 [*Brenguier et al.*, 2012], continuous displacement of this flank toward the east (up to 2.1 cm/yr with plate velocity removed) accompanied by subsidence (up to 3.5 cm/yr with plate velocity removed) has been observed (Figure 23.4a).

23.2.3. Shallow Magma Plumbing System

The first model of Piton de la Fournaise's magma plumbing system was proposed by *Lénat and Bachèlery* [1990] and consisted of an array of sills and dikes slowly cooling at 0.5–1.5 km beneath the Dolomieu crater. More recent geodetic and seismic data indicate a shallow magma reservoir located around 2.5 km below the summit craters (±1 km; *Battaglia et al.*, 2005; *Peltier et al.*, 2008, 2009a; *Massin et al.*, 2011; *Prôno et al.*, 2009), although its shape (single reservoir versus dense network of interconnected sills/dikes) is still debated (Figure 23.5a). Two deeper and larger reservoirs are suggested by seismicity, one at around 7.5 km depth [*Battaglia et al.*, 2005] and the other located west of Enclos Fouqué caldera at around 15 km depth [*Peltier et al.*, 2009a], but their exact locations are poorly constrained (Figure 23.5a).

Most of the recent eruptions at Piton de la Fournaise have been fed by dikes propagating from or at least passing through the shallow plumbing system centered below the summit cone. Dikes propagate at a rate of 0.2–2 m/s [*Peltier et al.*, 2005; *Roult et al.*, 2012] and are emplaced in two stages: an initial vertical propagation centered below the summit crater followed by lateral propagation in the case of eruptions that occur along the rift zones [*Toutain et al.*, 1992; *Peltier et al.*, 2005], with dike emplacement typically lasting a few minutes to hours [*Peltier et al.*, 2005; *Roult et al.*, 2012]. Three-dimensional (3D) surface wave tomography based on seismic noise correlation shows that the northeast and southeast rift zones extend to a depth of 0.5 km below sea level (~3 km beneath the surface [*Brenguier et al.*, 2007]); the N120 rift zone is not recognizable from this tomography, and there is no evidence for molten material below these depths beneath the rift zones.

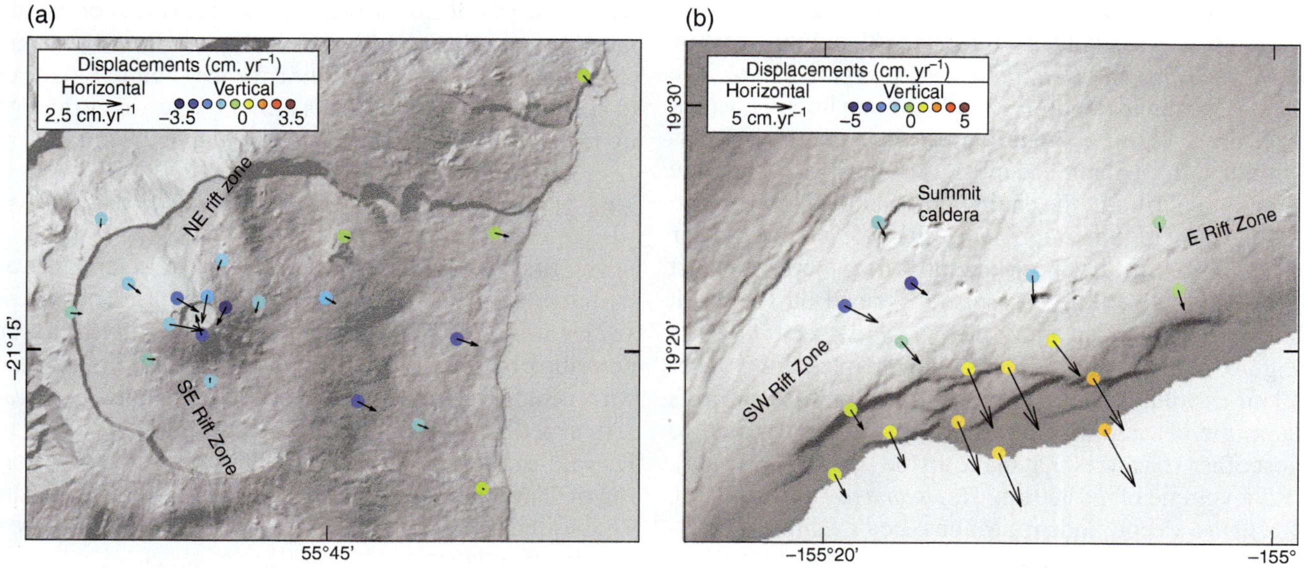

Figure 23.4 Velocities (in cm/yr) recorded on continuous GNSS stations at (a) Piton de la Fournaise (during 2012–2013, a time period that was characterized by no intrusion or other volcanic activity) and (b) Kīlauea (during 1997–2002, a time period that was characterized by mostly secular motion of the south flank and deflation of the summit, without major intrusion). Note that the scales of horizontal and vertical motion are different for each volcano.

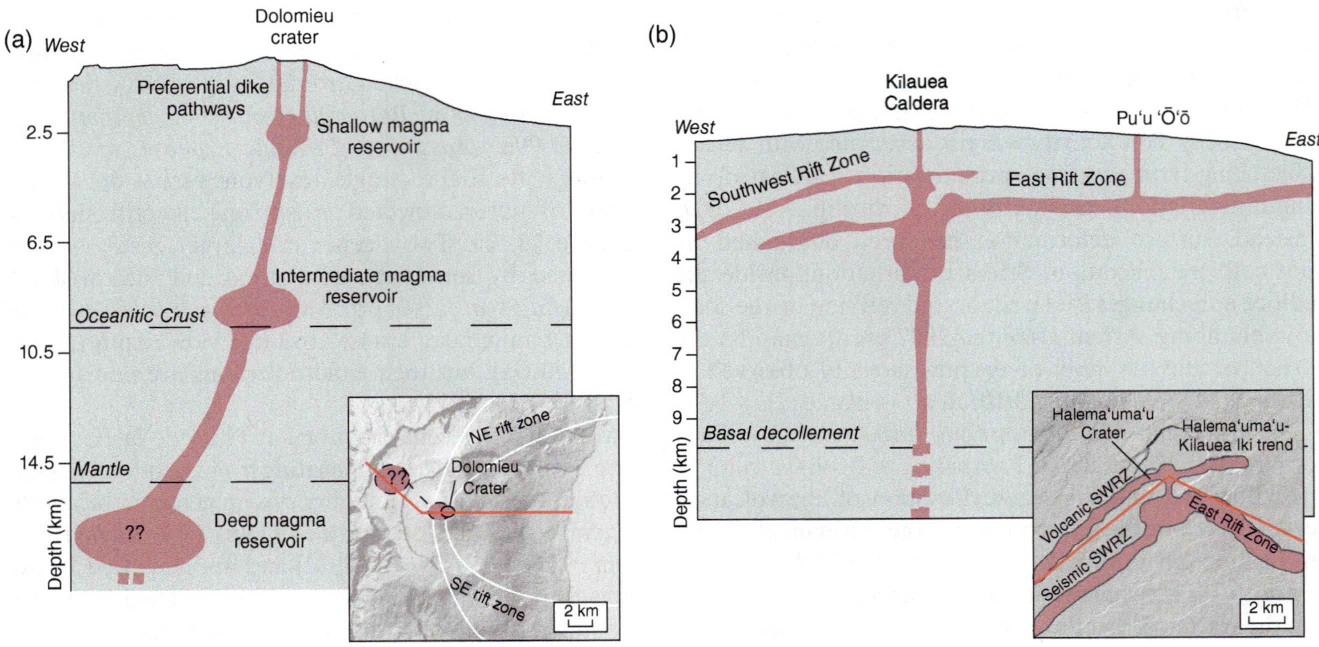

Figure 23.5 Schematic cross sections through the shallow magma plumbing systems of (a) Piton de la Fournaise [after *Battaglia et al.,* 2005; *Peltier et al.,* 2009a; *Prôno et al.,* 2009] and (b) Kīlauea [after *Poland et al.,* 2014]. Insets show plumbing in map view, with red lines giving the location of the cross sections.

23.3. KĪLAUEA VOLCANO

23.3.1. Geologic Setting

Kīlauea (19°25'00"N, 155°16'10"W) lies near the southeast end of the Hawaiian-Emperor chain of seamounts and islands, an age-progressive alignment of over 100 volcanoes that extends more than 6000 km across the North Pacific ocean basin and spans ~80 million years of activity [*Clague and Dalrymple*, 1987] (Figure 23.1b). The chain is the manifestation of the Hawaiian hotspot, which is fed by a plume of high-temperature material that is upwelling from mantle depths. Although it is not the youngest volcano in the chain (a distinction belonging to Lō'ihi, located about 35 km south of the Island of Hawai'i) Kīlauea has been the most active over the past 100 years [*Klein*, 1982]. Because it formed on the south flank of the larger and older Mauna Loa volcano (Figure 23.2b), there is considerable uncertainty over Kīlauea's volume. If Kīlauea is a veneer on Mauna Loa's flank, the volcano may be relatively small (compared to most other volcanoes that make up the Island of Hawai'i), with a volume of 11,000 km^3 [*Lipman and Calvert*, 2013]. If Kīlauea deeply indents Mauna Loa's flank [*Morgan*, 2006], however, the volume may be as much as 30,000 km^3 [*Robinson and Eakins*, 2006]. Submarine samples of preshield alkalic basalt suggest an initiation date of no later than about 280 ka [*Calvert and Lanphere*, 2006], but vigorous tholeiitic shield building did not commence until about 100–150 ka [*Lipman et al.*, 2006; *Lipman and Calvert*, 2013].

The gross structure of Kīlauea is defined by rift zones that radiate to the east and southwest from a summit caldera (Figure 23.2b). The East Rift Zone (ERZ) is by far the dominant of the two, extending for 130 km (75 km of which is offshore) [*Clague et al.*, 1995]. The Southwest Rift Zone (SWRZ), on the other hand, does not extend much beyond the coastline [*Holcomb*, 1987], about 30 km from the volcano's summit. The rift zones bound Kīlauea's unstable south flank, which is moving seaward at rates of up to ~10 cm/yr [*Owen et al.*, 1995] (Figure 23.4b).

23.3.2. Recent Activity

The first written description of Kīlauea's summit was by missionary William Ellis, who visited the volcano in 1823 [*Ellis*, 1963]. Eruptive activity continued to be described by missionaries for several decades thereafter, with geologic accounts supplied by occasional exploratory expeditions, including James Dwight Dana's visit to Hawai'i in 1840 [*Dana*, 1849, 1890]. Systematic monitoring of Hawaiian volcanoes began in 1911–1912 with the establishment of the Hawaiian Volcano Observatory on the rim of Kīlauea Caldera [*Tilling et al.*, 2014].

Kīlauea's eruptive activity prior to the visit of Ellis has been reconstructed from geologic deposits, from which it is clear that the volcano has experienced cycles of explosive and effusive behavior over the last few thousand years [*Fiske et al.*, 2009; *Swanson et al.*, 2012]. The current caldera formed in about A.D. 1500, after an ~60 year long effusive eruption that resurfaced much of the volcano's east flank and presaged an ~300 year long period of explosive activity from summit vents [*Swanson et al.*, 2012]. In addition to being recorded in geologic deposits, the caldera formation and preceding lava effusion may be referenced in Hawaiian oral traditions, which demonstrates another means of investigating pre-1823 volcanism, at least as far back as A.D. ~1000, when Polynesians began settling in Hawaiʻi [*Swanson*, 2008].

Following the postcaldera explosive period, which ended just prior to Ellis's visit, volcanism at Kīlauea was dominated by lava effusion and accumulation within the caldera, with occasional small eruptions from the ERZ and SWRZ [*Holcomb*, 1987]. The summit effusions ended abruptly in 1924, when lava withdrew from the summit, presumably to feed an ERZ intrusion, and a series of small explosions occurred as groundwater interacted with the hot, recently evacuated conduit [*Decker and Christiansen*, 1984; *Holcomb*, 1987]. Eruptive activity was minor after that time, with no eruption recorded between 1934 and 1952. Major ERZ eruptions in 1955 and 1960, the latter of which was preceded by the spectacular lava fountains of the Kīlauea Iki eruption at the volcano's summit, portended a new phase of Kīlauea activity that was dominated by eruptions from the ERZ. Between 1960 and 1983, numerous discrete eruptions occurred from the ERZ, including the nearly continuous 1969–1974 Mauna Ulu eruption, with occasional effusions from the summit and SWRZ [*Holcomb*, 1987] (Figure 23.3b). In 1983, high lava fountaining from a vent on the ERZ marked the opening of the currently ongoing (as of 2014) Puʻu ʻŌʻō eruption [*Heliker and Mattox,* 2003]. In 30+ years of activity, the eruption has shifted location and eruption style several times while effusing over 3 km^3 of lava (Chapter 18, this volume). In 2008, the ERZ activity was joined by the onset of the first summit eruption at Kīlauea in 26 years [*Wilson et al.*, 2008; *Houghton et al.*, 2011]. The summit eruption, also currently ongoing, is characterized by persistent emissions of gas and ash from a vent along the east margin of Halemaʻumaʻu Crater, within the Kīlauea Caldera (Figure 23.2b). The vent also hosts a circulating lava lake with a fluctuating level [*Patrick et al.*, 2011; *Orr et al.*, 2013], and the similar geochemistry of nearly coincident (in time) summit and Puʻu ʻŌʻō eruptive products indicates that the two vents are connected via Kīlauea's magma plumbing system (Chapter 20, this volume).

23.3.3. Shallow Magma Plumbing System

The first model of Kīlauea's magma plumbing system was developed by *Eaton and Murata* [1960], who used geophysical and geochemical data associated with eruptive activity in 1959 and 1960 to infer that magma rose from mantle depths into storage areas beneath the summit, from which it may erupt at the summit or move laterally along the rift zones to feed distal intrusions and eruptions. While this general view remains unchanged, the details of Kīlauea's presently active shallow plumbing have been refined as geophysical data have attained better sensitivity. The current view of Kīlauea's summit magma system includes multiple magma storage areas, with the main reservoir located beneath the south caldera at a depth of ~3 km [e.g., *Eaton*, 1962; *Thurber*, 1984; *Cervelli and Miklius*, 2003; *Baker and Amelung*, 2012] (Figure 23.5b). A smaller reservoir, which is a source of cyclic summit deformation and also deflates rapidly during rift zone dike emplacement, underlies the east margin of Halemaʻumaʻu Crater at 1–2 km depth [e.g., *Dawson et al.*, 1999; *Battaglia et al.*, 2003; *Cervelli and Miklius*, 2003; *Poland et al.*, 2009; *Baker and Amelung*, 2012] (see also Chapter 11, this volume). Magma may also be stored in a small reservoir at ~3 km beneath the southeast part of the caldera [*Dawson et al.*, 1999; *Battaglia et al.*, 2003; *Baker and Amelung*, 2012].

Kīlauea's rift zones are thought to contain molten or partially molten cores at depths of ~3 – 5 km through which magma may travel to distal sites of intrusion and eruption [*Johnson*, 1995] (Figure 23.5b). Beneath this depth, magma accumulation is suggested by a lack of seismicity [*Ryan*, 1988] and opening-mode deformation [e.g., *Delaney et al.*, 1990; *Owen et al.*, 1995; *Cayol et al.*, 2000], but high densities inferred from gravity surveys [*Kauahikaua et al.*, 2000] and elevated seismic velocities [*Okubo et al.*, 1997] argue for the presence of olivine cumulates, which may drive Kīlauea's south flank instability [*Clague and Denlinger*, 1994]. The character of the deep part of Kīlauea's rift zones therefore remains ambiguous.

23.4. COMPARING AND CONTRASTING PITON DE LA FOURNAISE AND KĪLAUEA

Due to their high levels of eruptive activity (Figure 23.3), which are especially well monitored with multidisciplinary sensor networks, Kīlauea and Piton de la Fournaise are ideal sites for detailed studies of eruptive mechanisms and associated phenomena, while ongoing research into the geology and geochemistry of the volcanoes is used to test models of how hotspot-related volcanic islands evolve. A better understanding of the similarities and differences between the two volcanoes can help support continued research into their activity by providing both complimentary and contradictory examples of various processes. Below, we explore the major elements of volcanism at Kīlauea and Piton de la Fournaise from the supply of magma at depth to its eruption at the surface.

23.4.1. Magma Supply and Tectonic Setting

Any comparison of Kīlauea and Piton de la Fournaise must begin with the broader context in which these volcanoes are situated. Both are basaltic ocean-island shields that are the products of hotspot volcanism, and both lie at or close to the presumed locations of the mantle plumes feeding the active ends of long hotspot chains (Figure 23.1). In addition, both volcanoes sit on old oceanic crust, i.e., ~60 Ma for La Réunion and ~90 Ma for Hawai'i [*Müller et al.*, 2008], implying crustal thicknesses of the same order [*Watts*, 1978; *White*, 1993]. The effective elastic thickness of crust beneath Hawai'i is 30–40 km [*Watts and ten Brink*, 1989; *Wessel*, 1993] and beneath La Réunion is ~15 − 20 km [*Driad*, 1997; *Tiwari et al.*, 2007]

The magma supply rates to the two volcanoes differ by an order of magnitude. The average supply to the shallow magma system at Kīlauea was steady from the 1950s through the early 2000s, with a rate of ~0.1 km^3/yr [*Swanson*, 1972]. This rate more than doubled during a short-lived surge in supply in 2003–2007 before returning to background levels [*Poland et al.*, 2012]. At Piton de la Fournaise, on the other hand, the supply rate is modeled as being 0.04 km^3/yr [*White*, 1993]. This difference in magma supply to the two volcanoes is reflected in the output of historical eruptions. During 1950–2010, La Réunion volcanism emitted ~1 km^3 of lava [*Staudacher et al.*, 2015], whereas Kīlauea erupted over 3 km^3 of lava in only half that time, between 1983 and 2013 (Chapter 18, this volume). Analysis of the hotspot chains indicates that the present rates of magma supply are not representative of the supply rates over the life of the hotspot. Melt production in Hawai'i is higher now than at any time in the past ~80 million years, the time over which the hotspot chain is currently exposed [*White*, 1993; *Van Ark and Lin*, 2004; *Vidal and Bonneville*, 2004]. In contrast, melt production at the La Réunion hotspot peaked in the early stages of that hotspot's life, during emplacement of the 65 Ma Deccan Traps flood basalt [*Courtillot et al.*, 1986; *White*, 1993]. Since the lithosphere beneath Hawai'i is slightly thicker (which should result in lower melt productions rates; *White*, 1993), the difference in melt production between the two hotspots suggests that the plume feeding Hawaiian volcanism is hotter and more buoyant than that of La Réunion.

Plume character exerts a primary control on the distribution and composition of volcanism at the surface. In Hawai'i, the last few million years of hotspot volcanism have produced a double chain of volcanoes, the Kea and Loa trends, that lie along parallel tracks and have distinct geochemical signatures [e.g., *Jackson et al.*, 1972; *Tatsumoto*, 1978]. The appearance of the dual chain occurred during the high melt production rates of the past few million years and also at the onset of strong manifestation of a geochemical component indicating recycled oceanic crust and sediments, one of several components being tapped by the melting region [*Weis et al.*, 2011]. How these components interact to form the dual chain is a subject of ongoing debate among geochemists [e.g., *Ren et al.*, 2005; *Pietruszka et al.*, 2013] and geodynamicists [e.g., *Farnetani and Hofmann*, 2010] (see also Chapter 3, this volume). In contrast to Hawai'i, there is only a single chain of volcanoes making up the La Réunion hotspot track (volcanic ridges located 100 km south southwest of La Réunion and oriented almost parallel to the Mauritius transform zone are probably controlled by structural alignments [*Deplus et al.*, 2009; *Saint-Ange*, 2009]). As therefore might be expected, the recent chemistry of La Réunion lavas reflects a more homogeneous source than that of Hawai'i, with a composition that includes either a plume with two distinct components or the entrainment of surrounding mantle by a uniform plume [e.g., *Pietruszka et al.*, 2009]. There is no chemical signature of subducted oceanic crust in the Réunion plume, indicating that the plume taps mantle that has not been influenced by recycled tectonic plates [*Vlastélic et al.*, 2006].

The higher supply rate in Hawai'i has implications for volcano size and duration and also for the behavior of the oceanic plate. The rate of Pacific Plate motion over the Hawaiian hotspot, about 10 cm/yr [*Gripp and Gordon*, 1990], is much faster than that of the African (Somalian) Plate over the La Réunion hotspot, about 2–2.5 cm/yr [*DeMets et al.*, 1994]. Hawaiian volcanoes, as a result, have shorter life spans than those of La Réunion. This shorter period of activity is compensated by a greater supply of magma in Hawai'i, allowing Hawaiian volcanoes to reach larger volumes than their counterparts from La Réunion (e.g., Piton de la Fournaise's volume represents less than 20%, and probably less than 10%, of Kīlauea's). The great volume of Hawaiian volcanoes emplaces a large load on the oceanic crust, causing the plate to subside beneath the island chain but creating flexural uplift away from the load. These flexural arches are sites of substantial off-axis volcanism [e.g., *Lipman et al.*, 1989; *Clague et al.*, 1990; *Frey et al.*, 2000], possibly driven by decompression of the asthenosphere beneath the uplift [*Bianco et al.*, 2005]. Such a model explains the alkalic nature of the arch volcanism and may also provide a mechanism for rejuvenated volcanism in Hawai'i. In stark contrast, plate flexure is not observed at La Réunion [*De Voogd et al.*, 1999]. According to *Michon et al.* [2007b], this lack of flexure may be due to thermal erosion of the slow-moving lithosphere by the plume feeding La Réunion volcanism and/or to strong underplating. The thermal erosion would lead to an isostatic disequilibrium and uplift of the lithosphere, balancing the load imposed by the volcanic edifice [*Michon et al.*,

2007b]. There is also no known well-developed off-axis volcanism at La Réunion, perhaps also a consequence of the lack of flexure. Only volcanic ridges (located 100 km south southwest of La Réunion and most probably related to the underlying oceanic crust [*Michon et al.*, 2007b]) and volcanic seamounts along the hotspot track (170–200 km southwest of La Réunion) have been detected by marine surveys [*Deplus et al.*, 2009].

The difference in the rate of plate motion between Hawai'i and La Réunion may also explain the discrepancy in the He isotopic ratio of the Mascarene and Hawaiian islands during their evolution. Most hotspot chains are characterized by He isotopic ratios (R/R_a; $R={}^3He/{}^4He$, R_a = atmospheric ratio of 1.384×10^{-6}) that range from close to midocean ridge basalt (MORB) (~8) to several times that value. *Graham et al.* [1993] measured R/R_a ratios between 8 and 23 in the Galápagos chain, similar to the range measured by *Farley et al.* [1993] in the Juan Fernandez archipelago (between 8 and 18). The highest and most primitive R/R_a values (up to 39) were measured in submarine rock samples from Lō'ihi (southeast of the Island of Hawai'i) [*Kurz and Jenkins*, 1982; *Hofmann et al.*, 2011]. A detailed study on Mauna Loa and Mauna Kea suggested that R/R_a ratios decline with time in Hawaiian volcanoes and that the highest R/R_a ratios occur in the first shield stages of the volcanoes, whereas these ratios approach typical MORB in the later shield stages [*Kurz et al.*, 1996; *Hofmann et al.*, 2011]. This trend led *Kurz et al.* [1995] to conclude that the upwelling mantle plume is heterogeneous and bends in response to the fast plate motion. The plume therefore erupts more primitive material during the early stage of shield growth and more asthenosphere-like material ($R/R_a = 8$) later in the shield stage as mantle material contaminates the plume over time. Alternatively, *Hofmann et al.* [2011] suggested that helium is partitioned into less dense and less viscous carbonatite melt, which becomes decoupled from a tilted plume (deflected due to plate-driven upper mantle flow) and rises more vertically to feed volcanoes at early stages of their growth.

The helium isotopic composition of La Réunion, in contrast, displays a surprisingly homogeneous R/R_a ratio of 13, which is the same for the most recent lava sampled on Piton de la Fournaise [*Staudacher et al.*, 1990], the oldest exposed rock (2 Ma) of Piton des Neiges, and rocks from Mauritius and the 65 million year old Deccan Traps [*Basu et al.*, 1993]. Slow plate motion at La Réunion may prevent significant tilting of the plume, which would reduce the potential for both contamination by asthenospheric material and separation of helium-rich carbonatite melts. These helium data, in addition to Sr and Nd systematics, reflect a lack of depletion in U and Th that accounts for the less primitive (lower) helium ratios [*Vlastélic et al.*, 2006] of La Réunion compared to Hawai'i, whereas the consistency of the helium ratio over time at La Réunion may be a consequence of the slow plate motion.

23.4.2. Island Evolution

Despite the difference in geochemical and geodynamic complexity of the source, the geochemical evolution of the volcanoes of Hawai'i and La Réunion is broadly similar. Hawaiian volcanoes progress through four stages of volcanism, each marked by a distinct chemical composition: alkalic preshield, tholeiitic shield, alkalic postshield, and alkalic rejuvenated [*Clague and Dalrymple*, 1987]. Kīlauea is currently in the midst of the tholeiitic shield-building stage, which marks the most vigorous period of volcano growth. The transitions from alkalic to tholeiitic and back to alkalic seem to be related to changes in the degree of partial melting [*Frey et al.*, 1991; *Garcia et al.*, 1995]. Piton de la Fournaise appears to be in the midst of a similar geochemical succession. Although all lavas erupted from the volcano are transitional between alkalic and tholeiitic, the progression over time has been from mildly alkalic to slightly tholeiitic, an evolution that has been ascribed to a shallowing depth of fractionation over time [*Albarède et al.*, 1997]. Late-stage volcanism in both Hawai'i and La Réunion includes differentiated magmas, as exemplified by Mauna Kea [*Frey et al.*, 1991] and Piton des Neiges [*Gillot and Nativel*, 1982], which suggests a decline in the supply rate of mantle-derived magma and that the volcanoes evolve in a closed system during their waning phases of activity [*Deniel et al.*, 1992].

The structural evolution of the volcanoes of Hawai'i and La Réunion share many similarities, e.g., flank instabilities (discussed further in Section 23.4.3), formation of calderas, and potentially the development of rift zones. The latter feature is of particular importance, given that rift zones can control both volcanic and local tectonic activity. The volcanoes of both hotspot chains grew sequentially, with younger volcanoes often forming on the flanks of older ones. Kīlauea grew on the flank of Mauna Loa [*Lipman and Calvert*, 2013], whereas Piton de la Fournaise was constructed either on the flank of Piton des Neiges [e.g., *Le Friant et al.*, 2011] or ProtoFournaise (Les Alizées) [e.g., *Malengreau et al.*, 1999] (Figure 23.2). This differs from some other hotspot chains, e.g., the Galápagos, where the volcanoes grew concurrently [*Simkin*, 1984]. Sequential development of volcanoes is thought to control the formation of rift zones in Hawai'i, with rift zone orientation in the younger volcano determined by the geometry of the older volcano [*Fiske and Jackson*, 1972]. As a result, rift zones are major structural elements of Hawaiian volcanoes, serving as both magma transport pathways and important tectonic elements (Figures 23.2b and 23.5b). Geophysical surveys reveal that the deep parts of Hawaiian rift zones are dense [*Kauahikaua* et al., 2000] and have high seismic velocities [*Okubo et al.*, 1997], suggesting the presence of cumulates that may have been left behind as the volcano and its magma pathways grew upward over time. Hawaiian rift zones are also aerially extensive, i.e., Kīlauea's ERZ extends more than 130 km from the summit, and can separate tectonic provinces, for example, at Kīlauea dividing the stable and unstable flanks [e.g., *Owen et al.*, 1995]. Kīlauea's rift zones have apparently migrated south with time due to seaward motion of the volcano's unstable south flank [*Swanson et al.*, 1976].

In contrast, the three rift zones of Piton de la Fournaise (northeast, southeast, and N120) are less extensive and less developed (Figure 23.2a and 23.5a). Surface wave tomography based on seismic noise correlation and deformation data suggests that they do not reach deeper than 3 km [*Brenguier et al.*, 2007; *Peltier et al.*, 2009a; *Fukushima et al.*, 2010], and they extend only a few kilometers offshore [*Bachèlery*, 1981; *Michon et al.*, 2007b]. The mechanism for rift zone formation at Piton de la Fournaise is ambiguous. The northeast and N120 rift zones align with crustal structures [*Michon et al.*, 2007b], and intrusions into the northeast rift zone have been shown to create a stress field that promotes intrusions along the N120 trend [*Michon et al.*, 2009]. Rift zone formation may therefore result from a combination of the regional stress field induced by uplift of the oceanic crust and the local stress field related to the dynamic behavior of the volcano, possibly including the stresses imparted by flank instability [*Walter and Troll*, 2003]. The Piton de la Fournaise rift zones may also be controlled by preexisting structures, as is the case at Kīlauea. Two-dimensional models of gravity measurements indicate two dense structures, with a root at ~3.5 km beneath the surface, that are interpreted to be hypovolcanic intrusive complexes, i.e., the Grand Brûlé complex (10 × 15 km in horizontal extent on the eastern flank of the volcano) and the Plaine des Sables-Enclos Fouqué complex (5 km in diameter located west of the Enclos Fouqué caldera), possibly associated with volcanism that predates Piton de la Fournaise [*Gailler et al.*, 2009]. These structures bracket the approximate sea-level depth northeast and southeast rift zones, suggesting that rift zone geometry would also be controlled by the configuration of the ancient volcanic forms. The N120 rift zone, which is not detectable in tomographic images [*Brenguier et al.*, 2007] and therefore probably not deeply rooted, may be solely a consequence of regional and/or local stresses.

23.4.3. Flank Instability

Similar to many large basaltic shield volcanoes, both Kīlauea and Piton de la Fournaise display varying styles of flank instability. Hawai'i is a type locality for instability of oceanic volcanoes, with evidence of catastrophic collapse [*Moore et al.*, 1989] and active creep [e.g., *Swanson et al.*, 1976; *Owen et al.*, 1995]. Large-scale debris avalanches are also evident offshore of La Réunion [e.g., *Lénat et al.*, 1990; *Oehler et al.*, 2005; *Lebas*, 2012], and recent deformation data indicate seaward motion of the east flank of Piton de la Fournaise [*Augier et al.*, 2008; *Clarke et al.*, 2013].

In the main Hawaiian Islands, seafloor mapping has revealed the presence of at least 17 distinct landslides with runout lengths and volumes of up to 200 km and 5000 km^3, respectively [*Moore et al.*, 1989, 1994a,b]. The largest landslides are comparable to the volumes of some individual volcanoes, for example, Māhukona volcano may have a volume of as little as 6000 km^3 [*Lipman and Calvert*, 2013]. Numerous avalanche-like deposits have similarly been identified offshore of Piton de la Fournaise from marine data and have been interpreted as resulting from flank collapses [*Lénat et al.*, 1990; *Labazuy*, 1996; *Ollier et al.*, 1998; *Bachèlery et al.*, 2003; *Oehler et al.*, 2008; *Le Friant et al.*, 2011] or gravitationally driven volcanic spreading [*Oehler et al.*, 2005; *Delcamp et al.*, 2008; *Michon and Saint-Ange*, 2008]. The eastern submarine flank of the volcano is covered by more than 500 km^3 of avalanche deposits; the maximum runout reaches 75 km [*Labazuy*, 1996; *Oehler et al.*, 2005; *Saint-Ange*, 2009; *Lebas*, 2012].

Kīlauea's active flank instability was first quantified in the 1960s [*Swanson et al.*, 1976] and is manifested by steady seaward motion and uplift of the south flank by several centimeters per year (Figure 23.4b). This pattern of deformation, combined with rift zone subsidence, indicates slip along a low-angle thrust (décollement) fault that lies at the boundary between the preexisting seafloor and volcanic pile at 7–10 km beneath the surface [e.g., *Delaney et al.*, 1990; *Denlinger and Okubo*, 1995; *Owen et al.*, 1995, 2000; *Cayol et al.*, 2000]. The décollement is indicated by seismicity that defines a plane at ~8 km depth [e.g., *Got and Okubo*, 2003] and is required by models of coupled flank and rift zone deformation [e.g., *Delaney et al.*, 1990; *Owen et al.*, 1995, 2000; *Cayol et al.*, 2000]. The submarine morphology of Kīlauea's south flank is characterized by compressional structures [e.g., *Morgan et al.*, 2000] that, together with uplift of the flank recorded by GNSS (Figure 23.4b), argue for active décollement faulting. Décollement slip occurs in three distinct styles: steady motion [e.g., *Owen et al.*, 1995], quasi-periodic slow slip events [*Brooks et al.*, 2006; *Montgomery-Brown et al.*, 2010] (see also Chapter 13, this volume), and strong earthquakes [*Lipman et al.*, 1985].

Although instability of Piton de la Fournaise's east flank, which is open to the sea, is implied by its morphology and was suggested by asymmetrical deformation that accompanied eruptions [*Cayol and Cornet*, 1998; *Peltier et al.*, 2007, 2008, 2009a], its active motion was not clearly confirmed until 2007, when a distal eruption was accompanied by strong displacement of the flank [*Tinard*, 2007; *Peltier et al.*, 2009a; *Fukushima et al.*, 2010; *Got et al.*, 2013]. Interferometric synthetic aperture radar (InSAR) measurements from the April 2007 distal eruption detected up to 1.4 m of eastward motion of the volcano's east flank associated with a large lava effusion and collapse of Dolomieu summit crater [*Augier et al.*, 2008; *Clarke et al.*, 2013]. Six permanent GNSS stations were subsequently established on the east flank in 2009–2010, and they have recorded continuous eastward motion of up to 2.1 cm/yr and subsidence of up to 3.5 cm/yr through 2009–2013 (Figure 23.4a). Although the continuous GNSS time series suggests steady flank motion, similar to that recorded at Kīlauea, the short time spanned by the measurements introduces uncertainty in such a conclusion, the extent to which lava flow cooling and contraction contribute to the deformation signal, for example, is ambiguous. GNSS surveys on the east flank of Piton de la Fournaise during the 1990s and early 2000s show no detectable motion [*Villeneuve*, 2000], although the measurement stations may have been located in areas (close to the sea) where deformation would not be resolvable (as inferred from not-continuous GNSS data recorded at the same locations during 2009–2013). InSAR similarly has shown no evidence for steady flank motion prior to 2007 [*Tinard*, 2007], but the displacement rate might have been below the detection threshold. As an alternative model of steady motion, flank deformation at Piton de la Fournaise may be episodic and controlled by flank eruptions, with high rates of coeruptive flank displacement that decay over time [*Got et al.*, 2013]. Available data are not sufficient to distinguish between these possibilities, although the April 2007 coeruptive flank motion clearly demonstrates the importance of magmatism to flank instability.

As at Kīlauea, a décollement has been proposed to underlie Piton de la Fournaise [*Hirn et al.*, 1991; *Battaglia et al.*, 2005], but the data needed to test the existence and location of such a plane are sparse. Indeed, a model of deformation associated with a flank eruption in February 2000 found that no slip occurred along a basal plane underlying the east flank of the volcano [*Fukushima et al.*, 2005], while flank displacements in 2007 have been modeled as the result of sill intrusion and shearing along a detachment [*Chaput et al.*, 2014a]. Although the ocean crust at La Réunion is located 7–8 km beneath the surface, similar to Kīlauea, Piton de la Fournaise lies on the flank of the older Piton des Neiges [e.g., *Le Friant et al.*, 2011] or ProtoFournaise (Les Alizées) [e.g., *Malengreau et al.*, 1999]. The interface between Piton de la Fournaise and the underlying edifice is a potential décollement plane, but current data argue that no décollement exists beneath the volcano. Seismicity beneath the volcano's east flank is rare, only a few events per year occur as part of two clusters between 1 and 4 km below sea level, and earthquakes are not aligned along a plane, as observed beneath Kīlauea's south flank. The Piton de la Fournaise earthquakes are probably related to brittle fracture due to magmatic and/or gravitational stresses [*Nercessian et al.*, 1996]. Moreover, *Le Friant et al.* [2011] have reported a lack of compressional structures (such as those present at Kīlauea) that would be expected if a décollement existed, an observation that led these authors to propose that Piton de la Fournaise does not spread on a décollement.

The data described above demonstrate that the geometries of the unstable flanks of Kīlauea and Piton de la Fournaise are quite different. Kīlauea's south flank is characterized by compressional structures and experiences persistent deformation, including uplift at the coast and horizontal displacements that increase toward the coast. The mobile east flank of Piton de la Fournaise, on the other hand, shows no structural evidence for compression, subsidence occurs along the coast, horizontal displacements do not increase toward the coast, and strong flank motions appear to be associated with magmatic activities. These disparities suggest that the driving forces of flank motion at the two volcanoes are also distinct. Kīlauea's south flank motion is thought to be driven by gravitational collapse caused by outward flow of olivine cumulates that underlie the summit and rift zones [*Clague and Denlinger*, 1994]. Modeling suggests that gravitational loading alone can explain the deformation [*Got et al.*, 2008; *Plattner et al.*, 2013], but there is evidence from the coincidence of rift zone intrusions and aseismic slip events on the décollement [*Montgomery-Brown et al.*, 2010] (see also Chapter 13, this volume) and from stress modeling of rift zone intrusions [*Dieterich et al.*, 2003] that magmatic activity can trigger episodes of flank motion. Both mechanisms therefore probably play a role, but the extent to which they interact or one dominates over the other is a subject of debate. At Piton de la Fournaise, steady flank deformation measured by GNSS during 2009–2013, a period with no significant magma injection, suggests a gravitational driving force, although such an influence is probably minor. In stark contrast to observations from Kīlauea samples [*Helz*, 1987], cumulates from beneath Piton de la Fournaise show no evidence of plastic deformation [*Welsch et al.*, 2012], which suggests that cumulates do not creep at Piton de la Fournaise (although some plastically deformed olivine has been described in Piton de la Fournaise picrites [*Albarède et al.*, 1997]). Evidence in favor of magmatism driving Piton de la Fournaise flank motion is provided by coeruptive deformation, particularly in 2007 [*Got et al.*, 2013]. The April 2007 eruption emitted 240 Mm^3 of lava, 100 times more than the mean historical volume for individual eruptions [*Bachèlery et al.*, 2010; *Roult et al.*, 2012]. This large volume of injected magma, following a few months of stress accumulation, may have caused the eastern flank to exceed a plastic threshold, leading to the large displacement [*Got et al.*, 2013]. The displacement may occur due to slip or sill intrusion along a shallow detachment surface [*Chaput et al.*, 2014a,b], as opposed a basal décollement fault such as that beneath Kīlauea. Whether the catastrophic landslides suggested by deposits offshore of Piton de la Fournaise were also driven by eruptive activity is still debated. That mechanism has been proposed for giant Hawaiian landslides, although magma pressure alone does not seem sufficient to drive such large mass movements [*Iverson*, 1995]. An alternative mechanism, suggested by *Michon et al.* [2007b], is that some of La Réunion's landslides could result from reactivation of preexisting basement faults enhanced by the presence of hydrothermally weakened material.

23.4.4. Magma Storage and Transport

Multitiered magma storage appears to be common to many basaltic volcanoes. At least two magma reservoirs at different depths are present at Etna, Italy [*Aloisi et al.*, 2011], Fernandina, Galápagos [*Chadwick et al.*, 2011; *Bagnardi and Amelung*, 2012], Kīlauea [*Cervelli and Miklius*, 2003; *Baker and Amelung*, 2012], and Piton de la Fournaise [*Battaglia et al.*, 2005; *Peltier et al.*, 2009a; *Prôno et al.*, 2009]. The major difference between Kīlauea and Piton de la Fournaise in this respect is the depth of storage (Figure 23.5). At Kīlauea, a shallow reservoir is present 1–2 km beneath the east margin of Halema'uma'u Crater, near the caldera center, whereas the deeper reservoir, which constitutes the main magma storage area for the volcano, is located 3 km beneath the south part of the caldera [*Cervelli and Miklius*, 2003] (Figure 23.5b). Other small, shallow storage areas are also episodically active in the summit area [e.g., *Poland et al.*, 2014]. According to geodetic and seismic data, Piton de la Fournaise's primary reservoir is ~2.5 km beneath the summit craters (±1 km; *Battaglia et al.*, 2005; *Peltier et al.*, 2008, 2009a; *Massin et al.*, 2011; *Prôno et al.*, 2009). There is no geophysical evidence of magma storage above that level, although melt inclusions [*Bureau et al.*, 1998a, 1998b; *Famin et al.*, 2009] and mineral textures [*Welsch et al.*, 2009] suggest that olivine in erupted basalts grows at depths less than 2 km. This condition changed following the 2007 summit collapse, after which shallow magma storage was suggested by deformation [*Peltier et al.*, 2010] and geochemistry (Chapter 9, this volume), although this is probably transient. At deeper levels, magma storage is suspected from seismicity beneath Piton de la Fournaise at the 7.5 km depth [*Battaglia et al.*, 2005], near the boundary between the volcanic pile and the oceanic crust, as well as the ~15 km depth beneath La Plaine des Sables [*Peltier et al.*, 2009a] (Figure 23.5a). No similarly deep storage has been recognized by deformation or seismic imaging at Kīlauea, although a crustal or upper mantle reservoir has been proposed to serve as a filter that prevents eruption of xenoliths [*Clague*, 1987].

The mechanism for the "stacked" reservoir geometry probably relates to the density contrast between magma and host rock. Neutral buoyancy is commonly given as a reason for the level of magma accumulation [e.g., *Ryan*, 1987; *Tilling and Dvorak*, 1993; *Battaglia et al.*, 2005; *Peltier et al.*, 2009a; *Fukushima et al.*, 2010] and may explain the potential for magma storage at the contact between volcanic rock and ocean crust beneath Piton de la Fournaise (i.e., at ~7.5 km depth; Figure 23.5a). The 1–2 km depth reservoir at Kīlauea might reflect a density contrast between bubble-rich magma stored at that depth and denser, bubble-poor magma stored 3 km beneath the surface [*Clague and Dixon*, 2000]. The shallow reservoir is also apparently high enough within the volcanic edifice to allow for degassing of sulfur [*Gerlach*, 1986], as evidenced by persistent SO_2 emissions even during periods of quiescence [e.g., *Poland et al.*, 2009]. Degassing at Piton de la Fournaise occurs mainly during and immediately following eruptions, probably because the primary reservoir is too deep for gas species other than CO_2 to exsolve.

At Kīlauea, summit eruptions are generally fed from the 1–2 km depth reservoir, which is located near Halema'uma'u Crater, a site of nearly continuous lava lake activity during the nineteenth and early twentieth centuries, sporadic eruptions during the latter half of the twentieth century, and a lava lake starting in 2008 [*Holcomb*, 1987; *Orr et al.*, 2013]. Lateral dikes initiate from both the shallow 1–2 km depth storage zone and the 3 km depth reservoir. From the deeper reservoir, the lateral magma pathways mark the major rift zones that define the structural boundaries of the volcano. Both the East Rift Zone and Southwest Rift Zone probably have 3–5 km depth molten cores that grew from repeated intrusions [*Johnson*, 1995]. Evidence for such a configuration is provided by the 1960 Kapoho eruption, which occurred at the eastern tip of the subaerial part of the ERZ. The eruption started on January 13, but the summit of the volcano, ~50 km uprift, did not begin to subside, an indication of magma transport to the eruption site, until four days later [*Eaton and Murata*, 1960]. The eruption was therefore initiated not by a dike that propagated from the summit but rather from magma stored within the rift zone that eventually tapped the summit through the molten conduit. Magma currently feeding the 30+ year long Pu'u 'Ō'ō eruption utilizes the same conduit, and dikes that have interrupted that activity have initiated from a depth of ~3 km [e.g., *Montgomery-Brown et al.*, 2010; *Lundgren et al.*, 2013]. Lateral dikes that propagate away from the shallower reservoir follow slightly different paths, as indicated by alignments of eruptive fissures and fractures [*Holcomb*, 1987], but they also trend to the east and southwest from the summit. The 1919–1920 Mauna Iki eruption, ~10 km southwest of Kīlauea's, caldera, followed one of these routes. Magma was observed flowing through cracks just beneath the surface between the eruption site and the summit [*Jaggar*, 1919], suggesting that these pathways lie within a few hundred meters of the surface.

As at Kīlauea, eruptions at Piton de la Fournaise mainly occur in the summit area or along the rift zones. Vertical dike propagation is more common, however, for eruptions of Piton de la Fournaise as indicated by intensive geodetic monitoring that started in the 1990s [e.g., *Toutain et al.*, 1992; *Peltier et al.*, 2005, 2008]. The Piton de la Fournaise monitoring networks show that all recent dikes initiated below the summit craters from, or at least passing through, the 2.5 km depth reservoir. All dikes propagate first vertically from that reservoir (at a velocity of 0.2–0.8 m/s) before feeding summit eruptions or transitioning to lateral propagation (probably due to a density boundary [*Fukushima et al.*, 2010]) along the rift zones (at a depth of ~1.5 km and a velocity of ~2 m/s [*Peltier et al.*, 2005, 2009a]) to feed lateral eruptions. Two main dike pathways have been highlighted by deformation data, one that initiates from the west side of Dolomieu Crater that feeds summit and proximal (i.e., near-summit) eruptions and the other east of Dolomieu Crater that feeds distal eruptions [*Peltier et al.*, 2008, 2009a] (Figure 23.5a). Unlike Kīlauea's rift zones, those of Piton de la Fournaise do not appear to contain molten cores, at least during the last few decades. This difference may explain the longer-lived recent eruptions at Kīlauea compared to Piton de la Fournaise.

23.4.5. Eruptive Activity

Historical eruptive activity (which we define as that which occurred since written records were kept) at both Kīlauea and Piton de la Fournaise has been dominated by effusive eruptions of basaltic lava. At Kīlauea, these eruptions can have durations ranging from hours to decades (Figure 23.3b). Between 1950 and 1969, a number of small to moderate (~10 Mm^3) eruptions occurred from vents on the volcano's summit and ERZ [*Dzurisin et al.*, 1984], culminating in the 1969–1974 nearly continuous effusion from the ERZ Mauna Ulu vent, the longest-duration historical eruption to that time [*Holcomb*, 1987]. Kīlauea's summit magma system drained into the ERZ in response to the 1975 M_w 7.7 earthquake beneath the volcano's south flank [*Jachens and Eaton*, 1980; *Dzurisin et al.*, 1980], and for several years thereafter intrusions dominated over eruptions as the magma storage system adjusted to the rift opening that accompanied the earthquake. The onset of the 30+ year long Pu'u 'Ō'ō eruption in 1983 may indicate the complete recovery of Kīlauea's plumbing system [*Dzurisin et al.*, 1984; *Cayol et al.*, 2000] and, since that time, the vast majority of all magma that has been supplied to the volcano has been erupted [e.g., *Poland et al.*, 2012]. The effusion rate of Kīlauea eruptions varies broadly, ranging from 300–400 m^3/s for high lava-fountaining eruptions [*Rowland and Walker*, 1990] to ~4 m^3/s for long-term effusive eruptions [*Swanson*, 1972; *Heliker and Mattox*, 2003]. Kīlauea magma is dominated by olivine fractionation, although plagioclase and clinopyroxene also fractionate in the summit magma storage system [*Garcia et al.*, 2003] and within the rift zones [*Wright and Fiske*, 1971; *Garcia et al.*, 1989, 1992; *Thornber et al.*, 2003]. Mixing between freshly supplied and stored, differentiated magma also occurs beneath the summit [*Garcia et al.*, 2003] and along the rift zones [e.g., *Wright*, 1971; *Wright and Fiske*, 1971; *Clague et al.*, 1995]. Olivine-rich lava (often referred to as picrites) erupted particularly along the subaerial and submarine ERZ probably as a consequence of ascending magma that entrained olivine crystals from the base of the summit magma complex [*Clague et al.*, 1995], similar to one of the mechanisms proposed for eruption of picrites at Piton de la Fournaise [*Albarède and Tamagnan*, 1988].

In contrast to Kīlauea's frequent periods of continuous eruptive activity, Piton de la Fournaise has been characterized by discrete eruptions over the past several decades (Figure 23.3a), including activity confined to the summit craters, within close proximity to the summit craters, and low on the flanks. From 2000 to 2007, cycles of eruptive activity were apparent, with each cycle beginning with a succession of summit/proximal eruptions and ending with a larger distal eruption accompanied by seaward displacements of the eastern flank [*Peltier et al.*, 2008, 2009a]. The summit and proximal eruptions have low effusion rates (0.5–20 m^3/s) and low erupted volumes (often less than 3×10^6 m^3), whereas distal eruptions are characterized by larger volumes (between 15 and 240×10^6 m^3) and high effusion rates (with a peak of 300 m^3/s recorded during April 2007 [*Roult et al.*, 2012; *Staudacher et al.*, 2015]). Eruption durations range from 12 hours to 196 days, with an average of 20 days. The volume of intruded magma is only about 10%–20% of that erupted (as deduced from deformation modeling and lava flow mapping), indicating that most magma supplied to Piton de la Fouranise also erupts and that volcano growth is dominantly exogenous (at least over the last few decades during which monitoring data are available) [*Fukushima et al.*, 2010; *Peltier et al.*, 2012].

Two main types of lavas have been emitted during recent Piton de la Fournaise eruptions: aphyric basalts with homogeneous bulk-rock composition (6–7 wt% MgO) and picrites (also called océanites on Piton de la Fournaise), which are enriched in olivine macrocrysts (8–29 wt% MgO). A third type of lava, an abnormal

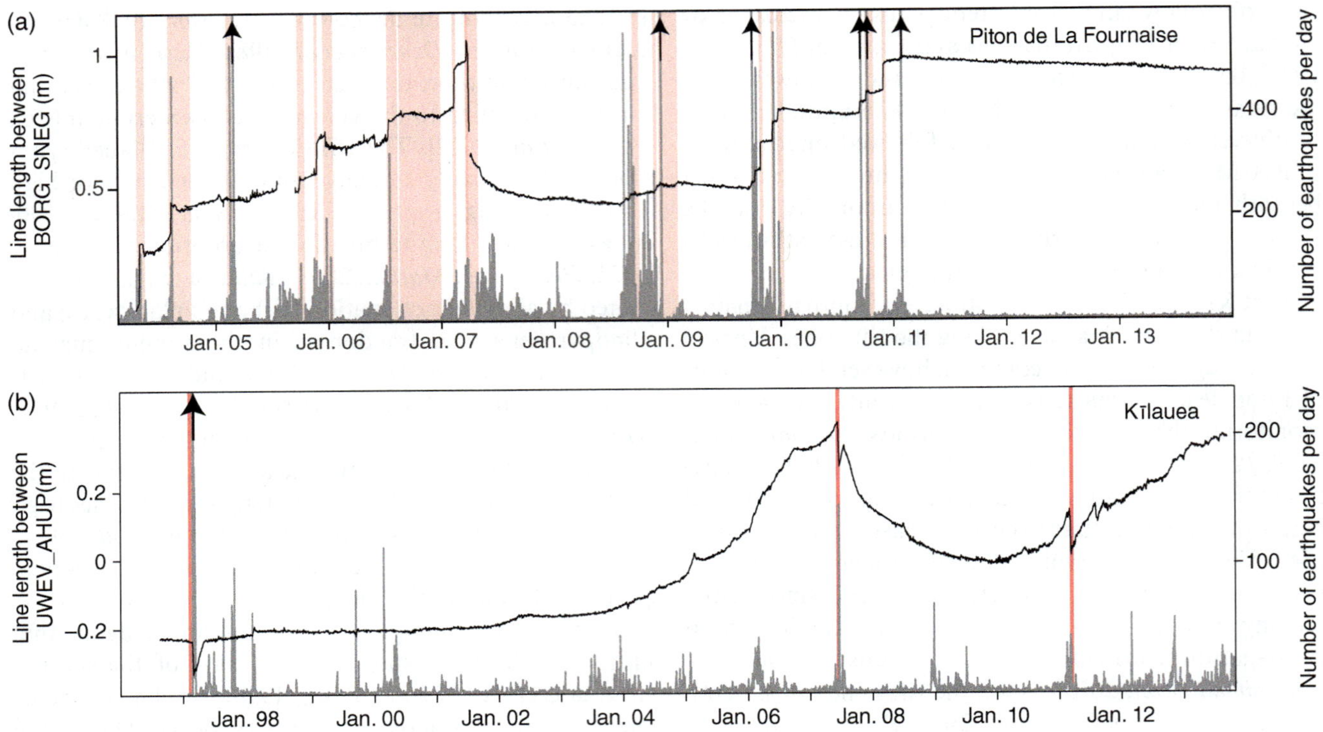

Figure 23.6 Examples of eruption precursors at (a) Piton de la Fournaise and (b) Kīlauea. (a) Time series plot covering 2004–2013 (the first continuous GNSS measurements were made in 2004) showing the daily number of earthquakes (in gray) and the distance change between GNSS stations BORG and SNEG (in black), which span Piton de la Fournaise's summit crater. Positive change is interpreted as inflation and negative as deflation. Red shaded areas represent eruptive periods. Gray arrows indicate days when more than 600 earthquakes were located (greater than the upper limit of the plot). (b) Time series plot covering 1997–2013 and showing the daily number of earthquakes (in gray) and the distance change between GNSS stations UWEV and AHUP (in black), which span Kīlauea's summit caldera. Increases in line length generally correlate with inflation and decreases with deflation, although seaward motion of the volcano's south flank causes a few centimeters per year of extension unrelated to inflation/deflation. Gray arrows indicate days when more than 200 earthquakes were located (greater than the upper limit of the plot). The Pu'u 'Ō'ō East Rift Zone eruption, which started in 1983, was active throughout the period, and persistent summit eruptive activity began in March 2008. Additional East Rift Zone fissure eruptions that interrupted the ongoing activity occurred in 1997, 2007, and 2011 and are indicated by red lines.

group (AbG) enriched in Mg, Fe, Ti, Na, and K and depleted in Ca and Si relative to aphyric basalts, was rarely emitted, mostly recently in 1998. This composition of magma may rise from a greater depth (>7.5 km [*Bureau et al.*, 1998a]) or might be associated with assimilation of olivine-rich cumulate xenoliths [*Salaün et al.*, 2010]. Most of the lava emitted during summit/proximal eruptions is largely aphyric, with small amounts of olivine, clinopyroxene, and plagioclase, while lava from distal eruptions is often picritic, with a high proportion (20%–40%) of olivine crystals. The olivine in distal flank eruptions may reflect incorporation of dunitic xenocrysts from a cumulate body or differentiation of a mafic magma by crystallization and settling of olivine crystals [e.g., *Albarède et al.*, 1997; *Famin et al.*, 2009; *Welsh*, 2010]. Crustal contamination is also occasionally an important factor in the chemistry of Piton de la Fournaise lava [*Pietruszka et al.*, 2009].

Episodes of flank motion exert control on eruptive activity at both Kīlauea and Piton de la Fournaise. The large-volume flank eruption and displacement at Piton de la Fournaise in April 2007 ended a period of eruption cyclicity and steady inflation of the summit. The major flank displacement associated with this event appears to have created a situation where intrusive events generally failed to culminate in eruptions during 2007–2013, a rare condition during the preceding 50 years at Piton de la Fournaise [*Roult et al.*, 2012]. Similarly, the ratio of eruptions to noneruptive intrusions changed from 4:1 to 1:4 in the years following Kīlauea's south flank earthquake in 1975, although this condition came to an end with the onset of the decades-long ERZ eruption in 1983 [*Dzurusin et al.*, 1980,1984; *Cayol et al.*, 2000].

While there is no written record of any eruption at Piton de la Fournaise with a years-long duration, there is geologic evidence for long-lasting periods of continuous eruptive activity (years to decades) between the mid-seventeenth and mid-nineteenth centuries [*Lénat et al*, 2001; *Peltier et al.*, 2012; *Michon et al.*, 2013]. During the eighteenth century, eruptions formed a large flow field that covered the Enclos Fouqué floor (Champ de Lave de l'Enclos Fouqué, $450-900 \times 10^6 m^3$ of lava flows [*Lénat et al.,* 2001]). Even if recorded eruptions have not persisted for years, such as those of Kīlauea, deformation evidence indicates quasi-continuous refilling of the shallow magma reservoir beneath Piton de la Fournaise between 2000 and 2007, probably resulting in high eruption frequency and cyclic eruption patterns during that time period and perhaps previously [*Peltier et al.*, 2009a]. Based on deformation data and numerical modeling, *Got et al.* [2013] showed that the eruptive cycles were linked to nonlinearity in the stress state of the edifice. Plastic displacement of the east flank during distal eruptions allows the stress accumulated since the beginning of a cycle (during summit and proximal eruptions) to relax, thus resetting the system and restarting the cycle. In the presence of a constant magma supply rate, therefore, the behavior of the edifice itself modulates magma transfer to the surface.

The dense geophysical monitoring networks installed on Kīlauea and Piton de la Fournaise show that many eruptions at the two volcanoes are preceded by significant seismic activity and surface deformation (Figure 23.6). Some episodes of fissure eruption that have taken place on Kīlauea's ERZ during the ongoing Pu'u 'Ō'ō eruption, for example, in June 2007 [*Poland et al.*, 2008, 2012; *Montgomery-Brown et al.*, 2010] and March 2011 [*Lundgren et al.*, 2013; (see also Chapter 18, this volume) have occurred at the culmination of weeks to months of summit inflation and elevated seismicity along the upper ERZ. The emplacement of a rift zone dike is accompanied by strong seismicity along with rapid extension of the rift zone and deflation of the summit as magma drains from the summit to feed the growing intrusion [e.g., *Poland et al.*, 2008] (Chapter 18). Eruptions have also occurred in the absence of strong precursory activity. For example, the March 2008 onset of eruptive activity at Kīlauea's summit took place during a period of deflation and without significant earthquake activity, although SO_2 emissions and seismic tremor did steadily increase in the four months prior to the explosion that opened the eruptive vent [*Poland et al.*, 2009]. Similarly, an ERZ fissure eruption in January 1997 and an intrusion in September 1999 both occurred during periods of summit deflation and nominal seismicity, although deformation and seismicity that accompanied the emplacement of dikes during these time periods were typical of other dike propagation events (Figure 23.6b). Such intrusions are thought to be a response to passive opening of the ERZ due to seaward motion of the volcano's south flank [*Owen et al.*, 2000; *Cervelli et al.*, 2002].

At Piton de la Fournaise, all eruptions monitored since the establishment of the observatory in 1980 have been preceded by activity occurring on two timescales [*Peltier et al.*, 2009a; *Roult et al.*, 2012] (Figure 23.6a): (1) a few weeks/months before an eruption, an increase in seismicity (up to 100 volcano-tectonic (VT) events per day) and, since 2000, slow inflation of the summit, indicating pressurization of the shallow magma reservoir, and (2) a few minutes/hours before the onset of the eruption, a seismic crisis (up to thousands of VT events) is accompanied by strong deformation (tens of centimeters) as a dike propagates toward the surface. The beginning of an eruption is always marked by the appearance of tremor on seismographs. Tremor not associated with any surface activity, but only with SO_2-rich degassing, has been recorded only twice since the founding of the Observatory, lasting only a few hours each time during emplacement of shallow noneruptive intrusions on 12 September 2008 and 18 October 2009 [*Roult et al.*, 2012].

While effusive activity has dominated the last few hundred years at both Kīlauea and Piton de la Fournaise, explosive eruptions have also been recorded both by historical observations and the geologic record. Since the first written account of Kīlauea's summit eruptive activity in 1823, two explosive phases have occurred, in 1924 and 2008 to present. The 1924 explosions consisted of several dozen events that took place during an 18 day period in response to drainage of magma from the summit lava lake in the preceding weeks. Groundwater subsequently infiltrated the conduit, came into contact with hot rock, and flashed to steam, driving explosions that ultimately resulted in one death [*Decker and Christiansen*, 1984]. The opening of Kīlauea's summit eruptive vent in 2008 was also associated with a small explosion but, instead of involving magma or water, was driven by a buildup of volatile pressure following blockage of a gas vent [*Houghton et al.*, 2011]. Subsequent small explosions during Kīlauea's ongoing summit activity were due to collapse of conduit rim and wall material into the active lava lake [*Orr et al.*, 2013]. Prior to 1823, base surge deposits associated with an explosive eruption in 1790 killed several hundred people. This eruption and several others that occurred during 1500–1800 are thought to have been driven by magma-water interactions, including both groundwater and surface water [*Decker and Christiansen*, 1984; *Mastin*, 1997; *Mastin et al.*, 2004; *Swanson et al.*, 2012]. Many of these explosions have been restricted to the region immediately surrounding the summit of the volcano. Much larger explosions, with eruption columns high enough to reach the jet stream (5–17 km in elevation), have also occurred during the past ~2000 years, as indicated by mapping of tephra fall deposits ~20 km from Kīlauea's summit [*Fiske et al.*, 2009]. The mechanism for such large explosions is uncertain but may involve exsolution of CO_2 as a driving force [*Fiske et al.*, 2009].

Phreatomagmatic and phreatic explosions have also been recognized at Piton de la Fournaise in the past several hundred years. During the eighteenth and the nineteenth centuries, frequent phreatomagmatic-to-phreatic explosions, associated with the formation of pit craters at the summit, alternated with periods of summit effusive activity lasting years to decades [*Bachèlery*, 1981; *Carter et al.*, 2007; *Peltier et al.*, 2012; *Michon et al.*, 2013]. The most violent explosions ejected lithic blocks tens of centimeters in size to a distance of 1 km from the summit crater. Since the start of the twentieth century, eruptive activity has been primarily effusive, with only rare and weak explosions. Interaction between water and magma is often cited to explain the occurrence of explosive eruptions, and geophysical studies at Piton de la Fournaise have identified a hydrothermal system at >300 m depth beneath the summit craters [*Barde-Cabusson et al.*, 2012; *Lénat et al.*, 2012]. *Michon et al.* [2013], however, suggested that phreatic and phreatomagmatic explosions during the past three centuries were triggered at shallower levels, resulting from interaction between magma and water coming from a shallow temporary aquifer and not from the deeper hydrothermal fluids. Mapping the locations and extents of subsurface water beneath volcanoes such as Piton de La Fournaise and Kīlauea is of obvious importance for assessing potential hazards from explosive magma-water interactions, all the more since a recent study proposed a model to explain the contribution of the hydrothermal system in triggering of magmatic intrusions at Piton de La Fournaise [*Lénat et al.*, 2012].

During the April 2007 summit caldera collapse at Piton de la Fournaise, no major explosive activity occurred. Presumably, there was no magma present at shallow levels beneath the summit cone at the time of the collapse, so there was no heat available to drive explosive interactions between water and hot rock [*Peltier et al.*, 2012]. The existence of magma in the collapsed block is therefore the most important factor for causing explosive eruptions during caldera collapse [*Geshi et al.*, 2002; *Peltier et al.*, 2012]. *Michon et al.* [2013] emphasized that magma-groundwater interaction producing explosive activity would be favored by long-lasting eruptions with abundant magma at very shallow depth below the summit, followed by rapid drainage to feed lateral intrusions. Kīlauea's 1924 explosions, which occurred after drainage of a lava lake from the summit, provide an example of such a process.

The largest well-known prehistoric explosions at Piton de la Fournaise date back to the Bellecombe ash episodes, spanning from 2000 to 5000 years ago [*Bachèlery*, 1981]. The ash deposits, which occur up to tens of kilometers from the summit, were related to major phreatomagmatic and phreatic explosions that were probably associated with the formation of the Enclos Fouqué caldera [*Bachèlery*, 1981; *Abchir et al.*, 1998]. Additional work on Piton de la Fournaise's explosive history is needed to further explore the mechanisms and impact of such volcanism on the island. Such work at Kīlauea [e.g., *Swanson et al.*, 2012] has helped to redefine the recent history of that volcano and offers similar promise at La Réunion.

23.5. DISCUSSION AND CONCLUSIONS

Kīlauea and Piton de la Fournaise are often cited as analog volcanoes, perhaps in part because their written records of activity extend back to a similar time (eighteenth century), and activity since then has been dominated by frequent effusive eruptions. In addition, both volcanoes are products of hotspot volcanism, and both are the sites of observatories that monitor eruptive activity with an advanced array of geophysical, geological, and geochemic measurements. These parallels are largely superficial, but probing deeper into how these volcanoes work supports the assumption of similarity when considering the development of the volcanoes and their plumbing systems and also their eruptive activity.

Analogous aspects between Kīlauea and Piton de la Fournaise include the following:

- Magma plumbing systems consist of multiple levels of magma storage, with the main storage at about 3 km depth.
- The geometry of the major structural rift zones is controlled in large part by preexisting structure: the influence of Mauna Loa in the case of Kīlauea and the presence of older volcanic shields and intrusive complexes beneath Piton de la Fournaise. Regional and local stresses may also control part of their formation. Rift zones at the two volcanoes provide important pathways for lateral magma transport and enable eruptions away from the summit.
- Both volcanoes display a compositional evolution from alkalic to tholeiitic, although the evolution is much milder at Piton de la Fournaise than it is at Kīlauea.
- The seaward-facing flanks of Kīlauea and Piton de la Fournaise are unstable, perhaps because they are not buttressed by preexisting, larger volcanoes. Rift zones divide the stable and unstable flanks of both volcanoes.
- Catastrophic landslides are evident in the offshore bathymetry of both volcanoes.
- Episodes of significant flank motion can fundamentally alter the style of volcanism. At Kīlauea following

the 1975 earthquake and at Piton de la Fournaise after the 2007 distal eruption/summit collapse/flank slip, both volcanoes experienced several years where intrusions dominated over eruptions, presumably because of stress perturbations inside the edifice and flank slip that opened space within the volcanoes that needed to be filled before frequent eruptions could resume.

• Volcanic activity is currently dominated by effusive eruptions at both volcanoes, but periods of explosive activity have also occurred in the past several thousand years, presumably driven in large part by magma-water interactions.

• During the past few decades (over which time monitoring data can be used to constrain the volumes of subsurface magma accumulation), most of the magma supplied to each volcano seems to have been erupted, although the proportion of magma that is stored within Kīlauea's rift zones and does not erupt is still a matter of debate.

Noteworthy differences also characterize Kīlauea and Piton de la Fournaise. For example, a shallow magma chamber at Kīlauea (1–2 km depth) allows for persistent degassing of SO_2 (even prior to the onset of summit eruptive activity in 2008). The lack of a similarly well-developed shallow reservoir at Piton de la Fournaise may explain why noneruptive degassing is not an important process at that volcano. Most differences between the two volcanoes, however, are related to plate tectonic processes and involve the scales of the two edifices:

• Kīlauea's volume is about an order of magnitude larger than that of Piton de la Fournaise.

• Hawaiian volcanoes have much shorter life spans than those of La Réunion.

• Plate flexure is an important process in Hawai'i and may control the occurrence of off-axis and rejuvenated volcanism, but no flexure has been recognized around La Réunion.

• Volcanism in Hawai'i since ~3 Ma has formed two parallel chains of compositionally distinct volcanoes, whereas a single chain characterizes La Réunion.

• Flank instability in Hawai'i is driven by a combination of magmatic and gravitational processes, while mainly magmatic forces control flank motion at Piton de la Fournaise. This is probably a result of the differing scales of the two magmatic systems. At Kīlauea, voluminous magmatism not only has caused some flank instability but also has resulted in the formation of a large, dense cumulate body at the base of the volcano that pushes the unbuttressed flank seaward along a décollement at the base of the volcanic pile. The combination of magmatic and gravitational forces causes steady motion, flank uplift, and the development of compressional structures. Such a cumulate body is less developed at Piton de la Fournaise and the edifice itself behaves in a stiffer manner than Kīlauea. Episodic large flank motion at Piton de la Fournaise is therefore driven mostly by occasional intrusions with slip and/or sill intrusions occurring along shallow detachments, even if a continuous eastern flank's sliding of a few centimeters per year is also probably driven by gravitational processes.

• The source of magma feeding hotspot volcanism at La Réunion appears to be geochemically more homogeneous than that feeding Hawai'i. Some of this chemical difference may be due to bending of the plume in Hawai'i, which does not appear to be significant at La Réunion.

• The transition between alkalic and tholeiitic compositions in Hawai'i and La Réunion may be caused by different processes, variations in the degree of partial melting in Hawai'i versus differences in the depth of fractionation at La Réunion.

Many of these differences can be ascribed to the greater magma supply to Hawaiian volcanoes and the faster plate motion relative to the Hawaiian hotspot compared to La Réunion. Magma supply to Kīlauea is an order of magnitude higher than that to Piton de la Fournaise, presumably because the plume feeding Hawaiian volcanism is hotter and more buoyant. The higher supply rate explains the large discrepancy in the volumes of the two volcanoes as well as the development of a large, gravitationally unstable cumulate pile at Kīlauea but not at Piton de la Fournaise. The distribution of volcanoes at the surface may be an additional consequence of differences in magma supply: Low supply yields a single hotspot track at La Réunion while high supply could promote a double chain of volcanoes in Hawai'i. Higher magma supply may also facilitate persistent eruptive activity at Kīlauea, whereas recorded eruptions at Piton de la Fournaise are generally discrete and often fall into groups separated by months/years of quiescence, suggesting that the lower supply rate may allow the mechanical behavior of the volcano itself to modulate eruptive activity. The difference in duration of volcanoes formed by the two hotspots is a consequence of plate motion relative to the hotspot, which is several times faster in Hawai'i than at La Réunion. Plate motion characteristics may also influence plate flexure, which is an important process in Hawai'i that may control off-axis and rejuvenated volcanism. Finally, faster plate motion might cause the Hawaiian plume to bend and introduce geochemical complexity to the melting process, whereas slower plate motion at La Réunion results in a more vertical plume with a more homogenous composition.

Understanding the similarities and differences between Kīlauea and Piton de la Fournaise, and Hawai'i and La Réunion in general, is central to any future work that might exploit the volcanoes as analogs, and there is clearly much to be gained by conducting complimentary studies on the two systems. For example, exploration of flank instability at Piton de la Fournaise could benefit from lessons learned at Kīlauea, particularly with regard to the mechanisms for flank motion and the potential occurrence of large earthquakes and episodic slow slip events. Likewise, joint analysis of explosive eruption deposits at Kīlauea and Piton de la Fournaise can shed light on the characteristics and causes of explosive basaltic volcanism.

Collaborative studies between Hawai'i and La Réunion researchers offer exceptional prospects for furthering knowledge about how each of the two systems works. We are hopeful that this review will provide motivation for such cooperative work.

ACKNOWLEDGEMENTS

We are grateful to our many colleagues of the Hawaiian Volcano Observatory (HVO) and the Observatoire Volcanologique du Piton de la Fournaise (OVPF) for discussions and comments that guided our analysis and for the dedication of the scientific and technical teams (and their collaborators) for installing and maintaining the monitoring networks on both volcanoes. This publication represents the first step toward a close collaboration between HVO and OVPF, as expressed in an agreement between the U.S. Geological Survey and the Institut de Physique du Globe de Paris. GNSS data from Kīlauea shown in Figures 23.4 and 23.6 were provided by Asta Miklius of HVO. Reviews were provided by Valérie Cayol, Asta Miklius, Aaron Pietruszka, and an anonymous reviewer, and we are grateful for discussions with Benoit Welsch. This is IPGP contribution number 3480.

REFERENCES

Abchir, A. M., M. P. Semet, G. Boudon, P. Ildefonse, P. Bachèlery, and R. Clocchiariatti (1998), Huge hydrothermal explosive activity on Piton de la Fournaise, Reunion Island: The Bellecombe Ash Member, 2700 BC, in *Proceedings, 2nd Workshop on European Laboratory Volcanoes*, pp. 447–455.

Albarède, F., and V. Tamagnan, (1988), Modelling the recent geochemical evolution of the Piton de la Fournaise volcano, Réunion Island, 1931–1986, *J. Petrol.*, *29*, 997–1030.

Albarède, F., B. Luais, G. Fitton, M. Semet, E. Kaminski, B. G. J. Upton, P. Bachèlery, and J. L. Cheminée (1997), The geochemical regimes of Piton de La Fournaise volcano (Reunion) during the last 530000 years, *J. Petrol.*, *38*(2), 171–201.

Aloisi, M., M. Mattia, C. Ferlito, M. Palano, V. Bruno, and F. Cannavò (2011), Imaging the multi-level magma reservoir at Mt. Etna volcano (Italy), *Geophys. Res. Lett.*, *38*(L16306), doi:10.1029/2011GL048488.

Augier, A., J. L. Froger, V. Cayol, Y. Fukushima, P. Tinard, T. Souriot, O. Mora, T. Staudacher, P. Durand, B. Fruneau, and N. Villeneuve (2008), The April 2007 eruption at Piton de la Fournaise, Réunion Island, imaged with ENVISAT-ASAR and ALOS-PALSAR data, USEReST workshop, Napoli, Italy.

Bachèlery, P. (1981), Le Piton de la Fournaise (Ile de la Réunion). Etude volcanologique, structurale et pétrologique, Ph.D. Thesis, Univ. of Clermont-Ferrand II, France.

Bachèlery, P., B. Robineau, M. Couteaud, and C. Savin (2003), Avalanches de débris sur le flanc Occidental du volcan bouclier Piton des Neiges (Réunion), *Bull. Soc. Geol. Fr.*, *174*, 125–140, doi:10.2113/174.2.12.

Bachèlery, P., F. Saint-Ange, N. Villeneuve, B. Savoye, A. Normand, E. Le Drezen, A. Barrère, J. P. Quod, and C. Deplus (2010), Huge lava flow into the sea and caldera collapse, April 2007. Piton de la Fournaise volcano, Workshop on Collapse Calderas, La Réunion 2010, pp. 73–74, Abstract.

Bagnardi, M., and F. Amelung (2012), Space-geodetic evidence for multiple magma reservoirs and subvolcanic lateral intrusions at Fernandina Volcano, Galápagos Islands, *J. Geophys. Res.*, *117*(B10406), doi:10.1029/2012JB009465.

Baker, S., and F. Amelung (2012), Top-down inflation and deflation at the summit of Kīlauea Volcano, Hawaii observed with InSAR, *J. Geophys. Res.*, *117*(B12406), doi:10.1029/2011JB009123.

Barde-Cabusson, S., et al. (2012), Structural control of collapse events inferred by self-potential mapping on the Piton de la Fournaise volcano (La Réunion Island), *J. Volcanol. Geotherm. Res.*, *209–210*, 9–18.

Barruol, G., and F. R. Fontaine (2013), Mantle flow beneath La Réunion hotspot track from SKS splitting, Earth Planet. *Sci. Lett.*, *362*, 108–121.

Basu, A. R., P. R. Renne, D. K. DasGupta, T. Feichman, and R. J. Poreda (1993), Early and late alkali igneous pulses and a high-3He plume origin for the Deccan flood basalts, *Science*, *261*, 902–906.

Battaglia, J., J.-L. Got, and P. Okubo (2003), Location of long-period events below Kilauea Volcano using seismic amplitudes and accurate relative relocation, *J. Geophys. Res.*, *108*(2553), doi:10.1029/2003JB002517.

Battaglia, J., V. Ferrazzini, Th. Staudacher, K. Aki, and J-L. Cheminée (2005), Pre-eruptive migration of earthquakes at Piton de la Fournaise volcano (Réunion Island), *Geophys. J. Int.*, *161*, 549–558.

Bianco, T. A., G. Ito, J. M. Becker, and M. O. Garcia (2005), Secondary Hawaiian volcanism formed by flexural arch decompression, *Geochem. Geophys., Geosyst.*, *6*(Q08009), doi:10.1029/2005GC000945.

Bonali, F., C. Corazzato, and A. Tibaldi (2011), Identifying rift zones on volcanoes: An example from La Réunion Island, Indian Ocean, Bull. *Volcanol.*, *73*(3), 347–366.

Bory de Saint-Vincent (1804), Voyages dans les quatre principales iles des mers d'Afrique, Reedition: Lafitte reprints, 1980, Marseille.

Brenguier, F., N. M. Shapiro, M. Campillo, A. Nercessian, and V., Ferrazzini (2007), 3-D surface wave tomography of the Piton de la Fournaise volcano using seismic noise correlation, *Geophys. Res. Lett.*, *34*, L02305, doi:10.1029/2006GL028586.

Brenguier, F., P. et al. (2012), First results from the undervol high resolution seismic and gps network deployed on Piton de La Fournaise Volcano, *Seismol. Res. Lett.*, *83*(7), doi:10.1785/gssrl.83.1.97.

Brooks, B. A., J. H. Foster, M. Bevis, L. N. Frazer, C. J. Wolfe, and M. Behn (2006), Periodic slow earthquakes on the flank of

Kilauea volcano, Hawaii, Earth Planet. *Sci. Lett.*, *246*(3–4), 207–216, doi:10.1016/j.epsl.2006.03.035.

Bureau, H., N. Métrich, F. Pineau, and M. P. Semet (1998a), Magma-conduit interaction at Piton de La Fournaise volcano (Réunion Island): A melt and fluid inclusion study, *J. Volcanol. Geotherm. Res.*, *84*, 39–60.

Bureau, H., F. Pineau, N. Métrich, M. P. Semet, and M. Javoy (1998b), A melt and fluid inclusion study of the gas phase at Piton de la Fournaise volcano (Réunion Island), *Chem. Geol.*, *147*(1–2), 115–130.

Calvert, A. T., and M. A. Lanphere (2006), Argon geochronology of Kilauea's early submarine history, *J. Volcanol. Geotherm. Res.*, *151*(1–3), 1–18, doi:10.1016/j.jvolgeores.2005.07.023.

Carter, A., B. van Wyk de Vries, K. Kelfoun, P. Bachèlery, and P. Briole (2007), Pits, rifts and slumps: The summit structure of Piton de La Fournaise, *Bull. Volcanol.*, *69*, 741–756.

Cayol, V., and F. H. Cornet (1998), Three-dimensional modeling of the 1983–1984 eruption at Piton de la Fournaise Volcano, Réunion Island, *J. Geophys. Res.*, *103*(B8), 18,025–18,037, doi:10.1029/98JB00201.

Cayol, V., J. H. Dieterich, A. T. Okamura, and A. Miklius (2000), High magma storage rates before the 1983 eruption of Kilauea, Hawaii, *Science*, *288*(5475), 2343–2346, doi:10.1126/science.288.5475.2343.

Cervelli, P., and A. Miklius (2003), The shallow magmatic system of Kīlauea Volcano, in *The Pu'u 'Ō'ō-Kupaianaha Eruption of Kīlauea Volcano, Hawai'i: The First 20 Years*, U.S. Geol. Surv. Prof. Pap. 1676, edited by C. Heliker, D. A. Swanson, and T. J. Takahashi, pp. 149–163.

Cervelli, P., P. Segall, F. Amelung, H. Garbeil, C. Meertens, S. Owen, A. Miklius, and M. Lisowski (2002), The 12 Septembre 1999 Upper East Rift Zone dike intrusion at Kilauea Volcano, Hawaii, *J. Geophys. Res.*, *107*(B7), 2150, doi:10.1029/2001JB000602.

Chadwick Jr, W. W., S. Jónsson, D. J. Geist, M. Poland, D. J. Johnson, S. Batt, K. S. Harpp, and A. Ruiz (2011), The May 2005 eruption of Fernandina volcano, Galápagos: The first circumferential dike intrusion observed by GPS and InSAR, *Bull. Volcanol.*, *73*(6), 679–697, doi:10.1007/s00445-010-0433-0.

Chaput, M., V. Pinel, V. Famin, L. Michon, and J.-L. Froger (2014a), Cointrusive shear displacement by sill intrusion in a detachment: A numerical approach, *Geophys. Res. Lett.*, *41*(6), 1937–1943, doi:10.1002/2013GL058813.

Chaput, M., V. Famin, and L. Michon (2014b), Deformation of basaltic shield volcanoes under cointrusive stress permutations, *J. Geophys. Res.*, *119*(1), 274–301, doi:10.1002/2013JB010623.

Clague, D. A. (1987), Hawaiian xenolith populations, magma supply rates, and development of magma chambers, *Bull. Volcanol.*, *49*(4), 577–587, doi:10.1007/BF01079963.

Clague, D. A., and G. B. Dalrymple (1987), The Hawaiian-Emperor volcanic chain, part I: Geologic evolution, in *Volcanism in Hawaii*, U. S. Geol. Surv. Prof. Pap. 1350, edited by R. W. Decker et al., pp. 5–54.

Clague, D. A., and R. P. Denlinger (1994), Role of olivine cumulates in destabilizing the flanks of Hawaiian volcanoes, *Bull. Volcanol.*, *56*(6), 425–434, doi:10.1007/BF00302824.

Clague, D. A., and J. E. Dixon (2000), Extrinsic controls on the evolution of Hawaiian ocean island volcanoes, *Geochem. Geophys. Geosys.*, *1*(4), doi:10.1029/1999GC000023.

Clague, D. A., R. T. Holcomb, J. M. Sinton, R. S. Detrick, and M. E. Torresan (1990), Pliocene and Pleistocene alkalic flood basalts on the seafloor north of the Hawaiian islands, *Earth Planet. Sci. Lett.*, *98*(2), 175–191, doi:10.1016/0012-821X(90)90058-6.

Clague, D. A., J. G. Moore, J. E. Dixon, and W. B. Friesen (1995), Petrology of submarine lavas from Kilauea's Puna Ridge, Hawaii, *J. Petrol.*, *36*(2), 299–349, doi:10.1093/petrology/36.2.299.

Clarke, D., F. Brenguier, J-L. Froger, N. M. Shapiro, A. Peltier, and T. Staudacher (2013), Timing of a large volcanic flank movement at Piton de la Fournaise volcano using noise-based seismic monitoring and ground deformation measurements, *Geophys. J. Int.*, doi:10.1093/gji/ggt276.

Courtillot, V., J. Besse, D. Vandamme, R. Montigny, J.-J. Jaeger, and H. Cappetta (1986), Deccan flood basalts at the Cretaceous/Tertiary boundary? *Earth Planet. Sci. Lett.*, *80*, 361–374.

Dana, J. D. (1849), Geology, vol. 10 of *United States Exploring* Expedition, George P. Putnam, New York.

Dana, J. D. (1890), Characteristics of Volcanoes, University Press, John Wilson, Cambridge, Mass.

Dawson, P. B., B. A. Chouet, P. G. Okubo, A. Villasenor, and H. M. Benz (1999), Three-dimensional velocity structure of the Kilauea caldera, Hawaii, *Geophys. Res. Lett.*, *26*(18), 2805–2808, doi:10.1029/1999GL005379.

Decker, R. W., and R. L. Christiansen (1984), Explosive eruptions of Kilauea volcano, Hawaii, in *Explosive Volcanism: Inception, Evolution, and Hazards*, edited by National Academy Press, pp. 122–132, Washington, D. C.

De Crémont (1770), Relation du premier voyage fait au volcan de l'île de Bourbon, in *L'année littéraire par M. Fréron, Année 1770, t. VII, Lettre IV*, reprint by Harvard College Library, 1922.

Delaney, P. T., R. S. Fiske, A. Miklius, A. T. Okamura, and M. K. Sako (1990), Deep magma body beneath the summit and rift zones of Kilauea Volcano, Hawaii, *Science*, *247*, 1311–1316.

Delcamp, A., B. van Wyk de Vries, and M. James (2008), The influence of edifice slope and substrata on volcano spreading, *J. Volcanol. Geotherm. Res.*, *177*, 925–943, doi:10.1016/j.jvolgeores.2008.07.014.

DeMets, C., R. Gordon, D. Argus, and S. Stein (1994), Effect of recent revisions to the geomagnetic reversal time scale on estimates of current plate motions, *Geophys. Res. Lett.*, *21*(20), doi:10.1029/94GL02118.issn: 0094-8276.

Deniel, C., G. Kieffer, and J. Lecointre (1992), New 230Th-238U and 14C age determinations from Piton des Neiges volcano, Réunion: A revised chronology for the differentiated series, *J. Volcanol. Geotherm. Res.*, *51*, 253–267.

Denlinger, R. P., and P. Okubo (1995), Structure of the mobile south flank of Kilauea Volcano, Hawaii, *J. Geophys. Res.*, *100*(B12), 24,499–24,507, doi:10.1029/95JB01479.

Deplus, C., B. de Voogd, J. Dyment, D. Bissessur, E. Sisavath, F. Depuiset, and M. Mercier (2009), New insights on the oceanic lithosphere at La Reunion hotspot volcano, Geophysical Research Abstracts, 11-5728, EGU 2009, Vienna.

De Voogd, B., S. Pou Palomé, A. Hirn, P. Charvis, J. Gallard, D. Rousset, J. Danobeitia, and H. Perroud (1999a), Vertical movements and material transport during hotspot activity: Seismic reflection profiling offshore La Réunion, *J. Geophys. Res.*, *104*, 2855–2874.

De Voogd, B., S. P. Palome, A. Hirn, P. Charvis, J. Gallart, D. Rousset, J. Danobeitia, and H. Perroud (1999b), Vertical movements and material transport during hotspot activity: Seismic reflection profiling offshore La Reunion, *J. Geophys. Res. Solid Earth*, *104*, 2855–2874.

Dieterich, J. H., V. Cayol, and P. Okubo (2003), Stress changes before and during the Pu'u'O'o-Kupaianaha eruption, in *The Pu'u 'Ō'ō-Kupaianaha Eruption of Kīlauea Volcano, Hawai'i: The First 20 Years*, U.S. Geol. Surv. Prof. Pap. 1676, edited by C. Heliker, et al., pp. 187–202.

Driad, L. (1997), Structure profonde de l'édifice volcanique de La Réunion (océan Indien) par sismique réfraction et grand angle, Ph.D. Thesis, Université de Paris VII.

Duncan, R. A. (1981), Hotspots in the southern oceans: An absolute frame of reference for motion of the Gondwana continents, *Tectonophysics*, *74*, 29–24.

Dzurisin, D., L. A. Anderson, G. P. Eaton, R. Y. Koyanagi, P. W. Lipman, J. P. Lockwood, R. T. Okamura, G. S. Puniwai, M. K. Sako, and K. M. Yamashita (1980), Geophysical observations of Kilauea Volcano, Hawaii; 2, Constraints on the magma supply during November 1975–September 1977, *J. Volcanol. Geotherm. Res.*, *7*(3–4), 241–269.

Dzurisin, D., R. Y. Koyanagi, and T. T. English (1984), Magma supply and storage at Kilauea Volcano, Hawaii, 1956–1983, *J. Volcanol. Geotherm. Res.*, *21*(3–4), 177–206, doi:10.1016/0377-0273(84)90022-2.

Eaton, J. P. (1962), Crustal structure and volcanism in Hawaii, in Crust of the Pacific Basin, Geophys. Monogr' 6, edited by G. A. Macdonald and H. Kuno, pp. 13–29, AGU, Washington, D. C., doi:10.1029/GM006p0013.

Eaton, J. P., and K. J. Murata (1960), How volcanoes grow, *Science*, *132*(3432), 925–938, doi:10.1126/science.132.3432.925.

Ellis, W. (1963), Journal of William Ellis; narrative of a tour of Hawaii, or Owhyhee (with an introduction by Thurston Twigg-Smith), Honolulu, Advertiser Publishing Company, (reprinted from the London 1827 edition and the Hawaii 1917 edition), 342 pp.

Famin, V., B. Welsch, S. Okumura, P. Bachèlery, and S. Nakashima (2009), Three differenciation stages of a single magma at Piton de la Fournaise volcano (Reunion hot spot), *Geochem. Geophys. Geosyst.*, *10*(1), Q01007, doi:10.1029/2008GC002015.

Farley, K. A., A. R. Basu, and H. Craig (1993), He, Sr and Nd isotopic variations in lavas from the Juan Fernandez archipelago, SE Pacific, *Contrib. Mineral Petrol*, *115*, 75–87.

Farnetani, C. G., and A. W. Hofmann (2010), Dynamics and internal structure of the Hawaiian plume, *Earth Planet. Sci. Lett.*, *295*(1–2), 231–240, doi:10.1016/j.epsl.2010.04.005.

Fiske, R. S., and E. D. Jackson (1972), Orientation and growth of Hawaiian volcanic rifts: The effect of regional structure and gravitational stresses, *Proc. Roy. Soc. A*, *329*(1578), 299–326, doi:10.1098/rspa.1972.0115.

Fiske, R. S., T. R. Rose, D. A. Swanson, D. E. Champion, and J. P. McGeehin (2009), Kulanaokuaiki Tephra (ca. A.D. 400–1000): Newly recognized evidence for highly explosive eruptions at Kilauea Volcano, Hawai'i, *Geol. Soc. Am. Bull.*, *121*(5–6), 712–728.

Frey, F. A., M. O. Garcia, W. S. Wise, A. Kennedy, P. Gurriet, and F. Albarede (1991), The evolution of Mauna Kea Volcano, Hawaii: Petrogenesis of tholeiitic and alkali basalts, *J. Geophys. Res. Solid Earth*, *96*(B9), 14,347–14,375.

Frey, F. A., D. Clague, J. J. Mahoney, and J. M. Sinton (2000), Volcanism at the edge of the Hawaiian Plume: Petrogenesis of submarine alkalic lavas from the North Arch volcanic field, *J. Petrol.*, *41*(5), 667–691. doi:10.1093/petrology/41.5.667.

Fukushima, Y., V. Cayol, and P. Durand (2005), Finding realistic dike models from interferometric synthetic aperture radar data: The February 2000 eruption at Piton de la Fournaise, *J. Geophys. Res.*, *110*, B03206, doi:10.1029/2004JB003268.

Fukushima, Y., V. Cayol, P. Durand, and D. Massonnet (2010), Evolution of magma conduits during the 1998–2000 eruptions of Piton de la Fournaise volcano, Réunion Island, *J. Geophys. Res.*, *115*, B10204, doi:10.1029/2009JB007023.

Gailler, L. S., J. F. Lénat, M. Lambert, G. Levieux, N. Villeneuve, and J. L. Froger (2009), Gravity structure of Piton de La Fournaise volcano and inferred mass transfer during the 2007 crisis, *J. Volcanol. Geotherm. Res.*, *184*, 31–48.

Garcia, M. O., R. A. Ho, J. M. Rhodes, and E. W. Wolfe (1989), Petrologic constraints on rift-zone processes, *Bull. Volcanol.*, *52*(2), 81–96, doi:10.1007/BF00301548.

Garcia, M. O., J. M. Rhodes, E. W. Wolfe, G. E. Ulrich, and R. A. Ho (1992), Petrology of lavas from episodes 2–47 of the Puu Oo eruption of Kilauea Volcano, Hawaii: Evaluation of magmatic processes, *Bull. Volcanol.*, *55*(1–2), 1–16, doi:10.1007/BF00301115.

Garcia, M. O., A. J. Pietruszka, and J. M. Rhodes (2003), A petrologic perspective of Kilauea volcano's summit magma reservoir, *J. Petrol.*, *44*(12), 2313–2339.

Garcia, M. O., D. J. P. Foss, H. B. West, and J. J. Mahoney (2005), Geochemical and isotopic evolution of Loihi volcano, Hawaii, *J. Petrol.*, *36*(6), 1647–1674.

Gerlach, T. M. (1986), Exsolution of H_2O, CO_2, and S during eruptive episodes at Kilauea Volcano, Hawaii, *J. Geophys. Res.*, *91*(B12), 12,177–12,185, doi:10.1029/JB091iB12p12177.

Geshi, N., T. Shimano, T. Chiba, and S. Nakada (2002), Caldera collapse during the 2000 eruption of Miyakejima Volcano, Japan, *Bull Volcanol.*, *64*, 55–68.

Gillot P.-Y., and P. Nativel (1982), K-Ar chronology of the ultimate activity of Piton des Neiges Volcano, Reunion Island, Indian Ocean, *J. Volcanol. Geotherm. Res.*, *13*, 131–146.

Gillot, P.-Y., and P. Nativel (1989), Eruption history of the Piton de la Fournaise volcano, Reunion Island, Indian Ocean, *J. Volcanol. Geotherm. Res.*, *36*, 53–55.

Gillot, P.-Y., J.-C. Lefèvre, and P. Nativel (1994), Model for the structural evolution of the volcanoes of Réunion island, *Earth Planet. Sci. Lett.*, *122*, 291–302.

Got, J. L., and P. Okubo (2003), New insights into Kilauea's volcano dynamics brought by large-scale relative relocation of microearthquakes, *J. Geophys. Res.*, *108*(B7), 2337, doi:10.1029/2002JB002060.

Got, J. L., V. Monteiller, J. Monteux, R. Hassani, and P. Okubo (2008), Deformation and rupture of the oceanic crust may control growth of Hawaiian volcanoes, *Nature*, *451*, 453–456, doi:10.1038/nature06481.

Got, J. L., A. Peltier, T. Staudacher, P. Kowalski, and P. Boissier (2013), Edifice strength and magma transfer modulation at Piton de la Fournaise volcano, *J. Geophys. Res. Solid Earth*, *118*(9), 5040–5057, doi: 10.1002/jgrb.50350.

Graham, D. W., D. M. Christie, K. S. Harpp, and J. E. Lupton (1993), Mantle plume helium in submarine basalts from the Galapagos Platform, *Science*, *262*, 2023–2026.

Gripp, A. E., and R. G. Gordon (1990), Current plate velocities relative to the hotspots incorporating the NUVEL-1 global plate motion model, *Geophys. Res. Lett.*, *17*(8), 1109–1112, doi:10.1029/GL017i008p01109.

Héguerty, D. (1754), Relation d'un voyage à l'île Bourbon, Mémoire de l'Acad. Royale de Nancy, pp. 73–91.

Heliker, C., and T. N. Mattox (2003), The first two decades of the Pu-u O'o-Kupaianaha eruption: chronology and selected bibliography, in *The Pu'u O'o-Kupaianaha Eruption of Kilauea Volcano, Hawaii: The First 20 Years*, U.S. Geol. Sur. Prof. Pap. 1676, edited by C. Heliker et al., pp. 1–27.

Helz, R. T. (1987), Diverse olivine types in lava of the 1959 eruption of Kilauea volcano and their bearing on eruption dynamics, U.S. Geol. Surv. Prof. Pap. 1350(1), pp. 691–722.

Hirn, A., J-C. Lépine, M. Sapin, and H. Delorme (1991), Episodes of pit-crater collapse documented by seismology at Piton de la Fournaise, *J. Volcanol. Geotherm. Res.*, *47*, 89–104.

Hofmann, W., C. G. Farnetani, M. Spiegelman, and C. Class (2011), Displaced helium and carbon in the Hawaiian plume, Earth Planet. *Sci. Lett.*, *312*, 226–236.

Holcomb, R. T. (1987), Eruptive history and long-term behavior of Kilauea volcano, in *Volcanism in Hawaii*, U. S. Geol. Surv. Prof. Pap. 1350, edited by R. W. Decker et al., pp. 261–350.

Houghton, B. F., D. A. Swanson, R. J. Carey, J. Rausch, and A. J. Sutton (2011), Pigeonholing pyroclasts: Insights from the 19 March 2008 explosive eruption of Kilauea volcano, *Geology*, *39*(3), 263–266.

Hugoulin (1860), Dernière éruption du volcan de l'Ile de La Réunion, dans "Bridet H. G., un cyclone à la Réunion," *Revue algérienne et coloniale*, Première semestre 1860, pp. 483–487.

Iverson, R. M. (1995), Can magma-injection and groundwater forces cause massive landslides on Hawaiian volcanoes? *J. Volcanol. Geotherm. Res.*, *66*(1), 295–308, doi:10.1016/0377-0273(94)00064-N.

Jachens, R. C., and G. P. Eaton (1980), Geophysical observations of Kilauea volcano, Hawaii, 1. Temporal gravity variations related to the 29 November, 1975, M = 7.2 earthquake and associated summit collapse, *J. Volcanol. Geotherm. Res.*, *7*(3–4), 225–240.

Jackson, E. D., E. A. Silver, and G. B. Dalrymple (1972), Hawaiian-Emperor chain and its relation to Cenozoic circumpacific tectonics, *Geol. Soc. Am. Bull.*, *83*, 601–617.

Jaggar, T. A., Jr. (1919), December 6, 1919, December 13, 1919, December 20, 1919, December 27, 1919, and January 3, 1920, *Month. Bull. Hawaiian Volcano Observatory*, *7*(12), 175–199. Reprinted in Bevens, D., T. J. Takahashi, and T. L. Wright, eds. (1988), The Early Serial Publications of the Hawaiian Volcano Observatory, vol. 2, pp. 1061–1086, Hawaii National Park, Hawaii Natural History Association.

Johnson, D. J. (1995), Molten core model for Hawaiian rift zones, *J. Volcanol. Geotherm. Res.*, *66*(1–4), 27–35.

Kauahikaua, J., T. Hildenbrand, and M. Webring (2000), Deep magmatic structures of Hawaiian volcanoes, imaged by three-dimensional gravity methods, *Geology*, *28*(10), 883–886, doi:10.1130/0091-7613(2000)28<883:DMSOHV>2.0.CO;2.

Klein, F. W. (1982), Patterns of historical eruptions at Hawaiian volcanoes, *J. Volcanol. Geotherm. Res.*, *12*(1–2), 1–35, doi:10.1016/0377-0273(82)90002-6.

Kurz, M. D., and W. J Jenkins (1982), Helium isotopic systematics of ocean island and mantle heterogeneity, *Nature*, *297*, 43–47.

Kurz, M. D., T. C. Kenna, D. P. Kammer, J. M. Rhodes, and M. O. Garcia (1995), Isotopic Evolution of Mauna Loa Volcano: A View From the Submarine Southwest Rift Zone, in *Mauna Loa Revealed: Structure, Composition, History, and Hazards*, edited by J.M. Rhodes and J.P. Lockwood, American Geophysical Union, Washington, D. C., doi: 10.1029/GM092p0289.

Kurz, M. D., T. C. Kenna, J. C. Lassiter, and D. J. DePaolo (1996), Helium isotopic evolution of Mauna Kea Volcano: First results from the 1-km drill core, *J. Geophys. Res. Solid Earth, 101*, 11,781–11,791, doi:10.1029/95JB03345.

Labazuy, P. (1996), Recurrent landslide events on the submarine flank of Piton de la Fournaise volcano (Reunion Island), *in Volcano Instability on the Earth and Other Planets*, edited by W. J. McGuire et al., pp. 295–306, Spec. Publ. 110, Geological Society of London, London.

Labazuy, P. (1996), Recurrent landslide events on the submarine flank of Piton de la Fournaise volcano (Reunion Island), *in Volcano Instability on the Earth and Other Planets*, edited by W. J. McGuire et al., pp. 295–306, Spec. Publ. 110, Geol. Soc.

Lacroix, A. (1939), Les transfomations récentes du sommetndu volcan actif (Piron de la Fournaise) de l'Ile de la Réunion, Notes, *Memoires et rapports de volcanologie,* in *Bulletin volcanologique*, Série II, Tome V, pp. 3–17.

Lebas, E. (2012), Processus de Démantelement des edifices volcaniques au cours de leur evolution; Application à La Réunion et Montserrat et comparaison avec d'autres edifices, Ph.D. Thesis. Université Paris-Diderot.

Le Friant, A., E. Lebas, V. Clément, G. Boudon, C. Deplus, B. de Voogd, and P. Bachèlery (2011), A new model for the evolution of la Réunion volcanic complex from complete marine geophysical surveys, *Geophys. Res. Lett.*, *38*(9), L09312, doi :10.1029/2011GL047489.

Lenat, J.-F., and P. Bachelery (1990), Structure et fonctionnement de la zone centrale du Piton de La Fournaise, in *Le volcanisme de La Reunion*, edited by J.-F. Lenat, pp. 257–96, Centre de Recherches Volcanologiques, Clermont-Ferrand-France.

Lénat, J.-F., P. Bachèlery, A. Bonneville, A. Galdéano, P. Labazuy, D. Rousset, and P. Vincent (1990), Structure and morphology of the submarine flank of an active basaltic volcano: Piton de La Fournaise (Réunion Island, Indian Ocean), Oceanol. *Acta*, *10*, 211–223.

Lénat, J.-F., P. Bachèlery, and F. Desmulier (2001), Genèse du champs de lave de l'Enclos Fouqué: Une éruption d'envergure exceptionnelle du piton de la Fournaise (Réunion) au 18e siècle, *Bull. Soc. Géol. France*, *172*(2), 177–199.

Lénat, J.-F., P. Bachèlery, and A. Peltier (2012), The interplay between collapse structures, hydrothermal systems, and magma intrusions: The case of the central area of Piton de la Fournaise volcano, Bull. *Volcanol*, *74*, 407–421.

Lipman, P. W., and A. T. Calvert (2013), Modeling volcano growth on the Island of Hawaii: Deep-water perspectives, *Geosphere*, *9*(5), 1348–1383, doi:10.1130/GES00935.1.

Lipman, P. W., J. P. Lockwood, R. T. Okamura, D. A. Swanson, and K. M. Yamashita (1985), Ground deformation associated with the 1975 magnitude-7.2 earthquake and resulting changes in activity of Kilauea Volcano, Hawaii, U.S. Geol. Surv. Prof. Pap. 1276, 45 pp.

Lipman, P. W., D. A. Clague, J. G. Moore, and R. T. Holcomb (1989), South Arch volcanic field: Newly identified young lava flows on the sea floor south of the Hawaiian Ridge, *Geology*, *17*(7), 611–614, doi:10.1130/0091-7613(1989) 017<0611:SAVFNI>2.3.CO;2.

Lipman, P. W., T. W. Sisson, M. L. Coombs, A. Calvert, and J.-I. Kimura (2006), Piggyback tectonics: Long-term growth of Kilauea on the south flank of Mauna Loa, *J. Volcanol. Geotherm. Res.*, *151*(1–3), 73–108, doi:10.1016/j.jvolgeores.2005.07.032.

Lundgren, P., M. Poland, A. Miklius, T. Orr, S.-H. Yun, E. Fielding, Z. Liu, A. Tanaka, W. Szeliga, S. Hensley, and S. Owen (2013), Evolution of dike opening during the March 2011 Kamoamoa fissure eruption, Kīlauea Volcano, Hawai'i, *J. Geophys. Res.*, *118*(3), 897–914, doi:10.1002/jgrb.50108.

Malengreau, B., J-F. Lénat, and J. L. Froger (1999), Structure of Reunion Island, Indian Ocean inferred from the interpretation of gravity anomalies, *J. Volcanol. Geotherm. Res.*, *88*, 131–146.

Massin, F., V. Ferrazzini, P. Bachèlery, A. Nercessian, Z. Duputel, and T. Staudacher (2011), Structures and evolution of the plumbing system of Piton de la Fournaise volcano inferred from clustering of 2007 eruptive cycle seismicity, *J. Volcanol. Geotherm. Res.*, *202*, 96–106.

Mastin, L. G. (1997), Evidence for water influx from a caldera lake during the explosive hydromagmatic eruption of 1790, Kilauea volcano, Hawai'i, *J. Geophys. Res.*, *102*(B9), 20,093–20,109, doi:10.1029/97JB01426.

Mastin, L. G., R. L. Christiansen, C. Thornber, J. Lowenstern, and M. Beeson (2004), What makes hydromagmatic eruptions violent? Some insights from the Keanakāko'i Ash, Kīlauea Volcano, Hawai'i, *J. Volcanol. Geotherm. Res.*, *137*(1–3), 15–31, doi:10.1016/j.jvolgeores.2004.05.015.

McDougall, I. (1971), The geochronology and evolution of the younger volcanic island of Reunion, Indian Ocean, *Geochim. Cosmochim. Acta*, *36*, p.261–288.

McDougall, I., and F. H. Chamalaun (1969), Isotopic dating and geomagnetic polarity studies on volcanic rocks from Mauritius, Indian Ocean, *Geol. Soc. Am. Bull.*, *82*, 1419–1442.

Michon, L., and F. Saint-Ange (2008), Morphology of Piton de la Fournaise basaltic shield volcano (La Réunion Island): Characterization and implication in the volcano evolution, *J. Geophys. Res.*, *113*, B03203, doi:10.1029/2005JB004118.

Michon, L., T. Staudacher, V. Ferrazzini, P. Bachèlery, and J. Marti (2007a), April 2007 collapse of Piton de La Fournaise: A new example of caldera formation, *Geophys. Res. Lett*, *34* (L21301), doi:1029/2007GL031248.

Michon, L., F. Saint-Ange, P. Bachèlery, N. Villeneuve, and T. Staudacher (2007b), Role of the structural inheritance of the oceanic lithosphere in the magmato-tectonic evolution of Piton de la Fournaise volcano (La Réunion Island), *J. Geophys. Res.*, *112*, B04205, doi:10.1029/2006JB004598.

Michon, L., V. Cayol, L. Letourneur, A. Peltier, N. Villeneuve, and T. Staudacher (2009), Edifice growth, deformation and rift zone development in basaltic setting: Insights from Piton de la Fournaise shield volcano (Réunion Island), *J. Volcanol. Geotherm. Res.*, *184*(1–2), 14–30.

Michon, L., A. Di Muro, N. Villeneuve, C. Saint-Marc, P. Fadda, and F. Manta (2013), Explosive activity of the summit cone of Piton de la Fournaise volcano (La Réunion island): A historical and geological review, *J. Volcanol. Geotherm. Res.*, *263*, 117–133.

Montgomery-Brown, E. K., D. K. Sinnett, M. Poland, P. Segall, T. Orr, H. Zebker, and A. Miklius (2010), Geodetic evidence for en echelon dike emplacement and concurrent slow-slip during the June 2007 intrusion and eruption at Kīlauea volcano, Hawaii, *J. Geophys. Res.*, *115*(B07405), doi:10.1029/2009JB006658.

Moore, J. G., D. A. Clague, R. T. Holcomb, P. W. Lipman, W. R. Normark, and M. E. Torresan (1989), Prodigious submarine landslides on the Hawaiian Ridge, *J. Geophys. Res.*, *94*(B12), 17,465–17,848, doi:10.1029/JB094iB12p17465.

Moore, J. G., W. R. Normark, and R. T. Holcomb (1994a), Giant Hawaiian landslides, *Ann. Rev. Earth Planet. Sci.*, *22*, 119–144, doi:10.1146/annurev.ea.22.050194.001003.

Moore, J. G., W. R. Normark, and R. T. Holcomb (1994b), Giant Hawaiian underwater landslides, *Science*, *264*(5155), 46–47, doi:10.1126/science.264.5155.46.

Morgan, J. K. (2006), Volcanotectonic interactions between Mauna Loa and Kilauea: Insights from 2-D discrete element simulations, *J. Volcanol. Geotherm. Res.*, *151*(1–3), 109–131, doi:10.1016/j.jvolgeores.2005.07.025.

Morgan, J. K., G. F. Moore, D. J. Hills, and S. Leslie (2000), Overthrusting and sediment accretion along Kilauea's mobile south flank, Hawaii: Evidence for volcanic spreading from marine seismic reflection data, *Geology*, *28*(7), 667–670.

Morgan, W. J. (1981), Hotspot tracks and the opening of the Atlantic and Indian Oceans, in *The Sea*, edited by C. Emiliani, pp. 443–487, Wiley, New York.

Müller, R. D., M. Sdrolias, C. Gaina, and W. R. Roest (2008), Age, spreading rates, and spreading asymmetry of the world's ocean crust, *Geochem. Geophys. Geosyst.*, *9*(Q04006), doi:10.1029/2007GC001743.

Nercessian, A., A. Hirn, J. C. Lépine, and M. Sapin (1996), Internal structure of Piton de La Fournaise volcano from seismic wave propagation and earthquake distribution, *J. Volcanol. Geotherm. Res.*, *70*, 123–143.

Oehler, J. F., B. van Wyk de Vries, and P. Labazuy (2005), Landslides and spreading of oceanic hotspot and arc shield volcanoes on Low Strength Layers (LSLs): An analogue modelling approach, *J. Volcanol. Geotherm. Res.*, *144*, 169–189, doi:10.1016/j.jvolgeores.2004.11.023.

Oehler, J. F., J. F. Lénat, and P. Labazuy (2008), Growth and collapse of the Reunion Island volcanoes, *Bull. Volcanol.*, *70*, 717–742, doi:10.1007/s00445-007-0163-0.

Okubo, P. G., H. M. Benz, and B. A. Chouet (1997), Imaging the crustal magma sources beneath Mauna Loa and

Kilauea volcanoes, Hawaii, *Geology*, *25*(10), 867–870, doi:10.1130/0091-7613(1997)025<0867:ITCMSB>2.3.CO;2.

Ollier, G., P. Cochonat, J. F. Lénat, and P. Labazuy (1998), Deep sea volcaniclastic sedimentary systems: An example from La Fournaise volcano, *Réunion Island, Indian Ocean, Sedimentology*, *45*, 293–330, doi:10.1046/j.1365-3091.1998.0152e.x.

Orr, T. R., W. A. Thelen, M. R. Patrick, D. A. Swanson, and D. C. Wilson (2013), Explosive eruptions triggered by rockfalls at Kīlauea volcano, Hawai'i, *Geology*, *41*(2), 207–210, doi:10.1130/G33564.1.

Owen, S., P. Segall, J. T. Freymueller, A. Miklius, R. P. Denlinger, T. Árnadóttir, M. K. Sako, and R. Bürgmann (1995), Rapid deformation of the south flank of Kilauea Volcano, Hawaii, *Science*, *267*(5202), 1328–1332, doi:10.1126/science.267.5202.1328.

Owen, S., P. Segall, M. Lisowski, A. Miklius, M. Muray, M. Bevis, and J. Foster (2000), January 30, 1997 eruptive event on Kilauea Volcano, Hawaii, as monitored by continuous GPS, *Geophys. Res. Lett.*, *27*:(17), 2757–2760.

Patrick, M., D. Wilson, D. Fee, T. Orr, and D. Swanson (2011), Shallow degassing events as a trigger for very-long-period seismicity at Kīlauea Volcano, Hawai'i, Bull. *Volcanol.*, *73*(9), 1179–1186, doi:10.1007/s00445-011-0475-y.

Peltier, A., V. Ferrazzini, T. Staudacher, and P. Bachelery (2005), Imaging the dynamics of dyke propagation prior to the 2000–2003 flank eruptions at Piton de la Fournaise, Reunion Island, *Geophys. Res. Lett.*, *32*, doi:10.1029/2005GL023720.

Peltier, A., T. Staudacher, and P. Bachèlery (2007), Constraints on magma transfers and structures involved in the 2003 activity at Piton de La Fournaise from displacement data, *J. Geophys. Res.*, *112*(B03207), doi:10.1029/2006JB004379.

Peltier, A., V. Famin, P. Bachèlery, V. Cayol, Y. Fukushima, and T. Staudacher (2008), Cyclic magma storages and transfers at Piton de La Fournaise volcano (La Réunion hotspot) inferred from deformation and geochemical data, Earth Planet. *Sci. Lett.*, *270*, 180–188.

Peltier, A., P. Bachèlery, and T. Staudacher (2009a), Magma transfer and storage at Piton de La Fournaise (La Réunion Island) between 1972 and 2007: A review of geophysical and geochemical data, *J. Volcanol. Geotherm. Res.*, *184*(1–2), 93–108.

Peltier, A., T. Staudacher, P. Bachèlery, and V. Cayol (2009b), Formation of the April 2007 caldera collapse at Piton de La Fournaise volcano: Insights from GPS data, *J. Volcanol. Geotherm. Res.*, *184*, 152–163.

Peltier, A., T. Staudacher, and P. Bachèlery (2010), New behaviour of the Piton de La Fournaise volcano feeding system (La Réunion Island) deduced from GPS data: Influence of the 2007 Dolomieu caldera collapse, *J. Volcanol. Geotherm. Res.*, *192*, 48–56.

Peltier, A., F. Massin, P. Bachèlery, and A. Finizola (2012), Internal structures and building of basaltic shield volcanoes: The example of the Piton de La Fournaise terminal cone (La Réunion), *Bull. Volcanol,*. doi:10.1007/s00445-012-0636-7.

Pietruszka, A., E. H. Hauri, and J. Blichert-Toft (2009), Crustal contamination of mantle-derived magmas within Piton de la Fournaise volcano, Réunion Island, *J. Petrol.*, *50*(4), 661–684, doi:10.1093/petrology/egp016.

Pietruszka, A., M. D. Norman, M. O. Garcia, J. P. Marske, and D. H. Burns (2013), Chemical heterogeneity in the Hawaiian mantle plume from the alteration and dehydration of recycled oceanic crust, *Earth. Planet. Sci. Lett.*, *361*, 298–309.

Plattner, C., F. Amelung, S. Baker, R. Govers, and M. P. Poland (2013), The role of viscous magma mush spreading in volcanic flank motion at Kīlauea Volcano, Hawai'i, *J. Geophys. Res.*, *118*(5), 2474–2487, doi:10.1002/jgrb.50194.

Poland, M. P., A. Miklius, T. Orr, A. J. Sutton, C. R. Thornber, and D. Wilson (2008), New episodes of volcanism at Kilauea Volcano, Hawaii, Eos Trans. *AGU*, *89*(5), 37–38, doi:10.1029/2008EO050001.

Poland, M. P., A. J. Sutton, and T. M. Gerlach (2009), Magma degassing triggered by static decompression at Kīlauea Volcano, Hawai'i, *Geophys. Res. Lett.*, *36*(L16306), doi:10.1029/2009GL039214.

Poland, M. P., A. Miklius, A. J. Sutton, and C. R. Thornber (2012), A mantle-driven surge in magma supply to Kīlauea Volcano during 2003–2007, *Nature Geosci.*, *5*(4), 295–300, doi:10.1038/ngeo1426.

Poland, M. P., A. Miklius, and E. K. Montgomery-Brown (2014), Magma supply, storage, and transport at shield-stage Hawaiian volcanoes, Chapter 5 of Poland, M. P., T. J. Takahashi, and C. M. Landowski, (eds.), *Characteristics of Hawaiian Volcanoes*, U.S. Geol. Surv. Prof. Pap. 1801, pp. 179–234, doi:10.3133/pp18015.

Prôno, E., J. Battaglia, V. Monteiller, J. L. Got, and V. Ferrazzini (2009), P-wave velocity structure of Piton de la Fournaise volcano deduced from seismic data recorded between 1996 and 1999, *J. Volcanol. Geotherm. Res.*, *184*, 49–62.

Ren, Z. Y., S. Ingle, E. Takahashi, N. Hirano, and T. Hirata (2005), The chemical structure of the Hawaiian mantle plume, *Nature*, *436*, doi:10.1038/nature03907.

Robinson, J. E., and B. W. Eakins (2006), Calculated volumes of individual shield volcanoes at the young end of the Hawaiian Ridge, *J. Volcanol. Geotherm. Res.*, *151*(1–3), 309–317, doi:10.1016/j.jvolgeores.2005.07.033.

Roult, G., A. Peltier, B. Taisne, T. Staudacher, V. Ferrazzini, A. Di Muro, and the OVPF group (2012), A new comprehensive classification of the Piton de la Fournaise eruptions spanning the 1986–2011 period, Search and analysis of eruption precursors from a broad-band seismological station, *J. Volcanol. Geotherm. Res.*, *241–242*, 78–104.

Rowland, S. K., and G. P. Walker (1990), Pahoehoe and aa in Hawaii: Volumetric flow rate controls the lava structure, *Bull. Volcanol.*, *52*(8), 615–628, doi:10.1007/BF00301212.

Ryan, M. P. (1987), Neutral buoyancy and the mechanical evolution of magmatic systems, in Magmatic Processes: Physiochemical Principles, edited by B. O. Mysen, pp. 259–287, The Geochemical Society, University Park, Pa.

Ryan, M. P. (1988), The mechanics and three-dimensional internal structure of active magmatic systems: Kilauea volcano, Hawaii, *J. Geophys. Res.*, *93*(B5), 4213–4248, doi:10.1029/JB093iB05p04213.

Saint-Ange, F. (2009), La sédimentation volcanoclastique en contexte de point chaud (île de La Réunion, Océan Indien), Ph.D Thesis, University of La Réunion, 378 pp.

Saint-Ange, F., P. Bachèlery, N. Babonneau, L. Michon, and S. J. Jorry (2013), Volcaniclastic sedimentation on the submarine slopes of a basaltic hotspot volcano: Piton de la Fournaise volcano (La Réunion Island, Indian Ocean), *Marine Geol.*, *337*, 35–52.

Salaün, A., B. Villemant, M. P. Semet, and T. Staudacher (2010), Cannibalism of olivine-rich cumulate xenoliths during the 1998 eruption of Piton de La Fournaise (La Réunion hotspot): Implications for the generation of magma diversity, *J. Volcanol. Geotherm. Res.*, *198*, 187–204.

Simkin, T. (1984), Geology of Galápagos Islands, in Galápagos, edited by R. Perry, pp. 15–41, Pergamon Press, New York.

Staudacher, T., and C. J. Allègre (1993), Ages of the second caldera of Piton de la Fournaise volcano (Réunion) determined by cosmic ray produced 3He and 21Ne, Earth Planet. *Sci. Lett.*, *119*, 395–404.

Staudacher, T., and A. Peltier (2015), Ground deformation at Piton de la Fournaise (La Réunion Island), a review from 20 years of GNSS monitoring, in *Active Volcanoes of the Southwest Indian Ocean: Piton de la Fournaise and Karthala. Active Volcanoes of the World*, edited by P. Bachèlery, J.-F. Lénat, A. Di Muro, and L. Michon, Springer-Verlag Berlin and Heidelberg.

Staudacher, T., P. Sarda, and C. J. Allègre (1990), Noble gas systematics on Réunion Island, Indien Ocean, *Chem. Geol.*, *89*, 1–17.

Staudacher, T., V. Ferrazzini, A. Peltier, P. Kowalski, P. Boissier, P. Catherine, F. Lauret, and F. Massin (2009), The April 2007 eruption and the Dolomieu crater collapse, two major events at Piton de la Fournaise, *J. Volcanol. Geotherm. Res.*, *184*(1–2), 126–137.

Staudacher, T., A. Peltier, V. Ferrazzini, A. Di Muro, P. Boissier, P. Catherine, P. Kowalski, and F. Lauret (2015), Fifteen years of intense eruptive activity (1998–2013) at Piton de La Fournaise volcano (La Réunion): A review, in *Active Volcanoes of the Southwest Indian Ocean: Piton de la Fournaise and Karthala. Active Volcanoes of the World*, edited by P. Bachèlery, J.-F. Lénat, A. Di Muro, and L. Michon, Springer-Verlag Berlin and Heidelberg.

Swanson, D. A. (1972), Magma supply rate at Kilauea volcano, 1952–1971, *Science*, *175*(4018), 169–170, doi:10.1126/science.175.4018.169.

Swanson, D. A. (2008), Hawaiian oral tradition describes 400 years of volcanic activity at Kīlauea, *J. Volcanol. Geotherm. Res.*, *176*(3), 427–431, doi:10.1016/j.jvolgeores.2008.01.033.

Swanson, D. A., W. A. Duffield, and R. S. Fiske (1976), Displacement of the south flank of Kilauea Volcano: The result of forceful intrusion of magma into the rift zones, U.S. Geol. Surv. Prof. Pap. 963, 39 pp.

Swanson, D. A., T. P. Rose, R. S. Fiske, and J. P. McGeehin (2012), Keanakāko'i Tephra produced by 300 years of explosive eruptions following collapse of Kīlauea's caldera in about 1500 CE, *J. Volcanol. Geotherm. Res.*, *215–216*, 8–25.

Tatsumoto, M. (1978), Isotopic composition of lead in oceanic basalt and its implication to mantle evolution, *Earth Planet. Sci. Lett.*, *38*(1), 63–87, doi:10.1016/0012-821X(78)90126-7.

Thornber, C. R., C. Heliker, D. R. Sherrod, J. P. Kauahikaua, A. Miklius, P. G. Okubo, F. A. Trusdell, J. R. Budahn, W. I. Ridley, and G. P. Meeker (2003), Kilauea east rift zone magmatism; an episode 54 perspective, *J. Petrol.*, *44*(9), 1525–1559, doi:10.1093/petrology/egg048.

Thurber, C. H. (1984), Seismic detection of the summit magma complex of Kilauea Volcano, Hawaii, *Science*, *223*(4632), 165–167, doi:10.1126/science.223.4632.165.

Tilling, R. I ., and J. Dvorak (1993), Anatomy of a basaltic volcano, *Nature*, *363*, 125–133.

Tilling, R. I., J. P. Kauahikaua, S. R. Brantley, and C. A. Neal (2014), The Hawaiian Volcano Observatory—a natural laboratory for studying basaltic volcanism, Chapter 1 of Poland, M. P., T. J. Takahashi, and C. M. Landowski, (eds.), *Characteristics of Hawaiian Volcanoes*, U.S. Geol. Surv. Prof. Pap. 1801, pp. 1–64, doi:10.3133/pp18011.

Tinard, P. (2007), Caractérisation et modélisation des déplacements du sol associés à l'activité volcanique du Piton de la Fournaise, île de La Réunion, à partir de données interférométriques, Août 2003–Avril 2007, Ph.D. Thesis, Université de Clermont Ferrand, 334 pp.

Tiwari, V. M., I. Grevemeyer, B. Singh, and J. Phipps Morgan (2007), Variation of effective elastic thickness and melt production along the Deccan–Reunion hotspot track, *Earth Planet. Sci. Lett.*, *264*(1–2), 9–21, doi:10.1016/j.epsl.2007.08.023.

Toutain, J.-P., P. Bachèlery, P. A. Blum, J. L. Cheminee, H. Delorme, L. Fontaine, P. Kowalski, and P. Taochy (1992), Real time monitoring of vertical ground deformations during eruptions at Piton de la Fournaise, *Geophys. Res. Lett.*, *19*(6), 553–556.

Van Ark, E., and J. Lin (2004), Time variation in igneous volume flux of the Hawaii-Emperor hot spot seamount chain, *J. Geophys. Res.*, *109*(B11401), doi:10.1029/2003JB00294.

Vidal, V., and A. Bonneville (2004), Variations of the Hawaiian hot spot activity revealed by variations in the magma production rate, *J. Geophys. Res.*, *109*(B03104), doi:10.1029/2003JB002559.

Villeneuve, N. (2000), Apports multi-sources à une meilleure compréhension de la mise en place des coulées de lave et des risques associés au Piton de la Fournaise: Géomorphologie quantitative en terrain volcanique, Ph.D. Thesis, Institut de Physique du Globe de Paris, 378 pp.

Villeneuve, N., P. Bachèlery, and J. Kemp (2014), La Réunion Island: A typical example of a basaltic shield volcano with rapid evolution, in *Landscapes and Landforms of France, World Geomorphological Landscapes*, edited by M. Fort and M.-F. André, Springer Science + Business Media Dordrecht doi:10.1007/978-94-007-7022-5_25.

Vlastélic, I., E. Lewin, and T. Staudacher (2006), Th/U and other geochemical evidence for the Réunion plume sampling a less differentiated mantle domain, *Earth Planet. Sci. Lett.*, *248*(1–2), 379–393, doi:10.1016/j.epsl.2006.06.003.

Walter, T. R., and V. R. Troll (2003), Experiments on rift zone evolution in unstable volcanic edifice, *J. Volcanol. Geotherm. Res.*, *127*, 107–120.

Watts, A. B. (1978), An analysis of isostasy in the world's oceans: 1. Hawaiian-Emperor seamount chain, *J. Geophys. Res.*, *83*(B12), 5989– 6004, doi:10.1029/JB083iB12p05989.

Watts, A. B., and U. S. ten Brink (1989), Crustal structure, flexure, and subsidence history of the Hawaiian Islands,

J. Geophys. Res., *94*(B8), 10,473–10,500, doi:10.1029/JB094iB08p10473.

Weis, D., M. O. Garcia, J. M. Rhodes, M. Jellinek, and J. S. Scoates (2011), Role of the deep mantle in generating the compositional asymmetry of the Hawaiian mantle plume, *Nature Geosci.*, *4*(12), 831–838, doi:10.1038/ngeo1328.

Welsch, B., F. Faure, P. Bachèlery, and V. Famin (2009), Microcrysts record transient convection at Piton de la Fournaise volcano (La Réunion hotspot), *J. Petrol.*, *50*(12), 2287–2305.

Welsch, B., F. Faure, V. Famin, A. Baronnet, and P. Bachelery (2012), Dendritic crystallization: A single process for all the textures of olivine in basalts? *J. Petrol.*, *54*(3), 539–574.

Welsh, B. (2010), Signification des Océanites dans le fonctionnement du Piton de La Fournaise, Ile de La Réunion, Ph.D. Thesis, University of La Réunion, 303 pp.

Wessel, P. (1993), A reexamination of the flexural deformation beneath the Hawaiian Islands, *J. Geophys. Res.*, *98*(B7), 12,177–12,190, doi:10.1029/93JB00523.

White, R. S. (1993), Melt production rates in mantle plumes, *Philos. Trans. R. Soc. Lond. Ser. A*, *342*(1663), 137–153, doi:10.1098/rsta.1993.0010.

Wilson, D., T. Elias, T. Orr, M. Patrick, J. Sutton, and D. Swanson (2008), Small explosion from new vent at Kilauea's summit, Eos Trans. *AGU*, *89*(22), 203.

Wright, T. L. (1971), Chemistry of Kilauea and Mauna Loa lava in space and time, U.S. Geol. Surv. Prof. Pap. 735, 40 pp.

Wright, T. L., and R. S. Fiske (1971), Origin of the differentiated and hybrid lavas of Kilauea Volcano, Hawaii, *J. Petrol.*, *12*(1), 1–65, doi:10.1093/petrology/12.1.1.

24

"Points Requiring Elucidation" About Hawaiian Volcanism

Michael P. Poland

ABSTRACT

Hawaiian volcanoes, which are easily accessed and observed at close range, are among the most studied on the planet and have spurred great advances in the geosciences, from understanding deep Earth processes to forecasting volcanic eruptions. More than a century of continuous observation and study of Hawai'i's volcanoes has also sharpened focus on those questions that remain unanswered. Although there is good evidence that volcanism in Hawai'i is the result of a high-temperature upwelling plume from the mantle, the source composition and dynamics of the plume are controversial. Eruptions at the surface build the volcanoes of Hawai'i, but important topics, including how the volcanoes grow and collapse and how magma is stored and transported, continue to be subjects of intense research. Forecasting volcanic activity is based mostly on pattern recognition, but determining and predicting the nature of eruptions, especially in serving the critical needs of hazards mitigation, require more realistic models and a greater understanding of what drives eruptive activity. These needs may be addressed by better integration among disciplines as well as by developing dynamic physics- and chemistry-based models that more thoroughly relate the physiochemical behavior of Hawaiian volcanism, from the deep Earth to the surface, to geological, geochemical, and geophysical data.

24.1. INTRODUCTION

James Dwight Dana provided one of the first comprehensive investigations of Hawaiian volcanism in his 1890 book *Characteristics of Volcanoes* [*Dana*, 1890]. He visited volcanoes across the globe but clearly held a deep appreciation for those of Hawai'i, which he visited twice (in 1840 and, at the age of 74, in 1887) and in the interim was kept appraised of by letters from Reverend Titus Coan, who lived on the Island of Hawai'i. In the preface to his book, Dana raised a number of "points requiring elucidation" about the volcanoes of Hawai'i, some of which have been thoroughly addressed in the intervening years (for instance, the temperature of lava and the composition of volcanic gases) while others remain relevant topics of inquiry to this day, including compositional variations among Hawaiian shields and the mechanism for magma ascent from depth.

Ocean islands have long been recognized as exceptional natural laboratories. Understanding of biological speciation, extinction, and ecosystem evolution was largely developed on islands, and Earth scientists gravitate to oceanic volcanoes to study processes ranging from the deep Earth to the surface [e.g., *Gillespie and Clague*, 2009; *d'Ozouville et al.*, 2014]. Studies in Hawai'i have been particularly important. The island chain inspired the theories of hotstpots [*Wilson*, 1963] and thermochemical mantle plumes [*Morgan*, 1971] and provides a window into Earth that would otherwise be inaccessible to geologists [e.g., *Stolper et al.*, 2009]. Frequent eruptive activity and ease of access have established Hawai'i as a type locality for basaltic volcanism. It is easy to understand why Massachusetts Institute of Technology professor

Hawaiian Volcano Observatory, U.S. Geological Survey, Hawai'i National Park, Hawaii, USA

Hawaiian Volcanoes: From Source to Surface, Geophysical Monograph 208. First Edition.
Edited by Rebecca Carey, Valérie Cayol, Michael Poland, and Dominique Weis.

Companion Website: www.wiley.com/go/Carey/Hawaiian_Volcanoes

Thomas A. Jaggar, Jr., chose to build a facility—today's Hawaiian Volcano Observatory (HVO)—for continuous volcano observation at the summit of Kīlauea Volcano in 1912 [*Tilling et al.*, 2014]. Since then, Hawaiian volcanoes have served as a testing ground for monitoring equipment and techniques that have been applied to develop models of magma origin, ascent, storage, transport, and eruption, all applied to other volcanoes on Earth and across the solar system.

In the spirit of Dana's "points," this concluding chapter addresses current questions that challenge scientists who are working toward a better understanding of how Hawaiian and similar volcanoes work. Major research themes are divided by depth from source to surface: (1) magma origin and ascent, (2) magma storage and island evolution, and (3) eruptions. Each depth range is characterized in terms of problems of fundamental importance that represent ripe targets for future research.

24.2. MAGMA ORIGIN AND ASCENT

The Hawaiian-Emperor chain of seamounts and islands (Figure 24.1) records volcanic activity over at least the past 80 Ma, with an age progression that decreases to the southeast due to motion of the Pacific Plate over the Hawaiian "hotspot" melting anomaly. The origin of the hotspot is a subject of debate [e.g., *Foulger*

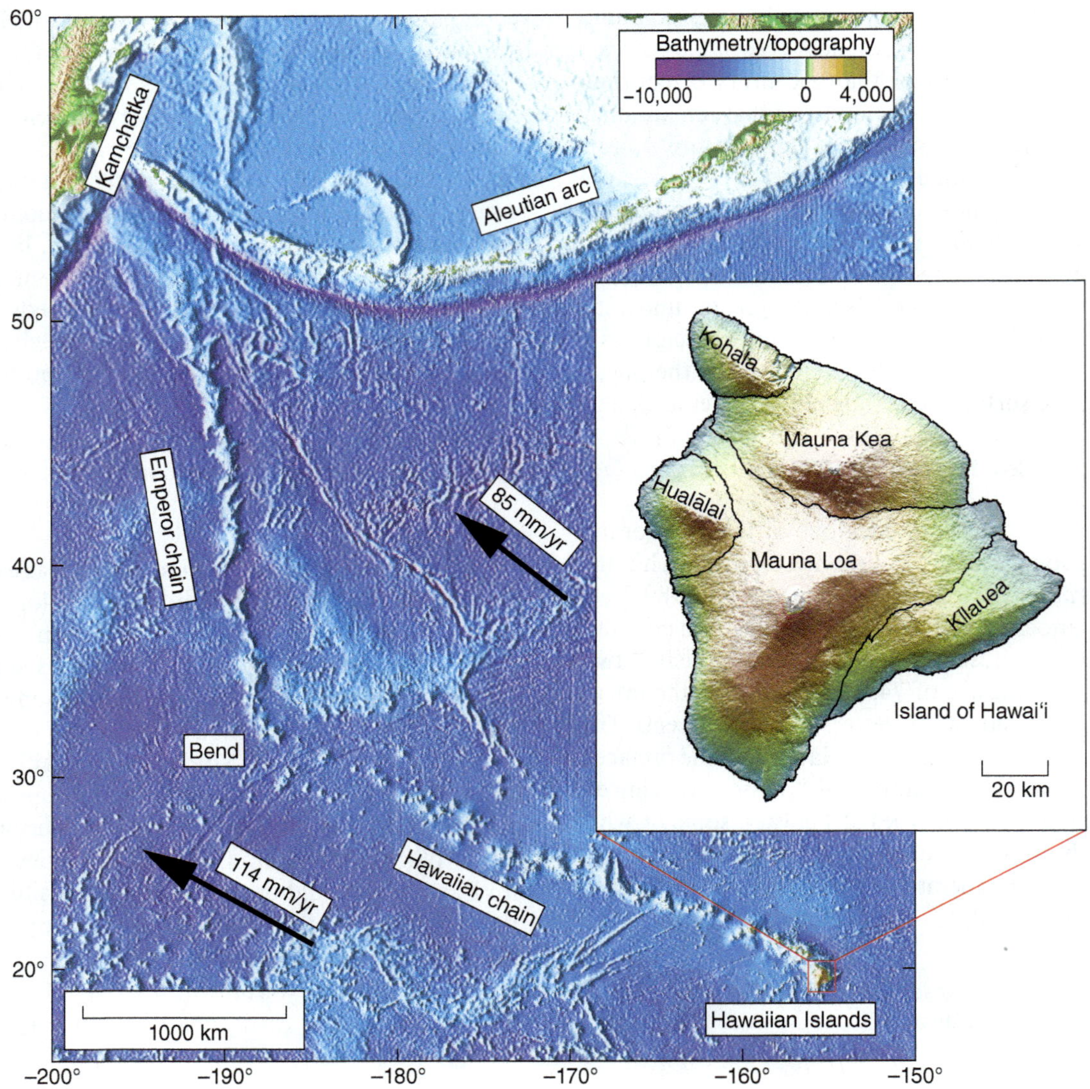

Figure 24.1 Bathymetric map showing the Hawaiian-Emperor chain of Island and seamounts in the northwest Pacific Ocean Basin. Inset shows relief map of the Island of Hawai'i, with the five volcanoes that make up the island outlined. Arrow indicates plate motion velocity in millimeters per year [*Gripp and Gordon*, 2002]. Figure taken from *Poland et al.* [2014].

et al., 2005], with some authors favoring a shallow (<150 km) depth of melting beneath Hawai'i [*Foulger*, 2007; *Anderson*, 2011]. The presence of a deep-seated mantle plume, however, is strongly suggested by primitive geochemical signatures (e.g., elevated helium isotopic ratios in erupted lavas [e.g., *Kurz et al.*, 1983]) and geophysical indications (especially from seismic imaging [e.g., *Wolfe et al.*, 2009]) of high-temperature upwelling from the lower mantle, although little is known about the plume itself or how it relates to surface volcanism.

24.2.1. Plume Origin, Structure, and Composition.

Seismic imaging of the Hawaiian plume indicates that it extends at least to the 660 km depth transition zone and possibly deeper [*Wolfe et al.*, 2009, 2011]. The specific depth of origin is difficult to constrain, due to a lack of sufficient resolution in seismic studies of the lowermost mantle, and the structure of the plume at depth is debatable [e.g., *Cao et al.*, 2011]. When projected vertically to the core-mantle boundary, the plume intersects the edge of a large low-shear-velocity province (LLSVP) that indicates the presence of a compositional boundary [*Weis et al.*, 2011]. Sampling of different sides of this boundary by the plume provides one explanation of the occurrence of parallel chains ("Loa" and "Kea") of geographically and geochemically distinct volcanoes at the surface.

The Loa (western) and Kea (eastern) volcanic trends, named for the largest volcanoes on each trend, were first recognized by *Dana* [1849] from the distribution of Hawaiian volcanoes (Figure 24.2a) [*Jackson et al.*, 1972]. The distinction is also manifested geochemically, most clearly in lead and other isotopes (Figure 24.2b) [*Tatsumoto*, 1978; *Abouchami et al.*, 2005; *Tanaka et al.*, 2008; *Weis et al.*, 2011] and also in major elements [e.g., *Hauri*, 1996; *Greene et al.*, 2010; *Jackson et al.*, 2012]. The occurrence of the geochemically distinct, parallel chains suggests that the plume may be divided into halves that preserve differing compositions over extreme length and timescales [*Abouchami et al.*, 2005]. Such a bilaterally asymmetric model is at odds with the common assumption of a concentrically zoned plume [e.g., *DePaolo et al.*, 2001; *DePaolo and Weis*, 2007] but is supported by numerical modeling of Hawaiian plume structure [e.g., *Farnetani and Hofmann*, 2010]. Other numerical models suggest that chemical zoning at the surface is a result, of spatial variations in temperature and pressure across the melting zone [*Bianco et al.*, 2008] or small-scale convection creating "topography" on the base of the lithosphere that impacts flow of material within the plume [*Ballmer et al.*, 2011]; chemical zonation of the mantle source would therefore be unnecessary.

A growing collection of geochemical data from shield-stage lava of several Hawaiian volcanoes emphasizes the complex nature of the Loa and Kea trends and argues for additional work to better understand the characteristics and source of the compositional and geographic distinction. Loa compositions have erupted from Kea volcanoes and vice versa, although Loa trend volcanoes tend to erupt more Loa-like compositions, with the same true of Kea trend volcanoes [e.g., *Garcia et al.*, 2012; *Xu et al.*, 2014]. This demonstrates that both compositions are

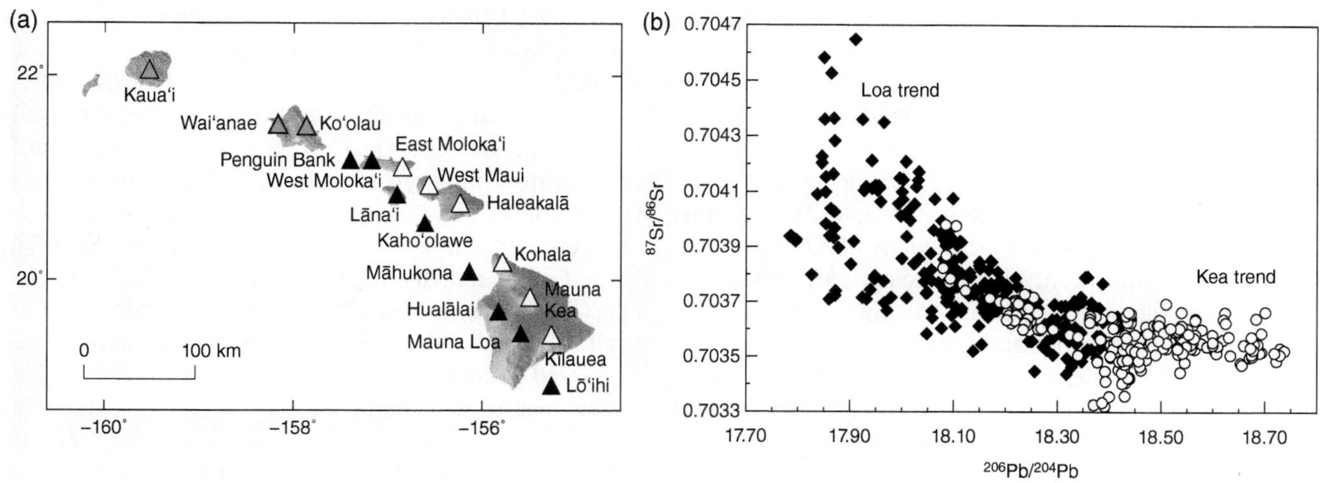

Figure 24.2 Parallel compositional trends that make up the recent islands of the Hawaiian chain. (a) Volcanoes formed over the last ~3 million years can be classified by their geochemistry as belonging to either the Kea (open triangles) or Loa (solid triangles) trends. Ko'olau marks the start of the Loa trend (according to *Tanaka et al.* [2008]). Before this time (gray triangles), volcanoes display characteristics of both trends. (b) Sr-Pb isotope plot of data from Hawaiian tholeiitic basalts highlighting the isotopic difference between Loa and Kea trend volcanoes. From Figure 1b inset of *Weis et al.* [2011].

available to volcanoes of both trends, prompting *Xu et al.* [2014] to suggest that the distinction may be caused by either chemical zoning in the plume or that the two components have different melting temperatures. Kea compositions melt at higher temperatures, meaning that they would dominate at volcanoes that are closer to the plume core, while Loa compositions would erupt from volcanoes on the plume margin. This interpretation is at odds with reconstructions of plume location based on geodynamics and isotopic evidence, which place the plume center closer to Loa trend volcanoes [*DePaolo and Stolper*, 1996; *DePaolo et al.*, 2001]. The fact that Loa and Kea compositions are coincident at several volcanoes argues that the plume is mixed with respect to both components and is highly heterogeneous [e.g., *Garcia et al.*, 2010, 2012; *Greene et al.*, 2010; *Pietruszka et al.*, 2013; *Xu et al.*, 2014].

The time of origin of the two components is also uncertain. The geographic distinction between Loa and Kea trends is most strongly manifested starting southeast of Koʻolau volcano, Oʻahu. Late-stage shield lavas there define the Loa end-member component [*Huang and Frey*, 2005], prompting *Tanaka et al.* [2008] to suggest Koʻolau as the start of the dual chain. The components are clearly present before Koʻolau, however, indicating that both components appear to be present in the plume. For example, West Kaʻena volcanism (between Kauaʻi and Oʻahu) has Loa-type composition [*Greene et al.*, 2010], and tholeiitic samples from Kauaʻi and Niʻihau straddle the Loa-Kea boundary, as defined by lead isotopes [*Garcia et al.*, 2010]. Northwest of the main Hawaiian Islands, Kea-type compositions appear to dominate in both the Hawaiian Ridge and Emperor Seamounts [*Regelous et al.*, 2003; *Tanaka et al.*, 2008; *Garcia et al.*, in press]. There is some evidence, however, for Loa-like compositions near the Hawaiian-Emperor bend, as suggested by lead isotope results [*Regelous et al.*, 2003; *Garcia et al.*, in press].

Parallel compositional chains occur at several hotspots that overlie the Pacific LLSVP [*Huang et al.*, 2011]. Multiple locations, in addition to Hawaiʻi, are thus available to test hypotheses related to plume structure, origin, and composition. These are clearly ripe topics for future investigation, especially using methods that account for dynamic interactions among a thermochemical plume, the mantle, and lithosphere, and that consider a variety of geophysical (e.g., bathymetry, gravity, and seismic velocity) and geochemical (e.g., major-element, isotope, and volatile) data.

The isotopic and mineralogical composition of the melting region for the Hawaiian plume is poorly constrained. Geochemical studies have identified a number of potential constituents, and there is general agreement that mixing among three components can explain most shield-stage lavas erupted during the past 5 m.y. [e.g., *Hauri*, 1996], with a fourth component needed to explain postshield and rejuvenated lava compositions [e.g., *Garcia et al.*, 2010]. These components are thought to reflect both primitive mantle and various compositions of recycled (i.e., subducted) oceanic crust [e.g., *Hauri*, 1996; *Ren et al.*, 2009; *Jackson et al.*, 2012; *Pietruszka et al.*, 2013]. Compositional changes in lava erupted from the Hawaiian hotspot over time reflect both a heterogeneous mantle source and variable degrees of melting. For example, the Emperor Seamounts preserve a systematic chemical evolution from 85 Ma to the time of the bend in the Hawaiian-Emperor chain (at ~47.5–49 Ma [*O'Connor et al.*, 2013; *Garcia et al.*, in press]) that suggests a gradual trend from compositions more akin to those seen at midocean ridges toward compositions that resemble those in lava erupted today, both chemically and isotopically. This progression reflects a change in source composition and varying degrees of melting beneath lithosphere of irregular thickness [*Lanphere et al.*, 1980; *Regelous et al.*, 2003].

Shields of the main Hawaiian Islands that formed during the past 4.5 m.y. span a range of compositions, which argues for geochemically distinct sources and extent of melting [*Frey and Rhodes*, 1993] or differences in the depth at which different materials begin to melt combined with spatial variability in the rate of melting due to plume-lithosphere interaction [*Bianco et al.*, 2011]. The small scale of some mantle heterogeneities is indicated by occasional departures from "normal" compositions at some volcanoes. For example, isotopic compositions of lavas erupted from Kīlauea and Mauna Loa over the last few thousand years moved away from values typical to each of those volcanoes to an intermediate value and back again [*Marske et al.*, 2007]. Geochemical changes over the last 200 years at Kīlauea span the entire compositional range from that volcano and likewise reflect variable source composition and changing magma supply [*Pietruszka and Garcia*, 1999a]. Heterogeneous mantle composition has even been proposed to explain short-term (months to years) variations in the composition of lavas erupted during Kīlauea's 1983-to-present East Rift Zone (ERZ) activity [e.g., *Garcia et al.*, 2000; *Marske et al.*, 2008; *Greene et al.*, 2013], although mixing between shallow-stored and freshly supplied magma has also been suggested [*Thornber*, 2003; *Thornber et al.*, 2003]. The debate over source composition is far from trivial, as the chemical make-up of the plume is a critical constraint on models of plume dynamics. For example, any estimate of plume temperature or rheology based on geophysical measurements is dependent on the assumed composition of the plume [e.g., *Herzberg*, 2006]. Such compositions should be constrained by petrologic data from the most primitive erupted products, including high-MgO glass in Kīlauea lava from the submarine Puna Ridge, which can

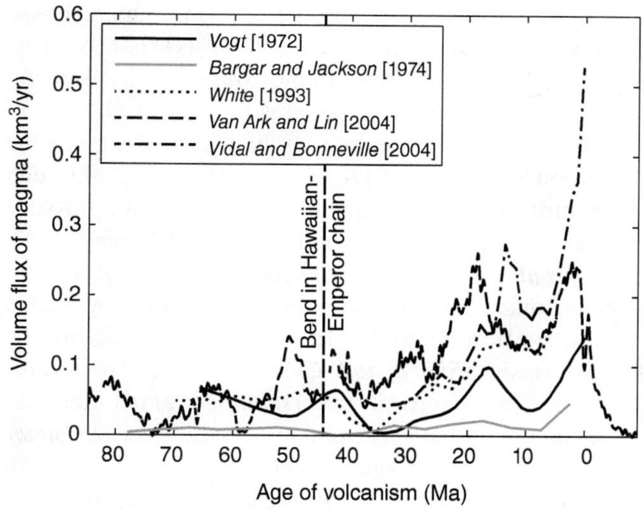

Figure 24.3 Volume flux of the Hawaiian hotspot over time as calculated by various authors. The estimate of *Bargar and Jackson* [1974] is based on the volume of seamounts and islands in the Hawaiian-Emperor chain (not including any estimate of the crustal root), with the timing established by *Robinson and Eakins* [2006]. *Vogt* [1972] used a similar volume-age relation but included an estimate for the volume of the crustal root. *White* [1993] based his volume flux curve on crustal thickening determined seismically, coupled with the volume of seamounts and volcanoes and their underplated roots (determined assuming Airy isostasy). *Vidal and Bonneville* [2004] distinguished topography/bathymetry created by the Hawaiian swell (caused by the mantle plume) from that caused by volcanism, which is a better measure of the magma production rate. Finally, *Van Ark and Lin* [2004] calculated crustal thicknesses from gravity data and incorporated a model of crustal flexure to compute hotspot volume flux over time. Dashed vertical line is the location of the bend in the Hawaiian-Emperor chain of seamounts and islands (see Figure 24.1). Taken from *Poland et al.* [2014].

be used to calculate parental melt characteristics [*Matzen et al.*, 2011]. There remains some question, however, as to whether or not geochemical trends observed at the surface are reflective of the composition of the lower mantle [*Cao et al.*, 2011].

The source composition, extent of melting, and dynamics of magma ascent control the composition of not only lavas erupted at the surface but also volatiles released to the atmosphere or trapped in crystals. Such information is critically important, given that volatile exsolution drives surface volcanism [e.g., *Sparks*, 1978]. With a plume that could tap as deep as the core-mantle boundary, Hawai'i provides an opportunity to investigate the volatile composition of the lower mantle, provided that surface volcanism preserves lower mantle chemistry (a point disputed by *Cao et al.* [2011]). Unfortunately, there is a dearth of samples that preserve the primary volatile compositions because lavas erupted at the surface will have mostly degassed by the time they crystalize and solidify. The best volatile quantifications therefore come from quenched seafloor lavas, where pressure from the overlying water column suppresses some volatile exsolution, and subaerial eruptive products (usually tephras) that come from magma that ascended rapidly enough that crystals with trapped volatiles reach the surface without being modified by shallow equilibration [*Wallace and Edmonds*, 2011]. Such studies are most useful for assessing volatile species like H_2O and S, which degas at low pressures. CO_2, in contrast, is poorly constrained, since it starts to exsolve at depths of ~30 km [*Gerlach*, 1986]. Water contents extracted from submarine glasses indicate differing source volatile compositions for lavas erupted at variable distances from the plume, supporting a view that the plume is zoned not just chemically or thermally but also with respect to volatiles [*Dixon and Clague*, 2001].

24.2.2. Hotspot Dynamics Over Time

The mass flux of a hotspot is a direct measure of the magma supply to the surface and provides a measure of hotspot vigor over time. A number of mass flux estimates have been made for the Hawaiian hotspot, initially based on bathymetry and topography [*Vogt*, 1972; *Bargar and Jackson*, 1974] and later incorporating seismic and gravity data, coupled with models of crustal loading [*White*, 1993; *Van Ark and Lin*, 2004; *Vidal and Bonneville*, 2004; *Robinson and Eakins*, 2006]. Although the rates calculated by these studies differ by about an order of magnitude (primarily because earlier studies had difficulty accounting for volumes obscured from bathymetry by flexure and isostatic compensation), they track one another in a gross sense and provide an ~80 million year record of mass flux that displays variations over timescales of millions of years (Figure 24.3). All flux estimates also show an increase in magma supply to the surface over the past 30 Ma, with the highest rates occurring at present. The flux variations probably reflect changes in lithospheric thickness (a factor that influences the extent of decompression melting) and pulsation of the plume [*Vogt*, 1972; *White*, 1993; *Regelous et al.*, 2003; *Van Ark and Lin*, 2004; *Vidal and Bonneville*, 2004] and may correlate with the composition of the source material as reflected by geochemical changes in erupted lava [e.g., *Greene et al.*, 2010]. For example, the strong manifestation of the Loa geochemical trend over the past few million years, from Ko'olau to the present, and the geographic manifestation of the dual volcanic chains occur during a time of increasing mass flux [*Weis et al.*, 2011] (although the Loa-type lava sampled at the bend in the Hawaiian-Emperor chain [*Regelous et al.*, 2003; *Garcia et al.*, in press] was during a period of comparatively low flux).

The volume of melt provided to a volcano by the hotspot also varies on timescales of less than a million years. The Hawai'i Scientific Drilling Project (HSDP) sampled nearly 500,000 years of Mauna Kea eruptive activity in a ~3.1 km deep drill hole. Geochemical diversity in retrieved samples is consistent with repeated waxing and waning supply during the volcano's shield-building stage, as opposed to a smoother increase and then decrease in supply as the volcano grew and was then carried away from the hotspot [*Stolper et al.*, 2004; *Rhodes et al.*, 2012]. Models of the growth of the Island of Hawai'i similarly suggest pulsed magma flow from the hotspot, based on volcano volumes and life spans [*Lipman and Calvert*, 2013] and lava accumulation rates [e.g., *Jicha et al.*, 2012], but even shorter-term changes in supply are evident from recent volcanism. A fluctuating supply to Kīlauea is indicated over centuries by cyclic transitions between explosive and effusive eruptions [*Trusdell*, 2011; *Swanson et al.*, 2014] and over years by monitoring data that indicate a more than doubling of supply during 2003–2007 compared to supply rates before and after that time period [*Poland et al.*, 2012; *Poland*, 2014]. Geochemical evidence suggests episodic intrusion of distinct magma batches into Kīlauea's summit reservoir system [e.g., *Wright*, 1971; *Pietruszka and Garcia*, 1999a,b; *Helz et al.*, 2014], which could be evidence for either pulsatory magma supply or fluctuations in source composition. While it is clear that supply to individual volcanoes can change over a variety of timescales, it is not known whether these changes indicate variations in melting and magma rise rates or, instead, changes in magma pathways, for instance, with melt being occasionally diverted into sublithospheric sills or other storage areas that cannot currently be detected geophysically. In other words, are supply rates to individual volcanoes reflective of mantle melting and magma ascent rates, or are melting and ascent rates more uniform over time, which would imply that changes in surface volcanism are reflective of pathway disturbances? Numerical models offer tantalizing indications that variations in magma supply to the surface are a normal consequence of the interaction between a thermochemical plume, the mantle, and the moving lithospheric plate [e.g., *Ballmer et al.*, 2011, 2013] (see also Chapter 3, this volume), but it is unlikely that such models can explain all timescales of magma supply changes.

In addition to variable mass flux through the hotspot, the prominent bend in the Hawaiian-Emperor chain (Figure 24.1) raises the possibility of variable plume motion over time. The bend was initially interpreted as reflecting a change in plate motion at the time of its 50 Ma formation [*Morgan*, 1971]. Paleomagnetic evidence, however, suggests that the hotspot was not fixed geographically but was moving south at tens of millimeters per year during formation of the Emperor Seamounts. The Suiko seamount, in the middle of the Emperor chain, has a paleolatitude that is farther north (~27°N) than present-day Hawai'i (19.5°N), for example [*Kono*, 1980]. A slowing in this southward motion, perhaps because of changes in mantle convection patterns, may have contributed to the formation of the bend [*Tarduno et al.*, 2003], although plate motion models suggest that Hawaiian hotspot motion may be relatively small [*Gripp and Gordon*, 2002; *Wessel et al.*, 2006; *Wessel and Kroenke*, 2008]. Subsequent plate tectonic analyses have returned to favoring a change in plate motion about 50 Ma due to changes across the Pacific Basin [*Sharp and Clague*, 2006], Australian-Antarctic plate reorganization [*Whittaker et al.*, 2007], or some combination of plate and hotspot motion [*O'Connor et al.*, 2013]. Refined ages will continue to shed light on this issue; for instance, *Garcia et al.* [in press] found an initiation age for the bend of 48–49 Ma, while the onset of arc volcanism in the western Pacific was 51–52 Ma, suggesting that the events were not synchronous.

24.2.3. Interactions Between Volcanoes

Numerous active volcanoes occur within close proximity of one another in Hawai'i, for instance, the summits of Kīlauea and Mauna Loa are less than 35 km apart. The geographic distribution of volcanoes has prompted speculation on the relation between adjacent shields, such as whether or not Kīlauea is a satellite vent of Mauna Loa [*Dana*, 1890]. Despite their close proximity, these volcanoes have distinct geochemical signatures and behave independently of each other, implying the presence of separate magma ascent pathways between the area of melt formation and the surface and effectively ruling out the presence of shallow (crustal) connections between volcanoes [*Frey and Rhodes*, 1993]—a conclusion that is corroborated by seismic evidence [e.g., *Okubo and Wolfe*, 2008]. How, then, is melt partitioned between volcanoes? There is some evidence that the volcanoes compete for magma supply. Statistical evidence from eruptive activity during the past ~200 years indicates that when Mauna Loa is highly active, Kīlauea is quiet or experiences persistent summit lava lake activity, for example, from the early 1800s to about 1950. Alternatively, when Kīlauea is experiencing heightened eruptive activity, which has been the case since about 1950, Mauna Loa is less active [*Moore*, 1970; *Klein*, 1982]. Over the past 4000 years, a similar trend is apparent. During times when Mauna Loa has had numerous eruptions, Kīlauea has experienced episodes of explosive volcanism that may be tied to decreases in magma supply to that volcano. Conversely, Kīlauea erupts more lava flows when supply shifts to favor it, while Mauna Loa has fewer overall eruptions [*Trusdell*, 2011; *Swanson et al.*, 2014].

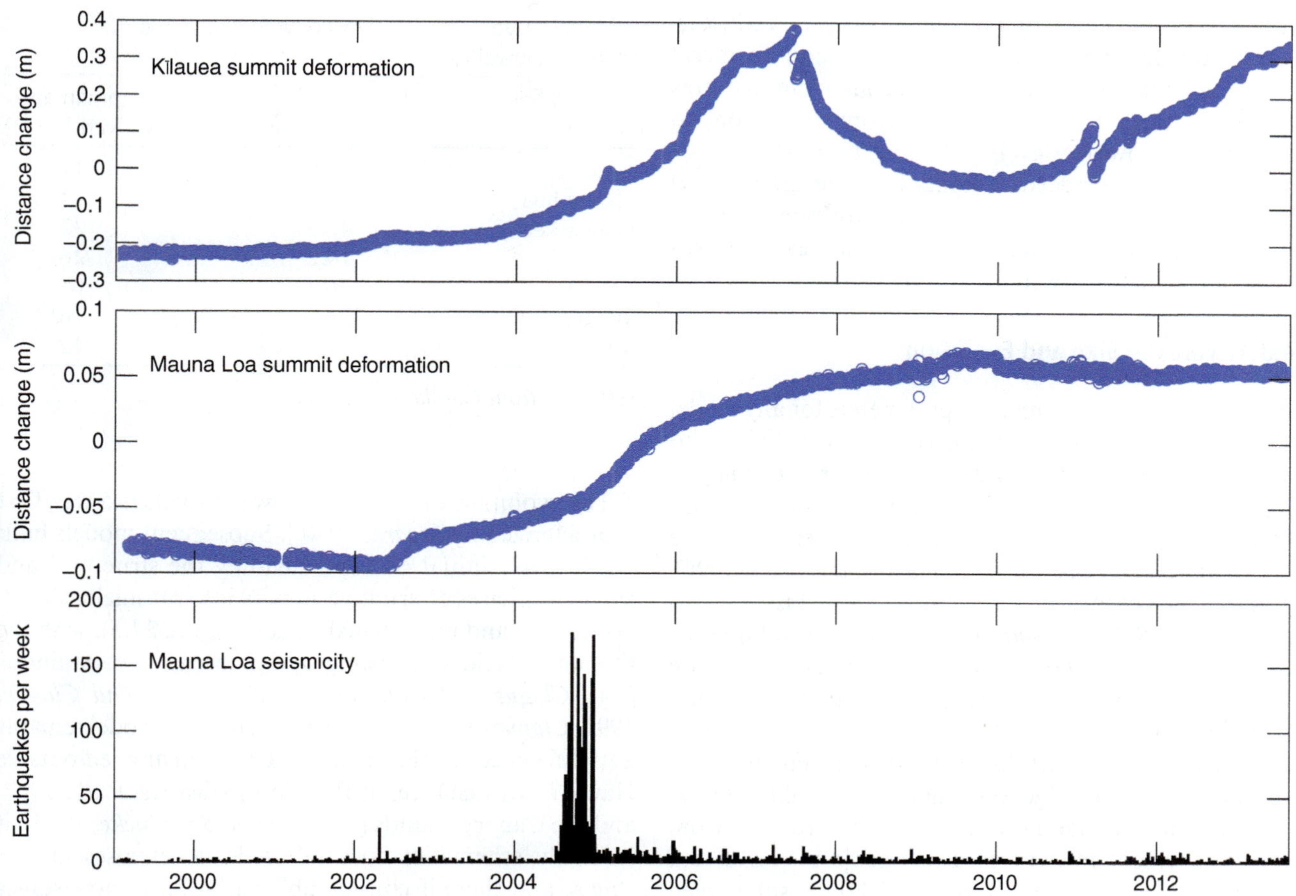

Figure 24.4 Time series of deformation at Kīlauea (top) and Mauna Loa (middle), and seismicity beneath Mauna Loa (bottom) during 1999–2013. Deformation time series are line lengths that cross the calderas of the two volcanoes. Increases in line length generally indicate inflation. Inflation at both volcanoes in the mid-2000s was caused by a surge in magma supply from the hotspot to the volcanoes [*Poland et al.*, 2012, 2014; *Gonnermann et al.*, 2012]. Seismicity time series is for the area beneath Mauna Loa's caldera. Most of the earthquakes during late 2004 occurred at 30–50 km depth [*Okubo and Wolfe*, 2008].

Direct correlations in volcanic activity between adjacent volcanoes have also been noted. During a 2003–2007 surge in magma supply to Kīlauea, Mauna Loa inflated (Figure 24.4) [*Poland et al.*, 2012] and experienced a swarm of deep, long-period earthquakes that were thought to be related to magma ascent [*Okubo and Wolfe*, 2008]. The coincidence in deformation can be modeled as magma influx into a broad asthenospheric melt layer that then supplied magma to both Kīlauea and Mauna Loa from spatially distinct areas, thus preserving their independent isotopic signatures [*Gonnermann et al.*, 2012].

Adjacent volcanoes may also interact chemically, but the mechanisms for such interactions are ambiguous and provide an opportune subject for future work. *Rhodes et al.* [1989] found that lava with a Mauna Loa-like geochemical signature that is less than 2000 years old is exposed in the walls of some pit craters on Kīlauea. If these flows erupted from Kīlauea, a model must be developed where magma from Mauna Loa can invade the plumbing system of Kīlauea, even though there is no evidence for a shallow connection between the two volcanoes. Eruption of lava from the "wrong" volcano has also been documented in the Galápagos, which is another locale with closely spaced, yet chemically distinct, basaltic volcanoes [*Geist et al.*, 1999].

24.3. MAGMA STORAGE AND VOLCANO EVOLUTION

Kīlauea and Mauna Loa are among the most active volcanoes in the world, and geophysical, geochemical, and geological data collected since the founding of the HVO in 1912 are well suited to support studies of magma storage and transport. Older volcanoes, e.g., Ko'olau, have portions of their magmatic systems that have been exposed by erosion, providing glimpses into the shallow (upper ~1 km) interiors of Hawaiian edifices. Finally, the large number of currently and formerly active volcanoes along the Hawaiian chain demonstrates how Hawaiian volcanoes evolve from inception to subsidence below sea level, including periods of catastrophic collapse. Indeed, insights into the growth and evolution of Hawaiian volcanoes have formed the basis for investigations at other basaltic volcanoes on Earth and other planets (especially Mars).

24.3.1. Volcano Size and Evolution

One of the most fundamental parameters for any model of volcano evolution is also one of the most difficult to quantify: Volume. Growth rates, age constraints, magma supply, and numerous other factors, even the modeled size of the hotspot, rely upon knowledge of volcano volume [e.g., *Moore and Clague*, 1992; *DePaolo and Stolper*, 1996]. Although the volume of the Island of Hawai'i is well constrained [e.g., *Robinson and Eakins*, 2006], the volumes of the individual volcanoes that make up the island are poorly known and have errors greater than 30% because of the interfingering and overlapping nature of the edifices [*Clague and Sherrod*, 2014] (Table 24.1). For example, the submarine Hilo Ridge was originally thought to represent a rift zone of Mauna Kea, which is the volcano from which the ridge appears to emanate [e.g., *Fiske and Jackson*, 1972; *Moore and Clague*, 1992]. A submerged shoreline terrace and isotopic compositions of dredged samples [*Holcomb et al.*, 2000], together with gravity evidence [*Kauahikaua et al.*, 2000], however, argue that the ridge is part of Kohala. If true, Mauna Kea is smaller than originally thought, and Kohala is much larger [*Lipman and Calvert*, 2013]. Similarly, there is disagreement over whether Kīlauea forms a veneer [*Coombs et al.*, 2006; *Lipman et al.*, 2006] or deeply indents [*Morgan*, 2006] the south flank of Mauna Loa—models that imply vastly different volumes (11×10^3 km^3 [*Lipman and Calvert*, 2013] versus 31.6×10^3 km^3 [*Robinson and Eakins*, 2006]) and, therefore, growth histories for Kīlauea.

Knowledge of volcano volumes and geometries opens the door to a variety of additional questions. Available evidence suggests substantial differences between volcanoes that form the Island of Hawai'i, for example, between Māhukona (6×10^3 km^3) [*Garcia et al.*, 2012] and Mauna Loa (74–83×10^3 km^3) [*Robinson and Eakins*, 2006; *Lipman and Calvert*, 2013]. This raises questions about the duration of volcanic activity and the rate of magma supply. Do smaller volcanoes have shorter life spans than larger ones or are they active for the same duration but with lower magma supply rates? Answers to such questions are central to understanding how the hotspot relates to surface volcanism [e.g., *DePaolo and Stolper*, 1996; *Lipman and Calvert*, 2013].

Table 24.1 Volumes (10^3 km^3) of volcanoes on the Island of Hawai'i.

Volcano	Robinson and Eakins [2006]	Lipman and Calvert [2013]
Kīlauea	31.6	11
Mauna Loa	74.0	83
Mauna Kea	41.9	22
Hualālai	14.2	26
Kohala	36.4	64
Māhukona	13.5	6[a]
Total	211.6	212

[a] *Estimate from Garcia et al. [2012].*

The evolution of a typical Hawaiian volcano was first characterized by *Stearns* [1946]. Subsequent models built on Stearns's initial concept, dividing the structural and chemical phases of eruptive activity into preshield, shield, postshield, and rejuvenated stages (Figure 24.5), with the tholeiitic shield stage being by far the most voluminous [e.g., *Clague and Dalrymple*, 1987; *Moore and Clague*, 1992; *Clague and Sherrod*, 2014]. While the model remains a useful generalization that has also been applied outside Hawai'i, for instance, at Réunion [*Albarède et al.*, 1997] and the Canary Islands [*Hoernle and Schmincke*, 1993], it is highly schematic; no single volcano exposes all the stages, and there is considerable variation in how stages are manifested. Some volcanoes lack entire stages, and the transition between stages can be gradual or rapid and may involve substantial gaps in time [e.g., *Clague and Sherrod*, 2014]. The mechanism for and significance of these differences in volcano evolution are unknown and are not currently well accounted for by models of plume dynamics or hotspot activity.

24.3.2. Magma Plumbing Systems

The first model of Kīlauea based on geophysical evidence was proposed by *Eaton and Murata* [1960], who used data associated with the paired 1959 Kīlauea Iki and 1960 Kapoho eruptions, at the summit and lower ERZ, respectively. The Eaton-Murata model included magma rising from >60 km depth in the mantle into storage reservoirs beneath the summit, from which it could erupt or move laterally through radial rift zones to intrude or erupt tens of kilometers from the summit (Figure 24.6a). The basic framework of this model has remained largely unchanged since it was proposed, with additional data providing increased resolution on the details of the plumbing system. *Ryan et al.* [1981] and *Ryan* [1987, 1988] offered a representation that mapped out magma storage areas based on hypocenter distributions and

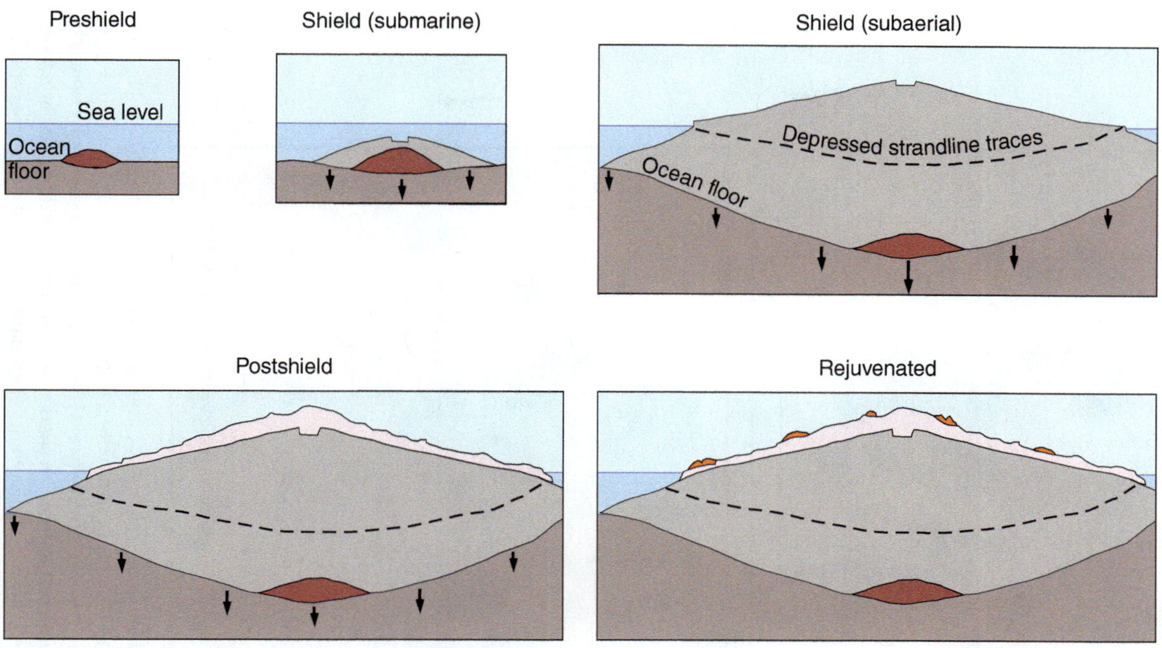

Figure 24.5 Generalized evolutionary stages of Hawaiian shield volcanoes. Arrows indicate subsidence of the ocean floor. Adapted from *Clague and Sherrod* [2014].

constrained by geological data (Figure 24.6b). *Poland et al.* [2014] used geophysical data collected since the late 1990s, coupled with geological data, to build upon these models (Figure 24.6c), identifying several discrete but interconnected magma storage areas beneath the summit (also identified by *Baker and Amelung* [2012]), with a small reservoir at ~1.5 km depth and a larger reservoir ~3 km beneath the surface. The mechanism behind the two-tiered magma system is ambiguous: How can the reservoirs act independent geophysically [*Baker and Amelung*, 2012; *Poland et al.*, 2014] yet show little petrologic distinction [e.g., *Pietruszka and Garcia*, 1999a,b] (see also Chapter 20, this volume)? Are the different levels caused by density contrasts between deeper, bubble-poor magma and shallow, bubble-rich magma that may occasionally overturn [*Clague and Dixon*, 2000], as suggested by episodic and cyclic deformation events at Kīlauea's summit (Chapter 11, this volume)? A similar magma storage geometry exists at other volcanoes, such as Piton de la Fournaise [*Peltier et al.*, 2009] (Chapter 23, this volume), Fernandina [*Chadwick et al.*, 2011; *Bagnardi and Amelung*, 2012], and Etna [*Aloisi et al.*, 2011], providing other opportunities to examine the problem. With regard to Kīlauea's rift zones, which branch to the east and southwest from the summit magma complex, geological and geophysical data suggest that each rift zone consists of paired arms that occupy different depths, with the northern arms ("HKIT" and "volcanic SWRZ" in Figure 24.6c) transporting magma within 1 km of the surface and the southern arms ("seismic SWRZ" and "East Rift Zone" in Figure 24.6c) representing the main structural boundaries of the volcano and having magma pathways at ~3 km depth [*Holcomb*, 1987; *Poland et al.*, 2014]. This geometry may reflect gradual southward migration of the rift zones over time due to motion of the volcano's south flank (see Section 24.3.3) [*Swanson et al.*, 1976]. While these models do not reflect the full complexity of magma storage and transport at Kīlauea (for example, the recent discovery that magma occasionally intrudes into and erupts from the central part of the Koaʻe Fault System [*Swanson et al.*, 2012a]), they provide a refined context for interpreting future data and magmatic events that may lead to further improvements.

A persistent issue related to shallow magma storage and transport at Kīlauea involves the connection between the ERZ and the summit. Some authors have proposed that magma that feeds ERZ eruptions and intrusions can bypass the summit reservoir complex, perhaps traveling through the deep part of the ERS (below the primary conduit at ~3 km depth) and therefore not mixing with shallow stored magma [e.g., *Garcia et al.*, 1996; *Vinet and Higgins*, 2010]. Opening of the deep rift zones has been modeled from deformation data [e.g., *Delaney et al.*, 1990; *Cayol et al.*, 2000; *Owen et al.*, 2000], and the region is mostly aseismic [*Ryan*, 1988], suggesting a plastic rheology that may indicate magma. High seismic velocities [*Okubo et al.*, 1997; *Park et al.*, 2007, 2009] and gravity anomalies indicative of dense rocks [*Kauahikaua et al.*, 2000], however, favor the presence of high-temperature cumulates in the deep rift zone that may even drive

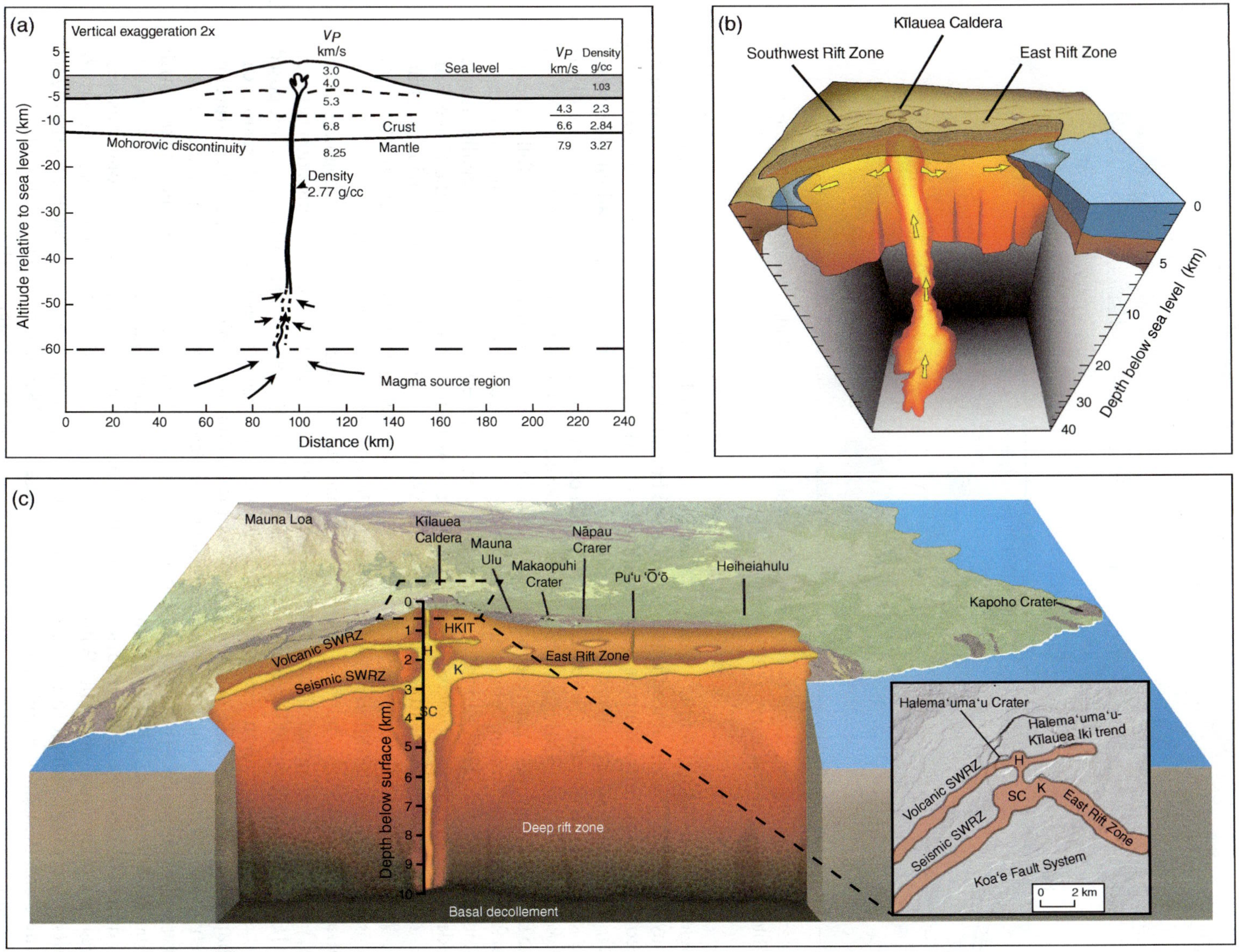

Figure 24.6 Models of Kīlauea's magma plumbing system. (a) Schematic cross section through an idealized Hawaiian volcano, as envisioned by *Eaton and Murata* [1960], based on seismic and deformation data. Magma ascends from about 60 km depth through open conduits and collects beneath the caldera. Seismic p-wave velocities (*Vp*) and representative densities are indicated for various depths. (b) Northward view of Kīlauea's internal structure modified from *Ryan* [1988] and based on seismicity beneath the volcano and subsurface density structure. Preferred magma pathway within the core region of the primary vertical conduit is shaded yellow, and lateral magma injection occurs along a level of neutral buoyancy 2–4 km beneath the surface (horizontal arrows). Magma occupies both the shallow and deep parts of the rift zones. Seismicity used by both *Eaton and Murata* [1960] and *Ryan* [1988] to define the vertical magma conduit below ~13 km depth was later shown to be caused by faulting in the lower crust and upper mantle and not related to any deep vertical feeder [*Wolfe et al.*, 2004]. (c) Proposed structure of Kīlauea's subsurface magma plumbing system from *Poland et al.* [2014], inferred from geophysical and geological data. Schematic cutaway shows a cross section through Kīlauea's summit and rift zones. Magma pathways and storage areas are exaggerated in size for clarity. H=Halema'uma'u reservoir, K=Keanakāko'i reservoir, SC=south caldera reservoir, SWRZ=Southwest Rift Zone, HKIT=Halema'uma'u-Kīlauea Iki trend. Plan view gives the relations of magma pathways to surface features and topography in the vicinity of Kīlauea Caldera.

gravitational instability of Kīlauea's south flank [*Clague and Denlinger*, 1994]; thus, the potential of the deep rift zone as a magma transport pathway is unclear.

The magma system of Mauna Loa appears simpler than that of Kīlauea in terms of geometry, probably because it is not as well constrained owing to the relative paucity of geophysically monitored eruptive and intrusive activity compared to Kīlauea. Inter- and co eruptive deformation data from Mauna Loa's summit area in the 1970s to the 1990s clearly indicated magma storage 3–4 km beneath the south part of the caldera [*Decker et al.*, 1983; *Lockwood and Lipman*, 1987; *Johnson*, 1995; *Miklius et al.*, 1995]. An episode of inflation in the 2000s was well monitored by Interferometric Synthetic Aperture Radar (InSAR) and Global Positioning System (GPS data) (Figure 24.4), and the deformation suggested that, in addition to the south caldera magma reservoir, a vertically elongated, tabular storage area underlies and follows the trend of the caldera below about 4 km depth [*Amelung et al.*, 2007]. A swarm of thousands of long-period earthquakes during Mauna Loa inflation in 2002–2004 was located between 30 and 50 km directly beneath the volcano's summit caldera, implying the presence of a focused magma feeder conduit that extends from these depths to shallow levels and that is independent of Kīlauea's feeder system [*Okubo and Wolfe*, 2008]. More detailed examinations of magma storage and transport await geophysical observations of future intrusive and eruptive activity at the volcano.

While there are some constraints on the geometry of magma storage at Kīlauea and Mauna Loa, the volume of storage is poorly known, and estimates are contradictory, especially between disciplines. Petrologic calculations for reservoir volume are generally smaller than geophysical ones, for example, *Pietruszka and Garcia* [1999b] found a volume for magma storage beneath Kīlauea's summit of ~2 km^3 based on isotope and trace-element ratios in historical lava, while *Segall et al.* [2001] modeled deformation data during a dike intrusion and favored a volume of ~20 km^3 (*Denlinger* [1997] modeled deformation and effusion rate data from 1991, during Kīlauea's ongoing ERZ eruption, to calculate a volume of 240 km^3, but this reflects the total volume of magma stored within the volcano rather than that within the summit reservoir system alone). This discrepancy may reflect that petrologic models consider the zone where batches fluid magma can mingle [*Pietruszka and Garcia*, 1999b], whereas geophysical models are based on physical and mechanical properties and behaviors [*Iyer*, 1984]. The existence of magma storage below the known summit reservoir complex is also ambiguous. A magma reservoir in the upper mantle (20–30 km depth) could explain the suggestion of high-pressure fractionation in Kīlauea lavas [*Wright*, 1971] and would also serve as a filter that removes mantle xenoliths from ascending magma during a volcano's shield stage [*Clague*, 1987], but these arguments provide indirect evidence only. Seismic swarms occur tens of kilometers beneath both Kīlauea [e.g., *Aki and Koyanagi*, 1981] and Mauna Loa [*Okubo and Wolfe*, 2008] but are not necessarily indicative of magma storage [*Wolfe et al.*, 2004]. Likewise, volume loss modeled at depths of ~25 km during Kīlauea's 1924 subsidence [*Mogi*, 1958] may be compromised by rod-scale error in leveling measurements of vertical deformation [*Eaton*, 1962; *Dvorak*, 1992].

24.3.3. Flank Instability

Hawai'i is the type locality for flank instability at oceanic volcanoes, displaying a range of behaviors from active creep to catastrophic collapse. The south flank of Kīlauea Volcano was first suspected of being unstable by *Moore and Krivoy* [1964] on the basis of their analysis of the 1962 ERZ eruption. A decade later these suspicions were confirmed by large-scale trilateration measurements [*Swanson et al.*, 1976], and, by the 1990s, GPS data were providing a comprehensive map of flank motion [*Owen et al.*, 1995] (Figure 24.7). Various workers have demonstrated that displacement of Kīlauea's south flank occurs along a landward-dipping décollement fault, which forms the boundary between the oceanic crust and volcanic pile [e.g., *Denlinger and Okubo*, 1995]. Motion on the décollement includes steady creep [e.g., *Owen et al.*, 1995], sudden slip associated with strong earthquakes [e.g., *Lipman et al.*, 1985], and episodic slow-slip events [e.g., *Cervelli et al.*, 2002; *Brooks et al.*, 2006] (see Chapter 13, Figure 13.11, of this volume). At least two forces have been proposed to drive this instability. *Swanson et al.* [1976] suggested that forcible intrusion of magma into the ERZ above about 3 km depth provided a mechanism for the displacement. *Dieterich* [1988] and *Delaney et al.* [1990] described the stress balance and feedback mechanism between pressurization of the entire rift zone, from the shallow magmatic system down to the base of the volcano, and motion along the basal décollement fault. Alternatively, gravitational collapse may drive seaward motion [*Moore and Krivoy*, 1964], perhaps caused by outward flow of olivine cumulates that underlie the summit area and rift zones [*Clague and Denlinger*, 1994]. Numerical modeling supports the hypothesis that a dense cumulate core alone can explain the instability [*Plattner et al.*, 2013], but the apparent triggering of a slow-slip event by a dike intrusion in 2007 [*Montgomery-Brown et al.*, 2010] (see also Chapter 13, this volume) suggests that shallow ERZ intrusions also play a role. The relative importance and temporal variability of the two mechanisms are therefore not well constrained.

The relation between the various types of flank motion at Kīlauea also remains poorly understood. It has been

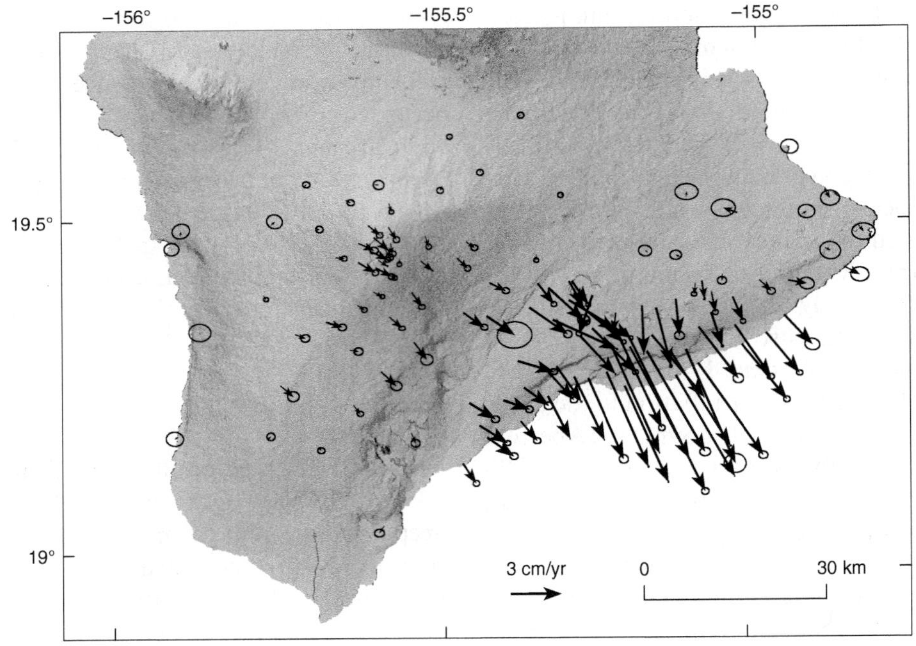

Figure 24.7 Horizontal displacements with respect to a fixed Pacific Plate that are measured at Global Positioning System (GPS) stations on the south half of the Island of Hawai'i during 1997–2002. Deformation is dominated by seaward motion of Kīlauea's south flank, which reaches magnitudes of ~8 cm/yr along the coast [*Cervelli and Miklius*, 2003]. Ellipses indicate 95% confidence bounds on station velocities.

suggested that at subduction zones a slow-slip event may accelerate into a destructive earthquake [*Segall and Bradley*, 2012], as apparently was the case with the 2011 M_w 9.0 Tohoku, Japan, earthquake [*Kato et al.*, 2012]. Whether M_w 7+ earthquakes beneath the south part of the Island of Hawai'i in 1975 or 1868 were associated with décollement slow slip is unknown. The conditions under which feedback occurs between the magmatic system and south flank (Chapter 13, this volume) are also unconstrained. These and other questions would benefit from a better understanding of the structure and dynamics of the south flank, especially the submarine environment. For example, uplift of the submarine portion of the Hilina slump has been documented by submarine geodesy [*Phillips et al.*, 2008] and also in seismic reflection and bathymetry data [*Morgan et al.*, 2003], but the rate, style, and spatial extent of the deformation, which are essential data for models of the geometry of flank motion, are poorly constrained. Likewise, the forces that control the occurrence of large earthquakes and the material and rheological properties of the south flank are not well understood, despite their clear importance for modeling earthquake hazard [*Denlinger and Morgan*, 2014].

Beyond Kīlauea, large landslides are ubiquitous throughout the Hawaiian Islands. At about the same time that Kīlauea's south flank instability was initially characterized, irregular topography apparent in preliminary bathymetric maps produced by the U.S. Navy in the vicinity of O'ahu and Moloka'i was interpreted as representing landslide deposits on the seafloor [*Moore*, 1964]. The landslide origin was debated for over two decades until GLORIA (Geologic LOng-Range Inclined Asdic) side-scan sonar surveys during 1986–1991 revealed their ubiquitous nature throughout the archipelago [*Moore et al.*, 1989]. Surveys of the entire Hawaiian chain, from the Island of Hawai'i to Midway Islands, revealed approximately 70 individual landslides that had runout distances of over 20 km. About 17 of these slides are located in the main Hawaiian islands (Figure 24.8). The largest runout lengths and volumes exceed 200 km and 5000 km^3, respectively, and the distribution of landslides suggests an average recurrence of ~350,000 years [*Moore et al.*, 1989, 1994a,b]. The volumes of the largest landslides are a significant fraction of the total volumes of their source volcanoes, implying a severe hazard. In fact, *Moore and Moore* [1984] identified tsunami deposits on several Hawaiian islands that suggest runup heights of several hundred meters. At Kohala volcano, the runup caused by the ~120 ka 'Ālika 2 slide, from Mauna Loa's west flank, may have exceeded 400 m in height and extended >6 km inland [*McMurtry et al.*, 2004]. Although the genesis of some of these deposits has been debated, the tsunami origin remains the most probable explanation [*Clague and Sherrod*, 2014].

While the subaerial and submarine extents of catastrophic landslides in Hawai'i have been mapped

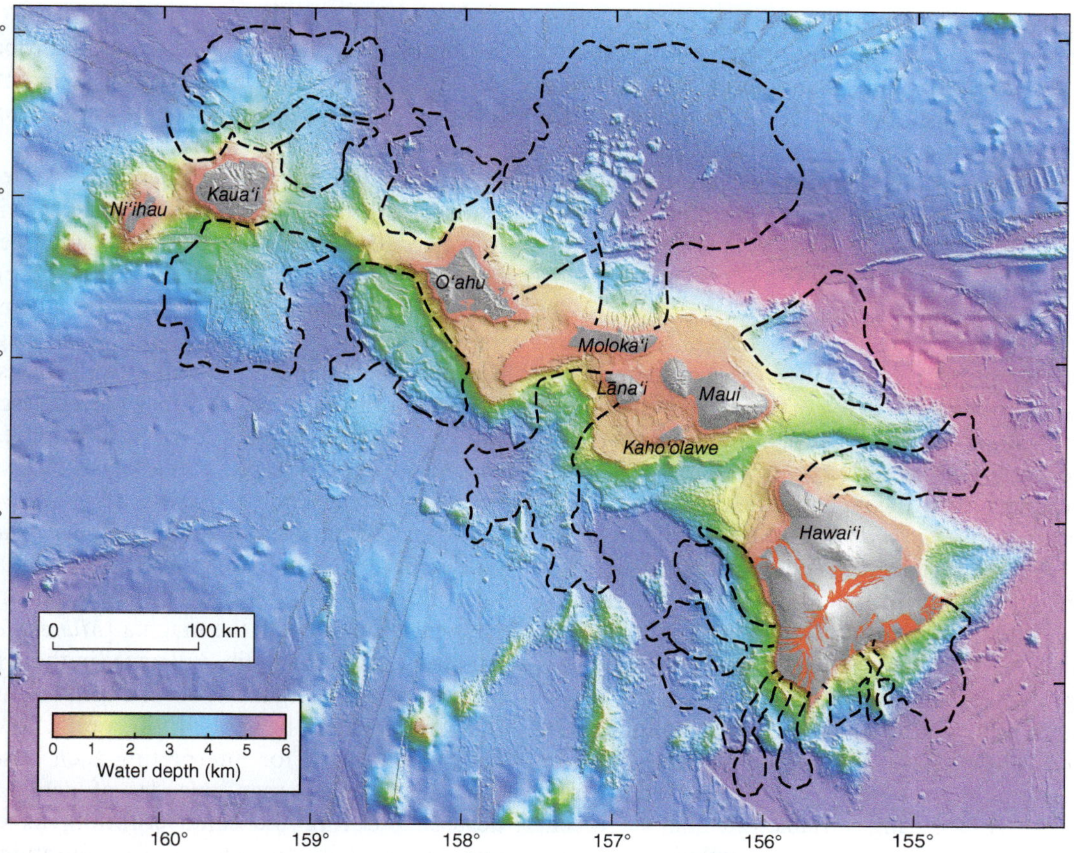

Figure 24.8 Bathymetric map of the region surrounding the Hawaiian Islands [modified from *Eakins et al.*, 2003] to show extent of mapped landslide deposits. Subaerial lava flows since about the year 1800 are shown in red. Mapped landslide deposits (based on Figure 2 of *Moore et al.* [1989]) are indicated by dashed lines.

extensively, the mechanisms and conditions under which such low-probability, yet high-impact, collapses take place are not well understood. Large landslides are known to occur during the periods of active volcano growth, including the preshield stage as demonstrated by the presence of landslide scars on Lō'ihi [*Fornari et al.*, 1988]. Whether or not collapse can also occur during postshield and rejuvenated stages is important for understanding the hazards posed by large landslides. The association between large slides and eruptive activity is also unknown, although magma pressure alone does not seem consistent with the magnitude of the largest collapses [*Iverson*, 1995]. Finally, can active slumps evolve into debris avalanches, as apparently occurred as part of the failure of Wai'anae volcano on O'ahu [*Coombs et al.*, 2004]? This question is particularly relevant when considering the hazard potential of Kīlauea's south flank.

Flank collapse hazards are not unique to the Hawaiian Islands. Indeed, the recognition of unstable zones and landslide deposits on Hawaiian volcanoes spurred the exploration for and recognition of flank instability at other large basaltic volcanoes. Large landslides appear to be common at oceanic shield volcanoes [e.g., *Holcomb and Searle*, 1991], as is creep of volcanic flanks, for instance, at Etna [e.g., *Rasa et al.*, 1996] and the Canary Islands [e.g., *González et al.*, 2010]. In some instances, perhaps most convincingly demonstrated at Etna, feedback between magmatism and flank slip is clear [*Solaro et al.*, 2010]. With so many examples worldwide of flank instability at basaltic volcanoes, comparative studies offer an excellent opportunity for developing a better understanding of the mechanisms for both catastrophic collapse and flank creep.

24.4. ERUPTIONS

The extent to which the timing, size, and duration of eruptions, including emissions of lava, tephra, and gas, can be forecast offers the ultimate affirmation of progress in the field of volcanology. Successful forecasts require not only monitoring data with sufficient temporal and spatial resolution and diversity between disciplines but also a good understanding of the physical and chemical processes that lead to and occur during eruptive activity.

Many important keys to the eruptive process take place in the subsurface and are beyond our ability to directly measure or observe. The characteristics, locations, amounts, and compositions of eruptive products and surficial changes, however, can be measured and mapped, and they provide fundamental insights into the internal state of a volcano, both leading up to and during an eruption.

24.4.1. Eruption Style

By the early twentieth century, gas exsolution had been recognized as a primary driver of volcanic activity (see discussion by *Mangan et al.* [2014]), although it took subsequent technological advancements before specific gas species and their relative concentrations could be definitively identified [*Sutton and Elias*, 2014]. Relating eruptive activity at the surface to gas exsolution requires not only knowledge of the volatile constituents of magma but also an understanding of the physical properties of the magma [e.g., *Wilson and Head*, 1981], bubble nucleation and growth [e.g., *Sparks*, 1978], and geometry of the conduit and vent (e.g., Chapter 17, this volume). For example, eruptive activity ranging from passive degassing to lava fountaining has been hypothesized to result from exsolution of H_2O during variable-rate magma ascent [*Wilson and Head*, 1981] or from accumulation and subsequent collapse of a CO_2-rich foam at the top of a magma reservoir [*Vergniolle and Jaupart*, 1990]. Lava fountaining at Pu‘u ‘Ō‘ō during 1983–1986 provided a test of these hypotheses, with geometric constraints [*Mangan et al.*, 2014] and a lack of significant CO_2 degassing from the ERZ [*Gerlach*, 1986] favoring the H_2O-based model for Kīlauea rift zone eruptions. Shallow lava lake fluctuations at Kīlauea where CO_2 degassing emissions are minor [*Orr and Rea*, 2012], however, may be driven by a combination of the models, i.e., accumulation and subsequent collapse of an H_2O-rich foam. CO_2 exsolution probably also drives some eruptive activity in Hawai‘i. For example, *Edmonds and Gerlach* [2007] measured a CO_2 degassing burst from Pu‘u ‘Ō‘ō, demonstrating that CO_2 bubbles can coalesce and rise through a shallow magma system. The spectacular lava fountains of the 1959 Kīlauea Iki eruption may also have involved CO_2. Magma feeding the Kīlauea Iki eruption rose rapidly from mantle depths, as inferred from a combination of geophysical and geochemical data [*Helz*, 1987; *Helz et al.*, 2014], and CO_2 bubbles may have provided the buoyancy needed for ascent of that olivine-rich magma [*Anderson and Brown*, 1993; *Stovall et al.*, 2011]. CO_2 exsolution has also been cited as a potential driver of explosive volcanism at Kīlauea, where ash columns have reached heights of >10 km [*Fiske et al.*, 2011], and may be associated with submarine pyroclastic eruptions at Lō‘ihi [*Schipper et al.*, 2011]. A combination of CO_2 and H_2O exsolution and foam collapse therefore probably plays a role in driving eruptive activity at different points on Kīlauea, from lava-lake-level fluctuations to lava fountaining to explosive eruptions.

While additional measurements of gas emissions in tandem with observations of eruptive activity are critically needed to assess the relation between degassing and eruption style (e.g., Chapter 14, this volume), models of gas exsolution and transport must also advance. Of chief importance is a better understanding of magma properties (e.g., surface tension) and the physics of bubble nucleation, growth, interaction, and coalescence, which can be advanced by both theoretical work [e.g., *Bottinga and Javoy*, 1991] and physical analysis of eruption products [e.g., *Mangan and Cashman*, 1996; *Carey et al.*, 2012] (see also, Chapter 19, this volume). The mechanics of two-phase (melt + gas) and even three-phase (melt + gas + crystals) flow, which controls eruptive activity and the physical properties of the magma [*Mangan et al.*, 2014], constitute a related topic. Advances in these fields, which are based on investigations occurring at volcanoes around the world and can be applied to Hawai‘i, could provide a better framework for interpreting such observations as (1) transitions between episodes of active and passive degassing during the same eruption, as emphatically demonstrated by episodes of lava fountaining at Pu‘u ‘Ō‘ō during 1983–1986 [*Wolfe et al.*, 1988] and at Mauna Ulu in 1969 [*Swanson et al.*, 1979] that were separated in time by passive degassing from a lava lake within the vent; (2) convection within lava lakes, which can be observed visually and is preserved by bubble populations in clasts erupted during explosions [e.g., *Carey et al.*, 2012]; (3) the causes of decades-long effusive activity, such as has been occurring from Kīlauea's ERZ since 1983 and occurred at Kīlauea's summit between at least 1823 until 1924; (4) ascent of bubbles independent of the magma from which those bubbles exsolved, as implied by an increase in CO_2 emissions that preceded increased lava effusion by more than a year during a magma supply surge to Kīlauea during 2003–2007 [*Poland et al.*, 2012]; and (5) many other processes related to magma ascent and eruption. Equally important is the geometry through which the two- or three-phase system flows, which can be elucidated at depth by geophysical measurements [e.g., *Lundgren et al.*, 2013] and at the surface by geological observations (e.g., Chapter 17, this volume).

Volatiles exert an obvious control on eruption style, but factors external to the magma also play an important role. Geological evidence suggests that explosive activity at Kīlauea over the last ~2000 years occurred when a deep caldera existed, parts of which were probably below the level of the water table [*Swanson et al.*, 2012b] (see also Chapter 16, this volume). Such a condition would facilitate magma-water interactions, which could drive large, exceptionally hazardous (in the summit region) hydrovolcanic explosions [*Mastin*, 1997; *Mastin et al.*, 2004]. Caldera collapse may also have driven high lava fountains by dynamically pressurizing the shallow magma storage system (e.g., Chapter 16, this volume). At a smaller scale, rockfalls into Kīlauea's summit lava lake have triggered minor explosions [*Orr et al.*, 2013].

24.4.2. Eruption Forecasting and Volcano Monitoring

A primary goal of volcano monitoring is to forecast eruptive activity, especially that which poses a hazard to society. Thus far, forecasting of volcanic eruptions has largely been empirical, that is, assessments of future activity are based on recognition of repeated patterns in monitoring data and geological deposits of prior events. The basis of numerous successful predictions of dome-building eruptions at Mount St. Helens, Washington, during 1980–1986 was recognition of repetitive preeruptive processes manifested in deformation, seismicity, and gas emissions [*Swanson et al.*, 1983]. Event trees provide an example of a quantitative probabilistic forecasting method, with the probability of future events assigned on the basis of, generally, past experience and geological evidence [*Newhall and Hoblitt*, 2002]. "Past experience" in Hawai'i, however, is limited by the short written record of eruptive activity, which extends only to the early nineteenth century [*Swanson*, 2008]. Examination of Hawaiian oral traditions can extend this record by several hundred years. *Swanson* [2008], for example, found excellent correspondence between the Pele-Hi'iaka chant and geological deposits that recorded a 60 year long effusive eruption at Kīlauea and subsequent formation of the caldera during the fifteenth and sixteenth centuries. Interpretation of other oral traditions by individuals or groups familiar with both volcanology and the Hawaiian language may yield other insights into past eruptive activity of Kīlauea, Mauna Loa, Hualālai, and Haleakalā, all of which have erupted since the time of Hawaiian settlement. Such work would be a good complement to geological studies, which provide critical insights into precontact volcanic activity (that is, volcanic activity that occurred prior to the 1778 contact between native Hawaiians and Europeans) based on mapping, stratigraphy, and geochronology. For example, mapping and dating of lava flows provide a basis for calculating lava resurfacing rates [e.g., *Trusdell*, 1995], while stratigraphic studies can reveal cycles of prehistoric explosive activity [*Hitchcock*, 1909; *McPhie et al.*, 1990; *Fiske et al.*, 2009; *Swanson et al.*, 2012b].

For current volcanic activity, geological, geophysical, and geochemical monitoring data may, in principle, be integrated in near-real time with models of the fundamental physical and chemical processes that are associated with volcanic unrest and eruptions [*Sparks*, 2003; *Segall*, 2013]. As increasingly realistic models of magma ascent and eruption become available [e.g., *Melnik and Sparks*, 1999] and as models of this type are quantitatively related to observations, such as ground deformation and rates of lava effusion [*Nishimura*, 2006; *Mastin et al.*, 2008; *Anderson and Segall*, 2011], probabilistic forecasts which incorporate all available information can be used to predict such critical information as the total duration or volume of an eruption [*Segall*, 2013]. Identifying the data that are most critical to advancing efforts to forecast volcanic activity should be an important factor in driving volcano-monitoring research.

Accepting that model development will require new types of monitoring data, what measurements will provide the most important input to models of Hawaiian volcanism? Two critical parameters are the rates of magma supply and lava effusion, since much eruptive and intrusive activity in Hawai'i can be ascribed to imbalance between magma input and output (e.g., Chapter 18, this volume). Magma supply is manifested in a variety of ways, but CO_2 emissions may constitute the most straightforward expression. CO_2 begins to exsolve from magma at tens of kilometers depth and, at Kīlauea, is degassed from the summit, making the CO_2 emission rate a good proxy for the magma supply rate [*Gerlach et al.*, 2002]. Measuring CO_2 emissions can be problematic, and new methods of continuous monitoring at high temporal resolution are needed, perhaps akin to those developed for SO_2 emission rate measurement [e.g., *Kern et al.*, 2013; *Sutton and Elias*, 2014]. Effusion rate monitoring is similarly challenged, since there are significant drawbacks to most techniques used to measure the volume of erupted lava. Changes in topography, measured either remotely or in the field, can provide an effusion rate if such changes are measured repeatedly over time [e.g., *Rowland et al.*, 1999; *Poland*, 2014] (see also Chapters 18 and 22, this volume). Topographic change will only capture subaerial lava flows, however, so derived effusion rates will not account for lava entering the ocean (as is often the case at basaltic ocean islands), and the most accurate topographic surveys (from Lidar) can be prohibitively expensive. The emission rate of SO_2 from Kīlauea's ERZ eruptive vent has been tied to effusion rate [*Sutton et al.*, 2001, 2003], but, since 2008, ERZ gas emissions have not been representative of eruption rate because of preeruptive degassing from the summit eruptive vent [*Elias and Sutton*, 2012]. Satellite measurements of thermal radiance have been tied to effusion rate, but the method requires cloud-free thermal imagery and field data to help constrain inputs to the thermal model [*Harris et al.*, 1998], and it breaks down when flows are confined to lava tubes [*Koeppen et al.*, 2013]. Very-low-frequency electromagnetic induction measurements over active lava tubes can give the cross-sectional area of the molten lava stream which, when combined with velocity measurements of lava within the tube (observed through a skylight), yields effusion rate but requires that all lava be carried through a single master tube and that the tube geometry be conducive to measurements [e.g., *Kauahikaua et al.*, 1996]. While conditions often exist when at least one of these techniques can be used, new methods are needed for more consistent quantification of effusion rate, especially on the same time scale as geophysical data are collected (i.e., seconds to minutes).

The growing accessibility of volcano-monitoring data, coupled with technological advancements that are providing new data streams, makes it challenging to manage and fully exploit the large volume of available measurements. Seismic monitoring in Hawai'i, for example, began with a few pendulum seismometers and evolved into a digital network of hundreds of telemetered components capable of tracking earthquake activity at a variety of temporal, spatial, and frequency scales [*Okubo et al.*, 2014] (Figure 24.9). HVO's modern seismic network yields data that are utilized for a range of studies, from interpreting processes based on large-scale earthquake relocations [*Matoza et al.*, 2013, 2014] to detecting subtle seismic precursors to eruptive activity [*Johnson and Poland*, 2013]. Deformation data have evolved from campaign measurements collected every few weeks to continuous measurements collected on a daily basis to subdaily point measurements [e.g., *Larson et al.*, 2010], complemented by high-spatial-resolution InSAR images, which have attained sufficient temporal density to be used to construct deformation time series [e.g., *Richter et al.*, 2013]. Geological observations using both visual and infrared cameras have also exploded in use [e.g., *Orr and Rea*, 2012; *Carbone et al.*, 2013], and remote-sensing data are ubiquitous in monitoring surface volcanism [e.g., *Harris et al.*, 1998]. Finally, gas emissions, once measured only by direct sampling, are now being tracked continuously by a variety of spectroscopic techniques that provide results on the same temporal and spatial scales as geophysical data [*Sutton and Elias*, 2014].

Efforts to forecast volcanic activity typically focus on the start time and style of an eruption but should not lose sight of the fact that eruption duration and changes in eruptive activity and style are also important, especially given the tremendous hazard implications of a short- versus long-duration eruption. As of 2015, Kīlauea is host to two long-term eruptions: one at the summit since

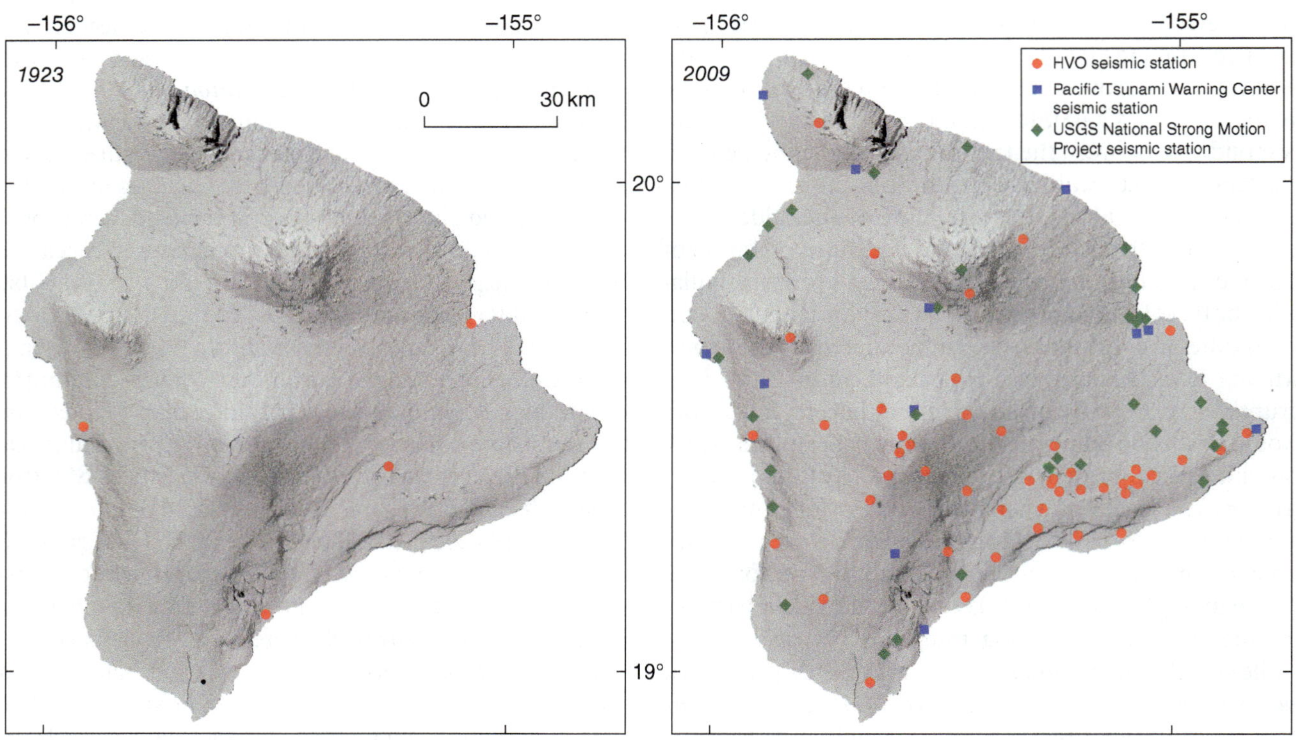

Figure 24.9 Seismic stations on the Island of Hawai'i in 1923 (left) and 2009 (right). Seismic network is a collaborative effort, with stations operated by the Hawaiian Volcano Observatory (red dots), National Oceanic and Atmospheric Administration's Pacific Tsunami Warning Center (blue squares), and U.S. Geological Survey's National Strong-Motion Project (green diamonds). Adapted from *Okubo et al.* [2014].

2008 and the other along the ERZ since 1983. The ERZ eruption has experienced several pauses and changes in vent location, none of which have resulted in an eruptive shutdown (Chapter 18, this volume]. Likewise, the summit eruption has persisted despite an episode of draining due to an ERZ fissure eruption in 2011 [*Carbone et al.*, 2013] as well as occlusion by debris on multiple occasions [*Johnson and Poland*, 2013]. What are the mechanisms that sustain long-term effusive eruptions and lava lakes? Do volcanoes grow endogenously during lava lake activity [*Francis et al.*, 1993]? For what reasons might a persistent eruption end? Perhaps only a large-scale and rapid episode of flank motion will be sufficient to stop one or both of Kīlauea's currently ongoing eruptions by diverting magma into newly created storage volumes within the volcano [e.g., *Denlinger and Morgan*, 2014], as apparently occurred as a result of the 1975 earthquake [*Dzurisin et al.*, 1980].

24.4.3. Volcanic Hazards

At its most basic level, assessment of volcanic hazards is based on the geological history of the volcano—especially eruptive style and frequency—which requires careful mapping of geological deposits. *Stearns and Macdonald* [1946] provided the first comprehensive geological map of the Island of Hawai'i, which became a foundation for subsequent work, similar to the more detailed mapping of *Wolfe and Morris* [1996] for the island, *Neal and Lockwood* [2003] for the summit of Kīlauea, and the statewide geological map compiled by *Sherrod et al.* [2007]. Such data are the basis for lava-flow-hazard zonation maps, both on the Island of Hawai'i (Figure 24.10) [*Mullineaux and Peterson*, 1974; *Mullineaux et al.*, 1987; *Wright et al.*, 1992] and Haleakalā volcano [*Sherrod et al.*, 2006]. These hazard maps are critical for civil defense and land use planning purposes, but quantitative probabilistic assessments are needed for charting the evolution of a volcanic hazard during an unfolding crisis situation. Future hazards research should therefore focus on developing realistic models of hazardous processes—for example, lava flows, gas emissions, and explosive eruptions—while continuing to refine the volcano's geological history through careful mapping and dating of deposits.

Lava flows are an obvious hazard requiring more research. While simple hazard zonation maps give a qualitative indication of the level of hazard in a certain area, lava flow pathway modeling [e.g., *Harris and Rowland*, 2001] (Chapter 21, this volume) would provide more useful information during eruptions when lava flows advance toward populated areas. *Kauahikaua et al.* [1998], for example, introduced the concept of a "lavashed," similar to a watershed, which defines an area in which a lava flow will be confined for a vent located upslope. An alternative approach is to calculate a steepest-descent path, which a gravitationally driven fluid will follow [*Kauahikaua et al.*, 2003]. Both methods require a digital elevation model (DEM); therefore, the accuracy of the flow-path forecast depends on the accuracy of the DEM. Updating the DEM over time thus becomes critical, especially because lava flows create new topography as they are emplaced [e.g., *Favalli et al.*, 2011]. Effusion rate exerts a primary control on lava flow emplacement [*Kauahikaua and Tilling*, 2014], so methods of monitoring the discharge of lava from a vent must be robust and useful in near-real time (see discussion in Section 24.4.2). Models of flow advance also need to account for lava cooling, degassing, crystallization, and other effects that will impact flow rheology and, therefore, advance rates and flow paths [*Cashman and Mangan*, 2014; *Kauahikaua and Tilling*, 2014] (see also Chapter 21, this volume). While improvements have been made in many of these fields (see discussion by *Cashman and Mangan* [2014]), most lava flow modeling does not account for the full physical complexity of the problem. The new implementation of the FLOWGO code (Chapter 21, this volume), however, is a step in this direction, and it offers the benefit of integrating models of lava flow extents and pathways. The code and future variants thus provide important tools for lava flow hazards assessment.

Kīlauea has long been known to have the potential for large-scale explosive eruptions that deposit tephra over broad areas [*Hitchcock*, 1909; *McPhie et al.*, 1990], and in 1790 an explosive eruption produced a pyroclastic density current that killed several hundred people [*Swanson and Christiansen*, 1973]. That eruption was much larger than the well-documented 1924 and 2008–present summit explosions [*Swanson et al.*, 2012b] and was thought by some workers to have produced the entirety of the meters-thick Keanakāko'i Tephra (Figure 24.11a) around Kīlauea's summit and to have coincided with caldera collapse. *Swanson et al.* [2012b], however, demonstrated that the tephra was a result of 300 years of explosive activity that began with caldera formation in about the year 1500. Other tephra deposits around Kīlauea [e.g., *Fiske et al.*, 2009] attest to a history of violent activity, and, in fact, about 60% of the last 2.5 k.y. at Kīlauea have been dominated by explosive eruptions [*Swanson et al.*, 2014], some of which probably resulted from interaction of groundwater with magma or hot wall rock. A few explosive eruptions reached the altitude of the jet stream (~5–17 km in elevation), based on the distribution of tephra east of the summit, which suggests powerful explosions that may have been driven in part by CO_2 exsolution [*Fiske et al.*, 2011]. Young explosive eruptions must also have occurred on Mauna Loa, given the presence of ballistic blocks on top of several-hundred-year-old lava flows near the volcano's

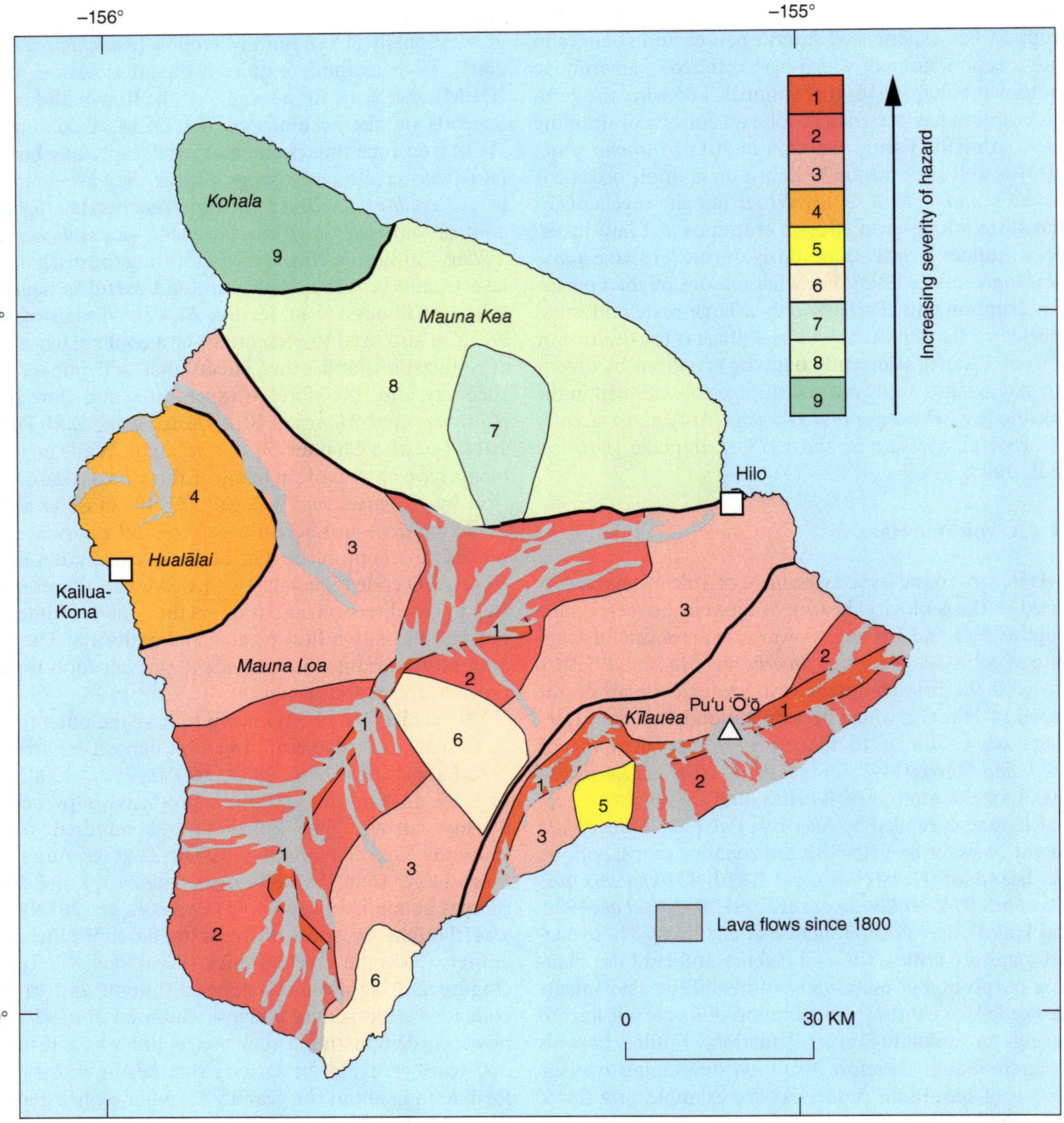

Figure 24.10 Lava flow hazards map for the Island of Hawai'i by *Wright et al.* [1992] [adapted from *Kauahikaua and Tilling*, 2014]. The greater the hazard, the lower the zone number. Thick black lines separate individual volcanoes on the island, and gray areas indicate lava flows emplaced since about the year 1800. Locations of Pu'u 'Ō'ō eruptive vent (white triangle) and cities of Hilo and Kailua-Kona (white squares) are shown.

summit (Figure 24.11b) [*Lockwood and Lipman*, 1987]. Several older ash layers were emplaced during the past several tens of thousands of years around Kīlauea and Mauna Loa, but these are incompletely studied [*Decker and Christiansen*, 1984; *Easton*, 1987]. Ash layers are also known from Hualālai and Mauna Kea, and thick ash beds have been recovered near Hilo from the HSDP core [*Beeson et al.*, 1996]. Clearly, fallout from explosive eruptions has impacted the entire Island of Hawai'i during the last several tens of thousands of years. Additional work defining the characteristics, mechanisms, and timing of explosive eruptions in Hawai'i is needed to better understand the potential hazards from explosive eruptive activity, including the areas that

Figure 24.11 Recent (since 1 ka) explosive deposits from two Hawaiian volcanoes. (a) Section through the Keanakāko'i Tephra [*Swanson et al.*, 2012b], located just outside the main building of the Hawaiian Volcano Observatory on the west rim of Kīlauea Caldera (lens cap for scale; base of deposit is obscured beneath loose rubble at the bottom of the photo). (b) Angular ballistics lie atop several-hundred-year-old lava flows near the summit caldera of Mauna Loa.

would be impacted, the recurrence intervals, and insights that might lead to recognition of potential future explosive eruptions before they occur.

Volcanic gases represent another hazard underappreciated until recently. The onset of continuous lava effusion in 1986 as part of Kīlauea's ongoing Pu'u 'Ō'ō eruption, in particular, led to the recognition that "vog" (volcanic smog) is a significant health hazard, due not only to direct inhalation but also indirectly, for example, by acidic water leaching lead into catchment drinking water [see discussion by *Sutton and Elias*, 2014]. The hazardous impact of gas emissions was compounded, starting in 2008, when a new eruptive vent opened at Kīlauea's summit. Elevated SO_2 emissions from that location affected a substantial swath of the Island of Hawai'i due to the vent's location farther inland than Pu'u 'Ō'ō, even leading to an agricultural disaster declaration [*Kauahikaua and Tilling*, 2014]. In response, HVO and University of Hawai'i scientists developed a vog measurement and prediction (VMAP) tool that provides gas emissions and aerosol data and forecasts via a publically available website [*Businger et al.*, 2011]. While VMAP represents an important first step in mapping and forecasting gas hazards, additional research on gas dispersion, long-term health effects, and other factors is needed to understand the impact of volcanic gases on society.

Earthquake activity, which is a direct consequence of volcanism, cannot be overlooked in any discussion of hazards in Hawai'i: Over 50 earthquakes greater than M6 have occurred in the archipelago since 1868 [*Klein et al.*, 2001]. The Island of Hawai'i is the focus of this activity, and although most earthquakes there are associated with magma movement and have magnitudes less than 3.5 [*Klein et al.*, 1987], slip on the décollement fault beneath the south flank of the island has resulted in M7+ earthquakes that caused heavy damage and some deaths in 1868 [*Wood*, 1914; *Wyss*, 1988] and 1975 [*Ando*, 1979; *Lipman et al.*, 1985]. The west side of the Island of Hawai'i is also subject to major ground shaking from décollement earthquakes, as demonstrated by M6+ events in 1929, 1950, 1951, and 1952 [*Wyss and Koyanagi*, 1992; *Klein et al.*, 2001]. Flexure of the lithosphere due to volcanic loading results in mantle earthquakes at tens of kilometers depths that are broadly felt and can reach M6+, as exemplified by damage on several islands due to M6.7 and M6.0 earthquakes northwest of the Island of Hawai'i on 15 October 2006 [*McGovern*, 2007; *Poland*, 2010; *Yamada et al.*, 2010]. Additional work on rupture characteristics of large crustal and mantle earthquakes, shaking associated with earthquakes of different types, and the relations between volcanic activity and strong earthquakes are ripe areas for future study [*Okubo et al.*, 2014].

The discussion above has focused on the direct volcanic hazards of lava flows, explosive eruptions, gas emissions, and earthquakes. Indirect hazards associated with volcanism in Hawai'i, such as ground subsidence and tsunami, also warrant additional research to better map their effects in space and time and forecast their occurrence. Effective communication to the public of hazards and risk information must be emphasized during the course of hazards research, especially given the lack of awareness of how to respond to volcanic events in some areas of Hawai'i [*Gregg et al.*, 2004].

24.5. INTEGRATED OCEAN ISLAND OBSERVATORY

The preceding discussion provides a sense of the problems that face scientists working on Hawaiian volcanism. As is typical for such large-scale issues, greater integration between disciplines and collaborations between researchers has been and will continue to be key to progress. Unfortunately, studies of oceanic hotspots are largely regional in nature, with little crossover between numerous investigators working separately to study a specific island chain or physical/chemical/biological process. Larger-scale research collaborations are needed to better integrate the community, following examples from other fields where interdisciplinary programs have done a good job of focusing a diverse group of investigators on key problems. The Deep Carbon Observatory, for instance, is a multidisciplinary, international, 10 year research effort begun in 2009 aimed at achieving a transformational understanding of the role of carbon in physical, chemical, and biological Earth processes [*Hazen et al.*, 2013]—a project with extraordinary vision given the importance of carbon to such questions as climate change, deep Earth composition, and the origin of life. Examples of research initiatives geared toward specific geological settings are the RIDGE and GEOPRISMS programs. RIDGE (Ridge Interdisciplinary Global Experiments) was started in 1999 with the goal of integrated multidisciplinary research at midocean ridges "from mantle to microbe" [*Fornari et al.*, 2012]. At continental margins, GeoPRISMS (Geodynamic Processes at Rifting and Subducting Margins), a continuation of the MARGINS initiative, aims to integrate geodynamics, Earth surface processes, and climate interactions over various timescales [*MARGINS Steering Committee*, 1998].

A similar community-wide effort directed at ocean islands is lacking. One of the recommendations stemming from a 2011 AGU Chapman Conference held in the Galápagos Islands was for an International Ocean Island Observatory (IOIO), which would represent a first-ever effort to bring together geoscientists and ecologists to study critical ocean island habitats [*Richards et al.*, 2012]. The approximately 180 scientists who participated in the 2012 Hawai'i AGU Chapman Conference [*Poland et al.*, 2013] also expressed a strong interest in an IOIO, demonstrating growing community support for cooperative research focused on ocean islands. Exploration of the potential for a future IOIO should be undertaken through collaboration between Earth and biological scientists studying ocean islands around the world and should involve research organizations from many countries. Greater collaboration and cooperation among the universities and research institutions located on islands or that focus on island research could form the nucleus of an IOIO: for example, the Darwin Research Station in the Galápagos; the Hawaiian Volcano Observatory and University of Hawai'i in Hawai'i; the Instituto Geografico Nacional for the Canary Islands; and the Observatoire Volcanologique du Piton de la Fournaise and the Université de la Réunion on Réunion Island. No other settings offer the opportunities to investigate deep Earth composition, biodiversity, and volcanism within such geographically restricted settings.

An IOIO could provide the organizational framework needed for supporting large-scale geoscience research efforts, e.g., drilling and submarine exploration—endeavors that have already demonstrated exceptional value in Hawai'i. The HSDP, which resulted in the recovery of more than 3 km of continuous core from the flank of Mauna Kea, spanning from about 240,000 to >650,000 years before present [*DePaolo and Stolper*, 1996; *DePaolo et al.*, 2001; *Garcia et al.*, 2007], provides an example of achievements made possible when fundamental questions are addressed by a coordinated group effort. Data from the HSDP have been used to interpret volcano sizes and life spans, mantle source compositions, plume dynamics, island subsidence, and other parameters of interest to the multi-disciplinary community of Hawai'i researchers. An IOIO would also facilitate better coordination between smaller-scale research efforts that, when integrated, can reach much farther than their individual components. For instance, geological mapping, age determinations, and geochemistry from locations away from the current locus of Hawaiian volcanism, such as Haleakalā [*Sherrod et al.*, 2003] and Kaua'i [*Garcia et al.*, 2010] (see also Chapter 6, this volume), provide constraints on the structure and composition of the Hawaiian plume over time, which can then be used in modeling plume dynamics [e.g., *Ballmer et al.*, 2011]. An area in which increased interdisciplinary collaboration would pay immediate benefits is the character of subvolcanic magma storage, as highlighted by studies of Kīlauea. Estimates of the volume of magma in reservoirs beneath Kīlauea's summit span several orders of magnitude, with low estimates based on petrology [e.g., *Pietruszka and Garcia*, 1999b] being impossible to reconcile with higher values required by geophysical models [e.g., *Segall et al.*, 2001]. What are the underlying hypotheses required for a model that can satisfy both data sets? HVO was founded to promote collaborative approaches to such problems by collecting and jointly interpreting a diversity of Earth science data: HVO's founding director, Thomas A. Jaggar, Jr., emphasized that the Observatory would "measure everything" [*Jaggar*, 1945]. HVO and other volcano observatories on ocean islands provide numerous examples of the achievements that are possible from greater integration between Earth science disciplines [*Tilling et al.*, 2014].

Increased interdisciplinary coordination through an IOIO will also require innovative modeling approaches to combine disparate data sets and to more realistically depict the physical and chemical processes associated with volcanic activity. For example, dike intrusion is generally explored using either deformation measurements or seismic data, but those two data types can be related by *Dieterich's* [1994] seismicity rate theory that links changes in stress rate to changes in seismicity rate (first applied to Kīlauea by *Dieterich et al.* [2003]). Using this model, *Segall et al.* [2013] demonstrated that seismicity and deformation could be jointly inverted to recover the propagation history of Kīlauea's 2007 ERZ dike intrusion. Similarly, by utilizing a physics-based model of magma ascent and eruption for 2004–2008 activity at Mount St. Helens, Washington [*Anderson and Segall*, 2011], *Anderson and Segall* [2013] were able to invert deformation and extrusion data to estimate the H_2O content of the primary melt, volume of the magma reservoir, and other properties of the volcano's magmatic system. Accounting for both uncertainty in model parameters and independent prior information can extend such models to probabilistic estimates of, for example, eruption duration [*Segall*, 2013], which is a constraint of obvious importance to volcanic hazards assessment and mitigation. The application of such techniques is far ranging. Physics- and chemistry-based models could, for instance, provide bounds on the probable range of CO_2 content of the primary magma (a problem discussed in Section 24.2.1) based on completely independent types of data, e.g., deformation and extrusion rate. Such models would provide a strong bridge within the community of IOIO geoscience researchers.

Incorporating biological investigations with these Earth science endeavors would open new realms of scientific inquiry [e.g., *Gillespie and Clague*, 2009]. An example of an IOIO focus might be the creation and degradation of ocean islands, which would require contributions from mantle dynamics, geomorphology, volcanology, and ecosystem evolution. On a smaller scale, the land-sea interface provides a focused zone where geology, chemistry, and biology are interconnected, even though the relations between these processes may not be clear.

24.6. SUMMARY AND CONCLUSIONS

Intense research and monitoring of Hawaiian volcanoes for over 100 years have highlighted a number of fundamental questions about how Hawaiian volcanoes work, from their source in the mantle to their eruptions at the surface. At the deepest levels of Hawaiian volcanoes, there is uncertainty regarding the depth of origin, source composition, and structure of the mantle plume, the causes for changes in magma supply over both short (years) and long (millions of years) timescales, and the changes in melt distribution between the various active volcanoes. The compositional and morphological evolution of the volcanoes themselves is poorly known, including the mechanisms behind and characteristics of flank instability, and the volume and geometry of magma storage within individual volcanoes are highly model and discipline dependent. Finally, a better grasp of the controls on eruption style will support volcano monitoring, forecasting efforts, and hazards assessment. These problems can be addressed by improved collaboration among researchers studying Hawaiian and similar volcanoes, especially between disciplines and physical depth ranges, and by the continued development of physics- and chemistry-based models, which will link traditionally independent data (for example, deformation and volatile content) and provide new tools for monitoring and forecasting volcanic activity. Ultimately, the community may seek to better organize the numerous researchers working on ocean island volcanism around the world through a collaborative IOIO. Such an effort would lead to a better understanding of basaltic hotspot volcanism as a whole, since studies of other ocean islands and seamounts will better inform research in Hawai'i and vice versa.

ACKNOWLEDGMENTS

This manuscript is derived from presentations and discussions at the 2012 AGU Chapman Conference, "Hawaiian Volcanoes, from Source to Surface." Discussions with conference planning committee members Rebecca Carey, Simon Carn, Valérie Cayol, Helge Gonnermann, Ken Hon, Paul Okubo, and Scott Rowland were critical. I am grateful to the staff of the Hawaiian Volcano Observatory for their input, particularly Don Swanson, Kyle Anderson, Tim Orr, Jim Kauahikaua, Jane Takahashi, Wes Thelen, Asta Miklius, and Carolyn Parcheta. My thanks also to Rebecca Carey, Dave Clague, Mike Garcia, Garrett Ito, Maggie Mangan, Dave Sherrod, and Dominique Weis for their insightful comments and reviews.

REFERENCES

Abouchami, W., A. W. Hofmann, S. J. G. Galer, F. A. Frey, J. Eisele, and M. Feigenson (2005), Lead isotopes reveal bilateral asymmetry and vertical continuity in the Hawaiian mantle plume, *Nature*, *434*(7035), 851–856, doi:10.1038/nature03402.

Aki, K., and R. Koyanagi (1981), Deep volcanic tremor and magma ascent mechanism under Kilauea, Hawaii, *J. Geophys. Res.*, *86*(B8), 7095–7109, doi:10.1029/JB086iB08p07095.

Albarède, F., B. Luais, G. Fitton, M. Semet, E. Kaminski, B. G. J. Upton, P. Bachèlery, and J.-L. Cheminee (1997), The geochemical regimes of Piton de la Fournaise volcano (Réunion) during the last 530,000 years, *J. Petrol.*, *38*(2), 171–201, doi:10.1093/petroj/38.2.171.

Aloisi, M., M. Mattia, C. Ferlito, M. Palano, V. Bruno, and F. Cannavò (2011), Imaging the multi-level magma reservoir at Mt. Etna volcano (Italy), *Geophys. Res. Lett.*, *38*(L16306), doi:10.1029/2011GL048488.

Amelung, F., S.-H. Yun, T. R. Walter, P. Segall, and S.-W. Kim (2007), Stress control of deep rift intrusion at Mauna Loa volcano, Hawaii, *Science*, *316*(5827), 1026–1030, doi:10.1126/science.1140035.

Anderson, A. T., and G. G. Brown (1993), CO_2 contents and formation pressures of some Kilauean melt inclusions, *Am. Mineral.*, *78*(7–8), 794–803.

Anderson, D. L. (2011), Hawaii, boundary layers and ambient mantle—Geophysical constraints, *J. Petrol.*, *52*(7–8), 1547–1577.

Anderson, K., and P. Segall (2011), Physics-based models of ground deformation and extrusion rate at effusively erupting volcanoes, *J. Geophys. Res.*, *116*(B07204), doi:10.1029/2010JB007939.

Anderson, K., and P. Segall (2013), Bayesian inversion of data from effusive volcanic eruptions using physics-based models: Application to Mount St. Helens 2004–2008, *J. Geophys. Res.*, *118*(5), 2017–2037, doi:10.1002/jgrb.50169.

Ando, M. (1979), The Hawaii earthquake of November 29, 1975: Low dip angle faulting due to forceful injection of magma, *J. Geophys. Res.*, *84*(B13), 7616–7626, doi:10.1029/JB084iB13p07616.

Bagnardi, M., and F. Amelung (2012), Space-geodetic evidence for multiple magma reservoirs and subvolcanic lateral intrusions at Fernandina Volcano, Galápagos Islands, *J. Geophys. Res.*, *117*(B10406), doi:10.1029/2012JB009465.

Baker, S., and F. Amelung (2012), Top-down inflation and deflation at the summit of Kīlauea Volcano, Hawaii observed with InSAR, *J. Geophys. Res.*, *117*(B12406), doi:10.1029/2011JB009123.

Ballmer, M. D., G. Ito, J. van Hunen, and P. J. Tackley (2011), Spatial and temporal variability in Hawaiian hotspot volcanism induced by small-scale convection, *Nature Geosci.*, *4*(7), 457–460, doi:10.1038/ngeo1187.

Ballmer, M. D., G. Ito, C. J. Wolfe, and S. C. Solomon (2013), Double layering of a thermochemical plume in the upper mantle beneath Hawaii, *Earth Planet. Sci. Lett.*, doi:10.1016/j.epsl.2013.06.022.

Bargar, K. E., and E. D. Jackson (1974), Calculated volumes of individual shield volcanoes along the Hawaiian-Emperor chain, *U.S. Geol. Surv. J. Res.*, *2*, 545–550.

Beeson, M. H., D. A. Clague, and J. P. Lockwood (1996), Origin and depositional environment of clastic deposits in the Hilo drill hole, Hawaii, *J. Geophys. Res.*, *101*(B5), 11,617–11,629, doi:10.1029/95JB03703.

Bianco, T. A., G. Ito , J. van Hunen, M. D. Ballmer, and J. J. Mahoney (2008), Geochemical variation at the Hawaiian hot spot caused by upper mantle dynamics and melting of a heterogeneous plume, *Geochem. Geophys. Geosyst.*, *9*(Q11003), doi:10.1029/2008GC002111.

Bianco, T. A., G. Ito, J. van Hunen, M. D. Ballmer, and J. J. Mahoney (2011), Geochemical variations at intraplate hot spots caused by variable melting of a veined mantle plume, *Geochem. Geophys. Geosyst.*, *12*(Q0AC13), doi:10.1029/2011GC003658.

Bottinga, Y., and M. Javoy (1991), The degassing of Hawaiian tholeiite, *Bull. Volcanol.*, *53*(2), 73–85, doi:10.1007/BF00265413.

Brooks, B. A., J. H. Foster, M. Bevis, L. N. Frazer, C. J. Wolfe, and M. Behn (2006), Periodic slow earthquakes on the flank of Kilauea volcano, Hawaii, *Earth Planet. Sci. Lett.*, *246*(3–4), 207–216, doi:10.1016/j.epsl.2006.03.035.

Businger, S., R. Huff, A. J. Sutton, T. Elias, and K. A. Horton (2011), The Vog Measurement and Prediction (VMAP) Project, Abstract V44C-05, presented at 2011 Fall Meet., AGU, San Francisco, Calif., 5–9 December.

Cao, Q., R. D. van der Hilst, M. V. de Hoop, and S.-H. Shim (2011), Seismic imaging of transition zone discontinuities suggests hot mantle west of Hawaii, *Science*, *332*(6033), 1068–1071, doi:10.1126/science.1202731.

Carbone, D., M. P. Poland, M. R. Patrick, and T. R. Orr (2013), Continuous gravity measurements reveal a low-density lava lake at Kīlauea Volcano, Hawai'i, *Earth Planet. Sci. Lett.*, *376*, 178–185, doi:10.1016/j.epsl.2013.06.024.

Carey, R. J., M. Manga, W. Degruyter, D. Swanson, B. Houghton, T. Orr, and M. Patrick (2012), Externally triggered renewed bubble nucleation in basaltic magma: The 12 October 2008 eruption at Halema'uma'u Overlook vent, Kīlauea, Hawai'i, USA, *J. Geophys. Res.*, *117*(B11202), doi:10.1029/2012JB009496.

Cashman, K. V., and M. T. Mangan (2014), A Century of Studying Effusive Eruptions in Hawai'i, Chapter 9 of Poland, M. P., T. J. Takahashi, and C. M. Landowski, (eds.), *Characteristics of Hawaiian Volcanoes*, U.S. Geol. Surv. Prof. Pap. 1801, pp. 357–394, doi:10.3133/pp18019.

Cayol, V., J. H. Dieterich, A. T. Okamura, and A. Miklius (2000), High magma storage rates before the 1983 eruption of Kilauea, Hawaii, *Science*, *288*(5475), 2343–2346, doi:10.1126/science.288.5475.2343.

Cervelli, P., and A. Miklius (2003), The shallow magmatic system of Kīlauea volcano, in *The Pu'u 'Ō'ō-Kūpaianaha Eruption of Kīlauea Volcano, Hawaii: The First 20 Years*, U.S. Geol. Surv. Prof. Pap. 1676, edited by C. Heliker et al., pp. 149–163.

Cervelli, P., P. Segall, K. Johnson, M. Lisowski, and A. Miklius (2002), Sudden aseismic fault slip on the south flank of Kilauea volcano, *Nature*, *415*(6875), 1014–1018, doi:10.1038/4151014a.

Chadwick, W. W., S. Jónsson, D. J. Geist, M. Poland, D. J. Johnson, S. Batt, H. K. S., and A. Ruiz (2011), The May 2005 eruption of Fernandina volcano, Galápagos: The first circumferential dike intrusion observed by GPS and InSAR, *Bull. Volcanol.*, *73*(6), 679–697, doi:10.1007/s00445-010-0433-0.

Clague, D. A. (1987), Hawaiian xenolith populations, magma supply rates, and development of magma chambers, *Bull. Volcanol.*, *49*(4), 577–587, doi:10.1007/BF01079963.

Clague, D. A., and G. B. Dalrymple (1987), The Hawaiian-Emperor volcanic chain, part I: Geologic evolution, in *Volcanism in Hawaii*, U.S. Geol. Surv. Prof. Pap. 1350, edited by R. W. Decker et al., pp. 5–54.

Clague, D. A., and R. P. Denlinger (1994), Role of olivine cumulates in destabilizing the flanks of Hawaiian volcanoes, *Bull. Volcanol.*, *56*(6–7), 425–434, doi:10.1007/BF00302824.

Clague, D. A., and J. E. Dixon (2000), Extrinsic controls on the evolution of Hawaiian ocean island volcanoes, *Geochem. Geophys. Geosys.*, *1*(4), doi:10.1029/1999GC000023.

Clague, D. A., and D. R. Sherrod (2014), Growth and degradation of Hawaiian volcanoes, Chapter 3 of Poland, M. P., T. J. Takahashi, and C. M. Landowski, (eds.), *Characteristics of Hawaiian Volcanoes*, U.S. Geol. Surv. Prof. Pap. 1801, pp. 97–146, doi:10.3133/pp18013.

Clague, D. A., J. G. Moore, J. E. Dixon, and W. B. Friesen (1995), Petrology of submarine lavas from Kilauea's Puna Ridge, Hawaii, *J. Petrol.*, *36*(2), 299–349, doi:10.1093/petrology/36.2.299.

Coombs, M. L., D. A. Clague, G. F. Moore, and B. L. Cousens (2004), Growth and collapse of Waianae Volcano, Hawaii, as revealed by exploration of its submarine flanks, *Geochem. Geophys. Geosyst.*, *5*(Q08006), doi:10.1029/2004GC000717.

Coombs. M. L., T. W. Sisson, and P. W. Lipman (2006), Growth history of Kilauea inferred from volatile concentrations in submarine-collected basalts, *J. Volcanol. Geotherm. Res.*, *151*(1–3), 19–49, doi:10.1016/j.jvolgeores.2005.07.037.

Dana, J. D. (1849), *Geology*, vol 10 of *United States Exploring Expedition*, George P. Putnam, New York.

Dana, J. D. (1890), *Characteristics of Volcanoes*, University Press, John Wilson, Cambridge, Mass.

Decker, R. W., and R. L. Christiansen (1984), Explosive eruptions of Kilauea volcano, Hawaii, in *Explosive Volcanism: Inception, Evolution, and Hazards*, pp. 122–132, National Academy Press, Washington, D. C.

Decker, R. W., R. Y. Koyanagi, J. J. Dvorak, J. P. Lockwood, A. T. Okamura, K. M. Yamashita, and W. R. Tanigawa (1983), Seismicity and surface deformation of Mauna Loa Volcano, Hawaii, *Eos Trans. AGU*, *64*(37), 545–547, doi:10.1029/EO064i037p00545-01.

Delaney, P. T., R. S. Fiske, A. Miklius, A. T. Okamura, and M. K. Sako (1990), Deep magma body beneath the summit and rift zones of Kilauea Volcano, Hawaii, *Science*, *247*(4948), 1311–1316, doi:10.1126/science.247.4948.1311.

Denlinger, R. P. (1997), A dynamic balance between magma supply and eruption rate at Kilauea volcano, Hawaii, *J. Geophys. Res.*, *102*(B8), 18,091–18,100, doi:10.1029/97JB01071.

Denlinger, R. P., and J. K. Morgan (2014), Instability of Hawaiian volcanoes, Chapter 4 of Poland, M. P., T. J. Takahashi, and C. M. Landowski, (eds.), *Characteristics of Hawaiian Volcanoes*, U.S. Geol. Surv. Prof. Pap. 1801, pp. 97–146, doi:10.3133/pp18014.

Denlinger, R. P., and P. Okubo (1995), Structure of the mobile south flank of Kilauea Volcano, Hawaii, *J. Geophys. Res.*, *100*(B12), 24,499–24,507, doi:10.1029/95JB01479.

DePaolo, D. J., and E. M. Stolper (1996), Models of Hawaiian volcano growth and plume structure: Implications of results from the Hawaii Scientific Drilling Project, *J. Geophys. Res.*, *101*(B5), 11,643–11,654, doi:10.1029/96JB00070.

DePaolo, D. J., and D. Weis (2007), Hotspot volcanoes and large igneous provinces, in *Continental Scientific Drilling*, edited by U. Harms et al., pp. 259–288, Springer, Berlin, doi:10.1007/978-3-540-68778-8_7.

DePaolo, D. J., E. Stolper, and D. M. Thomas (2001), Deep drilling into a Hawaiian volcano, *Eos Trans. AGU*, *82*(13), 149, 154–155, doi:10.1029/EO082i013p00149-02.

Dieterich, J. (1994), A constitutive law for rate of earthquake production and its application to earthquake clustering, *J. Geophys. Res.*, *99*(B2), 2601–2618, doi:10.1029/93JB02581.

Dieterich, J. H. (1988), Growth and persistence of Hawaiian volcanic rift zones, *J. Geophys. Res.*, *93*(B5), 4258–4270, doi:10.1029/JB093iB05p04258.

Dieterich, J. H., V. Cayol, and P. Okubo (2003), Stress changes before and during the Pu'u 'Ō'ō-Kūpaianaha eruption, in *The Pu'u 'Ō'ō-Kūpaianaha Eruption of Kīlauea Volcano, Hawaii: The First 20 Years*, U.S. Geol. Surv. Prof. Pap. 1676, edited by C. Heliker et al., pp. 187–202.

Dixon, J.E., and D. A. Clague (2001), Volatiles in basaltic glasses from Loihi Seamount, Hawaii: Evidence for a relatively dry plume component, *J. Petrol.*, *42*(3), 627–654. doi:10.1093/petrology/42.3.627.

d'Ozouville, N., D. Graham, K. Harpp, and E. Mittelstaedt, eds. (2014), *The Galápagos: A Laboratory for the Earth Sciences*, AGU, Monograph number 204, 427 pages.

Dvorak, J. J. (1992), Mechanism of explosive eruptions of Kilauea Volcano, Hawaii, *Bull. Volcanol.*, *54*(9), 638–645, doi:10.1007/BF00430777.

Dzurisin, D., L. A. Anderson, G. P. Eaton, R. Y. Koyanagi, P. W. Lipman, J. P. Lockwood, R. T. Okamura, G. S. Puniwai, M. K. Sako, and K. M. Yamashita (1980), Geophysical observations of Kilauea Volcano, Hawaii; 2, Constraints on the magma supply during November 1975–September 1977, *J. Volcanol. Geotherm. Res.*, *7*(3–4), 241–269, doi:10.1016/0377-0273(80)90032-3.

Eakins, B. W., J. E. Robinson, T. Kanamatsu, J. Naka, J. R. Smith, E. Takahashi, and D. A. Clague (2003), Hawaii's volcanoes revealed, U.S. Geol. Surv. Geol. Investig. Ser. Map I-2809, scale ~1:185,342.

Easton, M. (1987), Stratigraphy of Kilauea Volcano, in *Volcanism in Hawaii*, U.S. Geol. Surv. Prof. Pap. 1350, edited by R. W. Decker et al., pp. 243–260.

Eaton, J. P. (1962), Crustal structure and volcanism in Hawaii, in *Crust of the Pacific Basin*, Geophys. Monogr. 6, edited by G. A. Macdonald and H. Kuno, pp. 13–29, AGU, Washington, D. C. doi:10.1029/GM006p0013.

Eaton, J. P., and K. J. Murata (1960), How volcanoes grow, *Science*, *132*(3432), 925–938, doi:10.1126/science.132.3432.925.

Edmonds, M., and T. M. Gerlach (2007), Vapor segregation and loss in basaltic melts, *Geology*, *35*(8), 751–754, doi:10.1130/G23464A.1.

Elias, T., and A. J. Sutton (2012), Sulfur dioxide emission rates from Kīlauea Volcano, Hawai'i, 2007–2010, U.S. Geol. Surv. Open-File Rep. 2012–1107.

Farnetani, C. G., and A. W. Hofmann (2010), Dynamics and internal structure of the Hawaiian plume, *Earth Planet. Sci. Lett.*, *295*(1–2), 231–240, doi:10.1016/j.epsl.2010.04.005.

Favalli, M., S. Tarquini, and A. Fornaciai (2011), DOWNFLOW code and LIDAR technology for lava flow analysis and hazard assessment at Mount Etna, *Ann. Geophys.*, *54*(5), 552–566, doi:10.4401/ag-5339.

Fiske, R. S., and E. D. Jackson (1972), Orientation and growth of Hawaiian volcanic rifts: The effect of regional structure and gravitational stresses, *Proc. R. Soc. Lond., Ser. A*, *329*, 299–326, doi:10.1098/rspa.1972.0115.

Fiske, R. S., T. R. Rose, D. A. Swanson, D. E. Champion, and J. P. McGeehin (2009), Kulanaokuaiki Tephra (ca. A.D. 400–1000): Newly recognized evidence for highly explosive eruptions at Kilauea Volcano, Hawai'i, *Geol. Soc. Am. Bull.*, *121*(5–6), 712–728, doi:10.1130/B26327.1.

Fiske, R. S., T. R. Rose, and D. Swanson (2011), Diatreme-like eruption at Kilauea: The Kulanaokuaiki-3 Tephra (~900 CE), Abstract V33E-06 presented at 2011 Fall Meet., AGU, San Francisco, Calif., 5–9 December.

Fornari, D. J., M. O. Garcia, R. C. Tyce, and D. G. Gallo (1988), Morphology and structure of Loihi seamount based on seabeam sonar mapping, *J. Geophys. Res.*, *93*(B12), 15,227–15,238, doi:10.1029/JB093iB12p15227.

Fornari, D. J., S. E. Beaulieu, J. F. Holden, L. S. Mullineaux, and M. Tolstoy (2012), Introduction to the special issue: From RIDGE to Ridge 2000, *Oceanography*, *25*(1), 12–17, doi:10.5670/oceanog.2012.01.

Foulger, G. R. (2007), The "plate" model for the genesis of melting anomalies, in *Plates, Plumes and Planetary Processes*, Geol. Soc. Am. Spec. Pap., 430, pp. 1–28, doi:10.1130/2007.2430(01).

Foulger, G. R., J. H. Natland, D. C. Presnall, and D. L. Anderson (Eds.) (2005), *Plates, Plumes, and Paradigms*, Geol. Soc. Am. Spec. Pap. 388.

Francis, P., C. Oppenheimer, and D. Stevenson (1993), Endogenous growth of persistently active volcanoes, *Nature*, *366*(6455), 554–557, doi:10.1038/366554a0.

Frey, F. A., and J. M. Rhodes (1993), Intershield geochemical differences among Hawaiian Volcanoes: Implications for source compositions, melting process and magma ascent paths, *Philos. Trans. R. Soc. Lond., Ser. A*, *342*(1663), 121–136, doi:10.1098/rsta.1993.0009.

Garcia, M. O., J. M. Rhodes, F. A. Trusdell, and A. J. Pietruszka (1996), Petrology of lavas from the Puu Oo eruption of Kilauea Volcano; 3, The Kupaianaha Episode (1986–1992), *Bull. Volcanol.*, *58*(5), 359–379, doi:10.1007/s004450050145.

Garcia, M. O., A. J. Pietruszka, J. M. Rhodes, and K. Swanson (2000), Magmatic processes during the prolonged Pu'u 'O'o eruption of Kilauea Volcano, Hawaii, *J. Petrol.*, *41*(7), 967–990, doi:10.1093/petrology/41.7.967.

Garcia, M. O., E. H. Haskins, E. M. Stolper, and M. Baker (2007), Stratigraphy of the Hawai'i Scientific Drilling Project core (HSDP2): Anatomy of a Hawaiian shield volcano, *Geochem. Geophys. Geosyst.*, *8*(Q02G20), doi: 10.1029/2006GC001379.

Garcia, M. O., L. Swinnard, D. Weis, A. R. Greene, T. Tagami, H. Sano, and C. E. Gandy (2010), Petrology, geochemistry and geochronology of Kaua'i lavas over 4·5 Myr: Implications for the origin of rejuvenated volcanism and the evolution of the Hawaiian plume, *J. Petrol.*, *51*(7), 1507–1540, doi:10.1093/petrology/egq027.

Garcia, M. O., D. Hanano, A. Flinders, D. Weis, G. Ito, and M. D. Kurz (2012), Age, geology, geophysics, and geochemistry of Mahukona Volcano, Hawai`i, *Bull. Volcanol.*, *74*(6), 1445–1463, doi:10.1007/s00445-012-0602-4.

Garcia, M. O., J. R. Smith, J. P. Tree, D. Weis, L. Harrison, and B. R. Jicha (in press), Petrology, geochemistry and ages of lavas from Northwest Hawaiian Ridge volcanoes, in The Origin, Evolution, and Environmental Impact of Oceanic Large Igneous Provinces, *Geol. Soc. Am. Sp. Pap.* 511, edited by C. R. Neal, W. W. Sager, E. Erba, and T. Sano.

Geist, D., W. White, T. Naumann, and R. Reynolds (1999), Illegitimate magmas of the Galápagos: Insights into mantle mixing and magma transport, *Geology*, *27*(12), 1103–1106, doi:10.1130/0091-7613(1999)027<1103:IMOTGP>2.3.CO;2.

Gerlach, T. M. (1986), Exsolution of H_2O, CO_2, and S during eruptive episodes at Kilauea Volcano, Hawaii, *J. Geophys. Res.*, *91*(B12), 12,177–12,185, doi:10.1029/JB091iB12p12177.

Gerlach, T. M., K. A. McGee, T. Elias, A. J. Sutton, and M. P. Doukas (2002), Carbon dioxide emission rate of Kilauea Volcano: Implications for primary magma and the summit reservoir, *J. Geophys. Res.*, *107*(2189), doi:2110.1029/2001JB000407.

Gillespie, R., and D. A. Clague (Eds.) (2009), *Encyclopedia of Islands*, Univ. of Calif. Press, Berkeley, Calif.

Gonnermann, H., J. H. Foster, M. Poland, C. J. Wolfe, B. A. Brooks, and A. Miklius (2012), Coupling at Mauna Loa and Kīlauea via pore pressure in an asthenospheric melt layer, *Nature Geosci.*, *5*(11), 826–829, doi:10.1038/ngeo1612.

González, P. J., K. F. Tiampo, A. G. Camacho, and J. Fernández (2010), Shallow flank deformation at Cumbre Vieja volcano (Canary Islands): Implications on the stability of steep-sided volcano flanks at oceanic islands, *Earth Planet. Sci. Lett.*, *297*(3–4), 545–557, doi:10.1016/j.epsl.2010.07.006.

Greene, A. R., M. O. Garcia, D. Weis, G. Ito, M. Kuga, J. Robinson, and S. Yamasaki (2010), Low-productivity Hawaiian volcanism between Kaua'i and O'ahu, *Geochem. Geophys., Geosyst.*, *11*(Q0AC08), doi:10.1029/2010GC003233.

Greene, A. R., M. O. Garcia, A. J. Pietruszka, D. Weis, J. P. Marske, M. J. Vollinger, and J. Eiler (2013), Temporal geochemical variations in lavas from Kīlauea's Pu'u 'Ō'ō eruption (1983–2010): Cyclic variations from melting of source heterogeneities. *Geochem., Geophys., Geosyst.*, (*14*)11, 4849–4873 10.1002/ggge.20285.

Gregg, C. E., B. F. Houghton, D. M. Johnston, D. Paton, and D. A. Swanson (2004), The perception of volcanic risk in Kona communities from Mauna Loa and Hualālai volcanoes, Hawai'i, *J. Volcanol. Geotherm. Res.*, *130*(3–4), 179–196, doi:10.1016/S0377-0273(03)00288-9.

Gripp, A. E., and R. G. Gordon (2002), Young tracks of hotspots and current plate velocities, *Geophys. J. Int.*, *150*(2), 321–361, doi:10.1046/j.1365-246X.2002.01627.x.

Harris, A. J. L., and S. K. Rowland (2001), FLOWGO: a kinematic thermo-rheological model for lava flowing in a channel, *Bull. Volcanol.*, *63*(1), 20–44, doi:10.1007/s004450000120.

Harris, A. J. L., L. P. Flynn, L. Keszthelyi, P. J. Mouginis-Mark, S. K. Rowland, and J. A. Resing (1998), Calculation of lava effusion rates from Landsat TM data, *Bull. Volcanol.*, *60*(1), 52–71, doi:10.1007/s004450050216.

Hauri, E. H. (1996), Major-element variability in the Hawaiian mantle plume, *Nature*, *382*(6590), 415–419, doi:10.1038/382415a0.

Hazen, R. M., A. P. Jones, and J. A. Baross (Eds.) (2013), *Carbon in Earth*, *Reviews in Mineralogy and Geochemistry*, *75*. The Mineralogical Society of America, Chantilly, Virginia.

Helz, R.T. (1987), Diverse olivine types in lava of the 1959 eruption of Kilauea Volcano and their bearing on eruption dynamics, in *Volcanism in Hawaii*, U.S. Geol. Surv. Prof. Pap, 1350, edited by R. W. Decker et al., pp. 691–722.

Helz, R. T., D. A. Clague, T. W. Sisson, and C. R. Thornber (2014), Petrologic insights into basaltic volcanism at historically active Hawaiian volcanoes, Chapter 6 of Poland, M. P., T. J. Takahashi, and C. M. Landowski, (eds.), *Characteristics of Hawaiian Volcanoes*, U.S. Geol. Surv. Prof. Pap. 1801, pp. 237–292, doi:10.3133/pp18016.

Herzberg, C. (2006), Petrology and thermal structure of the Hawaiian plume from Mauna Kea volcano, *Nature*, *444*(7119), 605–609, doi:10.1038/nature05254.

Hitchcock, C. K. (1909), *Hawaii and Its Volcanoes*, Hawaiian Gazette Co., Honolulu.

Hoernle, K., and H.-U. Schmincke (1993), The petrology of the tholeiites through melilite nephelinites on Gran Canaria, Canary Islands: Crystal fractionation, accumulation, and depths of melting, *J. Petrol.*, *34*(3), 573–597, doi:10.1093/petrology/34.3.573.

Holcomb, R. T. (1987), Eruptive history and long-term behavior of Kilauea volcano, in *Volcanism in Hawaii*, U.S. Geol. Surv. Prof. Pap. 1350, edited by R. W. Decker et al., pp. 261–350.

Holcomb, R. T., and R. C. Searle (1991), Large landslides from oceanic volcanoes, *Mar. Geotechnol.*, *10*(1–2), 19–32, doi:10.1080/10641199109379880.

Holcomb, R. T., B. K. Nelson, P. W. Reiners, and N.-L. Sawyer (2000), Overlapping volcanoes: The origin of the Hilo Ridge, Hawaii, *Geology*, *28*(6), 547–550, doi:10.1130/0091-7613(2000)28<547:OVTOOH>2.0.CO;2.

Huang, S., and F. A. Frey (2005), Recycled oceanic crust in the Hawaiian Plume: Evidence from temporal geochemical variations within the Koolau Shield, *Contrib. Mineral. Petrol.*, *149*(5), 556–575, doi:10.1007/s00410-005-0664-9.

Huang, S., P. S. Hall, and M. G. Jackson (2011), Geochemical zoning of volcanic chains associated with Pacific hotspots, *Nature Geosci.*, *4*(12), 874–878, doi:10.1038/ngeo1263.

Iverson, R. M. (1995), Can magma-injection and groundwater forces cause massive landslides on Hawaiian volcanoes? *J. Volcanol. Geotherm. Res.*, *66*(1), 295–308, doi:10.1016/0377-0273(94)00064-N.

Iyer, H. M. (1984), Geophysical evidence for the locations, shapes and sizes, and internal structures of magma chambers beneath regions of Quaternary volcanism, *Philos. Trans. R. Soc. Lond. A*, *310*(1514), 473–510, doi:10.1098/rsta.1984.0005.

Jackson, E. D., E. A. Silver, and G. B. Dalrymple (1972), Hawaiian-Emperor chain and its relation to Cenozoic circumpacific tectonics, *Geol. Soc. Am. Bull.*, *83*(8), 601–618, doi:10.1130/0016-7606(1972)83[601:HCAIRT]2.0.CO;2.

Jackson, M. G., D. Weis, and S. Huang (2012), Major element variations in Hawaiian shield lavas: Source features and perspectives from global ocean island basalt (OIB) systematics, *Geochem. Geophys. Geosyst.*, *13*(Q09009), doi:10.1029/2012GC004268.

Jaggar, T. A., Jr. (1945), *Volcanoes Declare War; Logistics and Strategy of Pacific Volcano Science*, Paradise of the Pacific, Honolulu.

Jicha, B. R., J. M. Rhodes, B. S. Singer and M. O. Garcia (2012), ^{40}Ar/^{39}Ar geochronology of submarine Mauna Loa volcano, Hawaii, *J. Geophys. Res.*, *117*(B09204), doi:10.1029/2012JB009373.

Johnson, D. J. (1995), Gravity changes on Mauna Loa volcano, in *Mauna Loa Revealed: Structure, Composition, History, and Hazards*, Geophys. Monor. *92*, edited by J. M. Rhodes and J. P. Lockwood, pp. 127–143, AGU, Washington, D. C. doi:10.1029/GM092p0127.

Johnson, J. H., and M. P. Poland (2013), Seismic detection of increased degassing before Kīlauea's 2008 summit explosion, *Nature Comm.*, *4*, 1668, doi:10.1038/ncomms2703.

Kato, A., K. Obara, T. Igarashi, H. Tsuruoka, S. Nakagawa, and N. Hirata (2012), Propagation of slow slip leading up to the 2011 Mw 9.0 Tohoku-Oki earthquake, *Science*, *335*(6069), 705–708, doi:10.1126/science.1215141.

Kauahikaua, J., M. Mangan, C. Heliker, and T. Mattox (1996), A quantitative look at the demise of a basaltic vent: The death of Kupaianaha, Kilauea Volcano, Hawai'i, *Bull. Volcanol.*, *57*(8), 641–648, doi:10.1007/s004450050117.

Kauahikaua, J., F. Trusdell, and C. Heliker (1998), The probability of lava inundation at the proposed and existing Kulani prison sites, U.S. Geol. Surv. Open-File Rep. 98–794.

Kauahikaua, J., T. Hildenbrand, and M. Webring (2000), Deep magmatic structures of Hawaiian volcanoes, imaged by three-dimensional gravity methods, *Geology*, *28*(10), 883–886, doi:10.1130/0091-7613(2000)28<883:DMSOHV>2.0.CO;2.

Kauahikaua, J. P., and R. I. Tilling (2014), Natural Hazards and Risk Reduction in Hawai'i, Chapter 10 of Poland, M. P., T. J. Takahashi, and C. M. Landowski, (eds.), *Characteristics of Hawaiian Volcanoes*, U.S. Geol. Surv. Prof. Pap. 1801, pp. 397–427, doi:10.3133/pp180110.

Kauahikaua, J. P., D. R. Sherrod, K. V. Cashman, C. C. Heliker, K. Hon, T. N. Mattox, and J. A. Johnson (2003), Hawaiian lava-flow dynamics during the Pu'u 'Ō'ō-Kūpaianaha Eruption: A Tale of Two Decades, in *The Pu'u 'Ō'ō-Kūpaianaha Eruption of Kīlauea Volcano, Hawaii: The First 20 Years*, U.S. Geol. Surv. Prof. Pap. 1676, edited by C. Heliker et al., pp. 63–187.

Kern, C., C. Werner, T. Elias, A. J. Sutton, and P. Lübcke (2013), Applying UV cameras for SO_2 detection to distant or optically thick volcanic plumes, *J. Volcanol. Geotherm. Res.*, *262*, 80–89, doi:10.1016/j.jvolgeores.2013.06.009.

Klein, F. W. (1982), Patterns of historical eruptions at Hawaiian volcanoes, *J. Volcanol. Geotherm. Res.*, *12*(1–2), 1–35, doi:10.1016/0377-0273(82)90002-6.

Klein, F. W., R. Y. Koyanagi, J. S. Nakata, and W. R. Tanigawa (1987), The seismicity of Kilauea's magma system, in *Volcanism in Hawaii*, U.S. Geol. Surv. Prof. Pap. 1350, edited by R. W. Decker et al., pp. 1019–1185.

Klein, F. W., A. D. Frankel, C. S. Mueller, R. L. Wesson, and P. G. Okubo (2001), Seismic hazard in Hawaii: High rate of large earthquakes and probabilistic ground-motion maps, *Bull. Seismol. Soc. Am.*, *91*(3), 479–498, doi:10.1785/0120000060.

Koeppen, W. C., M. Patrick, T. Orr, A. J. Sutton, D. Dow, and R. Wright (2013), Constraints on the partitioning of Kīlauea's lavas between surface and tube flows, estimated from infrared satellite data, sulfur dioxide emission rates, and field observations, *Bull. Volcanol.*, *75*(5), 716, doi:10.1007/s00445-013-0716-3.

Kono, M. (1980), Paleomagnetism of DSDP Leg 55 basalts and implications for the tectonics of the Pacific plate, in *Initial Reports of the Deep Sea Drilling Project,* vol. 55, edited by E. D. Jackson et al., pp. 737–752, doi:10.2973/dsdp.proc.55.135.1980.

Kurz, M. D., W. J. Jenkins, S. Hart, and D. A. Clague (1983), Helium Isotopic variations in volcanic rocks from Loihi Seamount and the Island of Hawaii, *Earth Planet. Sci. Lett.*, *66*, 388–406, doi:10.1016/0012-821X(83)90154-1.

Lanphere, M. A., G. B. Dalrymple, and D. A. Clague (1980), Rb-Sr systematics of basalts from the Hawaiian-Emperor volcanic chain, in *Initial Reports of the Deep Sea Drilling Project,* vol. 55, edited by E. D. Jackson et al., pp. 695–706, doi:10.2973/dsdp.proc.55.131.1980.

Larson, K. M., M. Poland, and A. Miklius (2010), Volcano monitoring using GPS: Developing data analysis strategies based on the June 2007 Kīlauea Volcano intrusion and eruption, *J. Geophys. Res.*, *115*(B07406), doi:10.1029/2009JB007022.

Lipman, P. W., and A. T. Calvert (2013), Modeling volcano growth on the Island of Hawaii: Deep-water perspectives, *Geosphere*, *9*(5), 1348–1383, doi:10.1130/GES00935.1.

Lipman, P. W., J. P. Lockwood, R. T. Okamura, D. A. Swanson, and K. M. Yamashita (1985), Ground deformation associated with the 1975 magnitude-7.2 earthquake and resulting changes in activity of Kilauea Volcano, Hawaii, U.S. Geol. Surv. Prof. Pap. 1276.

Lipman, P. W., T. W. Sisson, M. L. Coombs, A. Calvert, and J.-I. Kimura (2006), Piggyback tectonics: Long-term growth of Kilauea on the south flank of Mauna Loa, *J. Volcanol. Geotherm. Res.*, *151*(1–3), 73–108, doi:10.1016/j.jvolgeores.2005.07.032.

Lockwood, J. P., and P. W. Lipman (1987), Holocene eruptive history of Mauna Loa volcano, in *Volcanism in Hawaii,* U.S. Geol. Surv. Prof. Pap. 1350, edited by R. W. Decker et al., pp. 509–535.

Lundgren, P., M. Poland, A. Miklius, T. Orr, S.-H. Yun, E. Fielding, Z. Liu, A. Tanaka, W. Szeliga, S. Hensley, and S. Owen (2013), Evolution of dike opening during the March 2011 Kamoamoa fissure eruption, Kīlauea Volcano, Hawai'i, *J. Geophys. Res.*, *118*(3), 897–914, doi:10.1002/jgrb.50108.

Mangan, M. T., and K. V. Cashman (1996), The structure of basaltic scoria and reticulite and inferences for vesiculation, foam formation, and fragmentation in lava fountains, *J. Volcanol. Geotherm. Res.*, *73*(1–2), 1–18, doi:10.1016/0377-0273(96)00018-2.

Mangan, M. T., K. V. Cashman, and D. A. Swanson (2014), The dynamics of Hawaiian-style eruptions: A century of study, Chapter 8 of Poland, M. P., T. J. Takahashi, and C. M. Landowski, (eds.), *Characteristics of Hawaiian Volcanoes*, U.S. Geol. Surv. Prof. Pap. 1801, pp. 323–354, doi:10.3133/pp18018.

MARGINS Steering Committee (1998), Program focuses attention on continental margins, *Eos Trans. AGU*, *79*(11), 137–143, doi:10.1029/98EO00097.

Marske, J. P., A. J. Pietruszka, D. Weis, M. O. Garcia, and J. M. Rhodes (2007), Rapid passage of a small-scale mantle heterogeneity through the melting regions of Kilauea and Mauna Loa Volcanoes, *Earth Planet. Sci. Lett.*, *259*(1–2), 34–50, doi:10.1016/j.epsl.2007.04.026.

Marske, J. P., M. O. Garcia, A. J. Pietruszka, J. M. Rhodes, and M. D. Norman (2008), Geochemical variations during Kīlauea's Pu'u 'Ō'ō eruption reveal a fine-scale mixture of mantle heterogeneities within the Hawaiian plume *J. Petrol.*, *49*(7), 1297–1318, doi:10.1093/petrology/egn025.

Mastin, L. G. (1997), Evidence for water influx from a caldera lake during the explosive hydromagmatic eruption of 1790, Kilauea volcano, Hawaii, *J. Geophys. Res.*, *102*(B9), 20,093–20,109, doi:10.1029/97JB01426.

Mastin, L. G., R. L. Christiansen, C. Thornber, J. Lowenstern, and M. Beeson (2004), What makes hydromagmatic eruptions violent? Some insights from the Keanakākoʻi Ash, Kīlauea Volcano, Hawaiʻi, *J. Volcanol. Geotherm. Res.*, *137*(1–3), 15–31, doi:10.1016/j.jvolgeores.2004.05.015.

Mastin, L. G., E. Roeloffs, N. M. Beeler, and J. E. Quick (2008), Constraints on the size, overpressure, and volatile content of the Mount St. Helens magma system from geodetic and dome-growth measurements during the 2004–2006+ eruption, in *A Volcano Rekindled: The Renewed Eruption of Mount St. Helens, 2004–2006*, U.S. Geol. Surv. Prof. Pap. 1750, pp. 461–488.

Matoza, R. S., P. M. Shearer, G. Lin, C. J. Wolfe, and P. G. Okubo (2013), Systematic relocation of seismicity on Hawaii Island from 1992 to 2009 using waveform cross-correlation and cluster analysis, *J. Geophys. Res.*, *118*(5), 2275–2288, doi:10.1002/jgrb.50189.

Matoza, R. S., P. M. Shearer, and P. G. Okubo (2014), High-precision relocation of long-period events beneath the summit region of Kīlauea Volcano, Hawaiʻi, from 1986 to 2009, *Geophys. Res. Lett.*, *41*(10), 3413–3421, doi:10.1002/2014GL059819.

Matzen, A. K., M. B. Baker, J. R. Beckett, and E. M. Stolper (2011), Fe–Mg partitioning between olivine and high-magnesian melts and the nature of Hawaiian parental liquids, *J. Petrol.*, *52*(7–8), 1243–1263, doi:10.1093/petrology/egq089.

McGovern, P. J. (2007), Flexural stresses beneath Hawaii: Implications for the October 15, 2006, earthquakes and magma ascent, *Geophys. Res. Lett.*, *34*(L23305), doi:10.1029/2007GL031305.

McMurtry, G. M., G. J. Fryer, D. R. Tappin, I. P. Wilkinson, M. Williams, J. Fietzke, D. Garbe-Schoenberg, and P. Watts (2004), Megatsunami deposits on Kohala volcano, Hawaii, from flank collapse of Mauna Loa, *Geology*, *32*(9), 741–744, doi:10.1130/G20642.1.

McPhie, J., G. P. L. Walder, and R. L. Christiansen (1990), Phreatomagmatic and phreatic fall and surge deposits from explosions at Kilauea Volcano, Hawaii, 1790 A.D.: Keanakakoi Ash Member, *Bull. Volcanol.* *52*(5), 334–354, doi:10.1007/BF00302047.

Melnik, O., and R. S. J. Sparks (1999), Nonlinear dynamics of lava dome extrusion, *Nature*, *402*(6757), 37–41, doi:10.1038/46950.

Miklius, A., M. Lisowski, P. T. Delaney, R. P. Denlinger, J. J. Dvorak, A. T. Okamura, and M. K. Sako (1995), Recent inflation and flank movement of Mauna Loa volcano, in *Mauna Loa Revealed: Structure, Composition, History, and Hazards*, Geophys. Monogr. 92, edited by J. M. Rhodes and J. P. Lockwood, pp. 199–205, AGU, Washington, D. C., doi:10.1029/GM092p0199.

Mogi, K. (1958), Relations between the eruptions of various volcanoes and the deformations of the ground surfaces around them, *Bull. Earthquake Res. Inst.*, *36*(2), 99–134.

Montgomery-Brown, E. K., D. K. Sinnett, M. Poland, P. Segall, T. Orr, H. Zebker, and A. Miklius (2010), Geodetic evidence for en echelon dike emplacement and concurrent slow-slip during the June 2007 intrusion and eruption at Kīlauea volcano, Hawaii, *J. Geophys. Res.*, *115*(B07405), doi: 10.1029/2009JB006658.

Moore, J. G., 1964, Giant submarine landslides on the Hawaiian Ridge, in *Geol. Surv. Res. 1964,* U.S. Geol. Surv. Prof. Pap. 501–D, pp. D95–D98.

Moore, J. G. (1970), Relationship between subsidence and volcanic load, Hawaii, *Bull. Volcanol.*, *34*(2), 562–576, doi:10.1007/BF02596771.

Moore, J. G., and D. A. Clague (1992), Volcano growth and evolution of the island of Hawaii, *Geol. Soc. Am. Bull.*, *104*(11), 1471–1484, doi:10.1130/0016-7606(1992)104<1471:VGAEOT>2.3.CO;2.

Moore, J. G., and H. L. Krivoy (1964), The 1962 flank eruption of Kilauea Volcano and structure of the east rift zone, *J. Geophys. Res.*, *69*(10), 2033–2045, doi:10.1029/JZ069i010p02033.

Moore, J. G., and G. W. Moore (1984), Deposit from a giant wave on the island of Lanai, Hawaii, *Science*, *226*(4680), 1312–1315, doi:10.1126/science.226.4680.1312.

Moore, J. G., D. A. Clague, R. T. Holcomb, P. W. Lipman, W. R. Normark, and M. E. Torresan (1989), Prodigious submarine landslides on the Hawaiian Ridge, *J. Geophys. Res.*, *94*(B12), 17,465–17,848, doi:10.1029/JB094iB12p17465.

Moore, J. G., W. R. Normark, and R. T. Holcomb (1994a), Giant Hawaiian landslides, *Annu. Rev. Earth Planet. Sci.*, *22*, 119–144, doi:10.1146/annurev.ea.22.050194.001003.

Moore, J. G., W. R. Normark, and R. T. Holcomb (1994b), Giant Hawaiian underwater landslides, *Science*, *264*(5155), 46–47, doi:10.1126/science.264.5155.46.

Morgan, J. K. (2006), Volcanotectonic interactions between Mauna Loa and Kilauea: Insights from 2-D discrete element simulations, *J. Volcanol. Geotherm. Res.*, *151*(1–3), 109–131, doi:10.1016/j.jvolgeores.2005.07.025.

Morgan, J. K., G. F. Moore, D. J. Hills, and S. Leslie (2000), Overthrusting and sediment accretion along Kilauea's mobile south flank, Hawaii: Evidence for volcanic spreading from marine seismic reflection data, *Geology*, *28*(7), 667–670, doi:10.1029/2003JB002411.

Morgan, J. K., G. F. Moore, and D. A. Clague (2003), Slope failure and volcanic spreading along the submarine south flank of Kilauea volcano, Hawaii, *J. Geophys. Res.*, *108*(2415), doi:10.1029/2003JB002411.

Morgan, W. J. (1971), Convection plumes in the lower mantle, *Nature*, *230*(5288), 42–43, doi:10.1038/230042a0.

Mullineaux, D. R., and D. W. Peterson (1974), Volcanic hazards on the Island of Hawaii, U.S. Geol. Surv. Open-File Rep. 74–239.

Mullineaux, D. R., D. W. Peterson, and D. R. Crandell (1987), Volcanic hazards in the Hawaiian Islands, in *Volcanism in Hawaii*, U.S. Geol. Surv. Prof. Pap. 1350, edited by R. W. Decker et al., pp. 599–621.

Neal, C. A., and J. P. Lockwood (2003), Geologic map of the summit region of Kīlauea Volcano, Hawaii, U.S. Geol. Surv. Geol. Invest. Ser. I-2759, scale 1:24,000.

Newhall, C., and R. Hoblitt (2002), Constructing event trees for volcanic crises, *Bull. Volcanol.*, *64*(1), 3–20, doi:10.1007/s004450100173.

Nishimura, T. (2006), Ground deformation due to magma ascent with and without degassing, *Geophys. Res. Lett.*, *33*(L23309), doi:10.21029/22006GL028101.

O'Connor, J. M., B. Steinberger, M. Regelous, A. A. P. Koppers, J. R. Wijbrans, K. M. Haase, P. Stoffers, W. Jokat, and D. Garbe-Schönberg (2013), Constraints on past plate and mantle motion from new ages for the Hawaiian-Emperor Seamount Chain, *Geochem. Geophys. Geosyst.*, doi:10.1002/ggge.20267.

Okubo, P. G., and C. J. Wolfe (2008), Swarms of similar long-period earthquakes in the mantle beneath Mauna Loa Volcano, *J. Volcanol. Geotherm. Res.*, *178*(4), 787–794, doi:10.1016/j.jvolgeores.2008.09.007.

Okubo, P. G., H. M. Benz, and B. A. Chouet (1997), Imaging the crustal magma sources beneath Mauna Loa and Kilauea volcanoes, Hawaii, *Geology*, *25*(10), 867–870, doi: 10.1130/0091-7613(1997)025<0867:ITCMSB>2.3.CO;2.

Okubo, P. G., J. S. Nakata, and R. Y. Koyanagi (2014), The Evolution of Seismic Monitoring Systems at the Hawaiian Volcano Observatory, Chapter 2 of Poland, M. P., T. J. Takahashi, and C. M. Landowski, (eds.), *Characteristics of Hawaiian Volcanoes*, U.S. Geol. Surv. Prof. Pap. 1801, pp. 67–94, doi:10.3133/pp18012.

Orr, T. R., and J. C. Rea (2012), Time-lapse camera observations of gas piston activity at Pu'u 'Ō'ō, Kīlauea volcano, Hawai'i, *Bull. Volcanol.*, *74*(10), 2353–2362, doi:10.1007/s00445-012-0667-0.

Orr, T. R., W. A. Thelen, M. R. Patrick, D. A. Swanson, and D. C. Wilson (2013), Explosive eruptions triggered by rockfalls at Kīlauea volcano, Hawai'i, *Geology*, *41*(2), 207–210, doi:10.1130/G33564.1.

Owen, S., P. Segall, J. T. Freymueller, A. Miklius, R. P. Denlinger, T. Árnadóttir, M. K. Sako, and R. Bürgmann (1995), Rapid deformation of the south flank of Kilauea Volcano, Hawaii, *Science*, *267*(5202), 1328–1332, doi:10.1126/science.267.5202.1328.

Owen, S., P. Segall, M. Lisowski, A. Miklius, R. Denlinger, and M. Sako (2000), Rapid deformation of Kilauea Volcano: Global Positioning System measurements between 1990 and 1996, *J. Geophys. Res.*, *105*(B8), 18,983–18,993, doi:10.1029/2000JB900109.

Park, J., J. K. Morgan, C. A. Zelt, P. G. Okubo, L. Peters, and N. Benesh (2007), Comparative velocity structure of active Hawaiian volcanoes from 3-D onshore–offshore seismic tomography, *Earth Planet. Sci. Lett.*, *259*(3–4), 500–516, doi:10.1016/j.epsl.2007.05.008.

Park, J., J. K. Morgan, C. A. Zelt, and P. G. Okubo (2009), Volcano-tectonic implications of 3-D velocity structures derived from joint active and passive source tomography of the island of Hawaii, *J. Geophys. Res.*, *114*(B09301), doi:10.1029/2008JB005929.

Peltier, A., P. Bachèlery, and T. Staudacher (2009), Magma transport and storage at Piton de La Fournaise (La Réunion) between 1972 and 2007: A review of geophysical and geochemical data, *J. Volcanol. Geotherm. Res.*, *184*(1–2), 93–108, doi:10.1016/j.jvolgeores.2008.12.008.

Phillips, K. A., C. D. Chadwell, and J. A. Hildebrand (2008), Vertical deformation measurements on the submerged south flank of Kilauea volcano, Hawai'i reveal seafloor motion associated with volcanic collapse, *J. Geophys. Res.*, *113*(B05106), doi:10.1029/2007JB005124.

Pietruszka, A. J., and M. O. Garcia (1999a), A rapid fluctuation in the mantle source and melting history of Kilauea Volcano inferred from the geochemistry of its historical summit lavas (1790–1982) *J. Petrol.*, *40*(8), 1321–1342, doi:10.1093/petroj/40.8.1321.

Pietruszka, A. J., and M. O. Garcia (1999b), The size and shape of Kilauea Volcano's summit magma storage reservoir: A geochemical probe, *Earth Planet. Sci. Lett.*, *167*(3–4), 311–320, doi:10.1016/S0012-821X(99)00036-9.

Pietruszka, A. J., M. D. Norman, M. O. Garcia, J. P. Marske, and D. H. Burns (2013), Chemical heterogeneity in the Hawaiian mantle plume from the alteration and dehydration of recycled oceanic crust, *Earth Planet. Sci. Lett.*, *361*, 198–309, doi:10.1016/j.epsl.2012.10.030.

Plattner, C., F. Amelung, S. Baker, R. Govers, and M. P. Poland (2013), The role of viscous magma mush spreading in volcanic flank motion at Kīlauea Volcano, Hawai'i, *J. Geophys. Res.*, *118*(5), 2474–2487, doi:10.1002/jgrb.50194.

Poland, M. P. (2010), Localized surface disruptions observed by InSAR during strong earthquakes in Java and Hawai'i, *Bull. Seismol. Soc. Am.*, *100*(2), 532–540, doi:10.1785/0120090175.

Poland, M. P., A. Miklius, A. J. Sutton, and C. R. Thornber (2012), A mantle-driven surge in magma supply to Kīlauea Volcano during 2003–2007, *Nature Geosci.*, *5*(4), 295–300, doi:10.1038/ngeo1426.

Poland, M. P., P. G. Okubo, and K. Hon (2013), Exploring Hawaiian volcanism, *EOS Trans. AGU*, *94*(7), 72, doi:10.1002/2013EO070005.

Poland, M. P. (2014), Time-averaged discharge rate of subaerial lava at Kīlauea Volcano, Hawai'i, measured from TanDEM-X interferometry: Implications for magma supply and storage during 2011–2013, *J. Geophys. Res.*, 119(7), 5464–5481, doi:10.1002/2014JB011132.

Poland, M. P., A. Miklius, and E. K. Montgomery-Brown (2014), Magma supply, storage, and transport at shield-stage Hawaiian volcanoes, Chapter 5 of Poland, M. P., T. J. Takahashi, and C. M. Landowski, (eds.), *Characteristics of Hawaiian Volcanoes*, U.S. Geol. Surv. Prof. Pap. 1801, pp. 179–234, doi:10.3133/pp18015.

Rasa, R., R. Azzaro, and O. Leonardi (1996), Aseismic creep on faults and flank instability at Mount Etna volcano, Sicily, in *Volcano Instability on the Earth and Other Planets*, Geol. Soc. Lond. Spec. Pap. 110, edited by W. J. McGuire et al., pp. 179–192.

Regelous, M., A. W. Hofmann, W. Abouchami, and S. J. G. Galer (2003), Geochemistry of lavas from the Emperor Seamounts, and the geochemical evolution of Hawaiian magmatism from 85 to 42 Ma, *J. Petrol*, *44*(1), 113–140, doi:10.1093/petrology/44.1.113.

Ren, Z. Y., T. Hanyu, T. Miyazaki, Q. Chang, H. Kawabata, T. Takahashi, Y. Hirahara, A. R. L. Nichols, and Y. Tatsumi (2009), Geochemical differences of the Hawaiian shield lavas: Implications for melting process in the heterogeneous Hawaiian plume, *J. Petrol.*, *50*(8), 1553–1573, doi:10.1093/petrology/egp041.

Rhodes, J. M., K. P. Wenz, C. A. Neal, J. W. Sparks, and J. P. Lockwood (1989), Geochemical evidence for invasion of Kilauea's plumbing system by Mauna Loa magma, *Nature*, *337*(6204), 257–260, doi:10.1038/337257a0.

Rhodes, J. M., S. Huang, F. A. Frey, M. Pringle, and G. Xu (2012), Compositional diversity of Mauna Kea shield lavas recovered by the Hawaii Scientific Drilling Project: Inferences on source lithology, magma supply, and the role of multiple volcanoes, *Geochem. Geophys. Geosyst.*, *13*(Q03014), doi:10.1029/2011GC003812.

Richards, M., D. Geist, C. Ebinger, G. Grant, K. Harpp, G. Ito, P. Ramon, and D. Toomey (2012), Report of the Organizing Committee for the July 2011 Chapman Conference on "The Galápagos as a Laboratory for the Earth Sciences," available at http://www.webpages.uidaho.edu/~dgeist/Chapman/GalapagosWhitePaperIllustrated.pdf.

Richter, N., M. P. Poland, and P. R. Lundgren (2013), TerraSAR-X interferometry reveals small-scale deformation associated with the summit eruption of Kīlauea Volcano, Hawai'i, *Geophys. Res. Lett.*, *40*(7), 1279–1283, doi:10.1002/grl.50286.

Robinson, J. E., and B. W. Eakins (2006), Calculated volumes of individual shield volcanoes at the young end of the Hawaiian Ridge, *J. Volcanol. Geotherm. Res.*, *151*(1–3), 309–317, doi:10.1016/j.jvolgeores.2005.07.033.

Rowland, S. K., M. E. MacKay, H. Garbeil, and P. J. Mouginis-Mark (1999), Topographic analyses of Kilauea Volcano, Hawai'i, from interferometric airborne radar, *Bull. Volcanol.*, *61*(1–2), 1–14, doi:10.1007/s004450050258.

Ryan, M. P. (1987), Neutral buoyancy and the mechanical evolution of magmatic systems, in *Magmatic Processes: Physiochemical Principles*, edited by B. O. Mysen, pp. 259–287, The Geochemical Society, University Park, Pa.

Ryan, M. P. (1988), The mechanics and three-dimensional internal structure of active magmatic systems: Kilauea volcano, Hawaii, *J. Geophys. Res.*, *93*(B5), 4213–4248, doi: 10.1029/JB093iB05p04213.

Ryan, M. P., R. Y. Koyanagi, and R. S. Fiske (1981), Modeling the three-dimensional structure of macroscopic magma transport systems: Application to Kilauea volcano, Hawaii, *J. Geophys. Res.*, *86*(B8), 7111–7129, doi:10.1029/JB086iB08p07111.

Schipper, C. I., J. D. L. White, and B. F. Houghton (2011), Textural, geochemical, and volatile evidence for a Strombolian-like eruption sequence at Lō`ihi Seamount, Hawai'i, *J. Volcanol. Geotherm. Res.*, *207*(1–2), 16–32, doi:10.1016/j.jvolgeores.2011.08.001.

Segall, P. (2013), Volcano deformation and eruption forecasting, in *Remote Sensing of Volcanoes and Volcanic Processes: Integrating Observation and Modelling*, Geol. Soc. Lond. Spec. Pap. 380, edited by D. M. Pyle et al., doi:10.1144/SP380.4.

Segall, P., and A. M. Bradley (2012), Slow-slip evolves into megathrust earthquakes in 2D numerical simulations, *Geophys. Res. Lett.*, *39*(18), doi:10.1029/2012GL052811.

Segall, P., P. Cervelli, S. Owen, M. Lisowski, and A. Miklius (2001), Constraints on dike propagation from continuous GPS measurements, *J. Geophys. Res.*, *106*(B9), 19,301–19,317, doi:10.1029/2001JB000229.

Segall, P., A. L. Llenos, S.-H. Yun, A. M. Bradley, and E. M. Syracuse (2013), Time-dependent dike propagation from joint inversion of seismicity and deformation data, *J. Geophys. Res.*, doi:10.1002/2013JB010251

Sharp, W. D., and D. A. Clague (2006), 50-Ma initiation of Hawaiian-Emperor bend records major change in Pacific plate motion, *Science*, *313*(5791), 1281–1284, doi:10.1126/science.1128489.

Sherrod, D. R., Y. Nishimitsu, and T. Tagami (2003), New K-Ar ages and the geologic evidence against rejuvenated-stage volcanism at Haleakala, East Maui, a postshield-stage volcano of the Hawaiian island chain, *Geol. Soc. Am. Bull.*, *115*(6), 683–694, doi:10.1130/0016-7606(2003)115<0683:NKAATG> 2.0.CO;2.

Sherrod, D. R., J. T. Hagstrum, J. P. McGeehin, D. E. Champion, and F. A. Trusdell (2006), Distribution, ^{14}C chronology, and paleomagnetism of latest Pleistocene and Holocene lava flows and Haleakalā volcano, Island of Maui, Hawai'i: A revision of lava flow hazard zones, *J. Geophys. Res.*, *111*(B05205), doi:05210.01029/02005JB003876.

Sherrod, D. R., J. M. Sinton, S. E. Watkins, and K. M. Brunt (2007), Geologic map of the State of Hawaii, U.S. Geol. Surv. Open-File Rep. 2007-1089.

Solaro, G., V. Acocella, S. Pepe, J. Ruch, M. Neri, and E. Sansosti (2010), Anatomy of an unstable volcano from InSAR: Multiple processes affecting flank instability at Mt. Etna, 1994–2008, *J. Geophys. Res.*, *115*(B10405), doi: 10.1029/2009JB000820.

Sparks, R. S. J. (1978), The dynamics of bubble formation and growth in magmas: A review and analysis, *J. Volcanol. Geotherm. Res.*, *3*(1–2), 1–37, doi:10.1016/0377-0273(78)90002-1.

Sparks, R. S. J. (2003), Forecasting volcanic eruptions, *Earth Planet. Sci. Lett.*, *210*(1–2), 1–15, doi:10.1016/S0012-821X(03)00124-9.

Stearns, H.T. (1946), Geology of the Hawaiian Islands, *Hawaii Div. Hydrogr. Bull.*, *8*, 106 pp.

Stearns, H. T., and G. D. Macdonald (1946), Geology and ground-water resources of the island of Hawaii, *Hawaii Div. Hydrogr. Bull.*, *9*, 363 pp.

Stolper, E., S. Sherman, M. Garcia, M. Baker, and C. Seaman (2004), Glass in the submarine section of the HSDP2 drill core, Hilo, Hawaii, *Geochem. Geophys. Geosyst.*, *5*(7), doi:10.1029/2003GC000553.

Stolper, E. M., D. J. DePaolo, and D. M. Thomas (2009), Deep drilling into a mantle plume volcano: The Hawaii Scientific Drilling Project, *Sci. Drilling*, *7*, 4–14, doi:10.2204/iodp.sd.7.02.2009.

Stovall, W. K., B. F. Houghton, H. Gonnermann, S. A. Fagents, and D. A. Swanson (2011), Eruption dynamics of Hawaiian-style fountains: The case study of episode 1 of the Kīlauea Iki 1959 eruption, *Bull. Volcanol.*, *73*(5), 511–529, doi:10.1007/s00445-010-0426-z.

Sutton, A. J., T. Elias, T. M. Gerlach, and J. B. Stokes (2001), Implications for eruptive processes as indicated by sulfur dioxide emissions from Kilauea Volcano, Hawai`i, 1979–1997, *J. Volcanol. Geotherm. Res.*, *108*(1–4), 283–302, doi:10.1016/S0377-0273(00)00291-2.

Sutton, A. J., T. Elias, and J. Kauahikaua (2003), Lava-effusion rates for the Pu'u 'Ō'ō-Kūpaianaha eruption derived from SO_2 emissions and very low frequency (VLF) measurements, in *The Pu'u 'Ō'ō-Kūpaianaha Eruption of Kīlauea Volcano, Hawaii: The First 20 Years*, U.S. Geol. Surv. Prof. Pap. 1676, edited by C. Heliker et al., pp. 137–148.

Sutton, A. J., and T. Elias (2014), One Hundred Volatile Years of Volcanic Gas Studies at the Hawaiian Volcano Observatory, Chapter 7 of Poland, M. P., T. J. Takahashi, and C. M. Landowski, (eds.), *Characteristics of Hawaiian Volcanoes*, U.S. Geol. Surv. Prof. Pap. 1801, pp. 295–320, doi:10.3133/pp18017.

Swanson, D. A. (2008), Hawaiian oral tradition describes 400 years of volcanic activity at Kīlauea, *J. Volcanol. Geotherm. Res.*, *176*(3), 427–431, doi:10.1016/j.jvolgeores.2008.01.033.

Swanson, D. A., and R. L. Christiansen (1973), Tragic base surge in 1790 at Kilauea Volcano, *Geology*, *1*(2), 83–86, doi:10.1130/0091-7613(1973)1<83:TBSIAK>2.0.CO;2.

Swanson, D. A., W. A. Duffield, and R. S. Fiske (1976), Displacement of the south flank of Kilauea Volcano: The result of forceful intrusion of magma into the rift zones, U.S. Geol. Surv. Prof. Pap. 963.

Swanson, D. A., W. A. Duffield, D. B. Jackson, and D. W. Peterson (1979), *Chronological narrative of the 1969–1971 Mauna Ulu eruption of Kilauea Volcano*, Hawaii, U.S. Geol. Surv. Prof. Pap. 1056.

Swanson, D. A., T. J. Casadevall, D. Dzurisin, S. D. Malone, C. G. Newhall, and C. S. Weaver (1983), Predicting eruptions at Mount St. Helens, June 1980 through December 1982, *Science*, *221*(4618), 1369–1376, doi:10.1126/science.221.4618.1369.

Swanson, D.A., R. S. Fiske, and C. R. Thornber (2012a), Vents and dikes in the heart of the Koa'e Fault System at Kilauea, Abstract V33B–2862, presented at 2012 Fall Meet., AGU, San Francisco, Calif., 3–7 December.

Swanson, D. A., T. P. Rose, R. S. Fiske, and J. P. McGeehin (2012b), Keanakāko'i Tephra produced by 300 years of explosive eruptions following collapse of Kīlauea's caldera in about 1500 CE, *J. Volcanol. Geotherm. Res.*, *215–216*, 8–25, doi:10.1016/j.jvolgeores.2011.11.009.

Swanson, D. A., T. R. Rose, A. E. Musek, M. O. Garcia, R. S. Fiske, and L. G. Mastin (2014), Cycles of explosive and effusive eruptions at Kīlauea Volcano, Hawai'i, *Geology*, *42*(7), 631–634, doi: 10.1130/G35701.1.

Tanaka, R., A. Makishima, and E. Nakamura (2008), Hawaiian double volcanic chain triggered by an episodic involvement of recycled material: Constraints from temporal Sr–Nd–Hf–Pb isotopic trend of the Loa-type volcanoes, *Earth Planet. Sci. Lett.*, *265*(3), 450–465, doi:10.1016/j.epsl.2007.10.035.

Tarduno, J. A., R. A. Duncan, D. W. Scholl, R. D. Cottrell, B. Steinberger, T. Thordarson, B. C. Kerr, C. R. Neal, F. A. Frey, M. Torii, and C. Carvallo (2003), The Emperor Seamounts: Southward motion of the Hawaiian hotspot plume in Earth's mantle, *Science*, *301*(5636), 1064–1069, doi:10.1126/science.1086442.

Tatsumoto, M. (1978), Isotopic composition of lead in oceanic basalt and its implication to mantle evolution, *Earth Planet. Sci. Lett.*, *38*(1), 63–87, doi:10.1016/0012-821X(78)90126-7.

Thornber, C. R. (2003), Magma-reservoir processes revealed by geochemistry of the Pu'u 'Ō'ō-Kūpaianaha eruption, in *The Pu'u 'Ō'ō-Kūpaianaha Eruption of Kīlauea Volcano, Hawaii: The First 20 Years*, U.S. Geol. Surv. Prof. Pap. 1676, edited by C. Heliker et al., pp. 121–136.

Thornber, C. R., C. Heliker, D. R. Sherrod, J. P. Kauahikaua, A. Miklius, P. G. Okubo, F. A. Trusdell, J. R. Budahn, W. I. Ridley, and G. P. Meeker (2003), Kilauea east rift zone magmatism: An episode 54 perspective, *J. Petrol.*, *44*(9), 1525–1559, doi:10.1093/petrology/egg048.

Tilling, R. I., J. P. Kauahikaua, S. R. Brantley, and C. A. Neal (2014), The Hawaiian Volcano Observatory—a natural laboratory for studying basaltic volcanism, Chapter 1 of Poland, M. P., T. J. Takahashi, and C. M. Landowski, (eds.), *Characteristics of Hawaiian Volcanoes*, U.S. Geol. Surv. Prof. Pap. 1801, pp. 1–64, doi:10.3133/pp18011.

Trusdell, F. (1995), Lava flow hazards and risk assessment on Mauna Loa volcano, Hawaii, in *Mauna Loa Revealed*, Geophys. Monogr. *92*, edited by J. M. Rhodes and J. P. Lockwood, pp. 327 – 336, AGU, Washington, D. C.

Trusdell, F. (2011), Dueling volcanoes: How activity levels at Kilauea influence eruptions at Mauna Loa, Abstract V33E-08, presented at 2011 Fall Meet., AGU, San Francisco, Calif., 5–9 December.

Van Ark, E., and J. Lin (2004), Time variation in igneous volume flux of the Hawaii-Emperor hot spot seamount chain, *J. Geophys. Res.*, *109*(B11401), doi:10.1029/2003JB002949.

Vergniolle, S., and C. Jaupart (1990), Dynamics of degassing at Kilauea Volcano, Hawaii, *J. Geophys. Res.*, *95*(B3), 2793–2809, doi:10.1029/JB095iB03p02793.

Vidal, V., and A. Bonneville (2004), Variations of the Hawaiian hot spot activity revealed by variations in the magma production rate, *J. Geophys. Res.*, *109*(B03104), doi: 10.1029/2003JB002559.

Vinet, N., and M. D. Higgins (2010), Magma solidification processes beneath Kilauea Volcano, Hawaii: A quantitative textural and geochemical study of the 1969–1974 Mauna Ulu lavas, *J. Petrol.*, *51*(6), 1297–1332, doi:10.1093/petrology/egq020.

Vogt, P. R. (1972), Evidence for global synchronism in mantle plume convection, and possible significance for geology, *Nature*, *240*(5380), 338–342, doi:10.1038/240338a0.

Wallace, P. J., and M. Edmonds (2011), The sulfur budget in magmas: Evidence from melt inclusions, submarine glasses, and volcanic gas emissions, *Rev. Mineral. Geochem.*, *73*(1), 215–246, doi:10.2138/rmg.2011.73.8.

Weis, D., M. O. Garcia, J. M. Rhodes, M. Jellinek, and J. S. Scoates (2011), Role of the deep mantle in generating the compositional asymmetry of the Hawaiian mantle plume, *Nature Geosci.*, *4*(12), 831–838, doi:10.1038/ngeo1328.

Wessel, P., and L. W. Kroenke (2008), Pacific absolute plate motion since 145 Ma: An assessment of the fixed hot spot hypothesis, *J. Geophys. Res.*, *113*(B06101), doi:10.1029/2007JB005499.

Wessel, P., Y. Harada, and L. W. Kroenke (2006), Toward a self-consistent, high-resolution absolute plate motion model for the Pacific, *Geochem. Geophys. Geosyst.*, *7*(Q03L12), doi:10.1029/2005GC001000.

White, R. S. (1993), Melt production rates in mantle plumes, *Philos. Trans. R. Soc. Lond, Ser. A*, *342*(1663), 137–153, doi:10.1098/rsta.1993.0010.

Whittaker, J. M., R. D. Müller, G. Leitchenkov, H. Stagg, M. Sdrolias, C. Gaina, and A. Goncharov (2007), Major Australian-Antarctic plate reorganization at Hawaiian-Emperor bend time, *Science*, *318*(5847), 83–86, doi:10.1126/science.1143769.

Wilson, J. T. (1963), A possible origin of the Hawaiian Islands, *Can. J. Phys.*, *41*(6), 863–868, doi:10.1139/p63-094.

Wilson, L., and J. W. Head III (1981), Ascent and eruption of basaltic magma on the Earth and Moon, *J. Geophys. Res.*, *86*(B4), 2971–3001, doi:10.1029/JB086iB04p02971.

Wolfe, C. J., P. G. Okubo, G. Ekström, M. Nettles, and P. M. Shearer (2004), Characteristics of deep (≥13 km) Hawaiian earthquakes and Hawaiian earthquakes west of 155.55°W, *Geochem., Geophys., Geosyst.*, *5*(Q04006), doi:10.1029/2003GC000618.

Wolfe, C. J., S. C. Solomon, G. Laske, J. A. Collins, R. S. Detrick, J. A. Orcutt, D. Bercovici, and E. H. Hauri (2009), Mantle shear-wave velocity structure beneath the Hawaiian hot spot, *Science*, *326*(5958), 1388–1390, doi:10.1126/science.1180165.

Wolfe, C. J., S. C. Solomon, G. Laske, J. A. Collins, R. S. Detrick, J. A. Orcutt, D. Bercovici, and E. H. Hauri (2011), Mantle P-wave velocity structure beneath the Hawaiian hotspot, *Earth Planet. Sci. Lett.*, *303*(3–4), 267–280, doi:10.1016/j.epsl.2011.01.004.

Wolfe, E.W., and J. Morris (1996), Geologic map of the Island of Hawaii, U.S. Geol. Surv. Misc. Invest. Ser. I-2524-A, scale 1:100,000.

Wolfe, E. W., C. A. Neal, N. G. Banks, and T. J. Duggan (1988), Geologic observations and chronology of eruptive events, in *The Puu Oo Eruption of Kilauea Volcano, Hawaii; Episodes 1 Through 20, January 3, 1983, Through June 8, 1984*, U.S. Geol. Surv. Prof. Pap. 1463, edited by E. W. Wolfe, pp. 1–97.

Wood, H, O. (1914), On the earthquakes of 1868 in Hawaii, *Bull. Seismol. Soc. Am.*, *4*(4), 169–203.

Wright, T. L. (1971), Chemistry of Kilauea and Mauna Loa lava in space and time, U.S. Geol. Surv. Prof. Pap. 735.

Wright, T. L., J. Y. F. Chun, J. Esposo, C. Heliker, J. Hodge, J. P. Lockwood, and S. Vogt (1992), Map showing lava-flow hazard zones, Island of Hawaii, U.S. Geol. Surv. Misc. Field Studies Map MF–2193, scale 1:250,000.

Wyss, M. (1988), A proposed source model for the great Kau, Hawaii, earthquake of 1868, *Bull. Seismol. Soc. Am.*, *78*(4), 1450–1462.

Wyss, M., and R. Y. Koyanagi (1992), Isoseismal maps, macroseismic epicenters, and estimated magnitudes of historical earthquakes in the Hawaiian Islands, U.S. Geol. Surv. Bull. 2006.

Xu, G., S. Huang, F. A. Frey, J. Blichert-Toft, W. Abouchami, D. A. Clague, B. Cousens, J. G. Moore, and M. H. Beeson (2014), The distribution of geochemical heterogeneities in the source of Hawaiian shield lavas as revealed by a transect across the strike of the Loa and Kea spatial trends: East Molokai to West Molokai to Penguin Bank, *Geochim. Cosmochim. Acta*, *132*, 214–237, doi:10.1016/j.gca.2014.02.002.

Yamada, T., P. G. Okubo, and C. J. Wolfe (2010), Kīholo Bay, Hawai'i, earthquake sequence of 2006: Relationship of the main shock slip with locations and source parameters of aftershocks, *J. Geophys. Res.*, *115*(B08304), doi:10.1029/2009JB006657.

INDEX

Hawaiian Volcanoes: From Source to Surface, Geophysical Monograph 208. First Edition.
Edited by Rebecca Carey, Valérie Cayol, Michael Poland, and Dominique Weis.

Companion Website: www.wiley.com/go/Carey/Hawaiian_Volcanoes